Fundamentals of Space Systems,
Second Edition

The Johns Hopkins University

Applied Physics Laboratory Series

In Science and Engineering

Series Editor
Kishin Moorjani

Renewable Energy from the Ocean: A Guide to OTEC
William H. Avery and Chih Wu

Software Engineering: A Holistic View
Bruce I. Blum

Beyond Programming: To a new Era of Design
Bruce I. Blum

Signal Processing Using Optics: Fundamentals, Devices, Architectures, and Applications
Bradley G. Boone

Flame Structure and Processes
Robert M. Fristrom

Desktop Computers: In Perspective
Richard A. Henle and Boris W. Kuvshinoff

Fundamentals of Space Systems, 2nd Edition
Vincent L. Pisacane (ed.)

Fundamentals of Space Systems

Second Edition

Edited by

Vincent L. Pisacane

2005

OXFORD
UNIVERSITY PRESS

Oxford University Press, Inc., publishes works that further
Oxford University's objective of excellence
in research, scholarship, and education.

Oxford New York
Auckland Cape Town Dar es Salaam Hong Kong Karachi
Kuala Lumpur Madrid Melbourne Mexico City Nairobi
New Delhi Shanghai Taipei Toronto

With offices in
Argentina Austria Brazil Chile Czech Republic France Greece
Guatemala Hungary Italy Japan Poland Portugal Singapore
South Korea Switzerland Thailand Turkey Ukraine Vietnam

Published by Oxford University Press, Inc.
198 Madison Avenue, New York, New York 10016

www.oup.com

Library of Congress Cataloging-in-Publication Data
Fundamentals of space systems / edited by Vincent L. Pisacane.—2nd ed.
p. cm.—(The John Hopkins University/Applied Physics Laboratory series in science and engineering)
Includes bibliographical references and index.
ISBN-13 978-0-19-516205-9
ISBN 0-19-516205-6
1. Astronautical instruments. 2. Astronautics—Systems engineering.
I. Pisacane, Vincent L. II. Series.

TL 1082 F86 2004
629.47—dc22 2003024375

9 8 7 6 5 4 3 2 1
Printed in the United States of America
on acid–free paper

Preface

This second edition of *Fundamentals of Space Systems* has undertaken to enhance the content of the first edition in two significant ways. First, the material has been updated to reflect changes in the technology and science of spacecraft systems engineering that have occurred since the publication of the first edition. Second, an attempt has been made to make the material more suitable for a space systems course as a prerequisite to a senior class project to design and perhaps build and launch a spacecraft or spacecraft instrument. To this end, the content has been revised on the basis of review and evaluation by the authors' students and their colleagues. In addition, more problems are provided (with the potential of a solutions manual) and a chapter pulling together the design concepts for a simplified spacecraft has been added. Since most of the authors also teach the material, this provides a text that is especially unique. As an academic offering, the material covered is suitable for a senior level or beginning level graduate course in any of the engineering disciplines.

The theme of the book is to expose the reader to the fundamentals of each of the subsystems in a spacecraft to a depth that should permit the reader to carry out a conceptual design. It should be noted that the authors of every chapter have had extensive experience and responsibility to develop actual spacecraft subsystems or systems as well as the responsibility to teach, in an academic program, how this is done. As a result, they are able to ignore material that may be of esoteric academic interest but not applicable to the development of actual space systems. The book should help scientists and systems or subsystem engineers working in the field to relate to the myriad of issues and compromises that are necessary to develop a system or subsystem to a set of requirements. With the maturing of the space industry, it is critical that practitioners have

	Two Semesters (hours)	One Semester (hours)
Space Systems Engineering	3	2
Space Environment	6	2
Astrodynamics	9	6
Propulsion and Flight Mechanics	6	3
Attitude Determination and Control	6	4
Space Power Systems	6	4
Thermal Control	6	3
Configuration and Structures	6	4
Communications	4.5	4
Command and Telemetry	3	1
Spacecraft Computer Systems and Software	4.5	1
Reliability, Quality Assurance, Radiation	6	2
Spacecraft Integration	3	1
Spacecraft Operations	3	1
Examinations	12	4
Total	84	42

a broader understanding of the issues and disciplines in order to enhance performance and reduce cost, schedule, and risk.

The material is based on lectures in courses presented in the Applied Physics Program of the Johns Hopkins University School of Engineering and in the Aerospace Engineering Department of the United States Naval Academy. A recommendation for an offering as either a two-semester or one-semester course is illustrated above, assuming a total of 42 contact hours per semester including a three-hour final examination.

To obtain a depth sufficient to carry out a conceptual design, it is recommended that the offering be a two-semester course. However, it can be offered as an introductory course over one semester by eliminating some material and restricting the remaining material to specific design concepts. For example, in astrodynamics, interplanetary trajectories may be ignored, in power systems nuclear power may be ignored, and in propulsion systems only spacecraft-based liquid or gaseous orbital propulsion systems may be addressed. To enhance the offering, it is recommended that the class be partitioned into a limited number of teams and provided with a set of mission requirements to which each team is required to design different or similar subsystems.

The book is edited solely by Dr. Pisacane, as Mr. Moore was unable to devote the time to the editing process because of commitments to ongoing space projects. The editor wishes to recognize his wife, Lois E. Wehren, M.D., for her inspiration, encouragement, and forbearance so that he was able to complete the task.

Finally, all of the authors wish to offer each reader, whether experienced practitioner, teacher, or student, a pleasant and profitable journey through the book.

Contents

Contributors

A brief description of the education and experience of the contributors, all of whom have had significant leadership roles in the design, development, test, and/or operation of space systems, follows.

Brian Anderson is a researcher in space physics and a member of the Principal Staff of the Johns Hopkins University Applied Physics Laboratory where he has been magnetometer instrument scientist for NASA's NEAR and MESSENGER missions. He earned his B.A. in physics, mathematics, and religion from Augsburg College in Minneapolis and a Ph.D. in physics from the University of Minnesota. His research interests include ultra-low frequency waves in the Earth's magnetosphere, wave–particle interactions in the magnetosphere and shocked solar wind of the magnetosheath, magnetic reconnection at the magnetopause, magnetosphere–ionosphere coupling, and magnetometer instrument development. Most recently he has used magnetic field measurements from the Iridium constellation of satellites to derive distributions of electric current coupling the magnetosphere and ionosphere. Dr. Anderson is a Member of the American Geophysical Union and served as Space Physics and Aeronomy Science Editor for *Geophysical* Research Letters from 2001 through 2003.

George Dakermanji received the B.S.E.E. degree from the University of Aleppo, Syria, and the M.S.E.E. and Ph.D. degrees from Duke University in Durham, North Carolina. He heads the Space Systems Section at the Johns Hopkins University Applied Physics Laboratory and was the lead power systems engineer on the MESSENGER and TIMED spacecrafts. Prior to joining APL in 1993, he headed the Power System Group at Fairchild Space Company and was also the lead engineer in the development of the NASA/GSFC Small Explorer/SAMPEX spacecraft power system and the XTE spacecraft power system electronics. His background and experience are in the areas of spacecraft power systems and power conditioning electronics.

Wayne F. Dellinger has B.S., M.S., and Ph.D. degrees in electrical engineering from Tennessee Technological University. He has worked in the Guidance and Control groups of Martin Marietta Denver Aerospace and Swales Aerospace. In 1997 he joined the Space Department at the Johns Hopkins University Applied Physics Laboratory where he is a member of the Principal Professional Staff and a section supervisor in the Mission Design, Guidance and Control Group. His background and experience are in control systems and estimation, and he has held the position of lead Guidance and Control engineer for the TIMED and CONTOUR missions.

Eric J. Hoffman received degrees in electrical engineering from M.I.T. and from Rice University, and joined the Johns Hopkins University Applied Physics Laboratory in 1964. He has performed systems engineering for APL's space communication and navigation systems as well as for entire spacecraft, supervised communications and navigation design activities, and led satellite conceptual designs. As Space Department Chief Engineer he provides technical guidance for space programs, promotes systems engineering, and sets standards for engineering design, design review, configuration management, and test. He has taught space systems design at the United States Naval Academy, Johns Hopkins University, National Taiwan University, in public courses, and on-site for NSA and at NASA centers. He has authored 60 papers and is a Fellow of the British Interplanetary Society and an Associate Fellow of the AIAA.

Mark Holdridge has served as Mission Operations Manager for three missions with the Johns Hopkins University Applied Physics Laboratory including the GEOSAT, NEAR, and CONTOUR missions. He is a member of the Principal Professional Staff and supervises a Mission Operations Section. He received his M.S. in astronautics from George Washington University and his B.S. in aerospace engineering from the University of Maryland. Mr. Holdridge has developed and refined APL's model for the conduct of low-cost planetary missions. He was responsible for planning the operations concepts for placing the NEAR spacecraft into orbit around the asteroid (Eros) that culminated in a soft landing on Eros's surface followed by surface operations. Prior to APL, Mr. Holdridge provided spacecraft mission analysis, control center software development, and operations management support to a variety of spacecraft missions for NASA, NOAA, U.S. NAVY, and commercial space missions, starting in 1983. He is currently the Deputy Mission Director for NASA's first mission to the Pluto/Charon system, the New Horizons mission.

Richard H. Maurer is a member of the Principal Professional Staff at the Johns Hopkins Applied Physics Laboratory. He received a B.S. and Ph.D. in physics from Long Island University and the University of Pittsburgh respectively. His experience includes being lead engineer/specialist on many APL space programs for radiation effects on spacecraft electronics (AMPTE, MSX, GEOSAT, TOPEX, NEAR, MESSENGER). He has used standard models of the Earth's radiation environment to predict total dose exposure and single-event upset rates for APL missions and has been instrumental in improving models of energetic solar particle events. Recent National Space Biomedical Research Institute (NSBRI) grants (1997–2004) supported Dr. Maurer and colleagues in the development and fabrication of a neutron energy spectrometer. Dr. Maurer has guided the effort that resulted in a successful high-altitude balloon flight in October 2003 and serves as the project manager. Allied research results on neutron production in thick targets of spacecraft structural and shielding materials have been obtained in accelerator experiments. Other research interests are in the areas of electronic packaging, gallium arsenide devices, lasers and laser diodes, and lithium battery primary cells. Related interests include statistical design and analysis of experiments, system reliability, accelerated environmental stress testing, and the physics of failure, including performance assurance activities for the NEAR and STEREO missions.

Douglas S. Mehoke received a B.S. and M.S. in mechanical engineering from the University of California at Santa Barbara and Stanford University, respectively. He is a member of the Senior Professional Staff of the Johns Hopkins University Applied Physics Laboratory and supervises the Thermal Design Section. He has also worked for the LMSC in Sunnyvale, California, where he was responsible for the thermal control systems in several satellite programs. He has contributed to a wide number of spacecraft thermal control systems including those on the Midcourse Space Experiment and the Contour Mission. His experience includes advanced thermal control methods, especially self-regulating thermal switches, composite radiators, and advanced radiator coatings.

Donald Grant Mitchell is a researcher in space physics and a member of the Principal Staff of the Johns Hopkins University Applied Physics Laboratory, where he has been a lead scientist on the HENA Instrument for the IMAGE MidEx Mission and Instrument Scientist on the Magnetospheric Imaging Instrument for the Cassini Saturn mission. He has a B.A. and a Ph.D. in physics from the University of Michigan and University of New Hampshire, respectively. Dr. Mitchell's research focuses on analysis and interpretation of energetic particles (both charged and neutral), plasma, and magnetic field observations in the Earth's magnetosphere, outer planet magnetospheres, and interplanetary space; design of very small, very low power energetic particle time of flight-energy analyzers, and design of instrumentation for the imaging of magnetospheres in energetic neutral atoms. Dr. Mitchell is a Member of the American Geophysical Union.

Robert C. Moore specializes in design of microprocessor-based hardware and software systems for embedded space flight applications. His education includes a B.S. in electrical engineering from Lafayette College and an M.S. in electrical engineering from the Johns Hopkins University. His experience includes lead engineer of the MESSENGER Safing & Fault Protection, lead engineer of the FUSE instrument data system, lead engineer of the digital data processor NEAR laser rangefinder, lead engineer of the baseband digital processing GPS telemetry transmitter, digital signal processor design of the Mars Observer Radar Altimeter, and lead engineer of the Energetic Particles Detector Data System of the GALILEO mission.

Max Peterson is retired from the Johns Hopkins University Applied Physics Laboratory (JHU/APL). He received a B.S. in electrical engineering from Kansas State University in 1961 and an M.S. in engineering from the Johns Hopkins University in 1968. Mr. Peterson joined JHU/APL in 1961 and worked on the Polaris Fleet Ballistic Missile Readiness Program. He supervised the Data Systems Design Section in the Space Telecommunications Group from 1969 until 1975. His work included data handling system design and test for several near-Earth spacecraft. He was the Assistant Program Manager for the AMPTE/CCE spacecraft and the System Engineer for the Polar BEAR Program. Mr. Peterson was Program Manager of the Midcourse Space Experiment (MSX) Program and NASA's MESSENGER mission to Mercury. Mr. Peterson has lectured at the US Naval Academy and the GWC Whiting School of Engineering on spacecraft integration and test and space communications.

Vincent L. Pisacane received a B.S. in mechanical engineering from Drexel, an M.S. in applied mechanics and mathematics, and a Ph.D. in applied mechanics and physics from Michigan State University, and was a post-doc in electrical Engineering at Johns Hopkins University. He is currently the Robert A. Heinlein Professor of Aerospace Engineering at the United States Naval Academy. At the Johns Hopkins University Applied Physics Laboratory he was Head of the Space Department, Assistant Director for Research and Exploratory Development, and Director of the Institute for Advanced Science and Technology in Medicine. His areas of expertise include systems engineering, astrodynamics, control systems, spacecraft propulsion systems, and space physiology. He has served as technology team leader for the National Space Biomedical Research Institute and on several NASA review committees. He has more than 60 publications and is a fellow of the AIAA.

Elliot H. Rodberg is a member of the Principal Professional Staff at the Johns Hopkins University Applied Physics Laboratory where he works in the Space Department's Integration, Test, and Mission Operations Group. He received a B.S. in physics from the University of Maryland and an M.S. in computer science from the Johns Hopkins University. He has designed hardware, software, and systems to support spacecraft testing and mission operations. He has worked on the integration, test, and launch teams for several NASA spacecraft, including AMPTE, TOPEX, ACE, TIMED, and MESSENGER, as well as Delta 180 for the Ballistic Missile Defense Organization. He was the Ground System Lead Engineer for the TOPEX Radar Altimeter, Advanced Composition Explorer (ACE) spacecraft, and TIMED spacecraft. He was also the Integration and Test Manager for the MESSENGER mission to Mercury.

Malcolm D. Shuster was educated at the Massachusetts Institute of Technology and the University of Maryland and is the author of many key papers on spacecraft attitude estimation. He is the originator of the QUEST algorithm for attitude determination, which is part of all our current missions to the other planets and a very large fraction of near-Earth missions. He has supported more than a dozen missions in a variety of mission analysis and planning, hardware acquisition, software development, and launch and early mission support. Dr. Shuster has been a senior astrodynamicist at the Johns Hopkins University Applied Physics Laboratory and a Professor of Aerospace Engineering at the University of Florida. In 2000 he received the Dirk Brouwer Award from the American Astronautical Society. Besides his work in astronautics, he spent a decade as a theoretical nuclear physicist and almost as many years working on defense systems. Currently, Dr. Shuster is director of research for the Acme Spacecraft Company.

William Skullney has a B.S. and M.S. degree in engineering mechanics from the Pennsylvania State University and over 25 years' professional experience. After three years of performing patrol analysis of the Navy's submarine fleet operations, Mr. Skullney switched to carrying out structural analysis for aerospace applications. His structural analysis efforts include lead engineer roles on the HILAT spacecraft program for the Defense Nuclear Agency, the Delta 180, 181, and 183 Sensor Module and the Midcourse Space Experiment Programs for the Strategic Defense Initiative Organization (currently the Missile Defense Agency), as well as a structural engineer role on the Hopkins Ultraviolet Telescope for the Johns Hopkins University. He was promoted to section supervisor of the Structural Analysis Section in 1990 and is currently the group supervisor of the Mechanical Systems Group in the Space Department of the Applied Physics Laboratory, a position he has held since 1991. Mr. Skullney is a member of AIAA (since 1993) and also an instructor on structural design and analysis for the Whiting School of Engineering of the Johns Hopkins University Evening College (since 1990) and for the Applied Technology Institute (since 1993).

Ralph M. Sullivan has a B.S. in physics from Boston College and an M.S. in applied science from the George Washington University. He has contributed to the design of numerous APL and NASA spacecraft power systems and was lead power system engineer for NASA's SAS-A, B & C and AMPTE/CCE spacecraft and SDIO's DELTA 180 and 181 spacecraft. From 1982 to 1991 he managed the Johns Hopkins University Applied Physics Laboratory's Space Power System Section, became Principal Professional Staff, and taught at the APL Evening College and various NASA centers. His experience includes the design of space power systems, determination of radiation effects on solar cells, and analysis of solar panel temperature and power balance. From 1991 to 1998 he was a space power system engineer at Swales Aerospace, Inc., and has since been a consultant.

Harry K. Utterback received an A.B. in mathematics from Gettysburg College and an M.S. in computer science from the Johns Hopkins University. He retired from the Johns Hopkins University Applied Physics Laboratory in 1999 after a 30-year career specializing in the design and implementation of real-time embedded software systems for a variety of spacecraft and their ground control systems. His most recent efforts were in the software quality assurance area. Mr. Utterback taught various computer science and software engineering courses in the GWC Whiting School of Engineering Continuing Professional Programs from 1979 to 1999.

Fundamentals of Space Systems,
Second Edition

1

Systems Engineering and Management

VINCENT L. PISACANE

It must be remembered that there is nothing more difficult to plan, more doubtful of success, more dangerous to manage, than the creation of a new system. For the initiator has the enmity of all who would profit by the preservation of the old institutions and merely lukewarm defenders in those who would gain by the new ones.

– Niccolo Machiavelli

When you come to a fork in the road, take it.

You've got to be very careful if you don't know where you are going because you might not get there.

– Yogi Berra

1.1 Introduction

There are compelling reasons to develop space instrumentation and systems: to employ them in space to contribute to our understanding of the space environment and the laws of physics, and to deploy them in space to perform functions to benefit society. Military, commercial, and research uses of space begin from the same basis, the ability to develop affordable and increasingly capable systems. The planning, developing, integrating, testing, launching, and operation of space systems constitute a systems engineering effort of profound proportions. Such an undertaking requires close coordination of disparate participants, whose requisite scientific, engineering, and management capabilities span a breadth of disciplines that continue to change at dramatic rates.

Space systems development is generally characterized by a broad range of requirements, procurement in small numbers, significant changes in the applicable technologies over the development cycle, launch costs that are a significant fraction of the total costs, and the inability to access the space environment to effect repairs or upgrades. As a consequence, space systems are generally unique, robust, and reliable, with minimal mass and power.

This chapter discusses the process and techniques utilized to concurrently develop the different subsystems of a sophisticated space system. Concurrent development poses significant engineering and management challenges. With the goal being the overall success of the system, the essence of systems engineering is compromise and trade-offs that are embodied in the often-quoted saying "The best is the enemy of the good" (Voltaire, 1764).

1.2 Fundamentals of Systems Engineering

A system is a collection of components that interact and work synergistically to satisfy specified needs or requirements. It is important to recognize that most systems do not exist independently and are a component of a supersystem or supersystems. Often this imposes unappreciated constraints on the development process. A taxonomy of the components of a system is as follows:

- A *system* is the group of segments deployed and operated to satisfy a mission objective, and usually consists of hardware, software, and operating personnel. Examples include the Global Positioning System, Defense Satellite Communication System, and NASA's Tracking and Data Relay Satellite System.
- A *segment* is a group of elements that together constitutes a major component or major function of a system. Examples include a constellation of satellites, a user community, and a tracking and control network.
- An *element* is a group of subsystems that together perform an important function, several of which integrate to form a segment. Examples are spacecraft of a constellation, a subset of users, and a tracking station of a network.
- A *subsystem* is a group of components that performs a function within an element, segment, or system. Examples include the attitude determination and control subsystem, the power subsystem, and the thermal control subsystem. Note that these are often called systems by their developers.
- An *assembly* is a group of items that performs a function that supports a subsystem. Examples relating to the attitude determination subsystem would be the solar attitude determination assembly, the magnetic field attitude assembly, the inertial measurement system assembly, and the attitude determination and control processor assembly.
- A *subassembly* is a functional subdivision of an assembly. Examples relating to the attitude determination and control system include the solar attitude detectors, solar attitude electronics, magnetometer sensors, and magnetometer electronics.
- A *part* is a hardware item that cannot be logically subdivided. Examples include a chip, a diode, a housing, and an attachment bolt.

The terms *segment* and *element* are employed less frequently and are usually not used for a single spacecraft or when there is a limited number of control stations or a limited number of direct users. In practice, the components of this taxonomy are not strictly followed. For example, most spacecraft subsystems are so complex that they are generally considered systems in their own right; e.g., attitude system and thermal control system.

The objective of systems engineering is to design, fabricate, operate, and dispose of a system that satisfies a set of specifications in a manner that is cost effective and conforms to a predetermined schedule with a defined acceptable risk. Consequently, the three major measures are performance, risk, and cost, with the latter including schedule—although sometimes schedule is considered separately. The systems engineering dilemma can be characterized by the fact that any of the two measures can be stipulated independently with the third dependent on them, as exemplified by the oft-quoted systems engineering dilemma:

- To reduce cost at constant risk, the performance must be reduced.
- To reduce cost at constant performance, the risk must be increased.
- To increase performance at constant cost, the risk must be increased.
- To increase performance at constant risk, the cost must be increased.
- To reduce risk at constant cost, the performance must be reduced.
- To reduce risk at constant performance, the cost must be increased.

Systems engineering addresses the concurrent development and operation of a system that must satisfy specific requirements, by recognizing the interrelationships among the components and the processes through which the system is developed and operated. The approach consists of identifying and quantifying the mission requirements, developing alternative concepts, carrying out design studies, identifying the best design, fabricating the design, integrating the subsystems and system, verifying that the system will meet the requirements, and assessing system performance. Systems engineering is an advisory function in that it provides to management an insight into the status of the system development and provides options for management decisions when these become advantageous or necessary. The International Council of Systems Engineering (INCOSE) defines systems engineering as follows:

- "Systems Engineering is an interdisciplinary approach and means to enable the realization of successful systems.
- It focuses on defining customer needs and required functionality early in the development cycle, documenting requirements, then proceeding with design synthesis and system validation while considering the complete problem: Operations, Performance, Test, Manufacturing, Cost & Schedule, Training & Support, Disposal.
- Systems Engineering integrates all the disciplines and specialty groups into a team effort forming a structured development process that proceeds from concept to production to operation.
- Systems Engineering addresses the business and technical needs of all customers with the goal of providing a quality product that meets the user needs."

Systems engineering is a systematic integrated approach to the concurrent development of products and services. As a result, systems engineering is an interdisciplinary, phased approach to develop and verify a system that meets a set of requirements or specifications. Inherent in systems engineering is that

- System level requirements are clearly established.
- The interval over which the system is to be developed and operated is established.

- The design that best satisfies the requirements may be complex, risky, or not obvious.
- The present state of a development is the integrated sum of the past and the basis of the future.

Systems engineering is evolving into a rigorous discipline, as evidenced by the ever-evolving standards published by a variety of organizations. As of this writing, several engineering standards have evolved from MIL-STD 499B, which was widely publicized but not officially released. Current standards are authored by the Electronic Industries Alliance (EIA), Institute of Electrical and Electronics Engineers (IEEE), European Cooperation for Space Standardization (ECSS), and the International Standards Organization (ISO). In addition, several other systems engineering standards are under development. These standards have attempted to focus on the processes and their related activities, avoiding specific approaches. Benefits associated with good systems engineering practices include:

- Reduced development time
- Improved satisfaction of requirements
- Reduced total life-cycle costs
- Reduced schedule
- Enhanced system quality, robustness, and reliability
- Reduced risks
- Enhanced ability to maintain and upgrade the system

An example of a systems engineering standard is the interim standard EIA/IS-731 systems engineering capability model (Electronic Industries Alliance, 1999). A *capabilities model* is a framework for designing and evaluating processes. Within this capabilities model, a hierarchical model of categories and focus areas is defined where each focus area consists of a set of practices and deliverables. The categories for EIA/IS-731 are technical (technical aspects involved in producing the deliverable), management (planning, control, and information management required to direct the project), and environment (the supporting infrastructure). The three capabilities and their nineteen focus areas are illustrated in table 1.1.

1.3 Concepts in Systems Engineering

Several concepts important to systems engineering follow below. These include functional analysis, verification and validation, technology readiness levels, mass margin, and trade analyses.

1.3.1 Functional Analysis

Functional analysis is one of the first steps to be carried out after the mission or system requirements have been determined. It is a top-down hierarchical decomposition in which all the functions and their interrelationships necessary to satisfy the mission requirements are identified. This provides insight into the system, its subsystems, and their interrelationships from which system and subsystem level requirements can be identified. Care must be taken to avoid specifying how the functions are to be carried out in favor of what functions are required.

Table 1.1 EIA/IS 731 systems engineering capability model with categories and focus areas

Systems Engineering Technical Category
Define stakeholder and system level requirements
Define technical problem
Define solsution
Assess and select
Integrate system
Verify system
Validate system
Systems Engineering Management Category
Plan and organize
Monitor and control
Integrate disciplines
Coordinate with suppliers
Manage risk
Manage data
Manage configurations
Ensure quality
Systems Engineering Environment Category
Define and improve the systems engineering process
Manage competency
Manage technology
Manage systems engineering support environment

1.3.2 Verification and Validation (V & V)

An important element of systems engineering is the verification and validation of the system prior to operation. *Verification* involves proving that the system satisfies the system level requirements; essentially, demonstrating that the system is built as specified. This can be accomplished by test, analysis, demonstration, and inspection. *Validation* consists of proof that the system accomplishes its intended purpose. It is generally more difficult to validate a system than to verify it. Validation can only occur at the system level, whereas verification is accomplished at the system and subsystem levels.

1.3.3 Technology Readiness Levels (TRLs)

One means of estimating the inherent risk in the development of a system is to identify the technology readiness level, that is, the maturity of each of the subsystems or their components. Categories of technology readiness levels from DOD 500.2R (Department of Defense, 2001) are provided in table 1.2. It is desirable not to start a large-scale development until critical items are at least at TRL four or higher.

1.3.4 Mass Margin

A critically important parameter for a space system is the spacecraft mass and the margin allowed for growth at the beginning of the project. Mass is the critical characteristic in

Table 1.2 Technology readiness levels (TRLs) and their definitions [from DOD 5000.2-R]

Technology Readiness Level	Description
1. Basic principles observed and reported	Lowest level of technology readiness. Scientific research begins to be translated into technology's basic properties.
2. Technology concept and/or application formulated	Invention begins. Once basic principles are observed, practical applications can be invented. The application is speculative and there is no proof or detailed analysis to support the assumption. Examples are still limited to paper studies.
3. Analytical and experimental critical function and/or characteristic proof of concept	Active research and development is initiated. This includes analytical studies and laboratory studies to physically validate analytical predictions of separate elements of the technology. Examples include components that are not yet integrated or representative.
4. Component and/or breadboard validation in laboratory environment	Basic technological components are integrated to establish that the pieces will work together. This is relatively "low fidelity" compared to the eventual system. Examples include integration of "ad hoc" hardware in a laboratory.
5. Component and/or breadboard validation in relevant environment	Fidelity of breadboard technology increases significantly. The basic technological components are integrated with reasonably realistic supporting elements so that the technology can be tested in simulated environment. Examples include "high fidelity" laboratory integration of components.
6. System/subsystem model or prototype demonstration in a relevant environment	Representative model or prototype system, which is well beyond the breadboard tested for level 5, is tested in a relevant environment. Represents a major step up in a technology's demonstrated readiness. Examples include testing a prototype in a high-fidelity laboratory environment or in simulated operational environment.
7. System prototype demonstration in an operational environment	Prototype near or at planned operational system. Represents a major step up from level 6, requiring the demonstration of an actual system prototype in an operational environment. Examples include testing the prototype in a test bed aircraft.
8. Actual system completed and qualified through test and demonstration	Technology has been proven to work in its final form and under expected conditions. In almost all cases, this level represents the end of true system development. Examples include developmental test and evaluation of the system in its intended system to determine if it meets design specifications.
9. Actual system proven through successful mission operations.	Actual application of the technology in its final form and under mission conditions, such as those encountered in operational test and evaluation. Examples include using the system under operational mission conditions.

the selection of many of the subsystems, especially the launch vehicle, the propulsion system, and the structure. As a result, there is typically a hard upper limit to mass. Violating the limit can have serious consequences, such as reducing the payload mass, modifying the trajectory, redesigning and rebuilding the structure, and perhaps changing

Table 1.3 Mass margin recommendations

Maturity	Recommended Growth Factor
Off the shelf or measured	1.05
Minor modifications of an existing device	1.07
Modifications of an existing device	1.10
New design, mass calculated	1.15
New design, thoughtful mass estimate	1.20
New design, uncertainty in mass estimate	1.30

the launch vehicle. There is a tendency for the mass of a system to increase during the development process, as discussed in chapter 8. This is due to uncertainty in the design and to the assumption that the mass of some items such as fasteners and harness components is allocated elsewhere. It is important that a hierarchical mass table be developed early in the development and be maintained under the control of the systems engineer. Table 1.3 gives recommended mass margins based on the maturity of the device or its design.

1.3.5 Trade Analyses

Trade analyses or *studies* are a critical element of systems engineering, employed to optimize performance at acceptable cost and risk with a documented rationale for the decisions. Trade studies are necessary when the system is complex and there is more than one selection criterion. It is important that the procedures used for the trade studies are consistent, although not necessarily identical, throughout the project. Sometimes it is sufficient to justify a decision by comparing the advantages and disadvantages of several alternatives. Another approach is to rank each alternative by each selection criterion and select the one that ranks higher overall.

A more formal procedure can be employed if the selection criteria can be quantified. One such approach is identified in figure 1.1. First the selection criteria, such as reliability, schedule, cost, power, mass, volume, complexity, safety, maintainability, accuracy, schedule, risk, familiarity, existence of infrastructure facilities, and equipment are identified. Next, relative weights for each criterion are selected to differentiate their relative importance. Utility functions are established that range from zero to one, with the least desirable attribute assigned a utility of zero and the most desirable attribute assigned a utility of one; intermediate attributes are assigned intermediate values. Then, the various alternatives are identified and a quantitative measure is determined for each by summing the weighted utilities for each to give relative utilities or costs. Finally, it is important to investigate the sensitivities of the outcome to small changes in the selection criteria, the weights, and the utility functions, prior to selecting the preferred configuration. By varying the utility functions, the effects of changes in the specifications can be evaluated in a consistent manner as sometimes small changes in specifications can provide significant benefits to enhance the overall development.

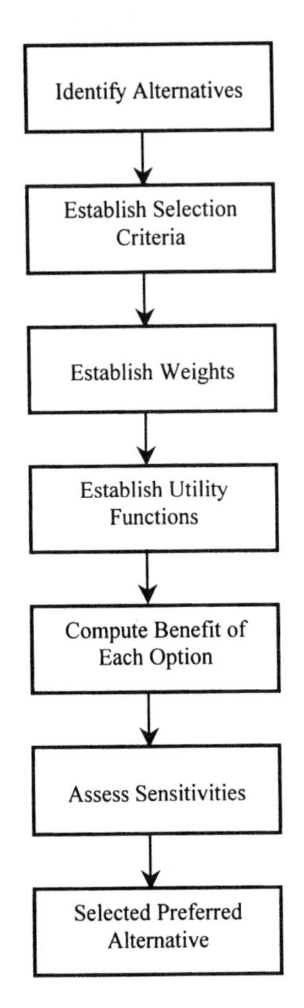

Figure 1.1 Trade analysis procedure.

1.4 Project Development Process

The system life-cycle model is a phased process involving the evolution of the system from requirements to development and operation and phase out. Decomposing the life cycle into phases organizes the project into more manageable components. This also provides management with improved visibility and a better quantitative assessment of progress. Each phase of a life-cycle model should have well-defined entrance and exit criteria that must be satisfied before it can be initiated or deemed complete. Life-cycle models vary by industry and complexity and magnitude of the undertaking. Inherent in each phase of the life cycle are four approaches that are iteratively applied to achieve the ultimate goals: *formulation*, *evaluation*, *approval*, and *implementation*, where

- *Formulation* results in a plan or concept to achieve a product.
- *Evaluation* provides an independent assessment of the ability to meet technical and/or programmatic objectives.
- *Approval* confirms the plan, indicating that a component of the project is ready to proceed to implementation.
- *Implementation* produces the desired product.
- *Evaluation* assures that the product is satisfactory and the phase completed.

Inherent in the life-cycle model are the following tasks:

- Identification and verification of system-level requirements
- Translation of system-level requirements into subsystem requirements or specifications
- Translation of subsystem requirements and specifications into subsystems that can be fabricated, integrated, tested, repaired, and have their performance assessed (it is critical that these elements be designed into the subsystem)
- Integration and test of the subsystems
- Integration of the subsystems into the system
- Test of the system, including an "end-to-end" test
- Integration of the space segment into the launch vehicle
- Operation of the system
- Post-mission evaluation
- Phase-out of the system

The life-cycle model used for development of space instrumentation and space systems generally consists of six phases:

- Pre-phase A—Advanced studies
  Identify new programs and projects
- Phase A—Conceptual design
  Conceptualize a system to satisfy needs or requirements
- Phase B—Definition
  Define the baseline system to detailed requirements
- Phase C—Design
  Carry out a detailed design
- Phase D—Development
  Build and test the system
- Phase E—Operations
  Operate and dispose of the system

This development approach can be viewed as a compromise between the spiral and the waterfall development methodologies. *Spiral development* is characterized by repeated iterations of the development cycle in which each cycle builds on the prior cycle (Boehm and Hansen, 2001). This is in contrast to the *waterfall development* approach which is characterized by a single linear cycle with *a priori* goals for each phase of development. The waterfall development approach can be defined in terms of phases of design, review, implement, and test. The spiral development would then be defined in terms of repeated cycles of design, review, implement, and test, with each cycle producing a prototype of an increasing level of complexity and functionality.

The duration of each phase to achieve a schedule of between 24 and 36 months for a small to medium satellite project is identified in table 1.4. These six phases, often utilized for space system development, are described next in more detail. For each phase, the purpose, major activities and their products, information baselines, and control gates are identified. These phases constitute a slightly modified version of the project life-cycle given in NASA SP-610S (NASA Headquarters, June 1995).

The *pre-phase A*—advanced studies—product is a set of mission goals and one or more concepts that can satisfy the goals. Approval to progress to the next stage is dependent on the goals, cost, and risk combining to offer a worthwhile opportunity

Table 1.4 Project phase schedule

Phase	Duration (Months)	Staffing (Staff–Years)
Pre-phase A—Advanced studies		
Phase A—Preliminary analysis	2–3	1–2
Phase B—Definition	3–4	3–4
Phase C—Design	6–14	*
Phase D—Development	13–27	*
Phase E—Operations	–	*
TOTAL (to delivery to launch site)	24–48	

*Dependent on specific mission requirements.

relative to other competing missions. Advanced studies are primarily carried out by organizations to develop program plans. Guidelines for this phase are given in table 1.5.

The *phase A*—conceptual design phase—product is the description of a mission that articulates the mission objectives, illustrates the benefits of satisfying the objectives, shows that the mission can be carried out with one or more credible concepts, and determines estimates of cost, schedule, and acceptable risks. This is accomplished by justifying the mission objectives; carrying out a conceptual design of the system and each subsystem; identifying launch dates, launch windows, and trajectories; describing

Table 1.5 Pre-phase A advanced studies guidelines

Purpose

To produce a broad spectrum of ideas and alternatives for missions from which new programs/projects can be selected

Major Activities and their Products

Identify missions consistent with charter
Identify and involve users
Perform preliminary evaluations of possible missions
Prepare program/project proposals, which include:

- Mission justification and objectives
- Possible operations concepts
- Possible system architectures
- Cost, schedule, and risk estimates

Develop master plans for existing program areas

Information Baselined

Program Master Plan

Control Gates

Mission concept review
Informal proposal reviews

Adapted from NASA SP-610S.

Table 1.6 Phase A conceptual design guidelines

Purpose

To determine the feasibility and desirability of a suggested new system, especially in preparation for the seeking of funding

Major Activities and their Products

Prepare mission needs statement
Identify project and system constraints
Identify payloads
Carry out a functional analysis
Develop top-level requirements
Develop subsystem-level requirements
Identify alternative operations and logistics concepts
Consider alternative design concepts, including:

- Feasibility and risk studies
- Cost and schedule estimates
- Advanced technology requirements

Identify system and subsystem characteristics
Identify top-level work breakdown structure
Demonstrate that credible mission concept(s) exist

Information Baselined

Program plan that identifies

- Benefits of the mission
- That (a) credible design(s) exists to carry out the mission
- Schedule and cost

Control Gates

Mission definition review
Conceptual design review
Conceptual non-advocate review
Conceptual program/project approval review

Adapted from NASA SP-610S.

a mission scenario; identifying make or buy decisions; defining necessary advanced technology development projects; determining a schedule and budget; and identifying risks and risk mitigation activities. A phase A study is generally employed to respond to a solicitation from a potential sponsor, such as a NASA announcement of opportunity or the development of an unsolicited proposal. Guidelines for this phase are given in table 1.6.

The *phase B*—definition phase—product is a description of a mission in sufficient detail to have a higher level of confidence in the projected performance, schedule, cost, and risks than was achieved in phase A. The detail must be sufficient so that a decision can be made to proceed to development. This is accomplished by sufficient analysis, design, studies, and limited prototyping of critical subsystems. Guidelines for this phase are given in table 1.7.

The *phase C*—design phase—product is the specification of a system in sufficient detail from which the system can be subsequently fabricated, integrated, and tested.

Table 1.7 Phase B definition guidelines

Purpose

To define the project in enough detail to establish an initial baseline capable of meeting mission needs so that a commitment to build can be made

Major Activities and their Products

Reaffirm the mission needs statement
Prepare a systems engineering management plan
Prepare a risk management plan
Prepare a configuration management plan
Prepare engineering specialty program plans
Restate mission needs as functional requirements
Reaffirm payloads
Reaffirm and enhance the functional analysis
Establish initial system requirements and verification requirements matrix
Perform and archive trade studies
Select a baseline design solution and a concept of operations
Define internal and external interface requirements
(Repeat the process of successive refinement to obtain specifications and drawings, verification plans, and interface documents to sufficient detail as appropriate)
Identify risks and risk mitigation plans
Define a more detailed work breakdown structure
Define verification approach end policies
Identify integrated logistics support requirements
Determine estimates of technical resources, life-cycle cost, and schedule
Initiate advanced technology developments
Revise and publish a project plan including schedule and cost

Information Baselined

System requirements and verification requirements matrix
System architecture and work breakdown structure
Concept of operations
Initial specifications at all levels
Project plans, including schedule, resources, acquisition strategies, and risk mitigation

Control Gates

Non-advocate review
Program/Project approval review
System requirements review(s)
System definition review
System-level design review
Lower-level design reviews
Safety review(s)

Adapted from NASA SP-610S.

At the conclusion of this phase, evidence should exist that all subsystems and the system will perform as expected. This supporting evidence can be obtained from studies, simulations, breadboards, brassboards for radio frequency devices, prior space flight experience, experiments, and tests. Guidelines for this phase are given in table 1.8.

Table 1.8 Phase C design guidelines

Purpose

To complete the detailed design of the system (and its associated subsystems, including its operations systems)

Major Activities and their Products

Add remaining lower-level design specifications to the system architecture
Refine requirements documents
Refine verification plans
Prepare interface documents
(Repeat the process of successive refinement to get "build-to" specifications and drawings, verification plans, and interface documents at all levels)
Augment baselined documents to reflect the growing maturity of the system:

- System architecture
- Verification requirements matrix
- Work breakdown structure
- Project plans

Monitor project progress against project plans
Develop the system integration plan and the system operation plan
Perform and archive trade studies
Complete manufacturing plan
Develop the end-to-end information system design
Refine the integrated logistics support plan
Identify opportunities for pre-planned product improvement
Confirm science payload selection

Information Baselined

All remaining lower-level requirements and designs, including traceability to higher levels
"Build-to" specifications at all levels

Control Gates

Subsystem (and lower level) preliminary and critical design reviews
System-level preliminary and critical design review

Adapted from NASA SP-610S.

The *phase D*—development phase or fabrication, integration, test, and certification phase—product is a system that has been fabricated, integrated, and tested to verify that the system and subsystems satisfy their specifications, and validated to achieve the overall objectives. Guidelines for this phase are given in table 1.9.

The *phase E*—deployment and operation—product is the operation of the system after deployment and its eventual decommissioning. Guidelines for this phase are given in table 1.10.

1.5 Management of the Development of Space Systems

This section describes the different procedures and tools that can be used to effectively and efficiently manage the development of complex space systems. The topics include the systems engineering management plan, program reviews, interface control documents,

Table 1.9 Phase D development guidelines

Purpose

To build the subsystems and integrate them to create the system, meanwhile developing confidence that will be met the system requirements through testing

Major Activities and their Products

Fabricate (or code) the parts (i.e., the lowest-level items in the system architecture)
Integrate those items according to the integration plan and perform verifications, yielding verified components and subsystems
(Repeat the process of successive integration to get a verified system)
Develop verification procedures at all levels
Perform system qualification verification(s)
Perform system acceptance verification(s)
Monitor project progress against project plans
Archive documentation for verifications performed
Audit "as-built" configurations
Document lessons learned
Prepare operator's manuals
Prepare maintenance manuals
Train initial system operators and maintainers
Finalize and implement integrated logistics support plan
Perform operational verification(s)

Information Baselined

"As-built" and "as-deployed" configuration data
Integrated logistics support plan
Command sequences for end-to-end command and telemetry validation and ground data processing
Operator's manuals
Maintenance manuals

Control Gates

Test readiness reviews (at all levels)
System acceptance review
Flight readiness review(s)
System functional and physical configuration audits
Operational readiness review
Safety reviews

Adapted from NASA SP-610S.

configuration management, work breakdown structure, scheduling, cost estimating, earned value management, and risk management.

1.5.1 Systems Engineering Management Plan

The *systems engineering management plan* (SEMP), sometimes called the *technical management plan*, is a document that describes the technical management of a project. While a generic plan may exist in an organization, it is important that the SEMP be tailored to the specific project. The plan generally consists of several sections, as illustrated in table 1.11 which is adapted from NASA Headquarters (1995); see also

Table 1.10 Phase E operations guidelines

Purpose
To operate the system in an expeditious manner and then to terminate operation in a responsible manner
Major Activities and their Products
Integrate with launch vehicle(s) and launch, perform orbit insertion, etc., to achieve a deployed system
Train replacement operators and maintainers
Conduct the mission(s)
Maintain and upgrade the system
Dispose of the system and supporting processes
Document lessons learned
Information Baselined
Mission outcomes, such as:
• Engineering data on system, subsystem and materials performance
• Science data returned
o High resolution photos from orbit
• Accomplishment records ("firsts")
o Discovery of the Van Allen belts
o Discovery of volcanoes on Io
Operations and maintenance logs
Problem/failure reports
Decommissioning procedures
Control Gates
Regular system operations readiness reviews
System upgrade reviews
Safety reviews
Decommissioning review

Adapted from NASA SP-610S.

Defense Systems Management College (1990). The *scope* section identifies the applicability of the document, its purpose, overall responsibility and authority for systems engineering management, and the document change process. The *technical planning and control* section identifies organizational responsibilities and authority; control processes for engineering data and documentation; and program assurance procedures. The *systems engineering process* section describes the procedures to implement the processes described in the previous section, including trade studies, modeling techniques, specification structures, and management of risk and cost. The section titled *engineering specialty integration* describes integration of the specialty engineering into the process to assure a cost-effective product.

1.5.2 Program Reviews

The development process described above involves a wide variety of personnel working concurrently to transform system-level requirements into a successful system. Reviews at the system and subsystem level are an effective means of communication, assuring

Table 1.11 Systems engineering management plan format

Title (specific to the project)

- Document Number
- Date
- Revision number

Table of Contents

1.0 Scope

This section should describe

- Applicability of the document
- Purpose of the document
- Relationship to other plans
- Overall responsibility and authority
- Document change process

2.0 References

This section should give applicable reference documents and standards

3.0 Technical Planning and Control

This section should describe

- Detailed responsibilities and authorities
- Baseline control process
- Change control process
- Interface control process
- Contracted (or subcontracted) engineering process
- Data control process
- Documentation control process
- Make-or-buy control process
- Parts, materials, and process control
- Quality control
- Safety control
- Contamination control
- Electromagnetic interference and electromagnetic compatibility (EMI/EMC) process
- Technical performance measurement process
- Control gates
- Internal technical reviews
- Integration control
- Verification control
- Validation control

4.0 Systems Engineering Process

This section should describe the

- System decomposition process
- System decomposition format
- System definition process
- System analysis and design process
- Requirements allocation process
- Trade study process
- System integration process
- System verification process
- System qualification process
- System acceptance process
- System validation process

Table 1.11 (Continued)

- Risk management process
- Life-cycle cost management process
- Specification and drawing structure
- Configuration management process
- Data management process
- Use of mathematical models
- Use of simulations
- Tools to be used

5.0 Engineering Specialty Integration

This section should contain, as needed, the project's approach to

- Concurrent engineering
- The activity phasing of specialty disciplines
- The participation of specialty disciplines
- The involvement of specialty disciplines
- The role and responsibility of specialty disciplines
- The participation of specialty disciplines in system decomposition and definition
- The role of specialty disciplines in verification and validation
- Reliability
- Maintainability
- Quality assurance
- Integrated logistics
- Human engineering
- Safety
- Producibility
- Survivability/vulnerability
- Environmental assessment
- Launch approval

6.0 Appendices

Adapted from NASA Headquarters, *Systems Engineering Handbook*, SP-610S, June 1995, edited 20 June 1999.

that the requirements are met and signifying that a milestone has been achieved. The benefits of a review are that it

- Provides an independent and critical assessment
- Identifies issues
- Assures that interfaces are well understood
- Promotes communication between participants
- Formalizes and documents progress
- Signifies that a milestone has been met
- Provides incentives to the participants

A successful review is not necessarily one that uncovers no issues. The success of a review can only be judged by the capabilities and independence of the panel and the subjective thoroughness of the review. To help assure a successful review, it is important that the

- Purpose of the review is understood.
- Date is known far enough in advance for the presenters and reviewers to prepare.
- Agenda allows sufficient time.

- Documentation is clear, concise, consistent, and distributed far enough in advance.
- Presenters are prepared.
- Review board is knowledgeable and independent.
- Chairperson controls the proceedings.
- Action items are identified but solutions not attempted.
- Forms are available to document potential action items in writing.
- Proceedings and action items are documented in writing.
- Meeting concludes with review of action items with responsible individual named and a date for closure.
- Dry run is held.

Reviews can be either informal or formal; small subsystem reviews tend to be less formal whereas system reviews tend to be formal. For the presenter, the review is an opportunity to demonstrate proficiency, get well-deserved recognition, and have peers approve of the work product. For management, a successful review is affirmation that the development is on course and that a milestone has been achieved. A variety of reviews may be undertaken for a space system.

A *red team review* is generally utilized to review a project that will be evaluated competitively. Thus it addresses a broad range of topics including needs of the potential sponsor, the goals and objectives of the mission, potential competitors, winning strategy, mission requirements, performance, risk, schedule, and budget.

Needs reviews are generally held if there are uncertainties in the needs, when changes have been proposed, or when it is necessary to promulgate the needs to a diverse audience. Generally, the needs are well enough understood so that needs reviews are bypassed in terms of requirements reviews.

Requirements reviews can resolve, finalize, and formalize the requirements of a system or its subsystems. It is important that there be no uncertainty in the requirements since they are the seminal point for the development. In a requirements review, it is important to identify the needs, including operational aspects if appropriate. The requirements, their sensitivities, and, if applicable, goals must be clearly and unambiguously defined in a manner that provides the maximum latitude in how the requirements may be satisfied. Goals are sometimes defined that are objectives above and beyond the requirements, which would enhance the system if attainable at no increase in cost, schedule, or risk. Critical is the development of a *requirements traceability matrix* that hierarchically relates system-level requirements to the sequentially derived requirements at the segment, element, subsystem, and assembly levels.

Design reviews are held at the subsystem and system level to assure that the design satisfies the engineering requirements. A typical agenda for a system-level design review is given in table 1.12. It is important that the purpose and the rules of order of the review be understood by all and the status of action items from prior reviews be addressed. Often it is convenient to partition a spacecraft into the primary payloads and the supporting subsystems. Primary payloads are those that are required to directly carry out the mission. The supporting subsystems or spacecraft bus is synthesized from subsystems that support the primary payloads as identified in table 1.12. The last task of the review is to identify each action item with a completion date and a responsible individual. Costs should be explicitly excluded in a design review.

Conceptual design reviews (CoDRs) are held as part of a phase A conceptual design, initially at the subsystem and then at the system level. The objectives are to assure that

Table 1.12 Typical agenda for a space system design review

Introduction

- Purpose of the review
- How the review is to be conducted
- Resolution of action items from previous reviews

Mission

- Mission purpose
- System-level requirements
- Description of the overall mission
- Launch description
- Reliability and quality assurance plan

Payloads

- Payload 1
- Payload 2
- ...
- Payload N

Spacecraft review

- Tracking, guidance, and orbit determination
- Mechanical and structure
- Telecommunications
- Command and data handling
- Flight processors and software
- Attitude determination and control
- Thermal subsystem
- Power subsystem

Ground support equipment
Integration and test
Electromagnetic compatibility
Safety
Launch site operations
Mission operations
Risks and risk mitigation plan
Documentation
Action items identified

the proper requirements are identified and that the overall design concept can satisfy the requirements.

Preliminary design reviews (PDRs) are generally held as part of phase C, definition, initially at the subsystem and then at the system level. Their purpose is to confirm that the approach for each subsystem and system is ready to proceed into the detailed design phase. These reviews are conducted when the system definition effort has proceeded to the point where a preliminary design exists for each subsystem. A PDR verifies that the

- Design approach for each subsystem or system satisfies the subsystem or system requirements.
- Design is validated by preliminary engineering results.

- Risks have been identified and mitigated to an acceptable level.
- Preliminary integration and test plans are completed.

Critical design reviews (CDRs) are held as part of the phase C definition phase following the system-level preliminary design review. They are initially held at the subsystem and then at the system level. Their purpose is to demonstrate that the design is completed and the project is ready to proceed to phase D, development. A CDR at the subsystem or system level verifies that the

- Detail design for each subsystem exists.
- Subsystem designs satisfy the subsystem requirements and/or the system design satisfies the system requirements.
- Design is validated by engineering results.
- Design can be fabricated, integrated, and tested.
- Appropriate documentation exists to fabricate the system.
- Final integration and test plans are complete.
- Risks are acceptable.

Engineering design reviews (EDRs) are sometimes held for less complicated or complex systems in place of the multiple design reviews discussed above. The example system-level design review agenda given in table 1.11 is applicable here as well.

Fabrication feasibility reviews are held to assure the feasibility of the manufacture and assembly of the design. The basis of these reviews, which occur primarily at the subsystem level, is the detailed specifications and drawings that were developed during the critical design. The participants should be the engineers and designers responsible for the design and the engineers and craftsmen responsible for manufacture and assembly. The latter should not be seeing the design for the first time at this review. It is important that they have been consulted and involved since the conceptual design to help avoid potential manufacturing difficulties. If there has been good communication during the design phases, these reviews should be without difficulty and simply consist of turning the design over to manufacturing and resolving minor manufacturing details. These reviews are best held prior to the CDR.

Design release reviews are to assure that the fabrication drawings and specifications for a subsystem are complete, checked, and approved and that fabrication should begin. The importance of complete and accurate fabrication specification cannot be overemphasized. Changes during fabrication can be costly and delay delivery.

Integration readiness reviews are important, prior to integration of the subsystems and the integration of subsystems into the system. These reviews address the proper integration sequence and the qualification tests necessary to assure that the subsystem or system can be integrated without risk.

Test reviews assure that the subsystems or system is ready for test, test plans and procedures are appropriate and thorough, and that test facilities and test personnel are qualified. These reviews are used for major subsystem and system tests, where inadequate or improper testing could be costly or damaging to significant components of the system. At the systems level, this review should involve all the principals.

Acceptance reviews are held at the completion of all acceptance testing to assure that the deliverable (subsystem or system) is ready for delivery and flight. It certifies that the deliverable has met all the test criteria, test anomalies have been explained, criteria in the interface control documents (ICDs) have been satisfied, alignments and calibrations have been verified, deliverable documentation and software are completed, and the field operations plan is acceptable.

Flight operations readiness reviews are held to approve the launch and post-launch flight operations. These reviews include the mission controllers and representatives of the different ground and space assets that will participate. Of prime importance is that the mission will satisfy its objectives, the range facilities and personnel are prepared, and that the launch can be safely undertaken. A space mission can involve many disparate ground and space assets to satisfy its objectives, and the coordination of these facilities into a cohesive unit is a nontrivial task.

1.5.3 Interface Control Documents

An *interface control document* (ICD) is a formal document that identifies and controls the physical, functional, environmental, operational, and procedural interfaces between a device and subsystem or subsystem and its system. The specifications in the ICD follow from the system or subsystem specifications and are generally established in an *interface requirements document* (IRD) or *interface requirements specification* (IRS). The ICD is generally a component of the contract with the organization to produce the deliverable and is signed by the developing and procuring organizations. The ICD includes minute detail, identifying, for example, for an electronic device the amount of heat generated, operating temperature, magnitude and frequency of radio frequency emissions, connector type with each pin specified, fastener types and location, and so on. The format of a typical ICD is given in table 1.13.

1.5.4 Configuration Management

Configuration management is the procedure for providing a disciplined approach to identify, control, account, and verify the requirements, specifications, and implementation documentation to formally document the configuration of the products. Its purpose is to establish that the contractual obligations have been satisfied, permit replication of an existing design, and be able to effectively address in-orbit anomalies. Configuration management is generally implemented at the conclusion of the critical design review. It consists of configuration identification, configuration control by the configuration control board, configuration accounting, and configuration verification.

Configuration identification is carried out by formally identifying documentation that describes the project requirements and specifications such as drawings, processes, and material lists that are used to fabricate the system. An important feature is a hierarchical use of part and drawing numbers to assure fidelity in tracking and coordinating the documentation during subsequent changes.

Configuration control is the formal process used to establish the baseline configuration and control changes. Changes are controlled by the formal action of a configuration control board established to review and approve changes to hardware, software, and how the system will be operated. The process to assure configuration control is illustrated in figure 1.2. *Engineering change requests* (ECRs) are generally designated as class I or class II. A class I change by contractual agreement requires approval by the sponsoring

Table 1.13 Typical interface control document (ICD) format

1.0 Summary

1.1 Purpose

Purpose and objectives

1.2 Scope

Describes the scope of the interface control document

Identifies the more critical aspects of the interfaces

1.3 Definitions, Acronyms, and Abbreviations

1.4 Overview

Simplified block diagram describing physical interfaces

Brief description of the interface

2.0 Applicable Documents

Identify only documents (drawings, sketches, specifications, etc.) referenced in subsequent sections

3.0 Interface Definition

This section contains text, drawings, tables, etc. required to completely define the interface requirements

3.1 Physical/Mechanical Interface

Envelope, volume, mass, moments of inertia, center of mass, principal axes, etc.

3.2 Radio Frequency Interface

Frequencies, power, modulation, signal-to-noise ratio, electromagnetic compatibility, connectors, etc.

3.3 Command and Data Interface

Formats, data rates, real time and delayed commands, etc.

3.4 Structural Interface

Static, dynamic, vibration, acoustic, and shock loads and bending and torsional moments, allowed deflections and strains, etc.

3.5 Mechanical Interface

Positioning tolerances, special supports, adjustability, attachment details, etc.

3.6 Electrical Requirements

Voltage, current, power, peak power, EMI, connectors with pin assignments, etc.

3.7 Environmental Interface

Thermal flux, power, and operating and survival temperatures, etc.

Magnetic flux density, rate of change, etc.

Radiation total dose, type, flux density

3.8 Hydraulic/Pneumatic Interface

Requirements for vacuum or non-conventional fluids, such as cryogenics, high-purity or toxic gases, etc. The oil in the modulator tank is an example

3.9 Other Interfaces

4.0 Verification

4.1 Quality Assurance

Quality assurance requirements, ICD requirements verification matrix

4.2 Tests

Test constraints, facility requirements, inspection requirements

4.3 Shipping Preparation

Appendices

Delivery Schedule

Command Lists

Data Formats

Etc.

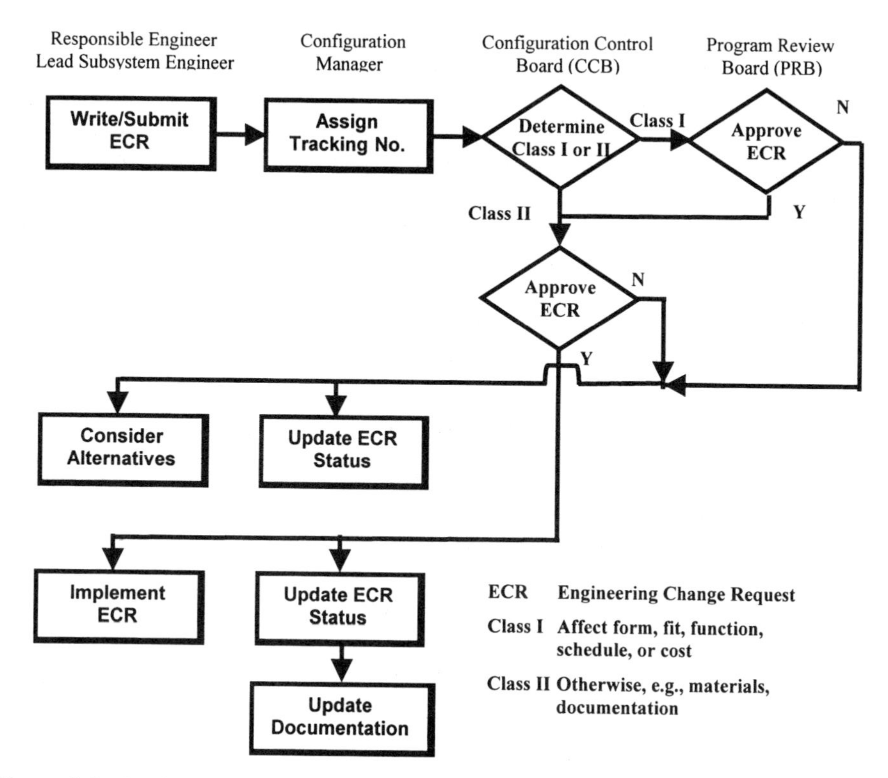

Figure 1.2 Configuration control process.

organization. Class I includes changes that affect performance, schedule, risk, or cost and generally result in additional documentation. A class II change is one that does not qualify as a class I change and generally results in a revision change to existing documentation.

The *configuration control board* (CCB) is generally populated by the lead subsystem engineers and chaired by the systems engineer. Representation of each of the subsystems on the CCB assures that a proposed change is communicated and that the potential impact on the other subsystems can be explicitly considered. The issues addressed in the CCB include

- Proposed change and justification
- Effect of not making the change
- Impact on design, performance, risk, schedule, and cost
- Impact on other subsystems
- Identification of affected documentation

Configuration accounting, under the responsibility of the configuration manager, has the responsibility to provide the single authoritative documentation of the system. Documentation, consisting of reports, parts lists, drawings, and specifications, is baselined after the design is frozen, generally at the conclusion of the critical design review, and is updated with an identifiable audit trail only after changes are approved by the configuration control board and project management. In principle, this means that modifications to the implementation documents must await completion of a formal change request, revision of the documentation, verification of the revised documentation, and final approval. An alternative that works well to reduce cost and schedule impacts is for the configuration manager to maintain approved changes by hand-written or computer-generated "red lined" modifications to the baseline documentation. This can eliminate a significant amount of the delay that may occur between approval by the configuration control board and project management and the formal implementation and verification of the changes.

Configuration verification insures that the documentation conforms to the system, that required changes have been incorporated into the contract, and that the system is fabricated and tested accordingly. Verification is carried out by reviews and audits.

1.5.5 Work Definition and Work Breakdown Structure

Work definition is the process of defining the specific tasks that are necessary to satisfy the system requirements. This necessitates identifying in a systematic manner all of the deliverables, the tasks necessary to produce the deliverables, and their interrelationships. Defining the work properly is critical to the success of a project. An omission or error may jeopardize the development if it is uncovered only after considerable resources have been expended. Work definition should be done top-down in a hierarchical manner to assure that a number of management functions can be supported, including

- Establishing clear project goals
- Planning and tracking the schedule
- Assessing technical performance
- Estimating cost and formulating and tracking budgets
- Identifying and leveling resources
- Defining statements of work for contracted efforts

A critical tool to define the work is the work breakdown structure (WBS). The *work breakdown structure* is a top-down hierarchical decomposition, breakdown, or family tree, of the work to be completed into successive levels with increasing detail. Each activity (also called work package, work element, element, or task) in a WBS should represent an identifiable work product traceable to one or more system-level requirements. Identifiable work products are composed of hardware, software, services, data, and facilities. The tasks within an activity may include functions such as requirements analysis, design engineering, materials, rework and retesting, software engineering and development, travel, computer support, systems engineering, program management, training, spares, and cost-saving efforts such as quality management. For each activity the following characteristics are identified:

- Activity identifier
- Dependencies between activities such as precedent and successor activities
- Products and milestones
- Duration
- Resources required
- Special facilities needed
- Significant subcontracts and purchases

On large projects, activities are sometimes collected into *work packages*. Consistency of the WBS structure across different products permits historical comparisons of costs and schedules to assist in making estimates on subsequent projects. The number of levels utilized in a WBS depends on the size and complexity of the undertaking and should be appropriate to the risks to be managed, resulting in manageable activities at the lowest level. A *manageable activity* is one in which the

- Product expected can be quantified.
- Completion can be determined.
- Milestones can be identified.
- Success can be measured.
- Resource requirements can be reasonably estimated.
- Duration can be reasonably estimated.

The number of levels is generally a compromise between management's desire to maintain visibility into schedule and costs and the additional cost of planning and reporting. The cost and preparation time increases exponentially for each layer added to the WBS. The level one activity consists of the system, or a subsystem of a very large system. Examples might be a spacecraft mission, spacecraft system, spacecraft bus, or spacecraft payload. Level two activities are the major subsystems that constitute the system. These include hardware and software development, aggregation of system-level services (such as test and evaluation, systems engineering, and program management), and data. Typically, level two activities are organized by subsystem. Level three activities are subordinate to level two activities and generally consist of smaller subsystems or assemblies, a type of service (for example, test and evaluation of a subsystem), or a type of data (for example, a test report). Lower levels follow the same pattern. A typical spacecraft development will generally employ a WBS of three levels with selected activities having one or two more levels as required. Level three of the WBS is the normal reporting level for external contractual information. It is helpful to construct a WBS dictionary that contains a description of the characteristics of each activity. Each activity should be numbered in such a manner that they are traceable across levels. For example, if the activity at level one is numbered by 10, 20, 30, then the numbers at level two traceable to activity 10 may be numbered by 10.1, 10.2, 10.3, and so on and activities at level three traceable to activity 10.1 may be numbered by 10.1.1, 10.1.2, 10.1.3. Each subcontractor effort is generally assigned to a single WBS activity. A generic WBS is given in figure 1.3.

To develop a WBS it is necessary to identify precedent or dependency relationships between activities. A *successor activity* is one that succeeds another activity. A *predecessor activity* is one that either entirely or partially precedes another activity.

There are three types of dependencies between predecessor and successor activities:

Mandatory dependencies are those that are inherent to the work to be done and are considered hard dependencies. As an example, an assembly cannot be integrated until the subassemblies are available, or a test cannot be undertaken until the test article is available.

Discretionary dependencies are those that are defined by management and are considered soft dependencies. They need to be introduced with care since they may preclude meeting or reducing cost and schedule.

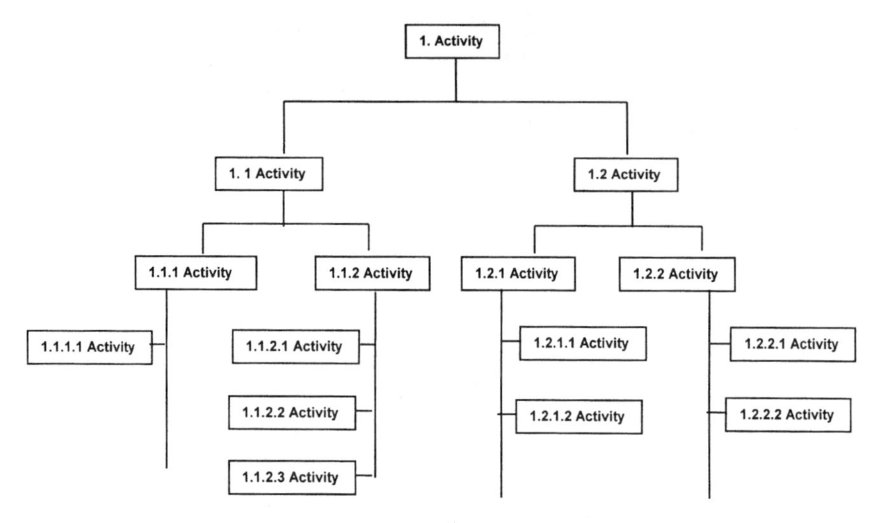

Figure 1.3 Sample Work Breakdown Structure (WBS).

External dependencies are those that are imposed by non-project activities. Examples may be the availability of an external facility or the delivery of a part from an external supplier.

For further information on constructing a WBS see MIL-HDBK-881 [Department of Defense, 1998) and SP-610S (NASA Headquarters, 1995). A representative work breakdown structure for a spacecraft development is illustrated in table 1.14.

1.5.6 Scheduling

Scheduling has the goal to organize a realistic sequence of activities that provides a basis for coordinating resources and monitoring project progress. Activities identified in the work breakdown structure represent the work to be performed. To complete the project on time requires scheduling the activities and identifying the resources required as a function of time. Scheduling generally involves identifying for each activity:

- Title, by its primary purpose or product
- Hierarchical number identifier
- Precedent activities
- Deliverables
- Milestones
- Resource required
- Equipment and facilities needed
- Purchases
- Duration

The benefits of scheduling are that it

- Promotes detailed and thoughtful planning
- Helps ensure all of the phases of the project have been addressed

Table 1.14 Representative work breakdown structure for spacecraft development

No.	Item
1.	Program management
1.01	Program manager
1.01	System engineer
1.01.1	Specifications
1.01.2	Interface control documents
1.01.3	Design reviews
1.01.4	Software engineering oversight
1.02	Program scientist
1.03	Financial accounting support
1.04	Programmatic documentation
1.05	Scheduling and status
1.06	Administrative support
1.07	Coordination with fabrication department
1.08	Security
2.	Reliability quality assurance
2.01	Reliability predictions
2.01	Failure modes effects and critically analysis
2.02	Subcontractor and vendor qualification
2.03	Safety
2.04	Reviews and documentation
3.	Structure
3.01	Vehicle mass properties list
3.01	Stress analysis
3.02	Launch vehicle interface
3.03	Deployment/separation mechanisms
3.04	Engineering mock-up
3.05	Test fixtures
3.06	Reviews and documentation
4.	Power subsystem
4.01	Load requirements
4.01	Solar arrays
4.02	Battery storage
4.03	Thermoelectric cells
4.04	Voltage conversion
4.05	Voltage regulation and monitoring
4.06	Test equipment
4.07	Reviews and documentation
5.	Thermal
5.01	Heat flow analysis
5.01	Active thermal management
5.02	Passive thermal management
5.03	Thermal blanket
5.04	Thermal vacuum test fixtures
5.05	Reviews and documentation
6.	RF Communications
6.01	Link requirements
6.01	Frequency allocation
6.02	Antennas
6.03	Receivers
6.04	Transmitters
6.05	Test equipment
6.06	Reviews and documentation
7.	Command and data handling
7.01	Command requirements
7.01	Telemetry requirements
7.02	Command processor
7.03	Software
7.04	Bulk-in test
7.05	Data storage
7.06	Power switching
7.07	Test equipment
7.08	Reviews and documentation
8.	Electromagnetic compatibility (EMC)
8.01	EMC environment requirements
8.01	Reviews and documentation
9.	Orbit/attitude control
9.01	Orbit/attitude analysis
9.01	Attitude determination
9.01.1	Sun detector
9.01.2	Magnetometer
9.01.3	Star camera
9.01.4	Horizon sensor
9.01.5	Gyros
9.02	Attitude adjustment
9.02.1	Reaction wheels
9.02.2	Magnetic torquers
9.03	Orbit determination
9.04	Orbit adjustment
9.05	Test equipment
9.06	Reviews and documentation
10.	Propulsion
10.01	Control
10.02	Thrusters
10.03	Test equipment
10.04	Reviews and documentation
11.	Experiments
11.01	Theoretical modeling
11.02	Sensor development
11.03	Experiment planning
11.04	Experiment interface
12.	Major subcontract (if appropriate)
13.	Integration and test
13.01	Harness
13.02	Test equipment
13.03	Environmental tests
14.	Ground support equipment (GSE)
14.01	Satellite tracking facility
14.02	GSE Computer and peripherals
14.03	Special-purpose hardware
14.04	Software
14.05	Encryption/decryption
14.06	Experiment interfaces
14.07	Mission control center

(Continued)

Table 1.14 (Continued)

14.08	Test equipment	15.02	Spacecraft checkout
14.09	Reviews and documentation	15.03	Range support coordination
14.10	User documentation	16.	Missions operations
15.	Launch field operations	16.01	Mission planning
15.01	Spacecraft/launch vehicle integration	16.02	Spacecraft operations
		16.03	Post launch data analysis

- Enhances visibility of major events and milestones
- Establishes the dates and duration for the execution of the project
- Establishes the basis for time-phased performance measurement
- Helps identify and manage risk
- Identifies necessary resources, material, facilities, and so on
- Identifies and assigns responsibilities
- Enhances communications
- Provides critical information for proactive decision making

Scheduling is initiated by addressing activities at the highest levels and developing a *master schedule*, the highest-level schedule, which shows the schedule for the entire project. From the master schedule, it is then possible to develop detailed hierarchical schedules for activities at each level of the WBS. It is generally necessary to iterate and adjust the schedules at all levels to assure that personnel resources, equipment, and facilities are available and that the planned delivery date can be met. Adjusting the schedule to assure that the required personnel resources match what is available is known as *resource leveling*. Resource leveling is based on the presumption that personnel costs may be reduced by eliminating large fluctuations in the personnel resources required.

As activities are completed, either earlier or later than planned, the schedule will need to be revised, consistent with constraints on personnel, costs, equipment, and facilities, with the revisions being shown on the schedule. It is important to maintain original and revised dates on a schedule until a formal replanning effort is undertaken. Having revised dates provides important information, as is illustrated in the extreme case when the completion date of an activity is "slipping a day per day."

Three parameters are important in scheduling: *free float*, *path float*, and the *critical path*:

- *Free float* is the length of time that an activity can be delayed without causing delay to any other activity.
- *Path float* is the length of time that interrelated or dependent activities can be delayed without affecting the delivery date of the final deliverable.
- *Critical path* is the series of dependent activities for which the path float is zero or near zero so that delay of any activity would affect the delivery date—unless there are modifications or "work arounds" to the schedule.

Several techniques available to systemize scheduling are discussed in turn: these include the Gantt or bar chart, milestone chart, combination of the two, and network diagrams.

1.5.6.1 Gantt or Bar Charts

A *Gantt* or *bar chart* is a horizontal bar chart developed as a production control tool by Henry L. Gantt in 1917. It is constructed with the horizontal axis representing the calendar times of the activities and the vertical axis representing different activities. Time can be in hours, days, weeks, or months. In assessing status, a vertical line is used to represent the report date, progress can be represented by darkened or colored horizontal bars, and important events indicated by tick marks. Arrows between activities can show precedents, but this does not clearly illustrate how delay of one activity affects subsequent activities. Several adaptations of the Gantt chart are in use, with variations in the colors employed and the identifiers of free and path floats and the critical path(s). Gantt charts can be constructed hierarchically from the master schedule to the level deemed sufficient for the project. A sample Gantt chart for a spacecraft command processor is shown in figure 1.4 where the free float, path float, and critical path are identified.

1.5.6.2 Milestone Charts

Milestone charts are used to represent milestones on the horizontal axis indicating calendar time, with different activities on the vertical axis. A *milestone* is defined as an important event that occurs during the project in order to achieve the project objective. These events occur at specific instances of time, such as completion of a test, occurrence of a review, delivery of an assembly, and so on. Milestone charts are also developed hierarchically from the master schedule to the level deemed sufficient for the project. A sample of a milestone chart is shown in figure 1.5. Milestone charts do not clearly show free and path floats or the interrelationships of the dependences of activities.

1.5.6.3 Network Diagrams or Work Flow Diagrams

Network diagrams or *work flow diagrams* are flowcharts that show the sequence of activities from an initial activity to a final activity. This configuration illustrates explicitly precedent or dependency relationships between activities and the beginning and start times of each activity. As with Gantt and milestone charts, network diagrams can be developed hierarchically. The elements of the network diagram are usually the elements of the work breakdown structure which provide a consistent approach to project management. In principle, there are four types of dependencies or precedent relationships:

- *Finish-to-start*: Start of the successor activity depends on the completion of the predecessor activity.
- *Start-to-start*: Start of the successor activity depends on the start of the predecessor activity.
- *Finish-to-finish*: Completion of the successor activity depends on the completion of the predecessor activity.
- *Start-to-finish*: Completion of the successor activity depends on the initiation of the predecessor activity.

The most commonly used type of logic relationship is the finish-to-start. Representation of the four possibilities is illustrated in figure 1.6.

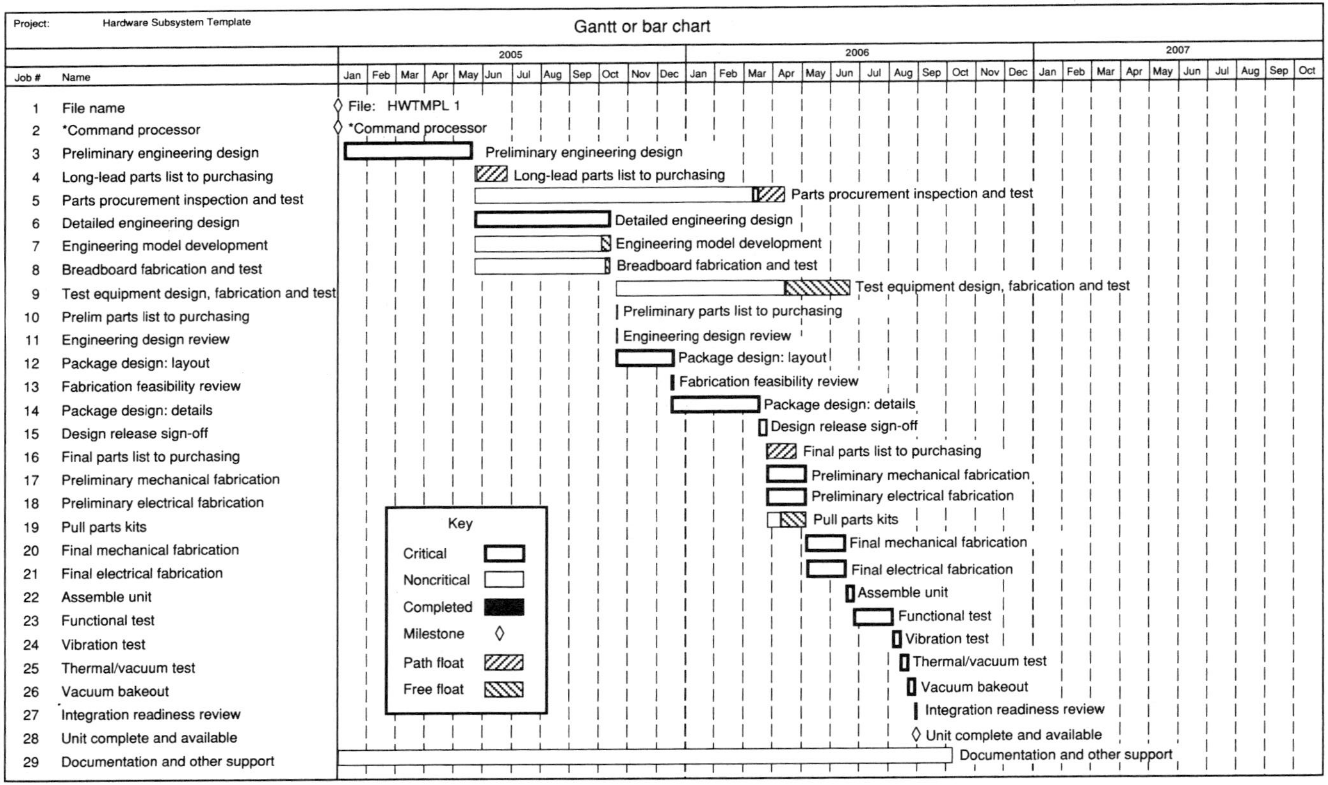

Project: Hardware Subsystem Template

Gantt or bar chart

2005: Jan | Feb | Mar | Apr | May | Jun | Jul | Aug | Sep | Oct | Nov | Dec
2006: Jan | Feb | Mar | Apr | May | Jun | Jul | Aug | Sep | Oct | Nov | Dec
2007: Jan | Feb | Mar | Apr | May | Jun | Jul | Aug | Sep | Oct

Job #	Name
1	File name
2	*Command processor
3	Preliminary engineering design
4	Long-lead parts list to purchasing
5	Parts procurement inspection and test
6	Detailed engineering design
7	Engineering model development
8	Breadboard fabrication and test
9	Test equipment design, fabrication and test
10	Prelim parts list to purchasing
11	Engineering design review
12	Package design: layout
13	Fabrication feasibility review
14	Package design: details
15	Design release sign-off
16	Final parts list to purchasing
17	Preliminary mechanical fabrication
18	Preliminary electrical fabrication
19	Pull parts kits
20	Final mechanical fabrication
21	Final electrical fabrication
22	Assemble unit
23	Functional test
24	Vibration test
25	Thermal/vacuum test
26	Vacuum bakeout
27	Integration readiness review
28	Unit complete and available
29	Documentation and other support

File: HWTMPL 1
*Command processor
Preliminary engineering design
Long-lead parts list to purchasing
Parts procurement inspection and test
Detailed engineering design
Engineering model development
Breadboard fabrication and test
Test equipment design, fabrication and test
Preliminary parts list to purchasing
Engineering design review
Package design: layout
Fabrication feasibility review
Package design: details
Design release sign-off
Final parts list to purchasing
Preliminary mechanical fabrication
Preliminary electrical fabrication
Pull parts kits
Final mechanical fabrication
Final electrical fabrication
Assemble unit
Functional test
Vibration test
Thermal/vacuum test
Vacuum bakeout
Integration readiness review
Unit complete and available
Documentation and other support

Key
Critical
Noncritical
Completed
Milestone
Path float
Free float

Figure 1.4 Sample Gantt or bar chart.

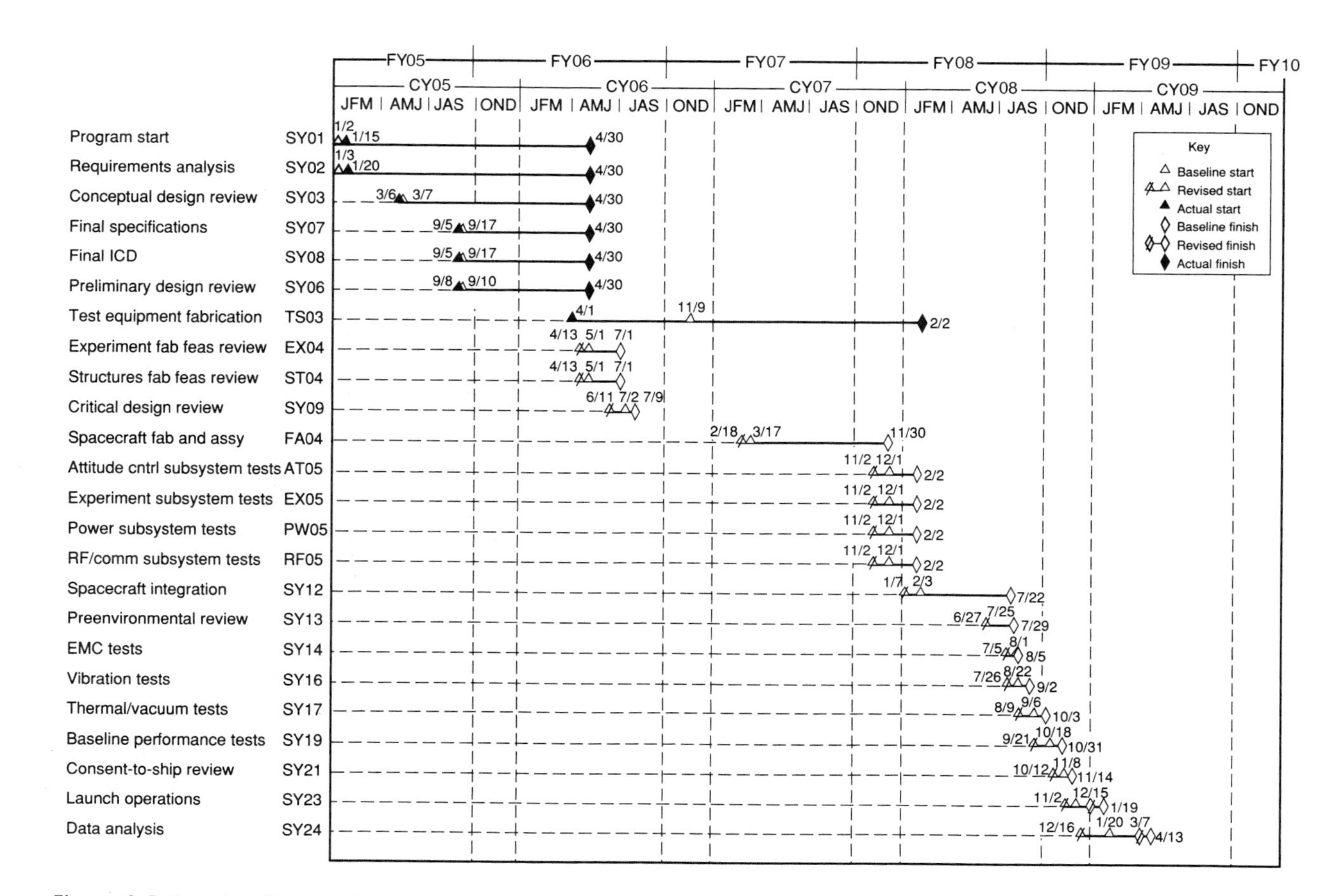

Figure 1.5 Sample milestone chart.

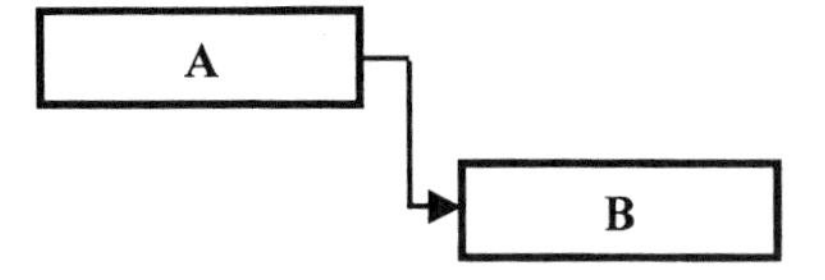

Finish-to-Start (FS)
Activity B cannot start until Activity A finishes

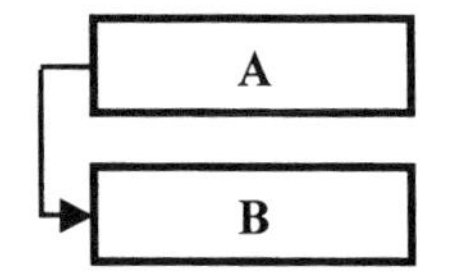

Start-to-Start (SS)
Activity B cannot start until Activity A starts

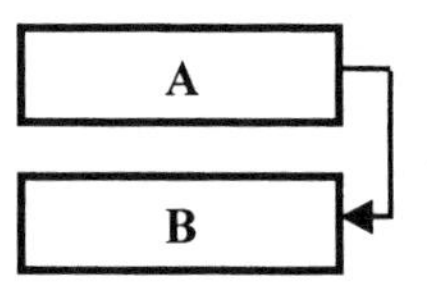

Finish-to-Finish (FF)
Activity B cannot finish until Activity A finishes

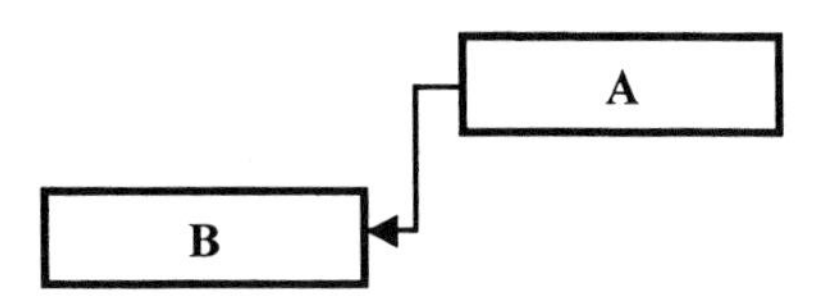

Start-to-Finish (SF)
Activity B cannot finish until Activity A starts

Figure 1.6 Four types of dependencies and precedents in work breakdown structures.

1.5.6.4 Arrow-Diagramming Method and Precedence-Diagramming Method

Network diagrams utilize either of the two graphical formats illustrated in figures 1.7 and 1.8, the arrow-diagramming method and the precedence-diagramming method. Both of these implementations are sometimes referred to, although inaccurately, as PERT charts.

The *arrow-diagramming method* (ADM), also called the a*ctivities-on-arrow* (AOA) method, uses arrows to represent activities and circles to represent nodes. The tail of the arrow represents the start of an activity and the head of the arrow represents its completion. The length of the arrow has no significance. An activity utilizes resources and time to produce a product while a node requires no resources or time and is the junction point for all dependencies, representing the earliest starting time or latest completion time of an activity. ADM sometimes requires the introduction of dummy activities with

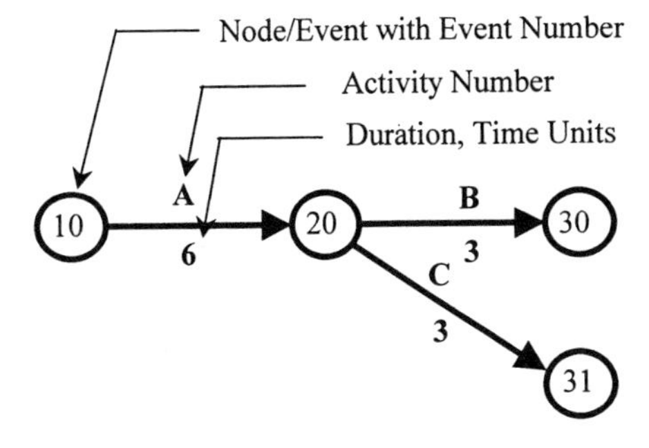

Figure 1.7 Arrow-diagramming-method (ADM) or activities-on-arrow (AOA) method

zero duration to enforce dependencies or constraints between nodes. Loops must be avoided—a loop occurs when activities are arranged in a way that an activity returns to an earlier activity. The ADM format is generally used along with the program evaluation and review technique (PERT), to be discussed later.

The *precedence-diagramming method* (PDM), also known as the *activity-on-node* (AON) method, uses rectangles as nodes to represent activities and arrows to connect the rectangles to show dependences. Because of its simpler visual format, no need for dummy activities, capability to easily show lag factors, and simpler depiction of the logical relationships, the precedence diagram has become more common in recent years. A *lag factor* means, for instance, that activity B cannot start until some defined time period after the completion of activity A.

1.5.6.5 Network Analysis, Program Evaluation and Review Technique (PERT), and Critical Path Management (CPM)

The two approaches generally used for network analysis are the program evaluation and review technique and the critical path method, with the latter the more prevalent.

The *program evaluation and review technique* (PERT) is a statistical approach that, although mentioned often, is not in significant use today. Consequently, only the principles are presented here. PERT uses three time estimates for the duration of each activity: the *most optimistic duration* t_o, the *most pessimistic duration* t_p, and the *most likely duration* t_l. From these, it is possible to determine for each activity the *mean duration* and the *duration variance,* where the *mean duration* of an activity, t_m, is

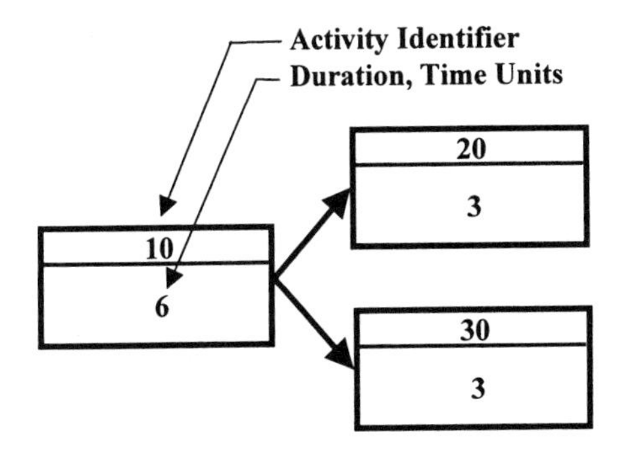

Figure 1.8 Precedence-diagramming method (PDM) or activity-on-node (AON) method.

given by

$$t_m = (t_0 + 4t_1 + t_p)/6, \tag{1.5.1}$$

and the *variance of the duration*, σ^2, of an activity is given by

$$\sigma^2 = [(t_p - t_0)/6]^2. \tag{1.5.2}$$

The mean durations are used through forward and backward passes through the network to determine two estimates of the completion date for each activity, the *earliest expected time* and the *latest allowed time,* and the *slack time* of an activity.

- A *forward pass* is a left to right (start to completion) calculation through the network (sequentially from the first activity to the last activity in the sequence) that establishes the earliest expected time a task can be completed.
- The *earliest expected time* (TE) for the completion of the activity is the maximum of the sums of the mean durations t_m of all precedent activities.
- A *backward pass* is a right to left (completion to start) calculation through the network that establishes the latest allowed time that a task can be completed and satisfy the schedule.
- The *latest allowed time* (TL) of the last activity is either set equal to its earliest expected time (TE) or to a later time of completion. From this, the latest allowed times of prior activities are obtained by subtracting the largest of the sums of the durations of subsequent activities from the TL of the last activity.
- The *slack time* (TS) for an activity is determined by subtracting the earliest expected time TE from the latest allowed time TL, where

$$\mathrm{TS} = \mathrm{TL} - \mathrm{TE}. \tag{1.5.3}$$

Slack time is the time the activity can be delayed without affecting subsequent activities.

A statistical completion date and its variance can also be determined by using the beta distribution, knowing the TE and TL of the delivery date (same as the TE and TL of the last activity) and the sum of the variances along the critical path.

The *critical path method* (CPM) is a deterministic approach that uses a single estimate for the duration of each activity and in many ways is similar to PERT. Four dates are computed in two steps for each activity: the earliest start, latest start, earliest finish, and latest finish dates. In the first step, *earliest start* and *earliest finish* dates are determined for each activity by a forward pass through the network, beginning at the first activity.

- A *forward pass* is a left-to-right calculation through the network that establishes the earliest a task can start and finish, based on precedent activities and the duration of each activity.
- The *earliest start* (ES) date for each activity is the latest of the earliest finish dates of precedent activities.
- The *earliest finish* (EF) date for each activity is the duration of the activity added to its earliest start date.

In the second step, *latest start* and *latest finish* dates are determined by a backward pass through the network, beginning at the last activity for which a date is specified as the latest finish date. This date can be selected to be either its earliest finish date or a later date.

- *Backward pass* is a right-to-left calculation through the network that establishes the latest start and finish dates for each activity.
- The *latest start* (LS) date of each activity is the duration of the activity subtracted from its latest finish date.
- The *latest finish* (LF) date of each activity, it is the earliest of the latest start dates of succeeding activities.

With the earliest and latest finish times determined for each activity, it is possible to determine the free float and path float for each activity and if any, the critical paths.

- The *free float* (FF) for each activity is the earliest finish date subtracted from the earliest of the earliest start dates of succeeding activities.

Free Float represents the length of time that the earliest finish date of an activity can be delayed without affecting the earliest start date of any succeeding activities.

- The *path float* (PF) for each activity, is the difference between the latest finish date and the earliest finish date or the latest start date and the earliest start date, i.e.,

$$\mathrm{PF} = \mathrm{LF} - \mathrm{EF} = \mathrm{LS} - \mathrm{ES}. \tag{1.5.4}$$

Activities or a combination of activities along a path can be delayed for the duration of the path float and, in principle, not delay the delivery of thc final product. However, utilization of the path float early in the development minimizes downstream flexibility. Also, delayed completion of activities might precipitate additional delays due to unavailability of facilities or personnel. The path through the network with the least float is the shortest time in which the project can be completed.

- The *critical path(s)* is the path through the network for activities with zero or near zero. path float.

Consequently, the critical path(s), if there are any, have activities that must be completed on time or the delivery date will be missed unless changes can be made to the schedule. In a dynamic project, critical paths can change frequently so it is prudent to monitor activities on paths that have small path floats as well. In fact, the path float for each activity should establishes the priorities for management attention.

An example of the precedence-diagramming method with critical path management is given in Figure 1.9.

1.5.7 Cost Estimating

A variety of techniques can be employed to develop cost estimates. It is important to select an estimating methodology that can reliably and economically estimate the costs of the project under consideration. This is dependent on having a well-defined technical baseline and schedule. When making a cost estimate the following criteria should be considered:

- Completeness: the estimate must be complete, providing all costs time-phased to sufficient detail.
- Consistency: the estimate should be internally consistent, with a clear format.
- Credibility: the technique used must be acceptable, with validated assumptions and historical costing data.
- Documentation: the estimating technique must be clearly described in sufficient detail that the costs can be independently verified.

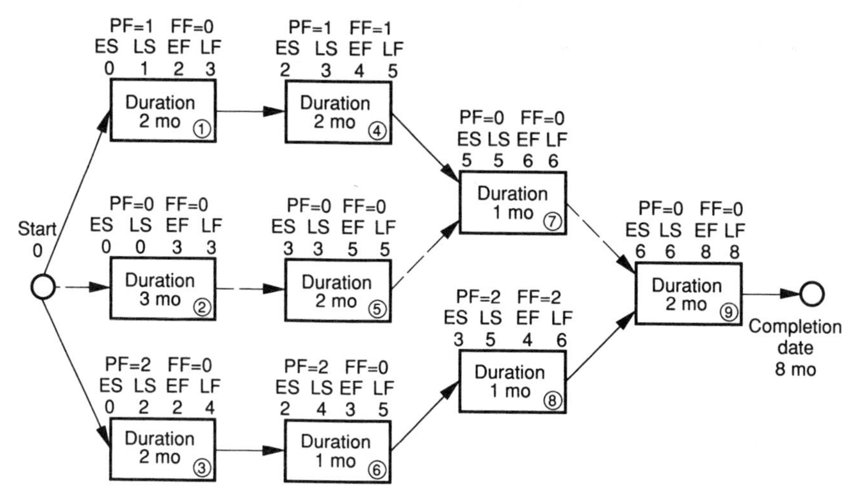

Figure 1.9 Network scheduling using the precedence-diagramming method and critical path method. The critical path method (CPM) involves both a forward pass and a backward pass. ES = earliest start; the latest of the earliest finish dates of precedent activities. EF = earliest finish; duration of an activity added to its earliest start date. LF = latest finish; earliest of the latest start dates of the immediately succeeding activities in time. LS = latest start; duration of an activity subtracted from its latest finish date. PF = path float; the earliest finish date subtracted from the latest finish date or the earliest start date subtracted from the latest start date. FF = free float; the earliest finish date of the given activity subtracted from the earliest of the earliest start dates of subsequent activities in time. n = activity number. Dashes represent the critical path.

Items to be included in a detailed cost estimate are:

- Description of the methodology
- Cost estimates, including the definition of the cost elements
- Quantitative risk assessment or confidence in the estimate
- Cost drivers
- Sensitivity analysis

Three of the more frequently employed cost-estimating techniques are the analogous estimate, bottoms-up or grassroots estimate, and parametric or top-down estimate. Each is described in turn.

1.5.7.1 Analogous

An *analogous* or *analog estimate* is based on the cost data of one or more past similar or reference projects that are technically representative of the project to be estimated. Differences in the project are then identified along with their concomitant cost differences, either positive or negative. Care must be emphasized to assure that the differences are properly identified. The cost estimate is then determined by adjusting the cost of the reference system by the differences and adjusting for changes in inflation, change in labor costs and indirect costs, and so on. The accuracy of this approach is dependent on the knowledge of the reference projects and their similarity to the project being estimated.

Advantages of this approach are that the estimate is based on related historical data, the estimate is easily defended, and it can be especially accurate if the deviations from the reference projects are minor. Disadvantages are that the estimate is based on a few, possibly one, instantiation; an appropriate analog may not exist; and there may be some hidden cost savings or hidden efficiencies not realizable by the current project. If several appropriate and recent analogs are available whose costs and program plans are sufficiently detailed, this may be the preferred approach.

1.5.7.2 Engineering Build-Up, Bottoms-Up, or Grassroots

The *engineering build-up*, *bottoms-up*, or *grassroots* approach is based on estimates by the lead individuals for each major activity of the work breakdown structure. A concern of this approach is the tendency to overestimate the cost by the addition of margin at each level of the work breakdown structure. The benefit of this approach is the buy-in of the lead participants since they are responsible for the estimates and delivery of the subsystem at cost and on schedule. This approach works especially well if the lead individuals have significant experience, an *a priori* cost limit is targeted, and the personnel doing the estimate are expected to carry out the activity. The key to this approach is the negotiation of costs with each lead engineer by technically knowledgeable management since there is generally no competition within an organization.

Advantages of this approach are that the estimates can be easily justified, there is technical visibility into each activity, the major cost contributors are clearly identified, and the estimate can be easily modified. Disadvantages are that it involves significant time and personnel, does not provide statistical confidence information, and requires experienced personnel. This approach can be the most time consuming and costly of the three approaches.

1.5.7.3 Parametric or Top Down

Parametric estimating can provide estimates at a lower cost and shorter cycle time than other traditional estimating techniques. It employs *cost estimating relationships* (CERs) that relate technical and programmatic characteristics of the system to costs based on historical data of resources utilized on similar projects. This approach is especially useful in the early phases of a project if little specific information is available. The CERs can be developed internally or obtained commercially. An example is the NASA/Air Force Cost Model (NAFCOM), based on a comprehensive set of historical cost and technical data for NASA spacecraft programs (NAFCOM, 2001). These data have been broken down to the subsystem level, normalized, and stratified by mission type, that is, launch vehicles, manned space vehicles, unmanned spacecraft, and scientific instruments.

Advantages of the parametric approach are that once the model is established it provides rapid answers to changes with statistical confidence levels, it eliminates subjectivity, the process is easily documented, cost ranges can be provided by varying the input parameters, and the results are easily defensible if the basic model is accepted or one of the standard models is used. A disadvantage is that the model must be applicable to the subject project and that its use at the fringe and outside of its relevant parameter range may give erroneous estimates. Costs unique to the project and outside the parameters of the model employed can be added to the estimate when appropriate.

In summary, the choice of estimating techniques depends on the level of detail known about the project, the experience of the personnel, availability of historical data bases, and access to cost models. For example, during phase A, conceptual definition, the parametric or analogous method can be effectively used. However, during phase B, concept definition, and phases C/D, design and development, the engineering build-up or grassroots approach is generally more appropriate.

1.5.8 Earned Value Management

The budget of a project is the plan by which costs are estimated and expenditures tracked to provide assessments of status. *Earned value management* (EVM) is an effective methodology for assessing both cost and schedule performance by comparing "planned" work, "accomplished" work, and the "actual expenditures" in terms of the budget assigned to each activity. This technique identifies and establishes, for each activity in the work breakdown structure, milestones for technical accomplishments, a schedule, and a budget. A most important issue for effective use of EVM is that each activity of the WBS has discrete milestones to which work accomplishments can be measured and that costs can be attributed to each activity by the organization's cost accounting system.

At the time of an assessment, three quantities form the basis for ERM: budgeted cost of work scheduled or *planned value;* budgeted cost of work performed or *earned value;* and actual cost of work performed or *actual cost* of work accomplished.

- The *budgeted cost of work scheduled* (BCWS) is the sum of the budgets for the activities that were scheduled to be completed by the assessment time.
- The *budgeted cost of work performed* (BCWP) is the sum of all of the budgets for the activities that were actually completed at the assessment time.
- The *actual cost of work performed* (ACWP) is the actual costs incurred in accomplishing work on each activity from the organization's cost accounting system.

The above definitions do not indicate how to quantify the BCWP if an activity is not completed when the assessment is made. Several options can be employed, two of which are to either use an estimate for the BCWP based on completed milestones for the activity or assign 50% of the BCWP of the activity if it has been started and not assign the other 50% until the activity is completed; this latter approach is known as the 50/50 rule. This approach avoids any disagreements that might arise in estimating the BCWP.

With the BCWS, BCWP, and the ACWP, it is possible to determine the cost variance, schedule variance, percent complete, cost performance index, and schedule performance index. These follow in turn.

The *cost variance* (CV) is the difference between the budget and the actual cost of the work performed at the time of the assessment, given by

$$\mathrm{CV} = \mathrm{BCWP} - \mathrm{ACWP}. \tag{1.5.5}$$

It represents the funds that have been expended in excess of the budget for the activities that are completed or partially completed and thus eliminates time as a factor.

The *schedule variance* (SV) is the difference between the budgeted cost of the work performed and the budgeted cost of the work scheduled to be performed at the time of the assessment:

$$SV = BCWP - BCWS. \tag{1.5.6}$$

It represents the difference between the number of activities that were scheduled and those completed, weighed by their individual budgets. Thus, it gives a measure of being ahead (positive values) or behind (negative values) in schedule in dollars. One way to interpret the SV is to divide it by the average expenditure per unit time to obtain a schedule difference in time units.

The *percent complete* (PC) gives the fraction of the work completed, that is, the budgeted cost of work performed (BCWP), in terms of the total budget at completion:

$$PC = BCWP/BAC, \tag{1.5.7}$$

where the *budget at completion* (BAC) is the total budget to complete the project; BAC equals the BCWS at the planned completion date.

The *cost performance index* (CPI) represents cost efficiency as given by the ratio of the budgeted cost of work performed (BCWP) to the actual costs of the work performed (ACWP), where

$$CPI = BCWP/ACWP. \tag{1.5.8}$$

Less than planned cost efficiencies are represented by $CPI < 1$ and greater than planned efficiencies are represented by $CPI > 1$.

The *schedule performance* index (SPI) is the ratio of the budgeted cost of work performed (BCWP) to the budgeted cost of the work scheduled (BCWS):

$$SPI = BCWP/BCWS. \tag{1.5.9}$$

where $SPI < 1$ indicates that a project is behind schedule, $SPI = 1$ indicates a project is on schedule, and $SPI > 1$ indicates a project is ahead of schedule.

It is also possible to estimate the cost at completion and the schedule at completion using respectively the cost performance index (CPI) and the schedule performance index (SPI). An *estimate (of budget) at completion* (EAC) can be determined by assuming that the same cost efficiency will prevail, so that the cost performance index (CPI) gives

$$EAC = BAC/CPI. \tag{1.5.10}$$

Just adding the cost variance (CV) to the budget at completion (BAC) will generally underestimate the *estimate at completion of cost* (EAC) unless significant cost savings can be realized in future activities.

An *estimated schedule at completion* (ESC) in time units can be obtained from assuming that the same schedule efficiency will prevail, so that the schedule performance index (SPI) gives

$$ESC = SAC/SPI, \tag{1.5.11}$$

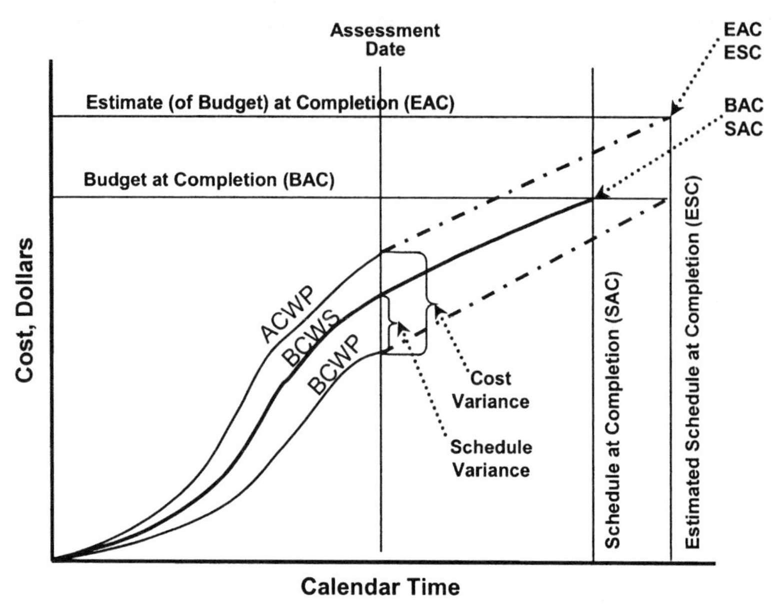

Figure 1.10 Earned value management example.

where the *schedule at completion* (SAC) is the budgeted duration of the project in time units. Just adding the duration of the time left in the schedule to the elapsed time will generally underestimate the delivery date.

The above-defined earned value status parameters are illustrated in figure 1.10. A variety of software programs exists to support project scheduling and budgeting and tracking performance and cost. Their capabilities range from addressing a single small project to large multi-project efforts.

1.5.9 Risk Management

Risk is the uncertainty of achieving a desired outcome and is a measure of both the likelihood of an adverse event and the severity of its consequence. Risk management is a systematic process to reduce risk to an acceptable level so as to enhance the probability of achieving the program objectives. Early detection and mitigation of potential risk can increase the probability of success of the project and enable more efficient use of resources. Potential adverse outcomes include schedule and cost in addition to performance. The steps in risk management include identify, analyze, plan, track, and control and document (figure 1.11). Each will be discussed in turn.

1.5.9.1 Identify

To manage risk, it is first necessary to *identify* the risks and their consequences that may impact program objectives. This can be accomplished from expert opinion, lessons learned on similar projects, hazard analysis, test results, engineering analyses, and simulations. Hazard analysis is a process used to determine how a system can cause faults or hazards to occur. The most common types of hazard analysis are the *fault tree*

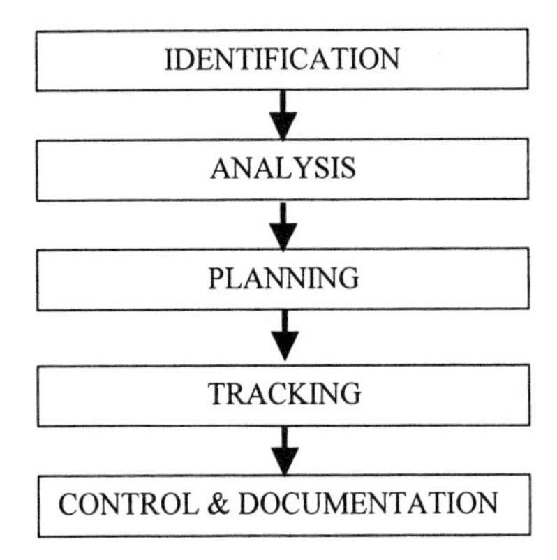

Figure 1.11 Risk managment process.

analysis (FTA) and the *failure modes and effects analysis* (FMEA), also known as *failure modes effects criticality analysis* (FMECA). The FTA is a deductive, top-down method of analyzing system design and performance given a particular failure mode. It involves specifying a top event to analyze (such as a fire, inadvertent propulsion burn, and so on), followed by identifying all of the associated elements in the system that could cause that top event to occur. The FMEA or FMECA is a bottoms-up approach to analyzing a design of a product or process in order to determine a list of potential failure modes and their high-level manifestations. The technique assumes a failure or fault at the lowest level of a piece part, subassembly, or assembly. Other risks may include reliability of a facility and lower productivity due to introduction of a new computer language. Through any approach, the identification process should result in a list of risks, the causes of each, and their consequences.

1.5.9.2 Analyze

Analysis of the risk should define its attributes by determining the expected severity of the consequence, its probability, and its timeframe. Several approaches can be used. The simplest is a tri-level attribution where the risk probability is identified as very likely, probable, improbable; the consequence as catastrophic, critical, and marginal; and the time frame as near-term, mid-term, and far-term. Another approach is to sequentially decompose the system into components for which the probability of risks can be assigned as a function of time and their consequences can be assigned, usually in dollars. A probabilistic analysis would then give the risk probability as a function of time and the potential cost in dollars. In any event, the analysis process should result in a list of prioritized risks, their probabilities, consequence, or cost, and time frame of occurrence.

1.5.9.3 Plan

Planning is the function of deciding what, if anything, should be done to mitigate a risk. It consists of classifying or grouping risks of a similar nature; ranking the risks; identifying the risk according to technical, schedule, or cost; assigning the risk to an individual; determining whether to accept or mitigate each risk; determining the approaches to be used to mitigate each risk; and specifying the level to which the risk should be reduced. Plans should be developed for the highest priority risks. Risks can be mitigated by adding redundancy, increased testing or analysis, redesign, changing procedures, or changing the way in which the system is operated. Risks identified in phase A can be mitigated by adding budget and schedule reserves and planning appropriate technology development or explicit risk mitigation activities.

1.5.9.4 Track

Formal *tracking* should be undertaken of each risk and the status or progress of its risk mitigation plan. This will identify if any changes occur to reduce or increase the risk and if the risk mitigation plan is effective, if the plan is being executed, and if it is on schedule.

1.5.9.5 Control and Document

Control and documentation ensures that the risks to the project continue to be managed effectively and efficiently and the actions undertaken are documented. It involves the execution of decisions to address risks by analyzing the tracking data and taking appropriate action such as replanning, closing the risk as an action item if mitigated, and invoking contingency plans if necessary. All plans, results, and status should be documented.

1.6 Organization

To successfully orchestrate the resources of a project requires a well-thought-out project organization with positions that have well-defined responsibilities and authority and an unambiguous reporting structure. The organization should be specifically tailored to the needs of the project. Figure 1.12 shows a typical organization for a spacecraft project. The identified positions are described subsequently.

The *project manager* has the overall responsibility and decision-making authority for the project, subject to review by the organization's management. The project manager is also the single point of contact with the sponsor or customer.

The *project "scientist"* reports to the project manager and is responsible for assuring the overall scientific or mission objectives are satisfied. Even in a non-scientific mission the equivalent of a project scientist is warranted to assure that the utility of the payloads are maximized. The scientist must understand the scientific or operational objectives of the mission and assure they are not compromised during the engineering development process. In a multi-payload mission, this person will work with the science or user teams to assure that the needs of each team are addressed.

The *systems engineer* reports to the project manager and is responsible for understanding and maintaining the system-level requirements, assuring the orderly flow-down of the system requirements to the subsystem requirements, and the overall design and test of the system. The systems engineer usually chairs the configuration control board, and provides management with alternatives to address any issues that arise.

The *project controller* or *scheduler* reports to the project manager and is responsible for developing the schedule and monitoring progress.

The *resource manager* reports to the project manager and is responsible for tracking the budget and cost accounting.

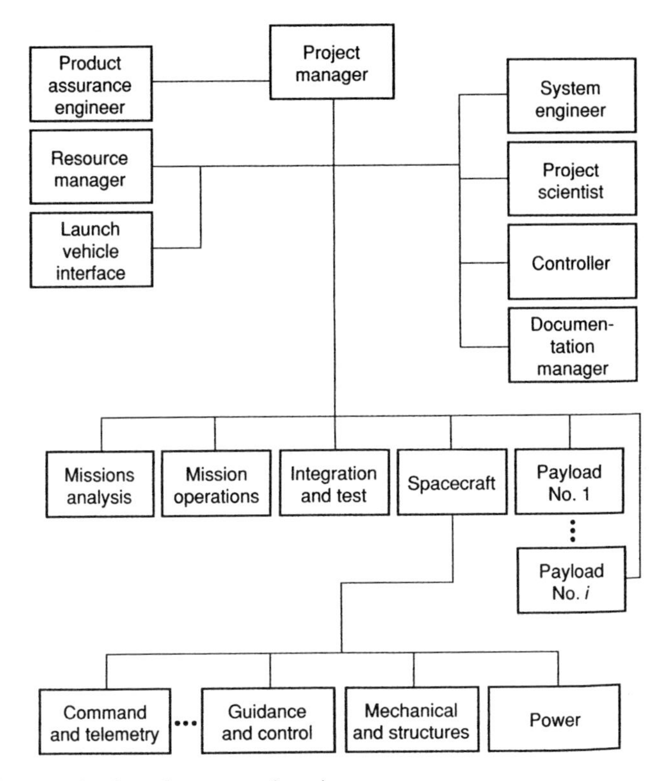

Figure 1.12 Organization of a spacecraft project.

The *product assurance engineer* generally has dual reporting responsibilities to the project manager and to higher-level management. The project responsibility is to implement the project quality assurance plan while the organizational responsibility is to assure that the project quality and assurance plan conforms to the institution's guidelines.

Lead engineers are responsible for providing the subsystems in accordance with subsystems requirements. For a small spacecraft development the lead engineer may report directly to the project manager, whereas in a large project there may be a payloads manager and spacecraft bus manager to whom the engineer reports.

Problems

Possible journals for some of the assignments below include the *IEEE Transactions on Engineering Management*, *Engineering Management Journal* published by the American Society of Engineering Management,and *The Journal of Management in Engineering* published by the American Society of Civil Engineers.

1.1 Find, read, and turn in a one-page review of a relevant article in the engineering literature on systems engineering.

1.2 Find, read, and turn in a one-page review of a relevant article in the engineering literature on engineering trade studies.

1.3 Find, read, and turn in a one-page review of a relevant article in the engineering literature on risk management.

1.4 Find, read, and turn in a one-page review of a relevant article in the engineering literature on configuration management.

1.5 Find, read, and turn in a one-page review of a relevant article in the engineering literature on scheduling projects.

1.6 Find, read, and turn in a one-page review of a relevant article in the engineering literature on the program evaluation and review technique (PERT).

1.7 Find, read, and turn in a one-page review of a relevant article in the engineering literature on the critical path method (CPM).

1.8 Find, read, and turn in a one-page review of a relevant article in the engineering literature on earned value management (EVM).

1.9 Find, read, and turn in a one-page review of a relevant article in the engineering literature on verification and validation (V & V).

1.10 Find, read, and turn in a one-page review of a relevant article in the engineering literature on organizations for a space system or subsystem development.

1.11 A power system consists of (1) an off-the-shelf solar array with a mass estimate of 10 kg, (2) batteries that are a modification of an existing design with a mass estimate of 15 kg, (3) a concept for a redundant battery charge regulator that promises a reduced mass of 1 kg, (4) electrical wiring with a thoughtful estimate of mass of 2 kg, and (5) and a redundant power switching device that is a minor modification of an existing device with an estimated mass of 3 kg. (a) Compute the total mass of the power system; (b) estimate the additional mass margin that should be retained; and (c) compute the average mass margin factor.
Answers: (a) 31 kg, (b) 2.9 kg, (c) 1.09.

1.12 Construct a Gantt chart for project A, described at the end of the section, showing for each activity (a) earliest start time, earliest finish time, path float, and free float, and (b) activities on the critical path.
Answers: (a) activity 3: 10,10,25,25,0,0; activity 9: 50,50,65,65,0,0; activity 12: 35,70,40,75,35,35; (b) activities 1,3,5,6,9,11,13.

1.13 Construct a Gantt chart for project B, described at the end of the section, showing for each activity (a) earliest start time, earliest finish time, path float, and free float, and (b) activities on the critical path.
Answers: (a) activity 4: 40,40,80,80,0,0; activity 5: 80,90,90,100,10,10; (b) activities 1,3,4,6,7.

1.14 Construct a milestone chart for project B if additional milestones are:

Activity number	1	2	3	4	5	6	7
Milestones (time units from ES)	5	10	15	5	2	10	5
Milestones (time units from ES)		15	20	35			10
Milestones (time units from ES)							25

where ES = earliest start

1.15 Construct a network diagram using the precedence-diagramming method for project A showing for each activity (a) earliest start time, latest start time, earliest finish time, latest finish time, path float, and free float, and (b) activities on the critical path.
Answers: (a) activity 3: 10,10,25,25,0,0; activity 9: 50,50,65,65,0,0; activity 12: 35,70,40,75,35,35; (b) activities 1,3,5,6,9,11,13.

1.16 Construct a network diagram using the precedence-diagramming method for project B showing for each activity (a) earliest start time, latest start time, earliest finish time, latest finish time, path float, and free float, and (b) activities on the critical path.
Answers: (a) activity 4: 40,40,80,80,0,0; activity 5: 80,90,90,100,10,10; (b) activities 1,3,4,6,7.

1.17 Given at the time of the assessment, a project has a BCWS = \$10,000, BCWP = \$7,000, ACWP = \$5,000. Find (a) cost variance, (b) schedule variance, (c) cost performance index, and (d) schedule performance index.
Answers: (a) CV = \$2,000, (b) SV = −\$3, 000, (c) CPI = 1.4, (d) SPI = 0.7.

1.18 At the time of an assessment of a project the BCWS = \$20,000, BCWP = \$15,000, ACWP = \$25,000, BAC = \$50,000, and the project is scheduled to be completed in one year. Find (a) cost variance, (b) schedule variance, (c) cost performance index, (d) schedule performance index, (e) percent complete, (f) estimated cost at completion, and (g) estimated schedule at completion.
Answers: (a) CV = −\$10, 000, (b) SV = −\$5, 000, (c) CPI = 0.6, (d) SPI = 0.75, (e) PC = 30%, (f) EAC = \$83,333, (g) ESC = 1.33 years.

1.19 Carry out an earned value analysis for project A at the 40th time unit showing (a) cost variance, (b) schedule variance, (c) percent complete, (d) estimated (budget at) time of completion, and (e) estimated schedule at completion.
Answers: (a) CV = −60 \$units, (b) SV = −400 \$units, (c) PC = 47%, (d) EAC = 3,527, (e) ESC = 106.

1.20 Carry out an earned value analysis for project B at the 70th time unit showing (a) cost variance, (b) schedule variance, (c) percent complete, (d) estimated budget at time of completion, and (e) estimated schedule at completion.
Answers: (a) CV = −100 \$units, (b) SV = −100 \$units, (c) PC = 53%, (d) EAC = 1,685, (e) ESC = 146.

Project A

Activity No.	1	2	3	4	5	6	7	8	9	10	11	12	13
Precedent(s)	–	1	1	1	2,3	5	4	5	6	7	8,9	10	11,12
Duration (time units)	10	10	15	10	20	5	5	5	15	10	10	5	10
Budget (\$ units)	100	200	500	200	400	100	400	200	500	200	300	100	200
Actual costs (\$ units)	110	190	510	210	0	0	420	0	0	220	0	0	0
Percent complete	100	100	100	100	0	0	100	0	0	100	0	0	0

Project B

Activity No.	1	2	3	4	5	6	7
Precedent(s)	–	1	1	2,3	4	4	5,6
Duration (time units)	10	20	30	40	10	20	30
Budget (\$ units)	100	200	300	400	100	200	200
Actual costs (\$ units)	110	200	290	300	0	0	0
Percent complete	100	100	100	50	0	0	0

References

Blanchard, B. S., and Fabrycky, W. J., 1998. *Systems Engineering and Analysis* (3rd Edition). Upper Saddle River, NJ: Prentice-Hall.

Boehm, B., and Hansen, W. J., 2001. Understanding the spiral model as a tool for evolutionary acquisition. *Crosstalk*, **14**(5) (May): 4–10.

Defense Systems Management College, January 1990. Systems Engineering Management Guide Washington, DC: U.S. Government Printing Office.

Department of Defense, 2 January 1998. MIL-HDBK-881, DoD Handbook—Work Breakdown Structure. Washington, DC.

Department of Defense, 10 June 2001. 5000.2-R, Mandatory Procedures for Major Defense Acquisition Programs (MDAPS) and Major Automated Information System (MAIS) Acquisition Programs. Washington, DC.

Electronic Industries Alliance, 17 January 1999. EIA/IS-731, Systems Engineering Capability Model (SECM). http://www.incose.org/lib/731-news.html

Gause, D. C., 1989. *Exploring Requirements: Quality Before Design*. New York: Dorset House Publishing.

NAFCOM (NASA Air Force Cost Model), 2001. http://nafcom.saic.com/

NASA Headquarters, June 1994. Risk Management Policy, NMI 8070.4A. Safety Division, Core QS. Washington, DC.

NASA Headquarters, June 1995 (edited June 1999). Systems Engineering Handbook, SP-610S. Washington, DC.

NASA Headquarters, 3 April 1998. NASA Program and Project Management Processes and Requirements. NASA Procedures and Guidelines, NPG: 7120.5A. Office of Chief Engineer, Code AE/HQ. Washington, DC.

NASA Headquarters, 19 January 2000. Strategic Management Handbook, NPG 1000.2. Washington, DC.

NASA Headquarters, Spring 2002. NASA Cost Estimating Handbook, released by S. J. Isakowitz, NASA Comptroller. Washington, DC.

Stevens, R., Jackson, K., Brook, P., and Arnold, S., 1998. *Systems Engineering—Coping with Complexities*. Upper Saddle River, NJ: Prentice Hall.

Weinberg, G. M., 1992. *Quality Software Management, Volume 1: Systems Thinking*. New York: Dorset House Publishing.

2

The Space Environment

BRIAN J. ANDERSON AND DONALD G. MITCHELL

2.1 Earth's Space Environment

This chapter provides a description of the phenomena in space that affect spacecraft and launch vehicles. The common impression is that space is empty and that the vacuum of space is the only engineering challenge to be met. Even though the gas density in space is orders of magnitude lower than that achieved in the best vacuum systems, the environment is not benign and spacecraft are exposed to a range of hazards including intense particle and electromagnetic radiation, dense plasma flows, highly reactive species, and variable neutral gas densities in low Earth orbit (LEO). Additionally, spacecraft communications and radio navigation must account for propagation of electromagnetic waves through the ionospheric plasma in the uppermost layer of the Earth's atmosphere. We introduce here the basic concepts, equations, and references needed to identify the features of the space environment that must be considered for operations in Earth orbit. Interplanetary conditions involve some of the same phenomena and are also mentioned.

The terrestrial space environment extends roughly from the upper reaches of the atmosphere to tens of R_E (earth radii) altitude where orbits are no longer stable due to lunar gravitational perturbations. Earth's atmosphere does not have a sharp upper boundary but extends above 1 R_E altitude, albeit very tenuously. The atmosphere exerts an appreciable drag force on spacecraft, especially during launch (and reentry) but also in orbits below ~600 km altitude. Above ~150 km the atmospheric density and composition are not fixed but vary with solar and geomagnetic activity. Solar extreme ultraviolet (EUV) radiation ionizes a fraction of the neutral gas in the upper atmosphere above ~100 km, creating the *ionosphere*. The ionization fraction is high enough that the ionosphere must be treated as a *plasma*. The influence of the ionosphere on radio wave propagation is its primary effect on spacecraft design, but there are also chemical and charging effects that must be considered. The ionosphere has no sharp cut-off with altitude but can be said to end at ~1000 km. From there to ~10 R_E the Earth's magnetic field forms a cavity called the *magnetosphere*. The magnetosphere is populated by low-density but often high-energy charged particles. The Van Allen radiation belts are in the inner portion of the magnetosphere. Beyond the magnetosphere is the *solar wind* of magnetized plasma that continuously flows from the Sun at speeds between 300 and 800 km/s.

Magnetic activity on the Sun, closely correlated with sunspots, influences the Earth's space environment via several mechanisms. Active regions near sunspots are copious sources of EUV and X-ray emissions and the solar EUV/X-ray intensity is strongly correlated with the ~11 year cycle in sunspot abundance, which in turn leads to solar cycle variations in the Earth's uppermost atmosphere. Solar flares, sudden bursts of EUV/X-ray emissions from active regions, lead to prompt effects in the ionosphere and atmosphere. Active regions also give rise to explosive ejections of magnetized plasma clouds that hurtle through the solar wind with supersonic speeds. Called coronal mass ejections (CMEs), these blasts of plasma create shock waves in the solar wind which are powerful particle accelerators, generating solar energetic particles (SEPs) that stream ahead of the CME along lines of magnetic force in the solar wind and reach Earth within hours of CME initiation. Up to several days later, when the CME arrives at Earth, the leading shock wave and/or the magnetic cloud of the CME can produce a major geomagnetic storm. Geomagnetic storms have numerous consequences for the space environment from intense heating of the atmosphere above 100 km, strong turbulence and density structure changes in the ionosphere, and astonishingly dynamic temporary reconfigurations of the magnetosphere which act to accelerate charged particles and can lead to intensification or repopulation of the Van Allen radiation belts.

Of all the ultimate consequences of the solar activity, the most hostile is particle radiation. This consists of energetic [~10 keV (kilo electron volts) to hundreds of MeV (million electron volts)] electrons, protons, and other ions, some trapped in the magnetospheric field (radiation belts), others accelerated in solar flares (also called solar energetic particles or SEPs). Radiation is dangerous both to people in space and to spacecraft electronic components (chapters 11, 13) and solar cells (chapter 6), which are degraded or even destroyed promptly by penetrating energetic particles. The radiation dose is the primary factor that limits the lifetime of most space electronics systems. Particle radiation is encountered in all regions of the space environment. The particle radiation environment of Jupiter is particularly harsh owing to the intense magnetic field of the planet (Dessler, 1983), and is one of the most daunting engineering challenges in designing probes to study the Jovian satellites.

A number of excellent texts on the physics of Earth's space environment are now available. Tascione (1988) gives a good descriptive introduction while Parks (1991), Kivelson and Russell (1995), and Gombosi (1998) give more extensive treatments and derivations of the physical principles. Chen (1984) and Baumjohann and Treumann (1997) are two of the texts on plasma physics that are accessible to the upper-level undergraduate or beginning graduate student. More extensive discussions of the topics discussed below can be found in these works. We begin our discussion by considering orbital dynamics and the effects of atmospheric drag.

2.2 Gravity

Gravity is the attractive force between two objects, proportional to the product of their masses and inversely proportional to the square of the distance between them. Relativistic gravitation is important for some regimes, but, for masses and accelerations associated with Earth-orbiting spacecraft, gravity can be described in the simple Newtonian form. Newton's law of gravitation is

$$\mathbf{f}_2 = -Gm_1m_2\frac{\mathbf{r}_{12}}{r_{12}^3} \tag{2.2.1}$$

which gives the force on object 2 due to object 1 where m_1 and m_2 are the masses of two point objects. The *gravitational constant* $G = 6.673 \times 10^{-11}$ m^3 s^{-1} kg^{-1}, $\mathbf{r}_{12}$ is the vector from object 1 to object 2. The minus sign appears because the force is attractive. This force can be written as the gradient of a scalar potential, $\mathbf{f}_2 = -\nabla_2 U$, where

$$U = -Gm_1m_2\frac{1}{r_{12}} \tag{2.2.2}$$

and the gradient is taken with respect to the coordinates of object 2. Physically, this means that gravity is a conservative force, so that if we denote the kinetic energy as $K = mv^2/2$, where $v = |d\mathbf{r}/dt|$, then the total energy, $W = U + K$, remains constant. Since the point of zero potential energy is taken at infinity, U is always negative.

The fact that the Earth is not truly spherical but slightly oblate and that the density is not uniform has important consequences for orbital dynamics. To treat these effects one must use equation 2.2.2, replacing one of the masses with the mass of a differential volume element, and consider the integral over the Earth's actual shape and density distribution. This is considered in chapter 3.

For our purposes, however, we can assume a spherical Earth with mass $M_E = 5.97 \times 10^{24}$ kg, and a point-mass satellite with mass m. It turns out that integrating equation 2.2.2 over the assumed spherically symmetric Earth yields exactly the same form regardless of the variation of mass density with distance from the Earth's center of mass. The force on the satellite is the same as if the Earth's mass were concentrated at the Earth's center. The gravitational acceleration, defined as the force per unit mass, directed toward the Earth's center of mass, is therefore approximately

$$a(r) = |\mathbf{f}/m| = \frac{GM_E}{r^2} \tag{2.2.3}$$

At the Earth's surface, $r = R_E \approx 6.4 \times 10^6$ m, $a(r)$ evaluates to 9.8 m/s^2 and is written g. For altitudes z much smaller than R_E, $z/R_E \ll 1$, equation 2.2.3 can be written as

$$a(z) \cong g(1 - 2z/R_E) \tag{2.2.4}$$

At 100 km and 300 km altitude the reduction of $a(r)$ from g is 3.1% and 9.4%, respectively. To estimate the basic structure of the atmosphere below 100 km assuming $a = g$ is a reasonable approximation. However, the error is unacceptable when computing orbital periods for which $z > 300$ km.

Atmospheric drag is a dissipative force that decreases the total energy of an object, causing its orbit to decay. Drag is the factor that determines the lower altitude limit for orbits about solar system objects with atmospheres. To estimate this limit, consider a satellite with mass m and cross-sectional area A moving with speed v in a circular orbit with radius r through an atmosphere with mass density $\rho(r)$. In a time dt the satellite moves a distance vdt and sweeps through a volume $Avdt$ corresponding to a mass of gas $\rho(r)Avdt$. The rate at which the satellite transfers momentum to the gas is equal and opposite to the drag force on the satellite, $\mathbf{F}_\mathrm{d}$. If we assume that the gas molecules acquire the satellite's velocity along its direction of motion and are deflected symmetrically transverse to the spacecraft motion, as a spray of water is deflected by a wall, then $\mathbf{F}_\mathrm{d}$ is the negative of the satellite velocity times mass of gas encountered per unit time

$$\mathbf{F}_\mathrm{d} = d\mathbf{P}/dt = -\rho(r)Av\mathbf{v} \tag{2.2.5}$$

It turns out that this estimate is exact for a spherical object and is an overestimate by about 10% for other shapes (see chapter 3).

In using equation 2.2.5 to determine the rate of orbital decay, we use the result that an object in circular motion experiences a centrifugal acceleration equal to v^2/r. The case of orbit decay due to drag for a highly elliptical orbit is discussed in chapter 3. For a circular orbit the centrifugal acceleration must be exactly balanced by a centripetal gravitational acceleration, equation 2.2.3, so a circular orbit radius and satellite speed are related by

$$v^2 = \frac{GM_\mathrm{E}}{r} \tag{2.2.6}$$

which gives for the total energy

$$W = -\frac{GM_\mathrm{E}m}{2r} \tag{2.2.7}$$

Note also that equation 2.2.6 also immediately yields the orbital period

$$\tau_0 = \frac{2\pi r}{v} = \frac{2\pi r^{3/2}}{\sqrt{GM_\mathrm{E}}} \tag{2.2.8}$$

corresponding to Kepler's relation between orbit period and radius. Now the rate at which $\mathbf{F}_\mathrm{d}$ does work on the satellite is $\mathbf{v}\cdot\mathbf{F}_d$, which is equal to dW/dt. Then we have

$$W\bigg/\left(\frac{dW}{dt}\right) = -\frac{m}{2\rho(r)A}\sqrt{\frac{r}{GM_\mathrm{E}}} \equiv -\tau_d \tag{2.2.9}$$

If $\tau_\mathrm{d} \gg \tau_0$, then r is approximately constant and equation 2.2.9 integrates to $W = W_0 \exp\{-t/\tau_\mathrm{d}\}$, so that τ_d is the instantaneous decay rate of the orbit energy and, by equation 2.2.7, also of the orbit radius. Note that τ_d is proportional to m but inversely proportional to $\rho(r)$ and A. As we shall see below, $\rho(r)$ increases exponentially with decreasing altitude. This implies that orbit decay by atmospheric drag is a rapidly accelerating runaway process because τ_d decreases exponentially as the orbit radius falls. Above 100 km altitude $\rho(r)$ is variable, leading to time-dependent orbit decays. Understanding the mass density profile of the uppermost atmosphere and its variability is therefore of great importance.

2.3 The Atmosphere

The atmosphere is made up of neutral gases whose temperature, density, and pressure profiles as a function of height can be understood in terms of fairly simple physical concepts. The basic structure of the atmosphere is well established; more expanded discussions may be found in Haymes (1971), Ratcliffe (1972), Wallace and Hobbs (1977), and the *Handbook of Geophysics and the Space Environment* (1985). The *COSPAR International Reference Atmosphere* (1985) remains a good model for the atmosphere's basic density and temperature structure.

2.3.1 Altitude Structure

The structure of the atmosphere is dominated by Earth's gravity and solar radiation. Near the ground, the atmosphere contains 78% N_2, 21% O_2, by volume, plus trace amounts of other gases, the more important of which are H_2O, CO_2, Ar, and O_3. Up to a height of ~120 km, the atmosphere is convective and turbulent and this region is called the *turbosphere*. The turbosphere is well mixed so that the relative abundance of atmospheric constituents remains nearly constant. Above the turbosphere, atmospheric constituents exhibit separate altitude density profiles according to molecular mass.

The temperature of the atmosphere varies with altitude and this is the basis for dividing the atmosphere into different regions. Figures 2.1a and b give the atmospheric temperature variation with altitude for different solar conditions. The lowest layer, the *troposphere*, is heated from below by the surface which in turn is warmed by the Sun. The temperature falls with altitude up to ~10 km where the temperature reaches a local minimum. This minimum is called the *tropopause*. Terrestrial weather systems occur in the troposphere, and the flat tops that can occur with strong thunderstorm clouds are indicative of the tropopause since the instability that drives thunderstorm development only occurs if the temperature falls with altitude.

In the layer above the tropopause, the *stratosphere*, the temperature rises with altitude due to heating by the absorption of solar EUV radiation by ozone, O_3. The stratospheric temperature rises to ~50 km where a local maximum occurs. This point is called the *stratopause*. The region immediately above the stratopause is the *mesosphere*. Here the O_3 density is too low to counteract infrared radiative cooling from other species, principally the triatomic molecules CO_2 and H_2O which are excellent infrared radiators, and the temperature falls with altitude, reaching its lowest value in the atmosphere, ~180 K (–90°C), at the *mesopause*, ~80 km altitude.

The *thermosphere* lies above the mesopause. In the thermosphere the atmosphere heats again due to absorption of solar EUV and concomitant exothermic chemical reactions. At thermospheric altitudes there are very few triatomic molecules to radiate heat away. This contributes to thermospheric heating because this local cooling mechanism does not operate. Above the *thermopause* (that is, above ~300–400 km) lies the *exosphere*, the last remnant of the atmosphere that extends into space. In the exosphere the density is so low that the mean free paths are comparable to or larger than the scale height (see below). As a result, particle motions are essentially ballistic trajectories and different gas species are decoupled. The thermosphere and exosphere are the regions of primary interest here.

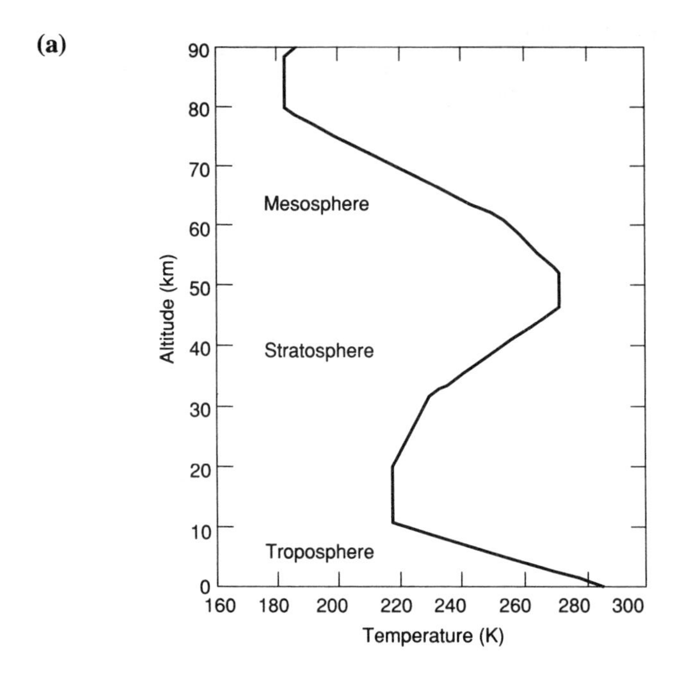

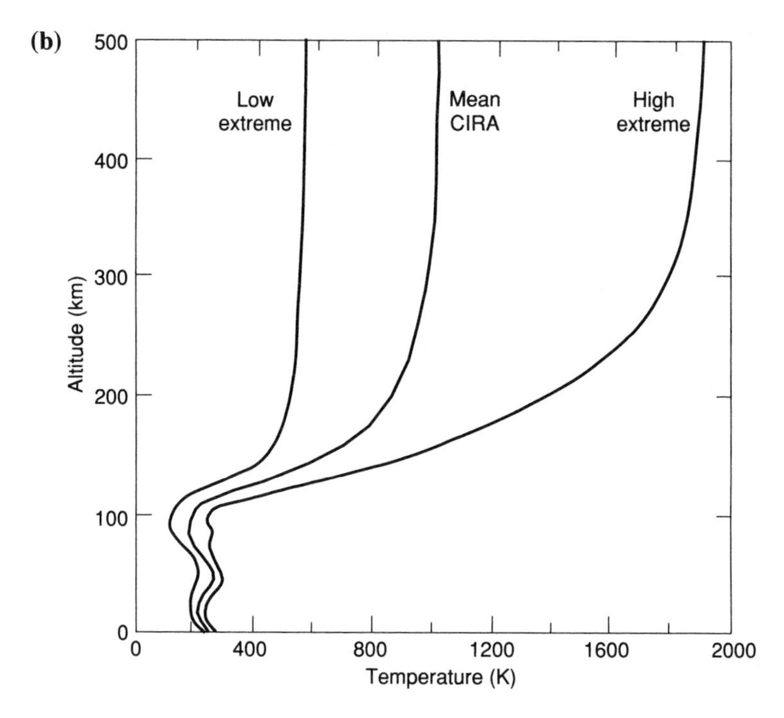

Figure 2.1 (a) The temperatures adopted for the "standard" atmosphere (adopted form the 1962 COESA report). (b) Mean CIRA temperatures and low extreme and high extreme temperatures (adopted from the *Handbook of Geophysics and the Space Environment*, 1985).

We can adequately describe the variation of atmospheric gas density with altitude by considering an ideal gas in Earth's gravity. For such a gas we can write

$$P = nkT \tag{2.3.1}$$

where P is the pressure (in N/m^2 or Pa), n is the number density (number/m^3), T is the temperature in Kelvins, and k is *Boltzmann's constant* (1.38×10^{-23} J/K). Consider an element volume of air of height dz with mass density ρ at rest in equilibrium under the downward force of gravity. Letting μ be the mean molecular mass, we have $\rho = \mu n$. For the volume to be at rest, its mass must be held up against gravity, which implies that there must be a larger pressure at the bottom of the volume element than at the top. The pressure difference dP between the upper and lower faces must balance the weight of the gas so that

$$\mu n a(r) dz + dP = 0 \tag{2.3.2}$$

In the thermosphere, above ~120 km, the temperature is approximately constant with altitude, so differentiating equation 2.3.1 with respect to z and using equation 2.3.2 gives

$$\frac{dn}{n} = -\frac{\mu a}{kT} dz \tag{2.3.3}$$

If we also ignore changes in $a(r)$ of a few percent, equation 2.3.3 integrates to

$$n = n_0 \exp\left(-\frac{z}{H}\right) \tag{2.3.4}$$

where n_0 is the density at a chosen reference altitude corresponding to r_0 and $H \equiv kT/\mu a(r_0)$ is the atmospheric scale height. The derivation of $P = P_0 \exp(-z/H)$ is left as an exercise.

Another factor that determines the behavior of the uppermost atmosphere is the mean distance an atom or molecule travels between collisions. The mean free path, l, is given by

$$l = \frac{1}{\sqrt{2}} \frac{1}{\sigma n} \tag{2.3.5}$$

where σ is the collision cross-sectional area. Since $n(z)$ falls exponentially with z, eventually the mean free path will become comparable to the scale height. Setting $H = l$ gives a rough estimate of where the gas makes the transition from fluid motions to ballistic trajectories of individual atoms or molecules. The result is $z_t = H \ln(\sqrt{2}\sigma n_0 H)$ relative to the reference altitude at which $n = n_0$. Using a typical cross-section of 10^{-18} cm^2 and the surface number density of 3×10^{19}/cm^3 gives z_t = 270 km. The transition is not sudden and the onset of diffusive rather than fluid transport begins much lower, at ~90 km. The dynamics of the mesosphere and lower thermosphere are therefore complicated by the fact that the dynamics of gas transport and flow change dramatically through this region.

The fact that H depends on μ is important for the thermosphere and exosphere. Below the turbopause, mixing ensures that $\mu \simeq 29 M_{\rm H}$ ($M_{\rm H}$ is the proton mass), corresponding to the average atmospheric composition. Using a temperature of 270 K, this gives a scale height of ~8 km. Corrections can be made for variations in temperature with height (see Ratcliffe, 1972, p. 5.) Above ~120 km, atmospheric mixing is ineffective and the scale heights for different gas species separate according to mass, so that, with base densities and temperatures determined by lower thermospheric density and temperature, the lightest elements (H, He) have much greater scale heights than heavier species (N, O). The top of the exosphere, called the *geochorona*, is dominated by atomic hydrogen and extends well into space, $z > 3\ R_E$ (Rairden et al., 1986). Although insignificant for satellite drag, the geochorona finds important application in the indirect imaging of the charged particle environment of the Earth and other planets (Roelof and Skinner, 2000; Mauk et al., 2003).

Figure 2.2a shows atmospheric mass density as a function of altitude. Note the increase in slope above ~120 km, reflecting much greater scale heights due to the high thermospheric and exospheric temperatures. Because the scale height is directly proportional to temperature, there is a strong diurnal variation in atmospheric density at altitudes above the turbosphere (figure 2.2b), where the temperature varies strongly between day and night (figure 2.2c).

Figure 2.3 gives the density/height profiles of the most common exospheric constituents. The slopes, which reflect the scale heights, are ordered by species mass. The scale heights depend strongly on solar activity due to the EUV emission from active regions on the Sun. When solar activity is high, more EUV radiation strikes the upper atmosphere and the thermospheric and exospheric scale heights increase. During geomagnetic storms, energetic particle precipitation from the magnetosphere and Joule heating due to magnetospheric currents closing in the ionosphere can dominate the heat input to the exosphere at middle and high latitudes. Figure 2.4 shows the strong correlation between solar and geomagnetic activity, and upper atmospheric density and temperature.

Using the mass density altitude profile in figure 2.2a and equation 2.2.9, one can readily estimate the lower altitude range at which orbits should remain stable to atmospheric drag. Taking typical values for a relatively large spacecraft, 3 m on a side and with a mass of 10^3 kg, one obtains $\tau_d \sim 6300$ s (two hours) at $z = 100$ km, 6×10^5 s (1 week) at 200 km, and 6×10^7 s (2 years) at 500 km. At 600 km and 800 km τ_d is ~20 years and ~200 years, respectively. The lowest useful altitude for long-term orbits is therefore about 800 km, and objects below 200 km altitude are removed by drag quite rapidly. As figure 2.3 indicates, the mass densities throughout this altitude range vary greatly, so proper estimates of atmospheric drag require a characterization of the uppermost atmosphere's variability.

2.3.2 Thermospheric and Exospheric Variability

The exospheric density changes in response to solar and geomagnetic activity, leading to large variations in drag forces in low Earth orbit (LEO). Measures of the solar and geomagnetic forcing are required to account for exospheric variability. A number of indices based on conveniently available, albeit indirect, measures of these driving effects have been developed and are widely used to develop numerous empirical models of atmospheric variability. Because solar X-ray and EUV radiation does not penetrate the

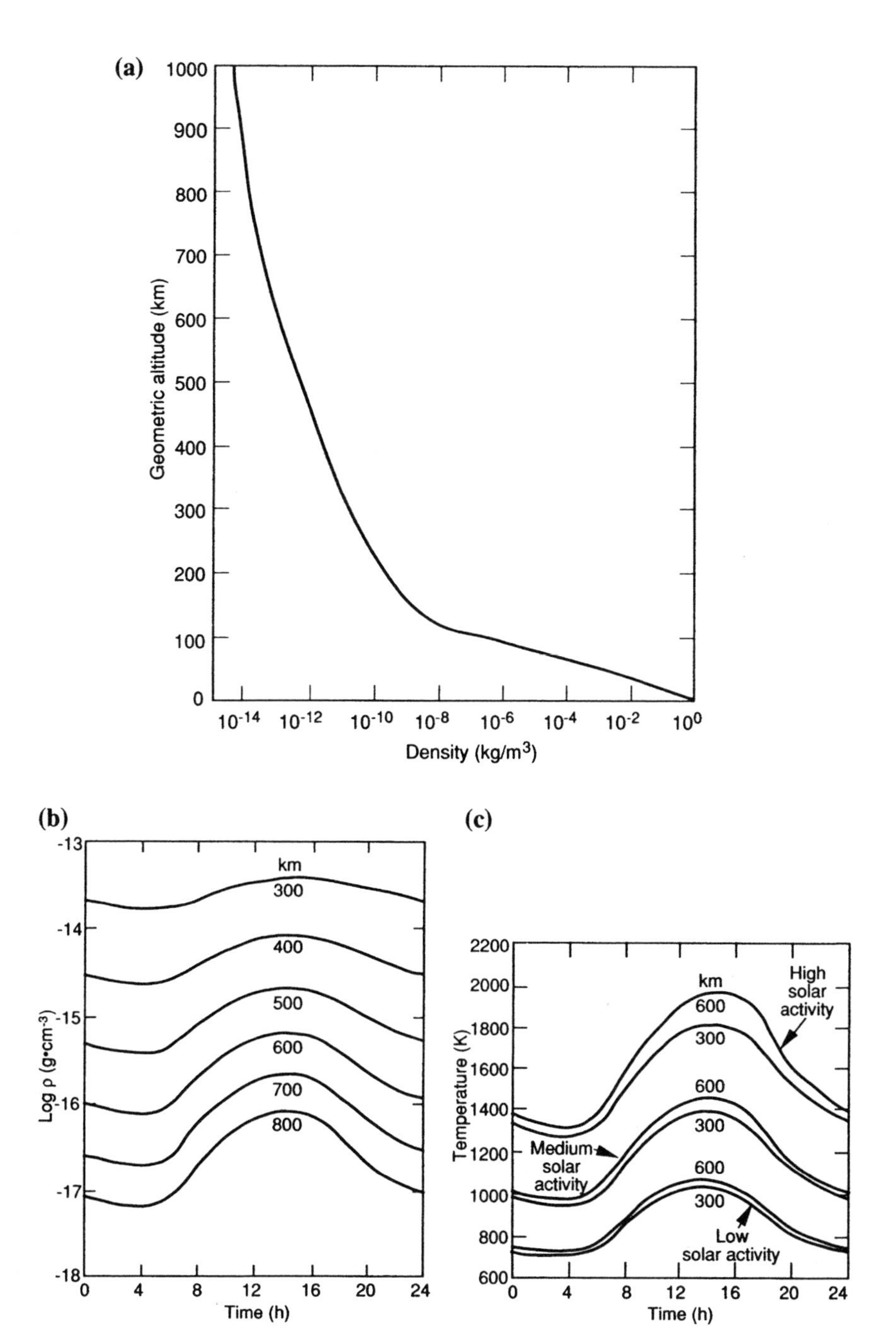

Figure 2.2 (a) Atmospheric mass density as a function of altitude (adapted from the *Handbook of Geophysics and the Space Environment*, 1985. (b) Diumal variation of the upper atmospheric dinsity for various altitudes for a medium level of solar activity (adapted from the *COSPAR International Reference Atmosphere*, 1964). (c) Diurnal variation of temperature for various altitudes for three levels of solar activity (adapted from *CIRA*, 1964)

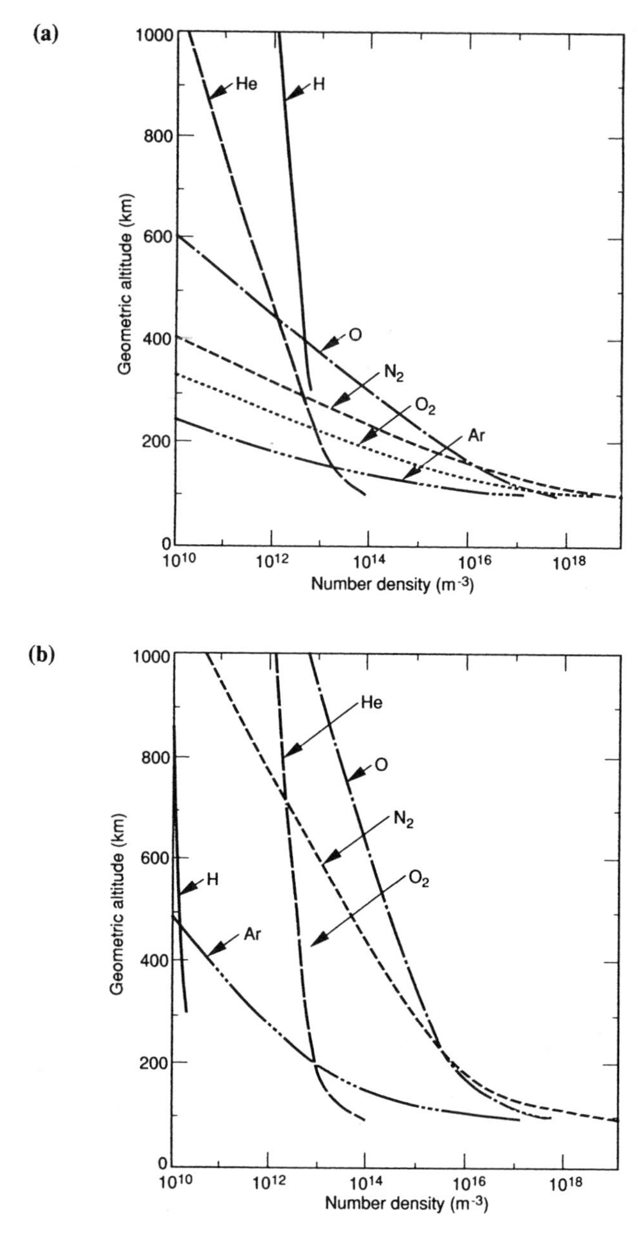

Figure 2.3 (a) Relative concentrations of atmospheric constituents during periods of minimum solar activity. (b) Relative concentrations atmospheric constituents during periods of maximum solar activity (adapted form *U.S. Standard Atmosphere*, 1976).

atmosphere to the ground, proxies for these emissions had to be used until fairly recently. Solar radio emission at 10.7 cm is highly correlated with sunspot number and flare activity, so an index based on the 10.7 cm intensity, F10.7, was developed. Direct space-based measures of solar X-ray and EUV intensity and distribution have been available

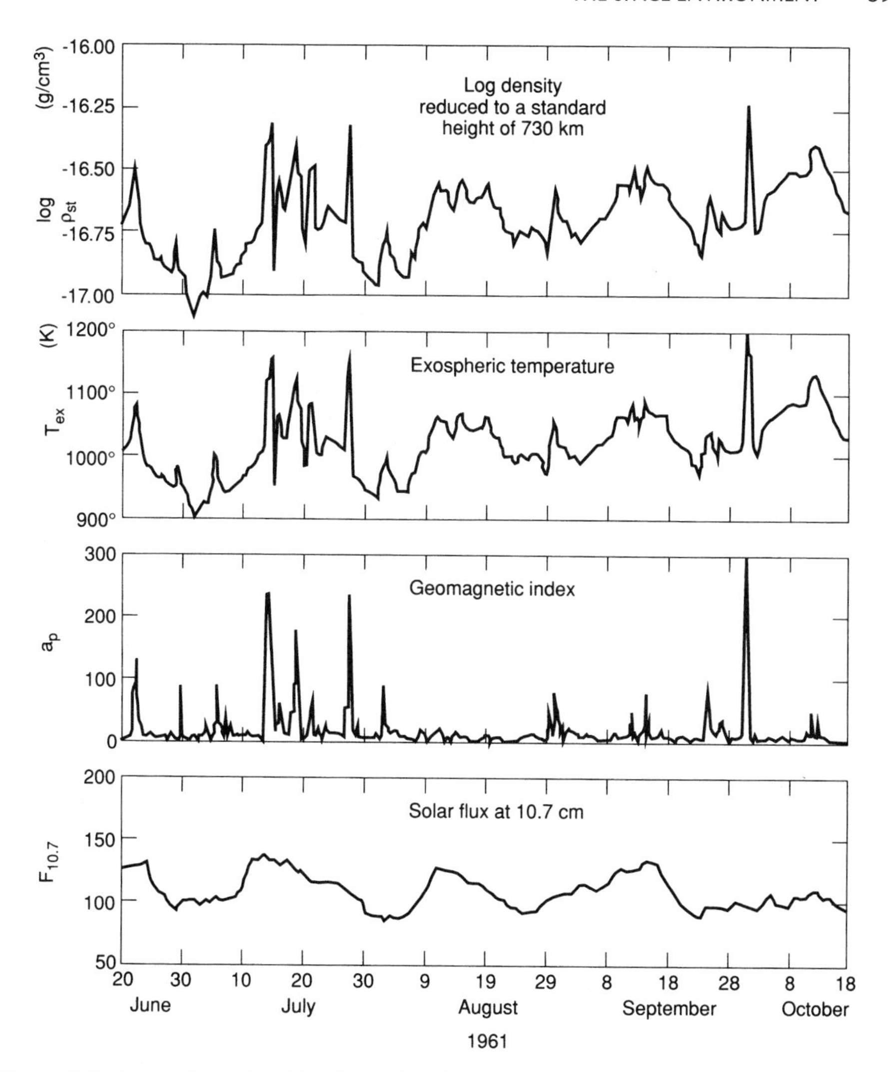

Figure 2.4 Atmospheric densities (log ρ_{st}) and temperature (T_{ex}) derived from the drag of the Explorer IX Satellite (1961–1981), compared with the geomagnetic index (a_p) and the 10.7 cm solar flux ($F_{10.7}$). The drag was determined from precise positional measurements on photographs taken with the Baker–Nunn cameras. Log density reduced to a standard height of 730 km.

only relatively recently with the advent of space-based X-ray and EUV sensors on the Yohkoh (1991–2001) and SOHO (1995 to the time of this writing) spacecraft and now, with the Solar X-ray Imager (SXI), on the latest generation of NOAA geostationary weather-monitoring satellites (GOES 12), the first of which became operational in September 2001. Because the F10.7 index provides coverage over a long time period covering several solar cycles, it remains extremely valuable and exospheric density models successfully employ it.

Geomagnetic variability has been known since the time of Gilbert, who first noted variations in the geomagnetic field associated with intense auroral storms. A variety of indices based on magnetic field measurements on the ground have been developed to indicate different features of the ionosphere and magnetospheric behavior. Two of the more reliable indices for monitoring geomagnetic storm dynamics, of importance for the exosphere, are the Kp and ap indices. The Kp index is a logarithmically scaled measure of the variability of the geomagnetic field at middle latitudes, while ap presents the same information but uses a linear scale (Mayaud, 1980). The F10.7, Kp, and ap indices are available through NOAA's Space Environment Center (SEC).

Figure 2.4 illustrates the correlation between exospheric densities and temperatures and indices of solar and geomagnetic activity for a five-month period. The density and temperature were inferred from drag effects on the orbit of the Explorer IX satellite. The density and temperature are highly correlated because it is the exospheric temperature that is responsible for increases in the scale height and consequently in the density. Note that while the inferred temperature variation is about 300 K or only ~30%, the density varies by a factor of 3 to 6. The short duration, ~1 day, spikes in density correlate one-to-one with spikes in the ap index while the longer term, ~30 day, variations in density correspond to the same period and maxima in the F10.7 index. The periodicity in F10.7 is indicative of active regions rotating quasi-rigidly with the Sun's 27-day rotation period. Because of these large density variations, low-altitude spacecraft may experience more than an order of magnitude increase in atmospheric drag during strong solar and/or geomagnetic activity, due to the increased density at orbit altitude.

Despite this apparent complexity in the exospheric density variations, it is possible to model the orbit-averaged exospheric density and temperature reasonably accurately. One commonly used model of the upper atmospheric state is the Jacchia (1970) model (see also Jacchia, 1972). Refined versions include more data, and updated model parameters (Jacchia, 1977) provide agreement with observations to within ~20% over the 500–1300 km altitude range (Marcos et al., 1978; Bass, 1980a, 1980b; Eisner, 1982; Liu et al., 1982; Gaposchkin and Coster, 1988; Marcos, 1988).

The model is based on a one-dimensional vertical diffusion equation for the number density, solved for each of four major constituents, N_2, O_2, O, and He. Hydrogen is treated differently since it is generated by dissociation of H_2O and CH_4 in the thermosphere and has a very large scale height owing to its low mass. To accommodate these issues the H density is taken as zero below 500 km. After the number densities are computed, the mass density is obtained by adding up the mass densities of all species. The basic equation for the density is (Jacchia, 1970)

$$\frac{1}{n_i}\frac{dn_i}{dz} = -\left(\frac{\mu_i g}{kT} + (1+\alpha_i)\frac{1}{T}\frac{dT}{dz}\right) \tag{2.3.6}$$

where α_i is the thermal diffusion coefficient, n_i is the number density, and μ_i is the molecular mass, all for the ith species. For $\alpha_i = 0$, equation 2.3.6 is what one obtains from equations 2.3.1 and 2.3.2 if one relaxes the constant temperature assumption, so departures in the thermodynamics from non-reactive ideal gas behavior are represented by α_i. The temperature is the same for all species and is expressed as

$$T = T_{\mathrm{EXO}} - (T_{\mathrm{EXO}} - T_0)\exp\{-\xi(z - z_0)\} \tag{2.3.7}$$

where T_0 is the temperature at the reference altitude z_0. The temperature asymptotes to the exospheric temperature, T_{EXO}, with a scale height of $1/\xi$. The reciprocal temperature scale height is determined empirically (Jacchia and Slowey, 1963; Jacchia, 1965) from

$$\xi = 0.0291 \exp\left\{-\frac{1}{2}\left[\frac{T_{EXO} - 800}{750 + 1.22 \times 10^{-4}(T_{EXO} - 800)^2}\right]\right\} \qquad (2.3.8)$$

Jacchia sets the boundary condition for equation 2.3.9 at $z_0 = 90$ km (or 120 km in some versions) and $T_0 = 183$ K.

The exospheric temperature is the parameter of the model through which solar and geomagnetic forcing are introduced. It accounts for the diurnal and semi-annual variation of solar illumination, the variation of solar EUV both with the solar cycle and shorter time scales via the F10.7 index, and geomagnetic activity via the Kp index. It is written

$$T_{EXO} = D_V(T_{SS} + \Delta T_{SS} + \Delta T_{SA}) + \Delta T_M \qquad (2.3.9)$$

where D_V is the diurnal variation applied to all solar effects, T_{SS} is the solar cycle effect, ΔT_{SS} is the effect of shorter time scale solar activity, ΔT_{SA} is the semi-annual variation due to the eccentricity of Earth's orbit, and ΔT_M is the contribution from geomagnetic activity. The diurnal variation is lagged two hours behind the maximum solar illumination. The variation with solar cycle is a linear function of the averaged F10.7, averaged over the previous three solar rotations. The shorter time scale variation, ΔT_{SS}, uses F10.7 lagged by one day. The geomagnetic response is lagged relative to Kp by six to seven hours. The expressions for the terms on the right-hand side of equation 2.3.9 are given in Jacchia's papers (Jacchia 1965, 1970, pp. 16–23; see also Jacchia and Slowey 1964, 1966, 1968). T_{EXO} values below 800 K are associated with "quiet" solar conditions while values above 1200 K are associated with "disturbed" solar conditions. The exospheric temperature T_{EXO} varies, approximately, between 600 K and 2080 K and the corresponding values of $1/\xi$, computed from equation 2.3.8, are ~35 to ~ 75 km, respectively.

The fact that this modeling approach yields results in agreement with observations to within 20% despite the large variations in exospheric density implies that its representation of the exospheric temperature as the primary factor governing neutral mass densities at orbit altitudes must be essentially correct. Other factors, including horizontal and vertical winds (Hedin et al., 1988), the spatial distribution of magnetospheric energy input, and thermodynamics of chemical and radiative reactions, are important in understanding thermospheric dynamics and represent important refinements to our understanding of atmospheric drag, but are not included in this model. A proper accounting of these effects requires a global, hydrodynamic circulation model such as the Thermosphere–Ionosphere General Circulation Model (TIGCM) (Fesen et al., 1993), developed at the National Center for Atmospheric Research. While providing critical insight into thermospheric dynamics and circulation, such fluid dynamics models are computationally intensive and are not suited to operational or engineering design uses. Moreover, their results are sensitive to the details of the thermospheric heating applied by the magnetosphere, which we are not yet able to specify with sufficient confidence to provide operationally useful inputs to thermospheric dynamics calculations. The empirical exospheric models therefore remain our best guides for estimating orbit dynamics.

Although we have not discussed the chemistry of the atmosphere, the presence of monoatomic species, positive ions, and free radicals under the ram pressure of an orbiting spacecraft can produce significant chemical interactions with exposed surfaces.

2.4 The Ionosphere

The same UV and X-ray solar radiation that heats the upper atmosphere also ionizes atoms and molecules in the upper atmosphere above ~80 km, creating a spherical shell of ionization embedded within the thermosphere that is called the ionosphere. The ionosphere extends from about 80 km to several thousand km, gradually increasing then decreasing in electron (and ion) density and smoothly joining the magnetospheric plasmas above. It is so named because it contains ions; that is, singly ionized atmospheric constituents and free electrons. No ionization is produced at low altitudes (below ~80 km), because the column density above 80 km is sufficient to absorb all of the ionizing wavelengths of the solar spectrum. Ionospheric ions and electrons have temperatures similar to those of their neutral source atoms, about 0.1 eV throughout most of the ionosphere. The ionosphere is important in the context of designing for space flight, primarily in the area of radio wave propagation. We shall describe the ionosphere and the basic physics behind its formation, structure, and its influence on radio waves. Reference material for this section may be found in Ratcliffe (1972), Haymes (1971), *the Handbook of Geophysics and the Space Environment* (1985), Tascione (1988), Parks (1991) and Kivelson and Russell (1995).

2.4.1 Ionization

Ionization takes place when air molecules, say O_2, are struck by UV or X-ray photons of sufficient energy $h\nu$ (where h is Planck's constant and ν is the frequency of the photon, $\nu = c/\lambda$, where λ is the wavelength and c the speed of light). Ionization proceeds at a rate determined by the local neutral density, the intensity of the ionizing radiation, and the cross-section of absorption of the radiation by the neutral species.

The basic structure of the ionosphere can be understood by considering the absorption of ionizing radiation as a function of altitude and the consequent altitude profile of ionization density. We begin by considering a slab of air of thickness dz at height z. If I is the intensity of the ionizing solar radiation entering the slab from above at an angle θ, and σ is the cross-section for absorption of the radiation (note that this is not the same cross-section that appears in equation 2.3.5), the amount of radiation absorbed in the slab, dI, can be written as

$$dI = n(z)dz\, \sigma I \sec\theta \tag{2.4.1}$$

Dividing by I and using equation 2.3.4, we get

$$dI/I = n_0 \exp\{-z/H\}\sigma\, dz \sec\theta \tag{2.4.2}$$

which we can integrate from z to ∞ to obtain

$$I = I_\infty \exp\{-\sigma H n_0 \sec\theta \exp(-z/H)\}. \tag{2.4.3}$$

Physically this expresses the important fact that the intensity of ionizing radiation is extinguished exponentially by the column density of gas between the source and z. A key feature of equation 2.4.3 is the nested exponential with z which arises because, while the intensity falls exponentially with the overlaying column density (the outer exponential), the density increases exponentially with decreasing z (the inner exponential). This means that eventually the intensity cuts off very abruptly.

Now the local rate of electron production, q, is directly proportional to the number density and the ionizing intensity, that is

$$q = \beta n I \tag{2.4.4}$$

where β is the cross-section for photoionization by the absorbed radiation. In general, $\beta \neq \sigma$ since radiation can be absorbed without producing ionization. Combining this with equation 2.3.7 gives the ionization rate as a function of altitude

$$q = \beta n_0 I_\infty \exp\{-(z/H + \sigma H n_0 \sec\theta \exp(-z/H))\} \tag{2.4.5}$$

For $z/H \gg 1$ the second term in the exponential argument is vanishingly small, so the ionization rate rises with decreasing z owing to the increasing density of the particles available for ionization. As the altitude continues to decrease, the rapid extinction of I takes over and the ionization rate plummets.

This behavior is best seen by evaluating equation 2.4.5 relative to the altitude of maximum ionization rate. To obtain the maximum ionization rate and the altitude where this occurs, consider the derivative of equation 2.4.4 along an oblique path, $s = z \sec\theta$,

$$\frac{dq}{ds} = \beta\left(I\frac{dn}{ds} + n\frac{dI}{ds}\right) \tag{2.4.6}$$

At the point along s where q is a maximum, dq/ds vanishes, so equation 2.4.6 gives

$$\left.\frac{1}{n}\frac{dn}{ds}\right|_M = -\left.\frac{1}{I}\frac{dI}{ds}\right|_M \tag{2.4.7}$$

where M indicates that equality holds at the maximum value of q. Substituting the variable z for s, recalling the definition of scale height, $(1/n)(dn/dz) = 1/H$, and using equation 2.4.1 in equation 2.4.7 gives

$$\sigma H n_M \sec\theta = 1 \tag{2.4.8}$$

Evaluating n_M at z_M using equation 2.3.7 gives

$$\sigma H n_0 \sec\theta = \exp(z_M/H) \tag{2.4.9}$$

which we substitute into equation 2.4.5 to obtain the maximum production rate

$$q_M = \beta n_0 I_\infty \exp\{-(z_M/H + 1)\} \tag{2.4.10}$$

Defining the new variable $y = (z - z_M)/H$, we finally have

$$q = q_M \exp\{1 - y - \exp(-y)\} \tag{2.4.11}$$

Equation 2.4.11 makes the altitude variation clear. For $z > z_M$ we have $y > 0$ and the ionization rate increases exponentially with decreasing altitude. Here, neutral densities are low enough that the incident solar radiation is not significantly attenuated by absorption, and the electron production is proportional to the neutral density. For $z < z_M$, however, $y < 0$, and the inner exponential term grows large, reflecting the attenuation of the incident radiation, so that the ionization rate falls precipitously.

The ionization rate is balanced either by electron–ion recombination or by transport out of the region of ionization, to yield a net ion/electron density. In equilibrium, electron loss will balance electron production. Here we consider only recombination loss, which is directly proportional to the probability that an electron and an ion will collide and recombine. Recombination can proceed by a variety of mechanisms that divide broadly into two categories, two-body and three-body recombination. The total loss rate may be written as

$$L = \alpha_2(z) n_\mathrm{e} n_\mathrm{i} + \alpha_3 n_\mathrm{e} n_\mathrm{i} n(z) = \alpha(z) n_\mathrm{e} n_\mathrm{i} \tag{2.4.12}$$

where n_e is the electron density, n_i is ion density, and the two- and three-body recombination coefficients are denoted $\alpha_2(z)$ and α_3, respectively. The third bodies are taken to be neutral gas molecules. Quantities other than n_e and n_i with the strongest altitude dependence are written as functions of z.

Two-body recombination is proportional to the collision of an electron and an ion, hence the product $n_\mathrm{e} n_\mathrm{i}$. Satisfying conservation of momentum and energy with only two bodies is not easy but is much faster if the ion is a molecule which dissociates into two atoms during the recombination. Radiative recombination proceeds by emission of a photon and is the only way atomic ions and electrons can recombine in two-body reactions. Radiative recombination is about four orders of magnitude slower than dissociative recombination. Since the molecular ion density falls more rapidly with altitude than the atomic ion density, $\alpha_2(z)$ falls with increasing altitude. With three-body recombination it is much easier to satisfy the conservation laws and there are several mechanisms by which recombination can occur, but the reactions all require a three-body collision, indicated by the additional factor $n(z)$. For purposes of the present discussion, we group the recombination processes together in the altitude-dependent net recombination coefficient $a(z)$. For altitudes below 200 km $a(z) \simeq 10^{-14}$ m^3/s and above 300 km $a(z) \simeq 10^{-16}$ m^3/s.

In general, the electron and ion densities are nearly equal, $n_e = n_i$, since any imbalance results in electric fields that act to restore overall charge neutrality. In equilibrium, $L = q$, and from equations 2.4.11 and 2.4.12 we obtain

$$n_\mathrm{e} = \sqrt{q_M/\alpha} \exp \tfrac{1}{2}\{1 - y - \exp(-y)\} \tag{2.4.13}$$

which means that the resulting electron density also maximizes at z_M and has an altitude profile closely related to that for q. This behavior results in a well-defined layer of peak electron (and ion) density; in other words, that layer where the ionizing UV radiation incident from above the atmosphere has not been absorbed because the column-integrated neutral density above the layer is too small, yet where the ambient neutral density has risen to the point that the radiation *is* likely to encounter and ionize many neutrals. Below this layer the neutral density is high, but the UV is too weak from absorption in the layers above to ionize many neutrals. Above this layer, the UV is strong

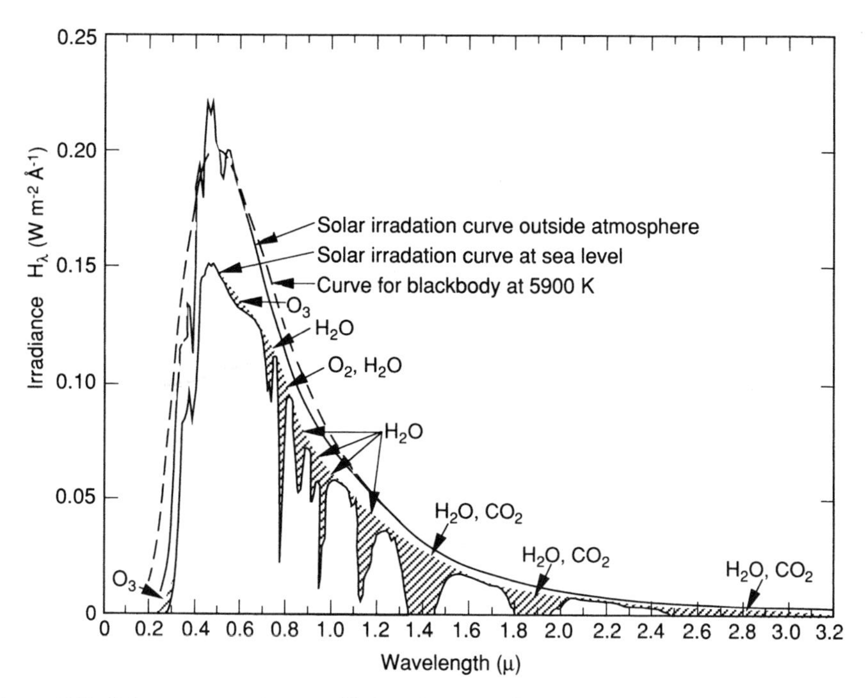

Figure 2.5 Solar spectrum above and below the atmosphere.

but there are very few neutrals available to ionize. This ionization layer is known as a Chapman layer after Sidney Chapman, a pioneering researcher in space physics who first explained the ionospheric density profile.

In fact, several ionization layers can form, since the absorption cross-section α is dependent both on the dominant atomic or molecular species and on the wavelength of the UV. In figure 2.5, the solar spectrum above the atmosphere is contrasted with the spectrum below the atmosphere. The spectrum is "notched out" at the frequencies where the indicated atomic or molecular absorption is strong, at various layers in the atmosphere. Most of the structure is in the infrared, although UV absorption by ozone appears below $\sim 0.3\mu$m. Table 2.1 gives the names of the major ionospheric layers, their altitudes, major ionic contributors, and the UV wavelength dominating that layer.

At night, the production from solar EUV and X-rays drops to zero, and we no longer attain an equilibrium state. Ignoring ionization associated with the aurora, the density profile in equation 2.4.13 gives the initial condition for decay. From equation 2.4.12 we have

$$\frac{dn_e}{dt} = -\alpha(z)n_e^2 \tag{2.4.14}$$

which integrates to give

$$n_e(t) = 1/[\alpha(z)(t - t_0) + 1/n_e(0)] \tag{2.4.15}$$

From this we see that the time for $n_e(0)$ to be reduced by half is given by

$$\tau_e = 1/\alpha(z)n_e(0) \tag{2.4.13}$$

Table 2.1 Layers of daytime midlatitude ionosphere

Layer	Altitude (km)	Major component	Production cause
D	70–90	NO^+, O_2^+	Lyman alpha, X-rays
E	95–140	O_2^+, NO^+	Lyman beta, soft X-rays, UV continuum
F1	140–200	O^+, NO^+	He II, UV continuum (100–800Å)
F2	200–400	O^+, N^+	He II, UV continuum (100–800Å)
Topside F	> 400	O^+	Transport from below
Plasmasphere	> 1200	H^+	Transport from below

Source: *Handbook of Geophysics and the Space Environment* (1985).

Note that the rate of ionization decay decreases as the electron density falls. Using a typical ionospheric density, $n_e(0) = 10^{11}/m^3$, and the values for $\alpha(z)$ quoted above, we have $\tau_e \simeq 10^3$ s below 200 km and $\tau_e \simeq 10^5$ s above 300 km. Below 120 km, three-body reactions are important and the decay time is even shorter. We therefore expect a strong diurnal variation in n_e below 200 km. Since the rate of recombination at some altitudes is so low, the electrons may move considerable distances before recombining. Because the mean free paths are large at these altitudes, electrons can be found in the greatest concentration far from the altitude where they are produced most rapidly.

Figure 2.6 gives the altitude profile of ionospheric electron density for different solar conditions. The D layer disappears through recombination at night. Electron and ion recombination and chemistry in the long mean-free-path region above ~120 km result in a diffusive coalescence of the F1 and F2 layers at night. These layers therefore appear as plateaus rather than as peaks. Above the F layer peak, the mean free paths are large and the ionization production rate is low, so that the electron density is determined by diffusion from below rather than by the local ionization.

The ion scale heights display an interesting property resulting from the dynamics of plasmas. One might naively expect the ions to have the same scale heights as their parent neutral atoms while the electrons, being a factor of 2,000 to 30,000 less massive, would have enormous scale heights. But the electrical attraction of the populations enforces quasi-neutrality and compels the ion and electron concentrations to be equal at all altitudes. This affects the scale heights of both species as follows. Because the ions and electrons must have the same altitude distribution, the ionized species scale height can be written as

$$H_i = \frac{kT_e + kT_i}{g(m_e + m_i)} \tag{2.4.14}$$

where T_e and m_e (T_i and m_i) are the electron (ion) temperature and mass. If $T_e = T_i = T$ and we take $m_e \simeq 0$ relative to m_i, this becomes

$$H_i = \frac{2kT}{gm_i} \tag{2.4.15}$$

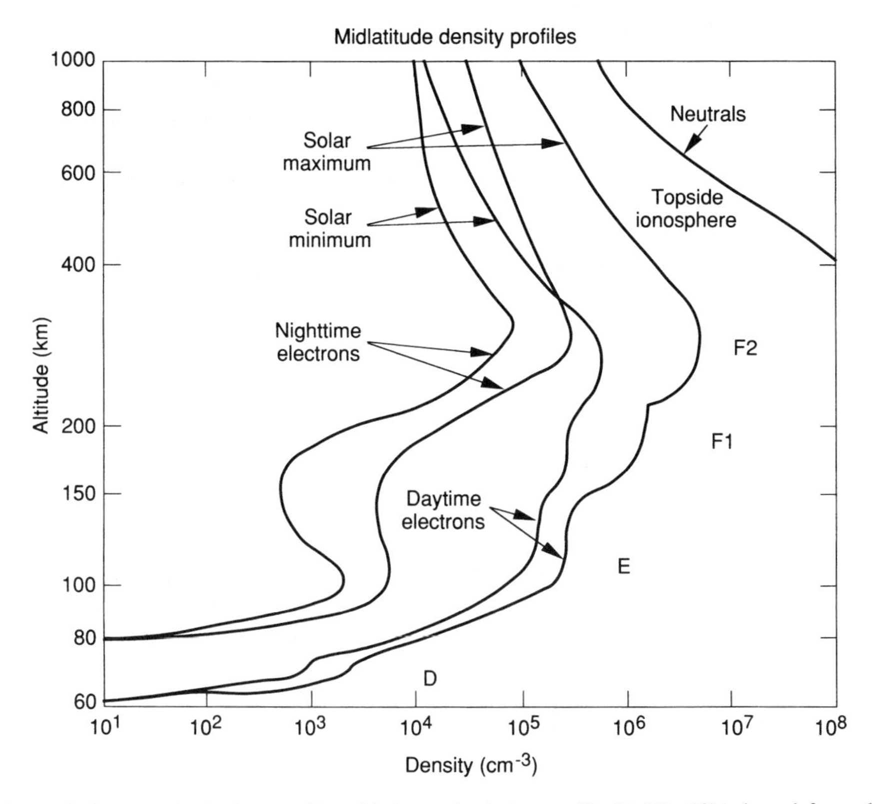

Figure 2.6 Total ionization profile with ionospheric layers (D, E, F1, F2)(adapted from the *Handbook of Geophysics and the Space Environment*, 1985)

which is twice the scale height of the parent neutral population. This doubling of the scale height is accompanied by an electric field that results because the electrons attempt to go to their natural, much greater scale height but are electrostatically constrained by the ions. A small charge separation results, with the electrons above the ions and pulling them upward. Since the scale height is doubled, the upward electric force must be half as strong as the gravitational force, so the electric field E is given simply by

$$E = \frac{g m_\mathrm{i}}{2e} \tag{2.4.16}$$

where e is the electronic charge. The electric field in this diffusive region above the F-layer peak is dominated by the ions with the highest density, namely O^+. (Monoatomic oxygen dominates the neutral density, and therefore the ion density, in this region above ~400 km; see figure 2.3.) However, hydrogen and helium, both of which are light atoms, are present at much lower densities. The ionospheric electric field, with a magnitude set by the mass of the dominant oxygen ions, is strong enough to completely dominate the force of gravity on the much lighter hydrogen and helium ions. The light ions essentially float above the oxygen ions at very great altitudes, and form the base populations for the plasmasphere and the magnetosphere.

2.4.2 Plasma Frequency

One of the most important aspects of the ionosphere to space instrumentation is its effect on the propagation of radio waves. The change in the propagation of radio waves at sunset that allows much better reception of distant AM stations is one familiar effect of the ionosphere. The ionosphere also affects spacecraft transmission and Global Positioning System (GPS) navigation signals. To understand these phenomena we need to consider electromagnetic waves in a plasma.

We begin with two of Maxwell's equations, Ampere's law

$$\nabla \times \mathbf{H} = \mathbf{j} + \frac{\partial \mathbf{D}}{\partial t} \tag{2.4.17}$$

and Faraday's law

$$\nabla \times \mathbf{E} = -\frac{\partial \mathbf{B}}{\partial t} \tag{2.4.18}$$

where $\mathbf{j}$ is the current density, $\mathbf{D}$ is the displacement vector ($= \varepsilon\mathbf{E}$), $\mathbf{B}$ is the magnetic induction, $\mathbf{H}$ is the magnetic field strength ($= \mu\mathbf{B}$), $\mathbf{E}$ is the electric field strength, ε is the dielectric constant or permitivity, and μ is the magnetic permeability (Jackson, 1972).

The current density in the plasma, under the approximation that the ions are too massive to move appreciably relative to the electrons in the radio-wave electric field, can be written

$$\mathbf{j} = -n_{\mathrm{e}}\, e\mathbf{V} \tag{2.4.19}$$

where $\mathbf{V}$ is the electron velocity. The acceleration $\mathbf{a}$ of the electrons by a sinusoidal radio wave is

$$\mathbf{a} = -(e/m_{\mathrm{e}})\mathbf{E}_0 \sin(\omega t) \tag{2.4.20}$$

where $\mathbf{E}_0$ is the amplitude of the wave, $\omega = 2\pi f$ is the wave angular frequency, and t is time. Integrating equation 2.4.20 to get $\mathbf{V}$ gives

$$\mathbf{j} = -\frac{n_{\mathrm{e}}\mathrm{e}^2}{m_{\mathrm{e}}\omega}\mathbf{E}_0 \cos(\omega t) \tag{2.4.21}$$

In free space $\varepsilon = \varepsilon_0$, so

$$\mathbf{D} = \varepsilon_0 E_0 \sin(\omega t) \tag{2.4.22}$$

so equation 2.4.17 becomes

$$\nabla \times \mathbf{H} = \left(\varepsilon_0\omega - \frac{n_{\mathrm{e}}e^2}{m_{\mathrm{e}}\omega}\right)\mathbf{E}_0 \cos(\omega t) \tag{2.4.23}$$

If the factor in brackets goes to zero, $\nabla \times \mathbf{H} = 0$, and the wave cannot propagate because, without a rotational magnetic field, there is no longer an induced electric field according to Faraday's law. (Mathematically, the time derivative term drops out if one takes the curl of equation 2.4.18, so the resulting equation does not have propagating wave solutions.) The frequency at which this occurs is defined as the electron plasma frequency

$$\omega_{pe} = \sqrt{\frac{n_e e^2}{\varepsilon_0 m_e}} \tag{2.4.24}$$

The remarkable feature of the plasma frequency is that it depends only on fundamental constants and the electron number density. Converting from angular frequency to frequency (cycles per second), equation 2.4.24 evaluates to

$$f_{pe} = 8.97 \text{ Hz– m}^{3/2}\sqrt{n_e} \tag{2.4.24}$$

so, for a typical F region electron density of 3×10^{12} m^{-3}, $f_{pe} = 16$ MHz.

The plasma frequency is a natural frequency of a plasma in which the electrons oscillate back and forth while the ions remain essentially motionless. An alternate derivation shows explicitly that the electron motion is simple harmonic oscillation with this frequency. The mode involves no magnetic field because the current density and displacement current cancel, consistent with equation 2.4.23. Recognizing that $\partial \mathbf{E}/\partial t = \omega \mathbf{E}_0 \cos(\omega t)$, we see that the factor in equation 2.4.23, is ω times the dielectric constant of the medium (compare equation 2.4.17):

$$\varepsilon = \left(1 - \frac{\omega_{pe}^2}{\omega^2}\right)\varepsilon_0 = N_R^2 \varepsilon_0 \tag{2.4.25}$$

where N_R is the index of refraction.

We can now discuss how the ionosphere affects radio transmission. At high frequencies, $\omega \gg \omega_{pe}$, $N_R \simeq 1$ and the effect on wave propagation is small. For GPS signals, however, extremely precise timing is critical and even small effects are important. The most accurate GPS navigation must account for the slight change in light travel time through the ionosphere (Klobuchar, 1987; Feess and Stephens, 1987). Electron density irregularities cause tiny erratic time delays and result in unpredictable and uncorrectable navigation errors. If the irregularities are strong, the received signal may appear to phase skip or scintillate, much as light "shimmers" through an exhaust plume or above a candle due to tiny irregularities in the index of refraction as the hot gases mix turbulently with the cooler ambient air. Severe ionospheric scintillation can result in loss of lock on the GPS satellite signal carrier.

More significant effects occur closer to ω_{pe}. For $\omega >\sim \omega_{pe}$, ε is significantly smaller than ε_0 and strongly dependent on n_e. Irregularities in n_e cause significant variation in N_R and an initially planar, coherent wave front will emerge with random delays and phase shifts. Exactly at ω_{pe}, N_R vanishes, meaning that the wave cannot propagate but will reflect (refract) into the lower density plasma from whence it came. For $\omega < \omega_{pe}$, N_R is imaginary, which means that the wave decays. A wave transmitted below ω_p from within the plasma will be absorbed and simply heat the electrons. For these reasons spacecraft communications must use frequencies well above the maximum f_{pe}.

The ionosphere has a profound effect on terrestrial radio propagation as well. Let $\omega_{pe,max}$ be the plasma frequency at the peak ionospheric density and consider a wave with $\omega < \omega_{pe,max}$ propagating up to the ionosphere from below. As the wave propagates up it encounters progressively higher n_e and eventually reaches an altitude where its frequency matches the local plasma frequency. The wave propagates no higher but is reflected approximately specularly back down. This explains why AM-band broadcasts reach a larger area at night. For any set of realistic conditions the AM band is well below $\omega_{pe,max}$, so AM signals bounce off the ionosphere. At night the D region disappears, so that the height of the reflection rises and the bounce distance increases. Moreover, the reflection degrades if the ionosphere is irregular because the wave is not coherently reflected. This is why HF communications can be disrupted by geomagnetic activity.

Ionospheric wave reflection is used to determine both the maximum n_e and the density profile below the maximum. This is done by sounding the ionosphere with a frequency-ramped signal and timing the return signal. The time delay between transmission and return uniquely specifies the altitude at which the wave and plasma frequencies match. The corresponding n_e is evaluated from equation 2.4.24. The maximum reflected frequency corresponds to $\omega_{pe,max}$.

Below the plasma frequency, a class of plasma waves known as electrostatic waves (waves with $\mathbf{H} = 0$) can propagate in a plasma, and can be generated by plasma instabilities. These waves do not exist in free space but are modes of a plasma. Since these waves are not relevant for spacecraft design issues they are not considered here. Treatments of other plasma waves can be found in Ratcliffe (1972), Chen (1984), and Swanson (1989).

2.4.3 Debye Length

We next consider the dynamics of plasmas in relation to electrostatic charges. Since a plasma is made up of ions and electrons, systems in space can acquire an electrostatic charge which depends on the materials involved, solar illumination, and the properties of the surrounding plasma. Because electrons and ions have very different masses but often similar temperatures, the average thermal speed of the electrons is typically a factor of at least fifty higher than that of the ions. As a result, systems in space do not generally remain at the same potential as the surrounding medium but tend to acquire a negative potential. In sunlight, however, positive potentials can occur if the photoelectron emission from spacecraft surfaces is high. The plasma in the vicinity of the spacecraft responds to the resulting electrostatic charge distributions. The limits of spacecraft charging and the distance range over which the plasma shields these charges are factors that can be important for spacecraft design.

We begin by considering the scale length for electrostatic shielding in a plasma. In equilibrium, a plasma with no external forces applied to it will remain quasi-neutral, with almost exactly equal numbers of positive and negative charges in a given volume. This condition is maintained because any inequality between charges within a volume will generate an electric field in the plasma to which the electrons (primarily) respond by moving, canceling the field on a time scale given by the plasma frequency. But since the electrons and ions have finite, non-zero, temperatures, their thermal velocities prevent the plasma from perfectly canceling the electrostatic field. Rather, the electric field penetrates into the plasma with a scale called the Debye length.

One can see that a finite temperature corresponds to finite electric field penetration by considering the following thought experiment. Imagine first a plasma of electrons and ions with a very low temperature. Place a positive test charge into this plasma. The electrons will be attracted to the charge, the ions repelled, and the charge will be neutralized by the inrushing electrons with the result that the electric field of the charge does not penetrate into the plasma. Now consider a second plasma, but let the electrons and ions be hot so that they are all moving rapidly. As before, place a test charge in the plasma. Although, as before, the electrons are attracted to the charge, the fact that they were already moving rapidly means that rather than rushing in to neutralize the charge they orbit the charge, creating a cloud of negative charge around the positive test charge. Far enough away (we will define "far enough" presently) the electric field of the test charge is shielded, but within the cloud of electrons the field of the test charge "penetrates" into the plasma and is closer to its vacuum value the closer one gets to the charge. If the electrons are very hot, the test charge does not affect the motions much and the shielding cloud does not cluster very tightly to the test charge. From this we see that a finite temperature will lead to finite penetration of electrostatic fields into a plasma and that the penetration length increases as the temperature increases. We now proceed to a quantitative derivation.

We want to solve for the electric field of a charge Q in a plasma with ion and electron temperatures T_{e} and T_{i}. For an electrostatic field we have $\partial \mathbf{B}/\partial t = 0$ so, from equation 2.4.18, $\nabla \times \mathbf{E} = 0$ which in turn implies that $\mathbf{E} = -\nabla\varphi$, where φ is the potential energy per unit charge. As with the gravitational force, this is equivalent to saying that the electrostatic field is conservative. We are free to add an arbitrary constant value to ϕ and, for the present problem, it is convenient to take $\varphi = 0$ at infinity. The equation governing electrostatic fields is Poisson's equation

$$\nabla \cdot \mathbf{E} = \frac{\rho_q}{\varepsilon_0} \tag{2.4.26}$$

where ρ_q is the charge density. Using $\mathbf{E} = -\nabla\varphi$, this becomes an equation for the electric potential

$$\nabla^2\varphi = -\frac{\rho_q}{\varepsilon_0} \tag{2.4.27}$$

The electron and ion densities together with the test charge give the charge density

$$\rho_q = e(n_{\mathrm{i}}(\mathbf{r}) - n_{\mathrm{e}}(\mathbf{r})) + Q\delta(\mathbf{r} - \mathbf{r}_Q) \tag{2.4.28}$$

where $\delta(\mathbf{r} - \mathbf{r}_Q)$ is the Dirac delta function and $\mathbf{r}_Q$ is the position of the test charge. For simplicity we have assumed that the ions are singly charged, but equation 2.4.28 is easily generalized. The Dirac delta function is defined as follows: $\delta(\mathbf{p}) = 0$ everywhere except at $\mathbf{p} = 0$; and $\int\delta(\mathbf{p})\mathrm{d}\mathbf{p}^3 = 1$ if the volume integral includes $\mathbf{p} = 0$. The delta function precisely describes the concept of a point charge.

To obtain an equation only in ϕ we need to express ρ_q, that is, n_{e} and n_{i}, in terms of the potential. From statistical mechanics (e.g. Reif, 1965) we define a conservative force field $\mathbf{F} = -\nabla U$, where U is the potential energy. In such a situation, the probability of finding a particle at a point $\mathbf{r}$ where the potential energy $U(\mathbf{r})$ is proportional to

$\exp(-U(\mathbf{r})/kT)$. This probability factor is called the Boltzmann factor. For the present problem, $U_e(\mathbf{r}) = -e\varphi(\mathbf{r})$ and $U_i(\mathbf{r}) = +e\varphi(\mathbf{r})$. At infinity $n_e = n_i \equiv n_0$ and $\varphi = 0$, so the electron and ion densities are given by the ratio of the Boltzmann factors at $\mathbf{r}$ divided by those at infinity (which are unity). We then have

$$\left.\begin{aligned} n_e(\mathbf{r}) &= n_0 \exp(e\varphi(\mathbf{r})/kT_e) \\ n_i(\mathbf{r}) &= n_0 \exp(-e\varphi(\mathbf{r})/kT_i) \end{aligned}\right\} \qquad (2.4.29)$$

Our equation for $\varphi(\mathbf{r})$ is obtained by substituting these in equation 2.4.28. For convenience we also let $\mathbf{r}_Q = 0$. We then have

$$\nabla^2\varphi(\mathbf{r}) = \frac{en_0}{\varepsilon_0}[\exp(e\varphi(\mathbf{r})/kT_e) - \exp(-e\varphi(\mathbf{r})/kT_i)] + \frac{Q}{\varepsilon_0}\delta(\mathbf{r}) \qquad (2.4.30)$$

Since this is a transcendental equation, we need to make an intelligent approximation to make further progress. Recall from our thought experiment that a cold plasma just neutralizes our test charge whereas a hot plasma leads to shielding. We can now quantify what we mean by cold and hot in terms of the ratio $e\varphi(\mathbf{r})/kT$. If $e\varphi(\mathbf{r})/kT \ll 1$ then the plasma is hot relative to the potential energy associated with the test charge, whereas if $e\varphi(\mathbf{r})/kT \simeq 1$ or >1 then the energy associated with the test charge is comparable to or larger than the thermal energy. Since we are interested in the hot plasma case we apply the approximation $e\varphi(\mathbf{r})/kT \ll 1$ in equation 2.4.30. Note that this approximation will apply only over a range of $\mathbf{r}$ which we will specify shortly. The Taylor expansion for $\exp(x)$ is $1 + x + x^2/2 + x^3/3! + \ldots$, so equation 2.4.30 becomes

$$\nabla^2\varphi(\mathbf{r}) = \frac{e^2 n_0}{\varepsilon_0}\left(\frac{1}{kT_e} + \frac{1}{kT_i}\right)\varphi(\mathbf{r}) + \frac{Q}{\varepsilon_0}\delta(\mathbf{r}) \qquad (2.4.31)$$

The solution to this equation is the potential of a shielded point charge

$$\varphi(r) = \frac{Q}{4\pi\varepsilon_0 r}\exp\left(-\frac{r}{L_D}\right) \qquad (2.4.32)$$

where the shielding scale length, L_D, is called the Debye length

$$L_D = \sqrt{\frac{\varepsilon_0 k}{e^2}} \Bigg/ \sqrt{\left(\frac{n_0}{T_e} + \frac{n_0}{T_i}\right)} \qquad (2.4.33a)$$

or

$$L_D = \sqrt{\frac{\varepsilon_0 k}{2e^2}}\sqrt{\left(\frac{T}{n_0}\right)} = 49\,\mathrm{K}^{-1/2}\mathrm{m}^{-3/2}\sqrt{\left(\frac{T}{n_0}\right)} \qquad (2.4.33b)$$

Equation 2.4.33b applies if $T_e = T_i$ and is evaluated for T in Kelvins, n_0 in particles/m^3 and L_D in meters. One can verify that equation 2.4.32 satisfies equation 2.4.31 and also, by using equation 2.4.32 and Gauss's law, that the total charge enclosed in a sphere at infinity is Q.

As expected, the influence of the test charge is shielded by the plasma and the shielding distance grows as the temperature increases and falls as the density increases. To examine the validity of the approximation $e\varphi(\mathbf{r})/kT \ll 1$, equation 2.4.32 can be rewritten (using $T_e = T_i$) as

$$\frac{e\varphi(r)}{kT} = \frac{1}{3N_D}\frac{Q}{e}\frac{L_D}{r}\exp\left(-\frac{r}{L_D}\right) \tag{2.4.34}$$

where N_D is the number of particles in a sphere of radius L_D

$$N_D = \tfrac{4}{3}\pi n_0 L_D^3 \tag{2.4.35}$$

From equation 2.4.34 we see that as long as $N_D \gg Q/e$ our approximation is valid even for $r \simeq L_D$. In the ionosphere (below 600 km), L_D ranges typically from 0.1 to 10 cm with $N_D \simeq 10^8$ whereas in the magnetosphere $L_D \simeq 0.1$ to 1 km and $N_D \sim 10^{15}$.

The phenomenon of Debye shielding has several implications for spacecraft design. If L_D is much smaller than a typical spacecraft dimension the plasma tends to keep the potential fairly uniform. The plasma also provides a conductive path between different portions of the spacecraft. Large potential differences can still occur in the plasma wake region of a satellite where the plasma density is evacuated by the passage of the satellite and L_D is locally large. In the opposite limit, where L_D is large relative to the spacecraft dimension, large relative potentials could develop at different points on the spacecraft. Dielectric surfaces on the spacecraft could accumulate charge, leading to a destructive discharge event. This problem is particularly relevant for solar arrays.

A large difference in electric potential between spacecraft and the space environment can also adversely affect measurements from particle detectors and electric field probes. Because the presence of the spacecraft itself modifies the local plasma environment, any measurement that will be affected by the relative potential of an instrument to the plasma must be made on an isolated, potential-controlled structure that protrudes beyond the spacecraft Debye sheath. Techniques to alleviate these problems include long (>100 m) electric field probes with voltage biasing, and active spacecraft potential control with electron guns to dissipate excess negative charge.

2.4.4 Spacecraft Charging

The previous section addressed the influence of a charged body on a plasma. One of the consequences of placing an initially uncharged spacecraft in a thermal plasma such as the ionosphere is that its equilibrium charge will not be zero, and a sheath will form about the spacecraft, characterized by the Debye length. This can be understood as follows. If a surface with no net charge is placed into a plasma, initially more electrons than ions will strike its surface due to the electrons' higher thermal speed. (High-energy electrons may penetrate several millimeters into spacecraft materials, thereby charging internal dielectrics; but lower energy ions and electrons deposit their charges directly onto the spacecraft surfaces.) Eventually, the surface potential becomes negative until the repulsion of electrons and attraction of ions adjusts the impinging ion and electron fluxes so that they cancel. A sheath is therefore formed, a region near the spacecraft within which positive ions outnumber electrons to such an extent that the potential due to the negative surface charge is canceled and the plasma outside the sheath does not feel a net potential. The behavior of the plasma in the vicinity of the (now charged) spacecraft is just that described in the previous section.

In addition to the mechanism mentioned above, photoelectron emission due to sunlight is an important effect. If the spacecraft is exposed to sunlight, ultraviolet photon bombardment of the spacecraft surface causes emission of electrons from the sunlit portions of the spacecraft. The photoelectron flux away from the spacecraft surface is usually greater than the plasma thermal electron flux to the surface, and so the spacecraft potential is driven positive. If the spacecraft exterior is not fabricated of a conducting material, potential differences will develop between the sunlit and the dark areas of the spacecraft. In the ionosphere, these potential differences will not amount to more than a few volts under normal conditions. However, in the magnetosphere, where the Debye length may be very large, and where the only significant ambient electron population may have characteristic energies up to 10 keV, the dark portions of a non-conducting spacecraft (or, in eclipse, the entire spacecraft, whether conducting or not) may rise to potentials of $\sim$10 kV due to the flux of these environmental electrons striking the surface (Koons et al., 1988). Large potential differences between spacecraft components can have very destructive results, such as a high-energy arc in which electrical discharge occurs between spacecraft surfaces that have been charged to different potentials. Avoiding such conditions is one of the goals of any good spacecraft design. This is particularly an issue in the design of solar cell power systems, where the use of semiconductor materials exposed to sunlight is required.

The P78-2 Spacecraft Charging at High Altitudes (SCATHA) satellite was specifically designed to study spacecraft charging and its effects. It was launched in January 1979. On 22 September 1982 the SCATHA satellite measured a -10 kV potential and detected twenty-nine current pulses from discharges. There was a two-minute loss of data from the spacecraft, which was attributed to the discharges (Koons et al., 1988). Gussenhoven et al. (1985) have reported short-lived charging events (with a duration of tens of seconds) with potentials of ~ -450 V on the DMSP satellites, which fly in 800 km (low-earth) polar orbits and are subject to charging from auroral activity. Garrett (1981) has written an extensive review of spacecraft charging, and Purvis et al. (1984) have published practical engineering guidelines for assessing and controlling spacecraft charging effects.

2.4.5 Ram-wake Effects

Another effect worth considering is the alteration of the ambient plasma caused by the motion of a body in a plasma, giving rise to ram and wake. Spacecraft motion through the ionosphere can be an important perturbation to the local plasma. Typical spacecraft velocities are 2 to 7 km/s. Ionospheric ions and neutrals typically have thermal velocities smaller than this, so densities build up on the side in the direction of motion and a low-density region forms behind the spacecraft. Density deviations can be several orders of magnitude from the ambient and the Debye length on the downstream side can be locally high, so that large differential potentials can arise. Since the plasma electrons have velocities much greater than those of the ions, the wake region tends to be charged negatively. Because some neutral constituents (monoatomic oxygen, in particular) may react with some surface materials, the high density and flux of neutrals on the surfaces facing the flow direction can cause erosion of surfaces and coatings. Chapter 7 covers surface chemistry effects.

2.5 The Magnetosphere

The region beyond the ionosphere, but earthward of interplanetary space, is called the magnetosphere because the Earth's magnetic field dominates and orders this region. The magnetosphere is bounded from interplanetary space by a current layer called the magnetopause. Inside this boundary, the magnetosphere consists of energetic low-density (10^5 to 10^6 ions and electrons per cubic meter) plasma trapped on geomagnetic field lines of force. Outside the magnetopause the solar wind flows nearly radially away from the sun. The solar wind is a completely ionized magnetized plasma of average density $\sim 10^7$ ions and electrons per cubic meter, flowing at 300 to 800 km/s. The solar wind magnetic field is typically 10 nT in magnitude at Earth's orbit. Useful references for properties of the magnetosphere can be found in Haymes (1971), Ratcliffe (1972), the *Handbook of Geophysics and the Space Environment* (1985), and Lyons and Williams (1984). Several recent texts including Parks (2004), Kivelson and Russell (1995), and Gombosi (1997) offer comprehensive treatments of space physics relevant to magnetospheric processes.

2.5.1 Earth's Intrinsic Magnetic Field

The Earth's magnetic field is, to first order, a dipole field with the north pole of the magnetic field near the southern geographic pole such that the lines of force point vertically upward in the south and downward in the north. The magnetic poles and the geographic poles are not co-located. The dipole field is not quite centered on the center of the Earth (see figure 2.7), but is displaced by $\sim$440 km toward the Pacific Ocean. In addition, it is tilted by $\sim 11°$ relative to the Earth's rotation axis.

Geophysical phenomena that are ordered by the magnetic field are most easily described using a coordinate system based on the magnetic field rather than geographic coordinates. Because the Earth's field is not a perfect dipole, the locations of the magnetic poles are not uniquely defined and a variety of magnetic coordinate systems have been developed, each with slightly different definitions optimized for the representation of particular types of data. For example, the poles called the *dip poles* are located where the surface magnetic field is vertical. These are the poles to which a compass points. In 1965 these were located at about 75.6°N, 101°W and 66.3°S, 141°E. By contrast, the geomagnetic poles, projected on the basis of the best approximation given by a dipole located at the center of mass of the earth, were at 78.6°N, 69.8°W and 78.6°S, 110.2°E in 1965. The invariant latitude poles, based on a more complicated model field, were in 1969 at 80.33° N, 279.09°E and 74.01°S, 126.3°E, indicating the extent of variation, several degrees or more, depending on the type of model being used.

It is important to use the model appropriate for a given application. The more accurate models are required for navigation and attitude control purposes and for estimating LEO radiation exposure and energetic particle precipitation into the ionosphere. They are also needed to order observations of the space environment at low altitudes or in the ionosphere. For altitudes greater than $\sim 1 R_E$ the dipole approximation is an adequate representation of the intrinsic field because the higher order terms of the main field become insignificant relative to magnetic fields from magnetospheric currents. We first consider the description of a dipole magnetic field.

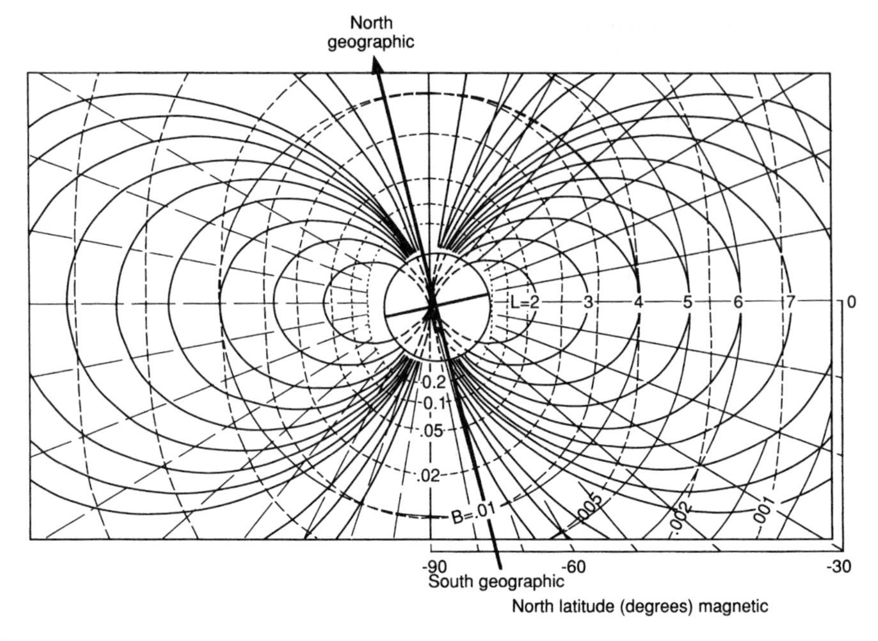

Figure 2.7 The eccentric-pole model of the Earth's magnetic field (B–L coordinates). The curves shown here are the intersection of a magnetic meridian plane with surfaces of constant B. The equivalent dipole is ~436 km distant from the center of the planet and is closest to the surface in the hemisphere that contains th Pacific. Hence, at a given altitude, the field is stronger over the Pacific than it is over Pacific than it is over the Atlantic. The geomagnetic axis is tilted 11.5° with respect to the Earth's rotational axis (the north–south line).

The equations for the dipole field in spherical coordinates (r, λ, Λ) are

$$B_r = (2B_0 \sin\lambda)/r^3 \tag{2.5.1}$$

$$B_\lambda = (B_0 \cos\lambda)/r^3 \tag{2.5.2}$$

$$B_\Lambda = 0 \tag{2.5.3}$$

where B_0 is the surface field at the equator, r is the geocentric radial distance in R_E, λ is the latitude, and Λ is azimuth. The dipole field lines are described by

$$r = R_0 \cos^2\lambda \tag{2.5.4}$$

where R_0 is the equatorial distance to the field line. This follows from integrating the differential relation $dr/B_r = rd\lambda/B_\lambda$, which in turn comes from considering the geometry of similar triangles resulting from a differential displacement $\mathbf{dr}$ along a dipole field line, using equations 2.5.1 and 2.5.2.

The magnitude of the dipole field may be written as

$$B = (B_0/r^3)(3\sin^2\lambda + 1)^{1/2} \tag{2.5.5}$$

These relations are used below to describe the magnetosphere.

More accurate representations of the Earth's main field are constructed using a potential formalism as follows. For a region outside the source currents of the magnetic field we have $\mathbf{j} = 0$. If we consider a time-independent system, $\partial/\partial t = 0$ and Ampere's law (equation 2.4.17) becomes simply

$$\nabla \times \mathbf{B} = 0 \tag{2.5.6}$$

which implies that $\mathbf{B}$ can be expressed as $-\nabla\phi_{\mathrm{B}}$, where ϕ_{B} is a scalar potential. Using Gauss's law, $\nabla \cdot \mathbf{B} = 0$, we obtain Laplace's equation for $\phi_{\mathbf{B}}$

$$\nabla^2 \phi_B = 0 \tag{2.5.7}$$

The general form for the solution to this is a spherical harmonic expansion (Arfken, 1970):

$$\phi_B = R_{\mathrm{E}} \sum_{n=1}^{\infty} \left(r^n T_n^{\mathrm{e}} + \frac{1}{r^{n+1}} T_n^{\mathrm{i}} \right) \tag{2.5.8}$$

where r is in R_{E} and the superscripts i and e refer to the field due to internal and external sources, respectively.

The T_n are given by

$$T_n = \sum_{m=0}^{n} (g_n^m \cos(m\phi) + h_n^m \sin(m\phi)) P_n^m(\cos\theta) \tag{2.5.9}$$

and

$$P_n^m(\cos\theta) = \sin^m(\theta) \frac{d^m}{d\cos\theta^m} P_n(\cos\theta), \quad m < n \tag{2.5.10}$$

and $P_n (\equiv P_n^0)$ are the associated Legendre polynomials. At the Earth's surface, ~99% of the field is described by the scalar potential due to internal sources and models typically use only the internal source terms. As we shall see, external sources become much more important at large geocentric distances ($>2R_{\mathrm{E}}$). The coefficients for the spherical harmonic expansion are determined from a combination of ground magnetic observatory data and space-based measurements. A widely used standard is the International Geophysical Reference Field (IGRF), which has been defined by the International Association of Geophysics and Aeronomy (IAGA) since 1945 in five-year increments (Barton, 1997) and is available from the National Space Science Data Center at NASA's Goddard Space Flight Center. Forward extrapolation from the most recent set of coefficients is performed in the IGRF model using estimates for the time rate of change of each coefficient which are provided with the most recent IGRF coefficients. The time rate of change is non-negligible and corresponds to a change of up to ~0.1% per year. The apparent dipole location moves, having shifted by 150 km over the past 100 years. Paleomagnetic evidence indicates that the dipole component of the internal field actually reverses direction over periods of $\sim 10^5$ years (higher moment terms typically change more rapidly).

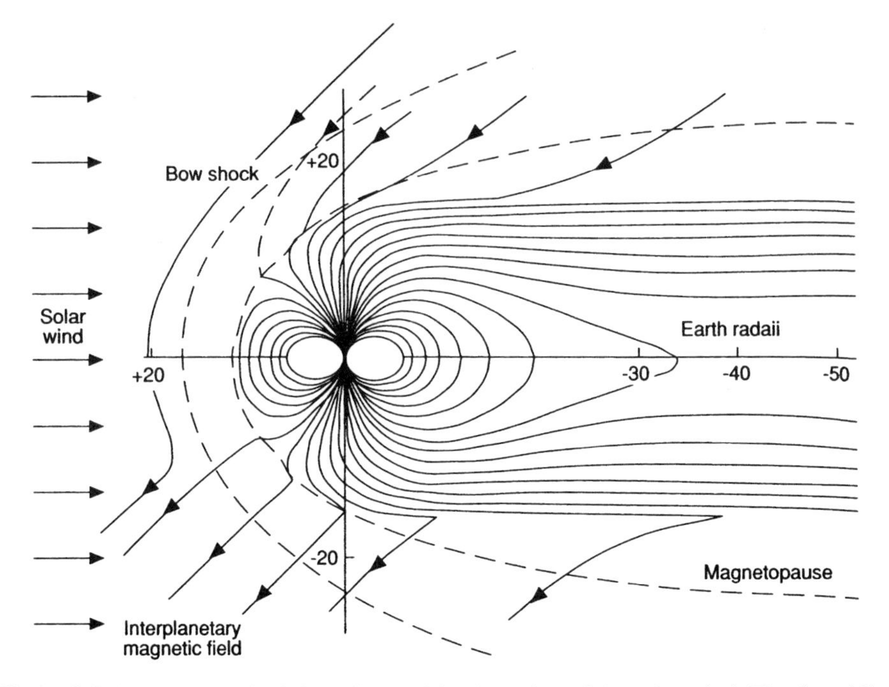

Figure 2.8 Geomagnetic field lines distorted by the action of the solar wind. The dotted line indicates the magnetopause, inside which the magnetic field is confined. Geocentric distances are indicated, in units of Earth radii, along the Sun–Earth line (adapted from Ratcliffe, 1972).

2.5.1 The Effect of the Solar Wind

The solar wind is a hot (10^5 to 10^6 K) plasma of density 5 to 20 $\times$ 10^6 ions and electrons per m^3. It flows radially from the sun at velocity $V = 300$ to 800 km/s and contains a weak (5 to 30 nT, or 5×10^{-5} to 3×10^{-4} gauss) magnetic field of no fixed direction but with general preferred directions. At some distance upstream of the Earth, the internal pressure of the Earth's magnetic field ($B^2/2\mu_0$) balances the solar wind dynamic pressure ($nmV^2/2$, where m is the mass of an average solar wind ion, about 1.05 M_H owing to the presence of He^{2+}, and n is the electron number density).

Inside 3 to 5 R_E the shape of the magnetosphere is little affected by the solar wind; however, the outer magnetosphere is greatly distorted by the ram pressure of the solar wind impacting the Earth's magnetic field. Figure 2.8 is a sketch of the magnetosphere, with the solar wind flowing from the dayside toward the nightside. Pressure balance between Earth's magnetic field and the solar wind ram pressure determines where the boundary between the solar wind and the magnetosphere lies. The distance from the center of the Earth sunward to the sub-solar magnetopause is called the stand-off distance. Since the solar wind magnetic field is very weak, a current layer, called the magnetopause current, is formed which separates the solar wind plasma from the Earth's magnetic field. The magnetopause current layer acts to exclude the Earth's magnetic field from the solar wind and, as a result, the field just inside the magnetopause on the Earth–Sun line is twice the dipole value at that distance. Pressure balance therefore gives

$$2B_{\mathrm{d}}(r)^2/\mu_0 = nmV^2/2 \tag{2.5.11}$$

where $B_{\mathrm{d}}(r)$ is the dipole field evaluated at a geocentric distance r. At the equator (where $\lambda = 0$), from equation 2.5.5, $B_{\mathrm{d}}(r) = B_0/r^3$. Using this in equation 2.5.11 and solving for r gives an estimate for the sub-solar magnetopause standoff distance

$$r_{\mathrm{mp}} = (4B_0^2/\mu_0 nmV^2)^{1/6}(\text{in } R_{\mathrm{E}}) \tag{2.5.12}$$

For nominal solar wind density $n \simeq 5\mathrm{cm}^{-3}$ and velocity 400 km/s, we get $r_{\mathrm{mp}} \simeq 10R_{\mathrm{E}}$. The most intense coronal mass ejections or high-speed streams associated with coronal holes can lead to velocities up to 900 km/s and densities of 40 cm^{-3} which would reduce r_{mp} to only 3 R_{E}. Although these extreme conditions occur rarely, perhaps once per solar cycle, the magnetopause is pushed within geosynchronous orbit from time to time. The variability of the solar wind properties implies that the magnetopause is almost always in motion and that the magnetosphere is a highly dynamic system which can be altered dramatically in size and magnetic topology in response to the solar with in 10 to 20 minutes.

The shape of the dayside magnetopause is well determined by simple gas-dynamic physics. The nominal distance to the magnetopause at 90° to the Earth-Sun line is $\sim$15 R_{E} whereas, on the nightside, the interaction draws the outermost magnetosphere out into a long *magnetotail.*

2.5.2 Motion of Charged Particles in a Dipole Field

Charged particle motion in a dipole magnetic field has certain unique properties that determine the makeup and distribution of the plasmas found in the magnetosphere. Both ions and electrons exhibit four fundamental types of motion, on three time scales, which we will examine below. These motions are:

1. Rapid gyration perpendicular to the magnetic field direction.
2. $\mathbf{E} \times B$ drift, an average motion of charged particles perpendicular to the magnetic and electric fields.
3. Mirroring—a behavior in which the motion parallel to the field is reversed (reflected) in strong field (low altitude) regions.
4. Azimuthal ∇B and curvature of $\mathbf{B}$ drifts, such that the particle (while gyrating and mirroring) moves slowly longitudinally about the Earth.

To understand these motions we consider the equation of motion of a charged particle in specified magnetic and electric fields.

2.5.2.1 Rapid Gyration

The force felt by a charged particle of charge q moving with velocity $\mathbf{V}$ in a magnetic field $\mathbf{B}$ is

$$\mathbf{F} = q\mathbf{V} \times \mathbf{B} \tag{2.5.13}$$

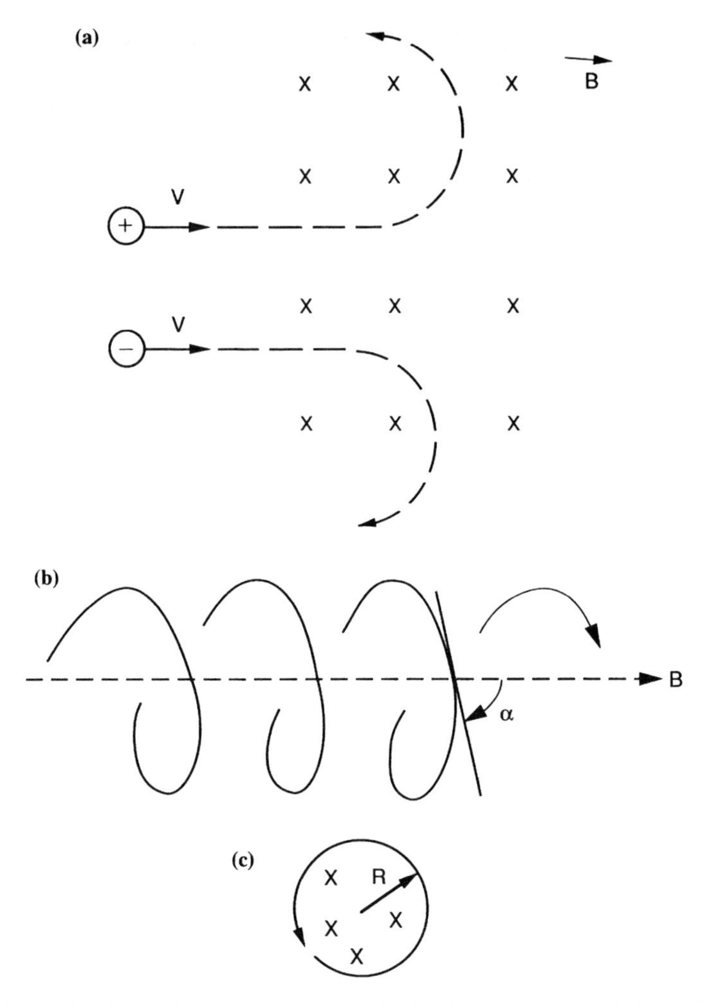

Figure 2.9 (a) Circular motion of a charged particle. (b) Helical motion. (c) Gyroradius.

This force is always perpendicular to both **V** and **B**, and implies circular motion in the plane perpendicular to **B** (figure 2.9a). Note also that $\mathbf{F} \cdot \mathbf{V} = 0$, so the magnetic field does no work on the particle. We are free to align the z direction with **B** so that equation 2.5.13 becomes

$$\left.\begin{aligned} m\frac{dV_x}{dt} &= qBV_y \\ m\frac{dV_y}{dt} &= -qBV_x \\ m\frac{dV_z}{dt} &= 0 \end{aligned}\right\} \tag{2.5.14}$$

If the magnetic field is constant and uniform, this has the following solution, given initial conditions $V_x = V_{x0}$, $V_y = 0$, and $V_z = V_z0$:

$$\left.\begin{aligned} V_x &= V_{x0}\cos\Omega t \\ V_y &= -V_{x0}\sin\Omega t \\ V_z &= V_{z0} \end{aligned}\right\} \tag{2.5.15}$$

where

$$\Omega = qB/m \tag{2.5.16}$$

is known as the *gyrofrequency*. Note that this frequency is independent of the particle velocity. The motion is helical (figure 2.9b), or circular if $V_{z0} = 0$.

To calculate the radius of the circular component of the motion (called the *gyroradius*; see figure 2.9c), we equate the magnetic force with the centrifugal force (which is required for circular motion), writing the velocity perpendicular to **B** as $V_\perp$:

$$qV_\perp B = mV_\perp^2/R_g \tag{2.5.17}$$

This yields

$$R_g = mV_\perp/qB = (2mW_\perp)^{1/2}/qB \tag{2.5.18}$$

where the perpendicular kinetic energy $W_\perp \equiv mV_\perp^2/2$. Typical values of R_g range from 3 cm for a low altitude, 0.1 eV electron to hundreds of kilometers for a high altitude, 100 keV proton; high energy ($\sim$ 1 GeV) cosmic ray ions have gyroradii larger than the entire magnetosphere.

2.5.2.2 Motion in E and B fields

For the motion of a charged particle in an electric and a magnetic field **B** along the z axis we add an electric field, **E**, to equations 2.5.14, obtaining

$$\left.\begin{aligned} m\frac{dV_x}{dt} &= qE_x + qBV_y \\ m\frac{dV_y}{dt} &= qE_y - qBV_x \\ m\frac{dV_z}{dt} &= qE_z \end{aligned}\right\} \tag{2.5.19}$$

Now we are also free to rotate our coordinates around the z direction so that $E_y = 0$. Then one solution would be $V_x = 0$, $V_y = -E_x/B$, $V_z = qE_zt/m$. If, however, the initial conditions do not correspond to that solution, then we must add a general solution for $V_{x0}, V_{y0}, V_{z0} \neq 0$:

$$\left.\begin{aligned} V_x &= V_{x0}\cos\Omega t + (V_{y0} + E_x/B)\sin\Omega t \\ V_y &= (V_{y0} + E_x/B)\cos\Omega t - V_{x0}\sin\Omega t - E_x/B \\ V_z &= V_{z0} + (qE_z/m)t \end{aligned}\right\} \tag{2.5.20}$$

In addition to the original gyratory motion derived for the case where $\mathbf{E} = 0$, these equations show that an applied electric field results in a constant velocity motion perpendicular to $\mathbf{E}$ and $\mathbf{B}$, independent of the sign or magnitude of the charge q. The general result for the electric field drift is

$$\mathbf{V}_{\text{DE}} = \mathbf{E} \times \mathbf{B}/B^2 \tag{2.5.21}$$

We could just as well have used any uniform, constant force $\mathbf{F}$ in this derivation and, since $\mathbf{E}$ is the force per unit charge, the general result for any uniform applied force is

$$\mathbf{V}_{\text{DF}} = \mathbf{F} \times \mathbf{B}/(qB^2) \tag{2.5.22}$$

The net motion can be viewed as gyration in circles of radius R_{g} about a point called the gyrocenter or guiding center, plus the new motion, called the $\mathbf{E} \times \mathbf{B}$ drift, which can be thought of as the motion of the guiding center perpendicular to $\mathbf{E}$ and $\mathbf{B}$. There will also be acceleration parallel (or anti-parallel) to $\mathbf{B}$ if $\mathbf{E}$ has a component parallel (or anti-parallel) to $\mathbf{B}$. The direction of the parallel acceleration depends on the sign of q.

If the mean free path of the particle for collisions with other particles is comparable to or smaller than the gyroradius, then a drift perpendicular to $\mathbf{B}$ and parallel (for q positive) or anti-parallel (for q negative) to $\mathbf{E}_\perp$ appears. This collisional drift is thus qualitatively different in that ions and electrons move in opposite directions whereas in the first drift (called the $\mathbf{E} \times \mathbf{B}$ drift) ions and electrons drift in the same direction at the same speed. The second (collisional) drift results in a current parallel to the electric field and the dissipation of energy by collisions with the neutral gas. This drift results in what is known as the Pederson conductivity in the ionosphere. At high altitude, collisions can be ignored, and this collisional drift is not significant. The Pedersen conductivity maximizes in the 110 km to 130 km altitude range. As a result, the closure of magnetospheric currents via Pedersen currents in the ionosphere results in neutral gas heating near the base of the thermosphere. This Joule dissipation therefore results in dramatic effects in the neutral density at higher altitudes because it changes the temperature at the base of the exosphere.

2.5.2.3 Mirroring in a Dipole Field

On a given line of magnetic force in a dipole field, the magnetic field intensity increases towards the ends of the field line near the poles and is a minimum at the equator. This feature acts like a magnetic bottle, trapping charged particles in the Earth's magnetic field. That this must happen can be seen by considering the magnetic moment of a particle's gyromotion, which turns out very significantly to be a constant of the particle motion to first order. Note first that a particle with perpendicular velocity $V_\perp$ travels in a circle whose radius is given by equation 2.5.18. This motion can be thought of as a circular current about an area πR_{g}^2, of magnitude $qV_\perp/2\pi R_{\text{g}}$. The magnetic moment μ of this current loop is the current times the area or

$$\mu = IA = \frac{qV_\perp}{2\pi R_{\text{g}}} \pi R_{\text{g}}^2 = \tfrac{1}{2} q V_\perp R_{\text{g}} \tag{2.5.23}$$

$$\mu = W_\perp/B \tag{2.5.24}$$

using equation 2.5.17 and $W_\perp = mV_\perp^2/2$. If the particle has a non-zero velocity parallel to **B**, it will move along the field and will experience a change in the field magnitude because |**B**| is not constant along a dipole field line. By Faraday's law this will result in an induced electromotive force in the particle frame of reference equal to $\pi R_g^2(\partial B/\partial t)$ so that in one gyration

$$\nabla W_\perp = q\pi R_g^2(\partial B/\partial t) \tag{2.5.25}$$

If the relative change in B in one gyroperiod is small, we can approximate $\partial B/\partial t$ as

$$\partial B/\partial t = \Delta B/\Delta t = V_\perp \Delta B/2\pi R_g \tag{2.5.26}$$

where the time for one gyration is given by $2\pi R_g/V_\perp$. Using equations 2.5.18 and 2.5.25, equation 2.5.26 becomes

$$\Delta W_\perp = mV_\perp^2 \, \Delta B/2B = W_\perp \Delta B/B \tag{2.5.27}$$

The change in μ in one gyration is

$$\Delta\mu = \Delta(W_\perp/B) = \Delta W_\perp/B - W_\perp \Delta B/B^2 \tag{2.5.28}$$

$$= 1/B(\Delta W_\perp - W_\perp \Delta B/B) \tag{2.5.29}$$

so

$$\Delta\mu = 0 \tag{2.5.30}$$

from equation 2.5.27. This says that even though B changes, μ does not, which implies that $W_\perp$ must change such that $W_\perp/B$ is constant. However, since the only force on the particle is the $q\mathbf{V} \times \mathbf{B}$ force, which does no work on the particle, the total kinetic energy of the particle must be conserved, in the absence of a force parallel to its motion. That is

$$W = mV^2/2 = \text{constant} \tag{2.5.31}$$

Now we let $V_\perp = V \sin\alpha$, $V_\parallel = V \cos\alpha$, where α is the angle between **V** and **B** and is called the pitch angle. Then we have

$$W_\perp = (mV^2 \sin^2\alpha)/2 \tag{2.5.32}$$

Since both V^2 and $W_\perp/B$ are constant, equation 2.5.32 implies

$$\sin^2\alpha/B = \text{constant} \tag{2.5.33}$$

At some point along the field, the field strength may be large enough that

$$B_m = B_1/\sin^2\alpha_1 \tag{2.5.34}$$

which implies that $\alpha_m = 90°$ and that all of the motion is perpendicular to **B**. That is, the parallel velocity is zero. What happens is that the gradient in B along **B**, $(\partial B/\partial z)$, acting on the magnetic moment μ, forces the particle toward weaker B, and so the particle reflects (mirrors) and travels in the opposite direction. This motion is illustrated

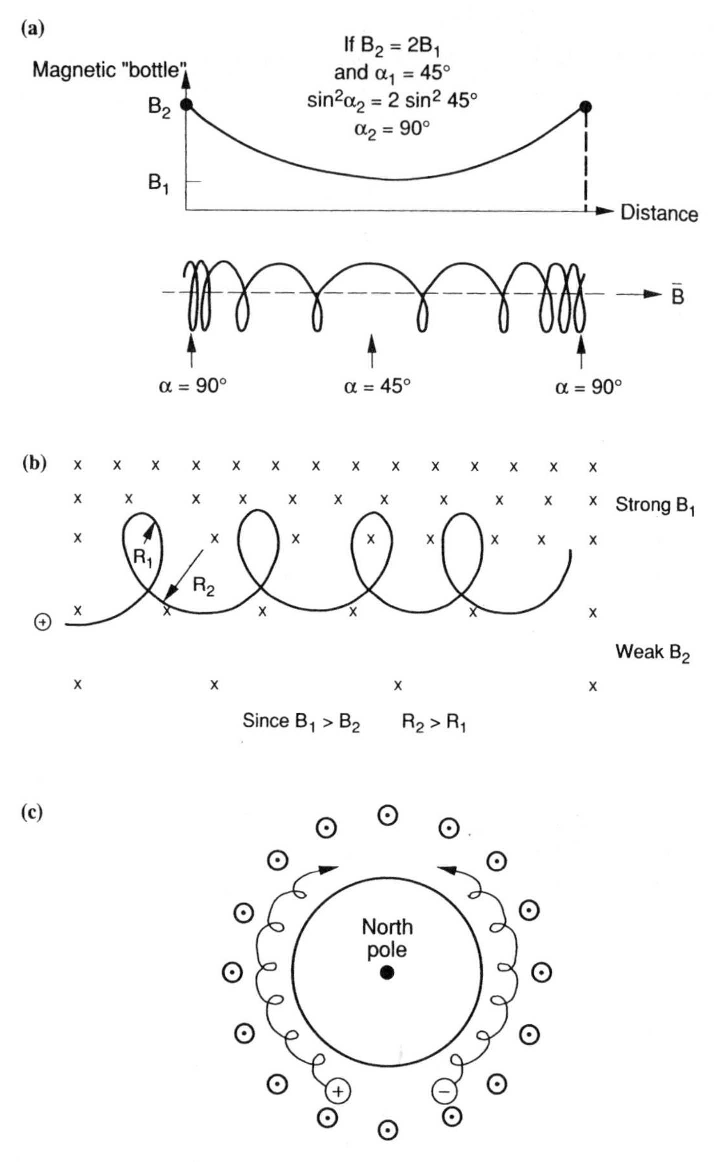

Figure 2.10 (a) Magnetic "bottle." (b) Gradient B drift. (c) Azimuthal direction in a dipolar geometry. B is directed upward and decreases as $1/r^3$.

in figure 2.10a. The point at which the particle reverses its parallel motion is called the mirror point, and is the same for all particles of the same pitch angle, regardless of charge, mass, or energy. The dipole field therefore acts like a natural magnetic bottle such that, once particles are loaded onto a field line and mirror above the atmosphere, where they would be lost due to collisions, they stay trapped by the magnetic field until something acts to change their equatorial pitch angle or, in the case of ions, they charge exchange with a hydrogen atom of the geochorona.

2.5.2.4 Azimuthal Drifts

The fourth type of motion is azimuthal drift of the particle guiding center (center of gyration motion). The azimuthal drift is a result of curvature and gradients in the magnetic field. These two effects result in guiding center velocities

$$\mathbf{v}_{\text{grad}} = \frac{mV_\perp^2}{2qB^2}\hat{\mathbf{B}} \times \nabla B \tag{2.5.35}$$

$$\mathbf{v}_{\text{curv}} = \frac{mV_\parallel^2}{qB^2}\hat{\mathbf{R}} \times \mathbf{B} \tag{2.5.36}$$

where $\mathbf{R}$ is the radius vector from the center of curvature of the field line to the field line, $V_\parallel$ is the velocity parallel to $\mathbf{B}$, and $V_\perp$ is the velocity perpendicular to $\mathbf{B}$. See Lyons and Williams (1984) for a derivation of these equations.

These drifts are always perpendicular to $\mathbf{B}$, as is seen from the cross-products in the equations. The gradient B drift is illustrated in figure 2.10b. In a dipole geometry, their direction is strictly azimuthal (figure 2.10c). In the real geomagnetic field, while the drift is not exactly azimuthal, over one drift orbit about the Earth the radial components cancel and the particle returns to its starting point. Other key features to note about these drifts are: they increase in proportion to particle energy; they increase linearly with radial distance (in a dipole field where equation 2.5.5 holds); and they are opposite for positive and negative charges.

2.5.2.5 Summary of Charged Particle Motion

Each of these motions (gyration, $\mathbf{E} \times \mathbf{B}$ drift, mirroring, and azimuthal gradient and curvature drifts) has a time scale. The relative importance of these different motions in the Earth's dipole field depends on the particle energy. For low-energy particles ($W \leq\sim 10$ keV), the $\mathbf{E} \times \mathbf{B}$ drift is more important in tracing the particles' trajectories than are the azimuthal gradient and curvature drifts. For higher energy particles, the $\mathbf{E} \times \mathbf{B}$ drift becomes insignificant relative to the azimuthal gradient and curvature drifts. In the following discussion we will concentrate on higher energy particles, because they are more important in terms of environmental effects on space systems. For these particles, the time periods for gyration, mirroring, and azimuthal drifts are of interest. For gyration, the period is (from equation 2.5.16)

$$T_g = 2\pi/\Omega = 2\pi m/qB \tag{2.5.37}$$

For mirroring, the period can be calculated by integrating the distance from a mirror point to the opposite hemisphere mirror point along the magnetic field (times two), divided by the velocity parallel to $\mathbf{B}$, $V_\parallel$:

$$T_m = 2\int_{m1}^{m2} \frac{ds}{V_\parallel} = \frac{4}{V}\int_{0}^{B2} \frac{ds}{\sqrt{1 - B(s)/B_{\text{m}}}} \tag{2.5.38}$$

$$= (4L/V)[1.3 - 0.56\sin\alpha_0] \tag{2.5.39}$$

where L is the radial distance in R_{E} at the equator and α_0 is the particle pitch angle at the equator.

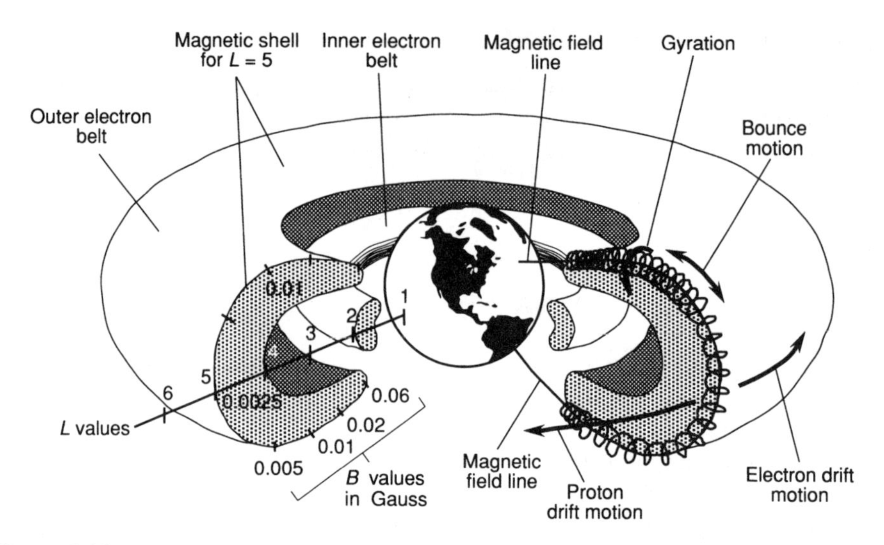

Figure 2.11 Trapping of energetic ions and electrons in the Earth's dipole field.

For azimuthal drift motion, the period is longer still and is evaluated to (Lyons and Williams, 1984)

$$T_d \sim 1.03 \times 10^4[\gamma/(1+\gamma)]F \text{ seconds,} \tag{2.5.40}$$

where $\gamma \equiv (1 - V^2/c^2)^{1/2}$ and F is a function that ranges from about 1.0 to 1.5. Schultz and Lanzerotti (1974) give a detailed discussion of these formulas.

For all of these motions, the periods and the motions themselves are valid only for collisionless motion and small field changes over the respective characteristic time scales. Figure 2.11 shows how these three types of motion combine in the Earth's dipole field in such a way that the energetic ions and electrons remain trapped for long periods of time, gyrating, mirroring, and drifting in "shells" about the Earth's dipole axis. Table 2.2 gives representative gyroradii, gyroperiods, mirroring periods, and azimuthal drift periods at 2000 km and 4 R_E equatorial altitudes.

As shown in Figure 2.11, particles executing these motions tend to stay on *shells*, or surfaces in the magnetospheric field traced out by the set of field lines on which the mirroring, drifting particles are trapped. In a dipole field, these surfaces can be uniquely identified by one parameter, L, defined as the radius (in R_E) of the equatorial crossing point of the field. Once a shell is described by L, any point along the field can be specified by its latitude and longitude. However, because (in a dipole field) $B_\phi = 0$ and $\partial B/\partial\phi = 0$, the longitude coordinate does not describe the shell in any meaningful way. From equations 2.5.4 and 2.5.5

$$r = L\cos^2\lambda \tag{2.5.41}$$

and

$$B = (B_0/r^3)(4 - 3r/L)^{1/2} \tag{2.5.42}$$

Table 2.2 Characteristic periods and equatorial gyroradii

Energy 2000 km altitude	Species	Gyroradius (km)	Gyroperiod (s)	Mirror period (s)	Drift period (min)
50 keV	electrons	0.005	2.5×10^{-6}	0.25	690
1 MeV	electrons	0.032	7×10^{-6}	0.1	53
1 MeV	protons	10	0.004	2.2	32
10 Mev	protons	30	0.0042	0.65	3.2
500 Mev	protons	250	0.006	0.11	0.084
4 R_E altitude					
1 keV	electrons	0.2	7.4×10^{-5}	4.0	11000
10 keV	electrons	0.63	7.5×10^{-5}	1.3	1100
100 keV	electrons	2	8.8×10^{-5}	0.46	90
1 MeV	electrons	6.3	2.2×10^{-4}	0.27	3.7
1 keV	protons	8.8	0.14	172	11000
10 keV	protons	28	0.14	54.5	1100
100 keV	protons	88	0.14	17.2	110
1 MeV	protons	280	0.14	5.45	11

so for any r and $|\lambda|$ there is a one-to-one functional relationship to B and L (figure 2.12). For some purposes, namely describing the trapped particle fluxes, this B–L coordinate system is the preferable one. Thus, on a given L shell, specifying B identifies two points that are equal distances from the equator on the field line, and should by symmetry carry identical energetic particle fluxes.

Since the Earth's field is distorted by magnetospheric currents, the dipole approximation will not be accurate enough to define the B–L coordinates in this manner. One approach to obtaining B–L coordinates is to use a good model field for **B** (including asymmetries introduced both by internal deviations from the dipole field and by external influences, that is, distortion in the solar wind). Using such a model of **B**, one then can invent a complicated expression for L. A reasonably accurate expression, good up to about $L \leq 3$, is

$$B_{\mathrm{m}} L_{\mathrm{m}}^3 / B_0 \simeq 1 + 1.350474\, r^{1/3} + 0.465380\, r^{2/3} + 0.047546\, r \tag{2.5.43}$$

This parameter, L_{m}, was originally defined by McIlwain (1966). A discussion of B–L space, its definition, and its approximations, may be found in Schultz and Lanzerotti (1974).

2.5.3 Regions of the Magnetosphere

The magnetosphere is divided into a number of regions that are differentiated on the basis of their topology and the characteristics of the particle populations found there. They are illustrated in figure 2.13. We will discuss them in order of increasing altitude.

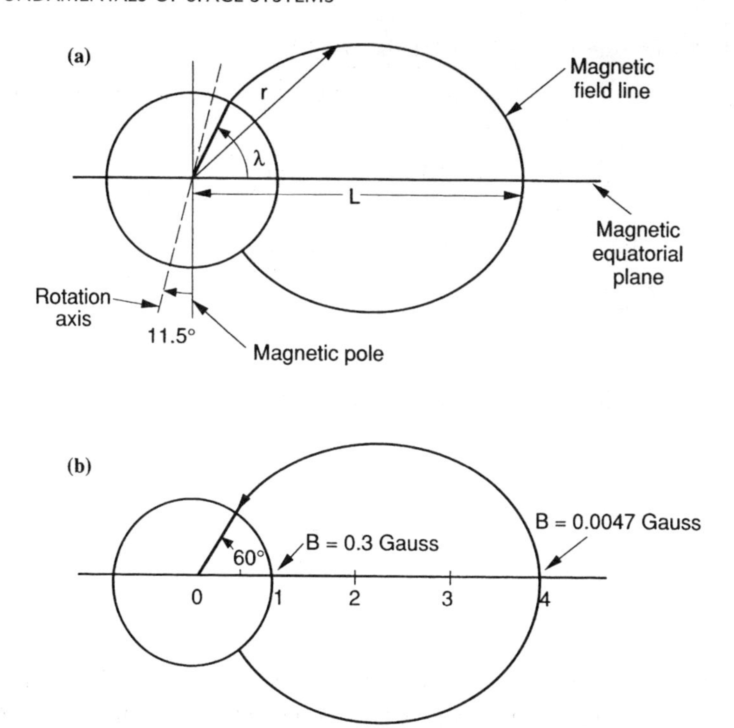

Figure 2.12 Direct functional relationship of dipoles B (a) and L (b) to geomagnetic r and γ.

2.5.3.1 The Plasmasphere

The *plasmasphere* is essentially a high-altitude extension of the ionosphere. Its primary constituents are H^+ and He^+ (and electrons), with typical ionospheric energies of 0.1 eV. At such low energies, the motions are dominated by electric field drift, the ion and electron gyroradii are very small, and their mirror motion and azimuthal drifts are slow compared with their $\mathbf{E} \times \mathbf{B}$ drift. A prototypical electron concentration profile with equatorial distance is shown in figure 2.14. The steep drop in density beyond about 4 R_E is the result of a change in the dominant electric fields in the regions. In the inner region, ions and electrons diffuse from the ionosphere along the field and remain on their original field lines as they co-rotate with the earth.

To understand co-rotation, consider a point in the magnetic equator above the ionosphere at radius r. Due to the Earth's rotation, the magnetic field sweeps past this point with a velocity $\mathbf{V}_\Omega = \Omega_E \times \mathbf{r}$, where Ω_E is the Earth's angular velocity vector whose direction is given by the right-hand rule. The motion of the magnetic field corresponds to an induced electric field experienced by a charged particle

$$\mathbf{E}_{ind} = -\mathbf{V}_\Omega \times \mathbf{B} \tag{2.5.44}$$

that leads to an induced drift given by $\mathbf{V}_{DE}$. Equation 2.5.44 can be obtained by evaluating the force per unit charge with a particle velocity of $-\mathbf{V}_\Omega$ in equation 2.5.13. Using

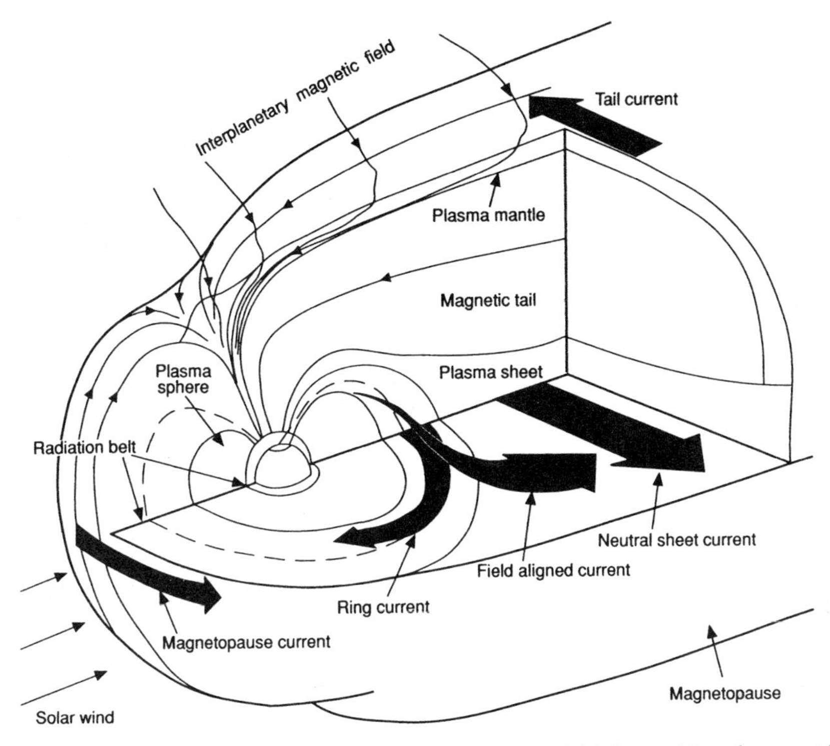

Figure 2.13 The configuration of the Earth's dipole magnetic field distorted into the comet-like shape called the magnetosphere. The various current systems that flow in this complicated plasma laboratory are labeled. The interplanetary magnetic field is the magnetic field of the Sun, which has a modulating effect on the processes that occur within the magnetosphere.

equation 2.5.21 with $\mathbf{E}_{\mathrm{ind}}$ and the vector identities $\mathbf{a} \times \mathbf{b} = -\mathbf{b} \times \mathbf{a}$ and $\mathbf{a} \times (\mathbf{b} \times \mathbf{c}) = \mathbf{b}(\mathbf{a} \cdot \mathbf{c}) - \mathbf{c}(\mathbf{a} \cdot \mathbf{b})$ gives

$$\mathbf{V}_{\mathrm{ind}} = \mathbf{V}_{\Omega} - \mathbf{B}(\mathbf{B}\cdot\mathbf{V}_{\Omega})/B^2 \tag{2.5.45}$$

The second term is the component of $\mathbf{V}_{\Omega}$ along $\mathbf{B}$, so subtracting means that $\mathbf{V}_{\mathrm{ind}}$ is the part of $\mathbf{V}_{\Omega}$ that is perpendicular to $\mathbf{B}$. In the case of the Earth, Ω_E is aligned within 11° to the dipole moment so that $\mathbf{B}$ and $\mathbf{V}_{\Omega}$ are nearly orthogonal. Hence $\mathbf{V}_{\mathrm{ind}} \simeq \mathbf{V}_{\Omega}$, so that plasma in regions where $\mathbf{V}_{\Omega}$ is larger than the other drifts will co-rotate with the magnetic field.

The region of cold plasma co-rotation extends only over the range in r for which $\mathbf{E}_{\mathrm{ind}}$ dominates. Note that $\mathbf{E}_{\mathrm{ind}}$ falls with radial distance as $1/r^2$. There is a dawn-to-dusk directed convection electric field $\mathbf{E}_{\mathrm{conv}}$ inside the magnetosphere, produced by interaction with the solar wind magnetic field. The resulting convection drift $\mathbf{V}_{\mathrm{conv}}$ is sunward since $\mathbf{B}$ is northward in the equatorial plane. $\mathbf{E}_{\mathrm{ind}}$ falls below $\mathbf{E}_{\mathrm{conv}}$ in the range $L = 3$ to $6\ R_{\mathrm{E}}$ exactly where depends on solar wind conditions. Beyond this transition L-shell, low-energy plasma no longer rotates with the Earth but convects sunward and is lost to the solar wind (figure 2.15). This leads to a boundary called the

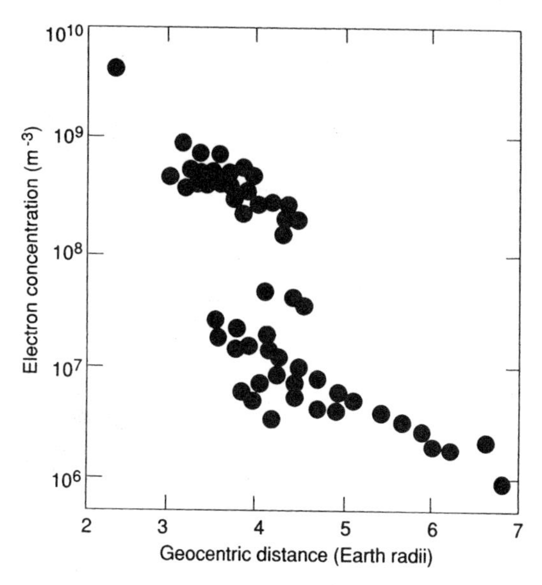

Figure 2.14 The electron concentration as a function of geocentric distance above the equator. The sharp gradient near four Earth radii represents the plasmapause (adapted from Ratcliffe, 1972).

plasmapause, corresponding to the sudden drop in density associated with the transition from co-rotation to sunward convection.

2.5.3.2 Trapped Radiation

High-energy particles are also found within $r \sim 4R_E$. The azimuthal gradient and curvature drifts in equations 2.5.35 and 2.5.36 are proportional to kinetic energy so that, for particles with energies greater than about 1 MeV, the particle motion is dominated by the gradient/curvature drift. As a result, these energetic particles drift azimuthally around the Earth on drift paths that completely encircle the Earth—they are trapped in the dipole field. These stable high-energy populations are known as the Van Allen radiation belts after James Van Allen who first discovered them in 1957 using balloon-borne Geiger tubes. Although the density of the more energetic population is small ($\sim 10^4$ m^{-3}), their energy per particle is large enough that this population generally dominates the plasma energy density in the magnetosphere. An extensive discussion of the radiation belts can be found in Shultz and Lanzerotti (1974), and a less comprehensive but more recent and accessible treatment is given by Vampola (1989b).

Figure 2.16 shows the distribution of very energetic ions and electrons in the magnetosphere. The greatest flux of very energetic ions is concentrated just outside the inner electron belt. Complete models of the distribution of energetic ions and electrons from tens of keV to hundreds of MeV energies, as a function of solar cycle epoch, can be found in publications of the National Space Science Data Center by Vette and by Sawyer and Vette, dated 1972 to 1976 (see NSSDC, 1985, and Bilitza, 1990). In fact, the distribution of particles is fairly continuous, although there are distinct regions of especially high flux at certain energies (hence the term *belts*). Because the collision cross-section of energetic particles with the low-energy plasma and neutrals decreases

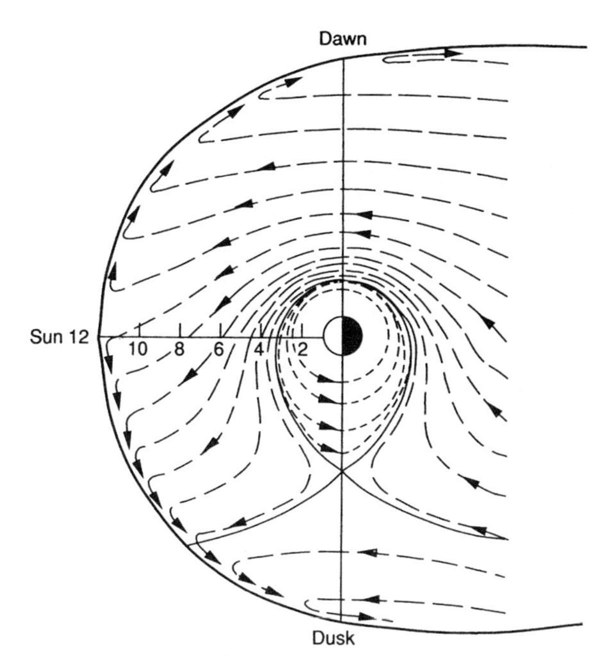

Figure 2.15 Equipotential contours for an electric field in the Earth's equatorial plane (dashed lines). These are also drift paths for very-low-energy particles. The electric field is a superposition of a co-rotational E-field due to the rotation of the Earth and its embedded magnetic field and a uniform dawn–dusk electric field (adapted from the *Handbook of Geophysics and the Space Environment*, 1985).

strongly with increasing energy, the more energetic particles have very long lifetimes (up to years) before they will be lost from the system.

The energetic particles in the radiation belts have two primary sources. One is the solar and galactic cosmic ray flux. Cosmic rays can either collide with the atmosphere, producing energetic fragments in the collisions, decay spontaneously (for example, a neutron into a proton and an electron), or enter directly and lose energy and become trapped by scattering. The other source is acceleration of lower energy particles in the magnetosphere by the electric fields that result from geomagnetic storms and substorms (discussed below) caused by solar wind interactions. The magnetospheric acceleration process can either be prompt, driven by the arrival of an interplanetary shock wave, or can occur over the period of several days following a geomagnetic storm.

Radiation belt loss mechanisms include scattering of the particles into the atmosphere or charge exchange with exospheric neutral hydrogen. Due to their low mass, the electrons are lost primarily by scattering either by collision with low-energy ions and neutrals or in wave disturbances of the magnetic field. The waves can be either low-frequency magnetohydromagnetic waves caused by interaction with the solar wind or plasma whistler waves. Scattering changes the particle pitch angles, thereby changing their mirror points, and those whose mirror points are shifted to points in or below the atmosphere are lost when they collide with atoms and molecules in the atmosphere. Protons are much less easily scattered, and, although the fluxes of protons into the atmosphere can be significant for LEO satellites, the radiation belt protons

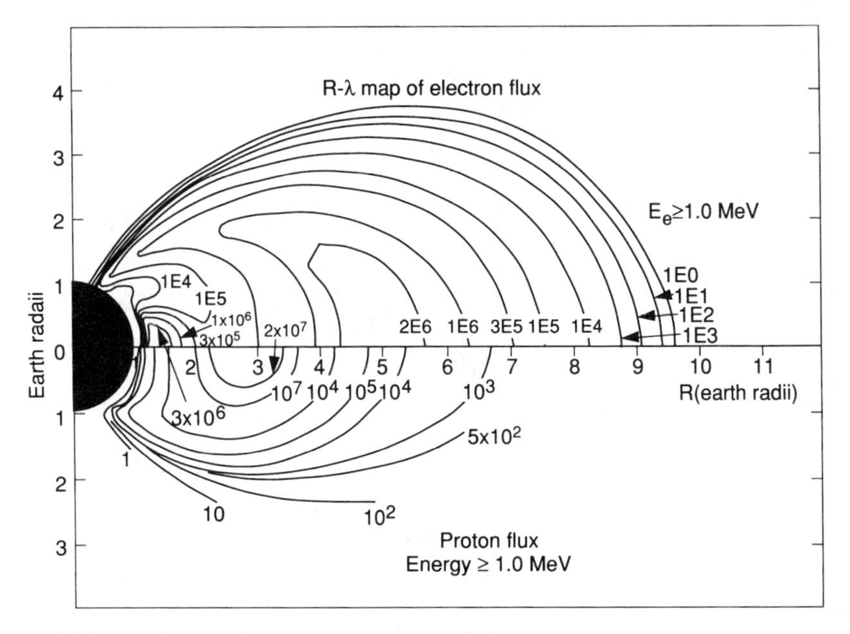

Figure 2.16 Distribution of very energetic ions and electrons.

are lost primarily by charge exchange with the neutral hydrogen of the geochorona. Because of their high energy, the energetic protons and electrons penetrate fairly deep into the atmosphere, causing D-region ionization. For this reason periods following geomagnetic storms are often associated with a sustained D-region at middle and high latitudes.

The deposition of radiation belt particles precipitated into the atmosphere is not uniformly distributed in longitude or latitude. Because the Earth's magnetic field is not a true dipole and is offset toward the Pacific Ocean, there is a region over the South Atlantic Ocean where the magnetic field is anomalously low compared to the same latitudes elsewhere. For a given particle pitch angle and L value, the mirror altitude is a surface of constant B. This surface is lowest in the South Atlantic anomaly. This has two related consequences. First, this is the region where particle mirror points first reach low enough in the atmosphere so that the particles are lost by collisions with the neutral atmosphere. The South Atlantic anomaly acts to scavenge the particles that reach closest to the Earth. Second, the radiation belt flux is lower for a constant altitude at other longitudes because the surface of constant B corresponding to the atmosphere in the South Atlantic is at higher altitudes everywhere else. The radiation exposure in LEO is therefore less intense elsewhere and the South Atlantic anomaly is the region of harshest radiation exposure. Figure 2.17 shows the precipitation pattern of > 1 MeV electrons at 500 km altitude.

2.5.3.3 Polar Cap

The magnetic field that connects with the polar cap is bent anti-sunward by interaction with the solar wind. This magnetic flux, which would otherwise connect with its

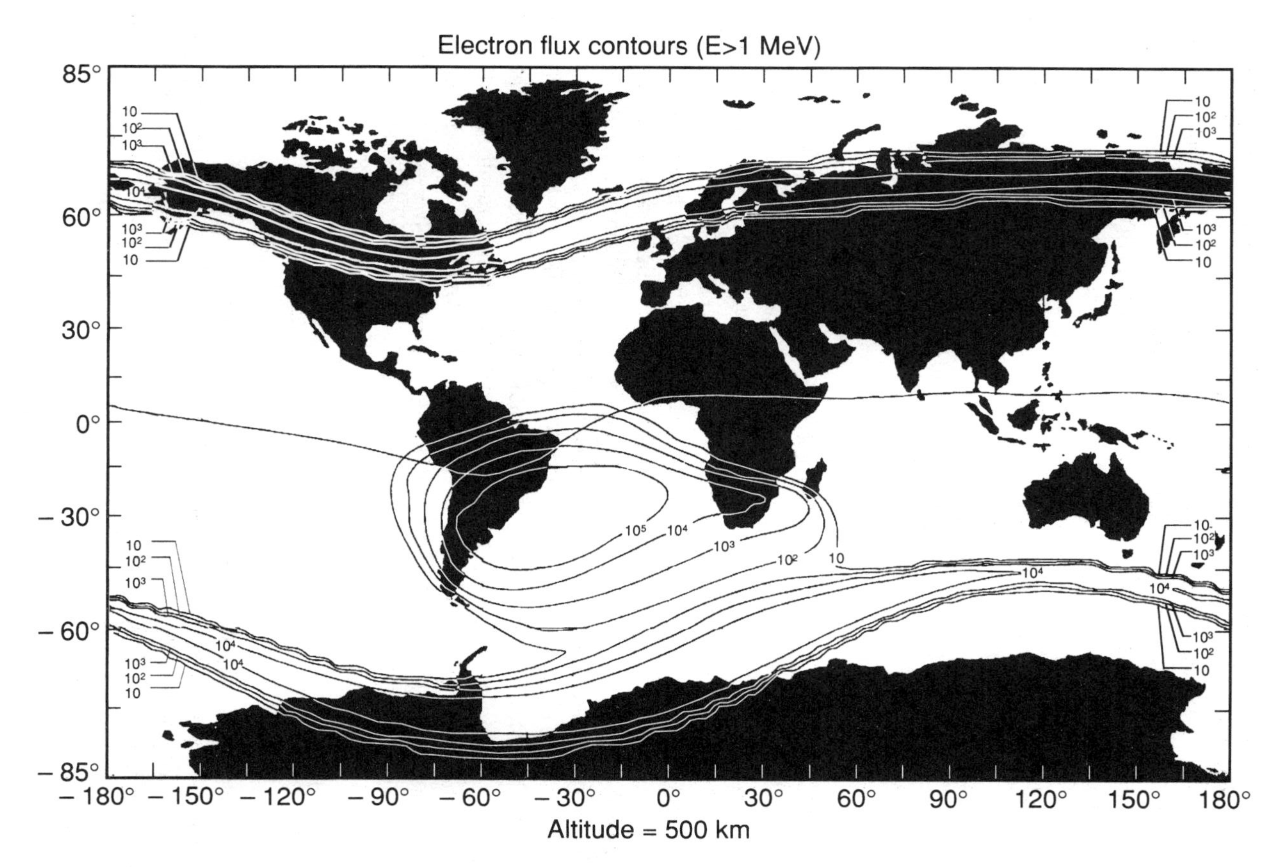

Figure 2.17 Precipitation pattern of > 1 MeV electrons at 500 km altitude.

counterpart in the opposite hemisphere, eventually transects the magnetopause and joins with the interplanetary magnetic field. Therefore, energetic particles cannot be trapped in the polar regions. However, energetic particles accelerated by coronal mass ejections (CMEs) travel along magnetic fields in the solar wind and have direct access to the polar caps. The polar cap is therefore a region of potential severe radiation exposure during SEP events in which protons are accelerated to relativistic energies and arrive at Earth within minutes to hours of flare initiation. Polar-orbiting satellites in LEO can therefore be more exposed to solar energetic particles than equatorially orbiting LEO satellites or even geostationary satellites. During geomagnetic storms the polar cap expands equatorward and, since storms are often correlated with SEP events, orbits with inclinations above about 45° geographic latitude can be at risk for radiation exposure during geomagnetic storms. In the *magnetotail*, the distended anti-sunward portion of the magnetosphere, the polar-cap-connected field regions are known as the "lobes" and form a pair of giant half-cylinders of magnetic field, similar to a pair of solenoids. The magnetotail extends hundreds of R_E anti-sunward, well beyond lunar orbit ($\sim 60\ R_E$).

2.5.3.4 The Plasma Sheet

Also in the magnetotail, but on closed magnetic flux tubes, is a region of moderately hot plasma called the *plasma sheet*. Although these flux tubes are closed, the drift paths for both low- and high-energy ions and electrons are not. Recall that in the trapping regions, energetic particles drift azimuthally in the Earth's magnetic field, such that they remain on an L-shell. To a first approximation the azimuthal gradient drift shell about the Earth is characterized by constant equatorial magnetic field magnitude. At the magnetic equator, the field strength is lower on the nightside than it is on the dayside for the same radial distance. This implies that drifting particles in the range 6 to 10 R_E at midnight move outward as they drift from the nightside to the dayside and encounter the magnetopause, where they are lost to the magnetosphere. Further tailward, the drift paths transit directly across the tail so that energetic particles drift across the magnetotail and into the adjacent solar wind. The plasma sheet is known as a quasi-trapping region because energetic particles are confined to magnetospheric field lines only so long as it takes them to execute one transit, either across the tail or from the nightside to the dayside, where they are then lost to the solar wind at the magnetopause.

2.5.4 Storms and Substorms

The magnetosphere and ionosphere are very dynamic. The greatest single influence on their state is the Sun. We have already discussed how solar activity, with the associated enhanced UV and X-ray flux, affects the ionosphere and upper atmosphere. Ionospheric electron and ion concentrations can be dramatically increased, and the neutral exospheric and thermospheric densities and scale heights, which depend strongly on thermospheric temperature, also increase greatly.

Solar activity affects the near-Earth space environment in two other ways. Flare activity can produce high-density, high-velocity solar wind streams. If such a stream strikes the magnetosphere, the magnetopause stand-off distance can be reduced from $\sim 10\ R_E$ to $\leq 6\ R_E$. This compression can generate magnetohydrodynamic (MHD) waves that can scatter trapped energetic particles such that they strike the upper atmosphere, depositing energy and heating and ionizing the neutral atmosphere.

The solar wind interaction with the magnetosphere is strongly modulated by the direction of the interplanetary magnetic field (IMF). When the IMF is parallel to the Earth's field at the magnetopause, there is little interaction. When the IMF is antiparallel (southward), a process known as merging or reconnection takes place. In reconnection the IMF and the Earth's field are interconnected and energy is transferred from the bulk flow of the solar wind, which acts as an electromotive dynamo, to the magnetosphere and ionosphere, which act as loads. This transfer is also dependent on solar wind velocity, so that high-speed flows with southward IMF transfer the most energy.

This takes place through two primary pathways. In the first path, energy is deposited in the thermosphere as Joule heating via electric currents resulting from the reconnection solar wind dynamo. This is a direct path in which electromagnetic energy from the solar wind is converted into electric currents whose closure via Pedersen currents in the ionosphere causes Joule heating at the base of the thermosphere. The power dissipated by this path is typically 50 GW but can rise to over 500 GW during geomagnetic storms. The second pathway proceeds by a complex chain of events called a substorm. Reconnection and convection add magnetic flux to the polar cap and then the lobe regions in the magnetotail, increasing the magnetic pressure ($B^2/2\mu_0$) there. This in effect charges the pair of solenoids comprising the magnetotail. This requires an increase in the plasma sheet currents, particularly the current sheet which separates the northern and southern lobes. This energy storage process is called the substorm growth phase. This cross-tail current eventually reaches the limit that the plasma in the plasma sheet can support and the current becomes unstable. Since the magnetic configuration cannot be sustained, the magnetic energy stored in the lobes is then released explosively and transferred to thermal energy of ions and electrons. The beginning of explosive energy release is known as substorm onset and is associated with sudden brightening of an aurora near midnight. The convulsive relaxation of the magnetotail drives these particles earthward where some are injected onto trapped drift paths. This progression of energy release is associated with intense and dynamic auroras that rapidly expand poleward and toward morning and evening along with their associated ionospheric currents. This stage, following substorm onset, is called auroral breakup and is associated with spectacular auroral displays.

If the energy input from the solar wind is especially strong and sustained, a magnetic storm results. Particularly intense substorms can occur repeatedly during geomagnetic storms, but geomagnetic storms display their own dynamics as well. During storms, particles are energized and injected deep inside the magnetosphere, within 3 R_E, where they remain for many days drifting in the stable trapping regions. These drifting ions and electrons carry a current, known as the ring current because it forms a ring about the earth near the equator. This current can significantly alter the magnetic field magnitude and direction, even at the Earth's surface. Substorm currents also affect the magnetic field, but these effects are localized to the auroral zones.

The third way in which solar activity affects the Earth environment is through flare-accelerated energetic particles. In a flare, ions and electrons can be accelerated to high energies (100 MeV to 1 GeV). If these energetic particles reach the earth, they can enter the magnetosphere and ionize the neutral upper atmosphere. This is most effective in the polar caps, where direct access along the open field lines is available.

2.6 Radiation

Penetrating radiation is a significant element of the environment, and space systems must be designed to withstand the radiation damage (see chapter 13). Whereas the photon radiation is very important for ionospheric effects, it is less destructive to materials in space than particle radiation. UV and X-ray fluxes must be considered for their photoionization effects on surfaces (as discussed earlier) and, along with all wavelengths, for their effects on sensitive instrumentation. For example, a telescope built for detecting faint objects might be damaged by direct sunlight, or blinded by light reflected or scattered from spacecraft structures.

Another important consideration is thermal influence. Most systems designed for space must operate within particular temperature limits. This operating range must be achieved by considering the balance between solar input and radiation to space. If passive control cannot be achieved by using either reflective or absorbing surface materials, or by other means, active heating and/or cooling may become necessary (see chapter 7). The total energy flux from the Sun is $\sim 1350\ \mathrm{W/m^2}$ at 1 AU and is nearly constant over time. The energy flux in the UV and X-ray energy range is $\sim 10^{-2}\ \mathrm{W/m^2}$, or less than 10^{-5} of the total. This is not important thermally, but it is quite variable and, as we have seen, important to the ionosphere.

Particle radiation, at energies greater than ~10 keV, can be damaging to materials. The effects of ions and electrons are different, with electrons penetrating more easily but ions in some ways being more destructive. The primary sources of destructive particle radiation are (1) the trapping regions, or Van Allen radiation belts; (2) the Sun, in solar flares and subsequent shock-accelerated SEPs; (3) cosmic rays, extremely energetic ions and electrons, as well as gamma rays, accelerated by unknown processes and entering the solar system with a nearly constant flux; (4) low-altitude beams of electrons and protons accelerated in intense auroras.

In designing systems for the space environment, models of the radiation as a function of time (solar cycle) and orbit location are used. For the radiation belts, the best existing models are the AE-8 and AP-8, as well as the AE-4 and AP-4 models (Vette, 1968). Figure 2.18 is a copy of a page from the *NSSDC News*, an article listing some of the atmospheric, ionospheric, geomagnetic, and radiation environment models available from the NSSDC on magnetic disk. These models give fluxes of ions and electrons at various energies as a function of position (in B–L coordinates) in the radiation belts. By using them and the known time history of the orbit, one can predict the total radiation dose and therefore the specification for radiation resistance and shielding of components.

The solar flare dose rate is extremely variable, so in calculating requirements one generally uses very conservative estimates; that is, one assumes flare activity modeled on worst-case previous experience for the portion of the solar cycle under consideration. The galactic cosmic ray flux is quite constant, so its effects can be modeled accurately.

2.7 The Interplanetary Medium

The interplanetary medium lies beyond the Earth's magnetosphere, and encompasses all the space in the heliosphere outside planetary magnetospheres. The solar wind is the major plasma in the interplanetary medium. Its density, magnetic field strength,

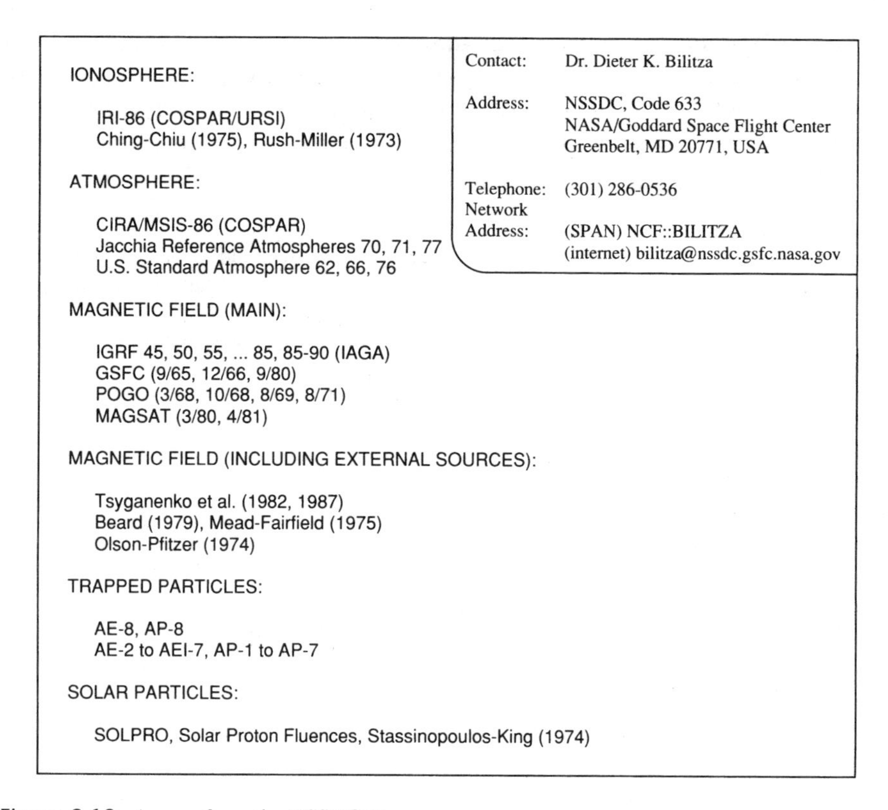

IONOSPHERE:

IRI-86 (COSPAR/URSI)
Ching-Chiu (1975), Rush-Miller (1973)

ATMOSPHERE:

CIRA/MSIS-86 (COSPAR)
Jacchia Reference Atmospheres 70, 71, 77
U.S. Standard Atmosphere 62, 66, 76

MAGNETIC FIELD (MAIN):

IGRF 45, 50, 55, ... 85, 85-90 (IAGA)
GSFC (9/65, 12/66, 9/80)
POGO (3/68, 10/68, 8/69, 8/71)
MAGSAT (3/80, 4/81)

MAGNETIC FIELD (INCLUDING EXTERNAL SOURCES):

Tsyganenko et al. (1982, 1987)
Beard (1979), Mead-Fairfield (1975)
Olson-Pfitzer (1974)

TRAPPED PARTICLES:

AE-8, AP-8
AE-2 to AEI-7, AP-1 to AP-7

SOLAR PARTICLES:

SOLPRO, Solar Proton Fluences, Stassinopoulos-King (1974)

Contact: Dr. Dieter K. Bilitza

Address: NSSDC, Code 633
NASA/Goddard Space Flight Center
Greenbelt, MD 20771, USA

Telephone: (301) 286-0536

Network Address: (SPAN) NCF::BILITZA
(internet) bilitza@nssdc.gsfc.nasa.gov

Figure 2.18 A page from the *NSSDC News*.

temperature, and velocity near Earth vary in time. Recall that the properties of the interplanetary plasma at 1 AU are typically as follows: density between about 5 and 20 protons/cm^3, velocity between about 300 and 800 km/s, magnetic field between 5 and 30 nT, temperature between 10^5 and 10^6 K or about 10 to 100 eV. As a function of distance from the Sun, the density drops as $1/r^2$; the radial component of the magnetic field dominates within about 1 AU and falls as $\sim 1/r^2$. The azimuthal magnetic field component dominates beyond about 1 AU and falls off only as $1/r$. The velocity increases gradually with r, but shows less variability beyond 1 AU than within 1 AU, and the temperature drops slowly with r.

The interplanetary medium is generally more benign than the near-Earth environment. Radiation is still present in solar photons, flare- and CME-accelerated particles, but as one travels to the outer planets these effects become less significant. Even near Earth, spacecraft whose orbits spend most or all of their time in the IPM generally have far longer life expectancy than those that orbit in the radiation belts. Degradation of solar voltaic cells due to radiation is often the ultimate cause of their demise. Some of the longest lived operational spacecraft are the Pioneer 10 and 11 spacecraft, powered by radioactive thermal generators and traveling far out into the outer heliosphere. Other magnetized planets, including Jupiter, Saturn, and Uranus, have magnetospheres similar to Earth's, complete with radiation belts. The Jovian intrinsic magnetic field is far stronger than Earth's and, as a result, the trapped radiation at Jupiter is far more energetic and dense than at Earth (Dessler, 1983), such that the radiation dose for one year in orbit at one of the Galilean satellites exceeds 1 Mrad, more than 30 times a multi-year dose in LEO. Missions to orbit other planets must therefore make careful assessment of the trapped radiation in the local magnetospheric environment, which may be quite different from that at Earth.

Acknowledgments The authors are grateful for the encouragement and suggestions offered by the late Dr. T. A. Potemra. His notes from a similar course that he taught formed the basis for some of the ideas and presentation here. We also thank Drs. D. J. Williams, A. T. Y. Lui, H. D. Black, R. H. Maurer, and especially Dr. A. L. Vampola for their critical readings of early drafts.

Problems

1. (a) Derive the expression for the altitude (h) dependence of atmospheric pressure (P) given a surface pressure of $P_0 = 1$ bar $= 1 \times 10^5$ Pa, and numerically evaluate the scale height at the Earth's surface ($k = 1.38 \times 10^{-23}$ J/K).
   (b) Perform the same exercise for Venus; $P_0 = 91$ bar, $T_{surface} = 730$ K, the atmosphere is predominantly CO_2, the mass of Venus = 0.82 that of the mass of Earth, and $R_V \simeq R_E$.
   (c) How deep would Earth's atmosphere be if it became liquid? (Hint: remember, once it is a liquid, it will not obey laws based on ideal gases.)
2. (a) What fraction of the neutral atomic hydrogen (H) at the base of the exosphere ($\sim$400 km) at solar quiet times will have escape velocities? (Assume a Maxwellian thermal distribution, $f(\mathbf{v}) = f_0 \exp[-m/2kT(v_x^2 + v_y^2 + v_z^2)]$.)
   (d) What fraction of atomic oxygen (O) at the base of the exosphere during very high solar activity will have escape velocities?
3. Assume the ionosphere is dominated by singly ionized oxygen, with trace amounts of ionized hydrogen. What is the proton scale height? (Remember to consider the *net* force on the hydrogen ions.)
4. An incident ionizing UV flux of I_0 falls vertically on an isothermal atmosphere of single constituent composition with mass m, surface density n_0, gravity g, absorption cross-section σ, ionization cross-section β, and recombination coefficient α. Find the net ion production rate and equilibrium ion number density at the altitude where the UV flux is $I_0 e^{-1}$.
5. (a) Assuming that density variations of 50% are randomly distributed in the F-layer of the ionosphere, what frequency range would be unacceptable for satellite communications?
   (b) What if the F-layer were smooth, but 50% fluctuations existed in the E-layer?
6. (a) What electric potential relative to the surrounding medium would a spacecraft in the nightside ionosphere assume?
   (b) In sunlight, photoionization of spacecraft surface materials may drive the spacecraft potential positive. Typical photoelectrons have energies of $\sim$3 eV. What equilibrium potential will a conductive surfaced spacecraft assume in the dayside ionosphere?

7. (a) A 1 MeV proton at 90° pitch angle at the equator at $L = 7$ in a dipole field is convected Earthward to $L = 4$ (causing the particle to gain energy), and the first adiabatic invariant (μ) is conserved. What is its final energy?
   (b) Assume the initial pitch angle was 45° at $L = 7$ at the equator, and the proton diffuses inward, conserving μ. What is the final energy and equatorial pitch angle? (Assume $W_{\parallel}$ does not change.)
8. Assume that all particles that mirror below 200 km in the atmosphere are precipitated, while those above 200 km mirror and return to high altitude. In a centered dipole field, what is the size of the angle (called the loss cone) about the magnetic field direction at the equator that will be empty of particles for a field line that intersects the Earth's surface at 50° latitude? (Hint: particles that mirror at higher altitudes will have larger pitch angles at any point along the field than those that mirror at 200 km.)
9. At times, the solar wind standoff distance (nominally 10 R_E) decreases to geosynchronous orbit (6.6 R_E). Derive and plot a curve that bounds the values of solar wind density and velocity for which $R_{\text{standoff}} \leq 6.6\, R_E$.
10. For a magnetic field line with L-value $L = 4\,\text{R}_E$, what is the magnitude of the co-rotation electric field (a) at the equator and (b) on the ground at the latitude where the $L = 4\, R_E$ field line intersects the surface?

References

Arfken, G., 1970. *Mathematical Methods for Physicists*, Academic Press, New York.

Barton, C. E., 1997. International geomagnetic reference field: the seventh generation, *J. Geomag. Geoelectr.*, **49**, 123–148.

Bass, J. N., 1980a. Analytical representation of the Jacchia 1977 Model Atmosphere, *AFGL-TR-80-0037. AD-A085781.*

Bass, J. N., 1980b. Condensed storage of diffusion equation solutions for atmospheric density model computations, *AFGL-TR-80-0038, AD-A086868.*

Baumjohann, W., and R. A. Treumann, 1997. *Basic Space Plasma Physics*, Imperial College Press, London, UK.

Bilitza, D., 1990. Solar-terrestrial models and application software, National Space Science Data Center/World Data Center A, *NSSDC/WDC-A-R&S 90-19.*

Chen, F. F., 1984. *Introduction to Plasma Physics and Controlled Fusion, Vol. 1: Plasma Physics*, Plenum Press, New York.

COSPAR International Reference Atmosphere (CIRA), 1986. National Space Science Data Center.

Dessler, A. (ed) 1983. *Physics of the Jovian Magnetosphere*, Cambridge University Press, Cambridge, MA.

Eisner, A., 1982. Evaluation of accuracy of the orbital lifetime program "LIFE", *Proc. Workshop on Satellite Drag*, Boulder, CO, March 18–19, Space Environment Services Center, Space Environment Lab., NOAA.

Feess, W. A., and S. G. Stephens, 1987. Evaluation of GPS ionospheric time-delay model, *IEEE Trans. Aerosp. Electron. Sys.*, ***AES-23***, 332–338.

Fesen, C. G., R. G. Roble, and E. C. Ridley, 1993. Thermospheric tides simulated by the National Center for Atmospheric Research Thermosphere–Ionosphere General Circulation Model at equinox, *J. Geophys. Res.*, **98**, 7805–7820.

Gaposchkin, E. M., and A. J. Coster, 1988. Analysis of drag, *Lincoln Lab. J.*, **1**, 203–224.

Garrett, H. B., 1981. The charging of spacecraft surfaces, *Rev. Geophys. Space Phys*, **19**(4), 577–616.

Gombosi, T. I., 1998. *Physics of the Space Environment*, Cambridge University Press, Cambridge, UK.

Gussenhoven, M. S., D. A. Hardy, F. Rich, W. J. Burke, and H.-C. Yeh, 1985. High-level spacecraft charging in the law-altitude polar auroral environment, *J. Geophys. Res.*, **90**, 11, 11,009–11,023.

Handbook of Geophysics and the Space Environment, 1985. Adolf S. Jura (ed.), Air Force Geophysical Laboratory, Document Accession Number ADA 167000, 1985.

Haymes, R. C., 1971. *Introduction to Space Sciences*, John Wiley & Sons, New York.

Hedin, A. E., N. W. Spencer, and T. L. Killeen, 1988. Empirical global model of upper thermosphere winds based on Atmosphere and Dynamics Explorer satellite data, *J. Geophys. Res.*, **93**, 9959–9978.

Jacchia, L. G., 1965. Static diffusion models of the upper atmosphere with empirical temperature profiles, *Smithson. Contrib. Astrophys.*, **8**, 215–257.

Jacchia, L. G., 1970. New static models of the thermosphere and exosphere with empirical temperature profiles, *Smithsonian Astrophys. Obs. Spec. Rep. 313*.

Jacchia, L. G., 1972. Atmospheric models in the region from 110 to 2000 km, in *CIRA 1972*, Akademie-Verlag, Berlin p. 227.

Jacchia, L. G., 1977. Thermospheric temperature, density and composition: New models, *SAO Special Report No. 375*, Smithsonian Astrophysical Observatory, Cambridge, MA.

Jacchia, L. G. and J. Slowey, 1964. Temperature variations in the upper atmosphere during geomagnetically quiet intervals, *J. Geophys. Res.*, **69**, 4145–4148.

Jacchia, L. G. and J. Slowey, 1966. The shape and location of the diurnal bulge in the upper atmosphere, *SAO Special Report. 207*.

Jackson, J. D., 1962. *Classical Electrodynamics*, John Wiley & Sons, New York.

King-Hele, D. G, 1975. The Bakerian Lecture, 1974; A view of Earth and air, *Phil. Trans. Roy. Soc. Lond.*, **A 278**, 67–109.

King-Hele, D. G. and D. M. C. Walker, 1988. Upper atmosphere zonal winds from satellite orbit analysis, *Planet. Space Sci.*, **36**, 1089–1093.

Kivelson, M. J., and C. T. Russell (eds.), 1995. *Introduction to Space Physics*, Cambridge University Press, Cambridge, UK.

Klobuchar, J. A., 1989. Ionospheric time-delay algorithm for single-frequency GPS users, *IEEE Trans. Aerosp. Electron. Sys.*, ***AES-23***, 325–331.

Koons, H. C. et al., 1988. Severe spacecraft charging event on SCATHA in September, 1982, *J. Spacecr. Rockets*, **25**(3) 239–243.

Liu J. F. F., et al., 1982. An analysis of the use of empirical atmospheric density models in orbital mechanics, *Proc. Workshop on Satellite Drag*, Boulder, CO, Mar 18–19, Space Environment Services Center, Space Environment Lab., NOAA.

Lyons, L. R., and D. J. Williams, 1984. *Quantitative Aspects of Magnetospheric Physics*, D. Reidel, Boston, MA.

Marcos, F. A., 1988. Accuracy of satellite drag models, *Proc. Atmospheric Neutral Density Specialist Conf.*, Colorado Springs, CO, 22–23 Mar., AD-A225 249.

Marcos, F. A., et al., 1978. Variability of the lower thermosphere determined from satellite accelerometer data, *AFGL Report TR-78-0134*.

Mauk, B. H., D. G. Mitchell, S. M. Krimigis, E. C. Roelof, and C. P. Paranicas, 2003. Energetic neutral atoms from a trans-Europa gas torus at Jupiter, *Nature*, **421**(6926), 920–922.

Mayaud, P. N., 1980. *Derivation, Meaning and Use of Geomagnetic Indices*, AGU Geophysical Monograph 22, American Geophysics Union, Washington, DC.

McIlwain, C. E., 1966. Magnetic coordinates, *Space Sci. Rev.*, **5**, 585.

NSSDC Data Listing, 1985. Supplementary Data Listing, National Space Science Data Center/World Data Center A, *NSSDC/WDC-A-R&S*, vol. 85–05.

Parks, G. K., 2004. *Physics of Space Plasmas: An Introduction*, Westview Press, Advanced Book Program, Bouldes, CO.

Purvis, C. K., et al., 1984. *Design guidelines for assessing and controlling spacecraft charging effects*, NASA Technical Paper 2361, Lewis Research Center, Cleveland, OH.

Rairden, R. L., L. A. Frank, and J. D. Craven, 1986, Geocoronal imaging with Dynamics Explorer, *J. Geophys. Res.*, **91**, 13613–13630.

Ratcliffe, J. A., 1972, *An Introduction to the Ionosphere and Magnetosphere*, Cambridge University Press, London.

Reif, F., 1965. *Fundamentals of Statistical and Thermal Physics*, McGraw-Hill, New York.

Roelof, E. C., and J. A. Skinner, 2000. Extraction of ion distributions from magnetospheric ENA and EUV images, *Space Sci. Rev.*, **91**(1/2) 437–459.

Schultz, M., and L. J. Lanzerotti, 1974. Particle diffusion in the radiation belts, Vol. 7 of *Physics and Chemistry in Space*, J. G. Roederer (ed.), Springer-Verlag, New York.

Swanson, D. G., 1989. *Plasma Waves*, Academic Press, San Diego, CA.

Tascione, T. F., 1988. *Introduction to the Space Environment*, Orbit, Malabar, FL.

U.S. Standard Atmosphere, 1976. National Oceanic and Atmospheric Administration, National Aeronautics and Space Administration, and U.S. Air Force, Washington, DC.

Vampola, A. L., 1989a. Solar cycle effects on trapped energetic particles, *J. Spacecr. Rockets, **26***, 416–427.

Vampola, A. L., 1989b. The space particle environment, *Report SD-TR-89-30*, Space Systems Division, Air Force Systems Command.

Vette, J. I., M. J. Teague, D. M. Sawger, and K. W. Chan, 1979, Modeling the Earth's radiation belts. A review of quantitative data based electron and proton models, in Marshall Space Flight Center, *Terest. Predictions Proc.*, **2**, 21–35 (SEE N80-24678).

Wallace, J. M., and P. V. Hobbs, 1977. *Atmospheric Science: an Introductory Survey*, Academic Press, New York, 1977.

Yionoulis, S., et al., 1972. A geopotential model (APL 5.0-1967) determined from satellite Doppler data at seven inclinations, *J. Geophys. Res.*, **77** (20), 3671–3677.

3

Astrodynamics

VINCENT L. PISACANE

3.1 Introduction

Astrodynamics, which includes the topics of orbital mechanics, celestial mechanics, and dynamical astronomy, is the study of the motion of natural and artificial bodies in space. Its fundamental objective is to determine and predict the position of celestial bodies on which the dominating force is gravity. Astrodynamics is experimentally based on the observations of the planets and moons and on Kepler's laws, and is theoretically based on the laws of classical mechanics and Newton's and Einstein's laws of motion. Its study is important to spacecraft missions because the trajectory of the spacecraft is a fundamental design parameter with major implications for all subsystems of the spacecraft. This chapter addresses only the fundamentals of astrodynamics that are important to spacecraft missions. For a more detailed exposition of the topic, appropriate references are identified.

The history of dynamics begins with Aristotle (384–322 BC) who stated that heavier objects fall faster than lighter ones and that the Earth is at the center of the universe. Claudius Ptolemy (~87–150 AD) stated that all heavenly bodies move in circles and epicycles about the Earth. Nicolaus Copernicus (1473–1543) proposed that all planetary bodies moved about the Sun instead of the Earth. This was rejected by many in the belief that the stars were close enough for motion of the Earth to cause stellar parallax. Tycho Brahe (1546–1601) collected and cataloged astronomical data on which his colleague, Kepler, based his findings. Johannes Kepler (1571–1634) used Brahe's data to conclude his three famous laws:

Lex I: The orbit of each planet is in the shape of an ellipse with the Sun at one of the foci (1609).

Lex II: The radius vector from the Sun sweeps out equal areas in equal times (1609).
Lex III: The squares of the periods of planets are proportional to the cubes of their semimajor axis (1619).

Galileo Galilei (1564–1642) stated that bodies fall at the same speed independent of their masses. He also proposed what is now known as the Galilean principle of relativity: that a sailor who drops a mass from the mast of a ship would see it fall vertically while a person on land would see it take a different trajectory due to the motion of the ship. Isaac Newton (1642–1727) proposed his three laws of motion and a law of gravitation that placed astrodynamics on a quantitative basis. Newton's three laws of motion are:

Lex 1: A particle remains at rest or moves with constant velocity (magnitude and direction) unless acted on by a force. This effectively defines an inertial reference system.
Lex 2: Relative to an inertial reference system, the motion of a particle is described by

$$\mathbf{F} = \frac{d\mathbf{p}}{dt} \tag{3.1.1}$$

where $\mathbf{F}$ is the applied force and $\mathbf{p} = m\mathbf{v}$ is the linear momentum, where m is the mass and $\mathbf{v}$ is the velocity.
Lex 3: For every action there is an equal and opposite reaction.

Newton's law of gravitation relates the force on a mass m in the vicinity of a mass M by

$$\mathbf{F}_{\mathrm{g}} = -\frac{GmM}{r^2}\hat{\mathbf{r}} = -\frac{GmM}{r^3}\mathbf{r} \tag{3.1.2}$$

where G is the universal gravitational constant ($6.67259 \times 10^{-11}\mathrm{m/s^2\ kg}$), $\mathbf{r}$ is the position vector of m with respect to M, and $\hat{\mathbf{r}}$ is the unit vector of $\mathbf{r}$. The minus sign indicates that the force is attractive.

Albert Einstein (1879–1955) proposed the special theory of relativity for bodies with velocities approaching the speed of light and the general theory of relativity for bodies acted on by gravitational forces. These theories need be considered only in the most precise developments in astrodynamics and are generally ignored for Earth satellites.

3.2 Fundamentals of Dynamics

Newton's three laws of motion and his law of universal gravitation are the central axioms of astrodynamics. Taking the cross-product of equation 3.1.1 with the position vector $\mathbf{r}$ from a point in the inertial reference system, it follows that

$$\mathbf{N} = \frac{d\mathbf{H}}{dt} \tag{3.2.1}$$

where $\mathbf{N}$ is the *moment of the force* and $\mathbf{H}$ is the *moment of the momentum* or *angular momentum* where

$$\mathbf{N} \equiv \mathbf{r} \times \mathbf{F} \quad \text{and} \quad \mathbf{H} \equiv \mathbf{r} \times \mathbf{p} \tag{3.2.2}$$

since $\dot{\mathbf{r}} \times \dot{\mathbf{r}} = 0$. Another useful result, named the *work–energy equation*, is obtained by taking the scalar product of Newton's second law with $d\mathbf{r}$ and integrating between positions 1 and 2 to yield

$$T_2 - T_1 = W_{1-2} \tag{3.2.3}$$

where the *kinetic energy* T_i is defined to be

$$T_i \equiv \tfrac{1}{2} m\dot{\mathbf{r}}(t) \cdot \dot{\mathbf{r}}(t), \quad i = 1, 2 \tag{3.2.4}$$

where $(\dot{\ })$ denotes derivative with respect to time and *work* W_{1-2} is defined by

$$W_{1-2} \equiv \int_1^2 \mathbf{F} \cdot d\mathbf{r}. \tag{3.2.5}$$

A force is defined to be *conservative* if the work done depends only on the endpoints and not on the path. This property is defined mathematically by

$$\int_{\text{closed path}} \mathbf{F} \cdot d\mathbf{r} = 0 \tag{3.2.6}$$

for all arbitrary paths. This equation will be satisfied if the force can be expressed solely in terms of the position vector and perhaps time, $\mathbf{F}(\mathbf{r}, t)$. Stokes's theorem states that for an arbitrary vector field $\mathbf{F}$ the relationship between the line integral and surface integral of an elemental area is

$$\int_{\text{closed path}} \mathbf{F} \cdot d\mathbf{r} = \int_{\text{surface area}} (\nabla \times \mathbf{F}) \cdot d\mathbf{A}. \tag{3.2.7}$$

For a conservative force, the expression on the left is zero, so for an arbitrary path the integrand on the right must be zero. Thus

$$\nabla \times -\mathbf{F}(\mathbf{r}, t) = 0 \tag{3.2.8}$$

where the minus sign is introduced by convention. Since the curl of the gradient of a scalar function is identically zero,

$$\nabla \times \nabla V = 0, \tag{3.2.9}$$

it follows that a conservative force can be expressed in terms of the scalar V by

$$\mathbf{F} = -\nabla V(\mathbf{r}, t). \tag{3.2.10}$$

Integrating this equation, holding time fixed, gives

$$V = -\int_{t=\text{constant}} \mathbf{F}(\mathbf{r}, t) \cdot d\mathbf{r}. \tag{3.2.11}$$

The advantage of treating conservative forces as a potential is that a scalar quantity has one value as a function of the independent variables **r** and t instead of the three of a vector.

The gravitational force given by equation 3.1.2 can also be written in the convenient form of the *acceleration of gravity* g_0 at a characteristic altitude r_0 where

$$\mathbf{F}_g = -m\left(\frac{GM}{r_0^2}\right)\left(\frac{r_0}{r}\right)^2\hat{\mathbf{r}} = -mg_0\left(\frac{r_0}{r}\right)^2\hat{\mathbf{r}}, \tag{3.2.12}$$

where

$$g_0 \equiv \left(\frac{GM}{r_0^2}\right) = \text{acceleration of gravity at the radius } r_0, \tag{3.2.13}$$

where r_0 is typically taken at the surface of the body. The gravitational potential for a point mass can be determined from equations 3.1.2 and 3.2.11 to be

$$V = -\int -\frac{GmM}{r^3}\mathbf{r}\cdot d\mathbf{r} = -\frac{GmM}{r}, \tag{3.2.14}$$

where the constant of integration can be assumed to be zero without loss of generality since the derivative of V determines the motion. Often, the *specific potential* U is employed if one of the masses is normalized, where

$$U \equiv \frac{V}{m} = -\frac{GM}{r}. \tag{3.2.15}$$

The specific potential for a finite body of mass M follows directly as

$$U = -G\int_M \frac{dM}{r}. \tag{3.2.16}$$

Example

Determine the specific potential of a shell of thickness t and radius p with constant mass density γ as illustrated in figure 3.1.

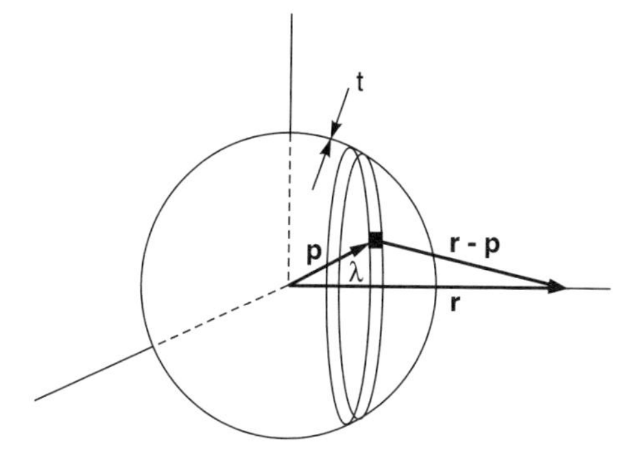

Figure 3.1 Gravitational potential of a spherical shell, t = shell thickness, $\mathbf{p}$ = position vector to elemental mass of shell, $\mathbf{r}$ = position vector to point where force is to be determined, λ = angle between $\mathbf{p}$ and $\mathbf{r}$.

Solution: The specific potential is given by

$$U = -G \int_m \frac{dM}{|\mathbf{r} - \mathbf{p}|} = -G \int_m \frac{2\pi\, p^2 \gamma t \sin\lambda\, d\lambda}{(r^2 + p^2 - 2rp\cos\lambda)^{1/2}}, \tag{a}$$

where $M = 4\pi p^2 t$ is the mass of the shell. Integration gives

$$U = -\frac{GM}{2pr}[(r + p) \pm (r - p)], \tag{b}$$

so the potential inside the shell is

$$U = -\frac{GM}{p}, \qquad r < p, \tag{c}$$

and outside the shell is

$$U = -\frac{GM}{r}, \qquad r > p. \tag{d}$$

The specific potential inside the shell is constant, so the gravitational force is zero as though the shell were not present. The specific potential outside the shell is the same as if the mass of the shell were concentrated at its center.

Since most celestial bodies are nearly spherical and their density varies to a large extent with radius, it is appropriate as an approximation to treat them as a point mass even for distances close to the body. For example the geopotential, the gravitational potential of the Earth, deviates from that of a point mass by about one part in a thousand.

For precise orbit determination, it is necessary to take into account the higher order variations in the gravitational potential of the Earth or any nearby celestial body. A convenient representation of the gravitational potential (Kaula, 1966) is

$$U = -\frac{GM}{r} \sum_{n=2}^{\infty} \sum_{m=0}^{n} \left(\frac{a}{r}\right)^n [\overline{C}_{n,m} \cos m\lambda + \overline{S}_{n,m} \sin m\lambda] \overline{P}_{n,m}(\sin\phi), \tag{3.2.17}$$

where

U = specific gravitational potential
GM = gravitational constant of the body
r = distance from center of mass of the body
a = semi-major axis of the average surface radius
n, m = degree and order, respectively
ϕ = geocentric latitude
λ = geocentric longitude
$\overline{C}_{n,m}, \overline{S}_{n,m}$ = normalized gravitational coefficients
$P_{n,m}(\sin\phi)$ = associated Legendre function of the first kind
$\overline{P}_{n,m}(\sin\phi)$ = normalized associative Legendre function of the first kind

$$= \left[\frac{(n - m)!(2n + 1)k}{(n + m)!}\right]^{1/2} P_{n,m}(\sin\phi) \tag{3.2.18}$$

where $k = 1$ when $m = 0$, $k = 2$ when $m \neq 0$, $P_{n,m}(\sin\phi)$ = associated Legendre function of the first kind

$$P_{n,m}(\sin\phi) = (\cos\phi)^m \frac{d^m}{d(\sin\phi)^m} P_n(\sin\phi) \tag{3.2.19}$$

$$= \frac{(\cos\phi)^m}{2^n n!} \frac{d^{n+m}(\sin^2\phi - 1)^n}{d(\sin\phi)^{n+m}} \tag{3.2.20}$$

$$P_n(\sin\phi) = \text{Legendre polynomial}$$

$$= \frac{1}{2^n n!} \frac{d^n(\sin^2\phi - 1)^n}{d(\sin\phi)^n} \tag{3.2.21}$$

The first few Legendre functions are

$$\left.\begin{aligned}
&P_{0,0} = 1 \\
&P_{1,0} = \sin\phi, \quad P_{1,1} = \cos\phi \\
&P_{2,0} = \tfrac{1}{2}(3\sin^2\phi - 1), \quad P_{2,1} = 3\sin\phi\cos\phi, \quad P_{2,2} = 3\cos^2\phi \\
&P_{3,0} = \tfrac{1}{2}(5\sin^3\phi - 3\sin\phi), \quad P_{3,1} = \tfrac{3}{2}(5\sin^2\phi - 1)\cos\phi \\
&P_{3,2} = 15\sin\phi\cos^2\phi, \quad P_{3,3} = 15\cos^3\phi \\
&\ldots
\end{aligned}\right\} \tag{3.2.22}$$

The origin of the reference system is generally selected by definition to be at the center of mass, so that

$$\overline{C}_{0,0} = 1, \overline{C}_{1,0} = \overline{C}_{1,1} = \overline{S}_{1,1} = 0 \tag{3.2.23}$$

and if an axis is along the principal moment of inertia, then

$$\overline{C}_{2,1} = \overline{S}_{2,1} = 0. \tag{3.2.24}$$

If the coefficients are given in the unnormalized format $C_{n,m}$, $S_{n,m}$ then

$$\begin{bmatrix} \overline{C}_{n,m} \\ \overline{S}_{n,m} \end{bmatrix} = \left[\frac{(n+m)}{(n-m)!(2n+1)k}\right]^{1/2} \begin{bmatrix} C_{n,m} \\ S_{n,m} \end{bmatrix}, \tag{3.2.25}$$

where $k = 1$ for $m = 0$ and $k = 2$ for $m \neq 0$. The first few coefficients for the Earth of the *World Geodetic System 1984* (Defence Mapping Agency, 1991), for which the coefficients are specified to $n = m = 41$, are

$$\left.\begin{aligned}
&\overline{C}_{2,0} = -0.484165371736 \times 10^{-3}, \quad \overline{S}_{2,0} = 0 \\
&\overline{C}_{2,1} = -0.186987635955 \times 10^{-9}, \quad \overline{S}_{2,1} = +0.119528012031 \times 10^{-8} \\
&\overline{C}_{2,2} = +0.243914352398 \times 10^{-5}, \quad \overline{S}_{2,2} = -0.140016683654 \times 10^{-5} \\
&\overline{C}_{3,0} = 0.957254173792 \times 10^{-6}, \quad \overline{S}_{3,0} = 0 \\
&\ldots \\
&\overline{C}_{4,0} = 0.539873863789 \times 10^{-6}, \quad \overline{S}_{4,0} = 0 \\
&\ldots \\
&\overline{C}_{5,0} = 0.685323475630 \times 10^{-7}, \quad \overline{S}_{5,0} = 0.
\end{aligned}\right\} \tag{3.2.26}$$

An alternative unnormalized expression often used for the gravitational potential is

$$U = -\frac{GM}{r}\left[1 + \sum_{n=2}^{\infty}\left(\frac{a}{r}\right)^n J_n P_n(\sin\phi) + \sum_{n=2}^{\infty}\sum_{m=1}^{n}\left(\frac{a}{r}\right)^n J_{n,m} P_{n,m}(\sin\phi)\cos m(\lambda - \lambda_{nm})\right] \tag{3.2.27}$$

where

$$J_n \equiv J_{n,0} = (2n+1)^{1/2}\overline{C}_{n,0}, \tag{3.2.28}$$

$$J_{n,m} = \left[\frac{2(n-m)!(2n+1)}{(n+m)!}\right]^{1/2}(\overline{C}^2_{n,m} + \overline{S}^2_{n,m})^{1/2}, \tag{3.2.29}$$

$$\tan\lambda_{n,m} = \overline{S}_{n,m}/\overline{C}_{n,m}. \tag{3.2.30}$$

The equivalent coefficients given by equation 3.2.26 are

$$\begin{aligned} J_2 &= -1.082626683 \times 10^{-3}, \\ J_3 &= +2.532665648 \times 10^{-6}, \\ J_4 &= +1.619621591 \times 10^{-6}, \\ J_5 &= +2.272960829 \times 10^{-6}. \end{aligned} \tag{3.2.31}$$

The specific potential U has $(n-m)$ sign changes over $-\pi/2 \leq \phi \leq \pi/2$ and $2m$ zeros over $0 \leq \lambda < 2\pi$. A *surface spherical harmonic function* is the portion of the gravitational expansion for a given degree n and order m that is a function on ϕ and λ. Three categories of surface spherical harmonic functions are realized:

- *Zonal* harmonic when $m = 0$.
- *Sectoral* harmonic when $m = n$.
- *Tesseral* harmonic when $m \neq n \neq 0$.

For zonal harmonics, the $P_{n,0}(\sin\phi)$ vanish at n parallels of latitude on the surface of a sphere and divide the surface into horizontal zones. The zonal harmonics are symmetric in longitude. For sectoral harmonics, the $\sin n\lambda P_{n,n}(\sin\phi)$ and $\cos n\lambda P_{n,0}(\sin\phi)$ vanish at lines of longitude and divide the surface of a sphere into vertical sectors. For tesseral harmonics, the $\sin m\lambda P_{n,m}(\sin\phi)$ and $\cos m\lambda P_{n,m}(\sin\phi)$ vanish at $n-m$ parallels of latitude and $2m$ lines of longitude that divide the surface of a sphere into quadrangles whose angles are right angles. Examples of each are given in figure 3.2. Models of the

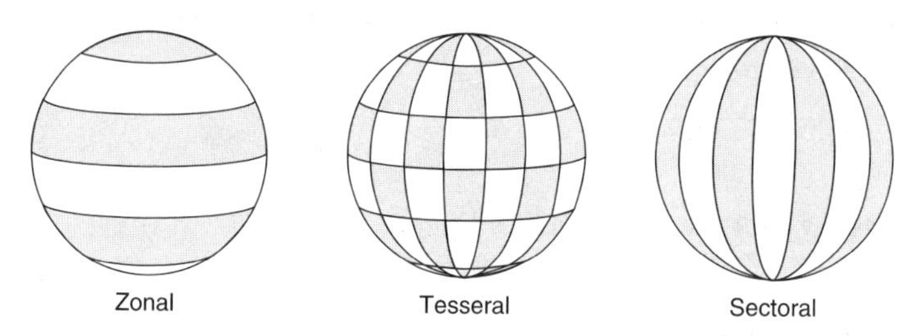

Figure 3.2 Zonal, tesseral, and sectoral surface spherical harmonics.

gravitational field of the Earth and planets have coefficients that number into the tens of thousands. The unclassified geoid height for the WGS 84 model is illustrated in figure 3.3. The geoid height is the distance of a geopotential surface from a reference mean ellipsoid.

3.3 Two-Body Central Force Motion

3.3.1 Equation of Motion

The equation of motion of a near-Earth spacecraft, developed here, includes the effects of the Sun and the Moon. The purpose is to show that the two-body central force formulation is a valid approximation for near-Earth spacecraft, and to estimate the relative perturbing influence of the Sun and Moon (Bate et al., 1971; Brower and Clemence, 1961; Danby, 1962; Geyling and Westerman, 1971; Smart, 1953; Tisserand, 1889; Vallado, 2001; Weisel, 1989). Consider a simplified solar system consisting of a central body or Sun, Earth, Moon, and spacecraft. The geometry is illustrated in figure 3.4, where $O(X, Y, Z)$ is an inertial reference system and $O(x, y, z)$ has its origin fixed at the center of mass of the Earth and its axes parallel to those of the inertial reference system. Although the axes of $O(x, y, z)$ remain parallel, this is not an inertial reference system since its origin moves with the Earth about the Sun. Note that in this development the motion of the Earth is unspecified. The equation of motion for the spacecraft follows from Newton's second law, equation 3.1.1, to be

$$m_0(\ddot{\mathbf{r}}_e + \ddot{\mathbf{r}}_0) = -\nabla\left(-\frac{Gm_e m_0}{r_0} - \frac{Gm_s m_0}{|\Delta_s|} - \frac{Gm_m m_0}{|\Delta_m|}\right), \tag{3.3.1}$$

where

$$\begin{aligned}
r_e &= \text{radius vector to Earth from the inertial reference system,} \\
r_s &= \text{radius vector from Earth to Sun,} \\
r_m &= \text{radius vector from Earth to Moon,} \\
r_0 &= \text{radius vector from Earth to spacecraft,} \\
m_e &= \text{mass of the Earth,} \\
m_s &= \text{mass of the Sun,} \\
m_m &= \text{mass of the Moon,} \\
m_0 &= \text{mass of the spacecraft,} \\
\Delta_s &= \mathbf{r}_0 - \mathbf{r}_s = \text{vector from the Sun to the spacecraft,} && (3.3.2) \\
\Delta_m &= \mathbf{r}_0 - \mathbf{r}_m = \text{vector from the Moon to the spacecraft.} && (3.3.3)
\end{aligned}$$

Carrying out the differentiation and dividing by m_0 gives

$$\ddot{\mathbf{r}}_e + \ddot{\mathbf{r}}_0 = -\frac{Gm_e}{r_0^2}\hat{\mathbf{r}}_0 - \frac{Gm_s}{|\Delta_s|^2}\hat{\Delta}_s - \frac{Gm_m}{|\Delta_m|^2}\hat{\Delta}_m, \tag{3.3.4}$$

where the caret denotes a unit vector. The analogous equation for the motion of the Earth is

$$\ddot{\mathbf{r}}_e = \frac{Gm_0}{r_0^2}\hat{\mathbf{r}}_0 + \frac{Gm_s}{r_s^2}\hat{\mathbf{r}}_s + \frac{Gm_m}{r_m^2}\hat{\mathbf{r}}_m \tag{3.3.5}$$

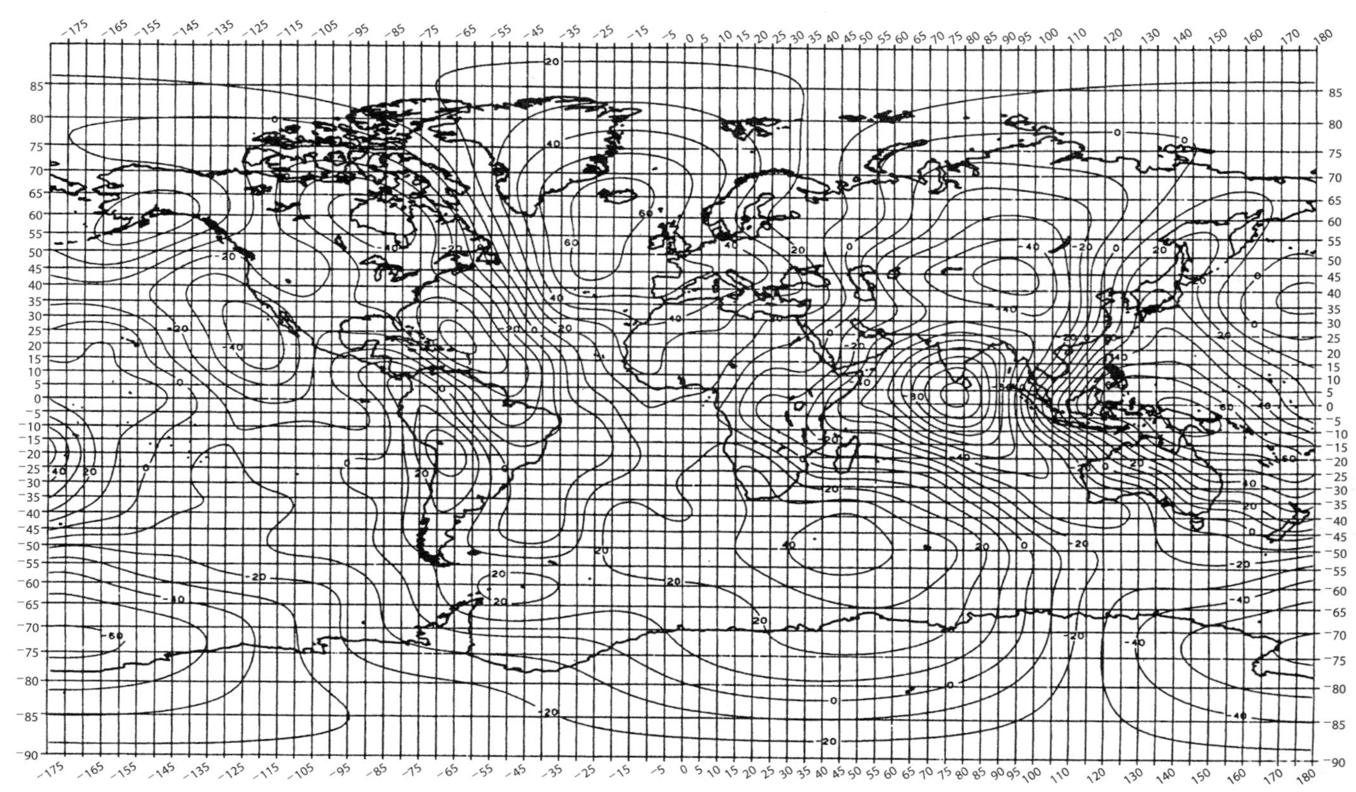

Figure 3.3 Geoid height in meters referenced to WGS 84 ellipsoid, *n* and *m* truncated at 18. (Copyright 1998 by the National Imagery and Mapping Service, U.S. Government. No domestic copyright claimed under Title 17 U.S.C. All rights reserved.)

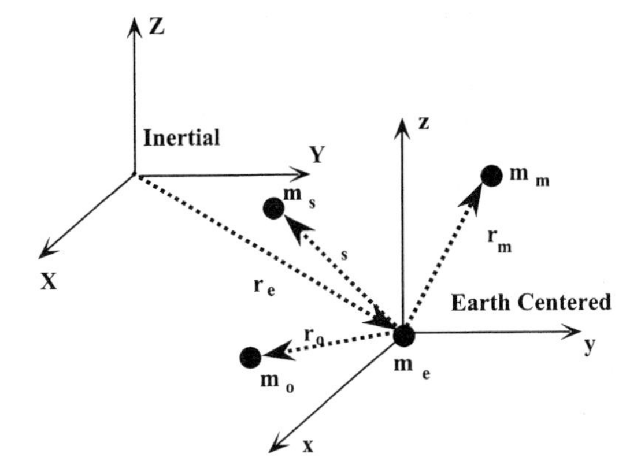

Figure 3.4 Simplified solar system. m_m = mass of Moon, m_s = mass of Sun, m_0 = mass of spacecraft, $O(X, Y, Z)$ = internal frame at center of mass, $O(x.y, z)$ = Earth-centered reference frame, $\mathbf{r}_0$ = spacecraft position vector, $\mathbf{r}_\mathrm{s}$ = Sun position vector, $\mathbf{r}_\mathrm{m}$ = Moon position vector, $\mathbf{r}_\mathrm{e}$ = Earth position vector.

which, when substituted into the prior equation, gives the motion of the spacecraft with respect to the Earth to be

$$\ddot{\mathbf{r}}_0 = -\frac{Gm_\mathrm{e}}{r_0^2}\left[\frac{(m_\mathrm{e}+m_0)}{m_\mathrm{e}}\hat{\mathbf{r}}_0 + \frac{m_\mathrm{s}r_0^2}{m_\mathrm{e}r_\mathrm{s}^2}\left(\hat{\mathbf{r}}_\mathrm{s} + \frac{r_\mathrm{s}^2}{\Delta_\mathrm{s}^2}\hat{\Delta}_\mathrm{s}\right) + \frac{m_\mathrm{m}r_0^2}{m_\mathrm{e}r_\mathrm{m}^2}\left(\hat{\mathbf{r}}_\mathrm{m} + \frac{r_\mathrm{m}^2}{\Delta_\mathrm{m}^2}\hat{\Delta}_\mathrm{m}\right)\right] \tag{3.3.6}$$

The relative orders of magnitude of the three terms in the bracket can be approximated as follows for a near-Earth satellite, assuming that $m_0 \ll m_\mathrm{e}$ and $r_0 \approx R_\mathrm{e}$ the radius of the Earth:

$$\frac{m_\mathrm{e}+m_0}{m_\mathrm{e}} \approx 1, \tag{3.3.7}$$

$$\frac{m_\mathrm{s}r_0^2}{m_\mathrm{e}r_\mathrm{s}^2} \approx \frac{m_\mathrm{s}R_\mathrm{e}^2}{m_\mathrm{e}r_\mathrm{s}^2} \approx 6\times 10^{-4}, \tag{3.3.8}$$

since $m_\mathrm{s}/m_\mathrm{e} = 3.3\times 10^5$ and $R_\mathrm{e}/r_\mathrm{s} = 4.3\times 10^{-5}$, and

$$\frac{m_\mathrm{m}r_0^2}{m_\mathrm{e}r_\mathrm{m}^2} \approx \frac{m_\mathrm{s}R_\mathrm{e}^2}{m_\mathrm{e}r_\mathrm{m}^2} \approx 3.4\times 10^{-6}, \tag{3.3.9}$$

since $m_\mathrm{m}/m_\mathrm{e} = 1/81.3$ and $R_\mathrm{e}/r_\mathrm{m} = 1/60.1$.

Equations 3.3.6 through 3.3.9 illustrate that the effects of the Sun and Moon are three to four orders of magnitude less than the effect of the Earth on the motion of near-Earth satellites. Thus, the presence of the Sun and Moon can be neglected to first order and treated as perturbations of the motion. As a consequence, the motion of the Earth satellite can be treated through the mechanism of the two-body central force problem. Note that the equations developed above can be generalized to a point mass body in the vicinity of any celestial body and include the perturbations of additional bodies, for example, Jupiter.

3.3.2 Solution to the Two-Body Central Force Problem

The equation of motion of a satellite about a primary body, for example a satellite about the Earth, ignoring the effects of the Sun and Moon, is given by equation 3.3.6 as

$$\ddot{\mathbf{r}} = -\frac{G(m_{\mathrm{e}} + m_0)}{r^2}\hat{\mathbf{r}}, \tag{3.3.10}$$

where the subscript 0 on r has been dropped. This equation can be written in several different forms

$$\ddot{\mathbf{r}} = -\frac{Gm_{\mathrm{e}}m_0}{r^2}\left(\frac{m_{\mathrm{e}} + m_0}{m_{\mathrm{e}}m_0}\right)\hat{\mathbf{r}} = -\frac{\mu}{r^2}\hat{\mathbf{r}} = -\frac{k/m}{r^2}\hat{\mathbf{r}}, \tag{3.3.11}$$

where

$$\mu \equiv G(m_{\mathrm{e}} + m_0) \approx Gm_{\mathrm{e}}, \tag{3.3.12}$$

$$k \equiv Gm_{\mathrm{e}}m_0, \tag{3.3.13}$$

$$m \equiv \frac{m_{\mathrm{e}}m_0}{m_{\mathrm{e}} + m_0} \approx m_0, \tag{3.3.14}$$

$$\mu = \frac{k}{m}, \tag{3.3.15}$$

and m is known as the *reduced* or *equivalent mass*. Multiplying equation 3.3.11 by the equivalent mass m and taking the cross-product of both sides with $\mathbf{r}$ gives

$$\mathbf{r} \times m\ddot{\mathbf{r}} = \mathbf{r} \times \left(-\frac{k}{r^2}\right)\hat{\mathbf{r}} = 0 \tag{3.3.16}$$

which can be integrated to yield

$$\mathbf{r} \times m\dot{\mathbf{r}} = \mathbf{H} = \text{constant} \tag{3.3.17}$$

where $\mathbf{H}$, the angular momentum of the satellite, is the constant of integration. This equation can be rewritten as

$$\mathbf{r} \times \dot{\mathbf{r}} = \frac{\mathbf{H}}{m} \equiv \mathbf{h} = \text{constant} \tag{3.3.18}$$

where $\mathbf{h}$ is the *specific angular momentum*, also constant.

The differential area swept out by the position vector in an increment of time dt, known as the *areal velocity*, is illustrated in figure 3.5 and is simply

$$dA = \tfrac{1}{2} r\,dr \sin\alpha, \tag{3.3.19}$$

which can be rewritten as

$$dA = \frac{1}{2}\left|\mathbf{r} \times d\mathbf{r}\right| = \frac{1}{2m}\left|\mathbf{r} \times m\frac{d\mathbf{r}}{dt}\right| dt, \tag{3.3.20}$$

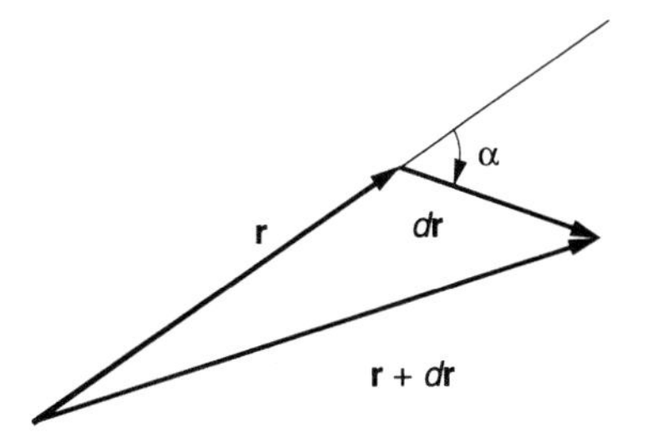

Figure 3.5 Areal velocity geometry. $\mathbf{r}$ = position vector; $d\mathbf{r}$ = change in position vector, α = included angle.

so that

$$\frac{dA}{dt} = \frac{H}{2m} = \frac{h}{2} = \text{constant} \tag{3.3.21}$$

where $H = |\mathbf{H}|$ and $h = |\mathbf{h}|$. Thus Kepler's second law has been proved to be valid and is illustrated in figure 3.6.

The equation for the energy follows directly from equation 3.3.11 by taking the scalar product with $d\mathbf{r}$ so that

$$m\ddot{\mathbf{r}} \cdot d\mathbf{r} - \frac{k}{r^2}\hat{\mathbf{r}} \cdot d\mathbf{r} = 0, \tag{3.3.22}$$

which can be integrated to yield

$$\frac{1}{2}m\dot{\mathbf{r}} \cdot \dot{\mathbf{r}} - \frac{k}{r} = E = \text{constant}, \tag{3.3.23}$$

where the constant of integration E is the energy. Using equation 3.3.15, equation 3.3.23 can be rewritten in terms of v, the magnitude of the velocity $\mathbf{v}$, as

$$\frac{1}{2}v^2 - \frac{\mu}{r} = \varepsilon = \text{constant} \equiv -\frac{\mu}{2a}, \tag{3.3.24}$$

where the *specific energy* is defined by

$$\varepsilon \equiv \frac{E}{m} \tag{3.3.25}$$

and the constant ε is expressed in terms of the parameter a, which is a distance. As a result, it is possible to express the magnitude of the velocity in terms of the parameter a and the distance r by

$$v = \sqrt{2\left(\varepsilon + \frac{\mu}{r}\right)} = \sqrt{\mu\left(\frac{2}{r} - \frac{1}{a}\right)}, \tag{3.3.26}$$

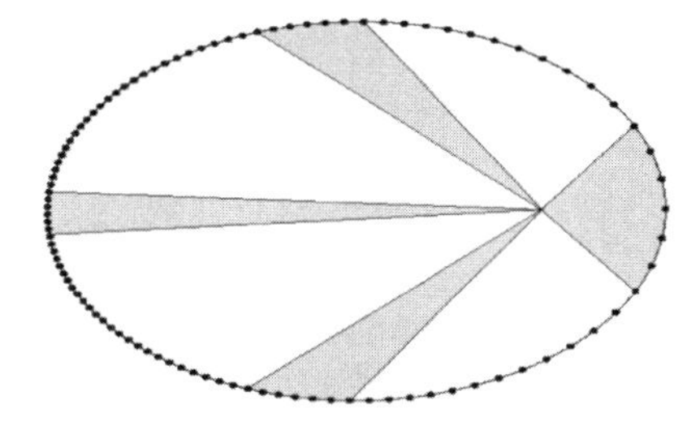

Figure 3.6 Areal velocity, illustrating constant areas.

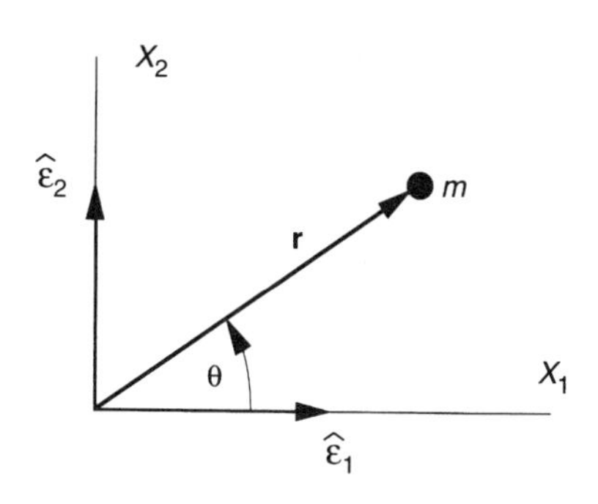

Figure 3.7 Position vector in polar coordinates, $\mathbf{r}$ = radius, r, θ = polar coordinates, $\boldsymbol{\varepsilon}_1$ = unit vectors ($i = 1, 2$).

which is known as the *vis-viva* (living force) *equation*. The term C3 (km^2/s^2), sometimes employed in interplanetary trajectories, is equal to twice the specific energy: $\text{C3} = 2\varepsilon$.

Since the angular momentum is constant, the motion must take place in a plane, so the position vector can be represented, using two-dimensional polar coordinates (r, θ) as illustrated in figure 3.7, by

$$\mathbf{r} = r \cos\theta \, \hat{\boldsymbol{\varepsilon}}_1 + r \sin\theta \, \hat{\boldsymbol{\varepsilon}}_2. \tag{3.3.27}$$

Substituting $\mathbf{r}$ into equation 3.3.11 gives two equations

$$m\ddot{r} - mr\dot{\theta}^2 = -\frac{k}{r^2} \quad \text{or} \quad \ddot{r} - r\dot{\theta}^2 = -\frac{\mu}{r^2}, \tag{3.3.28}$$

$$mr^2\dot{\theta} = H \quad \text{or} \quad r^2\dot{\theta} = h, \tag{3.3.29}$$

where H is the magnitude of the angular momentum and h is the magnitude of the specific angular momentum, both of which are constant (equations 3.3.17 and 3.3.18). It is possible to eliminate $\dot{\theta}$ in equation 3.3.28 by substituting equation 3.3.29, but the resulting nonlinear equation in terms of r as a function of time leads to elliptic functions. However, it is possible to eliminate the time dependency in equation 3.3.28 by expressing equation 3.3.29 as the operator

$$\frac{d}{dt} = \frac{h}{r^2}\frac{d}{d\theta}, \tag{3.3.30}$$

so that equation 3.3.28 can be written as

$$\frac{d^2u}{d\theta^2} + u = \frac{\mu}{h^2}, \tag{3.3.31}$$

where

$$u \equiv \frac{1}{r}. \tag{3.3.32}$$

Equation 3.3.31 is in the form of the undamped linear harmonic oscillator with a constant forcing function, whose solution is simply a constant, and a harmonic term written here as

$$u = \frac{\mu}{h^2}[1 + A\cos(\theta - \theta_0)] = \frac{\mu}{h^2}[1 + A\cos f], \tag{3.3.33}$$

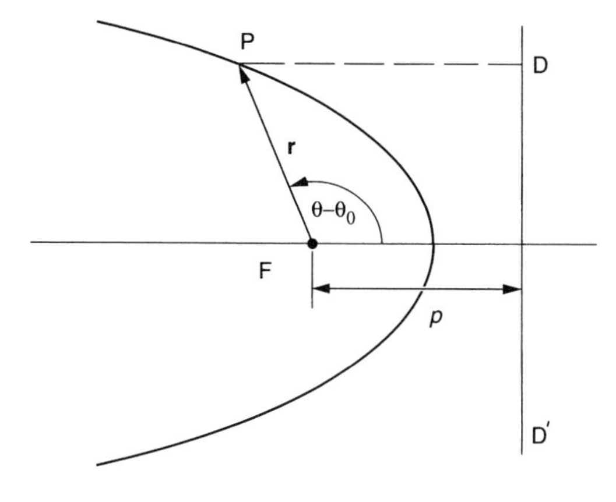

Figure 3.8 Geometry of a conic. D-D′ = directrix; F = focus, $f = \theta - \theta_0$.

so that

$$r = \frac{h^2/\mu}{1 + [1 + 2\varepsilon h^2/\mu^2]^{1/2} \cos f} \tag{3.3.34}$$

where f is introduced as the *true anomaly*, defined by

$$f \equiv \theta - \theta_0, \tag{3.3.35}$$

and the constant of integration A can be determined in terms of the energy. This calculation is left for an exercise at the end of the chapter.

Equation 3.3.34 has the form of the equation of a conic section in polar coordinates with the origin at one of the foci, which was studied by the Greeks long before its application to astrodynamics. A conic section is the locus of points in a plane such that the ratio of the distances from a fixed point, called the focus, and a fixed line, called the directrix, is a constant as illustrated in figure 3.8. Conic sections are also curves generated by the intersection of a plane with a cone, as illustrated in figure 3.9.

The general equation of a conic section is

$$r = \frac{pe}{1 + e \cos f} \tag{3.3.36}$$

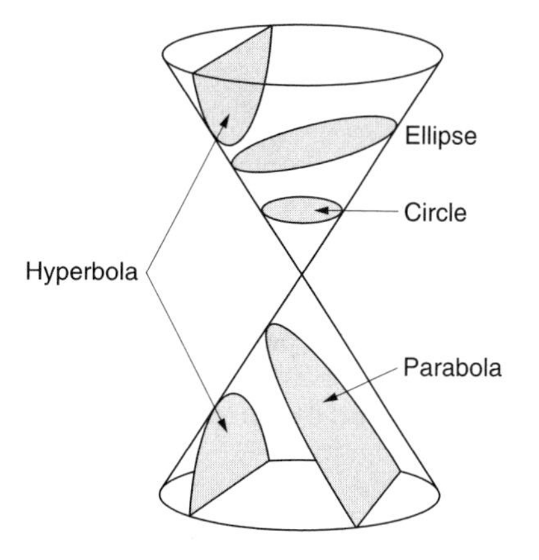

Figure 3.9 Conic sections.

Table 3.1 Relationships for conic sections

Conic	Specific Energy	Semimajor Axis	Eccentricity
Circle	$\varepsilon = -\mu^2/2h^2$	$a > 0$	$e = 0$
Ellipse	$-\mu^2/2h^2 < \varepsilon < 0$	$a > 0$	$0 < e < 1$
Parabola	$\varepsilon = 0$	$a = \infty$	$e = 1$
Hyperbola	$\varepsilon > 0$	$a < 0$	$e > 1$

where e, the eccentricity, is defined as the ratio of the length of the line marked by PD to the distance r and p is the distance to the directrix from the focus, as illustrated in figure 3.8. Comparing the terms in equations 3.3.34 and 3.3.36 gives, for p and e,

$$pe = \frac{h^2}{\mu}, \tag{3.3.37}$$

$$e = \left[1 + \frac{2\varepsilon h^2}{\mu^2}\right]^{1/2}. \tag{3.3.38}$$

Substituting equations 3.3.24 and 3.3.37 into equation 3.3.38 gives

$$pe = a(1 - e^2), \tag{3.3.39}$$

so that equation 3.3.36 can be rewritten in terms of a, e, and f as

$$r = \frac{a(1 - e^2)}{1 + e\cos f}. \tag{3.3.40}$$

The relationships between specific energy, semimajor axis, and eccentricity are shown in table 3.1 for the different conic sections, and the different trajectories are illustrated in figure 3.10.

3.3.3 Circular Orbit

If the specific energy is

$$\varepsilon = -\frac{\mu^2}{2h^2} \tag{3.3.41}$$

it follows from equations 3.3.38 and 3.3.40 that the *eccentricity* and *radius* are

$$e = 0, \quad r = a. \tag{3.3.42}$$

Consequently, the directrix is at infinity and the trajectory is a circle, as shown in figure 3.11.

The *magnitude of the velocity* can be found from the vis viva equation, 3.3.26, to be

$$v = \sqrt{\frac{\mu}{a}}, \tag{3.3.43}$$

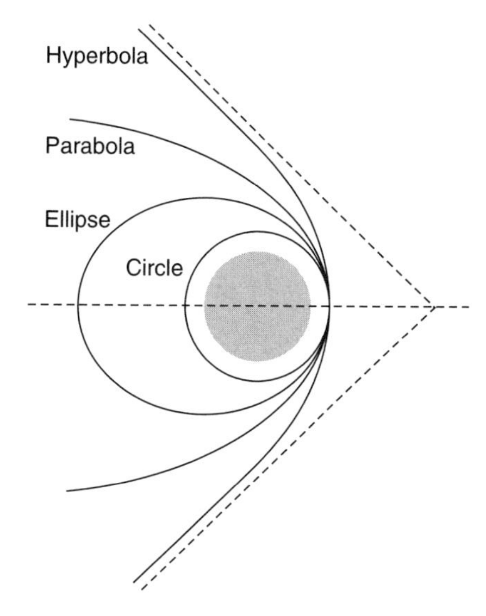

Figure 3.10 Relationships of conic sections.

and the *period* τ can be determined by dividing thc circumference by the velocity to obtain

$$\tau = \frac{2\pi a}{v} = 2\pi\sqrt{\frac{a^3}{\mu}}. \tag{3.3.44}$$

Note that the velocity decreases with increasing semimajor axis; if the semimajor axis increases by a factor of four the velocity decreases by a factor of two. The period increases with increasing semimajor axis; if the semimajor axis increases by a factor of four the period will increase by a factor of eight.

Example

Determine the Schuler period, which is defined as the period of a satellite around the Earth at zero altitude.

Solution: Using equation 3.3.44 and the fact that the equatorial radius of the Earth is 6 378 137 m gives

$$\tau = 2\pi\sqrt{\frac{a^3}{\mu}} = 2\pi\sqrt{\frac{6\ 378\ 137^3}{3.986\ 005 \times 10^{14}}} = 5\ 073.138\ \text{ s} = 84.55\ \text{ min}\,.$$

The Schuler period can be used to determine the period at any altitude from

$$\tau = 84.55 a^{3/2}\,\text{min}, \tag{3.3.45}$$

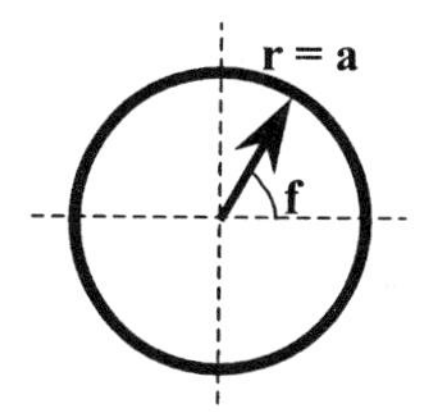

Figure 3.11 Circular orbit. a = semimajor axis, f = true anomaly.

with a in Earth radii. The *mean motion* n, which is the average angular rate of the position vector, follows from the period as

$$n = \frac{2\pi}{\tau} = \sqrt{\frac{\mu}{a^3}}, \tag{3.3.46}$$

and it follows that the *true anomaly* as a function of time is

$$f = (\theta - \theta_0) = n(t - t_0). \tag{3.3.47}$$

Example

Determine the radius of a satellite in an equatorial orbit whose position remains fixed with respect to the Earth so that its period is one sidereal day, equivalent to 1/1.002 737 909 35 of a solar day.

Solution: The duration of a sidereal day is

$$t = 86\,400/1.002\,737\,909\,35 = 86\,164.090\,54 \text{ seconds}$$

The equation for the semimajor axis in terms of the period can be determined from equation 3.3.44 as

$$a = \left[\left(\frac{\tau}{2\pi}\right)^2 \mu\right]^{1/3} = \left[\left(\frac{86\,140.090\,54}{2\pi}\text{s}\right)^2 3.986\,005 \times 10^{14}\ \text{m}^3/\text{s}^2\right]^{1/3},$$

$$a = 42\,156.34 \text{ km}.$$

3.3.4 Elliptical Orbit

If the specific energy is bounded by

$$-\frac{\mu^2}{2h^2} < \varepsilon < 0, \tag{3.3.48}$$

equation 3.3.38 gives the *eccentricity* to be

$$0 < e < 1 \tag{3.3.49}$$

and the trajectory is an ellipse, as illustrated in figure 3.12, with *semimajor axis* a and *semiminor axis* b, where

$$b = a\sqrt{1 - e^2}. \tag{3.3.50}$$

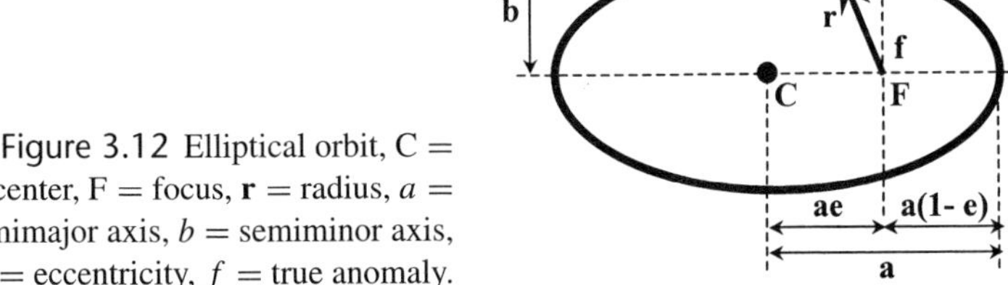

Figure 3.12 Elliptical orbit, C = center, F = focus, $\mathbf{r}$ = radius, a = semimajor axis, b = semiminor axis, e = eccentricity, f = true anomaly.

The solution of an elliptic trajectory confirms Kepler's first law. The radius at *periapsis* or, in the case of the Earth, *perigee*, r_{p}, is the minimum value of r and follows from equation 3.3.40 to be

$$r|_{f=0} \equiv r_{\mathrm{p}} = a(1-e), \tag{3.3.51}$$

and the radius at *apoapsis* or, in the case of the Earth, *apogee*, r_a, is the maximum value of r that follows from equation 3.3.40 to be

$$r|_{f=\pi} \equiv r_a = a(1+e). \tag{3.3.52}$$

The *period* is obtained from the constancy of the areal velocity and equations 3.3.21, 3.3.37, 3.3.39, and 3.3.50 to be

$$\tau = \frac{A}{\dot{A}} = \frac{\pi ab}{h/2} = 2\pi\sqrt{\frac{a^3}{\mu}}, \tag{3.3.53}$$

where the derivation is left as an exercise. The *mean motion*, or average angular velocity, follows as

$$n = \frac{2\pi}{\tau} = \sqrt{\frac{\mu}{a^3}}. \tag{3.3.54}$$

Note that this is the same expression as for a circular orbit. For two planets of masses m_1 and m_2 with semimajor axes a_1 and a_2 in motion about the Sun, whose mass is m_{s}, equation 3.3.53 gives the ratio of the periods of the planets to be

$$\left(\frac{\tau_1}{\tau_2}\right)^2 = \left(\frac{a_1}{a_2}\right)^3 \left(\frac{m_2+m_{\mathrm{s}}}{m_1+m_{\mathrm{s}}}\right)^2. \tag{3.3.55}$$

For $m_{\mathrm{s}} \gg m_i$ $(i = 1, 2)$, the expression reduces to

$$\left(\frac{\tau_1}{\tau_2}\right)^2 \approx \left(\frac{a_1}{a_2}\right)^3, \tag{3.3.56}$$

which is Kepler's third law.

Equation 3.3.40 gives the radius to the spacecraft as a function of the true anomaly. To find the motion as a function of time, it is possible to utilize the geometrical construct illustrated in figure 3.13 where the position of the body is projected onto the auxiliary circle and forms an angle called the *eccentric anomaly, E*. The eccentric anomaly can be expressed as

$$\cos E = \frac{ae + r\cos f}{a}, \tag{3.3.57}$$

which, if substituted for $\cos f$ in equation 3.3.40, yields for the radius

$$r = a(1 - e\cos E). \tag{3.3.58}$$

The relationship between the eccentric anomaly and the true anomaly is

$$\tan\frac{f}{2} = \left(\frac{1+e}{1-e}\right)^{1/2} \tan\frac{E}{2}, \tag{3.3.59}$$

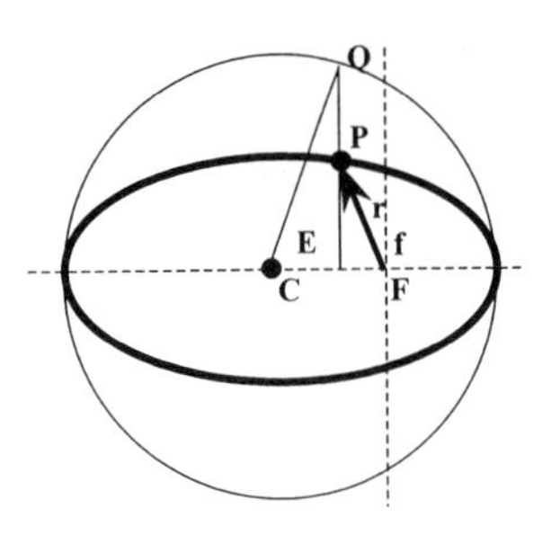

Figure 3.13 Eccentric anomaly for elliptic orbit. C = center, F = focus, **r** = position vector, E = eccentric anomaly, f = true anomaly, P = satellite position, Q = projection of P on auxillary circle.

which can be obtained directly from figure 3.13 and is left as an exercise. The specific energy given by equation 3.3.24 can be written in the form

$$\dot{r}^2 r^2 - \mu r^2 \left(\frac{2}{r} - \frac{1}{a} \right) + \mu a (1 - e^2) = 0, \tag{3.3.60}$$

using $v = (\dot{r}^2 + r^2\dot{\theta}^2)^{1/2}$ and equations 3.3.29, 3.3.37, and 3.3.39. Substituting equation 3.3.58 for r and $\dot{r}$and using equation 3.3.54 for the mean motion yields

$$\dot{E}(1 - e \cos E) = \sqrt{\frac{\mu}{a^3}} = n, \tag{3.3.61}$$

which upon integration gives

$$E - e \sin E = n(t - t_0) \equiv M, \tag{3.3.62}$$

where t_0 is the time of periapsis crossing and M is the *mean anomaly*. Equation 3.3.62 is known as *Kepler's equation* and is a transcendental relationship between the eccentric anomaly E and time t. Once the eccentric anomaly E is determined for a given time, the radius r and true anomaly $f$can be found from equations 3.3.58 and 3.3.59, respectively.

There is no known exact solution to Kepler's equation, but there are more than a hundred different approximate solutions. When the eccentricity is small, the eccentric anomaly can be easily found by successive approximations from

$$E_{i+1} = M + e \sin E_i, \quad i = 1, 2, 3, \ldots, \tag{3.3.63}$$

where $M = n(t - t_0)$ and the initial estimate for E can be zero, that is, $E_1 = 0$. The procedure is illustrated by the following example.

Example

From satellite tracking observations it is determined that a spacecraft has a semimajor axis of three Earth radii and an eccentricity of one-third. Find the radius and the true anomaly three hours after perigee passage.

Solution: The mean anomaly follows from the semimajor axis and the given time to be

$$M = n(t - t_0) = \left(\frac{\mu}{a^3} \right)^{1/2} (t - t_0) = \left(\frac{3.986\ 005 \times 10^{14}}{(3 \times 6\ 378\ 137)^3} \right)^{1/2} \times 3 \times 3\ 600$$
$$= 2.576\ 303\ 315 \text{ radians}.$$

Using equation 3.3.63, the iterations are seen to converge to 10 significant digits by 19 iterations.

i	E	i	E
1	0	13	2.714 407 258
2	2.576 303 315	14	2.714 406 907
3	2.754 856 704	15	2.714 407 014
4	2.702 025 797	16	2.714 406 981
5	2.718 152 495	17	2.714 406 991
6	2.713 269 717	18	2.714 406 988
7	2.714 751 923	19	2.714 406 989
8	2.714 302 335	20	2.714 406 989
9	2.714 438 738	21	2.714 406 989
10	2.714 397 357	22	2.714 406 989
11	2.714 409 911	23	2.714 406 989
12	2.714 406 102	24	2.714 406 989

The radius r follows directly from equation 3.3.58, given a and E, to be

$$r = a(1 - e\cos E) = 3\left(1 - \frac{1}{3}\cos 2.714\ 406\ 989\right)$$
$$= 3.910\ 135\ 363 \text{ Earth radii.}$$

The true anomaly follows from equation 3.3.59 to be

$$\tan\frac{f}{2} = \left(\frac{1+e}{1-e}\right)^{1/2}\tan\frac{E}{2} = \left(\frac{1+\frac{1}{3}}{1-\frac{1}{3}}\right)^{1/2}\tan\left(\frac{2.714\ 406\ 989}{2}\right),$$

which gives two answers for f:

$$f = 2.837\ 219\ 573 \text{ radians } (162.560\ 707\ 1 \text{ degrees}) \text{ or}$$
$$f = 5.978\ 812\ 227 \text{ radians } (342.560\ 707\ 1 \text{ degrees})$$

The ambiguity must be resolved, generally, by noting that f is in the same half-plane as E, so the first answer is correct. Thus the solution is:

At the time $t = 3$ h, $r = 3.910\ 135\ 363$ Earth radii and $f = 2.837\ 219\ 573$ radians or 162.560 707 1 degrees.

3.3.5 Parabolic Orbit

If the specific energy is

$$\varepsilon = 0, \tag{3.3.64}$$

equation 3.3.38 gives the *eccentricity* to be one,

$$e = 1, \tag{3.3.65}$$

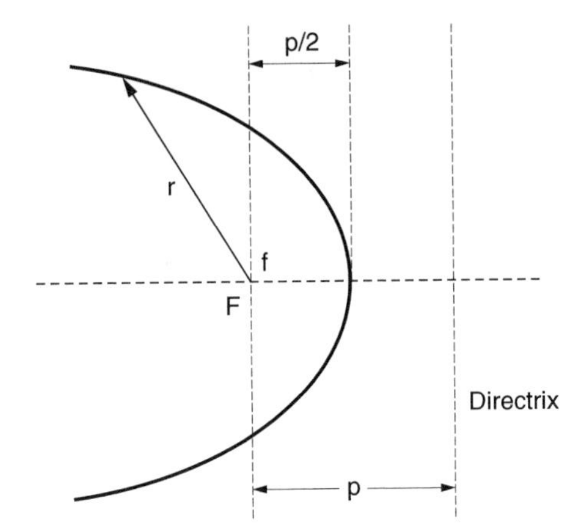

Figure 3.14 Parabolic trajectory. F = focus, r = radius, f = true anomaly, $p/2$ = distance to perlapsis.

and equation 3.3.24 gives the *semimajor axis* to be infinity,

$$a = \infty. \tag{3.3.66}$$

The *equation for the trajectory*, equation 3.3.40, cannot be used since the numerator is zero times infinity, so equations 3.3.34 and 3.3.36 must be used, where

$$r = \frac{h^2/\mu}{1 + \cos f} = \frac{p}{1 + \cos f}, \tag{3.3.67}$$

whose trajectory is illustrated in figure 3.14. The magnitude of the velocity of a body on a parabolic orbit is given by the vis viva equation, 3.3.26, where for $a \to \propto$

$$v = \sqrt{\mu\left(\frac{2}{r} - \frac{1}{a}\right)} = \sqrt{\frac{2\mu}{r}}, \tag{3.3.68}$$

which indicates that the magnitude of the velocity goes to zero as the radius approaches infinity. This equation also defines the *escape velocity*, v_{esc} for a body on a circular orbit of radius r. Comparison of equations 3.3.43 and 3.3.68 gives the escape velocity for a circular orbit to be

$$v_{\text{esc}} = \sqrt{2}\, v_{\text{circular}}. \tag{3.3.69}$$

The relationship between the true anomaly and time is given by the well-known Barker's equation

$$\tan\frac{f}{2} + \frac{1}{3}\tan^3\frac{f}{2} = 2\sqrt{\frac{\mu}{p^3}}(t - t_0), \tag{3.3.70}$$

where p is twice the distance of closest approach and t_0 is the time at closest approach. Once f is determined for a given p and time t, equation 3.3.67 can be used to determine the radius.

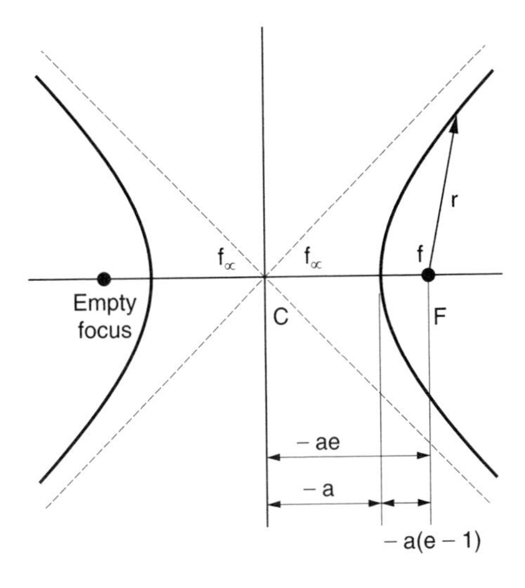

Figure 3.15 Hyperbolic trajectory. F = focus, C = center, $\mathbf{r}$ = position vector, f = true anomaly, a = semimajor axis, e = eccentricity, f_∞ = asymptote.

3.3.6 Hyperbolic Orbit

If the specific energy is greater than zero, $\varepsilon > 0$, it follows from equation 3.3.38 that the eccentricity is greater than one, $e > 1$, and from equation 3.3.24 that the semimajor axis must be negative, $a < 0$. The equation for the trajectory follows from equation 3.3.40:

$$r = \frac{a(1-e^2)}{1+e\cos f}, \qquad (3.3.40')$$

as illustrated in figure 3.15. When $f = 0$, the periapsis distance follows from equation 3.3.40 to be

$$r_\mathrm{p} = a(1-e), \qquad (3.3.71)$$

and the asymptotes are given by $r \to \infty$ as

$$f_\infty = \pm\cos^{-1}(-e^{-1}). \qquad (3.3.72)$$

The velocity is given by the vis viva equation, 3.3.26, which for $r \to \infty$ gives

$$v_\infty = \sqrt{\frac{-\mu}{a}}, \qquad (3.3.73)$$

which is known as the *hyperbolic excess speed.* The development of the analog of Kepler's equation for the hyperbolic orbit is beyond the scope of this presentation, but if the radius is represented in terms of an eccentric anomaly F by

$$r = a(1-e\cosh F), \qquad (3.3.74)$$

the relationship between the eccentric anomaly and the true anomaly follows as

$$\tan\frac{f}{2} = \sqrt{\frac{e+1}{e-1}}\tanh\frac{F}{2}, \qquad (3.3.75)$$

where

$$M = n(t - t_0) = e \sinh F - F, \tag{3.3.76}$$

and the mean motion is

$$n = \sqrt{\frac{\mu}{(-a)^3}}. \tag{3.3.77}$$

At a given eccentricity, semimajor axis, and time, equation 3.3.76 yields F, and then r and f follow from equations 3.3.74 and 3.3.75, respectively.

3.4 Reference Systems

3.4.1 Fundamentals

This section discusses the reference systems and the orbital elements used to represent the trajectories of artificial or natural celestial bodies in three-dimensional space. Reference systems used in astrodynamics and astronomy are generally classified by whether or not they are an inertial system, the location of the origin, and the orientation of two of the axes. Inertial, quasi-inertial, or near-inertial reference systems are convenient for integrating the equations of motion, and non-inertial reference systems are generally the basis of observations for orbit determination.

The motion of the Earth about the Sun is illustrated in figure 3.16. The plane traced out by the Earth is the *ecliptic plane* or *ecliptic* and the axis normal to it is the *ecliptic pole*. The Earth's spin axis is known as the *celestial equatorial pole* (CEP) or *celestial pole*, and the plane orthogonal to it is known as the *celestial equatorial plane* or *equatorial plane* whose intersection with the celestial sphere is known as the *equator*. The equatorial plane is inclined to the ecliptic plane by about 23.44 degrees, known as the *obliquity of the ecliptic*. The intersection of the equatorial plane with the ecliptic plane is known as

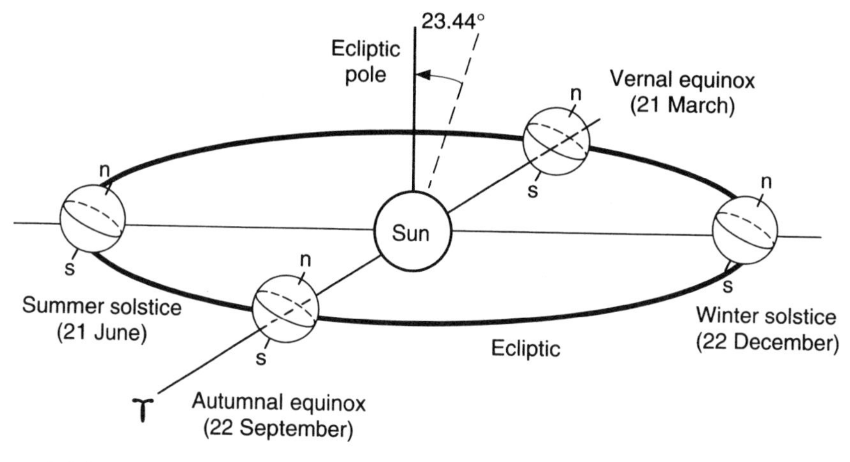

Figure 3.16 Orbital motion of the Earth. n = north, s = south, dates are typical.

the *line of equinox*, or the *equinox* (equal night). As the Earth moves around the Sun, the equinox moves on the plane of the ecliptic. Twice a year the Sun will be simultaneously in the ecliptic and equatorial planes. Between 19 and 21 March of each year, the Sun will appear to move from the southern hemisphere to the northern hemisphere, an event known as the *vernal equinox*. At this time, the line from the Earth to the Sun is known as the *first point of Aries* and is indicated by a ram's head in illustrations because when it was first defined about 3000 years ago it was pointing toward the constellation Aries. Today the equinox is pointing toward the constellation Pisces, moving slowly toward Aquarius, which it will reach about the year 2600.

The ecliptic plane, equatorial plane, ecliptic pole, and celestial pole are in motion with respect to the stars, so the equinox is not fixed in space. Instantaneous or actual positions of planes and directions are identified by the adjective *true* or *apparent* and are referenced to an instant of time identified as the *epoch* or *date*. These motions are irregular and are conveniently described in terms of two components, precession and nutation. *Precession* is the steady secular motion that remains after the effects of the *nutation* have been removed. *Nutation* is defined as the periodic and near-periodic effects of less than 300 years' duration. It is therefore convenient to define average positions given by the precession alone and identified as the *mean* referenced to an epoch or date. Consequently, at a given epoch the equinox and celestial pole can be specified as either the *true* (or *apparent*) *of date* or *mean of date*.

The two primary causes of perturbations are planetary perturbations and lunisolar perturbations. *Planetary perturbations* are a result of the gravitational forces of the other planets and bodies, such as Jupiter and the Moon, that perturb the Earth from traveling in a plane, since all celestial bodies do not lie in the ecliptic plane. *Lunisolar perturbations* result from the gravitational torques of the Sun and Moon on the spinning and oblate Earth. Together, these effects are identified as *general perturbations*. Both perturbations give rise to precession and nutation. The combined effects of planetary and lunisolar precession are known as *general precession*. The principal effect of planetary precession is a precession of the equinox of 12 seconds of arc per century and a decrease in the obliquity of about 47 seconds of arc per century. The principal effect of lunisolar precession is the precession of the Earth's celestial pole about the ecliptic pole, which causes the equinox to precess with a period of 25,765 years or 50 seconds of arc per year. The current theory for nutation consists of 106 harmonic terms due to planetary perturbations and 85 terms due to lunisolar perturbations. The four dominant periods for nutation are 18.6 years (precession period of lunar orbit), 183 days (half a year), 14 days (half a month), and 9.3 years (rotation period of the lunar perigee), and the nutation has a maximum amplitude of 20 seconds of arc. The precession of the ecliptic pole is much less, typically about 0.5 arc seconds per year. Figure 3.17 illustrates the presession and primary nutation of the International Earth Rotation Service (IERS) reference pole.

3.4.2 The International Celestial Reference System (ICRS)

The *International Celestial Reference System*, adopted in 1998, is the fundamental quasi-inertial, barycentric reference system for the measurement of positions and motions of celestial objects. Its orientation is identified with respect to the positions of 608 extragalactic radio sources. For objects with respect to the Earth, the axes are transferred to the center of mass of the Earth. The ICRS has been selected to be a good approximation

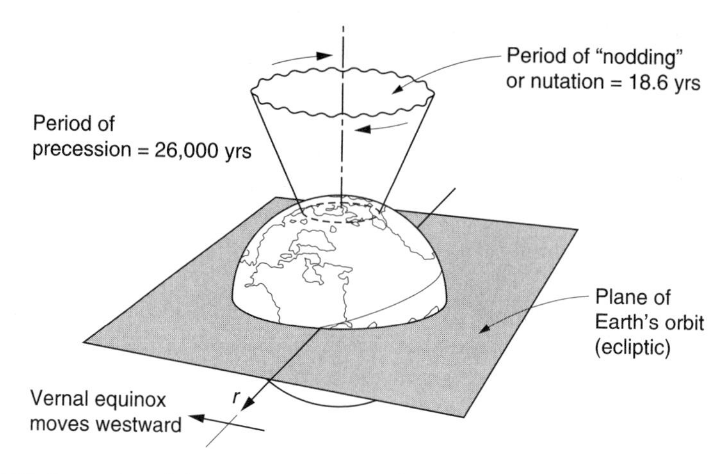

Figure 3.17 Precession of the equinox and the primary nutation of the IERS reference pole.

to the dynamic equator and equinox at the epoch of J2000.0 (12 h TT on 1 January 2000 or JD 2 451 545.0). As a result, at the epoch J2000, the Z-axis is nearly coincident with the *celestial (ephemeris) pole* and the X-axis is closely aligned with the vernal equinox. Positions of bodies are identified by *right ascension*, measured eastward in the equatorial plane from the vernal equinox, and *declination*, measured positive from the equator to the north and negative to the south. Relative to the Earth, the ICRS is often referred to as the Earth-centered inertial reference (ECI) system.

3.4.3 The International Terrestrial Reference System (ITRS)

The International Terrestrial Reference System, developed in the late 1980s, is the system used to locate positions with respect to the Earth. The polar axis is the *IERS reference pole* (IRP), which is essentially the Earth's rotational pole. The prime meridian is the *IERS reference meridian* (IRM), which is essentially the Greenwich meridian. As a result, the xy-plane is essentially the equator, the z-axis points along the IRP, and the x-axis is in the IRM. Positions are identified by *longitude* measured in the xy-plane with positive east from the IERS reference meridian and *latitude* measured positive from the equator to the north and negative to the south. Because of continental drift, positions in the IERS are time dependent and are defined relative to positions and velocities of a set of stations observed by very-long-baseline interferometry (VLBI), lunar laser ranging (LLR), global positioning system (GPS), and by Doppler orbitography and radio positioning integrated by satellites (DORIS) systems. These parameters were originally determined for 1997 and have been updated to J2000 to be consistent with the ICRS.

3.4.4 IERS Earth Orientation Parameters (EOP)

The *IERS Earth orientation parameters*, published by the International Earth Rotation Service, describe the relative orientation of the ITRS with respect to the ICRS (or vice versa) as a function of time. In principle the orientation of the ITRS with respect to the

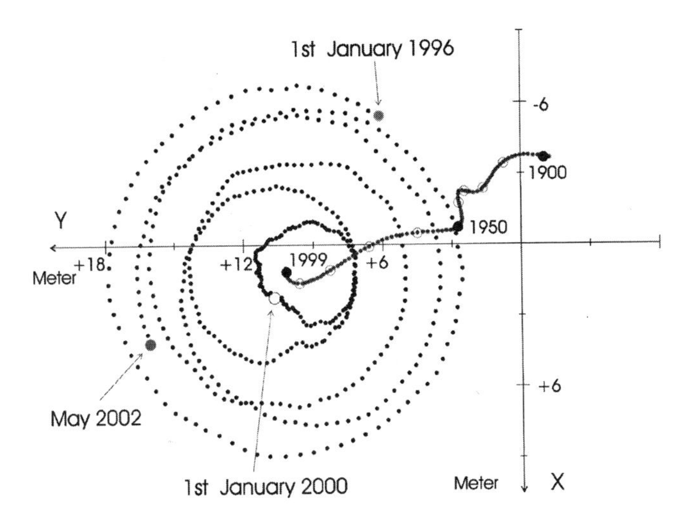

Figure 3.18 Polar motion at five-day intervals and mean path, 1900–2002. (Courtesy of Daniel Gambis, Earth Orientation Center of the International Earth Rotation Service.)

ICRS can be achieved by three parameters, but five are used for convenience. The earth orientation parameters are:

Polar motion (PM): two coordinates (x, y) that give the orientation of the CEP with respect to the IRP. The x-axis is in the direction of the IRM and the y-axis is in the direction 90 degrees west longitude. Illustrated in figure 3.18.

Length of day (LOD): the difference between the length of the day and 86 400 seconds gives the relative orientation of IRM with respect to the equinox when multiplied by the mean rotation rate of the Earth (360°/86 164.090 54s). This is specified as UT1–UTC or UT1–TAI, where UT1 is universal time, UTC is coordinated universal time, and TAI is International Atomic Time. UT1 is related to the Greenwich mean sidereal time (GMST) by an algorithmic relationship. TAI is the atomic time scale calculated by the Bureau International des Poids et Mesures (BIPM) such that UT1–TAI is approximately 0 on January 1st 1958. Illustrated in figure 3.19.

Celestial pole offsets $(d\Psi, d\varepsilon)$: these describe the difference in the observed position of the celestial pole from the predictions of the International Astronomical Union's precession and nutation models. These differences are attributable to atmospheric, oceanic, and Earth internal processes that are difficult to model. Illustrated in figure 3.20.

The Earth orientation parameters are published in daily, semi-weekly, and monthly bulletins by the IERS and are described more fully in the explanatory supplement to the IER bulletins A and B (International Earth Rotation Service, 2002).

The transformation of the position vector in the ITRS to the ICRS is given by successive matrix rotations

$$\mathbf{r}_{\mathrm{ICRS}} = \mathbf{P}(t)\mathbf{N}(t)\mathbf{\Theta}(t)\mathbf{\Pi}(t)\mathbf{r}_{\mathrm{ITRS}}, \tag{3.4.1}$$

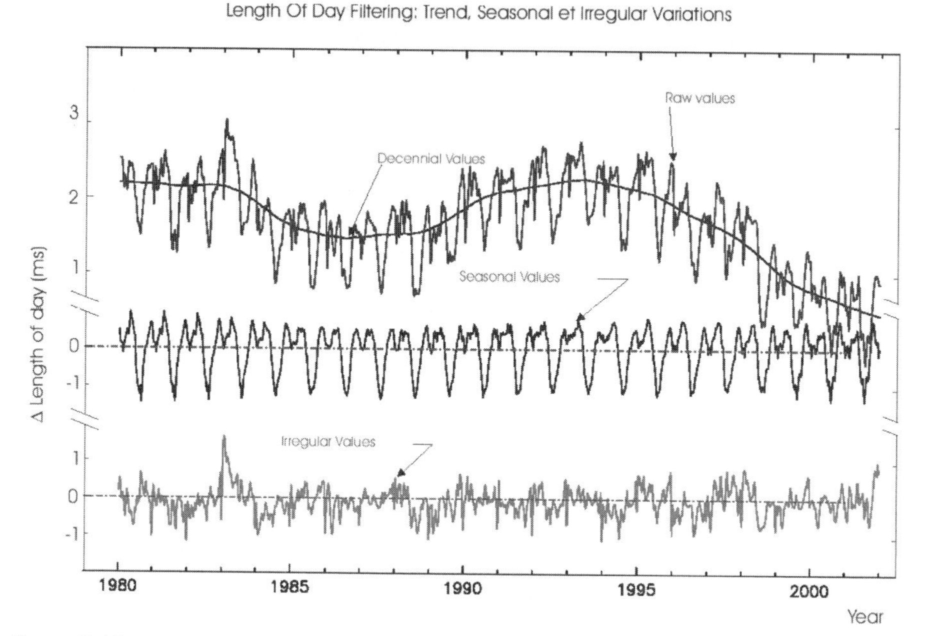

Figure 3.19 Excess of length of day over 86 400 s, 1980–2002. Raw values were split into various components; decennial, seasonal, and irregular terms. Bias of about 2ms per century since 1900. (Courtesy of Daniel Gambis, Earth Orientation Center of the International Earth Rotation Service.)

where

$$\begin{aligned}\mathbf{P} &= \text{precession transformation matrix,}\\ \mathbf{N} &= \text{nutation transformation matrix,}\\ \boldsymbol{\Theta} &= \text{Earth rotation matrix,}\\ \boldsymbol{\Pi} &= \text{polar motion transformation matrix.}\end{aligned}$$

The transformation matrices can be expressed in terms of the IERS Earth rotation parameters and the underlying precession and nutation theory, which is beyond the scope of this treatment but available in bulletins from the IERS EOP and ICRS product centers at the Paris Observatory.

3.4.5 Orbital Elements

Orbital elements are the set of six independent parameters that define the position and velocity of a body at a given time or as a function of time. Many different sets of parameters can be used including the cartesian components of position and velocity and, in Hamiltonian perturbation theory, the Delaunay or Poincare elements. However, there is a standard set in use, the *classical orbital elements* (COE), also known as the *Kepler elements,* illustrated in figure 3.21. These consist of:

- Size and shape of the orbit.

  Semimajor axis (a): describes the size of the orbit; dimensions of length, typically units of meters.

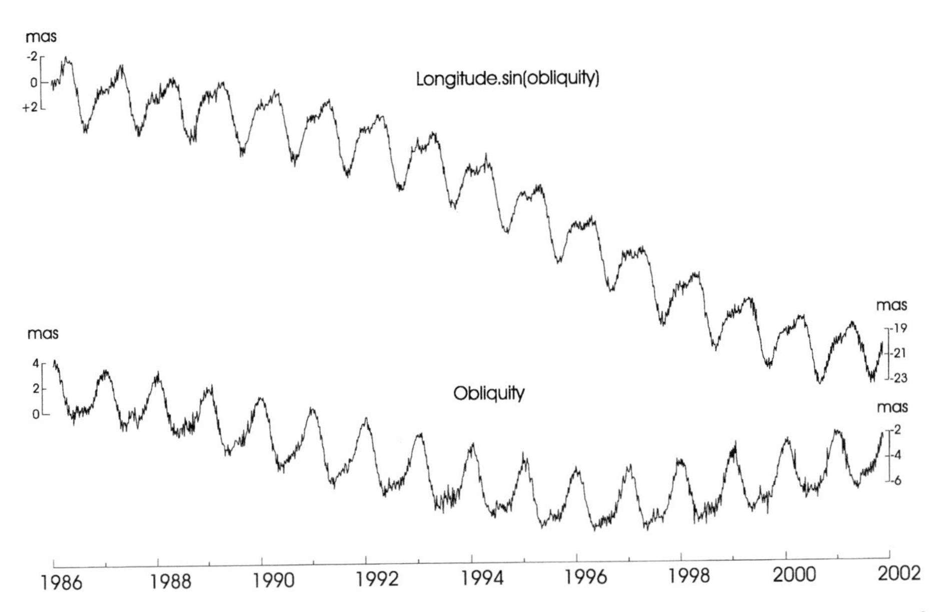

Figure 3.20 Celestial pole offsets: motion of the celestial pole relative to the IAU 1980 Theory of Nutation and the IAU 1976 Precession. Unit: 0.001 seconds of arc. (Courtesy of Daniel Gambis. Earth Orientation Center of the International Earth Rotation Service.)

Eccentricity (e): describes the shape of the orbit—circular, elliptic, parabolic, hyperbolic; dimensionless.

- Orientation of the plane in space

  Inclination (i): angle between the plane of the orbit and the equatorial plane or between the orbit angular momentum vector and the celestial pole. Inclinations less than 90° are *prograde* orbits those greater than 90° are *retrograde* orbits. Units are degrees or radians.

  Right ascension (or longitude) of the ascending node (Ω): angle from the equinox to the ascending node where the line from the origin to the ascending node is the *line of nodes.* Units are degrees or radians.

- Orientation of the orbit in the plane

  Argument of periapsis (ω): Angle from the line of nodes to the periapsis or point of closest approach of the orbit. Units are degrees or radians.

- Location of the body on the orbit

  True anomaly (f): angle relative to the origin from the periapsis, or point of closest approach, to the position of the body. Units are degrees or radians. Sometimes the *mean anomaly* (M) is specified in place of the true anomaly to expedite the solution of Kepler's equation. Units are degrees or radians.

- Epoch of the orbit

  Time (t_0): time that the orbital parameters are specified. Dimension is time with units generally the year, day, and seconds.

Consequently, the classical orbital elements are a, e, i, Ω, ω, f at t_0. An advantage of the classical orbital elements is that they can be interpreted intuitively and all, except the true anomaly, are constant for motion for the two-body problem. However,

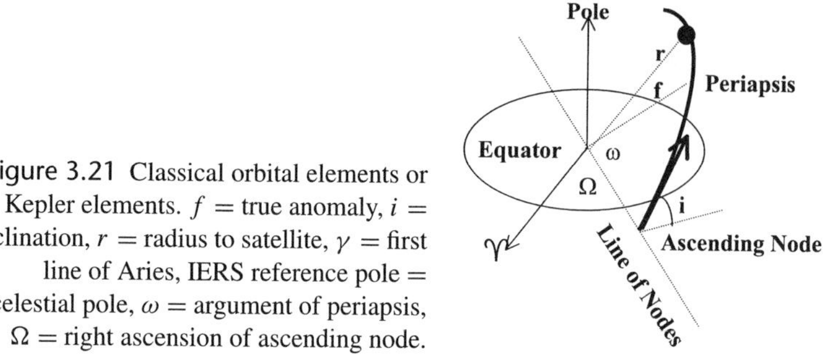

Figure 3.21 Classical orbital elements or Kepler elements. f = true anomaly, i = inclination, r = radius to satellite, γ = first line of Aries, IERS reference pole = celestial pole, ω = argument of periapsis, Ω = right ascension of ascending node.

sometimes alternative elements are useful. If the eccentricity is near zero, as for a near-circular orbit, the argument of periapsis is not well defined and the solutions for ω and f are ill conditioned. A useful alternative set of elements for near-circular orbits is: a, $e \cos\omega$, $e \sin\omega$, i, Ω, f, at t_0. For near-equatorial orbits, the inclination is near zero so the line of nodes is not well defined and the solutions for Ω and ω are ill conditioned. A useful alternative set of elements for near-equatorial orbits is: a, e, $\sin i \sin\Omega$, $\sin i \cos\Omega$, ω, f, at t_0. Other elements sometimes used include the mean motion n in place of the semimajor axis, the mean anomaly M in place of the true anomaly, the *longitude of periapsis* ϖ, where

$$\varpi \equiv \Omega + \omega, \tag{3.4.2}$$

in place of the argument of periapsis ω, and the *true longitude* L,

$$L \equiv \varpi + f = \Omega + \omega + f, \tag{3.4.3}$$

in place of the true anomaly. The classical orbital elements are illustrated in figure 3.21.

3.5 Time Systems

Time is the measurement of the interval between two events. The precise interpretation of time is critical to the determination of the ephemeris of a celestial body and extrapolation of its position forward or backward in time. Earth spacecraft have velocities on the order of 8 km/s, so that to determine its position to 10 cm or less will require precision in time to the order of 10 μs. Electromagnetic transmissions used in spacecraft observations, such as range or range rate, travel at the speed of light, 3×10^8 m/s, so that a precision of 0.3 ns is required to achieve a precision of 10 cm. In this section the different calendars and time systems employed in astrodynamics are described.

3.5.1 Julian Calendar

The *Julian calendar* was introduced by Julius Caesar in 45 BC and consisted of 12 months in a year of 365 days, with every year divisible by 4 being a leap year of 366 days. As a result, the average year is 365.2500 days, which is an approximation to the tropical year of 365.242 2 days. A *tropical year* is defined as the time it takes the

Table 3.2 Julian day number as function of calendar day

Year	Jan 0	Feb 0	Mar 0	Apr 0	May 0	Jun 0	Jul 0	Aug 0	Sep 0	Oct 0	Nov 0	Dec 0
1999	2451179	1210	1238	1299	1330	1360	1391	1422	1452	1452	1483	1513
2000	2451544	1574	1604	1635	1665	1696	1726	1757	1788	1818	1849	1879
2001	2451910	1941	1969	2000	2030	2061	2091	2122	2153	2183	2214	2240
2002	2452275	2306	2334	2365	2395	2426	2456	2478	2518	2548	2579	2609
2003	2452640	2671	2699	2730	2760	2791	2821	2852	2883	2913	2944	2974
2004	2453005	3036	3065	3096	3126	3157	3187	3218	3249	3279	3310	3340
2005	2453371	3402	3430	3461	3491	3522	3552	3583	3614	3644	3675	3705
2006	2453736	3767	3795	3826	3856	3487	3917	3948	3979	4009	4040	4070
2007	2454101	4132	4160	4191	4221	4252	4282	4313	4344	4374	4405	4435
2008	2454466	4497	4526	4557	4587	4618	4648	4679	4710	4740	4771	4801
2009	2454832	4863	4891	4922	4952	4983	5013	5044	5075	5105	5136	5166

Sun to appear to travel around the sky from a given point of the tropical zodiac back to that same point. It is generally approximated by the period for the Earth to travel from the equinox back to the equinox. Since the equinox is moving, the tropical year is different from the sidereal orbital period of 365.256 4 days, which is the time that it takes for the Earth to essentially complete one period with respect to the stars. The importance of correlating the calendar with the equinox is that it assures the repetitions of the seasons each year. Celestial observations are measured in *Julian Day Numbers* which are the number of days that have elapsed from the epoch of noon at Greenwich on 1 January 4713 BC (which denotes the year −4712 of the Julian proleptic calendar, where proleptic refers to a time prior to its adoption). The year 4713 was selected as the epoch because it was the year when three calendar cycles converged and it preceded any known historical astronomical data. It is important to note that the Julian day is from noon to noon, which is sometimes a source of confusion in translating between different time systems. The Julian day number for 1 January 2004 is 2 453 006 and the Julian day number for any given day is provided in the astronomical almanac published yearly (United States Naval Observatory, 2002). A table of Julian day numbers is given in table 3.2.

While the Julian day number refers to an integral number of days, the *Julian date* (JD) is an integer and decimal that accounts for the fraction of the day of the epoch since the preceding noon. It is often convenient to use, instead, the *Modified Julian date* (MJD), which is the number of days since midnight on 17 November 1858, which corresponds to a Julian date of 2 400 000.5 days. As a result, the MJD refers to midnight as opposed to noon for the Julian day or Julian date. A century of 36 525 days is used to convert Julian days to *Julian centuries*. Astronomical observations when specified by Julian date are denoted by the prefix J, so *J2000* denotes 12 hours universal time (UT) on 1 January 2000.

3.5.2 Gregorian Calendar

The average length of the year in the Julian calendar of 365.250 0 days differs from the tropical year of 365.242 2 by about 11 minutes a year. By the 16th century, this resulted in Easter occurring in the summer as it was based on astronomical observations. In

1582 Pope Gregory XIII issued a papal proclamation that established what is now called the *Gregorian calendar.* Ten days were omitted from the calendar so that Thursday 4 October 1582 would be followed by Friday 15 October 1582. The rule for leap year was also changed so that a year is a leap year if it is divisible by 4 except that century years must also be divisible by 400. Thus, 2000 was a leap year but 2100, 2200, 2300 are not. As a result, the average year is 365.242 5 days, which compares with the actual value of 365.242 2 by one day in 3300 years. The Gregorian calendar was adopted in the Catholic countries in 1582 but in other countries later, some in the 20th century; for example, Russia in 1918 and Greece in 1924. This is the current civil calendar now used worldwide.

3.5.3 International Atomic Time

International atomic time or *temps atomique international* (TAI) is derived from over 200 atomic oscillators located in more than 30 countries whose data are analyzed and published by the Bureau International des Poids et Mesures (BIPM) at Sèvres, France. The atomic standards are mostly cesium atomic clocks and hydrogen masers. The BIPM determines the official TAI epochs and distributes the offset of each clock by bulletin from intercomparisons between them. The basis of TAI is the International System of Units (SI) definition of the second. The SI second is defined as the duration of 9 192 631 770 cycles of the radiation emitted between two hyperfine ground states of cesium 133. This number of cycles was selected to be equal to 1/31 556 925.974 7 of the duration of the tropical year 1900 where the denominator is the number of seconds in that tropical year. The epoch for TAI was arbitrarily defined to be an offset of 32.184 seconds from terrestrial (dynamical) time (TT) for 1 January 1997 where

$$\mathrm{TT} = \mathrm{TAI} + 32.184\,\mathrm{s} \tag{3.5.1}$$

3.5.4 Dynamical Time

Dynamical time refers to the two time systems that are the independent argument in the ephemeredes of the Moon and the planets in the solar system. As a result, dynamical time is independent of the variations in the rotation rate of the Earth, polar motion, and nutation and precession, and is thus more uniform than the astronomical times. Its time base is the SI second. *Terrestrial time* (TT), the renamed terrestrial dynamical time (TDT), is the independent argument in the apparent geocentric ephemeredes and its relationship with TAI is given by equation 3.5.1. It is the uniform time that would be measured by an ideal clock on the surface of the geoid and measures in 86 400 SI seconds per day.

Barycentric dynamical time (TDB) is the independent argument of the ephemerides of the planetary bodies referred to the barycenter of the solar system and is more uniform than TT. TT differs from TDB by general relativistic periodic terms with amplitude less than 2 ms with annual and semiannual periods. Planetary ephmerides are published with TDB as the independent argument.

3.5.5 Coordinate Time

Barycentric coordinate time (TCB) is a relativistic time coordinate of the four-dimensional time barycentric frame, having its spatial origin at the barycenter of the solar system. The relationship between TCB and TDB in seconds is

$$\text{TCB} = \text{TDB} + 1.550\,505 \times 10^{-8}(\text{JD} - 2\,443\,144.5)86\,400, \tag{3.5.2}$$

where JD is the Julian date.

Geocentric coordinate time (TCG) is related to TCB and instead has its spatial origin at the center of mass of the Earth. The relation between TCG and TT in seconds is

$$\text{TCG} = \text{TT} + 6.969\,291 \times 10^{-10}(\text{JD} - 2\,443\,144.5)86\,400, \tag{3.5.3}$$

where JD is the Julian date.

3.5.6 Sidereal Time

Sidereal time is a measure of the rotational motion of the Earth about the celestial pole with respect to the equinox. It is measured in terms of the hour angle of the equinox given in hours, minutes, and seconds, where 15 degrees in arc length is equivalent to one hour of sidereal time with 24 sidereal hours in a sidereal day. Recall that the equinox is not fixed, undergoing precession and nutation, with the precession rate with respect to the stars of about 0.014 degrees per year (period of 25 765 years) or 9.2 milliseconds of sidereal time per day. *Greenwich apparent* (or *true*) *sidereal time* is the hour angle of the apparent or true equinox of date with the hour angle measured west from the IERS Reference Meridian or Greenwich meridian to the equinox. The Greenwich sidereal day begins when the Greenwich meridian coincides with the vernal equinox. *Greenwich mean sidereal time* is the hour angle of the mean vernal equinox of date with respect to the Greenwich meridian (see figure 3.22). The difference between apparent (or true) and mean sidereal times is due to the nutation of the equinox. The difference is given by the *equation of the equinoxes,* which is given in the Astronomical Almanac and has a maximum difference of about one second of time. Local apparent or mean sidereal time is determined from the Greenwich apparent or mean sidereal time plus the offset in longitude from Greenwich.

Figure 3.23 illustrates the difference between sidereal and universal (solar) time throughout the year. Sidereal and solar time agree only at the autumnal equinox and are 12 h apart at the vernal equinox.

3.5.7 Universal Time

Solar time, based on the position of the Sun, is the time that we prefer to use in our everyday lives. Universal time is not strictly solar time but is approximately related to the mean position of the Sun, so the term *mean solar time* is often used for universal time. Universal time is determined from astronomical observations of *Greenwich mean sidereal time* that specifies 0h UT1 from the formula

$$\begin{aligned}\text{GMST at } 0^{\text{h}}\text{UT1} = {} & [24\,110^{\text{s}}.548\,41 + 8\,640\,184.^{\text{s}}812\,866 T_{\text{U}} \\ & + 0^{\text{s}}.093\,104 T_{\text{U}}^2 - 6.^{\text{s}}2 \times 10^{-6} T_{\text{U}}^3]_{\text{modulo } 86400},\end{aligned} \tag{3.5.4}$$

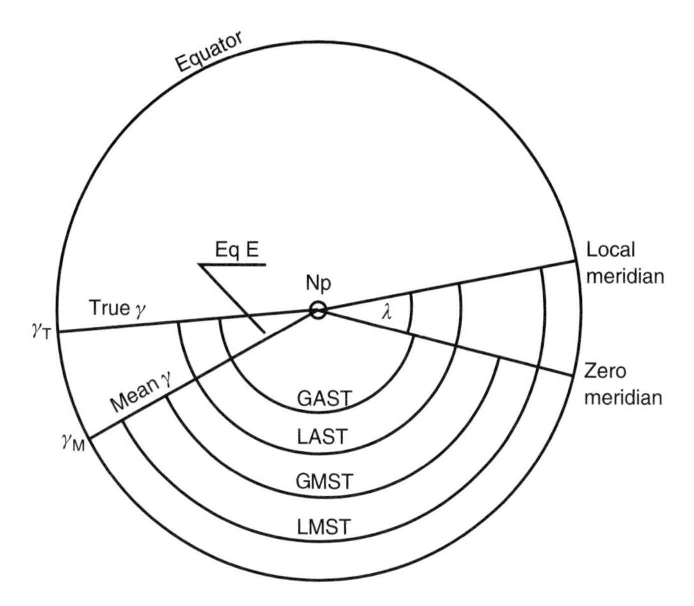

Figure 3.22 Mean and apparent sidereal time. E_qE = equation of the equinoxes, G = Greenwich, A = apparent, S = sidereal, T = time, Np = celestial pole, γ_T = true equinox, γ_M = mean equinox, Λ = longitude.

where

$$T_U = (\text{JD} - 2\,451\,545.0)/36\,525 \tag{3.5.5}$$

(Seidelmann, 1992). T_U is the interval of time measured in Julian centuries of 36 525 days from the epoch of J2000 at 12h UT1 (JD 2 451 545.0) on 1 January 2000 to the day of interest at 0h UTI; as a result it takes on values of ±0.5, ±1.5, ±2.5, ±3.5, etc. The ratio r of sidereal time to UT1 is given by

$$r = 1.002\,737\,909\,350\,795 + 5.900\,6 \times 10^{-11}\text{UT1} - 5.9 \times 10^{-15}\text{UT1}^2 \tag{3.5.6}$$

where the higher order terms can generally be neglected (Seidelmann, 1992).

Greenwich sidereal time agrees with universal time only at the autumnal equinox and is 12 hours different at the vernal equinox. At any other time, they differ. To compute the Greenwich mean sidereal time at an event other than at 0h UT1, it is necessary to add the elapse time. The orientation of Greenwich at a time UT1 beyond 0h UT1 follows from

$$\theta_m = (\text{GMST}(0^h\ \text{UT1}) + \text{UT1}) \times \omega_e \tag{3.5.7}$$

where

$$\omega_e = 360\text{deg}/86400\ \text{s} \tag{3.5.8}$$

for either UT1 or sidereal time. The Greenwich mean sidereal time θ_m in terms of the epoch of the observation follows from equation 3.5.4 by adding half a day to the first term to correct to the beginning of the day and a day per year to the second term

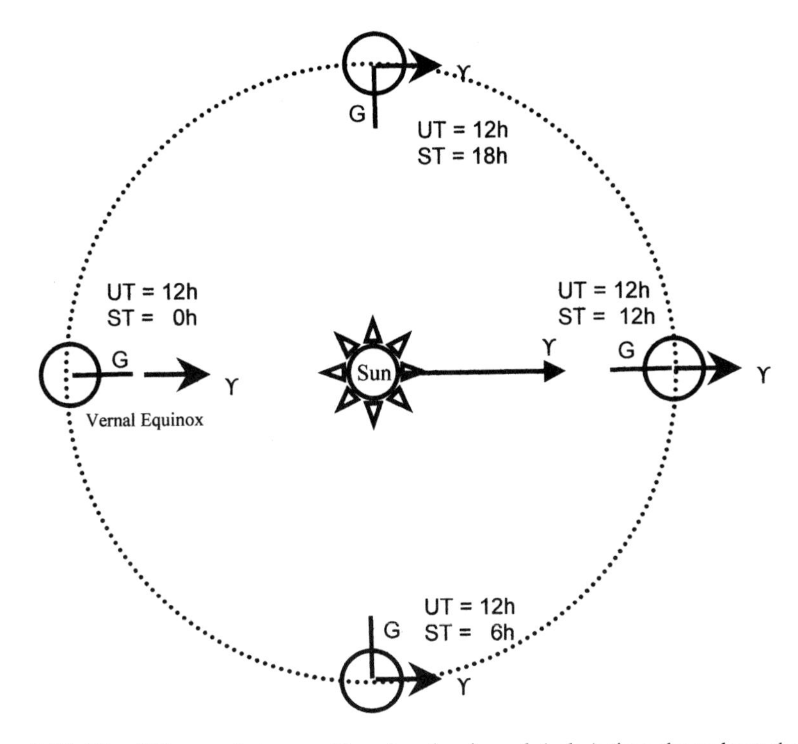

Figure 3.23 The difference between sidereal and universal (solar) time throughout the year. UT = universal (solar) time, ST = sidereal time, G = Greenwich meridian, γ = equinox.

to account for the difference between the number of sidereal and solar days, and is therefore

$$\begin{aligned}\theta_m &= 67\,310.^s 548\,41 + (876\,600^h + 8\,640\,184^s.812\,866) T_U \\ &\quad + 0^s.093104 T_U^2 - 6^s.2 \times 10^{-6} T_U^3 |_{\text{modulo}86400}\end{aligned} \tag{3.5.9}$$

where T_U is defined by equation 3.5.5 (Seidelmann, 1992).

Example

Determine the GMST and the right ascension of Greenwich at 0h UT on 1 January 2010.

Solution: At 0h UT on 1 Jan. 2010, equation 3.5.5 gives

$$\begin{aligned}T_U &= (2\,455\,197.5 - 2\,451\,545.0)/36\,525 \\ &= 0.100 \text{ exact}.\end{aligned}$$

Then equation 3.5.4 gives

$$\begin{aligned}\text{GMST at } 0^h \text{ UT1} &= 24\,110^s.548\,41 + 8\,640\,184.^s 812\,866 \times 0.1 \\ &\quad + 0^s.093\,104 \times 0.1^2 - 6.^s 2 \times 10^{-6} \times 0.1^3 \\ &= 8.^s 881\,290\,306\,276 \times 10^5.\end{aligned}$$

The hour angle of Greenwich follows as

$$\text{GMST at } 0^{\text{h}} \text{ UT} = \left[8^{\text{s}}.881\,290\,306\,276 \times 10^{5} \frac{15 \text{ deg}}{3600\text{s}} \right]_{\text{modulo } 360}$$

$$= 100.537\,627\,615 \text{ deg.}$$

Example

Determine the GMST of an event at 16 June 1994 at 18h UT1.

Solution: The Julian day number can be determined from the Astronomical Almanac, programs available from the Internet, applications programs, or computed from the elapse time from JD 2 451 545.0. For the time given, JD = 2 449 520.25, so that $T_U = -0.055\ 434\ 633\ 612$ follows from equation 3.5.5. The GMST follows from equation 3.5.9 to be 41 945.067 522 9.^{s} or $11^{\text{h}}\ 39^{\text{m}}\ 05^{\text{s}}.067\ 52$, or 174.771 114 679 deg.

For a site at east longitude λ relative to Greenwich, the Local Mean Sidereal Time (LMST) is then given by

$$\text{LMST} = \text{GMST} + \lambda \tag{3.5.10}$$

There are several variants of universal time. UT0 is the time that results from the precise observations referred to the prime meridian and the instantaneous pole and so is affected by polar motion. In order to properly compare observations at different sites, UT1 is determined by accounting for polar motion. UT2 is derived from UT1, with annual and semiannual variations in the Earth's rotation eliminated. The *equation of time* gives the difference between UT1 and apparent solar time, with a difference of up to 16 minutes (United States Naval Observatory, 2002).

Both sidereal and universal time use a 24-hour day. However, in a sidereal day, the meridian rotates from equinox to equinox and in a solar day the meridian rotates from pointing at the Sun to pointing back at the Sun, so the Earth must rotate almost 361 degrees per solar day as illustrated in figure 3.24. As a result, a sidereal day is 0.997 269 566 of a solar day (or universal day) or $3^{\text{m}}55.9^{\text{s}}$ less than a solar day.

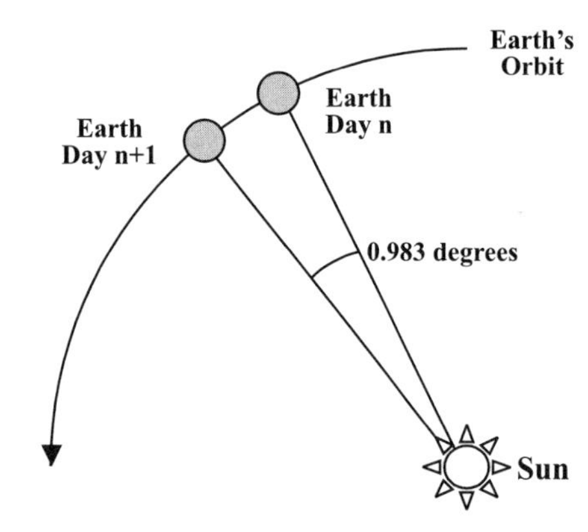

Figure 3.24 A solar day.

3.5.8 Coordinated Universal Time

Coordinated universal time (UTC), introduced in 1964, is a time scale that is maintained close to UT1 and is therefore an approximation to solar time. It is the modern successor to Greenwich Mean Time (GMT), which was used when the unit of time was the mean solar day and whose use has been discontinued. It is the basis for legal time worldwide and since 1972 it follows TAI exactly, except for being offset by a changing integral number of seconds called *leap seconds*. UTC is maintained to not deviate from UT1 by more than 0.9 s, so that

$$|\mathrm{UT1} - \mathrm{UTC}| < 0.9\ \mathrm{s}. \tag{3.5.11}$$

Recall that the SI second was defined by the rotation rate of the Earth in the year 1900. However, the Earth's rotation rate is undergoing a deceleration caused by the braking effects of the Earth tides, so that the period is currently about 2 ms per day less, as identified in figure 3.19. Consequently, on an average of 500 days, but with specific periods more variable due to short-term variations in the Earth's rotation rate, the difference between UTI and UTC would be about a second. To account for this, the IERS advises on the introduction of a *leap second* that may be inserted at midnight on the last day of June or December so that that day may have one more or less SI second. To date only additive seconds have been inserted. The difference between TAI and UTC is given by

$$\mathrm{TAI} - \mathrm{UTC} = \text{Number of integer leap seconds introduced.} \tag{3.5.12}$$

Bulletins are provided by the IERS and the USNO on whether or not a leap second will be introduced about six months in advance. A commission is now at work to recommend a replacement for the anomalistic leap second.

3.5.9 Global Positioning System Time

The *Global Positioning System* (GPS) is a constellation of at least 18 operational satellites whose primary purpose is to provide worldwide navigation information. It has also become an important means to calibrate clocks and to compare clocks at different locations. Three types of time are available from GPS: *GPS Time*, an estimate of UTC produced by the United States Naval Observatory, and the time on each of the spacecraft's two cesium and two rubidium time standards. GPS time, essentially atomic time, was synchronized to UTC on 6 January 1980 and, for operational reasons, has not incorporated the additional leap seconds since that time. Consequently, GPS differs from TAI time by a fixed offset of 19 s so that

$$\mathrm{GPS} = \mathrm{TAI} - 19\ \mathrm{s}, \tag{3.5.13}$$

and differs from UTC by the integral number of leap seconds introduced after January 1980.

A summary of the relationships between the time systems is illustrated in figure 3.25.

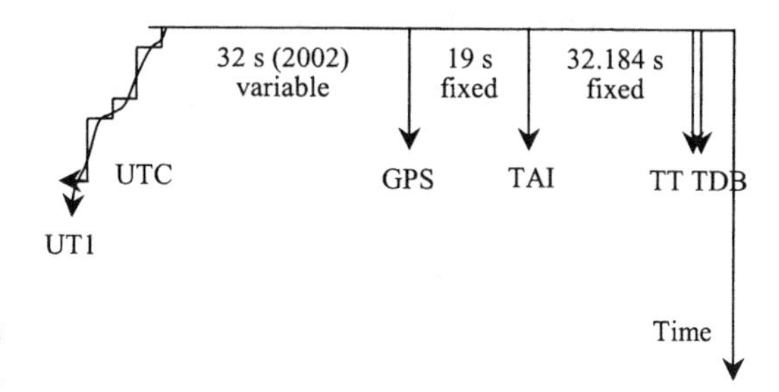

Figure 3.25 Relationships among time systems.

3.6 Trajectory Perturbation Theory Fundamentals

Earlier, the two-body central force problem was solved, giving the conic as the solution to motion in a Newtonian gravitational field. However, in practice other forces, *perturbing forces,* act on the spacecraft or celestial body that result in perturbation from the two-body solution. The trajectory determined by the primary gravitational force is known as the *unperturbed trajectory*. The trajectory resulting from all the forces is known as the *perturbed trajectory*. This section addresses the effects of these perturbing or disturbing forces by perturbation theory, which assumes the perturbing forces are small relative to the force of gravity. With the addition of these perturbing forces, the equation of motion, equation 3.3.11, becomes

$$\ddot{\mathbf{r}} = -\frac{G(m_{\mathrm{e}} + m_0)}{r^2}\hat{\mathbf{r}} + \mathbf{f} = -\frac{\mu}{r^2}\hat{\mathbf{r}} + \mathbf{f}, \tag{3.6.1}$$

where $\mathbf{f}$ represents the perturbing force divided by the mass of the body, which is the perturbing acceleration. There are two general techniques employed in treating perturbations, general and specific perturbation theory.

General perturbation theory involves solving for the perturbations analytically employing for example, the methods of variation of parameters. In traditional general perturbation theory, canonical transformations of the Hamilton–Jacobi equation are often employed, using Poincaré or Delaunay variables; this is beyond the scope of this treatment. It is also possible to treat the perturbations in terms of the classical orbital elements, as will be done here. In this case, the unperturbed trajectory can be specified in terms of the classical orbital elements $a_0, e_0, i_0, \omega_0, \Omega_0, M_0$ at time t_0. The trajectory perturbations can then be expressed by $\delta a_0, \delta e_0, \delta i_0, \delta\omega_0, \delta\Omega_0, \delta M_0$, which are functions of time. The perturbed trajectory is then given by $a = a_0 + \delta a_0, e = e_0 + \delta e_0, i = i_0 + \delta i_0, \omega = \omega_0 + \delta\omega_0, \Omega = \Omega_0 + \delta\Omega_0, M = M_0 + \delta M_0$, known as the *osculating elements*, as functions of time. Analytical solutions can provide information on the periods and amplitudes of the perturbation and the perturbed trajectory by evaluating analytical functions of time. However, depending on the characteristics of the forces being addressed, it is not always straightforward to derive a general perturbation theory.

Special perturbation theory involves numerically integrating the equations of motion. To minimize the effects of truncation and roundoff errors, special perturbation techniques usually involve numerically integrating only the components of the perturbations that cannot be handled analytically and adding the effects to that part of the problem that can be solved analytically. A number of sophisticated numerical integration routines are available that use predictor–corrector techniques to adjust the step size to minimize computational errors. The primary limit to special perturbation techniques is that the numerical errors accumulate with time. Unlike general perturbation theory, the equations must be integrated to a distant time.

3.6.1 Lagrange Planetary Equations

A result from general perturbation theory is the Lagrange planetary equations, which are presented in two forms. The perturbing force per unit mass of the body is first expressed in terms of a disturbing function R that is equal to the negative of the potential of the perturbing force, where

$$\left.\begin{aligned}
\frac{da}{dt} &= \frac{2}{na}\frac{\partial R}{\partial M},\\
\frac{de}{dt} &= \frac{1-e^2}{na^2e}\frac{\partial R}{\partial M} - \frac{\sqrt{1-e^2}}{na^2e}\frac{\partial R}{\partial \omega},\\
\frac{di}{dt} &= \frac{\cot i}{na^2\sqrt{1-e^2}}\frac{\partial R}{\partial \omega} - \frac{\csc i}{na^2\sqrt{1-e^2}}\frac{\partial R}{\partial \Omega},\\
\frac{d\Omega}{dt} &= \frac{\csc i}{na^2\sqrt{1-e^2}}\frac{\partial R}{\partial i},\\
\frac{d\omega}{dt} &= -\frac{\cot i}{na^2\sqrt{1-e^2}}\frac{\partial R}{\partial i} + \frac{\sqrt{1-e^2}}{na^2e}\frac{\partial R}{\partial e},\\
\frac{dM}{dt} &= n - \frac{1-e^2}{na^2e}\frac{\partial R}{\partial e} - \frac{2}{na}\frac{\partial R}{\partial a}.
\end{aligned}\right\} \tag{3.6.2}$$

They may also be expressed in terms of the components of the perturbing force per unit equivalent mass in a form known as the Gaussian–Lagrange planetary equations, given by

$$\left.\begin{aligned}
\frac{da}{dt} &= \frac{2}{n\sqrt{1-e^2}}[Re\sin f + T(1+e\cos f)]\\
\frac{de}{dt} &= \frac{\sqrt{1-e^2}}{na}\left[R\sin f + T\left(\frac{e+\cos f}{1+e\cos f}+\cos f\right)\right],\\
\frac{di}{dt} &= \frac{Wr\cos(f+\omega)}{na^2\sqrt{1-e^2}} = \frac{W\sqrt{1-e^2}\cos(f+\omega)}{na(1+e\cos f)},\\
\frac{d\Omega}{dt} &= \frac{Wr\sin(f+\omega)\csc i}{na^2\sqrt{1-e^2}} = \frac{W\sqrt{1-e^2}\sin(f+\omega)\csc i}{na(1+e\cos f)},\\
\frac{d\omega}{dt} &= \frac{\sqrt{1-e^2}}{nae}\left[-R\cos f + T\frac{2+e\cos f}{1+e\cos f}\sin f - \frac{e\sin(f+\omega)\cot i}{1+e\cos f}W\right],\\
\frac{dM}{dt} &= n - \frac{(1-e^2)^2}{na}\left[R\left(\frac{2}{1+e\cos f}-\frac{\cos f}{e}\right) + T\left(\frac{2+e\cos f}{1+e\cos f}\right)\frac{\sin f}{e}\right],\\
&= n - \frac{2Rr}{na^2} - \sqrt{1-e^2}(\dot\omega + \dot\Omega\cos i).
\end{aligned}\right\} \tag{3.6.3}$$

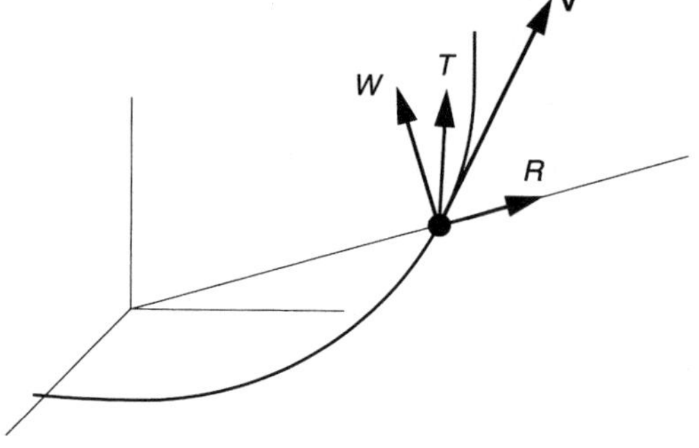

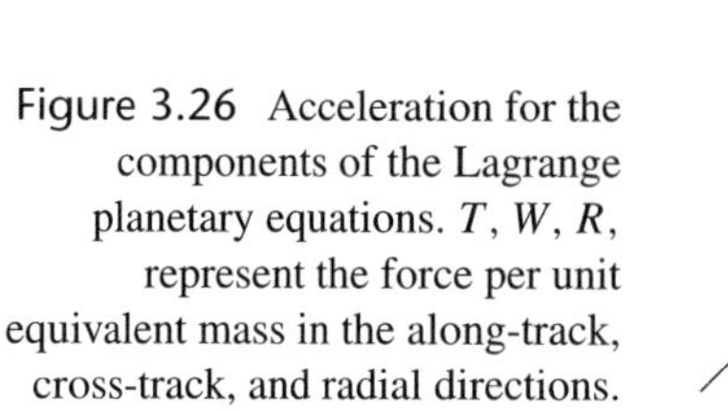

Figure 3.26 Acceleration for the components of the Lagrange planetary equations. T, W, R, represent the force per unit equivalent mass in the along-track, cross-track, and radial directions.

where T, W, R are the components of the accelerations or force per unit mass in the along-track, cross-track, and radial directions, as illustrated in figure 3.26. Note that the equations consist of secular and periodic terms that are functions of the true anomaly f and the sum of the true anomaly and argument of periapsis $(f + \omega)$. These equations can be averaged over these arguments to obtain secular, long period, and short period effects.

For small values of eccentricity or inclination, the Lagrange planetary equations, as presented, are ill conditioned. For small eccentricities, equations 3.6.2b,e and 3.6.3b,e can be reformulated in terms of $e \sin \omega$ and $e \cos \omega$. For small inclinations, equations 3.6.2c,d and 3.6.3c,d can be reformulated in terms of $\sin i \sin \Omega$ and $\sin i \cos \Omega$.

3.6.2 Euler–Hill Equations

The *Euler–Hill* or *Hill equations*, also known as the *Clohessy and Wiltshire equations*, are stated in terms of the components of the perturbed position and velocity and the components of the perturbing force per unit mass. The linearized form for an orbit with zero eccentricity is

$$\left.\begin{aligned} \ddot{H} - 2n\dot{L} - 3n^2 H &= f_H, \\ \ddot{L} + 2n\dot{H} &= f_L, \\ \ddot{C} + n^2 C &= f_C, \end{aligned}\right\} \tag{3.6.4}$$

where f_H, f_L, f_C are the components of the perturbing force per unit equivalent mass in the radial (H), along-track (L), and cross-track (C) directions as illustrated in figure 3.27. Analogous equations are available for nonzero eccentricity but are highly nonlinear (Henriksen, 1977). The complementary solution for L, C, and H can be determined by letting $f_L = f_C = f_H = 0$. The equation for C is decoupled from the others and the two coupled equations have repeated roots where the solution is

$$\left.\begin{aligned} C &= C_0 \cos nt + \frac{\dot{C}_0}{n} \sin nt, \\ H &= (4 - 3\cos nt) H_0 + \sin nt \frac{\dot{H}_0}{n} + 2(1 - \cos nt)\frac{\dot{L}_0}{n}, \\ L &= 6(\sin nt - nt) H_0 + L_0 - 2(1 - \cos nt)\frac{\dot{H}_0}{n} + (4 \sin nt - 3nt)\frac{\dot{L}_0}{n}, \end{aligned}\right\} \tag{3.6.5}$$

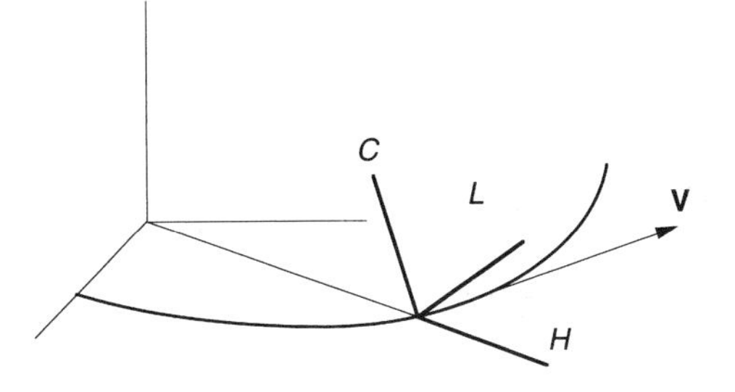

Figure 3.27 Euler-Hill coordinate system. Displacement components are H = radial, L = along track, C = cross-track.

where L_0, C_0, H_0, and $\dot{L}_0$, $\dot{C}_0$, $\dot{H}_0$ are the initial conditions of the components of position and velocity. The particular solutions depend on the specific functional form of the perturbing forces. These equations can also be used to study rendezvous problems where the L, C, H represent the differences in position of the two bodies.

3.7 Trajectory Perturbations

The primary potential perturbing forces of interest to satellite motion are the

- Inhomogeneous mass distribution of the primary body.
- Other celestial bodies, primarily Sun and moons.
- Solid body tides.
- Atmospheric drag.
- Radiation pressure (direct solar, reflected solar, and planet infra-red).

3.7.1 Inhomogeneous Mass Distribution

The inhomogeneous distribution of mass of the primary body, that is, its departure from a series of concentric shells of uniform radially symmetric mass density, is an important perturbation. The general form for the gravitational potential was given by equations 3.2.17 and 3.2.27. Note that the experimentally determined parameters $C_{n,m}$ and $S_{n,m}$ or $J_{n,m}$ are multiplied by $(a/r)^n < 1$ where a is the semimajor axis of the planetary ellipsoid and r is the radius to the satellite. Consequently, as the radius to the satellite and the degree of the expansion increase, the perturbation is reduced. For the Earth, the J_2 or C_{20} term, which represents the departure of the Earth from a sphere to an oblate spheroid, is three orders of magnitude smaller than the Newtonian term and three orders of magnitude larger than the other terms of the geopotential, where $J_2 = -0.001082626683$. The deviation of the geopotential surface from a reference ellipsoid is given in figure 3.3 and has a maximum deviation of 60 m. For small eccentricity, the first-order secular effect of J_2 can be derived from the Lagrange planetary equations by substituting equation 3.2.27 for the disturbing function R where, from spherical trigonometry, $\sin\phi = \sin(f + \omega)\sin i$, so that

$$\begin{aligned} R &= \frac{GM}{r}\frac{J_2}{2}\left(\frac{a_\mathrm{e}}{r}\right)^2(3\sin^2\phi - 1), \\ &= \frac{GM}{r}\frac{J_2}{2}\left(\frac{a_\mathrm{e}}{r}\right)^2[3\sin^2(f+\omega)\sin^2 i - 1]. \end{aligned} \tag{3.7.1}$$

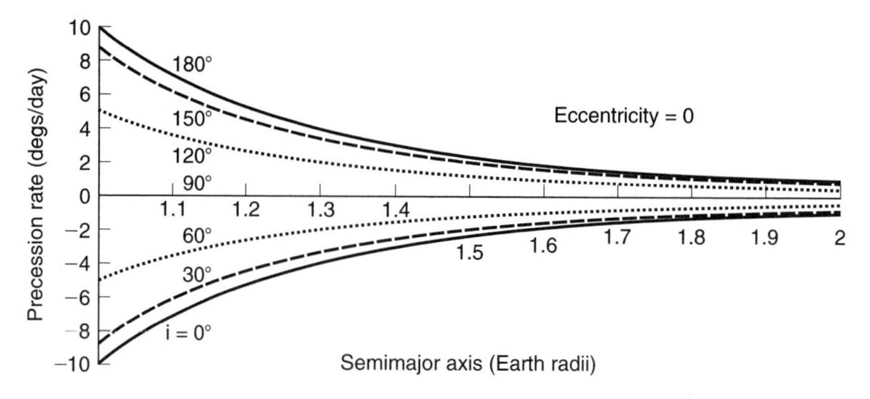

Figure 3.28 Precession rate of right ascension of the ascending node.

Averaging over one complete period of the perigee gives

$$\left.\begin{aligned}
\left\langle\frac{d\Omega}{dt}\right\rangle_{\text{secular}} &= \frac{3J_2na_{\text{e}}^2}{2a^2(1-e^2)^2}\cos i, \\
\left\langle\frac{d\omega}{dt}\right\rangle_{\text{secular}} &= -\frac{3J_2na_{\text{e}}^2}{4a^2(1-e^2)^2}(4-5\sin^2 i), \\
\left\langle\frac{dM}{dt}\right\rangle_{\text{secular}} &= n-\frac{3J_2na_{\text{e}}^2}{4a^2(1-e^2)^{3/2}}(2-3\sin^2 i),
\end{aligned}\right\} \tag{3.7.2}$$

where a, e, i, Ω, ω, M are the classical orbital elements of the spacecraft, a_{e} is the radius of the primary body, and J_2 is the coefficient of the surface spherical harmonic that represents the oblateness of the primary body.

Incidentally, all the even zonal harmonics (J_4, $J_6, \ldots$) produce the same type of result with significantly reduced magnitude. Note that secular effects, that is, a linear function of time in this case, arise only for the right ascension of the ascending node Ω, the argument of perigee ω, and the mean anomaly M. The secular precession of the ascending node, $\langle d\Omega/dt\rangle_{\text{secular}}$, is negative for a *prograde* orbit defined by $0 \leq i < \pi/2$, zero for a polar orbit with $i = \pi/2$, and positive for a *retrograde* orbit defined by $\pi/2 < i \leq \pi$ as illustrated in figure 3.28. As the semimajor axis increases, the precession rate of the ascending node reduces. When the precession rate of the ascending node equals the average apparent precession of the Sun, +0.983 degrees per day, the plane of the orbit will maintain its relative geometry with respect to the Sun; this is known as a *Sun-synchronous orbit*. Typical Sun-synchronous orbits have an altitude of about 800 km and an inclination of about 98.6 degrees. They are useful for observing the same sunlight conditions at their subsatellite point and for simplifying the design of the spacecraft since one side can be made to face the Sun 100% of the time, thus permitting maximum output of the solar array and a stable thermal environment. The differential precession of the right ascension of the ascending node can also be used to change the relative right ascension of the ascending node of a satellite in a constellation and

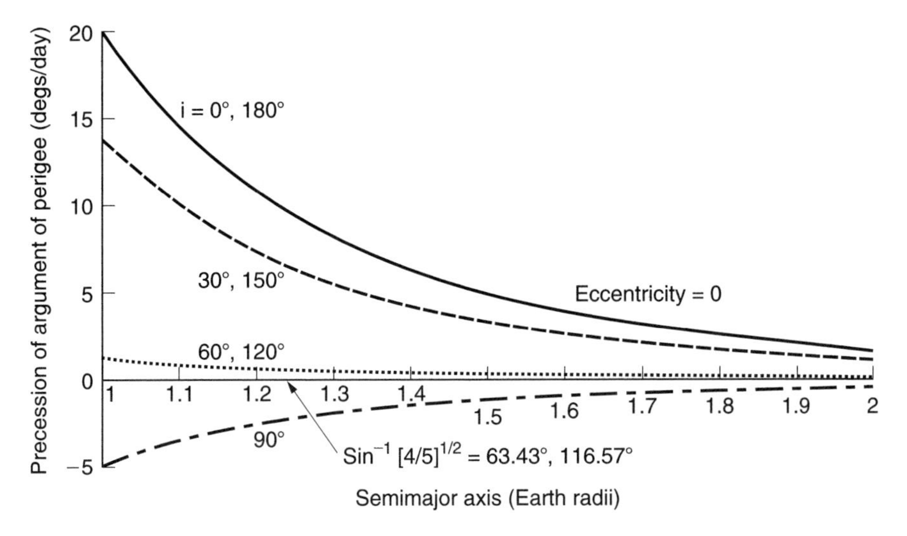

Figure 3.29 Precession rate of argument of perigee.

minimize fuel consumption. This is accomplished by changing the inclination a small amount and waiting until the correct relative orientation is attained, when the inclination is then returned to the original value. The differential rate can be determined from the following equation:

$$\left\langle \frac{\partial \dot{\Omega}}{\partial i} \right\rangle_{\text{secular}} = -\frac{3J_2 n a_\text{e}^2}{2a^2(1-e^2)^2} \sin i, \tag{3.7.3}$$

which is left as an exercise.

The inclinations at $i = \sin^{-1}[\pm(4/5)^{1/2}] = 63.4$ and 116.6 degrees, where the secular precession rates (figure 3.29) are zero, are known as the *critical inclinations*. This property is used to advantage in the Russian Molniya satellites that have an inclination of 63.4 degrees to maintain apogee over the northern hemisphere so as to provide communication to regions not accessible from geosynchronous communication satellites.

Example

Determine the secular precession rates for a satellite in a circular orbit with an altitude of 1000 km and an inclination of 45°.

Solution: Substituting

$$a = 6\,378\,137 + 1\,000\,000 = 7\,378\,137\ \text{m},$$

$$e = 0,$$

$$i = 45°,$$

$$J_2 = -0.001\,082\,626\,683,$$

it follows from equations 3.7.2 that

$$\left.\begin{aligned}\left\langle\frac{d\Omega}{dt}\right\rangle_{\text{secular}} &= -4.23^\circ/\text{day},\\ \left\langle\frac{d\omega}{dt}\right\rangle_{\text{secular}} &= 4.49^\circ/\text{day},\\ \left\langle\frac{dM}{dt}\right\rangle_{\text{secular}} - n &= 1.50^\circ/\text{day}.\end{aligned}\right\}$$

In addition to the secular effects, the zonal component of the geopotential also induces long-period perturbations that are functions of multiples of the period of the argument of periapsis and short-period perturbations that are functions of the period of the true anomaly. Perturbations due to some of the tesseral harmonics can result in a perturbation that is close to an integral number of orbits per sidereal day, which will amplify the effect for satellites with a period of an integral multiple of a sidereal day. This has been exploited to determine more precisely the values of selected coefficients in the geopotential field model (Yionoulis, 1965, 1967).

3.7.2 Sun and Moon

The equivalent force per unit mass that the Sun and Moon exert on a spacecraft is given by equation 3.3.6. These forces produce both period and secular effects, in which the latter will dominate in time. These perturbations are given in Giacaglia (1973), Taff (1985), Cook (1962), and Seeber (1993), and the secular perturbations are given in Deutsch (1963) as

$$\left.\begin{aligned}\left\langle\frac{d\Omega}{dt}\right\rangle_{\text{secular}} &= -\frac{3n_{\text{d}}^2\cos i}{4n\sqrt{1-e^2}}\mu_{\text{d}}\left(1+\tfrac{3}{2}e^2\right)\left(1-\tfrac{3}{2}\sin^2 i_{\text{d}}\right),\\ \left\langle\frac{d\omega}{dt}\right\rangle_{\text{secular}} &= \frac{3n_{\text{d}}^2}{4n\sqrt{1-e^2}}\mu_{\text{d}}\left(2-\tfrac{5}{2}\sin^2 I+\tfrac{1}{2}e^2\right)\left(1-\tfrac{3}{2}\sin^2 I\right),\end{aligned}\right\} \tag{3.7.4}$$

where

$$\left.\begin{aligned}\sin^2 I \equiv{}& \tfrac{1}{2}\sin^2 i_{\text{d}}(1+\cos^2\chi)+\sin^2\chi\cos^2 i_{\text{d}}\\ &+\tfrac{1}{2}\sin 2\chi\sin 2i_{\text{d}}\cos\Omega_{\text{d}}-\tfrac{1}{2}\sin^2 i_{\text{d}}\sin^2\chi\cos 2\Omega_{\text{d}}\end{aligned}\right\}$$

d = signifies disturbing body, Sun or Moon,
e = spacecraft eccentricity,
i = spacecraft inclination,
i_{d} = disturbing body inclination,
n = spacecraft mean motion,
n_{d} = disturbing body mean motion,
t = time,
Ω = spacecraft right ascension of the ascending node,
Ω_{d} = disturbing body right ascension of the ascending node,
χ = angle between the spacecraft and the disturbing body,
μ_{d} = mass of disturbing body divided by the mass of the disturbing body plus the mass of the Earth,
ω = spacecraft argument of perigee,

and the secular effects on the other classical orbital elements are essentially zero. For the solar perturbations, $I \approx 0$ and $\mu_d \approx 1$.

3.7.3 Solid Earth Body Tides

The gravitational forces of the Sun and Moon distort the shape of the Earth since it is not a rigid body; this phenomenon is called solid Earth body tides. These tides are in addition to the ocean tides. This distortion can be represented as a spherical harmonic expansion, which gives the specific force as (Seeber, 1993)

$$\mathbf{f}_e = \frac{k_2 G m_d a_e^5}{2 r_d^3 r^4} (3 - 15 \cos^2 \theta)\, \hat{\mathbf{r}} + 6 \cos\theta\, \hat{\mathbf{r}}_d, \tag{3.7.5}$$

where

$$\begin{aligned}
d &= \text{disturbing body, Sun or Moon,}\\
\mathbf{f}_e &= \text{specific perturbing force, units of force/mass;}\\
\hat{\mathbf{r}} &= \text{unit vector to satellite,}\\
\hat{\mathbf{r}}_d &= \text{unit vector to disturbing body,}\\
k_2 &= 0.3 = \text{Love's number, response of Earth to disturbing body,}\\
m_d &= \text{mass of disturbing body.}
\end{aligned}$$

The amplitude of body tide due to the Sun is about 26 cm and has a period of about 12 h. This effect influences the motion of the spacecraft directly and the radius of a ground station used to provide tracking data.

3.7.4 Atmospheric Forces

The atmosphere of a celestial body can introduce aerodynamic forces on a spacecraft whose principal component is drag, a force opposed to the motion with respect to the atmosphere. Lift and side slip are usually small and can be ignored. The specific drag force can be represented by

$$\mathbf{f}_d = -\frac{1}{2} C_d \rho(\mathbf{r}, t) \frac{A}{m} \mathbf{v}_r \cdot \mathbf{v}_r = -\frac{1}{2} \frac{\rho(\mathbf{r}, t)}{B} \mathbf{v}_r \cdot \mathbf{v}_r, \tag{3.7.6}$$

where

$$\begin{aligned}
A &= \text{cross-sectional area in the direction of the velocity,}\\
B &\equiv m/(C_d A), \text{ ballistic coefficient,}\\
C_d &= \text{coefficient of drag,}\\
\mathbf{f}_d &= \text{specific force due to drag,}\\
m &= \text{spacecraft mass,}\\
\mathbf{v}_r &= \text{relative velocity of the satellite with respect to the atmosphere,}\\
\rho(\mathbf{r}, t) &= \text{atmospheric density.}
\end{aligned}$$

The drag coefficient C_d is a parameter dependent on the shape of the body, the nature of the aerodynamic flow, and the accommodation coefficient. The *accommodation coefficient* is the temperature–dependent ratio of the average energy actually transferred between a scattering surface and impinging gas molecules to the average energy that would theoretically be transferred if the impinging molecules reached complete thermal equilibrium with the surface before leaving it. It has been shown that in a free-molecular and hyperthermal regime where the satellite speed exceeds the mean molecular speed by a large amount the drag coefficient assumes values between 2.0 and 2.5 for a wide spectrum of shapes and surface compositions (King-Hele, 1987; King-Hele and Walker, 1987). The theoretical value for a sphere in free molecular flow is 2.0. The typical value of C_d used for spacecraft is 2.2. Often, C_d is determined experimentally as part of the orbit determination process.

Aerodynamic forces are particularly difficult to model because of uncertainty or variability of the cross-sectional area and the drag coefficient and, more importantly, the variability of the atmospheric density (see chapter 2 for a more detailed discussion). Typical atmospheres have densities that vary approximately exponentially as a function of altitude. Several models are available, with the Jacchia (1972) model one of the preferred. To be useful, the model adopted should take into account the variation with density due to

- Solar activity.
- Magnetic activity.
- Latitudinal variations.
- Diurnal variations.
- Seasonal variations.
- Rotation of the atmosphere.
- Atmospheric tides.

For an atmosphere fixed with respect to the planet, the changes to the orbital elements are given by Fitzpatrick (1970) and Sterne (1960) as

$$\left.\begin{aligned}
\frac{da}{dE} &= -2\beta\rho a^2 \frac{(1+e\cos E)^{3/2}}{(1-e\cos E)^{1/2}}(1-\tau)^2, \\
\frac{de}{dE} &= -2\beta\rho a(1-e^2)\left(\frac{1+e\cos E}{1-e\cos E}\right)^{\frac{1}{2}}(1-\tau)^2\cos E \\
&\quad - \beta\rho a e d\frac{(1-e\cos E)^{3/2}}{(1+e\cos E)^{1/2}}(1-\tau)\sin^2 E, \\
\frac{di}{dE} &= -\frac{\beta\rho a\omega_\mathrm{p}}{2n(1-e^2)^{1/2}}\sin i(1+\cos 2u)(1-e\cos E)^{5/2}(1+e\cos E)^{1/2}(1-\tau), \\
\frac{d\Omega}{dE} &= -\frac{\beta\rho a\omega_\mathrm{p}}{2n(1-e^2)^{1/2}}\sin 2u(1-e\cos E)^{5/2}(1+e\cos E)^{1/2}(1-\tau), \\
\frac{d\omega}{dE} &= -\dot{\Omega}\cos i - \frac{2\beta\rho a}{e}(1-e^2)^{1/2}\sin E\left(\frac{1+e\cos E}{1-e\cos E}\right)(1-\tau) \\
&\quad \times\left[1-\frac{d}{2(1-e^2)}(1-e\cos E)(2-e^2-e\cos E)\right], \\
\frac{dM}{dE} &= n + \frac{2\beta\rho a}{e}(1-e^3\cos E)^{1/2}\sin E\left(\frac{1+e\cos E}{1-e\cos E}\right)^{1/2}(1-\tau) \\
&\quad \times\left[1-\frac{d(1-e\cos E)}{2(1-e^3\cos E)}(2-e^2-e\cos E)\right]
\end{aligned}\right\} \tag{3.7.7}$$

where

$$
\begin{aligned}
a &= \text{semimajor axis,}\\
A &= \text{spacecraft cross-sectional area,}\\
C_\mathrm{d} &= \text{spacecraft drag coefficient,}\\
d &= (\omega_\mathrm{p}/n)(1-e^2)^{1/2}\cos i,\\
e &= \text{eccentricity,}\\
E &= \text{eccentric anomaly,}\\
f &= \text{true anomaly,}\\
i &= \text{inclination,}\\
m &= \text{spacecraft mass,}\\
M &= \text{mean anomaly,}\\
n &= \text{mean motion,}\\
u &= \omega\tau f = \text{argument of latitude,}\\
\beta &= C_\mathrm{d}A/2m,\\
\rho &= \text{atmospheric density,}\\
\tau &= d(1-e\cos E)/(1+e\cos E),\\
\omega &= \text{argument of perigee,}\\
\omega_\mathrm{p} &= \text{planetry rotation,}\\
\Omega &= \text{right ascension of ascending node.}
\end{aligned}
$$

A more simplified result can be obtained directly from the Gaussian–Lagrange planetary equations when the rotation of the atmosphere is neglected so the drag force is along the negative velocity vector in inertial space. For the flight path angle γ, the specific force normal to the orbit W is zero and the other components of the specific drag force are

$$
\left.
\begin{aligned}
T &= f_\mathrm{d}\cos\gamma = -\frac{C_\mathrm{d}A}{2m_\mathrm{s}}\rho v_\mathrm{r}^2\frac{1+e\cos f}{(1+2e\cos f+e^2)^{1/2}},\\
R &= f_\mathrm{d}\sin\gamma = -\frac{C_\mathrm{d}A}{2m_\mathrm{s}}\rho v_\mathrm{r}^2\frac{e\sin f}{(1+2e\cos f+e^2)^{1/2}},
\end{aligned}
\right\}
\tag{3.7.8}
$$

where the expressions for the sine and cosine of the flight path angle are left as an exercise. Substituting for T and R in the Gaussian–Lagrange planetary equations and assuming that the coefficient of drag, atmospheric density, and cross-sectional area are constant, that the eccentricity is zero, and averaging over a period, gives the result that all the orbital elements are zero except the semimajor axis

$$
\left.
\begin{aligned}
\left\langle\left[\frac{da}{dt}\right]_{e=0}\right\rangle &= -C_\mathrm{d}\rho a^2 n\left(\frac{A}{m_\mathrm{s}}\right),\\
\left\langle\left[\frac{dM}{dt}\right]_{e=0}\right\rangle &= n,
\end{aligned}
\right\}
\tag{3.7.9}
$$

where this also is left as an exercise. This illustrates that a circular orbit will remain circular but the semimajor axis will decrease, linearly with time if the parameters on the right side of equation 3.7.9a are constants. The change in the period τ of a circular orbit can be determined by differentiating the equation for the period, equation 3.3.44, to obtain

$$\left.\begin{aligned} \frac{d\tau}{dt} &= \frac{3}{2}\frac{\tau}{a}\frac{da}{dt}, \\ \frac{d\tau}{da} &= \frac{3}{2}\frac{\tau}{a}. \end{aligned}\right\} \tag{3.7.10}$$

Substituting equation 3.7.9a into equation 3.7.10a yields the rate of change of the period for the circular orbit to be

$$\left\langle \left[\frac{d\tau}{dt}\right]_{e=0} \right\rangle = -\frac{3}{2}\tau C_{\mathrm{d}}\rho a n\left(\frac{A}{m}\right). \tag{3.7.11}$$

For orbits with large eccentricities, it can be assumed that the drag force acts only in the vicinity of periapsis. In this case, $\cos f \approx 1$and, if the velocity of the atmosphere is assumed to be so small that it can be ignored, the vis viva equation, equation 3.3.26, gives the velocity at periapsis to be

$$v_{\mathrm{r}}^2 = a^2 n^2 \left(\frac{1+e}{1-e}\right). \tag{3.7.12}$$

Substituting equation 3.7.12 into the expressions for the components of the specific force, equations 3.7.8, and substituting into the Lagrange planetary equations, equations 3.6.3, and averaging over the period yields

$$\left.\begin{aligned} \left\langle \left[\frac{da}{dt}\right]_{e=\text{large}} \right\rangle &= -C_{\mathrm{d}}\rho a^2 n\left(\frac{A}{m_{\mathrm{s}}}\right)\left(\frac{1+e}{1-e}\right)^{3/2}, \\ \left\langle \left[\frac{de}{dt}\right]_{e=\text{large}} \right\rangle &= -C_{\mathrm{d}}\rho a n\left(\frac{A}{m_{\mathrm{s}}}\right)\frac{(1+e)^{3/2}}{(1-e)^{1/2}}, \end{aligned}\right\} \tag{3.7.13}$$

where i, Ω, ω and M are unaffected. This derivation is also left as an exercise. Consequently, a highly circular orbit will experience reductions in both the semimajor axis and the eccentricity.

The effect on the radii at periapsis and apoapsis is determined as follows. Differentiating equations 3.3.51 and 3.3.52 give

$$\left.\begin{aligned} \frac{dr_{\mathrm{p}}}{dt} &= \frac{da}{dt}(1-e) - a\frac{de}{dt}, \\ \frac{dr_{\mathrm{a}}}{dt} &= \frac{da}{dt}(1+e) + a\frac{de}{dt}. \end{aligned}\right\} \tag{3.7.14}$$

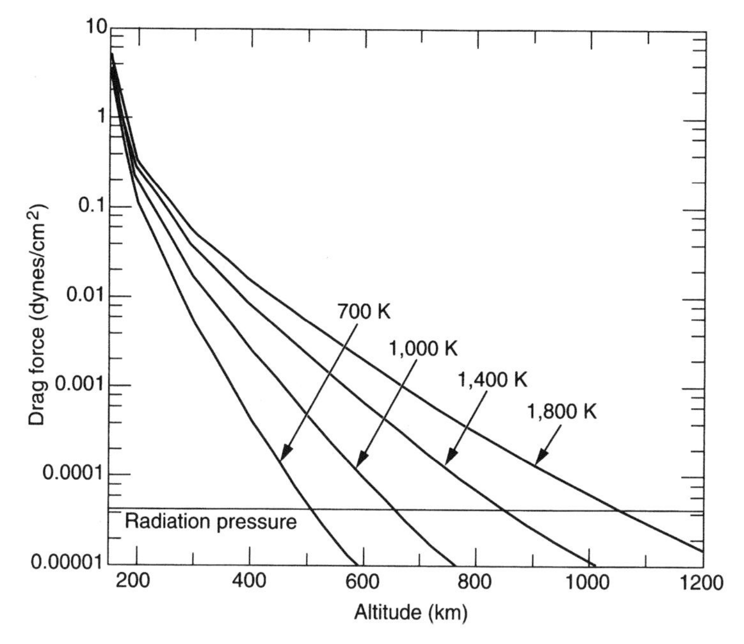

Figure 3.30 Drag force per unit area as a function of altitude and exospheric temperature; Jacchia model.

The term $\langle [de/dt]_{e=\text{large}} \rangle$ can be determined in terms of $\langle [da/dt]_{e=\text{large}} \rangle$ from equations 3.7.13 where

$$\left\langle \left[\frac{de}{dt}\right]_{e=\text{large}} \right\rangle = \left(\frac{1-e}{a}\right) \left\langle \left[\frac{da}{dt}\right]_{e=\text{large}} \right\rangle, \tag{3.7.15}$$

which when substituted into equations 3.7.14 yields

$$\left.\begin{aligned} \left\langle \left[\frac{dr_p}{dt}\right]_{e=\text{large}} \right\rangle &= 0, \\ \left\langle \left[\frac{dr_a}{dt}\right]_{e=\text{large}} \right\rangle &= 2 \left\langle \left[\frac{da}{dt}\right]_{e=\text{large}} \right\rangle. \end{aligned}\right\} \tag{3.7.16}$$

This first-order analysis shows that, for a highly elliptic orbit, the periapsis radius remains essentially constant while the apoapsis radius decreases at twice the rate of the semimajor axis.

The drag force per unit cross-sectional area, or drag pressure, as a function of orbit altitude and exospheric temperature is illustrated in figure 3.30 for the Jacchia density model. An exospheric temperature of 700 K corresponds to "quiet" solar conditions and a temperature of 1800 K corresponds to "disturbed" solar conditions. Also shown is the solar radiation pressure, which is approximately 4.7×10^{-5} dyn/cm^2. Even when the drag is significantly smaller than other perturbing forces, its integrated effect can dominate since it always acts in the same direction. Better models for the atmospheric

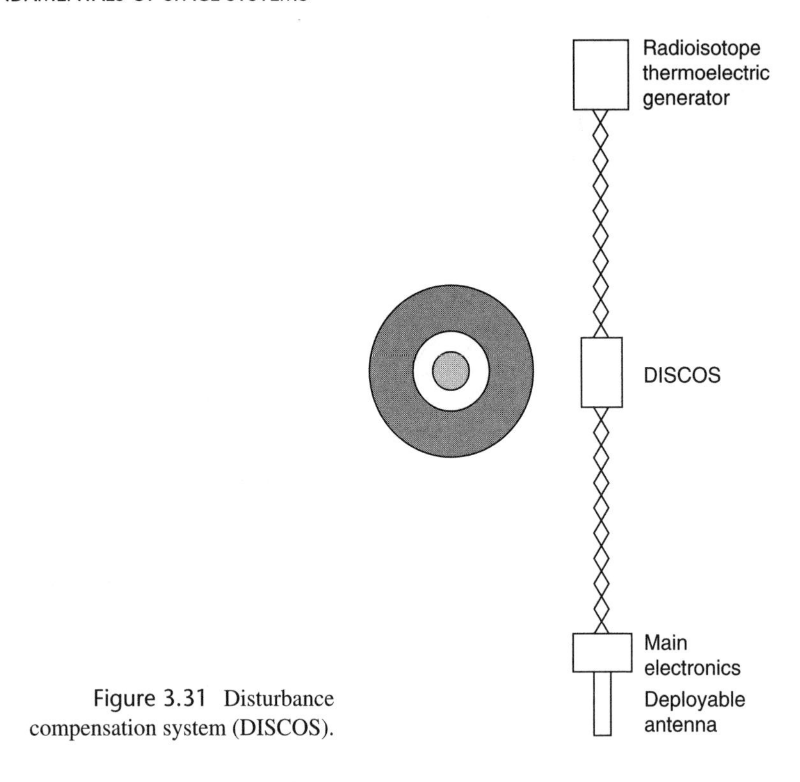

Figure 3.31 Disturbance compensation system (DISCOS).

density can often be in error by 10% to 20%, and up to 100% in periods of extreme solar activity. One technique to eliminate the effects of both drag and radiation pressure is the Disturbance Compensation System (DISCOS) described by the Staff of the Space Department et al. (1974) and by Black (1990). This device, illustrated in figure 3.31, consists of a small proof mass housed in a cavity and shielded from drag and radiation pressure. A propulsion system is used to assure that the spacecraft follows the trajectory of the proof mass, which is free from all atmospheric and radiation pressure forces.

Atmospheric forces limit the lifetime of a low-altitude satellite. The decrease in semimajor axis will cause the satellite to reach an altitude where it will experience sufficient aerodynamic force and heating that it may break up and then reenter. There is generally insufficient information to make accurate estimates of satellite lifetimes due to the variability of the atmospheric drag, the uncertainty in the coefficient of drag, and the cross-sectional area if the satellite's attitude is not known. Several estimations are given below, but at best these are gross approximations.

Perkins (1958) showed that if the atmospheric density is modeled by

$$\rho = \rho_0 \left(\frac{h}{h_0} \right)^{-k_1}$$

where ρ is the density at altitude h, ρ_0 is the density at altitude h_0, and k_1 is a scaling constant, the lifetime t_L of a satellite in a circular orbit is

$$t_L = -\frac{h_0}{k_1 + 1} \left\langle \frac{da}{dt} \right\rangle, \tag{3.7.17}$$

where $\langle da/dt \rangle$ is the initial time derivative of the semimajor axis, either measured or estimated by equation 3.7.9a. If, instead, it is believed that the density is better modeled by an exponential

$$\rho = \rho_0 e^{-k_2(h_0 - h)}, \tag{3.7.18}$$

where ρ_0 is the atmospheric density at altitude h_0 and k_2 is the inverse of the scale height, then the lifetime t_L is estimated to be

$$t_L = -\frac{h_0}{k_2} \left\langle \frac{da}{dt} \right\rangle, \tag{3.7.19}$$

where $\langle da/dt \rangle$ can be either measured or estimated by equation 3.7.9a.

For eccentric orbits, King-Hele (1964, 1978) showed that for initial eccentricities between 0.02 and 0.2 the lifetime t_L can be estimated by

$$t_L = -0.9e \bigg/ \frac{3}{2a} \left\langle \frac{da}{dt} \right\rangle, \tag{3.7.20}$$

and for initial eccentricities > 0.2

$$t_L = -e(0.62 + 1.4e) \bigg/ \frac{3}{2a} \left\langle \frac{da}{dt} \right\rangle, \tag{3.7.21}$$

where e, a, and $\langle da/dt \rangle$ are the eccentricity, semimajor axis, and the time derivative of the semimajor axis, either estimated by equation 3.7.9a or determined experimentally.

3.7.5 Radiation Pressure

Solar direct, solar albedo, and planet thermal radiation are the external radiation fluxes that can act on a spacecraft. In addition to thermal heating of the surfaces of the spacecraft, they result in forces that perturb the motion. Direct solar radiation pressure at the Earth's average distance from the Sun is approximately 0.47 dyn/m^2, with an annual variation of 3.5% due to the eccentricity of the Earth's orbit. Despite this being well known, estimates of the specific force of radiation suffer from uncertainties in the area illuminated because of complex spacecraft geometries; reflection off multiple surfaces; surface properties that degrade as a function of time due to the radiation, contamination, and abrasion; uncertainties in orientation with respect to the Sun; changing geometries due to deployment and steerable appendages; partial shadowing (penumbra) and total shadowing (umbra) due to the primary planet; and self shadowing of complex surfaces. In most treatments, only the direct radiation pressure is taken into account. The specific radiation force per unit area can be expressed in terms of the three terms that account for the specular and diffuse reflection and the absorption of the radiation as

$$d\mathbf{f} = -\frac{I_s dA\left(\hat{\mathbf{s}} \cdot \hat{\mathbf{n}}\right)}{mc} \left[2r_s\left(\hat{\mathbf{s}} \cdot \hat{\mathbf{n}}\right)\hat{\mathbf{n}} + r_a\hat{\mathbf{s}} + r_d\left(\hat{\mathbf{s}} \cdot \hat{\mathbf{n}}\right)\left(\hat{\mathbf{s}} + \frac{2}{3}\hat{\mathbf{n}}\right)\right], \tag{3.7.22}$$

where

$$
\begin{aligned}
c &= \text{speed of light in a vacuum,}\\
dA &= \text{differential surface area,}\\
\mathbf{f} &= \text{force per unit mass,}\\
I_s &= \text{energy flux from the source}(I_s(\text{Sun}) = 1360\text{W/m}^2),\\
m &= \text{spacecraft mass,}\\
\hat{\mathbf{n}} &= \text{unit vector normal to } dA,\\
r_a &= \text{percent absorbed, } 0 \le r_a \le 1,\\
r_d &= \text{percent diffusely reflected, } 0 \le r_d \le 1,\\
r_s &= \text{percent specularly reflected, } 0 \le r_s \le 1,\\
r_s &+ r_d + r_a = 1,\\
\hat{s} &= \text{unit vector toward the source.}
\end{aligned}
$$

The effect of radiation pressure is proportional to the area-to-mass ratio of the spacecraft, and thus is larger for smaller spacecraft, since the area-to-mass ratio is often proportional to r^{-1} where r is a characteristic dimension of the spacecraft. In the case of a constant radial radiation pressure force, for example due to the Earth's thermal radiation, the Gaussian–Lagrange planetary equations show that, when averaged over a revolution, the only element affected is the mean anomaly. In addition, a constant component of force normal to the orbit will result in no secular change in the inclination or right ascension of the ascending node. With shadowing, the analysis is more complex, with the major influence of solar radiation pressure being an increase in orbital eccentricity.

3.8 Orbit Determination

Orbit determination is the process of determining the ephemeris of a spacecraft, using theoretical models to relate observations obtained over an interval of time. The *ephemeris* is the position and velocity of the spacecraft as a function of time. The procedure usually results in a set of orbital elements that is tightly coupled to the unique model used in their determination. The orbital elements can then be used with the same model to extrapolate either forward or backward in time outside of the observation interval. The parameters determined consist of the six orbital elements and auxiliary parameters that represent uncertainties in the model. Typical auxiliary parameters include: drag coefficient, radiation pressure area, instrumentation biases, gravitational perturbations, polar motion, and corrections to the location of the tracking sites. Even with the utmost care, the equations of motion will be approximations because of modeling errors, truncation and rounding in digital procedures, and because the observations will contain measurement errors. Consequently, it is generally advantageous to determine the ephemeris over shorter rather than longer spans of time. The optimum time span and extrapolation span for orbit determination depend on the altitude of the satellite, the accuracy of the model, and the quantity, distribution, and accuracy of the observations.

Several tracking sites are generally required to obtain a good distribution of observations around the orbit to minimize errors correlated with a tracking site, such as location or calibration errors. At one time, there was a significant advantage from observations from a globally distributed network, but geopotential models are now sufficiently accurate that a global distribution is only necessary if high-precision ephemerides are required. Many different types of observations can be utilized for orbit determination, including:

- Two-way range and range rate.
- Angular positions from microwave radars.
- Range from optical laser sites.
- Doppler frequency measurements from one-way continuous-wave microwave transmissions.
- Positions against the star background from optical images.
- Images of landmarks obtained from the spacecraft.

Observations are generally obtained at subsecond intervals and transmitted to a central computing facility for processing. They are usually preprocessed to edit bad data that could result from transmission and formatting errors, noise, and instrumentation malfunction. When possible, a convenient means to edit the observations is to navigate the observing site with the observations. Deviation from the known station position and the character of the residuals has been found useful for editing and weighting the data. However, care must be used in the editing process to avoid biasing the solution. When high data rate observations are collected, it is sometimes efficient to perform local smoothing to reduce the number of data points to be subsequently processed. At this time it may be advantageous to correct the data for instrument calibration, clock drifts, timing biases, ionospheric and tropospheric effects, and so on (Black, 1978; Hopfield, 1980).

Estimation algorithms include weighted least-squares and Kalman filtering, with the latter generally used when the ephemeris is required in real-time. The weighted least-squares formulation in scalar form is

$$F[\mathbf{x}(\mathbf{r}_0, \dot{\mathbf{r}}_0; \mathbf{p})] = \frac{1}{n} \sum_{i=1}^{n} W_i [O_i^e - O_i(\mathbf{x})]^2, \tag{3.8.1}$$

where W_i is the weight for the ith observation: O_i^e is the ith observation; O_i is the calculated i^{th} observation; $\mathbf{x}(\mathbf{r}_0, \dot{\mathbf{r}}_0; \mathbf{p})$ represents a vector of parameters to be determined where $\mathbf{r}_0$ is the initial position, $\dot{\mathbf{r}}_0$ is the initial velocity, $\mathbf{p}$ is a vector of the auxiliary parameters; n is the number of observations; and F is a scalar to be minimized. The weights are employed to account for different types and quality of data. The set of parameters $\mathbf{x}$ that minimizes the function F is, by definition, the "best fit." However, the process is nonlinear, so a direct solution cannot be found.

It is sometimes more convenient to rewrite equation 3.8.1 in matrix form as

$$F[\mathbf{x}(\mathbf{r}_0, \dot{\mathbf{r}}_0; \mathbf{p})] = \frac{1}{n}[\mathbf{O}^e - \mathbf{O}(\mathbf{x})]^{\mathrm{T}} \mathbf{W} [\mathbf{O}^e - \mathbf{O}(\mathbf{x})], \tag{3.8.2}$$

where $\mathbf{W}$ is an $n \times n$ matrix of weights, $(\mathbf{O}^e - \mathbf{O})$ is an $n \times 1$ column vector of the observations $\mathbf{O}^e$ minus their calculated values $\mathbf{O}$, $\mathbf{x}(\mathbf{r}_0, \dot{\mathbf{r}}_0; \mathbf{p})$ is a vector of the s parameters to be determined, n is the number of observations, and F represents the scalar to be minimized. Differentiating the equation for F, with $\mathbf{x} = \mathbf{x}_0 + \Delta\mathbf{x}$, and setting $dF/d\mathbf{x}$ to zero, yields

$$\Delta\mathbf{x} = \left[\left(\frac{\partial\mathbf{O}}{\partial\mathbf{x}}\right)_{x0}^{\mathrm{T}} \mathbf{W} \left(\frac{\partial\mathbf{O}}{\partial\mathbf{x}}\right)_{x0}\right]^{-1} \left(\frac{\partial\mathbf{O}}{\partial\mathbf{x}}\right)_{x0}^{\mathrm{T}} \mathbf{W}(\mathbf{O}^{\mathrm{e}} - \mathbf{O}_{\mathbf{x}_0}), \tag{3.8.3}$$

where $(\partial\mathbf{O}/\partial\mathbf{x})_{x0}$ is the $n \times s$ matrix of partial derivatives evaluated for each observation that allows the solution to be determined iteratively by *differential correction*. To carry out the differential correction process it is necessary to compute the theoretical observation and their partial derivatives with respect to each of the parameters for each observation. The partial derivatives can be obtained either analytically by evaluating each observation or numerically by changing each parameter and reintegrating the ephemerides and numerically computing the partials for each observation. The matrix inversion in equation 3.8.3 involves a square matrix of dimension equal to the number of parameters being determined. The solution for $\Delta\mathbf{x}$ is repeated several times until the successive results for the parameters show little change. Some of the auxiliary parameters can be of independent interest. For instance, in the Navy Navigation Satellite System, the location of the daily coordinates of the celestial pole was determined (Anderle, 1972; Pisacane and Dillon, 1981).

For techniques to correct for atmospheric refraction in the orbit determination process, see Black (1978) and Hopfield (1980). For the fundamentals of orbit determination, see Escobal (1965), and for descriptions of the effect of perturbations on orbit determination see Pisacane et al. (1973).

3.8.1 Two-Line Elements

The United States Strategic Command (USSTRATCOM, formerly NORAD) maintains general perturbation elements for all resident space objects; these are called the two-line elements. These elements are periodically refined so as to maintain a reasonable prediction capability. The elements can be propagated in time to obtain the position and velocity of the body at any time in the past or future. It is important, as mentioned earlier, that the same theory be used to extrapolate the ephemerides as the one that was used to determine the parameters. The algorithmic software program can be obtained from the USSTRATCOM through the Internet in a variety of computer languages including Fortran, C, C++, and Pascal. The reference system used is an Earth-centered inertial (ECI) system. The two-line elements for each object are illustrated in table 3.3. These can be used to obtain the position to an accuracy of about a few kilometers routinely and with higher accuracy when required.

3.8.2 Global Positioning System

The Global Positioning System (GPS) followed the development of the first operational satellite system, the Navy Navigation Satellite System, which had pioneered satellite navigation and satellite-to-satellite navigation. GPS consists of at least 18 operational satellites, of which four are needed to carry out a navigation solution for a static or moving surface, air, or spacecraft object. The standard positioning system of GPS provides better than 20 m accuracy in position and 50 ns accuracy in time at least 95% of the time.

Table 3.3 Two-line orbital elements

	Field	Column	Description
Line 1	1.1	01–01	Line number
	1.2	03–07	Satellite number
	1.3	08–08	Classification (U = Unclassified)
	1.4	10–11	International designator (last two digits of launch year)
	1.5	12–14	International designator (launch number of the year)
	1.6	15–17	International designator (segment of launch)
	1.7	19–20	Epoch year (last two digits of year)
	1.8	21–32	Epoch (day number and fractional portion of the day)
	1.9	34–43	One half of the first time derivative of the mean motion (rev/day^2)
	1.10	45–52	One-sixth of the second time derivative of mean motion (decimal point assumed) (rev/day^3)
	1.11	54–61	BSTAR drag term (decimal point assumed, equal to $C_D A_\rho R/2m^*$)
	1.12	63–63	Ephemeris type
	1.13	65–68	Element number
	1.14	69–69	Check sum (modulo 10) (Letters, blanks, periods, plus signs = 0; minus signs = 1)
Line 2	2.1	01–01	Line number of element data
	2.2	03–07	Satellite number
	2.3	09–16	Inclination (degrees)
	2.4	18–25	Right ascension of the ascending node (degrees)
	2.5	27–33	Eccentricity (decimal point assumed)
	2.6	35–42	Argument of perigee (degrees)
	2.7	44–51	Mean anomaly (degrees)
	2.8	53–63	Mean motion (rev/day)
	2.9	64–68	Revolution number at epoch (rev)
	2.10	69–69	Check sum (modulo 10)

$^*C_D$ = coefficient of drag, m = satellite mass, ρ = atmospheric density, R = radius of the Earth, A = Area.

Example:

```
  123456789012345678901234567890123456789012345678901234567890123456789
 GOES 10
  1 24786U 97019A    98060.33155985 -.00000097 00000-0  00000+0  0  1678
  2 24786  0.0463  285.0622 0001527  70.5278 175.3261  1.00270131  3138
```

Spacecraft can determine their position reliably in real time using GPS if their radii are less than the semimajor axis of GPS, 26,400 km. This would preclude requirements for ground tracking and computationally intensive orbit determination.

3.9 Spacecraft Coverage

Spacecraft coverage is an important criterion for orbit selection for a variety of spacecraft missions including communications, navigation, weather, surveillance, search and rescue, Earth resource monitoring, and monitoring the space environment. Coverage connotes a broad range of criteria, for example, the degree of uniform sampling of the Earth, size of the instantaneous coverage area, time intervals between revisits of specific locations, duration that a satellite will be in view, and the fraction of a day that specific locations are in the coverage area. As a result, coverage specification is unique to the particular mission. Consequently, only fundamental concepts are considered here. Since

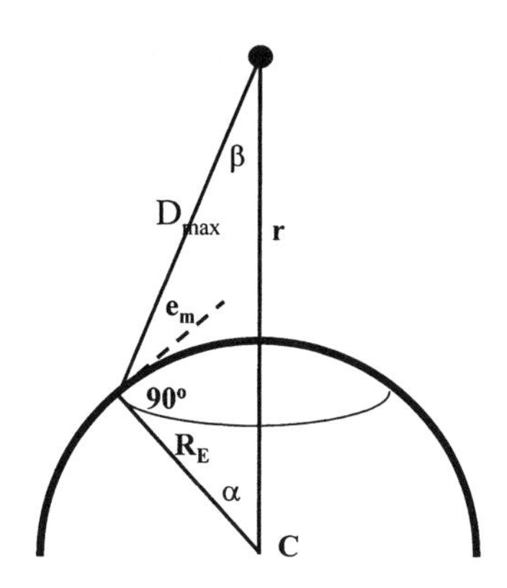

Figure 3.32 Instantaneous satellite coverage geometry. R_e = Earth radius, e_m = minimum elevation angle, r = radius to satellite, α = cone half-angle at C, C = center of Earth, β = half-cone angle at satellite, D_{max} = maximum distance to subsatellite point.

the Earth is oblate to one part in 300, it can be well approximated by a sphere in the geometrical calculations.

3.9.1 Coverage Formulae

The instantaneous area that a satellite is in view is illustrated in figure 3.32 where the largest coverage area is to the horizon. However, practical limits on minimum elevation angles generally limit the coverage area to elevations greater than 5 to 10 degrees. The half-cone angle at the spacecraft, β, can be determined from

$$\beta = \mathrm{Sin}^{-1}\left(\frac{R_E}{r}\cos e_m\right), \tag{3.9.1}$$

where β is the half-angle of the intersecting cone from the spacecraft, r is the radius to the spacecraft, e_m is the minimum elevation angle, and R_E is the radius of the Earth. The central angle α of the spherical cap is

$$\alpha = \frac{\pi}{2} - \beta - e_m = \cos^{-1}\left(\frac{R_E}{r}\cos e_m\right) - e_m. \tag{3.9.2}$$

The area of the instantaneous spherical cap is

$$A_t = 2\pi R_E^2(1 - \cos\alpha) = 2\pi R_E^2(1 - \sin(\beta + e_m)), \tag{3.9.3}$$

and the fraction of the Earth that is in view is

$$F_t = \frac{1}{2}(1 - \cos\alpha) = \frac{1}{2}[1 - \sin\alpha(\beta + e_m)], \tag{3.9.4}$$

as illustrated in figure 3.33. Not surprisingly, the percentage of coverage increases with increasing radius and decreases with increasing minimum elevation angle.

The maximum distance D_{max} to the coverage region is

$$D_{max} = R_E\frac{\sin\alpha}{\sin\beta} = R_E\frac{\cos(\beta + e_m)}{\sin\beta} = R_E\left[\left(\frac{r^2}{R_E^2} - \cos^2 e_m\right)^{1/2} - \sin e_m\right]. \tag{3.9.5}$$

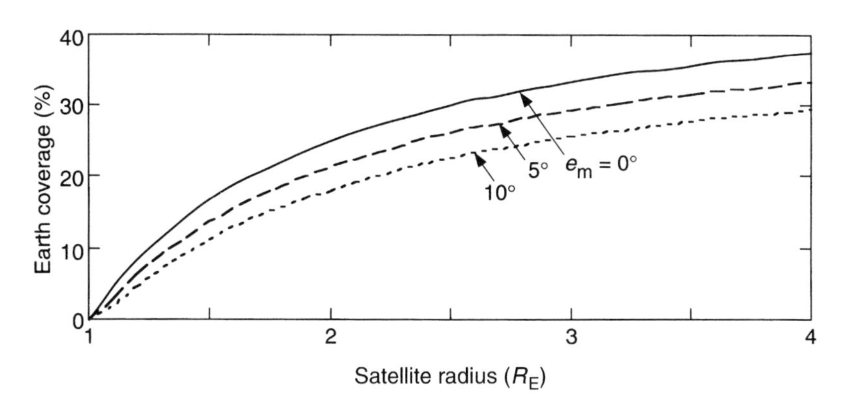

Figure 3.33 Percent Earth coverage for minimum elevation angle. R_e = Earth radius, e_m = minimum elevation angle.

The *average coverage rate,* which is the rate at which the area is swept out, can be approximated by

$$\frac{dA}{dt} = \frac{4\pi R_E^2}{\tau} \sin\alpha = \frac{4\pi R_E^2}{\tau} \cos(\beta + e_m) \tag{3.9.6}$$

where τ is the period, which is assumed large in comparison to the rotational period of the Earth.

3.9.2 Elevation and Azimuth Pointing

The elevation and azimuth, illustrated in figure 3.34, required to point an antenna or telescope to a satellite are defined by

- *Elevation* (El) = the angle from the horizon of the observer in the plane containing the spacecraft, observer, and center of the Earth.
- *Azimuth* (Az) = the angle measured in the horizontal plane of the observer from north to the intersection with the plane containing the spacecraft, observer, and center of the Earth.

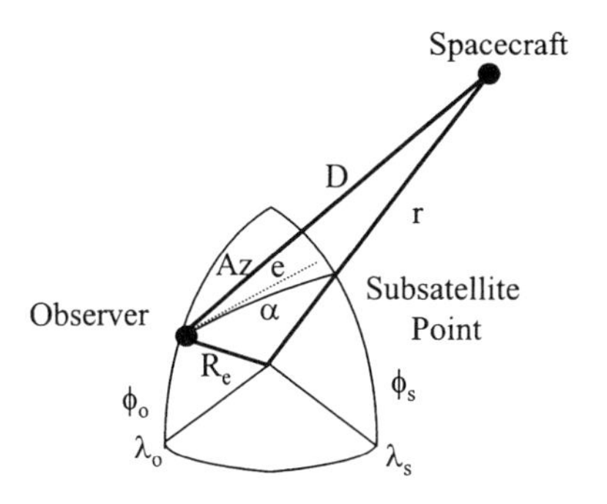

Figure 3.34 Geometry for azimuth (Az), elevation (El), and distance (D).

Given the Kepler elements of the spacecraft, the geographic latitude and longitude of the subspacecraft point is

$$\phi_s = \sin^{-1}[\sin i \sin(\omega + f)], \tag{3.9.7}$$

$$\lambda_s = \Omega - \lambda_g + \tan^{-1}\left[\frac{\cos i \sin(f + \omega)}{\cos(f + \omega)}\right] + C, \tag{3.9.8}$$

where

$$\begin{aligned} C &= 0 \text{ if } \cos(f + \omega) \geq 0, \\ &= \pi \text{ if } \cos(f + \omega) < 0 \text{ and } \cos i \sin(f + \omega) > 0, \\ &= -\pi \text{ if } \cos(f + \omega) < 0 \text{ and } \cos i \sin(f + \omega) < 0, \\ f &= \text{true anomaly}, \\ i &= \text{spacecraft inclination}, \\ \phi_s &= \text{latitude of subspacecraft point}, \\ \lambda_s &= \text{longitude of subspacecraft point}, \\ \lambda_g &= \text{hour angle of Greenwich}, \\ \omega &= \text{argument of periapsis}, \\ \Omega &= \text{right ascension of the ascending node.} \end{aligned}$$

The central angle or great circle angle between the observer and the subsatellite point is then given by

$$\left.\begin{aligned} \cos\alpha &= \cos\left(\frac{\pi}{2} - \phi_o\right)\cos\left(\frac{\pi}{2} - \phi_s\right) \\ &\quad + \sin\left(\frac{\pi}{2} - \phi_o\right)\sin\left(\frac{\pi}{2} - \phi_s\right)\cos(\lambda_s - \lambda_o) \\ \alpha &= \cos^{-1}[\sin\phi_o \sin\phi_s + \cos\phi_o \cos\phi_s \cos(\lambda_s - \lambda_o)] \end{aligned}\right\} \tag{3.9.9}$$

where

$$\begin{aligned} \alpha &= \text{central angle or great circle angle between observer and subsatellite point}, \\ \phi_0 &= \text{latitude of observer}, \\ \lambda_0 &= \text{longitude of observer.} \end{aligned}$$

The elevation, azimuth, and distance follow as

$$El = \tan^{-1}\left(\frac{\cos\alpha - R_e/\mathrm{r}}{\sin\alpha}\right), \tag{3.9.10}$$

$$Az = \pm\cos^{-1}\left(\frac{\sin\phi_s - \sin\phi_0\cos\alpha}{\cos\phi_0\sin\alpha}\right), \tag{3.9.11}$$

where Az is negative if $0 \leq \lambda_s - \lambda_0 \leq \pi$ and positive if $-\pi < \lambda_s - \lambda_s < 0$

$$D = r\frac{\sin\alpha}{\cos El} = R_E\frac{\sin\alpha}{\cos(El + \alpha)}, \tag{3.9.12}$$

where

$$El = \text{elevation},$$
$$Az = \text{azimuth},$$
$$D = \text{range between spacecraft and station},$$
$$r = \text{radius to the satellite},$$
$$R_E = \text{radius of the Earth}.$$

3.9.3 Doppler Shift

The received frequency f_r of a transmitted signal from a spacecraft with frequency f_t in motion with relative velocity $\mathbf{v}$ at position $\mathbf{r}$ is given by

$$f_r = f_t \frac{1 - \frac{\mathbf{v}\cdot\hat{\mathbf{r}}}{c}}{(1 - \frac{v^2}{c^2})^{\frac{1}{2}}} \tag{3.9.13}$$

where c is the speed of light in a vacuum. The change in frequency can be approximated by

$$\Delta f = f_r - f_t \approx -\frac{f_t}{c}\mathbf{v}\cdot\hat{\mathbf{r}}. \tag{3.9.14}$$

The Doppler shift must be taken into account in the design of spacecraft telecommunication systems and can be used as tracking data to determine the orbit.

3.9.4 Selected Orbit Configurations

Several popular orbit configurations will be discussed in turn. *Near-circular polar orbits* have been used to carry out many missions. Equation 3.7.2a shows that to, first order, the precession of the node of a polar satellite is zero so that a constellation of similar orbits will remain fixed in inertial space. Since each satellite passes over the poles, coverage at high latitudes is greater than for lower latitudes. A variant of this configuration is a near-circular near-polar orbit with inclination equal to $90°-\alpha$, where α is given by equation 3.9.2, so that the minimum elevation angle at the poles is specified. This would improve the coverage at lower latitudes relative to polar orbits. For example, a satellite at 1000 km altitude and with a minimum elevation angle of 10° at the poles can have the range of inclinations of between 68.4° and 111.6°.

The geostationary or *geosynchronous equatorial orbit* (GEO) has a period identical to the sidereal period of the Earth, which occurs at a semimajor axis of 42,156 km. The advantages of this orbit are the large coverage area, of about 42%, and that the satellite appears motionless to an observer on the Earth so that fixed high-gain antennas can be used in place of tracking antennas. Latitudes as high as 80 degrees can be served. If the eccentricity is not zero, the satellite will appear to oscillate in longitude. If the inclination is not zero, the satellite will trace out a figure eight as it moves from one side of the equator to the other.

The *Molniya orbit,* described in section 3.7, is a highly eccentric ($e \approx 0.75$) orbit with a period of half a sidereal day of $11^{\mathrm{h}}\ 58^{\mathrm{m}}$, and an inclination equal to the critical inclination of 63.4°. With the perigee in the southern hemisphere, the spacecraft permits long observation times of up to 8 h while in the vicinity of apogee in the northern hemisphere where a maximum of abut 43% of the Earth's surface is visible. This orbit is particularly useful for providing communications at the high latitudes not served by equatorial geosynchronous satellites.

Sun-synchronous orbits, described in section 3.7, have nodal recession rates the same as the apparent motion of the Sun, of about 0.983° per day. As a result the relative geometry between the spacecraft and the Sun will remain constant. This orbit is used to achieve the same sunlight conditions as on the Earth's surface for scientific purposes and its phase can be set anywhere between the dawn–dusk to noon–midnight configuration. This orbit also provides engineering advantages in that a solar array can be pointed to the Sun 100% of the time so that its area can be minimized, and the thermal conditions would be essentially constant, making its design easier with minimal temperature variations.

A *repeating ground track orbit* is one in which the subsatellite trace closes on itself after a number of revolutions. Thus, the period must be a submultiple of a sidereal day. This orbit is useful for obtaining repeated observations over the same region, or when it is necessary to specify the ground-track in advance.

A *Walker* (1973, 1984) *constellation* is composed of a near-uniform spacing of spacecraft characterized by (t, p, f), where

t is the number of satellites,
p is the number of orbital planes, and
f is the phase parameter that defines the spacing between spacecraft in adjacent planes.

The number of satellites per plane, s, is given by

$$s = t/p, \tag{3.9.15}$$

the in-plane spacing of spacecraft, ΔM, is

$$\Delta M = 360/s = 360\ p/t, \tag{3.9.16}$$

the separation of nodes $\Delta\Omega$ between orbital planes is

$$\Delta\Omega = 360/\mathrm{p}, \tag{3.9.17}$$

and the phase difference $\Delta\phi$ between spacecraft in adjacent orbital planes is

$$\Delta\phi = 360 f/t. \tag{3.9.18}$$

As an example, a constellation such as the GPS constellation of 18 operational and 6 spare satellites in 6 planes can be approximated by a Walker constellation of (18,6,–) where

$$s = t/p = 18/6 = 3,$$
$$\Delta M = 360/s = 360/3 = 120°,$$

$$\Delta\Omega = 360/p = 360/6 = 60^\circ,$$
$$\Delta\phi = 360f/t = 20f^\circ, \text{ where}$$
$$\Delta\phi|_{f=1} = 20^\circ,$$
$$\Delta\phi|_{f=2} = 40^\circ,$$
$$\Delta\phi|_{f=3} = 60^\circ.$$

The proposed European Space Agency Galileo constellation of navigation satellites is currently envisioned as a (27,3,2) Walker constellation.

For additional information on orbit selection and coverage see Luders (1961), Boehm (1964), Emara and Leondes (1977), Black et al. (1984), and Draim (1987).

3.10 Interplanetary Trajectories

Interplanetary missions differ from spacecraft orbiting a planet is that the spacecraft is exposed to more than one dominant force. For example, a spacecraft may leave the Earth, pass close to the Moon, travel on an interplanetary trajectory dominated by the Sun, and have close encounters with planets and asteroids before reaching its target planet whose gravitational force will dominate its approach. Parking orbits around the initial or final bodies add intermediate goals and well-defined phases that increase the flexibility in mission planning and scheduling. For example, reentry or landing can require a precise trajectory that is easier to obtain from a parking orbit than an interplanetary trajectory because landing sites can be surveyed, the status of the spacecraft determined, a precise orbit determined, and orbital corrections made.

3.10.1 Gravitational Assists

Significant changes to the spacecraft trajectory require a substantial mass of propellant. An alternative approach is to utilize a gravitational assist from an intermediate planet that can change the magnitude (increase or decrease) or direction of the velocity vector. This approach has proven essential for several missions including the Galileo mission that had two encounters with the Earth and one each with Venus and an asteroid to reach Jupiter more quickly, the Ulysses mission that used Jupiter to direct the trajectory out of the ecliptic plane so the spacecraft could sample the poles of the Sun, and the Cassini mission with encounters with Venus, Earth, and Jupiter to speed it on its way to Saturn. The mechanism of the gravitational assist for either increasing or decreasing the velocity of the satellite is illustrated in figure 3.35. While to first order the velocity with respect to the planet is constant, the velocity with respect to the Sun is either increased or decreased. Figure 3.36 illustrates the heliocentric velocity for Voyager 2 and the heliocentric escape velocity as a function of the distance from the Sun. The increases in heliocentric velocity achieved during the gravitational assists are apparent.

3.10.2 Patched Conics

One approach to approximate interplanetary trajectories is the technique of *patched conics*, based on the concept of the sphere of influence or activity sphere proposed

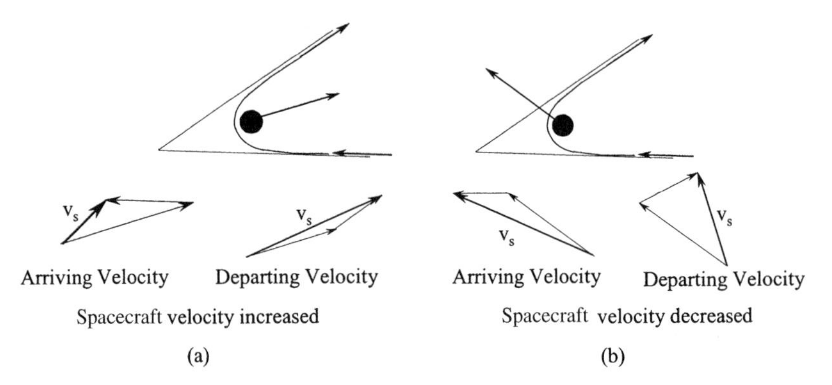

Figure 3.35 Examples of gravitational assist. (a) Spacecraft velocity increased. (b) Spacecraft velocity decreased.

by Lagrange. Consider a spacecraft acted upon by two gravitational bodies, one with mass m_a, the other with mass m_b with $m_a < m_b$, separated by distance d. The *sphere of influence* is the spherical region surrounding the smaller mass m_a where its gravity field dominates the motion of the spacecraft. Outside the sphere of influence, the larger mass will dominate its motion. The sphere of influence is determined by constructing the equations of motion first with respect to the larger mass, m_b, treating the smaller mass m_a, as a perturbation and then with respect to the smaller mass m_a, treating the larger mass m_b, as a perturbation. The radius around the smaller mass body where the primary and disturbing effects are the same is defined quantitatively as the *radius of the sphere of influence* ρ, where

$$\rho = d\left(\frac{m_a}{m_b}\right)^{2/5}. \tag{3.10.1}$$

Note that equation 3.10.1 is not simply the distance at which the gravitational forces are equal. Radii of the spheres of influence of the planets are given in table 3.4.

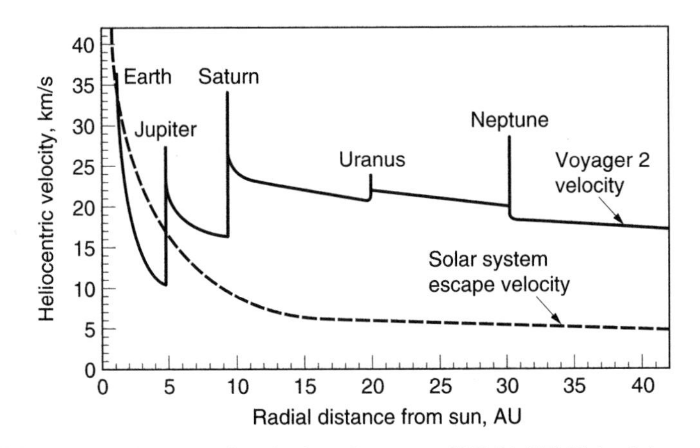

Figure 3.36 Voyager 2 heliocentric velocity. (Courtesy of NASA/JPL/Caltech.)

Table 3.4 Sphere of influence of celestical bodies

Celestial Body	$R_{\text{influence}}$ (10^9 m)
Mercury	0.1
Venus	0.6
Earth	1.0
Mars	0.6
Jupiter	50.0
Saturn	55.0
Uranus	52.0
Neptune	86.0
Moon	0.065

By defining the spheres of influence, the *method of patched conics* can be utilized, which notionally patches two conics together at the trajectory's intersection with the sphere of influence. The technique assumes the spacecraft is affected only by the primary body, the Sun, until it enters the sphere of influence of a secondary body where it is affected only by that body. The transition, at the boundary of the sphere of influence, is generally from an elliptical orbit around the Sun to a hyperbolic orbit around the body. As the body leaves the sphere of influence, the transition reverts to its orbit around the Sun. This is an effective and efficient technique to carry out a preliminary analysis. However, detailed analyses are required to take into account all the effects including radiation pressure, multiple gravity forces, thruster firings, and so on. For more information on the method of patched conics and interplanetary trajectories, see Breakwell et al. (1961) and Battin (1987).

3.10.3 Aerobraking

Aerobraking is a technique that can be employed in the capture process of a spacecraft by a planet with an atmosphere in order to reduce the amount of propellant required. As the spacecraft is at its closest point, the propulsion system is employed to capture the spacecraft into a highly elliptic orbit. From the earlier discussion on the effect of drag on an elliptic orbit in section 3.7, if the periapsis radius is sufficiently low it will, to first order, remain a constant while the apoapsis radius decays at a rate of twice the rate of change of the semimajor axis. When the apoapsis reaches the altitude of the target orbit, the propulsion system is used at apoapsis to raise the periapsis out of the atmosphere so that the target orbit will persist. Aerobraking was first used successfully at Venus during the Magellan missions and requires knowledge of the density of the planet's atmosphere.

As an example, consider the aerobraking of the Mars Global Surveyor spacecraft as illustrated in figure 3.37. Rather than being placed immediately into the mapping orbit at Mars orbit injection (MOI) on 12 September 1997, the spacecraft was placed into a highly elliptical orbit with a period of 48 hs. The plan was that 40 days after MOI, the period would be increased to about 34 Earth hours, after 60 days to 24 Earth hours, after 80 days to 12 Earth hours, after 100 days to 6 Earth hours, and after

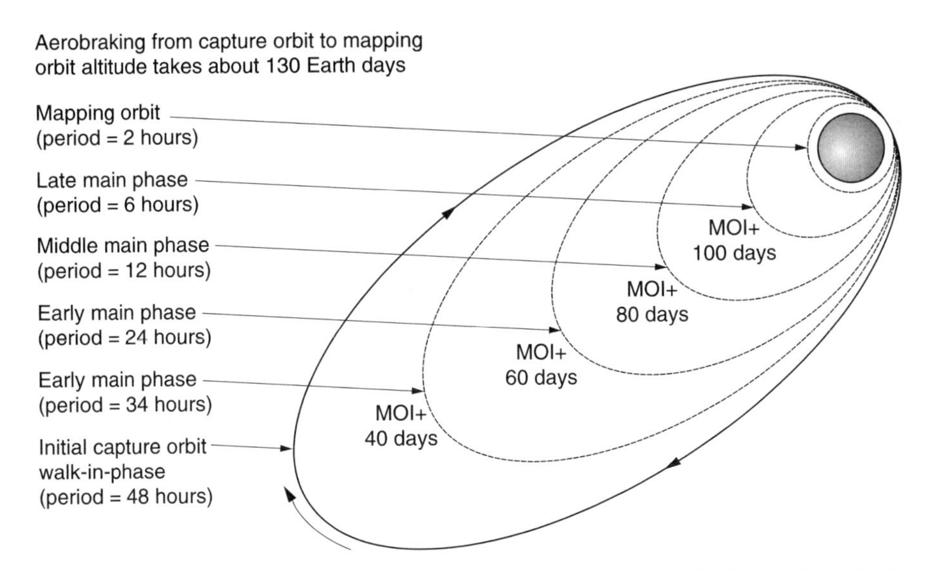

Figure 3.37 Mars Global Surveyor planned aerobraking scenario. MOI = Mars orbit injection. (Courtesy of NASA/JPL/Caltech.)

130 days it was to be placed in its final mapping orbit with a period of two Earth hours. It was estimated that a significant amount of propellant would be saved. Note, however, that the saving in propellant mass would be at the expense of an additional duration of 140 Earth days to achieve the final mapping orbit. Unfortunately, during deployment of the solar arrays, one of the panels did not indicate it was properly locked into place. During the 15th aerobraking pass, the spacecraft encountered a significant increase in drag such that the unlocked panel apparently suffered minor structural damage. The periapsis was raised and aerobraking temporarily suspended. Aerobraking was subsequently continued, but it took 17 months to achieve the final orbit instead of the planned 130 days.

NASA's Mars Odyssey spacecraft completed its successful aerobraking phase on 11 January 2002 after 332 revolutions, saving more than 220 kg of propellant. The spacecraft had been was launched on a DELTA II 7925 launch vehicle rather than a more capable but expensive launch vehicle.

3.10.4 Lagrange Libration Points

Euler and Lagrange studied the restricted three-body problem consisting of two large masses and one small mass in which the large masses are aligned and rotate in a plane such as the Sun and Earth or Earth and Moon. Euler found three points, L_1, L_2, and L_3, along the line connecting the two rotating masses where the small mass could be placed at a position of equilibrium. Lagrange found two additional points, L_4 and L_5, that form equilateral triangles in the plane of the motion. These five *Lagrange libration points* are illustrated for the Sun–Earth and Earth–Moon systems in figure 3.38. L_1, L_2, and L_3 are unstable while L_4 and L_5 are stable. However, the presence of the Moon in the Sun–Earth system and the presence of the Sun in the Earth–Moon system cause L_4

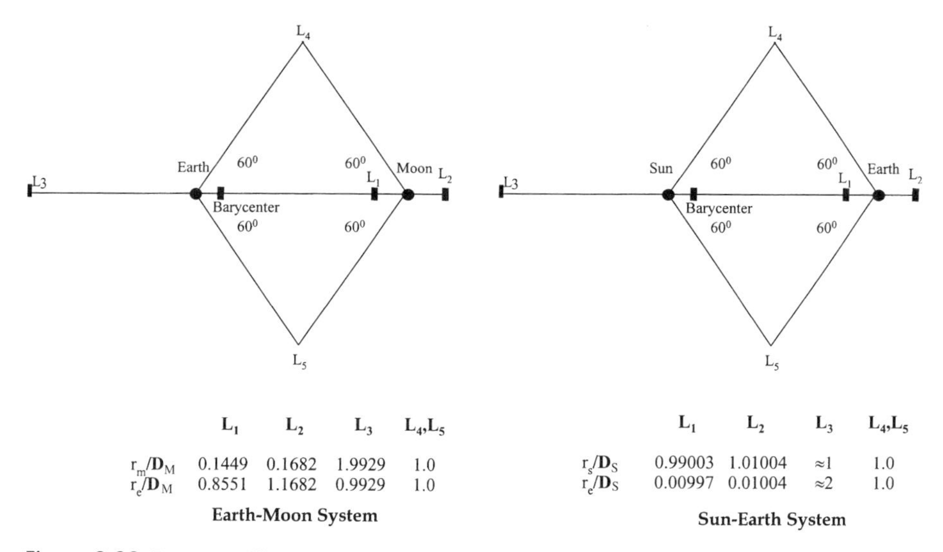

Figure 3.38 Lagrange libration points, r_{M} = distance from Moon, r_{E} = distance from Earth, D_{M} = distance between Earth and Moon = 384,000 km, r_{S} = distance to Sun, D_{S} = distance between Earth and Sun = 149.6 million km. Unstable: L_1, L_2, L_3. Stable: L_4, L_5.

and L_5 also to be unstable. Nevertheless, L_4 and L_5 for the Sun–Jupiter system is stable and there are twelve asteroids (the Trojans) in oscillation about them.

In the Sun–Earth system, L_1 is a convenient place to position a satellite to provide continuous monitoring of the Sun as it will require minimum fuel to keep it a fixed distance between the Earth and the Sun. L_2 is a convenient place for infrared satellites, such as the Next Generation Space Telescope (NGST), because the point is outside the Moon's orbit and the thermal and gravitational effects are therefore reduced.

Problems

1. Show that the constant of integration e in the solution $u = \frac{\mu}{h^2}[1 + e\cos f]$ of the differential equation $\frac{d^2u}{d\theta^2} + u = \frac{\mu}{h^2}$ can be expressed in terms of the specific energy ε as

$$e = \left[1 + \frac{2\varepsilon h^2}{\mu^2}\right]^{1/2}.$$

2. Find the expression for the period of an elliptical orbit based on the areal velocity being constant.

   Answer: $\tau = 2\pi\sqrt{\frac{a^3}{\mu}}$.
3. Derive Kepler's equation, $E - e\sin E = n(t - t_0) = M$, from the energy equation by introducing the eccentric anomaly E, where $r = a(1 - e\cos E)$.
4. Derive the relationship between the eccentric anomaly E and the true anomaly f.

   Answer: $\tan\frac{f}{2} = \left(\frac{1+e}{1-e}\right)^{1/2}\tan\frac{E}{2}$.

5. Determine the classical orbital elements given the position vector $\mathbf{r}$ and velocity vector $\mathbf{v}$ of a spacecraft with respect to the International Celestial Reference System.
   Answers: Let the unit vectors be $\boldsymbol{\varepsilon}_i (i = 1 \text{ to } 3)$ with $\boldsymbol{\varepsilon}_1$ along the equinox and $\boldsymbol{\varepsilon}_3$ along the IRM. With $\hat{\mathbf{h}} = (\mathbf{r} \times \mathbf{v})|\mathbf{r} \times \mathbf{v}|^{-1}$,

$$\varepsilon = \frac{v^2}{2} - \frac{\mu}{r}, \; a = \left(\frac{2}{r} - \frac{\mathbf{v} \cdot \mathbf{v}}{\mu}\right)^{-1/2},$$

$$e = \left(\frac{1 + 2\varepsilon h^2}{\mu^2}\right)^{1/2}, \quad \cos i = \mathbf{h} \cdot \boldsymbol{\varepsilon}_3,$$

$$\tan \Omega = (\mathbf{h} \cdot \boldsymbol{\varepsilon}_1)(-\mathbf{h} \cdot \boldsymbol{\varepsilon}_2)^{-1}, \cos f = \frac{1}{e}\left[\frac{a(1 - e^2)}{r} - 1\right],$$

$$\cos(f + \omega) = \frac{\mathbf{r}}{r} \cdot (\boldsymbol{\varepsilon}_1 \cos \Omega + \boldsymbol{\varepsilon}_2 \sin \Omega).$$

6. Determine the sine and cosine of the flight path angle of a spacecraft on an elliptical orbit.
   Answer: $\cos y = \dfrac{1 + e \cos f}{(1 + 2e \cos f + e^2)^{1/2}}, \sin \gamma = \dfrac{e \sin f}{(1 + 2e \cos f + e^2)^{1/2}}.$
7. On 9 February 1986, Halley's comet was at perihelion at a distance of 0.5872 AU and eccentricity 0.9673. (a) Calculate the period. (b) Determine the true anomaly and radius for 1 January 1992.
   Answer: (a) $\tau = 76.1$ years, (b) $f = 163.38°$, $r = 15.80$ AU.
8. A satellite is put into an Earth orbit at a height of one-quarter of the Earth's radius with a velocity normal to the radius. (a) Find the speed of the satellite if the orbit is to be circular. (b) Find the speed if the orbit just grazes the Earth at perigee. (c) Find the true anomaly where the satellite would intersect the Earth if its speed were 0.2, 0.6, 0.9 times the velocity required for a circular orbit.
   Answer: (a) 7071 m/s, (b) 6667 m/s, (c) f = 171.7°, 149.2°, 86.2°.
9. A scientific spacecraft is in an Earth orbit with perigee radius of 7000 km and apogee radius 10,000 km.
   (a) Find the altitude when the true anomaly is 90°.
   (b) Find the velocity when the true anomaly is 90°.
   (c) Determine the period of the spacecraft.
   Answer: (a) 1857.3 km, (b) 7.065 km/s, (c) 130 min.
10. Determine the secular rate of change with respect to time of the classical orbital elements for a satellite in a circular orbit subject to a constant drag force. Neglect the motion of the atmosphere relative to the spacecraft.
    Answers: see equations 3.7.9.
11. Determine the secular rate of change with respect to time of the classical orbital elements for a satellite in a highly elliptical orbit subject to drag at peripasis. Neglect the motion of the atmosphere relative to the spacecraft. Begin with the equations for the components of the specific force given by equations 3.7.6.
    Answers: see equations 3.7.13.
12. (a) Derive an expression for a mass of a celestial body if the period and semimajor axis of an orbiting spacecraft can be determined. (b) If a spacecraft orbits an asteroid

with a period of 10 days and a semimajor axis of 1000 km, determine the mass of the asteroid.

Answers: (a) $m_a = (4\pi^2 a^3 / G\tau^2) - m_s$, (b) 7.9×10^{17} kg.

13. An asteroid on a hyperbolic orbit passes by the planet Uranus at a minimum distance of 100 000 km and with an eccentricity of 2.0. Determine the time from periapsis that the asteroid is at a distance of 200 000 km.

    Answer: 14 008 s.

14. A spacecraft has a radius of 15 000 km, speed 8.8 km/s, and flight path angle 45°. Find: (a) eccentricity, (b) C_3, (c) semimajor axis, and (d) periapsis radius.

    Answer: (a) 1.5271, (b) 24.29 km^2/s^2, (c) -1.6408×10^7 m, (d) 8.6488×10^6 m.

15. The two-line elements provide the mean motion n in revolutions/day and the rate of change of the mean motion $\dot{n}/2$ in revolutions/day^2. Determine an expression for the change in radius in one day.

    Answer: da/dt (meters/day)= $-2a$ (meters) $\dot{n}$ (rev/day^2)/$3n$ (rev/day)

16. Determine the percentage coverage of the Earth for a geostationary satellite with a radius of 6.61 Earth radii for minimum elevation angles of 0°, 5°, 10°.

    Answers: (a) 42.4%, (b) 38.19%, (c) 34.08%.

17. For the two-line element that follows, find the position vector at epoch and one hour later, ignoring the effects of drag.

    MOLNIYA 3–48

```
1234567890123456789012345678901234567890123456789012345678901234567890123456789
1 24640U 96060A   01231.14299455 -.00000013  00000-0  10000-3 0  9194
2 24640  64.9085 268.1908 7053286 273.7677  15.2835  2.00648381 35299
```

Answer: (–383, –12 778, 43) km, (7041, –17 861, 16 233) km.

18. The altitude and true anomaly of an Earth satellite is determined at two instances where, at the first instance, the true anomaly is 30° and the altitude is 600 km and, at the second instance, the true anomaly is 120° and the altitude is 1800 km. Determine (a) the eccentricity, (b) the semimajor axis, and (c) the apogee altitude.

    Answer: (a) 0.118, (b) 7800 km, (c) 2342 km.

19. Determine an expression for the radius of a hyperbolic orbit in terms of the semimajor axis and the fraction k of the velocity at a radius of infinity.

    Answer: $r = -a/2(k^2-1)$.

20. Given the two-line elements for the International Space Station, find (a) year, day, time of the epoch in EDT (Eastern Daylight Time); (b) period; (c) semimajor axis, and perigee and apogee altitudes; and (d) expected change in altitude in one day.

```
1234567890123456789012345678901234567890123456789012345678901234567890123456789
1 25544U 98067A   02281.42009497  .00055235 00000-0  63516-0  0  2311
2 25544  51.6353 204.3265 0017279 348.2917 120.3454 15.61284021221696
```

Answers: (a) 1996, 8 October, 5h 4 min 56.205 41 s; (b) 92.232 min; (c) 6762.085 km, 372.263 km, 395.632 km; (d) –319.6 m/day.

21. Determine the geographic latitude and longitude of a spacecraft with period = 90 min, eccentricity = 0.0008, inclination = 28.5°, argument of perigee = 30°, right ascension of the ascending node = 30° when the hour angle of Greenwich = 10,000 s sidereal time.

    Answers: latitude = 24.4080°, longitude = 195.0299°.

22. Find an article in a technical journal that addresses the effect of perturbations on the trajectory of an artificial earth satellite. Turn in a one-page summary and include a copy of the article.
23. An Earth satellite is in an orbit with perigee altitude $h_p = 400$ km and eccentricity $e = 0.5$. Find (a) semimajor axis, (b) apogee radius, (c) apogee velocity, (d) perigee velocity, (e) period, (f) true anomaly when the radius to the satellite is equal to the semimajor axis, and (g) flight path angle when the radius to the satellite is equal to the semimajor axis.
    Answer: (a) 13 556 km, (b) 20 334 km, (c) 3.13 km/s, (d) 9.39 km/s, (e) 15 708 s, (f) $\pm 120°$, (g) 30°.
24. The distance, speed, and true anomaly of a meteoroid with respect to the Earth are 500 000 km, 2000 m/s, and 130 degrees. Find (a) the eccentricity, (b) the altitude, and (c) the speed at closest approach.
    Answer: (a) 1.2569, (b) $3.618\,26 \times 10^4$ km, (c) 4.597 km/s.
25. The periapsis of a spacecraft in a parabolic trajectory about the Earth is 8000 km. Find the distance between two points on the trajectory at distances from the center of the Earth of 10 000 km and 18 000 km.
    Answer: 12 720 km.
26. Determine the Greenwich Mean Sidereal Time for an event at 10 April 1987 at $19^h\,21^m$.
    Answer: $8^h\,34^m\,57.^s0896$.

References

Anderle, R. J., 1972. Pole position for 1971 based on Doppler satellite observations. *Tech. Rep. TR-2432*, Washington, DC: Nav. Weapons Lab.

Bate, R., D. Mueller, and J. White, 1971. *Fundamentals of Astrodynamics*. New York: Dover.

Battin, R. H., 1987. *An Introduction to the Mathematics and Methods of Astronautics*. Washington, DC: American Institute of Aeronautics and Astronautics (AIAA).

Black, H. D., 1978. An easily implemented algorithm for the tropospheric range correction. *J. Geophys. Res.*, **83**(B4): 1825–1828.

Black, H. D., 1990. Early development of Transit, the Navy navigation satellite system. *J. Guidance, Control, and Dynamics*, **13**(4): 577–585.

Black, H. D., A. Eisner, and J. Platt, 1984. Choosing a satellite constellation for the search and rescue satellite system. *NMEA* (Natl. Marine Electronics Assoc.) *News*, **17,** 82–90.

Boehm, B., 1964. Probabilistic evaluation of satellite missions involving ground coverage. *Celestial Mechanics and Astrodynamics*, vol. 14 of *Progress in Astronautics and Aeronautics,* V. G. Szebehely, ed. New York: Academic Press.

Breakwell, J. V., R. W. Gillespie, and S. Ross, 1961. Researches in interplanetary transfer. *ARS J.*, **31**: 201–208.

Brower, D., and G. M. Clemence, 1961. *Celestial Mechanics*. New York: Academic Press.

Cook, G. E., 1962. Luni-solar perturbations of the orbit of an Earth satellite. *Geophys. J. Roy. Astrono. Soc.*, **6**(3): 271–291.

Danby, J. M. A., 1962. *Foundations of Celestial Mechanics*. New York: Macmillan.

Defense Mapping Agency, 1 September 1991. *Department of Defense World Geodetic System 1984*, DMA TR 8350.2, second edition.

Deutsch, R., 1963. *Orbital Dynamics of Space Vehicles*. Englewood Cliffs, NJ: Prentice-Hall.

Draim, J., 1987. A common-period four-satellite continuous global coverage constellation. *J. Guidance, Control, and Dynamics*, **10**(5): 492–499.

Emara, E. T., and C. Leondes, 1977. Minimum number of satellites for three-dimensional world-wide coverage. *IEEE Trans. Aerosp. Electron. Syst.*, **AES-13**: 108–111.

Escobal, P. R., 1965. *Methods of Orbit Determination*. New York: John Wiley.

Fitzpatrick, P. M., 1970. *Principles of Celestial Mechanics*. New York: Academic Press.

Geyling, F. T., and H. R. Westerman, 1971. *Introduction to Orbital Mechanics*. Reading, MA: Addison-Wesley.

Giacaglia, G. E. O., 1973. *Lunar perturbations on artificial satellites of the Earth*. SAO Spec. Rep. 352, Cambridge, MA.

Henriksen, S., ed., 1977. *National Geodetic Satellite Program*. NASA SP 365, 2 vols.

Hopfield, H. S., 1980. Improvements in the tropospheric refraction correction for range measurements. *Phil. Trans. Roy. Soc. Lon. Ser. A*, **294**: 341–352.

International Earth Rotation Service, March 2002. *Explanatory Supplement to IERS Bulletins A and B*. Paris Service de la Rotation Terrestre Observatoire de Paris.

Jacchia, L. G., 1972. *Atmospheric models in the region from 110 to 2000 km*. COSPAR International Reference Atmosphere,1972. Berlin: Akademie-Verlag.

Kaula, W., 1966. *Theory of Satellite Geodesy; Applications of Satellites to Geodesy*, Waltham, MA: Blaisdell Publishing.

King-Hele, D. G., 1964. *Theory of Satellite Orbits in an Atmosphere*. London: Butterworths.

King-Hele, D. G., 1978. Methods for predicting satellite lifetimes. *J. Brit. Interplanet. Soc.*, **31**:181–196.

King-Hele, D., 1987. *Satellite Orbits in an Atmosphere. Theory and Applications*. Glasgow and London: Blackie.

King-Hele, D. G., and D. M. C. Walker, 1987. The contraction of satellite orbits under the influence of air drag. *Proc. Roy. Soc. Lond. A*, **414**: 271–295.

Luders, R. D., 1961. Satellite networks for continuous zonal coverage. *ARS J.*, 179–184.

National Imagery and Mapping Agency, 4 July 1997. *Department of Defense World Geodetic System 1984*. TR 8350.2, third edition. Bethesda, MD.

Perkins, F. M., 1958. An analytical solution for flight time of satellite in eccentric and circular orbits. *Astronaut. Acta,* **IV**: 113.

Pisacane, V. L., and S. C. Dillon, 1981. Determining coordinates of the rotational pole using satellite data from four sites. *J. Geophys. Res.*, **86**(2): 899–902.

Pisacane, V. L., B. B. Holland, and H. D. Black, 1973. Recent (1973) improvements in the Navy navigation satellite system. *Navigation, J. Ins. Navig.*, **20**(3): 224–229.

Seeber, G. 1993. *Satellite Geodesy, Foundations, Methods, and Applications*. Berlin: Walter deGruyter.

Seidelmann, P. K., ed., 1992. *Explanatory Supplement to the Astronomical Almanac*. Sansalite. CA: University Science Books.

Smart, W.M., 1953. *Celestial Mechanics*. London: Longmans, Green.

Staff of the Space Department, Johns Hopkins University, APL, and Staff of the Guidance and Control Laboratory, Stanford University, 1974. A satellite freed of all but gravitational forces: Triad I. *J. Spacecraft Rockets*, **11**: 637–644.

Sterne, T. E., 1960. *An Introduction to Celestial Mechanics*. New York: Interscience.

Taff, L. G., 1985. *Celestial Mechanics: a Computational Guide for the Practitioner*. New York: John Wiley and Sons.

Tisserand, F., 1889. *Traité de mechanique celeste*, vol. 1.

United States Naval Observatory, 2002. *The Astronomical Almanac* (published yearly). Washington, DC.

Vallado, D. A., 2001. *Fundamentals of Astrodynamics and Applications*, 2nd edition. El Segundo, CA: Microcosm/Kluwer.

Walker, J.G., 1973. Continuous whole Earth coverage by circular orbit satellites. *Proc. Satellite Systems for Mobile Communication and Surveillance, IEE Conference Public*, **95**: 35–38.

Walker J. G., 1984. Satellite constellations. *J. Brit. Interplanet. Soc.*, **37**: 559–572.

Weisel, W. E., 1989. *Spaceflight Dynamics*. New York: McGraw-Hill.

Yionoulis, S. M., 1965. Studies of the resonance effects due to the Earth's potential function. *J. Geophys. Res.*, **70**(4): 5991–5976. (See also **71**(4): 1289–1291.)

Yionoulis, S. M., 1967. The use of satellite Doppler data for geodetic purposes. *Les méthodes dynamiques de géodesie par satellites*. CNES Colloque International, Paris: 127–160.

4

Spacecraft Propulsion, Launch Systems, and Launch Mechanics

VINCENT L. PISACANE

4.1 Introduction

Propulsion systems are needed to place spacecraft into orbit, to maintain or change their trajectories, and to control their attitude. Rocket systems are self-contained in that they change stored energy into kinetic energy to produce thrust. Rocket systems are classified in a variety of ways: by their nature of their use (for example, launch, attitude control, orbit adjustment, apogee kick motor, station keeping), by the type of energy (chemical, electric, nuclear), and by the phase of the propellant (liquid, solid, hybrid).

This chapter first addresses the fundamental equations for the motion of a rocket and the thermodynamic processes that produce thrust. From these basic concepts, the motion of a rocket is addressed, including force-free motion with single and multiple stages, and then the effects of gravity are considered. Flight mechanics addresses the important issues associated with launch and transfer trajectories approximated by the Lagrange planetary equations, and the Hohmann and bielliptic transfers. This is followed by discussion of the characteristics of solid, liquid, electrical, and other rocket systems. The chapter concludes with a discussion of the sizing of a spacecraft's propellant system.

4.2 Fundamental Equations for Rocket Propulsion

Newton's second law of motion for a dynamical system in an inertial reference frame is

$$\mathbf{F} = \frac{d\mathbf{p}_I}{dt}, \tag{4.2.1}$$

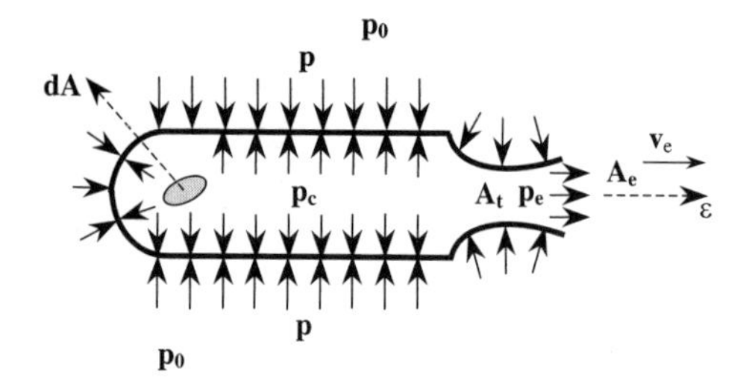

Figure 4.1 Force distribution on a rocket, ε = unit vector normal to exit cross-section, A_e = nozzle area, A_t = throat area, p_e = exhaust pressure, p_0 = ambient pressure, p = pressure on external surface, dA = differential area, A = surface area exclusive of nozzle exhaust area, $\mathbf{v}_e$ = exhaust velocity.

where $\mathbf{F}$ is total force, $\mathbf{p}_l$ is linear momentum, and t is time. The force distribution acting on the control volume of the rocket, or the exterior of a rocket in the atmosphere, is illustrated in figure 4.1, where

$$\mathbf{F} = -\int_{A_e} p_e d\mathbf{A} - \int_{A} p d\mathbf{A} + \mathbf{F}_b, \qquad (4.2.2)$$

which can be written as

$$\mathbf{F} = -(p_e - p_0)A_e\frac{\mathbf{v}_e}{v_e} - \int_{A} (p - p_0) d\mathbf{A} + \mathbf{F}_b, \qquad (4.2.3)$$

where p_e is the average nozzle exit pressure, p_0 is the ambient pressure, A_e is the nozzle exit area, $\mathbf{v}_e$ is the exhaust velocity with v_e its magnitude, $d\mathbf{A}$ is the differential surface area vector normal to the surface area, A is the surface area of the rocket exclusive of the nozzle exit area, $\mathbf{F}_b$ is a body force (typically, only gravity is assumed to act), and the identity $\int_{A+A_e} p_0 d\mathbf{A} = 0$ is used. Defining the aerodynamic force by

$$\mathbf{F}_a \equiv -\int_{A} (p - p_0) d\mathbf{A} \qquad (4.2.4)$$

permits equation 4.2.3 to be rewritten as

$$\mathbf{F} = -(p_e - p_0)A_e\frac{\mathbf{v}_e}{v_e} + \mathbf{F}_a + \mathbf{F}_b. \qquad (4.2.5)$$

This completes the determination of the forces that act on the rocket.

The time rate of change of momentum for the control volume, illustrated in figure 4.2, is

$$\frac{d\mathbf{p}_\ell}{dt} = \lim_{t\to 0}\frac{[(m + \Delta m)(\mathbf{v} + \Delta\mathbf{v}) + (-\Delta m)(\mathbf{v} + \mathbf{v}_e)] - m\mathbf{v}}{\Delta t} = m\dot{\mathbf{v}} - \dot{m}\mathbf{v}_e, \qquad (4.2.6)$$

where m is the mass of the rocket, $\dot{m}$ is the time derivative of the mass of the rocket, which is negative, $\mathbf{v}$ is the velocity of the rocket, and $\mathbf{v}_e$ is the exhaust velocity relative to the rocket. Substituting equations 4.2.5 and 4.2.6 into equation 4.2.1 gives

$$\mathbf{F}_a + \mathbf{F}_b - (p_e - p_0)A_e\frac{\mathbf{v}_e}{v_e} + \dot{m}\mathbf{v}_e = m\dot{\mathbf{v}}. \qquad (4.2.7)$$

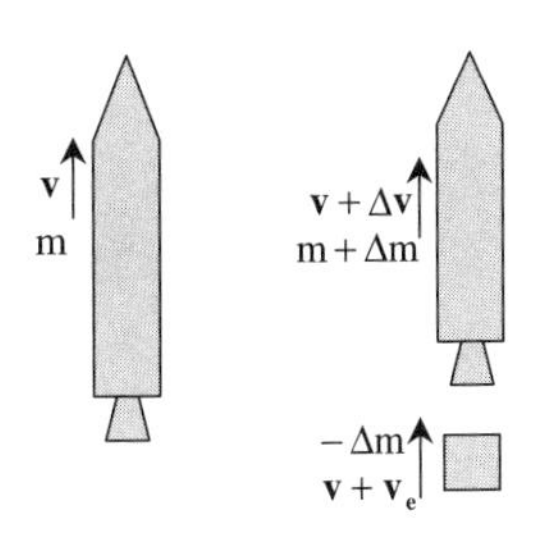

Figure 4.2 Rocket conservation of momentum. m = mass, $\mathbf{v}$ = rocket velocity, v_e = exhaust velocity relative to rocket, Δm = change in mass, Δv = change in velocity.

The propellant thrust $\mathbf{F}_p$ is defined to be

$$\mathbf{F}_p \equiv -(p_e - p_0)\, A_e \frac{\mathbf{v_e}}{v_e} + \dot{m}\mathbf{v}_e, \tag{4.2.8}$$

where the first term is known as the *pressure thrust* and the second term the *momentum thrust*. With $p_e > p_0$ and $\mathbf{v}_e$ being in the rearward or negative direction, both the pressure and momentum thrust are in the forward or positive direction. The majority of the thrust in actual rocket systems results from the momentum thrust. For v_e in the aft direction, the magnitude of the thrust in the forward direction follows simply as

$$F_p = (p_e - p_0)A_e - \dot{m}v_e = (p_e - p_0)A_e + \dot{m}_e v_e, \tag{4.2.9}$$

where the rate of change of the mass of the rocket $\dot{m}$ is equal to the negative of the change in the mass of the exhaust products or efflux $\dot{m}_e$ so that

$$\dot{m}_e = -\dot{m}. \tag{4.2.10}$$

It is convenient to express the propellant thrust as

$$\mathbf{F}_p = -(p_e - p_0)\, A_e \frac{\mathbf{v}_e}{v_e} - \dot{m}_e\mathbf{v}_e \equiv \dot{m}\mathbf{C} = -\dot{m}_e\mathbf{C}, \tag{4.2.11}$$

where $\mathbf{C}$, known as the *effective exhaust velocity*, is

$$\mathbf{C} = \mathbf{v}_e + (p_e - p_0)\frac{\mathbf{v}_e}{v_e}\frac{A_e}{\dot{m}_e}. \tag{4.2.12}$$

With the exhaust velocity in the rear direction, the magnitude of the thrust and the effective exhaust velocity follow as

$$F_p = (p_e - p_0)A_e + \dot{m}_e v_e = \dot{m}_e C, \tag{4.2.13}$$

$$C = v_e + (p_e - p_0)\frac{A_e}{\dot{m}_e}. \tag{4.2.14}$$

The equation of motion follows from equations 4.2.7 and 4.2.11 to be

$$\mathbf{F}_a + \mathbf{F}_b + \dot{m}\mathbf{C} = m\dot{\mathbf{v}}, \tag{4.2.15}$$

where $\dot{m}$ is negative.

Three situations arise in relation to the magnitude of the exhaust pressure, as illustrated in figure 4.3. If $p_e > p_0$, the efflux is *under-expanded* and expansion waves

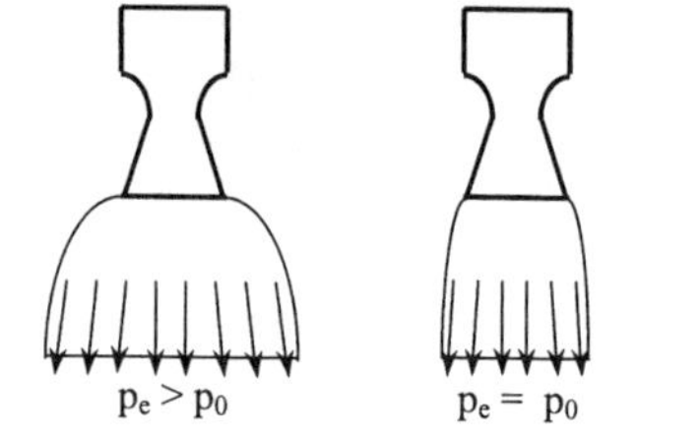

Figure 4.3 Effect of different exhaust velocities on exit flow.

are established at the nozzle exit to equalize the pressure. This results in an increase in the pressure thrust that only partially compensates for the decrease in momentum thrust. If $p_e = p_0$, the nozzle is said to be *correctly expanded* as the exhaust pressure of the nozzle matches the ambient pressure. In this case the thrust is a maximum. If $p_e < p_0$, the exhaust is said to be *over-expanded* and a negative pressure thrust results through compression shock waves that are established at the exit to the nozzle to equalize the pressure. When this occurs, the momentum thrust is increased but the increase only partially compensates for the decrease in the pressure thrust. Because of the tendency for self-compensation in both the under-expanded and over-expanded cases, the net force is relatively insensitive to small changes in pressure. In practice, rockets are usually designed to operate with the exhaust pressure equal to or slightly greater than the ambient pressure. In launch vehicles, whose rockets transcend different pressure environments with constant exhaust pressure, the thrust increases with increased altitude as the ambient pressure decreases. An adjustable nozzle, to correctly expand the flow and operate at the highest possible thrust, is not used in practice because of the complexities of the implementation. In space, the ambient pressure is zero so maximum thrust can be obtained. Rocket performance is often rated for operation at sea level and in a vacuum.

The thrust can also be expressed in terms of the *specific impulse* I_{sp}, usually in units of seconds, defined as the total impulse per unit weight of propellant consumed

$$I_{sp} = \frac{\int F_p dt}{g_0 \int \dot{m}_e dt}, \tag{4.2.16}$$

where g_0 is the standard acceleration of gravity, 9.806 65 m/s^2. Substituting equation 4.2.13 for F_p and assuming a constant propellant mass flow rate $\dot{m}_e$ yields a relationship between C and I_{sp}

$$I_{sp} = \frac{C \dot{m}_e \Delta t}{g_0 \dot{m}_e \Delta t} = \frac{C}{g_0}. \tag{4.2.17}$$

Table 4.1 provides effective exhaust velocities and specific impulses of typical spacecraft and launch rocket systems.

4.3 Thermodynamic Relations

This section develops the more important concepts for the thermodynamics of chemical propulsion. Consider the chamber in figure 4.1 in which a propellant is under pressure

Table 4.1 Typical effective velocities and specific impulses

Prepellant	Effective Exhaust Velocity(km/s)	Specific Impulse (a)
Cold gas	0.5–0.7	50–70
Monopropellants	2.0–2.2	200–225
Solids	2.5–2.9	250–300
Bipropellants	2.9–4.4	300–450
Electrothermal	2.9–4.9	300–500
Electrostatic	11.8–49.0	1200–5000
Electromagnetic	6.9–29.4	700–3000

and/or at elevated temperature so when the propellant is expanded through the nozzle it converts potential energy into kinetic energy. For this treatment, the assumption made is that the nozzle cross-section varies slowly so that the flow is essentially steady, axial, and has uniform properties over the cross-section. In addition, it is assumed that there is insufficient time for significant heat transfer, so the process can be treated as reversible adiabatic or isentropic for which the first law of thermodynamics, a generalization of the law of conservation of energy, is

$$h + \frac{1}{2}v^2 = \text{constant}, \tag{4.3.1}$$

where h is the specific enthalpy (energy/unit mass), equal to

$$h \equiv u + pV = u + \frac{p}{\rho}, \tag{4.3.2}$$

where v is the magnitude of the velocity, u is the specific internal energy (energy/unit mass), V is the specific volume (volume/unit mass), p is the pressure (force/unit area), and ρ is the mass density (mass/unit volume). For an ideal gas, the change in enthalpy between two states, with the initial state noted by i, can be expressed as

$$h_\text{i} - h = \frac{\gamma R(T_\text{i} - T)}{(\gamma - 1)M}, \tag{4.3.3}$$

where γ is the *specific heat ratio*, equal to the ratio of specific heat at constant pressure c_p to the specific heat at constant volume c_v, T is the absolute temperature, R is the universal gas constant (8314.472 J/kmol-K), and M is the molecular mass of the propellant (kg/kmol). A *kilomole* (kmol) is the mass of gas in kilograms equal to the molecular mass in atomic mass units (amu). The number of molecules in a kmol is equal to Avogadro's number $N_\text{A} = 6.022\ 141\ 99 \times 10^{26}$. For two points in the flow, equations 4.3.1 and 4.3.3 give

$$\frac{1}{2}v^2 - \frac{1}{2}v_\text{i}^2 = h_\text{i} - h = \frac{\gamma R(T_\text{i} - T)}{(\gamma - 1)M},$$

from which the velocity can be expressed as

$$v = \left[\frac{2\gamma R T_\text{i}}{(\gamma - 1)M}\left(1 - \frac{T}{T_\text{i}}\right) + v_\text{i}^2\right]^{1/2}. \tag{4.3.4}$$

For an isentropic process

$$\frac{T}{T_i} = \left(\frac{\rho}{\rho_i}\right)^{\gamma-1} = \left(\frac{p}{p_i}\right)^{\frac{\gamma-1}{\gamma}}, \quad p = \frac{\rho RT}{M}, \tag{4.3.5}$$

so equation 4.3.4 can be expressed as either

$$v = \left[\frac{2\gamma p_i}{(\gamma-1)\,\rho_i}\left(1-\left(\frac{p}{p_i}\right)^{\frac{\gamma-1}{\gamma}}\right) + v_i^2\right]^{1/2} \tag{4.3.6a}$$

or

$$v = \left[\frac{2\gamma}{\gamma-1}\frac{RT_i}{M}\left(1-\left(\frac{p}{p_i}\right)^{\frac{\gamma-1}{\gamma}}\right) + v_i^2\right]^{1/2} \tag{4.3.6b}$$

The velocity at any point in the nozzle, including at the exhaust, can be related to the conditions in the combustion chamber, with i = c, by neglecting v_c since it is small, so

$$v = \left[\frac{2\gamma}{\gamma-1}\frac{p_c}{\rho_c}\eta_c\right]^{1/2} = \left[\frac{2\gamma}{\gamma-1}\frac{RT_c}{M}\eta_c\right]^{1/2}, \tag{4.3.7}$$

where η_c, the *ideal cycle efficiency*, is defined by

$$\eta_c \equiv 1 - \frac{T}{T_c} = 1 - \left(\frac{p}{p_c}\right)^{\frac{\gamma-1}{\gamma}}. \tag{4.3.8}$$

Note that as $p/p_c \to 0$ or $T/T_c \to 0$ then $\eta_c \to 1$, indicating that the potential energy available is totally converted to kinetic energy. The exhaust velocity follows from equations 4.3.7 and 4.3.8 to be

$$v_e = \left[\frac{2\gamma}{\gamma-1}\frac{p_e}{\rho_c}\left(1-\left(\frac{p_e}{p_c}\right)^{\frac{\gamma-1}{\gamma}}\right)\right]^{1/2} \tag{4.3.9}$$

or

$$v_e = \left[\frac{2\gamma}{\gamma-1}\frac{RT_c}{M}\left(1-\left(\frac{p_e}{p_c}\right)^{\frac{\gamma-1}{\gamma}}\right)\right]^{1/2}. \tag{4.3.10}$$

Equations 4.3.9 and 4.3.10 illustrate that the exhaust velocity is a function of the pressure ratio p_e/p_c, the specific heat ratio γ, and either the ratio of the chamber pressure to mass density of the propellants or the ratio of the combustion chamber temperature to the molecular mass of the propellants. It follows from these equations that the exhaust velocity is:

(1) a maximum for a correctly expanded nozzle when the exhaust pressure and ambient pressure are nearly equal to zero, so the ideal cycle efficiency is close to one;

(2) increased by either increasing the combustion chamber temperature or decreasing the effective molecular mass of the propellant;
(3) increased by either increasing the combustion chamber pressure or decreasing the effective density of the propellant.

The specific impulse can be expressed in terms of the temperature in the combustion chamber and the molecular mass of the propellants from equations 4.2.14, 4.2.17, and 4.3.10 as

$$I_{sp} = \frac{1}{g_0}\left[v_e + (p_e - p_0)\frac{A_e}{\dot{m}_e}\right]$$
$$= \frac{1}{g_0}\left\{\left[\frac{2\gamma}{\gamma - 1}\frac{RT_c}{M}\left(1 - \left(\frac{p_e}{p_c}\right)^{\frac{\gamma-1}{\gamma}}\right)\right]^{1/2} + (p_e - p_0)\frac{A_e}{\dot{m}_e}\right\}. \tag{4.3.11}$$

The *Mach number* is a convenient dimensionless parameter defined as the ratio of the flow velocity v to the *local speed of sound* a by

$$\mathrm{M_a} \equiv \frac{v}{a}, \tag{4.3.12}$$

where the speed of sound in a gas is given by

$$a \equiv \left(\left[\frac{dp}{d\rho}\right]_s\right)^{1/2} = \left(\frac{\gamma RT}{M}\right)^{1/2}, \tag{4.3.13}$$

where the subscript s denotes an adiabatic process that signifies constant entropy. Consequently, the Mach number follows from equation 4.3.7 to be

$$\mathrm{M_a} = \left[\frac{2}{\gamma - 1}\frac{T_c}{T}\eta_c\right]^{1/2}, \tag{4.3.14}$$

which can be rewritten, using equation 4.3.5, to obtain

$$\frac{p}{p_c} = \left(\frac{\gamma - 1}{2}\mathrm{M_a^2} + 1\right)^{\frac{-\gamma}{\gamma-1}}. \tag{4.3.15}$$

The geometrical characteristics of the nozzle can be determined, using the Mach number, as follows. For a steady flow of velocity v, the mass flow rate $\dot{m}_e$ through any cross-section of area A of the nozzle must be constant and equal to

$$\dot{m}_e = \rho A v = \text{constant}, \tag{4.3.16}$$

where ρ is the density. Differentiating and dividing by $\rho A v$ gives

$$\frac{d\rho}{\rho} + \frac{dA}{A} + \frac{dv}{v} = 0. \tag{4.3.17}$$

The equation of *dynamic equilibrium* for an unsteady compressible flow, in the absence of external forces, is

$$\left.\begin{aligned} &\rho\frac{D\mathrm{v}}{Dt} + \nabla p = 0, \\ \text{where} \quad & \\ &\frac{D(\,)}{Dt} \equiv \nabla(\,)\cdot \mathrm{v} + \frac{\partial(\,)}{\partial t}. \end{aligned}\right\} \tag{4.3.18}$$

For steady one-dimensional adiabatic flow along a channel with distance ds it follows that

$$\left.\begin{aligned} &\mathbf{v} = v\boldsymbol{\varepsilon}, \quad d\mathbf{s} = ds\boldsymbol{\varepsilon}, \quad dv = (d\mathbf{s}/dt)\cdot\boldsymbol{\varepsilon}, \quad \nabla = \frac{\partial}{\partial s}\boldsymbol{\varepsilon}, \quad \frac{\partial v}{\partial t} = 0, \\ &\frac{D\mathbf{v}}{Dt}\cdot d\mathbf{s} = \left(\frac{\partial v}{\partial t}\boldsymbol{\varepsilon}\right)\cdot ds\ \boldsymbol{\varepsilon} = v dv, \\ &\nabla p \cdot d\mathbf{s} = \frac{\partial p}{\partial s}\boldsymbol{\varepsilon}\cdot ds\boldsymbol{\varepsilon} = dp, \end{aligned}\right\} \tag{4.3.19}$$

which, together with equation 4.3.13, permits equation 4.3.18a to be rewritten as

$$v dv + a^2\frac{d\rho}{\rho} = 0. \tag{4.3.20}$$

Substituting equations 4.3.13 and 4.3.20 into equation 4.3.17 gives

$$\frac{dv}{ds} = \frac{v}{A(\mathrm{M}_\mathrm{a}^2 - 1)}\frac{dA}{ds}. \tag{4.3.21}$$

This equation provides interesting results. The following cases are indicated, dependent on the Mach number $\mathrm{M_a}$:

- For $\mathrm{M_a} < 1$, the flow is subsonic and will decelerate ($dv/ds < 0$) in an expanding channel ($dA/ds > 0$) and accelerate ($dv/ds > 0$) in a converging channel ($dA/ds < 0$).
- For $\mathrm{M_a} = 1$, the flow is sonic and the change in area ($dA/ds = 0$) must be zero, so that sonic flow is not possible in either an expanding or a converging nozzle.
- For $\mathrm{M_a} > 1$, the flow is supersonic and will accelerate ($dv/ds > 0$) in an expanding channel ($dA/ds > 0$) and decelerate ($dv/ds < 0$) in a converging channel ($dA/ds < 0$).

Consequently, to produce supersonic flow from the combustion chamber, where the velocity is essentially zero, the nozzle must decrease in cross-sectional area to a minimum in the throat area where the flow velocity is sonic, and then increase in area. This type of nozzle is called the *de Laval nozzle* after the Swedish engineer who studied it extensively experimentally, producing supersonic flow velocities.

The pressure differential required for sonic flow to occur is obtained from equation 4.3.15, with $\mathrm{M_a} = 1$, to be

$$p_\mathrm{t} = p_\mathrm{c}\left(\frac{2}{\gamma + 1}\right)^{\frac{\gamma}{\gamma-1}}, \tag{4.3.22}$$

where p_t is known as the *critical pressure* and p_t/p_c is the *critical pressure ratio*. For $1.0 < \gamma < 1.4$, it follows that the critical pressure ratio is $0.6 > p_t/p_c > 0.5$. The velocity in the throat v_t is obtained by substituting this equation into equations 4.3.7 and 4.3.8 to obtain

$$v_t = \left[\frac{2\gamma}{\gamma+1}\frac{p_c}{\rho_c}\right]^{1/2} = \left[\frac{2\gamma}{\gamma+1}\frac{RT_c}{M}\right]^{1/2}. \tag{4.3.23}$$

The propellant mass flow rate at any point in the nozzle $\dot{m}_e$ can be obtained from equation 4.3.16, substituting equations 4.3.5 and 4.3.7, as

$$\dot{m}_e = \rho A v = p_c A\left(\frac{\rho}{p_c}\right) v = p_c A\left[\frac{2\gamma}{\gamma-1}\frac{M}{RT_c}\left(\frac{p}{p_c}\right)^{\frac{2}{\gamma}}\left(1-\left(\frac{p}{p_c}\right)^{\frac{\gamma-1}{\gamma}}\right)\right]^{1/2} \tag{4.3.24}$$

or

$$\dot{m}_e = \rho A v = \rho_c A v\left(\frac{\rho}{\rho_c}\right) = A\left[\frac{2\gamma\rho_c p_c}{\gamma-1}\left(\frac{p}{p_c}\right)^{\frac{2}{\gamma}}\left(1-\left(\frac{p}{p_c}\right)^{\frac{\gamma-1}{\gamma}}\right)\right]^{1/2}, \tag{4.3.25}$$

which are known as *Saint Venant's equation*. These equations show that changes in area A and pressure p must occur in a manner to keep $\dot{m}_e$ constant. It is convenient to write equation 4.3.25 in terms of conditions in the throat area A_t, where the Mach number is unity, $\rho = \rho_t$, $p = p_t$, and, using equation 4.3.22, to obtain

$$\dot{m}_e = A_t\left[\gamma\rho_c p_c\left(\frac{2}{\gamma+1}\right)^{\frac{\gamma+1}{\gamma-1}}\right]^{1/2}, \tag{4.3.26a}$$

or

$$\dot{m}_e = A_t p_c\left[\frac{\gamma M}{RT_c}\left(\frac{2}{\gamma+1}\right)^{\frac{\gamma+1}{\gamma-1}}\right]^{1/2}. \tag{4.3.26b}$$

Note that the propellant flow rate for sonic speed at the throat does not depend on conditions beyond the throat since it is impossible for disturbances to travel upstream faster than the speed of sound. The relative shape of the nozzle can be obtained from the ratio of the nozzle area at any location in the channel with respect to the throat area, by dividing equations 4.3.25 by 4.3.26a to obtain

$$\left(\frac{A}{A_t}\right) = \left(\frac{2}{\gamma+1}\right)^{\frac{\gamma+1}{2(\gamma-1)}}\left(\frac{p}{p_c}\right)^{\frac{1}{\gamma}}\left[\frac{2}{\gamma-1}\left(1-\left(\frac{p}{p_c}\right)^{\frac{\gamma-1}{\gamma}}\right)\right]^{\frac{1}{2}} \tag{4.3.27}$$

The area ratio can be determined solely in terms of the Mach mumber and the specific heat ratio by substituting equation 4.3.15 for the pressure ratio p/p_c to obtain

$$\left(\frac{A}{A_t}\right) = \frac{1}{\mathrm{M_a}}\left[\frac{(\gamma-1)\mathrm{M_a^2}+2}{\gamma+1}\right]^{\frac{\gamma+1}{2(\gamma-1)}}. \tag{4.3.28}$$

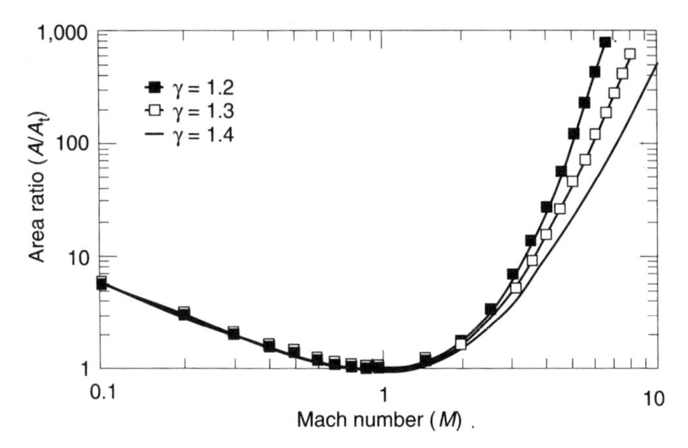

Figure 4.4 Area ratio as a function of Mach number. M = Mach number, A_t = throat area, A = area, γ = ratio of specific heats.

The area ratio as a function of Mach number with specific heat ratio a parameter is illustrated in figure 4.4.

The *expansion ratio* ε of a nozzle, defined as the ratio of the exhaust area to the throat area, is determined from equation 4.3.27 to be

$$\begin{aligned}\varepsilon &\equiv \left(\frac{A_e}{A_t}\right) \\ &= \left(\frac{2}{\gamma+1}\right)^{\frac{\gamma+1}{2(\gamma-1)}} \left(\frac{p_e}{p_c}\right)^{\frac{1}{\gamma}} \left[\frac{2}{\gamma-1}\left(1-\left(\frac{p_e}{p_c}\right)^{\frac{\gamma-1}{\gamma}}\right)\right]^{-1/2}, \qquad (4.3.29)\end{aligned}$$

where, for a properly expanded nozzle, the pressure p_e at the exit equals the ambient pressure p_0. The expansion ratio as a function of exit to combustion pressure ratio is illustrated in figure 4.5. Note that for space application $p_e \to 0$, and it follows from equation 4.3.29 that $A_e/A_t \to \infty$.

The thrust produced by the rocket, given by equation 4.2.9,

$$F_p = (p_e - p_0)A_e + \dot{m}_e v_e,$$

can be rewritten by substituting equation 4.3.26 for $\dot{m}_e$ and equation 4.3.9 for v_e to obtain

$$F_p = p_c A_t \left[\frac{2\gamma^2}{(\gamma-1)}\left(\frac{2}{\gamma+1}\right)^{\frac{\gamma+1}{\gamma-1}}\left(1-\left(\frac{p_e}{p_c}\right)^{\frac{\gamma-1}{\gamma}}\right)\right]^{1/2} + (p_e - p_0)A_e. \qquad (4.3.30)$$

It is sometimes convenient to express the thrust in terms of the *thrust coefficient* C_F, defined by

$$C_F \equiv \frac{F_p}{p_c A_t}, \qquad (4.3.31)$$

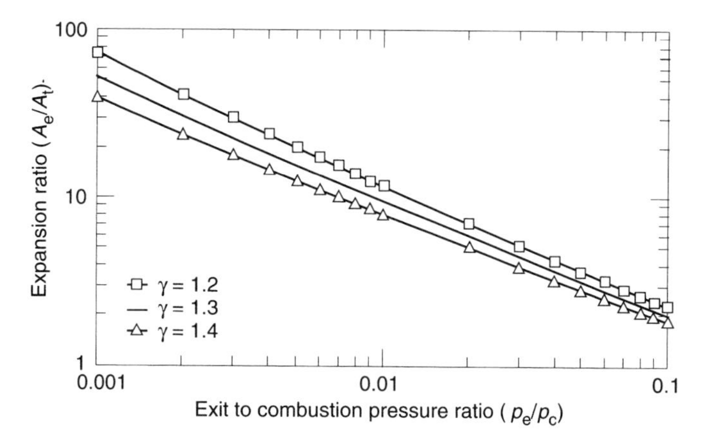

Figure 4.5 Expansion ratio as a function of exit to combustion pressure ratio, A_e = exit area, A_t = throat area, p_e = exit pressure, p_c = combustion chamber pressure, γ = ratio of specific heats.

and it follows from equation 4.3.30 that

$$C_F = \left[\frac{2\gamma^2}{(\gamma-1)} \left(\frac{2}{\gamma+1} \right)^{\frac{\gamma+1}{\gamma-1}} \left(1 - \left(\frac{p_e}{p_c} \right)^{\frac{\gamma-1}{\gamma}} \right) \right]^{1/2} + \left(\frac{p_e - p_0}{p_c} \right) \frac{A_e}{A_t}. \quad (4.3.32)$$

This equation gives $C_F(\gamma, p_e/p_c, p_0/p_c, A_e/A_t)$; however, equation 4.3.29 gives a relationship between A_e/A_t, p_e/p_c, and γ, so the parameters are not independent. Equation 4.3.32 can be expressed in terms of independent variables as $C_F(\gamma, p_e/p_c, p_0/p_c)$ or $C_F(\gamma, p_0/p_c, A_e/A_t)$. The thrust coefficient as a function of the expansion ratio for selected values of p_0/p_c for a given value of γ is illustrated in figure 4.6.

For a specified pressure ratio p_0/p_c, the thrust coefficient has a maximum value known as the *optimum thrust coefficient,* as illustrated in figure 4.6 as the maximum C_F of each of the individual curves for p_0/p_c. This occurs when $p_e = p_0$ which can be determined by differentiating equation 4.3.32 with respect to p_0/p_c and setting $dC_F/d(p_0/p_c)$ to zero. For a correctly expanded nozzle $p_e = p_0$ and, for small p_e/p_c, equation 4.3.32 gives for the maximum value of C_F

$$C_F|_{max} \approx \left[\frac{2\gamma^2}{(\gamma-1)} \left(\frac{2}{\gamma+1} \right)^{\frac{\gamma+1}{\gamma-1}} \right]^{1/2}. \quad (4.3.33)$$

The thrust coefficient is useful since it can be determined experimentally by measuring the thrust F_p, combustion chamber pressure p_c, and throat area A_t, since p_e/p_c is small.

Another useful parameter is the *characteristic exhaust velocity* C^*, verbalized as C-star, defined as the ratio of the effective exhaust velocity C to the thrust coefficient C_F, where

$$C^* \equiv \frac{C}{C_F} = \frac{F_p/\dot{m}_e}{F_p/p_c A_t} = \frac{p_c A_t}{\dot{m}_e}, \quad (4.3.34)$$

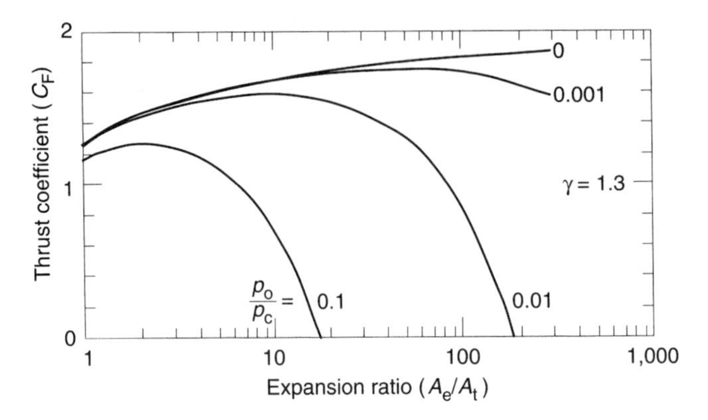

Figure 4.6 Thrust coefficient versus expansion ratio. A_e = exit area, A_t = throat area, γ = ratio of specific heats, p_0 = ambient pressure, p_c = combination pressure.

utilizing equations 4.2.13 and 4.3.30. C^*, like C_F, can be determined experimentally by measuring the combustion pressure, throat area, and propellant mass flow rate. By measuring both C^* and C_F, the effective exhaust velocity C can be determined from equation 4.3.34 and, from C, the specific impulse can be determined from equation 4.2.17. C^* expressed in terms of pressure or temperature follows from its definition by equation 4.3.34 and, substituting equation 4.3.26 for $\dot{m}_e$,

$$C^* = p_c^{1/2}\left[\gamma \rho_c \left(\frac{2}{\gamma+1}\right)^{\frac{\gamma+1}{\gamma-1}}\right]^{-1/2} = \left(\frac{\gamma R T_c}{M}\right)^{1/2}\left[\gamma^2 \left(\frac{2}{\gamma+1}\right)^{\frac{\gamma+1}{\gamma-1}}\right]^{-1/2}. \tag{4.3.35}$$

It is important to recall the assumptions that were made in obtaining the results of this section, these include:

(1) The mass flow is homogeneous, steady, and directed axially.
(2) Properties across a cross-section are constant.
(3) Perfect gas laws are valid and the process is reversible adiabatic, that is, isentropic.
(4) Specific heat ratio remains constant.
(5) Velocity in the combustion chamber is small enough that it can be ignored.

With these assumptions, the results predicted are generally accurate to better than 10%. Two examples follow that illustrate the use of the above developments.

Example

Rockets used to control the attitude of a spacecraft produce a force of 1 N at a duty cycle of 0.1%. Molecular hydrogen is used as the working fluid and is heated to 2000 K in the combustion chamber. If a one-year supply of propellant is required, what is the mass of hydrogen needed for each thruster? The specific heat of molecular hydrogen is 1.4.

Solution: The mass m_e(kg) required for each thruster is

$$m_e = \dot{m}_e t f, \tag{a}$$

where $\dot{m}_{\mathrm{e}}(\mathrm{kg/s})$ is the average mass flow rate during an impulse, $t(\mathrm{s})$ is the time duration of the mission, and $f(\%)$ is the duty cycle. Since $p_{\mathrm{e}} \ll p_{\mathrm{c}}$, the propellant mass flow rate is determined from equations 4.2.13, 4.3.33, 4.3.34, and 4.3.35 to be

$$\dot{m}_{\mathrm{e}} = \frac{F_{\mathrm{p}}}{C} = \frac{F_{\mathrm{p}}}{C_{\mathrm{F}}C^*} = F_{\mathrm{p}}\left[\frac{M}{\gamma R T_{\mathrm{c}}}\frac{\gamma-1}{2}\right]^{1/2}, \tag{b}$$

where for molecular hydrogen $M = 2.016$ kg/kmol, so that

$$\dot{m}_{\mathrm{e}} = 1\mathrm{N}\left[\frac{2.016\ \mathrm{kg/kmol}}{1.4 \times 8314.472\ \mathrm{J/K\ kmol} \times 2000\ \mathrm{K}} \times \frac{1.4-1}{2}\right]^{1/2}$$

$$= 1.3276 \times 10^{-4}\ \mathrm{kg/s}. \tag{c}$$

For the mission duration of $t = 1$ year and a duty cycle of $f = 0.1\%$, equation (a) gives the mass required for each thruster to be

$$m_{\mathrm{e}} = \dot{m}_{\mathrm{e}} t f = \left(1.3276 \times 10^{-4}\mathrm{kg/s}\right)(365.25 \times 86\,400\ \mathrm{s})(0.001) = 4.19\ \mathrm{kg}. \tag{d}$$

Example

Determine the (1) throat velocity, (2) exhaust velocity, (3) propellant mass flow rate, (4) throat area, (5) exit area, and (6) exit temperature for an ideal rocket that is to operate at an altitude of 30 km where the ambient pressure is $0.02 \times 10^6\ \mathrm{N/m^2}$. The rocket generates a thrust of 5000 N with a propellant having a ratio of specific heats of 1.3, a mean molecular mass of 24 kg/kmol, and combustion chamber pressure and temperature of $2 \times 10^6\ \mathrm{N/m^2}$ and 3000 K respectively.

Solution:

(1) *Throat velocity*: equation 4.3.23 gives the throat velocity where the Mach number is unity to be

$$v_{\mathrm{t}} = \left[\frac{2\gamma}{\gamma+1}\frac{RT_{\mathrm{c}}}{M}\right]^{1/2}$$

$$= \left[\frac{2 \times 1.3}{1.3+1}\frac{8314.472\ \mathrm{J/(K\ kmol)} \times 3000\ \mathrm{K}}{24\ \mathrm{kg/kmol}}\right]^{1/2}$$

$$= 1083.9\ \mathrm{m/s} \tag{a}$$

(2) *Exhaust velocity* is given by equation 4.3.10 where

$$v_{\mathrm{e}} = \left\{\frac{2\gamma}{\gamma-1}\frac{RT_{\mathrm{c}}}{M}\left[1-\left(\frac{p_{\mathrm{e}}}{p_{\mathrm{c}}}\right)^{\frac{\gamma-1}{\gamma}}\right]\right\}^{1/2}$$

$$= \left\{\frac{2 \times 1.3}{1.3-1}\frac{8314.472\ \mathrm{J/(K\ kmol)} \times 3000\ \mathrm{K}}{24\ \mathrm{kg/kmol}}\left[1-\left(\frac{0.02}{2}\right)^{\frac{1.3-1}{1.3}}\right]\right\}$$

$$= 2428\ \mathrm{m/s} \tag{b}$$

(3) *Propellant mass flow rate* is given by equation 4.2.13 which, for a correctly expanded nozzle ($p_e = p_0$), reduces to

$$\dot{m}_e = \frac{F_p}{(p_e - p_0)A_e + v_e} = \frac{F_p}{v_e} = \frac{5000 \text{ N}}{2428 \text{ m/s}} = 2.059 \text{ kg/s} \tag{c}$$

(4) *Throat area* is given by equation 4.3.30 with $p_e = p_0$, as

$$\begin{aligned} A_t &= \frac{F_p}{p_c}\left[\frac{2\gamma^2}{(\gamma-1)}\left(\frac{2}{\gamma+1}\right)^{\frac{\gamma+1}{\gamma-1}}\left(1-\left(\frac{p_e}{p_c}\right)^{\frac{\gamma-1}{\gamma}}\right)\right]^{-1/2} \\ &= \frac{5000 \text{ N}}{2 \times 10^6 \text{ N/m}^2}\left[\frac{2 \times 1.3^2}{(1.3-1)}\left(\frac{2}{1.3+1}\right)^{\frac{2.3}{0.3}}\left(1-\left(\frac{0.02}{2}\right)^{\frac{0.3}{1.3}}\right)\right]^{-1/2} \\ &= 0.1573 \times 10^{-2} \text{ m}^2 (15.73 \text{ cm}^2) \end{aligned} \tag{d}$$

(5) *Exit area*: the expansion ratio area can be determined from equation 4.3.29 as

$$\begin{aligned} \varepsilon &= \left(\frac{A_e}{A_t}\right) = \left(\frac{2}{\gamma+1}\right)^{\frac{\gamma+1}{2(\gamma-1)}}\left(\frac{p_e}{p_c}\right)^{\frac{1}{\gamma}}\left[\frac{2}{\gamma-1}\left(1-\left(\frac{p_e}{p_c}\right)^{\frac{\gamma-1}{\gamma}}\right)\right]^{\frac{1}{2}} \\ &= \left(\frac{2}{1.3+1}\right)^{\frac{1+1.3}{2(1.3-1)}}\left(\frac{0.02}{2}\right)^{\frac{1}{1.3}}\left[\frac{2}{1.3-1}\left(1-\left(\frac{0.02}{2}\right)^{\frac{1.3-1}{1.3}}\right)\right]^{\frac{1}{2}} \\ &= 9.68. \end{aligned}$$

from which it follows from equations (d) and 4.3.29 that the nozzle exit area is

$$\begin{aligned} A_e &= 9.68 A_t = 9.68 \times 0.1573 \times 10^{-2} \text{ m}^2 \\ &= 1.5227 \times 10^{-2} \text{ m}^2 (152.27 \text{ cm}^2) \end{aligned} \tag{e}$$

(6) *Exit temperature* is given by equation 4.3.5 as

$$T_e = T_c\left(\frac{p_e}{p_c}\right)^{\frac{\gamma-1}{\gamma}} = 3000 \text{ K}\left(\frac{0.02}{2}\right)^{\frac{1.3-1}{1.3}} = 1036.5 \text{ K} \tag{f}$$

4.4 Nozzles

Nozzles typically have a circular or rectangular cross-section with a variety of longitudinal characteristics. The actual nozzle length is determined by many considerations, such as constraints on size and mass and the required nozzle efficiency. The efficiency or effectiveness of a nozzle increases for smaller divergent angles following the throat because less of the propellant is directed sideward.

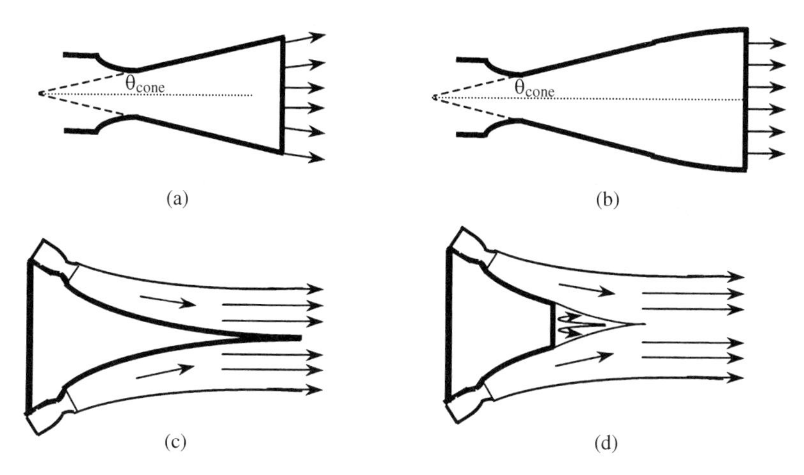

Figure 4.7 Nozzle schematics: (a) conical; (b) contour; (c) spike; (d) aerospike.

Cone or *conical* nozzles have dimensions following the throat that increase linearly with length. Typical divergence angles are between 10°and 20°, with about 15° the typical value. The length of a circular conical nozzle can be approximated directly from its geometry, as illustrated in figure 4.7a, where the length L is given by

$$L = \frac{D_e - D_t}{2 \tan \theta_{cone}}, \tag{4.4.1}$$

where D_e is the diameter of the exit, D_t is the diameter of the throat, and θ_{cone} is the divergent angle. A correction factor of less than one for the thrust can be used to account for the flow orthogonal to the axis of the nozzle. The correction factor, being the ratio of the actual thrust to the ideal thrust, is given by

$$\lambda \equiv \frac{\text{actual thrust}}{\text{ideal thrust}} = \frac{1}{2}(1 + \cos \theta_{cone}), \tag{4.4.2}$$

where θ_{cone} is the half-angle of the conical nozzle. This is left to be derived in the problems at the end of the chapter. As θ_{cone} increases, the efficiency decreases. For a nozzle with a divergence angle of 30°, so that θ_{cone} = 15°, the correction factor for the momentum thrust is 0.983. Although not adopted in practice, the conical nozzle can be utilized as a *two-step nozzle* where another cone, collinear with the basic nozzle, is initially retracted and is extended when appropriate to increase the exit area and decrease the exit pressure. Issues with this approach are the seal between the two nozzles, the added mass, and the complication of the extension mechanism.

The popular *bell* or *contour* shaped nozzle has a diameter that initially increases more rapidly than that of a conical nozzle and then increases less rapidly, so that at the end of the nozzle the diameter is almost constant with a half-angle of just a few degrees, as illustrated in figure 4.7b. This directs more of the flow aft, at the expense of a nozzle that is longer, heavier, and more costly to fabricate. In principle, the contour nozzle can be extended into a two-step nozzle like the conical nozzle, but this is not employed in practice. Most spacecraft and launch propulsion systems use the contour nozzle.

The *spike nozzle* is configured as either annular or rectangular with the inner surface a real boundary and the outer surface an aerodynamic boundary, as illustrated in figure 4.7c. This allows the aerodynamic boundary to expand as the pressure decreases and contract as the pressure increases to form a nozzle with varying geometry. The disadvantage of the spike nozzle is the long spike that protrudes for some distance and is relatively massive. The *plug nozzle* is a spike nozzle with a portion of the spike cut off. The *aerospike nozzle* (figure 4.7d) is a plug nozzle in which lower temperature gases are expelled where the spike is terminated to aerodynamically extend the inner surface of the nozzle.

4.5 Force-Free Rocket Motion

In this section, the force-free motion of both single-stage and multiple-stage or multiple-step rocket systems is considered in the absence of external forces. Although restrictive, this treatment does provide an understanding of the fundamental concepts. The equation of motion and the equation for the effective exhaust velocity, equations 4.2.15 and 4.2.17, give

$$m\dot{v} = -C\dot{m} \quad \text{where} \quad \mathrm{C} = \mathrm{I_{sp}g_0}. \tag{4.5.1}$$

This can be rewritten as

$$m\frac{dv}{dt} = -C\frac{dm}{dt} \quad \text{or} \quad dv = -C\frac{dm}{m}, \tag{4.5.2}$$

and if the effective exhaust velocity C is assumed constant, integration gives

$$\begin{aligned} \Delta v = v(t) - v_i &= -C \ln \frac{m(t)}{m_\mathrm{i}} \\ &= C \ln \frac{m_\mathrm{i}}{m(t)}, \end{aligned} \tag{4.5.3}$$

where Δv is the increase in velocity, $v(t)$ is the velocity at time t, v_i is the initial velocity, $m(t)$ the mass at time t, and m_i the initial mass. Recall that the logarithm of a number less than one is negative. When the propellant is expended, with the final mass of the rocket m_f and the final velocity v_f, equation 4.5.3 gives

$$\Delta v_\mathrm{f} = v_\mathrm{f} - v_\mathrm{i} = -C \ln \frac{m_\mathrm{f}}{m_\mathrm{i}} = -C \ln r_\mathrm{m} = C \ln r_\mathrm{m}^{-1}, \tag{4.5.4}$$

where r_m is the *mass fraction*, defined by

$$r_\mathrm{m} \equiv \frac{m_\mathrm{f}}{m_\mathrm{i}}. \tag{4.5.5}$$

Equation 4.5.4 is known as the *force-free rocket equation*. It shows that the change in velocity depends on the ratio of the final to the initial mass, not on the individual magnitude of either. The initial mass m_i and final mass m_f can be further partitioned into the propellant mass m_p, structural mass m_s, and payload mass m_ℓ, where

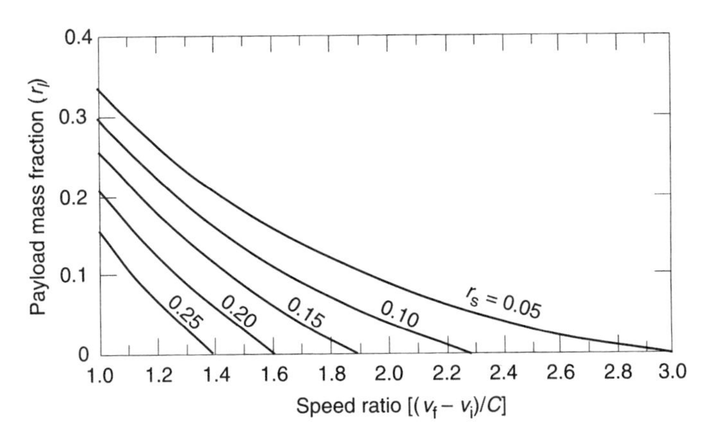

Figure 4.8 Payload mass fraction r_l as function of $\Delta v/C$. Δv = change in velocity, C = effective exhaust velocity, r_s = structural mass fraction.

$$m_i = m_p + m_s + m_l, \quad m_f = m_s + m_l, \tag{4.5.6}$$

so the mass fraction r_m can be expressed by

$$r_m = r_s(1 - r_l) + r_l = r_l(1 - r_s) + r_s, \tag{4.5.7}$$

where

$$r_s \equiv \frac{m_s}{m_s + m_p} \quad \text{and} \quad r_l \equiv \frac{m_l}{m_i}, \tag{4.5.8}$$

where r_s is the *structural mass fraction* and r_l is the *payload mass fraction*. This development is left as an exercise in the problems at the end of the chapter. The change in velocity, given by equation 4.5.4, can then be expressed as

$$\begin{aligned} \Delta v_f &= v_f - v_i = -C \ln r_m \\ &= -C \ln[r_s(1 - r_l) + r_l] = -C \ln[r_l(1 - r_s) + r_s]. \end{aligned} \tag{4.5.9}$$

The maximum Δv_f occurs when $r_l \to 0$ and $r_s \to 0$, which is not practical. A typical structural mass fraction of 10% with no payload gives an upper bound of

$$\Delta v_f|_{r_l=0, r_s=0.1} = -C \ln 0.1 = 2.3\, C. \tag{4.5.10}$$

The payload mass capability follows from equation 4.5.9 as

$$r_l = \frac{\exp(-\frac{\Delta v}{C}) - r_s}{1 - r_s}, \tag{4.5.11}$$

and is illustrated in figure 4.8.

Multiple-stage or multiple-step rocket systems are employed to reduce the deleterious effects of the structural mass. Dividing the propulsion system into distinct stages allows the structural masses of the spent stages to be separated, reducing the mass that must be accelerated by the remaining stages. To analyze the performance of multiple-stage rockets, let the stages be numbered in increasing order, according to the sequence in which they are used, by $k(k = 1, 2, 3\ldots)$. It is important to note that, as defined, a stage includes all subsequent stages. Similar to that for a single stage, let $m_{i,k}$ and $m_{f,k}$ be the initial and final mass of the kth stage so that, by definition,

$$m_{\mathrm{i},k} = m_{\mathrm{p},k} + m_{\mathrm{f},k}, \quad m_{\mathrm{f},k} = m_{\mathrm{s},k} + m_{\mathrm{i},k+1}, \quad k = 1, 2, \ldots, n, \tag{4.5.12}$$

where $m_{\mathrm{p},k}$ and $m_{\mathrm{s},k}$ are the masses of the propellant and structure, respectively, of the kth stage. The mass fractions for multiple stages follow as

$$\left.\begin{aligned} r_{\mathrm{m},k} &\equiv \frac{m_{\mathrm{f},k}}{m_{\mathrm{i},k}} = \frac{m_{l,k} + m_{\mathrm{s},k}}{m_{\mathrm{i},k}}, \\ r_{\mathrm{s},k} &\equiv \frac{m_{\mathrm{s},k}}{m_{\mathrm{s},k} + m_{\mathrm{p},k}}, \\ r_{l,k} &\equiv \frac{m_{l,k}}{m_{\mathrm{i},k}} = \frac{m_{\mathrm{i},k+1}}{m_{\mathrm{i},k}}, \end{aligned}\right\} \tag{4.5.13}$$

where $r_{\mathrm{m},k}$, $r_{\mathrm{s},k}$ and $r_{l,k}$ are the mass fraction, structural mass fraction, and payload mass fraction, respectively, of the kth stage, so that

$$r_{\mathrm{m}.k} = r_{\mathrm{s},k}(1 - r_{l,k}) + r_{l,k} = r_{l,k}(1 - r_{\mathrm{s},k}) + r_{\mathrm{s},k}. \tag{4.5.14}$$

The change in velocity of the k^{th}stage is then

$$\Delta v_k = v_{\mathrm{f},k} - v_{\mathrm{i},k} = -C_k \ln r_{\mathrm{m},k}, \tag{4.5.15}$$

$$\Delta v_k = -C_k \ln[r_{\mathrm{s},k}(1 - r_{l,k}) + r_{l,k}] = -C_k \ln[r_{l,k}(1 - r_{\mathrm{s},k}) + r_{\mathrm{s},k}], \tag{4.5.16}$$

where C_k is the effective exhaust velocity of the kth stage. The final velocity of the rocket, after all the propellant stages have been expended, can be determined by recognizing that the final velocity of the kth stage is equal to the initial velocity of the $(k+1)^{\mathrm{th}}$ stage,

$$v_{\mathrm{f},k} = v_{\mathrm{i},k+1}. \tag{4.5.17}$$

The final velocity $\Delta v_{\mathrm{f},n}$ when the last stage has been expended follows as

$$\begin{aligned} \Delta v_{\mathrm{f},n} &= v_{\mathrm{f},n} - v_{\mathrm{i},1} = (v_{\mathrm{f},n} - v_{\mathrm{i},n}) + (v_{\mathrm{f},n-1} - v_{\mathrm{i},n-1}) + \cdots + (v_{\mathrm{f},1} - v_{\mathrm{i},1}) \\ &= \sum_{k=1}^{n} (v_{\mathrm{f},k} - v_{\mathrm{i},k}), \end{aligned} \tag{4.5.18}$$

so that, substituting equations 4.5.15 and 4.5.16,

$$\begin{aligned} \Delta v_{\mathrm{f},n} &= -\sum_{k=1}^{n} C_k \ln r_{\mathrm{m},k} \\ &= -\sum_{k=1}^{n} C_k \ln[r_{\mathrm{s},k}(1 - r_{l,k}) + r_{l,k}] \\ &= -\sum_{k=1}^{n} C_k \ln[r_{l,k}(1 - r_{\mathrm{s},k}) + r_{\mathrm{s},k}]. \end{aligned} \tag{4.5.19}$$

In the unique case that the various mass fractions and the effective exhaust velocities are the same for each stage, then

$$r_s = r_{s,k} \quad r_l = r_{l,k} \quad C = C_k, \tag{4.5.20}$$

the total change in velocity is

$$\begin{aligned} \Delta v_{f,n} &= v_{f,n} - v_{i,1} = -nC \ln r_m, \\ &= -nC \ln[r_s(1 - r_l) + r_l] = -nC \ln[r_l(1 - r_s) + r_s], \end{aligned} \tag{4.5.21}$$

and the payload mass fraction r_l for each stage follows as

$$r_l = \frac{\exp\left(-\dfrac{\Delta v_{f,n}}{nC}\right) - r_s}{1 - r_s}. \tag{4.5.22}$$

The *overall payload mass fraction*, denoted by r_l^* and defined as the ratio of the payload of the nth stage $m_{l,n}$ to the initial mass of the rocket m_I, is

$$\begin{aligned} r_l^* &= \frac{m_{l,n}}{m_{i,1}} = \frac{m_{i,n+1}}{m_{i,n}} \frac{m_{i,n}}{m_{i,n-1}} \frac{m_{i,n-1}}{m_{i,n-2}} \cdots \frac{m_{i,3}}{m_{i,2}} \frac{m_{i,2}}{m_{i,1}} \\ &= r_{l,n} r_{l,n-1} r_{l,n-2} r_{l,n-3} \cdots r_{l,1} = \prod_{k=1}^{n} r_{l,k}. \end{aligned} \tag{4.5.23}$$

From equations 4.5.22 and 4.5.23, the overall payload mass fraction follows as

$$r_l^* = r_l^n = \left[\frac{\exp\left(-\dfrac{\Delta v_{f,n}}{nC}\right) - r_s}{1 - r_s} \right]^n. \tag{4.5.24}$$

If the quantity in the bracket, the individual payload mass fraction, is negative, it represents an unrealistic case. The value for r_l^* for an infinite number of stages can be determined from equation 4.5.24 to be

$$r_l^*|_{n\to\infty} = \exp\left(-\frac{\Delta v_{f,n}}{C(1 - r_s)} \right), \tag{4.5.25}$$

so that

$$\Delta v_{f,n}|_{n\to\infty} = -C(1 - r_s) \ln r_l^*. \tag{4.5.26}$$

This derivation is left as an exercise in the problems set at the end of the chapter. The last two equations give the ultimate potential of staging when the payload mass fraction is the same for each stage. The benefit of staging to increase the payload mass fraction is illustrated in figure 4.9. An example that follows indicates the advantages of staging.

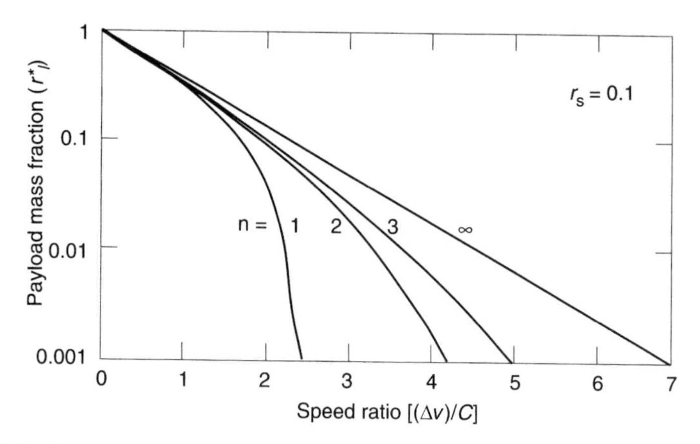

Figure 4.9 Payload mass fraction for an n-stage rocket, assuming that characteristics of each stage are identical and structural mass fraction $r_{\mathrm{S}} = 0.1$. Δv = change in velocity, C = effective exhaust velocity, n = number of stages.

Example

(a) A multiple-stage rocket system has identical characteristics for each stage with the structural mass fraction 0.1. Starting from rest, it is desired to achieve an increase in velocity of $\Delta v_n / C = 2$. Determine the ratio of the overall payload mass fraction for multiple stages. (b) Then repeat for $\Delta v_{\mathrm{n}} / C = 4$.

Solution: Equations 4.5.24 and 4.5.25 give:

(a)

n	r^*_l	Ratio
1	0.0393	1.00
2	0.0886	2.25
3	0.0969	2.47
4	0.1003	2.55
5	0.1022	2.60
⋮	⋮	⋮
∞	0.1084	2.76

(b)

n	r^*_l	Ratio
1	–	–
2	0.0015	1.00
3	0.0060	4.00
4	0.0078	5.20
5	0.0088	5.87
⋮	⋮	⋮
∞	0.0117	7.80

The ratio is with respect to the payload mass fraction of the minimum number of stages. This example illustrates that there is little advantage in utilizing more than two or three stages in case (a) whereas multiple stages are required to satisfy the conditions of case (b) but the gain diminishes with increasing number of stages.

Because of the complexity of staging, most launch systems utilize three to four stages. The retired Scout launch vehicle uses four solid rocket stages with the option of a fifth stage.

4.6 Rocket Motion with Gravity

Launch vehicles must act against the force of gravity. This can be approximated by an average value for the narrow range of altitudes over which the stage of a launch vehicle generally operates. For vertical motion, the one-dimensional rocket equation given by equation 4.5.1 becomes

$$m\dot{v} = -C\dot{m} - mg, \tag{4.6.1}$$

where m is the mass of the rocket, v its velocity, $\dot{m}$ is the rate of change of the rocket mass and is negative, C is the effective exhaust velocity, and g is the acceleration of gravity. Multiplication by dt and division by m yields

$$dv = -C\frac{dm}{m} - gdt, \tag{4.6.2}$$

and integration, assuming C and g to be constant, yields

$$\Delta v = v(t) - v_\mathrm{i} = -C \ln \frac{m}{m_\mathrm{i}} - gt, \tag{4.6.3}$$

where t is the elapse time of powered fight, v_i is the initial velocity, and m_i is the initial mass. When all of the propellant is expended

$$\Delta v_\mathrm{f} = v_\mathrm{f} - v_\mathrm{i} = -C \ln \frac{m_\mathrm{f}}{m_\mathrm{i}} - gt_\mathrm{f}, \tag{4.6.4}$$

where v_f is the final velocity, m_f is the final mass, and t_f is the duration of powered flight. In terms of the mass fractions defined by equations 4.5.5 and 4.5.7, this equation can be written as

$$\Delta v_\mathrm{f} = v_\mathrm{f} - v_\mathrm{i} = -C \ln r_\mathrm{m} - gt_\mathrm{f} \tag{4.6.5}$$

$$= -C \ln[r_l(1 - r_\mathrm{s}) + r_\mathrm{s}] - gt_\mathrm{f} = -C \ln[r_\mathrm{s}(1 - r_l) + r_l] - gt_\mathrm{f}. \tag{4.6.6}$$

To find the height at the end of powered flight, it is necessary to specify the fuel consumption in order to determine the time of powered flight t_f. If the propellant mass flow rate $\dot{m}_e$ is assumed constant, an analytical solution can be found from equations 4.2.13 and 4.5.5 to be

$$t_f = \frac{m_p}{\dot{m}_e} = \frac{C}{g_0 r_t}(1 - r_m), \tag{4.6.7}$$

where the *thrust-to-initial-weight ratio* r_t is defined by

$$r_t \equiv \frac{F_p}{m_i g_0}. \tag{4.6.8}$$

With this, equation 4.6.5 can be rewritten as

$$\Delta v_f = v_f - v_i = -C \ln r_m - \frac{gC}{g_0 r_t}(1 - r_m) \tag{4.6.9}$$

$$= -C \ln[r_l(1 - r_s) + r_s] - \frac{gC}{g_0 r_t}(1 - r_s)(1 - r_l) \tag{4.6.10}$$

Example

Find the increase in velocity for a rocket that takes off vertically from the Earth with $r_l = 0$, $r_s = 0.1$, $g = g_0$, for different values of r_t.

Solution: It follows from the problem statement and equation 4.6.10 that

$r_t =$ 0.5	1.0	2.0	3.0	4.0	...	∞
$\delta v/C =$ 0.5	1.40	1.85	2.00	2.08	...	2.30

As r_t increases, the burn time decreases and the change in velocity increases. If $r_t < 1$, the rocket will expend propellant without moving until $r_t = 1$, which signifies that the thrust is equal to the gravitational force on the rocket.

The height h_f that a rocket travels during powered flight from time 0 to t_f is given by integrating equation 4.6.3 to obtain

$$h_f = -C \int_0^{t_f} \ln \frac{m}{m_i} dt - \frac{1}{2} g t^2 + v_i t, \tag{4.6.11}$$

where it is again assumed that C and g are constants. The integral in equation 4.6.11 can be written for constant rate of change of mass $\dot{m}$ as

$$\int_0^{t_f} \ln \frac{m}{m_i} dt = \frac{m_i}{\dot{m}} \int_{m_i}^{m_f} \ln \frac{m}{m_i} d\left(\frac{m}{m_i}\right)$$

$$= \frac{m_i}{\dot{m}} \left(\frac{m_f}{m_i} \ln \frac{m_f}{m_i} - \frac{m_f}{m_i} + 1 \right), \tag{4.6.12}$$

and using

$$\dot{m} = \frac{(m_\mathrm{f} - m_\mathrm{i})}{t_\mathrm{f}}, \tag{4.6.13}$$

the height at the end of powered flight is

$$h_\mathrm{f} = -Ct_\mathrm{f}\left(\frac{r_\mathrm{m}\ln r_\mathrm{m}}{r_\mathrm{m}-1} - 1\right) - \frac{1}{2}gt_\mathrm{f}^2 + v_\mathrm{i}t_\mathrm{f}. \tag{4.6.14}$$

The vehicle will coast above this height until the gravitational force reduces the velocity to zero. The coast height of the rocket can be obtained by conservation of energy

$$\frac{1}{2}m_\mathrm{f}v_\mathrm{f}^2 - \frac{m_\mathrm{f}gR^2}{R+h_\mathrm{f}} = -\frac{m_\mathrm{f}gR^2}{R+h_\mathrm{f}+h_\mathrm{c}}, \tag{4.6.15}$$

where R is the radius of the planet, g is the acceleration of gravity at the distance $R+h_\mathrm{f}$ from the center of the planet, and h_c is the coast height that can be expressed from equation 4.6.15 as

$$h_\mathrm{c} = \frac{\frac{1}{2}\frac{v_\mathrm{f}^2}{gR^2}(R+h_\mathrm{f})^2}{1 - \frac{1}{2}\frac{v_\mathrm{f}^2}{gR^2}(R+h_\mathrm{f})}. \tag{4.6.16}$$

If it can be assumed that the height of powered flight $h_\mathrm{f}$is small relative to the radius from the center of the body R, then $h_\mathrm{f} \ll R$ and equation 4.6.16 can be approximated by

$$h_\mathrm{c} \approx \frac{\frac{1}{2}\frac{v_\mathrm{f}^2}{g}}{1 - \frac{1}{2}\frac{v_\mathrm{f}^2}{gR}}. \tag{4.6.17}$$

If it can further be assumed that the specific kinetic energy at the end of powered flight $\frac{1}{2}v_\mathrm{f}^2$ is small relative to the specific potential energy gR, then $\frac{1}{2}v_\mathrm{f}^2 \ll gR$, so that a further approximation is that

$$h_\mathrm{c} \approx \frac{1}{2}\frac{v_\mathrm{f}^2}{g} \tag{4.6.18}$$

which is applicable when the height of powered flight and the coast height relative to the planetary radius are small, so that the gravitational force can be considered constant and independent of height. Generally, the powered flight may be short enough for equation 4.6.17 to be valid, but the coast height is generally too large for the approximation given by equation 4.6.18 to be sufficiently accurate.

The effect of gravity on a single-stage rocket can be generalized to the multiple-stage rocket as follows. Equation 4.6.5 for the k^{th}stage is

$$\Delta v_{\mathrm{f},k} = v_{\mathrm{f},k} - v_{\mathrm{i},k} = -C_k \ln r_{\mathrm{m},k} - g_k t_{\mathrm{f},k}, \tag{4.6.19}$$

where, for the kth stage, $v_{\text{f},k}$ and $v_{\text{i},k}$ are the final and initial velocities, C_k is the effective exhaust velocity, $r_{\text{m},k}$ is the mass fraction, g_k is the acceleration of gravity, and $t_{\text{f},k}$ is the burn time. The total change in velocity for n stages follows, as before, as

$$\begin{aligned}\Delta v_{\text{f},n} &= \sum_{k=1}^{n} \Delta v_{\text{f},k} = -\sum_{k=1}^{n} C_k \ln r_{\text{m},k} - \sum_{k=1}^{n} g_k t_{\text{f},k} \\ &= -\sum_{k=1}^{n} C_k \ln[r_{l,k}(1 - r_{\text{s},k}) + r_{\text{s},k}] - \sum_{k=1}^{n} g_k t_{\text{f},k} \\ &= -\sum_{k=1}^{n} C_k \ln[r_{\text{s},k}(1 - r_{l,k}) + r_{l,k}] - \sum_{k=1}^{n} g_k t_{\text{f},k}.\end{aligned} \tag{4.6.20}$$

These equations reduce to equations 4.5.19 in the absence of the gravitational field.

The results of this section permit the characteristics of the vertical motion of a multistage rocket in a gravitational field to be determined. Gravity acts to oppose the motion and consequently to reduce the maximum velocity and therefore determines the height the rocket can achieve. The longer the burn time, the lower the final velocity at the end of powered flight, as indicated by equations 4.6.5 and 4.6.20. The loss in velocity is only partially compensated by the higher altitude achieved while the rocket is burning. A real loss of energy of the rocket occurs for longer burn times. This can be explained by noting that the energy expended in raising the propellant to a higher altitude is not recovered and is lost to the rocket. The increase in potential energy of the expended propellant is equal to the loss in energy of the rocket. Consequently, in the absence of an atmosphere a short burn time, so that maximum velocity of the rocket is achieved at a lower altitude, will result in a higher altitude at the end of the coast phase. This can be demonstrated by the following example.

Example

Compare the altitude achieved by a two-stage rocket in a vertical trajectory in a uniform gravitational field, first by a burn–coast–burn–coast scenario and then by a burn–burn–coast scenario, where the burnout velocities of the two stages are v_1 and v_2 respectively. Assume the burns are instantaneous.

Solution: With instantaneous burn times the height of powered flight h_f is zero, and with a constant gravitational field equation 4.6.18 is applicable and gives, for the burn-coast–burn–coast scenario, the sum of the two coast heights as

$$h_\text{c}|_{\text{burn-coast-burn-coast}} = \frac{1}{2}\frac{v_1^2}{g} + \frac{1}{2}\frac{v_2^2}{g} = \frac{v_1^2 + v_2^2}{2g}. \tag{a}$$

For the burn–burn–coast scenario the coast height follows as

$$h_\text{c}|_{\text{burn-burn-coast}} = \frac{(v_1 + v_2)^2}{2g}. \tag{b}$$

The burn–burn–coast scenario will result in a higher altitude than the burn—coast–burn–coast scenario, since

$$h_c|_{\text{burn-burn-coast}} = h_c|_{\text{burn-coast-burn-coast}} + \frac{v_1 v_2}{g}. \tag{c}$$

4.7 Launch Flight Mechanics

4.7.1 Introduction

Launch vehicles are usually accelerated from a vertical position and must overcome aerodynamic and gravitational forces to place a payload into a specified orbit. The launch trajectory in three-dimensional space depends on the launch site, safety constraints on the trajectory, and the initial conditions of the spacecraft orbit that the launch vehicles try to achieve. The launch trajectory is governed by equation 4.2.15:

$$\mathrm{F_a} + \mathrm{F_b} + \dot{m}\mathrm{C} = m\dot{v}$$

This equation for the acceleration $\dot{v}$ can be integrated numerically, requiring the aerodynamic force F_a to be specified as a function of position, velocity, geometry, and attitude; the body force F_b, in this case the gravitational force, is to be specified as a function of the mass and the position with respect to the Earth; the magnitude and direction of the propulsive thrust $\dot{m}C$; also required are the mass of the rocket m, and the boundary conditions, in other words, the initial position and velocity at launch and the position and velocity at burnout which are the initial condition of the final orbit.

The determination of the ascent trajectory is an optimization problem of determining the propellant thrust as a function of time, given the initial and final boundary conditions and subjected to a variety of constraints. These constraints may include a limit on the propellant mass flow rate, number of stages allowed, mass of each stage, final payload mass, trajectory restrictions for safety, and angle of attack to minimize aerodynamic torques. Determination of the optimal solution requires sophisticated computer simulations well beyond the intended scope of this treatment. In the case of solid rocket motors, the technique includes the use of energy-wasting maneuvers to achieve the orbit injection parameters.

4.7.2 Gravity-Turn Trajectory

To minimize the structural mass required for launch from a celestial body with an atmosphere, the launch vehicle generally flies a *gravity-turn, zero-angle-of-attack,* or *zero-lift* trajectory while in the atmosphere. In this case, the thrust vector is programmed to be in the opposite direction to the velocity vector except immediately after lift-off when the rocket is moving slowly and a small-angle adjustment is employed

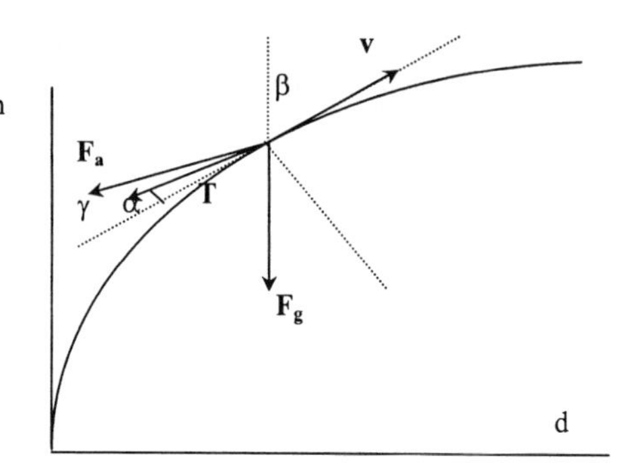

Figure 4.10 Gravity-turn trajectory geometry, $\mathbf{F}_\mathrm{a}$ = aerodynamic force, γ = deviation of aerodynamic force from aft, $\mathbf{F}_\mathrm{g}$ = gravitational force, $\mathbf{v}$ = velocity, β = deviation of velocity vector from vertical, $\mathbf{T}$ = thrust vector, α = deviation of thrust vector from aft, d = downrange, h = altitude.

to produce a small horizontal component of velocity. For motion in a plane, illustrated in figure 4.10, the equation of motion, 4.2.15, can be written using

$$\begin{aligned} v &= v \sin\beta\, \boldsymbol{\varepsilon}_\mathrm{d} + v\cos\beta\, \boldsymbol{\varepsilon}_\mathrm{h}, \\ \mathbf{F_a} &= -F_\mathrm{a}\sin(\beta+\gamma)\boldsymbol{\varepsilon}_\mathrm{d} - F_\mathrm{a}\cos(\beta+\gamma)\boldsymbol{\varepsilon}_\mathrm{h}, \\ \mathbf{C} &= -C\sin(\beta+\alpha)\boldsymbol{\varepsilon}_\mathrm{d} - C\cos(\beta+\alpha)\boldsymbol{\varepsilon}_\mathrm{h}, \\ \mathbf{F}_\mathrm{g} &= -mg\boldsymbol{\varepsilon}_\mathrm{h}, \end{aligned} \tag{4.7.1}$$

where

$\mathbf{C}$ = effective exhaust velocity,
$\boldsymbol{\varepsilon}_\mathrm{d}, \boldsymbol{\varepsilon}_\mathrm{h}$ = downrange and altitude unit vectors,
$\mathbf{F}_\mathrm{a}$ = aerodynamic force,
g = acceleration of gravity,
m = mass of rocket,
v = velocity of the rocket,
α = angle of effective exhaust velocity from the antivelocity vector,
β = deviation of the velocity vector from the vertical,
γ = angle of aerodynamic force from the antivelocity vector,

so that

$$\left.\begin{aligned} \dot{v} &= -\frac{\dot{m}}{m}C\cos\alpha - \frac{F_\mathrm{a}}{m}\cos\gamma - g\cos\beta, \\ \dot{\beta} &= \frac{\dot{m}}{m}\frac{C}{v}\sin\alpha - \frac{F_\mathrm{a}}{mv}\sin\gamma + \frac{g}{v}\sin\beta. \end{aligned}\right\} \tag{4.7.2}$$

For a gravity-turn trajectory, $\alpha = \gamma \approx 0$, it follows that

$$\dot{v} = -\frac{\dot{m}C + F_\mathrm{a}}{m} - g\cos\beta, \quad \dot{\beta} = \frac{g}{v}\sin\beta. \tag{4.7.3}$$

The downrange distance and altitude follow from

$$d = \int_0^t v\sin\beta\, dt, \quad h = \int_0^t v\cos\beta\, dt. \tag{4.7.4}$$

The second equation of 4.7.3 shows that if the rocket is initially in a vertical position, so that the initial value of β is zero, $d\beta/dt$ is also zero and β will remain zero for all time. To achieve downrange distance, there must be a deviation from vertical at liftoff or soon thereafter so that $\dot{\beta}$ is nonzero. The initial angle, $\beta(t \approx 0)$, taken to be small, is known as the *initial kick angle* and is accomplished by a pitch maneuver soon after the launch vehicle clears the platform and is moving with a low velocity. Since $\dot{\beta}$ is inversely proportional to the velocity and directly proportional to sin β, $\dot{\beta}$ will increase more slowly as the velocity increases and faster as β increases.

Even if the drag term is ignored, equations 4.7.3 are nonlinear and no general solution is available. An analytical solution can be obtained in two cases, when: (a) the ratio of the acceleration of gravity g to the pitch rate is constant, and (b) the acceleration of gravity and the ratio of the sum of the aerodynamic and propulsive force to the mass are constant. Each case is considered in turn.

In the first case, when the ratio of the acceleration of gravity to the pitch rate is constant, the pitch rate $\beta/\dot{\beta}$ is constant. Differentiating the second of equations 4.7.3 gives

$$\dot{v} = g \cos \beta, \tag{4.7.5}$$

and substituting this in the first equation of 4.7.3 to eliminate β gives

$$\dot{v} = -\frac{\dot{m}C + F_a}{2m}. \tag{4.7.6}$$

With $v = v_i$ at t = 0, integration gives

$$v - v_i = -\int_0^t \frac{\dot{m}C + F_a}{2m} dt. \tag{4.7.7}$$

This is half of the velocity that would be achieved in the absence of gravity, see equation 4.7.3.

In the second case, when the ratio of the sum of the aerodynamic and propulsive force to the mass $(\dot{m}C + F_a/2m)$ and the acceleration of gravity g are constant, it is convenient to represent the ratio in equation 4.7.3 as

$$-\frac{\dot{m}C + F_a}{2m} = \text{constant} \equiv fg. \tag{4.7.8}$$

Dividing the first equation of 4.7.3 by the second gives

$$\frac{dv}{v} = (f \csc \beta - \cot \beta) d\beta,$$

which can be integrated to give the velocity as a function of the angle β as

$$v = k \csc \beta \left(\tan \frac{\beta}{2}\right)^f, \tag{4.7.9}$$

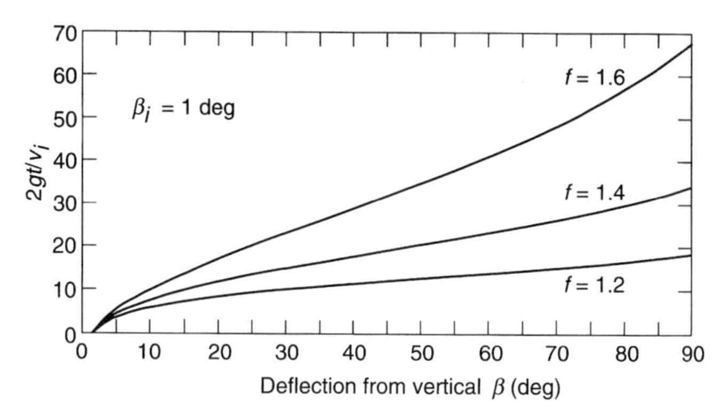

Figure 4.11 Gravity-turn trajectory for constant net specific thrust, g = acceleration of gravity, t = time, $\mathbf{v}_i$ = initial velocity, f = dimensionless parameter defining net specific thrust.

where k is the constant of integration. Note that when $\beta = \pi/2$, v is equal to k. The constant k can be determined from the initial conditions at $t = 0$, denoted by $(\ )_i$, to be

$$k = v_i \sin\beta_i \left(\cot\frac{\beta_i}{2}\right)^f. \tag{4.7.10}$$

Substituting equation 4.7.9 into the second equation of 4.7.3 and integrating gives the elapse time as a function of the angle β as

$$t = \frac{k}{2g}\left[\frac{(\tan\frac{\beta}{2})^{f+1}}{f+1} + \frac{(\tan\frac{\beta}{2})^{f-1}}{f-1}\right]_{\beta_i}^{\beta}. \tag{4.7.11}$$

With β a function of time in equation 4.7.11, equation 4.7.9 can be used to determine the velocity v as a function of time. Combining equations 4.7.10 and 4.7.11 eliminates k to give time as

$$\frac{2gt}{v_i} = \sin\beta_i \left(\cot\frac{\beta_i}{2}\right)^f \left[\frac{(\tan\frac{\beta}{2})^{f+1}}{f+1} + \frac{(\tan\frac{\beta}{2})^{f-1}}{f-1}\right]_{\beta_i}^{\beta}. \tag{4.7.12}$$

This equation is illustrated in figure 4.11 where it can be seen that, to achieve a given value of β, a longer time is required for larger values of the net specific force parameter f.

4.7.3 Ascent into Orbit

There are several sequences in the trajectory of a launch vehicle. The initial rise of the vehicle occurs at a fixed angle, usually zero. Typically, about 15s or so after launch the initial guidance or pitch maneuver is initiated, with a rate depending on the intended orbit injection parameters. This puts the vehicle into a gravity-turn trajectory until the vehicle is out of the atmosphere. The final guidance depends on the nature of the upper stages. Liquid propellant engines use variable burn times and solid rocket motors use energy-wasting maneuvers to achieve the injection parameters of the target orbit.

The point during ascent when the launch vehicle is subjected to maximum dynamic pressure is known as Max Q and produces the largest loads on the launch vehicle. A deployable fairing is used to protect payloads from the aerodynamic and thermal environments during launch and is separated when the aerodynamic force is at an acceptable magnitude, generally above 100 km altitude. For example, the three-piece Titan IV fairing is made of a composite material, 5.08 m in diameter, ranging from 17.1 to 26.2 m in length and from 3600 to 6300 kg in mass. The fairing is generally deployed at the lowest possible safe altitude to maximize the payload mass. However, spacecraft failures have occurred when the fairing was deployed at too low an altitude, resulting in excessive aerodynamic heating.

4.7.4 Launch Site

Launch sites are selected to provide a field of view over which launch vehicles can be safely aborted without damage to life or property. For example, the Eastern Space and Missile Center (ESMC) at Cape Canaveral Air Force Station, Florida, at latitude 28.5°, has over 40 launch complexes. It has an easterly field of view ranging in azimuth from 37° to 114°, resulting in direct (without plane change) launch inclinations from 28.5° to 57° (figure 4.12). Smaller or larger azimuth angles would launch a spacecraft over habitable land masses, adversely affect safety options for abort or vehicle separation conditions, or present the undesirable possibility that expendable components could land on foreign land or sea space. Other inclinations require additional maneuvers and the use of propellant mass, reducing the payload or orbital capability. The maximum payload mass is obtained by launching directly east to take advantage of the rotation rate of the Earth, resulting in an inclination of 28.5°. The eastward rotation rate of the Earth as a function of latitude ϕ provides a velocity of

$$v_{\text{east}} = 0.46 \cos\phi \text{ km/s}. \tag{4.7.13}$$

Higher inclination launches in the United States take place from the Western Space and Missile Center (WSMC) at Vandenberg Air Force Base, California, at a latitude of 34.6°; this center has approximately 52 launch complexes. Safe launches are to the south with azimuths from 158° to 201°, resulting in direct orbit inclinations from 70° to 104°. A launch due south would produce an inclination of about 94°, due to the rotation of the Earth.

4.7.5 Launch Window

The launch window is the interval of time during which a launch may take place that will achieve the target orbit. Launch windows generally range from a few seconds to minutes. The capability to launch in a narrow window is achieved by building into the countdown *holds,* which are periods of time that provide buffers to the launch countdown. For Earth satellites, the launch time principally determines the right ascension of the ascending node, which is the angle between the equinox and the line of nodes of the orbit defined by the intersection of the spacecraft with the equatorial plane as it travels north. If the orientation of the line of nodes is to be maintained with respect to the Sun, then the launch time would be the same universal (civil) time each day if there is a delay in

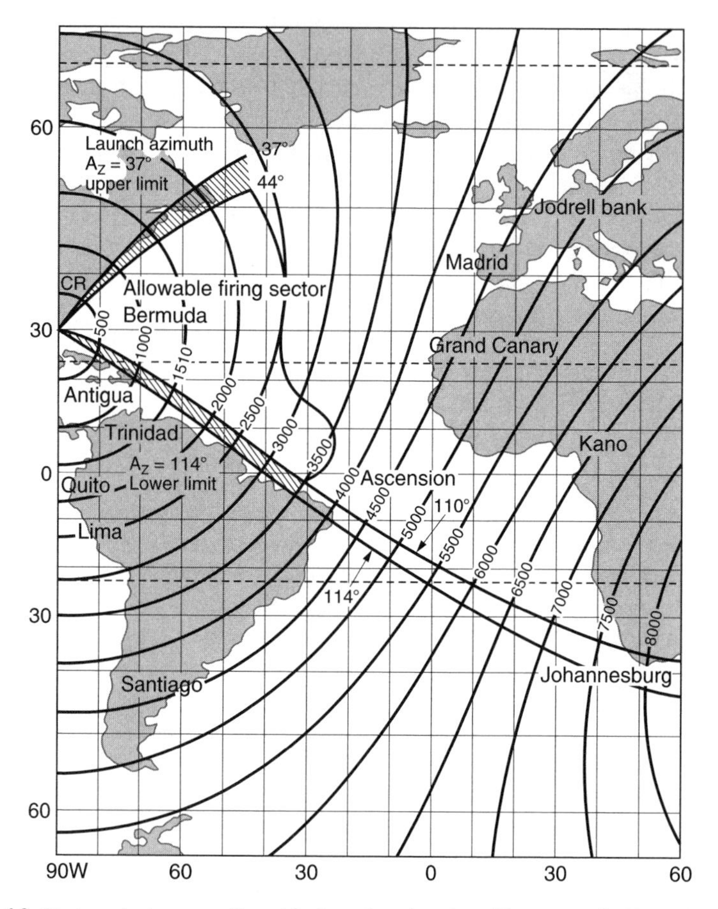

Figure 4.12 Eastern test range allowable launch azimuths. (Courtesy of USAF 45 SW/Safety Office, Chapter 2 EWR 127-1.)

the launch. If the line of nodes is to be maintained with respect to the equinox, such as in establishing or replacing spacecraft in a constellation, the launch time would be at the same sidereal time each day, so that solar time would slip by 3 minutes 56 seconds each day that a launch is delayed. If a rendezvous is planned with another spacecraft, the precession of the orbit of the other spacecraft must also be taken into account. For an interplanetary launch, the window is constrained first to a number of weeks by the relative location of the Earth and the targeted celestial body or bodies and within an hour or so to take advantage of the Earth's rotational motion.

4.8 Transfer Trajectories

4.8.1 Introduction

To change the orbital elements, the Lagrange planetary equations discussed in chapter 3 can be used to determine in- and out-of-plane orbital specific impulses, which are equivalent to a change in velocity, Δv. The assumption of an impulse is generally accurate as most propellant burns are for short periods of time compared to the period of the orbit. For changes in the trajectory relative to a local vertical reference system, which is especially important for rendezvous, the Euler–Hill equations, also provided in chapter 3, can be integrated to determine the specific impulses required. For changes to several of the classical orbital elements, it is generally not possible to determine an optimum solution except by trial and error. At the same time as the amount of propellant is determined, the minimum and maximum thrust must be determined to establish the nozzle(s) mass flow rates so that the thrusters can be sized. The maximum required thrust determines the throat area of the nozzle. The minimum Δv determines the precision to which the thrust must be controlled. When these requirements are incompatible in a single nozzle, then multiple nozzles are employed: a nozzle with a large throat area to achieve the maximum thrust and one with a small throat area to satisfy the precision to which the final velocity is required. The margins used for the equivalent Δv are a function of the characteristics of the specific mission but range from 10–20% for orbit corrections to 100% for attitude control systems. For in-plane changes from one circular orbit to another, both the Hohmann transfer and the bielliptic transfer can be utilized with the minimum required Δv. For out-of-plane or a combination of in- and out-of-plane transfer the Lagrange planetary equations are utilized to determine the required $\Delta \nu$.

The specific energy of a spacecraft is given by equation 3.3.24 of chapter 3:

$$\varepsilon = \frac{1}{2}v^2 - \frac{\mu}{r} = -\frac{\mu}{2a} \approx \frac{1}{2}v^2 - \frac{Gm_e}{r}, \tag{4.8.1}$$

from which the vis viva equation, 3.3.26, follows as

$$v = \sqrt{\mu\left(\frac{2}{r} - \frac{1}{a}\right)}, \tag{4.8.2}$$

where

$$\begin{aligned}
a &= \text{semimajor axis,}\\
G &= \text{universal gravitational constant,}\\
m_e &= \text{mass of Earth,}\\
m_0 &= \text{mass of satellite,}\\
r &= \text{radius from center of Earth,}\\
v &= \text{velocity,}\\
\varepsilon &= \text{specific energy,}\\
\mu &= G(m_e + m_0) \approx G\, m_e.
\end{aligned}$$

It follows from equation 4.8.1 that the change in the specific energy $\delta\varepsilon$ of a spacecraft as the result of an applied $\Delta\nu$ at any place in the orbit defined by position r is

$$\delta\varepsilon = \nu\delta\nu, \tag{4.8.3}$$

which indicates that the increase in specific energy is the greatest when the change in velocity is applied at the moment the spacecraft has its maximum velocity. For example, for an elliptical orbit the change in specific energy will be a maximum if the change in velocity occurs at periapsis. This can be explained by conservation of energy in that the propellant will reside at the radius at which it is released which would have lower specific potential energy at periapsis, so the specific kinetic energy of the spacecraft must be greater than if expended elsewhere. As a result, most interplanetary trajectories carry out a major thrusting maneuver relatively close to the Earth rather than at a further distance where the velocity would be reduced.

4.8.2 Lagrange Planetary Equations

The Lagrange planetary equations, given as equations 3.6.3 in chapter 3, constitute the method of choice for determining the magnitude and direction of the specific thrust impulses and where in the orbit to best apply them. These equations can be rewritten as

$$\left.\begin{aligned}
\Delta a &= \frac{2}{n\sqrt{1-e^2}}[R\Delta t\, e \sin f + T\Delta t(1+e\cos f)],\\
\Delta e &= \frac{\sqrt{1-e^2}}{na}\left[R\Delta t \sin f + T\Delta t\left(\frac{e+\cos f}{1+e\cos f}+\cos f\right)\right],\\
\Delta i &= \frac{W\Delta t\, r\cos(f+\omega)}{na^2\sqrt{1-e^2}} = \frac{W\Delta t\sqrt{1-e^2}\cos(f+\omega)}{na(1+e\cos f)},\\
\Delta\Omega &= \frac{W\Delta t\, r\sin(f+\omega)\csc i}{na^2\sqrt{1-e^2}} = \frac{W\Delta t\sqrt{1-e^2}\sin(f+\omega)\csc i}{na(1+e\cos f)},\\
\Delta\omega &= \frac{\sqrt{1-e^2}}{nae}\left[-R\Delta t\cos f + T\Delta t\frac{2+e\cos f}{1+e\cos f}\sin f\right.\\
&\quad\left. - W\Delta t\frac{e\sin(f+\omega)\cot i}{(1+e\cos f)}\right],\\
\Delta M &= n - \frac{(1-e^2)}{na}\left[R\Delta t\left(\frac{2}{1+e\cos f}-\frac{\cos f}{e}\right)\right.\\
&\quad\left. + T\Delta t\left(\frac{2+e\cos f}{1+e\cos f}\right)\frac{\sin f}{e}\right],
\end{aligned}\right\} \tag{4.8.4}$$

where a is the semimajor axis, e the eccentricity, i the inclination, Ω the right ascension of the ascending node, ω the argument of periapsis, M the mean anomaly, f the true anomaly, n the mean motion, and R, T, and W are the specific forces (force divided by the mass of the spacecraft) in the radial, along-track, and cross-track directions respectively. As can be seen from the equations, there are specific locations in the orbit and directions of the specific impulse that are most effective for changing a particular orbital element. For example, a change in inclination can only be accomplished by a specific impulse in the cross-track direction, it is best accomplished at the equator when $f+\omega = 0$ and this will leave Ω and ω unchanged. It cannot be accomplished when the spacecraft reaches its extreme latitudes where $f+\omega = \pm\pi/2$.

For specific impulses that occur over intervals that are not small relative to the orbital period, as with some electrical thrusters, it is necessary to integrate the equations. As an example, the change in semimajor axis for a specific impulse in the radial direction

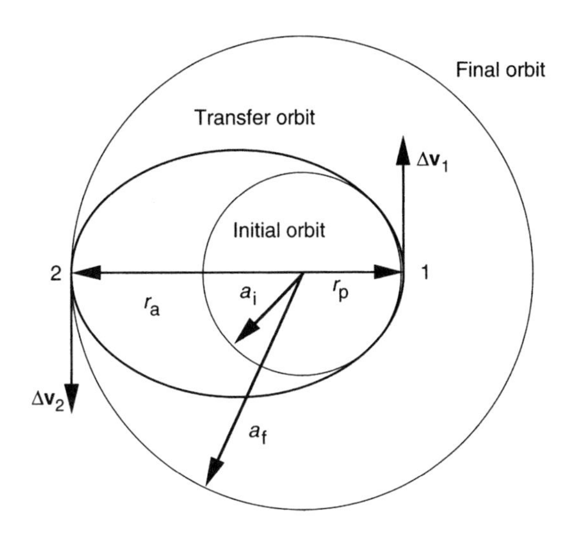

Figure 4.13 Hohmann transfer orbit, a_f = semimajor axis of final orbit, r_p = periapsis radius of transfer orbit, r_a = apoapsis radius of transfer orbit, a_i = semimajor axis of original orbit, Δv_i = change in velocity.

integrated over one full revolution gives zero change since the average value of sin f is zero.

4.8.3 Hohmann Transfer

Transfer between two coplanar circular orbits can be accomplished by a two-impulse maneuver known as the *Hohmann transfer*. The Hohmann transfer orbit is an elliptical orbit that is tangent to both the initial orbit and the final orbit. Consider the outward transfer shown in figure 4.13 where impulses are applied at the points 1 and 2. The periapsis, r_p, and apoapsis, r_a, radii of the transfer orbit follow from equations 3.3.51 and 3.3.52 of chapter 3 to give

$$r_\mathrm{a} = a(1+e) = a_\mathrm{f}, \tag{4.8.5}$$

$$r_\mathrm{p} = a(1-e) = a_\mathrm{i}, \tag{4.8.6}$$

where a_i, a, and a_f are the semimajor axes of the initial, transfer, and final orbits respectively. The eccentricity e of the transfer orbit can be determined from equations 4.8.5 and 4.8.6 as

$$e = \frac{r_\mathrm{a} - r_\mathrm{p}}{r_\mathrm{a} + r_\mathrm{p}} = \frac{a_\mathrm{f} - a_\mathrm{i}}{a_\mathrm{f} + a_\mathrm{i}}. \tag{4.8.7}$$

The velocity at periapsis v_p for the transfer orbit follows from equations 4.8.2, 4.8.6, and 4.8.7 to be

$$v_\mathrm{p}^2 = 2\mu\left(\frac{1}{a_\mathrm{i}} - \frac{1}{2a}\right) = \frac{\mu}{a_\mathrm{i}}(1+e) = \frac{2\mu}{a_\mathrm{i}}\left(\frac{a_\mathrm{f}/a_\mathrm{i}}{1 + a_\mathrm{f}/a_\mathrm{i}}\right). \tag{4.8.8}$$

The velocity v_i on the initial circular orbit follows from equation 4.8.2 to be

$$v_\mathrm{i}^2 = 2\mu\left(\frac{1}{a_\mathrm{i}} - \frac{1}{2a_\mathrm{i}}\right) = \frac{\mu}{a_\mathrm{i}}. \tag{4.8.9}$$

The change in velocity Δv_1 required to achieve the transfer from the initial circular orbit follows from equations 4.8.8 and 4.8.9 to be

$$\Delta v_1 = v_{\rm p} - v_{\rm i} = \left(\frac{\mu}{a_{\rm i}}\right)^{1/2} \left[\left(\frac{2a_{\rm f}/a_{\rm i}}{1 + a_{\rm f}/a_{\rm i}}\right)^{1/2} - 1\right]. \tag{4.8.10}$$

The velocity v_2 required at point 2 to achieve the final circular orbit follows from equation 4.8.2 to be

$$v_{\rm f} = \left[2\mu\left(\frac{1}{a_{\rm f}} - \frac{1}{2a_{\rm f}}\right)\right]^{1/2} = \left(\frac{\mu}{a_f}\right)^{1/2}. \tag{4.8.11}$$

The velocity on the transfer orbit at apoapsis can be determined from conservation of the specific angular momentum h

$$h = a_{\rm i} v_{\rm p} = a_{\rm f} v_{\rm a}, \tag{4.8.12}$$

and equation 4.8.8 to be

$$v_{\rm a} = v_{\rm p}\frac{a_{\rm i}}{a_{\rm f}} = \left[\frac{2\mu}{a_{\rm f}}\left(\frac{1}{1 + a_{\rm f}/a_{\rm i}}\right)\right]^{1/2}. \tag{4.8.13}$$

The change in velocity Δv_2 required to achieve the final circular orbit follows from the difference between equations 4.8.11 and 4.8.13, which is

$$\Delta v_2 = v_{\rm f} - v_{\rm a} = \left(\frac{\mu}{a_{\rm f}}\right)^{1/2} \left[1 - \left(\frac{2}{1 + a_{\rm f}/a_{\rm i}}\right)\right]^{1/2}. \tag{4.8.14}$$

The total Δv required for the Hohmann transfer is then the sum of equations 4.8.10 and 4.8.14. This derivation was for an outward transfer from a circular orbit with a small semimajor axis to a circular orbit with a larger semimajor axis. The equations for the change in velocity for an inward transfer, from a circular orbit with a larger semimajor axis to a circular orbit with a smaller semimajor axis, are the same if the directions of the velocities are reversed. Typical changes in velocity required for the Hohmann transfer are given in figure 4.14.

4.8.4 Bielliptic Transfer

The bielliptic transfer, illustrated in figure 4.15, consists of three impulses and two elliptical transfer orbits to transfer between the initial and final circular orbit. The first impulse produces the initial transfer ellipse with an apoapsis distance that is greater than the target orbit. The second impulse, applied at the apoapsis of the first transfer ellipse, increases the velocity to produce the second transfer ellipse whose periapsis radius is equal to the semimajor axis of the final orbit. Thc third impulse is applied at the periapsis of the second transfer ellipse to circularize the orbit.

Let $a_{\rm i}$ be the semimajor axis of the initial orbit, $a_{\rm f}$ the semimajor axis of the final orbit, and r the apoapsis distance of the first transfer ellipse, which is arbitrary. The optimum conditions for the Hohmann and bielliptic transfers follow:

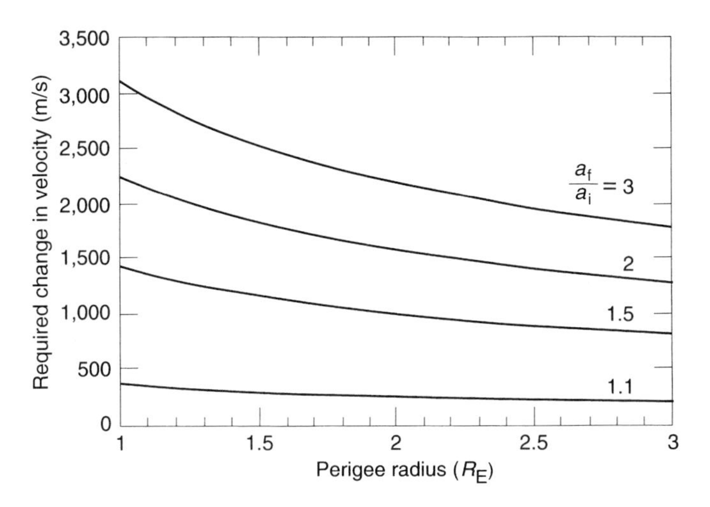

Figure 4.14 Velocity required for Hohmann transfer.

- For transfers for $1 < a_f/a_i < 11.94$, the Hohmann transfer is optimum.
- For transfers for $a_f/a_i = 11.94$, a bielliptic transfer, where the intermediate impulse is applied at $r = \infty$, requires less Δv than the Hohmann transfer but takes an infinite time.
- For transfers for $11.94 < a_f/a_i < 15.58$, the bielliptic transfer is optimum only if r, the apoapsis of the first transfer orbit where the intermediate impulse is applied, is greater than a certain value that approaches infinity as $a_f/a_i \rightarrow 11.94$.
- For transfers for $a_f/a_i > 15.58$ and $a_f < r < \infty$, the bielliptic transfer is optimum.

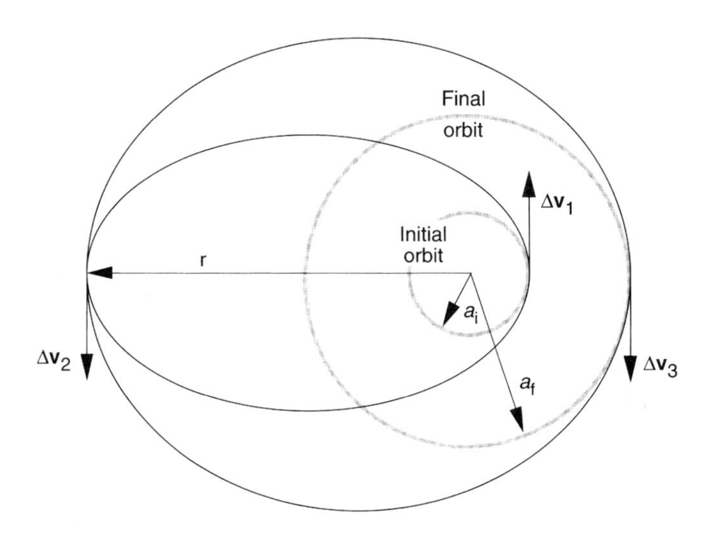

Figure 4.15 Bielliptic transfer. Δv = change in velocity, a_f = semimajor axis of final orbit, a_i, = semimajor axis of initial orbit, r = radius of Δv_2.

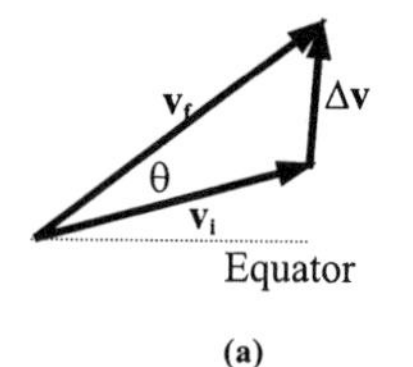

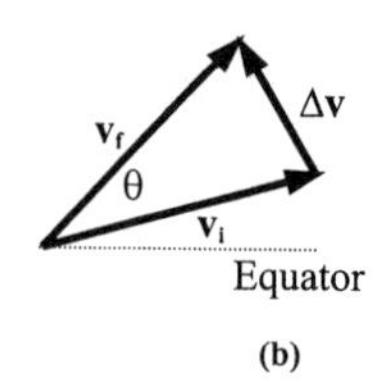

Figure 4.16 Change to inclination; (a) with change to magnitude of velocity; (b) without change to magnitude of velocity, v_{i} = initial velocity, v_{f} = final velocity, Δv = velocity increment, θ = change in inclination.

When the bielliptic transfer is more efficient than the Hohmann transfer, it takes more time to accomplish. The reason behind the efficiency of the bielliptic orbit is the benefit of applying a Δv at a lower altitude and higher velocity. A small Δv applied at the reduced velocity at apoapsis can have a significant effect on the periapsis that more than compensates for the larger expenditure of Δv at the lower altitude.

4.8.5 Selected Transfers

A change to the inclination for an impulse applied at the Equator is illustrated in figure 4.16 for the cases of (a) a combined maneuver with change in inclination and magnitude of the velocity, and (b) one with only a change in the inclination. It follows directly from figure 4.16a and the law of cosines that the Δv for a combined maneuver that changes the inclination by the angle θ and the magnitude of the velocity from v_{i} to v_{f} is given by

$$\Delta v = [v_{\mathrm{i}}^2 + v_{\mathrm{f}}^2 - 2v_{\mathrm{i}}v_{\mathrm{f}}\cos\theta]^{1/2}. \qquad (4.8.15)$$

For a change in inclination only it is necessary that $v_{\mathrm{f}} = v_{\mathrm{I}}$, from which it follows directly from either figure 4.16b or equation 4.8.15 that the required Δv is

$$\Delta v = 2v\sin(\theta/2). \qquad (4.8.16)$$

It would seem that the Δv for the simple change in inclination given by equation 4.8.16 would be optimum. However, it can be shown that a three-impulse maneuver utilizing the bielliptic transfer may be more efficient. Of course, this increase in efficiency comes at the cost of an increased time to carry out the transfer. A change in inclination and an in-orbit change in parameters can be accomplished by a Hohmann or bielliptic transfer, either preceded or followed by the plane change. It should be noted that it is more efficient to make the plane change while in the orbit in which the velocity is the smallest. This is illustrated by equation 4.8.16, where the required change in velocity is proportional to the velocity.

4.9 Solid Propulsion Systems

4.9.1 Introduction

The first rockets with solid propellants are attributed to the Chinese and are the earliest known types of rocket. The "rockets' red glare" from the "Star Spangled Banner" of Francis Scott Key identifies their use in the War of 1812 between the United States and Britain.

Significant advances in the development of solid propellants occurred during World War II and have continued. Today solid propellant rockets are used as launch vehicles, boosters to assist other types of launch vehicle, and to change orbits and trajectories.

Solid rockets are traditionally identified as *motors*. A solid rocket motor is simply a container filled with a solid propellant that, when ignited, expels hot gases through a nozzle. The propellant must burn at the proper rate to maintain the desired chamber pressure and have the structural rigidity to withstand the rigors of ignition and the subsequent dynamic loads. Solid propellants combine both the fuel and the oxidizer in a solid block called a *grain*. As a result, the motor case needs to be less massive than for other forms of propellant.

The primary advantages of solid propellant rockets are:

- High reliability, resulting from few if any moving parts
- Limited preparation required prior to ignition
- Favorable structural mass ratio when compared to liquid rockets
- Ease of storage and long storage life
- Lower cost than liquid rockets

The primary disadvantages are:

- Low specific thrust compared to liquid rockets
- Cannot be throttled
- Cannot be restarted
- Grain has a potential for combustion instability
- Propellants may be toxic and unsafe to handle

Solid rocket motors consist of:

- Motor case
- Insulator liner
- Propellant
- Igniter
- Nozzle
- Thrust vector control

Each is discussed in turn.

4.9.2 Motor Case

The motor case of a solid propellant system can be made from steel, titanium alloy, or filaments. Filament motor cases consist of continuous inorganic or organic strands that are wound into shape and held by a matrix material. Filaments enjoy significant strength-to-weight ratio advantages over other materials. Length-to-diameter ratios for cylindrical motors are typically between 2 and 6 and have structural mass fractions between 0.05 and 0.2. Longer motors are made from segmented sections of motor cases.

4.9.3 Insulator Liner

Between the motor case and the propellant grain is a rubber-like organic insulator liner that assures bonding of the grain to the motor case and acts as a thermal insulator.

Generally the grain is geometrically constructed so that the propellant is consumed as the liner is exposed to the process of combustion, so that its mass is minimized.

4.9.4 Propellants

Solid propellants are classified as single-base, double-base, composite, or composite-modified double-base.

Single-base propellants have a single active constituent such as nitrocellulose (gun cotton) or nitroglycerin. These are unstable and are not used for modern-day rockets.

Double-base propellants are a homogeneous mixture of two active ingredients, typically nitrocellulose (guncotton) and an energetic nitrated plasticizer, such as nitroglycerin, that causes it to dissolve and harden into a uniform solid.

Composite propellants contain a heterogeneous mixture of fuel and oxidizer as separate compounds suspended together in a hydrocarbon binder in the proper proportions to support combustion. The fuels are typically metal powders. The binder forms the constituents into a rubbery mass that because of its elastic property, reduces potential burning instabilities due to pressure variations. The chemical oxidizers and inorganic fuels are mixed in powder form with a liquid binder that holds them in suspension until the mixture is cured in an oven to a solid. Binders are long-chained molecules that form a matrix to constrain the homogeneity of the mixture of fuel and oxidizer and also act as a fuel. Darkening agents, such as carbon black or lampblack, are added to suppress heat radiation so as to prevent below-surface ignition that would result in uncontrolled burning. A *composite-modified double-base propellant* is a combination of double-base and composite propellants.

Composite propellants make up the majority of the modern rockets. The principal mass constituents include 20–40% fuel, 50–70% oxidizer, 10–20% binder that acts as a fuel, and 0–5% plasticizer that may be added to improve flexibility and assist in the processing, also acting as a fuel.

Examples of constituents of solid propellant follow from Sutton and Biblarz (2001) and others:

- *Fuels*
  - Aluminum powder
  - Beryllium powder
  - Magnesium powder
  - Sodium powder
  - Hydrocarbons
  - Polymers
  - Plastics
  - Rubber
- *Oxidizers*
  - Ammonium perchlorate (NH_4ClO_4)
  - Lithium perchlorate ($LiClO_4$)
  - Potassium perchlorate ($KClO_4$)
  - Sodium perchlorate ($NaClO_4$)
  - Ammonium nitrate (NH_4NO_3)
  - Potassium nitrate (KNO_3)
  - Sodium nitrate ($NaNO_3$)

Cyclotetramethylene tetranitramine
Cyclotrimethylene trinitramine

- *Binders*
  Carboxy-terminated polybutadiene (CTPB)
  Epoxides
  Hydroxy-terminated polybutadiene (HTPB)
  Nitrocellulose
  Polybutadiene acrylic acid (PBAA)
  Polybutadiene acrylic acid acrylonitrile (PBAN)
  Polyurethane
  PVC plastisol
- *Curing agents*
  Aziridines
  Epoxy resins
  Hexamethylene diisocyanate (HMDI)
- *Plasticizers*
  Nitroglycerin
  Triethyleneglycol dinitrate
  Trimethylolethane trinitrate

As an example, the Shuttle solid rocket motor consists of 69.93% ammonium perchlorate as the oxidizer, 16% powdered aluminum as the fuel, 0.07% iron oxide powder as a catalyst, 12.04% polybutadiene acrylic acid acrylonitrile as a binder, and 1.96% epoxy-curing agent, with the binder and epoxy also acting as fuel. The solid rocket provides a specific impulse of 242 at sea level and 268.6 in a vacuum.

Specific impulses of solid propellants are generally between 220 and 270s. The thrust level, as a function of time, is determined by the change in surface area of the propellant as it burns. The shape of the grain can be configured to produce a thrust that is *progressive* (increases as function of time), *neutral* (near-constant as a function of time), *regressive* (decreases as a function of time), or various combinations of these. A near-constant thrust is generally preferred so that the maximum dynamic loads are minimized.

A solid core, or end-burner, that burns from back to front, maximizes the amount of propellant in the motor case and produces a near-constant thrust. However, this configuration has several disadvantages, including an increasing volume in the combustion chamber with a constant exposed area of the propellant, combustion occurring at the liner for a sustained period of time so that it need be more massive, and a significant change in the center of mass, complicating attitude control. A propellant grain with a longitudinal void exposes a larger area so the thrust will be greater than that of an end-burner, will extinguish as the combustion reaches the liner, and will minimize any changes to the center of mass. However, as the grain burns outward the surface area increases, so the thrust increases, resulting in a progressive grain. To produce a near-constant thrust, that is, a neutral burn, a star- or fin-shaped longitudinal central cavity is generally employed. As the propellant burns, the area remains nearly constant, resulting in a near-constant, burn. Several different grain configurations are illustrated in figure 4.17.

The ideal, or best possible, solid propellant has many of the following characteristics:

(1) *Performance*

  a. High chemical energy density that produces high combustion temperatures and pressures
  b. Low molecular mass of the products of combustion

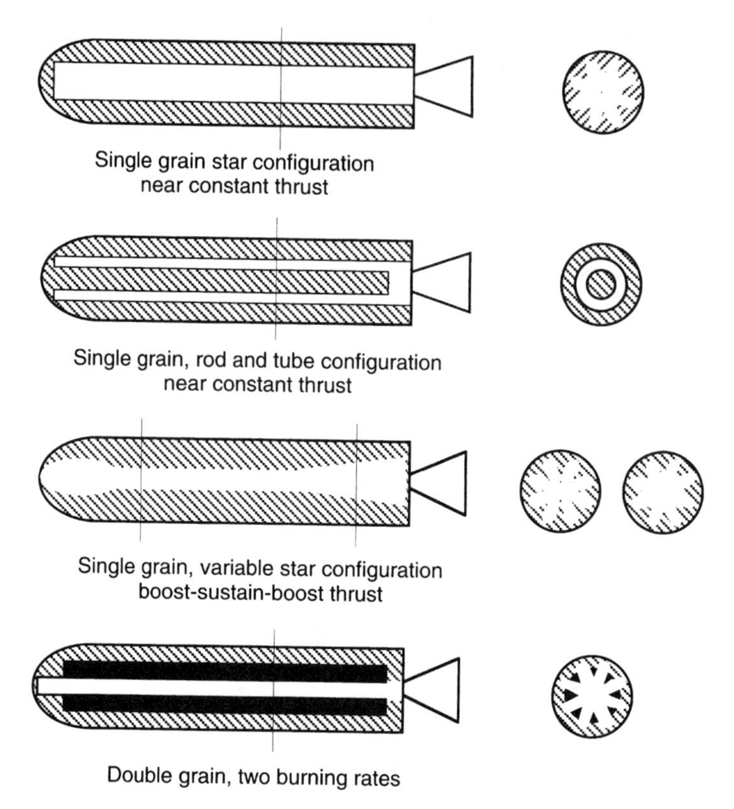

Figure 4.17 Solid propellant grain configurations.

c. Stable performance from rocket to rocket
d. Burns at a steady predictable rate

(2) *Operating conditions*

a. Limited preparation prior to firing
b. Properties are stable, with a long shelf life
c. Unaffected by environmental conditions such as humidity, temperature, and impact
d. Propellants are non-corrosive and non-toxic
e. Resistant to accidental ignition from heat, impact, and stray electric fields
f. Easy to handle and transport

(3) *Manufacturability*

a. Desirable fabrication properties
b. Performance insensitive to minor variations in tolerances
c. Materials and manufacturing processes are inexpensive

4.9.5 Igniter

An igniter performs the function that the name implies and generally consists of a propellant with a high burn rate that produces high pressure and high temperature gases, igniting the solid propellant of the rocket. A *pyrotechnic igniter* generally consists of a sensitive primer ignited by an electrical impulse. It is designed to burn for a very short interval and can take on different shapes, such as a small container, perforated tube, stringer, or sheet. The *pyrogen igniter* is essentially a small rocket motor that is designed to produce high pressure and hot gases, but not thrust. The electrical initiator of an igniter is known as a squib, glow plug, or primer.

Unintended ignition of an igniter can take place, caused by (a) static electricity, (b) induced currents from electromagnetic fields produced by electrical equipment, (c) currents from test equipment, or (d) heat, vibration, or shock from inappropriate handling. There are two approaches to safeguard against unintended ignition: the classical safe-and-arm device, and/or the use of safeguards built into the igniter. The safe-and-arm approach uses an electrical switch to safe the igniter by grounding the electrical circuit that initiates ignition, sometimes mechanically inhibiting the ignition even if the igniter should fire. The igniter is safed until just before ignition, when it is armed.

4.9.6 Nozzle

The nozzle of a solid propellant rocket is typically of the contour configuration described earlier. Severe conditions exist in the throat area where the heat transfer is a maximum, so that erosion is an issue. To provide control, the nozzle is often gimbaled with sliding parts. As a consequence, some of the few solid rocket failures are due to problems with the nozzle. Severe erosion in the throat can cause a reduction in thrust and, if asymmetric, can divert the direction of the thrust vector, possibly overwhelming the attitude control system. An increase in the throat area of more than 5% as a result of ablation is severe.

Several approaches are used to minimize nozzle erosion. Throat inserts are often used in the design; these are primarily made from pyrolytic graphite. The surface of the nozzle can be protected by locating a cooler-burning propellant in the combustion chamber near the nozzle to form a boundary layer of cooler gases. Special materials can be used in nozzles to provide either insulation or enhanced heat transfer. Graphites, tungston, and molybdenum are used to enhance heat conduction, and pyrolytic graphite and refractory materials are used to provide insulation. Ablative materials such as refrasil, asbestos phenolic, and graphite can be selectively used to carry heat away.

4.9.7 Thrust Vector Control

Thrust vector control of a solid rocket motor by a gimbaled nozzle is problematic because of the conditions to which the nozzle throat is exposed. Consequently, alternatives are available to replace or augment a gimbaled nozzle. *Jet vanes* are longitudinal heat-resistant fins mounted on the inside of the nozzle near the exit that rotate to direct the thrust and produce a sideward force. Typically four are used at 90° spacing with opposing pairs used to provide pitch and yaw control and all four used to provide roll control. Their disadvantage is that they disrupt the flow, reducing thrust, and may erode prematurely. A *Jetavator* is a circular ring positioned perpendicular to the flow just in front of the nozzle exit. It is rotated into the flow to deflect it and so provide a sideward thrust. A Jetavator can provide pitch and yaw control but not roll control. A *reaction control system* that consists of a separate propulsion system with thrusters orthogonal to the axis of the rocket can be used to provide a torque to control its orientation. Hydrazine, or a cold gas such as nitrogen, is often used. If the rocket is in the atmosphere, *aerodynamic fins* can be used to control its orientation. *Jet tabs* or *spoilers* are small blunt bodies attached to the nozzle exit perpendicular to the axis of the nozzle that can be rotated into or out of the flow to provide a sideward control thrust. Inserting the tab in the nozzle exit induces a shock wave and increased pressure upstream of the tab, producing a sideward thrust. Jet tabs can provide pitch and yaw control but not roll control. Useful for larger solid rocket motors is a *secondary injection thrust vector control* system that injects fluid, either gas or liquid, into the exhaust through ports in the expansion section of the nozzle. The sideward thrust arises from the imbalance in pressure caused by the reaction of the injected fluid. This approach requires the complication of a storage tank, piping, injectors, control valves, and control logic, but it can produce large control thrusts.

Examples of thrust vector control follow. The Titan III Solid Rocket Motor (SRM) uses fluid injection through 24 equidistant injectors in the exhaust nozzle. The Minuteman stage II uses a gimbaled nozzle augmented by a fluid injection system. The Pegasus first stage uses three aerodynamic fins while the second and third stages use a cold nitrogen reaction control system. The Delta I employs a reaction control system consisting of two vernier engines. The Atlas E and I use a hydraulically gimbaled nozzle assisted by two reaction control vernier engines. The Atlas II, IIA, and IIAS use a hydraulically gimbaled nozzle augmented by a hydrazine reaction control system. The U.S. Navy's Polaris missile uses the Jetavator. The Taurus first, third, and fourth stages use a gimbaled nozzle with 3° deflection and a cold nitrogen reaction control system while the second stage uses hydraulic jet vanes. The Titan Solid Rocket Motor Upgrade (SRMU) uses a gimbaled nozzle with a maximum deflection of 6°. The nozzle in a spacecraft is generally fixed, with the spacecraft attitude control system maintaining the desired orientation.

Control of the magnitude of the thrust is accomplished by the geometrical configuration of the grain, as described earlier. Once a solid rocket is ignited it burns to completion and cannot be throttled or restarted. Tests have shown that a slug of water can quench the flame and that a sudden decrease in combustion pressure, perhaps by opening a large hole in the rocket, can cause a flame-out, but these have not been used in an operational rocket. A typical solid rocket motor and its performance are illustrated in figure 4.18.

4.10 Liquid Propellant Systems

4.10.1 Introduction

The liquid-propellant rocket was first launched by Robert H. Goddard in 1926 and was perfected into the V-2 rocket of World War II. Since then, the state of the art of liquid propellant rockets has been further advanced so that they now produce the greatest thrust of all current operational propulsion systems. As a result, they are preferred for launch systems. A liquid propellant system is typically identified as an *engine* as opposed to a solid rocket that is identified as a motor. In a typical liquid engine, the fuel and oxidizer are stored in tanks and utilize a system of pipes, valves, and combustion chamber where they are combined and burned to produce thrust through a nozzle. In a launch vehicle turbopumps can be added to obtain high mass flow rates. A schematic of a bipropellant liquid engine from a launch vehicle is given in figure 4.19.

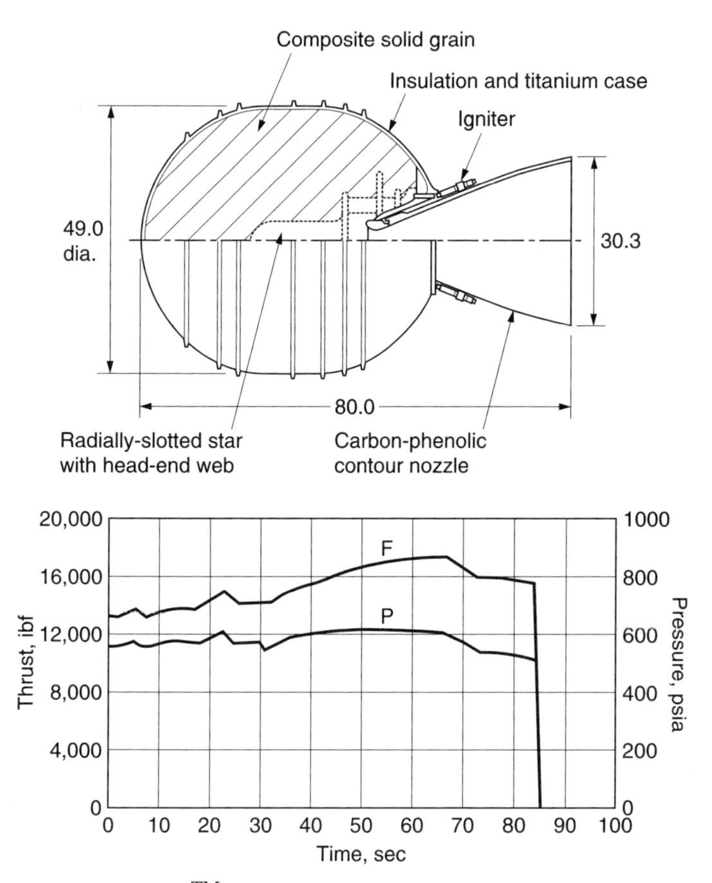

Figure 4.18 ATK-Elkton Star^{TM}-48B solid rocket motor, Propellants: AP 71%, Al 18%, HTBP binder 11%. Throat insert: 3D carbon–carbon. Throat diameter: 10.11 cm. Average expansion ratio: 47.2; initial expansion ratio: 54.8. Structural mass fraction: 0.061. Effective specific impulse: 292.4s. Total impulse: 5 799 169 N s. Total mass: 2141.4 kg. (Permission granted by ATK-Elkton LLC, Elkton MD.)

These engines are more complex then solid propellant motors but they offer several advantages:

- Higher specific impulse than solids
- Greater thrust than solids
- Thrust can be easily throttled
- Engine can be restarted many times
- Whole thrust chamber can be gimbaled, resulting in simpler nozzle design
- Flow of propellants can be precisely monitored to control the magnitude of the thrust

The primary disadvantages are:

- Lower reliability because of the multiplicity of parts
- Potential for propellants to rupture lines if excessive cold causes freezing
- Potential for propellants to burst containers if excessively hot
- Prelaunch preparation can be extensive

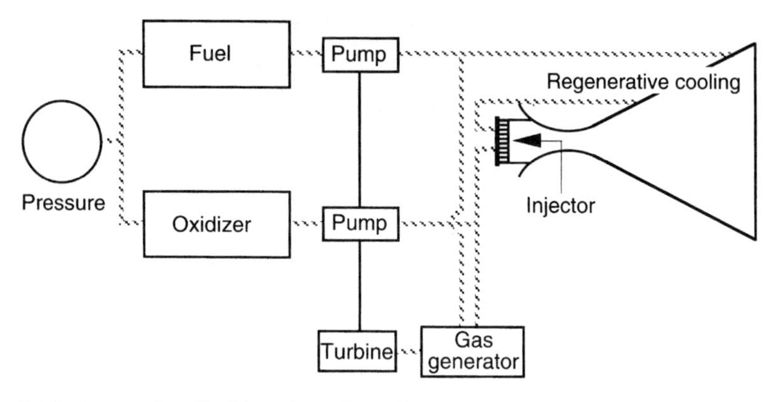

Figure 4.19 Bipropellant liquid engine schematic.

- Less favorable structural mass ratios, especially for smaller rockets
- Potential danger in handling some fuels

Liquid rocket engines for spacecraft consist of

- Propellant(s)
- Propellant flow controls
- Injector
- Igniter
- Thrust chamber
- Thrust control

Each is discussed in turn.

4.10.2 Propellants

Liquid propellants are classified as either a monopropellant or a bipropellant. A *monopropellant* inherently contains an oxidizing agent and a combustible substance in either a single chemical substance or a mixture of several substances. It decomposes or reacts exothermically by being either heated or exposed to a chemical catalyst. Examples are hydrogen peroxide (H_2O_2), and hydrazine (N_2H_4) and its related organic compounds: momomethylhydrazine (MMH, CH_3NHNH_2) and unsymmetrical dimethylhydrazine (UDMH, $(CH_3)_2NNH_2$). UDMH is more stable than hydrazine and both UDMH and MMH have wider liquid temperature ranges, thus tolerating larger spacecraft temperature variations. While hydrogen peroxide was used in the early developments of monopropellant systems, the three forms of hydrazine are now used extensively.

Bipropellants consist of a fuel and oxidizer stored separately and mixed in the combustion chamber. They produce the largest thrust of any of the current propulsion systems. Liquid bipropellants are also classified as either storable liquids at room temperature, or cryogenic liquids that must be maintained significantly below room temperature. Examples of a cryogenic bipropellant is liquid hydrogen (LH, H_2), which is a liquid below 20.4 K, and liquid oxygen (LOX, O_2), which is a liquid below 90.0 K. Liquid oxygen is made by liquefying air and boiling off the nitrogen and other gases. Such a low boiling point causes an extremely high rate of evaporation. For this reason, storage without refrigeration, results in appreciable loss.

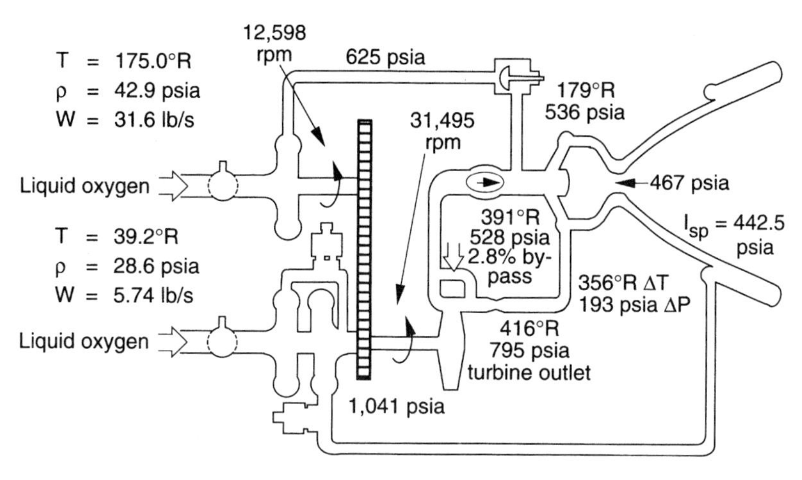

Figure 4.20 Centaur liquid oxygen/liquid hydrogen engine. RL-10A-3-3A simplified propellant flow schematic; 16.5 K thrust, 5.5:1 mixture ratio. (Courtesy of General Dynamics, now Lockheed Martin.)

Examples of bipropellant systems for launch systems include:

- Liquid hydrogen and oxygen
  - Space shuttle main engine (SSME)
  - RL-10 engine on the Titan III stage, Atlas Centaur second stage, Delta III and IV second stage
  - Vulcain engine on Ariane V first and second stage
- Liquid oxygen and kerosene
  - RS-27 engine on Delta II, III, and IV first stage
  - MA-5 engine on Atlas first stage
  - RD-170 engine on the liquid Energia booster
- Nitrogen tetroxide and monomethylhydrazine
  - Gemini main engine for rendezvous and docking
  - Shuttle Orbiting Maneuvering System (OMS)
  - RS-72 engine on Ariane V second stage
- Nitrogen tetroxide and aerozine-50
  - AJ 23 engine on Titan III and IV first and second stage
  - AJ10 engine on Delta II second stage
  - Lunar Module
- Liquid oxygen and RP1 (refined petroleum)
  - AJ-26 engine on Russian N1 first and second stage
  - RD-180 engine on Atlas III booster
  - F1 engine on Saturn V
- Nitrogen tetroxide and unsymmetrical dimethylhydrazine
  - Viking 5 engine on Ariane IV first and second stage
  - Viking 2 engine on Ariane II first stage
  - Proton first stage

The RL-10 liquid bipropellant engine is illustrated in figure 4.20 and a schematic of the Space Shuttle Main engine (SSME) is given in figure 4.21.

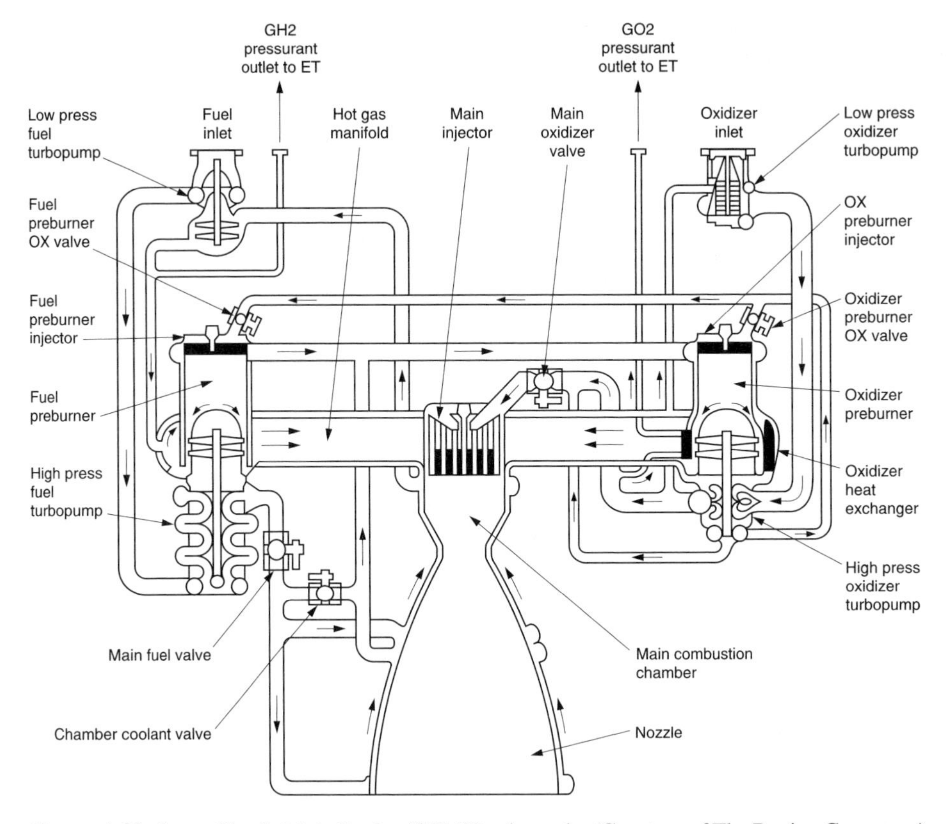

Figure 4.21 Space Shuttle Main Engine (SSME) schematic. (Courtesy of The Boeing Company.)

Hydrazine is unique in that it can be used as a monopropellant or a bipropellant. The choice for most spacecraft propulsion systems today is one of the three forms of hydrazine, with or without nitrogen tetroxide as the oxidizer.

Desired characteristics of liquid propellants are as follows:

(1) *Performance*
   a. High chemical energy density that produces high combustion temperatures and pressures; that is, high specific impulse
   b. Low molecular mass of the products of combustion
   c. Burns at a steady predictable rate
   d. Low freezing point

(2) *Operating conditions*
   a. Low toxicity and corrosiveness
   b. Stable constituents
   c. Properties are stable, with a long shelf life
   d. Unaffected by environmental conditions such as humidity, temperature, and impact
   e. Propellants are non-corrosive and non-toxic
   f. Easy to handle and transport

(3) *Manufacturability*
   a. Availability of raw materials
   b. Ease of manufacture
   c. Materials and manufacturing processes are inexpensive

4.10.3 Propellant Flow Controls

For proper combustion to occur, the propellants must be forced under pressure into the combustion chamber at controlled rates. There are two basic means to achieve the proper flow, a pressure-fed system and a pump-fed system. A simple pressure-fed system consists of a high-pressure supply tank of inert gas, tanks of propellants, piping, control valves, one-way check valves, pressure regulators, orifices, filters, and fill/drain service ports. Rubberized bladders in the propellant tanks separate the inert gas from the propellant. The inert gas pressure forces the propellants into the combustion chamber. A pump-fed system is used to increase the mass flow rate in the majority of liquid engines in launch systems and retains all the above features, including the inert gas to prime the pumps. The pumps, usually centrifugal pumps, are driven by turbines which are in turn driven by a gas generator that generally uses propellants that produce high-pressure gas at much lower temperatures than the propellants. Sometimes the primary propellants are diverted to drive the pumps, to simplify the overall system. The liquid bipropellant engines in Figures 4.19 and 4.20 illustrate pump-fed systems.

The ratio of the mass flow rates of oxidizer to the fuel is called the *mixture ratio* and must be carefully controlled for proper combustion. As examples, the mixture ratio for nitrogen tetroxide to the various forms of hydrazine is approximately 1.7, while the mixture ratio for liquid oxygen to liquid hydrogen is approximately 6 to 1.

4.10.4 Injectors

The injector meters the flow of propellant to the combustion chamber, atomizes the propellants into a vapor, and uniformly distributes the propellants in the proper mixture ratio for ignition and for uniform and efficient burning. Injectors come in a variety of configurations with two basic types: impinging and non-impinging. The *impinging injector* is used for bipropellants where the fuel and oxidizer are kept separate until injected through a large number of separate orifices in such a manner that they impinge on each other to provide proper mixing and atomization. The impinging pattern can be like impinging (fuel-on-fuel and oxidizer-on-oxidizer) or unlike impinging (fuel on oxidizer). The *non-impinging injector* or shower-head type of injector employs non-impinging streams of propellant and oxidizer and relies on turbulence and diffusion to provide mixing. The size of the combustion chamber is generally larger for non-impinging injectors, to achieve good mixing, so they are rarely used today. The throttling range of the engine is limited primarily by the ability to achieve a vaporized mixture at low-pressure drops across the injector. To achieve a large variation in thrust, which is required in some launch engines to be able to achieve a high degree of precision in the resulting velocity, requires a variable area injector. A sleeve assembly can be used that varies the area so near-optimum injector pressure drops and propellant velocities can be maintained at each thrust level. In spacecraft systems, fixed injector thrusters are

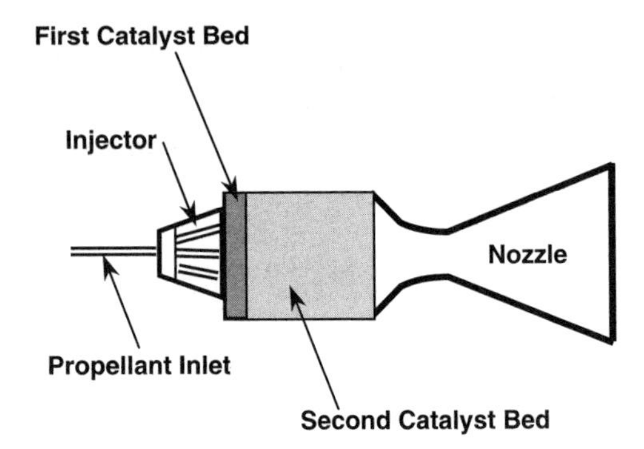

Figure 4.22 Monopropellant catalytic bed thruster.

generally employed, with thrusters of different sizes used to achieve both large thrust levels and high precision in the resulting velocity.

4.10.5 Igniters

Ignition of monopropellants can be produced thermally or by a catalytic material. For hydrazine and its derivatives, a catalytic bed of iridium is preferred as it is an effective catalyst at room temperatures while iron, nickel, and cobalt can be used at higher temperatures on the order of 400 K. As a consequence, hydrazines are the propellant of choice for most spacecraft monopropellant systems. To enhance the chemical process, sometimes the catalyst bed is preheated. The usual catalyst for the various forms of hydrazine consists of iridium deposited on a porous bed of aluminum oxide. Hydrazine decomposes into gaseous ammonia (NH_3) and nitrogen (N_2), releasing heat, and the ammonia decomposes further into nitrogen and hydrogen gases which absorb some of the heat. A schematic of a catalytic bed monopropellant nozzle is illustrated in figure 4.22.

Hypergolic propellants that combust spontaneously on contact do not need an igniter. These propellants include hydrazine, monomethylhydrazine, and unsymmetrical dimethylhydrazine with nitrogen tetroxide (N_2O_4), nitric acid (HNO_3), or hydrogen peroxide (H_2O_2); kerosene with hydrogen peroxide; and liquid hydrogen with hydrogen peroxide. *Spark plug igniters* are used with liquid hydrogen, kerosene, and RP-1 fuels.

4.10.6 Thrust Chambers

Thrust chambers consist of the combustion chamber, the exhaust nozzle, and the thrust chamber cooling system as one unit. The combustion chamber must be of the correct size to assure complete combustion of the propellants. Typically, the chamber is smaller than the nozzle to which it is attached. The nozzle is usually of the contour configuration discussed earlier. Temperature control of the thrust chamber, especially in the throat area of the nozzle, can be accomplished by radiation and regenerative cooling. Radiation cooling may be used in bipropellant systems when the burn times are short, or in monopropellant systems that burn at a lower temperature, as in spacecraft systems.

Launch systems often use *regenerative cooling* whereby liquid propellant is circulated around the combustion chamber and the nozzle to absorb heat prior to its introduction into the combustion chamber.

4.10.7 Thrust Control

Thrust vector control maintains the correct orientation during the application of the thrust. Launch engines often gimbal the whole thrust chamber by connecting it to the propellant supply by flexible hoses. This provides pitch and yaw control. Roll control can be provided by the techniques discussed for solid propellants. In spacecraft systems the spacecraft can be oriented to point the nozzle, so gimbaling is not ordinarily employed. However, there is an advantage that a static gimbaled nozzle is able to orient the thrust vector through the center of mass to minimize attitude disturbances. As propellant is consumed, the center of mass of the spacecraft can change. Liquid propellants can be easily throttled, shut off, and restarted by controlling the propellant flow, which, as pointed out earlier, is a significant advantage.

4.10.8 Examples

Schematics of a monopropellant and a bipropellant propulsion system are illustrated in figures 4.23 and 4.24.

An example of a typical long-duration spacecraft propulsion system is that of NASA's Cassini spacecraft. Cassini is a mission to study the planet Saturn and its moons, especially its largest, Titan. It was launched on 15 October 1997 and arrived in July 2004. It has separate monopropellant and bipropellant systems. The monopropellant system is used to remove momentum from the reaction wheels of the attitude stabilization system and to make small trajectory corrections. The bipropellant system is to carry out the deep space maneuvers, to make large changes to the trajectory, and to inject the spacecraft into orbit at Saturn. A top-level schematic is given in figure 4.25 and a more detailed schematic is given in figure 4.26.

4.11 Cold Gas System

Generally classified as liquid systems, cold gas systems are used in spacecraft and launch systems for attitude control and in spacecraft for station keeping and small changes to orbit parameters. These systems are characterized by their simplicity, as illustrated in figure 4.27. Typical cold gas propellants include hydrogen, helium, nitrogen, argon, air, krypton, and freon. Their advantages include simplicity, reliability, relatively lower cost, non-toxicity, handling safety, and some do not contaminate sensitive spacecraft surfaces such as optics. Their disadvantages include a tendency to leak since the systems are under high pressure, relatively poor mass fractions, and low specific impulses when compared to solid propellant, monopropellant, or bipropellant systems.

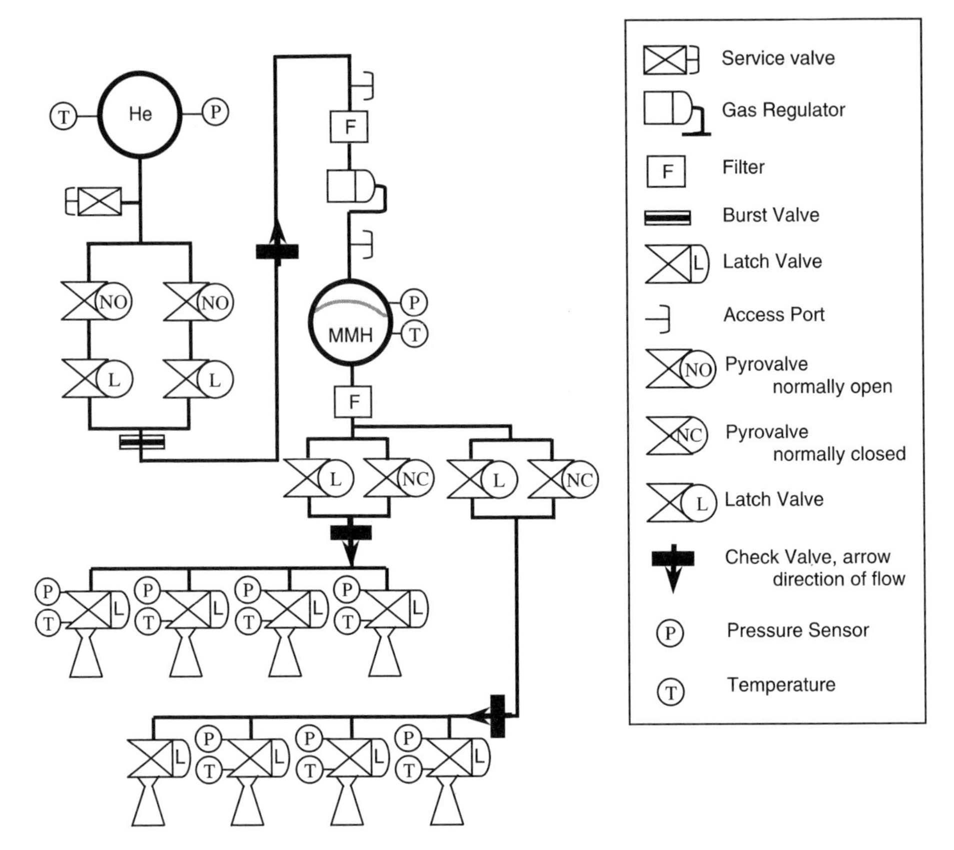

Figure 4.23 Monopropellant propulsion schematic.

4.12 Hybrid Rockets

Hybrid rockets are those that have one component of the propellant in a solid phase and the other in a liquid phase. The usual configuration is that of a solid fuel with liquid oxidizer.

The primary advantages of hybrid rockets are:

- Can be throttled and restarted
- Higher specific impulse than solid propellants
- High reliability than liquid engines, fewer moving parts
- Less preparation required prior to ignition than liquid engines
- Ease of storage and long storage life
- Lower cost than a liquid engine

The primary disadvantages are:

- Lower specific thrust than liquid rockets
- Some fuels are toxic

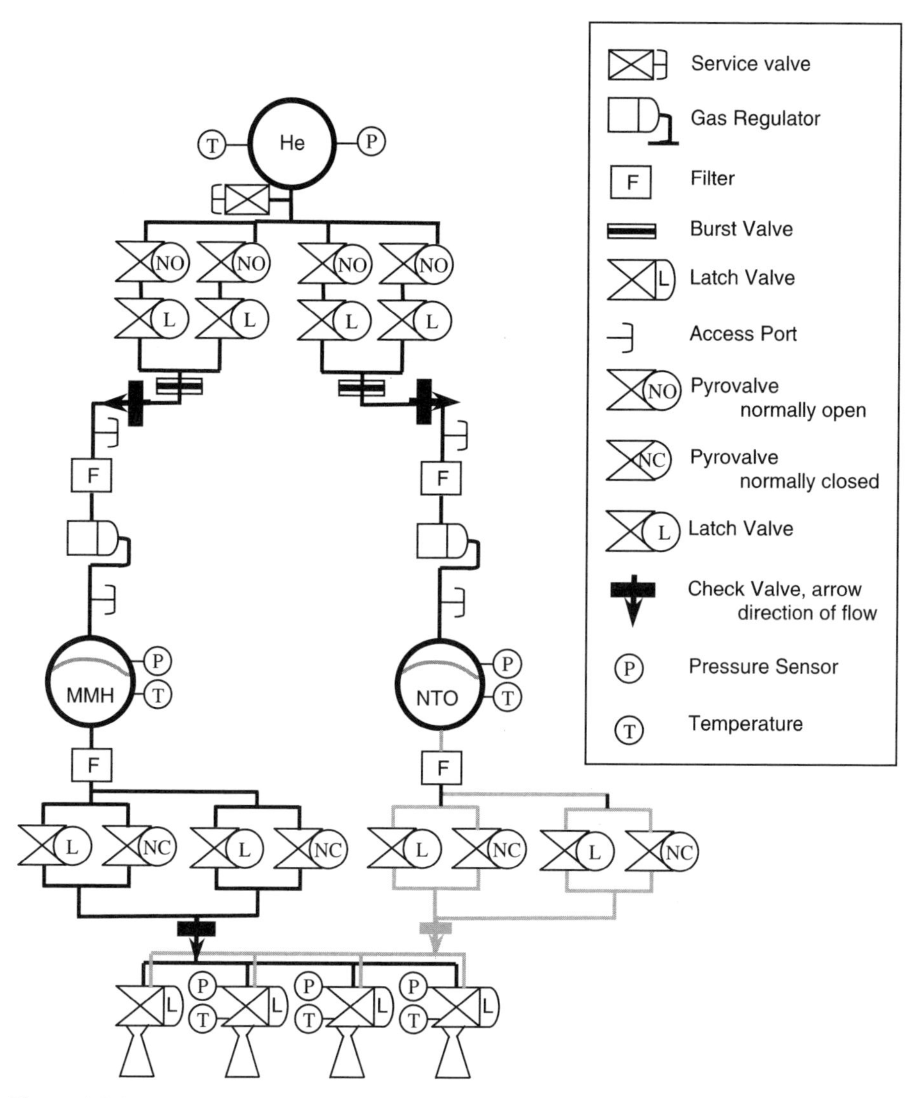

Figure 4.24 Bipropellant propulsion schematic.

Components of both the solid rocket motor and the liquid rocket engine are used in the hybrid rocket. Typical hybrid fuels include polyethylene (PE), polymethyl methacrylate (PMMA, plexiglas), polyvinyl chloride (PVC), and hydroxyl-terminated polybutadiene (HTPB). Complementary fuels that increase specific impulse include light metals such as beryllium, lithium, and aluminum. Typical hybrid oxidizers include nitrous oxide (NOX, N_2O), gaseous oxygen (GO_2), hydrogen peroxide (H_2O_2), liquid oxygen (LOX, O_2), nitrogen tetroxide (N_2O_4), nitric acid (HNO_3), and fluorine/liquid oxygen (FLOX). The propellant of choice for large hybrid rockets is a liquid oxygen oxidizer and HTPB fuel.

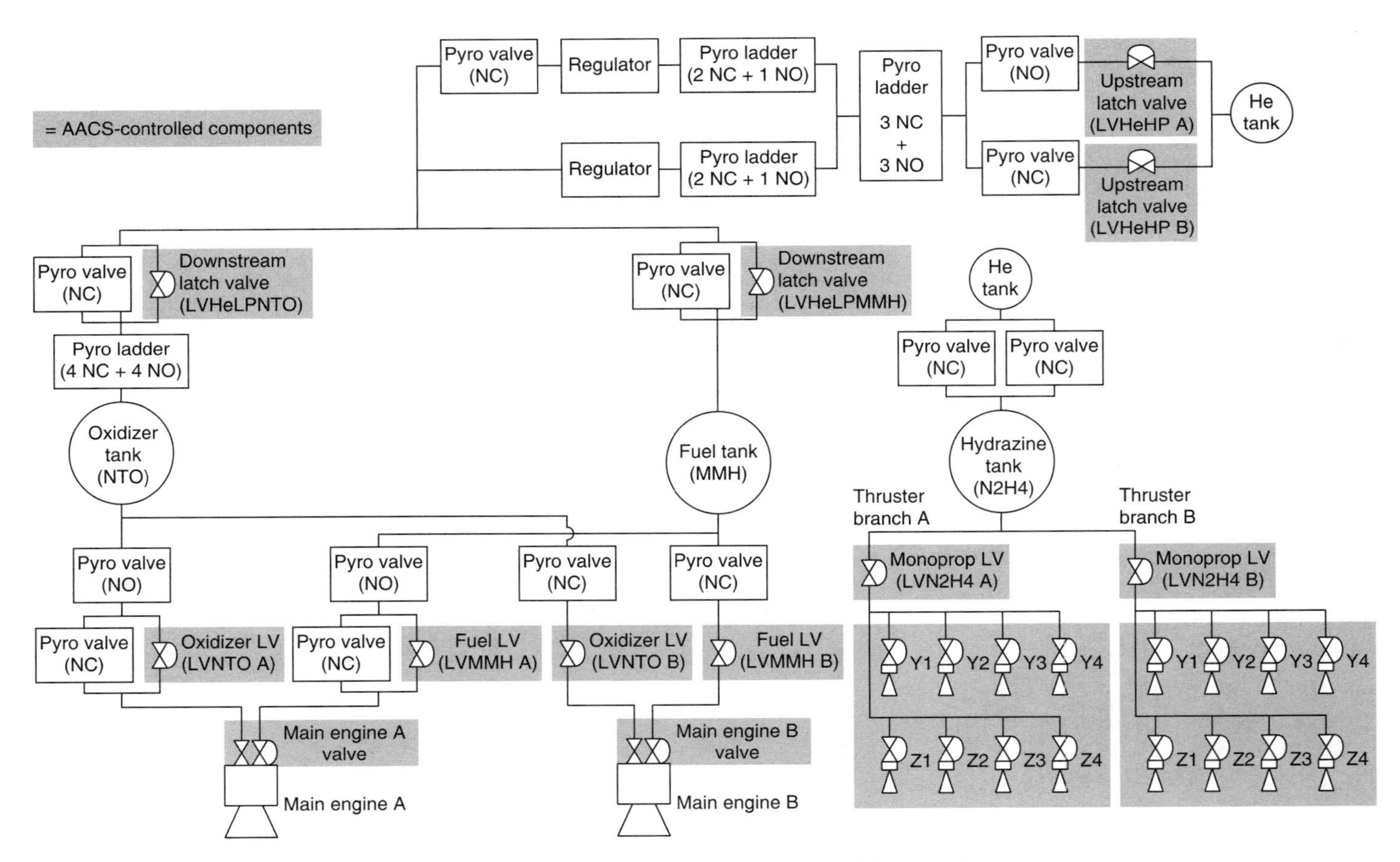

Figure 4.25 Schematic overview of Cassini spacecraft propulsion system. (Courtesy of NASA/JPL/Caltech.)

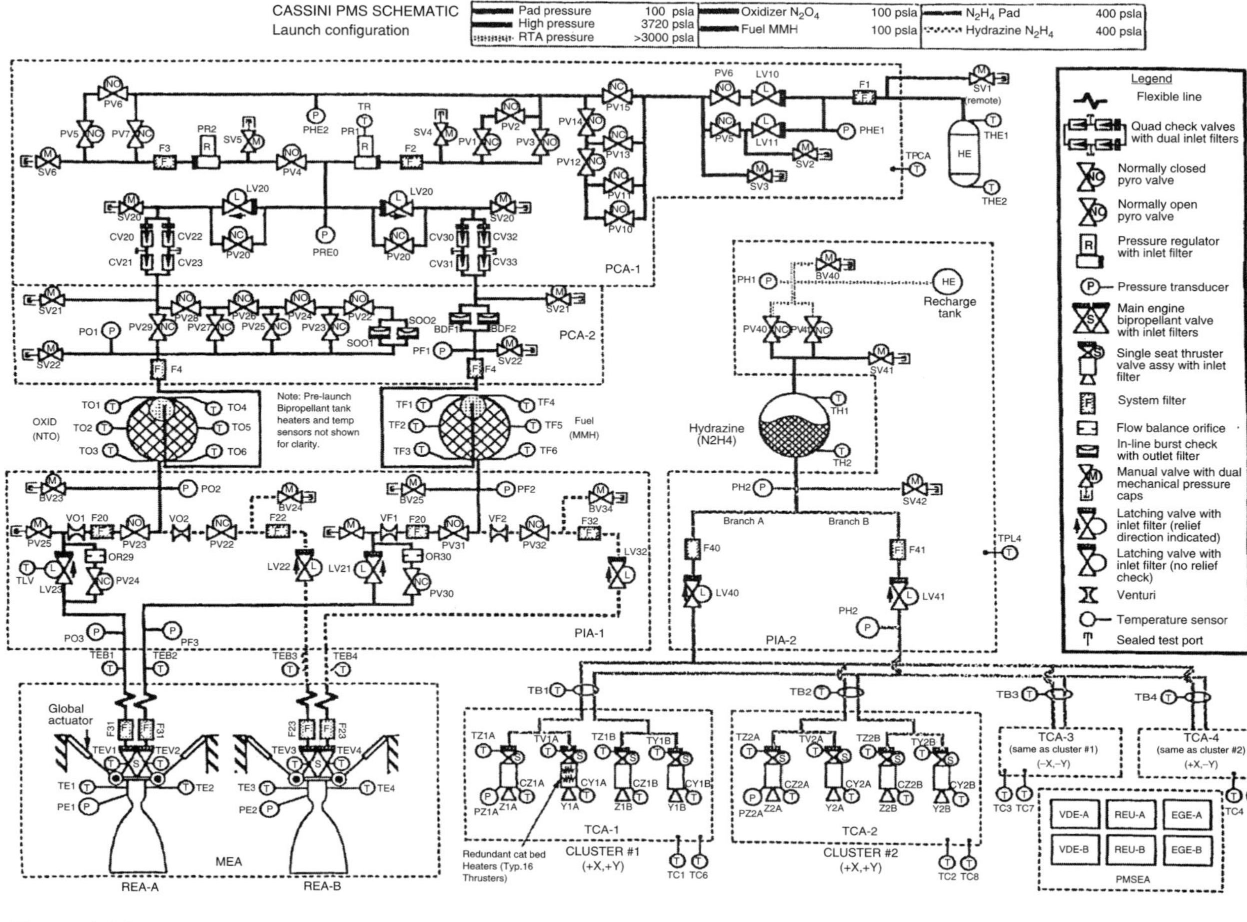

Figure 4.26 Detailed schematic of Cassini spacecraft propulsion system. (Courtesy of NASA/JPL/Caltech.)

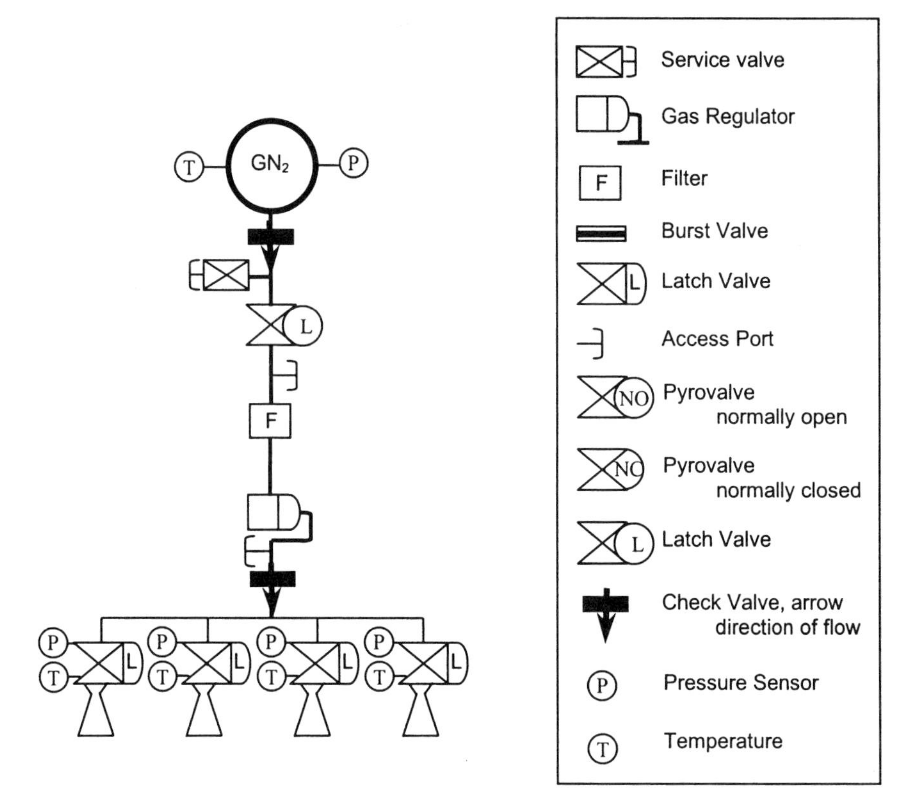

Figure 4.27 Cold gas system schematic.

Ignition is accomplished by either a heat source that causes some vaporization of the solid fuel and injection of a hypergolic substance, or electrical ignition with injection of the oxidizer. To assure good mixing to maximize thrust a multiple-chamber grain configuration is generally used, as illustrated in figure 4.28. To further assure good mixing and maximum thrust, a mixing chamber is located just before the nozzle.

4.13 Electrical Propulsion Systems

4.13.1 Introduction

Electric propulsion involves the acceleration of propellants by electrical heating or electromagnetic fields to provide thrust, and involves the conversion of electrical energy into the kinetic energy of the propellants. Advancements in the technology of electrical propulsion systems, based on ground and in-flight testing, has brought them to a high level of maturity. These areas of improvements include thruster design, materials, propellants, and a better understanding of the basic physics.

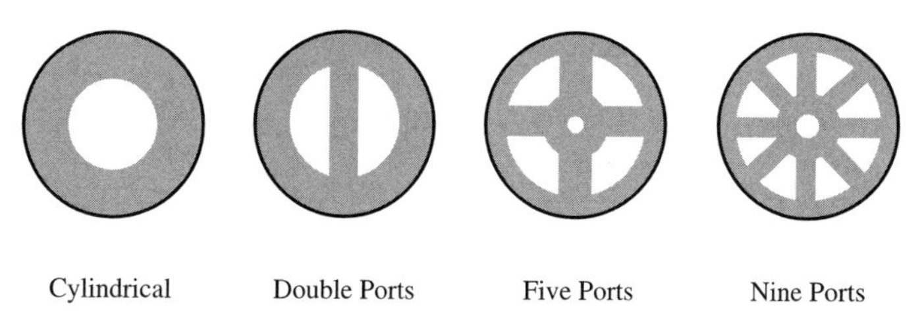

Figure 4.28 Port configurations for hybrid rocket.

Electrical thrusters can outperform other types of thruster in selected circumstances because of their generally higher specific impulse. The total efficiency η_t of an electrical propulsion thruster is given by the power in the rocket efflux divided by the input electrical power, where

$$\begin{aligned}\eta_t &= \frac{\text{kinetic power}}{\text{electrical power}} = \frac{\frac{d}{dt}(\frac{1}{2}m_e C^2)}{P_e} \\ &= \frac{\dot{m}_e C^2}{2P_e} = \frac{F_p C}{2P_e} = \frac{F_p g_0 I_{sp}}{2P_e}, \qquad (4.13.1)\end{aligned}$$

using equations 4.2.13 and 4.2.17, where

$$\begin{aligned}\dot{m}_e &= \text{mass flow rate,} \\ C &= \text{effective exhaust velocity,} \\ F_p &\equiv \dot{m}_e C = \text{propulsive thrust,} \\ P_e &= \text{electrical power input into thruster,} \\ g_0 &= \text{standard surface gravity, } 9.80665 \text{ m/s}^2, \\ I_{sp} &= \text{specific impulse.}\end{aligned}$$

There are three basic types of electrical thruster: electrothermal, electrostatic, and electromagnetic, although some thrusters can fall into more than one category.

4.13.2 Electrothermal Propulsion

In an electrothermal thruster the propellant is heated electrically and is thermodynamically expanded to supersonic speeds, as in a chemical rocket. There are two principal types of electrothermal thruster, the resistojet and the arcjet.

The *resistojet* is the simplest of electrical thrusts, employing the principle that the enthalpy of a propellant increases when it flows over a heated surface. As in a chemical thruster, the thermal energy is transformed into kinetic energy by the nozzle that produces thrust. The heating element is generally a coil, not too dissimilar to those found in electrical heaters. The gaseous propellant is fed into the area of the heating coil where it is heated and expanded through the nozzle. Resistojets have operated at DC and AC in steady and pulse modes, with input power levels ranging from watts to kilowatts. They

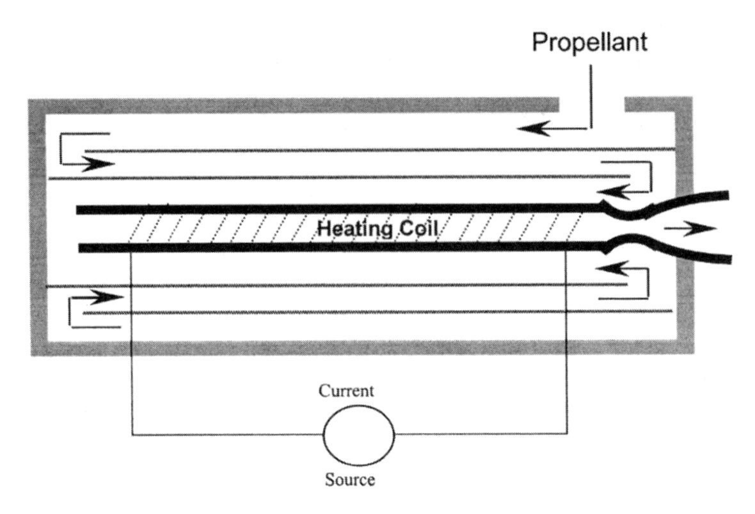

Figure 4.29 Resistojet schematic.

generate thrusts from several millinewtons to several newtons with specific impulses ranging from 200 to 400 seconds. Many different propellants have been employed including hydrogen (H_2), carbon dioxide (CO_2), methane (CH_4), hydrazine (N_2H_4), ammonia (NH_3), nitrogen (N_2), helium (He), and argon (Ar).

The first resistojet was developed for the Vela-III satellite in 1965 and used 90 W of power to generate a thrust of 0.187 N with a specific impulse of 123 s, using nitrogen as the working medium. Resistojets are also used to augment the propulsive thrust of the typical hydrazine propulsion system by increasing the temperature of the propellant. Resistojets are often used for station keeping, with hydrazine-augmented resistojets employed on several geostationary communication satellite systems. Four resistojets are used on the INTELSAT V communication satellites for station keeping. These produce a thrust of 0.22 to 0.49 N with a specific impulse of 300 s, requiring 250 to 550 W of power. A schematic of a resistojet is illustrated in figure 4.29.

An *arcjet* utilizes an arc discharge between a central cathode and a coaxial anode that acts as the thruster's nozzle. The gaseous propellant is fed into the thruster and is heated while flowing through and around the discharge. Stagnation temperatures of 5000–10 000 K can be obtained. Propellants employed include those identified above for the resistojet. An example is the hydrazine arcjet on the Lockheed Martin 7000 series communication satellites used for north–south station keeping, which require electrical power of 1.5–2.0 kW with a specific impulse of 520 s and a thrust of 0.4 N. A schematic of an arcjet is illustrated in figure 4.30.

4.13.3 Electrostatic Propulsion (Ion Propulsion)

Electrostatic thrusters use a high-voltage static electric field to accelerate ions to produce thrust and are secondarily identified by the ionization method. In electrostatic thrusters ions are accelerated, rather than the electrons, because of their heavier mass. As a result,

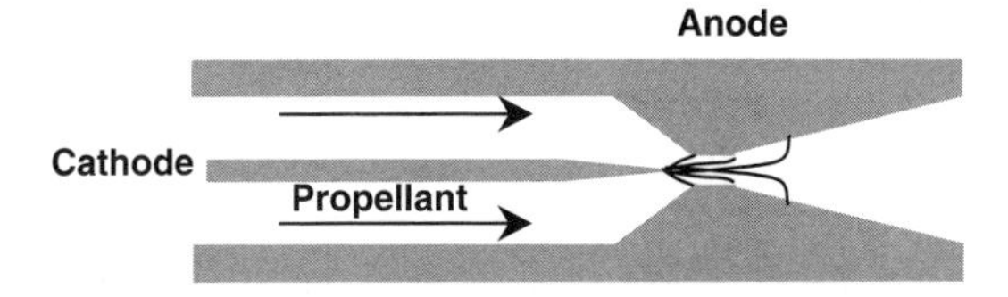

Figure 4.30 Arcjet Schematic.

gases used are those with heavy molecular masses such as xenon, krypton, cesium, mercury, and argon.

Electrostatic thrusters typically employ five stages to produce thrust. First, a low-pressure gas propellant is introduced into the ionization chamber. Second, the gas propellant is ionized. This can be accomplished by several different techniques including radio-frequency excitation, electron bombardment, electron–cyclotron resonance, and pulsed induction. Third, the plasma must be contained, generally by a magnetic field. Fourth, positive ions are accelerated by an electrostatic field, typically with a potential difference of 1–10 kV. Fifth, an electron current is discharged from the spacecraft to prevent a charge build-up. Since the ion beam exhaust is positively charged, the spacecraft would otherwise become negatively charged, resulting in a retarding force between the spacecraft and the exhaust. Radio frequency excitation produces a plasma by pumping a cavity of neutral gas with high-frequency electromagnetic fields. Electron bombardment produces a plasma by bombarding a neutral gas with energetic thermionic electrons. Electron–cyclotron resonance (ECR) produces a plasma through microwave-frequency electromagnetic radiation. Pulsed inductive excitation produces a plasma by a current pulse from high-capacity capacitors that creates a time-varying electromagnetic field. *Hall-effect thrusters* utilize orthogonal electric and magnetic fields to produce a Hall current that interacts with the magnetic field to produce a force that accelerates ions in the axial direction. Hall thrusters in use generally employ hydrogen and argon.

A schematic of an ion thruster is given in figure 4.31. An example of an ion propulsion thruster is the NSTAR (NASA Solar electric propulsion Technology Application Readiness) xenon thruster on the interplanetary probe Deep Space 1. This 30 cm diameter thruster operates over an input power range from 0.5 to 2.3 kW and produces a range of thrusts from 19 to 92 mN. The specific impulse ranges from 1900 s at 0.5 kW to 3100 s at 2.3 kW and the propellant mass flow rate is a few milligrams per second. The service life requirement of the engine is 8000 hours, with a qualification life of 12 000 hours.

4.13.4 Electromagnetic Thrusters

Electromagnetic thrusters utilize time-varying electrical and magnetic fields to accelerate highly ionized plasmas. Unlike ion thrusters, electromagnetic thrusters accelerate both the ions and electrons in a plasma so that the exhaust is electrically neutral. There are several types of electromagnetic thruster, including magnetoplasmadynamic thrusters, pulsed plasma thrusters, helicon thrusters, and pulsed inductive thrusters.

Magnetoplasmadynamic (MPD) thrusters, either pulsed or stationary, first ionize a propellant gas with a large current. Interaction between the current and the azimuthal

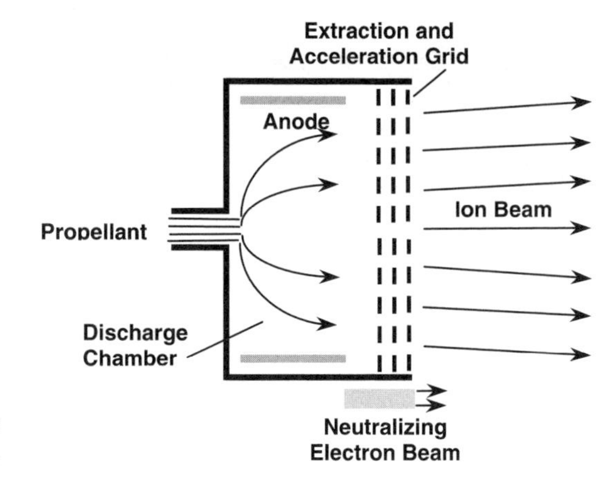

Figure 4.31 Electrostatic thruster schematic.

magnetic field that it produces then accelerates the plasma (ions and electrons) in the axial direction by a Lorentz force. *Pulsed plasma thrusters* (PPTs) utilize electrical power to ablate, ionize, and electromagnetically accelerate ions by the Lorentz force from a block of solid propellant material, typically Teflon. State-of-the-art pulse plasma thrusters have a specific impulse of 800–1500 s, a thrust of 220–1100 mN, and an efficiency of 5–15%. *Helicon thrusters* utilize electromagnetic waves, often by a helical coil, that interact with a current to create an electromagnetic field that accelerates plasma in the axial direction. *Pulsed inductive thrusters* (PITs) use capacitors to generate a magnetic pulse that induces an electric field to ionize a gas in which the plasma is accelerated by the Lorenz force in the magnetic field. These systems can be operated employing a variety of propellants including hydrazine, ammonia, argon, and carbon dioxide at specific impulses ranging from 1000 to 5000 s. The Hall thruster identified above is also classified as an electromagnetic thruster when the ion motion is primarily axial and the electron motion is primarily spiral.

4.14 Other Propellant Systems

4.14.1 Nuclear

Nuclear propulsion systems have the ability to overcome thrust limitations of chemical rockets because the source of energy and the propellant are independent of each other. The energy source is a critical nuclear reactor in which neutrons split fissile isotopes, for example, uranium (92-U-235) or plutonium (94-Pu-239). The reactor releases fission products, gamma rays, and enough extra neutrons to maintain the nuclear reaction. The energy density of nuclear fuel is significant in that one kilogram of fissile material has as much energy as ten million kilograms of chemical propellant. There are two approaches to utilizing heat from a nuclear thermal rocket. The first is to use the heat from the reactor to expand a low-molecular mass propellant, such as hydrogen, through a conventional nozzle. The other is to use the heat to produce electrical power which then powers an electrical thruster.

Direct heating of the propellant is generally accomplished by passing the propellant through the reactor. A *solid core nuclear thermal rocket* uses a solid reactor core with channels through which the propellant can flow to pick up heat. Temperatures can reach 3000 K, heating hydrogen to achieve a specific impulse of 500–1100 s. A *liquid core nuclear thermal rocket* can reach a temperature of 6000 K, heating hydrogen to achieve specific impulses in excess of 1500–2000 s. A *gaseous core nuclear thermal rocket* can reach a temperature of 25 000 K, heating hydrogen to achieve a specific impulse of 3000–7000 s.

To convert nuclear thermal power into electrical power in order to operate electrical thrusters requires energy conversion, either through a static or a dynamic converter. Static conversion is accomplished by the thermoelectric or thermionic effect that involves non-moving devices. Dynamic conversion is accomplished by thermodynamic processes such as the Rankine, Brayton, or Stirling cycles, which involve moving parts such as a turbine or reciprocating engine.

An advantage of the static converters is their simplicity, resulting from their lack of moving parts, so they are highly reliable. Their disadvantage is that they are relatively inefficient. In a *thermionic converter,* electricity is produced by electrons emitted from an hot emitter or cathode at temperatures of 1500–2000 K across a small inter-electrode gap of less than 1 mm, enclosed in a container with conductive vapors that enhance current flow, to a cooler surface or anode at 800–1200 K. Thermionic systems typically have a theoretical efficiency of around 10–15%, defined as the ratio of the electrical power produced to the thermal power available, with actual efficiencies ranging from 3 to 8%. The *thermoelectric converter* is based on the principle that, when two dissimilar metals are joined in a closed circuit and the two junctions are kept at different temperatures, an electrical voltage, and subsequent electrical current, is produced. This is known as the Seebeck effect, after the individual who discovered the principle in 1826. Thermoelectric converters operate typically at a source temperature of 1500 K. Thermoelectric electric conversion has a theoretical efficiency of 10–15%, with actual systems achieving efficiencies of about 5%. All the United States space radioisotope thermoelectric generators deployed employ thermoelectric converters.

In *nuclear dynamic systems* the heat generated from the radioactive decay of radioisotopes is used to warm a gas or a liquid to rotate a turbine or reciprocating engine that in turn drives a generator that produces electricity. Rankine cycle conversion has an efficiency of about 20%, whereas Stirling or Brayton cycle conversion systems have efficiencies of about 40%. The Rankine cycle utilizes a working fluid that changes phase during the operation of the conversion cycle. The Stirling cycle is a closed-cycle reciprocating engine that uses a high-pressure single-phase gaseous working fluid. The Brayton cycle utilizes a single-phase gaseous working fluid that absorbs energy at constant pressure through a rise in temperature.

4.14.2 Solar Sailing

A *solar sail* utilizes a large, lightweight surface to reflect light, either electromagnetic energy from the Sun or an external laser, to provide a propulsive force. The advantages of a solar sail are that it does not require expendables for propulsion and has few moving parts. The disadvantages are that it cannot function while in the eclipse, the force from the Sun decreases as the distance squared, it requires very large sail areas that are subject to damage from micrometeoroids, and they generate low thrust levels. By proper orientation of the sail with respect to the energy source, the spacecraft can maneuver to its target. For example, a cargo-carrying spacecraft can travel out from the Earth to Mars and then back to the Earth without expenditure of fuel for propulsion. The gravitational field acts as the equivalent of the keel of a sailboat.

The relativistic momentum p of a photon is given in terms of the energy E by $p = E/c$, where c is the speed of light. The force resulting from a photon striking a surface is due to its change in momentum and is equal to the momentum if the surface is non-reflecting and equal to twice the momentum if the surface is a perfect reflector. At the Earth's distance from the Sun, sunlight has an energy of about 1368 $\mathrm{W/m^2}$ so that the radiation force on an orthogonal non-reflecting surface would be $4.7 \times 10^{-6}\ \mathrm{N/m^2}$. A solar sail with an area of 1 $\mathrm{m^2}$ produces a force between 4.7 $\cos\beta\ \mathrm{N/m^2}$ and 9.4 $\cos\beta\ \mathrm{N/m^2}$, depending on the sail's reflecting properties, where β is the angle between the normal to the surface and the direction of the energy source. A 10 metric ton spacecraft with a sail area of 1 $\mathrm{km^2}$ would take about 400 days to transit from Earth to Mars. Designs of solar sails have included circular, square, and blade configurations using a strong polymer coated with aluminum for high reflectivity. Applications for solar sails include exploration of the solar system and beyond, maintenance of special artificial orbits, and delivery and return of large cargos.

4.15 Propellant System Sizing

In-orbit propulsion systems are employed on spacecraft to maintain them on a particular trajectory, change their trajectory, rendezvous with a target, or provide thrust for the attitude control system. The design of the propulsion system is generally partitioned into several steps. For changes to the orbit, the required linear impulse or change in spacecraft velocity is determined, and for attitude control systems the required angular impulse is determined. Next, the type of system, such as cold gas, monopropellant, bipropellant, or electrical, is selected by specifying for a vacuum the effective exhaust velocity or the specific impulse. From these, it is then possible to determine the amount of propellant required and the mechanical and electrical characteristics of the propulsion system, such as: the size, number, and types of thrusters; number and size of tanks; control valves and logic; the total mass; volume; and cost. A spacecraft propulsion system generally requires an attitude determination system and an active control system to point the thruster in the correct direction and maintain its direction when there is an offset between the propulsive thrust vector and the center of mass of the spacecraft.

To change the trajectory of the spacecraft, the Lagrange planetary equations are used to determine the required in- and out-of-plane total impulse, which is equivalent to the total change in velocity, Δv. The assumption of an applied impulse is generally accurate enough as most propulsive maneuvers are for short periods of time compared to the period of the orbit. Alternatively, the Euler-Hill equations, discussed earlier in this chapter and in chapter 3, can be integrated to determine the total impulse required and the total change in velocity. This approach is generally employed for rendezvous. For changes to several of the orbital elements, it is generally not possible to determine an optimum solution except by trial and error. At the same time the minimum and maximum thrust levels must be specified to determine propellant mass flow rates. The margins used for the equivalent Δv are a function of the characteristics of the specific mission but range from 10–20% for orbit corrections to 100% for attitude control systems.

Once the Δv is determined and the specific impulse for candidate propulsion systems is selected, equations 4.5.4 and 4.2.17 can be used to determine the mass of propellants when written in the form

$$m_{\mathrm{p}} = m_{\mathrm{i}}[1 - \mathrm{e}^{-\Delta v/g_0 I_{\mathrm{sp}}}], \tag{4.15.1}$$

where

m_{p} = mass of propellant,
m_{i} = initial mass of the spacecraft,
Δv = change in velocity,
g_0 = standard acceleration of gravity, $9.80665 \mathrm{m/s^2}$,
I_{sp} = propellant specific impulse.

Given a structural mass fraction r_{s} for the propulsion system, the total mass of the propellant system is

$$m_{\mathrm{total}} = \frac{m_{\mathrm{p}}}{1 - r_{\mathrm{s}}}. \tag{4.15.2}$$

Note that the specific impulse is dependent on the nature of the propellant, nozzle design, ambient pressure, and combustion efficiency. The structural mass fraction of most propulsion systems ranges between 5–10% for solids to 25% for a liquid bipropellant. An example to determine the mass of propellant required to satisfy a specified change in velocity Δv follows.

Example

Determine the mass of propellant needed to change the velocity of a spacecraft by 500 m/s if its initial mass is 1000 kg.

Solution: Equation 4.15.1 gives that

$$m_{\mathrm{e}} = 1000\ \mathrm{kg}\lceil 1 - \mathrm{e}^{-500/9.80665 I_{\mathrm{sp}}}\rceil.$$

A table of propellant mass for different candidate propulsion systems follows:

Type of System	Specific Impulse I_{sp} (s)	Propellant Mass (kg)
Cold gas	50	639.3
Monopropellant	230	198.8
Bipropellant	300	156.3
Ion thruster	3000	16.9

If a large thrust is required that the the ion thruster cannot provide, then it may be necessary to choose the bipropellant system. If the structural mass fraction is estimated to be 0.20, the total mass of the system follows from equation 4.15.2 to be

$$m_{\text{total}} = \frac{m_{\text{p}}}{1 - r_{\text{s}}} = \frac{156.3}{1 - 0.2} = 195.4 \text{ kg}.$$

Problems

1. During propulsion testing, a rocket produces a total impulse of 3×10^6 N s over 100 s during which the average propellant mass flow rate is 10 kg/s. Find the specific impulse and the effective exhaust velocity.
   Answer: 306.1 s and 3000 m/s.
2. A rocket operates at sea level, using propellants whose products of combustion have a specific heat ratio of 1.3. Find (a) the required chamber pressure and (b) the nozzle expansion ratio if the exit velocity has a Mach number of 3.0. The nozzle inlet velocity can be assumed to be small.
   Answer: (a) 40.55 atm, (b) 5.16.
3. An ideal rocket has a thrust coefficient of 1.5, an effective exhaust velocity of 2000 m/s, a mass flow rate of 75 kg/s, and a nozzle throat area of 0.025 m^2. Compute (a) the thrust, (b) the chamber pressure, and (c) the specific impulse.
   Answer: (a) 1.5×10^5 N, (b) 4×10^6 N/m^2, (c) 204.1 s.
4. Find (a) the nozzle throat area, (b) the nozzle exit area, (c) the exhaust speed, and (d) the specific impulse of a rocket engine that has a chamber pressure of 20 atmospheres, a correctly expanded nozzle with an exhaust pressure of 1 atmosphere, a chamber temperature of 2800 K, a mean molecular mass of the products of combustion of 22 kg/kmol, a specific heat ratio of 1.3, and generates a thrust of 1500 N.
   Answer: (a) 5.32 cm^2, (b) 17.12 cm^2, (c) 2139.5 m/s, (d) 218.3 s.
5. A rocket launch system has the following characteristics: initial mass of 10 000 kg, a propellant flow-rate of 150 kg/s, a nozzle exit velocity of 2000 m/s, a burn time of 50 s, a nozzle exit area of 0.25 m^2, and a nozzle exit pressure of 0.0345×10^6 N/m. For operation in a 1 atmosphere environment ($0.101\,325 \times 10^6$ N/m^2), determine (a) the thrust, (b) the effective exhaust velocity, (c) the initial thrust-to-weight ratio, (d) the initial vertical acceleration, and (e) the mass fraction.
   Answer: (a) 283 300 N, (b) 1888.7 m/s, (c) 2.89, (d) 18.5 m/s^2, (e) 0.25.
6. A propulsion system has a mass flow rate of 100 kg/s, products of combustion with a molecular mass of 24 kg/kmol, a ratio of specific heats of 1.3, combustion chamber temperature and pressure of 6000 K and 5 MPa respectively, and operates in an environment with an ambient pressure of 50 kPa. Determine the thrust for a nozzle whose exhaust pressure is (a) such that the nozzle is correctly expanded, (b) twice ambient, and (c) one-half ambient.
   Answer: (a) 0.3434×10^6 N, (b) 0.3401×10^6 N, (c) 0.3393×10^6 N.
7. Determine the percent variation in thrust between sea level and the altitudes of 5 and 50 km for a rocket with a chamber pressure of 20 atmospheres, a nozzle expansion ratio of 6, and propellants with a specific heat ratio of 1.3. The atmospheric pressures

at 5 and 50 km altitude on the Earth are 0.533 13 and 0.000 787 atmospheres, respectively, where 1 atmosphere is $0.101\ 325 \times 10^6$ N/m^2. Note that an iteration must be performed.
Answer: 10.5 % and 22.5 %.

8. Derive the mass fraction r_m in the form $r_m = r_s(1 - r_l) + r_l = r_l(1 - r_s) + r_s$ where r_s and r_l are the structural and payload mass fractions respectively.
9. Derive the theoretical correction factor for the thrust of a conical nozzle of half-angle θ_{cone}.
Answer: $(1 + \cos\theta_{cone})/2$.
10. A single-stage rocket in a force-free environment has a change in velocity of 1800 m/s caused by a propulsion system with a specific impulse of 300 s. Determine the mass fraction.
Answer: 0.542.
11. A 20 000 kg spacecraft is in a circular orbit around the Earth at an altitude of 1000 km. Determine the length of the burn required to escape from the Earth if the effective exhaust velocity is 3000 m/s and the propellant mass flow rate is 30 kg/s.
Answer: 425.2 s.
12. A rocket with a total mass of 10 000 kg is vertical on its Earth-based launch pad. For a propellant flow rate of 10 kg/s, find the critical value of the effective exhaust velocity for which the rocket will just begin to rise.
Answer: 9.8×10^3 m/s.
13. A rocket is tested at sea level and the following are measured: burn duration of 50 s, initial mass of 1500 kg, final mass of 300 kg, average thrust of 60 000 N, chamber pressure of 6 MPa, nozzle exit pressure of 0.08 MPa, nozzle throat diameter of 0.1 m, and nozzle exit diameter of 0.3 m. Determine the (a) mass flow rate, (b) exhaust velocity, (c) characteristic velocity, (d) specific impulse, (e) effective exhaust velocity, and (f) thrust coefficient.
Answers: (a) 24 kg/s, (b) 2562.7 m/s, (c) 1963.5 m/s, (d) 254.9 s, (e) 2500 m/s, (f) 1.273.
14. It is required to achieve the escape velocity of 12.5 km/s in a force-free environment using a multiple-stage rocket with each stage having a structural mass fraction of 0.10 and an effective exhaust velocity of 2.5 km/s. Determine the ratio of the payloads for a three-stage rocket and a four-stage rocket. Would a two-stage rocket meet the velocity requirements?
Answer: 0.001 844/0.000 963 = 1.91; no.
15. A rocket is launched vertically from the surface of the Earth and expels propellant at the rate of 0.005 times the initial mass of the rocket per second. For an effective exhaust velocity of 5000 m/s and ignoring the variation in the gravitational force with altitude, determine (a) the speed and (b) the height of the rocket at 10 s into the flight.
Answer: (a) 158 m/s, (b) 781 m.
16. Show that the overall payload mass fraction for a multiple-stage rocket given by equation 4.5.24,

$$r_l^* = r_l^n = \left[\left(\exp\left(-\frac{\Delta v_n}{nC}\right) - r_s\right) \Big/ (1 - r_s)\right]^n,$$

can be approximated for an infinite number of stages by $r_l^*|_{n\to\infty} = \exp(-\Delta v_{f,n}/(C(1-r_s)))$. Hint: take the logarithm of both sides and approximate the logarithm and exponential by a series expansion.

17. Ignoring the effects of aerodynamic forces, determine (a) the burnout velocity and (b) the burnout height of a rocket that is launched vertically from the Earth. The mass of the propellant is 60% of the rocket, the burn time is 5 s, and the effective exhaust velocity is 3000 m/s.
Answer: (a) 2700 m/s, (b) 5715 m.
18. A rocket with an initial mass of 40 000 kg, launched vertically upward, has a propellant mass of 30 000 kg, a propellant mass flow rate of 200 kg/s, and an effective exhaust velocity of 2500 m/s. Find the velocity and height at burnout.
Answer: 1995 m/s and 91 362 m
19. A two-stage rocket launched vertically from the Earth consists of two separate rockets. The first to ignite has a structural mass of 200 kg and a propellant mass of 1200 kg. The second to ignite has a structural mass of 50 kg and a propellant mass of 500 kg. The payload has a mass of 200 kg. The specific impulse of the propellant of each rocket is 250 s. (a) Calculate the maximum height of the rocket system when the two rockets ignite in series, ignoring the effects of burn time (i.e., burn–burn–coast strategy). (b) Then calculate the maximum height of the rocket system when the second rocket ignites when the maximum height is achieved after the first rocket has ignited, again ignoring the effects of burn time (i.e., burn–coast–burn–coast strategy).
Answer: (a) 1364 km, (b) 631 km.
20. Five similar rockets, each with a propellant mass of 10000 kg, a structural mass of 500 kg, and a specific impulse of 350 s, are to be used to launch a payload of 500 kg. Ignore the effect of burn time. Find the terminal velocity if (a) all the rockets are ignited at one time, (b) four are ignited initially and the fifth is ignited after the four burn out and are separated, (c) three are ignited initially and the fourth and fifth are ignited in series after the prior rockets burn out and are separated, (d) two are ignited initially and the other three are ignited in series after the prior rockets burn out and are separated, and (e) the first rocket is ignited and all subsequent rockets are ignited in series after the prior rocket burns out and is separated.
Answer: (a) 9856 m/s, (b) 13 054 m/s, (c) 13 243 m/s, (d) 13 290 m/s, (e) 13 303 m/s.
21. A launch vehicle first stage rocket that operates at sea level has a combustion chamber pressure of 66 atmospheres, a combustion temperature of 4000 K, and uses kerosene as the fuel and oxygen as the oxidizer with an oxygen-to-fuel mixture ratio of 2.8235; the products of combustion have a specific heat ratio of 1.3, and the nozzle is correctly expanded. (a) Determine the theoretical exhaust velocity. If the thrust produced in the rocket at sea level is measured to be 30×10^6 N and the propellant mass flow rate is 13 000 kg/s, calculate (b) the theoretical specific impulse, (c) the measured specific impulse, and (d) the efficiency of the rocket, defined as the ratio of the measured to the theoretical specific impulse. The chemical combustion process is $C_{12}H_{26} + nO_2 \rightarrow nCO + nH_2O + H_2$, where n is the number of moles of oxygen.
Answer: (a) 2645 m/s for $M = 25.52$ kg/kmol and $n = 15$, (b) 270.6 s, (c) 235.3 s, (d) 87.0%.

22. The altitude of a spacecraft in a circular obit around the Earth is 1000 km. It is necessary to change the orbit to a circular orbit of 400 km, so that it can be serviced by a Shuttle mission. Find (a) the specific energy of the Hohmann transfer orbit, (b) the velocity change required to achieve the transfer orbit, (c) the velocity change required to circularize into the final orbit, and (d) the minimum time of flight. (a) –28.158 km^2/s^2, (b) 157.4 m/s, (c) 160.9 m/s, (d) 49.4 min.
23. A satellite is launched from the orbit of the Earth into a Hohmann transfer orbit to the planet Jupiter with which it is to rendezvous. Given the gravitational constant for the Sun is $GM_{sun} = 1.327\,1244 \times 10^{20}\ m^3/s^2$, the radius of the Earth's orbit is $R_e = 1.495\,9789 \times 10^{11}$m, and the radius of Jupiter's orbit is $5.2028R_e$, determine (a) the increase in velocity needed and (b) its percentage of the velocity needed to escape from the solar system.
Answer: 14 436 km/s and 34.3%.
24. A rocket is launched vertically into a gravity-turn trajectory, during which the drag force and propulsive thrust divided by the mass, $-(\dot{m}C + F_d)/m$, is equal to the constant fg with $f = 1.2$ and $g = 9.800\,665 m/s^2$. It is desired to achieve a trajectory that is parallel to the Earth's surface 500 s after launch. (a) Find the velocity that must be developed at the instant of the pitch-over maneuver of 0.1 degrees. (b) Then find the velocity of the rocket when its trajectory is parallel to the Earth's surface.
Answer: 219.6 m/s, (b) 1797 m/s.

References

Agrawal, B. N., 1986. *Design of Synchronous Spacecraft.* Englewood Cliffs, NJ: Prentice-Hall.

Culler, G. J., and B. D. Fried, 1957. Universal gravity turn trajectories. *J. Appl. Phys.,* **28**(6): 672–676.

Griffin, M. D., and J. R. French, 1991. *Space Vehicle Design.* Washington, DC: American Institute of Aeronautics and Astronautics.

Humble, R. W., G. N. Henry, and W. J. Larson, 1995. *Space Propulsion Analysis and Design.* New York: McGraw-Hill.

Jensen, J., G. Townsend, J. Kork, and D. Kraft, 1962. *Design Guide to Orbital Flight.* New York: McGraw-Hill.

Prussing, J. E., and B. A. Conway, 1993. *Orbital Mechanics.* New York: Oxford University Press.

Sutton, G. P., and O. Biblarz, 2001. *Rocket Propulsion Elements*, 7th edition. New York: J. Wiley.

Turner, M. J. L., 2000. *Rocket and Spacecraft Propulsion.* Chichester, UK: Springer-Praxis.

Wertz, J. R., and W. J. Larson, 1999. *Space Mission Analysis and Design.* Dordrecht: Kluwer.

5

Spacecraft Attitude Determination and Control

MALCOLM D. SHUSTER AND WAYNE F. DELLINGER

5.1 Introduction

By spacecraft attitude, we mean how a spacecraft is oriented in space. Every spacecraft carries a complement of instruments, usually called a *payload*, that must be directed in some way, and the performance of this payload depends in a fundamental way on the attitude. Thus, the ability to know the spacecraft attitude (*attitude determination*) and the ability to command a desired attitude (*attitude control*) are indispensable to the performance of the spacecraft.

The attitude needs of spacecraft differ greatly. A scientific spacecraft designed to study the Earth, such as the *EOS* spacecraft or a weather spacecraft, may require an attitude in which the spacecraft payload is always pointed to the Earth. A communications spacecraft will have to point an antenna toward the Earth. The spacecraft may be spinning, have a fixed inertial orientation, or, in the case of Earth-pointing spacecraft, be rotating very slowly.

Attitude-sensing requirements are also varied. If a high level of accuracy is not needed, the attitude may be sensed from measurements of the Earth's horizon or the magnetic field. For very high accuracy, the spacecraft may measure the directions of stars. There are even degrees of sensitivity within types of sensors. For example, a wide variety of sensors exist that measure the direction of the Sun with accuracies slightly better than one degree or as precisely as a few arc-seconds.

Attitude control also takes a number of forms. It may be passive, relying on its natural response to the environment to maintain the attitude, it may be active and rely on computed actuator torques to achieve and maintain a desired attitude, or it may combine these two control strategies. Depending on the needs of the mission, attitude control requirements may be to control within only a few degrees of some nominal attitude, or the spacecraft may be controlled to within a small fraction of an arcsecond, as for the Hubble Space Telescope.

The subject of the present chapter falls into three areas: attitude representations (that is, the parameters that describe the attitude), attitude determination, and attitude dynamics and control. The subject of attitude determination and control is vast; therefore, this chapter is necessarily selective in the material it presents. Additional material can be found in several other sources, and detailed bibliographies are listed in each section and at the end of the chapter.

The attitude determination specialist must—on the basis of the mission requirements, cost, and time constraints—select a suite of attitude sensors of the appropriate accuracy and an attitude determination method that uses the data from those sensors. The attitude control specialist must determine the effect of environmental disturbance torques on the spacecraft and—also on the basis of the mission requirements, cost, and time constraints—select a suite of control hardware and an attitude control method. The control hardware may include attitude sensors, and the control method may use the attitude determined by the attitude determination system. Although generally one speaks of an "attitude determination and control system," very often the two are separate. The accuracy requirements for attitude determination and attitude control may be totally unrelated. It is not uncommon for the definitive attitude to be computed long after the mission, and attitude control is frequently effected on board the spacecraft without the simultaneous computation of the most accurate attitude. This chapter addresses the main functions of attitude determination and attitude control without regard to their possible interrelation.

The attitude specialist is burdened by the fact that attitude is somewhat complicated to describe. The description of position is fairly straightforward and generally mastered long before the beginning of university studies. Describing attitude is a very different matter, however. In some situations, it is easier to describe attitude by a 3×3 matrix, whereas in others it is easier to describe it in terms of three angles, by a four-dimensional vector, or by any of several other fundamentally different and often exotic choices. The attitude equations of motion are also more complicated than the equations of motion for spacecraft position. Even in the general case of a free rigid body (that is, a body not acted on by torques), the equations of motion are coupled, nonlinear, and not solvable in terms of elementary functions. The "simple" special case of a free rotating symmetric body is solved in this chapter, but even that appears forbiddingly complicated. In the words of one monograph (Griffin and French, 1991), attitude determination and control "is generally considered [to be] the most complex and least intuitive of the spacecraft vehicle design disciplines." With that dire warning, we proceed.

The present chapter contains more material than can be presented comfortably in a few lectures, and some material will certainly seem unnecessarily abstruse to readers for whom this chapter will be their final contact with spacecraft attitude determination and control. In particular, the reader who will not continue with attitude studies may wish to skip some sections (in particular, sections 5.2.5, 5.3.4, 5.5.3, 5.5.4, 5.6.3, 5.6.9), at least on a first reading.

The single most extensive general reference to date on spacecraft attitude is by Wertz (1978), totaling more than 800 pages. A more recent reference can be found in Wie (1998). Kaplan (1976) has written an advanced undergraduate textbook on this subject, biased heavily toward dynamics and control. Much shorter treatments than the present chapter can be found in the books by Carrou (1984), Agrawal (1986), Fortescue and Stark (1991), Griffin and French (1991), and Larson and Wertz (1992). Portions of this chapter have been adapted from Shuster (1989). A good source for mathematics background is the book by Wylie and Barrett (1982).

5.2 Attitude Representations

We begin by addressing the problem of how to describe or, to use the more sophisticated term, *represent* attitude. Attitude expresses a relation between two coordinate systems. Thus, a proper understanding of attitude will benefit from a review of some of the general properties of vectors and coordinate systems, especially the relation between vectors as abstract objects (that is, the vector as it can be understood without reference to a coordinate system) and the numerical representation of vectors (that is, the set of its components in a particular coordinate system).

The attitude of a body has much in common with the position of a body, which can be described by giving the displacement (or translation) of a coordinate system fixed in that body from some reference coordinate system, for example, a coordinate system fixed in the Earth or with respect to a celestial reference. Likewise, the attitude or orientation of an extended object can be described as a *rotation*. Imagine a coordinate system with the same origin as the body's coordinate system but whose axes are parallel to those of the reference coordinate system. The body's attitude is then expressible as the rotation that carries these translated and rotated reference coordinate axes into the body's coordinate axes. The study of attitude is therefore the study of rotations.

Whereas the study of translations is fairly simple, the study of attitude is much more complicated. There are no solitary point masses in attitude problems, and the equations of motion are nonlinear and coupled. For this reason, the baggage of attitude studies is very extensive and not particularly easy to acquire. Only the most important parts of that baggage are developed in the present section.

Of the many ways to represent attitude, the present section presents only the most useful: (1) the axis and angle of rotation; (2) the rotation matrix; (3) the Euler angles; and (4) the quaternion. Because attitude is more complicated to describe than position, one resorts to one or more representations in order to take advantage of a particular property that simplifies a particular part of a problem. It is not uncommon to make use of a half-dozen different representations in treating a single attitude problem.

Of the representations treated here, the axis and angle of rotation appeal most to our geometrical intuition of what a rotation is and help us to express mathematically what the other rotations mean geometrically; otherwise, they are of little practical use. The rotation matrix is the matrix that in general we must construct when we wish to transform vectors from one frame to another. The Euler angles are convenient for treating spinning spacecraft and gimbaled gyros and for archiving attitude, since there are only three variables to record. The Euler angles have the psychological advantage of appealing to our need to visualize, but otherwise they are not very practical. The quaternion is the most convenient representation to use in dynamical simulation of attitude, because it makes the best compromise between simplicity of the dynamical equations and the dimension of the system. There are, in fact, a dozen (or even three dozen, depending on how one counts) different attitude representations, each with its special niche. The four treated here are the most important.

5.2.1 Right-Handed Orthonormal Coordinate Systems

An *orthonormal basis* $\{\boldsymbol{i}, \boldsymbol{j}, \boldsymbol{k}\}$ is simply a set of three unit vectors that are mutually orthogonal. Thus, the vectors of an orthonormal basis satisfy the scalar-product relations

$$\boldsymbol{i} \cdot \boldsymbol{j} = \boldsymbol{i} \cdot \boldsymbol{k} = \boldsymbol{j} \cdot \boldsymbol{k} = 0, \tag{5.2.1}$$

$$\boldsymbol{i} \cdot \boldsymbol{i} = \boldsymbol{j} \cdot \boldsymbol{j} = \boldsymbol{k} \cdot \boldsymbol{k} = 1. \tag{5.2.2}$$

Generally, the bases that we encounter most often are not only orthonormal but also *right-handed*; that is, their vector products satisfy

$$\boldsymbol{i} \times \boldsymbol{j} = -\boldsymbol{j} \times \boldsymbol{i} = \boldsymbol{k}, \tag{5.2.3}$$

$$\boldsymbol{j} \times \boldsymbol{k} = -\boldsymbol{k} \times \boldsymbol{j} = \boldsymbol{i}, \tag{5.2.4}$$

$$\boldsymbol{k} \times \boldsymbol{i} = -\boldsymbol{i} \times \boldsymbol{k} = \boldsymbol{j}. \tag{5.2.5}$$

Bases in this chapter will generally be at least orthonormal.

If we are given a physical vector $\boldsymbol{r}$ in three-dimensional space and an orthonormal basis $\{\boldsymbol{i}, \boldsymbol{j}, \boldsymbol{k}\}$, we can find coefficients x, y, and z, also called *components* or *coordinates*, such that

$$\boldsymbol{r} = x\boldsymbol{i} + y\boldsymbol{j} + z\boldsymbol{k}, \tag{5.2.6}$$

and the components are given simply by

$$x = \boldsymbol{i} \cdot \boldsymbol{r}, \quad y = \boldsymbol{j} \cdot \boldsymbol{r}, \quad z = \boldsymbol{k} \cdot \boldsymbol{r}. \tag{5.2.7}$$

This result follows directly from equations 5.2.1 and 5.2.2.

We can arrange the components of r in a column vector $\mathbf{r}$, which we define to be

$$\mathbf{r} \equiv \begin{bmatrix} x \\ y \\ z \end{bmatrix}. \tag{5.2.8}$$

This column vector $\mathbf{r}$ is in fact a 3×1 matrix. If we know the basis, then knowing the column-vector representation completely specifies the vector. A column-vector representation is not the same as an abstract (physical) vector, however. We can talk about vectors in the abstract without reference to a coordinate system, but a column vector gives no information about the physical vector unless the coordinate system (that is, the basis) is also indicated. In general, in order to distinguish the two types of vectors in this chapter, we will write physical vectors using boldface italic characters ($\boldsymbol{r}$) and the associated column vector using nonitalicized boldface characters ($\mathbf{r}$). It is important to keep in mind that although there is only one physical vector, $\boldsymbol{r}$, there are infinitely many corresponding column vectors, $\mathbf{r}$, depending on the choice of basis.

Abstract operations with vectors (abstract because they are not expressed in terms of numerical representations of the vectors), as in equations 5.2.1 and 5.2.2, can be expressed equally well in terms of column-vector representations. We note that if we write the physical vectors $\boldsymbol{u}$ and $\boldsymbol{v}$ in terms of an orthonormal basis as

$$\boldsymbol{u} = u_1\boldsymbol{i} + u_2\boldsymbol{j} + u_3\boldsymbol{k} \tag{5.2.9}$$

and

$$\boldsymbol{v} = v_1\boldsymbol{i} + v_2\boldsymbol{j} + v_3\boldsymbol{k}, \tag{5.2.10}$$

then their column-vector representations are simply

$$\mathbf{u} = \begin{bmatrix} u_1 \\ u_2 \\ u_3 \end{bmatrix} \quad \text{and} \quad \mathbf{v} = \begin{bmatrix} v_1 \\ v_2 \\ v_3 \end{bmatrix}. \tag{5.2.11}$$

From the definition of an orthonormal basis, equations 5.2.1 and 5.2.2, it follows that the scalar product of two physical vectors may be expressed in terms of their components with respect to a common basis as

$$\boldsymbol{u} \cdot \boldsymbol{v} = \sum_{k=1}^{3} u_k v_k. \tag{5.2.12}$$

We can also write the scalar product more compactly in terms of column-vector representations as

$$\boldsymbol{u} \cdot \boldsymbol{v} = \mathbf{u}^{\mathrm{T}}\mathbf{v} \equiv \mathbf{u} \cdot \mathbf{v}, \tag{5.2.13}$$

where $\mathbf{u}^{\mathrm{T}}$ is the matrix transpose of $\mathbf{u}$,

$$\mathbf{u}^{\mathrm{T}} \equiv [u_1 \quad u_2 \quad u_3], \tag{5.2.14}$$

that is, a row vector, which is a 1×3 matrix, and the implied operation in the central member of equation 5.2.13 is simply matrix multiplication. (The product of a 1×3 matrix and a 3×1 matrix is a 1×1 matrix, in other words, a scalar, as required.)

Likewise, from equations 5.2.3 through 5.2.5, we can write the vector product in terms of physical vectors as

$$\boldsymbol{u} \times \boldsymbol{v} = (u_2v_3 - u_3v_2)\boldsymbol{i} + (u_3v_1 - u_1v_3)\boldsymbol{j} + (u_1v_2 - u_2v_1)\boldsymbol{k}, \tag{5.2.15}$$

and, from equation 5.2.15, the column-vector representation of $\boldsymbol{u} \times \boldsymbol{v}$ is just

$$\mathbf{u} \times \mathbf{v} \equiv \begin{bmatrix} u_2v_3 - u_3v_2 \\ u_3v_1 - u_1v_3 \\ u_1v_2 - u_2v_1 \end{bmatrix}. \tag{5.2.16}$$

The vector product of two column vectors can be written as a matrix operation, namely,

$$\mathbf{u} \times \mathbf{v} = [\mathbf{u}\times]\mathbf{v} \tag{5.2.17}$$

where $[\mathbf{u}\times]$ is defined to be the 3×3 matrix

$$[\mathbf{u}\times] = \begin{bmatrix} 0 & -u_3 & u_2 \\ u_3 & 0 & -u_1 \\ -u_2 & u_1 & 0 \end{bmatrix}. \tag{5.2.18}$$

Thus, if we work in terms of column-vector representations, vector operations can be replaced by matrix operations. The matrix $[\mathbf{u}\times]$ turns up quite frequently in attitude work.

The distinction between physical vectors and column vectors (or vector representations) may seem artificial. However, in attitude studies, especially in attitude estimation, we often have to deal with the representation of the same vector in two or more bases, and without this distinction things very quickly become confusing.

It should be noted that every relation that is true for physical vectors is also true for their column-vector representations with respect to a common basis. The inverse is not true. However, if a relation holds between column-vector representations with respect to a common basis for every basis, then it must also hold true for the physical vectors.

5.2.2 Orthogonal Transformations

We now answer the question of how to describe attitude in mathematical terms. We have said that attitude can be understood as the rotation of one coordinate basis into another. Attitude is expressible, therefore, as the matrix of the transformation coefficients, which we describe below. This matrix, as we shall see, is orthogonal, and the transformation is also referred to therefore as an orthogonal transformation.

Suppose that we are given two orthonormal bases, $\{\boldsymbol{I}, \boldsymbol{J}, \boldsymbol{K}\}$ and $\{\boldsymbol{i}, \boldsymbol{j}, \boldsymbol{k}\}$. Most often, the first basis will consist of three orthonormal axes that are fixed inertially, and the second will consist of three orthonormal axes fixed in the spacecraft. What is the nature of the transformation connecting these two bases?

Since $\{\boldsymbol{I}, \boldsymbol{J}, \boldsymbol{K}\}$ and $\{\boldsymbol{i}, \boldsymbol{j}, \boldsymbol{k}\}$ are both orthonormal bases, the vectors of each basis may each be expressed in terms of the vectors of the other basis as

$$\left.\begin{aligned} \boldsymbol{i} &= C_{11}\boldsymbol{I} + C_{12}\boldsymbol{J} + C_{13}\boldsymbol{K}, \\ \boldsymbol{j} &= C_{21}\boldsymbol{I} + C_{22}\boldsymbol{J} + C_{23}\boldsymbol{K}, \\ \boldsymbol{k} &= C_{31}\boldsymbol{I} + C_{32}\boldsymbol{J} + C_{33}\boldsymbol{K}, \end{aligned}\right\} \tag{5.2.19}$$

and

$$\left.\begin{aligned} \boldsymbol{I} &= C'_{11}\boldsymbol{i} + C'_{12}\boldsymbol{j} + C'_{13}\boldsymbol{k}, \\ \boldsymbol{J} &= C'_{21}\boldsymbol{i} + C'_{22}\boldsymbol{j} + C'_{23}\boldsymbol{k}, \\ \boldsymbol{K} &= C'_{31}\boldsymbol{i} + C'_{32}\boldsymbol{j} + C'_{33}\boldsymbol{k}, \end{aligned}\right\} \tag{5.2.20}$$

for some sets of coefficients C_{ij} and C'_{ij}. Since the bases are each orthonormal, it follows from inspection of equations 5.2.19 and 5.2.20 that

$$\left.\begin{aligned} C_{11} &= \boldsymbol{I}\cdot\boldsymbol{i} = C'_{11}, \\ C_{12} &= \boldsymbol{J}\cdot\boldsymbol{i} = C'_{21}, \\ C_{13} &= \boldsymbol{K}\cdot\boldsymbol{i} = C'_{31}, \end{aligned}\right\} \tag{5.2.21}$$

and similarly for the other matrix elements. Thus the coefficients of C_{ij} and C'_{ij} are not independent but satisfy

$$C'_{ij} = C_{ji}. \tag{5.2.22}$$

If we interpret C_{ij} and C'_{ij} as elements of 3×3 matrices C and C', respectively, then these two matrices satisfy

$$C' = C^{\mathrm{T}}, \tag{5.2.23}$$

that is, one matrix is the transpose of the other. The coefficients C_{ij} and C'_{ij} are usually called the *direction cosines*, because $\boldsymbol{I} \cdot \boldsymbol{i}$, for example, is the cosine of the angle between the two directions $\boldsymbol{I}$ and $\boldsymbol{i}$. Likewise, C and C' are called *direction-cosine matrices*.

Let us explore the properties of the direction-cosine matrices more fully. If we substitute equation 5.2.20 into 5.2.19, we find that

$$\begin{aligned} \boldsymbol{i} = {} & (C_{11}C'_{11} + C_{12}C'_{21} + C_{13}C'_{31})\boldsymbol{i} \\ & + (C_{11}C'_{12} + C_{12}C'_{22} + C_{13}C'_{32})\boldsymbol{j} \\ & + (C_{11}C'_{13} + C_{12}C'_{23} + C_{13}C'_{33})\boldsymbol{k}, \end{aligned} \tag{5.2.24}$$

and similarly for $\boldsymbol{j}$ and $\boldsymbol{k}$. We recognize immediately that the coefficients of equation 5.2.24 are simply the elements of a matrix product, so that we may rewrite equation 5.2.24 as

$$\boldsymbol{i} = (CC')_{11}\boldsymbol{i} + (CC')_{12}\boldsymbol{j} + (CC')_{13}\boldsymbol{k}. \tag{5.2.25}$$

Since the vectors in any single basis are linearly independent, it follows that the first coefficient in equation 5.2.24 or 5.2.25 must be unity while the other two vanish, and similarly for the equivalent relations for $\boldsymbol{j}$ and $\boldsymbol{k}$. Thus,

$$\sum_{k=1}^{3} C_{ik}C'_{kj} = \sum_{k=1}^{3} C_{ik}C_{jk} = \delta_{ij}, \tag{5.2.26}$$

where δ_{ij} is the Kronecker symbol defined as

$$\delta_{ij} = \left\{ \begin{matrix} 1, & \text{if} \quad i = j, \\ 0, & \text{if} \quad i \neq j. \end{matrix} \right\}. \tag{5.2.27}$$

Had we instead substituted equation 5.2.19 into 5.2.20, the result would have been

$$\sum_{k=1}^{3} C'_{ik}C_{kj} = \sum_{k=1}^{3} C_{ki}C_{kj} = \delta_{ij}. \tag{5.2.28}$$

We may write equations (5.2.26) and (5.2.28) equivalently as matrix equations

$$CC^{\mathrm{T}} = C^{\mathrm{T}}C = I, \tag{5.2.29}$$

where I is a 3×3 identity matrix

$$I = \begin{bmatrix} 1 & 0 & 0 \\ 0 & 1 & 0 \\ 0 & 0 & 1 \end{bmatrix}. \tag{5.2.30}$$

The Kronecker symbol δ_{ij} is just the (i, j) element of the 3×3 identity matrix I. The matrix C^T is thus the matrix inverse of C. A square matrix for which

$$C^{-1} = C^{\mathrm{T}} \tag{5.2.31}$$

is said to be *orthogonal*. Only matrices that satisfy equation 5.2.29 or 5.2.31 can relate two orthonormal bases.

Having discovered how to describe the relation between two orthonormal bases, we now address the problem of relating column-vector representations of vectors with respect to these two bases. An arbitrary vector can be represented with respect to either one basis or the other as

$$\boldsymbol{r} = X\boldsymbol{I} + Y\boldsymbol{J} + Z\boldsymbol{K} \tag{5.2.32}$$

or as

$$\boldsymbol{r} = x\boldsymbol{i} + y\boldsymbol{j} + z\boldsymbol{k}. \tag{5.2.33}$$

We wish to determine the relation between the two representations of $\boldsymbol{r}$, each with respect to a different basis. We write this, indicating the bases explicitly, as

$$\boldsymbol{r}_{\{I,J,K\}} \equiv \begin{bmatrix} X \\ Y \\ Z \end{bmatrix} \quad \text{and} \quad \boldsymbol{r}_{\{i,j,k\}} \equiv \begin{bmatrix} x \\ y \\ z \end{bmatrix}. \tag{5.2.34}$$

Examine now the two representations in equation 5.2.34. If we use equation 5.2.7 to compute the scalar product of $\boldsymbol{r}$ with the basis vectors $\boldsymbol{i}$, $\boldsymbol{j}$, and $\boldsymbol{k}$, as given by equation 5.2.19, we will find that

$$\begin{aligned} x = \boldsymbol{i} \cdot \boldsymbol{r} &= C_{11}\boldsymbol{I} \cdot \boldsymbol{r} + C_{12}\boldsymbol{J} \cdot \boldsymbol{r} + C_{13}\boldsymbol{K} \cdot \boldsymbol{r} \\ &= C_{11}X + C_{12}Y + C_{13}Z, \end{aligned} \tag{5.2.35}$$

and similarly for the other two components. In matrix notation this becomes

$$\begin{bmatrix} x \\ y \\ z \end{bmatrix} = \begin{bmatrix} C_{11} & C_{12} & C_{13} \\ C_{21} & C_{22} & C_{23} \\ C_{31} & C_{32} & C_{33} \end{bmatrix} \begin{bmatrix} X \\ Y \\ Z \end{bmatrix}. \tag{5.2.36}$$

Thus, the same coefficients that transform the bases into one another also transform the representations.

Despite the fact that direction cosines appear in both, equations 5.2.19 and 5.2.36 are very different in character. The multiplication operations in equation 5.2.19 are multiplication of an abstract vector by a scalar, whereas the operation in equation 5.2.36 is matrix multiplication.

Equations 5.2.29 and 5.2.31 have important consequences for the determinant of C. From the general properties of the determinant, we have that

$$\det(CC^{\mathrm{T}}) = (\det C)(\det C^{\mathrm{T}}) = (\det C)^2 = 1, \tag{5.2.37}$$

from which it follows that

$$\det C = \pm 1. \tag{5.2.38}$$

When $\det C = +1$, we say that the orthogonal matrix is *proper*. Otherwise, we say that it is *improper*. For any orthogonal matrix, it can be shown that

$$(C\mathbf{x}) \cdot (C\mathbf{y}) = \mathbf{x} \cdot \mathbf{y}, \tag{5.2.39}$$

and

$$(C\mathbf{x}) \times (C\mathbf{y}) = (\det C)C(\mathbf{x} \times \mathbf{y}). \tag{5.2.40}$$

We leave the proof of equation 5.2.39 as an exercise to the reader. The proof of equation 5.2.40 is much more difficult. We see that every orthogonal transformation preserves scalar products, but only proper orthogonal transformations preserve vector products. (We shall see that proper orthogonal transformations are synonymous with rotations.) As a consequence of equation 5.2.39, if $\mathbf{x}$ and $\mathbf{y}$ are orthogonal, so are $C\mathbf{x}$ and $C\mathbf{y}$. The orthogonal transformations are thus *right-angle preserving*. The root meaning of orthogonal in Greek is, in fact, "right angle."

Rotations are limited to proper orthogonal transformations. An example of an improper orthogonal transformation is that described by the matrix

$$\begin{bmatrix} -1 & 0 & 0 \\ 0 & -1 & 0 \\ 0 & 0 & -1 \end{bmatrix}. \tag{5.2.41}$$

Since every rotation leaves at least one axis unchanged (the axis of rotation), such a transformation, which changes the direction of every vector, is impossible by means of a rotation.

5.2.3 The Rotation Matrix

Up to now we have considered transformations between orthonormal bases in the most general form without paying attention to their geometrical significance. We now remedy that situation. We shall see that not all orthogonal transformations are rotations, but only the proper orthogonal transformations.

Consider a rotation about the $\boldsymbol{K}$-axis (figure 5.1). Mathematically, this rotation is described by the equations

$$\left.\begin{aligned} \boldsymbol{i} &= \cos\phi \boldsymbol{I} + \sin\phi \boldsymbol{J}, \\ \boldsymbol{j} &= -\sin\phi \boldsymbol{I} + \cos\phi \boldsymbol{J}, \\ \boldsymbol{k} &= \boldsymbol{K}. \end{aligned}\right\} \tag{5.2.42}$$

Alternatively, following the example of equation 5.2.19, we write the transformation as

$$\left.\begin{aligned} \boldsymbol{i} &= R_{11}(\mathbf{K}, \phi)\boldsymbol{I} + R_{12}(\mathbf{K}, \phi)\boldsymbol{J} + R_{13}(\mathbf{K}, \phi)\boldsymbol{K}, \\ \boldsymbol{j} &= R_{12}(\mathbf{K}, \phi)\boldsymbol{I} + R_{22}(\mathbf{K}, \phi)\boldsymbol{J} + R_{23}(\mathbf{K}, \phi)\boldsymbol{K}, \\ \boldsymbol{k} &= R_{13}(\mathbf{K}, \phi)\boldsymbol{I} + R_{23}(\mathbf{K}, \phi)\boldsymbol{J} + R_{33}(\mathbf{K}, \phi)\boldsymbol{K}, \end{aligned}\right\} \tag{5.2.43}$$

where we now write $R(\mathbf{K}, \phi)$ to designate the direction-cosine matrix corresponding to a rotation about an axis $\mathbf{K}$ through an angle ϕ.

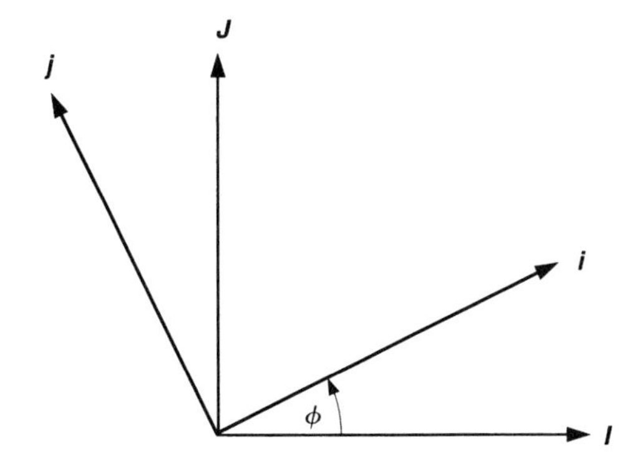

Figure 5.1 Rotation about the $\boldsymbol{K}$-axis. Here $\boldsymbol{I}$ and $\boldsymbol{J}$ are the initial coordinate axes and $\boldsymbol{i}$ and $\boldsymbol{j}$ are the final coordinate axes.

If we compare equations 5.2.42 and 5.2.43, we must have

$$R(\mathbf{K}, \phi) = \begin{bmatrix} \cos\phi & \sin\phi & 0 \\ -\sin\phi & \cos\phi & 0 \\ 0 & 0 & 1 \end{bmatrix} \tag{5.2.44}$$

and

$$\begin{bmatrix} x \\ y \\ z \end{bmatrix} = R(\mathbf{K}, \phi) \begin{bmatrix} X \\ Y \\ Z \end{bmatrix}, \tag{5.2.45}$$

where the indicated operation in equation 5.2.45 is matrix multiplication. It is a simple matter to derive the rotation matrices for rotations about the other two axes. These are given by

$$R(\mathbf{I}, \phi) = \begin{bmatrix} 1 & 0 & 0 \\ 0 & \cos\phi & \sin\phi \\ 0 & -\sin\phi & \cos\phi \end{bmatrix}, \tag{5.2.46}$$

$$R(\mathbf{J}, \phi) = \begin{bmatrix} \cos\phi & 0 & -\sin\phi \\ 0 & 1 & 0 \\ \sin\phi & 0 & \cos\phi \end{bmatrix}. \tag{5.2.47}$$

Note that we have written the rotation matrix in equations 5.2.43 through 5.2.45 not as a function of $\boldsymbol{K}$, a physical vector, but of $\mathbf{K}$, a vector representation, because the rotation matrix, being a numerical quantity, is written logically as a function of numerical vectors, that is, vector representations. This might cause some alarm because it is unclear whether the representation should be chosen with respect to the initial or the final basis. It makes no difference, however, because from equation 5.2.42 it follows that $\boldsymbol{K}$ and $\boldsymbol{k}$ will have the same representation with respect to either basis.

For an arbitrary rotation through an arbitrary angle ϕ about an axis $\hat{\mathbf{a}}$ (the circumflex of $\hat{\mathbf{a}}$ serves to remind us that $\hat{\mathbf{a}} \cdot \hat{\mathbf{a}} = 1$), the general formula is

$$R(\hat{\mathbf{a}}, \phi)\mathbf{v} = \cos\phi\mathbf{v} + (1 - \cos\phi)(\hat{\mathbf{a}} \cdot \mathbf{v})\hat{\mathbf{a}} - \sin\phi\hat{\mathbf{a}} \times \mathbf{v}, \tag{5.2.48}$$

which is equivalent to

$$R(\hat{\mathbf{a}}, \phi) = \cos\phi I + (1 - \cos\phi)\hat{\mathbf{a}}\hat{\mathbf{a}}^{\mathrm{T}} - \sin\phi[\hat{\mathbf{a}}\times], \tag{5.2.49}$$

where the matrix $[\hat{\mathbf{a}}\times]$ is given by equation 5.2.18. Either equation 5.2.48 or 5.2.49 is known as *Euler's formula*. If we write

$$\hat{\mathbf{a}} = \begin{bmatrix} a_1 \\ a_2 \\ a_3 \end{bmatrix}, \tag{5.2.50}$$

then we can write the individual elements of $R(\hat{\mathbf{a}}, \phi)$ as

$$R(\hat{\mathbf{a}}, \phi) = \begin{bmatrix} c + a_1^2(1-c) & a_1 a_2(1-c) + a_3 s & a_1 a_3(1-c) - a_2 s \\ a_2 a_1(1-c) - a_3 s & c + a_2^2(1-c) & a_2 a_3(1-c) + a_1 s \\ a_3 a_1(1-c) + a_2 s & a_3 a_2(1-c) - a_1 s & c + a_3^2(1-c) \end{bmatrix}, \tag{5.2.51}$$

where $c = \cos\phi$ and $s = \sin\phi$. One can verify directly that the determinant of the expression in equation 5.2.51 is $+1$. Thus, rotation matrices are proper orthogonal.

Equation 5.2.49 or 5.2.51 permits us to write out the rotation matrix for a rotation through an arbitrary angle of rotation about an arbitrary rotation axis. Euler showed that the converse is also true, namely, that every rotation can be expressed as a rotation about a single axis. This last result is known as *Euler's theorem*. Note that the axis of rotation has two free parameters, and that the angle of rotation is a third parameter. Hence, rotations are characterized by three parameters, which means that the nine elements of C must be subject to six constraints, expressed by equation 5.2.29 or 5.2.31.

5.2.4 The Euler Angles

Consider a sequence of three consecutive rotations about body axes,

$$C = R(\hat{\mathbf{a}}_3, \phi_3) R(\hat{\mathbf{a}}_2, \phi_2) R(\hat{\mathbf{a}}_1, \phi_1), \tag{5.2.52}$$

where the three rotation-axis column vectors, $\hat{\mathbf{a}}_1$, $\hat{\mathbf{a}}_2$, and $\hat{\mathbf{a}}_3$, must be chosen from the set consisting of the three unit column vectors

$$\hat{\mathbf{u}}_1 \equiv \begin{bmatrix} 1 \\ 0 \\ 0 \end{bmatrix}, \quad \hat{\mathbf{u}}_2 \equiv \begin{bmatrix} 0 \\ 1 \\ 0 \end{bmatrix}, \quad \text{and} \quad \hat{\mathbf{u}}_3 \equiv \begin{bmatrix} 0 \\ 0 \\ 1 \end{bmatrix}. \tag{5.2.53}$$

This section focuses on one possible choice, $\hat{\mathbf{a}}_1 = \hat{\mathbf{u}}_3$, $\hat{\mathbf{a}}_2 = \hat{\mathbf{u}}_1$, and $\hat{\mathbf{a}}_3 = \hat{\mathbf{u}}_3$, called a 3-1-3 sequence.

From the discussion of the last subsection, we can interpret this sequence of rotations as follows. The first rotation $R(\hat{\mathbf{a}}_1, \phi_1)$ connects the initial (typically, the inertial) basis $\{\boldsymbol{I}, \boldsymbol{J}, \boldsymbol{K}\}$ to an intermediate basis $\{\boldsymbol{I}', \boldsymbol{J}', \boldsymbol{K}'\}$. For the 3-1-3 sequence, since the column-vector representation of the axis of rotation is chosen to be $\hat{\mathbf{u}}_3$, the *physical axis* of rotation $\hat{\mathbf{a}}_1$ can only be the basis vector $\boldsymbol{K}$ (or, equivalently, $\boldsymbol{K}'$, since they are the same). Perhaps better said, the first rotation is about the body z-axis.

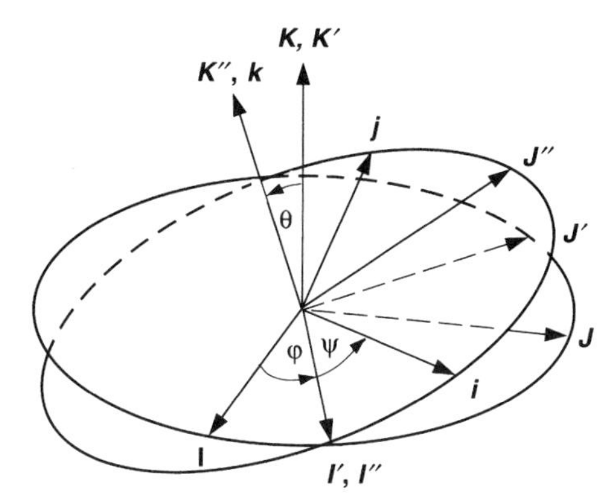

Figure 5.2 Rotation in three dimensions using 3-1-3 Euler angles.

The second rotation $R(\hat{\mathbf{a}}_2, \phi_2)$ connects the first intermediate basis $\{\boldsymbol{I}', \boldsymbol{J}', \boldsymbol{K}'\}$ to the second intermediate basis $\{\boldsymbol{I}'', \boldsymbol{J}'', \boldsymbol{K}''\}$. For the 3-1-3 sequence, the representation of the second axis of rotation is chosen to be $\hat{\mathbf{u}}_1$, and this representation must be with respect to either $\{\boldsymbol{I}\prime, \boldsymbol{J}', \boldsymbol{K}'\}$ or $\{\boldsymbol{I}'', \boldsymbol{J}'', \boldsymbol{K}''\}$. Thus, the second physical axis of rotation, $\hat{\mathbf{a}}_2$, is $\boldsymbol{I}'$ (or, equivalently, $\boldsymbol{I}''$, since they are the same). In either case, to someone seated on the spacecraft $\boldsymbol{I}'$ and $\boldsymbol{I}''$ are just the body x-axis.

In the same way, the third and final rotation for the 3-1-3 sequence must be about the body z-axis again. In space, the third axis is not necessarily identical to the first axis, but it is the direction toward which the body z-axis now points after being reoriented by the previous rotations. The rotation-axis column vectors in equation 5.2.52, if they are taken from the set given in equation 5.2.53, always correspond, therefore, to rotations about the current body coordinate axes. We say "current" because our point of view is typically that of an external observer. If we were seated on the spacecraft, we would think that the axes of rotations were just the body coordinate axes and the first rotation axis was the same as the last. A pictorial representation of the 3-1-3 sequence is shown in figure 5.2.

Every rotation matrix can be written as a 3-1-3 Euler sequence. (One could also write the rotation matrix as the product of four or more rotations, but, since rotations can be completely characterized by three parameters, three rotations about an appropriately chosen set of body axes are sufficient.) The axes, $\hat{\mathbf{a}}_1$, $\hat{\mathbf{a}}_2$, and $\hat{\mathbf{a}}_3$, are called the *Euler axes* and the angles, ϕ_1, ϕ_2, and ϕ_3, are called the *Euler angles*. The Euler angles are more commonly written as ϕ, θ, and ψ rather than ϕ_1, ϕ_2, and ϕ_3. Note that ϕ here does not have the same significance as previously, where it denoted the angle of rotation of the *entire* rotation. For a 3-1-2 Euler sequence (the third rotation is now about the body y-axis) or, in fact, for any sequence in which the three indices are distinct, the Euler angles are often called roll, pitch, and yaw, in analogy with aeronautics terminology. For a spacecraft that is Earth-locked, the *roll axis* is typically the body axis nominally closest in direction to the velocity, the *yaw axis* is the body axis nominally closest to the zenith, and the *pitch axis* is the body axis nominally closest to the orbit normal (the direction of the orbital angular velocity). When the spacecraft is not Earth-locked, assignment of these names may be quite arbitrary. Except when the Euler angles are very small, the values of roll, pitch, and yaw are not unambiguous because the values of the Euler angles corresponding to a given attitude depend in general on the order in which the three rotations are performed. The reader is cautioned, therefore, that even though the roll, yaw, and pitch axes of two spacecraft may be the same, the roll, yaw, and pitch angles do not necessarily have the same meaning if the order of the Euler angles is defined differently in each case.

Not every sequence of Euler axes can be used to parameterize any attitude. Rotations, as we have seen, have three degrees of freedom. In order for the representation in terms of the Euler angles to have three degrees of freedom, it is necessary that

$$\hat{\mathbf{a}}_1 \neq \hat{\mathbf{a}}_2 \quad \text{and} \quad \hat{\mathbf{a}}_2 \neq \hat{\mathbf{a}}_3. \tag{5.2.54}$$

If this were not true, we could combine the first two or the last two rotations by adding their angles, and there would be no more than two degrees of freedom (only one degree of freedom if we chose all three axes to be the same). Thus, once we choose the first axis, there are only two possible choices for the second axis; after choosing the second axis, there are only two possible choices for the third axis. It follows that there are twelve possible sets of axes that satisfy these conditions and, therefore, twelve possible sets of Euler angles. The 3-1-3 sequence of Euler angles has been particularly popular for the description of spacecraft attitude motion.

The dependence of the direction-cosine matrix on the Euler angles is somewhat forbidding. Consider the 3-1-3 Euler angles. If we carry out the two matrix multiplications, we find that the attitude matrix is given by

$$\begin{aligned} C(\phi, \theta, \psi) &= \begin{bmatrix} \cos\psi & \sin\psi & 0 \\ -\sin\psi & \cos\psi & 0 \\ 0 & 0 & 1 \end{bmatrix} \begin{bmatrix} 1 & 0 & 0 \\ 0 & \cos\theta & \sin\theta \\ 0 & -\sin\theta & \cos\theta \end{bmatrix} \begin{bmatrix} \cos\phi & \sin\phi & 0 \\ -\sin\phi & \cos\phi & 0 \\ 0 & 0 & 1 \end{bmatrix} \\ &= \begin{bmatrix} c\psi c\phi - s\psi c\theta s\phi & c\psi s\phi + s\psi c\theta c\phi & s\psi s\theta \\ -s\psi c\phi - c\psi c\theta s\phi & -s\psi s\phi + c\psi c\theta c\phi & c\psi s\theta \\ s\theta s\phi & -s\theta c\phi & c\theta \end{bmatrix}, \end{aligned} \tag{5.2.55}$$

where we have written $s\psi$ for $\sin\psi$, etc. In machine computation, it is generally advisable to have the computer carry out the two matrix multiplications rather than compute the individual element from the final formula, which increases the possibility of programming errors. Every rotation matrix can be written in terms of 3-1-3 Euler angles (or in terms of any of the twelve sets).

The Euler angles are not unique. In fact, it is easy to show by direct substitution for a 3-1-3 set of Euler angles that the two sets (ϕ, θ, ψ) and $(\phi + \pi, -\theta, \psi - \pi)$ generate identical attitude matrices. (The axis-angle representation is also double-valued, since $(\hat{\mathbf{a}}, \phi)$ and $(-\hat{\mathbf{a}}, -\phi)$ generate the same rotation matrix.) To make the Euler angles unique, one usually limits the 3-1-3 Euler angles to the values

$$0 \leq \phi < 2\pi, \quad 0 \leq \theta \leq \pi, \quad 0 \leq \psi < 2\pi. \tag{5.2.56}$$

The great advantage of the Euler angles is that they contain only three parameters as opposed to four for an axis and angle or nine for the direction-cosine matrix.

The situation often arises that we have computed the attitude matrix by some method and wish to determine the Euler angles in order to store the attitude more efficiently. To determine the Euler angles from a given rotation matrix, we note first from equation 5.2.55 that

$$C_{33} = \cos\theta. \tag{5.2.57}$$

Hence,

$$\theta = \arccos(C_{33}). \tag{5.2.58}$$

The principal value of the arccosine is consistent with equation 5.2.56. We note also that

$$C_{31} = \sin\theta \sin\phi \quad \text{and} \quad C_{32} = -\sin\theta \cos\phi, \tag{5.2.59}$$

from which we can write, for $\sin\theta > 0$,

$$\phi = \text{ATAN2}(C_{31}, -C_{32}), \tag{5.2.60}$$

where ATAN2(y, x) is a double-entry function that gives the angle whose tangent is y/x and which is in the correct quadrant. In equation 5.2.60 we have used the Fortran and MATLAB name for this function. In the same way, from

$$C_{13} = \sin\theta \sin\psi \quad \text{and} \quad C_{23} = \sin\theta \cos\psi, \tag{5.2.61}$$

it follows, for $\sin\theta > 0$, that

$$\psi = \text{ATAN2}(C_{13}, C_{23}). \tag{5.2.62}$$

For $\sin\theta = 0$, the situation becomes more complicated. In this case, the rotation matrix for the 3-1-3 set of Euler angles, evaluated from equation 5.2.55, has the form

$$C = \begin{bmatrix} \cos(\phi \pm \psi) & \sin(\phi \pm \psi) & 0 \\ -\sin(\phi \pm \psi) & \pm\cos(\phi \pm \psi) & 0 \\ 0 & 0 & \pm 1 \end{bmatrix}, \tag{5.2.63}$$

where the plus or minus sign is chosen according to the sign of $\cos\theta$. Thus, ϕ and ψ are not uniquely determined. One usually chooses arbitrarily as a solution

$$\psi = 0 \quad \text{and} \quad \phi = \text{ATAN2}(C_{12}, C_{11}). \tag{5.2.64}$$

It should be pointed out that only convenience and a distaste for ambiguity drive us to the restrictions of equations 5.2.56 and 5.2.64. Nothing is lost by relaxing these constraints, and in some cases (for example, when attempting to fit a smooth curve to the Euler angles) violation of these restrictions may be unavoidable.

Unfortunately, the Euler angles are not a well-behaved representation of attitude. In particular, they are singular because of ambiguity in the 3-1-3 Euler angles when $\theta = 0$ or π. If, instead, we had used 3-1-2 Euler angles, these would have been well behaved at $\theta = 0$ or π but would have shown singular behavior when $\theta = \pm\pi/2$. Thus, to avoid the singularity, we must resort to two sets of Euler angles and occasionally switch from one set to the other. We will see what the singularity difficulty means in practice when we consider attitude kinematics. Because of the singularity problem and the large number of trigonometric functions to be computed, the Euler angles are used less frequently nowadays, although their use was very frequent in the past. In one case, however, the Euler angles have a genuine advantage: in the support of spinning spacecraft. In this case, two of the Euler angles can be chosen to be the spherical angles of the spin axis and the remaining Euler angle the angle of rotation about the spin axis. We see, in fact, from equation 5.2.55 that if we had chosen the body z-axis to be the spin axis, the Euler angles

θ and ψ would be its spherical angles. For spinning spacecraft, the spin axis is generally stable and confined to some small region about a nominal direction. Two of the Euler angles then have limited variation, whereas the third tends to have an almost constant rate. Thus, for spinning spacecraft, the singularity can be avoided and the angles are well behaved.

5.2.5 The Quaternion

Because of the singularity difficulty of the Euler angles and the large number of elements in the direction-cosine matrix, one commonly uses another representation, the quaternion, in numerical calculations. The quaternion is based in part on Chasle's theorem, which states that "every rigid body motion can be realized by a rotation about an axis combined with a translation parallel to that axis." The quaternion is free of the analytical complexity that plagues the Euler angles and has only one addition component.

We define the *quaternion* $\bar{q}$ as the 4×1 matrix

$$\bar{q} \equiv \begin{bmatrix} q_1 \\ q_2 \\ q_3 \\ q_4 \end{bmatrix} = \begin{bmatrix} \mathbf{q} \\ q_4 \end{bmatrix}, \tag{5.2.65}$$

where

$$\mathbf{q} \equiv \begin{bmatrix} q_1 \\ q_2 \\ q_3 \end{bmatrix} \equiv \sin(\phi/2)\hat{\mathbf{a}}, \quad q_4 \equiv \cos(\phi/2), \tag{5.2.66}$$

the unit-column vector $\hat{\mathbf{a}}$ is, as usual, the representation of the rotation axis, and ϕ is the rotation angle. We call $\mathbf{q}$ the vector component of the quaternion and q_4 the scalar component. The quaternion thus has four components and, because this is one more component than it needs, it must satisfy a constraint, namely

$$\bar{q}^{\mathrm{T}}\bar{q} = q_1^2 + q_2^2 + q_3^2 + q_4^2 = 1, \tag{5.2.67}$$

as can be verified from equation 5.2.66.

In terms of the quaternion, the rotation matrix as given by equation 5.2.49 can be rewritten as

$$R = (q_4^2 - |\mathbf{q}|^2)I_{4\times 4} + 2\mathbf{q}\mathbf{q}^{\mathrm{T}} - 2q_4[\mathbf{q}\times], \tag{5.2.68}$$

where the matrix $[\mathbf{q}\times]$ is given by equation 5.2.18. For the individual elements, this expression is equivalent to

$$R = \begin{bmatrix} q_1^2 - q_2^2 - q_3^2 + q_4^2 & 2(q_1q_2 + q_4q_3) & 2(q_1q_3 - q_4q_2) \\ 2(q_2q_1 - q_4q_3) & -q_1^2 + q_2^2 - q_3^2 + q_4^2 & 2(q_2q_3 + q_4q_1) \\ 2(q_3q_1 + q_4q_2) & 2(q_3q_2 - q_4q_1) & -q_1^2 - q_2^2 + q_3^2 + q_4^2 \end{bmatrix}. \tag{5.2.69}$$

The derivation of equations 5.2.68 and 5.2.69 from equation 5.2.49 is left as an exercise for the reader.

The advantage of equation 5.2.68 or 5.2.69, which expresses the rotation matrix in terms of the quaternion, over equation 5.2.52 or 5.2.55, which expresses the rotation matrix in terms of the Euler angles, is that, instead of trigonometric functions, only a few simple multiplications or additions appear. There is a price to be paid, however, namely, that four parameters now specify attitude rather than three. In practice, this is a small price.

The above equations show us how to calculate the rotation matrix from the quaternion. We can calculate the quaternion from the rotation matrix equally readily. The result is

$$q_4 = \tfrac{1}{2}\sqrt{1 + R_{11} + R_{22} + R_{33}}, \tag{5.2.70}$$

and, if $q_4 \neq 0$, the remaining components can be calculated from the off-diagonal elements as

$$q_1 = \frac{1}{4q_4}(R_{23} - R_{32}), \quad q_2 = \frac{1}{4q_4}(R_{31} - R_{13}), \quad q_3 = \frac{1}{4q_4}(R_{12} - R_{21}). \tag{5.2.71}$$

These results follow directly from equations 5.2.67 and 5.2.69. Note that the sign of $\overline{q}$ is not determined by equations 5.2.52 and 5.2.55. However, since R is a quadratic function of $\overline{q}$, this does not matter; $\overline{q}$ and $-\overline{q}$ generate the same rotation matrix.

If q_4 is close to zero, equations 5.2.70 and 5.2.71 will not be very accurate. However, since the sum of the squares of the components of the quaternion is unity, at least one component must have a magnitude greater than or equal to $\frac{1}{2}$. Thus, we get around the problem of having $q_4 = 0$ by simply calculating one of the other components first and then calculating the other components from it. A simple, efficient method for calculating the quaternion in this way has been published by Shepperd (1978).

Unlike the case of the Euler angles where, depending on the attitude, two different sets of angles are needed to avoid the singularity, only one quaternion is needed. (Apart from the overall sign, only one quaternion characterizes a given attitude.) This is not the quaternion's only advantage. Unlike the Euler angles, which cannot be combined easily when one combines rotations, the quaternion has a very simple composition rule. We denote quaternion multiplication by simple juxtaposition of the two quaternions. Thus, if $\overline{q}$ is the quaternion of the first rotation and $\overline{q}'$ is the quaternion of the second quaternion, the combined rotation is represented by $\overline{q}''$, where

$$\overline{q}'' = \overline{q}'\overline{q}. \tag{5.2.72}$$

For this to be true, we must have also that

$$R(\overline{q}'') = R(\overline{q}')R(\overline{q}). \tag{5.2.73}$$

$R(\overline{q}'')$ is computed from $R(\overline{q})$ and $R(\overline{q}')$ by simple matrix multiplication. For quaternions, the multiplication rule, consistent with equations 5.2.72 and 5.2.73, is

$$\overline{q}'' = \overline{q}'\overline{q} \equiv \begin{bmatrix} q_4'\mathbf{q} + q_4\mathbf{q}' - \mathbf{q}' \times \mathbf{q} \\ q_4'q_4 - \mathbf{q}' \cdot \mathbf{q} \end{bmatrix}, \tag{5.2.74}$$

where the upper part of the expression on the right of equation 5.2.74 gives the vector component $\mathbf{q}''$, while the lower part gives q_4''. The composition rule for the quaternions is not unique, either, because we could have changed the sign of the rightmost member of equation 5.2.74. The sign convention of equation 5.2.74 is the one generally accepted and most convenient.

The quaternion has several advantages over the rotation matrix as a representation of attitude. First, it has fewer elements (four instead of nine), so it requires less storage. Second, there are fewer constraints (one instead of six). Third, the composition rule is simpler (16 multiplications instead of 27). Finally, the constraint is very simple to enforce. If, because of accumulated numerical roundoff error, the rotation matrix that we calculate turns out not to be orthogonal, it is a complicated process to make it orthogonal again. For the quaternion, however, we simply replace $\overline{q}$ by

$$\frac{1}{\sqrt{\overline{q}^{\mathrm{T}}\overline{q}}}\,\overline{q}. \tag{5.2.75}$$

5.2.6 Further References

The variety of attitude representations is considerably greater than has been presented in the present section. The short contributions by Markley (1978) and Duchon (1984) provide a quick summary of the more important representations. The books by Hughes (1986) and Junkins and Turner (1986) contain long chapters devoted to attitude representations. The first chapter of Kane et al. (1983) also treats representations. The reader is warned, however, that these last authors use a rather different convention for the rotations, which leads to many differences in sign in the equations from those presented here. The books by Wertz (1978) and Hughes (1986) contain tables of the Euler angles for all twelve sets. An exhaustive survey of the attitude representations has been prepared by Shuster (1993).

5.3 Attitude Kinematics

It is convenient to separate the study of attitude motion into two parts: attitude kinematics and attitude dynamics. By *attitude dynamics*, which we treat later in this chapter, we mean the response of a body to an applied torque. By *attitude kinematics*, on the other hand, we mean the fundamental description of how a change in attitude with time is characterized or, equivalently, what we mean by an *angular velocity*. To appreciate the distinction between kinematics and dynamics, we recall that for translational mechanics the kinematic equation is simply

$$\frac{d}{dt}\mathbf{r} = \mathbf{v}, \tag{5.3.1}$$

which has only geometric content. The dynamical equation is Newton's second law of motion,

$$\frac{d}{dt}\mathbf{v} = \frac{1}{m}\mathbf{F}(\mathbf{r}). \tag{5.3.2}$$

The corresponding equations for attitude, as we shall see, are much more complicated.

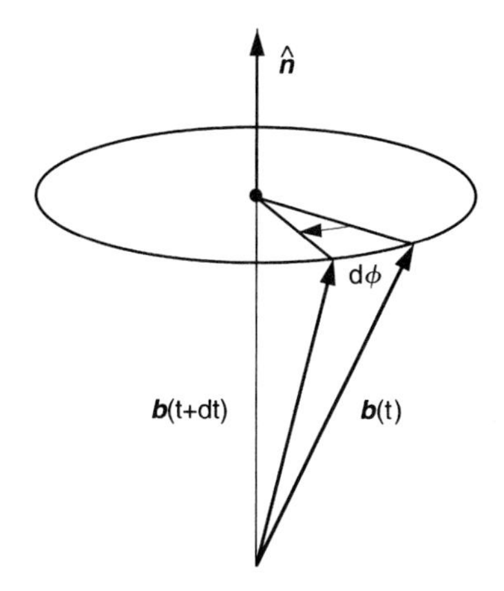

Figure 5.3 Temporal change of a rotating vector.

5.3.1 Rates of Change in Moving Frames

Consider a body rotating about an axis $\hat{\boldsymbol{n}}$ and with a vector $\boldsymbol{b}$ that is fixed in the rotating body, and is otherwise unchanging, as shown in figure 5.3. The change of the vector $\boldsymbol{b}$ over a time interval Δt is given by

$$\Delta \boldsymbol{b} = \boldsymbol{b}(t + \Delta t) - \boldsymbol{b}(t) = \Delta\phi(\hat{\boldsymbol{n}} \times \boldsymbol{b}), \tag{5.3.3}$$

where $\Delta\phi$ is the angle through which the body has rotated about the (instantaneous) axis $\hat{\boldsymbol{n}}$. The temporal derivative of $\boldsymbol{b}$ is simply

$$\frac{d}{dt}\boldsymbol{b} = \lim_{\Delta t \to 0} \frac{\Delta \boldsymbol{b}}{\Delta t} = \lim_{\Delta t \to 0} \frac{\Delta\phi}{\Delta t}(\hat{\boldsymbol{n}} \times \boldsymbol{b}) = \frac{d\phi}{dt}(\hat{\boldsymbol{n}} \times \boldsymbol{b}). \tag{5.3.4}$$

The physical vector defined by

$$\overline{\boldsymbol{\omega}} \equiv \frac{d\phi}{dt}\hat{\boldsymbol{n}} \tag{5.3.5}$$

is called the *angular velocity* of the body, and we write more simply

$$\frac{d}{dt}\boldsymbol{b} = \overline{\boldsymbol{\omega}} \times \boldsymbol{b}. \tag{5.3.6}$$

Suppose now that $\{\boldsymbol{I}, \boldsymbol{J}, \boldsymbol{K}\}$ is a right-handed orthonormal basis fixed in the inertial frame. Then

$$\frac{d}{dt}\boldsymbol{I} = \frac{d}{dt}\boldsymbol{J} = \frac{d}{dt}\boldsymbol{K} = \boldsymbol{0}, \tag{5.3.7}$$

since the inertial axes do not change with time. Thus,

$$\frac{d}{dt}(\boldsymbol{I} \cdot \boldsymbol{b}) = \boldsymbol{I} \cdot \frac{d}{dt}\boldsymbol{b}, \tag{5.3.8}$$

and similarly for $\boldsymbol{J}$ and $\boldsymbol{K}$, so that the temporal derivative of the inertial components of $\boldsymbol{b}$ are just the inertial components of the temporal derivative of $\boldsymbol{b}$. Hence, from equations 5.2.40 and 5.3.6,

$$\frac{d}{dt}\mathbf{b}_{\mathrm{I}} = \boldsymbol{\omega}_{\mathrm{I}} \times \mathbf{b}_{\mathrm{I}}, \tag{5.3.9}$$

where the subscript $_{\mathrm{I}}$ denotes the column-vector representation with respect to inertial coordinates.

What about the components of $\boldsymbol{b}$ with respect to a basis $\{\boldsymbol{i}, \boldsymbol{j}, \boldsymbol{k}\}$ fixed in the body? Since the vector is rotating with the body, its components with respect to body axes must be constant. Hence,

$$\frac{d}{dt}\mathbf{b}_{\mathrm{B}} = \mathbf{0}, \tag{5.3.10}$$

where the subscript $_{\mathrm{B}}$ denotes the column-vector representation with respect to the body-fixed coordinate system.

5.3.2 Kinematic Equation for the Direction-Cosine Matrix

If the body is rotating, the direction-cosine matrix must be changing with time. What do these relations tell us about the change in the attitude? Suppose that the physical vector $\boldsymbol{b}$ (still fixed in the body) has no intrinsic time dependence and consider equation 5.3.6 again. From our discussion in the last section we know that for any (physical) vector $\boldsymbol{r}$

$$\mathbf{r}_{\mathrm{B}}(t) = C(t)\mathbf{r}_{\mathrm{I}}(t). \tag{5.3.11}$$

Differentiating both sides with $\boldsymbol{r} = \boldsymbol{b}$ (and suppressing the explicit time dependence in our expression) leads to

$$\begin{aligned} \mathbf{0} = \frac{d}{dt}\mathbf{b}_{\mathrm{B}} &= \left(\frac{d}{dt}C\right)\mathbf{b}_{\mathrm{I}} + C\frac{d}{dt}\mathbf{b}_{\mathrm{I}} \\ &= \left(\frac{d}{dt}C\right)\mathbf{b}_{\mathrm{I}} + C(\boldsymbol{\omega}_{\mathrm{I}} \times \mathbf{b}_{\mathrm{I}}). \end{aligned} \tag{5.3.12}$$

Since C is a proper orthogonal matrix, we know from equation 5.2.40 that we can write the last line above as

$$\left(\frac{d}{dt}C\right)\mathbf{b}_{\mathrm{I}} + (C\boldsymbol{\omega}_{\mathrm{I}}) \times (C\mathbf{b}_{\mathrm{I}}) = \mathbf{0}. \tag{5.3.13}$$

Now, the expression $(C\boldsymbol{\omega}_{\mathrm{I}})$ is just $\boldsymbol{\omega}_{\mathrm{B}}$, the representation of $\overline{\boldsymbol{\omega}}$ with respect to body-fixed coordinates. Recalling also the definition of the cross-product matrix, we may recast equation 5.3.13 as

$$\left(\frac{d}{dt}C\right)\mathbf{b}_{\mathrm{I}} + [\boldsymbol{\omega}_{\mathrm{B}}\times](C\mathbf{b}_{\mathrm{I}}) = \mathbf{0}, \tag{5.3.14}$$

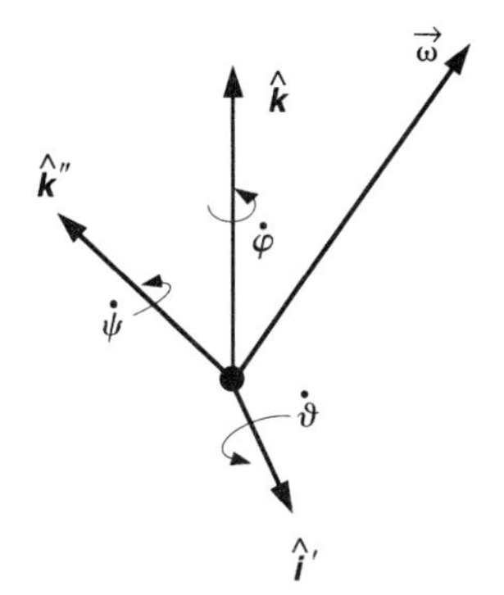

Figure 5.4 Angular velocity vector from Euler rates.

and, noting that matrix multiplication is associative, this becomes

$$\left(\frac{d}{dt}C\right)\mathbf{b}_{\mathrm{I}} + ([\boldsymbol{\omega}_{\mathrm{B}}\times]C)\mathbf{b}_{\mathrm{I}} = \mathbf{0}. \tag{5.3.15}$$

We now note that the 3×1 matrix $\mathbf{b}_{\mathrm{I}}$ that appears in both terms of equation 5.3.15 is arbitrary. Therefore, we may cancel it from the equation to obtain

$$\frac{d}{dt}C = -[\boldsymbol{\omega}_{\mathrm{B}}\times]C. \tag{5.3.16}$$

This is the kinematic equation for the direction-cosine matrix.

From equation 5.3.16 we may now write the general expression relating the time derivatives of the representations of vectors in two frames. Consider equation 5.3.11 and let $\boldsymbol{r}$ have any arbitrary time dependence. Differentiating equation 5.3.11, with $\boldsymbol{r}$ now arbitrary, leads directly to

$$\begin{aligned}\frac{d}{dt}\mathbf{r}_{\mathrm{B}} &= \frac{dC}{dt}\mathbf{r}_{\mathrm{I}} + C\frac{d}{dt}\mathbf{r}_{\mathrm{I}}\\ &= -[\boldsymbol{\omega}_{\mathrm{B}}\times]C\mathbf{r}_{\mathrm{I}} + C\frac{d}{dt}\mathbf{r}_{\mathrm{I}}\\ &= -\boldsymbol{\omega}_{\mathrm{B}} \times \mathbf{r}_{\mathrm{B}} + C\frac{d}{dt}\mathbf{r}_{\mathrm{I}}.\end{aligned} \tag{5.3.17}$$

Equation 5.3.17 is true no matter what the nature of the time dependence of $\boldsymbol{r}$. Note the difference in sign between equations 5.3.17 and 5.3.9.

5.3.3 The Kinematic Equation for the Euler Angles

The kinematic relationship for the Euler angles is more complicated. The angular velocity in terms of the Euler angle rates is obtained by adding up the angular velocities from each of the degrees of freedom. Thus, in terms of (abstract) physical vectors,

$$\overline{\omega} = \dot{\phi}\hat{\boldsymbol{a}}_1 + \dot{\theta}\hat{\boldsymbol{a}}_2 + \dot{\psi}\hat{\boldsymbol{a}}_3, \tag{5.3.18}$$

as shown in figure 5.4 for the 3-1-3 sequence.

The body-axis representation of the angular velocity vector is then simply

$$\boldsymbol{\omega}_{\mathrm{B}} = \dot{\phi}(\hat{\mathbf{a}}_1)_{\text{final body}} + \dot{\theta}(\hat{\mathbf{a}}_2)_{\text{final body}} + \dot{\psi}(\hat{\mathbf{a}}_3)_{\text{final body}}. \tag{5.3.19}$$

We have written explicitly that the representation of the Euler axis vectors in equation 5.3.19 is now with respect to the final body axes $\{\boldsymbol{i}, \boldsymbol{j}, \boldsymbol{k}\}$ for *all three* Euler axis vectors, since we desire a representation of $\overline{\omega}$ with respect to that basis. The values of the column-vector representations of the Euler axes that are used in evaluating the rotation matrix, however, are not the representations with respect to the final body axes but the representations with respect to the intermediate (that is, the "current") body axes, which are not identical to representations with respect to the final body bases (except for $\hat{\mathbf{a}}_3$). Thus we must transform the simple representations of the Euler axes with respect to the intermediate bases into representations with respect to the final body axes. To do this we note that the physical vector $\hat{\mathbf{a}}_2$ is one of the coordinate axes of $\{\boldsymbol{I}', \boldsymbol{J}', \boldsymbol{K}'\}$ or $\{\boldsymbol{I}'', \boldsymbol{J}'', \boldsymbol{K}''\}$ and its representation with respect to either of these two bases will be the same. Thus we may assume the set of intermediate axes to be the latter one and we need only multiply the representation $\hat{\mathbf{a}}_2$ by the last rotation $R(\hat{\mathbf{a}}_3, \phi)$ to obtain the representation with respect to the final body axes. For $\hat{\mathbf{a}}_1$ we must multiply by the last two rotations. Thus, carrying out the necessary transformations, equation 5.3.19 becomes

$$\boldsymbol{\omega}_{\mathrm{B}} = \dot{\psi}\hat{\mathbf{a}}_3 + \dot{\theta}R(\hat{\mathbf{a}}_3, \psi)\hat{\mathbf{a}}_2 + \dot{\phi}R(\hat{\mathbf{a}}_3, \psi)R(\hat{\mathbf{a}}_2, \theta)\hat{\mathbf{a}}_1, \tag{5.3.20}$$

where the unit column vectors $\hat{\mathbf{a}}_1$, $\hat{\mathbf{a}}_2$, and $\hat{\mathbf{a}}_3$ are the representations of the three Euler axes with respect to the intermediate bases and take on any of the three values given by equation 5.2.53.

For the case of a 3-1-3 sequence of Euler angles, if we substitute for the rotation matrices and carry out the indicated matrix multiplications, equation 5.3.20 becomes

$$\begin{aligned}\begin{bmatrix}\omega_1\\ \omega_2\\ \omega_3\end{bmatrix}_{\mathrm{B}} &= \dot{\psi}\hat{\mathbf{u}}_3 + \dot{\theta}R(\hat{\mathbf{u}}_3, \psi)\hat{\mathbf{u}}_1 + \dot{\phi}R(\hat{\mathbf{u}}_3, \psi)R(\hat{\mathbf{u}}_1, \theta)\hat{\mathbf{u}}_3\\ &= \dot{\psi}\begin{bmatrix}0\\0\\1\end{bmatrix} + \dot{\theta}\begin{bmatrix}\cos\psi\\ -\sin\psi\\ 0\end{bmatrix} + \dot{\phi}\begin{bmatrix}\sin\theta\sin\psi\\ \sin\theta\cos\psi\\ \cos\psi\end{bmatrix}\\ &= \begin{bmatrix}\dot{\phi}\sin\theta\sin\psi + \dot{\theta}\cos\psi\\ \dot{\phi}\sin\theta\cos\psi - \dot{\theta}\sin\psi\\ \dot{\phi}\cos\theta + \dot{\psi}\end{bmatrix}.\end{aligned} \tag{5.3.21}$$

These can be solved for the Euler angle rates as a function of the angular velocity vector to yield

$$\begin{bmatrix}\dot{\phi}\\ \dot{\theta}\\ \dot{\psi}\end{bmatrix} = \frac{1}{\sin\theta}\begin{bmatrix}\sin\psi & \cos\psi & 0\\ \sin\theta\cos\psi & -\sin\theta\sin\psi & 0\\ -\cos\theta\sin\psi & -\cos\theta\sin\psi & \sin\theta\end{bmatrix}\boldsymbol{\omega}_{\mathrm{B}}. \tag{5.3.22}$$

We now make an unhappy discovery: the Euler angle rates become infinite when $\sin\theta = 0$ even if the angular velocity vector is finite. This is another indication that the Euler angles are singular. This singularity is not only a mathematical phenomenon. In real gyros, where the Euler angles correspond to actual gimbal angles, this situation forces two of the gimbals to have infinite angular velocities, which is impossible since the gimbals are not massless. Instead of traveling with infinite velocity the gimbals seize, causing a phenomenon known as *gimbal lock*.

This problem is remedied by using a different sequence of Euler angles, say, 1-3-2, in our calculations when this condition is approached. For physical gyros, one adds an extra gimbal, which permits the system to change mechanically from, say, 3-1-3 to 1-3-2 (or to 2-3-1, depending on whether the fourth gimbal has been added to the inside or outside of our 3-1-3 sequence). The 1-3-2 and 2-3-1 sequences of Euler angles are not trouble-free either but show gimbal-lock effects when $\cos\theta = 0$. Consequently, one must alternate between two sequences of Euler angles to represent the attitude as it changes. This singularity problem is inherent in the Euler angles but absent in the rotation matrix and the quaternion, which are well behaved at all attitudes.

5.3.4 The Kinematic Equation for the Quaternion

Like that for the rotation matrix, the kinematic equation for the quaternion is very simple and has the form

$$\left.\begin{aligned} \frac{d}{dt}\mathbf{q} &= \tfrac{1}{2}(q_4\boldsymbol{\omega}_\mathrm{B} - \boldsymbol{\omega}_\mathrm{B}\times\mathbf{q}), \\ \frac{d}{dt}q_4 &= -\tfrac{1}{2}\boldsymbol{\omega}_\mathrm{B}\cdot\mathbf{q}. \end{aligned}\right\} \tag{5.3.23}$$

Of prime importance is the fact that the kinematic equation for the quaternion is linear in the elements of both $\boldsymbol{\omega}_B$ and $\overline{q}$. This simplifies numerical integration of the kinematic equation. For this reason and the simple method of correcting normalization errors, the quaternion is the favorite representation for simulation studies.

5.3.5 Further References

Short summaries of attitude kinematics are presented in Wertz (1978) and Duchon (1984). The books by Hughes (1986) and Junkins and Turner (1986) treat these topics in more detail, as does the book of Kane et al. (1983). The reader is again cautioned that these last authors use a rather different convention for the rotations from the present chapter and from the other cited works. A complete account of the kinematic equations for the different attitude representations has been given by Shuster (1993) in his survey of attitude representations.

5.4 Attitude Measurements

By an attitude measurement, we mean generally the measurement of any quantity sensitive to the attitude, for example, the magnetic field vector; the direction of the geocenter, the Sun, a star, or some other body; the measurement of an angle (such as solar aspect, or nadir angle); or the measurement of angles or angular rates from gyros. The character of gyro measurements is very different from that of the other measurements, because the attitude information they provide is not absolute.

Let us consider first absolute attitude measurements. The sensors may be divided into two groups: coarse attitude sensors and precise attitude sensors. In the first group are primarily coarse magnetometers, Sun sensors, and Earth horizon sensors; in the second are primarily fine Sun sensors and star sensors. Each of these sensors measures effectively the three components of a vector. The magnetometer actually measures three separate components; the other sensors measure the direction of a vector and, hence, measure a unit vector.

5.4.1 Magnetometers

In the most general terms, a magnetometer measures the strength of a magnetic field in one direction. A 3-axis magnetometer, consisting of three orthogonal magnetometers, measures the three components of the magnetic field. If $\mathbf{B}_\mathrm{B}$ is the value of the magnetic field in the spacecraft body coordinate system as determined from the magnetometer, and $\mathbf{B}_\mathrm{I}$ is the known value of the magnetic field in inertial coordinates, then

$$\mathbf{B}_\mathrm{B} = C\mathbf{B}_\mathrm{I}, \tag{5.4.1}$$

where C is the direction-cosine matrix [rotation matrix, (proper) orthogonal matrix, attitude matrix] describing the attitude. Thus the measurement of the magnetic field provides a measurement of the attitude (in this case relative to inertial coordinates). The vector $\mathbf{B}_\mathrm{I}$ is generally given by the International Geomagnetic Reference Field (IGRF) model, which requires knowledge of the spacecraft position.

The magnetometer does not measure the magnetic field in the body frame but in a frame fixed in the magnetometer. The measured magnetic field in the body frame is obtained from the measured field in the sensor frame according to

$$\mathbf{B}_\mathrm{B} = S_\mathrm{mag}\mathbf{B}_\mathrm{mag}, \tag{5.4.2}$$

where S_mag is the magnetometer alignment matrix, a proper orthogonal matrix that transforms representations from the magnetometer frame to the body frame. This matrix is determined before launch but is often re-estimated in space.

Measurements are never perfect but are corrupted by noise. Thus, we should write more accurately

$$\mathbf{B}_\mathrm{B} = C\mathbf{B}_\mathrm{I} + \Delta\mathbf{B}_\mathrm{B}, \tag{5.4.3}$$

where $\Delta\mathbf{B}_\mathrm{B}$ is the random noise, resolved along body axes. The random noise can be made up of alignment errors, biases, scale factor errors, spacecraft electrical activity, and the like. The largest component of this noise may arise not from the sensor itself but from the magnetic field model, that is, our value for $\mathbf{B}_\mathrm{I}$. For near-Earth orbits, the error in the modeled field direction may vary from 0.5° near the Equator to 3° near the magnetic poles, where erratic auroral currents play a large role. Far from the Earth, the magnetic field is too weak to be an acceptable attitude reference and, in fact, is dominated not by the geomagnetic field but by the interplanetary field, whose time dependence is complicated and largely unpredictable.

Measurement of the magnetic field alone cannot be sufficient to determine the attitude completely, because the magnitudes of the vectors in equation 5.4.1 provide no information about the attitude (if we multiply both vectors by the same factor, the equation

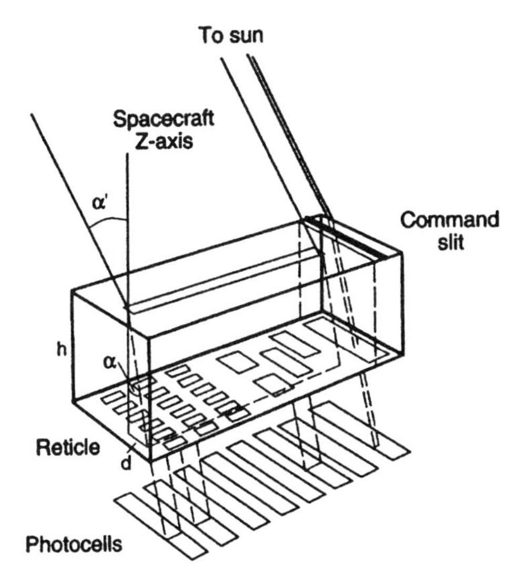

Figure 5.5 Single-slit Sun sensor.

will still be true). Thus the magnetic field measurement, although a vector, has only two degrees of freedom that are sensitive to the attitude (namely, the direction). Since we need three parameters to specify the attitude, one vector measurement is not enough. The other vector measurements listed above are all directions and, therefore, unit vectors with only two degrees of freedom. Thus, measurements of at least two vectors are needed to determine the attitude. (It is possible, over time and with appropriate dynamics modeling, to estimate the attitude with a magnetometer alone, using measurements separated significantly in time.)

5.4.2 Sun Sensors

The vector Sun sensor works on a different principle. In fact, a vector Sun sensor consists of two sensors, each consisting essentially of a rectangular chamber with a thin slit at the top (figure 5.5). Light entering the chamber through the slit casts an image of a thin line on the bottom of the chamber, which is lined with a network of light-sensitive cells that effectively measures the distance d of the image from a centerline. If h is the height of the chamber, then the angle of refraction α is given by

$$\tan \alpha = \frac{d}{h}. \tag{5.4.4}$$

The angle of incidence to the sensor, α', is related to the angle of refraction by Snell's law, $\sin \alpha' = n \sin \alpha$, where n is the index of refraction of the sensor block material, and the index of refraction of space is unity. By placing two sensors perpendicular to each other (figure 5.6), one can measure the complete direction of the Sun with respect to the sensor axes. The geometry for interpreting these two measurements is shown in figure 5.7.

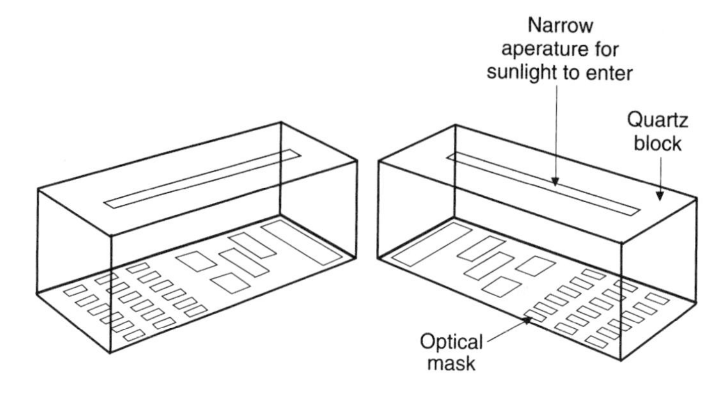

Figure 5.6 Two-slit Sun sensor.

To determine the Sun direction in the sensor frame from the two angle measurements α and β, we note that these are related to the Sun direction by

$$\tan\alpha = \frac{\hat{\mathbf{x}} \cdot \hat{\mathbf{S}}}{\hat{\mathbf{z}} \cdot \hat{\mathbf{S}}}, \quad \tan\beta = \frac{\hat{\mathbf{y}} \cdot \hat{\mathbf{S}}}{\hat{\mathbf{z}} \cdot \hat{\mathbf{S}}}, \tag{5.4.5}$$

where $\hat{\mathbf{x}}$, $\hat{\mathbf{y}}$, and $\hat{\mathbf{z}}$ are the axes of the sensor coordinate system. The z-axis of the sensor is usually defined as the outwardly directed normal. From these measurements we construct the unit vector $\hat{\mathbf{S}}_{\text{Sun}}$ in sensor coordinates as

$$\hat{\mathbf{S}}_{\text{Sun}} = \frac{1}{\sqrt{1 + \tan^2\alpha + \tan^2\beta}} \begin{bmatrix} \tan\alpha \\ \tan\beta \\ 1 \end{bmatrix}. \tag{5.4.6}$$

This measurement, of course, is of the Sun direction in the Sun sensor frame. The measurement in the body frame is obtained from

$$\hat{\mathbf{S}}_{\text{B}} = S_{\text{Sun}}\hat{\mathbf{S}}_{\text{Sun}} \tag{5.4.7}$$

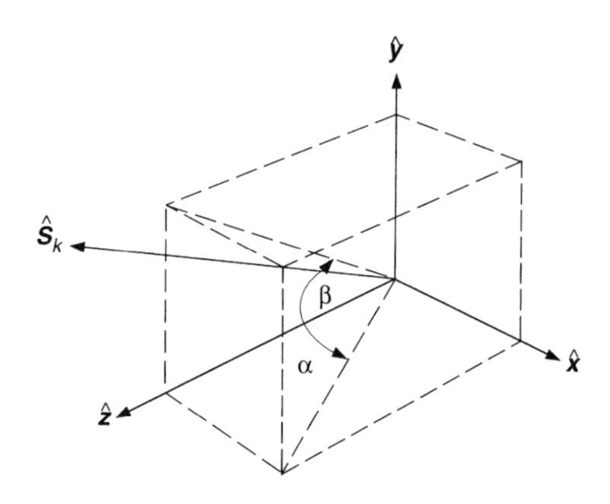

Figure 5.7 Projected angles for direction sensors.

and, allowing for instrument noise,

$$\hat{\mathbf{S}}_{\mathrm{B}} = C\hat{\mathbf{S}}_{\mathrm{I}} + \Delta\hat{\mathbf{S}}_{\mathrm{B}}, \tag{5.4.8}$$

in analogy with similar relations for the vector magnetometer.

Purely digital Sun sensors (sometimes called *digital solar aspect detectors*, or DSADs) determine the angles of the Sun by determining which of the light-sensitive cells in the sensor is the most strongly illuminated. Thus the accuracy of this sensor is limited by the angular diameter of the Sun, which is approximately 0.5° as seen from the Earth. Fine Sun sensors use analog information in addition to digital information. By knowing the intensity of sunlight striking neighboring pixels, the direction of the centroid of the Sun can be computed to within a few arcseconds. Less accurate measurements can be had by simply exposing an unobstructed solar cell to the Sun. Modeling of the cell output as a function of the Sun angle relative to the cell normal (as will be seen in the next chapter) provides a coarse measurement of the Sun angle. Use of multiple cells, generally in a pyramid configuration, can provide a measurement of the Sun vector.

Vector Sun sensors generally have fields of view of ±60°. Therefore, if the spacecraft is not inertially stabilized at one attitude throughout the mission, it will be necessary to use more than one sensor head to ensure that the Sun is always visible in one head. Generally, for agile spacecraft equipped with purely digital Sun sensors (that is, of low accuracy), the sensor assembly has five separate heads, usually sharing the same electronics unit. A suite of five precise Sun sensor heads may be uneconomical, however. Like the magnetometer, a Sun sensor alone (or even five Sun sensors) is not sufficient to determine the attitude.

In addition to the "fixed" Sun sensors described, there are also "spinning" Sun sensors. It is often not the Sun sensor that actually spins in this case, but the spacecraft, which then carries the Sun sensor with it. The sensor is often similar to the single-slit Sun sensor discussed earlier but is accompanied by a bright-object sensor so that the Sun angle is read into a buffer only when the Sun line crosses this bright-object sensor. Resolution of these sensors is typically 1°. Since the orientation of the sensor in the spacecraft body frame is known, this Sun crossing provides, in fact, the complete direction (that is, both spherical angles) of the Sun at the time of crossing, not just its angle from the spin axis.

Another way in which this spinning Sun sensor can be implemented is as a V-slit sensor. Such a sensor consists of two narrow slits with a field of view up to 160°. The two slits are not parallel but make some small angle. If the angular velocity of the spacecraft is known, the time interval between the two Sun crossings is a measure of the angle between the Sun and the spin axis. The accuracy of this type of sensor can be better than 0.1°.

5.4.3 Earth Horizon Sensors

Earth horizon sensors tend to be of two types: scanning and static. Static horizon sensors have a field of view slightly larger than the Earth and contain a number of sensors that sense infrared radiation from the Earth's surface (figure 5.8). The signal from each of the individual sensing elements is proportional to the fraction of its field of view on which the Earth intrudes. From these signals the direction of the geocenter, that is, the nadir, can be determined with an accuracy of from 0.1° in near-Earth orbit to 0.01° at

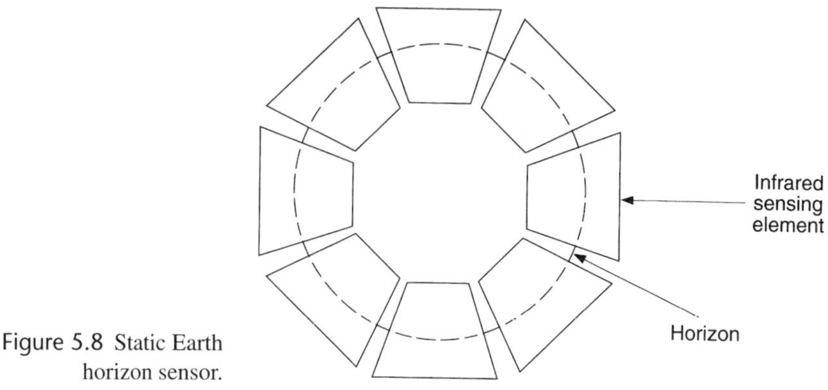

Figure 5.8 Static Earth horizon sensor.

geosynchronous altitudes. Their use is generally restricted to spacecraft with circular orbits.

Earth horizon scanners employ a spinning mirror or prism to focus a narrow pencil of light onto a sensing element that is usually called a *bolometer* (figure 5.9). Spinning of the mirror may be a result of the spacecraft spinning, or of a motor spinning the mirror while the spacecraft is rotating only enough to remain nadir pointing. The latter case can be accomplished with a dedicated motor, or the mirror could be attached to a momentum wheel. As the mirror or prism rotates (typically at 120 or 240 rpm for a dedicated motor), its field of view sweeps out a cone with half-cone angle typically around 45°. Electronics in the sensor detect when the infrared (IR) signal from the Earth is first received or finally lost during each sweep of the scan cone (figure 5.10). The time between the arrival and loss of signal (AOS and LOS) determines the Earth width. From this, the roll angle (the angle about the axis perpendicular to both the scan-cone axis and the vertical) can be inferred. (The nomenclature is somewhat loose here since the horizon scanner roll angle is not necessarily the same as the spacecraft attitude roll

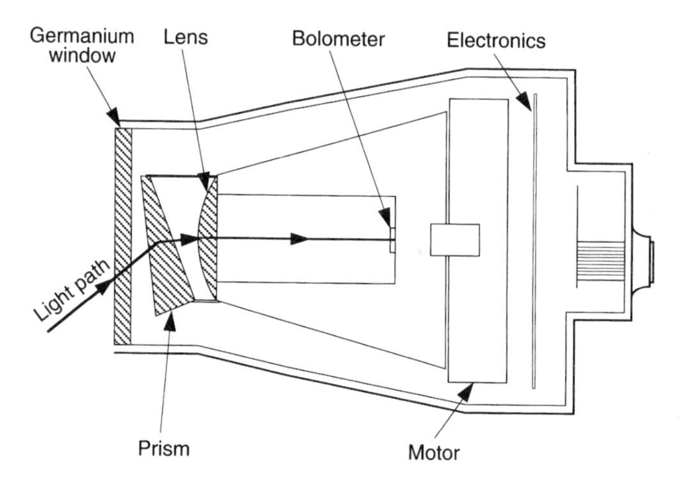

Figure 5.9 Scanning Earth horizon sensor.

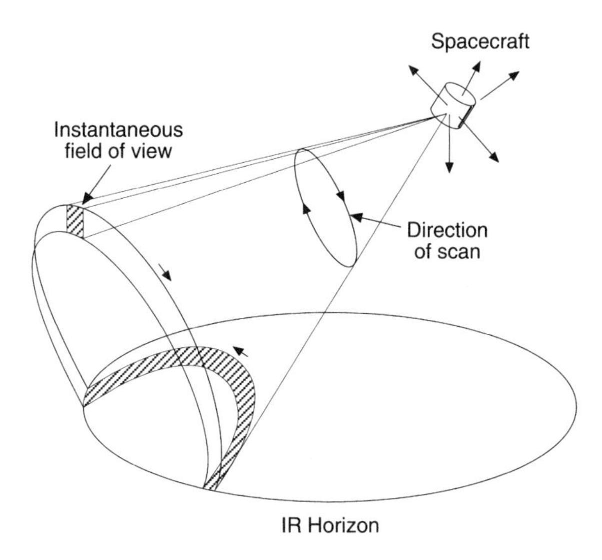

Figure 5.10 Scan cone geometry.

angle defined in section 5.2. Unfortunately this practice is ingrained.) The Earth width measured by the horizon scanner is denoted by Ω,

$$\Omega = \omega_{\text{scanner}}(t_{\text{LOS}} - t_{\text{AOS}}), \tag{5.4.9}$$

with ω_{scanner} the scanner spin rate. If the angular radius of the Earth as seen from the spacecraft is denoted by ρ and the half-cone angle of the scan by γ, then the scanner roll angle (sometimes also called the *nadir angle*) η is a solution of

$$\cos\rho = \cos\gamma\cos\eta + \sin\gamma\sin\eta\cos(\Omega/2). \tag{5.4.10}$$

For altitudes such that $\rho \geq \gamma$, only a single value of η, $0 \leq \eta \leq \pi$, satisfies equation 5.4.10. However, for altitudes such that $\rho < \gamma$ there will, in general, be two solutions to equation 5.4.10, only one of which is correct. Other information must then be used to eliminate the false solution. Another approach is to use two horizon scanners to eliminate the ambiguity, since the false nadir angle will generally be different for the two scanners. The angular radius of the Earth ρ decreases with altitude, being $\pi/2$ at zero altitude and 0 infinitely far from the Earth. For a typical scanner half-cone angle of 45 degrees, the altitude at which $\rho = \gamma$ is roughly 1700 km. Thus, for near-Earth spacecraft generally there is only a single solution for η.

If there is a "pickoff" element in the sensor whose crossing can be detected with each sensor pass, called the pitch reference point, then the scanner pitch angle (the angle about the scan-cone axis, which frequently *but not always* corresponds to the spacecraft pitch) can also be determined. If t_{pickoff} is the time at which the scanner field of view crosses the pitch reference point, then the scanner pitch angle is given by

$$p = \omega_{\text{scanner}}\left[\tfrac{1}{2}(t_{\text{LOS}} - t_{\text{AOS}}) - t_{\text{pickoff}}\right]. \tag{5.4.11}$$

The term in parentheses in equation 5.4.11, the midpoint between the AOS and LOS crossings, is often called the "split" point; hence, the frequently used alternative terminology for p, the "split-to-index angle."

If the spin axis in body frame is the $\boldsymbol{k}$-axis, and pitch is measured from the $\boldsymbol{i}$-axis, then the nadir vector $\hat{\mathbf{E}}$ in sensor coordinates is

$$\hat{\mathbf{E}}_{\mathrm{HS}} = \begin{bmatrix} \sin\eta\cos p \\ \sin\eta\sin p \\ \cos\eta \end{bmatrix}, \tag{5.4.12}$$

and the nadir vector in body coordinates is

$$\hat{\mathbf{E}}_{\mathrm{B}} = S_{\mathrm{HS}}\hat{\mathbf{E}}_{\mathrm{HS}}. \tag{5.4.13}$$

As for the vector magnetometer and Sun sensor,

$$\hat{\mathbf{E}}_{\mathrm{B}} = C\hat{\mathbf{E}}_{\mathrm{I}} + \Delta\hat{\mathbf{E}}_{\mathrm{B}}. \tag{5.4.14}$$

Typically, horizon sensor accuracies are from 0.1° to 1.0°. The accuracy of the horizon scanner or static horizon sensor depends to some degree on the amount of data processing done. In addition to random electronic noise, account must be taken of the fact that the Earth's horizon is not perfectly circular (because the Earth is oblate) and that the horizon sensor detects not the first contact with land or ocean but the point in the atmosphere at which the 16 μm infrared radiation reaches a certain intensity. This last effect varies as a function of both the season and the latitude. In near-Earth orbit, the neglect of these oblateness and horizon radiance corrections can decrease the accuracy from 0.3° to 1.0°.

5.4.4 Star Sensors

From the perspective of data processing, the two-axis star sensor is very similar to that of the fine Sun sensors. Although very different technologies are in use, virtually all two-axis star sensors function in the same way: starlight strikes a light-sensitive surface, the point of impact on the surface is determined, and the plane angles α and β are determined from this point. From α and β one determines the unit vector of the star in the sensor frame.

Great advances have occurred in the technology of star sensors, both in how they sense starlight and in their processing capabilities. In the past, star sensors used cathode ray tubes (CRTs) and photomultipliers to measure the point of impact. These required large amounts of power and generated large magnetic fields, which sometimes interfered with other attitude sensors or with the spacecraft payload. The current line of star sensors uses charged-coupled devices (CCDs), similar to (and sometimes identical to) the optical element in a video camera, to measure the point of impact. These solid-state devices have much smaller power requirements and hence smaller currents, and they therefore generate much smaller magnetic disturbances than CRTs and photomultipliers. The two-axis star sensors are called *star trackers* if they are able to lock onto and follow a star. Star sensors that can determine the directions of several stars simultaneously are sometimes called *star cameras*. This nomenclature is overlapping, and its usage is by no means consistent. It is even more confusing now since most, if not all, of the new sensors lock onto and follow multiple stars. In general we now almost exclusively use the term star tracker.

Up until the last five years or so, the attitude determination from star data was left to spacecraft onboard computer, or ground, processing. Increased processing capabilities now allow star trackers to track multiple stars (some track 30 or more) and to process this star data using an on-camera star catalog and algorithm such as QUEST to calculate the attitude. Hence the output of many new trackers is an inertial-to-sensor quaternion, making them not a two-axis star sensor but a true three-axis attitude sensor.

Star trackers are used typically only on three-axis stabilized spacecraft, because the sensors generally do not function accurately at angular velocities above a few degrees per second without additional motion compensation. Performance (attitude determination) of the new trackers is better about the axes orthogonal to the boresight. This is due to the relatively small field-of-view of the sensor, which limits the separation between the various measured star vectors. The next section will elaborate on this point. Typical accuracies for the current trackers are 1–3 arcseconds off boresight, and 10–30 arcseconds about the boresight. These accuracies are generally quoted as 1-sigma values, implying a nice statistical Gaussian noise distribution. The complexity of these sensors makes this not the case, and for very fine attitude determination many error effects must be taken into account. These include noise equivalent angle (NEA, which includes pixel nonuniformity, photon noise, dark current), centroiding errors, optical errors, star catalog uncertainties, and velocity aberration, to name a few. The accumulation of these errors tends to inflate the vendor specifications to some extent.

Depending on the angular velocity, time delay integration (TDI) compensation can be used to allow for higher body rates. In general this is used for spacecraft spinning about a single axis at a constant rate, since the TDI algorithm requires knowledge of the rate and performance degrades with variation in that rate (unless the TDI algorithm is furnished the information). Also, for use on spinning spacecraft, there exist V-slit star trackers which function in a manner similar to the V-slit Sun sensors.

5.4.5 Gyroscopes

Gyroscopes or gyros exist in many forms, and it would be impossible to describe here all the different technologies in current use. Functionally, gyros may be divided into two classes: gimbaled and strap-down. A four-axis gimbaled platform is depicted in figure 5.11. This gyro consists of an inertially stabilized platform and four gimbals, whose angles can be read. The outermost gimbal is attached to the spacecraft. As the spacecraft orientation changes, the gimbals adjust to the changing attitude. Usually only three of the four gimbals are free at any one time. Clearly, the gimbal angles in this case are simply Euler angles, and, by selecting either the outer three or the inner three gimbals, gimbal lock can be avoided. If the gyro gimbals are set so that the gimbal angles

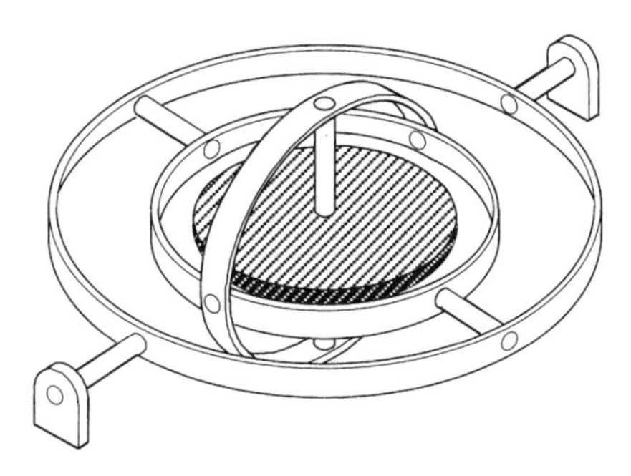

Figure 5.11 Gimbaled platform.

are zero at the initial attitude, then these gimbal angles will provide a direct measure of the attitude relative to the initial gyro frame. The use of three or four gimbals in an individual gyro assembly is not required. Some systems use two or more sets of two-axis gyros (that is, equipped with two gimbals) or three or more sets of single-axis gyros (that is, each with only a single degree of freedom). Gyros can also be constructed to give a direct measure of the angular velocity.

Strap-down attitude systems work on a different principle (they can, however, be implemented using gimbaled gyros). Typically, a strap-down system measures a small change in the attitude of the platform from its previous value and applies small torques to return the platform to that attitude. The output of the gimbaled gyro consists of these small attitude adjustments. Generally, these adjustments are performed at high frequency, often more than 1 kHz, so that by dividing these small adjustments by the time interval we obtain a measure of the angular velocity. By integrating the kinematic equations, we can obtain the relative attitude between any two times; that is, the attitude of the spacecraft at one time with respect to its attitude at the other time. Gyros also compute their outputs with respect to a gyro frame, so that these outputs must be transformed appropriately. Depending on the type of gyro, angular velocity accuracies from 1 degree per hour to as little as 10 arcseconds per hour are attainable.

A general problem with gyros is that of *drift*. Gyro measurements are not absolute but relative to some frame, either physical or mathematical. For gimbaled gyros this frame is that of the physical gyro platform, whose orientation can change due to friction in the gimbal bearings. For strap-down gyros this frame is the computed inertial frame, whose orientation will differ from that of the true inertial frame because of measurement noise, the discretization of the gyro measurements in the computer, and the accumulation of roundoff errors.

The great value of gyros is that they provide continuous output and are not constrained by the need for external references as for, say, the star tracker, which can provide attitude information only when it has located identifiable stars. However, because of the problem of gyro drift, the gyro is generally coupled with an inertially referenced sensor, like the star tracker, whose measurements provide periodic corrections.

Gyro construction is generally of three types: mechanical, laser, and electrical. *Mechanical gyros* consist generally of a spinning mass in some housing, commonly gimbals, although the spinning mass may also be electrostatically suspended. Another type of gyro that can be considered mechanical in nature is the hemispherically resonating gyro (HRG). The HRG operates under the principle that sound waves emanating from a body set up a beat frequency when the body is rotated. This is similar to one rubbing one's finger around the rim of a wine glass and causing it to "sing." HRGs are very low noise devices with exceptional life expectancies. *Laser gyros* use the interference of two laser beams to determine the angular rate about the gyro axis. (The effect behind the operation of these laser gyros has its origins in general relativity.) Laser gyros in which the beam is contained by mirrors are called *ring-laser gyros* (RLGs). Those that use optical fibers to constrain the beam are called *fiber-optic gyros* (FOGs). The fiber-optic gyros demand less power than RLGs but, from the standpoint of attitude accuracy, FOG technology has lagged behind. The gap is slowly narrowing. Finally, the term *electrical gyros* is used to denote those gyros whose technology is found in *microelectromechanical systems* (MEMS). These devices employ vibrating beams, for example, on a microscopic scale.

While requiring very little power, and even less space, they are not very accurate and are typically used on low-cost microsatellite missions.

5.4.6 Spacecraft Sensor Configuration

The choice of sensors to support a spacecraft will be a function of the mission requirements. (Remember that one requires measurements of at least two distinct objects to determine attitude, hence potentially two different sensors.) Scientific satellites performing precise experiments in space will generally require the most accurate sensors possible. At present these are star trackers. Recall that the largest error of a star tracker is about the boresight axis. Many times, to attain very good accuracy about all axes, two trackers will be used and mounted such that their boresights are 90° apart. In this manner the boresight error of one tracker is compensated for by the other tracker's off-boresight axes. The attainable attitude accuracy of this configuration in 2004 is better than 3 arcsec. Such spacecraft may be equipped with gyros as well. The gyro not only provides a direct measurement of the angular velocity, which may be important to the mission in its own right, but also permits us to combine attitude data from different times to sharpen the attitude estimate and provide attitude information during the data gaps of the other sensors. (Dynamical models are frequently not sufficiently reliable to accomplish this.) Very often the scientific payload may function as an attitude sensor. For example, the Hubble Space Telescope may use star sightings made by the telescope itself to help determine the attitude.

Note that whereas *some* star sensors are accurate to 0.001 arcsec, the reference directions of stars generally are not known to better than 0.1 arcsec, which currently sets the accuracy limit on attitude determination. Attitude control may be more precise, but spacecraft attitude is then controlled with respect to a reference frame that is not as precisely known as the accuracy level of the control law. This is the case, for example, with the Hubble Space Telescope, for which pointing is controlled to 0.01 arcsec but the absolute attitude is not known to much better than a few seconds of arc.

Spacecraft with more modest accuracy requirements will rely chiefly on coarse digital Sun sensors, horizon sensors, and magnetometers. For an inertially stabilized or Earth-locked spacecraft, the canonical choice has been a two-axis (that is, vector) digital Sun sensor, a three-axis (vector) magnetometer, and one or two infrared horizon scanners. The accuracy of the digital Sun sensor and magnetometer is typically 1° per axis. The accuracy of the horizon scanner will depend on the complexity of the data processing employed, but will be no worse than 1°. There is thus a large gap in accuracies between 10 arcsec and 1 degree (a factor of 360), which are the typical accuracies of precise and course attitude determination hardware, respectively. Spacecraft with very demanding attitude accuracy requirements will generally be equipped with both the coarse and the precise sensor sets, because the precise sensors may not function well when the spacecraft motion is great at the beginning of the mission. One usually depends on the less accurate coarse sensors for "safe" modes, for example, pointing solar arrays directly at the Sun when an anomaly occurs.

For spinning spacecraft one generally relies on single-axis Sun sensors, magnetometers, and horizon sensors. At high altitude, the horizon sensor is usually a slit sensor, which functions much like the V-slit Sun sensor. Thus, the attitude accuracy of a spinning spacecraft is typically 1° in any single set of measurements. Since the spin-axis direction generally varies very slowly in these cases, it may be determined with much greater accuracy by combining many measurements. The CONTOUR spacecraft, which used a V-slit Sun sensor and a dual-beam horizon sensor, obtained spin axis attitude knowledge to around 0.1° through the use of large data sets.

For geosynchronous satellites the situation is much simpler. The magnetic field is too weak to be useful as an attitude reference. Therefore, one is left with only Earth, Sun, and star references. For communication satellites, accuracies of 1° are generally sufficient, and precise sensors are not required. Thus, the general sensor configuration here may be a static horizon sensor and a digital Sun sensor. During part of the orbit, the Sun will be occulted by the Earth. To ensure continuous knowledge of the attitude, one generally includes a gyro. Another solution is to include a sensor that measures the direction of the polar star (Maute and Defonte, 1990). The next generation (beyond 2004) Geostationary Operational Environmental Satellite (GOES) will employ a star tracker to satisfy its tight attitude knowledge requirements.

5.4.7 Further References

More detailed descriptions of most of the sensors may be found in Wertz (1978). Descriptions of some of the sensors and attitude determination methods can be found in Agrawal (1986), Foliard (1984), Griffin and French (1991), Mobley (1988), Larson and Wertz (1992), and Shuster (1989). The most detailed work on gyroscopes is Radix (1978) (in French—there is no comparable work in English). In addition to the material in chapter 2 of the present work, more detailed information on the Earth and the spacecraft environment can be found in Stacey (1977), Jacobs (1987), and Bertotti and Farinella (1990).

5.5 Attitude Estimation

By *attitude estimation*, we mean how we determine the attitude given the attitude measurements. We may seek to estimate the full direction-cosine matrix or its equivalent, in which case we speak of *three-axis attitude determination*, or we may wish to estimate only the direction of a single axis (usually the spin axis of a spinning spacecraft), in which case we speak of *single-axis attitude* (or *spin-axis attitude*). The estimation method may be *deterministic* and use only the minimum amount of data necessary to determine the attitude uniquely (but not necessarily most accurately); or the method may be *optimal*, using more than the minimal amount of data in order to find the attitude that optimizes some criterion.

5.5.1 Deterministic Three-Axis Attitude Determination

As an example of deterministic three-axis attitude determination, consider the determination of the spacecraft attitude from simultaneous measurement of two directions, say, the direction of the magnetic field vector and the Sun vector, in the spacecraft body frame.

Denote the two unit-vector measurements in the spacecraft body frame by $\hat{\mathbf{W}}_1 = \hat{\mathbf{S}}_\mathrm{B}$ and $\hat{\mathbf{W}}_2 = \hat{\mathbf{B}}_\mathbf{B}$. The corresponding representations in the inertial reference frame are given by $\hat{\mathbf{V}}_1 = \hat{\mathbf{S}}_\mathrm{I}$ and $\hat{\mathbf{V}}_2 = \hat{\mathbf{B}}_\mathrm{I}$. The representations in the inertial reference frame are given by mathematical models which were themselves inferred from earlier measurements, often terrestrial. Ideally,

$$\hat{\mathbf{W}}_1 = C\hat{\mathbf{V}}_1, \quad \hat{\mathbf{W}}_2 = C\hat{\mathbf{V}}_2, \tag{5.5.1}$$

where C is the direction-cosine (that is, the attitude) matrix. More realistically, however, these two equations cannot be satisfied, because they require that

$$\hat{\mathbf{W}}_1 \cdot \hat{\mathbf{W}}_2 = \hat{\mathbf{V}}_1 \cdot \hat{\mathbf{V}}_2. \tag{5.5.2}$$

Deviations from this equation will occur because of measurement noise (or from errors in the reference vectors, which at some point were "measured"). Thus, in general, a solution does not exist for the attitude because of inconsistencies in the data. If we are willing to ignore these inconsistencies, a very simple and elegant solution for the attitude is available. The algorithm, often called the *triad algorithm*, is said to be *deterministic* because it uses a minimal subset of the data from which the attitude can be determined.

To construct this deterministic attitude, define two orthonormal triads of vector representations according to

$$\hat{\mathbf{r}}_1 = \hat{\mathbf{V}}_1, \quad \hat{\mathbf{r}}_2 = \frac{\hat{\mathbf{V}}_1 \times \hat{\mathbf{V}}_2}{|\hat{\mathbf{V}}_1 \times \hat{\mathbf{V}}_2|}, \quad \hat{\mathbf{r}}_3 = \hat{\mathbf{r}}_1 \times \hat{\mathbf{r}}_2, \tag{5.5.3}$$

$$\hat{\mathbf{s}}_1 = \hat{\mathbf{W}}_1, \quad \hat{\mathbf{s}}_2 = \frac{\hat{\mathbf{W}}_1 \times \hat{\mathbf{W}}_2}{|\hat{\mathbf{W}}_1 \times \hat{\mathbf{W}}_2|}, \quad \hat{\mathbf{s}}_3 = \hat{\mathbf{s}}_1 \times \hat{\mathbf{s}}_2. \tag{5.5.4}$$

Let us assume (incorrectly) that the spacecraft measurements contain no errors, so that equation 5.5.1 is satisfied. Then, by equation 5.2.40 and the hypothesis of equation 5.5.1,

$$\hat{\mathbf{s}}_i = C\hat{\mathbf{r}}_i, \quad i = 1, 2, 3. \tag{5.5.5}$$

We now note that even if a proper orthogonal matrix C could not be found that satisfied equation 5.5.1, we can nonetheless find such a matrix that satisfies equation 5.5.5 because, by explicit construction, $\hat{\mathbf{s}}_i \cdot \hat{\mathbf{s}}_j = \hat{\mathbf{r}}_i \cdot \hat{\mathbf{r}}_j = \delta_{ij}$.

If we now define two matrices according to their columns as

$$A \equiv [\hat{\mathbf{s}}_1 \ \hat{\mathbf{s}}_2 \ \hat{\mathbf{s}}_3], \quad B \equiv [\hat{\mathbf{r}}_1 \ \hat{\mathbf{r}}_2 \ \hat{\mathbf{r}}_3], \tag{5.5.6}$$

which are both proper orthogonal, it follows that

$$A = CB \tag{5.5.7}$$

which is just equation 5.5.5 gathered into matrix form, as can be easily verified by carrying out the multiplications explicitly. Since B is proper orthogonal, its inverse is given by its transpose. Thus, if we multiply both members of equation 5.5.7 on the right by B^T, we can write

$$C = AB^\mathrm{T}, \tag{5.5.8}$$

which is the desired attitude matrix. This algorithm is used frequently.

As we have said, equation 5.5.1 is not satisfied in practice due to various errors; therefore, equation 5.5.8 cannot provide the *correct* attitude. The attitude calculated from real data always contains some error. However, even though the data (and perhaps the reference vectors) contain errors, the construction always leads to a proper orthogonal matrix for C, because the construction of the attitude matrix effectively removes those components from $\hat{\mathbf{W}}_2$ and $\hat{\mathbf{V}}_2$ that could possibly violate equation 5.5.2.

Note also that although the procedure leads to a unique solution, it is not the only unique solution. Had we allowed the first vector to be the magnetic-field vector and the second vector to be the Sun vector, the algorithm would have led to a different (but, we hope, very close) value for C. Thus, uniqueness occurs only when we specify which vector is to be regarded as the first vector. Since data is deleted from the second vector, the more accurate attitude estimate is usually obtained by choosing the more accurate vector to be the first vector.

In general, the measurements are never exactly simultaneous. If the spacecraft is inertially stabilized, the error caused by the lack of simultaneity may be small enough to be ignored. If not, some correction must be applied by using an estimate of the spacecraft angular velocity to correct for the fact that the spacecraft is rotating between measurements.

5.5.2 Optimal Three-Axis Attitude Determination

Instead of a somewhat naive (but excellent) deterministic method, one could try to make use of all the information provided by the two measurements, and perhaps also those from an Earth horizon sensor as well, by finding the attitude that minimized the cost function

$$J(C) = \tfrac{1}{2}a_{\mathrm{S}}|\hat{\mathbf{S}}_{\mathrm{B}} - C\hat{\mathbf{S}}_{\mathrm{I}}|^2 + \tfrac{1}{2}a_{\mathrm{B}}|\hat{\mathbf{B}}_{\mathrm{B}} - C\hat{\mathbf{B}}_{\mathrm{I}}|^2 + \tfrac{1}{2}a_{\mathrm{E}}|\hat{\mathbf{E}}_{\mathrm{B}} - C\hat{\mathbf{E}}_{\mathrm{I}}|^2, \tag{5.5.9}$$

where a_{S}, a_{B}, and a_{E} are three non-negative weights. We can generalize equation 5.5.9 for an arbitrary number of measurements by writing the cost function as

$$J(C) = \tfrac{1}{2}\sum_{k=1}^{N} a_k|\hat{\mathbf{W}}_k - C\hat{\mathbf{V}}_k|^2, \tag{5.5.10}$$

where $\hat{\mathbf{W}}_k$ is a vector measurement (with respect to the spacecraft body coordinates) and $\hat{\mathbf{V}}_k$ is the corresponding direction in inertial coordinates. Equation 5.5.10 assumes implicitly that all measurements are simultaneous.

If the data were perfect (in other words, contained no errors), it would be possible to find a proper orthogonal matrix C for which every term in equation 5.5.10 vanished. As we know, this is not possible generally; so we seek instead the optimal value of C that minimizes the cost function. In choosing the cost function, we wish to give more emphasis to the more accurate measurements and less emphasis to the less accurate measurements. We control this in our calculation by choosing appropriate values for the weights, giving values to these in some inverse proportion to the accuracy of the measurements.

We now come to the central and frequent difficulty in estimating attitude. We would not wish to minimize $J(C)$ over the nine elements of C, because only three are independent. Nor would we wish to take the six constraints into account explicitly in our calculations, which would greatly increase their complexity. The recommended procedure, therefore, is to minimize $J(C)$ over some minimum parameter set such as the Euler angles. Thus, we define

$$J(\phi, \theta, \psi) \equiv J(C(\phi, \theta, \psi)), \tag{5.5.11}$$

and $J(\phi, \theta, \psi)$ now depends only on independent variables.

We now come to the second important difficulty in attitude determination. While J is a simple function of C (except for the constraint, of course), it is a very complicated and unappealing function of the Euler angles. This will be obvious if we substitute equation 5.2.55 into 5.5.11. Minimization of $J(\phi, \theta, \psi)$ is generally not possible by closed-form analytical methods, and therefore must be carried out by an iterative numerical procedure. The most common instrument for this is the Newton–Raphson method.

5.5.3 The Newton–Raphson Method

The *Newton–Raphson method* is a method for finding an extremum of a function. To simplify the notation of equation 5.5.11, define a 3×1 column vector $\mathbf{x}$ according to

$$\mathbf{x} = \begin{bmatrix} \phi \\ \theta \\ \psi \end{bmatrix}. \tag{5.5.12}$$

The column vector $\mathbf{x}$ has no interesting connection to any physical vector. However, it is simpler to write J as a function of a column vector $\mathbf{x}$ than of the three Euler angles separately. The problem now is to minimize an arbitrary non-negative multivariate function $J(\mathbf{x})$. If this minimum does not lie on the boundary over which $\mathbf{x}$ is defined and the derivative of J with respect to its arguments is defined, then $\mathbf{x}^*$, the value of the argument of J for which J is a minimum, is a solution of

$$\frac{\partial J}{\partial \mathbf{x}}(\mathbf{x}^*) \equiv \begin{bmatrix} \frac{\partial J}{\partial \mathbf{x}_1} & \frac{\partial J}{\partial \mathbf{x}_2} & \frac{\partial J}{\partial \mathbf{x}_3} \end{bmatrix}^{\mathrm{T}} = \mathbf{0}. \tag{5.5.13}$$

To find this optimal value, we develop an algorithm that, given an approximate value for $\mathbf{x}^*$, yields a better approximation for this quantity.

To develop this algorithm, suppose that we have an earlier estimate, $\mathbf{x}(i)$, for which this equation holds approximately. The argument i indicates that we have applied our (still undetermined) iterative algorithm i times and we are now about to apply it an $(i+1)^{\text{th}}$ time. To do this we begin by expanding $\partial J / \partial \mathbf{x}$ in a Taylor series about the value $\mathbf{x}(i)$,

$$\begin{aligned} \frac{\partial J}{\partial x_m}(\mathbf{x}) = {} & \frac{\partial J}{\partial x_m}(\mathbf{x}(i)) \\ & + \sum_{n=1}^{3} \left(\frac{\partial^2 J}{\partial x_m \partial x_n}(\mathbf{x}(i)) \right) (x_n - x_n(i)) + \cdots, \end{aligned} \tag{5.5.14}$$

or, in less cumbersome matrix notation,

$$\frac{\partial J}{\partial \mathbf{x}}(\mathbf{x}) = \frac{\partial J}{\partial \mathbf{x}}(\mathbf{x}(i)) + \left(\frac{\partial^2 J}{\partial \mathbf{x} \partial \mathbf{x}^{\mathrm{T}}}(\mathbf{x}(i))\right)(\mathbf{x} - \mathbf{x}(i)) + \cdots. \tag{5.5.15}$$

Here, $\partial J/\partial \mathbf{x}$ stands for the N-dimensional column vector (that is, the $N \times 1$ matrix) whose m^{th} element is $\partial J/\partial x_m$, and $\partial^2 J/\partial \mathbf{x}\partial \mathbf{x}^{\mathrm{T}}$ stands for the matrix whose (m, n) element is equal to $\partial^2 J/\partial x_m \partial x_n$. If we keep only terms through linear in equation 5.5.15 and set this truncated series to zero, we can solve the resulting linear equation for $\mathbf{x}$, which gives the next approximation, $\mathbf{x}(i + 1)$. Thus,

$$\mathbf{x}(i + 1) = \mathbf{x}(i) - \left(\frac{\partial^2 J}{\partial \mathbf{x} \partial \mathbf{x}^{\mathrm{T}}}(\mathbf{x}(i))\right)^{-1} \frac{\partial J}{\partial \mathbf{x}}(\mathbf{x}(i)). \tag{5.5.16}$$

This is our algorithm.

If we continue this procedure forever, then

$$\lim_{i \to \infty} \mathbf{x}(i) = \mathbf{x}^*. \tag{5.5.17}$$

Fortunately, we seldom need to do this. After typically a half-dozen iterations, a good solution is obtained. If a good starting value can be obtained, say, from a deterministic algorithm like the triad method treated earlier in this section, sometimes even a single iteration is sufficient. (In that case one often speaks of a *differential corrector*.)

The Newton–Raphson method is burdensome when applied to the Euler angles because of the large number of trigonometric functions that must be differentiated. In addition, the method assumes that we have some starting value, $\mathbf{x}(0)$, which frequently is not available. A poor choice of the starting value may cause the algorithm to converge to a minimum that is only a local minimum and not the value of $\mathbf{x}$ that achieves the absolutely smallest value of J. Thus, if the data are such that a simple deterministic method is not available, the fact that the starting value $\mathbf{x}(0)$ is not provided by this iterative method itself is a serious drawback.

5.5.4 Spin-Axis Attitude Determination

Sometimes we do not wish to know the attitude of the spacecraft but only the direction in space of one spacecraft axis, which customarily we call a *single-axis attitude*. (In contrast, the attitude is often called the *three-axis attitude* or even the *two-axis attitude*.) This situation occurs generally only for spinning spacecraft, in which frequently only the direction of the spin axis is of interest. For this reason, spin-axis attitude and single-axis attitude are often used interchangeably.

The same methods that were employed for optimal three-axis attitude determination can be employed also for spin-axis attitude determination. Consider a spacecraft that is equipped with a solar aspect sensor and a vector magnetometer. We suppose that the magnetometer is so mounted on the spacecraft that one component of the measured magnetic field is along the spin axis. To simplify the discussion, we will assume that the Sun sensor output is simply the component of the Sun's direction along the spin axis; that is, that we have a solar cosine detector rather than a solar aspect detector (the transformation from solar cosine detector to solar aspect detector being made in our computer).

We thus have two sets of measurements, which we write as a function of the spin-axis $\hat{\mathbf{n}}$ as

$$z_{S,k} \equiv \cos \beta_k = \hat{\mathbf{n}} \cdot \hat{\mathbf{S}}_{I,k} + \Delta z_{S,k}, \tag{5.5.18}$$

and

$$z_{B,k} \equiv \cos \gamma_k = \hat{\mathbf{n}} \cdot \hat{\mathbf{B}}_{I,k} + \Delta z_{B,k}, \tag{5.5.19}$$

where $\hat{\mathbf{S}}_{I,k}$ and $\hat{\mathbf{B}}_{I,k}$ are the representations with respect to inertial axes of the directions of the Sun and geomagnetic field vectors, respectively, at time t_k. Generally, a very large number of measurements is used in estimating the spin-axis attitude.

The cost function that we minimize to find the optimal estimation of the spin-axis attitude is therefore

$$J(\mathbf{n}) = \frac{1}{2} \sum_{k=1}^{N} \{a_{S,k} |z_{S,k} - \hat{\mathbf{n}} \cdot \hat{\mathbf{S}}_{I,k}|^2 + a_{B,k} |z_{B,k} - \hat{\mathbf{n}} \cdot \hat{\mathbf{B}}_{I,k}|^2\}, \tag{5.5.20}$$

where again $a_{S,k}$ and $a_{B,k}$ are non-negative weights. Since $\hat{\mathbf{n}}$ has only two degrees of freedom, we do not wish to differentiate with respect to its three components as if they were independent. Thus, we write $\hat{\mathbf{n}}$ as a function of the two spherical angles,

$$\hat{\mathbf{n}} = \begin{bmatrix} \sin\theta \cos\phi \\ \sin\theta \sin\phi \\ \cos\theta \end{bmatrix}, \tag{5.5.21}$$

and minimize instead

$$J(\theta, \phi) \equiv J(\hat{\mathbf{n}}(\theta, \phi)) \tag{5.5.22}$$

with respect to θ and ϕ, using the Newton–Raphson method, as for optimal three-axis determination.

5.5.5 Kalman Filtering

A discussion of attitude estimation would not be complete without mention of one of the most widely used techniques for onboard attitude estimation, that being what has been termed "Kalman filtering." The algorithm presented in section 5.5.2 is considered a weighted least-squares algorithm. That is, it determines the attitude matrix C by means of minimizing the weighted square of the error over an entire batch of data points. This is generally not useful for on-orbit calculations where the attitude is required immediately. A recursive least-squares process allows the estimate to be updated at each measurement time, thereby avoiding the need to maintain an extensive backlog of previous measurements. The Kalman filter, developed by R. E. Kalman, goes a step further than a recursive least-squares algorithm by employing the statistical knowledge of the modeling process and the attitude measurements. In fact, the linear least-squares process can be viewed as a special case of the Kalman filter.

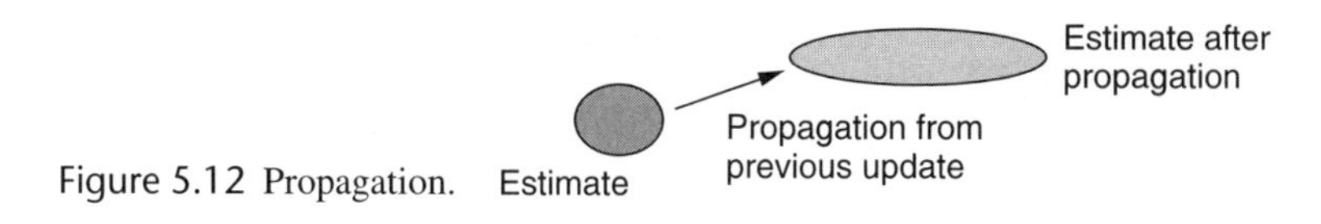

Figure 5.12 Propagation.

In very simplistic terms, the Kalman filter performs an endless series of propagations and updates. A model is developed of a system, and in spacecraft that model is generally the kinematics of the attitude. The propagation step involves calculating the attitude of the spacecraft at some time t_i in the future through the kinematic equations. Gyro measurements usually provide the rate measurements necessary for this calculation. Due to gyro noise and modeling errors, together termed process noise, the propagated, or predicted, attitude will be different from the achieved attitude. At time t_i the propagated attitude is corrected, or updated, with attitude measurements. These attitude measurements are themselves noisy; this is termed the measurement noise. The Kalman filter statistically combines attitude measurements with various noise characteristics, along with the model and process noise, to arrive at the statistically optimal estimate for a linear system.

As an illustration of this process, consider figures 5.12 and 5.13. At the start of the propagation step the attitude is estimated to within a certain accuracy, depicted in figure 5.12 as a small ellipse. After propagation the attitude estimate is less accurate, due to process noise effects, and is portrayed by the larger ellipse. Figure 5.13 illustrates the update process. Here, the attitude estimate after propagation and the measurement of the attitude by the attitude sensors are statistically combined to result in an estimate that is generally more accurate than the individual entities.

5.5.6 Further References

There are many books on estimation theory. Junkins (1978) introduces estimation theory in a manner that may be most approachable to readers of this chapter. Wertz (1978) contains much useful information on attitude determination, somewhat biased, however, toward spin-axis attitude determination and the use of horizon scanners. Very brief introductions can be found in Kaplan (1976), Foliard (1984), and Griffin and French (1991). There are many intricacies to the design and implementation of a Kalman filter, and the interested reader is directed to Gelb (1989) and Anderson and Moore (1979) for more information on this topic.

The triad method was first published by Black (1964) and a covariance analysis was developed by Shuster and Oh (1981). This last paper also presents the QUEST method, one of the more popular methods for solving the particular optimal estimation

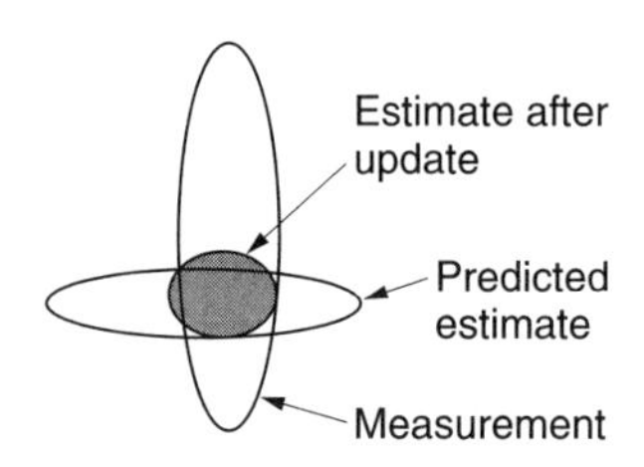

Figure 5.13 Propagation update.

problem treated in this chapter. Simple vectorial methods for solving spin-axis attitude determination problems have been developed by Shuster (1983). A very complete discussion of the geometry of attitude measurements, and of arc length and dihedral angles in particular, can be found in Wertz (1978).

This section has omitted a number of related topics that are important and often crucial for mission support and have been of particular concern to the primary author. These topics include postlaunch bias determination (Lerner and Shuster, 1981; Lerner, 1990), alignment calibration (Shuster et al., 1991), and sequential attitude estimation techniques, especially those using the Kalman filter (Lefferts et al., 1982). Additionally, not included in this chapter is a discussion of attitude estimation using the Global Positioning System (GPS). Parkinson et al. (1996) provide extensive detail of the GPS, while Cohen (1992), Melvin and Hope (1993), and Crassidis et al. (1998) focus on attitude estimation with GPS.

5.6 Attitude Dynamics

We understand *attitude dynamics* to mean how the motion of a body is affected by applied torques. The fundamentals of attitude kinematics were treated in section 5.3. The description of the attitude motion of a body depends on the treatment of both the dynamical and kinematical aspects.

Attitude motion, as we shall see, is much more complicated than translational motion. In translational motion, the force-free case leads to motion with constant linear velocity. For a totally asymmetric rigid body, even torque-free attitude motion requires elliptical integrals for its description. Even for a body with an axis of symmetry, the attitude motion is nontrivial but manageable in terms of simple trigonometric functions.

5.6.1 Angular Momentum and Euler's Equation

Consider the angular momentum of a rigid body with respect to its center of mass. The angular momentum vector $\boldsymbol{L}$ is given by

$$\boldsymbol{L} = \sum_{i=1}^{N} m_i \boldsymbol{r}_i \times \boldsymbol{v}_i , \tag{5.6.1}$$

where m_i, $i = 1, \ldots, N$, are the component point masses making up the rigid body, which for the moment we will assume to be finite in number. Since we are considering the motion of the body with respect to its center of mass, only rotational degrees of freedom are available, and we can write as previously

$$\boldsymbol{v}_i = \overline{\boldsymbol{\omega}} \times \boldsymbol{r}_i . \tag{5.6.2}$$

Thus the angular momentum of the rigid body becomes

$$\boldsymbol{L} = \sum_{i=1}^{N} m_i \boldsymbol{r}_i \times (\overline{\boldsymbol{\omega}} \times \boldsymbol{r}_i). \tag{5.6.3}$$

Applying the vector triple product, this expression becomes

$$L = \sum_{i=1}^{N} m_i[(\boldsymbol{r}_i \cdot \boldsymbol{r}_i)\overline{\omega} - (\boldsymbol{r}_i \cdot \overline{\omega})\boldsymbol{r}_i]. \tag{5.6.4}$$

Through the use of dyadics, the angular momentum in the inertial frame can be written as

$$L = \mathbf{I}_\mathrm{I} \cdot \overline{\omega}, \tag{5.6.5}$$

where the inertia dyadic $\mathbf{I}_\mathrm{I}$ (a symmetric matrix called the inertia tensor) is given by

$$\mathbf{I}_\mathrm{I} = \int [(\boldsymbol{r}_\mathrm{I} \cdot \boldsymbol{r}_\mathrm{I})I - \boldsymbol{r}_\mathrm{I}\boldsymbol{r}_\mathrm{I}^\mathrm{T}]dm. \tag{5.6.6}$$

Note that the angular momentum of a rigid body about its center of mass depends on the mass distribution only through the inertia dyadic. Thus, we need only know the inertia dyadic of a rigid body to study its attitude dynamics.

Because $\boldsymbol{r}$ is generally time dependent, the inertia dyadic is not generally constant with time. However, for a rigid body, the mass density in the body frame is constant. Therefore, the representation of the inertia dyadic with respect to body coordinates, the body-referenced inertia tensor, will be constant. Thus, for studies of spacecraft rotational dynamics, it is preferable to work in body coordinates, where the inertia tensor $\mathbf{I}_B$, the 3×3 matrix representation of the inertia dyadic, is given by

$$\begin{aligned} \mathbf{I}_\mathrm{B} &= \int [(\boldsymbol{r}_\mathrm{B} \cdot \boldsymbol{r}_\mathrm{B})I - \boldsymbol{r}_\mathrm{B}\boldsymbol{r}_\mathrm{B}^\mathrm{T}]dm \\ &= \int \begin{bmatrix} y^2 + z^2 & -xy & -xz \\ -yx & x^2 + z^2 & -yz \\ -zx & -zy & x^2 + y^2 \end{bmatrix}_\mathrm{B} dm \end{aligned} \tag{5.6.7}$$

and is constant in time for a rigid body. (In the second line of equation 5.6.7 we have written B as a subscript on the entire matrix rather than on every individual component of $\boldsymbol{r}$ in order to avoid a very cumbersome expression.) The diagonal elements of $\mathbf{I}_\mathrm{B}$ are called the *moments of inertia*; the off-diagonal elements are called the *products of inertia*. The products of inertia are sometimes defined without the minus sign, thus it is very important when working with the different engineering disciplines that everyone agrees on the definition, lest there be a lurking sign error in one's analysis.

In general, if the inertia tensor $\mathbf{I}$ is known in one right-hand orthonormal basis, its value $\mathbf{I}'$ with respect to another right-hand orthonormal basis having the same origin is given by

$$\mathbf{I}' = R\mathbf{I}R^\mathrm{T}, \tag{5.6.8}$$

where R is the proper orthogonal matrix connecting the two bases. Similarly, if the inertia tensor of the rigid body is $\mathbf{I}$ with respect to some coordinate system whose origin is at the center of mass of the body, and if the coordinate system is displaced by a translation $\mathbf{a}$ while keeping the axes parallel, then the inertia tensor with respect to the new coordinate system is given by

$$\mathbf{I}' = \mathbf{I} + M[(\mathbf{a} \cdot \mathbf{a})I - \mathbf{a}\mathbf{a}^\mathrm{T}], \tag{5.6.9}$$

where M is the mass of the rigid body. This last result is known as the *parallel-axis theorem*.

The inertia tensor is a symmetric matrix. Therefore, there exists a basis in which it is diagonal and has the form

$$\mathbf{I} = \begin{bmatrix} I_1 & 0 & 0 \\ 0 & I_2 & 0 \\ 0 & 0 & I_3 \end{bmatrix}. \tag{5.6.10}$$

The axes of this basis are called the *principal axes*, and the values I_1, I_2, I_3 are called the principal moments of inertia. It is easy to show from equation 5.6.7 that the inertia tensor is positive definite. Hence, the principal moments of inertia will always be positive.

The (physical) angular momentum vector $\boldsymbol{L}$ satisfies

$$\frac{d}{dt}\boldsymbol{L} = \boldsymbol{N}, \tag{5.6.11}$$

where

$$\boldsymbol{N} \equiv \sum_{i=1}^{N} \boldsymbol{r}_i \times \boldsymbol{F}_i \tag{5.6.12}$$

is the torque acting on the rigid body. As earlier, the sum is over the elements of the body. The representation of equation 5.6.11 in the inertial reference frame, following equation 5.3.8, is simply

$$\frac{d}{dt}\mathbf{L}_{\mathrm{I}} = \mathbf{N}_{\mathrm{I}}. \tag{5.6.13}$$

In body coordinates at the center of mass, on the other hand, and recalling equation 5.3.17, this becomes

$$\frac{d}{dt}\mathbf{L}_{\mathrm{B}} = -\boldsymbol{\omega}_{\mathrm{B}} \times \mathbf{L}_{\mathrm{B}} + C\mathbf{N}_{\mathrm{I}}, \tag{5.6.14}$$

or, absorbing the attitude matrix C,

$$\frac{d}{dt}\mathbf{L}_{\mathrm{B}} + \boldsymbol{\omega}_{\mathrm{B}} \times \mathbf{L}_{\mathrm{B}} = \mathbf{N}_{\mathrm{B}}, \tag{5.6.15}$$

which is *Euler's equation.*

Euler's equation is the basis of all our dynamical studies. In this work we will study only rigid bodies, for which

$$\mathbf{L}_{\mathrm{B}} = \mathbf{I}_{\mathrm{B}}\omega. \tag{5.6.16}$$

If we substitute this into equation 5.6.15, we obtain

$$\mathbf{I}_{\mathrm{B}}\frac{d}{dt}\omega + \omega \times (\mathbf{I}_{\mathrm{B}}\omega) = \mathbf{N}_{\mathrm{B}}. \tag{5.6.17}$$

Equation 5.6.17, however, is only half the picture and must be supplemented by one of the kinematical relationships such as, for the 3-1-3 Euler angles,

$$\begin{bmatrix} \omega_1 \\ \omega_2 \\ \omega_3 \end{bmatrix} = \begin{bmatrix} \dot{\phi} \sin\theta \sin\psi + \dot{\theta}\cos\theta \\ \dot{\phi} \sin\theta \cos\psi - \dot{\theta}\sin\psi \\ \dot{\phi}\cos\theta + \dot{\psi} \end{bmatrix}, \tag{5.6.18}$$

or similar equations for the attitude matrix, the quaternion, or any other attitude representation. (Henceforth, when there is no subscript or other indication of the representation, it will be assumed that representations are with respect to a body-fixed axis.)

For torque-free motion, the right-hand member of equation 5.6.17 vanishes. Even in this case, the equation is nonlinear and has no simple solution. Thus, in almost all instances, Euler's equations can be solved only numerically.

5.6.2 Torque-Free Motion of a Rigid Body

The environmental torques that act on the spacecraft are generally quite small. They cannot be ignored, however, because they can act for a very long time. All the same, a great deal can be learned about spacecraft motion by studying the torque-free case.

If we assume that the spacecraft body axes are aligned with the principal axes of the inertia tensor and set $\mathbf{N}_\mathrm{B} = \mathbf{0}$, then we may write equation 5.6.17 in component form as

$$\left.\begin{aligned} I_1\dot{\omega}_1 + (I_3 - I_2)\omega_2\omega_3 &= 0, \\ I_2\dot{\omega}_2 + (I_1 - I_3)\omega_3\omega_1 &= 0, \\ I_3\dot{\omega}_3 + (I_2 - I_1)\omega_1\omega_2 &= 0. \end{aligned}\right\} \tag{5.6.19}$$

These equations are solvable only in terms of the Jacobian elliptic integrals (Thomson, 1986). For the special case of a symmetric spacecraft ($I_1 = I_2$), these become

$$\left.\begin{aligned} I_1\dot{\omega}_1 + (I_3 - I_1)\omega_2\omega_3 &= 0, \\ I_1\dot{\omega}_2 + (I_1 - I_3)\omega_3\omega_1 &= 0, \\ I_3\dot{\omega}_3 &= 0, \end{aligned}\right\} \tag{5.6.20}$$

which can be solved in terms of simple trigonometric functions. In general, spacecraft are never exactly symmetric, even for the second-order moments that make up the inertia tensor. However, it is often true that the difference between two principal moments of inertia is small compared with the difference of these from the third, so this assumption is useful for understanding the approximate motion of the spacecraft. In realistic simulations, however, these equations must always be integrated numerically, especially when considering nonvanishing spacecraft torques.

Note first from equation 5.6.20 that $\omega_3 =$ constant, so that the first two equations may be written as

$$\left.\begin{aligned} \dot{\omega}_1 - \lambda_\mathrm{b}\omega_2 &= 0, \\ \dot{\omega}_2 + \lambda_\mathrm{b}\omega_1 &= 0, \end{aligned}\right\} \tag{5.6.21}$$

where

$$\lambda_\mathrm{b} = \left(\frac{I_1 - I_3}{I_1}\right)\omega_3. \tag{5.6.22}$$

Euler's equation thus reduces in this case to two coupled first-order linear differential equations. Substituting each of the equations of 5.6.21 into the other yields two uncoupled second-order linear differential equations,

$$\left.\begin{aligned} \ddot{\omega}_1 + \lambda_\mathrm{b}^2\omega_1 &= 0, \\ \ddot{\omega}_2 + \lambda_\mathrm{b}^2\omega_2 &= 0. \end{aligned}\right\} \tag{5.6.23}$$

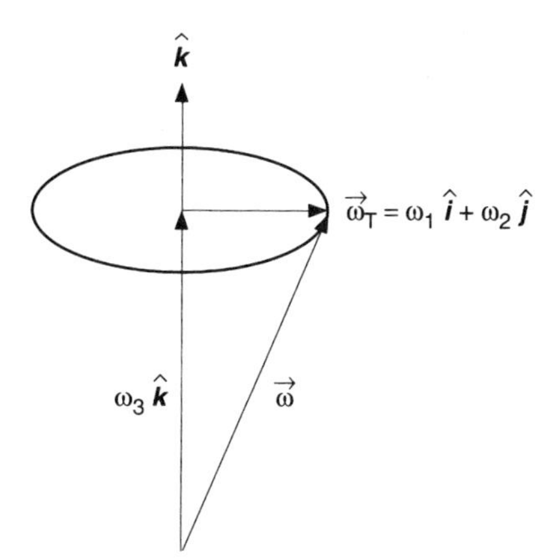

Figure 5.14 Torque-free motion of the angular velocity vector.

Thus, ω_1 has the general solution

$$\omega_1(t) = \omega_T \cos[\lambda_b(t - t_0)], \tag{5.6.24}$$

and, from equation 5.6.21,

$$\omega_2(t) = -\omega_T \sin[\lambda_b(t - t_0)]. \tag{5.6.25}$$

In the body frame, then, the angular velocity vector performs a coning motion about the spacecraft symmetry axis with angular rate λ_b and half-cone angle γ, with

$$\tan \gamma = \frac{\omega_T}{\omega_3}. \tag{5.6.26}$$

This motion is illustrated in figure 5.14.

The important lesson of this subsection is that the attitude motion of a torque-free rigid body is circular. Thus, in the body-fixed reference frame, the angular velocity and the angular momentum vectors (really, their representations) cone about the spacecraft symmetry axis. In the inertial reference frame, on the other hand, the direction of the angular momentum remains fixed and the spin axis and the angular velocity vector cone about it.

5.6.3 Detailed Attitude Motion of a Symmetric Spacecraft

After determining the solution for the angular velocity, it remains to determine the attitude motion. To simplify the notation, we define without loss of generality an inertial coordinate system with $\mathbf{L}_I$ along the inertial z-axis. Then, if we use a 3-1-3 set of Euler angles whose rotation matrix is given by equation 5.2.55, the body-reference angular momentum is

$$\begin{bmatrix} L_1 \\ L_2 \\ L_3 \end{bmatrix} = \begin{bmatrix} I_1\omega_1 \\ I_2\omega_2 \\ I_3\omega_3 \end{bmatrix} = L \begin{bmatrix} \sin\theta \sin\psi \\ \sin\theta \cos\psi \\ \cos\theta \end{bmatrix}, \tag{5.6.27}$$

and we have used the fact that $L_I = L$. Thus, two of the Euler angles become the spherical angles of the angular momentum vector in body coordinates. (Recall in this equation that $I_1 = I_2$.) Since ω_3 is a constant, it follows that θ is a constant. Noting then $\dot{\theta} = 0$, we can write the kinematic equations for the Euler angles as (see equation 5.3.21)

$$\left.\begin{aligned} \omega_1 &= \dot{\phi} \sin\theta \sin\psi, \\ \omega_2 &= \dot{\phi} \sin\theta \cos\psi, \\ \omega_3 &= \dot{\phi} \cos\theta + \dot{\psi}. \end{aligned}\right\} \tag{5.6.28}$$

If we solve the first line of equation 5.6.27 for $\sin\theta \sin\psi$ and substitute it into the first line of equation 5.6.28, we obtain

$$\dot{\phi} = L/I_1 \equiv \lambda_i. \tag{5.6.29}$$

Finally, substituting this result into the last line of equation 5.6.28 (and recalling equation 5.6.27) leads to

$$\dot{\psi} = \lambda_b. \tag{5.6.30}$$

Thus, for the torque-free motion of a symmetric rigid body, the Euler angle rates appropriately defined are all constant (and for $\dot{\theta}$ that constant is 0). The solution for the attitude motion in terms of the Euler angles is

$$\left.\begin{aligned} \phi &= \lambda_i t + \phi_0, \\ \theta &= \theta_0, \\ \psi &= \lambda_b t + \psi_0. \end{aligned}\right\} \tag{5.6.31}$$

Although the Euler angles have some unpleasant analytical properties, we see from equation 5.6.31 that they are ideally suited to some problems.

To understand the geometric significance of these results, return to equation 5.6.27 and note that in the body frame

$$\left.\begin{aligned} L_1 &= L \sin\theta \sin(\lambda_b t + \psi_0), \\ L_2 &= L \sin\theta \cos(\lambda_b t + \psi_0), \\ L_3 &= L \cos\theta. \end{aligned}\right\} \tag{5.6.32}$$

Thus, the angular momentum vector rotates about the body axis $\mathbf{k}_B$ with angular rate λ_b and half-cone angle θ, with

$$\tan\theta = \frac{L_T}{L_3} \equiv \frac{I_1 \omega_T}{I_3 \omega_3}. \tag{5.6.33}$$

The body-referenced angular velocity rotates about $\mathbf{k}_B$ with the same angular rate but with half-cone angle γ. In fact, it can be shown that $\mathbf{L}_B$, $\boldsymbol{\omega}_B$, and $\mathbf{k}_B$ are coplanar, with

$$\mathbf{L}_B = \mathbf{I}_I(\boldsymbol{\omega}_B + \lambda_b \mathbf{k}_B). \tag{5.6.34}$$

In the inertial frame, on the other hand, the representation of the body $\boldsymbol{k}$ axis is given by

$$\mathbf{k}_I = C_{313}^{T}(\phi, \theta, \psi) \begin{bmatrix} 0 \\ 0 \\ 1 \end{bmatrix} = \begin{bmatrix} \sin\theta \sin(\lambda_i t + \phi_0) \\ -\sin\theta \cos(\lambda_i t + \phi_0) \\ \cos\theta \end{bmatrix}, \tag{5.6.35}$$

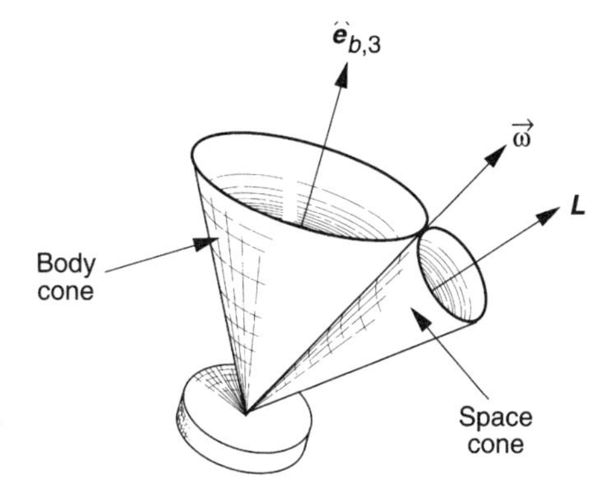

Figure 5.15 Torque-free attitude motion of a symmetric rigid body.

so that in the inertial coordinate system $\mathbf{k}_I$ rotates about $\mathbf{L}_I$ with angular rate λ_i. The angular velocities λ_b and λ_i are referred to as the body-referenced and inertial-referenced *nutation rates*. (The use of the term nutation here differs from that in top theory, where it indicates a wobble about the horizontal axis.)

Note also that the angular velocity vector is the axis of rotation. Thus, this axis must be instantaneously at rest in both the body frame and the space (that is, the inertial) frame. If we regard the body and space cones as the cones swept out by the angular velocity vector with respect to the body axis $\boldsymbol{k}$ and the angular momentum vector, respectively, then one cone must roll over (or within) the other, with the angular velocity vector the line of contact. Since the line of contact is instantaneously at rest, one cone therefore rolls over the other without slipping. An illustration of the coning motion is given in figure 5.15 for an oblate symmetrical rigid body ($I_1 < I_3$).

The most important point of this discussion is that, for a symmetrical spacecraft in a torque-free environment, the motion of the spin axis in inertial space is a cone about the total angular momentum vector. Thus, a point on the spin axis describes a circle in inertial space. We shall see this graphically when we study spacecraft control systems.

5.6.4 Spacecraft Torques

Having discussed the torque-free motion in some detail, we turn to a discussion of spacecraft torques. There are principally five: magnetic, reaction, gravity-gradient, aerodynamic, and solar radiation torques. The relative importance of these torques is a function of the spacecraft size, mass, mass distribution, and altitude. We can divide these torques into two overlapping groups: *environmental torques*, to which the spacecraft is subjected independent of any action it may take: and *control torques*, which are generated on the spacecraft specifically to control the attitude motion. The environmental torques include magnetic, gravity-gradient, aerodynamics, and solar radiation torques. Control torques could be said to include all of these, but the gravity-gradient, magnetic, and reaction torques are most important, and the most widely used. While some mention of attitude control torques is impossible to avoid in discussing spacecraft torques, this subsection will be devoted principally to environmental torques.

5.6.5 Magnetic Torques

Magnetic torques on the spacecraft result either from a commanded magnetic dipole moment as part of an attitude control strategy or from spurious uncontrolled magnetization of the spacecraft. This magnetization may result from: permanent magnets used in instruments; poor magnetic design of the spacecraft; residual fields from space-charge effects, which occur as the spacecraft passes through the atmosphere; other magnetic stabilization devices, such as eddy-current nutation dampers; or from active magnetic control devices. For near-Earth spacecraft at altitudes greater than 1000 km, magnetic disturbance torques can be the principal disturbance affecting spacecraft attitude. At some point, depending on altitude and spacecraft configuration, solar radiation pressure will become the dominant disturbance.

If $\boldsymbol{m}$ is the magnetic moment of the spacecraft (in A m^2) and $\boldsymbol{B}$ is the magnetic flux density (in Wb/m^2), the magnetic torque (in N m) is given by

$$\boldsymbol{N}_{\text{mag}} = \boldsymbol{m} \times \boldsymbol{B}. \tag{5.6.36}$$

The spurious magnetic moment is not known a priori, but an effective value can be estimated by studying the detailed dynamics of the spacecraft. To appreciate the size of these environmental magnetic torques, we note that typical values of the Earth's magnetic field at low altitude at the equator are 3×10^{-5} Wb/m^2. (One Wb/m^2 is also written one T (= tesla); in c.g.s. units one tesla = 10^4 gauss.) A spurious magnetic moment on the spacecraft may have an order of magnitude of 1 A m^2. Thus, the "environmental" magnetic torques at low altitude may have a typical magnitude of 3×10^{-5} N m. Since the magnetic field magnitude decreases with increasing distance from the geocenter roughly as R^{-3}, the magnetic torque is several orders of magnitude smaller at geosynchronous altitudes. (The reader should be warned that magnetic dipoles are often given in electrostatic units [esu] as pole-cm; 1000 pole-cm corresponds to 1 A m^2.)

5.6.6 Gravity-Gradient Torques

The magnitude of the gravitational force from the Earth is not constant but varies roughly as R^{-2}, where R here is the distance from the geocenter. Thus the gravitational force on one part of the spacecraft is different from that on another, and this difference results in a net torque.

To derive an expression for the gravity-gradient torque, consider the force $d\boldsymbol{F}$ on a mass element dm located at a position $\boldsymbol{R}$ with respect to the geocenter, as shown in figure 5.16.

$$d\boldsymbol{F} = -\frac{\mu \boldsymbol{R} dm}{R^3}, \tag{5.6.37}$$

where $\mu = GM_{\text{e}}$ is the Earth's gravitational constant. If we write

$$\boldsymbol{R} = \boldsymbol{R}_{\text{cm}} + \boldsymbol{r}, \tag{5.6.38}$$

where $\boldsymbol{R}_{\text{cm}}$ is the position of the center of mass of the spacecraft with respect to the Earth, the torque about the center of mass is then

$$N_{\text{GG}} = \int \boldsymbol{r} \times d\boldsymbol{F} = -\mu \int \frac{\boldsymbol{r} \times \boldsymbol{R}_{\text{cm}}}{|\boldsymbol{R}_{\text{cm}} + \boldsymbol{r}|^3} dm. \tag{5.6.39}$$

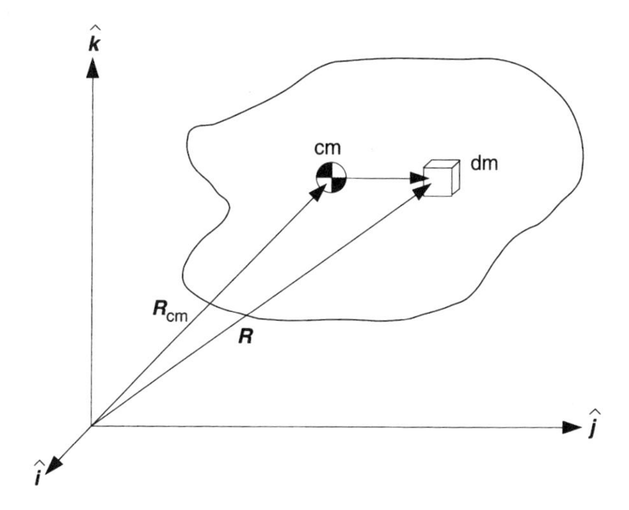

Figure 5.16 Gravity-gradient torque calculation.

After much mathematical manipulation one arrives at

$$N_{\text{GG}} = \frac{3\mu}{R_{\text{cm}}^3} \hat{\boldsymbol{R}}_{\text{cm}} \times (\mathbf{I} \cdot \hat{\boldsymbol{R}}_{\text{cm}}), \tag{5.6.40}$$

where $\mathbf{I}$ is the inertia tensor.

We can eliminate the appearance of the gravitational constant in our equations by noting that the mean motion for any elliptical orbit is

$$\omega_0^2 = \frac{\mu}{R_{\text{cm}}^3}. \tag{5.6.41}$$

Note that for circular orbits ω_0 , the orbital angular velocity of a body at radius R_{cm},is constant. Thus, equation 5.6.40 may be rewritten as

$$N_{\text{GG}} = 3\omega_0^2 \hat{\boldsymbol{R}}_{\text{cm}} \times (\boldsymbol{I} \cdot \hat{\boldsymbol{R}}_{\text{cm}}). \tag{5.6.42}$$

We must bear in mind that ω_0 is a function of R_{cm}^3.

The form of the gravity-gradient torque is very similar to that of the kinematic term in Euler's equation (equation 5.6.17), $\omega \times (\mathbf{I}\omega)$. In gravity-gradient stabilization, described in section 5.8, this term, not the gravity gradient, provides yaw stabilization. Note that if the inertia tensor is a multiple of the identity matrix, that is, if all three principal moments are equal, the gravity-gradient torque vanishes. Generally, however, this is seldom the case, and the difference in the principal moments is often of the same order of magnitude as the moments themselves. Note that we write this kinematic term as a function of matrices and column vectors rather than dyadics and physical vectors since it is specific to one frame, the body frame.

In low-Earth orbit the orbital period is typically 100 min, corresponding to an orbital angular velocity of roughly 10^{-3} s^{-1}. If the difference in the principal moments is on the order of 1000 kg m^2, the gravity-gradient torque could be as large as 3×10^{-3} N m.

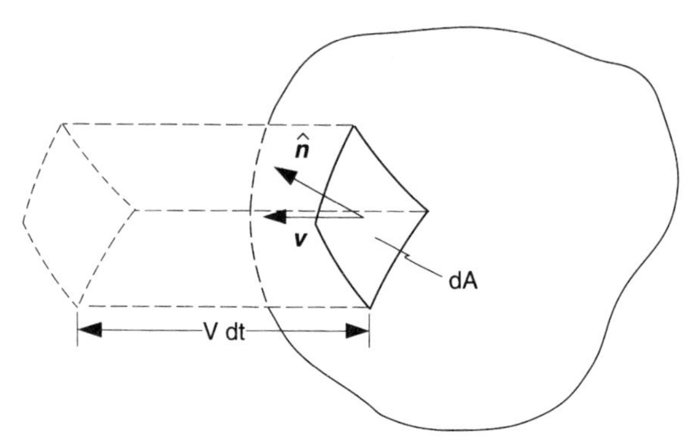

Figure 5.17 Geometry of the aerodynamic force.

5.6.7 Aerodynamic Torques

At altitudes below about 800 km the aerodynamic torque is generally the most important of the environmental torques. The end of a spacecraft mission often occurs when the aerodynamic torques become so great as the spacecraft reenters the Earth's atmosphere that the attitude control system ceases to remain effective and the spacecraft tumbles.

At low atmospheric densities, the collision of a spacecraft surface can generally be modeled as an inelastic impact, with the spacecraft surface absorbing some fraction of the momentum of the air particles. Let dA be the area element and $\hat{\boldsymbol{n}}$ denote the normal to the area element (figure 5.17). If $\boldsymbol{v}$ is the spacecraft velocity, then in a time interval dt the spacecraft surface element sweeps through a volume dV, where

$$dV = (\hat{\boldsymbol{n}} \cdot \boldsymbol{v}) dA dt. \tag{5.6.43}$$

If ρ is the atmospheric density, then the mass of the volume element is $\rho\ dV$, and its momentum $d\boldsymbol{p}$ relative to the spacecraft is $\rho \boldsymbol{v} dV$. The impulse imparted to the surface element dA in time dt is then

$$d\boldsymbol{p} = -\tfrac{1}{2} C_D \rho \boldsymbol{v} dV, \tag{5.6.44}$$

where C_D is the drag coefficient, which for spacecraft generally has a value between 2 and 2.2. These values correspond to the elastic deflection of the atmospheric molecules from the spacecraft and are most appropriate in regions of low atmospheric density. Combining equations 5.6.43 and 5.6.44 and dividing by dt, the aerodynamic force on the spacecraft surface element dA is

$$d\boldsymbol{F}_{\text{aero}} = \frac{d\boldsymbol{p}}{dt} = -\tfrac{1}{2} C_D \rho (\hat{\boldsymbol{n}} \cdot \boldsymbol{v}) \boldsymbol{v} dA. \tag{5.6.45}$$

If the surface element has a moment arm $\boldsymbol{r}$ from the spacecraft center of mass, the aerodynamic torque becomes

$$\boldsymbol{N}_{\text{aero}} = -\int \tfrac{1}{2} C_D \rho (\hat{\boldsymbol{n}} \cdot \boldsymbol{v})(\boldsymbol{r} \times \boldsymbol{v}) dA, \tag{5.6.46}$$

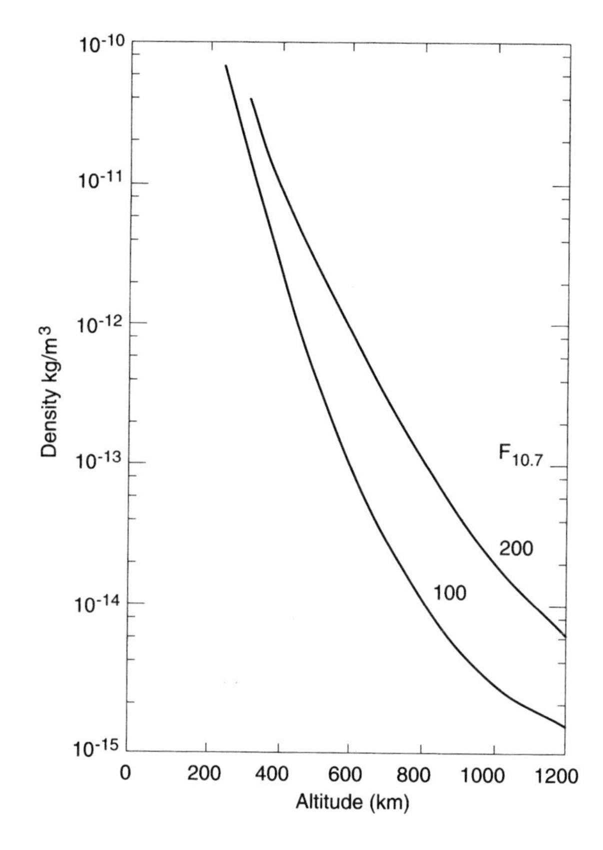

Figure 5.18 Atmospheric density as a function of altitude.

and the integral is taken over that part of the surface for which $\hat{\boldsymbol{n}} \cdot \boldsymbol{v} > 0$, that is, the portion of the spacecraft surface facing into the wind.

Calculation of aerodynamic torques for realistic spacecraft is not very accurate in general because of large uncertainties in the atmospheric density (as much as 100%), uncertainties in the drag coefficient, and shadowing effects. In addition, at high altitudes (say, 1000 km), the atmospheric density is sensitive to solar activity, which may cause the lower atmosphere to boil up, with sometimes drastic consequences. In the case of the Solar Maximum Mission spacecraft, an unexpected period of high solar activity led to increased atmospheric drag and a reduction of the spacecraft lifetime to the point where it was no longer possible either to retrieve it or to boost it to a higher orbit before reentry. A table of nominal atmospheric density is shown in figure 5.18. The two curves correspond to two different levels of solar activity.

A useful concept in computing aerodynamic torques is the center-of-mass–center-of-pressure offset. If we can characterize the spacecraft as being a finite collection of surfaces, then equation 5.6.46 can be written as

$$\boldsymbol{N}_{\text{aero}} = -\tfrac{1}{2} C_{\text{D}} \rho \sum_{i=1}^{0} A_i (\hat{\boldsymbol{n}}_i \cdot \boldsymbol{v})(\boldsymbol{r}_i \times \boldsymbol{v}), \tag{5.6.47}$$

where $\boldsymbol{r}_i$ is the position to the center of pressure of the ith surface, measured from the center of mass. For a flat surface, if there are no shadowing effects, the center of

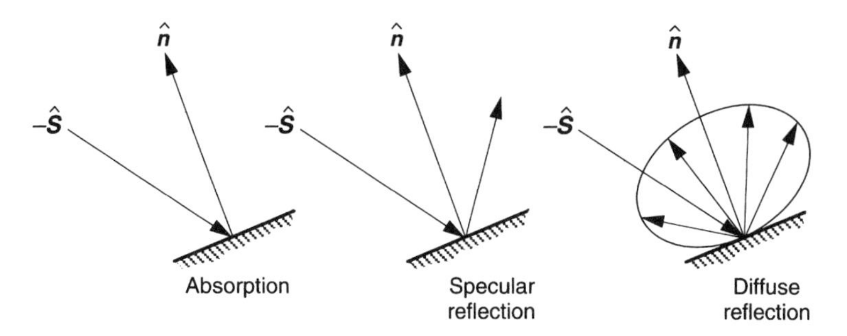

Figure 5.19 Absorption and reflection of solar radiation.

pressure is located at the geometric center of the body. If shadowing is present, the center of pressure is at the center of the unshadowed portion of the surface, at least in regions of low atmospheric density. Such a model provides a simple means of estimating aerodynamic torques.

As a numerical example, consider a spacecraft in a circular orbit with an altitude of 300 km. If we assume a high solar activity, the atmospheric density will be 5×10^{-11} kg m^{-3}. The spacecraft velocity is roughly 7.7 km/s. Assuming that the most significant surface has an area of 1 m^2 and its center of pressure is located at a point 1 m from the center of mass along a direction normal to the velocity, the aerodynamic torque will have the magnitude (assuming $C_\mathrm{D} = 2$, that is, perfectly elastic collisions with the atmospheric molecules)

$$\tfrac{1}{2}C_\mathrm{D}A\rho v^2 r_{\mathrm{cp-cm}} = 3\times10^{-3}\mathrm{N\ m}$$

At an altitude of 1000 km, the aerodynamic torque will be smaller by several orders of magnitude.

5.6.8 Solar Radiation Torques

The solar radiation torque, from a geometrical standpoint, is very similar to the aerodynamic torque, except that the incident particles are photons rather than air molecules. The mean momentum flux from the Sun, $\boldsymbol{P}$, is given by

$$\boldsymbol{P} = -\frac{F_\mathrm{e}}{c}\hat{\boldsymbol{S}} \equiv -P\hat{\boldsymbol{S}}, \tag{5.6.48}$$

where F_e is the solar constant (that is, the energy flux coming from the Sun), and c is the speed of light. $\hat{\boldsymbol{S}}$ is the unit vector from the spacecraft to the Sun.

Radiation from the Sun may be either completely absorbed, specularly reflected, or diffusely reflected, as shown in figure 5.19. The probabilities of each of these occurrences are called the coefficients of absorption, specular reflection, and diffuse reflection, and satisfy

$$C_\mathrm{a} + C_\mathrm{s} + C_\mathrm{d} = 1. \tag{5.6.49}$$

For complete absorption of the solar radiation on a surface element dA, the solar radiation force is

$$d\boldsymbol{F}_\mathrm{a} = -P(\hat{\boldsymbol{n}}\cdot\hat{\boldsymbol{S}})\hat{\boldsymbol{S}}dA, \tag{5.6.50}$$

where again $\hat{\boldsymbol{n}}$ is the normal to the surface element. Equation 5.6.50 is just the incident momentum flux multiplied by the area element.

For pure specular reflection, the incoming momentum flux is $-P(\hat{\boldsymbol{n}} \cdot \hat{\boldsymbol{S}})\,\hat{\boldsymbol{S}}$, and the outgoing momentum flux is $-P(\hat{\boldsymbol{n}} \cdot \hat{\boldsymbol{S}})\hat{\boldsymbol{S}}'$, where $\hat{\boldsymbol{S}}'$ is the outgoing direction. From the law of reflection, it follows that the momentum transfer will be

$$d\boldsymbol{F}_{\text{s}} = -2P(\hat{\boldsymbol{n}} \cdot \hat{\boldsymbol{S}})^2 \hat{\boldsymbol{n}}\, dA. \tag{5.6.51}$$

For pure diffuse reflection, we must average the momentum transfer over the angle of reflection. A common model is to assume that the reflected intensity is proportional to $\cos\varphi$, where φ is the angle of the reflected light from $\hat{\boldsymbol{n}}$. This leads to an expression for purely diffusely reflected radiation:

$$d\boldsymbol{F}_{\text{d}} = -P(\hat{\boldsymbol{n}} \cdot \hat{\boldsymbol{S}})\left(\frac{2}{3}\hat{\boldsymbol{n}} + \hat{\boldsymbol{S}}\right) dA. \tag{5.6.52}$$

Adding these three expressions, appropriately weighted, and recalling equation 5.6.49, leads to

$$d\boldsymbol{F}_{\text{solar}} = -P\left\{(1 - C_{\text{s}})\hat{\boldsymbol{S}} + 2\left[C_{\text{s}}(\hat{\boldsymbol{n}} \cdot \hat{\boldsymbol{S}}) + \frac{1}{3}C_{\text{d}}\right]\hat{\boldsymbol{n}}\right\}(\hat{\boldsymbol{n}} \cdot \hat{\boldsymbol{S}})\,dA, \tag{5.6.53}$$

and the total solar torque is

$$\boldsymbol{N}_{\text{solar}} = \int \boldsymbol{r} \times d\boldsymbol{F}_{\text{solar}}, \tag{5.6.54}$$

where $\boldsymbol{r}$ is the moment arm from the center of mass. Again, the integral is limited to that part of the surface for which $\hat{\boldsymbol{n}} \cdot \hat{\boldsymbol{S}} > 0$.

That value of the solar energy flux reaching the Earth is $F_e = 1400\ \text{W/m}^2$, and the value of the speed of light is $c = 3 \times 10^8$ m/s. If we assume unit reflectivity and pure specular reflection ($C_{\text{s}} = 1$) for a 1 m^2 surface with its normal in the Sun direction, the force on that surface is 9.33×10^{-6} N. Thus, if the offset of the solar center of pressure from the center of mass is 0.5 m and is perpendicular to the direction of the solar flux, the solar radiation torque will have a magnitude of 4.67×10^{-6} N m.

5.6.9 Attitude Simulation

When torques are present, analytical solutions of the attitude dynamics equations are generally not possible, in which case one must carry out computer simulations to study the spacecraft attitude dynamics. The question then arises as to which attitude representation to use. The attitude matrix, while having a simple kinematic equation, has six more elements than the minimum number of parameters needed to parameterize the attitude, and six constraints that must be satisfied. Numerical roundoff errors in the simulation will cause these constraints to become violated, and it is not a simple task to orthogonalize a matrix that is nearly orthogonal. The Euler angles have minimal dimension but, as we have seen, they have unpleasant analytical properties at some attitudes, and the computation of many trigonometric functions is burdensome. The best choice, therefore, is the quaternion, whose kinematic equation, given by equation 5.3.23, is linear and which satisfies only a single constraint that is easy to enforce.

For a rigid body subject to torques, the dynamical equations are given by equations 5.3.23 and 5.6.17. The equations of motion are therefore seven-dimensional and have the form

$$\left.\begin{aligned} \frac{d}{dt}\mathbf{q} &= \tfrac{1}{2}(q_4\omega - \omega \times \mathbf{q}), \\ \frac{d}{dt}q_4 &= -\tfrac{1}{2}\omega \cdot \mathbf{q}, \\ \frac{d}{dt}\omega &= \mathbf{I}_{\mathrm{B}}^{-1}[-\omega \times (\mathbf{I}_{\mathrm{B}}\omega) + \mathbf{N}_{\mathrm{B}}(\overline{q})], \end{aligned}\right\} \tag{5.6.55}$$

and we have noted the dependence of the torque on the attitude. For a dual-spin spacecraft, or a spacecraft with multiple reaction or momentum wheels, the dimensionality of these equations will become much larger. For obvious reasons, computer simulation is an important part of attitude dynamics and control studies.

5.6.10 Further References

The literature on rigid-body mechanics and spacecraft dynamics is very large. Among the general advanced books on classical mechanics, we can single out the books by Goldstein et al. (2002) and Meirovitch (1970), which are extremely rewarding but not easy reading. Dyadics are used extensively in Goldstein et al. (2002). The most complete treatment of dyadics is still that of Gibbs (1901). Among the books on spacecraft dynamics and control are the works of Kaplan (1976), Thomson (1986), and Wiesel (1989), who treat the material at an introductory level. The materiel in Wertz (1978) is both practical and useful. The information in Singer (1964) is now somewhat dated but still very useful. In addition to the two books on classical mechanics, very detailed and advanced treatments can be found in the books of Kane et al. (1983), Hughes (1986), Junkins and Turner (1986), and Rimrott (1989). A short discussion of environmental torques can be found in Duchon (1984).

In addition to the material in chapter 2 of the present work, more detailed information on the spacecraft environment can be found in Stacey (1977), Wertz (1978), Jacobs (1987), and Bertotti and Farinella (1990).

5.7 Attitude Actuators

Attitude actuators are devices that impact the dynamics of the spacecraft, thereby changing its attitude and/or velocity. There are numerous types of actuators with varying degrees of authority (that is, magnitude of capability). In addition, these devices can create either internal forces or torques, whereby the actuator acts against the spacecraft directly, or external forces or torques, where the dynamic change is a result of a reaction external to the spacecraft. These distinctions will become clearer in the following sections.

While attitude sensors are used in attitude determination, attitude actuators are used in attitude control. In some cases an actuator may be used passively, where no intervention by the ground or onboard computer is required, or actively, where commands are actively and routinely sent to the actuator.

5.7.1 Momentum/Reaction Wheels

Momentum wheels and *reaction wheels* are simply motors with a flywheel mounted on the motor shaft, the difference in terminology resulting primarily from the speed at which they operate. A momentum wheel typically operates at a relatively high constant speed, providing a means of momentum storage which in turn provides stability. Reaction wheels generally operate at low, and varying, speeds. Both types of wheel provide active control of the spacecraft by generating a reaction torque according to Newton's third law. As a torque is electrically/magnetically exerted on the motor shaft to cause the wheel to accelerate, an equal and opposite torque is generated on the spacecraft, causing the attitude to change.

Momentum wheels are commonly used singly or in pairs to provide spin stability and active control about, at most, two body axes; their use in control is detailed in section 5.8. Reaction wheels are used in a three-wheel triad formation, with the spin axes of the wheels mounted orthogonally, or in a four-wheel pyramid scheme, with the spin axes pointing normal to the faces of a pyramid. Either of these configurations is used to control the three-axis attitude of a spacecraft. The four-wheel pyramid configuration provides redundancy as well as greater flexibility in the distribution of wheel angular momentum. Very often a fourth wheel, whose axis does not coincide with any of the others, is added to the three-wheel triad formation to provide redundancy. (Reaction wheels are often biased to some small momentum—too small to provide sensible gyroscopic stabilization—in order to minimize the problem of stiction.)

A momentum wheel mounted in a one- or two-axis gimbal is generally called a control moment gyro (CMG). A motor spins a wheel at a (typically) constant speed, creating a momentum vector in the direction of the spin axis, while one or more motors are used to torque the momentum vector to change its direction. As the momentum direction is changed, a reaction torque is created on the spacecraft. CMGs can be made much more responsive than momentum wheel systems (that is, they provide much greater torques on the spacecraft) and therefore are used on very agile spacecraft, but at much greater financial cost and weight.

Momentum and reaction wheels have the advantage of providing quick and accurate attitude control. Also, they can be used at any altitude. Their disadvantage is that they can be costly, massive, and require large amounts of power. Also, as indicated previously, wheels may saturate because of the accumulated action of disturbance torques on the spacecraft, which are transferred to the wheels by the wheel control law. Thus, some additional means must be available to remove energy from (or desaturate) the wheels.

5.7.2 Magnetic Coils

In section 5.6.5 the effects of magnetic torque on a spacecraft were discussed. Of particular interest here is the *intentional* creation of a magnetic moment. The control of such a device is discussed in section 5.8.

A magnetic moment, $\mathbf{m}$, is created by current flowing through a wire loop and is given by

$$\mathbf{m} = NIA\hat{\mathbf{n}}, \tag{5.7.1}$$

where N is the number of loops (or turns), I is the current, A is the area of the planar loop, and $\hat{\mathbf{n}}$ is the unit vector normal to the plane of the loop. The direction of the magnetic moment is given by the right-hand rule; if one curls the fingers of one's right hand in the direction of the current flow, the thumb will point in the positive direction of the moment.

Typically, coils are wound around specially designed rods that provide near optimum magnetic moment for a given power and weight. These devices are called *torque rods* or *torquer bars*. There have been instances, however, where a wire loop was created simply by wrapping wire around the exterior of the spacecraft, in effect making the spacecraft a giant torque rod. Obviously this has implications with which other spacecraft subsystems and instruments would very likely be concerned.

5.7.3 Thrusters

Thrusters exert a force or torque on the spacecraft by expelling mass. In attitude control we are interested in the torque produced, since the torque changes the attitude. Generally speaking, the torque created by a force exerted in a direction that does not pass through the center of mass is not desirable because of the resulting change in the linear velocity. For this reason thrusters are commonly used in pairs for attitude control in order to generate a torque, but no net force, within the variability of the thrust from each jet (figure 5.20). In this way the spacecraft's orbital motion will not be altered by an attitude maneuver.

If a thruster has a moment arm $\boldsymbol{r}$ with respect to the body system centered at the center of mass of the spacecraft, and if it expels mass at a rate $\dot{m}$ with effective exhaust velocity $\boldsymbol{v}$, then the thruster torque will be

$$\boldsymbol{N}_{\text{thruster}} = \boldsymbol{r} \times (\dot{m}\boldsymbol{v}). \tag{5.7.2}$$

The mass expelled may be a fairly inert gas, such as freon or nitrogen, or it may be a chemically active gas such as hydrazine. Forces up to 5 N are possible with inert gases, while forces from roughly 1 to 10,000 N are possible with hydrazine. The tremendous force possible with hydrazine thrusters makes it an extremely dangerous substance for human interaction. Hydrazine charging on the ground is always an unwelcome exercise.

A possibility for greatly extending the lifetime of thrusters is through the use of electric propulsion, in which the expelled mass consists of ions accelerated in an electric field. The momentum added by the acceleration reduces the rate at which the mass need be expended. Examples of these are the Pulsed-Plasma Thruster (PPT) and the Xenon Ion Propulsion System (XIPS). The thrust produced by these devices is on the order of millinewtons, quite small compared to inert or active gas systems. Electric thrusters are a reasonable alternative for geosynchronous and deep-space spacecraft,

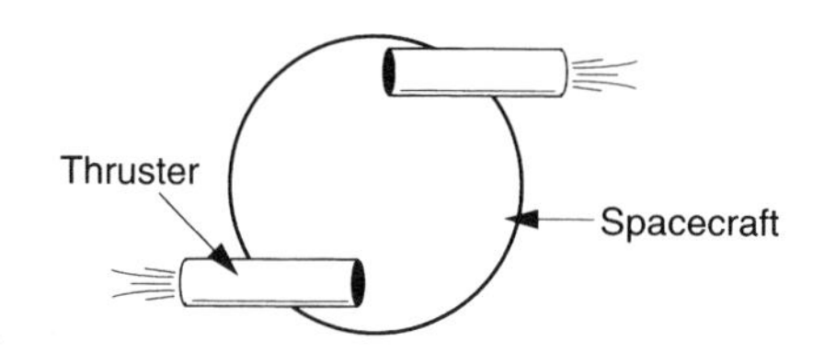

Figure 5.20 Thruster torques.

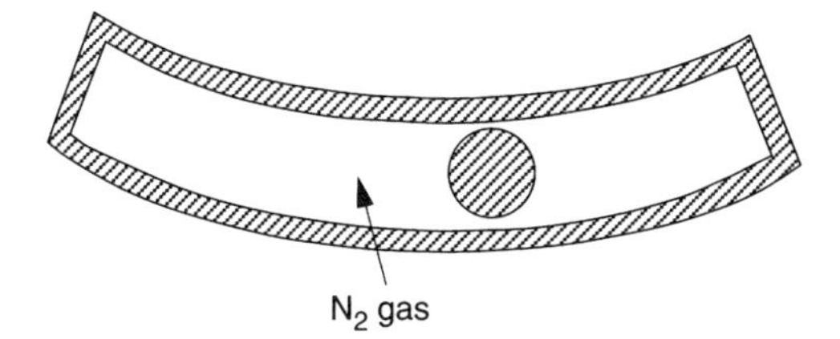

Figure 5.21 Ball-in-tube nutation damper.

where the gravity-gradient or geomagnetic field is too weak to be used for attitude control. Chapter 4 provides more information regarding electric propulsion.

5.7.4 Nutation Dampers

When a spacecraft is disturbed from a stable equilibrium it may exhibit unwanted oscillations about that equilibrium point. For a spinning spacecraft this is called nutation; for gravity-gradient stabilized spacecraft it is called libration. Nutation (or libration) control, discussed in section 5.8, can be accomplished in several ways. Here we present hardware used to reduce nutation. This reduction, or damping, is accomplished with various devices that dissipate energy passively by providing friction.

One example is a tube that contains a ball or a viscous liquid or both (figure 5.21). As the spacecraft oscillates, the ball, the liquid, or both will oscillate and dissipate part of the energy as friction on the side of the tube. Another, more efficient, damping device uses a conducting pendulum that is free to swing in the field of a permanent magnet. As the conductor accelerates in the field of the magnet it produces eddy currents, which quickly dissipate the energy. Eddy current nutation dampers are generally much more efficient than mechanical nutation dampers. The transfer of energy is enhanced by tuning the pendulum so that its natural frequency is the nutation frequency of the spacecraft motion.

5.7.5 Yo-yo Despin

Yo-yo despin masses are often released at the beginning of the attitude acquisition sequence to spin down the spacecraft. (While the name comes from their resemblance to a child's toy, some refer to them as yo despin devices since they do not return to their original position.) The two yo-yos are generally wrapped around the spacecraft, if it is circular, or around a circular flange. When released, they fly out roughly tangentially from the spacecraft (figure 5.22) and take up the angular momentum in much the same way as a figure skater slows herself from a spin by extending her arms. If m is the total mass of the two yo-yos, l is the length of the cord, R is the radius of the spacecraft (or circular flange), and I_z is the moment of inertia about the spin axis (assumed to be a principal moment of inertia), then the angular velocity about the spin axis when the yo-yos have reached the maximum length of the cord and are detached from the spacecraft is

$$\omega_{\text{final}} = \omega_{\text{initial}} \left(\frac{cR^2 - l^2}{cR^2 + l^2} \right), \tag{5.7.3}$$

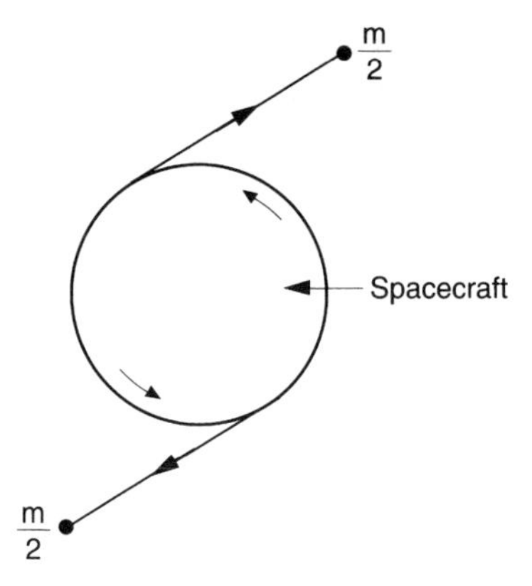

Figure 5.22 Deployment of yo-yo despin weights.

where

$$c = 1 + \frac{I_z}{mR^2}. \tag{5.7.4}$$

This formula assumes that the cord is massless, always tangential to the spacecraft, unstretched, and that the two masses fly out in exactly opposite directions. Thus, if

$$l = R\sqrt{c}, \tag{5.7.5}$$

the final angular momentum of the spacecraft will be zero. In general, this is undesirable, since for the sake of stability the spacecraft should have some spin during the attitude acquisition sequence. Since m is generally much smaller than the mass of the spacecraft, while R typically is only about twice the root-mean-square radius of the spacecraft mass from the spin axis, it follows that l is generally quite large compared with spacecraft dimensions.

Yo-yo despin weights have the advantage of being very inexpensive and lightweight. They have the disadvantage that they leave very lethal hockey pucks in space, where they pose a threat to future spacecraft. The tracking of space debris is a major task of NASA, NORAD (the North American Air Defense Command), and other agencies, and the ever-increasing amount of space debris is of growing concern.

5.7.6 Further References

While being a little dated, Wertz (1978) is probably still the best reference for information regarding the descriptions and mathematical modeling of much of the hardware discussed in this section. The control-moment gyro is described in O'Connor and Morine (1967). Chapter 4 of this book gives descriptions of thruster systems and their components. Electric propulsion (PPT, XIPS) is currently attracting great attention as a result of improvements in the mass and power of these systems. Good descriptions of electric propulsion systems can be had by simply performing a search on the Internet, using one's favorite search engine.

5.8 Attitude Control

By attitude control, we mean how we cause the spacecraft to have a certain attitude. Attitude control is generally of two kinds: *attitude stabilization*, in which case the purpose is to maintain the spacecraft at a given attitude; and *attitude maneuvers*, whose purpose it is to change the spacecraft attitude to a new orientation. Attitude maneuvers may be either for the initial acquisition of attitude at the beginning of the mission or simply to direct the payload at a new target, which may occur many times during the mission. In addition, attitude control may be *passive*, in which case the control torques operate without the need to compute the control response on the basis of attitude sensors, or controls may be *active*, in which case they are commanded on the basis of the (not necessarily simultaneous) output of the attitude sensors. In addition, controls may be *open loop*, by which we mean that a control law is calculated before it is activated, and is not altered by any information received during the action of the control, or controls may be *closed loop*, which means that the response of the control law is determined in real time by the state of the system. *Feedback control* is synonymous with closed loop control.

5.8.1 Reaction Control

By reaction control, we mean control of the spacecraft module by moving some other mass. This may be accomplished in two specific ways: (1) by expelling mass from the spacecraft, or (2) changing the position or orientation within the spacecraft. The first type includes thrusters and despin yo-yos; the second type includes momentum wheels, reaction wheels, and CMGs. Devices of the first type have the advantage of being capable of very large and rapid changes in the angular velocity of the spacecraft, but they are limited by the fact that the mass that is expelled is nonrenewable. The second type, since it does not expel mass, is renewable, but it suffers from the fact that wheels are limited in the maximum angular momentum they may store and can therefore saturate unless some other means is available to desaturate them.

Use of the term "control" typically implies the desire to impact a system in a positive way; we often speak of something uncontrolled in a negative way. The hardware or software used to generate the torque commands, whether they be to wheels or thrusters, is called the *controller*. The collection of the controller, actuators, and sensors is called the *control system*. There are numerous ways in which to control a spacecraft, with various types of attitude control hardware, and we will focus here on simple controllers using reaction wheels and thrusters.

Attitude control with reaction wheels is described easiest in a classical control manner. The discipline of control theory is vast and well beyond the scope of this book, but it is important to understand a few basics regarding the workings of a simple controller. To do this we will examine a simple spring–damper mechanical system, shown in figure 5.23. While not a spacecraft, this system is intuitively easier to understand.

The position $x(t)$ of the mass m is controlled by the forcing function $f(t)$ (that which we command) and is described mathematically as

$$m\ddot{x} = f(t) - B\dot{x} - Kx, \tag{5.8.1}$$

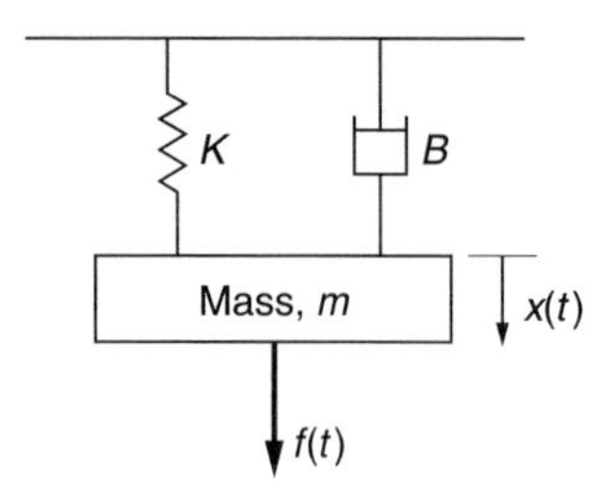

Figure 5.23 Spring–damper system.

where K is the spring stiffness and B is the viscous damper coefficient. Suppose that one wishes to move the mass in discrete steps such that $x(t)$ changes in increments of 1 cm, and assume for the time being that the damper does not exist ($B = 0$). One could envision the difficulty in performing this operation if the spring stiffness were very light; if one were not careful in choosing $f(t)$ the mass would begin to bounce back and forth, and could do so uncontrollably, that is, no $f(t)$ would exist such that the desired motion is achieved. On the other hand, the addition of the damper ($B \neq 0$) could, depending on the value of B, slow the motion to a sufficient degree to maintain the desired position of the mass. While the objective is to control the *position* of the mass, the damper provides a means of controlling the *rate* of the mass. In attitude control we are controlling angles rather than positions, but the concepts are the same.

An illustration of the effects of damping on this type of system is given in figure 5.24. In this figure are plotted several responses of the system to a step response,

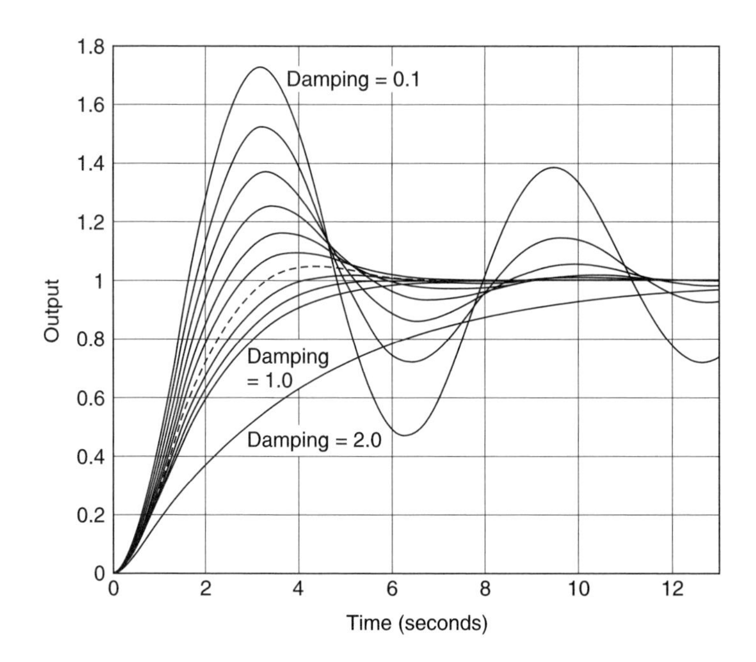

Figure 5.24 Step responses with various damping ratios.

that is, a step from one position to another. In these cases the step has been normalized to 1, and the damping values listed are damping ratios and do not equal the value of B. (As we will see below, the damping ratios in figure 5.24 are given by $B/2\sqrt{Km}$.) Several characteristics are worth noting. The greater the damping the slower the response in getting to the final destination. On the other hand, the smaller the damping the greater the overshoot, or going beyond the desired position and having to come back. Depending upon the requirements placed on the controller, overshoot may be acceptable in order to achieve the end position quickly. Similarly, greater damping may be required if the desire is to achieve the final position with no overshoot without regard to the amount of time taken. One speaks of systems with the damping ratios greater than one as being *overdamped*, while damping ratios less than one are considered *underdamped*. A system with a ratio of one is called *critically damped*; this value achieves the desired position in the fastest time with no overshoot.

In spacecraft attitude control we are interested in determining a control torque such that the prescribed spacecraft motion is obtained, and in this case we control angles rather than positions. In the above example of a mechanical system the control was on position, and rate was effectively controlled by the damper. This is not always the case; one must also actively control rate. Generally one selects a control law of the form

$$N_{\text{wheel}} = -k_1(\theta - \theta_0) - k_2(\omega - \omega_0), \tag{5.8.2}$$

and the constants k_1 and k_2 are chosen to make the control system slightly underdamped. Here θ and ω are the sensed angle and angular velocity of the spacecraft about a given spacecraft axis, and θ_0 and ω_0 are the desired angle and angular velocity about that axis. The first term (in k_1) provides the restoring torque, and the second term (in k_2) provides damping so that this control law will not lead simply to undamped oscillatory motion. Implementation of such a system for three-axis control requires that complete attitude and angular velocity be computed on board. A controller of the form given in equation 5.8.2 is called a *proportional-derivative* (PD) *controller* because the control torque is proportional to the attitude error and its derivative (rate error). Another type of controller typically used in attitude control includes a term that involves the integral of the angular error, and is called a *proportional-integral-derivative* (PID) *controller*. A PD controller will allow a residual error bias to remain, thus resulting in less accurate pointing. The inclusion of the integral term causes the controller ideally, over time, to reduce the bias error to zero.

Let us examine a single-axis momentum wheel case where the spin axis coincides with the pitch axis. The equations of motion of the spacecraft pitch motion and the motion of the momentum wheel are given by

$$\left.\begin{aligned} I_p \ddot{p} &= N_{\text{wheel}} + N_{\text{ext}}, \\ I_{\text{wheel}}(\dot{\omega}_{\text{wheel}} + \ddot{p}) &= -N_{\text{wheel}}, \end{aligned}\right\} \tag{5.8.3}$$

where p is the pitch angle of the spacecraft and I_p is the moment of inertia about the pitch axis. The first equation describes the dynamics of the spacecraft base module, the second the dynamics of the wheel. The angle and angular velocity errors are now just the pitch and pitch rate, so that the pitch feedback control (for an Earth-locked spacecraft) is given by

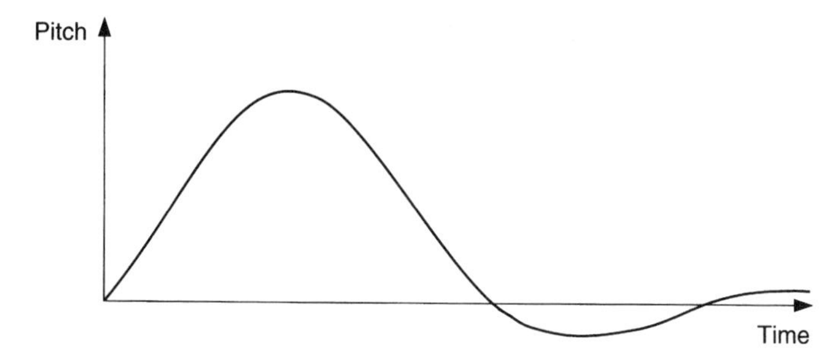

Figure 5.25 Typical temporal response of a momentum wheel system.

$$N_{\text{wheel}} = -K_p p - K_{\dot{p}}(\dot{p} - \omega_0), \tag{5.8.4}$$

where we have written $\theta = p$ and $\theta_0 = 0$. For a circular orbit $\omega_0 = \sqrt{\mu/R^3}$. This control law leads to an equation of motion

$$\ddot{p} + 2(\zeta\omega_p(\dot{p} - \omega_0)) + \omega_p^2 p = I_p^{-1} N_{\text{ext}}, \tag{5.8.5}$$

with

$$\omega_p \equiv \sqrt{K_p/I_p} \quad \text{and} \quad \zeta = \frac{K_{\dot{p}}}{2\sqrt{K_p I_p}}. \tag{5.8.6}$$

(Notice the similarity between equations 5.8.5 and 5.8.1.) Generally, the damping constant ζ is chosen to have the value 0.707 so that the system is slightly underdamped. This leads to some additional oscillation over the above critical damping, but also to a control law that is much faster from a practical standpoint. The choice of the value for ω_0 is usually determined by hardware and system constraints. The typical response of this type of system is shown in figure 5.25.

This system provides gyroscopic stabilization in roll and yaw and direct control of pitch through momentum exchange. The pitch and yaw will necessarily drift with time and therefore must be adjusted, using either thrusters or magnetic control. A three-axis system with three reaction wheels could be similarly designed.

Reaction control with thrusters is managed differently. The on-off nature of a thruster, versus the relatively continuous control provided by wheels, does not lend itself to the typical PD or PID control described above, due to its nonlinear nature. Thruster control is commonly implemented using *phase planes*. The use of phase planes comes out of nonlinear control theory, and involves plotting angular rate versus angle as time progresses. In the case of thruster control we plot angular rate error versus angular error. Thruster firing depends upon the values of the angular and rate errors at a given time. Figure 5.26 illustrates a phase plane thruster control logic.

Suppose the attitude angular error has a large negative value and the rate error has a large positive value. This would place the location in the upper left corner of the figure. In this situation the appropriate thruster would be fired so as to reduce the rate error without necessarily reducing the attitude error. The firing would continue until the "+ Rate limit" switch line is crossed and the system enters the "Coast zone." At this

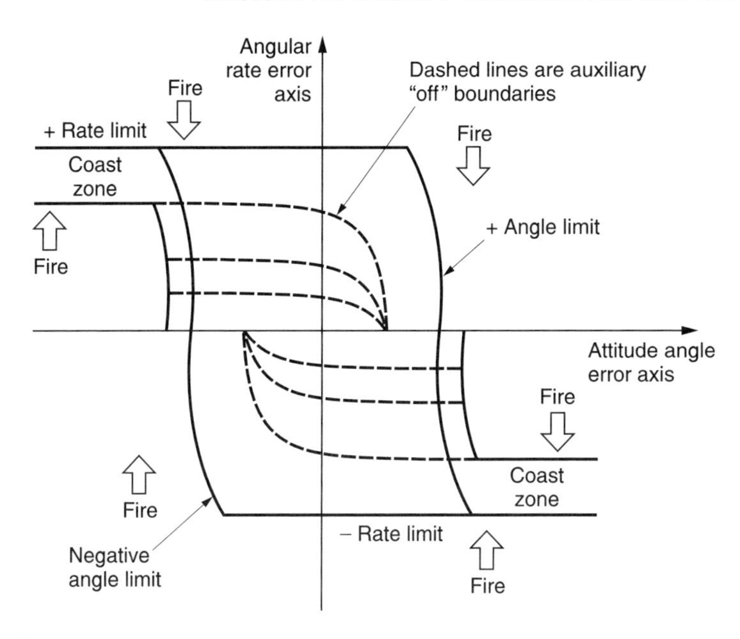

Figure 5.26 Phase plane thruster control logic.

point the thruster firing would cease. The positive rate error will cause the angular error to decrease without intervention by the thrusters. (This corresponds to a drifting to the right in the figure.) Eventually the angular error will become positive, and continue to grow, due to the positive rate error. At some point in time the angular error will cross the "+ Angle limit" switch line, causing the appropriate thrusters to fire such that the rate error is decreased and then forced negative. The firing continues until the "Coast zone" in the lower right hand corner is reached. The process then repeats, causing a drifting and firing from coast zone to coast zone. One can see that the spacecraft attitude and rate errors are defined by, and maintained within, the switch lines.

The definition of the switch lines determines the accuracy of the controller, and their selection is driven by system requirements. It must be remembered that thrusters use a non-replenishable propellant, and a phase plane designed to hold the attitude and rate errors tightly to zero will result in vociferous thruster firing, thus quickly consuming the available propellant.

5.8.2 Magnetic Control

Generally, momentum/reaction wheels, whether for active attitude control or for spin stabilization, are coupled with electromagnets. For spin-stabilized spacecraft, the electromagnets generally have an axis aligned with the axis of the momentum wheel and the other two axes aligned perpendicular to it. The electromagnets have two functions. One is *spin-axis maneuvers*, that is, to reorient the spin axis of the spacecraft, which may have moved because of the accumulated action of disturbance torques. The other is to spin up or down the momentum or reaction wheels, called *momentum control*. Momentum and reaction wheels can operate only within certain limits. Controller

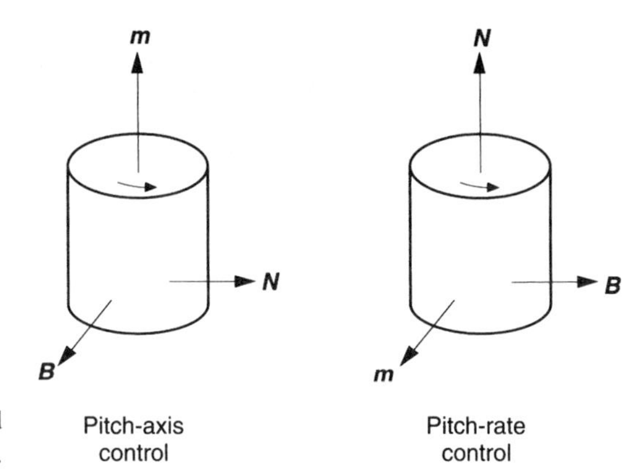

Figure 5.27 Pitch-axis control and pitch-rate control.

compensation for disturbance torques will generally cause these wheels to accumulate angular momentum, leading them to saturate eventually or to lose angular momentum to the extent that they no longer provide stabilization. It is necessary, therefore, to be able to exchange the momentum in the wheels with the environment. Since the momentum in the wheels is coupled to the spacecraft body angular momentum via the control laws, it is sufficient that it be possible to exchange momentum between the spacecraft body and the environment. This may be accomplished by either thrusters or magnetic torquing.

Consider equation 5.6.36. Suppose that the spin axis of the spacecraft is the pitch axis and one of the magnetic torquers is aligned with the pitch axis; that is, it will produce a magnetic moment in the direction of the pitch axis. If the magnetic field has a component that is perpendicular to the pitch axis, then activating the pitch-axis coil will cause the spacecraft spin axis to precess about the direction of the magnetic field. In this manner, the attitude of the pitch axis can be changed in the plane perpendicular to the magnetic field. If the spacecraft is in an equatorial orbit (more exactly, if the orbit lies in the magnetic equatorial plane) so that the magnetic field always points north, then the pitch axis can be precessed magnetically only about the north direction, and the attitude cannot be controlled completely. If, however, the orbit has a magnetic inclination that is significantly different from zero, then the direction of the magnetic field along the orbit will be constantly changing and, depending on the position of the spacecraft in the orbit, the pitch axis attitude can be altered.

The remaining two magnetic torquers are mutually perpendicular and perpendicular to the pitch axis. Hence, by controlling the relative strength of the two coils, a magnetic moment can be generated in any direction perpendicular to the pitch axis. In particular, it can be generated in the direction perpendicular to both the pitch axis and the external magnetic field. In this case the applied magnetic torque, $\mathbf{m} \times \mathbf{B}$, will be parallel or antiparallel to the pitch axis and will cause the spacecraft attitude to change about the pitch axis, or, equivalently, to cause the spacecraft spin rate about the pitch axis to increase or decrease. The different configurations between pitch-axis control and pitch-rate control are shown in figure 5.27.

For a non-spinning spacecraft, the distinction between the pitch axis and the other two axes disappears. It is clearly possible to control all three axes of the spacecraft using magnetic control alone (provided the orbit does not lie in the magnetic equatorial plane). Such a method is postulated by Musser and Ebert (1989).

5.8.3 Spin Stabilization

Spin stabilization provides an example of open-loop passive control. Suppose a spacecraft has a momentum wheel that has a very large angular momentum **h**. The angular momentum of the spacecraft is then (all representations in this section are with respect to body coordinates unless indicated otherwise)

$$\mathbf{L} = \mathbf{I}\boldsymbol{\omega} + \mathbf{h}, \tag{5.8.7}$$

and Euler's equation becomes, in the inertial frame,

$$\frac{d}{dt}(\mathbf{I}_\mathrm{I}\boldsymbol{\omega}_\mathrm{I}) + \frac{d}{dt}\mathbf{h}_\mathrm{I} = \mathbf{N}_\mathrm{I}. \tag{5.8.8}$$

In general, for spin stabilization, it is also true that $|\mathbf{h}| \gg |\mathbf{I}\omega|$ and the temporal derivatives of the first term of equation 5.8.8 can be neglected if the torque is not along **h**. Then we can approximate equation 5.8.8 by

$$\frac{d}{dt}\mathbf{h}_\mathrm{I} = \mathbf{N}_\mathrm{I}. \tag{5.8.9}$$

If we can ignore friction in the bearings of the momentum wheel, then the magnitude of $\mathbf{h}_\mathrm{I}$ will not change. Therefore, the angular momentum of the wheel can change only by changing its orientation, and the disturbance torque will tend to precess the spin axis of the momentum wheel (and with it the spacecraft) by some angle ϕ. From equation 5.8.9, this precession will be on the order of

$$\phi \approx \frac{1}{h}\left|\int \mathbf{N}_\mathrm{I} dt\right|. \tag{5.8.10}$$

If the angular momentum of the momentum wheel h is very large compared with the environmental torque, then ϕ will remain very small. Equation 5.8.10 can also be used to determine the amount of momentum required to maintain a gyroscopic stiffness such that the error remains below ϕ over a given period of time. For example, suppose one desires the attitude error to remain below 0.5 over a 24-hour period, and that the disturbance torque on the spacecraft over this time is a constant 1×10^{-5} N m. Integrating this torque over 24 hours results in an increase in momentum of 0.864 N m s. To maintain an error less than 0.5° (0.0087 radians) one would need a momentum wheel with a constant 99 N m s of stored momentum.

The same gyroscopic stabilization occurs for a rapidly spinning spacecraft, although the spin rate may not be as constant. However, if the spacecraft is spinning quickly enough the average torque per revolution will be very small, and the analysis for momentum wheels may also be applied in this case.

Spin stabilization of a spinning spacecraft can only be effective if the spacecraft is spinning about the major principal axis, that is, the principal axis with the largest moment of inertia. To see this, examine the energy of the rigid body,

$$E = \tfrac{1}{2}\boldsymbol{\omega}^\mathrm{T}\mathbf{I}\boldsymbol{\omega} = \tfrac{1}{2}\mathbf{L}^\mathrm{T}\mathbf{I}^{-1}\mathbf{L}. \tag{5.8.11}$$

If we choose the body-fixed coordinate axes to lie along the principal axes of the inertia tensor, then the (body-referenced) inertia tensor will be diagonal and equation 5.8.11 becomes

$$E = \frac{1}{2I_1}L_1^2 + \frac{1}{2I_2}L_2^2 + \frac{1}{2I_3}L_3^2. \tag{5.8.12}$$

Since the magnitude of $\mathbf{L}$ is nearly constant if the torque is small, the energy can be a true minimum only if $\mathbf{L}$ is parallel to the major principal axis. However, for the body-referenced spin axis to migrate from its original direction in the body to the major principal axis, the energy must change. Suppose the spinning spacecraft is prolate, so that the symmetry axis, which is supposedly also the spin axis, is the minor principal axis. Without some means of dissipating energy, the spacecraft would be at equilibrium (actually, unstable equilibrium) in this case. However, if there is any energy-dissipating device on board the spacecraft (for example, a nutation damper, propulsion propellant, or even a heat pipe), the spacecraft, after it dissipates as much energy as possible, will eventually come to spin about the major principal axis. Thus, it will eventually go into a "flat spin" unless some active control device is on the spacecraft to prevent this. In the past, this has occurred with disastrous effects.

Pure spin stabilization, being open-loop, cannot maintain the spacecraft attitude indefinitely because of the drift caused by environmental torques. Although the angular momentum of the spacecraft may be large, even a small torque, given sufficient time, will cause large attitude deviations. Thus, spin stabilization may be accompanied by some feedback mechanism, which need not be operating continuously, in order to correct for these drift effects. These spin-stabilized systems may be reoriented by ground command, or a reorientation maneuver may be activated automatically by a computer on board the spacecraft.

5.8.4 Gravity-Gradient Stabilization

The inertia tensor of a spacecraft may be made fairly large by the use of long booms. In this way the magnitude of the gravity-gradient torque acting on the spacecraft can be increased and used to control the attitude motion of the spacecraft. In general, gravity-gradient stabilization techniques are used to keep a spacecraft Earth pointing. To see that this is indeed possible, consider Euler's equation again. Let us assume that the spacecraft is in a circular orbit. If the angular velocity of the spacecraft is constant, as it must be if the spacecraft is truly Earth pointing, then, in the absence of other torques, Euler's equation becomes in the body frame

$$\begin{aligned} \frac{d}{dt}\mathbf{L} &= -\boldsymbol{\omega} \times (\mathbf{I}\boldsymbol{\omega}) + \mathbf{N}_{\mathrm{GG}} \\ &= -\boldsymbol{\omega} \times (\mathbf{I}\boldsymbol{\omega}) + \frac{3\mu}{R^3}\hat{\mathbf{R}} \times (\mathbf{I}\hat{\mathbf{R}}). \end{aligned} \tag{5.8.13}$$

The first term of the right-hand side of equation 5.8.13 gives the effective torque due to the gradient of the centrifugal force, which has much the same form as the gravity-gradient torque. If we assume that $\mathbf{R}$ is exactly along the local vertical, which we call the z-axis, and that $\boldsymbol{\omega}$, which is now identical with the orbital angular velocity, is exactly along the negative x-axis, then equation 5.8.13 becomes

$$\frac{d}{dt}\mathbf{L} = \omega_0^2(I_{31}\hat{\mathbf{y}} - I_{21}\hat{\mathbf{z}}) - 3\omega_0^2(I_{23}\hat{\mathbf{x}} - I_{13}\hat{\mathbf{y}}), \tag{5.8.14}$$

where we have used the fact that the orbital angular velocity for a circular orbit is given by $\omega_0^2 = \mu/R^3$. The angular momentum will remain constant, therefore, provided the products of inertia all vanish. Thus, a necessary condition for gravity-gradient stabilization is that the principal moments of inertia be aligned with the local vertical and the orbit normal. Gravity-gradient stabilized spacecraft, therefore, are always Earth pointing.

Now let us consider what happens when there is an additional torque. Consider the case of a spacecraft in a circular orbit with the vertical (yaw axis) and the negative orbit normal (pitch axis) aligned with principal axes. Let us suppose that this torque is about the orbit normal, so that the tendency of the torque is to rotate the spacecraft about the orbit normal. In this case, the pitch axis will not change (that is, the direction of the orbit normal will not change in the spacecraft body frame), so that the first term on the right-hand side of equation 5.8.13 will remain zero. In the body frame, $\hat{\mathbf{R}}$ now becomes

$$\hat{\mathbf{R}} = \begin{bmatrix} 0 \\ \sin\theta \\ \cos\theta \end{bmatrix}, \tag{5.8.15}$$

for some angle θ. If we compute the right-hand side of equation 5.8.13 now, we obtain

$$\begin{aligned} \frac{d}{dt}\mathbf{L} &= 3\omega_0^2(I_3 - I_2)\sin\theta\cos\theta\hat{\mathbf{x}} + N_x\hat{\mathbf{x}} \\ &= -\frac{3}{2}\omega_0^2(I_2 - I_3)\sin 2\theta\hat{\mathbf{x}} + N_x\hat{\mathbf{x}}. \end{aligned} \tag{5.8.16}$$

The spacecraft will find an equilibrium orientation at the value θ_e for which the derivative of $\mathbf{L}$ vanishes. This is given by

$$\theta_e = \tfrac{1}{2}\sin^{-1}\left(\frac{2N_x}{3\omega_0^2(I_2 - I_3)}\right). \tag{5.8.17}$$

It is not enough for this to be an equilibrium point. It must also be a point of stable equilibrium. Thus, the torque must be in the opposite direction to the angular displacement. From equation 5.8.16, this means that we must have $I_2 > I_3$.

This, however, only guarantees stability about one axis. Examine now the other two axes. If we consider a small torque about the vertical, then the zenith will not change direction in the spacecraft frame, and the second term of equation 5.8.13 will not contribute to the change of angular momentum. The gravity-gradient torque, therefore, cannot provide attitude control about the vertical. The angular velocity in this case, however, is

$$\boldsymbol{\omega} = \omega_0\begin{bmatrix} -\cos\theta \\ \sin\theta \\ 0 \end{bmatrix}. \tag{5.8.18}$$

Substituting this into equation 5.8.13 leads to

$$\frac{d}{dt}\mathbf{L} = -\tfrac{1}{2}\omega_0^2(I_1 - I_2)\sin 2\theta\hat{\mathbf{z}} + N_z\hat{\mathbf{z}}. \tag{5.8.19}$$

The torque in this case has not come from the gravity gradient but from the centrifugal-force gradient. Stability about the z-axis thus requires that $I_1 > I_2$.

If we consider finally a small torque acting about the y-axis, then both the zenith direction and the orbit normal, as viewed from the spacecraft, will change direction. In this case, responding to a small rotation about the y-axis,

$$\hat{\mathbf{R}} = \begin{bmatrix} -\sin\theta \\ 0 \\ \cos\theta \end{bmatrix} \quad \text{and} \quad \boldsymbol{\omega} = \omega_0 \begin{bmatrix} -\cos\theta \\ 0 \\ -\sin\theta \end{bmatrix}, \tag{5.8.20}$$

which leads to

$$\frac{d}{dt}\mathbf{L} = -2\omega_0^2(I_1 - I_3)\sin 2\theta\hat{\mathbf{y}} + N_y\hat{\mathbf{y}}. \tag{5.8.21}$$

Stability now requires that $I_1 > I_3$. Therefore, the spacecraft motion will be stable about all three axes, provided that

$$I_1 > I_2 > I_3. \tag{5.8.22}$$

Thus, three-axis stability by pure gravity-gradient stabilization requires that nominally the major principal axis be aligned with the orbit normal (or negative orbit normal, generally the pitch axis) and the minor principal axis be aligned with the nadir (or zenith, generally the yaw axis). Although this stability condition is valid, the actual situation is slightly more complicated, particularly for non-circular orbits. Quarter-orbit coupling, which we have not considered in the derivation, permits a second possible orientation of the rigid body, which is $I_2 > I_3 > I_1$. The reader is referred to Kaplan (1976) or Hughes (1986) for details. Gravity-gradient stabilization, although passive, is closed loop since a restoring torque is generated, based on the size of the error.

Gravity-gradient stabilization, although feasible, is limited because of the weakness of the gravity-gradient torque. Thus it cannot be used at very high altitudes, where this torque is very weak, and it cannot be used at very low altitudes, where it will be overwhelmed by aerodynamic torques. Also, because the gravity-gradient torque is so weak, attitude control by this means cannot be very precise. In general, it is difficult to achieve stability to better than a few degrees by gravity-gradient techniques alone. Also, the mission requirements of the spacecraft may not be consistent with the mass distribution required for gravity-gradient stabilization. This can be overcome by attaching a proof mass, typically around 1 kg, at the end of a long boom, which tends to increase the size of the other moments of inertia relative to the moment about the vertical, which for gravity-gradient stabilization must be the smallest. This is illustrated in figure 5.28, which depicts a typical gravity-gradient stabilized spacecraft. The long boom, however, may have significant flexure modes, typically of low frequency, which can couple to the attitude motion of the base module of the satellite. These flexure

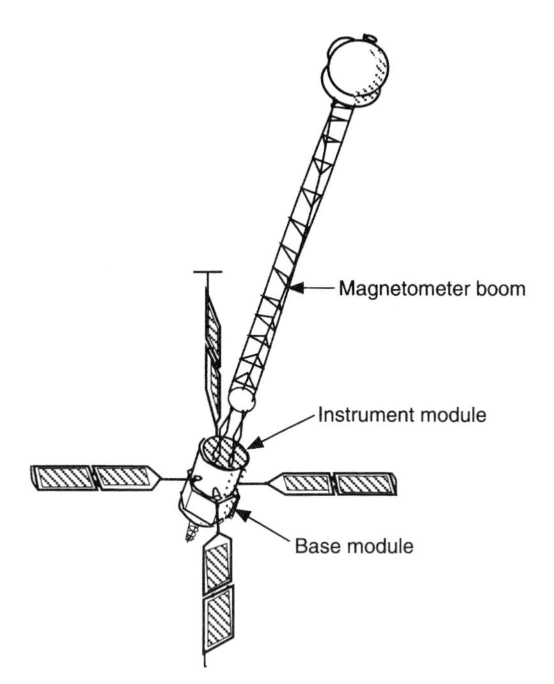

Figure 5.28 A typical gravity-gradient stabilized spacecraft.

modes can be attenuated by the use of momentum dampers (see section 5.8.6), but the remaining oscillation still leads to a degradation in control performance. Furthermore, for an elliptical orbit, oscillations in the gravitational field at the spacecraft, due to variations in the orbit radius, can lead to a destabilizing torque. These effects limit the usefulness of gravity-gradient stabilization for spacecraft with tight control requirements.

5.8.5 Combined Gravity-Gradient and Spin Stabilization

Many spacecraft use both gravity-gradient stabilization and a single-momentum wheel for three-axis attitude stability. Since most spacecraft are nearly symmetric, it is extremely difficult to obtain gravity-gradient stability about the axis of symmetry. Thus, some other means is necessary.

For an Earth-locked spacecraft, one method is to have a passive momentum wheel whose axis is aligned with the pitch axis. Thus, while the gravity-gradient torque provides roll and pitch stability, the momentum wheel provides roll and yaw stability. Generally some means, such as magnetic torquers or thrusters, is necessary to correct the direction of the momentum wheel for drift.

A great danger of combined gravity-gradient and spin stabilization is that changes in thermal flexure in the boom may lead to destabilizing torques, which may even cause the spacecraft to invert its attitude. In cases like this, the three-axis attitude must be reacquired.

5.8.6 Nutation Damping

In the discussion of gravity-gradient stabilization we saw that for small disturbance torques it was possible to find an equilibrium orientation of the spacecraft. However, the spacecraft will not come to rest at that orientation unless there is some means of dissipating the energy of oscillation about that equilibrium point. In fact, without some means of energy dumping, the spacecraft oscillations may eventually build up and drive it into an unstable attitude.

Passive nutation damping is accomplished by using various devices that dissipate energy by providing friction. These devices were discussed previously in section 5.7.4. Active nutation damping can be accomplished by using thrusters, reaction wheels, or electromagnets to torque the spacecraft in the direction opposite to the one in which it is nutating. Such a method requires an onboard computer to determine the direction of the nutational motion and to compute the opposing torque.

5.8.7 Typical Attitude Control Systems

The choice of a spacecraft attitude control system is usually determined by the spacecraft attitude and accuracy requirements. For low-Earth spacecraft with low perigee that will be subject to large aerodynamic torques, some sort of gyroscopic stabilization via spacecraft spin or a momentum wheel will often be used. Corrections for spin-axis drift can be accomplished via thrusters or magnetic torquing. For low attitude accuracy requirements, no additional control hardware may be needed. If attitude accuracy requirements are high, however, a set of three-axis momentum wheels is often the best solution to maintaining accurate three-axis attitude. At altitudes greater than 600 km the effect of aerodynamic torques is much reduced. Above this altitude, gravity-gradient stabilization, perhaps augmented by a momentum wheel, becomes practical, although attitude accuracies better than a few degrees are hard to achieve. Somewhat better accuracies can be delivered by a bias momentum wheel and magnetic torquers. Higher accuracies can be achieved by a set of three-axis thrusters (low-thrust, six total) or a set of three-axis momentum wheels. If wheels are chosen, then momentum management of the wheels will require thrusters or magnetic torquers as well. At geosynchronous altitudes, magnetic or gravity-gradient control is no longer feasible. Thus, attitude control can be accomplished only via thrusters or using momentum wheels, with thrusters providing momentum management.

The terms momentum bias and zero momentum are used to describe attitude control systems with momentum or reaction wheels. A *momentum bias system* contains one or more momentum wheels such that the total system momentum is non-zero. Rather than spinning the spacecraft, a momentum wheel is used to provide the spinning gyroscopic stiffness for attitude stability. These systems are generally used in nadir pointing spacecraft. A *zero momentum system* uses reaction wheels, letting the wheel speeds vary, and the total system momentum is controlled to zero (allowing for momentum due to orbital rate for nadir-pointing spacecraft). Zero momentum systems provide tighter control of the spacecraft, and they allow greater agility. Reorienting the attitude of a momentum bias system requires precession of the momentum vector, which can be a slow process. Zero momentum systems, because they normally do not have a momentum vector, are able to slew quickly to different attitudes. Whether momentum bias or zero momentum, external disturbance torques will cause the wheel speeds to increase, thus requiring a means to dump excessive momentum.

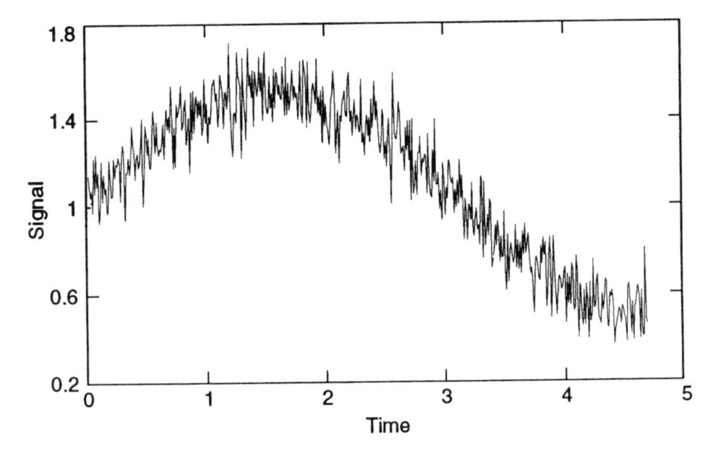

Figure 5.29 Jitter.

5.8.8 Jitter

An important consideration in attitude control is jitter, the very-high-frequency random motion that inevitably occurs in any maneuver (and that is easily distinguishable from the maneuver motion) or that interferes with science imaging (figure 5.29).

Jitter can have many sources. Attitude maneuvers are generally realized via thrusters, magnetic torquers, or momentum wheels. Thus, irregularities in the expulsion of propellant, in the currents of the magnetic torquers, random friction effects, or unbalanced masses in the momentum wheels can cause jitter. In manned spacecraft, movements of the crew are a source of jitter. In addition, in a feedback attitude control system, in which the sensed attitude error is provided by attitude sensors, random errors in the attitude sensors will lead to random errors in the attitude estimates, which will then also lead to jitter.

The effects of jitter can be significant. For large structures jitter can lead to the excitation of flexure modes and large errors in attitude control. The excitation of the flexible modes may come from motion of the spacecraft (slewing) or from some source within the spacecraft (for example, instruments, or the control system itself). Likewise, in precise attitude pointing, which with the Hubble Space Telescope has reached the milliarcsecond level (Taff, 1991), jitter must be very tightly controlled. Unfortunately, there is no easy means to control jitter except by demanding higher manufacturing tolerances and by limiting the speed of spacecraft maneuvers.

5.8.9 Further References

Attitude control is treated in detail in the books of Kaplan (1976), Wertz (1978), Hughes (1986), and Junkins and Turner (1986). Kaplan in particular gives very illuminating examples of the design of several attitude control systems. Hughes focuses on attitude stabilization, whereas Junkins and Turner focus on attitude maneuvers. Singer (1964) contains much useful material. Mobley (1988) presents an elementary but useful introduction with much practical information. A great deal of interest has been given to the control of flexible structures. For more on this topic, the reader is referred to the books by Joshi (1989) and Junkins (1990).

For early work on gravity-gradient stabilization the reader is referred to Fischell and Mobley (1964) and Pisacane et al. (1967). For the effects of thermal flexure on gravity-gradient stabilization, see Goldman (1975). The problem of instability of dual-spin gravity-gradient spacecraft has been discussed by Hunt and Williams (1987) and Williams and Hunt (1989).

Early studies of dual-spin spacecraft that led to our present understanding may be found in Landon and Stewart (1964), Likins (1964), Perkel (1966), Slafer and Marbach (1975), and Thomson (1962). Kaplan presents two interesting control design examples: the design of a bias-momentum system using a gimbaled reaction wheel (1974, 1976), and the design of a solar-electric thruster system (1975, 1976).

5.9 Illustrations of Actual Missions

5.9.1 Magsat

The Magsat spacecraft, launched in 1979, is a good example to illustrate the material presented in this chapter because it carried most of the sensors we have mentioned, and during its mission history had several different control phases. It is also an example of a momentum bias control system.

The goal of the Magsat mission was to measure the Earth's magnetic field with an accuracy of $6\,\gamma$ ($1\,\gamma = 10^{-9}$ Wb/m^2 $= 1$ nT). Because complete Earth coverage was important, the Magsat spacecraft was placed in a nearly polar orbit. To ensure also that the Sun would never come within the field of view of the star trackers and that there would always be adequate solar power, the orbit was chosen to be Sun-synchronous. This guaranteed that the Sun would always be available as an attitude reference as well.

The magnetic field generated by a multipole moment of order l decreases with distance as $r^{-(l+2)}$, where r is the distance from the source of the multipole field (see chapter 2). The sources of these multipole fields are largely the local regions of magnetization in the Earth's mantle. Thus, in order to observe the detailed structure of the Earth's magnetic field, the perigee of the Magsat spacecraft had to be quite low. As a result, the spacecraft would be subject to large atmospheric drag forces, which would severely limit its lifetime and also lead to large aerodynamic torques that would make stabilization of the spacecraft a problem. Had the only purpose of Magsat been to investigate magnetic anomalies in the Earth's mantle, a short lifetime might have been acceptable; but it was desired to obtain detailed information as well for the main field, which originates in the Earth's core. For this reason, the apogee was chosen to be much higher so that the varying altitude would permit separation of main and anomalous magnetic fields and prolong the life of the spacecraft. Actual launch values of the orbital parameters were: apogee 578 km, perigee 352 km, and inclination 96.8°. At an inclination of $\sim$97°, perturbations from the nonsphericity of the Earth would cause the spacecraft orbit plane to precess at approximately one degree per day to keep the orbit normal aligned with the Sun direction. (With respect to an inertially oriented frame fixed in the Earth, the Sun direction appears to rotate once per year, or $\sim$1° per day.) Chapter 3 provides more detail on this subject.

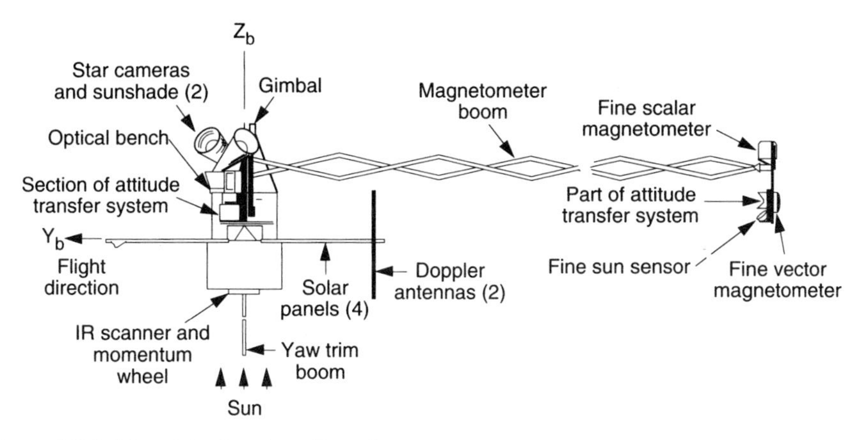

Figure 5.30 Magsat deployed configuration.

A pictorial view of the Magsat spacecraft is shown in figure 5.30. In order to isolate the scientific payload, which consisted of a very accurate vector magnetometer and an even more accurate scalar magnetometer (which measured only the magnitude of the magnetic field), from the base module (where the star trackers would generate large magnetic disturbances [they were old trackers indeed!]), these were separated by a 6 m long boom. Long booms bring as many problems as they solve since they are susceptible to flexure. Thus, the attitude of the payload at the end of the boom may not be the same as the attitude of the main spacecraft bus where the attitude sensors are located.

The attitude-determination hardware was extensive. Definitive attitude determination was provided by two fixed-head star trackers and a precise Sun sensor, each with an accuracy on the order of 10 arcsec. At the time, these were the most accurate attitude sensors available, and they determined the accuracy of the mission. Data from the star trackers were usable only at angular velocities below 200 arcsec/s. To provide coarse attitude during early mission, and as a backup to the definitive system, the spacecraft was also supplied with a single-axis "spinning" digital Sun sensor, a coarse (two-axis) Sun sensor, a coarse vector magnetometer, and an infrared horizon scanner. These sensors had an accuracy on the order of one degree. Since the boom was somewhat flexible and susceptible to distortion by solar heating, it was necessary to have a separate measure of the attitude on the experiment module relative to the base module. This was provided by an attitude transfer system consisting of a light-emitting diode (LED) source on the base module and a system of mirrors located on the base module and the experiment module. By observing the reflections of the light signals, the attitude of the experiment module relative to the base module could be inferred. Partial attitude rate measurements were provided by redundant pitch gyros. These gyros were not needed for attitude determination, but provided rate information for the pitch feedback control loop. Given the instrumental errors on the spacecraft in the magnetic field measurement, the total error for the attitude determination system was 20 arcsec/axis (rms).

Attitude control was provided by three-axis electromagnetic torquers, a bias momentum wheel (which was part of a combined horizon-scanner–momentum-wheel assembly), a nutation damper, despin yo-yos, and an aerodynamic trim surface. Attitude

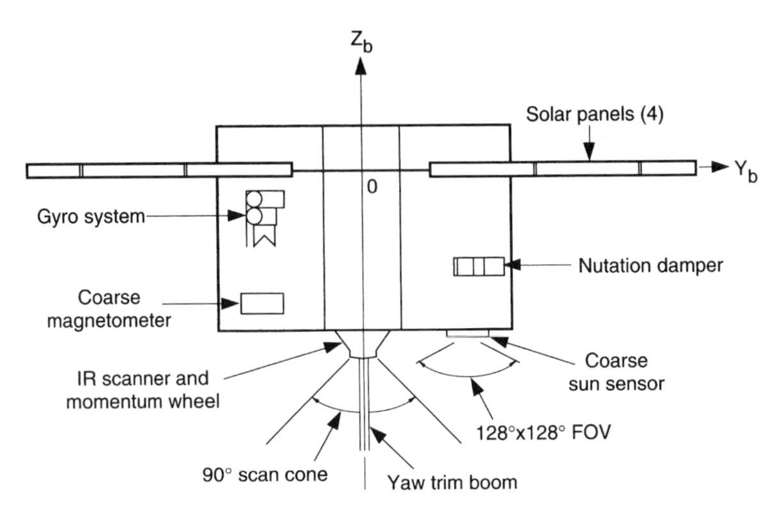

Figure 5.31 Base module attitude determination and control system (ADCS) hardware locations (side view).

control on the spacecraft was directed by an attitude signal processor, which consisted of an RCA 1802 microprocessor with 4094 bytes of programmable read-only memory and 1024 bytes of random access memory. The processor was programmed in MICRO-FORTH. By today's standards this microprocessor must seem primitive indeed, but at the time it was an important development in onboard attitude control. The long boom led to a large displacement of the center of mass from the center of pressure. Given the large surface area of the solar panels and the low perigee, the aerodynamic torques at perigee would have been great. For that reason, it was necessary to control the spacecraft attitude to be within a few degrees of the orientation at which the aerodynamic torques at perigee were minimized. The need to manage the aerodynamic torque led to the Earth-locked configuration of Magsat. The magnetic field vector could be observed from any spacecraft orientation. The disposition of attitude sensors and control hardware is shown in figures 5.30 and 5.31.

The design and operation of the attitude determination and control system (ADCS) can best be appreciated by reviewing the Magsat launch sequence. Magsat was launched by a four-stage solid-fuel Scout rocket from the Western Test Range at Vandenberg Air Force Base in California on October 30, 1979. The mass of the spacecraft was 183 kg. The launch time was selected so that the right ascension of the node of the orbit plane would be 90° from the right ascension of the Sun, thus putting Magsat into a "dawn–dusk" orbit. During launch, the momentum wheel in the spacecraft was driven at a fixed speed of 1500 rpm in order to provide passive gyroscopic stabilization following launch about the spacecraft pitch axis (the approximate symmetry axis of Magsat). On the fourth stage of the rocket the spacecraft was also spin-stabilized at a spin rate of approximately 150 rpm.

After injection into orbit, the attitude acquisition sequence began. The yo-yo despin masses were released to bring the spacecraft spin rate down to between 0 and −4 rpm. The solar panels were then deployed, which increased the moment of inertia about the spin axis, bringing the spin rate down to between 0 and −0.5 rpm. At this point, spin-rate gyroscopic stabilization is provided chiefly by the momentum wheel. The first telemetry was then transmitted to Earth.

To decrease the spacecraft nutation, the nutation damper was then uncaged. This consisted of a magnetically damped torsion pendulum whose natural frequency was matched to the expected nutation frequency of the spacecraft. Next, the aerodynamic trim boom was extended to a length of approximately 4.63 m. This trim boom offset the otherwise large aerodynamic torque on the spacecraft at perigee.

Magsat could not have met its power requirements unless the symmetry axis of the spacecraft (that is, the pitch axis) were aligned to within 60° of the Sun, which in this case was also the orbit normal. At injection the angle between the orbit normal and the pitch axis was nearly 80° (approximately 90° was anticipated). The spin-axis attitude of Magsat was determined from measurements of the magnetic field along the spin axis and the angle between the Sun and the spin axis, as described in section 5.5. The coarse vector Sun sensor was also activated at this time to verify the sign of the spacecraft spin rate (by observing the progression of quadrants in which the Sun line fell). The coarse vector Sun sensor unfortunately did not perform as planned (not every sensor survives launch), and its use was soon abandoned (with no loss to mission objective, since the precise Sun sensor performed flawlessly). To accomplish the necessary maneuver, a magnetic dipole of variable magnitude and sign (as determined by the onboard computer) was commanded that was parallel to the spacecraft pitch axis (see section 5.8.2).

Magsat was to operate in an Earth-locked configuration (hence with a spin rate of approximately 0.001 rev/s). The magnetic spin/despin mode was therefore activated to bring the spacecraft spin to within 0.05 rev/s. At this point, the IR horizon scanner was activated along with the pitch-loop control law. The latter was a feedback control law of the pitch and the pitch rate (from one of the two redundant pitch-axis gyros) of the type described in section 5.8.1. The pitch-loop control law quickly brought the spacecraft into an Earth-locked configuration. Had this control law been activated without the preceding magnetic despin, the larger change in angular velocity could have caused the pitch momentum wheel to saturate.

At the end of the third day of the mission the magnetometer boom was extended. This was the most difficult part of the launch sequence because the field of view of the attitude transfer system was only ±6 arcmin. Fortunately, largely because of experience gained from extensive simulations before launch, the boom deployment was executed flawlessly. The aerodynamic trim boom was next extended to its planned final length of 5.30 m. (It was determined much later from in-flight experience that an aerodynamic trim boom length of 4.48 m was optimum, and the boom length was so adjusted.) The star trackers and precise Sun sensor were now activated. By November 3, 1979, the spacecraft was fully operational, and scientific data collection began. Magsat was now in its final mission configuration.

The Magsat spacecraft was equipped with three commandable, mutually perpendicular electromagnets (torque rods). One was aligned with the pitch axis. The other two lay in the plane of the roll and yaw axes. The magnetic control modes were of two kinds: spin/despin and pitch control.

To control pitch, a magnetic moment is generated along the spacecraft pitch axis. As before, the magnetic torque is given by

$$\boldsymbol{N}_{\text{mag}} = \boldsymbol{m} \times \boldsymbol{B}. \tag{5.9.1}$$

At orbital latitudes near the magnetic equator, the magnetic field vector will lie in the orbit plane roughly parallel to the Earth's magnetic axis. If the sign of the magnetic moment is chosen properly at these latitudes, the torque will point toward the geocenter in a direction that will decrease the magnitude of the right ascension of the spacecraft pitch axis. At high magnetic latitudes, the magnetic field direction will be nearly parallel to the equator, and activating the pitch coils will now change the declination or the pitch axis. Thus, by activating the magnetic control law near the magnetic equator or at ± 45 degrees magnetic latitude, both the roll and the yaw could be controlled. The roll/yaw control law was activated only during four short segments (about 150 s) of the orbit. This time was selected to be an integral number of nutation periods (in fact, two periods) so that no net nutational motion would be excited. To determine the roll and yaw errors on board, it was noted that at the equator the IR scanner roll angle was, in fact, a measure of the pitch-axis right ascension. At the poles, however, it was a measure of spacecraft declination because of a phenomenon called quarter-orbit coupling. Thus, for a stable pitch axis, the IR scanner alone was able to determine the pitch-axis attitude by using measurements at different points on the spacecraft orbit. The value of the pitch-axis coil current was calculated by the onboard computer, using a simple field model and the value of the horizon-scanner roll.

The effect of a roll/yaw control sequence is shown in figure 5.32, which depicts a prelaunch simulation. Here right ascension and declination (of the spacecraft spin axis) are defined not with respect to inertial coordinates, but with respect to orbital coordinates. Thus, declination was measured from the horizontal plane, and right ascension was measured in the horizontal plane from the orbit normal. For a circular orbit (or at perigee and apogee for a general orbit), vanishing orbital right ascension and declination correspond to the spacecraft having the boom along the velocity vector, which is the desired orientation for minimizing the aerodynamic torque.

At the beginning of the simulation shown in figure 5.32 the roll/yaw error is less than 2°, the chosen threshold for the control. At high altitudes, where the torques are negligible compared with wheel momentum, the locus of the pitch-axis right ascension and declination is circular, as expected. (This would be true only near the origin. Far from the origin these loci would be altered by "trigonometric distortion.") At perigee, because of action of the aerodynamic torque, the attitude changes more quickly, leading to a tighter loop. Near the end of the fourth orbit the threshold of 2° is exceeded, and the magnetic control law is activated. Until this time (about 400 min into the simulation), the spacecraft has been controlled passively. Within an orbit, roughly, the desired pitch-axis attitude has been achieved, and the spacecraft pitch axis describes almost conical motion again. The loci are closer to being circular and closer together because the aerodynamic torque at perigee is now much smaller than before. Note also that the center of the motion is not roll $=$ yaw $=0$ but a bias point, which reflects the optimal pitch-axis attitude for minimizing the average aerodynamic torque. This attitude offset also decreased the aerodynamic drag in general and led to a longer mission lifetime.

The pitch-control law, which maintains the desired pitch angle by exchanging momentum between the spacecraft body and the pitch-axis momentum wheel, may lead to an increase or decrease in momentum in the wheel. Thus, either the momentum wheel will eventually saturate (in which case the pitch control law can no longer be effected) or the wheel momentum may decrease to the point where gyroscopic stability is no longer

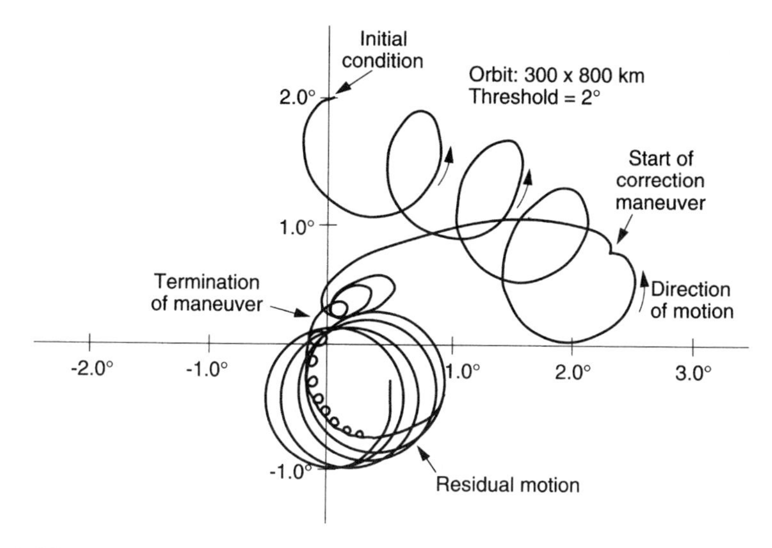

Figure 5.32 Simulated roll/yaw maneuver. The roll/yaw threshold is exceeded during the fourth orbit of this simulation. This relatively rapid maneuver excites nutational motion in the spacecraft that, due to the nutation dampers on the spacecraft, is damped out within an orbit (see progressively smaller sub-loops).

maintained. Thus it is necessary to spin up or despin the momentum wheel during the mission to maintain the wheel momentum within proper limits.

To accomplish this, as we noted in section 5.8.2, if a magnetic moment is commanded that lies in the plane of the roll and yaw axes and is also perpendicular to the magnetic field, the resulting torque is parallel to the spacecraft pitch axis and will cause the spacecraft pitch rate to increase or decrease, depending on the sign of the magnetic moment. The pitch loop control law, however, will transfer the spacecraft pitch rate to the momentum wheel. Thus, by commanding the appropriate control magnetic moment, the wheel can be spun up or down. To avoid conflict between the spin–despin and the roll/yaw control activities, the control laws were constructed so that spin–despin would occur only in regions where the roll/yaw control law was never active.

Attitude determination activities were equally varied. Three distinct attitude determination systems were developed. We have already described the determination of spin-axis attitude during the early part of the mission. This was determined by an attitude ground support system dedicated to near-real-time operation, designated MAGNRT. In addition to this system, two other attitude determination systems were developed. One of these, MAGFINE, would provide definitive processing of the attitude data from the fine attitude sensors (star trackers and precise Sun sensor), which arrived at a rate of four times per second. Another software system, MAGINT, would determine a less accurate attitude using only the coarse sensors (coarse Sun sensor—replaced by the fine Sun sensor—as well as the IR horizon scanner and coarse vector magnetometer) at a much slower rate (as slow as once per 30 s) to provide an intermediate definitive solution to the project scientists in much less time than that required to process the fine-attitude data.

MAGNRT would use the triad algorithm based preferentially on the Sun and IR horizon scanner data, if both were available, but using the magnetometer data if one was absent. The MAGINT system would use the triad algorithm as well but could also compute the attitude from a batch optimal algorithm of the type described in section 5.5, using all three coarse sensors. Because of the many trigonometric functions to be differentiated and the iterative nature of the algorithm, this method would be very slow and, because of the accuracy of the sensors, could yield attitude accuracies only on the order of 1.0°. It was originally planned that the MAGFINE system would be a similar optimal batch algorithm based on data from the fine-attitude sensors, but processing times would have been very long; hence the need for an intermediate definitive system, MAGINT.

Fortunately, a fast algorithm, QUEST (Shuster and Oh, 1981), was developed which minimized the cost function of equation 5.5.9 and was much faster than the intermediate definitive method in MAGINT. Therefore, after the first few weeks of the mission, the MAGINT software was permanently retired, replaced by the MAGFINE software. As a result, definitive data became available to the project scientists six months earlier than had been anticipated.

In addition to attitude determination activities, sensor alignments had to be reestimated regularly during the mission. This was done by time-consuming optimal batch methods, using the Newton–Raphson algorithm. Fortunately, these were computed relatively infrequently compared with the rate at which the attitude was computed. Similar methods were applied to the occasional calculation of bias offsets for the IR horizon scanner and the magnetometer. The experimenter was also able to determine alignments from the data, and computed them as a part of his own data-reduction procedures (Langel et al., 1981).

5.9.2 TIMED

The Thermosphere Ionosphere Mesosphere Energetics and Dynamics (TIMED) spacecraft is a good example of a 3-axis zero momentum controlled spacecraft. Like Magsat, TIMED carries many of the attitude determination and attitude control hardware mentioned in this chapter.

As part of NASA's Solar Connections program, the TIMED mission has the primary objective of investigating and understanding the energetics and dynamics of the mesosphere and lower thermosphere/ionosphere (MLTI) region. Encompassing altitudes from 60 to 180 km, this region has been the least explored and least understood of the Earth's atmosphere. The TIMED mission goals include determining the temperature, density, and wind structure of the MLTI, and determining the relative importance of various radiative, chemical, electrodynamical, and dynamical sources and sinks of energy in the MLTI region. TIMED was launched on December 7, 2001.

The TIMED G&C subsystem consists of two star trackers, redundant 3-axis inertial reference units (IRUs), redundant 3-axis magnetometers, redundant Sun sensors on each of the $+y$ and $-y$ sides, three torque rods containing redundant coils, four reaction wheel assemblies, and a pair of redundant processors. TIMED also carries a redundant Guidance and Navigation System (GNS). GNS uses Global Positioning Satellite (GPS) data to calculate a position, velocity, and time state vector at a rate of once per second. These data are passed to the G&C subsystem for use in calculating its commanded attitude. See figure 5.33 for a pictorial view of the spacecraft.

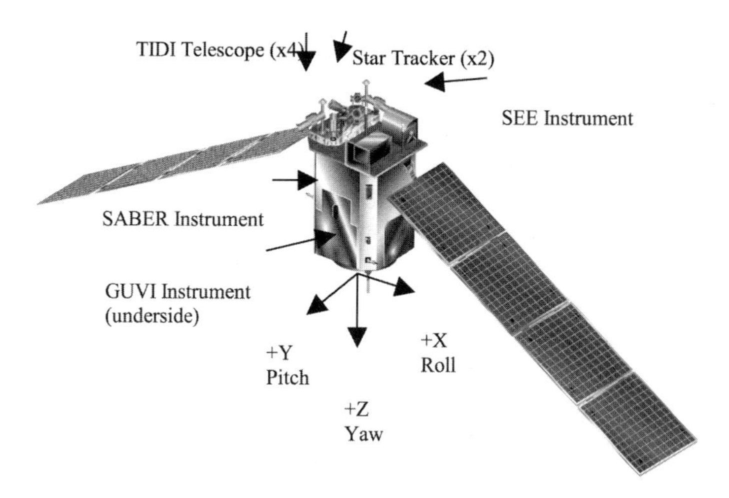

Figure 5.33 TIMED spacecraft.

During normal operation TIMED is required to maintain a nadir-pointing attitude with the $+z$ axis in the direction of the Earth's geocentric center, the $+y$ axis in the direction of the orbit normal, and the $+x$-axis generally in the velocity or anti-velocity direction. The Sun beta-angle, defined as the angle between the Sun and orbit normal vectors, dictates the x-axis direction as follows. The SABER instrument contains a thermal cooler for instrument temperature regulation and requires the pointing of the $+y$ vehicle axis away from the Sun. Orbit precession thus necessitates a 180-degree rotation, called the "yaw maneuver," about the z-axis approximately once every sixty days to avoid solar illumination of SABER.

Attitude determination is accomplished onboard with a Kalman filter (see section 5.5), using measurements from the star trackers and the gyros (that is, the IRU). The filter estimates attitude estimate error (to correct the attitude used in the controller) and gyro biases (used to correct the rate measurements from the gyros). Star tracker attitude measurements provide "updates" to the attitude filter once per second, and between these updates the filter propagates the attitude with gyro measurements. To reduce attitude estimation errors due to errors about the boresight, the two trackers are mounted with their boresights 90° apart.

The job of the controller is to maintain the nadir-pointing attitude by commanding the reaction wheels. On TIMED the wheels are configured in a pyramid with all four wheels working simultaneously; however, the configuration will maintain acceptable control of the spacecraft with three wheels should one wheel fail. A PID controller is used to control the spacecraft on the basis of attitude and rate errors. The attitude error is the difference between the commanded attitude and the estimated attitude from the filter. The rate error is the difference between the commanded rate and the measured rate from the gyros, appropriately corrected for the bias. The commanded attitude and rate are derived from the GNS-provided position and velocity vectors by calculating the desired body frame with respect to the inertial frame as

$$\mathbf{z}_B = -\frac{\mathbf{p}_I}{|\mathbf{p}_I|}, \quad \mathbf{y}_B = \frac{\mathbf{p}_I \times \mathbf{v}_I}{|\mathbf{p}_I \times \mathbf{v}_I|}, \quad \mathbf{x}_B = \mathbf{y}_B \times \mathbf{z}_B, \tag{5.9.2}$$

$$C_{\mathrm{I}} = \begin{bmatrix} \mathbf{x}_{\mathrm{B}}^{\mathrm{T}} \\ \mathbf{y}_{\mathrm{B}}^{\mathrm{T}} \\ \mathbf{z}_{\mathrm{B}}^{\mathrm{T}} \end{bmatrix}, \tag{5.9.3}$$

where $\mathbf{p}_{\mathrm{I}}$ and $\mathbf{v}_{\mathrm{I}}$ are the spacecraft position and velocity vectors in the inertial frame, and C_{I} is the direction cosine matrix describing the rotation from inertial coordinates to spacecraft body coordinates. The commanded rate in inertial coordinates is given by

$$\boldsymbol{\omega}_{\mathrm{I}} = \frac{\mathbf{p}_{\mathrm{I}} \times \mathbf{v}_{\mathrm{I}}}{|\mathbf{p}_{\mathrm{I}}|^2}. \tag{5.9.4}$$

This rate is then transformed into body coordinates through the use of equation 5.9.3 (actually through the quaternion representation of 5.9.3).

The PID controller gains were chosen according to Bauer et al. (1992) as

$$\left.\begin{aligned} K_{\mathrm{p}} &= \omega_{\mathrm{c}}^2, \\ K_{\mathrm{d}} &= 2\zeta\omega_{\mathrm{c}}, \\ K_{\mathrm{i}} &= \frac{\zeta\omega_{\mathrm{c}}^3}{10}, \end{aligned}\right\} \tag{5.9.5}$$

where ζ is the desired damping ratio and ω_{c} is the desired bandwidth of the controller. The selection of ω_{c} is governed by many variables, such as the speed of response and the amount of overshoot. There is also a practical limit to the setting of ω_{c}; while one can set its value arbitrarily large, the system itself (the spacecraft) has a limited response, due to actuator limitations. The selection of ζ and ω_{c} directly affects the stability of the system, as discussed in section 5.8.1, and much analysis goes into the proper selection of them.

Various environmental torques, principally aerodynamic, create a secular, or constant, torque on the spacecraft. This torque is relatively fixed in the body frame and causes the controller to command the wheels with a small bias torque. This bias torque will, over time, cause one or more of the wheels to saturate in wheel speed, thus causing the wheel to no longer be useful for control. The momentum controller prevents this situation. TIMED has a set of three torque rods mounted orthogonally to one another and aligned with the spacecraft body axes. When the system momentum error, that is, the momentum in excess of the normal momentum required about the y body axis to maintain nadir pointing, exceeds a threshold, the momentum controller commands one or more of the rods to cause the wheel controller to reduce the wheel speeds. (That is, the rods are commanded in such a way that the magnetic torque created causes the wheel controller to command wheel torques that, in order to compensate for the magnetic torque, cause the wheel speeds to decrease.) The commanded dipole is calculated as orthogonal to the momentum error and the measured magnetic field, both written in body coordinates, and in the opposite direction of the momentum error, or mathematically

$$\mathbf{m}_{\mathrm{cmd}} = \mathbf{h}_{\mathrm{err}} \times \mathbf{B}_{\mathrm{mag}}. \tag{5.9.6}$$

The rods can be commanded positively or negatively, thus providing 27 combinations between the three rods. The set of rods, and their polarity, are chosen such that the

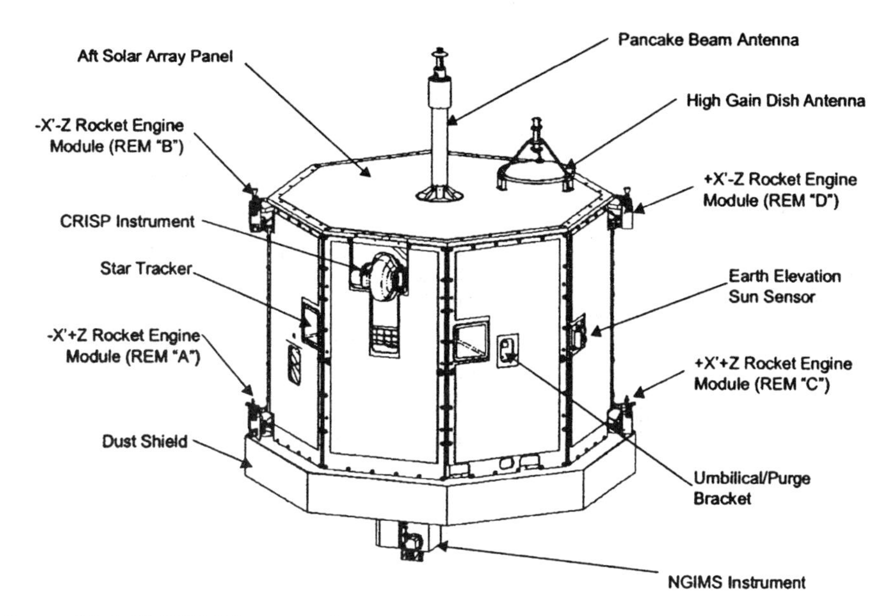

Figure 5.34 CONTOUR spacecraft.

resulting dipole is approximately in the same direction as that given in equation 5.9.6. In order to reduce chatter on the rod commands, that is, commanding the rods on and off too frequently, hysterisis is added to the decision logic. The momentum error must be above a particular threshold before a momentum unload is commanded. Once an unload is in progress it will not stop until the momentum error is reduced to a second threshold, set significantly below the original threshold. Once the second threshold is reached, momentum unloading will cease and not restart until the momentum error rises above the first threshold, in which case the process is repeated.

5.9.3 CONTOUR

While unfortunately short lived, the Comet Nucleus Tour (CONTOUR) spacecraft, shown in figure 5.34, is an excellent example of a spin-stabilized spacecraft employing only a liquid propulsion system (thrusters). The spacecraft mission was to collect various types of comet science data, ranging from images to dust analysis, during flybys of two comets. In order to reach the comets, a heliocentric orbital dance was to be performed, involving several Earth swingbys between the various comet encounters. During the time periods between encounters the spacecraft was spin stabilized, spinning about its z-axis (the axis containing the pancake beam antenna in figure 5.34). Comet encounters, during which the science data were to be collected, were to be performed in a 3-axis controlled mode. Unfortunately, six weeks after launch, during the firing of the internal solid rocket motor to escape Earth orbit, the spacecraft disintegrated, and no science data were ever collected. However, during the first six weeks the spin-stabilized vehicle performed flawlessly.

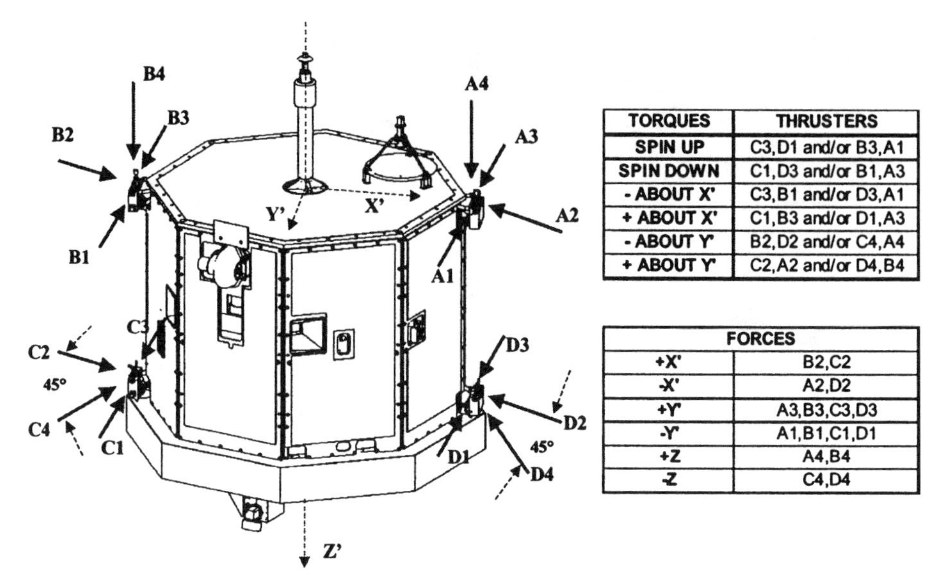

TORQUES	THRUSTERS
SPIN UP	C3,D1 and/or B3,A1
SPIN DOWN	C1,D3 and/or B1,A3
- ABOUT X'	C3,B1 and/or D3,A1
+ ABOUT X'	C1,B3 and/or D1,A3
- ABOUT Y'	B2,D2 and/or C4,A4
+ ABOUT Y'	C2,A2 and/or D4,B4

FORCES	
+X'	B2,C2
-X'	A2,D2
+Y'	A3,B3,C3,D3
-Y'	A1,B1,C1,D1
+Z	A4,B4
-Z	C4,D4

Figure 5.35 CONTOUR thruster layout.

Of particular interest here is a discussion of the spin-axis attitude estimation and the precess maneuvers. Briefly, attitude estimation was performed on the ground, using data from an Earth–Sun sensor. This sensor is specifically made for spinning spacecraft. It consists of a v-slit Sun sensor with one slit parallel to the spin axis of the vehicle (called the meridian slit), and two Earth sensors with beams separated by 5°. This angular separation was chosen for this particular mission on the basis of orbital geometry. The sensor outputs a pulse each time the sun crosses each of the Sun sensors, and a pulse each time the Earth sensors detect a space-to-Earth or Earth-to-space crossing. The timing of the various pulses with respect to the meridian slit was measured on board the spacecraft and telemetered to the ground. Processing of these data through various algorithms provided the spin axis direction. See van der Ha et al. (2003).

The CONTOUR spacecraft was controlled with thrusters located on four rocket engine modules (REMs), with each REM containing four thrusters. The thruster configuration provided the means to spin up or spin down, as well as to precess (change the direction of the spin axis) and perform orbital maneuvers. Figure 5.35 shows the thruster configuration, as well as the thrusters chosen to perform the various maneuvers. Note that the arrows denote the direction of the force on the spacecraft, not the direction of the nozzle.

In order to effect an orbital maneuver, it was generally required that the spacecraft spin axis be moved from its current direction, since its current direction was undoubtedly different than where the mission design team needed it to be. For stability purposes the spin rate was maintained at approximately 20 rpm about the z body axis. A precess maneuver was then a maneuver about an axis perpendicular to this spin axis. Figure 5.35 lists the combinations of thrusters required to accomplish maneuvers about the x- and y-axes. One must remember, however, that since the spacecraft is spinning about the

z-axis, a torque about the x-axis will produce a motion of the spin axis about the y-axis (recall Euler's equation, 5.6.15).

Since the thruster configuration only allows torques about the x and y body axes, how does one generate a torque such that the spin axis is driven in any direction? The answer is simply timing. By appropriately timing the firing of the thrusters within a revolution one can cause the spin axis to precess in any desired direction. The timing in this case is provided by the Sun pulse from the Earth–Sun sensor.

The process for determining the thruster commands for a precess maneuver was as follows. The navigation team would provide the orbit of the spacecraft. Knowing the time of the maneuver and the spacecraft position at that time, one determines the Sun position relative to the spacecraft and thus when the Sun sensor would produce a Sun pulse. Knowledge of the current and desired spin axis directions provides the direction in which the spin axis must be moved. With this information, and choosing a particular set of thrusters, the delay time from the Sun pulse occurrence to the time of thruster firing is calculated. Thruster on-times were chosen to be of a length corresponding to one-quarter of a revolution.

All precess maneuvers were performed open-loop by performing these calculations on the ground and then providing the thruster set, on-time, and delay time to the spacecraft. In all, twelve precess maneuvers were performed, all complete successes.

5.9.4 Further References

Almost every aspect of the Magsat mission is described in a special issue on Magsat of the *Johns Hopkins APL Technical Digest* (Potemra et al., 1980). The articles on attitude determination and attitude control systems are of special interest. The performance of the attitude control system is presented in Tossman et al. (1980). Additional data on the Magsat spacecraft are available in De Amicis (1987). Attitude ground operations for Magsat, with emphasis on the fine attitude system, are described in Abshire et al. (1981). The fine-attitude determination algorithm, QUEST, is developed in detail in Shuster and Oh (1981). The first scientific results of the Magsat mission are summarized in the Magsat special issue referred to above. Fuller accounts are given in Langel (1982, 1985). The TIMED mission, from a guidance and control perspective, is described in more detail in Dellinger (1999), and Dellinger et al. (2003). The CONTOUR mission is described in Rogers et al. (2001), Bunn and Rogers (2002), and van der Ha et al. (2003).

Problems

1. Show that the matrices $\hat{\mathbf{a}}\hat{\mathbf{a}}^{\mathrm{T}}$ and $[\hat{\mathbf{a}}\times]$ satisfy

$$\begin{aligned}
(\hat{\mathbf{a}}\hat{\mathbf{a}}^{\mathrm{T}})^2 &= \hat{\mathbf{a}}\hat{\mathbf{a}}^{\mathrm{T}}, \\
[\hat{\mathbf{a}}\times]\hat{\mathbf{a}}\hat{\mathbf{a}}^{\mathrm{T}} &= 0, \\
\hat{\mathbf{a}}\hat{\mathbf{a}}^{\mathrm{T}}[\hat{\mathbf{a}}\times] &= 0, \\
[\hat{\mathbf{a}}\times]^2 &= \hat{\mathbf{a}}\hat{\mathbf{a}}^{\mathrm{T}} - I.
\end{aligned}$$

Use these results and Euler's formula, equation 5.2.49, to show that

$$R^{\mathrm{T}}(\hat{\mathbf{a}}, \phi)R(\hat{\mathbf{a}}, \phi) = I$$

and

$$R(\hat{\mathbf{a}}, \phi_1)R(\hat{\mathbf{a}}, \phi_2) = R(\hat{\mathbf{a}}, \phi_1 + \phi_2).$$

2. Verify equations 5.2.12, 5.2.13, 5.2.15, and 5.2.17.
3. Let $\mathbf{x}$ and $\mathbf{y}$ be any two 3×1 column vectors and let C be any orthogonal matrix. Show that

$$(C\mathbf{x}) \cdot (C\mathbf{y}) = \mathbf{x} \cdot \mathbf{y}.$$

4. (a) Show that the direction-cosine matrix can be written as

$$C = \begin{bmatrix} i \cdot I & i \cdot J & i \cdot K \\ j \cdot I & j \cdot J & j \cdot K \\ k \cdot I & k \cdot J & k \cdot K \end{bmatrix}.$$

   (b) Give the physical interpretation of the columns and rows of C.
5. An orbital coordinate system is defined in terms of the spacecraft position and velocity, which we will assume have been given in inertial coordinates for all times. If the orbital x-axis has been defined as the direction of the spacecraft position $\hat{\mathbf{r}}$, and the orbital z-axis as the orbit normal (the direction of the spacecraft orbital angular momentum $\hat{\mathbf{L}}$), show that the proper orthogonal transformation that transforms column vectors from inertial coordinates to orbital coordinates is given by

$$C = [\hat{\mathbf{r}} \quad (\hat{\mathbf{L}} \times \hat{\mathbf{r}}) \quad \hat{\mathbf{L}}]^{\mathrm{T}},$$

   where

$$[\mathbf{a} \quad \mathbf{b} \quad \mathbf{c}]$$

   denotes a matrix whose columns are given by $\mathbf{a}$, $\mathbf{b}$, and $\mathbf{c}$, respectively.
6. Verify, using Euler's formula, that

$$R(\hat{\mathbf{a}}, \phi)\hat{\mathbf{a}} = \hat{\mathbf{a}}.$$

7. Carry out the necessary matrix multiplications to obtain an expression for the rotation matrix in terms of 2-1-3 Euler angles.
8. For the 3-1-3 Euler angles, verify that (ϕ, θ, ψ) and $(\phi + \pi, -\theta, \psi - \pi)$ generate identical rotation matrices.
9. Show that Euler's formula may be written in the form

$$R(\hat{\mathbf{a}}, \phi)\mathbf{v} = \mathbf{v} - \sin\phi \hat{\mathbf{a}} \times \mathbf{v} + (1 - \cos\phi)\hat{\mathbf{a}} \times (\hat{\mathbf{a}} \times \mathbf{v}),$$

   and in matrix form

$$R(\hat{\mathbf{a}}, \phi) = I - \sin\phi[\hat{\mathbf{a}}\times] + (1 - \cos\phi)[\hat{\mathbf{a}}\times]^2.$$

10. (a) If R is a rotation matrix, show that ϕ, the angle of rotation, is given by

$$\cos\phi = \tfrac{1}{2}(\mathrm{tr}C - 1),$$

   where $\mathrm{tr}C \equiv C_{11} + C_{22} + C_{33}$ is the trace operation.

(b) For $\sin\phi \neq 0$ show that the three components of the axis of rotation are given by

$$a_1 = \frac{1}{2\sin\phi}(C_{23} - C_{32}),$$

$$a_2 = \frac{1}{2\sin\phi}(C_{31} - C_{13}),$$

$$a_3 = \frac{1}{2\sin\phi}(C_{12} - C_{21}).$$

(c) Develop an expression for finding $\hat{\mathbf{a}}$ when $\sin\phi = 0$.

11. (a) Compute the attitude matrix for a 3-1-3 set of Euler angles with the values $\phi = \pi/4, \theta = \pi/3, \psi = -\pi/4$. (Do these values satisfy our normal conventions? Does it really matter?)
(b) Find the axis and angle of rotation corresponding to this attitude matrix.
12. Derive equations 5.2.68 and 5.2.69 from equation 5.2.49.
13. The kinematic equation for the Euler angles given by equation 5.3.22 is for the body-referenced angular velocity. Derive equivalent expressions from the Euler angle rates in terms of the inertially referenced angular velocity. (Hint: $\omega_{\text{body}} = C\omega_{\text{inertial}}$.)
14. (a) Suppose that the rotation matrix is given on some time interval by

$$C(t) = R(\hat{\mathbf{a}}, \omega_0 t),$$

with ω_0 a constant vector on the interval. Demonstrate by explicit differentiation of Euler's formula, equation 5.2.49, that

$$\frac{d}{dt}C(t) = -[\omega_0 \times]C(t).$$

(b) What is the most general rotation matrix that satisfies this equation?
15. Solve equation 5.4.10,

$$\cos\rho = \cos\gamma\cos\eta + \sin\gamma\sin\eta\cos(\Omega/2),$$

for the nadir angle. Show that the result has two solutions, which may be written

$$\cos\eta = \frac{1}{A}[\cos\rho\cos\gamma \pm \sin\gamma\cos(\Omega/2)(A - \cos^2\rho)^{1/2}],$$

where

$$A = \cos^2\gamma + \sin^2\gamma\cos^2(\Omega/2).$$

16. Suppose a spacecraft is equipped with two identical Earth horizon scanners with their scan cones oriented in opposite directions. Let the spacecraft roll axis be designated as the roll axis of the first scanner. Show that the spacecraft roll angle is given by

$$\cot\eta = \tfrac{1}{2}\tan\gamma(\cos(\Omega_2/2) - \cos(\Omega_1/2)).$$

17. Show that the angular radius of the Earth as seen from a spacecraft at an altitude h is given by

$$\rho = \arcsin\left(\frac{R_e}{R_e + h}\right),$$

where R_e is the radius of the Earth. For $R_e = 6378$ km, calculate the observed angular radius of the Earth at (a) the Earth's surface, (b) $h = 100$ km, (c) $h = 1000$ km, and (d) $h = 36{,}000$ km (approximately geosynchronous orbit).

18. The directions of the Sun and the magnetic field as observed by sensors on the spacecraft are

$$\hat{\mathbf{S}}_{S/C} = \frac{1}{\sqrt{2}}\begin{bmatrix} 1 \\ -1 \\ 0 \end{bmatrix}, \quad \hat{\mathbf{B}}_{S/C} = \frac{1}{\sqrt{2}}\begin{bmatrix} 0 \\ -1 \\ 1 \end{bmatrix}.$$

The direction of these vectors in the geocentric inertial coordinates system is

$$\hat{\mathbf{S}}_{GCI} = \frac{1}{\sqrt{2}}\begin{bmatrix} 1 \\ 0 \\ -1 \end{bmatrix}, \quad \hat{\mathbf{B}}_{GCI} = \frac{1}{\sqrt{2}}\begin{bmatrix} 1 \\ -1 \\ 0 \end{bmatrix}.$$

(a) Does there exist a well-defined orthogonal matrix C which satisfies

$$\hat{\mathbf{S}}_{S/C} = C\hat{\mathbf{S}}_{GCI}, \quad \hat{\mathbf{B}}_{S/C} = C\hat{\mathbf{B}}_{GCI}?$$

(b) Use the triad algorithm to calculate an attitude matrix C which satisfies the first of these equations exactly.
(c) Is the second question also satisfied exactly?

19. Suppose that the Sun and nadir directions in inertial coordinates are known to be

$$\hat{\mathbf{S}}_I = \begin{bmatrix} 0 \\ 0 \\ 1 \end{bmatrix}, \quad \hat{\mathbf{E}}_I = \begin{bmatrix} 1 \\ 0 \\ 0 \end{bmatrix}.$$

Measurements of these same directions made by the spacecraft yield values of

$$\hat{\mathbf{S}}_B = \begin{bmatrix} 0 \\ 1 \\ 0 \end{bmatrix}, \quad \hat{\mathbf{E}}_B = \begin{bmatrix} 0 \\ 0 \\ 1 \end{bmatrix}.$$

(a) Use the triad method to determine the attitude of the spacecraft with respect to inertial coordinates.
(b) Show explicitly (that is, based only on its numerical value) that the attitude matrix you have computed is proper orthogonal.
(c) What are the axis and angle of rotation that characterize this attitude?

20. Repeat problem 19, with

$$\hat{\mathbf{E}}_B = \frac{1}{\sqrt{2}}\begin{bmatrix} 1 \\ 0 \\ 1 \end{bmatrix}$$

first letting the Sun vector be the first vector and then repeating the computation with the nadir vector as the first vector. Are both matrices proper orthogonal? What do you say about the two estimates?

21. Denote the two direction-cosine matrices in problems 19 and 20 by C_1 and C_2. Define the relative attitude of these two estimates by

$$\delta C = C_2 C_1^{\mathrm{T}}.$$

How large is the angle of rotation characterizing δC? Is this reasonable?

22. Determine the inertia tensor for an ellipsoid of mass M and semiaxes a, b, and c. Assume that the axes of the ellipsoid pass through the center of mass and are aligned with the body axes, and that the density is uniform.
23. Suppose the ellipsoid of problem 22 is rotated 45° about the c-axis. What is the resulting inertia tensor?
24. Determine the inertia tensor for a rectangular solid of mass M and sides a, b, and c. Assume that the sides are perpendicular to the body axes, the center of mass coincides with the origin of the body-coordinate system, and that the density is uniform.
25. Determine the inertia tensor for a rectangular plate of mass M and with sides a, b, but zero thickness. Assume that the body axes are parallel to the sides of the plate, the center of mass coincides with the origin of the body coordinate system, and the density is uniform.
26. Determine the inertia tensor for a circular plate of mass M, radius a, and zero thickness. Assume that the z-axis is perpendicular to the plate and passes through the center, and that the density is uniform. What is the inertia tensor with respect to the coordinate axes if the center of the plate is located at $(a, 0, 0)$?
27. It is desired to maintain the attitude of a 3-axis stabilized spacecraft so that its symmetry axis makes an angle of 10° with the vertical. Suppose the spacecraft can be modeled as a dumbbell of total mass M with the two equal masses separated by a distance l. For $M = 200$ kg and $l = 3$ m, compute the magnitude of the gravity-gradient torque on the spacecraft at the Earth's surface and at geosynchronous altitude.
28. A small, very compact spacecraft in a circular orbit of altitude 500 km has a long flat boom with square cross-section and negligible mass. Assume the length of the boom is 10 m and the width is 5 cm, and that the boom is directed toward the nadir with the normal to one of the sides in the direction of the spacecraft velocity. What is the aerodynamic torque on the spacecraft? Use a typical value for the atmospheric density at this altitude.
29. It is desired to stabilize a near-Earth spacecraft against random disturbance torques using gravity-gradient stabilization. This is accomplished by attaching a long massless boom to the spacecraft bus, at the end of which is a proof mass. If the spacecraft bus (in this case all the spacecraft except for the boom) is spherical with mass 200 kg and radius 1 m, what is the inertia tensor of the spacecraft bus? If the proof mass is 5 kg and the (random) disturbance torques are of typical magnitude 10^{-4} N m, how long must the boom be to maintain a constant attitude to within 10°, 1°, or 0.1°? Does the result depend on the radius of the spacecraft bus? What are the relative contributions of the spacecraft bus and the proof mass to the gravity-gradient torque?
30. A cylindrical spacecraft of length 3 m, radius 0.25 m, and mass 500 km is spinning about the axis of symmetry with a spin rate of 1000 rpm. If the spacecraft is subjected

to a torque of 10^{-5} N m perpendicular to the symmetry axis, at what rate will the symmetry axis precess?

31. At the Equator at sea level the magnetic field is 3×10^{-5} Wb/m^2 and directed northward. The inertia matrix is given by

$$\mathbf{I} = \begin{bmatrix} 1000 & 0 & 0 \\ 0 & 800 & 0 \\ 0 & 0 & 2000 \end{bmatrix} \text{kg/m}^2.$$

If the spacecraft is equipped with three electromagnets that can generate a magnetic moment of 10 A m^2 in any direction, how quickly can the attitude be changed about the nadir at low Earth orbit and at geosynchronous altitude? How quickly can it change about the north direction? Assume the z-axis is pointed at nadir and the x-axis is pointed north.

32. A spacecraft is subjected to a constant pitch torque of 5×10^{-5} N m. It is desired to maintain the pitch attitude by the use of momentum wheels, which we will assume to be cylindrical with radius 25 cm. The allowed wheel speeds are between 200 and 1000 rpm. The nominal wheel speed is 600 rpm. If the wheels are to be despun no more than once per day, what must be the mass of the momentum wheel?

33. A cylindrical spacecraft with a mass of 200 kg and a radius of 1 m is initially spinning at 100 rpm. It is desired to despin the spacecraft completely by releasing two despin masses with a total mass of 200 g. What is the desired length of the cord by which the despin masses are attached to the spacecraft? What is the desired length if the final spin rate is to be 5 rpm?

34. Suppose that a spacecraft has a (body-referenced) inertia tensor of

$$\mathbf{I} = I_0 \begin{bmatrix} 1.0 & 0 & 0 \\ 0 & 0.8 & 0 \\ 0 & 0 & 0.2 \end{bmatrix}$$

with $I_0 = 1000$ kg m^2. Suppose also that the spacecraft is spinning about the body z-axis in a torque-free vacuum with an initial spin rate of 0.1 s^{-1}.

(a) What are the initial angular velocity vector and angular momentum vector of the spacecraft in body coordinates?

(b) What is the initial rate of change of the angular momentum vector in body coordinates?

(c) Suppose now that the spacecraft loses energy by very nearly isotropic heat radiation (that is, the radiant energy flux is very nearly the same in all directions, leading to an infinitesimally small net torque). About what axis will the spacecraft be spinning finally?

(d) What will be the final angular velocity vector and angular momentum vector of the spacecraft in body coordinates?

(e) What is the rotational kinetic energy of the spacecraft initially and finally?

35. A spinning spacecraft is required to maintain pointing of the spin axis to within 0.3°. If the constant external disturbance torque operating on the spacecraft perpendicular to the pitch axis is 1×10^{-4} N m, how much momentum would be required to have to maintain the pointing control over a four-day period? How fast would the spacecraft have to be spinning if the spin-axis inertia was 1000 kg m^2?

References

Abshire, G., R. McCutcheon, G. Summers, and F. G. VanLandingham, 1981. High precision attitude determination for Magsat. *Proceedings of ESA International Symposium on Spacecraft Flight Dynamics,* Darmstadt, Germany.

Agrawal, B. N., 1986. *Design of Geosynchronous Spacecraft.* Englewood Cliffs, N.J.: Prentice-Hall.

Anderson, B. D. O., and J. B. Moore, 1979. *Optimal Filtering.* Englewood Cliffs, N.J.: Prentice-Hall.

Bauer, F. H., M. D. Femiano, and G. E. Moser, 1992. Attitude control system conceptual design for the X-ray timing explorer. *Proceedings of the Guidance, Navigation, and Control Conference*, AIAA, Reston, VA, Aug. 1992.

Bertotti, B., and P. Farinella, 1990. *Physics of the Earth and the Solar System.* Dordrecht, The Netherlands: Kluwer.

Black, H. D., 1964. A passive system for determining the attitude of a satellite. *AIAA J.* **2**:1350.

Bunn, J. C., and G. D. Rogers, 2002. Influence of hardware selection on the design of the CONTOUR estimation and control software. *Proceedings, 5th International Conference on Dynamics and Control of Systems and Structures in Space 2002*, King's College, Cambridge.

Carrou, J. P., ed., 1984. *Mathématiques spatiales/Space mathematics.* Toulouse, France: CEPADUES Editions.

Carrou, J. P., ed., 1990. *Mécaniques spatiales/Space mechanics.* Toulouse, France: CEPADUES Editions.

Cohen, C. E., 1992. Attitude determination using GPS. Ph D Dissertation, Stanford University.

Crassidis, J. L., E. G. Lightsey, and F. L. Markley, 1998. Efficient and optimal attitude determination using recursive global positioning system signal operations. *Proceedings of the Guidance, Navigation, and Control Conference* (Boston, MA), AIAA, Reston, Aug. 1998 (AIAA Paper 98-4496).

De Amicis, S. J., 1987. *Artificial earth satellites designed and fabricated by the Johns Hopkins University Applied Physics Laboratory.* Laurel, MD. JHU/APL SDO 1600 (revised).

Dellinger, W. F., 1999. Attitude estimation and control for the TIMED spacecraft. *Proceedings, 14th World Congress of International Federation of Automatic Control (IFAC)*, Beijing, China.

Dellinger, W. F., H. S. Shapiro, J. C. Ray, and T. E. Strikwerda, 2003. Recent G&C experiences of the TIMED spacecraft. *Proceedings, 26th Annual American Astronautical Society (AAS) Guidance and Control Conference*, Breckenridge, CO.

Duchon, P., 1984. Modelisations nécessaries á la conception et la définition des systèmes de commande, d'attitude et d'orbite des véhicules spatiaux, in J.-P. Carrou, 1984, op. cit.

Fischell, R. E., and F. F. Mobley, 1964. A system for passive gravity-gradient stabilization of Earth satellites. *Progress in Astronautics and Aeronautics* **17**.

Foliard, J., 1984. Mesures et restitution d'attitude, in J.-P. Carrou, 1984, op. cit.

Fortescue, P., and J. Stark, eds., 1984, 1991. *Spacecraft System Engineering*. New York: John Wiley & Sons.

Gelb, A., 1989. *Applied Optimal Estimation.* Cambridge, MA: MIT Press.

Gibbs, J. W., 1901. *Vector Analysis.* E. B. Wilson, ed. New York: Dover Publications.

Goldman, R. L., 1975. Influence of thermal distortion on gravity-gradient stabilization. *J. Spacecraft and Rockets* **12**:406.

Goldstein, H., C. P. Poole, C. P. Poole, Jr., and J. L. Safko, 2002. *Classical Mechanics*. Englewood Cliffs, NJ: Prentice-Hall.

Griffin, M. D., and J. R. French, 1991. *Space Vehicle Design.* Washington, DC: American Institute of Aeronautics and Astronautics.

Hughes, P. C., 1986. *Spacecraft Attitude Dynamics.* New York: John Wiley & Sons.

Hunt, J. W., Jr., and C. E. Williams, 1987. Anomalous attitude motion of the Polar BEAR satellite. *Johns Hopkins APL Tech. Dig.* **8**:324.

Jacobs, J. A., 1987. *Geomagnetism*, vols. 1 and 2. Orlando, FL: Academic Press.

Joshi, S. M., 1989. *Control of Large Flexible Space Structures.* New York: Springer-Verlag.

Junkins, J. L., 1978. *An Introduction to Optimal Estimation of Dynamical Systems.* Alphen aan den Rijn, The Netherlands: Sijthoff & Nordhoff.

Junkins, J. L., ed., 1990. *The Mechanics and Control of Large Flexible Structures.* Washington, DC: American Institute of Aeronautics and Astronautics.

Junkins, J. L., and J. D. Turner, 1986. *Optimal Spacecraft Rotational Maneuvers.* Amsterdam: Elsevier.

Kane, T. R., P. W. Likins, and D. A. Levinson, 1983. *Spacecraft Dynamics.* New York: McGraw-Hill.

Kaplan, M. H., 1974. Active attitude and orbit control of body-oriented geostationary communications satellites. *Progress in Astronautics and Aeronautics* **33**:29.

Kaplan, M. H., 1975. Design and operational aspects of an all electric thruster system for geostationary satellites. *J. Spacecraft and Rockets* **12**:682.

Kaplan, M. H., 1976. *Modern Spacecraft Dynamics and Control.* New York: John Wiley & Sons.

Landon, R. V., and B. Stewart, 1964. Nutational stability of an axisymmetric body containing a rotor. *J. Spacecraft and Rockets* **1**:682.

Langel, R. A., ed., 1982. Magsat preliminary results. *Geophys. Res. Lett.* **9**.

Langel, R. A., ed., 1985. Magsat. *J. Geophys. Res.* **90**.

Langel, R. A., J. Herbert, T. Jennings, and R. Horner, 1981. Magsat data processing, a report for investigators. *NASA Technical Memorandum 81260.* Greenbelt, MD: NASA Goddard Space Flight Center.

Larson, W. J., and J. R. Wertz, eds., 1992. *Space Mission Analysis and Design.* Dordrecht, The Netherlands: Kluwer.

Lefferts, E. J., F. L. Markley, and M. D. Shuster, 1982. Kalman filtering for spacecraft attitude estimation. *J. Guidance Control and Dynamics* **5**:417.

Lerner, G. M., 1990. Attitude sensor calibration using scalar observations. *J. Astronaut. Sci.* **38**:201.

Lerner. G. M., and M. D. Shuster, 1981. In-flight magnetometer calibration and attitude determination for near-Earth spacecraft. *J. Guidance and Control* **4**:518.

Likins, P. W., 1964. Attitude stability for dual-spin spacecraft. *J. Spacecraft and Rockets* **4**:1638.

Markley, F. L., 1978. Parameterization of the attitude, in J. R. Wertz, 1978, op. cit.

Maute, P., and O. Defonte, 1990. A system for autonomous navigation and attitude determination in geostationary orbit, in J.-P. Carrou, 1990, op. cit.

Meirovitch, L., 1970. *Methods of Analytical Dynamics.* New York: McGraw-Hill.

Melvin, P. J., and A. S. Hope, 1993. Satellite attitude determination with GPS. *Advances in the Astronautical Sciences,* vol. 85, Part 1. AAS #93-556, pp. 59–78.

Mobley, F. F., ed., 1988. *Space Systems Course.* Laurel, MD: The Johns Hopkins University Applied Physics Laboratory.

Musser, K. L., and W. L. Ebert, 1989. Autonomous spacecraft attitude control using magnetic torquing only. *Proceedings, Flight Mechanics/Estimation Theory Symposium.* Greenbelt, MD: NASA Goddard Space Flight Center.

O'Connor, B. J., and L. A. Morine, 1967. A description of a CMG and its application to space vehicle control. *J. Spacecraft and Rockets* **6**:225.

Parkinson, B., et. al., eds., 1996. *Global Positioning System: Theory and Applications*, Vols. I, II. Washington, DC: American Institute of Aeronautics and Astronautics.

Perkel, H., 1966. Stabilite–a three-axis attitude control system utilizing a single reaction wheel. *Progress in Astronautics and Aeronautics* **19**:375.

Pisacane, V. L., P. P. Pardoe, and B. J. Hook, 1967. Stabilization system analysis and performance of the GEOS-A gravity-gradient satellite (Explorer XXIX). *J. Spacecraft and Rockets* **4**:1623.

Potemra, T. A., F. F. Mobley, and L. D. Echard, eds., 1980. *Johns Hopkins APL Tech. Dig.* **1**(3):162–248, *Special Issue on Magsat* 1.

Radix, J. C., 1978. *Gyroscopes et gyromètres.* Toulouse, France: CEPADUES Editions.

Rimrott, F. P. J., 1989. *Introductory Attitude Dynamics*. New York: Springer-Verlag.

Rogers, G. D., J. C. Bunn, and W. F. Dellinger, 2001. CONTOUR guidance and control system algorithm design and development. *Proceedings, 16th International Symposium on Space Flight Dynamics*, Pasadena, CA.

Shepperd, S. W., 1978. Quaternion from rotation matrix. *J. Guidance and Control* **1**:223.

Shuster, M. D., 1983. Efficient algorithms for spin-axis attitude estimation. *J. Astronautical Sciences* **31**:237.

Shuster, M. D., 1989. *Restitution d'attitude des véhicules spatiaux*, lecture notes, © 1989 by Malcolm D. Shuster.

Shuster, M. D., 1993. A survey of attitude representations. *J. Astronautical Sciences* **31**:439.

Shuster, M. D., and S. D. Oh, 1981. Three-axis attitude determination from vector observations. *J. Guidance and Control* **4**:70.

Shuster, M. D., D. S. Pitone, and G. J. Bierman, 1991. Batch estimation of spacecraft sensor alignments. *J. Astronaut. Sci.* **39**:519 (Part I) and 547 (Part II).

Singer, S. F., ed., 1964. *Torques and Attitude Sensing in Earth Satellites.* New York: Academic Press.

Slafer, L. and H. Marbach, 1975. Active control of the dynamics of a dual-spin spacecraft. *J. Spacecraft and Rockets* **12**:287.

Stacey, F. D., 1977. *Physics of the Earth.* New York: John Wiley and Sons.

Taff, L. G., 1991. An analysis of the Hubble Space Telescope fine guidance sensor fine lock mode. *Proceedings, Flight Mechanics/Estimation Theory Symposium.* NASA Goddard Space Flight Center, Greenbelt, MD.

Thomson, W. T., 1962. Spin stabilization of attitude against gravity torque. *J. Astronaut. Sci.* **9**:31.

Thomson, W. T., 1986. *Introduction to Space Dynamics.* New York: Dover.

Tossman, B. E., F. F. Mobley, G. H. Fountain, J. J. Heffernan, J. C. Ray, and C. E. Williams, 1980. *MAGSAT attitude control system design and performance.* Paper 80-1730, AIAA Guidance and Control Conference, Danvers, MA.

van der Ha, J., W. F. Dellinger, G. D. Rogers, and J. Stratton, 2003. CONTOUR's phasing orbits: attitude determination & control concepts and flight results. *Proceedings, 13th AAS/AIAA Space Flight Mechanics Conference*, Ponce, Puerto Rico.

Wertz, J. R., ed., 1978. *Spacecraft Attitude Determination and Control.* Dordrecht, The Netherlands: Kluwer.

Wie, B., 1998. *Space Vehicle Dynamics and Control.* Reston, VA: American Institute of Aeronautics and Astronautics.

Wiesel, W. E., 1989. *Spaceflight Dynamics.* New York: McGraw-Hill.

Williams, C. E., and J. W. Hunt, Jr., 1989. Spacecraft inversion using a momentum wheel. *Proceedings, First Pan-American Conference of Applied Mechanics,* Rio de Janeiro.

Wylie, C. R., and L. C. Barrett, 1982. *Advanced Engineering Mathematics.* New York: McGraw-Hill.

6

Space Power Systems

GEORGE DAKERMANJI AND RALPH SULLIVAN

6.1 Introduction

Reliable, continuous operation of the power system is essential to the successful fulfillment of a spacecraft mission. A failure, even a brief interruption in the source of power, can have catastrophic consequences for the spacecraft's attitude and thermal control as well as its electrical systems. Therefore, the power system and its components must be designed and fabricated with reliability as a primary requirement. The challenge for the power system designer is to accomplish this in the face of ever-increasing costs, demands for improved performance, and the usual competition with other systems for spacecraft resources.

The above can only be accomplished by a thorough understanding of the power system, its basic components, and the environment in which it must operate. There have been many advances in the decade since the first edition that have led to new space power system technologies and improvements in our understanding of the operating environment. This is a trend that promises to continue. The following is an attempt to describe the present state-of-the-art in power systems and to introduce a basic understanding that will enable one to keep abreast of future developments.

6.2 The Space Environment

6.2.1 Solar Energy

The spectral irradiance of the Sun at the Earth's mean distance, one astronomical unit (AU), is shown in chapter 2 (figure 2.5). It presents a visual picture of the solar spectrum outside the Earth's atmosphere (Air Mass Zero) and on the Earth's surface (Air Mass

One). The solar constant is the integrated energy under the Air Mass Zero (AM0) curve and is defined as the total solar energy incident on a unit area perpendicular to the Sun's rays at the mean Earth–Sun distance outside the Earth's atmosphere. The value accepted through the 1990s was 135.3 mW/cm^2. But more recent space measurements indicate a value of 136.6 mW/cm^2.

The value of this solar constant varies annually due to the eccentricity of the Earth's orbit about the Sun ($e = 0.016$). The difference in Earth–Sun distance between perihelion and aphelion is approximately 5,560,000 km, a very small annual change in the mean distance of 1.4961×10^8 km. Since the Sun's intensity varies inversely with the square of the distance (for large distances), this results in a solar constant variation of $\pm$ 3.5% about the mean of 136.6 mW/cm^2.

6.2.2 The Earth's Radiation Environment

As described in chapter 2 (section 2.5), the radiation environment near the Earth consists of electrons and protons trapped in the geomagnetic field, corpuscular radiation associated with large solar flare activity, and, to a lesser extent, galactic cosmic ray radiation. However, the most damaging environment for solar cells is that of the trapped electrons and protons within the Earth's magnetosphere and the solar flare protons, which peak sporadically at times near maximum solar activity. Within the magnetosphere, the trapped particles are the most damaging, particularly the low-energy protons that predominate at relatively low altitudes. Outside the magnetosphere, the solar flare protons predominate.

Figure 2.16 shows a cross section of the Earth's magnetosphere, illustrating the trapped electron and proton flux. The heart of the proton belt, approximately two to three Earth radii or 6000 to 12,000 km altitude, is so damaging to solar cell arrays that they become impractical for extended missions in this region. Therefore, most solar powered spacecraft are located in low Earth orbit (LEO), usually below 1100 km, or in high Earth orbit (HEO), above 20,000 km in altitude. Solar powered spacecraft that must experience the environment near the center of the proton belt are typically placed in highly elliptical orbits so that the length of time spent in this region is minimized.

Much less damaging radiation exists below an altitude of 1100 km, except for the accumulation of particles in the South Atlantic Anomaly, a depression in the magnetic field that causes an increase in the radiation density at lower altitudes off the coast of Brazil. Most of the solar cell radiation damage that occurs below this altitude is due to the low-energy protons that populate this region. However, there is another kind of damage that occurs to all spacecraft surfaces, including the solar cell array, at very low altitudes (below approximately 650 km): a surface abrasion of the coverglass and exposed electrical contacts caused by atomic oxygen.

Trapped radiation significantly impacts solar array lifetimes almost as far out as geosynchronous Earth orbit (GEO). But damage due to protons from solar flares becomes more significant at higher altitudes and inclinations, and is the predominant life-limiting concern in interplanetary space. J. Feynman and co-workers have developed the JPL 91 model to estimate the solar flare fluences (Feynman et al., 1990). This model uses a data set spanning three solar cycles and is used to predict the integral fluence as a function of confidence level and time. A computer program "Solar Proton Estimator (SPE)" was developed by Spitale and Feynman (1992) to estimate the proton integral fluences at

1 AU or in interplanetary space. The program calculates the radiation fluences for typically used confidence levels of 75, 90, 95, and 99%.

6.3 Orbital Considerations

6.3.1 The Geocentric Equatorial Coordinate System

Several coordinate systems are used to determine the position of a body in space. The galactic coordinate system is convenient for determining the position of stars and star clusters of our galaxy. A heliocentric system would be employed for orbits about the Sun, whereas a geocentric system is typically used to describe Earth orbits.

Figure 6.1 shows the celestial sphere with infinite radius and an origin that coincides with the center of the Earth. The celestial poles of this sphere are the extension of the Earth's North and South Poles, which define the Z-axis. The celestial equator is the great circle equidistant from these poles, and is co-planar with the Earth's equator. The apparent yearly path of the Sun as seen from the Earth is projected onto the celestial sphere, and is called the ecliptic. Its corresponding plane is the ecliptic plane. The equatorial and ecliptic planes intersect in a line that becomes the X-axis (also called the line of equinoxes or the line of nodes). The positive X-axis points to the vernal equinox, usually called the First Point of Aries. The Y-axis is defined as the line that passes through the origin and is normal to the Z–X plane.

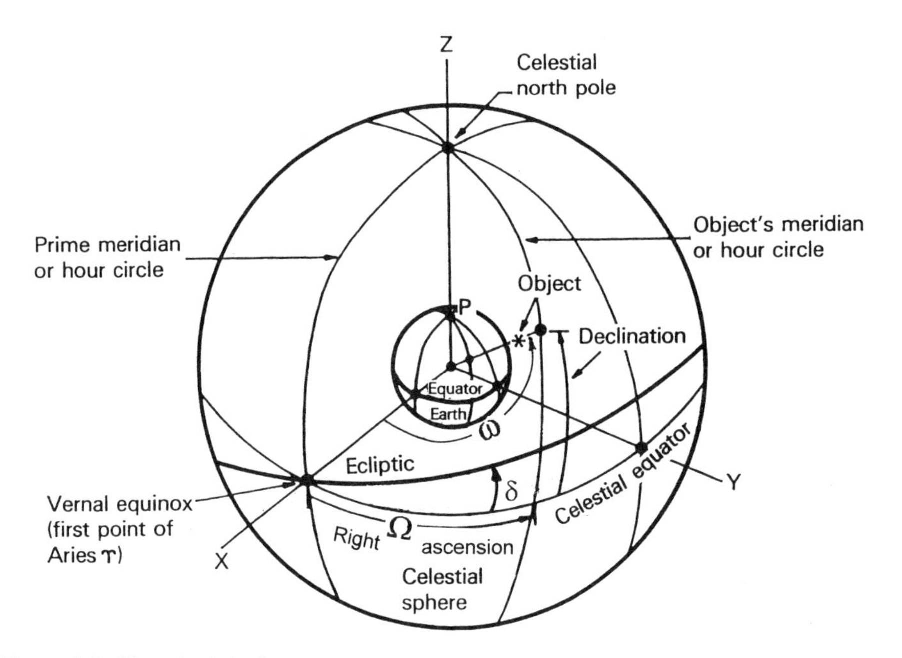

Figure 6.1 The celestial sphere.

In this coordinate system, the position of an object is determined by its right ascension and declination angles. The right ascension angle is measured from the First Point of Aries, eastward along the celestial equator. The declination angle is measured along the object's hour circle (meridian) from the celestial equator, ±90 degrees in either direction, with north declination taken as positive.

6.3.2 Spacecraft Position

The Kepler elements, also known as the classical orbital elements (see chapter 3), are most often used to determine the position of a spacecraft and its orbit in three-dimensional space because, to the first order, five of the six elements are constant. The classical orbital or Kepler elements are:

Orbit inclination (degrees), i
Orbit semimajor axis (km), a
Orbit eccentricity, e
Right ascension of the ascending node, Ω
Epoch time, t_0(the time when the orbital elements are defined, usually the time of perigee)
Argument of perigee, ω

The precession of perigee ($\dot{\omega}$) and the precession of the right ascension of the ascending node ($\dot{\Omega}$) are given by the following expressions:

$$\dot{\omega} = \frac{1.036 \times 10^{14}(5\cos^2 i - 1)}{a^{7/2}(1 - e^2)^2} \quad \text{deg/day} \tag{6.3.1}$$

and

$$\dot{\Omega} = -\frac{2.072 \times 10^{14}\cos i}{a^{7/2}(1 - e^2)^2} \quad \text{deg/day} \tag{6.3.2}$$

where, for circular orbits:

$a = R_{\mathrm{E}}+$ orbit altitude (km)
$R_{\mathrm{E}} =$ radius of the Earth, generally accepted as around 6378.137 km

6.3.3 Sun Position

To an observer on Earth, the Sun appears to circle the inertial Earth frame once per year. The apparent motion of its right ascension angle is approximately 1° per day eastward. The angle at which the celestial equator intersects the ecliptic is called the *obliquity of the ecliptic* and is treated as a constant 23.5° for the purpose of power system design. The Sun's declination (δ) is zero at the equinoxes and varies throughout the year from +23.5° at the summer solstice on June 21 to −23.5° at the winter solstice on December 21. (Note that the maximum and minimum solar distances (aphelion and perihelion) follow the solstices by almost two weeks on approximately July 4 and January 3, respectively.)

6.3.4 Orbital, Eclipse, and Sunlight Periods

The orbital period and the fraction of time that a spacecraft is in sunlight (and eclipse) are of fundamental importance to the design of both the thermal system and the power system. These determine the number of battery discharge cycles, a major determinant of battery lifetime. The orbital period T is developed in chapter 3 (equation 3.3.53) from Newton's formulation of Kepler's third law. It is reproduced here as equation 6.3.3. This formula is independent of the orbit eccentricity, and is therefore useful for both circular and elliptical orbits.

Orbital period:

$$T \cong 2\pi\sqrt{\frac{a^3}{\mu}} \cong 1.6585 \times 10^{-4} a^{3/2} \text{ minutes} \tag{6.3.3}$$

where:

$\mu = Gm_E = 3.986005 \times 10^{14}\ \text{m}^3/\text{s}^2 = 14.3496 \times 10^8\ \text{km}^3/\text{min}^2$
$a =$ the orbit semimajor axis (km) $= R_E +$ orbit altitude for circular orbits
$R_E =$ Radius of the Earth, around 6378.137 km

For all Earth orbits except those that are Sun synchronous, the Sun will lie in the orbit plane typically twice per year, causing the spacecraft to experience the maximum time in eclipse and the minimum time in sunlight. The maximum eclipse period determines the longest time that the battery must sustain the load, a major factor in defining the size of the battery. The minimum sunlight period determines the shortest time available for battery recharge, and therefore influences the solar array geometry. Since this so-called "minimum Sun" case usually drives both the battery and solar array size, it is typically the first case that the power system designer will analyze.

Figure 6.2 shows, for circular orbits, the geometry for calculating the fraction of time in sunlight and eclipse for this "minimum Sun" case. Consider that the Sun is in the orbit plane, its rays are essentially parallel, and the terminal rays are tangent to the Earth. A right triangle is then formed with the Earth's radius R_E as one leg and a hypotenuse of length equal to the sum of R_E and the orbit altitude A. These observations lead to the equations for the fraction of time in sunlight and eclipse, which are shown in the figure.

The result of a more general analysis where the Sun is not confined to the orbit plane is shown in figure 6.3 for a range that includes low Earth orbits (LEOs). Note that all orbits (except those that are Sun synchronous) are subject to the periods of maximum eclipse, which is very close to 36 minutes for LEO. Equatorial and other low inclination orbits are close to this condition all the time. Also, the minimum eclipse times get shorter as the inclination is increased until, for angles of high inclination, the spacecraft experiences periods of 100% Sun.

6.3.4.1 *Periodic Orbital Variations*

The annual variation in the eclipse and sunlight periods is fundamentally important to the power system design. If a solar power system is designed for the maximum eclipse and minimum sunlight times at the end-of-life (EOL), then it is overdesigned for all

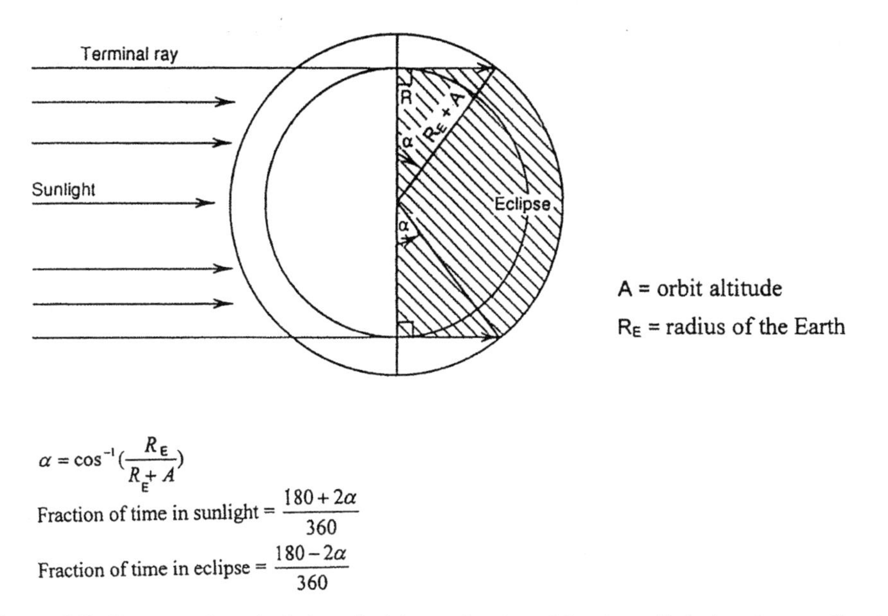

Figure 6.2 Geometry for calculation of minimum fraction of time in sunlight (maximum eclipse) for circular orbits.

other conditions, especially the minimum eclipse and maximum sunlight periods at the beginning-of-life (BOL). Therefore, these other conditions must also be examined to ensure that the power electronics is properly designed to process the solar array power and that the design will maintain energy balance throughout the mission.

Variations in the eclipse and sunlight periods due to the time of year are best expressed as a function of the angle between the Sun and orbit normal vectors, as shown in figure 6.4. The orbit normal vector's direction cosines are expressed in terms of the orbit's right ascension and inclination. The direction cosines of the Sun vector are expressed as functions of the Sun's right ascension and declination, which are functions of day number.

Figure 6.5 shows the percentage of time that a typical spacecraft in a circular, polar, low Earth orbit spends in sunlight during the course of one year. The curves are primarily a function of altitude and inclination, but also depend upon the launch date and initial right ascension angle (chosen arbitrarily for purposes of illustration). A different selection of launch date and/or initial right ascension angle would yield curves of a similar shape, but with a shift in phase. It can be seen that the lower altitude spacecraft not only have longer eclipse times, but dwell near the maximum eclipse for a much longer fraction of the year.

6.3.5 Solar Array Analysis

Fundamental to all solar array analysis is the determination of the angle between a solar panel normal vector and the Sun vector. The situation is shown in figure 6.6, in which same expression is used that was previously used for the cosine of the angle between two vectors. The coordinates are those of the spacecraft body coordinates and the two

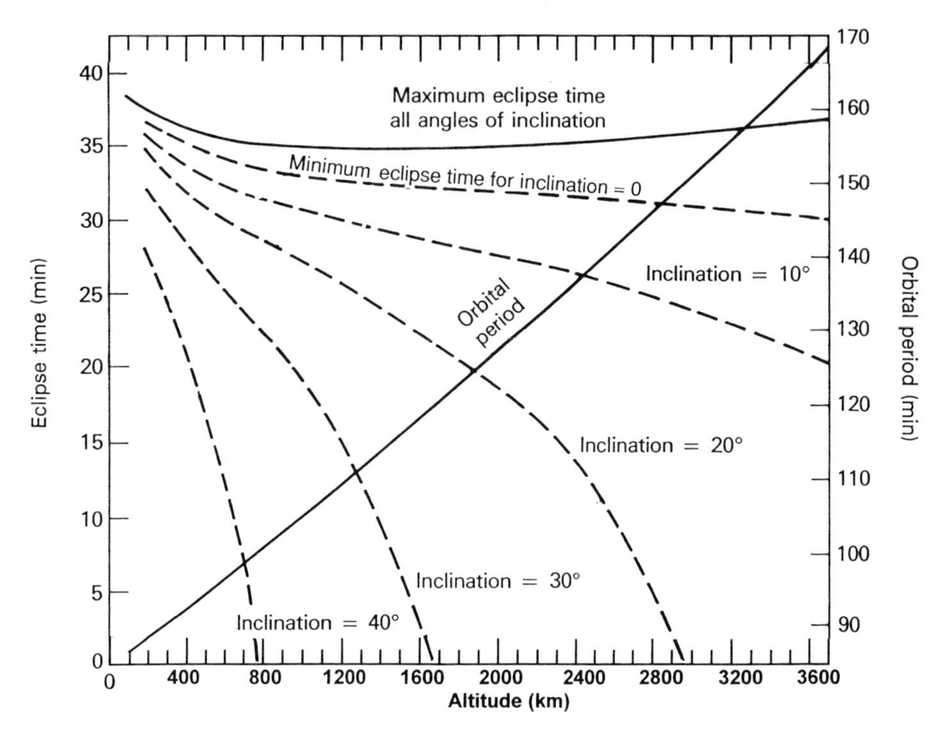

Figure 6.3 Orbital period, maximum and minimum eclipse time for circular orbits.

vectors are the solar panel normal and the spacecraft–Sun line. The cosine of the angle between the two vectors is once again the sum of the products of their direction cosines. Since a solar panel surface will be illuminated only when the angle is between 0° and 90°, negative values of the angle's cosine are excluded for solar panels with cells on only one side.

Using the formulas given, the output or effective area of any number of solar panel surfaces can be calculated and summed to obtain the useful solar cell area of an array for a single spacecraft position relative to the Sun. This calculation is then repeated for several positions of the spacecraft throughout the sunlight period. The same calculations can then be repeated for subsequent orbit–Sun positions as they change with time throughout the mission. Approximate analyses of this type are often used for the preliminary design until the configuration is finalized.

Once the basic configuration is determined, the array is then more precisely analyzed. A shadow study is sometimes necessary if the spacecraft and/or its appendages can sometimes shade the solar cell strings. The analysis is typically performed using a computer-generated three-dimensional model of the spacecraft and its solar panels with a simulated distant light source. Projections showing the shadows on the solar array are analyzed for relevant orientations of the spacecraft and solar array. The analysis can also be accomplished with the use of a physical three-dimensional scaled replica of the spacecraft and solar array, illuminated by a distant light source. In either case, the objective of the study is to show the size and shape of the shaded solar cell area for a representative number of spacecraft–Sun positions so that, for each position, the effect

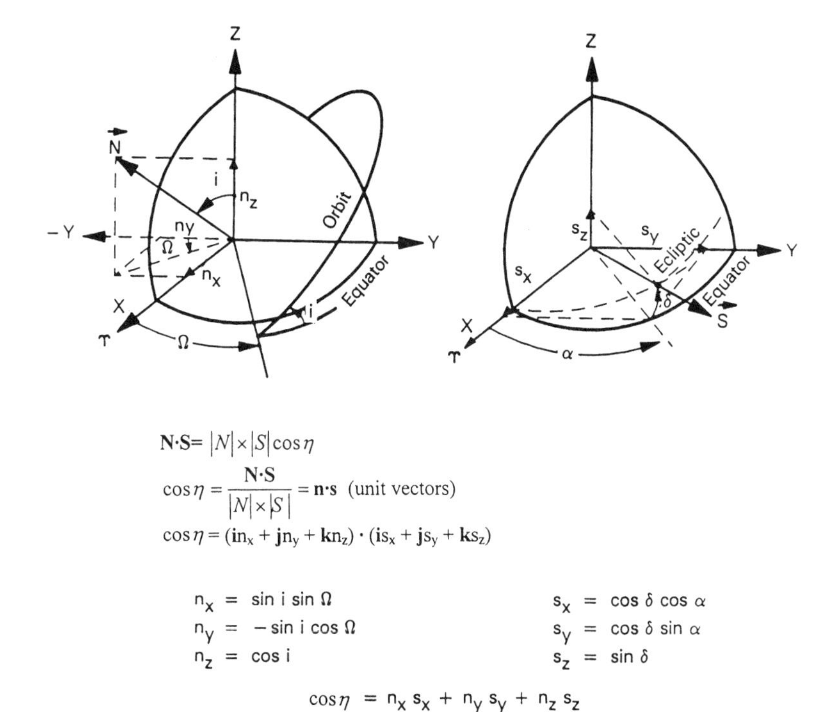

Figure 6.4 Direction cosines to determine the angle η between the orbit normal and the Earth–Sun line.

of shadows on the solar array's power generation can be determined. Then the analysis must be repeated, including the results of the shadow study, the estimated temperature, and the actual angular positions of the solar panels.

At larger angles of incidence, the electrical output of a solar panel is slightly less than would be predicted from cosine dependence. This is due to the increased reflection experienced at these angles, partly because of the increased reflection between optical media at larger angles of incidence[1]. Often, this angular response of the solar cell is empirically determined. One such determination shows that the solar cell response follows the cosine law for angles less than or equal to 50° and is approximated by the following relationship for larger angles:

$$\rho = -0.369\cos^3\phi + 0.637\cos^2\phi + 0.750\cos\phi - 0.015 \tag{6.3.5}$$

for $\phi > 50°$, where ϕ is the angle between the Sun vector and the panel normal.

6.4 Energy Sources

Every electrical power system source consists of a *primary energy source*, an energy *converter*, *energy storage* and *regulation and distribution*. The converter transforms

[1]For a more complete discussion of the solar cell's optical system and illumination at larger angles of incidence, see Rauschenbach (1980), p. 245.

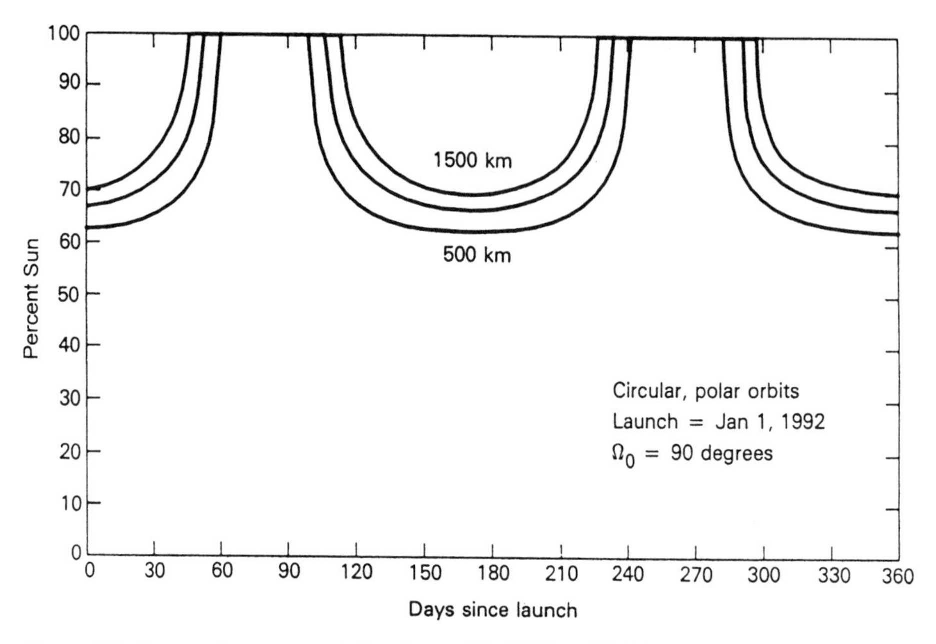

Figure 6.5 Percent Sun versus mission day at 500, 1000, and 1500 km.

energy from an available form into the more useful electrical form. Three basic conversion types that are used for the generation of power in space are solar, nuclear, and electrochemical. Depending on the size, duration and nature of the application, a wide range of conversion and energy storage systems has evolved that use one or more of these three primary energy sources.

6.4.1 Types of Energy Source

In some cases, the primary source and the converter are a single device. Solar cells and some electrochemical sources, such as batteries and fuel cells, are energy sources that produce electricity *directly* without requiring an electrical converter. By contrast, nuclear sources produce their energy in the form of heat and radiation. The heat from nuclear sources must then be transformed to electricity using a separate converter, with the radiation treated as a waste product.

The nuclear source can be either a radioisotope or a reactor. A radioisotope is a material that produces heat and radiation as a natural by-product of its decay. That is, its heat generation rate will be nearly constant, decreasing very gradually over a typical 5-to 20-year mission life. And, since it cannot be controlled, it must be either used or wasted. By contrast, a reactor's rate is actively controlled in order to regulate the electrical output of its converter. This control adds to the complexity of the reactor, but is one of the reasons why the nuclear reactor is usually more suitable than a radioisotope source for larger systems. (For reference, table 6.1 is a glossary of relevant space power terms.)

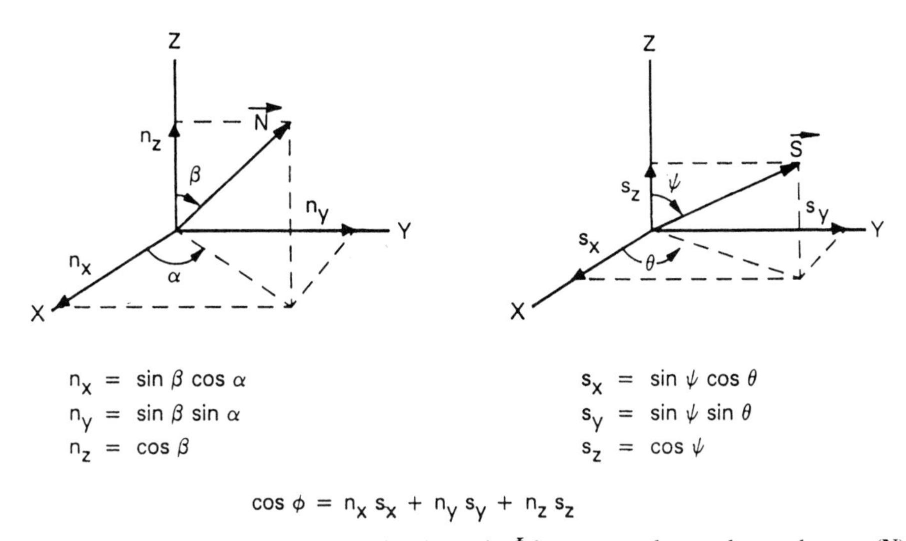

Figure 6.6 Direction cosines to determine the angle Φ between a solar panel normal vector (**N**) and the spacecraft–Sun line (**S**).

6.4.2 Radioisotope Thermoelectric Generators (RTGs)

Generally, RTGs are compact, relatively lightweight, and reliable, and offer significant advantages over photovoltaic systems for certain applications. Although the RTG is a distant second to the solar array–battery system in number of missions flown, it is the only practical alternative for U.S. missions to the outer solar system, missions very close to the Sun, or those missions that are subject to a very high radiation environment. For many planetary exploration missions there are advantages to the continuous power generated by an RTG, which minimizes the need for onboard energy storage and eliminates the problems of solar array–Sun orientation. And their performance is not degraded by particle irradiation or dust and debris. Together, the solar array–battery and RTG systems account for all of the U.S. long-duration spacecraft missions.

Since the size of the heat source and the electrical output of a nuclear power source are both measured in watts, the heat source is usually depicted in thermal watts (W_t) and the output in electrical watts (W_e) to avoid confusion. The size and specific power of RTGs used for space applications have grown considerably since the first SNAP-3 unit was launched by JHU/APL on the Naval Navigation Satellite, TRANSIT 4A, in June 1961.

In 1977 General Electric (GE) provided the RTGs for both the Voyager and Galileo missions. Three 150 W_e, 4 W_e/kg generators were used to power Voyager II. Larger RTGs of similar design were then developed for Galileo, which was launched in October 1989, powered by two such units. At the time of their manufacture, each produced 300 W_e at a specific power of 5.3 W/kg. A number of these units, General Purpose Heat Source RTGs (GPHS-RTGs), were produced in expectation of near-term use for future missions. In addition to the two that were used for Galileo, one of these was used for Ulysses and three were used for the Cassini–Huygens mission, and a similar RTG will be used to power the New Horizons spacecraft, which is slated for launch in

Table 6.1 Glossary of space power sources and related terms

Primary battery: An electrochemical cell or group of cells that is activated (charged) with electrolyte once and is used as a primary energy source on demand until depletion. A primary battery is not designed to be recharged.

Secondary battery: An electrochemical cell or group of cells recharged after use by passing current in the opposite direction to the discharge current. A secondary battery is a reusable energy storage device and is secondary to another primary source of energy.

Fuel cell: Similar to a *primary battery* except that the reactants (fuel) are stored externally and fed into the cell. The cell can operate until the fuel is depleted.

Regenerative fuel cell: A fuel cell that is similar to a *secondary battery* in that the reactants are restored by the application of energy to the cell's by-products. It is an energy storage device that is secondary to another primary source of energy.

Chemical dynamic: An electrical generator that uses ignited fuel as the basic heat source and a thermodynamic engine for conversion to electricity.

Dynamic: Used to describe power sources that employ a *thermodynamic engine* in converting to electricity from an energy source.

Nuclear: A power system that uses a *nuclear reactor* as its heat source (a radioactive fissionable material). It employs an *active control system* to regulate the fission rate, matching the electrical power generation to the load requirement. The conversion from thermal to electrical energy can be accomplished with a thermoelectric generator (*nuclear thermoelectric*), a thermionic converter (*nuclear thermionic*), or an engine-driven generator (*nuclear dynamic*).

Radioisotope: An energy source that is derived from the heat generated from the decay of a *radioactive isotope*. The isotope decays at its natural, predictable rate *without any active control*. When used as the heat source for a thermoelectric generator, the system is called a radioisotope thermoelectric generator (RTG).

Thermionic converter: A device consisting of two electrodes, an emitter maintained at high temperature and a collector maintained at a lower temperature, separated by a vacuum or plasma. The thermal energy of the emitter is sufficient to cause electrons to overcome the surface work function, traverse to the collector, and supply electrical power to the load. Any heat source can be used, but a nuclear reactor is usually considered for space applications.

Thermoelectric converter: A device consisting of two dissimilar materials (semiconductors) to form a *couple*, one connected to the heat source, the other to the heat sink or radiator. The heat flow through the couple generates electrical power due to the *Seebeck effect*. Any heat source can be used. For space applications, radioisotopes have been used for systems of modest size ($< 1\ kW_e$).

Photovoltaic (solar): The production of power across the junction of two dissimilar materials (n and p doped semiconductors) when exposed to light or electromagnetic radiation.

Photovoltaic (thermal): Similar to solar photovoltaic but uses semiconductor materials with proper band-gap energies to convert thermal infrared radiation directly to electricity. The heat source may be nuclear.

Solar dynamic: An electrical generator that uses sunlight as the basic heat source and a *thermodynamic engine* for the conversion to electricity.

January 2006. These RTGs have an exceptional performance record. The Voyager I and II grand tours of our planetary system, the Galileo mission to Jupiter, the Ulysses exploration of the polar regions of the Sun, and the Cassini–Huygens mission to Saturn and its moon, Titan, are all still functioning as expected.

However, the RTG does have special problems. Materials problems, caused by high generator operating temperatures, limit its lifetime. An even more serious problem is the fact that the fuel source is an expensive radioactive material that could become hazardous in the event of an accident. This leads to complex safety requirements and expensive ground handling procedures. Also, a post-launch mishap could have high-profile international (political) implications, which frequently introduce delays in the schedule. To minimize this hazard for mishaps during launch or on-orbit, the fuel capsules are designed to be recovered intact for reentry at nominal orbital velocities, thus precluding the uncontrolled spread of radioactive particles. Such considerations drive the cost of an RTG to many times that of a solar array–battery system, limiting their application to only those missions where a solar array is not practical.

6.4.2.1 RTG Description

All thermoelectric generators use n-doped and p-doped semiconductor elements (so-called N and P legs) that are electrically connected in series to form a couple. The legs are typically cylindrical in shape and are bonded or pressed to hot and cold electrically isolated thermal conductors (shoes). A radioisotope thermoelectric generator (RTG) includes an isotopic fuel source that provides heat to the semiconductor elements, causing the conversion of heat to electricity through the thermoelectric (or Seebeck) effect.

Figure 6.7 shows a configuration with the semiconductor elements arranged so that the heat flows uniformly through all of them in parallel but they are connected electrically in series. The heat flows from the source through the N and P legs and out to the radiator where it is rejected. The temperature difference across the thermocouple elements produces a proportional voltage difference due to the Seebeck effect. The polarity of the voltage difference is opposite for the N and P legs, so that the voltages add and part of the heat flow is converted into electrical power. These P–N elements are electrically connected in series to form a string at a useful system voltage. Such series strings are then connected in parallel to obtain the necessary current or power.

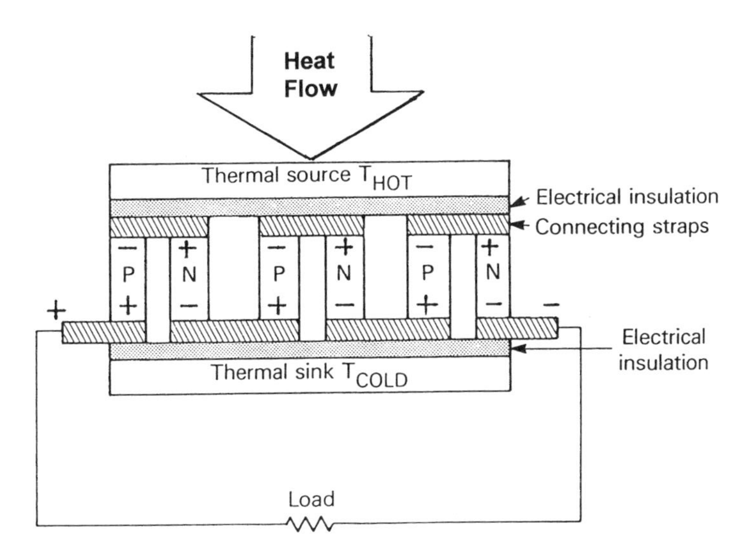

Figure 6.7 Thermoelectric generator.

6.4.2.2 Radioisotopes

It is essential that a radioisotope energy source should have a lifetime compatible with mission requirements and possess good engineering properties at high temperatures; that is, it should have high power density, low thermal conductivity, low gas evolution, and low weight. Its material properties should include dimensional stability, structural compatibility, high melting point, and resistance to corrosion. Although approximately 1300 radioisotopes are known to exist, only a relatively small number are useful as heat sources for an RTG. For some, the half-life is too short for a typical spacecraft mission life that is measured in years. Also, some isotopes are unacceptable because their emission of alpha and beta particles is accompanied by the release of electromagnetic energy in the form of gamma rays, an unwanted by-product, requiring heavy shielding to minimize the radiation hazard. Pu-238 has been selected for use on nearly all spacecraft, because its primary emissions (alpha particles) are easily shielded and it has a half-life of 86.8 years, a length that is acceptable to both systems and safety engineers.

6.4.2.3 RTG Sizing

The RTG must be sized to supply the load at the end-of-life (EOL). Therefore, it must be designed with excess power at the beginning-of-life (BOL) to provide for the loss of generator output power due to estimated couple degradation, changes in thermal radiator characteristics, and the gradual cooling of the heat source as the plutonium fuel decays throughout the mission. The RTG's power degradation is due mainly to the decay of the plutonium fuel at a rate that is easily predicted. Since the heat source is a continuous power generator that cannot be interrupted, this excess power must be diverted to shunt dissipaters, on the spacecraft, which effectively match the spacecraft load and generator source impedance to effect near-maximum power transfer continuously throughout the mission. Large mismatches between the load and source impedance cannot be tolerated for extended periods since they would cause the source to overheat, degrade the performance of the thermoelectric generator, and shorten its life.

6.4.2.4 The General Purpose Heat Source RTG (GPHS-RTG)

Figure 6.8 shows a cutaway view of the general purpose heat source radioisotope thermoelectric generator (GPHS-RTG) that was developed by a division of General Electric (now part of Lockheed Martin Corporation) in Valley Forge, Pennsylvania, under the sponsorship of JPL in the mid 1990s. It was designed to accommodate a modular heat source that could be sized to supply the power needs of future missions that were not yet conceived (hence the term: "General Purpose"). This was a very important aspect of the design because the costly effort of qualifying the GPHS to meet the launch-range safety requirements only had to be done once on a single GPHS unit (a fraction of a single heat source).

The original design was to provide 290 to 300 W_e (BOL), which was expected to degrade to about 250 to 260 W_e for a typical mission life of 10 years. For safety considerations, the fuel modules are designed to survive reentry into the Earth's atmosphere and remain intact after Earth impact in the event of an accident.

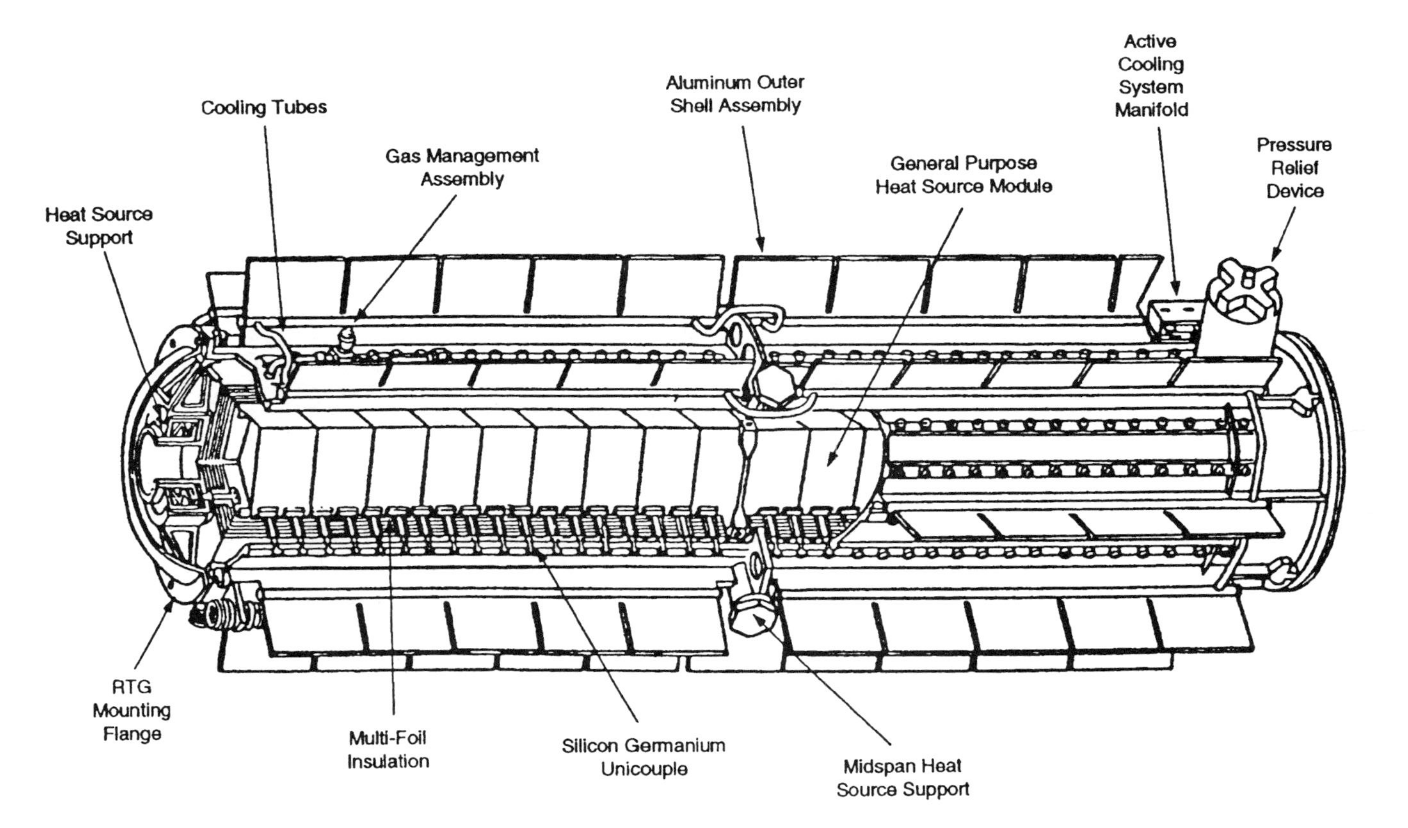

Figure 6.8 General purpose heat source (GPHS) RTG, cut-away illustration.

The heat source consists of a stack of 18 GPHS modules, eleven of which are shown as rectangular blocks, aligned along the central axis of the cylindrical unit. The silicon–germanium (SiGe) thermoelectric couples (so-called unicouples) are thermally connected to and sandwiched between the heat source hot shoe and the cold shoe at the radiator. [A unicouple comprises one electron-conduction (n-type) leg and one hole-conduction (p-type) leg thermally connected in parallel and electrically connected in series.] The hot junction temperature is approximately 1000°C (1832°F) and the cold junction temperature is 300°C(572°F). Both temperatures are higher than those of previous designs. Each GPHS module has eight SiGe radioisotope unicouples associated with it to form the RTG module.

Since RTGs produce approximately nine times more heat than electrical power, they have abundant waste heat. The GPHS-RTG has a specific power of 7.7 W_e/kg (not considering that any waste heat might be used for thermal control). One of the advantages of using an RTG instead of a solar array is that a battery is not required to support the load during eclipse. But, without a battery to supplement the RTG, transient loads must be examined and managed very carefully. A transient load demand that equals or exceeds the RTG's short circuit current would overload the bus and drive the system toward zero volts. Such an under-voltage condition could precipitate a reset of the spacecraft's logic circuits and its computer.

RTGs have no moving parts. They have an excellent proven record of very reliable operation over very long missions (Voyager spacecraft RTGs have been operating for over 25 years). However, RTGs have low efficiencies. Research, development, and life tests are ongoing for more efficient thermal-to-electric converters. Stirling engines with linear electric alternators are being considered as a replacement for Si–Ge thermoelectric generators to increase the conversion efficiencies to over 20%, thus providing more power to the spacecraft for the same nuclear heat source. Other thermoelectric generation concepts being studied include thermal photovoltaics, Brayton cycle engines, and alkali metal thermoelectric converters (AMTECs).

6.4.3 Solar Cells

6.4.3.1 Solar Cell Theory

Theory of operation: the photovoltaic effect. Since the silicon cell (particularly the n/p silicon cell) has been used almost since 1964, its theory of operation is the most highly developed. But the basic p–n junction theory is the same for GaAs and the individual junctions in multijunction cells. To make a silicon n–p junction (referring to figure 6.9), a crystal wafer is doped with small quantities of p-type impurities such as aluminum, boron, or gallium. Then, in a shallow diffusion process, the very top or front surface ($<1/4\ \mu m$) is injected with a higher concentration of n-type impurities such as antimony, arsenic, or phosphorus. The result is a very thin n-doped layer on the front surface that forms a junction with the base region, which remains p-doped. The excess n-type atoms ionize and supply electrons for conduction in the front layer. The excess p-type atoms in the base region take electrons from the lattice, leaving "holes" that can conduct current by moving from one atom to the next as an adjacent electron moves in. A p–n junction is developed in a similar manner with different materials and the roles of p-doped and n-doped materials reversed.

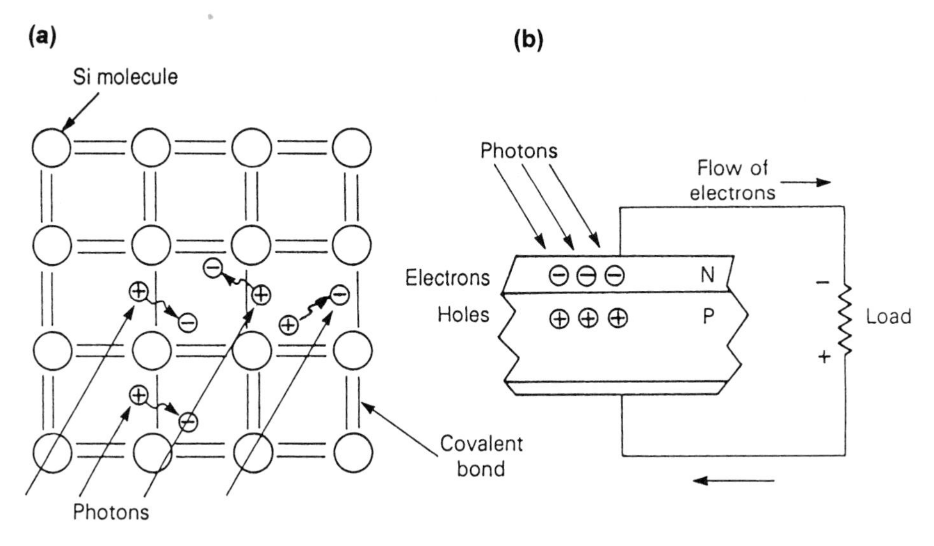

Figure 6.9 Illustration of photovoltaic effect: (a) crystal lattice structure; (b) n–p junction

When a p–n junction is produced in a crystal, electrons from the n-side thermally diffuse into the p-region, where they become minority carriers. Similarly, holes from the p-side become minority carriers in the n-region. Recombining with the majority carriers rapidly neutralizes these minority carriers. Since each region was initially electrically neutral, the loss of electrons from the n-region and the loss of holes from the p-region contribute to establishing a potential difference across the junction. This built-in electric field is used for photovoltaic power generation.

When a photon of light is absorbed in a crystal, it will ionize an atom and release an electron, thereby producing an electron–hole pair. After a short burst of illumination, the electrons and holes that have been injected both diffuse until they find an opposite number and recombine. The time for a minority carrier to recombine (the lifetime) is related to the diffusion length and depends on the density of recombination centers, which are crystal defects that provide sites where minority carriers are captured and then recombined readily with majority carriers.

When light is absorbed in the region near the p–n junction of a solar cell, some of the electrons and holes will diffuse to the junction. At this point, the charges will be separated by the built-in electric field at the junction, thereby providing a current that can flow through an external load. This method converts photon energy directly to electrical power and is known as the photovoltaic effect.

When a solar cell is illuminated, electron–hole pairs are produced throughout the base region below the junction to a depth that is dependent on the light's wavelength. Electron–hole pairs in single junction cells that originate close to the junction are mostly due to short wavelength light, whereas the longer wavelength, more penetrating light produces carriers further from the junction. Red light or infrared radiation produces electron-hole pairs more deeply below the surface than blue light, with current output dependent on the magnitude of the diffusion length for minority carriers. Particle radiation produces recombination centers throughout the base region, with greater effect on carriers that are further from the junction (it reduces the diffusion length). Therefore, radiation damage has a greater effect on the solar cell's long wavelength (red) response.

Also, a back surface field (BSF) layer is usually designed into the cells, behind the junction, to reflect minority carriers that are diffusing away from the junction. And, in silicon cells, back metallization is applied as a mirror to reflect photons that have not been absorbed and return them into the active region for another chance to be absorbed. Such cells are called back surface reflector (BSR) cells. Both of these techniques are commonly utilized in silicon cells, which are then designated as BSFR cells. In multijunction cells, the top junction materials are selected with the highest energy band-gap material that absorbs the shorter wavelength photons. The longer wavelength photons then pass to the second junction, which is made of material of lower band-gap energy. The remaining unabsorbed higher wavelength photons pass to the bottom junction in a triple junction cell. The longer wavelength energy that is not absorbed passes through the cell to be absorbed by the substrate as waste heat and is then radiated to space.

6.4.3.2 Solar Cell Types

Single crystal silicon cells were the main technology used on spacecraft until the mid-1980s. The efficiency of silicon cells improved slowly. From a little over 10% in the early 1960s, the efficiency grew to about 13% by the mid-1970s and is today available at about 16%. The primary advantages of using silicon as a solar cell material are that it is readily available, relatively inexpensive to purchase and process, and is rigid and rugged, enabling the manufacture of the complete cell from a single crystal. Single junction gallium arsenide (GaAs) cells were always an attractive alternative technology because the larger band gap energy of GaAs made it more efficient than silicon. Also, it was less susceptible to radiation damage than silicon. But the gallium arsenide was a more expensive material and it lacked physical integrity. The finished cell was brittle and fragile, making it very difficult to work with. In early attempts, a thin film of active GaAs was grown on an inert gallium arsenide wafer (the substrate) and was prone to breakage. Technologies were then developed to grow the GaAs on crystal lattice matched germanium (Ge) crystal wafers. Single junction GaAs/Ge cells achieved around 19% efficiency.

Substantial improvement in efficiency was achieved in the late 1990s with the development of multijunction cells. Dual junction (DJ) cells were developed by adding a GaInP junction on top of the GaAs junction on the inactive germanium wafer. DJ cells were used on numerous geosynchronous and scientific satellites including Deep Space 1 concentrator panels. Triple junction cells quickly followed, formed by activating the interface with the germanium wafer to form the third junction. Triple junction cells with efficiencies of over 28% are commonly used on spacecraft. Research is being performed on cells of four and more junctions in the quest for still higher cell efficiencies.

The multijunction cells consist of several junctions with different energy band gaps that are stacked in series, each converting a slightly different part of the solar spectrum, broadening the range of the cell's spectral response as shown in figure 6.10. This converts more of the solar spectrum, making more efficient use of the cell's area. The junctions are connected in series such that their generated voltages add, resulting in cells of lower current density than silicon or single junction GaAs, but of much higher voltage and power.

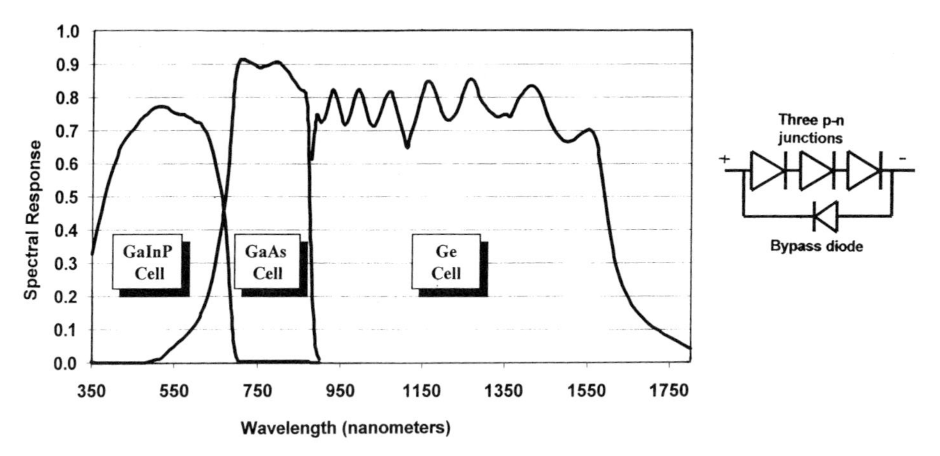

Figure 6.10 Triple junction solar cell spectral response and equivalent circuit, showing bypass diode.

As previously mentioned, silicon cells are made by doping with p- and n-type impurities in thin-cut and polished silicon wafers sliced from crystal ingots (5.5 to 7 mil [137–175 μm] thick), a relatively inexpensive process. By contrast, gallium arsenide and multijunction (MJ) cells are made by growing crystals of Ga- and As-based compounds in semiconductor reactors and adding the p- and n-type impurities, with the individual junction materials, as part of growing the crystal of each junction in a metal–organic vapor phase epitaxy (MOVPE) crystal growth process. These are much more expensive processes, with more steps in the operation. Therefore, silicon cells will always cost substantially less than GaAs and multijunction cells. All space silicon and multijunction cells are n–p junctions because they are less susceptible to damage from charged particle radiation than p–n junctions. The improved resistance of n-on-p cells to radiation can be attributed partly to the fact that electrons, the minority carriers in p-type material, have a diffusion constant about three-times greater than holes, the minority carriers in n-type material. Since radiation primarily damages the base region by reducing the diffusion length of its minority carriers, it is advantageous to use p-type for the base material due to its inherently longer diffusion length. Another contributing factor is that the types of recombination centers produced by radiation in p-type material are less effective in shortening the minority carrier diffusion length than those produced in n-type material.

Silicon solar cell thickness and weight are due primarily to the bulk material in the base region. For all gallium arsenide based cells (single, double, or triple junction), the germanium substrate material contributes nearly all of the cell thickness and weight. The actual thickness of the active p–n junction on a gallium arsenide cell is only a few micrometers. Therefore, the additional junctions in a multijunction solar cell are added with negligible increase in weight. So although they cost more to manufacture than silicon cells, their higher efficiency per unit area of gallium arsenide based cells often results in a cost advantage over silicon cells, measured in dollars per watt.

The compounds in the MJ solar cell are monolithically grown on top of the germanium wafer after activating the germanium junction. Figure 6.11 shows a simplified diagram of the various layers for a typical triple junction (MJ) solar cell. Since the three junctions

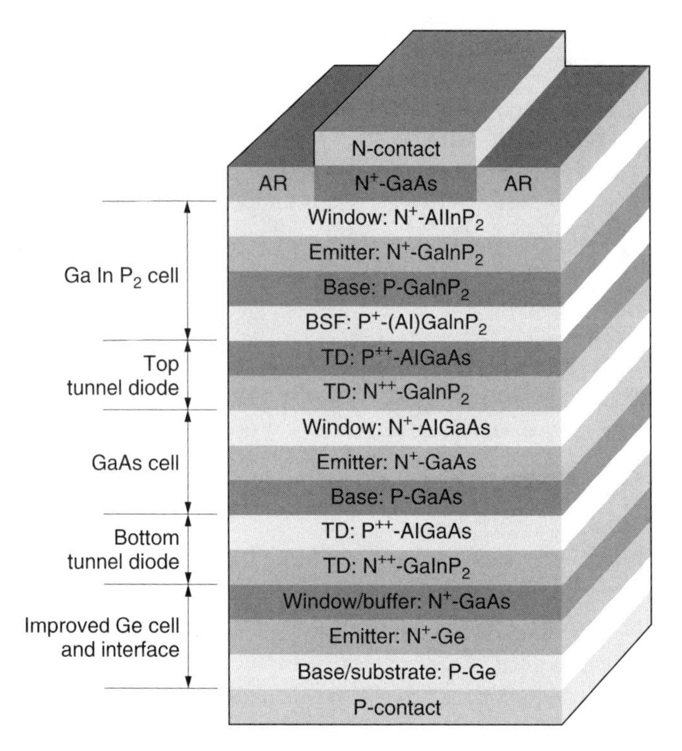

Figure 6.11 Illustration of n–p multijunction (MJ) solar cell construction.

are in series, their voltages add, but the junction with the lowest current density is the one that limits the current. In the example shown, at the beginning of life, the top junction is current limiting. After radiation, the middle (GaAs) junction degrades more than the top (GaInP) junction and limits the current. The active junctions in an MJ cell are electrically connected with tunnel junctions. Presently, the companies in the U.S. that manufacture solar cells for space applications are Spectrolab Inc. and EMCORE Corp.

6.4.3.3 Solar Cell Design

A typical solar cell consists of a rectangular wafer of high-purity semiconductor crystal with p–n junction (or junctions) grown on the top of the silicon base 5.5 to 8 mil thick or, for gallium arsenide and MJ cells, on the germanium substrate typically 5.5 to 7 mil thick. Electrical connections are applied to the front (top) layer and to the base or substrate region. Finger-like extensions of the front surface contact (called grids) are added to decrease the cell's internal resistance. These conducting grids are applied to the very thin front surface to reduce its resistance to lateral current flow, a significant contributor to the cell's internal resistance.

Germanium crystal substrates are a very expensive component that must be procured by solar cell manufacturers as thin, very pure, cylindrical wafers to produce high-quality GaAs or MJ cells. Therefore, the utilization of the germanium wafer is optimized to

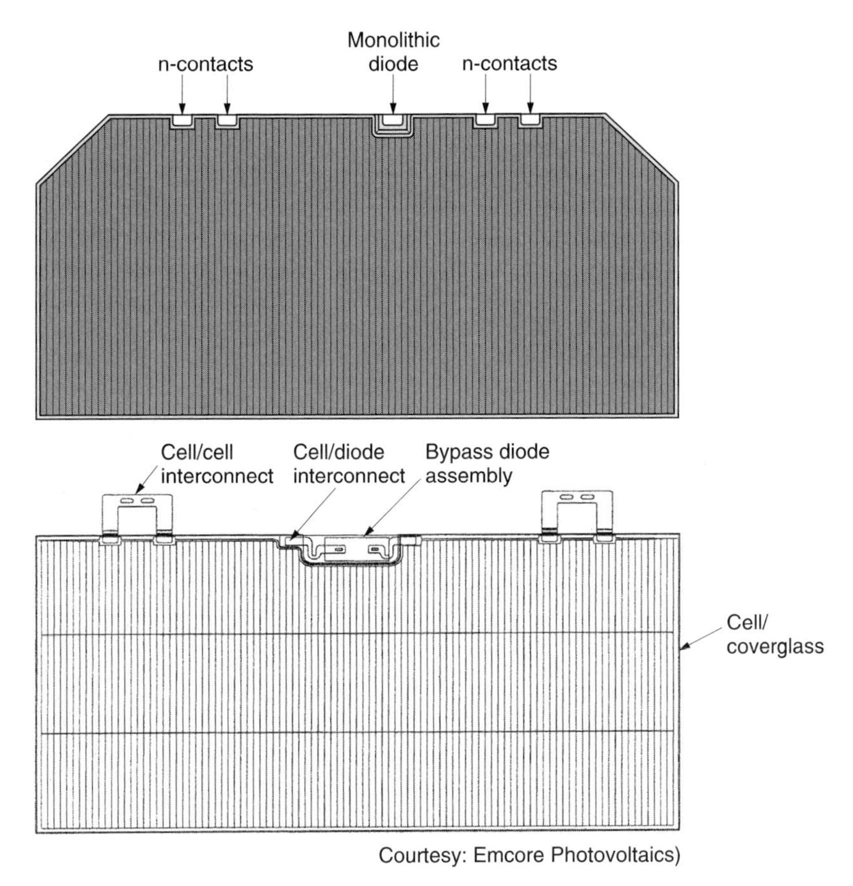

Figure 6.12 Multijunction (MJ) solar cells, showing selected bypass diode and grid configurations (front view). (Courtesy of Emcore Photovoltaics.)

reduce cost. Obtaining the largest size semi-rectangular shaped cells from a circular wafer led to the cropped corner cell shown in figure 6.12.

6.4.3.4 Solar Cell Electrical Characteristics

Table 6.2 summarizes the characteristics of selected solar cell technologies under standard measurement conditions of 28°C and AM0 sunlight. The five cell types range in efficiency from 14.6% for silicon to 27.2% for triple junction (multijunction) cells. All of these are currently available and have been used on spacecraft. However, most spacecraft before 1990 used the silicon cells, and most future spacecraft are likely to use the newer multijunction cells. Therefore, the other three cell types (high-efficiency silicon, gallium arsenide, and dual junction gallium arsenide) have experienced modest use as transition technologies. In addition to the electrical characteristics, the table also includes pertinent thermal properties of the cells, the effects of a typical radiation exposure, and information on the front surface thermal properties of these cells when covered with a typical coverglass and coating. Note that cell efficiencies are traditionally

Table 6.2 Solar cell characteristics at 28°C, AM0

Solar Cell Parameter[1]	Si (10 Ω-cm)	Hi-Eff-Si (10 Ω-cm)	GaAs/Ge Single Junction	Dual Junction	Triple Junction (Multijunction)
J_{sc} (mA/cm^2)	42.5	46.87	30	15.05	17.0
J_{mp} (mA/cm^2)	39.6	43.42	28.5	14.15	16.2
V_{oc} (V)	0.605	0.625	1.02	2.36	2.66
V_{mp} (V)	0.5	0.52	0.9	2.085	2.345
P_{mp} (mW/cm^2)	19.8	22.58	25.5	29	37.9
C_{ff} (fill factor)	0.77	0.77	0.82	0.83	0.83
E_{ff} (%)	14.6	16.7	19	21.8	28.0
Temperature coefficient @ 1×10^{15} 1-MeV[2] electrons/cm^2					
J_{sc} (μA/cm^2/°C)	22*	45	20	12	12.0
J_{mp} (μA/cm^2/°C)	22*	45	20	13	9.0
V_{mp} (mV/°C)	−2.15*	−1.97	−1.9	−5	−6.8
V_{oc} (mV/°C)	−1.96*	−1.9	−1.8	−4.8	−6.3
Radiation degradation ratio @ 1×10^{15} 1-MeV[2] electrons/cm^2					
J_{mp}	0.86	0.905	0.83	0.91	0.95
V_{mp}	0.83	0.79	0.9	0.91	0.90
P_{mp}	0.71	0.75	0.75	0.83	0.86
Thermal properties after coverglass[3] installation:					
Emittance of Microsheet**	0.85	0.85	0.85	0.85	0.85
Absorbtance (coating)**	0.79(AR)	0.78(IRR)	0.89(AR)	0.92 (AR)	0.92(AR)

(1) Solar cell parameters defined on figure 6.13.
(2) MeV = million electron volts (see section 6.4.3.5).
(3) Coverglass terms defined on figure 6.24.
*Pre-radiation.

**Ceria-doped Microsheet coverglass.

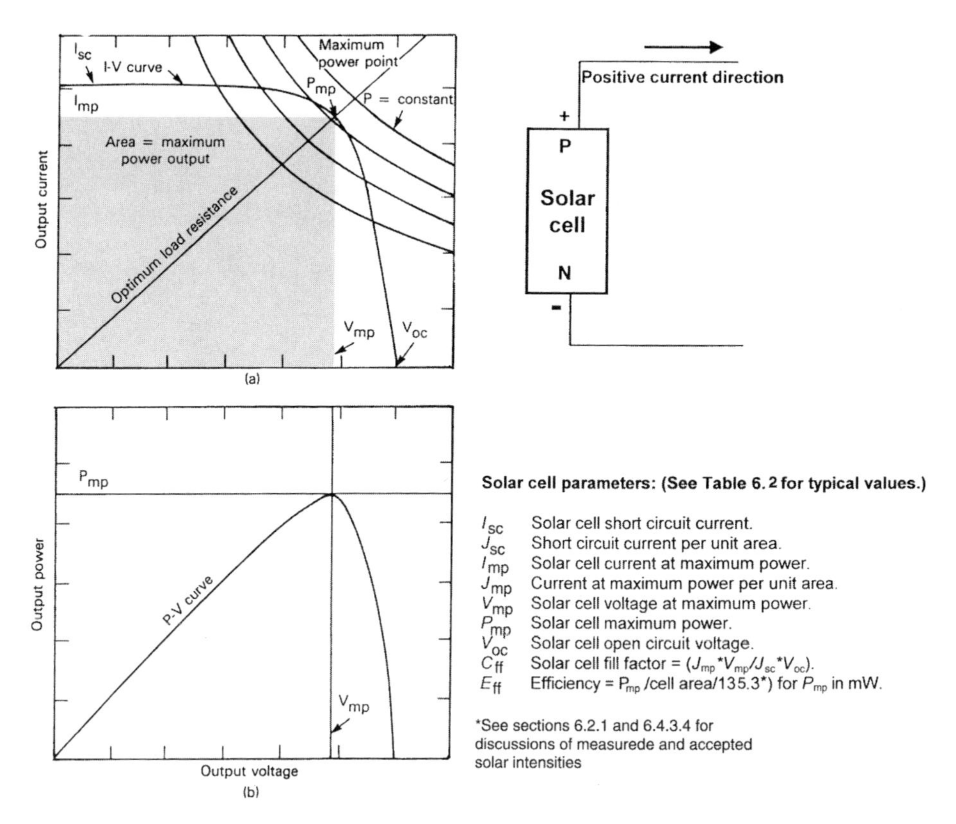

Figure 6.13 Solar cell electrical output characteristics: (a) I–V curve; (b) P–V curve.

referenced to 135.3 mW/cm2. This is done for historical reasons. The efficiency does not affect the power analysis and measurements, since the manufacturer uses reference cells tested at AM0 in calibrating and testing all cells and panels.

A typical solar cell current–voltage (I–V) characteristic and resulting power curve are shown in figure 6.13a. The defined points of interest are: the short-circuit current (I_{sc}), the open-circuit voltage (V_{oc}), and the current, voltage, and power at the maximum power point (I_{mp}, V_{mp}, and P_{mp}). Maximum power occurs at the point of maximum cell efficiency, where the largest area, also called the fill factor, can be drawn inside the cell's I–V curve (shaded area in figure 6.13b).

Figure 6.14 shows a simplified equation for a solar cell, neglecting the cell's shunt resistance, with the corresponding current–voltage characteristic and equivalent circuit. It is based on the well-established equation for an ordinary diode that represents the cell's n–p junction. Under short circuit conditions, $V = 0$ and the second term in the diode equation becomes zero. The influence of the diode current is eliminated, and the load current (I_L) is equal to the short-circuit current. The second term in the equation is the current lost through the diode, which increases with increasing voltage until, at open circuit voltage, all of the current goes through the diode and the load current is reduced to zero.

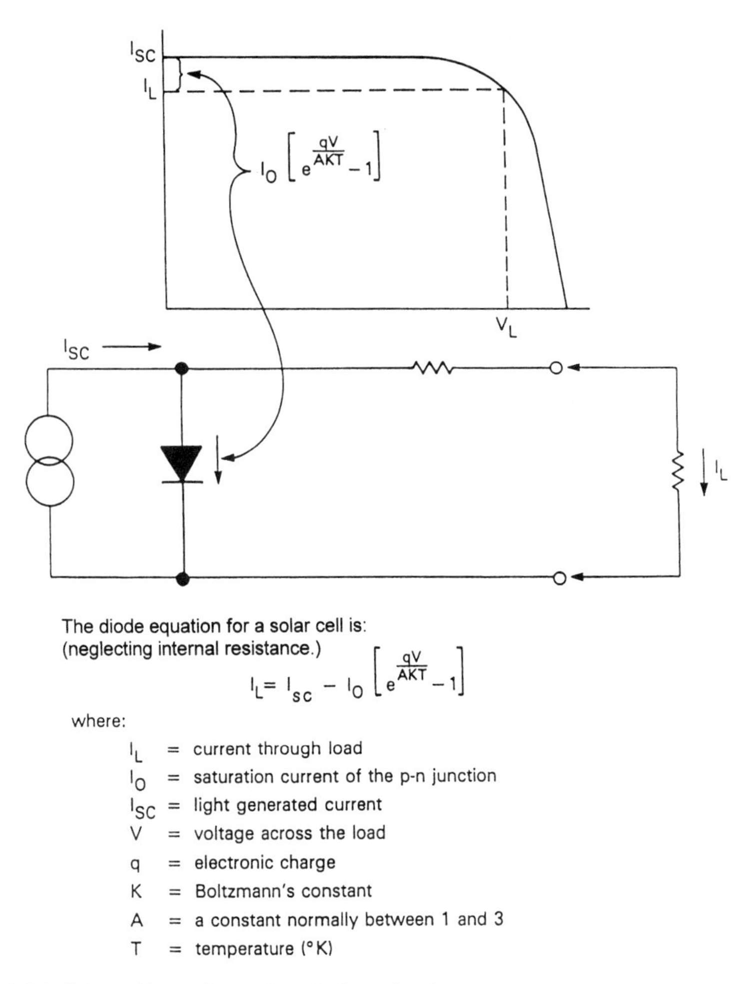

Figure 6.14 Solar cell equation and equivalent circuit.

Therefore, each p–n junction in the solar cell is basically a diode with both illuminated and dark characteristics, as shown in figure 6.15 for a typical solar cell. The power-producing part of the curve (shown here in the fourth quadrant) is the typical I–V curve that is of primary interest. The reverse characteristics resemble those of a diode with high reverse (leakage) current, and they are not well controlled during manufacture because they are not directly related to the production of power. However, these characteristics become important when the cell is driven into reverse by a solar array that is generating sufficient power to cause a reverse voltage breakdown. The most common occurrence of solar cell voltage reversal is for a cell to become shaded, which sharply reduces its current-carrying capacity and causes its voltage to reverse. The illuminated cells in its series string drive current through it. In other words, the shaded cell becomes part of the load. All p–n junctions have a breakdown voltage that causes them to conduct significantly increased current if their reverse voltage is sufficient.

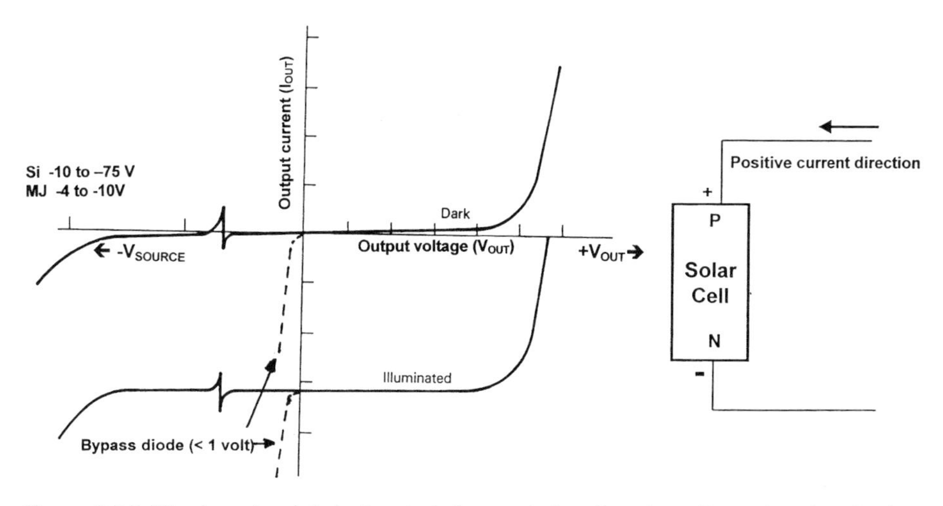

Figure 6.15 Illuminated and dark electrical characteristics of a solar cell p–n junction. Broken lines show the effect of a bypass diode on the reverse voltage.

For silicon cells, this breakdown voltage is large enough to cause local overheating (or hot spots) if the current is large enough and could even eventually damage the cell (or any cells that may be in parallel with it). For strings at modest voltages, the local power dissipation is limited to levels that do not pose a serious concern. At higher voltages for silicon cells, this potential problem can be prevented by the judicious use of bypass diodes, connected in parallel, across a cell or groups of cells in series.

However, GaAs and MJ cells are much more fragile and will be permanently damaged if they are reverse biased at voltages exceeding their breakdown voltage (which is much lower than for silicon cells). All GaAs and MJ cells require a bypass diode across each cell to limit the reverse voltage. The effect of the bypass diode on the reverse voltage breakdown is shown in figure 6.15. One manufacturer places the bypass diode under the cell in a specially etched cavity. Another manufacturer places the bypass diode in one of cropped edges of the cell. Also, multijunction cells with a monolithically grown bypass diode are currently in production. The diode is monolithically attached to the back contact of the MJ cell during cell fabrication, while the top connection is made during cell interconnect installation. The design reduces the amount of handling and work required to install and weld (or solder) the discrete diode across the cell, thus improving its reliability and reducing somewhat the panel fabrication cost with minimal reduction of the active solar cell area.

The equations below can be used to generate I–V curves from the estimated or measured solar cell parameters of V_{oc}, I_{sc}, V_{mp}, and I_{mp} as defined in figure 6.13. The equations were developed by TRW semi-empirically and have become accepted as the "TRW" solar cell I–V curve model. They have proven to be useful in power system solar array design and simulation.

"TRW" solar cell I–V curve model:

$$I = I_{sc}\lfloor 1 - C_1(e^{[V/(C2V_{oc})]} - 1)\rfloor \tag{6.4.1}$$

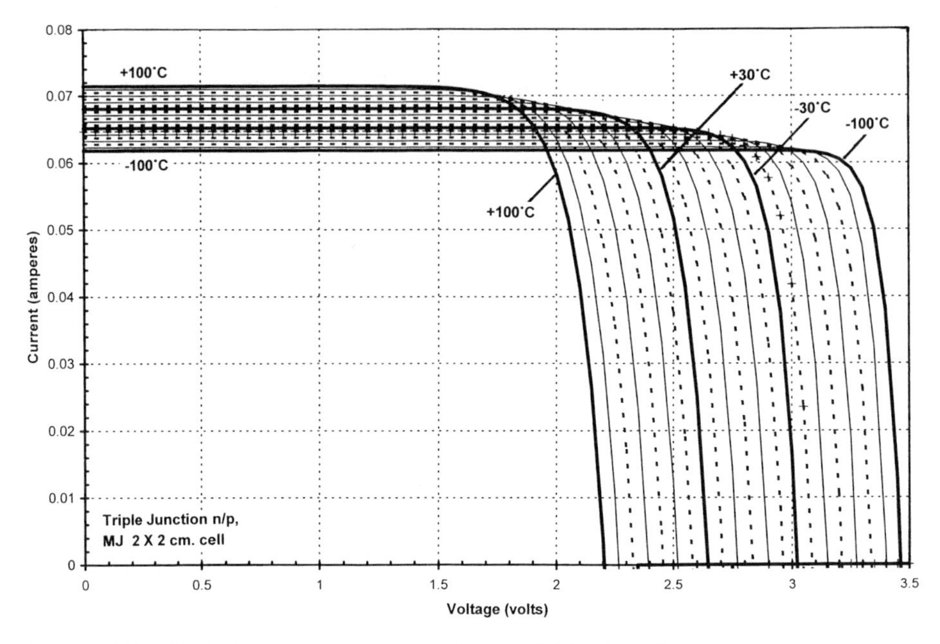

Figure 6.16 Effect of temperature on a multijunction (MJ) solar cell.

where

$$C_1 = \left(1 - \frac{I_{\mathrm{mp}}}{I_{\mathrm{sc}}}\right) e^{\frac{-V_{\mathrm{mp}}}{C_2 V_{\mathrm{oc}}}} \tag{6.4.2}$$

$$C_2 = \frac{\frac{V_{\mathrm{mp}}}{V_{\mathrm{oc}}} - 1}{\ln(1 - \frac{I_{\mathrm{mp}}}{I_{\mathrm{sc}}})} \tag{6.4.3}$$

Temperature is a very important parameter in the design of the solar cell array. Since efficiency decreases with increasing temperature, every precaution is taken to keep the operating temperature low. The current and power density of solar cells are directly proportional to normal incident sunlight intensity, which is inversely proportional to the square of the solar distance. The cell voltage varies slightly with changes in the Sun's intensity, with higher voltage at higher Sun intensity. Figures 6.16 and 6.17 show the effect of temperature and normal incident light intensity variations respectively on the voltage–current characteristics of a triple junction solar cell.

Measurements of this type are typically made using a solar simulator for MJ cells. This consists of a xenon lamp, filtered to match the solar spectral distribution in the wavelength range that these cells respond to. Silicon and GaAs cells are calibrated with a standard solar cell, which is referenced to one that is calibrated near the top of the Earth's atmosphere (AM0) via a balloon or high-altitude aircraft flight. This standard cell is used to adjust the intensity of the lamp to obtain 1 solar constant. However, MJ solar cells require calibrated individual junction versions of the MJ cell in order to calibrate the light source to the correct spectrum for each individual junction of the MJ cell.

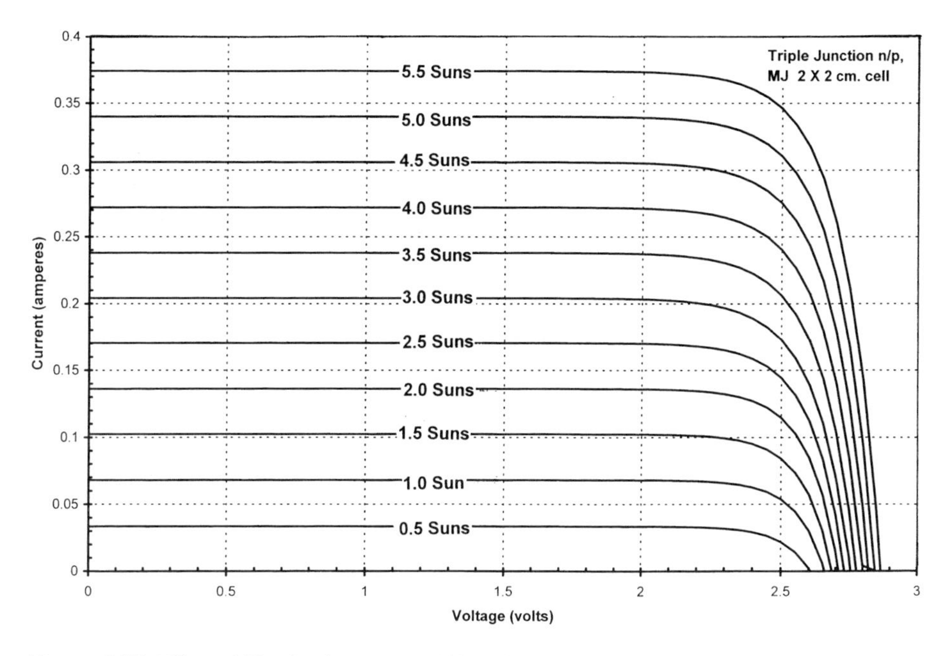

Figure 6.17 Effect of illumination on a multijunction (MJ) solar cell.

6.4.3.5 Effects of Radiation on the Solar Cell

Particle radiation, protons and electrons, cause trapping centers in the base region, decreasing the minority carrier lifetime and diffusion length. This affects the solar cell's spectral response and degrades its I–V curve. To determine the end-of-life (EOL) solar cell performance for a mission, the beginning-of-life (BOL) I–V curve must be measured and the effect of the orbital radiation environment must be calculated. (Typically, the influence of temperature is also calculated to obtain the hot case EOL I–V curve for worst-case analysis.)

Determination of the effect of radiation on a solar cell requires two sets of expensive, specialized equipment: one set to measure the solar cell output and another set to generate the damaging energetic particles. This is generally done by irradiating cells to a given set of particle fluences over a number of energies that mimic the space environment, and then performing electrical measurements after each exposure. Typical exposure times are on the order of hours. It also takes significant time to accumulate the particle dose (the fluence) expected in a typical mission. It is not practical to do this for a large number of energy levels for both electrons and protons. Due to the ready availability of 1 MeV electron accelerators, it has become common practice to use them to perform most of the solar cell radiation damage testing. Typically, these ground-based tests inject these *unidirectional* 1 MeV electrons at normal incidence into the front surface of a bare cell. Therefore, it has been the practice to empirically determine the damage coefficients that are the equivalent number of unidirectional 1 MeV electrons to produce the same solar cell damage as an omnidirectional particle (proton or electron) for each energy level

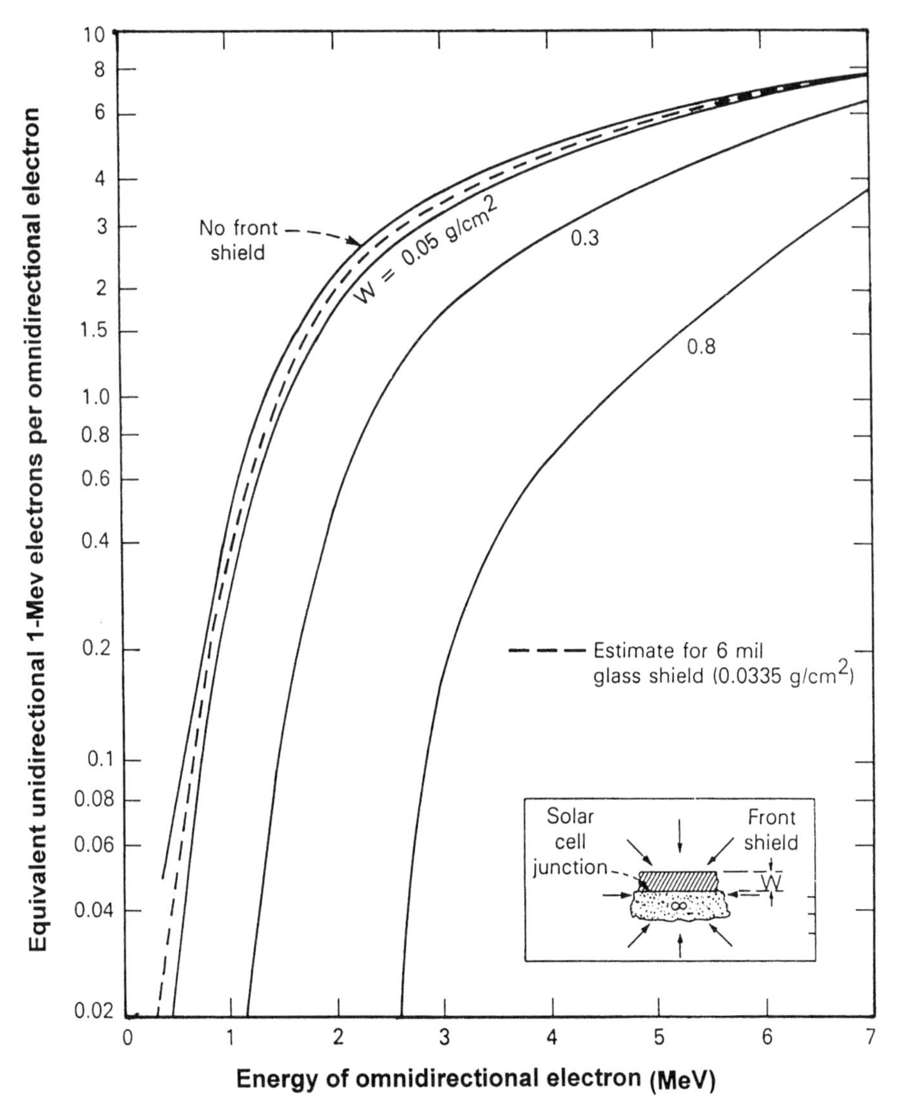

Figure 6.18 Equivalent damage for monoenergetic isotropic electrons on shielded *n–p* silicon solar cells. (Reprinted with permission; AT&T, 1963.)

and for each coverglass thickness for each type of solar cell. These factors are known as damage equivalence coefficients or relative damage coefficients (RDCs).

Measurements of this type were first made at Bell Laboratories for the Telstar satellite (Brown et al., 1963). Similar tests were continued by the Jet Propulsion Laboratory (Anspaugh et al., 1982; Anspaugh, 1996). More recently, the Aerospace Corporation has performed these characterization tests on multijunction cells (Marvin, 2000a, b). In the original Telstar tests, the omnidirectional effect was obtained by rotating the test cells in the beam of energetic particles and shielding them on the backside to obtain the effect of an infinite back shield. Figure 6.18 shows the damage coefficients for electrons on silicon cells (the number of unidirectional 1 MeV electrons that are equivalent to an omnidirectional electron) of energies up to 7 MeV for different front shielding

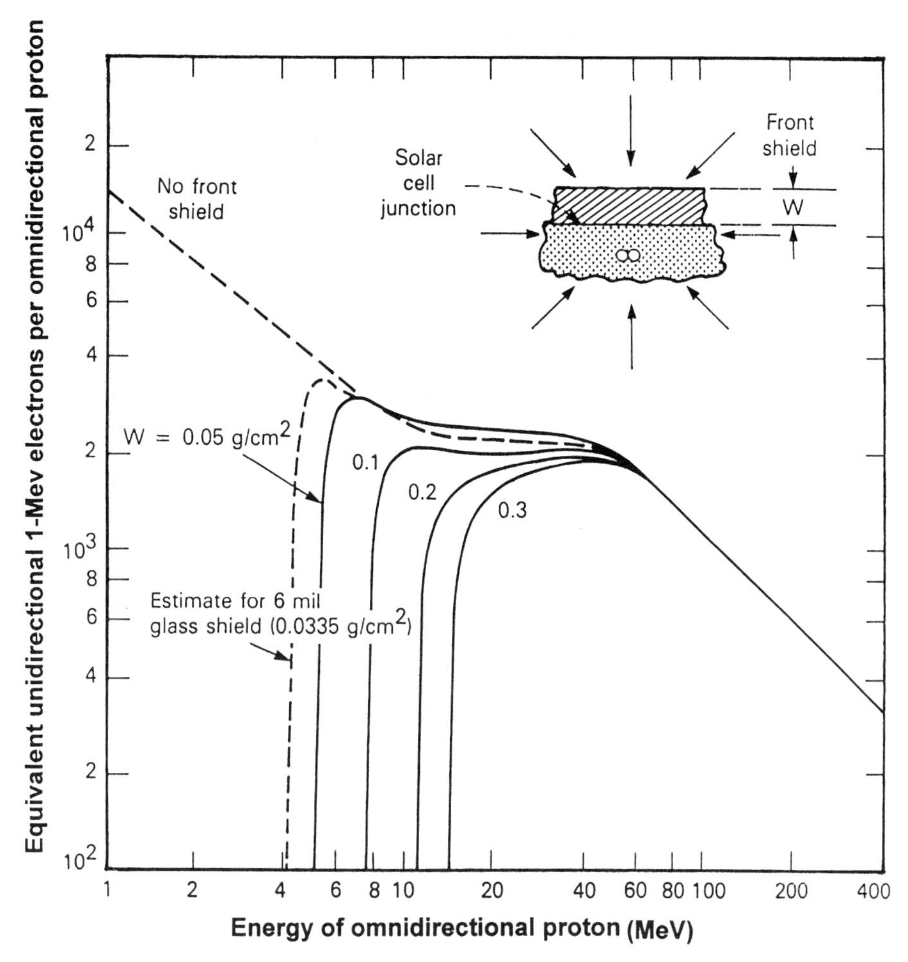

Figure 6.19 Equivalent damage for monoenergetic isotropic protons on shielded n–p silicon solar cells. (Reprinted with permission; AT&T, 1963.)

thicknesses. As one would intuitively expect, the damage increases with increasing electron energy for all shield thicknesses.

The damage coefficients that were obtained for protons on silicon cells are shown in figure 6.19 for different front shielding thicknesses. The equivalent number of unidirectional 1 MeV electrons per omnidirectional proton is shown up to energies of 400 MeV. Notice that approximately 3000 electrons are needed to damage a cell as much as one proton (protons are much more damaging than electrons) and the curves for protons are of opposite slope to those of the electrons. Low-energy protons are more damaging than high-energy protons. Note the sharp cutoff for cover densities 0.3 g/cm^2, suggesting that protons below the corresponding energy cannot penetrate the coverglass. For example, protons of less than 4.2 MeV energy cannot penetrate a 6 mil glass cover (where 1 mil = 0.001 inch). Unless otherwise specified, the coverglass material assumed in solar cell radiation damage analysis is fused silica with a density of 2.20 g/cm^3.

The following steps illustrate the method of calculating the effect of radiation on a solar cell assembly or solar array. For simplicity, the influence of radiation from the

Table 6.3 Calculation of equivalent 1-MeV electron flux resulting from the proton environment

Energy Range (MeV)					
E_k	E_{k+1}	NSSDC Proton map	Omnidirectional Protons/(cm^2 d) $E_{k+1} - E_k$	Damage Coefficient (figure 6.19)	Equivalent unidirectional 1-MeV electrons (e/(cm^2 d))
1	2		0.482E 08	0.0*	0.0*
2	3		0.413E 08	0.0*	0.0*
3	4		0.354E 08	0.0*	0.0*
4	5		0.304E 08	2.0E 03	0.608E 11
5	6		0.261E 08	3.5E 03	0.914E 11
6	7		0.224E 08	3.2E 03	0.717E 11
7	8		0.193E 08	3.2E 03	0.618E 11
8	9		0.166E 08	2.9E 03	0.481E 11
9	10	AP–4	0.143E 08	2.8E 03	0.400E 11
10	11		0.123E 08	2.7E 03	0.332E 11
11	12		0.107E 08	2.6E 03	0.278E 11
12	13		0.925E 07	2.5E 03	0.231E 11
13	14		0.800E 07	2.5E 03	0.200E 11
14	15		0.693E 07	2.5E 03	0.173E 11
15	20		0.810E 07	2.4E 03	0.194E 11
20	25	AP–2	0.649E 07	2.4E 03	0.156E 11
25	30		0.522E 07	2.4E 03	0.125E 11
30	35		0.443E 07	2.4E 03	0.106E 11
35	40	AP–1	0.350E 07	2.35E 03	0.082E 11
40	45		0.282E 07	2.2E 03	0.062E 11
45	50		0.229E 07	2.1E 03	0.048E 11
50	60		0.135E 07	1.9E 03	0.026E 11
60		AP–3	0.115E 08	6.0E 02	0.069E 11
					Total = 5.82E 11

Note: orbit altitude: 1100 km (600 n-mi) circular; orbit inclination: 90°; silicon solar cells.
Damage coefficient is for 150 μm (6-mil) cover of 0.0335 g/cm^2
*Protons < 4.2 MeV are stopped by 6 mil coverglass.

backside will be ignored (the substrate will be considered to be a perfect shield). Also, we will calculate only the 1 MeV electron equivalence of trapped protons, since their treatment is similar to that of trapped electrons and solar protons. (Note that there are other influences on the solar array that are not considered here: the second-order effect of the darkening of the coverglass bonding materials due to both ultraviolet and particle radiation, which is treated in the *JPL Radiation Handbook* (Anspaugh et al., 1982); and the mission-dependent effects of atomic oxygen and thermal cycling.)

1. Table 6.3 demonstrates the calculation of the equivalent unidirectional 1 MeV electron fluence for protons in an 1100 km circular polar orbit for silicon solar cells with 6 mil thick coverglass. The first step is to obtain the estimated omnidirectional protons per square centimeter per day for various energy ranges for this orbit from standard computer models of the environment such as NASA's AP-8 and AE-8. The first column lists the proton energy ranges and the second defines the proton environment (AP) map that was used to obtain the omnidirectional proton estimates for each range in column 3.

2. The second step is to transform the number of protons for each energy range into the equivalent number of 1 MeV electrons and add them. For this example, a coverglass thickness of 150 μm (6 mil) of microsheet was used. The damage coefficient for protons is determined by visual inspection of figure 6.19 and listed in column 4 (in practice, the data is available as a file for computer entry). Then, for each energy range, the number of omnidirectional particles (electrons or protons) is multiplied by the damage coefficient to obtain the equivalent number of unidirectional 1 MeV electrons per cm^2 per day. Since these units are now the same for each energy range, they can be added to obtain the total unidirectional 1 MeV electron fluence due to all omnidirectional protons. This same process must then be repeated for the electron environment, using the damage coefficients from figure 6.18 to obtain the total unidirectional 1 MeV electron fluence due to all omnidirectional electrons.
3. Since low-energy protons contribute most of the damage for this orbit, the final value of 5.82×10^{11} equivalent unidirectional 1 MeV electrons represents most of the radiation environment. However, a complete solution would require:
   (a) a similar calculation for the equivalent 1 MeV electrons due to trapped electrons;
   (b) a similar calculation for the equivalent 1 MeV electrons due to solar flare protons;
   (c) similar calculation for the equivalent 1 MeV electrons due to the trapped electrons and protons penetrating into the backside of the array through the substrate.

All of these contributions are then added to obtain the total number of equivalent unidirectional1 MeV electron/cm^2 d. This number is then multiplied by the appropriate mission length to obtain the total fluence. The resulting degradation is then obtained from the 1 MeV ground test data, available from the cell manufacturer and/or the JPL or Aerospace Corp. publications. (For illustration, this type of data is shown later in this chapter in figure 6.22 for MJ cells.)

The foregoing analysis of the proton environment on a silicon cell was intended to provide a simplified illustration of the process. Software tools exist for automating the solution to problems of this type. NSSDC's SOFIP (Short Orbital Flux Integration Program) integrates the omnidirectional particle environment for a defined orbit (Stassinopoulis, 1979). The output data is then entered into JPL's EQFLUX (Equivalent Flux Program) to obtain the 1 MeV electron equivalent flux for a defined coverglass (Anspaugh et al., 1982, p. D-6). Also, the European Space Agency (ESA) has assembled SPENVIS (SPace ENVironment Information System), a web-based tool available on (www.spenvis.oma.be) at no cost. It incorporates SOFIP and EQFLUX for different cell technologies. However, as will be explained in the following discussion of multijunction cell degradation, the original process used for Telstar has been modified.

Figures 6.20 and 6.21 show the *relative* damage coefficients that have been developed for multijunction cells (Marvin, 2000 a). While they are analogous to the data presented above for Telsar, there are a few differences:

- Only the unshielded, unidirectional (normal incidence) curves are based on data. The shielded (even the 0 mil cover) curves are derived from this data for the omnidirectional environment.
- The unshielded, unidirectional (normal incidence) curves are normalized so that the damage coefficient for 1 MeV electrons and 10 MeV protons is unity. To use

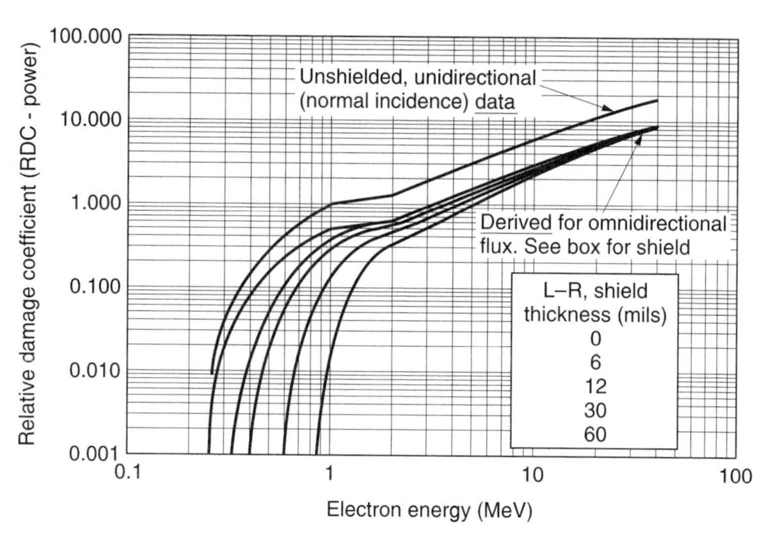

Figure 6.20 Relative damage coefficient (RDC) for power: omnidirectional electrons on shielded, Spectrolab multijunction (3J) solar cells. (Source: Marrin, 2000a, p. 44.)

this data, a cell-specific parameter (the *equivalent fluence*) must be known. The *equivalent fluence* parameter is the fluence of 1 MeV electrons needed to produce the same degradation as a single 10 MeV proton. It is dependent on the cell type and also whether the data is for the solar cell current, voltage or power. A number of these factors are listed in table 6 of the Aerospace Report, but for the cell corresponding to figures 6.20 and 6.21 this *eqivalent fluence* is 870. That is, 870 1-MeV electrons are required to produce the same degradation to the MJ cell's power as one 10 MeV proton.

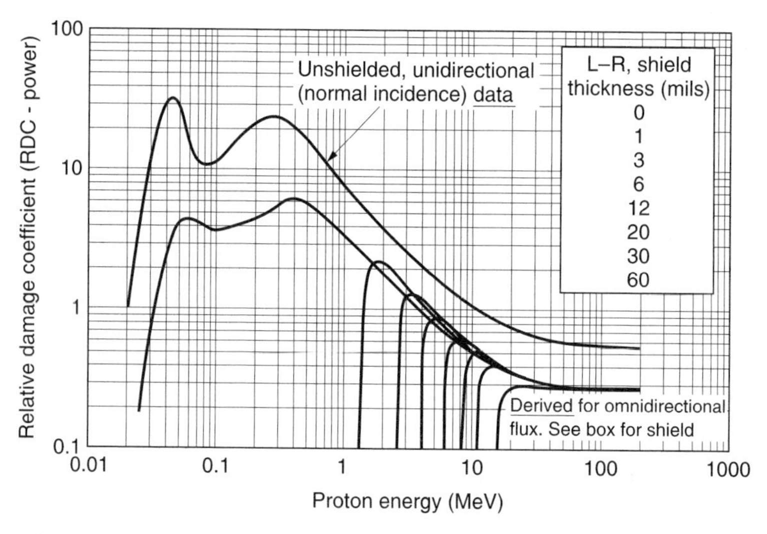

Figure 6.21 Relative damage coefficient (RDC) for power: omnidirectional protons on shielded spectrolab multijunction (3J) solar cells. (Source: Marvin, 2000a, p. 44.)

- The double maximum for the low energy protons is due, in part, to the multiple junctions, which degrade differently.
- Using this information, the 1 MeV equivalent damage to MJ cells can be calculated by the same process as shown above for silicon cells. MJ cells are more resistant to damage from radiation than either silicon or gallium arsenide single junction cells, and low-energy protons are not as damaging to any of the GaAs-based cells as they are to silicon cells.

Figure 6.22 shows the effect of 1 MeV electron radiation on the parameters of a bare (uncovered) MJ solar cell (source: EMCORE Report EWR P047). There is negligible damage to any of the parameters until after 1 E + 12 1-MeV electrons. Thereafter, the voltage parameters degrade somewhat more than the current parameters. After the equivalent number of 1 MeV electrons has been calculated as discussed above for silicon cells, the degradation of the MJ parameters can be determined from this figure.

The power available from solar arrays will vary over the mission life. Figure 6.23 shows the effect of radiation and temperature on the MJ I–V curves (the effects are similar for GaAs and silicon cells). One group of curves (solid lines) is for the beginning-of-life with no degradation. The other group (broken lines) is for the cell after it has been subjected to a fluence of 3.0 E + 15 1-MeV electrons per cm^2. (Note: in discussing ground test data, the "unidirectional" aspect of the 1 MeV electron unit is generally understood, making its restatement unnecessary.) Although the short-circuit current increases, its voltage and power both decrease with increasing temperature. However, radiation exposure degrades the entire I–V curve in both its current and voltage regions. The power system engineer is forced to design to the lower curve for the end-of-life available power, but the system must be designed so that it can accept the additional

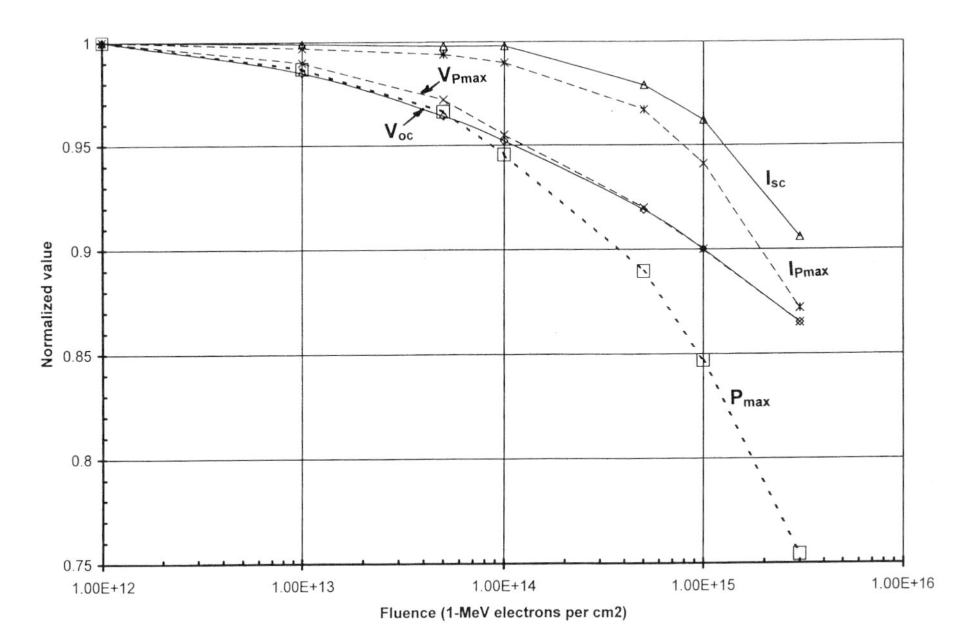

Figure 6.22 Normalized multijunction (MJ) solar cell parameters versus fluence. (Source: Emcore report EWRP047, May 2003, with permission.)

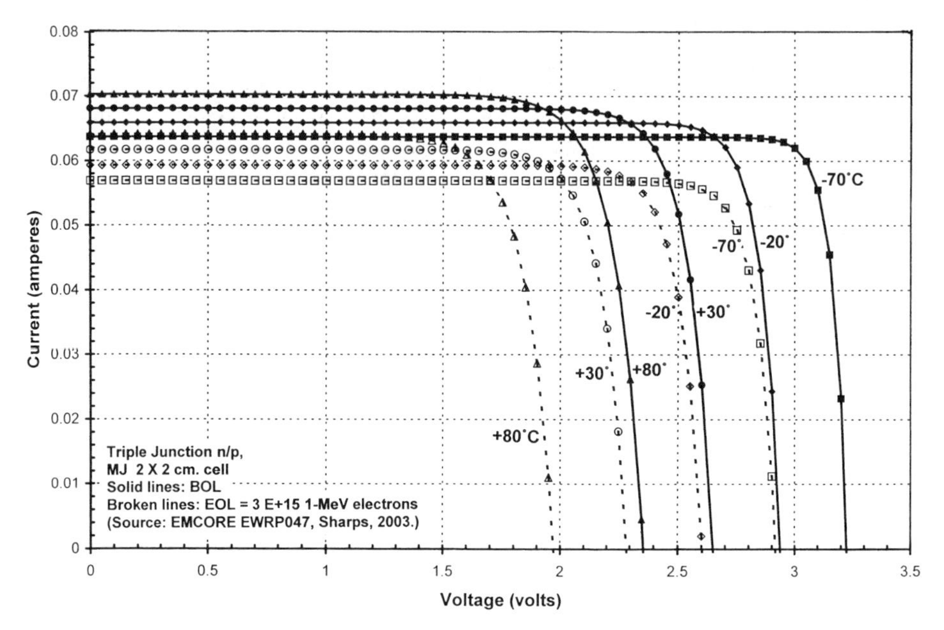

Figure 6.23 Effect of temperature and radiation on a multijunction solar cell.

power of the higher curve at the beginning-of-life. Therefore, it is important to start with an accurate prediction of accumulated damaging particle fluence for the expected lifetime of the spacecraft in its particular orbit.

6.5 Solar Cell Arrays

6.5.1 Array Construction

Figure 6.24 is a cross-section showing the typical elements in a solar cell array. In addition to the proper control of photovoltaic, optical, and mechanical properties, a significant amount of attention is given to its thermal properties. To maximize the cell efficiency, the system must be designed to absorb as much useful solar energy as possible and to run as cool as possible. Therefore, the coverglass front surface should have high emissivity and low reflectivity. To minimize its reflectivity, the front surfaces of both the coverglass and solar cell are covered with anti-reflection (AR) coatings. The ultraviolet (or multilayer blue) reflecting filter is often used on silicon panels and placed on the cell side of the coverglass (as shown). It is intended to reflect ultraviolet light that the cell cannot convert to electricity. It will reduce the absorptance of the cells and allow the panel to run cooler. For MJ cells, capable of converting light of shorter wavelength, the UV reflective coating is not used if its attenuation range reduces the usable light spectrum. To facilitate the heat rejection, the thermal conductance from the top to the bottom surface of the panel should be as high as possible, and the thermal control coating on the back of the panel is selected for maximum emissivity consistent with durability. For some missions it may also be required that the thermal control coating should have

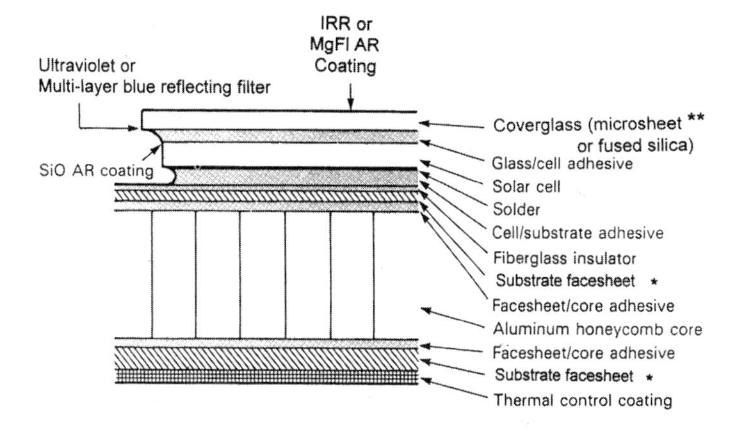

* **Substrate facesheet material** = aluminum or carbon fiber composite
** **Ceria doped microsheet coverglass:**
A borosilicate glass with 5% cerium dioxide for high infrared emittance and low solar absorptance, tailored to maximize transmitted solar energy at the peak response wavelength of each solar cell type (silicon, gallium arsenide, DJ and MJ). Ceria stabilizes the glass, preventing the formation of color centers under electron and proton irradiation.
AR Antirelection (coating) to increase absorption.
IRR Infra-red rejection multilayer coverglass coatings to reduce cell temperature.

Figure 6.24 Typical solar cell array cross-section. (After Rauschenbach, 1976.)

low outgassing properties and sufficient electrical conductivity to inhibit static charge accumulation on the panel surface. When, on occasion, this same requirement is imposed on the front surface of the coverglass, indium tin oxide is applied to its front surface (not shown) as described below in the discussion on high voltage considerations.

Because of its favorable strength-to-weight ratio and its low cost, aluminum was the material typically used for facesheets and for the aluminum honeycomb core of the solar panels for approximately the first thirty years of the space industry. However, a carbon fiber composite is an alternate material that has become more commonly used for the facesheets in recent years. Although somewhat more expensive, carbon fiber composite facesheets can be manufactured with very low thermal expansion coefficients so that the large temperature changes experienced between eclipse and sunlight cause minimal change in the dimensions of the array. This reduces the relative motion between interconnected solar cells and significantly reduces the stresses induced in the cells and on their electrical connections.

Long-term thermal cycling of panels with aluminum face sheets has been the primary cause of the solar array's ultimate failure. So, for lowEarth orbit spacecraft and other missions requiring high reliability, carbon fiber composite facesheets are preferred. For most missions the weight difference between these two choices is insignificant.

6.5.2 Series–Parallel Effects

The power from a solar array is equal to the power output from an individual solar cell multiplied by the number of solar cells in the array. The current from a single

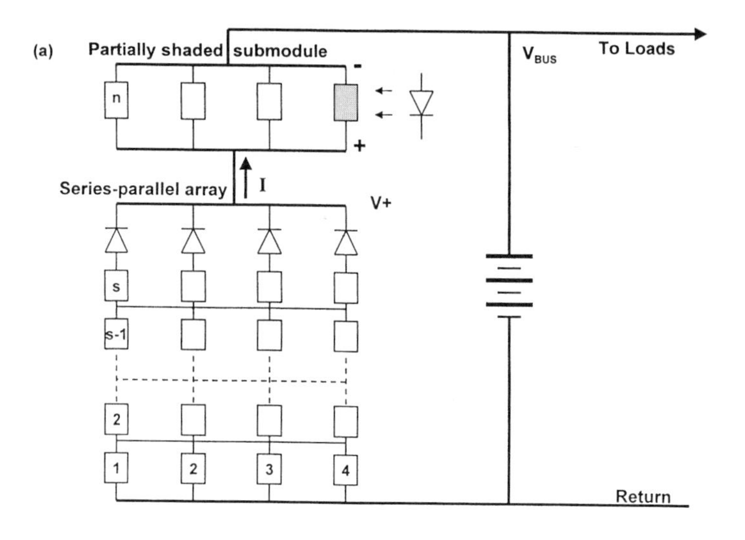

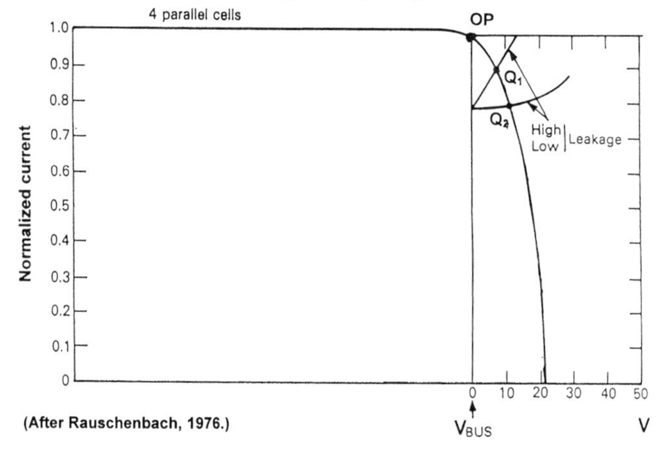

Figure 6.25 Overheating due to partial shading of a series–parallel solar array.

cell is multiplied by the number of cells connected in parallel and its voltage by the number of cells connected in series. Figure 6.25a illustrates the concept of a series–parallel array, an arrangement that was more commonly used with small silicon cells early in the space program. In the example shown, four cells are connected in parallel to form one submodule in part of a solar array (or module). Then the proper number of these submodules is connected in series to obtain the desired solar array voltage. This approach has advantages for low voltage systems using silicon cells, since the power loss is minimized when more than one cell is shaded.[2] Each illuminated cell is equivalent to a constant current source in parallel with a forward-biased diode. But a darkened

[2]The following concepts are nearly identical for shaded solar cells and those that have failed open circuit. For simplicity, we will develop the concepts related to "shaded cells" with the understanding that they are equally valid for cells that have failed in the "open circuit" mode.

cell is equivalent to a "reverse-biased" diode that opposes the flow of current. Because there are four parallel cells in this submodule, a single shaded cell will cause a loss of approximately one-fourth of the current. But a second such cell (that is not in the same submodule) has very little further effect. The same is true for any additional shaded cells as long as no two of them are in parallel (in the same submodule).

Rauschenbach (1980) gives an extensive treatment to the subject of series–parallel strings and to the resulting "hot spot" problem. As the demand for more powerful silicon solar arrays drove designers to use higher voltages, local overheating in the vicinity of an "affected" cell (one that has failed open circuit or is shaded) became a noticeable problem. Consider the series–parallel solar array shown in figure 6.25a. There are four parallel cells in a submodule and s cells in series. One cell in the partially-shaded (nth) submodule is shaded. This submodule is now effectively part of the load because the illuminated solar array is driving current through it (and reversing its voltage).

Figure 6.25b is a graphical illustration of the power dissipation in the cells of the partially shaded submodule. The voltage difference between the array with s submodules in series and with $s-1$ is neglected because s in silicon panels is usually larger than 100, causing only a small voltage shift. If there were no solar cell shading, the operating point for each solar array would be at the intersection of its $I–V$ curve and the bus voltage (point OP in this example), neglecting the voltage drop due to blocking diodes, line losses, etc. However, shading a single cell adds impedance to the flow of current through the submodule. Since this increased resistance is in series between the array and the bus, it increases the voltage of the solar array operating point. This new operating point will occur on the solar cell $I–V$ curve over a range from Q_1 to Q_2, depending on the reverse characteristics of the submodule which contains the shaded cell. The current reaching the bus is reduced from its value at OP to the value at a new operating point somewhere in the range between Q_1 and Q_2.

Since the array voltage, V_{oc}, is higher than that of the bus, the submodule containing the shaded silicon cell may be driven into reverse, dissipating more power than it was designed for, resulting in an increased local temperature or "hot spot." The reverse characteristics are shown for submodules with cells of both high (Q_1) and low (Q_2) leakage. The power dissipation (Q) in the reversed submodule is the product of its current and voltage, where its voltage is the difference between the voltage at its operating point (between Q_1 and Q_2) and V_{BUS}. For the case shown, this is between 10 and 15 V; more than ten times higher than the usual cell voltage of less than 1 volt. This is the so-called "hot spot" problem. A precise analysis is difficult because of the uncertainty in the reverse characteristics of the solar cells, which are typically not controlled by the manufacturer because they do not directly contribute to their power output. However, reasonable bounds on the power dissipation can be obtained by this graphic approach if the extreme array $I–V$ curves and submodule reverse characteristics can be determined. Note that the current through the submodule with the shaded cell is reduced by only about 20%, while the voltage drop can increase by orders of magnitude. But the significant point is that *this increased power is dissipated in the illuminated cells in parallel with the shaded cell*. Therefore, for higher voltage silicon solar arrays, the parallel interconnection of the cells in the submodule *increases their susceptibility* to overheating and damage, thus *decreasing the reliability* of the entire solar array. The increased power dissipation in the submodule (especially those illuminated cells in parallel with the shaded cell) can be high enough to cause their destructive overheating.

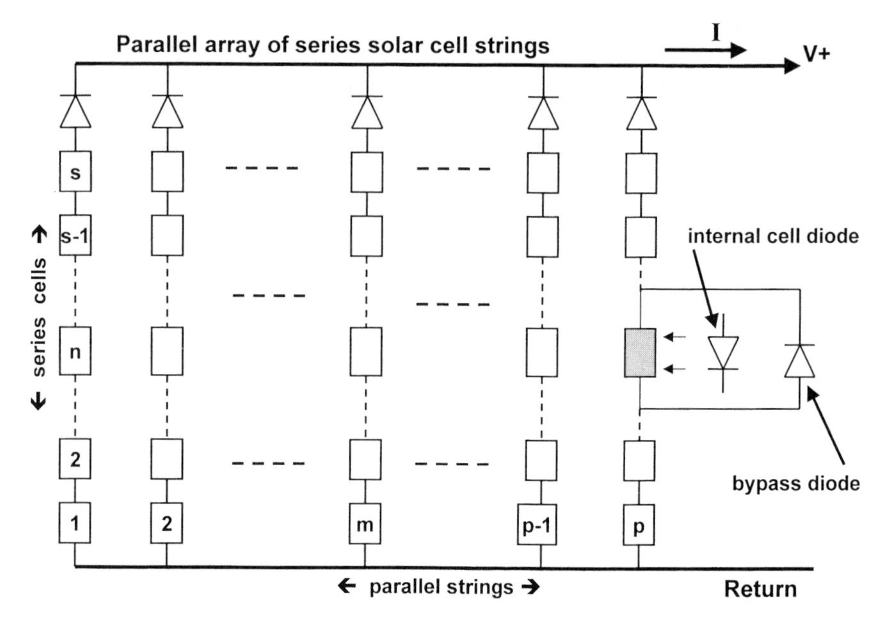

Figure 6.26 Typical solar array, illustrating bypass diode concept.

Also, the silicon cell's reverse characteristics can vary widely, so the power dissipation may not be uniformly distributed over all of the cells in the partially shaded submodule. The cell dissipating the most power becomes the most susceptible to overheating and failure. If another cell in parallel with the first affected cell fails open circuit, then the current is shifted to the remaining (now fewer) cells in the submodule, resulting in increased power dissipation in each cell. Obviously this problem is likely to cascade, causing the overheating and failure of all of the cells in the submodule with the catastrophic loss of power from the entire series–parallel array of the module.

To eliminate this concern, two changes have been made in solar array design. These are illustrated in figure 6.26. In this array, there are no parallel connections at the cell level. Instead, individual strings of series-connected solar cells are connected in parallel at the spacecraft ground and bus. With this arrangement, the illuminated cells in its string will reverse the single shaded cell. But its resulting large reverse voltage will not affect the cells in other strings, threatening them with damage. For some systems with silicon cells that can tolerate modest reverse voltages, this one change will eliminate the "hot spot" problem.

The use of bypass diodes eliminates the hot spot problem. Bypass diodes are used across groups of series-connected silicon cells, typically groups of five to nine cells. Whenever the bypass diode conducts current, the corresponding power is lost but the shadowed cells are not damaged. GaAs and multijunction cells, which are very susceptible to reverse voltage damage, require bypass diodes across each cell. Although the diode is conceptually external to the cell, the cell manufacturer must add the diode as an integral component of the cell to ensure that the cells are protected during the fabrication and test of the solar panels. For arrays of GaAs or multijunction solar cells, where each cell is individually shunted by a diode, the I–V curve, in the example above, remains close to its original current with a loss in voltage corresponding to the number of shadowed cells. (For each shadowed cell, there is the loss of that cell's voltage *plus* the additional voltage drop of its bypass diode.) So an individual bypass diode on each cell eliminates the potential cell reversal problems discussed above and improves the array in several ways: it provides protection for the cell, preventing its reverse bias at damaging voltages; and it optimizes the power contribution of strings with partially shadowed cells by providing a current path for the string. It also minimizes the reduction in string current that results from partial mechanical cracks in a cell.

In recent years, as the power of large geosynchronous spacecraft increased to over 10 kW, the bus voltage had to be raised to reduce the current and line losses. Today, most large spacecraft operate at voltages in the range of 100 V. The radiation environment contributes to electrostatic charge accumulation on the dielectric surfaces of the solar cell coverglasses. This has become a serious concern. Normally, the small charges will accumulate and then dissipate in a breakdown arc to ground without serious damage. However, once the arc is started, the high solar panel voltage may continue to feed the arc with the energy from the solar array section. Having numerous cells and strings in parallel without isolation diodes between strings will aggravate this situation since the energy of more than one string will feed the arc. This can lead to serious damage to the panel.

The approaches taken to minimize this problem include the use of single cell strings with diode isolation between the strings, as shown in figure 6.26, the use of more conductive coverglass materials, and precautions taken in designing the arrangement of the cells to ensure that the inter-cell voltage differences of adjacent strings are less than approximately 70 V. Some geosynchronous spacecraft manufacturers coat the cell coverglass with conductive material, typically indium-tin oxide, and electrically interconnect each of the coverglass conductive coatings so that they can be grounded.

6.5.3 Magnetic Considerations

The current from solar cell arrays will generate magnetic fields that can sometimes interfere with sensitive magnetometers. Even very small magnetic fields from the solar array can induce large errors in the magnetometer measurements.

An often-used technique is to run the wiring for each string directly beneath the string on the backside of the panel to minimize the magnetic size of the current loop and its magnetic field. In addition, adjacent strings are wired alternately clockwise and counterclockwise so that they generate opposite fields that tend to cancel one another. And non-magnetic materials, such as silver or silver-plated molybdenum, are often used for the interconnections and end terminations.

6.6 Energy Storage Devices

Batteries have experienced significant improvement during the past decade. The demands of countless new applications have driven research to increase the energy density and the range of available battery shapes and sizes as well as to extend their lifetimes.

This section will address the batteries and fuel cells available for space applications. Both are devices that convert the chemical energy contained in their active materials directly into electrical energy by means of an electrochemical oxidation–reduction reaction involving the ionic transfer of charge between electrodes internally and the external transfer of electrons from one material to another through an electrical circuit or load. However, in a battery the reactants are all contained within the cell, whereas in a fuel cell they are stored externally and delivered to the cell on demand.

6.6.1 The Electrochemical Cell

In this section, we will use the terms *cell, battery cell,* or *fuel cell* for the basic electrochemical unit and reserve the term *battery* for the device, usually consisting of two or more electrically connected cells, used to store electrical energy in a power system. The cell consists of four major components: the anode, cathode, electrolyte, and a fourth, very important component, not usually discussed: the separator.

1. The electrolyte gives up electrons to the *anode* during the electrochemical reaction, a process defined as *oxidation*. During discharge the anode is the electrode connected to the negative terminal and supplies electrons to the external circuit (a role that is reversed during charge).
2. The electrolyte accepts (gains) electrons from the *cathode* during the electrochemical reaction, a process defined *reduction*.[3] During discharge, the cathode is connected to the positive terminal and accepts electrons from the external circuit (a role that is reversed during charge).
3. The *electrolyte*is the ionic conductor that provides the medium for transfer of electrons, as ions, inside the cell between the anode and the cathode. The electrolyte is typically a liquid, such as water or another solvent, with dissolved salts, acids, or alkalis, whose molecules break up into positive and negative ions to carry electric charge between the electrodes. Examples are sulfuric acid and potassium hydroxide. Some batteries use solid electrolytes that are ionic conductors at the operating temperature of the cell.
4. In a typical cell, the electrodes are plates, with the anode and cathode plates alternating and arranged facing one another. The *separator* maintains equal spacing across the plates' surface and uniform spacing between adjacent pairs of plates. In "electrolyte starved" cells, such as nickel–cadmium and nickel–hydrogen, the electrolyte is limited to only the amount that will saturate the separator, which is the medium that holds a uniform concentration of the limited volume of electrolyte between the plates. Commonly used separator materials are nylon and zirconium. In response to environmental and hazardous concerns, asbestos is no longer used as a separator material.

A cell that can be only discharged by the user and cannot be recharged by reversing the discharge current flow is called a *primary* cell. It is charged by the manufacturer in a special activation process, used only once, and discarded. Since the user will never recharge it with another energy source, it will always be the *primary* source of energy for its application.

[3]The complete *reduction–oxidation* reaction in a cell is often abbreviated as a *redox* reaction and the cells are sometimes called *redox* cells.

A cell that can be both discharged and then recharged by the user is called a *secondary* cell. Simply reversing the discharge current flow with a power supply or some other energy source recharges it. In an application, this source of recharge becomes primary and the battery will be *secondary* to it. As noted above, the terms anode and cathode apply to different plates during the charge than they do for discharge—a source of confusion with secondary cells. Therefore, the terms *positive plate* (or *electrode*) and *negative plate* (or *electrode*) will be used to avoid this confusion since these names do not change meaning with a change from charge to discharge.

Using this terminology, when the cell is connected to an external load, electrons flow from the negative electrode (or plate) through the external load to the positive electrode (or plate). The electrical circuit is completed in the electrolyte by the flow of anions (negative ions) and cations (positive ions) between the positive and negative electrodes (or plates). During the recharge of a rechargeable or secondary battery, the external load is replaced by an energy source and the current flow is reversed.

The best plate materials are those that will be the lightest weight, rugged, and provide high cell voltage and capacity. Such combinations may not always be practical, however, due to reactivity with other cell components, polarization (the increased drop in voltage with current flow due to ohmic and electrode losses), difficulty in handling, high cost, and other such obstacles.

In a practical system, the plates are selected with the following properties: efficient as a reducing and/or oxidizing agent, good conductivity, stability (long life in a caustic environment), ease of fabrication, and low cost. Metals such as zinc and cadmium have these favorable properties and have often been selected as the negative plate material. Lithium, the lightest metal and the one with the highest energy density, is a very attractive negative plate material and is becoming more practical as suitable electrolytes and cell designs are developed to control its reactivity. The positive plate materials must have similar properties. Most of the positive plate materials are metallic oxides, such as silver oxide or nickel oxide. When an electrode is a gas, such as hydrogen in a nickel–hydrogen cell, Teflonated platinum-black serves as a catalytic gas electrode.

The electrolyte must have good ionic conductivity but not be electrically conductive since this would cause internal short-circuiting. Other important characteristics are that it should have minimum chemical reaction with the electrode and separator materials, little change in properties over the operating temperature range, safety in handling, long life, and low cost. Most electrolytes are aqueous solutions such as a 30% concentration of potassium hydroxide. However, there are important exceptions: for example, in lithium cells, non-aqueous electrolytes are typically used to avoid the reaction of the lithium metal with the electrolyte because of lithium's tendency to react with water.

The separator does not enter the electrochemical reaction, but its properties are very important for the proper operation of the cell. It must have the porosity to absorb and retain a uniform distribution of the electrolyte, a stable and constant thickness to maintain a uniform distance between adjacent electrodes, and must provide the insulation to electrically isolate the plates from one another and from the cell case. It must also maintain its permeability to the electrolyte to provide an even distribution of ionic motion. Obviously, any separator material, saturated with a caustic electrolyte and subjected to abrasion by the relative motion of the plates, will wear and eventually deteriorate. This is especially true at elevated temperatures, where the separator material is more susceptible to abrasion by plate motion and is less resistant to the penetration of metallic fibers that eventually grow from the plates. But even when the cell is controlled to its optimum operating temperature, separator wear will eventually degrade the cell's performance. The best that one can hope to achieve is to maximize performance and lifetime by taking a great deal of care in the selection, screening, and processing of the separator. The point is that even though the separator does not enter the electrochemical reaction, its properties and their quality control are very important for the consistent performance of the cell.

Underscoring this point is an incident that occurred in the spacecraft nickel–cadmium industry. In the late 1970s, to comply with local environmental standards, the Pellon Division of Freudenberg changed the nylon material that it supplied to General Electric in Gainesville, Florida, for use as the separator in the manufacture of spacecraft NiCd batteries. This seemingly simple change in material and the necessary cell redesign resulted in spacecraft batteries which exhibited inconsistent performance that was not evident until the completed spacecraft batteries were in or near their final flight qualification tests in the late 1980s. This experience caused the battery users to seek alternatives. The most obvious alternatives available at the time were the further development of nickel–hydrogen cells, with added emphasis on smaller sizes, and the development of the so-called super nickel–cadmium (SNiCd) cells from Hughes Aerospace Co. (HAC), which were similar in size and shape to the conventional nickel–cadmium cells.

A battery cell can be built in many sizes and shapes, with cylindrical and prismatic the most common configurations used in space flight applications. (Due to their high internal pressures, the nickel–hydrogen cells have different configurations that are discussed later.) In a cylindrical cell the plates are spirally wound (Swiss roll fashion). While this shape is better suited to containing high internal cell pressures, the spiral nesting of the plates induces an uneven pressure on the separator, resulting in uneven wear. The rectangular shape of prismatic cells results in more uniform abrasion of the separator, but the relatively large flat surface area of the external cell wall must usually be constrained by external "end-plates" to prevent buckling with increased internal pressure. The cell's internal components are designed to accommodate the cell shape. The cells are sealed in a variety of ways to prevent leakage and dry-out of the electrolyte. Some cells are provided with venting devices or other means to allow accumulated gases to escape, and some are fused to limit severe overcurrent. Suitable containers and means for terminal connection are added to complete the cell. Cell cases may be plastic, but are usually metal (stainless steel or inconel), for strength and to facilitate the sealing processes required for most aerospace cells.

Self-discharge. Self-discharge of battery cells is the electrical capacity lost duc to undesirable electrochemical processes within the cell. It is equivalent to the application of a small external electrical load: a continual drain, whether or not the cell is in use. For rechargeable batteries in orbit, this is not a significant consideration because the cyclical discharge and recharge rate are many orders of magnitude larger than the self-discharge, and dwarf its effect. However, it will decrease the available capacity of any charged battery in storage if not recharged periodically. And it must be considered as an additional load on non-rechargeable batteries, especially those involving extended storage and/or mission times. Self-discharge can usually be minimized by storing the cells at a lower temperature, where electrochemical activity is reduced, and then warming them before use.

Nickel–cadmium and nickel–hydrogen cells suffer relatively high self-discharge rates, presenting the user with a major logistical problem since it necessitates charging batteries of this type at the launch site, just before liftoff. Lithium-based cells typically have low self-discharge rates, although, as with all batteries, their self-discharge rate increases with temperature. While in storage, their rate of loss declines after a time due to the build-up of the passivation film on the surface of the lithium anode surface at open circuit. Lithium primary batteries have particularly good storage characteristics, due to this phenomenon.

Lifetime. There are three different battery cell lifetimes: *dry life*, *wet life,* and *cycle life*. The usual sequence in the fabrication of most battery cells is first to install all of the components and seal the container except for the "fill-tube," which is temporarily capped with a removable cover. At this point, before the electrolyte is added, the cell is said to be in the "dry" state. For extended storage periods, it is preferable to store cells in this state, filled with dry nitrogen. In this environment, the cells should have a *dry lifetime* that is much longer than after their exposure to the electrolyte.

The process of adding the electrolyte and sealing the fill-tube is called *activation.* This is the start of the cell's *wet life*. For an aerospace cell, the remaining wet life is a significant parameter in estimating its potential usefulness. Therefore, the usual practice is to stamp the activation date and serial number on each cell.

Once secondary batteries are put to use, there are a limited number of charge–discharge cycles that they can withstand before wearing out. This expected *cycle life* is typically estimated on the basis of ground testing of cells of the same design and often from the same lot. Most tests indicate that deeper depths of discharge will shorten a cell's cycle lifetime. Cycle life is usually the limiting lifetime for secondary cells.

6.6.2 Spacecraft Batteries

Battery cells are divided into two groups, *primary* and *secondary,* for the purpose of this discussion. *Primary* batteries are used in aerospace applications to activate pyrotechnic devices and for other single-use purposes. They may also be used as the main power source for short-term missions, especially for sub-orbital missions where a solar array would be an unnecessary expense, or for very low-power experiments where a solar array is impractical. The properties that one usually looks for in primary batteries are good shelf life, high energy density, non-hazardous attributes, and a wide range of operating temperature. It is also desirable that the probability of venting gases be low during the operating life of the spacecraft since these gases can sometimes contaminate optically sensitive surfaces such as sensor lenses and solar arrays. *Secondary* cells are capable of being recharged. They are most often used as part of the main power system to supply the load during eclipse, or whenever the load exceeds the solar array capability. Although one looks for the same properties in a secondary cell as those listed above for a primary cell, usually the most important characteristic is that it should have a long cycle life.

6.6.2.1 Primary Batteries

Fortunately, there is a wide variety of choices available to satisfy the diverse applications that exist for primary batteries. But selection of the optimum primary battery for a particular application can only be achieved by a careful review and specification of all the requirements. Some of these requirements are:

- Total energy required to provide the load–time profile, including ground testing and estimated overdesign, which is typically higher than for secondary batteries because primary battery capacity cannot be measured directly. It can only be estimated, on the basis of measurement of other batteries of the same design.
- An operating range within the expected ground and orbital temperature limits.
- An acceptable voltage range.
- A tolerable risk of hazard and gas evolution.
- Ability to satisfy the maximum discharge rate.
- An acceptable shelf life (tolerable self-discharge rate) and the required operating life.
- Ability to meet all regulations for hazardous equipment for the mission and for transportation to and storage at the launch site.

Table 6.4 lists the basic characteristics of selected primary batteries that might be considered for use on spacecraft.

Silver–zinc batteries and cells have been widely used in a variety of shapes and sizes for aerospace applications since the inception of the space industry in the early 1960s. They have been used extensively for sub-orbital flights. Silver–zinc cells are second only to lithium in energy density, but can be discharged at a much higher rate than any of the lithium systems. Their main advantages are:

- High gravimetric energy density (energy per unit weight) and volumetric energy density (energy per unit volume), second only to the lithium systems.
- High discharge rate capability. They have the lowest source impedance of any system, providing them with the highest discharge rate ability which is limited mainly by their ability to dissipate the heat.
- Cells discharge at a respectable 1.5 V with a flat discharge voltage characteristic.
- Fast response time.

Their main weaknesses are their short lifetime and relatively high cost. Although they can be stored for over 5 years in the dry condition, after electrolyte is added (activation), their wet life is measured in months. Recommended pre-launch storage time for spacecraft silver–zinc batteries varies from 30 to 90 days, depending on battery type and anticipated mission length. For this reason, they are typically stored dry and activated as late as possible before use. Even so, uncertainty in the launch schedule may impose a need for additional spare batteries or additional means for remote battery activation after launch, thus increasing the complexity and cost of a relatively expensive spacecraft component.

Although they can be designed as either a primary or secondary cell, their maximum cycle life is so low (fewer than 200 cycles) that their main use has been as a primary battery. Their need to vent gas products to the environment during discharge may be a disadvantage for some spacecraft where re-deposited gases could contaminate nearby equipment. For some applications, the narrow operating temperature range from 0 to 40°C is restrictive. And, as with all primary batteries, those that have not been used must be disposed of in a responsible manner.

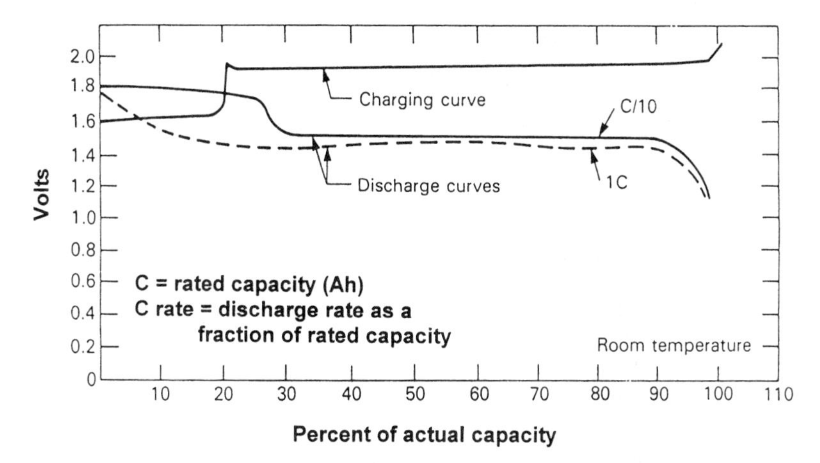

Figure 6.27 Typical change and discharge of Yardney low-rate, silver–zinc cells. (Courtesy of Yardney Technical Products, Inc.)

All silver–zinc cells can be discharged at a high rate relative to those of other chemistries. But some have been designed with lower discharge rate capability than others, either because weight restrictions have caused them to be designed with reduced heat conduction paths or because a longer lifetime requirement has dictated increased separator thickness, resulting in slightly higher source impedance. Figure 6.27 shows the charge and discharge characteristics for low-rate cells from Yardney Technical Products. Notice that, although the symbol C is used to denote the so-called *name-plate* or *rated capacity* (the rated capacity determined by the cell manufacturer), the abscissa refers to *actual capacity*. Therefore, the cells are completely discharged at 100%.

Silver–zinc cells have unique charge/discharge characteristics. Two plateaus are evident in figure 6.27 when the cell is discharged at a low rate of $C/10$; but the higher plateau nearly disappears at the increased discharge rate of $1C$. If this higher initial voltage plateau is undesirable, it can be suppressed by pre-discharging. Also, the charge voltage rises sharply twice, once early in the charge cycle and the second time at full charge. This second voltage rise is so consistent that it is used as a convenient indicator of the full charge condition. Note that the difference in plateau voltage between the discharge rates of $C/10$ and $1C$ is very small, because of the very low source impedance of these cells. Silver–zinc batteries are ideal sources of energy for very high pulsed loads.

Figure 6.28 shows the discharge characteristics for silver–zinc cells designed to discharge at higher rates. However, the higher rates are at the expense of significantly reduced capacity (but only slightly smaller voltage). By contrast, cells with low discharge rates will have significantly higher capacities. Silver–zinc cells can be designed to accept pulsed loads in excess of $50C$, the primary limitation being the heat dissipation capability of the combined battery and its environment.

Notice that the abscissa of this chart is labeled *rated capacity*, so that 100% is equal to $1C$. The chart illustrates the conceptual difference between *actual* and *rated* capacities. Since the supplier wishes to guarantee the *rated* capacity under all foreseeable condi tions, the *actual* capacity will usually exceed the rated capacity for most (but not all)

Table 6.4 Characteristics of selected primary battery cells for spacecraft use

	Liquid Cathode					Solid Cathode
				Lithium–oxyhalides		
	Silver–zinc	Lithium–sulfur dioxide	Lithium–thionyl chloride	Lithium–bromine complex (BCX)[1]	Lithium–sulfuryl chloride (CSC)[2]	Lithium–polycarbon monofluoride[3]
Applications	Hi pwr (pulse)	Hi pwr (pulse) Hi init. pressure	Hi pwr, low rate, battery backup	Low temperature continuous uses	Intermittent, pulse & high rate	Implantable uses ⟨& Defense⟩[6]
Chemical designation	AgZn	Li/SO_2	$Li/SOCl_2$	$Li/SOCl_2$, BrCl	Li/SO_2Cl_2	Li/CF_x or $Li(CF)_n$
Configuration	prismatic[4]	cylindrical	cylindrical	cylindrical	cylindrical	cylindrical
Capacity range (A h)	0.3 to 675	0.45 to 34	0.06 to ⟨24⟩[6]	3/4 to 40	3/4 to 30	⟨18 – 1200⟩[6]
Gravimetric energy density (W h/kg)	90 to 230	170 to 280	290 to 456 ⟨550⟩[6]	276 to 472	284 to 479	360 ⟨462 – 820⟩[6]
Volumetric energy density (W h/dm^3)	150 to 700	350 to 510	670 ⟨1100⟩[6]	770	720	680 ⟨760 – 1184⟩[6]
Operating temperature range (°C)	0 to 40	−60/−40 to 55/70	−60/−40 to 85[5]	−55/−40 to 70/85	−32 to 93	−40 to 85
Storage temperature range (°C)	0 to 30	0 to 50	−40 to 30	−40 to +25	−32 to +30	−40 to +50
Dry storage life	5 yr	>10 yr	>10 yr	>10 yr	>10 yr	>10 yr
Wet storage life	30 to 90 d	>10 yr	~10 yr	long life	long life	>10 yr
Open circuit voltage (V/cell)	1.6 to 1.86	3.0	3.65 to 3.67	3.9	3.9	3.0

Discharge voltage (V/cell)	1.5	2.3 to 2.9	3.0 to 3.6	3.0 to 3.7	3.2 to 3.7	2.5 to 2.8
Room temperature self-discharge (% 1st yr/ % per yr after 1st yr)	N.A.[8]/0.1[7]	N.A.[8]/ < 3.0	5.5/3.0	5.7/2.0	4.5/2.5	N.A.[8]/ < 1.0
Source impedance	very low	high	high	high	high	moderate
Discharge profile	flat	flat	flat	initially sloping	flat	flat
Manufacturer(s) (See key below)	EPT, SAFT, YTP	WGT, SAFT EPT	WGT, SAFT ⟨EPT⟩[6]	WGT	WGT	WGT ⟨EPT⟩[6]

(1) WGT introduced 1979, contains additive for ‘low temperature operation.’
(2) WGT introduced 1979, contains additive for ‘no voltage delay.’
(3) WGT’s cells used in drug infusion pumps, neurostimulators, and pacemakers. EPT’s typically larger cells used in aerospace (primarily missiles), and military.
(4) Some (usually smaller) sizes also available as cylindrical.
(5) Up to 200°C for special high-temperature, low-rate applications.
(6) The difference in energy densities between WGT’s medical and EPT’s military lithium cells is primarily due to the differences in their size and application.
(7) Safety considerations require that silver zinc batteries have a limited ‘maximum wet stand’ life of typically 1 to 12 months after activation.
(8) N.A. = not available

Key: Eagle Picher Technologies (EPT)
Electrochem (Division of Wilson Greatbatch) (WGT)
SAFT (Division of Alcatel)
Yardney Technical Products (YTP)

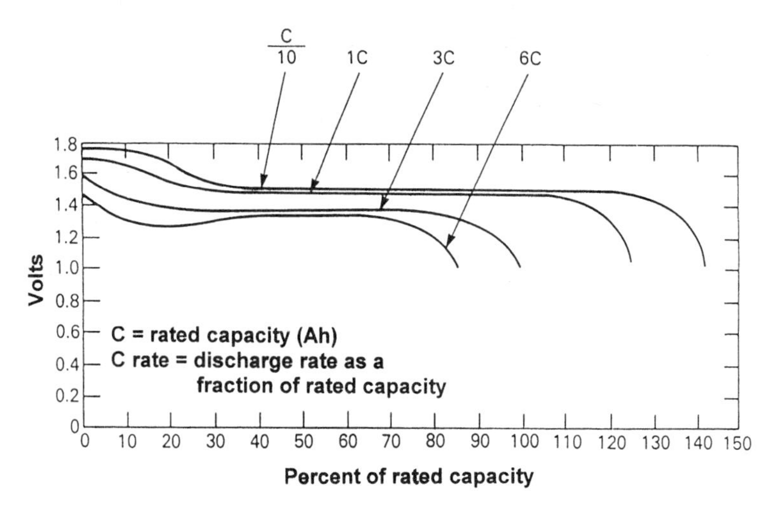

Figure 6.28 Typical discharge curves for silver-zinc cells at various rates. (Courtesy of Yardney Technical Products, Inc.)

applications and will typically be much larger for lower discharge rates. This "overdesign" is common practice in the battery industry and is typically larger for the older battery chemistries.

Lithium ion rechargeable batteries are replacing silver–zinc batteries in some applications, such as in launch vehicles and missiles, because of their long shelf life, low cost, and low self-discharge with the capability to be recharged to full state of charge before use.

Lithium cells. There is a large number of different lithium-based chemistries. Lithium, the lightest of all metals, has the greatest electrochemical potential and the largest energy content. For this reason, lithium primary cells have found extensive application in consumer, industrial, medical, military, and aerospace products from the early 1970s. However, safety concerns have limited their use, especially in large sizes for consumer applications. While there were a number of reasons for this, the one common problem was that they all used lithium metal as the negative electrode and depended on the movement of this metal to transport the ions between electrodes during discharge and recharge. Since lithium metal is violently reactive with water and even the nitrogen in the air, non-aqueous electrolytes and sealed containers were required. A wide range of potential positive electrode materials was explored, each in an effort to reduce the risk under the conditions imposed by a particular application. But, given the interest in lithium's potentially higher energy density, it is not too surprising that a multiplicity of lithium cell types has emerged, each with unique characteristics suitable for a particular application. The characteristics of a selected group of these are listed in table 6.4. In general, the development of lithium cells is a relatively new technology that offers the following advantages:

- The highest energy density
- Longest shelf life (lowest self-discharge rate)
- Widest range of operating temperature

- Highest discharge voltage, ranging from 2.8 to 3.6 V, depending primarily on the chemistry

Its disadvantages are:

- The cells and batteries can be hazardous, especially under abuse conditions. They cannot be discharged at rates higher than their design limits without risk of producing excessive gas products that could lead to venting. To minimize this risk, some applications may require the cells to include special safety devices such as fuses, vents, etc. Other safety devices, such as resettable circuit breakers or bypass circuits, are often installed in the battery or in the system by the user.
- Certain applications may require external circuitry to control the discharge. One concern is cell reversal. When lithium cells are completely discharged, their voltage drops rapidly and, since battery cells cannot be manufactured to have identical performance characteristics, one cell will be drained first. The still-active cells, in series with the depleted cell, will then drive it in the reverse direction. This is a hazardous condition for a lithium cell, especially a lithium thionyl chloride cell.
- The cell's high source impedance limits it to relatively low drain rates for most lithium chemistries.
- Delayed voltage response at "turn-on".

Since lithium cells are sealed units, they do not normally interact with or vent gas to their surroundings. So, with proper treatment in normal usage, they are safe and self-contained. But the battery electrodes are both very reactive and, since the cells are hermetically sealed, rapid gas evolution could result in destructive cell venting with the release of a caustic and reactive electrolyte to the surroundings. Fuses and vents are therefore an important part of the design, especially for the higher power lithium cells, and the development of inexpensive, reliable vents would greatly enhance the safety and acceptability of these cells. Given the hazardous nature of this chemistry, there is good reason to develop appropriate handling procedures and improve user understanding of the procedures and the technology to avoid abuse of these cells.

In table 6.4 four lithium cells are listed as having liquid cathodes and only one (the lithium polycarbon monofluoride) has a solid cathode. Cells with a solid cathode are inherently less hazardous, even when subjected to abuse conditions, but they are also lower-rate devices. The discharge of a solid cathode involves diffusion of lithium ions into the bulk of the cathode, which is a relatively slow process. By contrast, with a liquid cathode the discharge occurs at the surface of the carbon (which substitutes as the charge collector for the liquid). This is a much faster process and enables the liquid cathode cells to deliver higher rates of current than those with a solid cathode. For commercial applications, the lower-rate cells sometimes use solid electrolytes, especially in the smaller sizes. But most spacecraft applications require larger sizes and demand higher rates, obtainable only by using liquid electrolytes which have lower resistivity.

After activation, and when not in use, the metallic surface of the lithium negative electrode becomes passivated by reacting with the electrolyte. This is a salt crystal buildup on the surface, only a few micrometers thick, which is an insoluble electrical insulator capable of conducting lithium ions. It has the positive effect of reducing self-discharge to only about 10% capacity loss over five years and contributes to the long shelf life of lithium cells, which are typically rated as having a lifetime of over ten years. However, the downside is that the passivation layer also contributes to an unwanted voltage delay on startup, as shown in figure 6.29. The film increases the cell's internal

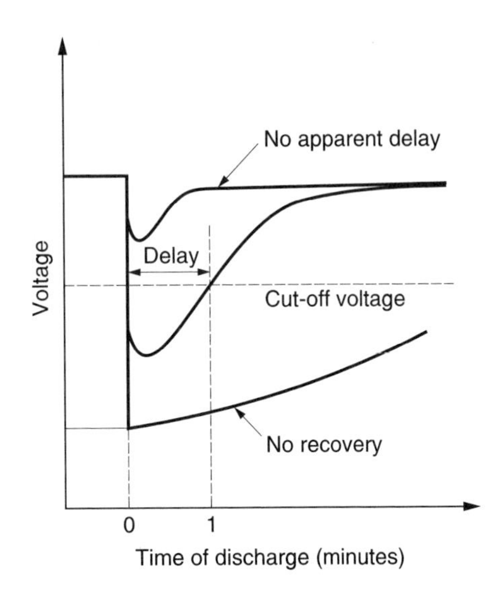

Figure 6.29 Voltage delay of lithium cells at "power on."

resistance until it has been broken up and removed by the discharge reaction. This phenomenon, present in all battery cells, is much more noticeable in lithium cells. It is typically longer for liquid cathode systems, especially lithium thionyl chloride, because the lithium is in direct contact with the cathode material. It is worse at low temperatures following prolonged storage at high temperature. At room temperature the delay may be less than 1 second, but at −20°C it can be as long as 1 minute.

Lithium–sulfur dioxide (Li/SO_2) cells are particularly good for low temperature and high current applications. Their construction is typically cylindrical and their discharge voltage is about 2.8 V, with a flat discharge curve. Shelf life is good, even at high temperatures, and expected lifetime is about 10 years at or below room temperature. But the cells must be designed to contain relatively high internal pressures up to 60 psi (0.4 MPa) and must be vented to safely accommodate excessive pressures.

Most *lithium–thionyl chloride* cells are spiral wound, cylindrical devices. The carbon electrode substitutes as the collector for the thionyl chloride ($SOCl_2$) positive electrode (a corrosive liquid that reacts with lithium to produce lithium chloride, sulfur, and sulfur dioxide). Although the cell is considered to be relatively low pressure, at high depths of discharge the sulfur dioxide is not entirely absorbed in the electrolyte and can cause the cell internal pressure to increase. (An increased cell pressure will also result if a reverse voltage is applied to it.) If this increased pressure were to rupture the cell, its surroundings would be exposed to a very caustic electrolyte. Therefore, these cells are used mainly at low rates unless modified with an additive (such as BrCl or Cl_2) to decrease their propensity for vigorous reaction.

Figure 6.30 compares the discharge characteristics of two types of lithium battery to those of two commercial D-size cells. Lithium cells clearly offer significant improvements in both voltage and available energy over commercial alkaline and carbon–zinc cells. Lithium–thionyl chloride cells, manufactured by Wilson Greatbach Technologies' Electrochem Division, have been used to power the lamps imbedded in the astronaut's headgear. Also, a very large lithium–thionyl chloride battery was used by JHUAPL

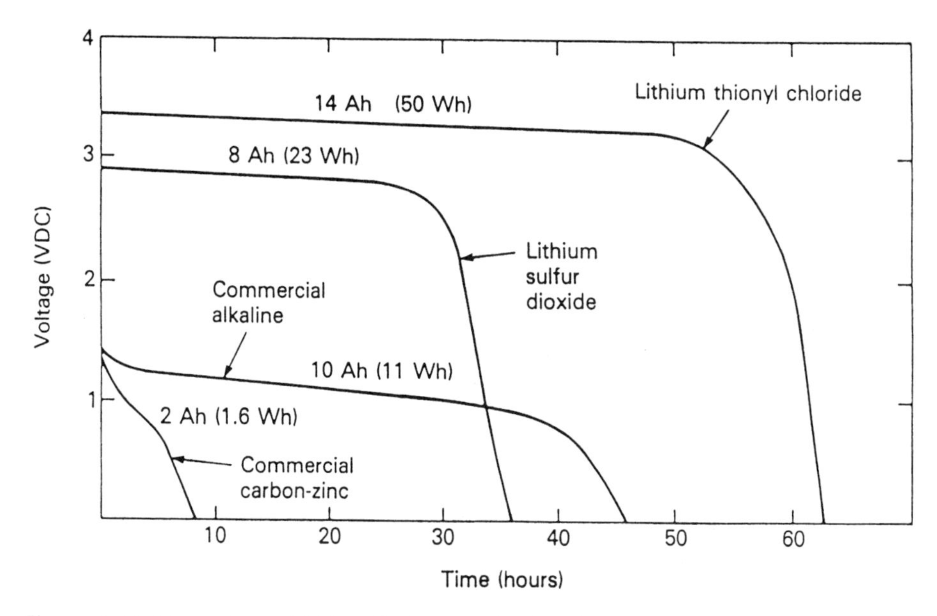

Figure 6.30 Comparison of lithium discharge with other chemistries.

for the main power system on the DELTA 181 spacecraft, a short-duration mission of 55 days, using about 64 kW h of energy.

Lithium–oxyhalide cells were developed as a less hazardous alternative to lithium–thionyl chloride cells. They are similar to the latter in energy, power density, and wide operating temperature range, but are somewhat safer. There are two kinds of oxyhalide cells:

- *Lithium–thionyl chloride plus bromine chloride,* (Li/$SOCl_2$, BrCl) is also referred to as *lithium–bromine complex (Li/BCX).* The addition of BrCl to the electrolyte makes the cells less hazardous under abusive conditions. It also leads to a higher open circuit and initial discharge voltage, which disappears after the first 10 to 20% of discharge, resulting in an initially sloping discharge profile.
- The *lithium–sulfuryl chloride* (Li/SO_2Cl_2) cell has a flat discharge profile and is used for high current applications. However, the cell is not as suitable for high currents as the lithium–bromine complex cell and the electrolyte does react aggressively with water.

Lithium–polycarbon monofluoride (Li/CFx) cells have a solid cathode and good moderate and low-rate capability at temperatures above −30°C.

The *lithium ion cell* is a significant advance in lithium battery technology, both for primary and secondary cells. However, most of the spacecraft applications and the lion's share of the funding, research, and test are directed toward the secondary cells. This technology will therefore be discussed along with the secondary cells.

6.6.2.2 Secondary Batteries

Secondary cells can be electrically recharged to restore their original capacity. They are formed into a battery for use in the main power system to supply the load whenever it exceeds the capability of the primary energy source. For a solar array powered system, this typically means that the battery supplies power to the electrical load during eclipse or whenever the load exceeds the capability of the solar array.

The basic characteristics of typical spacecraft secondary cells are listed in table 6.5. For these cells, long cycle life is usually more important than energy density. In fact, extended cycle life is achieved at the expense of energy density in secondary cells. For example, the separator in a silver–zinc secondary battery is significantly thicker than one in a silver– zinc primary battery, to prolong its life. But this thicker separator increases the weight of the secondary cell without contributing to its energy, resulting in a lower energy density. Secondary cells thus have lower energy densities than primary cells because they are generally designed and used to maximize cycle life as a primary consideration. In usage, deeper depths of discharge will increase the energy density achieved but will decrease the cell's cycle life. This typically results in the user curtailing the discharge depth to increase the probability of achieving the desired lifetime.

Due to their very low cycle lifetime, *silver–zinc batteries* are seldom used as secondary batteries on a spacecraft. Possible applications might be for short-duration missions of only a few orbits. The recent advances in rechargeable lithium ion cells make them a more attractive alternative for most of these types of application.

Secondary battery performance characteristics. Since secondary battery cells are typically used for long-life applications commensurate with their capability, they must be hermetically sealed. Therefore, their charge and discharge voltages and rates must be controlled to levels that ensure acceptable rates of gas evolution in order to avoid excessive internal pressures. Charge and discharge rates must also be controlled to levels that allow cell heat generation to dissipate at a rate that prevents overheating. For these reasons, a charge control system is required for rechargeable batteries. Also, load control and protective devices are typically used to limit the discharge.

The larger purpose of the battery charge control system is to place boundaries on the pertinent battery parameters of current, voltage, temperature, and pressure that will enhance the battery's performance and extend its life. Therefore the design requirements of the charge control system are derived from knowledge of the battery performance characteristics, which are different for each battery chemistry. For each of the following battery chemistries, we will present a brief history and those features necessary to compare one battery system to another. We will also attempt to describe the parameters and their boundaries that must be constrained by the chemistry's charge control system.

Conventional nickel–cadmium (NiCd) cells. Using NASA/GSFC's terminology, we will refer to the nickel–cadmium cells with nylon separators as *conventional* to distinguish them from the so-called super nickel–cadmium (SNiCd) cells developed by Hughes (HAC)—a later development using a zircar separator and adding special proprietary HAC additives to the electrolyte. Conventional nickel–cadmium cells were used for energy storage on the majority of spacecraft during the first three decades of the space era. They also had reasonably high energy density, and required relatively simple charge control systems. The original cylindrical cells, with spirally wound plates, were phased out by the mid-1960s in favor of the more easily packaged prismatic configuration.

Until the 1980s, the prismatic configuration of nickel–cadmium battery cells had been generally accepted as the qualified spacecraft rechargeable battery due to its long cycle life, relative design simplicity, and consistent performance. In spite of the disruption caused by the separator problem (previously discussed), they are still used today. The primary manufacturer that produces spacecraft cells and batteries using conventional nickel–cadmium technology is SAFT, in France. SAFT nylon separator is different from the type used in the U.S. Saft was therefore not affected by the problems that plagued U.S. sources, and has continued to provide spacecraft NiCd batteries of high reliability.

In a prismatic nickel–cadmium cell, the rectangular positive and negative plates are arranged alternately in the cell's stainless steel container. A separator maintains uniform intercell spacing and holds the electrolyte. Approximately 3.0 ml of potassium hydroxide electrolyte per ampere hour of rated capacity is added, and the container is hermetically sealed. There is no internal fuse or vent; the internal gas evolution rates and battery charge control methods are so well known and controlled that these protective devices are not needed.

In a typical battery with prismatic cells, the cells are stacked face to face, separated by insulators, and usually wired in such a way as to minimize the magnetic field resulting from the flow of current. To reduce the battery temperature and the temperature difference between cells, thin aluminum plates are sandwiched between the broad surfaces of adjacent cells and thermally connected to the base plate to create a thermal path for efficient heat removal. Coating the cell surfaces with a space-qualified silicone enhances this heat transfer.

Typical charge and discharge characteristics are shown in figure 6.31. Nickel–cadmium battery cells commonly have actual capacities 20% higher than the manufacturer's rated capacity and, to ensure full charge, it is necessary to charge the cells at a low rate beyond their actual capacity, allowing any excess charge to be dissipated as heat. Therefore, the curves show the battery being charged and discharged beyond the rated capacity.

Figure 6.31c shows that the discharge voltage and capacity are a function of discharge rate. Pulse discharge rates up to $10C$ are possible, depending upon the length of the pulse and the temperature. The charge–voltage curves (figure 6.31a and b) increase at a moderate slope until at or near full charge, when the slope noticeably increases. This effect is more pronounced at higher charge rates. At lower temperatures the level of the voltage changes, and the effect of increased overcharge rate on voltage becomes more dramatic. This property of nickel–cadmium batteries allows the power system designer to control the battery charge by limiting its voltage to a preset maximum that is a function of battery temperature. This is a commonly used way to limit the oxygen gas evolution rate during charge and overcharge so as to avoid excessive cell pressures. This approach, of temperature-compensated voltage limiting, also prevents hydrogen evolution, excessive oxygen evolution, and, if properly selected, limits the overcharge rate to a level that is acceptable to the thermal design of the battery.

The voltage versus temperature limit is empirically determined in a series of parametric tests, where the cells are run through typical charge/discharge cycles at the expected charge, discharge, and overcharge rates throughout the temperature range. The correct limit voltage as a function of temperature is determined from the results of these parametric tests.

The NiCd V/T curves most commonly used were those developed by NASA/GSFC, shown in figure 6.32. The negative slope of each curve results from the decreased battery voltage with increased temperature. There are eight parallel V/T limit levels, separated

Table 6.5 Characteristics of secondary batteries for spacecraft use

	Silver–zinc	Nickel–cadmium[1]	Super nickel–cadmium[1]	Nickel–hydrogen[1] IPV	Nickel–hydrogen[1] CPV	Nickel–hydrogen[1] SPV	Lithium ion (liquid electrolyte)[6]
Chemical designation	AgZn	NiCd	SNiCd	NiH_2			Li ion
Configuration	prismatic	prismatic	prismatic	cylindrical[2] pressure vessel			cylind/prismatic
Capacity range (A h)		10 to 40	4.8 to 50	10 to 400	4 to 100	30 to 120	0.5 to 100[6,3]
Typical DOD(%)	50	10 to 25	30	LEO = 35%		GEO = 70%	LEO = 20% GEO = 70%
Gravimetric energy density (W h/kg)	90 to 130	30 to 35	20 to 25	40 to 65	27 to 63	54	75 to 145
Volumetric energy density (W h/dm^3)	150 to 300	75 to 90	65 to 75	30 to 101	30 to 102	47 to 68	160 to 380
Energy efficiency	75	75 to 85	75 to 85	70 to 80	70 to 80	70 to 80	93 to 96
Operating temperature range: (°C)	0 to 20	0 to 20	0 to 20	–10 to 30	–10 to 30	–10 to 30	–10 to +40
Storage temperature range (°C)	0 to 30	0 to 30	10[4]	0 to 30	0 to 30	0 to 30	–40 to +50
Self discharge (%/day)	0.1	1	1	10	10	10	0.3
Dry storage life	~5 yr	~5 yr	~5 yr	~5 yr	~5 yr	~5 yr	N.A.
Wet storage life	30 to 90 days[5]	2 yr	>2 yr	>2 yr	>2 yr	>2 yr	>5 yr[6]
Cycle life	200 max. 20 to 50 typ.	20,000 (20% DOD)[6]	30,000 (25% DOD)[6]	40,000 LEO (35% DOD)[6] 15 yr GEO (70% DOD)			>10,000 LEO (20% DOD)[6] >7 yr GEO (70% DOD)[6]

Peak charge voltage (V/cell)	2.0	1.45	1.48	1.56	3.12	34.32	4.2
Charge cell voltage (V/cell)	1.86 to 2.0	1.45	1.45	1.50	3.0	33.0	4.1
Discharge voltage (V/cell)	1.8 to 1.45	1.25 nom. 1.0 min.	1.22 nom. 1.0 min.	1.25 nom. 1.0 min.	2.5 nom. 2.0 min.	27.5 nom. 22.0 min.	3.75 nom. 3.0 min.
Manufacturer(s) (see key below)	EPT, YTP	SAFT	HSC, EPT	EPT, SAFT,	EPT		AEA, EPT, JSB, MECO, SAFT, SONY, WGC, YTP

(1) Rechargeable nickel-based batteries use an aqueous electrolyte solution of concentrated KOH, which is still liquid at −40°C.
(2) EPT nickel-hydrogen cells are cylindrical pressure vessels with flat, toro-spherical or hemispherical ends. Pressures range from 400 to 1225 psig.
(3) Lithium ion cells for spacecraft up to 200 Ah under development.
(4) SNiCd cells are typically trickle-charged when stored at 10°C.
(5) Safety considerations require that silver–zinc batteries have a limited 'maximum wet stand' life, typically 1 to 12 months after activation.
(6) All data is as of mid-2003. Technology improvements in NiH_2 and particularly Li ion cells are continuing, leading to increases in cycle life and DOD. Consult manufacturers for the latest information.

Key: AEA Technology (AEA)
Eagle Picher Technologies (EPT)
Hughes Space and Communications Co. (HSC)
Japan Storage Batteries (JSB)
Mitsubishi Electric Corporation (MECO)
SAFT
SONY
Wilson Greatbatch Technologies (WGT)
Yardney Technical Products (YTP)

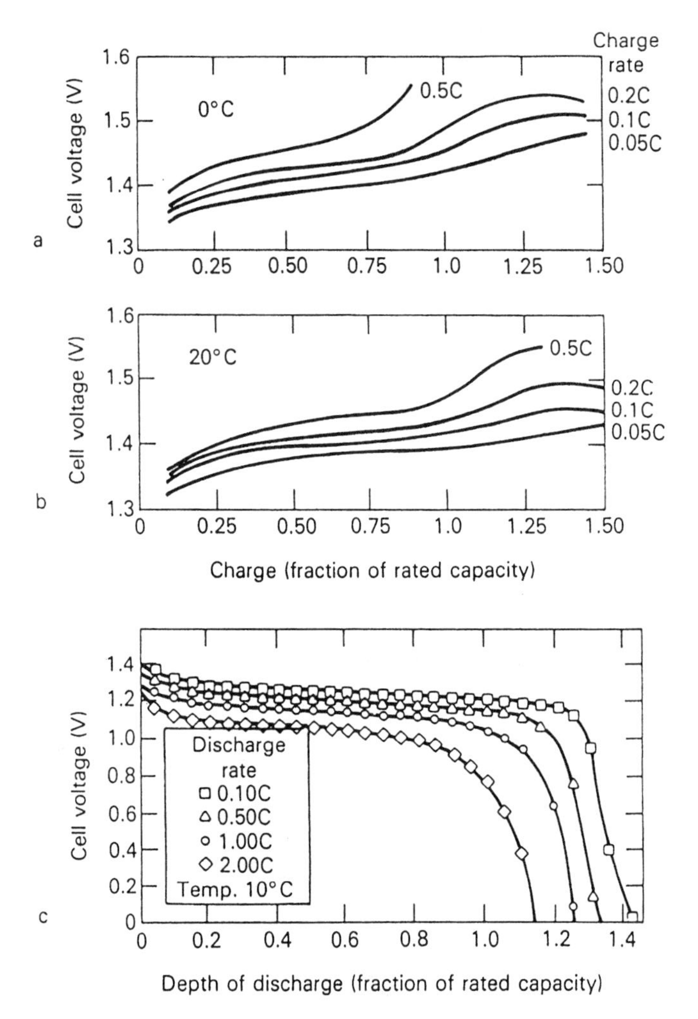

Figure 6.31 Charge and discharge voltages for NICd cells at different temperatures and rates. (Source: NSSDC.)

by 0.020 ± 0.20 V per cell, with a negative slope of −2.33 ±0.20 mV per °C and with level 8 voltage set to 1.520 ± 0.015 V at 0°C (Ford et al., 1994). In practice, *V/T* levels 5 or 6 were found to be suitable voltage limits for most nickel–cadmium batteries at the beginning of life (BOL). If the charge voltage increased after a few years in orbit, then the voltage limit might be increased to a higher, more suitable, level. Or, on those few occasions when a shorted cell developed, *V/T* level 1 or 2 might be selected to limit the charge rate at the dramatically reduced battery charge voltage. Note that the curves extend well beyond the expected range of battery operating temperature (0 to 10°C), so that battery charge control is not lost in the event of a spacecraft thermal problem.

This multilevel voltage limiter is a simple, reliable method of battery charge control for nickel–cadmium cells that is still in use today. However, it is now most often

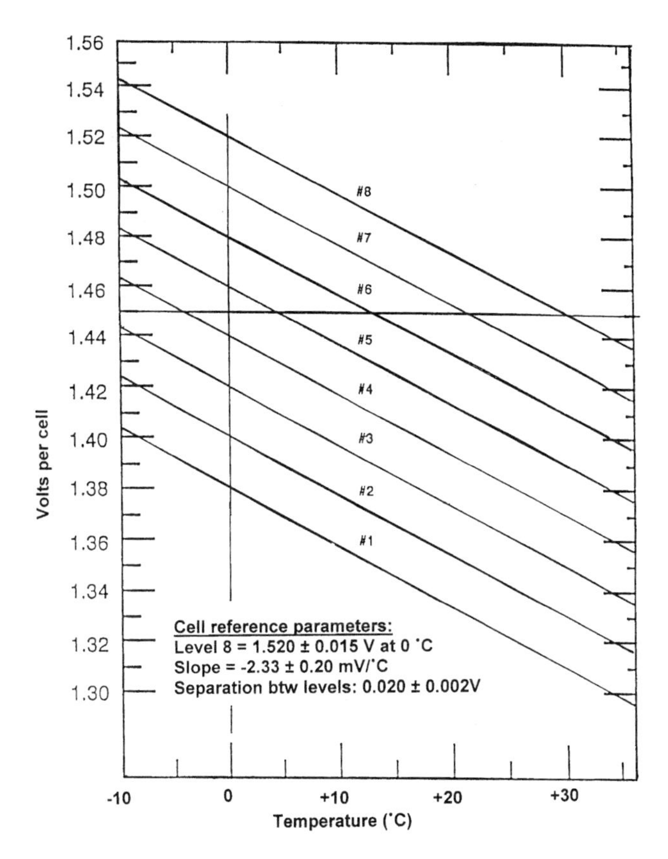

Figure 6.32 Voltage–temperature (V/T) limits for NiCd cells. (Source: GSFC.)

used in parallel with an electronic coulometer, which counts the discharge in ampere minutes and allows a recharge of the same amount, increased slightly to allow for battery inefficiency. Combining these two very different charge control methods has become very popular because the coulometer offers precise knowledge of the battery state of charge (SOC) and its control, but one that is subject to error if the spacecraft data system has to be reset. But the voltage limiter is autonomous and provides a reliable limit to severe battery overcharge. Another refinement is the digital control of the V/T curve level and slope, allowing a complete redefinition of the V/T curves in orbit.

The primary mode of nickel–cadmium cell degradation is that they develop a second (lower) voltage plateau with repeated charge–discharge cycling. This is commonly called the "memory effect." Left uncorrected, this voltage degradation will continue to result in capacity loss. This typical aging problem is shown in figure 6.33. It shows the results of a pack of twenty two 20 A h cells that were subjected to 25% depth of discharge every 90 min for about 2 years or 12,000 cycles (Baer, 1983). At the beginning-of-life (BOL) the discharge curve is reasonably flat at about 1.25 V per cell. After 3000 cycles the plateau has degraded to 1.2 V and a second plateau has begun to develop. This second plateau would not yet be noticed at 25 to 50% depth of discharge. However, after 12,000

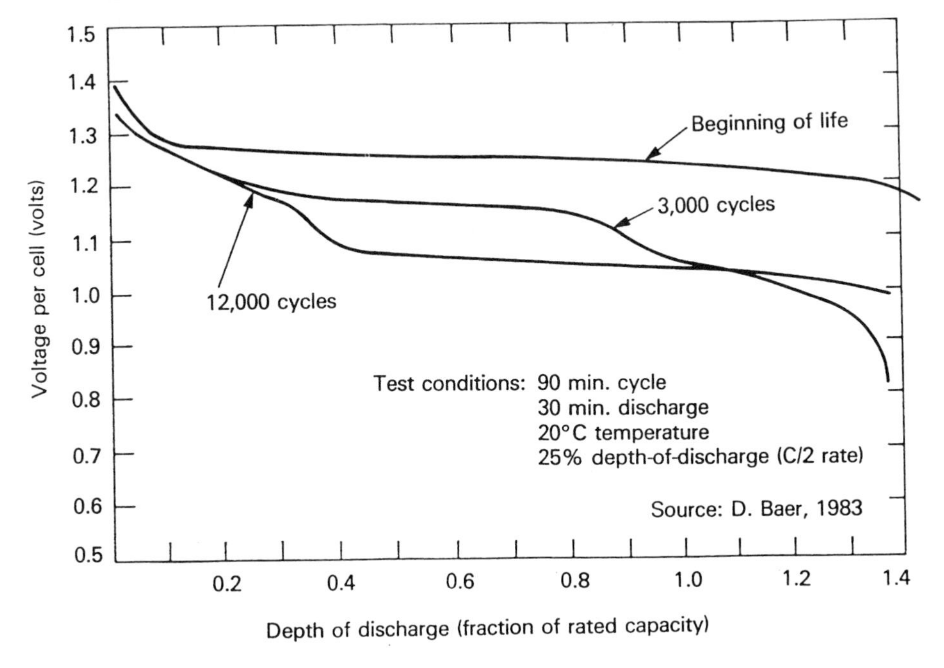

Figure 6.33 Effect of orbital cycling on discharge voltage of 20 A h NiCd cells.

cycles the second plateau has been established at about 1.05 V per cell and the abrupt drop in voltage would be experienced by a spacecraft discharging to 35% depth. This degradation will continue until, at some point, the battery will experience a drop to the second plateau. Notice that at 12,000 cycles the capacity, measured in ampere hours (A h), is more than at 3000 cycles. In fact it can be more than at the beginning-of-life, but it is at a reduced voltage.

This memory effect is at least partially reversible. The most effective method is to completely discharge the individual cells to zero volts. But this is usually not possible with a battery on an orbiting spacecraft where the minimum discharge voltage is often limited by spacecraft requirements. Even when a battery can be isolated from the spacecraft bus, its minimum discharge voltage is usually limited to something close to 1.0 V per cell due to concern that a lower voltage might result in reversing one or more cells. Fortunately, it is not absolutely necessary to discharge the cells completely because deep discharging is somewhat effective in erasing the battery's memory effect.

The problem can therefore be dealt with in two ways. One is to design the system so that each cell in the battery can be completely discharged in orbit and then to allow the battery to be recharged at a controlled rate. This "reconditioning" process provides a partial cure for a limited time and, once started, must be periodically repeated. The second solution to the problem is to design the spacecraft to operate down to a low voltage, corresponding to about 1.0 V per cell. This second approach is the most commonly implemented and, since it entails less risk, it is more acceptable to spacecraft operations managers.

The Naval Surface Weapons Center in Crane, Indiana (NSWC/Crane), has performed life-cycle tests on batteries since the early 1960s, accumulating the largest database of battery cycling history. But the change in separator material in the 1980s, discussed previously, caused many other changes in NiCd cell design to be made by its manufacturer. Some NSWC historical data on NiCd is therefore no longer valid for predicting the lifetime of newer cells. However, the facility continues to be used by GSFC and others to perform life-cycle tests in existing programs to qualify cells of different battery technologies in production for use on spacecraft.

Hughes Aerospace Corporation (HAC) (now part of Boeing) developed super nickel–cadmium (SNiCd) cells, using some ideas and components borrowed from the nickel–hydrogen technology. While these cells are similar to conventional nickel–cadmium cells in both configuration and performance, there are important differences. Internally, zirconium has replaced the nylon separator material, and there is an additive to the electrolyte that is proprietary to Hughes. The cells are heavier than conventional cells and are more expensive. They use the same charge control system as used by conventional nickel–cadmium batteries. However, the stability of their performance characteristics and lifetime has proven to be better than that of conventional, U.S. manufactured, nickel–cadmium cells.

Nickel–hydrogen cells. Because nickel–hydrogen (NiH_2) battery cells could be discharged to greater depths than nickel–cadmium batteries, they offered a significant weight improvement. For this reason, they were first developed for use on spacecraft in geosynchronous Earth orbit (GEO), where the weight to achieve orbit was a significant driver of launch vehicle cost. Comsat Corporation (now part of Lockheed Martin Corporation) and Hughes were responsible for most of this development as part of their responsibility for the communications satellites. These were large spacecraft which required large batteries. For such systems, the weight advantage overcame the disadvantages of increased battery cost and more complex thermal and mechanical design of the nickel–hydrogen batteries. The primary suppliers of these batteries are EaglePicher Technologies, SAFT, and Boeing.

A nickel–hydrogen cell from SAFT is shown in figure 6.34a. It reveals a distributed arrangement of the plate stack, which is located in the center of the cell, leaving the dome ends occupied by the hydrogen gas. Locating the two terminals on one end of the cell decouples the wiring from mounting considerations. This particular design was developed for geosynchronous applications. A detailed description of the available internal designs from battery manufacturers (much of which is proprietary) is beyond the scope of this book. But each manufacturer has developed unique internal designs in its quest to improve the performance of this technology.

A nickel–hydrogen individual pressure vessel (IPV) from Eagle Picher Technologies, LLC, is shown in figure 6.34b. This is the EPT ManTech cell design, developed under the Department of Defense (DOD) Manufacturing Technology (ManTech) Program. The configuration shown is "axial," with the terminals extending from opposite dome ends. "Rabbit-ear" configurations, with both terminals protruding from one end (as shown in the SAFT cell of figure 6.34a) are also available from EPT. The internal design makes extensive use of fuel cell technology. It is essentially a pressure vessel, typically manufactured from Inconel (or stainless steel). Nickel–hydrogen cells are typically sealed to withstand high internal pressure during overcharge, with an

a. SAFT Design for GEO, (Source: SAFT, Borthomieu, 2000.)

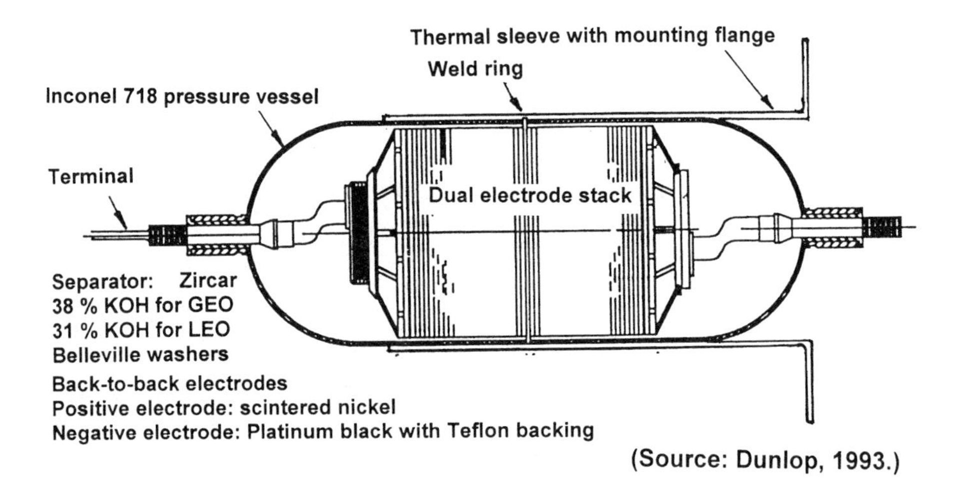

(Source: Dunlop, 1993.)

b. EPT IPV Mantech Design

Figure 6.34 Nickel-hydrogen battery cells: cross-sectional views.

operating range typically within 500 to 900 psig (3.4–6.2 MPa) and the maximum expected operating pressure (MEOP) less than 1100 psig (7.5 MPa). The stack of disc-shaped electrodes consists of slurry-sintered nickel positive plates, spaces for hydrogen gas (the negative electrode), Teflonated platinum-black catalyst-sintered negative plate, and zirconium separators. The electrolyte is an aqueous solution of potassium hydroxide with, typically, 31% concentration by weight. Since the negative electrode is hydrogen gas, the inert Teflonated platinum functions as a collector for the negative electrode.

The principal source of the cell's heat is the plate stack, which is concentrated in the center of the vessel, leaving the volume at the dome-shaped ends available for gases. The vessel is partially encased in a thermal sleeve that is bonded to the outside wall, covering at least 40 to 50% of the cell's external surface. This thermal sleeve is typically fabricated with a flange that will protrude, after bonding, from either the midpoint or the base of the cell for mechanical attachment and heat conduction to the structure. The sleeve improves the thermal conductivity along the wall, and its flange provides a means of attachment to the spacecraft and an effective heat path to the structure. The sleeve has typically been made from aluminum or magnesium, but increasingly fiber composite is being used.

Throughout the 1980s, a number of factors converged to cause nickel–hydrogen technology to advance in development and widen its application. The advantages of lower weight and longer lifetime became more apparent with continued use of the technology. Environmental restrictions made it increasingly difficult to use cadmium in any production processes.

Eagle Picher Technologies (EPT) developed their common pressure vessel (CPV)—a pressure vessel containing two nickel–hydrogen electrochemical cells, connected in series, to more directly compete with the smaller sized nickel–cadmium cells. Each of these two electrochemical cells was half the size of a single cell that would have fit in the same volume. The finished CPV had half the capacity and twice the voltage (the same total energy) as an IPV of the same size. The CPV is more suitable for smaller systems because only half as many cells (vessels) were required for a given voltage (simplifying the packaging) and the cells are available in smaller capacity sizes than with the IPV configuration. For example, a nominal 28 V system requires 22 IPV cells in its battery, whereas only 11 CPV cells are needed in a battery of the same voltage.

In January 1994, the Naval Research Laboratory (NRL) launched a scientific spacecraft, Clementine, which featured the first use of the single pressure vessel (SPV) in orbit. The SPV, developed by Johnson Controls Incorporated (JCI) and Comsat (Halpert and Surampudi, 1997), is a single pressure vessel containing an entire battery of 22 electrochemical cells connected in series (the number typically used to support an unregulated 28 V bus). Later, Eagle Picher acquired this division of JCI and continued the development of the SPV technology. Motorola[4] was the first to make extensive use of these batteries on the company's Iridium constellation of communication satellites (Sterz et al., 1997). The first twenty Iridium spacecraft each used a single 50 A h SPV, and the remaining 75 satellites used 60 A h sizes (Toft, 2000). All of the nickel–hydrogen technologies are continuing to be developed, with both larger and smaller sizes becoming available.

Nickel–hydrogen cells also exhibit a double plateau on discharge after extended cycling. However, this effect is not as prominent as it is with nickel–cadmium cells. (Zimmerman and Weber (1997) attribute the lower voltage of the second plateau to the intrinsic semiconducting characteristics of the active materials in the nickel electrodes.) Also, its occurrence can be both delayed and diminished by onboard reconditioning. Toward this end, reconditioning is periodically performed on the Hubble Space

[4]This division has since been acquired by Boeing.

Telescope (HST) NiH_2 cells. Some GEO satellites have a reconditioning capability with their NiH_2 cells, and use it as required during the mission prior to the start of eclipse seasons.

Nickel–hydrogen batteries have proven to have significantly longer life than nickel–cadmium systems. Lifetimes in excess of 40,000 cycles at 40% depth of discharge (DOD) have been demonstrated in ground tests, and projected lifetimes of 50,000 cycles are being made on the basis of accelerated testing. One reason for this is that the primary mode of nickel–cadmium battery failure, the gradual migration of cadmium metal through the separator, short-circuiting the plates, has been eliminated. The nickel–hydrogen cell uses hydrogen gas for the negative electrode, eliminating the cadmium metal. Also, hydrogen gas does not fade or become coated with metallic oxides, as do cadmium plates. Furthermore, the oxygen, produced in the nickel–hydrogen cell during overcharge, combines with the hydrogen to form water (a reversible reaction). This gives the nickel–hydrogen cell a much greater tolerance to high overcharge rates. The basic limitation to the charge rate is the heat dissipation limitations of the system, rather than an excessive rate of oxygen gas evolution.

Figure 6.35 shows the nickel–hydrogen charge–discharge voltage profiles and pressure at different temperatures for a group of 120 A h IPV cells from the Intelsat program. The cells were charged at 12 A ($C/10$ rate) for 16 h and then discharged at 60 A ($C/2$ rate) to 0.1 V (Dunlop et al., 1993). During charge, the internal pressure of these cells increases linearly with their state of charge until the battery approaches full charge. Then the pressure tapers to the cell's steady-state full charge pressure. During discharge the pressure decreases linearly with charge state. The battery pressure telemetry can therefore be calibrated and used to indicate the nickel–hydrogen battery's state of charge (SOC). In a battery consisting of IPV or CPV cells, strain gages to monitor the pressure are mounted on a number of cells, which are assumed to be representative of the battery. In an SPV battery, special pressure transducers are used which indicate the pressure of the complete 22-cell battery assembly.

Parametric tests on similar Intelsat cells showed that their charge efficiency, like that of nickel–cadmium cells, drops rapidly on approaching full charge, as shown in figure 6.36. Energy going into the cell after this drop in charge efficiency is mostly dissipated as heat. This is of particular concern in low Earth orbit where it is usually necessary to charge at a high rate due to the limited time in sunlight. A high charge rate must be reduced as full charge is approached, to prevent battery overheating and possible damage.

Figure 6.37 shows the available capacity as a function of cell temperature. At higher temperatures, the available capacity is significantly reduced. There is some reduction at lower temperatures as the electrolyte approaches freezing. Temperatures below–10°C should be avoided, with optimum operating temperature being in the range from −5 to 10°C.

Some of the methods of battery charge control for nickel–hydrogen cells are listed below. The first two methods are also used to determine the battery state of charge and so can be used passively in the monitor mode as well as for active charge control. The parallel combination of the coulometer and voltage V/T limiter with an appropriate slope of about −4.5 mV/°C is similar to the performance of nickel–cadmium battery controllers, which have a slope of −2.33 mV/°C.

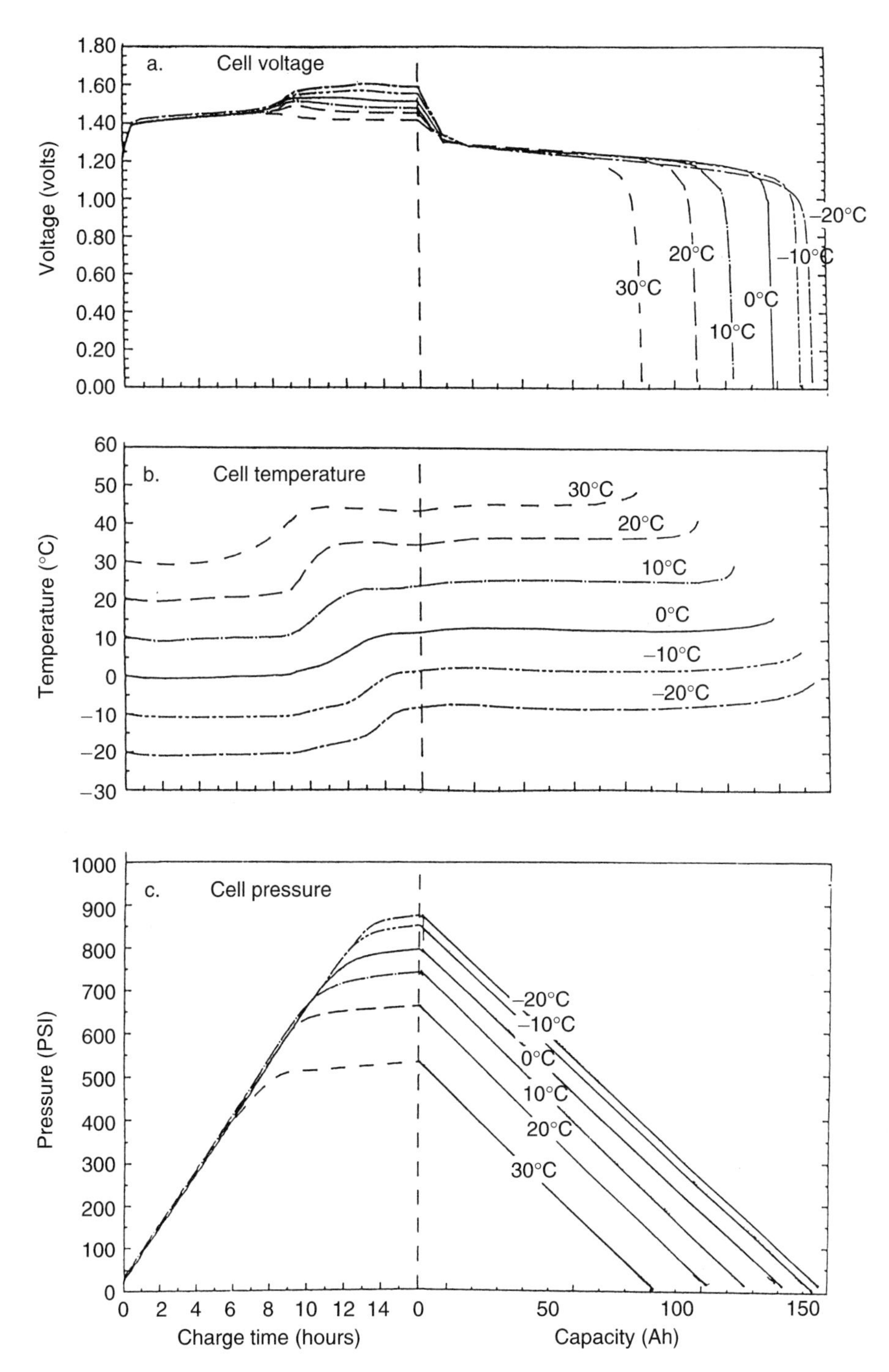

Figure 6.35 Charge–discharge characteristics versus temperature for Intelsat VII A Lot L01-004, 120 Ah nickel–hydrogen cells, $C/10$ charge rate. (Source: NASA Ref. Pub. 1314, *NASA Handbook for Nickel Hydrogen Batteries,* Sept. 1993, p. 5–9 & 5–14.)

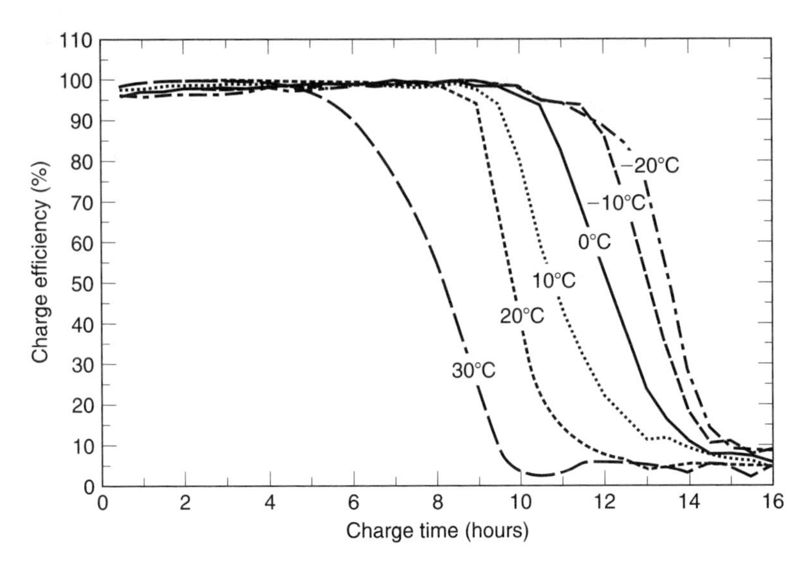

Figure 6.36 Charge efficienty versus temperature for Intelsat VII A Lot L01-004, 120 Ah nickel-hydrogen cells, $C/10$ charge rate. (Source: NASA Ref. Pub. 1314, *NASA Handbook for Nickel Hydrogen Batteries,* Sept, 1993, p. 5–15.)

- The coulometer is used to count the charge removed and returned, with an additional amount to account for battery inefficiencies and to transmit a signal to place the battery in trickle charge when full charge has been reached. On some spacecraft, the charge rate is reduced from a high rate to a lower charge rate (typically $C/10$) when the coulometer count reaches around 90% of full state of charge (SOC) to compensate for the lower battery charge efficiency as it approaches full charge, and to reduce the thermal dissipation and stresses in the battery.
- The use of cell pressure as an indicator of full charge. With the battery at a low state of charge, the battery internal losses are low and the relationship between pressure and temperature is given by the ideal gas law[5] $PV = NRT$. However, as the battery gets closer to full charge, the losses are large and the ideal gas law relationship gradually becomes less representative. At full charge, the pressure actually *decreases* with higher temperature. The effects of temperature on the full state of charge pressure value must be taken into account. For those systems where a single pressure limit value is used to indicate "full charge," a pressure value corresponding to the hottest predicted battery temperature at full state of charge is selected. Temperature-dependent full charge pressure limits can be derived from the battery test data and used in charge limit control.

The relationship between battery pressure and capacity also depends on charge rate and lifetime. Very low charge rates lead to denser crystalline growth, which causes lower pressures and capacity. Therefore, when using the pressure as an indication of charge state, it is necessary to periodically recalibrate their relationship.

[5]The ideal gas law is derived from Charle's law, Boyle's law and Avogadro's hypothesis. N is the number of moles of gas and R is the universal gas constant $= 8.3114 \times 10^7$ erg mol^{-1} K^{-1}.

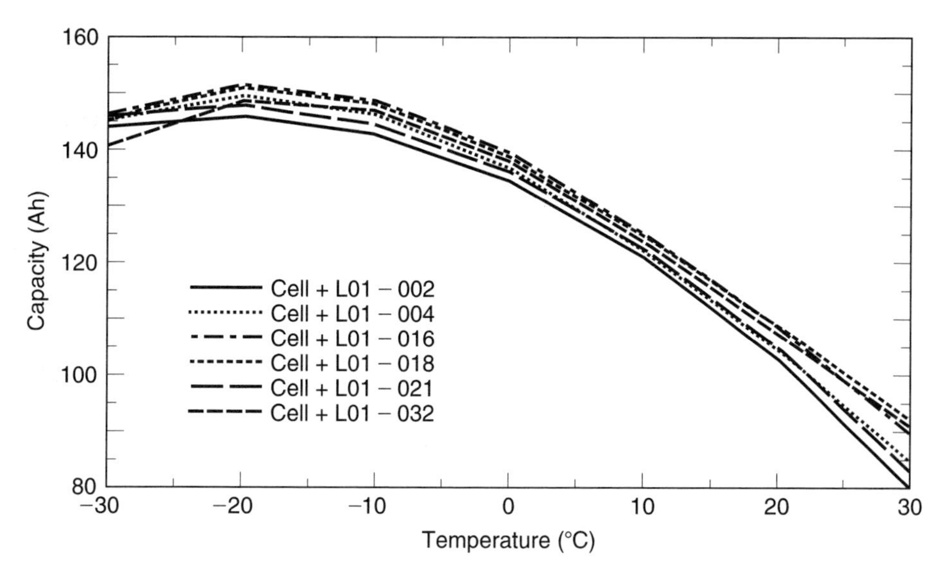

Figure 6.37 Capacity to 1.0 Volt versus temperature for selected Intelsat VII A, 120 Ah nickel–hydrogen cells. (Source: NASA Ref. Pub. 1314, *NASA Handbook for Nickel Hydrogen Batteries,* Sept. 1993, p. 5–10.)

- Voltage limit control V/T is sometimes used in NiH_2 charge control as a full charge voltage limiter, in parallel with another control method.
- Because the slope of the battery time-rate of change of pressure (dP/dt) changes during charge as the battery approaches full state of charge, it can also be used as an indicator of when full charge is reached. This technique has not often been used because a computer is required to track the very slow rate of pressure change and perform its time differentiation. Dependence on this computer raises concern about the possibility of anomalous operation of the processor under spacecraft abnormal fault conditions.
- The battery time-rate of change of temperature (dT/dt) changes rapidly as full charge is approached. This is true for both nickel–cadmium and nickel–hydrogen cells. Until the battery is about 80% charged, its charge efficiency if very high (over 95%), so the heat exchange between the battery and its surroundings is very small and the battery's temperature change is also small. As the battery nears full charge, its charge efficiency is reduced and an increasing fraction of the charge energy is dissipated as heat. At full charge the battery charge efficiency is zero and all charge energy is converted into heat. This causes the battery temperature to rise very fast. By detecting the battery temperature and calculating its dT/dt, the battery full state of charge can be determined.

  This technique, although theoretically correct, has not been used in an operational spacecraft, primarily because, if the spacecraft pointing becomes anomalous, the battery temperature may increase due to this incorrect spacecraft attitude, causing an erroneous "full charge" signal that terminates the battery charge when it is critically needed. Another potential problem with this control technique is that it requires a computer to perform a slow differentiation, a process that could yield erroneous results caused by memory resets during anomalies.

Lithium ion secondary cells: as previously noted in the discussion of primary lithium cells, safety concerns limited their use for consumer applications, especially in large sizes. The first breakthrough came in 1991 when Sony introduced cells in which the lithium metal was replaced by a layered-structure carbon (a lithiated carbon) as the negative electrode, which allowed lithium ions to pass into and out of it in large quantity.

Then a breakthrough in positive electrodes came from a program sponsored by the UK Atomic Energy Authority (now AEA Technology). J. Goodenough at Oxford University devised a new class of positive solid-solution electrode materials that could be produced with the lithium already synthesized into them. Upon charging, lithium ions are extracted from the positive electrode material and inserted into the negative electrode material. Upon discharging, the reverse process takes place. Most manufacturers use lithium cobalt oxide ($LiCoO_2$) for this positive electrode. But since cobalt is an expensive metal, a cheaper alternative, such as lithium manganese oxide ($LiMn_2O_4$) or lithium nickel oxide ($LiNiO_2$), is sometimes used for commercial applications. All of these oxides are less hazardous because they are stable in normal air, unlike lithium metal, which reacts with moisture and also with nitrogen in the air.

In the early 1980s, a lithium-ion-conducting polymeric was developed to replace the liquid organic electrolyte. This new polymer cell enabled the development of all-solid-state batteries. The polymer electrolyte is a very thin solid material, sandwiched between a lithium metal film and a metal film. But the polymer electrolyte currently has increased resistance, reducing the cell's rate capability compared to a cell of the same geometry with a liquid electrolyte, especially at lower temperatures. Since higher rates are usually required for batteries used in spacecraft power systems, cells with liquid electrolyte are currently favored for research and development. (However, research is continuing on cells with a polymer electrolyte since they promise a greater cycle life to be achieved, which would be an important advantage.)

There is intensive interest in rechargeable lithium ion cells for spacecraft in GEO and LEO orbits. Extensive ground testing by manufacturers, spacecraft developers, government and other aerospace organizations are continuing in order to understand and characterize lithium ion batteries for various applications. A number of spacecraft have been launched using lithium ion batteries and others are in various stages of development. In addition to the increased energy per unit weight and volume (more than 125 W h/kg and 350 W h/dm^3), there would be many other advantages to lithium ion secondary cells for spacecraft:

- The discharge voltage is higher than for nickel-based chemistries (an average 3.5 V compared to about 1.28 V for NiH_2 and NiCd), thus reducing the number of cells required per battery.
- A very good storage life (high charge retention) would also make lithium attractive for powering certain deep-space probes and planetary landers where there is a requirement for a limited number of recharges and long mission life. It also translates into requiring less launch pad management. Lithium ion batteries are baselined for use on several Mars missions including landers/rovers. Yardney's Lithion cells are used on the Mars Express Rovers.
- Simpler thermal control: a nickel-based battery is typically a major source of heat on the spacecraft and has a very narrow operating temperature range. The higher

efficiency (the useful energy recoverable on discharge divided by the energy required to charge) is typically >90% for lithium ion compared to <80% for nickel-based systems, resulting in substantially less heat dissipation. This reduction in wasted heat, together with the somewhat higher maximum temperature of operation, means that the thermal design of lithium ion batteries is less critical and will require minimal heater power for thermal control, so there should be additional weight savings in reduced battery thermal control elements such as thermal conductors and battery radiator plates and possibly heat pipes and louvers.
- No memory effect, as with nickel-based systems.
- The cell voltage is a very good indicator of the state of charge and is minimally dependent on cell temperature. Battery charge is terminated when the battery voltage reaches its full charge voltage (around 4.2 V). The discharge is terminated when the minimum allowed voltage is reached (around 3 to 3.5 V). The battery is kept at a specific state of charge by maintaining a fixed voltage across it.
- Low standby power requirements during long ground storage or long flight durations when the battery is in the standby mode. For long life, the battery should be kept cold at a low standby state of charge with a corresponding lower voltage (around 3.5 to 3.8 V, depending on cell manufacturer). The battery can then be recharged fully just before the anticipated need.
- Simpler battery mechanical and thermal design, compared to the cylindrical pressure vessels of nickel–hydrogen batteries.
- Since most lithium ion cells contain no ferromagnetic materials, they are "magnetically clean" compared to nickel-based chemistries. HEOS, GEOS, Giotto, and, most recently, the Cluster spacecraft all had to use silver–cadmium batteries, which have poor storage and cycle life performances. Lithium ion would allow much longer life and increase the scientific return for these missions.
- Lithium ion cells are relatively less expensive than NiH_2 or NiCd cells.

However, there are disadvantages to the lithium ion technology:

- No tolerance for overcharge. While overcharging a battery wastes energy, in most systems a limited amount of overcharge can be beneficial because it ensures that all of the cells, connected in series within the battery, reach full charge at the end of the charging period. This extra overcharge eliminates the risk that some cells that have less capacity or are operating at a slightly higher temperature, or have a greater self-discharge or leakage current than others, can "run down" over a large number of cycles, eventually reducing the useful capacity of the battery to a level insufficient to meet the needs of the payload during eclipse. Lithium ion secondary cells, in contrast, do not possess such a "safe" overcharge reaction mechanism. Small amounts of overcharge can convert the lithium oxide to metallic lithium, causing irreversible damage to a cell's subsequent performance, and excessive overcharging can lead to the buildup of internal pressure, leading to cell venting. Cells are carefully selected and matched before being assembled in a battery. With charge/discharge cycling, small fabrication differences between individual cells, and imperfect thermal gradient between cells, divergences between cells can occur. It is therefore not certain that it will be possible to maintain all series-connected cells in a battery at the same state of charge without additional circuitry to adjust the state of charge of individual cells.
- Some degradation of cell capacity with calender life. Oxidation of the electrolyte can occur, which leads to a reduction in cell life in addition to the effect of

charge/discharge cycling. The rate of oxidation increases with increased cell voltage. The exact mechanism of this degradation is not well understood.

- More complex charge control electronics may be necessary to monitor and control individual cells.
- Cell voltages are high, requiring fewer cells in a battery. The loss of a shorted cell will lower the average battery voltage at full charge by about 4 V. The power system must be designed to handle this reduction in voltage.
- Poor performance at low temperature: the capacity of lithium ion cells is lower at colder temperatures, especially at higher discharge rates, which is the reverse of nickel–hydrogen batteries. There are ongoing efforts to improve the low temperature capacity. This is very important in applications such as Mars rovers where the ability to operate at very cold temperatures during the Martian nights is important.
- Cells cannot be assembled in the discharged state. Cells are stored and shipped in the partially charged state (about 50%). Discharging the cells completely will damage them. Special care must be exercised during battery assembly.
- Cells are easily degraded or damaged outside narrow voltage limits for both charging (about 4.2 V) and discharging (below about 3 V).
- The maximum charge and discharge current must be limited.
- Special storage requirements: lithium ion batteries should be stored charged, preferably at between 50 and 75% of full capacity.
- Some inefficiency in solar array power utilization with some power system topologies, due to the gradual voltage rise in lithium ion cells during recharge. For nickel-based systems, the voltage increases to the maximum charge voltage after a relatively short period of charging. However, the charge voltage of lithium ion cells increases to its maximum more gradually (see figure 6.38).

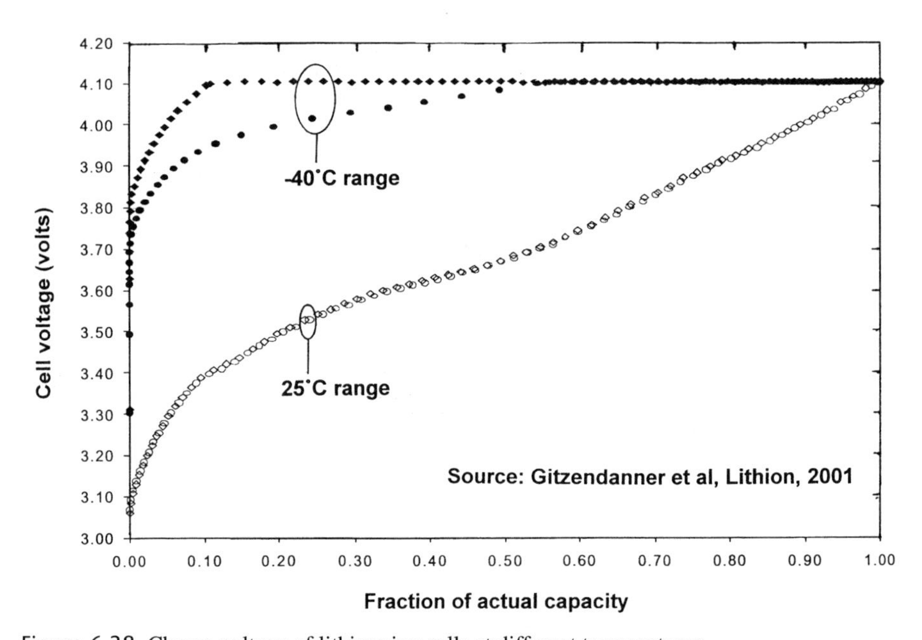

Figure 6.38 Charge voltage of lithium ion cells at different temperatures.

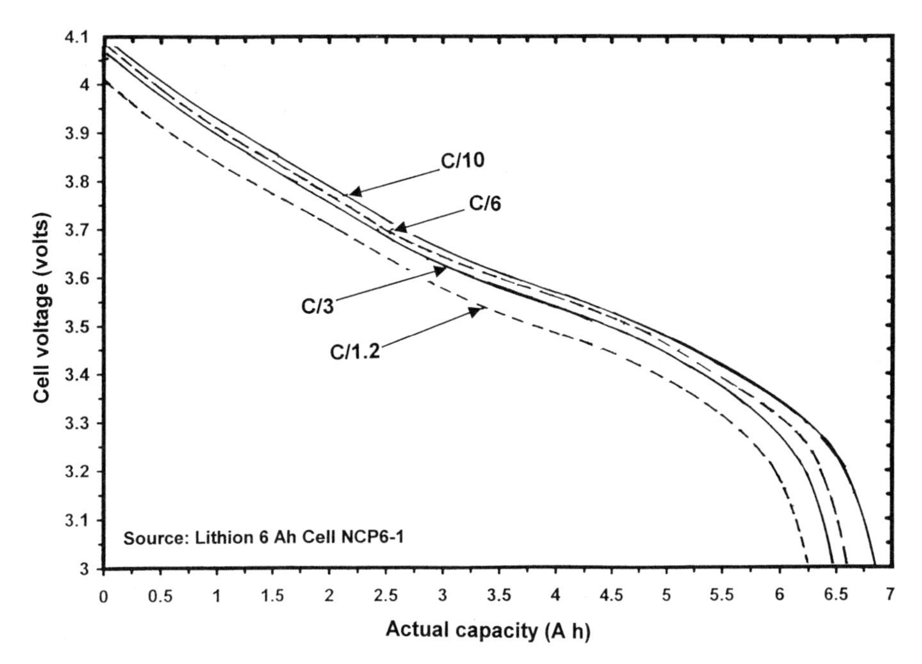

Figure 6.39 Discharge voltage of lithium ion cells at different rates.

Figure 6.38 also illustrates the typical charge characteristics of lithium ion cells as a function of temperature. While lithium ion cells perform well at room and higher temperatures, their performance is poor at cold temperatures. One experimental study indicates that this reduced performance at lower temperatures is related to increased impedance of the electrolyte and both plates. But the strongest influence seems to be the increase in positive plate impedance (Baker et al., 1999).

All of the manufacturers are continually evaluating this technology by conducting their own in-house tests and analyses. Figure 6.39 shows the discharge voltage of lithium ion cells for different rates. Most manufacturers recommend a minimum discharge voltage of 3.0 V. The preferred recharge method is to use constant current charging until full state of charge occurs at about 4.1 V, followed by constant voltage charging at that level (float voltage).

AEA in the UK supplies modular spacecraft batteries based on the Sony 18650 commercial type hard carbon (HC) cell. These small 1.5 A h cells have been in mass production since 1992. AEA screens each cell to a standard that has been in effect since 1995, allowing the accumulation of a database of statistical and performance data. Each Sony cell is individually equipped with a fuse to limit excessive current, and they are arranged in parallel groups to minimize the effect of individual cell failures through redundancy. The cells are connected in series strings which are then connected in parallel to form a circuit. These circuits are assembled into trays and trays are stacked into modules, which are connected in parallel to form the battery (Lizius et al., 2000). If a cell fails, the internal fuse will blow and isolate the string from the battery. The use of this approach in large batteries will lead to battery packs containing a large number of series/parallel cells.

Other manufacturers (notably SAFT, EaglePicher Technologies, LLC, Yardney, Lithlon, Japan Storage Batteries, Mining Safety Appliances, MECO, and others) are developing and testing large lithium ion battery cells with capacities from a few ampere hours to over 100 A h. Their development effort is supported by parametric and life cycle test programs. Life-cycling tests are an important part of battery development. They provide information not only on the "graceful" degradation in performance that inevitably accompanies cell ageing, but also on possible premature failure modes. Another important outcome from life-cycling tests is the results that allow the determination of the proper environmental and electrical limits and optimization of the required charge control parameters. These tests, together with abuse testing, provide the information to evaluate the effectiveness of the safety devices.

Part of any space-battery design is a failure mode and effects criticality analysis (FMECA), which has to include all conceivable ways in which a battery and its cells and other components may fail and to assess the impact of each of these failure modes on the battery's subsequent ability to support the spacecraft operation. To meet reliability requirements, a battery design may have to include appropriate means for minimizing the impact of a failed cell, whether failed as an open or a short circuit. In case of a short-circuit failure, the system design may have to meet the mission requirement with a lower battery voltage and less energy. In the case of nickel–hydrogen IPV and CPV cells, it is usually necessary to provide means for bypassing a cell that fails in the open circuit condition. This is often done by incorporating a circuit which detects an open circuit of a pressure vessel and activates a switch to bypass the failed vessel. Clearly this is not applicable in the case of the NiH_2 SPV battery. Open circuit failures in NiCd cells are possible but exceedingly unlikely and are usually not considered in the design. An open cell failure in a lithium ion cell is possible, so some lithium ion battery manufacturers place bypass circuits in parallel with the cells to isolate and bypass a badly degraded or an open circuit cell.

Battery sizing. The number of cells in a battery is determined by dividing the average battery bus voltage by the average cell discharge voltage:

$$N = \frac{V_{\text{ave}}}{V_{\text{cellave}}} \tag{6.6.1}$$

where

V_{ave} = average battery bus voltage (determined from system requirements),
N = number of cells in series,
V_{cellave} = average battery cell voltage during discharge (determined from cell tests).

The maximum battery bus voltage is equal to the number of cells in series times the maximum cell voltage (the open circuit voltage for primary cells). For secondary cells, the maximum cell voltage is at the maximum expected charge rate at the coldest battery temperature. The minimum bus voltage is similarly calculated from the minimum cell discharge voltage (measured at the maximum discharge and rate) at the coldest temperature.

The minimum battery capacity required is obtained by calculating the sum of the ampere hours required during the battery discharge. For secondary batteries, the capacity is sized taking into account the depth of discharge (DOD) that is allowed for the specific cell chemistry selected to satisfy the mission requirements:

$$C = \sum \frac{P_n T_n}{N V_{\text{cellave}} D_{\text{dod}}} \tag{6.6.2}$$

where

P_{n} = average battery load power during T_{n}, W,
T_{n} = duration of P_{n}, h,
D_{dod} = battery depth of discharge, % (100% for primary batteries),
N = number of cells in series,
V_{cellave} = average battery cell voltage during P_{n} discharge, V,
C = battery capacity, A h.

6.6.3 Fuel Cells

A fuel cell is an electrochemical device in which the chemical energy of a conventional fuel and oxidizer are stored external to the battery and are fed to it as needed. Because they consume both fuel and oxidizer, fuel cells are not usually considered for spacecraft missions of long duration. Typically, fuel cells designed for spacecraft applications produce power in the 1 kW to 5 kW range and may be required to operate for a time period ranging from a few minutes to approximately one month.

Fuel cells may be designed to utilize a wide range of fuels and oxidizers. However, the hydrogen–oxygen fuel cell is the only type that has been developed to a significant degree for spacecraft application, having been used successfully on both the Gemini and Apollo manned spacecraft programs. The Space Shuttle's Orbiter uses three fuel

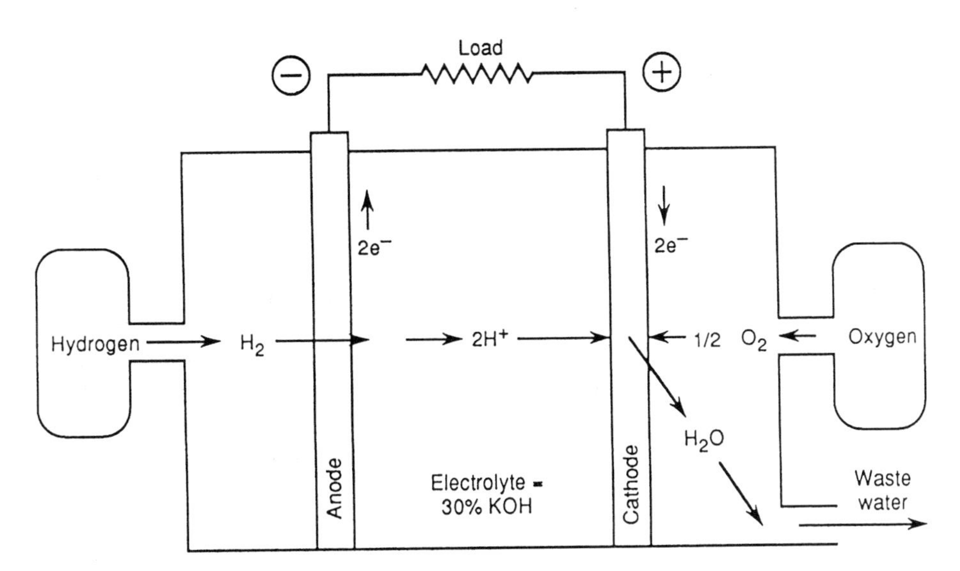

Figure 6.40 Simplified hydrogen–oxygen fuel cell.

cells. A simplified concept of a hydrogen–oxygen fuel cell is provided in figure 6.40 to illustrate its function. Gaseous hydrogen and oxygen are fed at controlled rates into the cell. At the anode–electrolyte interface, hydrogen passes through a porous anode and dissociates into hydrogen ions and electrons. The hydrogen ions migrate through the electrolyte to the cathode interface where they combine with the electrons, which have traversed the load, and with oxygen to form water as a by-product. The theoretical voltage generated by one such cell is approximately 1.23 V. In actual practice, a voltage of approximately 0.8 V is usually realized. Each cell has a fixed voltage output based on the electrochemical reaction that is taking place. To obtain higher voltages, cells are "stacked" or connected in series, similarly to the cells in a battery.

The design of a hydrogen–oxygen fuel cell for a spacecraft application is a significant challenge; many materials problems must be overcome and a mechanism must be devised to remove water from the cell in a zero-gravity environment. Screen wicks are invariably utilized in such designs to collect water by capillary action and to conduct it to some reservoir external to the generator. For manned missions, this can be used as a source of potable water.

The ion exchange membrane (IEM) for the hydrogen–oxygen fuel cell developed by the General Electric Company was an important breakthrough in early fuel cell technology. The ion exchange membrane is a quasi-solid electrolyte, which obviates the requirement for porous electrodes; thus, the selection and manufacture of electrode materials becomes a less complex task. The IEM fuel cell used on the Gemini spacecraft utilized platinum-coated titanium wire screen for electrodes, and this permitted a highly compact structure; the stacking factor of these cells was over 6 cells to the inch, and a power density of about 3.2×10^{-2} kW/kg (68 lb/kW) was achieved for the system. Approximately a pound of reactants was consumed, and a pint of water produced, per kW h of energy generated.

6.7 Solar Array Power Control Techniques

Over the past four decades of the space industry, a large number of solar array power control techniques have been employed. They fall into two broad categories: the series type that operates between the solar array and the bus, adjusting the solar array operating point voltage to control its power; and the shunt type that functions in parallel between the solar array and its load, diverting the current from the bus to accomplish the same result. A chart of these control techniques is shown in figure 6.41.

All of these techniques can be used with either a regulated or an unregulated bus. In general, the linear types provide the simplest control schemes and contribute less noise to the bus. However, they have inherently higher power dissipation. Both the digital and switching types have the advantage of minimizing the spacecraft internal power dissipation, but generate more noise on the bus. The switching-type regulators, driven by pulse width modulators (PWMs), tend to be more complex and heavier than the digital types. However, they (especially the buck and/or boost type) have the advantage of being useful with peak power trackers.

A number of these solar array power control techniques are presented. The control signals to these regulators are typically the bus voltage or the battery voltage and/or

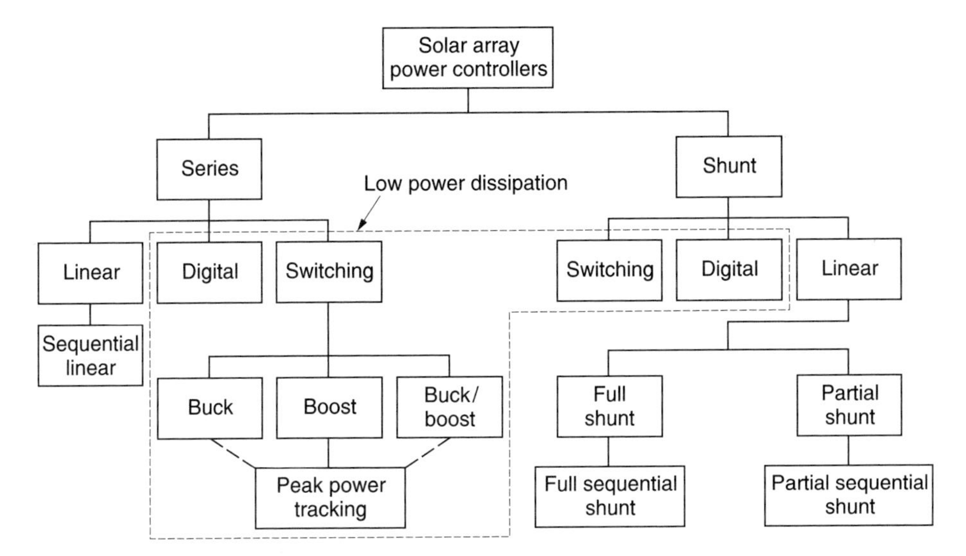

Figure 6.41 Classification of solar array power control techniques.

the battery current, battery temperature, and pressure. Error amplifiers compare the bus voltage, battery voltage, battery pressure, or current to a preset reference. The control signals from the error amplifiers drive the solar array power control circuits. The series pulse width modulated "buck" type peak power tracker controller is discussed in section 6.8.

6.7.1 Shunt Regulators

6.7.1.1 Linear Shunt Regulator

A circuit diagram and operating characteristics of a full linear shunt regulator is shown in figure 6.42. This has been a commonly used system, especially for small, low power (100 W) satellite applications. In this system, any power from the solar array not used by the loads and battery is dissipated in the shunt transistor(s) and resistors. The control signals to the shunt transistors control the proper amount of the array current to be diverted from the bus. In an actual system, a number of these units could be used in parallel to reduce the power dissipation in each unit.

Figure 6.42a shows the intersection of the solar cell I–V curve and the bus voltage during battery charge at P_1. This represents the current that can be delivered from the solar array to the bus. If the total demand of the spacecraft load and battery is defined as I_{bus}, then the remaining current (I_{sh}) must be accepted by the shunt. The total power dissipation in the shunt is simply $V_{bus}\, I_{sh}$ and is shown as the shaded rectangle in figure 6.42. Figure 6.42b shows how this power is distributed between the shunt transistor and resistor as a function of shunt current. The transistor power is low at low current, and it is also low at high current due to the low voltage drop of the saturated transistor. The maximum transistor power dissipation occurs at $I_m/2$, where I_m is the

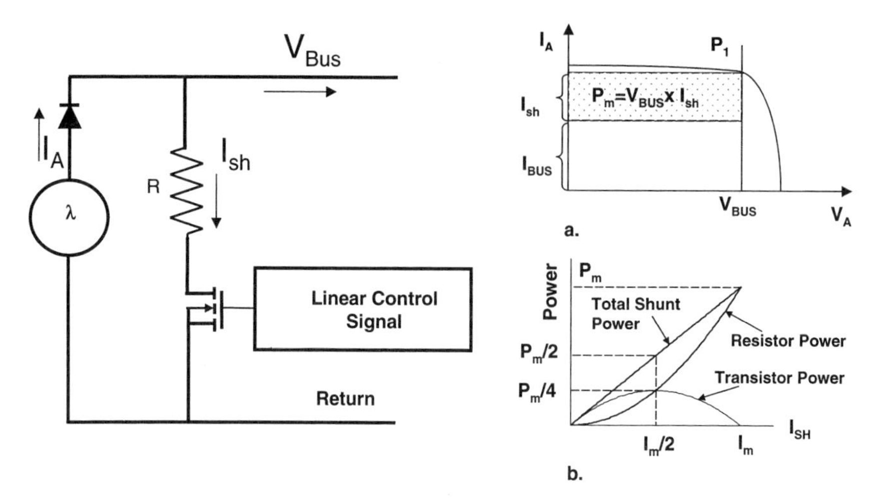

Figure 6.42 Solar array controller with full linear shunt.

maximum I_{sh}. At this point the total shunt is dissipating one-half of its maximum power, which is divided equally between the transistor and the resistor. Therefore, the transistor has a peak power of $P_{\text{m}}/4$.

Often shunt power levels are a significant problem to the thermal design, especially when there are large variations in the array current throughout the mission. One way to minimize this problem is to locate the shunt external to the spacecraft. The rugged construction of the passive resistor makes it an easy element to mount outside the spacecraft. It can also be manufactured as a distributed component that can be bonded to a radiative surface on the solar array, where its heat can be easily radiated to space. The transistor, however, is more susceptible to the effects of space radiation and the wide temperature excursions that are experienced outside the spacecraft. Therefore, the usual practice is to locate the shunt transistor inside and the resistor outside the spacecraft.

6.7.1.2 Partial Linear Shunt

The other ways of reducing a shunt's internal spacecraft power dissipation all employ techniques for not allowing unwanted solar cell array energy into the spacecraft. Rather, excess energy is left on the solar array for re-radiation to space. Figure 6.43 shows a regulator with a so-called *partial linear shunt*. In this configuration, the solar cell array is divided into upper and lower sections. The partial shunt regulator thus acts as a linear voltage regulator, reducing the array voltage to that point on the $I-V$ curve that supplies the required current. That is, the highest voltage at which the upper section can operate is $V_{\text{B}} - V_{\text{L}}$ (neglecting the diode voltage). By controlling V_{L} with the transistor shunt Q, the array I–V curve can be tailored to supply only the energy required by the system. Except for the energy dissipated in the shunt transistor Q, excess solar array energy is not transferred into the spacecraft.

Figure 6.43 can offer some clarification. The shaded area in figure 6.43a represents $P_{\text{sh}},$ the power dissipated in the shunt transistor Q. By comparison, the shaded area in

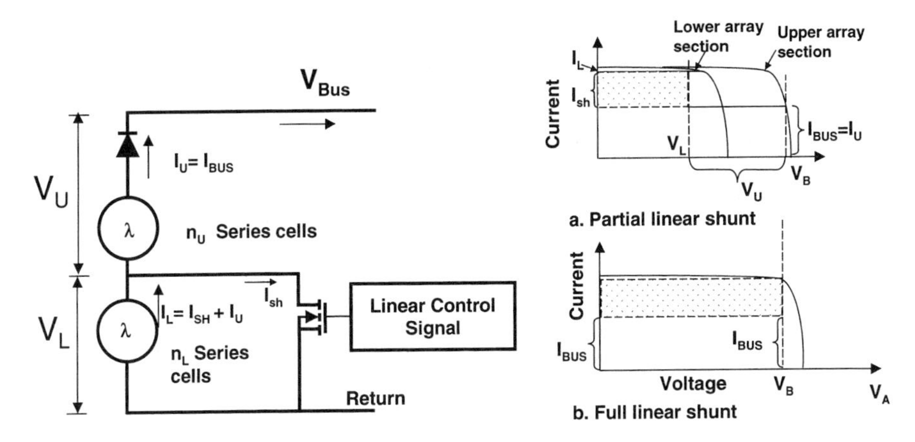

Figure 6.43 Solar array controller with partial linear shunt.

figure 6.43b, a similar graph for a full linear shunt regulator drawn to the same scale, is much larger. Notice that the partial linear shunt carries nearly the same current as the full linear shunt, but at a much lower voltage.

The number of cells in the upper section, n_U, is selected such that at the coldest predicted solar array temperature its voltage is not high enough to conduct any current if the shunt transistor is fully saturated, indicating that no power is needed from that section. Notice also that the partial shunt regulator, unlike the full shunt regulator, does not have a dissipator resistor in series with the control transistor. In the absence of this current-limiting device, it is common practice to make the transistor, Q in figure 6.43, quad-redundant to protect against a transistor short circuit. This will increase the reliability and reduce the power dissipation in each transistor since it distributes the same power over more devices. Note also that the solar array operating point on the I–V curve is on the short circuit (I_{sc}) side for the full shunt system whereas it is on the open circuit voltage (V_{oc}) side for the partial shunt.

6.7.1.3 Sequential Linear Shunt

Since the power dissipation in the transistor is low when it is either off or driven to saturation, it is possible to minimize the total transistor power dissipation in a group of parallel shunts by ensuring that only one transistor can operate in the linear range at any given time. All of the other transistors in the group would be either open circuit or driven to saturation. Figure 6.44 shows a full sequential linear shunt (a partial sequential type is another possibility). A sequencer is driven by the control signals. It ensures that all shunts are operated sequentially, where shunt m is operating through the linear range. All preceding shunts are saturated and all shunts following shunt m are off. This approach minimizes the total transistor power dissipation. It effectively ensures that most of the heat will be dissipated in the resistors, which can be located on an external radiator at the discretion of the thermal design engineer.

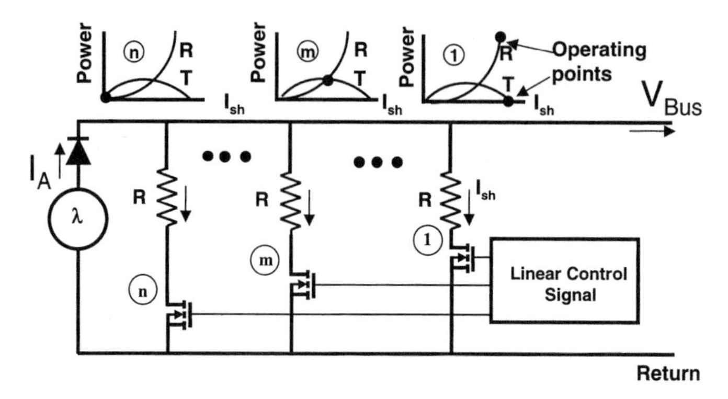

Figure 6.44 Solar array controller with full sequential linear shunt.

6.7.1.4 Digital Shunts

Still another way of limiting the spacecraft internal power dissipation due to excess energy from the array is with the use of digital shunts. These shunts have been used on many spacecraft to inhibit unwanted solar array energy from reaching the bus and going into the spacecraft, leaving it out on the array where it can easily be re-radiated to space. The solar cell array is divided into a number of parallel circuits, each of which is controlled by its own transistor switch. Each switch is connected in parallel across an array circuit and is either on or off, depending on the error-sensing signal(s). When the switch is on, it is saturated and effectively short-circuits the solar array circuit, dissipating very little power. When the switch is open it dissipates no power.

In practice, the digital shunt regulator is often augmented with the small, dissipative, linear shunt regulator shown in figure 6.45. In this system coarse control is provided by the digital shunts and fine control by the linear shunt. Instead of the small linear regulator, there are many systems, usually higher power systems, where a relatively small switching pulse width modulator (PWM) regulator is used to perform the fine control such as is used on French SPOT satellites and the NASA/GSFC EO-1.

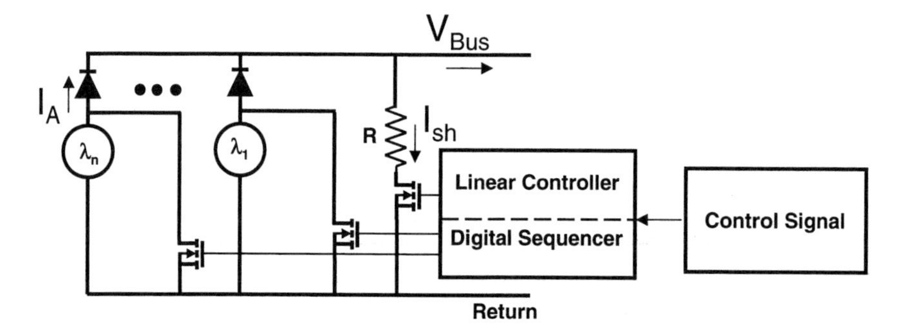

Figure 6.45 Solar array controller with combined digital and linear shunts.

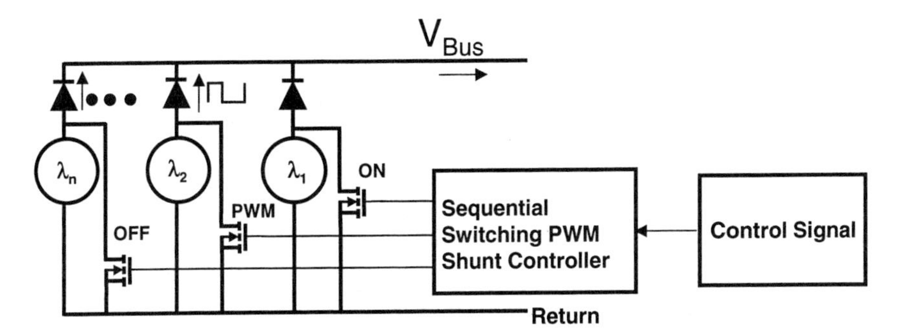

Figure 6.46 Solar array controller with sequentially switching shunt regulator (S3R).

This design, in its simplest form, is in a topology with the battery connected directly across the solar array. The digital shunt regulator does not use the linear shunt or a PWM regulator. Instead, only the digital shunts are used. After the battery is fully charged it will be continuously charged and discharged by very small amounts at a low repetition rate (about once per second). A controller of this type operated successfully for nearly five years on the APL-designed AMPTE/CCE (Active Magnetosphere Particle Tracer Experiment/Charge Composition Explorer) spacecraft, which was in a highly elliptical orbit and used a nickel–cadmium battery.

6.7.1.5 Sequentially Switching Shunt Regulator (S3R)

The simplified diagram of a sequentially switching shunt regulator, also known as an S3R, is shown in figure 6.46. The solar array is divided into segments, each with its own bypass-switching transistor. As with the sequential linear regulator shunt system, a sequencing function must be performed by the control system. On the basis of the error signal value, each segment switch transistor is pulse width modulated (PWM) at a high frequency between zero and 100% duty cycle. Only one segment transistor is operating in its PWM range; all others are either in saturation or open circuit. This approach further reduces the power dissipation in any one transistor compared to the linear regulator. More EMI is generated in this approach than in the linear control designs. This design has been used in very high power systems, especially in European geosynchronous satellites.

6.7.2 Series Regulators

Solar array power is controlled by transistor(s) placed between the solar array and the power bus. Whenever the pass transistor is driven to saturation, all of the current available from the solar array is provided to supply the load and recharge the battery. When the loads and the battery do not require as much power, the controller will begin to turn off the pass transistor, limiting the current from the solar array. Power limitation is achieved by forcing the solar array operating point out toward open-circuit voltage. This concept may be expanded to a sequential linear regulator where the solar array is divided into n parallel circuits, each with its own series pass transistor between its array circuit and the bus. As with the shunt system, a sequencing function must be performed by the control system.

The concept can be extended to series digital control, where the solar cell array is divided into a number of circuits, each of which is controlled by its own transistor switch connected in series between the array section and the bus. The switch is either on or off, depending on the error-sensing signal(s). When the switch is on it is saturated, dissipating very little power. When the switch is open it dissipates no power. This design is used on the HST spacecraft.

Series PWM controllers can also be implemented to perform linear or precision control, similar to the PWMs in the shunt controllers described above.

6.8 Space Power Systems

When there are no active components between the solar array and the load, their connection is said to be "direct" and the system belongs to a class called direct energy transfer (DET) systems. All other systems are labeled as non-DET. Most of the relatively simple systems are DET systems and would typically employ one of the shunt regulators described in the preceding section. However, those systems employing series regulators would belong to the non-DET class. Since peak power trackers all employ some form of series regulation, they fall into the non-DET class.

6.8.1 Direct Energy Transfer (DET) Systems

DET systems can be regulated or unregulated. The choice of whether to use a regulated or unregulated system is a complex issue and must be made early in the program. Prior experience of the participating engineers and the availability of existing hardware are two of the factors influencing this decision. Unregulated busses are simpler, but require more power regulation at the loads. For spacecraft with a large number of small loads, or in the case of GEO communication satellites with numerous identical transponders, regulated busses are often used. In addition, regulated busses decouple the battery from the solar array where the number of cells in series in the battery is independent of the bus voltage.

6.8.1.1 Unregulated DET System with Solar Array Clamped to Battery

A simplified block diagram of this type is shown in figure 6.47. This topology is used in low Earth orbit satellites where the battery charge power requirements are substantial and the recharge time is short, and also in small inexpensive satellites. The solar array is clamped to the battery. Therefore, the solar array power is directly supplied to the loads and battery recharge with no electronic processing. Very large transient loads are well accommodated in this design because of its low source impedance and minimal

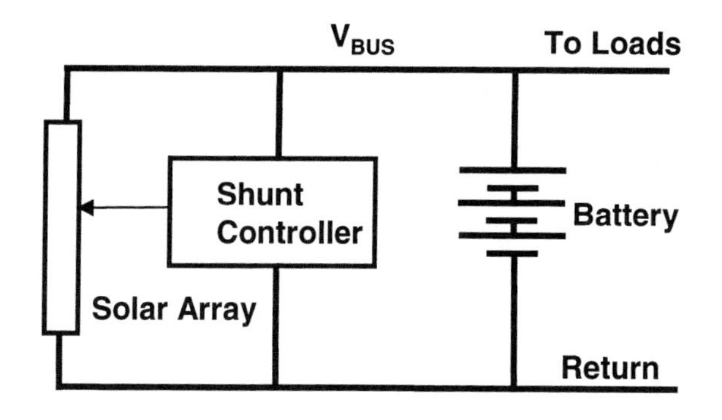

Figure 6.47 Unregulated direct energy transfer (DET) power system with solar array clamped to battery.

power-dissipating series elements. The simple design compensates somewhat for its inability to maximize the solar array power utilization.

The voltage of the battery determines the usable power from the solar array. Similarly to other DET systems, the number of solar cells in series must be selected such that the battery can be fully charged at minimum solar array voltage, which occurs at end-of-life and hot solar array conditions. As in any shunt system, the solar array operation is on I_{sc} side of the I–V curve (refer to figure 6.13). When the battery begins to recharge, following the end of discharge, its voltage (and the solar array power) will be low. Its voltage will increase as it charges, allowing the solar array power to increase. Considering that most spacecraft loads have high efficiency PWM converters regulating their power, they will present a constant power load to the bus. Since $I = P/V$, the required load current will decrease, allowing the battery charge rate to increase. But the battery charge efficiency decreases as it gets closer to full charge. Therefore, the charge rate must be decreased in order to not overcharge or overheat the battery, further reducing the solar array power utilization. For a nickel–hydrogen or nickel–cadmium battery, this reduction of the charge rate is normally accomplished by reaching a limit defined by the V/T or coulometer controllers (or the pressure controller for nickel–hydrogen), commanding the charge electronics to a lower charge rate. The lithium ion battery charge voltage rises more gradually than that of either of the nickel-based batteries, resulting in somewhat lower solar array utilization, a factor that must be considered in sizing the solar array. Also, a short-circuit failure of a battery cell will further reduce the operating voltage and hence the power from the solar array. The effect of this type of failure will be especially severe with lithium ion batteries since their cells have a full charge voltage of over 4 V.

The simplified relationship between the solar array power required to recharge the battery and that required to power the spacecraft loads is derived by calculating the average energy discharged during eclipse and the power required to return the discharged energy back to the battery. Taking into account the losses and ampere hour inefficiency of the battery, this is given by

$$P_{sa} = \frac{P_\ell}{\eta_u}\left[1 + \frac{T_{dis} V_{ch}}{T_{ch} V_{dis}} RF\right] \qquad (6.8.1)$$

where

P_{sa} = solar array power at full state of charge voltage, W,
(Constraint: The voltage of the solar array maximum power point for the hot case solar array at EOL must equal or exceed the maximum cold battery charge voltage required.)
P_ℓ = the load power (considered constant throughout the orbit), W,
T_{ch} = battery charge time, min,
T_{dis} = battery discharge time, min,
V_{ch} = average battery charge voltage, V,
V_{dis} = average battery discharge voltage, V,
η_u = solar array power utilization factor (approximately the ratio of nominal battery voltage to maximum battery voltage), dimensionless,
RF = battery ampere hour recharge fraction, which accounts for battery ampere hour inefficiency, dimensionless.

To perform an accurate solar array sizing, an energy balance program is required, which incorporates solar array and battery models and includes the panel thermal and orbital information for Sun–angle variations on the panels during the various phases of the orbits.

This topology has also been used on very low power small satellites. A linear full shunt regulator (section 6.7.1.1) is often used to control the solar array power. V/T, voltage, current, or pressure can be used to perform the battery charge control, depending on the battery technology selected. The SAS-A and B spacecraft were part of a series of Small Astronomy Spacecraft, developed by JHU/APL. Each had four deployed solar panels with cells on both sides. A small, 6 Ah, eight-cell NiCd battery was used to provide power to the spacecraft during launch and the eclipse periods. The eclipse power was about 25 W.

A single linear shunt, as shown in figure 6.42, was sized to dissipate the maximum solar array power, which could be provided by the solar array at maximum battery voltage for any illumination angle of the solar panels at beginning-of-life at winter solstice where the solar intensity is highest. The control unit was designed to employ a V/T limit to protect the NiCd battery from over-voltage and an electronic coulometer (ampere hour counter). The coulometer can be commanded to passively monitor battery charge state or can be selected to control the shunts to maintain proper battery current, reducing it when the battery is full and allowing higher charge rates when it is not.

Figure 6.48a shows the operation of this system as a function of time, over one discharge–recharge cycle, using just the voltage (V/T) limiter (i.e., no coulometer control). After discharge, the battery charge current increases to the amount available from the array. Then the voltage limit maintains a constant bus voltage for the remainder of the sunlight period as the battery current decays to about $C/10$ at room temperature. With knowledge of the NiCd battery characteristics, the V/T limit voltage was designed to reduce the overcharge rate to an acceptable level for that battery temperature. If the voltage level is set too low the battery will not be completely recharged. If too high, the battery cells will experience high internal pressures and overheating, shifting the operating point on the V/T curve (figure 6.32) to a higher battery temperature. If still higher overcharge rates result, it will cause even higher temperatures and a condition known as *thermal runaway* can occur.

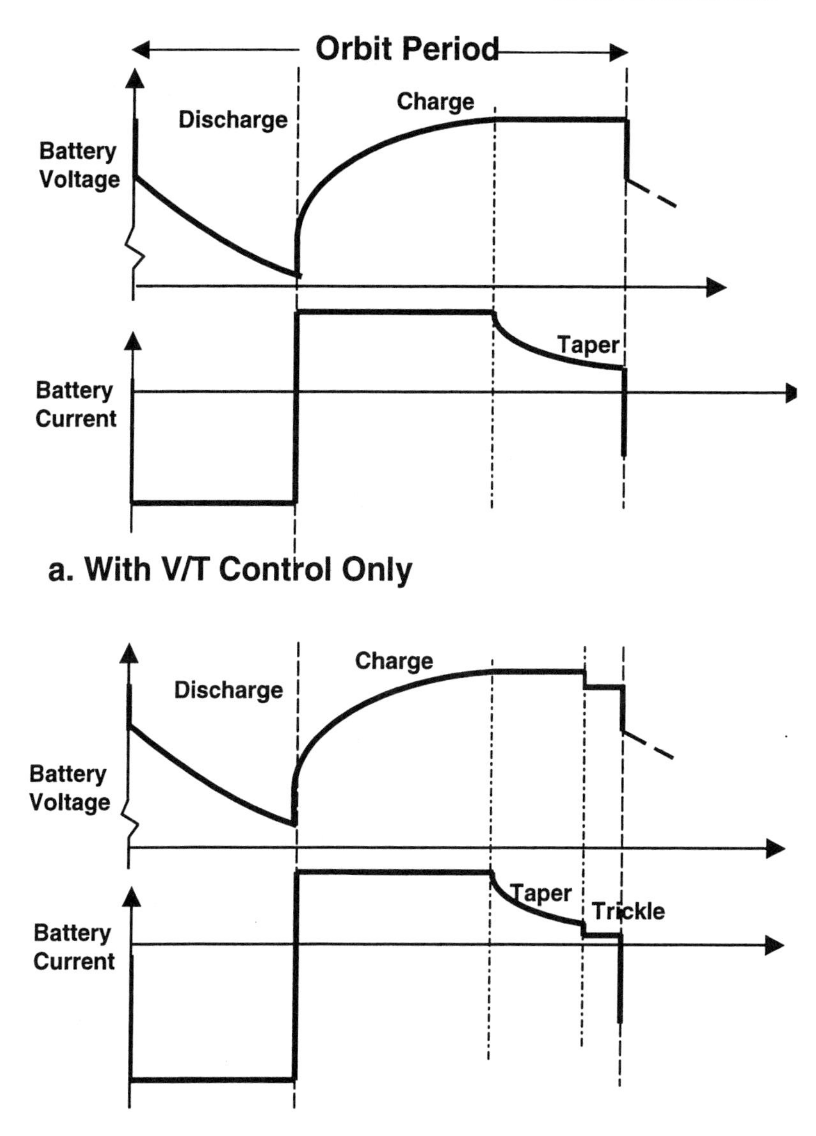

Figure 6.48 Discharge–recharge cycle of SAS-A power system with coulometer and/or V/T control. (Source: JHU/APL.)

It is prudent to design options into the system that reduce the system's dependence on a precise match between the battery and the voltage limiter characteristics. One approach to this for nickel-based batteries is to have multiple V/T curves that can be selected by ground command as shown in figure 6.32. NASA/GSFC uses eight such curves. Another approach is to use coulometer control, either instead of or in parallel with the V/T limiter.

Figure 6.48b shows the effect of using the electronic coulometer in parallel with the voltage limiter. The coulometer counts the number of ampere hours of discharge. However, during recharge, the counter is offset to count back more energy to compensate for battery inefficiency. The system behaves the same as with the voltage limiter until the coulometer indicates full charge. At that point, the battery current is reduced to a very low "trickle charge" rate of $C/100$, with a consequent slight drop in the charge voltage, an indication of a reduced rate of gas evolution within the battery cells.

If lithium ion batteries were used in this small satellite example, the V/T and coulometer controls would be modified to accommodate the lithium ion battery characteristics. The voltage is a good indicator of the lithium ion battery state of charge and the full charge voltage does not depend on battery temperature. The battery would be allowed to charge with available solar array current to a maximum limit. For short-duration missions, with a limited number of shallow discharges, individual lithium ion cell charge control is not needed. The battery voltage will be monitored and the charging would be terminated when the maximum voltage limit is reached. The battery will be maintained at full state of charge at this voltage.

6.8.1.2 DET Power System Regulated during Sunlight and Unregulated during Eclipse

In this DET system a battery charger decouples the solar array from the battery. Diodes provide a discharge path from the battery to the spacecraft power bus as shown in figure 6.49. The bus voltage is regulated during operation in sunlight to a fixed voltage when the solar array power exceeds the load and battery charge power. The charger is designed to support the maximum recharge rate required. When the load demand exceeds the solar array power, the bus voltage drops to the battery voltage, which provides the power difference. The battery charger is current limited. After discharge, when the solar array power exceeds the charger current limit level, the bus voltage recovers to the regulation level. Whenever the load demand exceeds the solar array power available, the bus voltage drops to the battery voltage. The maximum amplitude of power bus transients can be as high as the difference between the regulated bus voltage and the battery minimum discharge voltage.

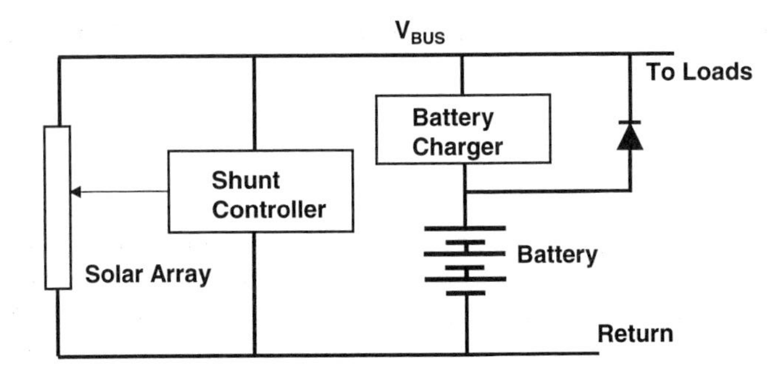

Figure 6.49 DET Power system regulated in sunlight and unregulated in eclipse.

This topology is used on spacecraft with relatively low load power (less than 1500 W), low bus voltage (typically 22 to 35 V), and missions with a relatively long battery recharge time as in geosynchronous, highly elliptical, or Sun-orbiting missions. It is prone to power system bus lock-up, which can occur when a transient load exceeds the solar array power capabilities, causing the solar array voltage to drop to the battery voltage while in sunlight. Although the array has ample power available at the higher regulation level voltage, its output power is lower at the lower battery voltage. The supply of power to loads will continue uninterrupted, with the battery supplying the difference in power between the load requirement and the solar panel capabilities at this low battery voltage. As the battery continues to discharge, its voltage will continue to drop and the portion of power supplied by the solar array will drop also. With constant power loads ($P = VI$), the load current will increase as the bus/battery voltage decreases. This will further aggravate the battery discharge rate. When the load transient ends, the bus voltage will return to its regulated level if the available solar array power at *that* battery voltage is more than the load power.

However, the bus voltage will remain at the battery voltage if the load power after the transient is larger than the solar panel power at that voltage; the battery will then continue to provide the difference in power. This situation will continue with further reduction in battery voltage and subsequent reduction in available solar array power at the lowered battery voltage. The system is said to be in "lock-up." This condition will continue until the spacecraft power system protections are activated which remove most of the loads from the power bus such that the remaining loads are less than the available solar panel power, thus allowing the power bus voltage to recover and the battery to recharge. Autonomy rules must be incorporated in the spacecraft to detect any possible lock-up and shed enough loads to ensure bus recovery with minimum solar array power at minimum battery voltage.

For LEO spacecraft, the simplified relationship between the solar array power required to recharge the battery and that required to power the spacecraft loads is derived by calculating the average energy discharged during eclipse and the power required to return the discharged energy to the battery, taking into account the losses of the battery, charge regulator, and the discharge diode inefficiencies;

$$P_{sa} = P_l \left[1 + \frac{V_{ch} T_{dis}}{V_{dis} T_{ch}} \times \frac{RF}{\eta_{dd} \eta_{cr}} \right] \tag{6.8.2}$$

where

P_{sa} = average solar array power during charge at regulated voltage value, W,
P_l = the load power (considered constant throughout the orbit), W,
T_{ch} = battery charge time, h,
T_{dis} = battery discharge time, h,
V_{ch} = average battery charge voltage, V,
V_{dis} = average battery discharge voltage, V,
η_{dd} = discharge diode circuit efficiency, dimensionless,
η_{cr} = charger circuit efficiency, dimensionless,
RF = battery ampere hour recharge fraction, dimensionless.

The solar array shunt limiter must be designed to handle the BOL solar array power not required by the minimum load. For GEO, MEO, and missions where the battery recharge duration is very long, a fixed charge rate is usually used (between $C/10$ and $C/20$, depending on the battery technology). The above equation can be simplified to:

$$P_{sa} = P_l + V_{bat} \times \frac{C}{K\eta_{cr}} \tag{6.8.3}$$

where C is the battery capacity, K is the charge rate selected, V_{bat} is the average battery voltage during charge, and η_{cr} is as defined above. This topology was used on the APL-designed CONTOUR and NEAR spacecraft, using digital shunts for coarse solar array power control and full sequential linear shunts for fine control.

6.8.1.3 Regulated DET bus power system

In this DET system the power bus is regulated at all times by charge and discharge regulators, as shown in figure 6.50. This design has more electronics than in the other two DET topologies described. However, it provides the most flexibility. Since the bus voltage does not vary during normal operation, it does not cause variations in the solar array power. It eliminates the lock-up concern, the power bus does not have large transients, the spacecraft load power converters are simpler, and the battery design does

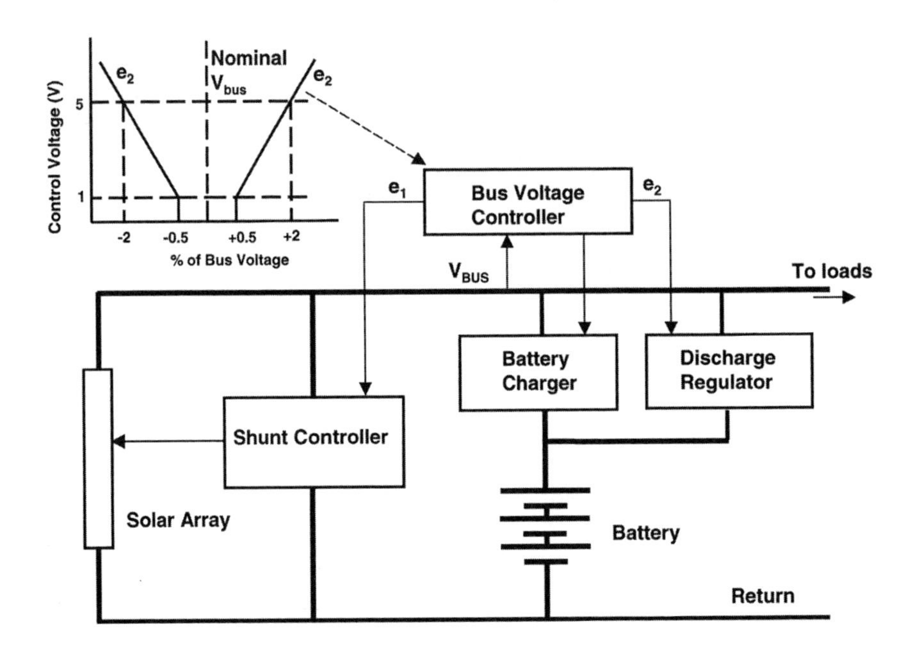

Figure 6.50 Simplified block diagram of a regulated direct energy transfer (DET) power system.

not depend on power bus voltage. This design is used in very high power spacecraft where high bus voltages (over 100 V) are required to reduce the wiring size and associated power distribution power losses and harness weight. The number of cells in the battery is made independent of the selected bus voltage. Practically all high power geosynchronous communication satellites, the Space Station, and some LEO spacecraft have this topology. The charge and discharge regulators on these high power systems are high efficiency PWM switching regulators.

The relationship between the solar array power required to recharge the battery and power the spacecraft loads is derived similarly to the previous system topologies, taking into account the discharge regulator efficiency.

$$P_{sa} = P_l \left[1 + \frac{V_{ch} T_{dis}}{V_{dis} T_{ch}} \times \frac{RF}{\eta_{dr} \eta_{cr}} \right] \tag{6.8.4}$$

where

P_{sa} = solar array power at regulated bus voltage, V,
P_l = the load power (considered constant throughout the orbit), W,
T_{ch} = battery charge time, min,
T_{dis} = battery discharge time, min,
V_{ch} = average battery charge voltage, V,
V_{dis} = average battery discharge voltage, V,
η_{dr} = discharge regulator circuit efficiency, dimensionless,
η_{cr} = charger circuit efficiency, dimensionless,
RF = battery ampere hour recharge fraction, dimensionless.

The shunt limiter must be designed to handle the BOL solar array power that is not required by the minimum load. The battery discharge regulator must be designed to support the total load power. The power system could use any of the shunt-type regulators described above in section 6.7.1.

Similarly to the system in section 6.8.1.2, for GEO, MEO, and missions where the battery recharge duration is very long, we can again use equation 6.8.3 to calculate the solar array power.

In the simplified block diagram of figure 6.50, the bus controller senses the bus voltage V_{bus}. The transfer function (inset a) indicates that when the bus is within $\pm$ 0.5% of the nominal V_{bus}, then both output signal lines (e_1 and e_2) from the bus voltage controller are less than the threshold that would activate either the shunt limiter or the discharge regulator (a boost regulator), and no action is taken.

If the bus voltage increases more than 0.5% above the nominal V_{bus}, then the bus voltage controller will produce a linear signal at e_1 that will cause the shunt limiter to maintain the bus voltage to within the specified range of $\pm$ 2% of its nominal V_{bus} value. If the bus voltage were to drop more than 0.5% below V_{bus}, as it certainly would during discharge, then the bus voltage controller would deliver a signal e_2 to the discharge (boost) regulator, enabling it. The signal e_2 will then cause the discharge regulator to deliver sufficient current from the battery to the bus to hold the bus voltage within 2% of V_{bus}.

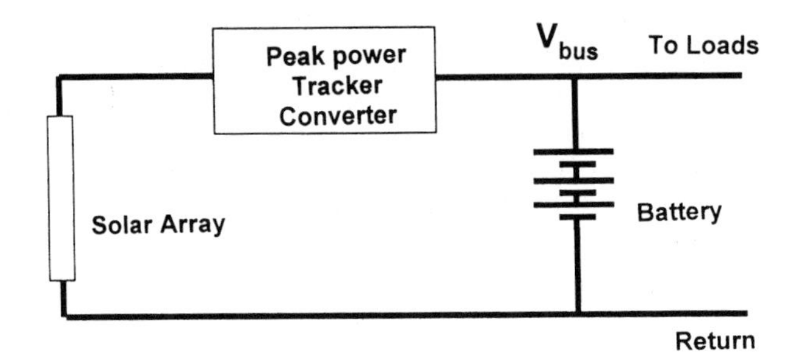

Figure 6.51 Peak power tracking (PPT) power system concept.

6.8.2 Non-DET Systems

6.8.2.1 Peak Power Tracker Power System

A simplified block diagram of a peak power tracking system is shown in figure 6.51. The peak power tracker (PPT) is capable of continuously extracting the maximum power from the solar array by moving the solar array operating voltage to the maximum power point voltage. This peak power tracking will continue as long as the load and the battery recharge requirements exceed the maximum power available from the solar array. When this maximum power is no longer required (as when the battery recharge current is reduced), the PPT controller automatically moves the solar array operating point away from the maximum power point until energy balance is reached (the generated energy is reduced to that which is required). With a typical step down (buck) PWM converter(s), the solar array operating point is moved off the maximum power point in the direction of V_{oc}.

The simplified relationship for the solar array power (P_{sa}) required for recharging the battery and powering the spacecraft loads (P_ℓ) is derived, as for the other topologies discussed above, by calculating the average energy discharged during eclipse and the solar array power required to return that discharged energy and power the loads taking in consideration the peak power point tracking accuracy (η_{tr}), the efficiency of the peak power point tracker converter electronics (η_{ppt}), and the ampere hour battery recharge fraction (RF) to compensate for the battery inefficiency:

$$P_{sa} = \frac{P_l}{\eta_{ppt}\eta_{tr}}\left[1 + \frac{T_{dis}V_{ch}}{T_{ch}V_{dis}}RF\right] \tag{6.8.5}$$

where

P_{sa} = solar array power required, W,
P_l = the load power (considered constant throughout the orbit), W,
T_{ch} = battery charge time, min,
T_{dis} = battery discharge time, min,
V_{ch} = average battery charge voltage, V,
V_{dis} = average battery discharge voltage, V,
RF = battery ampere hour recharge fraction ratio, dimensionless,
η_{ppt} = peak power tracker converter efficiency, dimensionless,
η_{tr} = peak power tracking accuracy, dimensionless.

The peak power tracking system is most advantageous for those applications where the solar array I–V curve varies widely throughout the mission. It is therefore often used in low Earth orbit missions and in planetary missions where the solar array operating temperature varies substantially during the mission. The PPT system decouples battery voltage variations from those of the solar array. The power system is tolerant of variations in the battery voltage caused by degradation over life (even a short-circuited battery cell) or unexpectedly large voltage degradation of the solar array. However, the power electronics is complicated and is usually heavy due to its relatively large thermal dissipation; this is a byproduct of its function to process all of the solar array power in order to recharge the battery and power the loads.

The power system for the Messenger spacecraft, built by JHU/APL, is a peak power tracking system. The spacecraft is designed to orbit the planet Mercury after a six-year cruise. During the mission the temperature on the solar array can vary between–150°C and 270°C. MESSENGER uses eight step-down (also called "buck type") regulating PPT converter modules, operating in parallel. The outputs of the PPT converters are connected to the battery. They are designed to process about 750 W of solar array power. The spacecraft computer performs the peak power calculations and sets the converter reference voltage. The control loop of the buck converters varies the duty cycle to maintain the input voltage from the solar array wings to a reference value set by the spacecraft computer.

The MESSENGER solar array current is monitored by the data acquisition system. The spacecraft computer sets the solar array reference voltage and calculates the instantaneous power of the solar array. The power value is compared to a previously stored power value. If the new value is larger than the previous value by a preset amount, then the computer control signal is moved in the same direction as the previous change. Otherwise it is moved in the opposite direction, with a smaller step voltage value.

Two step sizes are used to speed up the conversion. A large step is used initially. If the direction change is made, then the smaller step size is used. The step size reverts to the larger size if more than a predetermined number of steps are taken without a change in direction. The solar array's operating voltage range is bounded by maximum and minimum voltage values. The maximum limit is the predicted maximum solar array open-circuit voltage (V_{OC}) and the minimum is the minimum battery voltage at end of discharge. If these limit values are reached, then the direction of the solar array voltage change is reversed.

The TIMED spacecraft, developed at APL and launched in December 2001, is also a peak power tracker system. TIMED PPT converters were designed to process over 2000 W of solar array power.

6.8.3 A Power System Design Example

The starting point for any subsystem design is the mission and spacecraft requirements, which define the objective of the entire program and the specific objectives of each spacecraft. The first step in any power system design is therefore to develop the power system requirements from the mission/spacecraft requirements. The power system configuration, size, weight, and cost can then be determined. Once the requirements are defined, the design should proceed in a logical sequence. To illustrate: the following example is for the preliminary design of a power system for one spacecraft in a constellation of scientific satellites in low-Earth circular orbit.

6.8.3.1 Mission Requirements

A constellation of near-polar (high inclination) operational spacecraft is required to perform a search and rescue function. To minimize the cost of design, fabrication, and test, the spacecraft will all be designed and fabricated as nearly identical as possible. Each spacecraft will be Earth-pointing in a Sun-synchronous, low Earth orbit. The orbit planes (or their nodal crossings) will be separated by angles that will result in complete coverage of the Earth's surface. The primary payload of each spacecraft will be high-powered, Earth-pointing transmitters and receivers with wide-angle fields of view. The power level of the transmitters and the sensitivity of the receivers will be such that they will be able to communicate with low cost, low power, hand held, user-owned devices. The performance (reception and transmission) of these instruments will be insensitive to the spacecraft's rotation angle around its nadir-pointing vector. Therefore, as long as the primary instruments are aligned with the nadir vector, the spacecraft can be rotated about this vector to point the solar array toward the Sun. The vehicles must be designed for a minimum operational lifetime of four years.

Note: Sun-synchronous orbits are orbits with the secular rate of the right ascension of the ascending node equal to the right ascension rate of the mean Sun, so that the angle between the Sun vector and the orbit plane are nearly constant.

6.8.3.2 Spacecraft Requirements

The particular spacecraft for this example will be for a nominal twelve o'clock nodal crossing. Individual designs may use redundant components to enhance reliability, at the discretion of the appropriate designers and system engineers. But, to minimize costs, there is no spacecraft system requirement for redundancy of the major subsystems, such as the battery or power system electronics. Proven design and hardware will be used wherever possible. A circular orbit at low altitude has been selected to keep launch costs and instrument power level to a minimum. Also, the orbit altitude will be below 1000 km to minimize particle radiation and above 650 km to minimize the damaging effects of atomic oxygen. A preliminary study yields additional requirements, all of which are summarized as follows:

- Launch date: January 2006.
- Lifetime: 4 years (required minimum).
- Electrical load: 825 W (steady-state operational, constant power).
- Orbit: Sun synchronous, 700 km circular, 98.2° inclination, noon–midnight (12 am–12 pm) orbit, where the sun is always in the orbit plane.
- Attitude control: three axes stabilized: antenna and receiver are both nadir pointing.
- Solar array will be a single wing with single axis rotation.
- The rotation axis of the solar array will be held perpendicular to the Sun-line by controlling the rotation of the spacecraft about its nadir vector.
- The spacecraft guidance and control subsystem will have the responsibility to maintain the solar panel normal pointing toward the Sun.

6.8.3.3 Power System Topology

The power system topology selected for this LEO spacecraft is a non-DET peak power tracker with a single battery connected to the power bus.

6.8.3.4 Solar Array Configuration

Although there are many factors that influence the specific design of the solar array, the three primary drivers in deciding its basic configuration are:

- The power requirement.
- The tracking/pointing timeline in the orbital configuration.
- The volume constraints in the launch configuration.

As a result of trading off all these factors between the various spacecraft subsystems, a single wing with two interhinged, deployable panels was selected with a single axis of rotation. The two panels will be folded and stowed along the side of the spacecraft, inside the fairing (the heat shield), so that it has no direct interface with the launch vehicle at launch. After launch, when the heat shield has been rejected and the spacecraft separated from the launch vehicle, the panels will be deployed. Then, after the spacecraft attitude control has stabilized, the solar array wing will be rotated to track the Sun during the sunlit portion of the orbit. The spacecraft's orientation about its nadir-pointing vector will be adjusted periodically by the guidance and control (G&C) system to allow the panel rotation control to maintain the solar panel normal to the Sun-line. Triple junction MJ cells are used to minimize the size of the solar panels.

6.8.3.5 Radiation Effects

The 1 MeV electron/cm^2 equivalent radiation is calculated for different coverglass thicknesses. Four-mil-thick cerium-doped microsheet coverglass is selected to shield the cells from radiation. Thicker coverglass will make the overall panel requirements smaller but will increase the weight. The selection is normally a compromise between the increased weight of the thicker coverglass and that of the larger panels resulting from the use of thinner covers. Coverglass with a thickness of less than 3 mils is not used because it is fragile, leading to handling difficulties and increased manufacturing costs. The total dosage (fluence) received by the cell also takes into account the shielding provided by the mass of the solar panel substrate in calculating the impinging radiation from the back of the solar cells.

The total radiation dosage received by the solar cells over the 4-year mission, including solar flare protons for this orbit, will be taken in this example to be 1.0×10^{15} 1-MeV electrons/cm^2 (for the convenience of using the parameter values provided in table 6.2).

6.8.3.6 Effects of Orbit Parameters

Design parameters, determined for a 700 km circular polar orbit, are listed in table 6.6. They have been calculated from equation 6.3.3 and the three equations of figure 6.2.

Table 6.6 Orbit-driven design parameters

	Parameter	Value
1.	Orbital period	98.76 min
2.	Maximum eclipse time (35.72 %)	35.3 min (nearly constant*)
3.	Minimum time in Sun (64.28 %)	63.5 min (nearly constant*)
5.	Number of orbits per day	14,581
6.	Number of orbits per year	5322
7.	Number of orbits in four years	21,288

* Due to Sun synchronization of orbit plane.

6.8.3.7 Determination of the Nominal Bus Voltage

Often the bus voltage level is chosen to satisfy existing load hardware. For instance, a spacecraft power system nominal bus voltage may be selected as 28 V because existing subsystems and instruments have been previously designed to this level and now require this voltage. However, the choice of voltage is basically a tradeoff between the heritage of existing hardware, the availability and reliability of circuit components, and overall system power losses. At higher voltages there are fewer high-reliability relays, capacitors, and power transistors to select from. But at lower voltages the power system current must be higher to deliver the required load, leading to enlarged switching relay sizes and increased power losses in diodes, transistors, and the harness. Considering these factors, many high-power geosynchronous satellites operate at 100 V; for example, the Space Station bus voltage is 120 V. Most lower-powered spacecraft of less than 1500 W will elect to use a nominal 28 V bus. Since we have defined a lower power spacecraft in this design example, a nominal 28 V bus is selected.

6.8.3.8 Battery Type, Bus Voltage, and Current

For a mission of this lifetime (~20,000 charge–discharge cycles), the present options of batteries with flight heritage are nickel–cadmium and nickel–hydrogen cells. For this example, nickel–hydrogen IPV cells are selected because of their proven capability to meet the large number of charge–discharge cycles required. The manufacturer's test data provides us with the following battery cell parameters:

- The *maximum battery cell voltage,* which occurs during charge at the coldest operating temperature (−5°C) at the high charge rate, is about 1.6 V.
- The *minimum cell voltage*, which occurs at the end of discharge, typically taken to be the "knee" of the discharge curve, is about 1 V.
- The *average cell charge voltage* is 1.5 V.
- The *nominal cell discharge voltage* is about 1.25 to 1.28 V.

Cell voltages are limited to a maximum value because higher voltages can shorten the cell's lifetime. Similarly, they are limited in their minimum voltage. The "knee" is the point where further discharge will rapidly run the voltage to zero. Since the battery cells in a stack do not have identical capacities or operate at identical temperatures, they will not reach this point simultaneously. Therefore, a cell that has been discharged to less than 1 V is at risk of being driven to voltage reversal by the other cells that have more capacity. Since this would damage the cell or activate the cell bypass switch if it is being used, it is typically defined as the lower allowable limit voltage (the voltage at which the discharge current is interrupted during ground test or significantly reduced by load shedding in an orbital situation).

For an unregulated, nominal 28 V system, twenty-two cells ($\sim 28/1.28$) in series are required. Therefore, the expected power bus voltage extremes for the 22-cell battery are about 22 to 35 V. To account for possible loss of one failed short battery cell, the minimum voltage typically specified for spacecraft critical loads is 21 V.

6.8.3.9 Battery Capacity

From life test data on NiH_2 cells and discussions with the battery manufacturer (EaglePicher Technologies, LLC, http://www.epi-tech.com), the recommended depth of discharge (DOD) is 35%:

$$\text{Battery capacity} = \frac{(\text{eclipse load power}) \times (\text{longest eclipse time in hours})}{(\text{average battery discharge voltage}) \times (\text{battery DOD})}$$
$$= \frac{825 \times (35.28/60)}{28 \times 0.35} = 49.5 \text{ Ah}$$

From the manufacturer's existing cell designs, a 50 A h individual pressure vessel (IPV) cell is selected.

6.8.3.10 Solar Array Design

The energy balance equation (6.8.5) for a peak power tracking (PPT) system, from section 6.8.2.1, is:

Solar array power during the sunlight period:

$$P_{sa} = \frac{P_l}{\eta_{ppt}\eta_{tr}}\left[1 + \frac{T_{dis}V_{ch}}{T_{ch}V_{dis}}RF\right] \tag{6.8.6}$$

where

load power $P_l = 825$ W,
battery charge time $T_{ch} = 63.5$ min,
battery discharge time $T_{dis} = 35.3$ min,
battery average charge voltage $V_{ch} = 33$ V,
EOL battery average discharge voltage $V_{dis} = 28$ V,
battery recharge fraction RF (between 1.03 to 1.1 for NiH_2 to account for battery ampere hour inefficiency over life) = 1.1 (conservative value),
PPT converter efficiency $\eta_{ppt} = 0.95$,
PPT accuracy $\eta_{tr} = 0.99$.

The required solar array power at PPT converter input $= 825 \times [1 + (35.3/63.5)g \times (33/28) \times 1.1]/0.95 \times 0.99 = 825 \times 1.8296 \simeq 1510$ W.

The solar array sizing is strongly impacted by the predicted solar array operational temperature for the hot case at the end-of-life because this represents the lowest voltage from the solar cell array. Under these conditions, the solar array voltage must equal or exceed the maximum charge voltage of the battery plus the voltage drop of any components between the array and the battery. This is the case that determines the number of solar cells that are required in series. The hottest solar array temperature occurs at maximum solar intensity, or minimum distance between the Sun and the Earth (perihelion). This occurs annually on approximately January 3rd, a few weeks after winter solstice. The calculated maximum temperature of the solar array (SA) at this point is 90°C.

The maximum battery charge voltage $\leqslant$ 35.0 V
Voltage drop across the PPT converters = 3.0 V
Blocking diodes (to decouple the SA strings) = 0.7 V
Wiring between the battery and SA = 0.3 V
Minimum SA output voltage at the voltage of its maximum power point (V_{mp}) at EOL, hot case $= 35.0 + 3.0 + 0.7 + 0.3 = 39.0$ V

The two multijunction solar cell manufacturers in the US are Spectrolab Inc. and EMCORE Corporation.[6] The cell manufacturer's data sheets provide information on the primary solar cell parameters I_{sc}, V_{oc}, V_{mp}, and I_{mp}. They also provide information on the variation of these parameters with temperature and 1 MeV electron equivalent radiation. The maximum power and efficiency are then derived from these parameters and their temperature and damage coefficients. The solar array power calculations are typically done in a spreadsheet, using the solar cell parameters. The losses due to cell mismatch, welding/soldering of interconnects, UV radiation, coverglass/adhesive/cell spectral mismatch, spectrum calibration, micrometeorites in addition to charged particle radiation, array temperature, Sun distance variations, and changes in Sun angle to panel normal are considered in the spreadsheet. Only the main losses are considered in this example, namely the temperature, Sun distance variations during the year, and radiation. The other losses are relatively small and have been neglected for simplification. (Note: the array pointing is maintained normal by the spacecraft guidance and control system).

The power at the maximum power point is obtained in this example by calculating the EOL voltage and current, V_{mp} and I_{mp} respectively. It can also be calculated using the parameters for the maximum power (P_{mp}). From the manufacturer's data sheets (available from their websites), or using the typical cell data in table 6.2, the solar cell parameters are:

$V_{mp} = 2.345$ volts before radiation degradation

After radiation degradation of 1 E15 1-MeV electrons/cm^2, the maximum power point voltage becomes:

$2.345 \times 0.90 = 2.111V$. V_{mp} temperature coefficient $= -6.8\text{mV}/°\text{C}$

Therefore, the MJ cell voltage after radiation (at EOL) at maximum temperature is

$2.111 - [6.8 \times (90 - 28)/1000] = 1.689$ V

[6]Spectrolab Inc., Division of Boeing Co. (www.spectrolab.com), and EMCORE (www.emcore.com).

The number of cells in series in a string:

$39.0/1.689 = 23.09$; use 23 cells in series

The minimum available solar array power condition, which determines the size of the solar array, occurs at end-of-life (EOL) at that time of year when the solar intensity is at a minimum. Minimum solar intensity occurs when the distance between the Sun and the Earth is a maximum (aphelion). This happens annually on approximately July 4 th, a few weeks after summer solstice (SS). The worst-case power available in this "minimum power" case occurs when the temperature is a maximum during these orbits. Thermal analysis of the solar array indicates that at EOL and aphelion the maximum array temperature is 80°C.

The minimum EOL values for V_{mp} and I_{mp} of a string are calculated next. The solar cell string voltage under these conditions is:

$$23 \times (2.111 - 6.8(80 - 28)/1000) = 40.5 \text{ V}$$

In this preliminary design, it is convenient to express the diode and solar array harness voltage as a percentage of the solar array voltage. This approximates the percentage of lost power: $[(0.7 + 0.3)/40.5] \times 100 = 2.47\%$

The triple junction solar cell current density at maximum power point (J_{mp}) at the beginning-of-life (BOL) can be obtained from the manufacturer's data sheets. For our purposes, we will use table 6.2, which gives typical values for triple junction cells:

$\text{J}_{mp} = 16.2 \text{ mAcm}^2$
Temperature coefficient of $J_{mp} = 9.0\,\mu\text{A}/(°\text{Ccm}^2)$
J_{mp} after radiation degradation: $16.2 \times 0.95 = 15.39 \text{ mA/cm}^2$
J_{mp} at 80°C (aphelion) $= 15.39 + 9.0 \times (80 - 28)/1000 = 15.86 \text{ mA/cm}^2$
At aphelion, the current density will be less by a factor of 0.965 to account for the maximum solar distance.
J_{mp} at aphelion hot $= 15.86 \times 0.965 = 15.30 \text{ mA/cm}^2$

For a cell size of 24 cm^2, the current at the maximum power point at aphelion hot at EOL $= 24 \times J_{mp} = 24 \times 15.30 = 367.2$ mA.

Therefore the minimum power from a 23-cell string for the EOL (aphelion) hot case at the maximum power point is

$$(\text{Cell current}) \times (\text{string voltage}) = 367.2 \times 40.5/1000 = 14.87 \text{ W}$$

Taking into account the solar array line and diode losses, the solar cell string power at the input of the PPT converters is

$$14.87 \times (1 - 0.0247) = 14.50 \text{ W}$$

The total number of strings required is the solar array power at input of PPT converters divided by the power at PPT input per string:

$$1510/14.50 = 104.14; \quad \text{we will use 104 strings}$$

Therefore the net solar cell area required = (cell area)×(number of cells in a string)×(number of strings) $= 24 \times 104 \times 23 = 57,408 \text{ cm}^2 = 5.74 \text{ m}^2$.

For the preliminary design, assume a packaging factor of 0.85 to account for the spaces between cells, the stay-out areas for hinges, mechanisms, and other mechanical areas where cells can not be placed.

$$\text{The total solar array area required} = 5.74/0.85 = 6.75\ \text{m}^2$$

Assuming the panels to be nearly identical, the area of each of the two panels = $6.75/2 = 3.375\ \text{m}^2$

The solar array will deliver its maximum power and voltage at the beginning- of-life at the coldest array temperature (which will occur at the exit from the longest eclipses during those periods when the spacecraft is at its minimum distance from the Sun (perihelion)). For these conditions, the solar array thermal analysis indicates that the array's coldest temperature will be −70°C.

Next, we must calculate the maximum solar array voltage and power for the same conditions. The solar cell parameters are calculated as follows (as before, BOL values at 28°C and temperature coefficients are taken from table 6.2):

$$V_{oc}\text{@BOL, cold} = 23 \times [2.66 - 6.3 \times (-70 - 28)/1000] = 75.38\ \text{V}$$

$$V_{mp}\text{@BOL, cold} = 23 \times [2.345 - 6.8 \times (-70 - 28)/1000] = 69.26\ \text{V}$$

$$I_{mp}\text{@BOL, cold, perihelion} = 24 \times [17.0 + 9.0 \times (-70 - 28)/1000]/1000 = 0.387\ A \qquad (6.8.7)$$

BOL maximum power from the solar array = (number of strings) × (string voltage at V_{mp}) × (string current at V_{mp}):

$$\text{BOL maximum solar array power:} 105 \times 69.26 \times 0.387 = 2814\ \text{W}$$

The PPT electronics must handle these maximum solar array powers and voltages, even though they may only be transitory.

The above, simplified design assumes that the battery is charged with available solar power until the recharge ratio is reached. For each battery type (NiCd, NiH_2, or lithium ion), the solar array charge current is normally reduced to accommodate the reduction in battery charge efficiency as the battery approaches full charge. This reduction of the battery charge current will reduce the solar array power utilization and increase the required battery charge time, contributing to the need to increase the size of the solar array beyond the minimum size that we have calculated in this example.

To be complete, the simplified analysis described above needs to be refined using an energy balance program, which incorporates the appropriate solar array and battery models and includes the panel thermal and orbital information for Sun-angle variations on the panels during the various phases of the orbits.

The design of a solar cell array is an iterative process, in which the approximate array area is first determined by a solar array analysis of the type described. From this the approximate area is determined, and the mechanical designer then determines the best method of obtaining the required area. For this example, a single wing consisting of two solar panels is selected.

6.9 Conclusion

The power system is one of the major spacecraft subsystems, with interfaces to all other subsystems. However, it has particularly important, interdependent interfaces with the guidance and control, thermal, and mechanical subsystems. During the preliminary design phase, the system engineers must work closely together, sharing their ideas openly and efficiently. The power system designer must also be sensitive to the effect of that system on others, such as electro magnetic interference (EMI), electrostatic discharge (ESD), induced and static magnetic fields, contamination, and a host of possible spacecraft sensors and experiments. Failure to do this can result in expensive design changes after the program is in progress. The challenge for the spacecraft system designer is to be certain that the mission requirements for these systems are well defined and understood.

Problems

1. A Sun-synchronous orbit is one where the secular rate of right ascension of the ascending node is equal to the right ascension rate of the mean Sun. Calculate the following orbit parameters for a Sun-synchronous Earth orbit.
(a) What is the precession rate of a Sun-synchronous Earth orbit in degrees per day?
(b) What is the inclination angle of a Sun-synchronous circular orbit of 600 km altitude?
(c) What is the period (T) of this orbit?
(d) What is the maximum eclipse time for this orbit if it is defined as a "noon–midnight" orbit *where the Sun is always in the orbit plane*?
(e) What is the battery depth-of-discharge (DOD) for a spacecraft in this orbit using a 22 cell, 50 A h nickel–hydrogen battery with the loads connected directly to the battery, assuming the average eclipse power is 1500 W and the average battery cell discharge voltage is 1.25 V?
2. Calculate the following rates of orbit precession, using 6378 km for the Earth's radius. Assume that the Sun moves at an angular rate of 360/365.24 = +0.98565 degrees per day, which can be rounded to 0.9856°/d.
(a) What is the precession of the ascending node (precession of right ascension) for the inclinations listed below for a circular, 1000 km altitude orbit?
(b) To an observer on Earth, what is the rate of precession of the ascending node relative to the Sun for each inclination?

$$\text{Inclination } (^\circ): 110, 100, 90, 80, 60, 30, 0.$$

(c) At what inclination is the orbit plane synchronous with the Sun?
(d) What is the precession of the ascending node (precession of right ascension) for the altitudes listed below for a circular, 100° inclined orbit?
(e) What is the rate of precession of the ascending node relative to the Sun for each altitude?

$$\text{Altitude (km)}: 500, 1000, 1500.$$

(f) At what altitude is the orbit plane synchronous with the Sun?

3. Corning 0213 and 0214 glasses are ceria-doped borosilicates that are manufactured into drawn sheets, commonly called microsheet, that are typically used as the solar cell coverglass. In addition to providing protection from particle radiation, they absorb UV radiation. Microsheet is a very pure glass that is available at reasonable cost. There are different variations in transmission to match solar cells of different technology with different spectral response. Sapphire and fused silica are both purer, but are more expensive. The manufacturer's product data sheet lists its density as $2.6\ \mathrm{g/cm^3}$.
   (a) What is the density (in $\mathrm{g/cm^2}$) of a 1 mil thick microsheet coverglass? (1 mil = 0.001 in.)
   (b) Using figure 6.19 for a silicon cell, what thickness of coverglass (in mils) is required to stop a 4.2 MeV proton?
   (c) Approximately what thickness of coverglass is required to stop a 10 MeV proton?
4. (a) If the estimated radiation dose for a spacecraft is 8.20×10^{13} 1-MeV electrons per year as an estimate for the sum of the electron and solar proton environments, what is the equivalent number of (unidirectional) 1-MeV electrons for a 10-year lifetime?
   (b) Using the ground test data for MJ cells in figure 6.22, and their BOL values from table 6.2, what would the solar cell parameters be at the end of 10 years?
5. Calculate the orbital period for circular orbits of the following altitudes. Also, for these same orbits during minimum Sun, find the percent of the orbit time in sunlight, the percent of the orbit time in eclipse, and the maximum eclipse period.

   $$\text{Altitude: } 500\ \text{km},\ 1500\ \text{km},\ 15{,}000\ \text{km},\ 36{,}000\ \text{km}.$$

   For a 1000 W spacecraft with the loads connected directly to the battery and with a nominal bus voltage of 28 V, what is the battery capacity necessary to limit the average battery DOD to 50% in each of the above orbits?
6. For a balloon experiment of 24 h duration using a lithium thionyl chloride primary battery on the bus, what is the number of battery cells in series for a nominal bus voltage of 32 V? What is the maximum battery/bus voltage? What battery capacity is required for a 200 W load? Assume a nominal lithium thionyl chloride cell discharge voltage of 3.2 V and an open-circuit cell voltage of 3.5 V, and that the battery is designed to deliver 100% of its capacity.
7. For a 50 V, regulated GEO spacecraft power system with a voltage step-down battery charge regulator and a voltage boost-type discharge regulator as shown in figure 6.50, calculate the number of battery cells in series, using NiH_2 and Li ion cells. Assume that a minimum of 4 V is needed across the charge and discharge regulators. Calculate the battery capacity for a 5000 W load using the number of cells in series selected for each cell technology, assuming 70% battery DOD for the maximum eclipse time of 72 minutes and that the discharge regulator efficiency is 94%. The nominal discharge and peak charge voltages for NiH_2 and lithium ion cells are given in table 6.5.
8. Referring to paragraph 6.8.3, the power system design example:
   (a) If lithium ion cells were used, what would be the number of cells required for a nominal 28 V bus? What would be the maximum allowed battery/bus voltage? Use the battery cell characteristics given in table 6.5.

(b) Assuming the same temperatures and the same 1 MeV electrons/cm^2 equivalent charged particle radiation dosage on the solar panels of the design example, and using the solar cell characteristics given in table 6.2:

(i) Calculate the number of solar cells in series required using silicon cells.

(ii) Calculate the maximum voltage from the solar panels at the exit from the longest eclipse.

(Refer to sections 6.8.3.5 (Radiation Effects) and 6.8.3.10 (Solar Array Design).)

9. Referring to the solar array controller of figure 6.42, which uses a full linear shunt, assume that the solar array will supply a maximum of 10 A and that the maximum bus voltage is 35 V. The minimum load is 5 A, and the minimum battery current is 0.2 A.

(a) How much power must the linear shunt be designed to dissipate?

(b) What is the peak power that will be dissipated in the resistor?

(c) What is the peak power that will be dissipated in the transistor?

Suppose that you wish to design for the condition that someone may inadvertently disconnect the load with the solar array simulator during spacecraft integration and tast (I&T), which may cause the battery to charge at a high rate. Now what power should the shunt be designed to dissipate?

10. Using the TRW model for a solar cell I–V curve (equations 6.4.1–6.4.3):

(a) Develop a spreadsheet to plot the I–V curves of a 4 cm^2 MJ solar cell at the BOL for the nominal, hot, and cold temperatures of +28°C, +80°C, and −80°C. Use the MJ BOL solar cell parameters of table 6.2.

(b) Plot the I–V curves at the EOL for the nominal, hot, and cold temperatures of +28°C, +80°C, and −80°C. Assume an EOL radiation fluence of 1.0×10^{15} equivalent 1 MeV electrons/cm^2. Use the MJ BOL solar cell parameters of table 6.2, degraded by the proper factor determined from the curves of figure 6.22.

11. Assuming a power system with the battery on the bus as shown in figure 6.47. Li ion cells are used and the nominal battery/bus voltage is 28 V. For simplicity, ignore the effects of radiation over the spacecraft's lifetime and assume no voltage drop between the solar array and the bus and that the solar cell temperature coefficients are the same for BOL and EOL = 1.0 E15 1-MeV electrons/cm^2. Refer to table 6.5.

(a) What is the number of battery cells in series?

(b) What is the maximum battery voltage at full charge?

(c) How many series-connected silicon solar cells would be required for their maximum temperature of +85°C?

12. In problem 11, assume a non-DET series digital control solar array regulator is used where transistor or relay switches are placed between the solar array string and the battery to control the solar array power. Assume that the solar cell temperature coefficients are the same for BOL and EOL = 1.0 E15 1-MeV electrons/cm^2.

(a) How many series-connected silicon solar cells would be required at a maximum temperature of +85°C?

(b) What would be the maximum voltage from the array as it enters sunlight if its minimum temperature got as low as −120°C by the end of eclipse? (Note: if the switches are open circuit the solar array will be at open-circuit voltage (V_{oc}).)

(c) How many series-connected MJ solar cells would be required for a Sun-pointing solar array in low Earth orbit (LEO) with a maximum temperature of +85°C?

References

Anspaugh, B. E., 1996. *GaAS Solar Cell Radiation Handbook.* JPL Publication 96–9, July.

Anspaugh, B. E., R. G. Downing, H. Y. Tada, and J. R. Carter, 1982. *Solar Cell Radiation Handbook*, third edition. JPL Publication 82–69 (NASA-CR-169662), November.

Baker J., P. Shah, Baer, D., 1983. Internal GSFC Memorandum. (Mine Safety Appliances Co.), G. Nagasubramanian, and D. Doughty (Sandia National Laboratories), 1999. Impedance studies on lithium ion cells. Electrochemical Society Conference.

Borthomieu, Y. and M. Fabre, 2000. High specific energy NiH_2 batteries for GEO satellites. SAFT Defense and Space Division, Alcatel Space Industries, NASA Aerospace Battery Workshop, November 14–16.

Brown, W. L., J. D. Gabbe, and W. Rosenzweig, 1963. Results of the Telstar radiation experiments. *The Bell System Technical Journal*, **XLII** (July), 1505.

Chetty, P. R. K., 1991. *Satellite Technology and Its Applications*, 2nd edition. TAB Books.

Dunlop, James D., G. M. Rao, and T. Y. Yi, 1993. *NASA Handbook for Nickel Hydrogen Batteries.* NASA Ref. Pub. 1314, September.

Eagle-Picher Corporation, Battery Product Data Information, 2003.

Feynman, J., T. P. Armstrong, L. Dao-Gibner, and S. Silverman, 1990. A new interplanetary proton fluence model. *Journal of Spacecraft and Rockets*, **27**, 403.

Ford, F. E., G. M. Rao, and T. Y. Yi, 1994. *Handbook for Handling and Storage of Nickel-Cadmium Batteries: Lessons Learned.* NASA Ref. Pub. 1326.

Gitzendanner, R., F. Puglia, and C. Marsh (Lithion, Inc), 2001. Low temperature and high rate performance of lithium-ion systems for space applications. NASA Aerospace Battery Workshop, November 27–29.

Halpert, G., and S. Surampudi, 1997. An historical summary and prospects for the future of spacecraft batteries. NASA Battery Workshop, Huntsville, Al., November 18–20.

Lizius, D., Cowels, P., Spurrett, R., and Thwaite, C., 2000. Lithium-ion satellite batteries using small cells. NASA Aerospace Battery Workshop, Nov. 14–16.

Marvin, D. C., 2000a, Assessment of multijunction solar cell performance in radiation environment. Aerospace Corporation Report TOR-2000(1210)-1, 29 February.

Marvin, D. C., 2000b, Degradation prediction for multijunction solar cells on earth-orbiting spacecraft. Aerospace Corporation Report TOR-2000(1210)-2, 15 June.

Rauschenbach, H. S., 1976. *Solar Cell Array Design Handbook.* JPL SP 43–38, vols. 1 & 2.

Rauschenbach, H. S., 1980. *Solar Cell Array Design Handbook.* Van Nostrand Reinhold C.

Sharps, P. R., 2003. Results of radiation testing of EMCORE Photovoltaics advanced triple junction solar cell. EMCORE Photovoltaics, reports EWRP047 and EWRP036.

Spitale, G., and J. Feynman, 1992. Program SPE. Jet Propulsion Laboratory, California Institute of Technology, C.

Stassinopoulis, E. G., 1979. *SOFIP—A Short Orbital Flux Integration Program, GSC-12554.* Available from COSMIC, NASA's Computer Software Management Information Center.

Sterz, S., B. Parmele, D. Caldwell, and J. Bennett (Eagle-Picher Industries, Inc.), 1997. Nickel-hydrogen (NiH_2) single pressure vessel (SPV) battery development update. NASA Battery Workshop, Huntsville, AL, Nov. 18–20.

Toft, M., 2003. Private communication, July.

Yardney Technical Products, Inc., 1988. *Battery Energy Data Manual.*

Zimmerman, A., and N. Weber, 1997. *Cause for second plateau discharge in nickel Electrodes.* NASA Battery Workshop, Huntsville, AL, Nov. 18–20.

7

Spacecraft Thermal Control

DOUGLAS MEHOKE

7.1 Introduction

The function of the thermal control subsystem in a space flight program is to control the temperatures of all the individual components throughout the entire mission including ground, launch, and flight operations. Temperature control is an important part of the operation of most spacecraft systems. Spacecraft thermal designs balance the benefit gained by having equipment operate in a specific temperature range with the resources required to keep the equipment in that range. This balancing of requirements and resources is accomplished through the generation of a specific mission concept with detailed thermal requirements. Thermal requirements are a consideration in the design of electronic components, scientific sensors, precise alignment-controlled structures, high power applications, and extreme environments. Typically, these requirements come in the form of temperature limits. Table 7.1 shows typical temperature limits for a variety of commonly used spacecraft components. Other thermal requirements that influence the design of a spacecraft are the external environments to which the craft is exposed, the power and mass constraints imposed by the system, and the limits imposed on thermal hardware options. Heritage and experience are important considerations for the use of any mission-critical hardware.

Most electronic equipment is designed to operate over a specified temperature range. Normally, there is a connection between the life of an electronic component and its operating temperature. Generally, the hot temperature limit ensures the equipment will not fail due to the overheating of any individual electronic part. The cold limit is usually based on the minimum operating design temperature of the part. Hot and cold limits must cover the range of temperatures the unit is expected to experience during its life. In space applications, this range is applicable only to the specific mission and is the

Table 7.1 Typical spacecraft component temperature limits

Component or Subsystem	Operating Temperature (°C)	Survival Temperature (°C)
General electronics	−10 to 45	−30 to 60
Batteries	0 to 10	−5 to 20
Infrared detectors	−269 to −173	−269 to 35
Solid-state particle detectors	−35 to 0	−35 to 35
Motors	0 to 50	−20 to 70
Solar panels	−100 to 125	−100 to 125

basis of the proposed design approach. To cover uncertainties in the design and to ensure proper performance in flight, a temperature margin is included in all temperature requirements.

There is a variety of other impacts that temperature has on different types of systems. Some scientific sensors require specified temperature ranges, or operate more effectively over them. IR wavelength optical sensors require cryogenic temperatures to achieve the necessary signal strength. Particle detectors tend to operate better at colder temperatures, and lose sensitivity as the temperature increases. Some equipment, such as batteries and solar cells, operates best at a specific temperature. The predicted performance of these items needs to be adjusted to include the expected operating temperature. Some components require precise alignment between different parts of the system. Strict alignment control is tied directly to the temperatures and temperature gradients within the affected structure.

7.2 Design Process Summary

The design process for a space program involves the definition of requirements, the development of a design concept to support these requirements, the fabrication and integration of the various subsystems into the flight hardware, the qualification testing necessary to verify that the design meets the mission objectives, and the launch and flight operations. The details of the process represent an attempt to control the development of a complex and possibly unique system with a very small production run. The process stresses sponsor participation, external reviews, clearly defined requirements, and as thorough verification as is possible on the ground, as described in chapter 1.

Generally there are two levels of requirements on any flight program. Some requirements are at the system level; they directly support the primary mission objectives. These requirements cross design disciplines and may not be changed without agreement of the program management. Typical system-level requirements include mission lifetime, orbit trajectory, attitude, and any special requirements specific to the individual mission. Other requirements are derived from these higher-level requirements, and are developed to support the design of a particular subsystem. They affect the operation or hardware of particular subsystems and may be changed as the design progresses, as long as they do not conflict with any system-level requirements. Generally, temperature limits are derived requirements.

A proposed spaceflight program combines specific mission objectives with a particular hardware design. All programs have a system design that represents a negotiated compromise between the desired mission return products and the available resources. The development of the system concept includes the detailed design of the various subsystems and their interactions with other subsystems. The thermal subsystem has direct interfaces with most subsystems. The mechanical design is important in defining box locations (heat sources and sinks), heat paths, and the external surfaces where heat can be radiated to space. There are always interactions with the power subsystem, due to the fact that heater power is typically a significant part of the system power budget. Propulsion systems have temperature limits that require a separate thermal approach. Attitude and Control defines where the spacecraft points and the external environment. Command & Data Handling controls temperature sensors and automated heater control. The thermal subsystem covers most of the external surfaces of the spacecraft, which makes it a significant part of the Contamination and Electromagnetic Compatibility (EMC) designs. Missions Operations defines how the spacecraft operates, orbital precession, and eclipse durations.

The integration process brings together the separate subsystems into a unified whole. Before integration, each subsystem shows that it has satisfied its individual requirements and its interface agreements with the spacecraft. Typical integration activities for the thermal subsystem are the fabrication of multilayer insulation (MLI) blankets, installation of the heater system, incorporation of any thermal hardware, and coordination with the developing mission operations plans. Also, a major activity during integration is the planning for the upcoming spacecraft thermal vacuum test.

The environmental test program is aimed at verifying that the system will operate as expected over the predicted environmental extremes. For the thermal subsystem, the final verification test is the spacecraft thermal vacuum test. For this test, the spacecraft is installed into a large thermal vacuum chamber with a test fixture that duplicates the space environment as nearly as practicable. The system is operated while the spacecraft temperatures are cycled beyond those expected in flight. After successful completion of the system-level thermal vacuum test, the thermal subsystem is considered to be flightworthy.

7.3 Analysis

A major part of the thermal design task is the analytical modeling used to predict the operation of the subsystem. Modeling allows the hardware performance to be predicted before integration, eliminating the test and rework cycle. The goal of a modeling program is to allow the configuration to be refined and design options to be traded off before they become fixed by the hardware. The success of that modeling program is dependent on the speed and accuracy with which the system can be analyzed.

Spacecraft thermal analysis involves describing the physical structure and components in terms of analytical expressions that can be solved to yield the expected performance of the real hardware. The modeling process requires that the physical system be approximated by a mathematical representation that can be quantified. There are many ways that physical systems can be simulated. An important part of the modeling effort is to decide what techniques are best used for the particular situation. Both exact and numerical solutions are used.

Exact solutions can be used where the geometry is such that simple shapes can be used to approximate the system. An analytically exact solution, if one exists, can be used to predict the behavior of the system. The benefit of this approach is that the credibility of the solution is the greatest. The disadvantages of exact solutions are the limited geometries that can be analyzed, and the time required for the solution. While exact solutions are not able to describe complex shapes, they are useful in providing checks on other solution methods.

As systems become more complex, numerical solutions are required. Numerical solutions break the structure of interest into a discrete grid of smaller surfaces with interdependent responses. The system of equations defining the behavior of the individual surfaces is solved simultaneously to produce a global solution describing the entire structure. The approach has the benefits of being applicable to any geometry and there is a variety of software tools available to support the effort. The major snag is that the accuracy is dependent on the grid approximation and the solution method. It is up to the analyst to understand the limitations of the solution method used and to make the appropriate checks to ensure its accuracy.

The goal of the analysis effort is to create an approach that can be used reliably to model arbitrarily shaped systems quickly and accurately. To be useful in the design cycle, the analysis must be able to predict the results for proposed changes in the same timeframe as when the design decisions need to be made. Speed is typically a function of the detail required in the task. As modern spacecraft become more and more complex, more is required from the analytical effort. Performance margins are narrowed and systems are optimized to produce more with less. During this process it is the analytical effort that defines what is practical and how much more may be possible.

The review process is a key part of the analytical cycle. Design reviews are normally held at various points in the programs to ensure that the evolving design is meeting its requirements. Thermal presentations at most reviews present analytical results. It is important that the review process includes not only the predicted results but also the models and methods used to achieve them.

7.4 Thermal Analysis

All thermal analysis begins with the first law of thermodynamics, which states that energy in a defined system is conserved. That law is represented in equation 7.4.1. The rate at which heat is added to the system, Q, minus the rate of work production by the system, W, is equal to the change in the internal energy U of the system.

$$Q - W = dU/dt \tag{7.4.1}$$

To be practical, spacecraft thermal design requires an analytical method where the thermal performance of arbitrary shapes and materials can be predicted. The goal is then to develop the appropriate equations to predict the temperatures of a given system in terms of easily definable quantities. Therefore, equation 7.4.1 must be put into a form that relates the temperatures of the system with its geometry, physical characteristics, and boundary conditions.

Prior to the succeeding discussion, it is worthwhile to look at the units of energy and power. In equation 7.4.1, U refers to the energy of the system and has the units of joules or BTUs. Work and heat have the units of energy. Rates of heat and work production are energy flows, and have the units of power. Power units are energy/time (W, J/s, or BTU/s). In the following discussions, Q will be used to represent the heat flux and will have the units of power, while U will be used to represent energy.

For a spacecraft in flight, the work done by the system is zero so the net heat flux added to the system is equal to the change in its internal energy. The change in the internal energy U of a material can be described in terms of its shape, physical properties, and temperature. For a uniform solid with cross-sectional area A and length dx, the change in internal energy is shown in equation 7.4.2, assuming constant properties.

$$dU/dt = A\,dx\,\rho c_{\mathrm{p}}\,dT/dt \tag{7.4.2}$$

In this equation, ρ is the material's density and c_{p} is its specific heat at constant pressure. T is the temperature of the material, and t is time. The units of density are mass per volume (kg/m^3) and the units of specific heat are energy per mass per change in temperature (J/kg°C). For normal applications, material properties are not strongly dependent on temperature and may be assumed to be constant. This approximation is not valid if the temperatures vary significantly from ambient conditions. Combining equations 7.4.1 and 7.4.2 for an isolated system that does no work gives the basic heat equation in terms of the geometry, material properties, and temperature of the system.

$$Q = A\,dx\,\rho c_{\mathrm{p}}\,dT/dt \tag{7.4.3}$$

The net heat flux into a system is the summation of all the heat flows entering the system minus all those leaving it. This summing of all the heat flows into and out of a system is referred to as the heat balance on the system, and shown simply as

$$Q_{\mathrm{net}} = Q_{\mathrm{in}} - Q_{\mathrm{out}} \tag{7.4.4}$$

There is a variety of ways that heat can flow into and out of a system. The heat balance must include all the forms.

7.5 Conduction

The process by which heat moves through a solid is defined as conduction. The one-dimensional rate equation, published by Fourier in 1822, is

$$Q_x = -kA\,dT/dx \tag{7.5.1}$$

where Q_x is the heat flow in the x direction.

Heat flow in a solid is proportional to the temperature gradient in the material multiplied by the material conductivity k and the cross-sectional area through which the heat flows, A. Using equation 7.5.1, the heat movement and storage in a solid can be defined. The net heat flux added to a one-dimensional solid is then the heat flowing in one side minus the heat flowing out the other side:

$$Q_{\mathrm{in}} = -kA\frac{dT}{dx} \quad \text{and} \quad Q_{\mathrm{out}} = -kA\left[\frac{dT}{dx} + \frac{d(dT/dx)}{dx}dx\right] \tag{7.5.2}$$

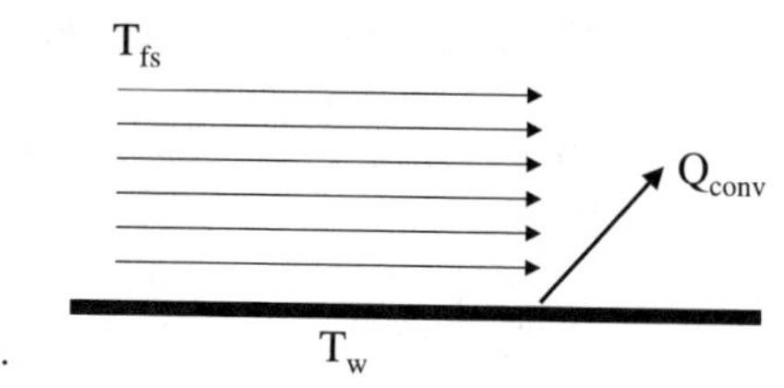

Figure 7.1 Convection.

so,

$$Q_{\text{net}} = kA\, d^2T/dx^2 dx \tag{7.5.3}$$

Combining equations 7.4.3 and 7.5.3 allows the thermal behavior of a solid to be described entirely in terms of the geometry and material properties of the system:

$$d^2T/dx^2 = \rho c_{\text{p}}/k\; dT/dt \tag{7.5.4}$$

The term $k/\rho c_{\text{p}}$ is called the thermal diffusivity of a material and is a measure of how fast heat propagates through a substance. Materials with a high diffusivity will see temperatures propagate much faster through the material.

As with the material properties mentioned before, conductivity is not strongly affected by temperature. It can be used as a constant, though we must remember that significant changes do occur at very high or low temperatures.

7.6 Convection

Convection is a term used to quantify the microscopic conduction that occurs at the interface of a moving fluid over a solid surface into a macroscopic value. The general convection problem is shown in figure 7.1. The heat transfer between the plate and moving fluid is defined as

$$Q_{\text{conv}} = h(T_{\text{w}} - T_{\text{fs}}) \tag{7.6.1}$$

where T_{w} is the surface temperature, T_{fs} is the free stream temperature, and h is the convective heat transfer coefficient. The coefficient h is a function of the fluid properties, speed, and mixing conditions. Values for h are available in most heat transfer texts. Convection is not typically involved in space applications. However, it may be important in planetary lander applications.

7.7 Nodal Approximation

Numerical analyses require a structure to be broken up into smaller pieces that can be analyzed as a group. The modeling method defines how these pieces are defined and how they interact with their neighbors. Nodal approximation is used, as part of a numerical solution technique, to represent an arbitrarily shaped structure in terms of discrete surfaces that can be modeled individually. The two types of nodal analytical method in widespread use are finite difference and finite element. Most of the work in spacecraft thermal analysis is done using finite difference techniques.

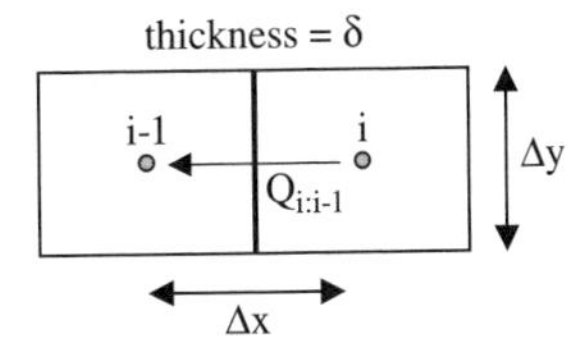

Figure 7.2 One-dimensional nodal heat flow.

The finite difference method divides the surface into smaller sections called nodes. Each node is assumed to be a point at which all the associated material properties of the section are concentrated. The nodes are connected by conductors that allow heat to flow between them on the basis of their temperature differences.

A one-dimensional grid structure is shown in figure 7.2. The points in the centers of each rectangle are the nodes. The conductive heat transfer between the two nodes can be defined using equation 7.5.1 and the approximation

$$dT/dX \simeq \Delta T/\Delta x \tag{7.7.1}$$

as

$$Q_{i:i-1} = -k\Delta y\delta[T_i - T_{i-1}]/\Delta x \tag{7.7.2}$$

for a nodal spacing of Δx, a width of Δy, and a thickness of δ. The two-dimensional generalized grid structure is shown in figure 7.3. Allowing for heat generation $g_{i,j}$ per unit volume and time, a heat balance calculated for the center node yields the following:

$$\begin{aligned} &\frac{\delta\Delta y}{\Delta x}(T_{i-1,j} - T_{i,j}) + \frac{\delta\Delta y}{\Delta x}(T_{i+1,j} - T_{i,j}) \\ &\quad + \frac{\delta\Delta x}{\Delta y}(T_{i,j-1} - T_{i,j}) + \frac{\delta\Delta x}{\Delta y}(T_{i,j+1} - T_{i,j}) \\ &\quad + g_{i,j}\frac{\delta\Delta x\Delta y}{l} = \frac{\rho c_p}{l}\delta\Delta x\Delta y\frac{(T^*_{i,j} - T_{i,j})}{l} \end{aligned} \tag{7.7.3}$$

where all the terms are evaluated time $= n - 1$, except for T^*, which is evaluated at time $= n$.

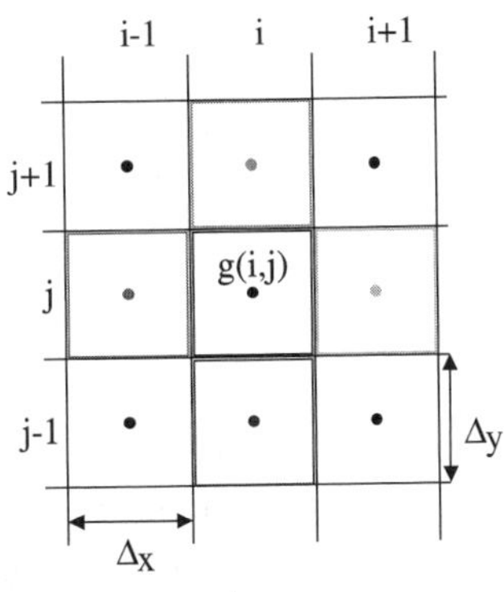

Figure 7.3 Two-dimensional nodal heat balance.

Equation 7.7.3 can be simplified by substituting a thermal resistance term R that defines the heat paths between two nodes, a bulk thermal capacitance C for the node, and a heat input Q. With the substitutions

$$\left.\begin{aligned} R_{i-1,j:i,j} = R_{i+1,j:i,j} = \Delta x/k\delta\Delta y \; R_{i,j-1:i,j} = R_{i,j+1:i,j} = \Delta y/k\delta\Delta x \\ C_{i,j} = \rho c_{\mathrm{p}}\Delta x\Delta y\delta \; Q_{i,j} = g_{i,j}\Delta x\Delta y\delta \end{aligned}\right\} \tag{7.7.4}$$

equation 7.7.3 becomes

$$\begin{aligned} &\frac{T_{i-1,j} - T_{i,j}}{R_{i-1,j:i,j}} + \frac{T_{i+1,j} - T_{i,j}}{R_{i+1,j:i,j}} + \frac{T_{i,j-1} - T_{i,j}}{R_{i,j-1:i,j}} \\ &+ \frac{T_{i,j+1} - T_{i,j}}{R_{i,j+1:i,j}} + Q_{i,j} = C_{i,j}\frac{T^{*}_{i,j} - T_{i,j}}{\Delta t} \end{aligned} \tag{7.7.5}$$

Equation 7.7.5 is the conductive energy balance equation that defines the temperature of node i,j with the geometry and materials used in the structure. A similar equation could be written for every node in the system. Equation 7.7.5 expresses the heat flow in a rectangular grid pattern in terms of the nodal quantities ($T_{i,j}$, $C_{i,j}$, and $Q_{i,j}$) and the connections between the nodes $R_{i,j:k,l}$. The equation can be generalized to non-regular nodal geometries and can include other types of heat transfer. For the general case the i,j notation is replaced with a single subscript s to represent a particular node with no specific geometrical orientation. Including convection and internal heat generation, equation 7.7.5 becomes

$$\sum_{i \neq s=1}^{n} \frac{(T_i - T_s)}{R_{i:s}} + \sum_{i \neq s=1}^{n} hA_s(T_i - T_s) + Q_s = C_s\frac{(T^{*}_s - T_s)}{\Delta t}, \quad k = 1 \text{ to } n \tag{7.7.6}$$

When completed for all nodes, equation 7.7.6 forms a system of linear equations that can then be solved to produce a set of nodal temperatures predicting the thermal response over the period of interest, based on the details of the system and a set of initial and boundary conditions assumed. Initial conditions are the state of the system at the start of the period of interest. Boundary conditions are the constraints on heat flow at the boundaries of the model. Typical boundary conditions are fixed temperature or adiabatic (no heat flow).

The thermal nodal approximation is similar to the method used for electrical circuits, with the following analogies:

Temperature	is equivalent to	Voltage
Heat flow	is equivalent to	Current
Thermal resistance	is equivalent to	Electrical resistance

7.8 Direction-Dependent Material Properties

An important part of the analysis process is the material property that defines how heat moves through a solid. That property, introduced earlier as the thermal conductivity k, is usually treated as a constant, but in reality it can be dependent on both temperature and the direction of the heat flow. A requirement for the use of any material property is to know its variation over the temperature range in question, and whether the material property is independent of direction.

Materials where the property is independent of direction are termed isotropic. The definition for thermal conductivity k, given in equation 7.5.1, relates the heat flow with the change in temperature in a given direction. Assuming a uniform solid, the relationship between heat flow and temperature gradient would be the same in any direction. Therefore, the thermal conductivity would be independent of direction throughout the material.

There are materials where the thermal conductivity is directionally dependent. These materials are termed non-isotropic. Sometimes this dependency is inherent in the material because of its construction, as in a graphite epoxy laminate. Other times, directional dependency occurs when the structure is made of different parts but is treated as a single material for analytical simplicity. An example of a non-isotropic material is a honeycomb panel with aluminum facesheets. When treated as a single material there is a different conductivity in the plane of the panel as opposed to that perpendicular to the plane. For non-isotropic materials, a thermal conductivity is defined for each axis.

The same approach is used to calculate thermal resistances in non-isotropic materials as in isotropic materials. However, in non-isotropic situations the directional conductivities are used. In graphite fiber/epoxy composites there is a thermal conductivity in the direction of the fibers and one perpendicular to it. The effective thermal conductivity is found by calculating separate thermal resistances for the fibers and epoxy base using the separate volumetric fractions the two materials occupy in the composite. The effective conductivity is thus found using

$$k_{\text{eff}} = (k_{\text{fiber}} A_{\text{fiber}} + k_{\text{base}} A_{\text{base}})/(A_{\text{fiber}} + A_{\text{base}}) \tag{7.8.1}$$

When layers are laid up into a laminate, the arrangement can eliminate the directional dependency of the in-plane conductivity. However, there will still be a difference between the in-plane and cross-plane conductivities.

7.9 Radiation

A third type of heat transfer, which is very important in spacecraft thermal analysis, is thermal electromagnetic radiation. All heat transfer to and from the spacecraft, and much that occurs within the structure, takes place through radiation. Radiation is governed by Planck's law, which states that

$$e_{\text{b}\lambda}(T) = \frac{C_1 \lambda^{-5}}{e^{c_2/\lambda T} - 1} \tag{7.9.1}$$

where $C_1 = 3.74 \times 10^8$ W μm^4/m^2, $C_2 = 1.44 \times 10^4$ μm K, and $\lambda\ e_\text{b}(T)$ is the blackbody thermal radiation, which is defined as the energy emitted per second, per wavelength, and per area by a perfect emitter at the given wavelength and temperature.

When equation 7.9.1 is integrated over all wavelengths, the total heat flux emitted is proportional to the absolute temperature to the fourth power,

$$Q_\text{b} = \sigma T^4 \tag{7.9.2}$$

where b refers to a blackbody or ideal radiator, and $\sigma = 5.67051 \times 10^{-8}$ W/m^2 K^4 and is the Stefan–Boltzmann constant.

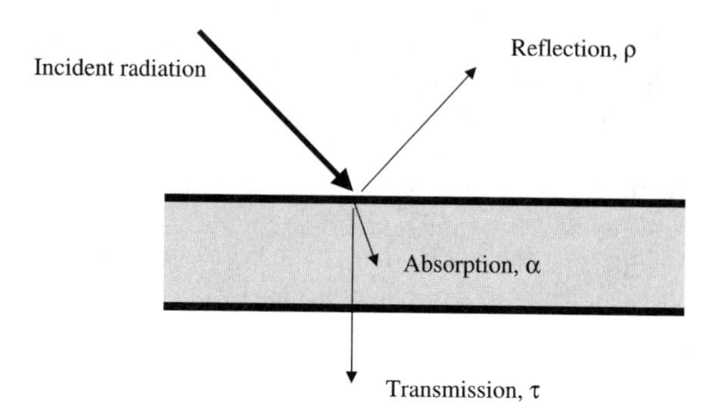

Figure 7.4 Radiation properties.

7.9.1 Radiation Properties

Working with radiation heat transfer requires some special constants that define how radiation is transmitted from surface to surface. A generic material is shown in figure 7.4. Radiation striking this surface must either be absorbed into the material, be reflected from the surface, or pass through the material. Three terms are defined to account for how the incoming ray reacts with the surface.

$\alpha =$ absorptance
$\rho =$ reflectance
$\tau =$ transmittance

and to conserve energy

$$\alpha + \rho + \tau = 1 \tag{7.9.3}$$

In addition, as shown in figure 7.5, reflected radiation comes off in either a specular or a diffuse manner. The specular portion is reflected with $\phi_1 = \phi_2$, while the direction of the diffuse portion is considered to be independent of the incident angle. The reflectance ρ is therefore refined into specular, ρ_s, and diffuse, ρ_d, components. The amount of heat radiating from a real surface is defined as the amount that would leave a blackbody at the same temperature times the emissivity ε:

$$Q_{\text{actual}} = \varepsilon Q_b \tag{7.9.4}$$

For bodies in thermal equilibrium at the same temperature, the energy being absorbed must equal the energy emitted, and so

$$\varepsilon = \alpha \tag{7.9.5}$$

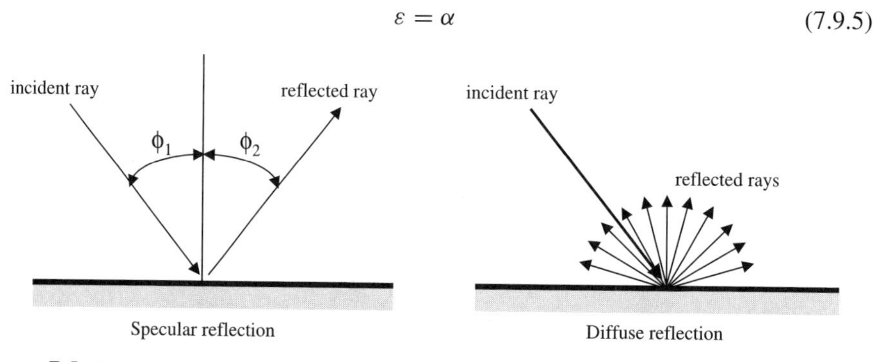

Figure 7.5 Specular and diffuse reflections.

The reflectance, absorptance, transmittance, and emissivity discussed so far are total. They represent the wavelength-dependent values integrated over all wavelengths. Therefore, the total properties will vary depending on the incident radiation, which will vary according to the temperature of the source.

To simplify the radiative heat transfer calculations, it is important to remove the temperature dependency from the radiative properties. This simplification is accomplished by looking at the radiation from real surfaces in terms of the monochromatic properties. For a real surface, the emitted power can be defined in terms of wavelength-dependent properties integrated over all wavelengths λ:

$$Q = \int_{\lambda=0}^{\infty} \varepsilon_\lambda Q_{b\lambda} d\lambda \tag{7.9.6}$$

Similarly, for the other radiative properties,

$$Q_{\text{absorbed}} = \int_{\lambda=0}^{\infty} \alpha_\lambda Q_{\lambda \text{ incident}}\, d\lambda \tag{7.9.7}$$

$$Q_{\text{reflected}} = \int_{\lambda=0}^{\infty} \rho_\lambda Q_{\lambda \text{ incident}}\, d\lambda \tag{7.9.8}$$

$$Q_{\text{transmitted}} = \int_{\lambda=0}^{\infty} \tau_\lambda Q_{\lambda \text{ incident}}\, d\lambda \tag{7.9.9}$$

Next, it is noted that in spacecraft analyses there are two general wavelength regions of interest. In space, radiation sources are limited to two broad categories: physical surfaces and the Sun. Most real surfaces spend most of their time at or near room temperatures and very few will spend any time above 100°C or 200°C. In contrast the Sun's radiation temperature is over 5000°C. The wavelength-dependent emissive power from equation 7.9.1 is plotted in figure 7.6 for different source temperatures. The plot shows that most of the energy from solar radiation is in the visible wavelength region, while virtually all of the energy for a 200°C object is in the IR region. Keeping in mind that radiative heat transfer is an integrated product of the wavelength-dependent radiation properties and the wavelength-dependent radiation source, from equations 7.9.7 to 7.9.9, total radiation properties can be defined that are more useful analytically.

Two terms that are widely used are the solar absorptance and the IR emittance. The solar absorptivity α_s is defined as the fraction of solar flux absorbed by a particular surface. It is found by integrating the solar flux times the absorptance of the surfaces over all wavelengths, divided by the total solar flux. The IR emmitance ε_{IR} is the energy absorbed through radiation exchange with surfaces at normal temperatures. It is found in a similar manner to α_s, except that it uses a heat flux generated by a source near room temperature, or an IR source.

$$\alpha_s = \frac{\int_{\lambda=0}^{\infty} \alpha_\lambda Q_{\lambda(T_{\text{solar}})} d\lambda}{\int_{\lambda=0}^{\infty} Q_{\lambda(T_{\text{solar}})} d\lambda} \tag{7.9.10}$$

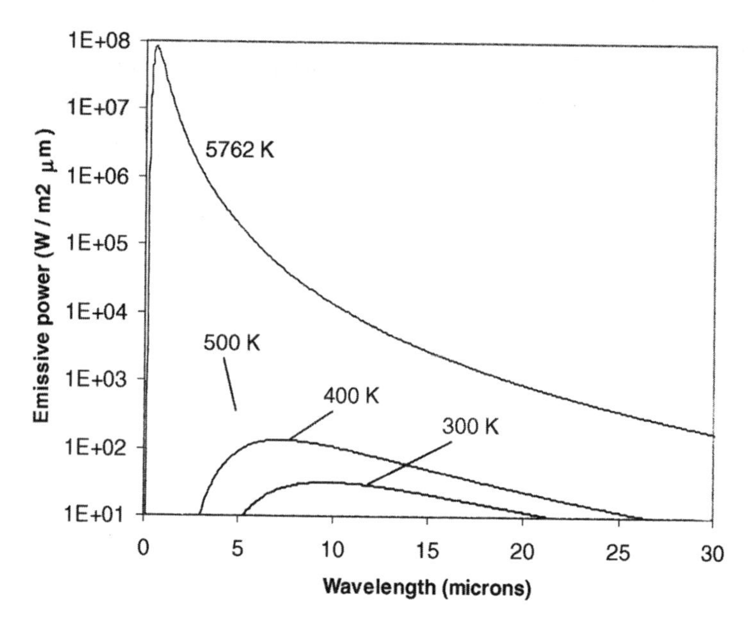

Figure 7.6 Wavelength variation of sources at different temperatures.

$$\varepsilon_{\mathrm{IR}} = \frac{\int_{\lambda=0}^{\infty} \alpha_\lambda Q_{\lambda(T_{\mathrm{IR}})} d\lambda}{\int_{\lambda=0}^{\infty} Q_{\lambda(T_{\mathrm{IR}})} d\lambda} \tag{7.9.11}$$

From equation 7.9.5, the energy absorbed and emitted by surfaces in thermal equilibrium at the same temperature are equal. Therefore, the IR emittance is also the property described in equation 7.9.4, which defines the energy actually emitted by a real surface.

The absorptance values used, needed for equations 7.9.10 and 7.9.11, are based on measured data. The integration needs to be performed numerically as

$$\alpha_{\mathrm{s}} = \frac{\sum Q_{\lambda(T\,\mathrm{solar})} \alpha_{\mathrm{mat}} \Delta\lambda}{\sum Q_{\lambda(T\,\mathrm{solar})} \Delta\lambda} \tag{7.9.12}$$

A typical plot of measured reflectance is shown in figure 7.7. Since most surfaces used in spacecraft design applications are opaque, the transmittance is zero. Therefore, from equation 7.9.3, at a given wavelength λ,

$$\alpha_\lambda = 1 - \rho_\lambda \tag{7.9.13}$$

Surface absorptance values are typically obtained from measured reflectance data.

7.9.2 Radiation Exchange Between Real Surfaces

The next step in the description of radiation heat transfer is a discussion of how radiation is exchanged between real surfaces. The details of how the power is absorbed or emitted by a particular surface are discussed above. The remaining topic to be covered is how much power moves from one surface to another.

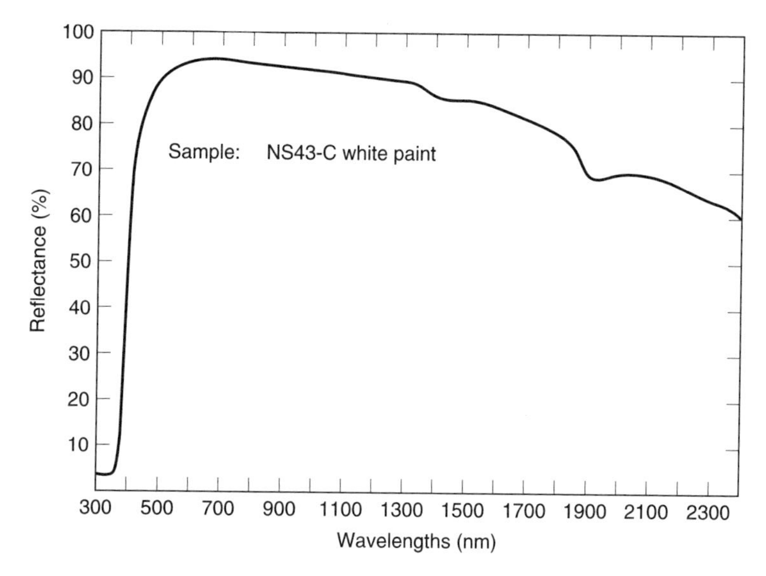

Figure 7.7 Measured reflectance data for NS 43C white paint.

A energy flux balance between two generic surfaces is given in figure 7.8. Surface 1 has a temperature of T_1(K), an area of $A_1(\mathrm{m}^2)$, and an emissivity of ε_1. Surface 2 has the properties T_2, A_2, and ε_2. The net heat exchange between the two surfaces is the heat flux leaving the first surface that is absorbed by the second surface minus the heat flux emitted by the second surface that is absorbed by the first surface.

		heat flux available to leave ideal surface	fraction of flux actually leaving surface	fraction of flux reaching other surface	heat flux absorbed by other surface		
$Q_{1\text{to}2}$	=	$Q_{b1}A_1$	ε_1	F_{1-2}	ε_2	=	$A_1F_{1-2}\varepsilon_1\varepsilon_2\sigma T_1^4$
$Q_{2\text{to}1}$	=	$Q_{b2}A_2$	ε_2	F_{2-1}	ε_1	=	$A_2F_{2-1}\varepsilon_2\varepsilon_1\sigma T_2^4$

(7.9.14)

The net heat flux exchanged is then

$$Q_{1-2} = Q_{1\text{to}2} - Q_{2\text{to}1} \tag{7.9.15}$$

To reduce equation 7.9.14 further requires a discussion of the new term, F_{n-m}, introduced there. F_{n-m} is the view factor between surfaces n and m, and is defined as the fraction of the heat flux leaving surface n that strikes surface m. Note that view factors describe the heat flux actually leaving one surface that is incident on a second

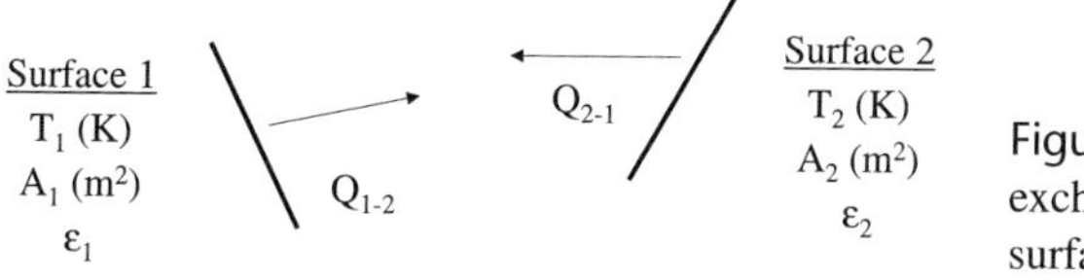

Figure 7.8 Radiation exchange between real surfaces.

surface. The details of what takes place at the surface are covered by the radiation terms already discussed. This new term is concerned only with the geometric view that one surface has of another. View factors can be calculated analytically for certain basic shapes. Some basic shapes and expressions are given in most texts on heat transfer. Two basic properties of view factors are given below:

A_1 A_2

$$A_1 F_{1-2} = A_2 F_{2-1} \tag{7.9.16}$$

A_1 A_2 A_3

$$F_{1-2,3} = F_{1-2} + F_{1-3} \tag{7.9.17}$$

Combining equations 7.9.14, 7.9.15, and 7.9.16 gives a general expression for the radiation heat exchange between two surfaces:

$$Q_{1-2} = A_1 \boldsymbol{F}_{1-2} \varepsilon_1 \varepsilon_2 \sigma (T_1^4 - T_2^4) \tag{7.9.18}$$

A complete radiation energy flux exchange for a general system of surfaces requires the geometry of the surfaces, surface properties, and temperatures. As part of a general analytical method, view factors between arbitrary shapes must be available. These are typically calculated numerically with specially designed computer codes. The codes allow for the fact that radiation can reach from one surface to another by reflections off other surfaces in the system. The reflections off the other surfaces will be influenced by the radiation properties of those surfaces. A new factor, $\boldsymbol{F}_{1-2}$, is used to account for the total energy leaving an ideal surface 1 that reaches and is absorbed by surface 2, including any interactions with other surfaces in the system. With this substitution, equation 7.9.18 becomes

$$Q_{1-2} = A_1 \boldsymbol{F}_{1-2} \sigma (T_1^4 - T_2^4) \tag{7.9.19}$$

and the general heat balance equation, neglecting convection, from equation 7.7.6 can be extended to include radiation:

$$\sum_{i \neq \mathrm{s}}^{n} \frac{(T_i - T_\mathrm{s})}{R_{i:\mathrm{s}}} + \sum_{i \neq \mathrm{s}}^{n} \sigma \boldsymbol{F}_{\mathrm{s:i}} A_\mathrm{s} (T_i^4 - T_\mathrm{s}^4) + Q_\mathrm{s} = C_\mathrm{s} \frac{(T_\mathrm{s}^* - T_\mathrm{s})}{\Delta t} \tag{7.9.20}$$

The system of equations created by equation 7.9.20 can be solved to produce the nodal temperatures at all points for an arbitrary mechanical structure. The constants are based on the materials and the geometry of the structure.

7.10 Steady-State and Transient Solutions

Equation 7.9.20 can be used to solve problems with time-varying or constant boundary conditions. The former are known as transient problems, and the latter are called steady-state.

In a steady-state problem, there is no change in temperature with time so the right-hand side of equation 7.9.20 goes to zero. The equation becomes

$$\sum_{i \neq s}^{n} \frac{(T_i - T_s)}{R_{i:s}} + \sum_{i \neq s}^{n} \sigma \boldsymbol{F}_{s:i} A_s (T_i^4 - T_s^4) + Q_s = 0 \qquad (7.10.1)$$

Note that as the temperature variation between time steps disappears, the effects from the material properties of density and specific heat are lost. This fact means that steady-state problems can be solved as simple heat flux balances. The final temperatures are the ones necessary for the heat inputs into any particular node to match the heat lost.

Solution routines for transient problems are more complex. The left side of equation 7.9.20 is a heat balance on node s:

$$\sum (\text{heat flows in/out of node s}) = C_s(T_s^* - T_s)/\Delta t \qquad (7.10.2)$$

where T^* denotes the temperature at the end of the next time step. The simplest way to propagate the solution forward in time is to evaluate all the properties of the model at the present time step, and then use equation 7.10.2 to find the next value.

$$T_s^* = \Delta t \left(\sum (\text{heat flows})_s / C_s\right) + T_s \qquad (7.10.3)$$

The transient solution technique used in equation 7.10.3 is called forward-differencing, in that the properties at one instant in time are used to propagate the solution forward. Forward-differencing is an easy method to implement. Its major drawback is that the method can be unstable. If the time step is too large, the nodal temperatures can over-predict the change during the time step. To control the solution, it is necessary to relate the time step to the properties of the model.

Like the electrical analogy discussed earlier, transient thermal models have an inherent time constant. An estimate of the time constant can be obtained from

$$\text{Time const}_s = R_s C_s, \text{ where } 1/R_s = \sum (1/R_{s:i}) \qquad (7.10.4)$$

To maintain accuracy and stability, the time step in a forward-differencing method should be a fraction of the smallest time constant in the system. The thermal capacitance C_s is related to the mass of the node, and the thermal resistances are related to how well it is thermally tied to the other nodes. For nodes of small mass that are well thermally coupled, the system time constant can get quite small, making the run time for the system extremely long. To attack the run time/accuracy problem, other types of solution techniques are used.

Backward-differencing refers to a method where the new values are calculated by looking backward in time. Equation 7.9.20 can be written as a system of linear equations in matrix form if the radiation term is not included. By inverting the constants matrix the new temperatures can be found, using the heat flows at the new time step. This method has the benefit of being unconditionally stable, but it has the drawback of requiring a lot of computer storage for the matrices required and the inversion process. Additionally, if the radiation term is included, the problem is nonlinear and other approximations need to be included.

In commercial heat transfer programs typically a variety of transient solution methods is included. Predictor–corrector algorithms, intermediate loops, relaxation criteria are used as a way of achieving an acceptable level of accuracy while keeping the run time reasonable. Also, typically included is some mechanism to provide a check on the system by assessing the overall heat balance during the transient calculations. A requirement of the analytical process is to understand the limitations of the solution technique used and to provide the needed checks to make sure the solution is correct.

7.11 Environmental Heat Inputs

In most space applications, there are other surfaces besides those on the spacecraft that need to be considered in the radiative heat balance. Environmental heating is the term used to express the radiation heat transfer received on a spacecraft surface from the Sun or a planetary body. Equations 7.9.20 can be used to predict the thermal performance for any system of surfaces. Using the nodal method described above, it would be possible to create nodes representing the heavenly bodies and to treat these surfaces like ordinary nodes in the analysis. Additionally, heat lost to space must be accounted for in the energy balance. The main problem with using this approach is that the external bodies are then treated like ordinary parts of the model and the solution technique gets bogged down waiting for them to reach their steady-state temperatures.

Environmental inputs can be included into a model without significantly complicating the calculation of the solution. A way to include these external inputs can be found by examining in detail all the radiation connections for a particular surface. For surface s, the complete radiation connections to node s would have the form

$$\begin{aligned}\sum \sigma A_s \boldsymbol{F}_{s-i}(T_i^4 - T_s^4) + \sigma A_s \boldsymbol{F}_{\text{s-earth}}(T_{\text{earth}}^4 - T_s^4) + \sigma A_s \boldsymbol{F}_{\text{s-sun}}(T_{\text{sun}}^4 - T_s^4)\\ + \sigma A_s \boldsymbol{F}_{\text{s-space}}(T_{\text{space}}^4 - T_s^4) = Q_{\text{rad}}\end{aligned} \tag{7.11.1}$$

Separating terms gives

$$\begin{aligned}\sigma As \sum \boldsymbol{F}_{s-i} T_i^4 + \sigma A_s \boldsymbol{F}_{\text{s-earth}} T_{\text{earth}}^4 + \sigma A_s \boldsymbol{F}_{\text{s-sun}} T_{\text{sun}}^4\\ + \sigma A_s \boldsymbol{F}_{\text{s-space}} T_{\text{space}}^4 - \sigma A_s \sum \boldsymbol{F}_{s-i} T_s^4 - \sigma A_s \boldsymbol{F}_{\text{s-earth}} T_s^4\\ - \sigma A_s \boldsymbol{F}_{\text{s-sun}} T_s^4 - \sigma A_s \boldsymbol{F}_{\text{s-space}} T_s^4 = Q_{\text{rad}}\end{aligned} \tag{7.11.2}$$

But since, from view factor algebra,

$$\left(\sum \boldsymbol{F}_{s-i} + \boldsymbol{F}_{\text{s-earth}} + \boldsymbol{F}_{\text{s-sun}} + \boldsymbol{F}_{\text{s-space}}\right) = 1 \tag{7.11.3}$$

and the radiative heat fluxes from the earth and sun are

$$\left.\begin{aligned} Q_{\text{earth}} &= \sigma A_s \boldsymbol{F}_{\text{s-earth}} T_{\text{earth}}^4 \\ Q_{\text{sun}} &= \sigma A_s \boldsymbol{F}_{\text{s-sun}} T_{\text{sun}}^4 \end{aligned}\right\} \tag{7.11.4}$$

and noting that the temperature of space is essentially zero, equation 7.11.2 becomes

$$\sum \sigma A_s \boldsymbol{F}_{s-i}(T_i^4 - T_s^4) + Q_{\text{s-earth}} + Q_{\text{s-sun}} - \sigma A_s \left(1 - \sum \boldsymbol{F}_{s-i}\right) T_s^4 = Q_{\text{rad}} \tag{7.11.5}$$

If $\boldsymbol{F}_{\text{s-space}}$ is redefined as

$$\boldsymbol{F}_{\text{s-space}} = \left(1 - \sum \boldsymbol{F}_{s-i}\right) \tag{7.11.6}$$

and T_{space} is treated as a fixed temperature node at 0 K, then the $\sigma A_s \boldsymbol{F}_{\text{s-space}} T_s^4$ term in equation 7.11.5 can be brought inside the summation over all the nodes and equation 7.11.5 becomes

$$\sum \sigma A_s \boldsymbol{F}_{s-i}(T_i^4 - T_s^4) + Q_{\text{s-earth}} + Q_{\text{s-sun}} = Q_{\text{rad}} \tag{7.11.7}$$

where space is included as one of the nodes of the system. When equation 7.11.7 is inserted into 7.9.20, the heat fluxes from the Earth and Sun can be treated in the same way as internal heat generations, and the general form becomes

$$\sum_{i \neq s}^{n} \frac{(T_i - T_s)}{R_{i:s}} + \sum_{i \neq s}^{n} \sigma \boldsymbol{F}_{s:i} A_s (T_i^4 - T_s^4) + Q_s + Q_{\text{env}} = C_s \frac{(T_s^* - T_s)}{\Delta t} \tag{7.11.8}$$

The general spacecraft thermal control problem becomes the generation of the appropriate resistances, viewfactors, and environmental heat fluxes from the mechanical design and orbital parameters. System temperatures are then predicted on the basis of those inputs. The calculation of environmental heat fluxes for anything other than a very simple surface geometry is better left to the computer.

Environmental heat fluxes come in three forms: direct solar, albedo (reflected solar), and planetshine. Direct solar and albedo are radiation that is in the visible wavelengths originating at the Sun. Direct solar is treated as a collimated source, with all the incident radiation assumed to be coming in at the same angle. Albedo, being reflected off the Earth, is a diffuse source, so that the incident radiation is coming in at a variety of angles. Planetshine is radiation in the IR wavelengths coming from the warm surface of the planet. It is also treated as a diffuse source. The difference between collimated and diffuse radiation affects how the surface heat fluxes are calculated.

Direct solar, Q_{sun}, is treated as constant collimated light with an intensity at the Earth of 1353 W/m^2. As shown in figure 7.9, there is a variation of $\pm 3.5\%$ in the solar constant for Earth-orbiting spacecraft because of the change in the Earth–Sun distance over the year. Heat fluxes are calculated on a projected area basis. The projected area is the value of the cosine of the angle between the sun–spacecraft vector and the surface normal integrated over the surface. More simply, it is the effective area a surface has when viewed from the Sun. The general heat flux onto a surface s is

$$Q_{\text{sun-s}} = Q_{\text{sun}} A_{\text{proj-s}} \alpha_{\text{Ss}} \tag{7.11.9}$$

where $A_{\text{proj-s}}$ is the projected surface of surface s and α_{Ss} is the solar absorptance of the surface s.

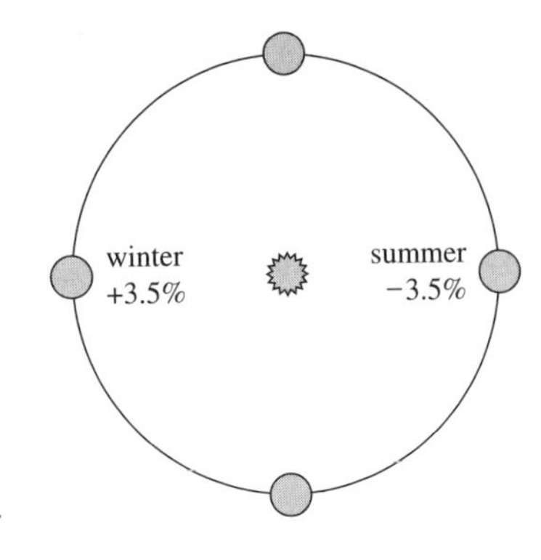

Figure 7.9 Solar constant variation.

The albedo factor, ρ_{alb}, is the fraction of the solar energy incident on a planet's surface that is reflected into space. Albedo surface heat fluxes are more difficult to compute because of the diffuse nature of the reflecting surface. The energy reflecting off the Earth varies with location around the Earth due to the variation in the local solar input and the surface properties, which can vary with the time of the year. The general trend of the surface reflectivity or albedo factor is shown in figure 7.10. It has a nominal value in the equatorial regions but rises to almost 1 at the poles due to the presence of the ice caps. However, albedo is not typically modeled this way. It is simpler to model the albedo heat flux as the product of an albedo factor times the cosines of the angles defining the position of the surface relative to the sub-solar point (the closest point on the Earth to the Sun), as shown in figure 7.11. The average albedo factor used should include the time-averaged specific distribution in the local albedo factor for the specific orbit. Typically, values for ρ_{alb} between 0.28 and 0.4 are used. The range allows for uncertainty in the environmental input. Since radiation from the planet is diffuse, the farther the surface is away from the plate the lower the incident flux. To account for this dependency on geometry the concept of the view factor, discussed earlier, is used. For

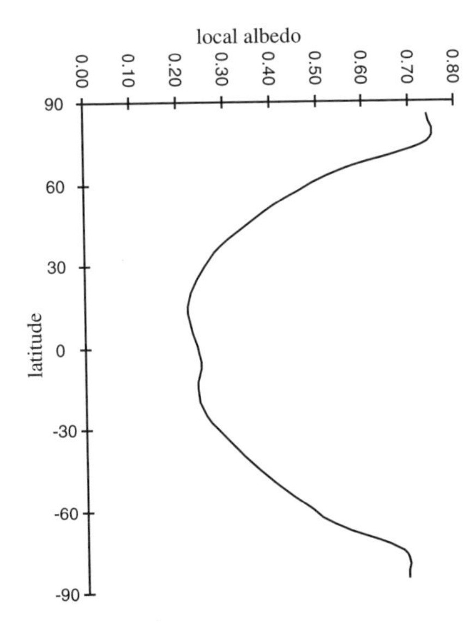

Figure 7.10 Typical albedo as a function of latitude.

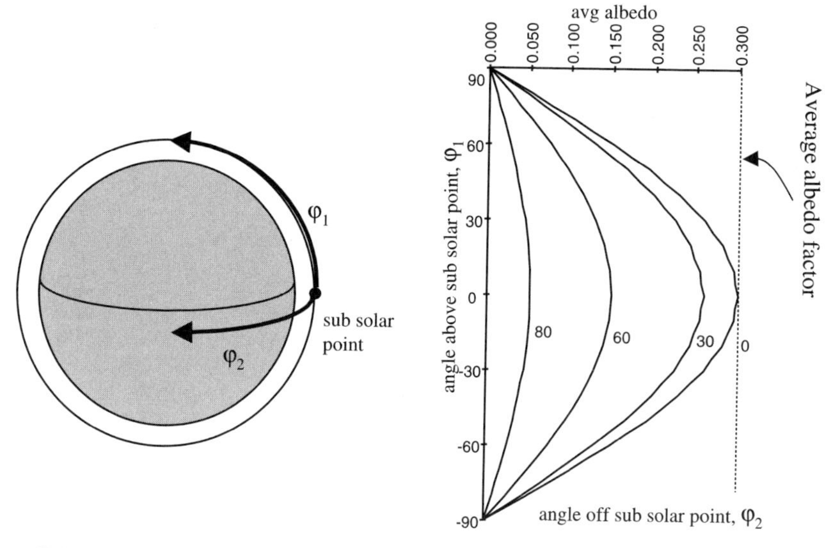

Figure 7.11 Modeled albedo variation with deviation from sub-solar point.

situations like balloon experiments, where the payload spends most of its time over one location, it is important to use the local albedo factor.

A surface above a planetary body will have some view factor to the planet below it. For a planar surface with its normal pointing toward the center of a planet, $F_{\text{s-planet}}$ can be approximated as $(R_{\text{planet}}/R_{\text{SC}})^2$. Graphs are available for other geometries (see figure 7.12), but a more useful source is the commercially available computer codes developed for these purposes. The albedo hitting a surface s becomes

$$Q_{\text{alb}} = Q_{\text{sun}}\, \rho_{\text{alb}} \cos\phi_1 \cos\phi_2\, A_{\text{s}} F_{\text{s-planet}}\, \alpha_{\text{Ss}} \tag{7.11.10}$$

where ϕ_1 and ϕ_2 are as defined in figure 7.11, A_{s} is surface area, $F_{\text{s-earth}}$ is the view factor of the surface to the Earth, and α_{Ss} is the surface's solar absorptance. Equation 7.11.10 is valid when ϕ_1 and ϕ_2 are between 0° and 90°, $Q_{\text{alb}} = 0$ for all other angles.

The last environmental heat flux to be discussed is planetshine. Planetshine is the normal radiation emitted from the surface of the planet due to its own temperature. As with the albedo discussion, view factors from the surface to the planet need to be included. If the planet has regions with significantly different temperatures, the heat flux calculation needs to include the view factors of the spacecraft surfaces to the different regions. Unlike the albedo, the radiation is in the IR wavelengths, so the IR emissivity needs to be used in calculating the heat absorbed by the surface. The Earth is modeled as a constant temperature source. A typical range used for Earthshine Q_{EarthIR} is between 220 W/m^2 and 270 W/m^2; this range covers seasonal variation. As with albedo, the specific planetshine values used should be checked on the basis of the specific orbit track.

The general equation for plansetshine to a surface s is

$$Q_{\text{planet}} = Q_{\text{planetIR}}\, A_{\text{S}}\, F_{\text{s-planet}} \varepsilon_{\text{IRs}} \tag{7.11.11}$$

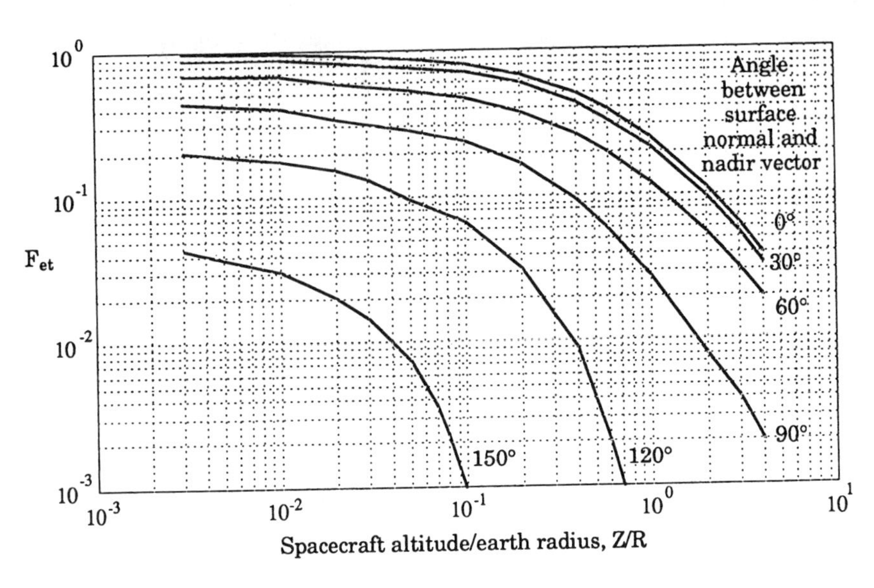

Figure 7.12 Earth albedo view factors.

where all the terms are described as before. Note that the view factor is the same as the one used in the albedo calculation because it is a function of the geometry, not the wavelength of the energy involved.

7.12 Orbit Definition

The calculation of environmental heat fluxes requires the spacecraft position to be defined in terms of the planet and the Sun. This information can be given in terms of an orbit for planetary missions or distance from the Sun for interplanetary ones. In either case, the analysis needs to identify the different types of environmental heating that will occur over the mission, and develop bounding cases.

For orbiting spacecraft, the β angle is introduced as a way of defining an orbit with a specific angle to the sun. The definition of β angle is shown in figure 7.13. It is the angle between the orbit plane and the spacecraft–Sun line. The β angle has the useful property of simply defining a specific orbit plane relative to the sun and the related showing characteristics. Nominally, Earth orbits are defined in terms of semimajor axis, eccentricity, inclination, right ascension of the ascending node, and argument of periapsis. Mission planning predicts the range of orbit angle path the spacecraft will encounter throughout its life. Rather than duplicating the actual orbits, the β angle is varied between the minimum and maximum values. Environmental heating fluxes are calculated to determine the hot and cold cases for the specific design. These bounding cases bracket the worst case hot and cold temperatures the spacecraft will encounter.

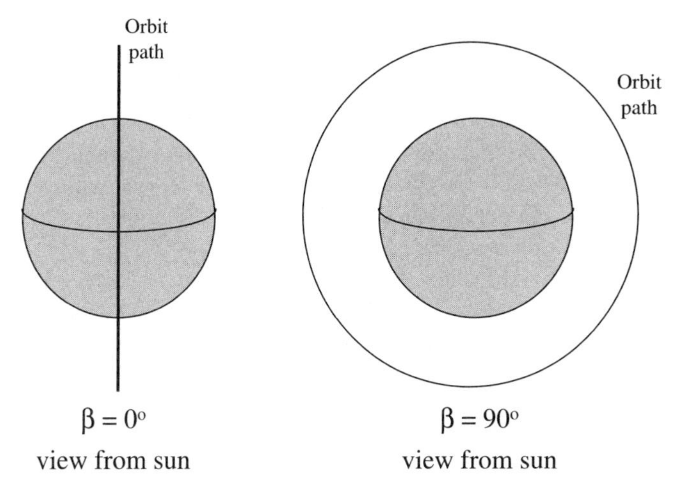

Figure 7.13 β angle definition.

7.13 Analytical Codes

In the earlier discussion it was noted that spacecraft thermal analyses are broken into two main categories: the calculation of the thermal resistances and heat fluxes appropriate to model the design under consideration; and the solution of the system of equations resulting from that model. Hand calculations are important to provide checks during the analysis, but they are restricted to extremely simple problems. Environmental heat fluxes can be estimated, using the approximations discussed in the preceding paragraphs, for early calculations and as checks on the computer simulations. However, anything consisting of more than the simplest problem requires a specialized computer code.

Since modern spacecraft thermal analyses are tied to specialized computer codes, using those codes is an important part of the effort. Historically, two programs, TRASYS and SINDA, were used for the bulk of the modeling work done. TRASYS was used to calculate the radiation interchange factors $\boldsymbol{F}_{i\text{-}j}$ and the environmental heat fluxes. SINDA was the equation solver that took those and other inputs and calculated temperatures. There were generic formats for both codes, and models in the proper format could be passed back and forth rather easily. Now there is a wide variety of programs available that have, at the same time, simplified and complicated the thermal analysis process. Complicated models can be built and checked out much faster than before. However, the variety of different codes leads to a variety of formats, and the transfer of models between organizations is becoming a more serious issue. Some new programs commercially available are:

- TSS (Thermal Synthesizer System) from Spacedesign Corporation.
- Thermal Desktop from C&R Technologies.
- IDEAS TMG from the Structural Dynamics Research Corporation.
- Thermal Analysis Kit from K&K Associates.
- Thermal Analysis System from Harvard Thermal Inc.

7.14 Design Considerations

The thermal design is closely coupled with most of the other spacecraft design areas. An important part of the thermal design task is coordination with the other design disciplines to optimize the return from the proposed mission. The basic thermal design approach revolves around the spacecraft's payload, orbital environment, mechanical configuration, attitude control, power generation, and internal dissipation. The approach is always based on controlling the heat flows into and out of the system.

The process begins with the desire to place a particular payload in a specific place in space. Generally, the spacecraft is divided into the payload (scientific instruments or other components directly involved in the primary mission) and the bus (spacecraft subsystems required to support the payload). The payload is usually unique to the particular mission; the bus is more likely to be similar between different missions. The payload usually has certain basic thermal requirements: temperature limits, configuration, and power dissipation. Temperature limits determine whether the payload can be thermally coupled to the spacecraft bus or needs to be thermally isolated. Typical spacecraft bus temperatures are designed to range between $-10°C$ and $40°C$. Thermally coupling the payload to the bus can simplify the overall design of the bus and payload, but may complicate the individual designs. Isolating the payload from the bus allows the two designs to proceed in parallel with simpler interfaces. The simpler interfaces of an isolated design may be beneficial whether the payload and bus have different temperature ranges or not. The payload configuration includes any external surfaces which affect heat fluxes from external sources and provide potential heat rejection surfaces. The configuration also determines whether there are any alignment issues that may require special thermal gradient requirements. Power dissipation and the variation in that dissipation determine how the particular area will need to be controlled.

The next step in the design process is the analysis of the prospective environmental heat fluxes. The mission design includes a definition of the design orbits, including the launch and any transfer orbits. The desired orbits are analyzed to determine the range of angles the spacecraft will see relative to the Sun due to orbital precession. Rather than look over the predicted mission timeline, a set of thermal design cases is generated that bounds the extremes of the environmental inputs the spacecraft is expected to see. The desire is to define areas on the spacecraft where the environmental heat fluxes are small and use those surfaces to reject the internal power dissipated by the spacecraft or subsystem.

A payload and bus configuration is then developed around the mission design. The bus includes the major spacecraft support electronics. Generally, these components are thermally tied to the spacecraft structure. A power budget is developed that captures the operational requirements of the mission. The mechanical design needs to include enough radiator area to reject the internal power dissipation of the spacecraft, including the environmental heat fluxes. Heat paths inside the bus are estimated in the early description of the bus structural configuration, and then revised as the design and analytical efforts progress. If external organizations are responsible for parts of the spacecraft, specific interface requirements are negotiated that allow both designs to proceed without being affected by the ongoing design evolutions.

Once the basic configuration has been defined, the spacecraft is divided into thermally similar zones, and a thermal design approach is developed for each zone. Component temperature limits are reviewed to determine whether items can be grouped into zones. The spacecraft bus is usually the largest zone. Other typical zones include the solar array, battery, propulsion system, and star camera. A thermal design approach is defined for each zone that controls the heat flows, which allows the temperature to be controlled. Hot and cold boundary conditions (conductive and radiative) and environmental heat fluxes are developed for each zone.

The most basic thermal control approach uses radiators to control temperatures in the hot design condition and heater power to maintain them in the cold condition. The maximum power dissipation is combined with the warmest boundary conditions and highest environmental fluxes to produce the maximum power inputs to the system. Radiators are added to the configuration to reject the required heat into the warmest environment. That radiator size is then used with the coldest boundary conditions and lowest internal power dissipation to determine the heater power required. Heater power is added to make up the difference between the heat lost by the given radiator size and that required to maintain the component temperatures above their minimum levels. This approach is usually the simplest to implement, but requires the most heater power. Approximately a 25% variation in internal power dissipation can be accommodated by a heater/radiator design.

Other approaches maintain temperatures by adjusting the heat lost by the spacecraft. These designs require special thermal hardware like louvers or heat pipes that can modify the heat flows out of the spacecraft. The complexity of the added hardware has to be traded off against the heater power saved.

7.15 Thermal Control Hardware

Thermal control hardware is classified as either passive or active. Passive components have no changing element associated with them. Examples of passive components are radiators, thermal control coatings, fixed-conductance heat pipes, and multilayer insulation (MLI). Passive designs balance the variations in environmental heat fluxes with changing spacecraft temperatures. Active components include something that changes the spacecraft heat balance during the mission. Examples of active components are thermostatically controlled heaters, louvers, and variable-conductance heat pipes. Trade-offs between active and passive designs include subsystem cost, control range, and reliability. The following sections describe some of the more common thermal control components.

7.15.1 Radiators

A radiator can be any part of the spacecraft that is allowed to directly radiate energy to space. A radiator typically uses an existing part of the structure that is exposed to space. Heat flows are conducted to the radiator from internal parts of the spacecraft. The location of power-dissipating components relative to the radiators is a major part of the ongoing discussion between the thermal and mechanical subsystems. To maximize the heat lost by the radiator surface, a coating is applied to it that has a high emissivity. If the surface has environmental heat fluxes, those heat flows need to be included.

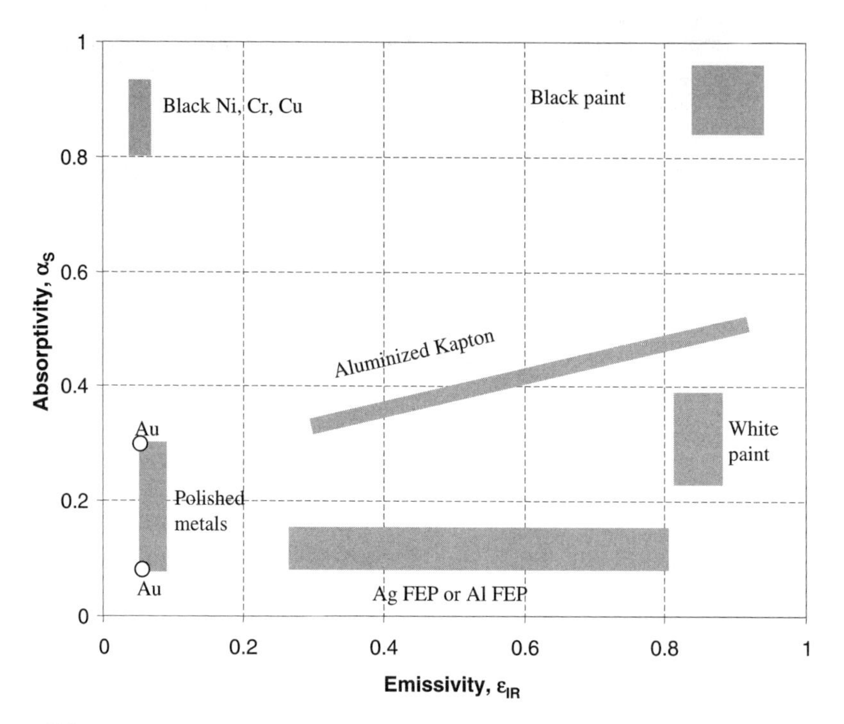

Figure 7.14 Typical radiation properties.

Coatings designed to have specific radiative properties are called thermal control coatings. Ideally, a radiator surface would have a very low absorptivity but a high emissivity. An isolating surface would have low values for both properties. A heat-absorbing surface would have a high absorptivity and a low emissivity. Unfortunately, in real surfaces there is a limited range of radiative properties. The ratio of absorptivity to emissivity, $\alpha_s/\varepsilon_{IR}$, is used to classify surfaces. Surfaces with a low $\alpha_s/\varepsilon_{IR}$ ratio, <0.4, are ideal for radiator surfaces that experience some solar inputs. Surfaces with high ratios, >1.0, will get hot when exposed to solar heating. Figure 7.14 gives the ranges of solar absorptivities and IR emissivities for real surfaces. Generally, the IR emissivity of a surface does not vary significantly with environmental effects, but the solar absorptivity can vary a great deal.

White paints are one of the most common thermal control coatings. These coatings have a low $\alpha_s/\varepsilon_{IR}$ ratio, but are subject to degradation due to environmental effects. Typically, UV radiation and local contamination both act to increase the absorptivity of the paint. UV causes the pigment to darken. Contamination coats the surface with a film that has its own radiative property, which is typically higher than that of the base surface.

Another option for a low $\alpha_s/\varepsilon_{IR}$ is the use of second surface reflectors. The idea behind first and second surface thermal control is shown in figure 7.15. A second surface reflector combines the properties of two separate layers to produce a surface with a low and stable $\alpha_s/\varepsilon_{IR}$ ratio. The top layer of the reflector has a high emissivity, but is transparent in the solar wavelengths. The solar radiation is transmitted through this layer and

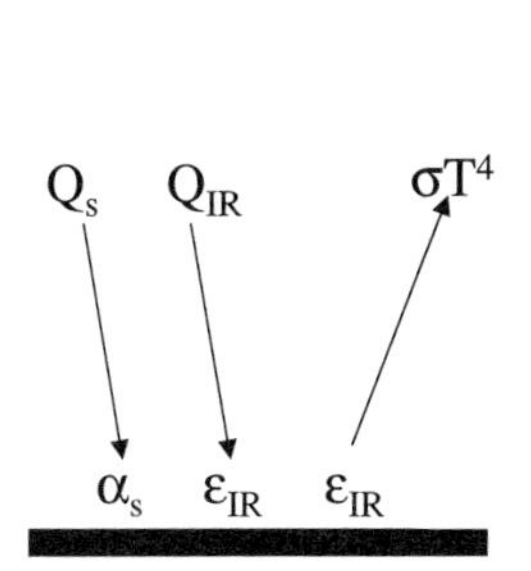

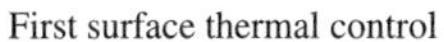
First surface thermal control

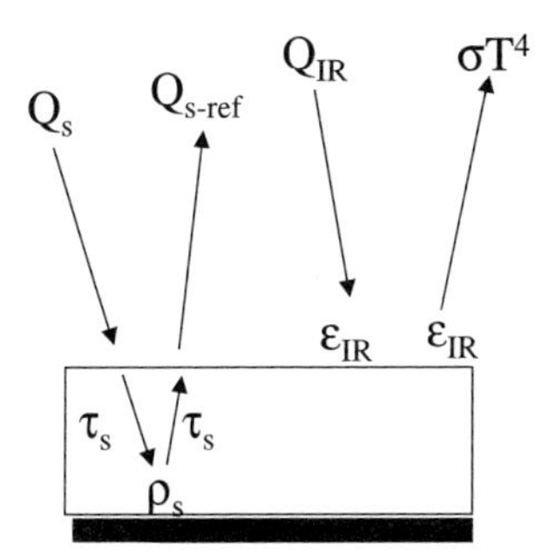

Second surface thermal control

Figure 7.15 Second surface mirror.

reflected off a shiny rear layer, making the net solar reflectance high (and the absorptivity low). The stability of the surface is a function of how well the outside layer maintains its transmissivity. Second surface reflectors come in two general forms: metalized fluorinated ethylene propylene (FEP) Teflon films, and optical solar reflectors (OSRs). Metalized Teflon surfaces come as either aluminized Teflon (Al/FEP) or silverized Teflon (Ag/FEP) films, and have an outer layer of transparent FEP Teflon. The films come in thicknesses between 0.002 and 0.010 inches (50 and 250 μm) thick, and are flexible. They are applied to a surface using an acrylic transfer adhesive. Silver-backed films have a better $\alpha_S/\varepsilon_{IR}$ ratio then the aluminum films, but are subject to significant corrosion in the presence of water. OSR tiles use an external layer of quartz with a silver backing. The quartz is insensitive to radiation and UV darkening, making the radiative properties very stable. The tiles are applied in much the same manner as solar cells. The additional cost of the tiles needs to be traded off against the lower operating temperatures resulting from the stability of the OSRs.

Knowing the stability of thermal control coatings is an important factor in the use of these coatings. The use of these surfaces should include both beginning-of-life (BOL) and end-of-life (EOL) values. BOL values are measured and so are relatively easy to obtain. Determination of the EOL properties is based on UV exposure testing and flight experience. The cumulative solar exposure for the radiator surface over the mission should be calculated and used in the determination of the EOL surface degradation. Typical values for common thermal control surfaces are given in table 7.2.

Table 7.2

Surface	α_S BOL	α_S EOL	ε IR
OSR	0.08	0.20	0.80
Ag/FEP	0.09	0.30	0.78
Al/FEP	0.13	0.40	0.78
Z93 white paint	0.13	0.30	0.85
A276 white paint	0.26	0.50	0.85
0.001 Kapton	0.38	0.45	0.67
Z306 black paint	0.95	0.95	0.85
Bare aluminium	0.2	0.2	0.1

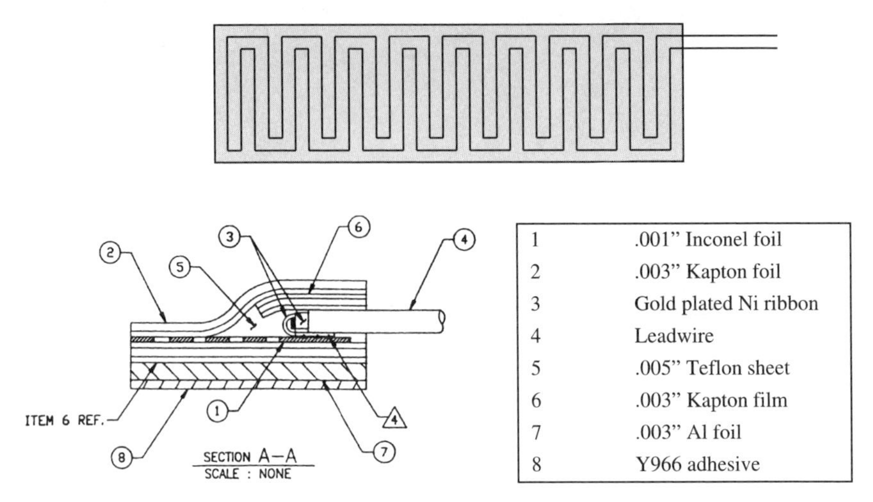

1	.001" Inconel foil
2	.003" Kapton foil
3	Gold plated Ni ribbon
4	Leadwire
5	.005" Teflon sheet
6	.003" Kapton film
7	.003" Al foil
8	Y966 adhesive

Figure 7.16 Kapton foil heater construction.

7.15.2 Heaters

Heaters are used to support the spacecraft during cold periods. The most common type of heater uses a resistance element laminated between two Kapton film layers, as shown in figure 7.16. The heaters are attached to a substrate using an acrylic film adhesive. The heaters come in standard sizes, but can also be made for specific applications. Key features for using Kapton heaters are the substrate temperature, surface topology, power density, resistance density, and system voltage. Typically, the transfer adhesive used to bond the heater to the substrate has a temperature limit of 150°C. Use above this limit may result in debonding and failure of the heater. There are some adhesives that allow operation at a higher temperature.

The heaters have a distributed resistive element that is spread over the underlying substrate. Surface irregularities that cause the heater not to contact the underlying deck risk local overheating and possible failure. The power density is the amount of heat flux output by the heater divided by its surface area. On a good heat sink, Kapton heaters have a maximum power density of about 9 W/in^2. In a practical design, the maximum power density used should be about 2 W/in^2. Since the heater element is an etched Inconel foil, there is a limit on the length of foil that can be put into a certain area. Typical limits for area resistances are 100 Ω/in^2 for foil elements and 400 Ω/in^2 for wire elements.

Sizing a heater requires the ranges of the input voltage to be defined. Both the average and peak power of heaters are important. Spacecraft bus voltages may either be controlled to a narrow range (28 V ± a few percent) or vary over a wide range (22 V to 36 V). For a controlled bus, the voltage variation should include some allowance for line losses (∼ 0.5 V). For an uncontrolled bus, the heater needs to be set on the basis of the lowest possible voltage the bus may see for an extended period (a few seconds). This lower design voltage may be significantly higher than the minimum instantaneous voltage.

The power dissipated in a heater is given by

$$Q_{\text{htr}} = V_{\text{bus}}^2 / R_{\text{htr}} \tag{7.15.1}$$

Typically, a 25% margin is used for heater power margin. For a nominally sized heater, requiring 10 W, the heater would be sized to produce 12.5 W at the minimum voltage. If 22 V is used as the minimum voltage and 36 V as the maximum, the peak power drawn by the heater would be 33.5 W. Clearly, for a system with many heaters, the instantaneous current drawn, being more than three times the average value, could be a design issue.

Heaters can be controlled either through commands or thermostatically. For very simple systems, the heater commands can be sent from the ground in response to slowly changing criteria. More often, heaters are controlled automatically by the onboard computer. Depending on the spacecraft telemetry, the computer can power specific heaters on the basis of particular system temperatures. Heaters can also be automatically controlled with thermostats. Thermostats include a bimetallic switch that either makes or breaks the electrical connection between the two contacts, depending on the temperature of the device. Thermostats are typically rated at 100,000 contacts and can carry 5 A current. Practical designs should limit the cycles to about 50,000 and a current of 1 A.

7.15.3 Multilayer Insulation (MLI)

MLI is used in all spacecraft applications to control the heat lost from the spacecraft body by radiation. MLI consists of multiple layers of low-emissivity surfaces arranged in parallel. Key design aspects of MLI are the external and internal layer materials, the expected temperature ranges, number of internal layers, shape of the item covered, surface conductivity, and contamination requirements.

The theory behind MLI as insulation comes from the analysis of parallel radiation shields. For two parallel planes with emissivities ε_1 and ε_2, the heat flux exchanged between the surfaces has to allow for the fact that the radiation can bounce back and forth between the shields. The heat balance for two arbitrary surfaces is given in equation 7.9.18. For two parallel surfaces that are close together, the view factor between the surfaces is effectively 1. The two surface emissivities are replaced by a single effective emittance term that accounts for the multiple radiation bounces between the two surfaces,

$$Q_{1-2} = A_1\,\varepsilon_{\text{eff}}\,\sigma\,(T_1^4 - T_2^4) \tag{7.15.2}$$

where

$$\varepsilon_{\text{eff}} = 1/[1/\varepsilon_1 + 1/\varepsilon_2 - 1] \tag{7.15.3}$$

This approach can be extended for N radiation shields in series, all having the same emissivity, by the equation

$$Q_{1-N} = A_1\,\varepsilon_{\text{eff}}\,\sigma(T_1^4 - T_N^4) \tag{7.15.4}$$

where

$$\varepsilon_{\text{eff}} = (N + 1)(2/\varepsilon_1 - 1) \tag{7.15.5}$$

Clearly, as N gets to be a big number, the insulating value should become very high. However, in real life the effectiveness of the insulation is limited by localized contact between the surfaces. Figure 7.17 compares the real and analytical insulating performance of MLI for various numbers of layers. For design purposes, analysis of MLI

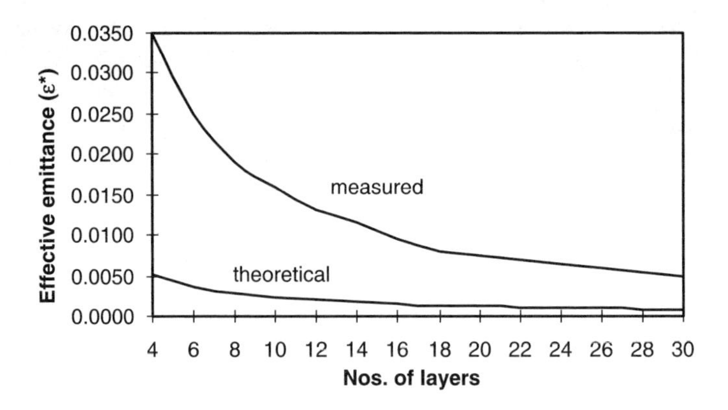

Figure 7.17 Comparison between real and analytical MLI performance.

blankets assumes one-dimensional heat transfer between the internal and external layers, with an effective emissivity defined as ε^*. Typical values for ε^* are between 0.01 and 0.03 for large flat blankets. For smaller three-dimensional shapes, ε^* values as high as 0.05 or 0.1 may be used. The variation in ε^* bounds the expected performance of the MLI blankets. The actual performance needs to be verified in the system-level thermal vacuum testing of the article.

MLI consists of an outer layer facing the external environment, multiple inside layers providing the insulating performance, and an internal protective layer. The outer layer material is chosen on the basis of temperature, surface erosion, or electrical conductivity requirements. Kapton, Al/FEP, and Beta cloth are typical outer later materials. Atomic erosion at lower altitudes, < 500 km, needs to be considered. Kapton is especially sensitive to damage from atomic oxygen. If electrical conductivity is required the Kapton or Al/FEP can be coated with indium titanium oxide (ITO), which is a thin transparent coating that increases the surface conductivity of the layer.

The role of the internal layers is to provide the maximum insulating performance for the minimum mass. Materials used for the inside layers are either 0.0025 in. (62.5 μm) thick Mylar or 0.003 in (75 μm) thick Kapton. To achieve the desired low emissivity, the surfaces are coated with vacuum deposited aluminum (VDA). As mentioned before, the performance of the isulation is limited by the local conductance between the layers. Several different methods are used to reduce the effects of these local contacts. One approach uses a fine netting (Dacron or silk) to separate the layers. Another approach uses crinkled material to minimize contact points. A third method ignores the contacts and uses the just the VDA sheets.

An important part of MLI design is the venting of the material. There is a very large surface area ratio (about 40:1) between the internal MLI surface area and the actual area covered by the blanket. Any surface contamination and trapped air need to make their way out of the blanket during the transition from ambient pressure to vacuum. Again, different methods are used. One method uses perforated layers that let the blanket vent either through the external or internal surface. Another approach uses unperforated layers and lets the blankets vent out through the edges. The internal layer is generally a handling or protective layer. Its role is to keep the inside layers from being damaged during assembly or handling. Either Kapton or a Dacron sail cloth can be used.

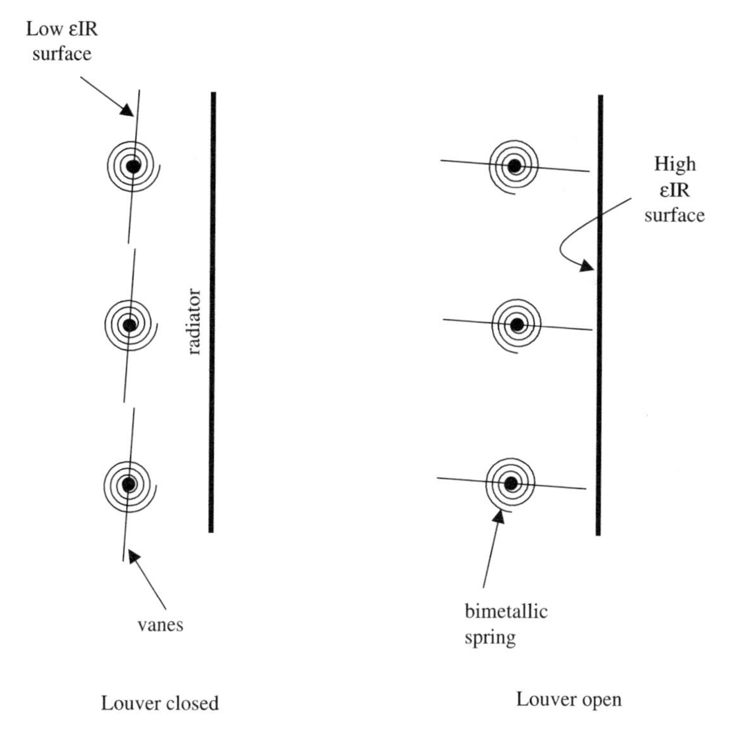

Figure 7.18 Louver construction.

The specific MLI blanket construction techniques used are based on the experience and heritage of the particular institution building them.

7.15.4 Louvers

Louvers are used to vary the heat flowing out from a surface through an effective change of its surface emissivity. They consist of movable low-emissivity surfaces that can cover or expose a high-emissivity underlying surface. Key design features of louvers are the actuation method, actuation time, and the possibility of solar exposure.

A typical louver construction is illustrated in figure 7.18. The louvers consist of several thin vanes covering a radiator surface. The vanes are made of thin VDA-coated Kapton layers laminated to an internal axis. One end of the axis is attached to an actuator that can rotate the vanes through 90°. In the closed position, the surface exposed to the external environment is the VDA coating on the vanes, resulting in a low radiative heat loss from the system. As the vanes open, the high-emissivity radiator surface is exposed, allowing more radiative heat loss. For analytical purposes, a louvered radiator is replaced with a flat surface, which is given an effective emissivity that includes the change in radiative heat loss as a function of vane positions. A typical performance curve for a vane louver is given in figure 7.19. In the fully open position, blockage from the vanes results in a lower emissivity than could be expected from the underlying radiator surface alone.

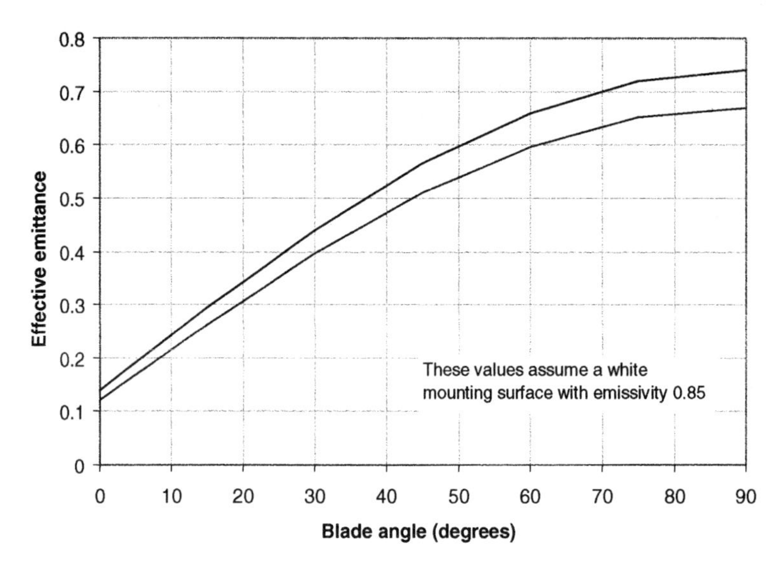

Figure 7.19 Louver performance.

Louvers are generally controlled on the basis of temperature. Bimetallic elements are used that rotate as their temperature changes. The temperature of the bimetallic element is tied to that of the louver frame, which is coupled to the surface on which it is mounted. Louvers can also be controlled somewhat remotely through the use of heaters mounted near the actuator housings. The heaters allow a tighter temperature deadband through computer control of the louvers, at the price of additional power. Actuation time for louvers is typically several minutes.

Most louvers are not designed to operate while exposed to the sun. The low-emissivity coating on the vanes has a high $\alpha_S/\varepsilon_{IR}$ ratio, making it a hot surface. Too high a temperature will cause a failure in the laminating adhesive holding the vane together. Furthermore, the action of the louver is to open as the temperature rises. When exposed to the sun, an open louver would allow more solar heating in and could result in a further temperature increase.

7.15.5 Heat Pipes

Heat pipe technology uses the latent heat associated with the vapor–gas phase change to carry heat very effectively over reasonably long distances. A heat pipe consists of a sealed tube with an internal working fluid. Key design features are the internal construction, operating temperature range, and thermal interfaces of the pipe.

Heat pipe operation is based on a working fluid that remains in both liquid and vapor states over its normal operating temperature range. As shown in figure 7.20, heat is added to one part of the tube, the evaporator, and removed at another, the condenser. The heat added at the evaporator vaporizes liquid, which is transported to the condenser where it reliquifies. Heat is transferred between the two areas, without an associated temperature drop, through the movement of the vapor. The liquid is returned from the condenser to the evaporator by means of a capillary wick structure surrounding the vapor

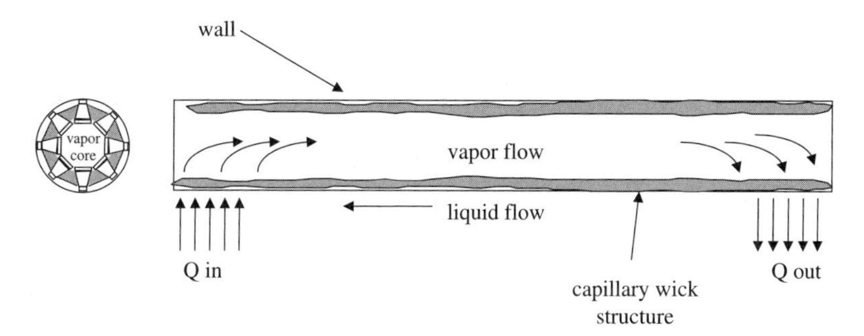

Figure 7.20 Heat pipe operation.

core. Because of the large amount of heat associated with the liquid–gas phase change, heat pipes are very effective at transporting heat.

The operating temperature range of the pipe determines the choice of the working fluid. Figure 7.21 shows the liquid–vapor curves for several typical working fluids. Furthermore, the fluid's latent heat of vaporization signals how much energy the vapor state can carry. The working fluid must also be compatible with the pipe material. The most common heat pipe construction uses an aluminum extrusion, with an axial groove wick structure, and an ammonia working fluid. The heat pipe will continue to operate as long as the liquid return rate through the wick structure is enough to keep the evaporator filled. As the heat flux rises past the allowable range, the liquid will vaporize faster than it can be replenished and "burn-out" occurs. The ability of the capillary wick to move fluid is based on its design and length. The pumping ability of the wick structure needs

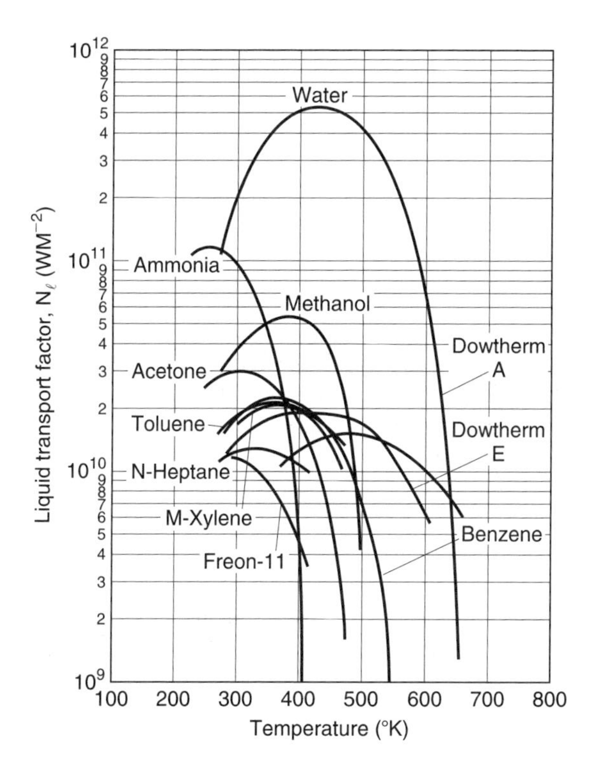

Figure 7.21 Heat pipe vapor curves.

to be larger then the frictionally induced pressure drop in the fluid flowing through that structure.

The design of the heat pipe interface is an important part of the overall system. The pipe material is required to effectively move the heat from the pipe interface to the vapor core at the evaporator and then move it back at the condenser. Since spacecraft surfaces are typically flat, saddles may be added to make the transition between the heat transfer surfaces and the pipe wall. Additionally, as heat is added to one side of a pipe, the resulting temperature variation around the evaporator will cause the liquid grooves on the hot side to burn out earlier than the grooves on the cooler side, even though the pipe may still operate.

7.15.6 Doublers

Doublers are layers of material added to an existing structure to increase its conductivity locally. In spacecraft designs, a standard structural material used is an aluminum honeycomb panel. The facesheets for these panels are typically between 0.005 in. and 0.025 in. (125 and 625 μm) thick, and they provide most of the thermal conductivity in the panels. The key design issue with the use of a doubler is to match the analytical model with the actual geometry.

When power-dissipating components are attached to a honeycomb panel, the heat is conducted radially out from the interface joint into the panel before it is conducted or radiated out of the system. The local heat flow around the interface results in a temperature rise between the panel average and the contact point. If this temperature rise is too large, additional material can be used to increase the material thickness locally around the interface point. Doublers are typically bonded to the underlying surface and use the same material to eliminate CTE mismatch concerns. The design of doublers is tied closely to both the geometry involved and the analytical modeling process. Since doublers are added to an existing structure, they have no inherent size or shape but are specially designed for the specific situation. Special attention needs to be paid to the nodal configuration around the interface point to ensure that the heat flow is accurately reflected in the local area around that point. Typically, the node size should be reduced in areas with large temperature gradients.

7.16 Testing

Thermal testing is used to provide information needed for the analytical efforts, to verify the system is properly designed, and to validate the analyses used in the design. Testing is divided into three general areas: engineering testing, component-level qualification testing, and system-level qualification testing. All three areas are important parts of the design process.

Thermal testing requirements are typically based on specific guidelines. Examples of testing requirements are Mil-Std 1540, "Product Verification Requirements for Launch, Upper Stage, and Space Vehicles"; JPL D-22011, "System Thermal Testing Standard"; GSFC, "General Environmental Verification Specification (GEVS) for STS and ELV Payloads, Subsystems, and Components, Revision A"; and APL SDO-11225, "Space Flight System Performance Verification Requirements."

7.16.1 Engineering Testing

Engineering tests are designed to provide empirical information about the thermal behavior of key parts of the flight hardware. As mentioned above, thermal analysis involves calculating thermal resistances between specified nodes. When that resistance is dominated by a complicated geometry or ill-defined contact between surfaces, the analysis must either include a large variation for the uncertainty or rely on better information derived by test. As an example, MLI performance is almost entirely based on empirical testing.

Engineering tests also provide confidence in the survivability of hardware exposed to severe environments. Designs operating in corrosive environments or close to the material failure point require detailed information of how the material in the specific configuration behaves in the regime. Environments requiring engineering testing include material radiative property degradation due to UV exposure or atomic oxygen exposure.

As the results of these tests are collected and correlated, general design guidelines emerge that replace the need for specific testing. The material and radiative properties discussed earlier in this chapter are all based on engineering testing. It is important to understand the applicability of the design guidelines to the specific application. As an example, the conductivity and specific heat of most materials are used as if constant, but they change significantly at cryogenic temperatures. MLI performance is affected by the size, configuration, and how it is modeled.

7.16.2 Component-level Testing

Component-level testing is focused on finding design and/or workmanship faults. Thermal vacuum testing consists of exposing the component to both temperature and vacuum exposures. Thermal testing typically involves either thermal balance or thermal cycling. Thermal balance testing is aimed at verifying the thermal analysis by simulating specific heat paths in and out of the component. The important part of any balance test is to define the test conditions so that the component and its heat paths can be analyzed. Thermal cycling verifies the workmanship of the component and its proper operation over its required temperature range. The important parts 3 of the cycling testing are the temperature limits, the temperature dwell periods, and the number of cycles. Component temperature ranges capture the basic requirements of the thermal control subsystem. The dwell periods and number of cycles are based on the specific mission and the institutional preferences of the design organization.

7.16.3 System-level Testing

System level-testing is performed at the highest level of assembly prior to delivery of the flight hardware. The test verifies that the spacecraft or instrument operates as required when subjected to thermal conditions more severe than those expected in flight, and that the thermal control design performs correctly. Since all the components should have been previously qualified at the component level, the purpose of the system-level test is not to duplicate those tests but rather to verify the performance of the system as a whole, including the interfaces between the components. Since the thermal control system is part of the overall system, its verification includes the analysis supporting the design. The system test is typically divided into two parts: balance and cycling testing.

The balance test needs to collect sufficient data to allow correlation of the system analytical thermal model, and to detect deficiencies in the thermal design. The proper correlation analytical model is important because that model is used to predict untestable conditions. The thermal margin existing in the design should be demonstrated during the test. The test also allows operation of any function that cannot be fully exercised except in a simulated mission-like environment.

The cycling part of the thermal vacuum (TV) test is the only time during integration that the entire system is tested in flight-like, on-orbit conditions. Nominally, as much of the system is in a flight-like condition as is practicable. Exceptions are made for items like solar arrays, antennas, and booms that do not fit or cannot be adequately tested at the system level. Subsystem performance testing is repeated to ensure that no anomalies are introduced from either the temperature or vacuum conditions. For some instruments involving high-voltage or cold sensors, the system-level TV test allows the instrument to be fully tested. The length of the test also provides time for the system to be operated in a flight condition over several weeks.

7.17 Design Example

As an illustration of the preceding discussion, the thermal control system for a 1 m cube is developed and analyzed. The cube is assumed to be in a circular orbit around the earth with a 1000 km altitude. The power dissipated inside the cube varies from 150 W to 200 W. The attitude of the cube in the orbit is earth oriented. The cube mass is 500 kg. The cube flies such that one face is always pointed toward the Earth ($-Z$) and one face away from the Earth ($+Z$). The $+X$ face always points into the velocity direction. The design example geometry is shown in figure 7.22. The orbit precession is such that all orbit beta angles will be seen over the life of the mission. It has been decided that the hot orbit design case is $\beta = 90°$ (full sun) and the cold orbit design case is $\beta = 0°$. The orbital design cases are shown in figure 7.23.

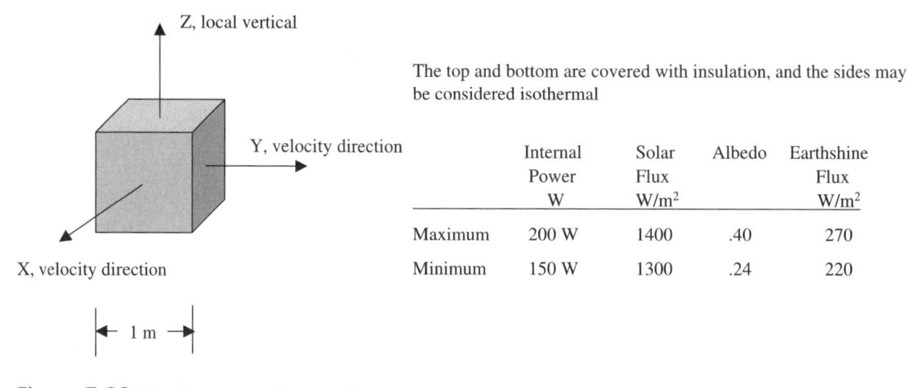

	Internal Power W	Solar Flux W/m²	Albedo	Earthshine Flux W/m²
Maximum	200 W	1400	.40	270
Minimum	150 W	1300	.24	220

Figure 7.22 Design example—problem definition.

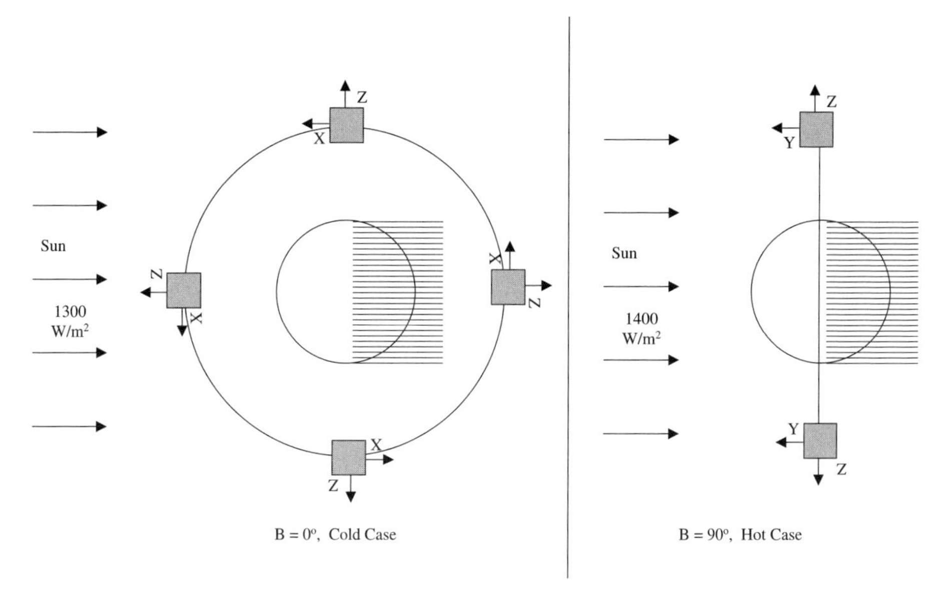

Figure 7.23 Design example—orbital design cases.

7.17.1 Direct Solar Heating—Hot Case

The direct solar heating is simply the projected areas of the cube face with respect to the Sun (equation 7.11.9, with $Q_{\text{sun}} = 1400\ \text{W/m}^2$ and $\alpha_s = 1$). For the hot case the cube is always in the same position with respect to the Sun, so the solar heating on all faces is constant (Table 7.3).

7.17.2 Direct Solar Heating—Cold Case

For the cold case, the motion of the cube around the planet exposes four sides to the sun in varying amounts. The angles at which the spacecraft enters and exits the eclipse are 150° and 210°, respectively. The orbit angle is measured from the subsolar point (the point in the orbit nearest the Sun). They can be calculated using the Earth radius of 6378 km and the 1000 km altitude. The period is about 100 minutes. Taking the cosine of the angle between the normal of each face and the Sun line, equation 7.11.9 gives, for $Q_{\text{sun}} = 1300\ \text{W/m}^2$ and $\alpha_s = 1$, the incident fluxes as shown in table 7.4.

Note that the points just before and after the eclipse entry are added to allow for correct averaging.

Table 7.3 Incident direct solar energy on each cube face (W/m^2)—Hot case

$+X$	$-X$	$+Y$	$-Y$	$+Z$	$-Z$
0	0	1400	0	0	0

Table 7.4 Incident direct solar energy on each cube face (W/m²)—Cold case

Orbit Angle (°)	+X	−X	+Y	−Y	+Z	−Z
0	0	0	0	0	1300	0
30	0	650	0	0	1126	0
60	0	1126	0	0	650	0
90	0	1300	0	0	0	0
120	0	1126	0	0	0	650
149.9	0	652	0	0	0	1125
150 } Eclipse	0	0	0	0	0	0
180 } Eclipse	0	0	0	0	0	0
210 } Eclipse	0	0	0	0	0	0
210.1	652	0	0	0	0	1125
240	1126	0	0	0	0	650
270	1300	0	0	0	0	0
300	1126	0	0	0	650	0
330	650	0	0	0	1126	0
360	0	0	0	0	1300	0

7.17.3 Albedo Heating—Hot Case

Albedo heating rates are calculated using equation 7.11.10. For the $\beta = 90°$ orbit, the angle φ_2 is 90° and so the albedo heating disappears (table 7.5).

7.17.4 Albedo Heating—Cold Case

Albedo heating (equation 7.11.10) included a term, $F_{\text{s-planet}}$, that describes the view the surface has to the Earth. These view factors are given in figure 7.12 for flat plates at various angles. For the Earth-facing surface, $-Z$, of the cube,

$$F_{\text{s-planet}} \simeq 0.7$$

Using the approximation given in the text, we get

$$[R_e/(Z + R_e]^2 = (6378/(6378 + 1000))^2 = 0.75$$

The view factor to each side of the cube, from figure 7.12, is

$$F_{\text{s-planet}} \simeq 0.2$$

Using the minimum albedo factor, $\rho_{\text{alb}} = 0.24$, allows the incident fluxes to be calculated around the orbit from equation 7.11.10 with $\alpha_s = 1$ (table 7.6).

Table 7.5 Incident albedo heating on each cube face (W/m²)

$+X$	$-X$	$+Y$	$-Y$	$+Z$	$-Z$
0	0	0	0	0	0

Table 7.6 Incident albedo heating on each cube face (W/m^2)

Orbit Angle (°)	$+X$	$-X$	$+Y$	$-Y$	$+Z$	$-Z$
0	62	62	62	62	0	218
30	54	54	54	54	0	189
60	31	31	31	31	0	109
90	0	0	0	0	0	0
120	0	0	0	0	0	0
149.9	0	0	0	0	0	0
150 } Eclipse	0	0	0	0	0	0
180 } Eclipse	0	0	0	0	0	0
210 } Eclipse	0	0	0	0	0	0
210.1	0	0	0	0	0	0
240	0	0	0	0	0	0
270	0	0	0	0	0	0
300	31	31	31	31	0	109
330	54	54	54	54	0	189
360	62	62	62	62	0	218

7.17.5 Earth IR Heating—Cold Case

Earth IR heating is calculated using equation 7.11.11, with $\varepsilon_{IR} = 1$. The IR heating from the planet is not affected by the orbital position because the Earth is assumed to be at a constant temperature. The same view factors as were used in the albedo case apply here. The minimum and maximum Earthshine values are 220 W/m^2 and 270 W/m^2 respectively (table 7.7).

7.17.6 Earth IR Heating—Hot Case

Similarly, using the maximum Earth IR flux, we get the values listed in table 7.8.

Table 7.7 Incident Earth IR heating on each cube face (W/m^2)

Orbit Angle (°)	$+X$	$-X$	$+Y$	$-Y$	$+Z$	$-Z$
0	44	44	44	44	0	154
30	44	44	44	44	0	154
60	44	44	44	44	0	154
90	44	44	44	44	0	154
120	44	44	44	44	0	154
149.9	44	44	44	44	0	154
150 } Eclipse	44	44	44	44	0	154
180 } Eclipse	44	44	44	44	0	154
210 } Eclipse	44	44	44	44	0	154
210.1	44	44	44	44	0	154
240	44	44	44	44	0	154
270	44	44	44	44	0	154
300	44	44	44	44	0	154

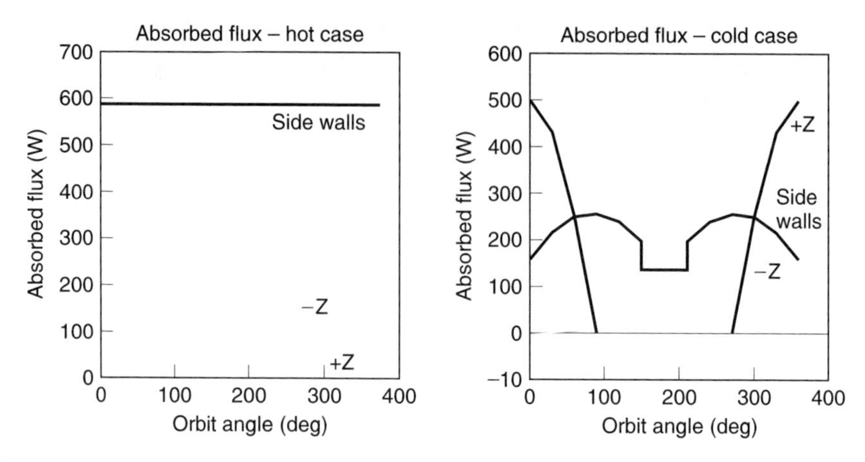

Figure 7.24 Design example—absorbed energy.

Putting all the heating rates together requires the radiative properties of the surfaces to be defined. Assume that the $+Z$ and $-Z$ faces are covered with MLI insulation with a 0.001 inch (25 μm) Kapton external layer. Effectively, the heat lost from these two surfaces is eliminated. It is further assumed that the four remaining side faces are covered with an Ag/FEP radiator surface. From table 7.2, we get the radiative properties for the two surfaces. BOL properties are used for the cold case and EOL properties for the hot case.

For BOL,	Kapton	$\alpha_s/\varepsilon_{IR} = 0.38/0.67$
	Ag/FEP	$\alpha_s/\varepsilon_{IR} = 0.09/0.78$
For EOL,	Kapton	$\alpha_s/\varepsilon_{IR} = 0.45/0.67$
	Ag/FEP	$\alpha_s/\varepsilon_{IR} = 0.30/0.78$

Remembering that both direct solar and albedo use the solar absorptivity and that Earthshine uses the IR emissivity allows total absorbed heating rates to be calculated for the three surfaces of interest: the top insulation surface, the bottom insulation surface, and the combined four side walls. All the areas are 1 m^2. The resultant absorbed energies on the three surfaces are shown in figure 7.24.

Once the absorbed fluxes are calculated, either transient or steady-state temperatures can be found. Since the heat inputs for the $\beta = 0°$ case vary, but not for the $\beta = 90°$ case, it is difficult to compare the two cases directly. To allow such a comparison, the concept of orbit averages is introduced. If a time-weighted average of the absorbed energy fluxes is made for the two design orbits, the resulting orbit-average heat inputs are as given in table 7.9. The side-wall heat flux includes the individual heating rates for the four side walls added together.

The hot case does indeed have the largest heat input for the side walls, but not for the $+Z$ and $-Z$ blanket surfaces. There can be different hot and cold orbits for different parts of the structure. Making sure the design cases adequately bound the design is an important part of the process.

Table 7.8 Incident Earth IR heating on each cube face (W/m^2)

Orbit Angle (°)	$+X$	$-X$	$+Y$	$-Y$	$+Z$	$-Z$
0	54	54	54	54	0	189
30	54	54	54	54	0	189
60	54	54	54	54	0	189
90	54	54	54	54	0	189
120	54	54	54	54	0	189
149.9	54	54	54	54	0	189
150 } Eclipse	54	54	54	54	0	189
180 } Eclipse	54	54	54	54	0	189
210 } Eclipse	54	54	54	54	0	189
210.1	54	54	54	54	0	189
240	54	54	54	54	0	189
270	54	54	54	54	0	189
300	54	54	54	54	0	189
330	54	54	54	54	0	189
360	54	54	54	54	0	189

7.17.7 Putting it All Together

Finding the temperature of the cube requires the energy balance equation for a steady-state condition. Equation 7.11.8 is simplified for a single-node, steady-state system, where the spacecraft surfaces see only space. The resulting equation is

$$A_{sc}\, \varepsilon\, \sigma\, T_{sc}^4 = Q_{env} + Q_{int}$$

In the hot case, the maximum environmental orbit-average heat inputs are combined with the maximum internal dissipation to produce the hot case temperature

$$\begin{aligned} T_{sc} &= [(588.5\text{ W} + 200\text{ W})/(4\text{m}^2 \times 0.78 \times 5.669 \times 10^{-8}\text{ W/m}^2\text{ K}^4)]^{0.25} \\ &= 258.4\text{ K} = -14.6^\circ\text{C} \end{aligned}$$

The cold case uses the minimum values

$$\begin{aligned} T_{sc} &= [(212.2\text{ W} + 150\text{ W})/(4\text{m}^2 \times 0.78 \times 5.669 \times 10^{-8}\text{ W/m}^2\text{ K}^4)]^{0.25} \\ &= 211.4\text{ K} = -61.6^\circ\text{C} \end{aligned}$$

The temperature referred to represents that of the radiating surface of the spacecraft. Internal temperature distributions will be based on this radiating surface. In this case,

Table 7.9 Orbit-averaged absorbed energy (W)

Hot Case			Cold Case		
Side Walls	$+Z$	$-Z$	Side Walls	$+Z$	$-Z$
588.5	0.0	126.6	212.2	153.6	205.7

while the hot temperature may be acceptable, the lower one is probably too cold. The system temperature can be raised by reducing some of the side-wall radiator area. For example, if the $+Y$ and $-Y$ surfaces were blanketed, the heat input for the hot case would be reduced substantially.

Problems

1. A flat plate is exposed to the sun in space. Which of the following plate properties are primarily responsible for determining the equilibrium temperature it will reach?

density, ρ	solar absorptivity, α_s
specific heat, c_p	IR emissivity, ε_{IR}
conductivity, k	transparence, τ

   Answers: α_s, ε_{IR}, and τ.
2. An IR detector is cooled to 80 K. The surrounding instrument is at 200 K. Four thin-walled G-10 tubes mechanically support the detector. Neglecting any radiative heat transfer, which of the tubes, properties are primarily responsible for determining the heat leak into the detector?

density, ρ	solar absorptivity, α_s
specific heat, c_p	IR emissivity, ε_{IR}
conductivity, k	transparence, τ

   Answers: k.
3. Find the solar heat input to the following geometrical shapes. Assume $Q_{Sun} = 1353$ W/m^2 and $\alpha_s = 1$.
   (a) A sphere with 1 m diameter.
   (b) A cylinder with 0.5 m diameter and 2 m long (with length perpendicular to the Sun vector).
   (c) Find the minimum and maximum solar input to a cube with 1 m sides and the Sun in any orientation.
   Answers: (a) 1062.6 W, (b) 1353.0 W, (c) min. 1353.0 W, max. 2343.5 W.
4. A satellite uses two solid aluminum spheres as calibration sources for optical measurements. The spacecraft ejects the spheres and images them after their temperatures have stabilized. Assume that the spheres are 5 cm in diameter. One is painted black, and the other is vapor-deposited aluminum. The spheres are ejected at 20°C, and each is continually exposed to the Sun afterwards. Find the equilibrium temperature each sphere will reach. Assume $Q_{Sun} = 1353$ W/m^2, $\alpha_s = 0.15$ and $\varepsilon_{IR} = 0.05$ for vapor-deposited aluminum, and $\alpha_s = 0.95$ and $\varepsilon_{IR} = 0.85$ for black paint
   Answers: 244°C for vap. dep. sphere, 131°C for black sphere.
5. A black plate, 1 foot square, is suspended in a vacuum chamber with a liquid nitrogen cold wall (90 K). The plate temperature is maintained at 300 K with heaters. Find the heater power required to maintain the plate's temperature. (Assume $\sigma = 5.67 \times 10^{-8}$ W/m^2 K^4, $\varepsilon_{IR} = 1.0$ both sides.)
   Answers: 84.6 W.

6. A mechanical cooler keeps the detector in problem 2 at 80 K. The sensor is mounted to a frame that runs at 200 K by four thin-walled hollow tubes that are 2 in. long by 0.5 in. outer diameter. The total heat leak through the four tubes into the detector has to be kept at less than 200 mW. Assuming the tubes are made of G-10 with a thermal conductivity of 0.3 W/m K, what wall thickness is required?
   Answers: 0.084 in.
7. A solar telescope looks at the Sun continually while in orbit. The dimensions of the optical elements are given in table 7.10. Using the given solar flux spectrum, calculate (a) the solar constant at 1 AU, (b) the heat flux absorbed on the primary mirror, (c) the heat flux absorbed on the secondary, and (d) the heat flux that remains in the beam. Assume all the light that hits the primary mirror is reflected onto the secondary. The primary and secondary mirror diameters are 80 cm and 10 cm respectively. Use the information in the table for both mirrors and the solar spectrum.

Table 7.10

Wavelength (μm)	Mirror reflectivity ρ(%)	Solar flux spectrum (W/m$^2\mu$)	Wavelength (μm)	Mirror reflectivity ρ(%)	Solar flux spectrum (W/m$^2\mu$)
0.20	0.0%	12.0	1.20	97.5%	500.0
0.30	16.7%	590.0	1.40	97.6%	330.0
0.32	20.0%	792.0	1.60	98.0%	223.0
0.40	87.0%	1600.0	1.80	98.2%	148.0
0.50	93.0%	2040.0	2.00	98.4%	102.0
0.55	95.9%	1980.0	2.50	98.4%	49.7
0.60	98.9%	1870.0	3.00	98.4%	26.3
0.65	98.8%	1670.0	4.00	98.4%	9.3
0.70	98.8%	1490.0	5.00	98.4%	4.1
0.75	98.7%	1290.0	6.00	98.4%	2.1
0.80	98.6%	1140.0	8.00	98.4%	0.6
0.90	98.3%	900.0	10.00	98.4%	0.2
1.00	98.0%	740.0	12.00	98.4%	0.1
1.10	97.8%	610.0			

   Answers: (a) 1412 W/m^2, (b) 41.3 W, (c) 24.9 W, (d) 643.5 W.
8. Two plates are suspended in space so that the first plate faces the sun and the second is completely shaded by the first. The first plate's solar reflectivity is 30% and its solar transmissivity is 40%. Its IR reflectivity and transmissivity are 0%. The second plate's solar and IR reflectivities are 0% and its solar and IR transmissivities are 0%. Assume no heat transfer from the backside of the second plate or the area between the plates. Find the temperatures of the two plates.
   Answers: $T_{\text{plate1}} = 359.5$ K, $T_{\text{plate2}} = 402.5$ K.
9. A battery design calls for a temperature range of 5°C to 15°C. The battery internal heat dissipation varies from 0 W to 20 W. The battery absorbs 0 W to 30 W from the external environment. Find the (a) radiator size ($\varepsilon_{\text{IR}} = 0.85$) needed to keep the battery below its maximum temperature with the maximum internal dissipation, and

(b) the heater power required to keep it above its minimum temperature limit with the minimum internal dissipation.
Answers: $A_{batt} = 0.151 \text{ m}^2$, $Q_{htr} = 43.4$ W.

10. A cube with 1 m sides is in a low Earth orbit with an 800 km altitude. The orbit beta angle is 0°. The cube's $+X$ face is in the direction of the velocity vector and the $-Z$ axis always points toward the Earth. Find the orbit-average heat fluxes on the six cube faces.
Answer: +X face = $-X$ face = 314 W,
+Y face = $-Y$ face = 0 W,
+Z face = 430 W, $-Z$ face = 48 W.

References

Campbell, W. A., and J. J. Scialdone, 1993. *Outgassing Data for Selecting Spacecraft Materials.* NASA reference publication 1124, Rev. 3, September.

Carslaw, H. S., and J. C. Jaeger, 1959. *Conduction of Heat in Solids*, 2nd ed. Oxford: Oxford University Press.

Doenecke, J., 1993. Survey and evaluation of multilayer insulation heat transfer measurements. SAE Technical Paper No. 932117, presented at the 23rd International Conference on Environmental Systems, Colorado Springs, Co.

Henninger, J. H., 1984. *Solar Absorptance and Thermal Emittance of Some Common Spacecraft Thermal-Control Coatings.* NASA reference publication 1121, April.

Holman, J. P., 1976. *Heat Transfer.* 4th ed. New York: McGraw-Hill.

Jursa, A. S., 1985. *Handbook of Geophysics and the Space Environment.* Air Force Geophysics Laboratory, NTIS Doc. No. ADA 167000.

Lin, E. L., and J. W. Stultz, 1996. Effective emittance for Cassini multilayer insulation blankets and heat loss near seams. *Journal of Thermophysics and Heat Transfer*, Vol. 10, No. 2.

Minkowycz, W. J., E. M. Sparrow, G. E. Schneider, and R. H. Pletcher, 1988. *Handbook of Numerical Heat Transfer.* New York: John Wiley & Sons.

O'Neil, R. F., et al., 1983. A vector approach to numerical computation of view factors and application to space heating. Paper No. AiAA-83-0157, presented at the 21st AIAA Aerospaces Sciences Meeting, Reno, NV, January 10–13.

Pisacane, V. L., and R. C. Moore, 1994. *Fundamental of Space Systems*, 1st ed. New York: Oxford Universtiy Press.

Siegel, R., and J. R. Howell, 1981. *Thermal Radiation Heat Transfer*, 2nd ed. New York: McGraw-Hill.

Silverman, E. M., 1996. Composite spacecraft structures design guide. NASA Contractor Report 4708, March.

Sparrow, E. M., and R. D. Cess. 1978. *Radiation Heat Transfer*, Augmented edition, New York: McGraw-Hill.

Wiedemann, C., et al., 1997. Effective emissivity determination of multilayer insulation for practical spacecraft application. SAE Paper No. 972354, presented at the 27th International Conference on Environmental Systems, Lake Tahoe, NV.

8

Spacecraft Configuration and Structural Design

W. E. SKULLNEY

8.1 Introduction

In considering missions of opportunity, very few investigators or program managers are interested in flying just the mechanical or structural system. The role of the mechanical system and structure is, however, an important one. The primary function of the structural subsystem is to provide the mechanical interface with the launch vehicle and structural support to all spacecraft subsystems. This structure must be sturdy enough to withstand the severe launch environment, yet have minimum mass and provide adequate protection of the sensitive payload so that the mission objectives can be successfully completed. Since many of the spacecraft subsystems may provide conflicting requirements, one of the primary tasks of the mechanical designer is to carefully coordinate inputs from each area and, after careful consideration, arrive at the most optimum configuration that will satisfy the mission requirements. In addition, the designer must package each subsystem's hardware such that it fits into the launch vehicle envelope while allowing as much access as possible and providing the most optimum mass distribution.

As indicated in chapter 1, five phases in the system development process are:

(1) conceptual design (phase A);
(2) preliminary design/analysis (phase B);
(3) detailed design/analysis (phase C);
(4) development (fabrication, assembly, integration, and test), (phase D); and
(5) operation (phase E).

During each of these phases, the mechanical designer and analyst have various responsibilities. A brief description of these responsibilities is presented for phases B, C, and D in the following paragraphs of this section.

During the preliminary design/analysis phase, a general spacecraft configuration is developed using many of the basic fundamentals of mechanical engineering (statics, dynamics, strength of materials, etc.). Preliminary estimates of mass, volume, and power are made and initial estimates of worst-case loads are used to size the major structural elements of the spacecraft. Based on the configuration developed and the preliminary mass properties, fundamental spacecraft frequencies (that is, the spacecraft stiffness) are estimated. Adequate stiffness of the spacecraft structure is necessary to ensure that dynamic coupling does not occur between the spacecraft and the launch vehicle. Undesired dynamic coupling of the two systems could result in loads that are much higher than the initial estimates. A primary goal of the mechanical system team in this preliminary phase is to establish a configuration that packages all of the required hardware in the available envelope, and to develop a structure that provides the best possible primary load path through which structural loading can be transferred. A number of iterations are typically required to arrive at the "final" design. During the preliminary phase, the appropriate structural materials that will best fit the application are defined and deployable mechanisms are identified.

Some programs may require the development of an engineering, or prototype, structure to evaluate the design concept being used. These entities, described in chapter 1, are non-flight components built prior to the flight hardware and used primarily for testing at levels that significantly exceed those predicted in launch and flight. The primary goal is to demonstrate an adequate design margin and confirm predicted performance or provide data where an analytical model may fall short of reality. In addition, this hardware can be used to identify any major difficulties that may occur in the mechanical assembly process. Hardware of this nature is typically built in the early stages of the program so that the results can be used to improve the flight design.

In the detailed design/analysis phase, detailed parts are developed and analyzed using updated information on structural loading. With the definition of more detail, finite element models are generated to more accurately determine loads, stresses, and displacements (static analysis), and fundamental frequencies and mode shapes (modal analysis). The spacecraft system finite element model is used by the launch vehicle contractor to conduct a coupling analysis. This analysis provides a more accurate prediction of the flight loads (bending moments, shear forces, displacements, accelerations) that are generated throughout the structural system. The updated loads information is used to verify the previous analyses and possibly to further optimize the system design. During the detailed design phase, interaction between program management, subsystem engineers, and fabrication personnel ensures that the structure being designed can indeed be fabricated and that it will satisfy all program requirements.

The fabrication phase represents a time when the mechanical team completes the last details of the design, including mechanical ground support equipment, resolves difficulties that arise during the fabrication process, and makes final plans and preparation for the upcoming test and integration phase. During the integration and test phase, described in chapter 14, the flight spacecraft and components are subjected to acceptance tests to provide screening from manufacturing defects and to verify the hardware can withstand the worst-case expected loading environments. In some organizations, the structural engineer "changes hats" and assumes the role of the test engineer during this time. Testing is typically conducted at the component level, prior to system- or subsystem-level tests, to establish a sufficient degree of confidence that the hardware will survive the next level of tests. Early discovery and elimination of difficulties could save considerable time and effort once the spacecraft has been assembled. Spacecraft-level tests are conducted to ensure that once all hardware is assembled, the system can withstand the load conditions expected in flight. Testing areas applicable to the mechanical/structural system are defined later in this chapter.

8.2 Spacecraft System Requirements

In developing the spacecraft configuration, the first task is to determine the requirements. These requirements can come from several sources such as launch vehicle design manuals, instrument or payload requirements, and subsystem or component requirements. The engineer or designer responsible for the development of the configuration must consider all of these requirements and make the necessary tradeoffs to arrive at the final design. This development effort is a highly iterative process and requires a great deal of interaction with engineers developing all flight components. In many cases it is necessary to establish a priority of these requirements to arrive at the best configuration. A number of requirements that the mechanical system must satisfy are the following:

(a) Provide the interface with the launch vehicle
(b) Satisfy launch vehicle mass and center of gravity requirements
(c) Satisfy launch vehicle strength requirements; in other words, provide support of the total spacecraft with adequate strength
(d) Satisfy launch vehicle stiffness requirements; that is, provide sufficient stiffness to preclude dynamic interactions (dynamic coupling) with the launch vehicle
(e) Provide the mounting platform for payload and electronics (including stable pointing platforms for instruments and critical components)
(f) Protect hardware from launch vehicle environments
(g) Accommodate and provide the interface for all instruments, electronics, and subsystems
(h) Package system hardware in the launch vehicle fairing
(i) Provide a platform for the stowage and deployment of any spacecraft appendages (for example, antennae, solar arrays, booms, etc.)
(j) Accommodate static and dynamic mass balance requirements, if appropriate
(k) Satisfy component pointing co-alignment and alignment requirements
(l) Provide clear fields of view (FOVs) for instruments and appropriate subsystem components
(m) Provide the necessary mechanisms for deployments, separation, etc.
(n) Allow for proper control of the spacecraft's attitude via proper design of system deployment dynamics
(o) Satisfy system jitter requirements
(p) Satisfy thermal requirements by
    (a) providing the necessary thermal paths to maintain proper temperature control of system components; and
    (b) providing sufficient radiator surfaces to deep space
(q) Satisfy system purge and contamination requirements
(r) Satisfy electromagnetic compatibility requirements regarding necessary mounting surface treatments, structural bond resistance, and surface resistivity of external exposed surfaces
(s) Provide ease of assembly and disassembly

(t) Provide hardware accessibility during integration operations
(u) Accommodate test and integration handling
(v) Satisfy the necessary requirements for mission duration

These numerous requirements must be satisfied on the ground, during lift-off, and throughout the mission on orbit. Many of them will be addressed in further detail throughout the remaining parts of this chapter.

The designer faces many areas of concern common to any choice of launch vehicle. The total interface of every package and experimental instrument must be understood, identified, and documented in words and drawings with great clarity. These interfaces include not only weight, volume, bolt hole patterns, mounting surface flatness, and tolerances, but also fields of view, temperature restrictions, and access requirements. As the detailed information becomes available, package locations are defined. The designer collaborates with the structural engineer to ensure that the major structure is defined, the primary load path identified, and the spacecraft structural concept is formed. This conceptual approach is then presented to the other specialists in the areas of thermal control, power generation and storage, attitude guidance and control, command and telemetry, and communications. Based on inputs from these areas, the spacecraft structural concept is modified until concurrence is obtained from all disciplines.

Agreement on the spacecraft configuration as described above is not always as simple as presented. Many times incompatible design requirements among the spacecraft disciplines lead to conflicting views on the design, or the requirements placed on the system are over-constrained. Some examples of this are:

(1) instrument FOV requirements interfere with the deployment of the solar panels; or
(2) the structural engineer and thermal engineer may have difficulty choosing a structural joint configuration that provides adequate strength with sufficient heat flow restriction; or
(3) difficulties in equipment location may arise because the attitude control lead engineer may be unable to accept a magnetic field generated by the power distribution circuit design, or the choice of high-strength magnetically permeable fasteners; or
(4) the lead power engineer responsible for solar array design may have a conflict with communications engineers because of an antenna which shadows the array.

When this occurs the designer must propose a variety of possible compromises until a solution that is acceptable to all engineers is reached. The ideal structure is thus one that provides the most benign environment, utilizes no space, provides total accessibility until launch, provides both infinite and zero heat transfer as required, and finally has zero mass! In most applications, the solution is less than ideal in all respects but is a viable compromise. After many iterations and compromises, the conceptual idea emerges into a preliminary design and finally a detailed design that can be fabricated.

8.3 System Configuration Development

The selection of the launch vehicle is also an important factor in developing a spacecraft configuration for a particular mission. A launch vehicle imposes numerous constraints on

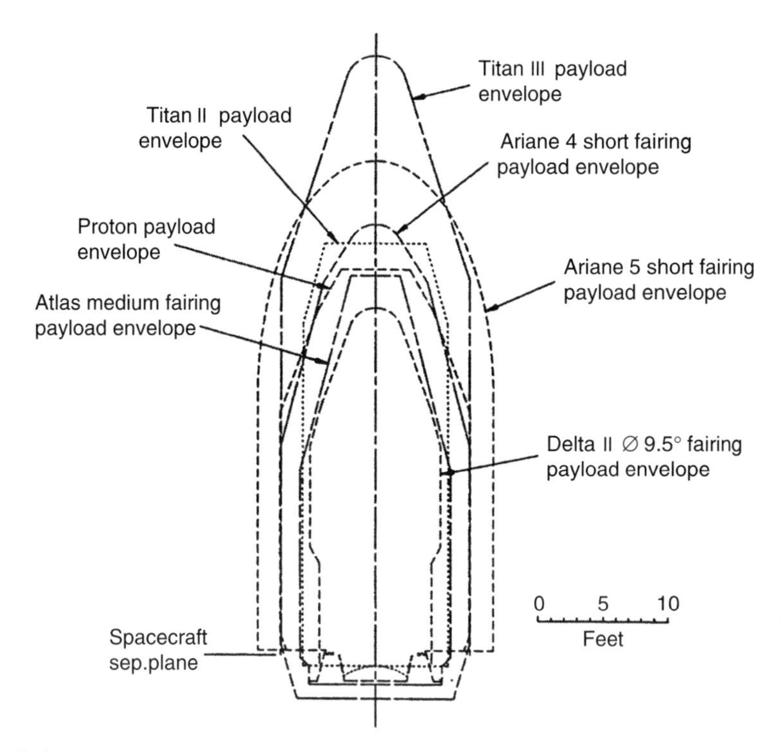

Figure 8.1 Launch vehicle fairing comparison.

the spacecraft design, with some of the primary ones being mass, size (volume), electrical and mechanical interfaces, and dynamic and thermal environments during launch.

Typically, for each mission, there are a number of possible launch vehicle candidates. The specific vehicle selected is based on a number of factors—not the least of which is cost—and is not always defined at the onset of the program. When such is the case, the designer and design team must address the requirements of each potential launch vehicle candidate and establish requirements that allow the design to proceed. Typically, requirements are "enveloped" to give a worst-case scenario.

One such area that the designer must start with is the amount of space, or envelope, available in the fairing for the applicable launch vehicles. Figure 8.1 illustrates, in simple form, a number of the fairing envelopes available for a variety of launch vehicles including Delta, Atlas, and Titan (American), and Proton and Ariane (International). As one can see, a variety of comparable "sizes" is available which could satisfy various mission requirements. For each of the launch vehicles, a user's design manual is available that provides more detail on the basic requirements to be used in developing the initial spacecraft configuration. As an example, detailed fairing information for the Atlas family of launch vehicles is given in figure 8.2.

In considering the available envelope for each spacecraft fairing, the designer must identify whether the envelope is specified as static or dynamic. A static envelope is exactly as indicated. The designer creates the configuration from the available subsystem information and, as long as the equipment outline falls within the envelope,

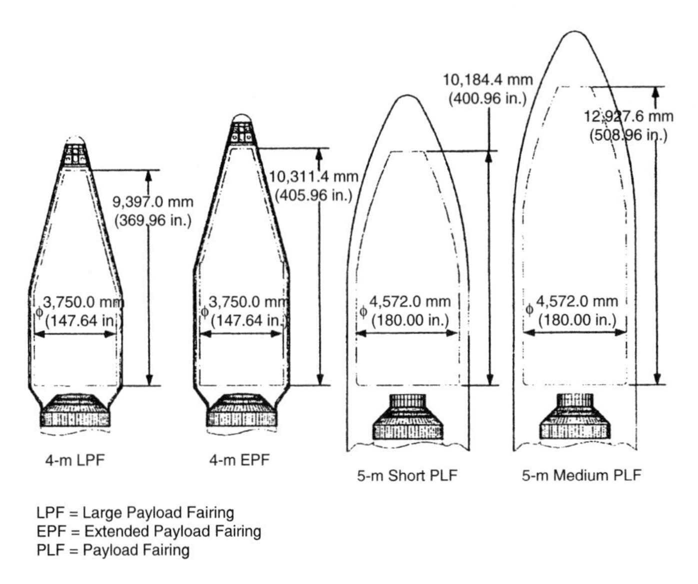

Figure 8.2 Atlas static payload envelope. (Courtesy of Lockheed Martin Corporation and International Launch Services.)

then the requirement is met. For a dynamic envelope, an additional factor comes into play. Here, in developing the configuration, the designer must also allow room for any displacement that may occur when the system experiences disturbances due to the launch environments. From a worst-case perspective this is typically on the order of 1–2 inches of displacement. An example of such a configuration layout, which uses a Delta II 10-foot fairing, is shown in figure 8.3a. This mission, the STEREO mission, is a stack of two spacecraft which are nearly identical systems. A view of one of the spacecraft in the orbital configuration is shown in figure 8.3b.

In some situations, the design engineer may discover that the spacecraft equipment outline slightly exceeds the available envelope in one or more places. All is not lost in this situation as the launch vehicle contractor may, after a closer look, find that that indeed there is available space to accommodate this slight violation. This is possible since in most cases the launch vehicle contractor typically has additional "margin" available which may be used when needed. In cases where the violation is somewhat large, or in particular regions where launch vehicle equipment is located, this may not always be possible. An example where the outline exceeded the envelope and was acceptable is shown in figure 8.4 for NASA's Advanced Explorer Composition (ACE) spacecraft. Here the solar array and magnetometer boom exceed the available space (indicated by the cross-hatched area). This violation is of particular concern in this circumstance as it occurs in the vicinity of the launch vehicle/spacecraft separation plane. In addition to the static perspective, the dynamics of what might happen during the separation sequence must be considered. As one can see from the figure, the solar array extends below the separation plane. At separation, the spacecraft would be "pushed" away at the observatory attach fitting (OAF) from the launch vehicle via the separation springs (typically 3 or 4) at the interface. Since each of these springs will not be exactly identical, slightly varying forces will be exerted at each location, giving rise to a possible tipping of

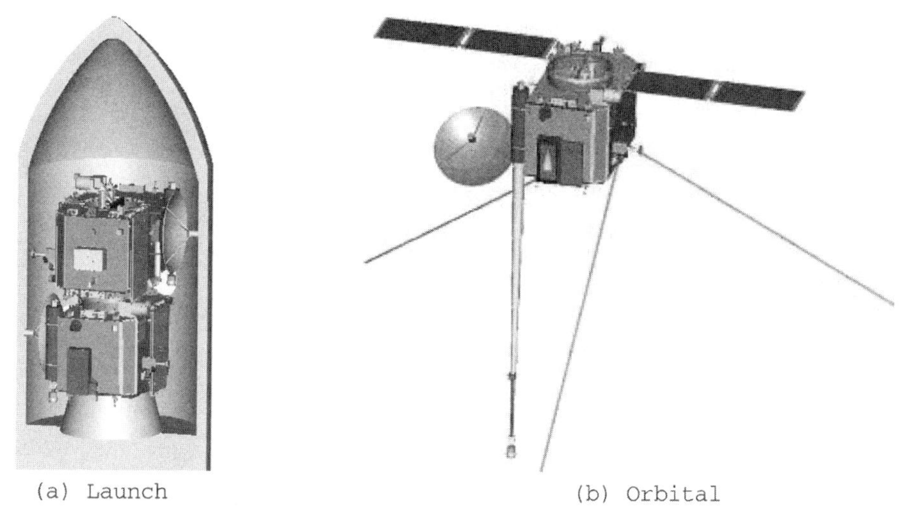

Figure 8.3 STEREO spacecraft configurations; (a) launch; (b) orbital.

the spacecraft during separation. Depending on the extent of the tipping, the solar array could come into contact with the payload attach fitting (PAF), thus possibly causing damage to the array or causing the spacecraft to tumble uncontrollably away from the launch vehicle.

As indicated previously, the design engineer has the responsibility of packaging all of the hardware required for the mission. This includes not only the spacecraft bus hardware but also the payload. A layout of equipment for the top mounting deck of the ACE spacecraft is shown in figure 8.5. In developing the layout of hardware, numerous factors must be considered. Some of the items, indicated in the figure, are:

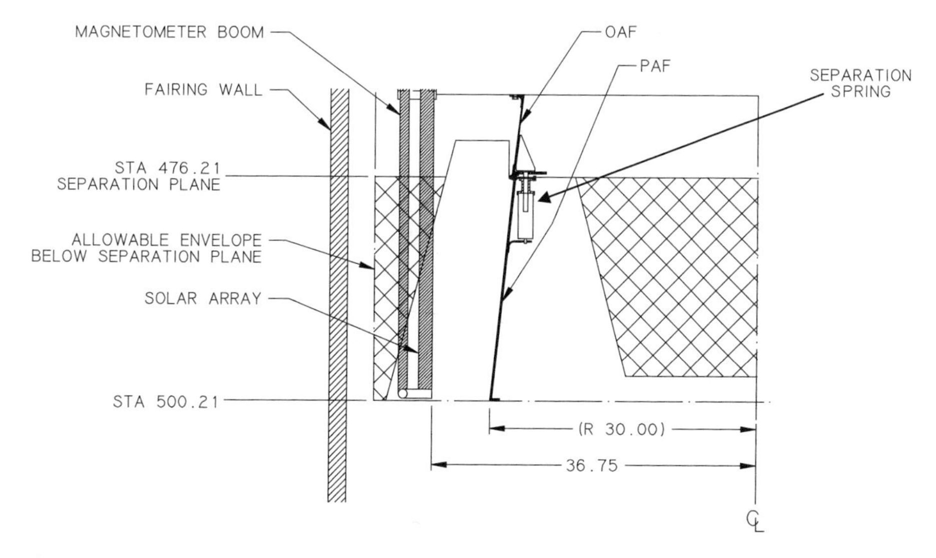

Figure 8.4 Envelope considerations near the separation plane.

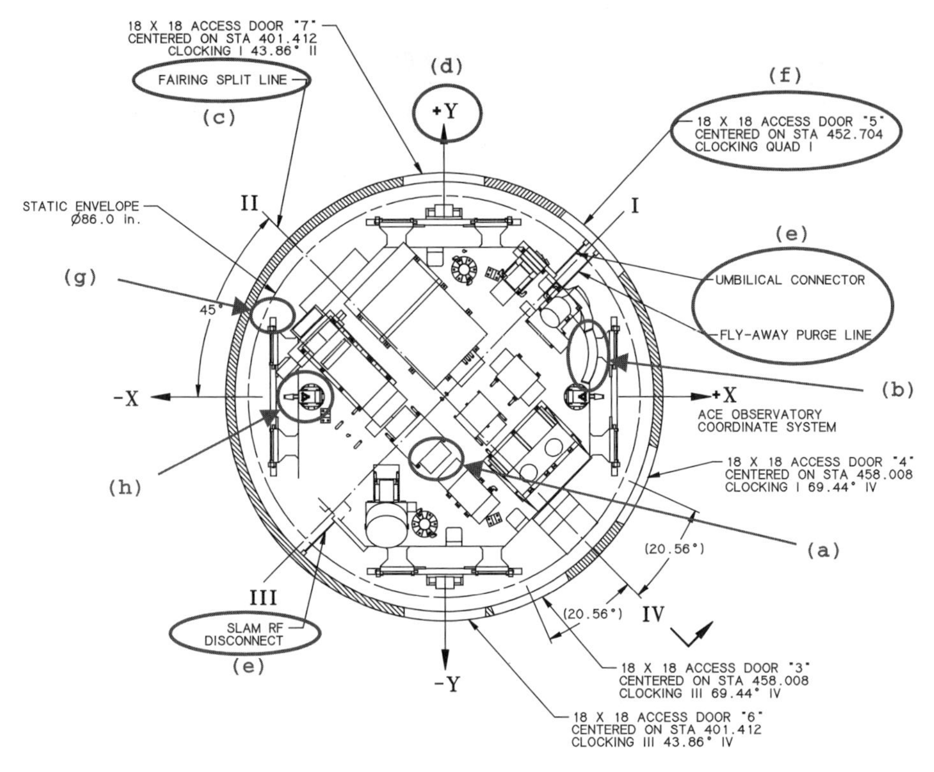

Figure 8.5 ACE spacecraft deck configuration.

(a) allowing space for harness routing and connector locations;
(b) allowing clearance for the deployment of items such as instrument covers;
(c) location of fairing split line, which can define spacecraft clocking;
(d) launch vehicle versus spacecraft coordinates;
(e) location of disconnects relative to spacecraft clocking and access (spacecraft clocking refers to the orientation of the spacecraft system relative to the fairing and, specifically, the fairing split or separation line);
(f) location of access doors which provide "last minute" access for final arming and removal of non-flight items;
(g) proximity of spacecraft allowable fairing envelope;
(h) thruster locations and concerns regarding contamination and plume impingement.

A typical configuration at the launch vehicle/spacecraft separation plane is shown in figure 8.6. Here the spacecraft and launch vehicle are joined by a clamp band, or V-band, configuration. This configuration is frequently used because a circular interface adapter (cylinder or conical frustrum) is commonly used. This kind of joint is desirable in that it provides a more uniform distribution of load at the interface, a condition typically desirable from both a launch vehicle and spacecraft perspective. For these types of interfaces not-to-exceed values are specified by the launch vehicle contractor. The clamp band typically has two halves, with a number of individual segments attached to each section. The band is properly preloaded via instrumented bolts to withstand the loading environment during launch. Upon command, pyrotechnic bolt cutters cut the

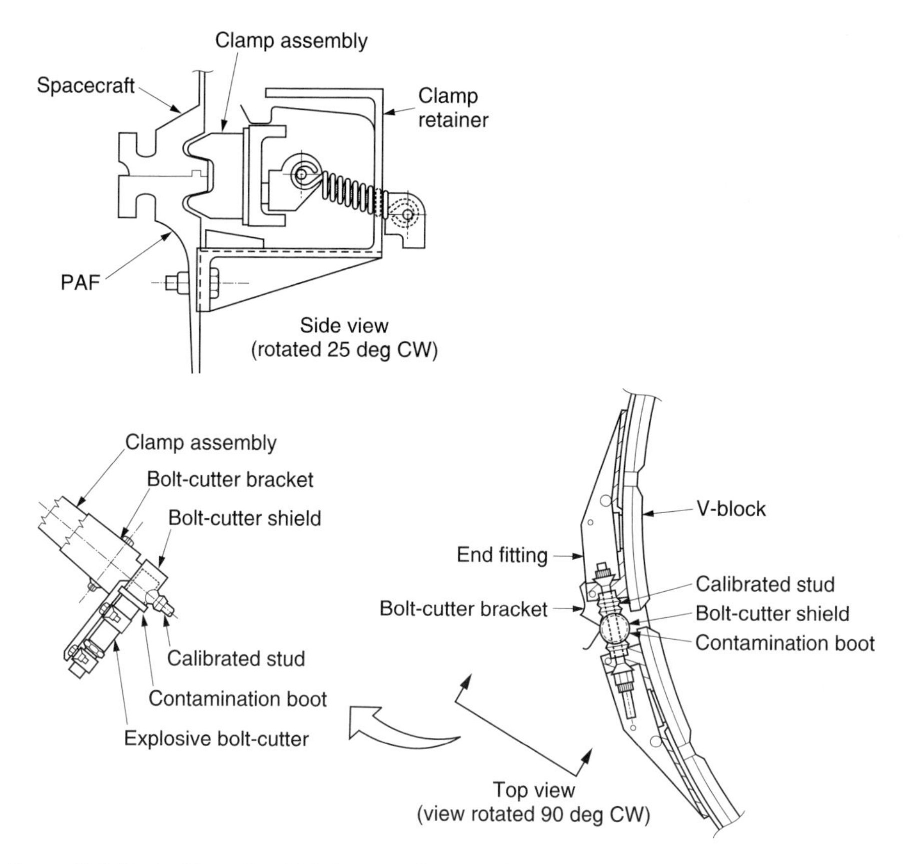

Figure 8.6 Typical launch vehicle/spacecraft separation mechanism.

bolt and release the band. Stored energy from preloading the V-band as well as the spring-loaded mechanism shown will force the clamp band back into the clamp retainer. The separation springs shown in figure 8.4 then push the spacecraft away from the launch vehicle, usually at a predetermined separation velocity.

One other aspect of the design layout which can be a significant driver in the development of the configuration is the orientation of the payload, in this case instruments, to satisfy the field of view (FOV) requirements. Figure 8.7 illustrates a couple of the FOV conditions for a single STEREO spacecraft configuration. Examining the FOV requirements is aided by the use of software such as Pro/Engineer, SDRC IDEAS, or a number of other computer-aided drafting (CAD) packages. In figure 8.7a an interference between the solar arrays and an instrument FOV is identified, while figure 8.7b illustrates a clear FOV for the deployable high-gain antenna (shown in a number of positions). To illustrate the level of complexity for the STEREO spacecraft system, table 8.1 indicates all of the system FOV constraints. This particular configuration was severely constrained not only by the large number of requirements but also by the magnitude of some of the individual requirements and the impact of trying to configure them all within one system. As is indicated by the figure and table, not all of the requirements could be completely satisfied. The table indicates the close-to-final negotiated FOV configuration.

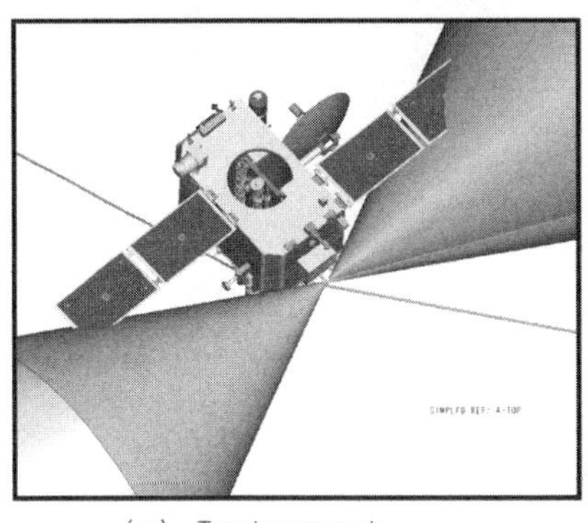

(a) Instrument

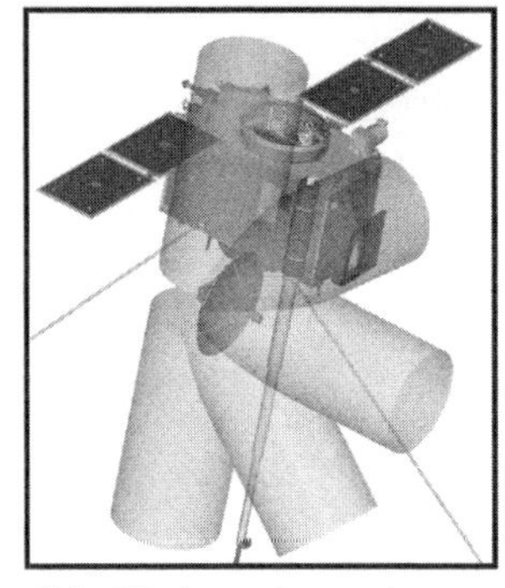

(b) High gain antenna

Figure 8.7 Field of view requirements: (a) instrument; (b) high-gain antenna.

8.4 Mass Properties Constraints and Estimation

8.4.1 Mass Properties Constraints

Launch vehicle selection also places constraints on the lift mass and mass properties of the spacecraft configuration. For missions with spinning upper-stage vehicles, the location of the center of gravity and the mass balance are important factors in determining the amount of weight that must be added to balance the spacecraft system. This process is much like that of balancing the tires on an automobile in that additional mass is added to the tire rim to prevent "wobbling". Significant imbalance of the spacecraft during the spinning sequence can greatly impact launch vehicle performance.

The spacecraft mass, combined with the location of the center of mass above the separation plane, must also meet certain requirements for the given launch vehicle adapter selected. Figure 8.8 depicts the mass versus center of mass requirements for a number of Delta launch vehicle payload attach fittings (PAF). These mass/center of mass requirements are critical as they represent limitations on the structural capability of the PAF. The break in the graph occurs because the 3724 PAF is used only for smaller payloads.

8.4.2 Mass Estimation

Estimation of mass starts in the early stages of the program and continues up through the final stages of integration. Since there are many unknowns early on, the process of estimating mass must be based predominantly on historical percentages. Table 8.2 presents typical subsystem mass percentages that can be used for early-on estimation of the mass for spacecraft missions. The breakdown is based on systems that do not require propulsion systems. For systems requiring propulsion systems, the mass can vary substantially and is based on the mission needs. For example, propulsion systems for orbit maintenance may be only 15–20% of the total spacecraft mass whereas interplanetary spacecraft propulsion systems could be as much as 60–65%. Note that in some circumstances subsystem masses can vary significantly as program needs vary.

Table 8.1 STEREO spacecraft FOV requirements

Instruments	FOV Requirement	Aim	Comment
SECCHI–SCIP suite			
Cor1	170 UFOV cone	$+X$ Sun	Clear FOV
Cor2	150 UFOV cone	$+X$ Sun	Clear FOV
EUVI	8 UFOV cone	$+X$ Sun	Clear FOV
GT	20 UFOV cone	$+X$ Sun	Clear FOV
SECCHI–HI	183 × 183 square	$-Z$ Earth	Clear FOV
IMPACT suite			
Plastic WAP	14 × 360 cone–sweep	Y-axis	Impingement—several
Plastic SWS	55 wedge × 40 FOV angle	$+X$ Sun	Clear FOV
SEP suite			
SEPT–E	Twin 52 fwd/aft cone	P-spiral	Clear FOV
SEPT–NS	Twin 52 fwd/aft cone	Y-axis	Impingement—solar panels
HET	60 fwd cone	P-spiral	Clear FOV
LET	30 × 130 fwd/aft complex	P-spiral	Clear FOV
SIT	17 × 44 fwd rectangle	P-spiral	Clear FOV
BOOM suite			
STE-D	80 × 80 pyramid (fwd-aft)	P-spiral	Clear FOV
STE-U	80 × 80 pyramid (aft-fwd)	P-spiral	Clear FOV
SWEA	120 × 360 cone-sweep	X-axis	Impingement—solar panels, SWAVES
SWAVES	3 orthogonal antennas 120° apart	X-axis	
Spacecraft			
High gain ant.	1.2m beam × 180° sweep	$+X$ to $-X$	Clear FOV
Low gain ant.	2 × 2π steridian	$+Z$ to $-Z$	$+Z$ = clear FOV, $-Z$ impinged by HI
DSAD	5 × 128 × 128 pyramid	All axis	Impinged by solar panels
Star camera	25 cone	$+Z$-axis	Clear FOV
Solar panels	4 × 180 direct Sun	$+X$-axis	Clear FOV
Thrusters	12 × 100 cone	$+X$, $-X$, $+Z$	10 × clear FOV, 2 × SWAVES

UFOV = Unobstructed field of view.

The subsystem percentage the program decides to use is based on the hardware requirements. In addition to historical percentages, mass can be estimated by use of approximate densities that have been established throughout the years. Some of these densities are:

- "Black" boxes: 0.025–0.05 lb/in^3
- DC/DC converters: 0.06 lb/in^3
- PC boards: 0.015–0.03 lb/in^2
- Thermal blankets: 0.2 lb/ft^2

Using the above techniques and other calculations, a total spacecraft mass is developed. Estimates early in the program are based on limited details about the subsystem components and, as a result, the total system mass commonly increases throughout the program. Figure 8.9 depicts a typical mass growth curve for programs.

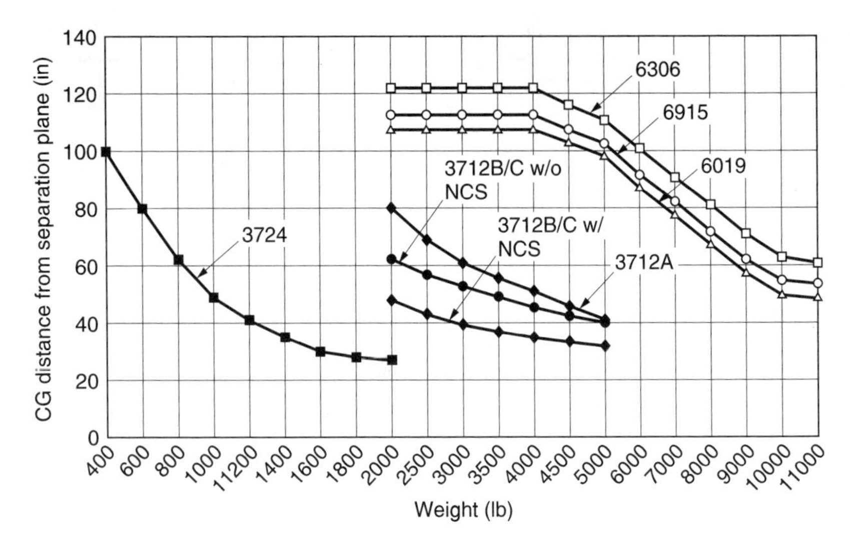

Figure 8.8 Delta PAF capability.

The amount of mass margin carried during the program is based on the maturity of the design and the hardware. At the onset of the program, since there are many uncertainties regarding the program, a higher percentage of mass margin should be retained. As the program evolves and the design matures, less margin is needed. Typical margins at certain phases of the program are shown in table 8.3. In calculating mass margin, two methods are utilized:

$$\left.\begin{aligned} \%\text{margin} &= \frac{\text{Maxlift} - \text{CBE}}{\text{Maxlift}} = 1 - \frac{\text{CBE}}{\text{Maxlift}} \\ \%\text{margin} &= \frac{\text{Maxlift} - \text{CBE}}{\text{Maxlift}} = \frac{\text{Maxlift}}{\text{CBE}} - 1 \end{aligned}\right\} \tag{8.4.1}$$

where Maxlift = maximum launch vehicle lift capacity for the mission and CBE = current best estimate. The method used is often dictated by the organization building the hardware or the sponsor for whom the hardware is being built.

In table 8.2, structure mass percentages are presented as primary and secondary structure. Understanding the differences between these structural categories will be

Table 8.2 Typical Subsystem Mass Percentages

Item	Range (%)
Structure (total)	15–22
Primary structure	12–15
Secondary structure	2–5
Fasteners	1–2
Power	12–30
Thermal control	4–8
Harness	4–10
Avionics	3–7
Guidance & control	5–10
Communication	2–6
Payload	7–55

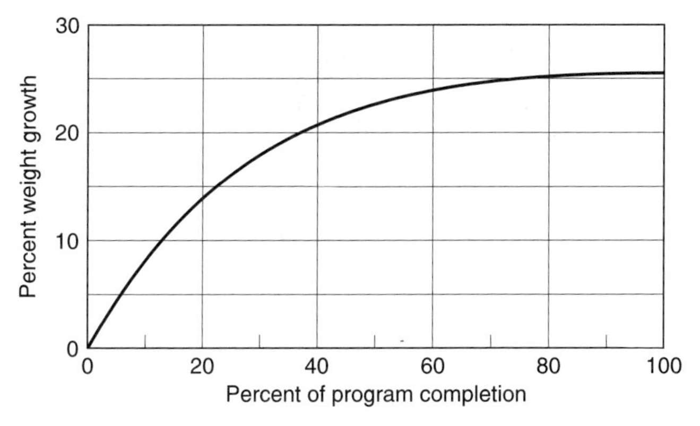

Figure 8.9 Average program mass growth.

important in the following sections. *Primary structure* is defined as the collection of structural elements that transfer load from one part of a structure to another. The transfer of loading is referred to as the primary load path. Elements in the primary load path experience loading in excess of that created by their own mass. Examples of primary structure are main support cylinders or payload adapters. *Secondary structure* is defined as the collection of structural elements which provide support for components or transfer loading which is not in excess of their own created mass and that of the hardware mounted thereon. Examples of secondary structure include support brackets sometimes required for instruments, antennae, or thrusters. Figure 8.10 illustrates the primary structure for one of the STEREO spacecraft systems depicted in figure 8.3.

8.5 Structural Design Criteria—Launch Environments

In order to proceed with the spacecraft structural analysis, the analyst must first understand the predominant environments the spacecraft structure must survive. Attention must be directed towards not only the magnitudes of these environments but also the type of the environment. Certain environments associated with the launch phase are depicted in figure 8.11. Although various expendable launch vehicles may vary slightly, most are quite similar to this typical profile. Through experience with a particular launch vehicle or early discussions with the structural dynamicist from a launch vehicle contractor, certain critical launch events and their dynamic characteristics are obtained. This information is utilized in the early design phase. It should also be noted that, in the design of a spacecraft, "stronger" does not always imply a better structural subsystem. Generally

Table 8.3 System mass margins

Program start	20% minimum
Conceptual design review	20%
Preliminary design review	15%
Critical design review	10%

NOTE: Program start should be with other three concepted design review, etc.

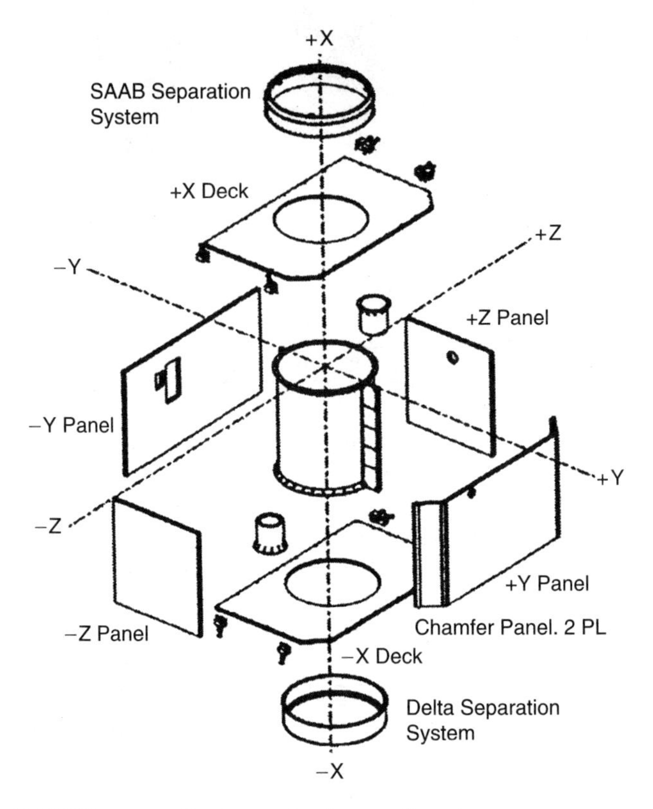

Figure 8.10 STEREO spacecraft primary structure configuration.

stronger implies stiffer, and with the environment being a function of frequency, the environment (that is, dynamic response) may change at a greater rate than the corresponding change in strength of the structural members. In fact, many spacecraft are designed to have fundamental structural frequencies such that the structure becomes an isolator for the payloads, thus causing the inputs to be reduced. Prior to more specific discussions regarding some of these environments, a brief description of the events shown in figure 8.11 is given in the following paragraphs.

8.5.1 Liftoff

Liftoff usually provides the most significant spacecraft *acoustic* environment, the primary reason being the reverberant effect the ground provides in addition to the larger engines associated with the first stage. The effect of the acoustic environment is to generate a random vibration due to noise pressure acting on the surfaces of the spacecraft. Typically, acoustic environments inside heat shields can reach levels of 145 dB[1] (reference pressure of 2×10^{-5} N/m^2) with a frequency content of concern up to several

[1] dB = 10 $\log_{10}$ (power ratio), where power ≡ acceleration density (g^2/ Hz), or dB = 20 $\log_{10}$ (voltage ratio), where voltage ≡ acceleration (g).

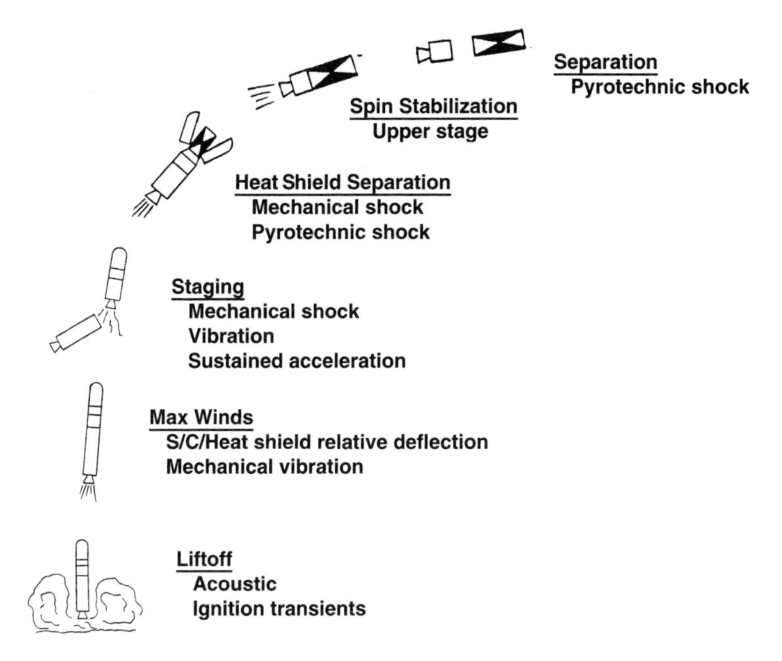

Figure 8.11 Events during typical launch scenario.

thousand hertz (Hz). These environments are derived from microphones located within the heat shield during early flights. Acoustic environments are presented as "dB versus frequency" plots with a reference sound pressure level of 0 dB, which is equivalent to an rms pressure of 2×10^{-5} N/m^2 (2.9×10^{-9} psi). Frequency is typically defined in either octave band or 1/3 octave band center frequencies. The acoustic environment is generally not a concern for the primary support structure, but must be considered in the design of structures having a large surface area and relatively low mass density (less than 158 N/m^2 or 3.3 lb/ft^2). An example of such a structural element is the solar arrays.

In addition, at liftoff, primary structural loading can be developed by the engine ignition *transients*. These transients usually excite dynamic modes of less than 100 Hz, a primary area of concern for structural components since most spacecraft have fundamental structural modes that are less than 100 Hz. During these events vibration disturbances in combination with any sustained acceleration generally provide the most significant structural loadings.

8.5.2 Max Winds and Transonic Buffeting

There are two mid-launch phenomena, max winds and transonic buffeting, which are usually a more severe design condition for the heat shield than for the spacecraft. The max winds phenomena describes the structural loading due to high-altitude winds (100 mph), while transonic buffeting is a loading condition that occurs (shock waves impacting the heat shield) in going from subsonic to sonic speeds. Loading during the max winds and

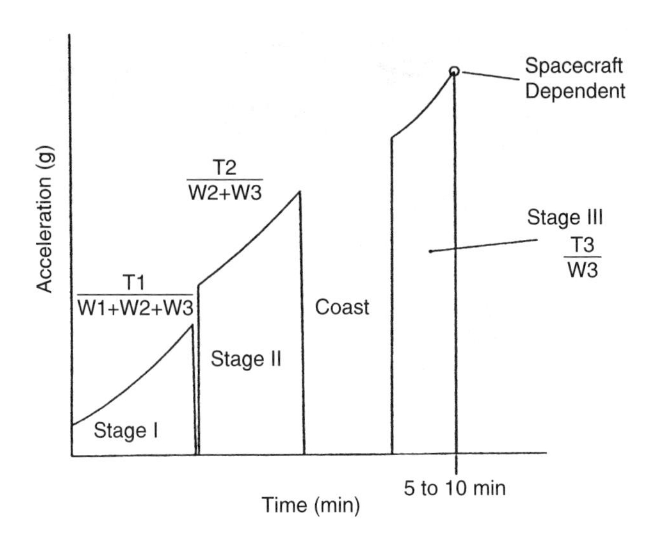

Figure 8.12 Sustained acceleration for three-stage vehicle.

transonic buffeting conditions cause the clearance between the spacecraft and the heat shield to be reduced; thus the relative clearance between the components located on the spacecraft perimeter and the heat shield must be a design consideration.

8.5.3 Staging

During pre-staging, that is, just prior to the separation of two stages of the launch vehicle or between the last stage of the launch vehicle and the spacecraft, the maximum sustained acceleration for a particular stage configuration is achieved. The sustained (steady) acceleration is simply the constant thrust divided by the decreasing weight of the launch vehicle, and it therefore peaks at the end of each stage. For the final stage, this sustained acceleration is heavily dependent on the mass of the spacecraft as expended upper stages often weigh less than the attached spacecraft. A general time versus acceleration curve for a three-stage launch vehicle is shown in figure 8.12.

Acceleration levels during these events satisfy the relationship

$$a = \frac{T}{W} = \frac{T}{W/g} \tag{8.5.1}$$

where a = maximum acceleration, T = thrust, g = acceleration due to gravity, and W = weight of the spacecraft. The staging event is similar to the engine ignition at liftoff. Again, development of primary modal response of the structure is the concern.

Also, during pre-staging or staging events, some level of vibration or transient environment can exist during periods of high sustained acceleration. These transients are typically smaller in magnitude than the sustained accelerations. With a properly designed spacecraft configuration these components of acceleration can be combined and analyzed by static analysis. The combination of sustained and transient accelerations is more commonly referred to as *quasi-static acceleration*. Figure 8.13 shows a typical launch vehicle stage shutdown. From this it can be seen that the transient

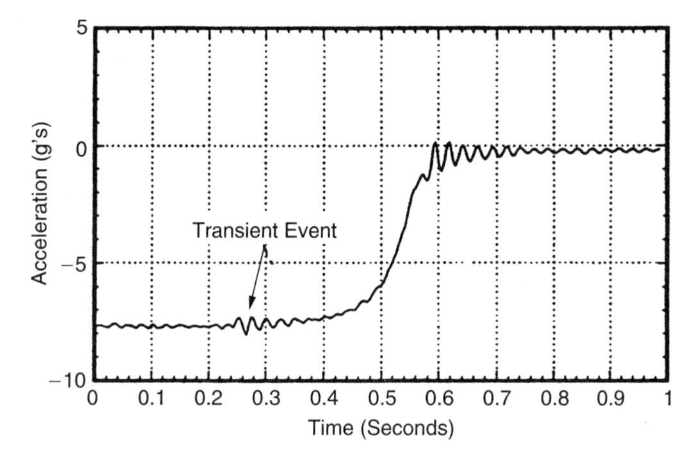

Figure 8.13 Acceleration during launch vehicle stage shutdown.

event adds approximately $0.5g$ to the steady-state thrusting of $7.7g$, giving a quasi-static acceleration of $8.2g$ to be used in the analysis.

8.5.4 Heat Shield Separation

Usually a pyrotechnically activated mechanism initiates the separation of the heat shield. Pyrotechnics typically imply an extremely high frequency content and, although not a great concern for structures, can be quite damaging to electronics. The physical separation that follows usually imparts a lower frequency mechanical shock, which may have a structural impact.

8.5.5 Spin Stabilization

The final stages of an expendable launch vehicle are often pointed, spin imparted, and thrusted. Small solid rockets or gas thrusters are often used to spin up the spacecraft/upper stage to rates that range from 50 to 200 rpm. The effects generated are the tangential acceleration associated with spin-up and despin, and the centripetal acceleration induced by the steady-state rotation. These are defined as follows:

$$\left.\begin{aligned} a_t &= r\alpha \\ a_n &= r\omega^2 \end{aligned}\right\} \tag{8.5.2}$$

where a_t = tangential acceleration, a_n = normal acceleration, α = angular acceleration, ω = angular velocity, and r = radius.

8.5.6 Separation

Separation of the spacecraft from the launch vehicle is usually accomplished using a pyrotechnic initiator to sever bolts and free a clamp strap. During this event a high impulsive force of very short duration, or shock, is generated. The impact to equipment mounted in the close proximity of this environment can be significant, whereas hardware mounted a considerable distance away or removed by many joints will not experience much of an effect. The environment generated during this event is again, as in heat shield separation, more of a concern for electronic hardware due to its high-frequency content.

8.6 Operational Environments

One would think that after an exhausting five- to fifteen-minute launch phase the structural subsystem has fulfilled all its requirements. Generally, this is not the case. A sample of the requirements often imposed on the structure and subsystems during the functional lifetime of a spacecraft is briefly described below.

8.6.1 Propulsion

A transtage, small solid or liquid rocket, often provides more severe vibration and larger sustained acceleration than those of the primary rocket. The proximity to the spacecraft and the lesser spacecraft mass are the main reasons for this level of severity.

8.6.2 Deployment Dynamics

The deployment of structures to an orbital configuration is usually the critical design environment for the deploying structure as well as for local areas of the spacecraft. Typically, the environments which must be addressed include shock (pyrotechnic and/or mechanical) due to the initiation of the event and tangential acceleration due to the actual deployment kinematics.

8.6.3 Ultra-stable Structures

Many astronomical measurements require the structural support of instruments within arc-second angular accuracy and stability. The design of such "optical benches" is generally a combined thermal and structural responsibility. The material sciences side of the structures engineer is brought to bear on these problems as well as the ability to predict distortions from a transient thermal environment.

8.6.4 Flexible Body Control System

Spacecraft "control laws" must take into account flexible body characteristics which cannot be adequately separated from control system frequencies. Dynamic disturbances due to spacecraft components (for example, movement of tape recorder reels, scanning of mirrors, thermal impact on flexible structures) as well as the response of the total structure to these disturbances must be identified.

8.6.5 Reentry/Reuse

With the advent of the Space Transportation Systems (Shuttle), evaluation of fatigue and crack propagation in the design process has become a necessity due to the reuse of equipment. Shuttle launched systems must be designed for reentry and landing without incurring danger to the shuttle and its equipment. In some cases, the shuttle landing can provide the most significant load on the spacecraft if the deployment is aborted and the system is returned to earth. For expendable launch vehicles, the combined ground test and launch lifetime generally does not cause such a concern.

8.7 Structural Design and Test Criteria

The design of spacecraft structures involves tradeoffs between material strength and structural stiffness. Strength can be defined as a member's load-carrying capability. In considering the strength of the structure to withstand launch loading, the governing criterion is the material's capability. This capability is defined in terms of the material's yield and/or ultimate strength. The yield point is the point beyond which the material exhibits substantial permanent deformation. The ultimate allowable is the maximum stress the material can withstand prior to failure.

Stiffness is a system's inherent resistance to deformation. In considering a system's stiffness, its natural, or fundamental, frequency is examined. This is the frequency at which a system will oscillate naturally when an external disturbing force is placed on the structural element and then removed. This cyclic motion results from the transfer of energy within the structural member and represents a time at which the system can experience substantial loading environments.

As previously mentioned, the spacecraft is subjected to major mechanical loads during the launch sequence. The structure must not only be designed with strength to survive these loads but also provide protection to the other systems (for example, instruments on the spacecraft) from the launch environment. Stiffnesses of spacecraft structures and subsystems must be properly designed to ensure that this protection exists during the launch sequence. Once the spacecraft is in orbit, the loads experienced will, in most circumstances, be lower than those that occurred during launch. For this phase, structural stiffness can again play an important role. Stiffness must be properly defined to avoid interaction between any appendage vibrations and the attitude control system. Sufficient structural stiffness will also help minimize thermal distortion, thus enabling more accurate pointing of any onboard instruments.

While a great many programs today implement the "fly what you test" approach, in some programs it may be necessary to follow slightly different approaches. The hardware categories that could be encountered are discussed in the following paragraphs. The philosophy followed, in conjunction with the number of systems being produced, generally dictates the level of margin that must be carried in the final design.

Engineering model, or development, hardware is not intended for flight and is sufficiently representative of a proposed flight item to make test results a valid indication of flight behavior. This hardware is primarily used for developmental testing of new systems and may be tested to load levels that significantly exceed flight levels. Development tests are performed on engineering model hardware for the purpose of obtaining data for evaluation and with no specific success or failure criteria. An example of such hardware was developed on the Messenger program to test a propulsion system tank support structure. This structure was embedded in a composite mounting panel and was designed to carry a significant load during the launch. A representative panel structure with the tank support structure was fabricated and tested to evaluate the concept that would be used for the flight hardware. Load and deflection data were collected, and ultimately the panel was loaded to failure. Engineering model hardware and testing precede prototype and/or flight hardware testing.

Prototype, or qualification, hardware is equipment that is not intended for flight use. Prototype hardware is built and tested to demonstrate that the test item will function within performance specifications under simulated conditions more severe than those expected during ground handling, launch, and orbital operations. The purpose of such hardware is to uncover deficiencies in the design and method of manufacture. The hardware is not intended to exceed design safety margins or to introduce unrealistic modes of failure. In the absence of any deficiencies, prototype hardware is still considered non-flight due to the severity of testing that has been conducted.

Protoflight hardware, encapsulating the "fly what you test" approach, is equipment designated to serve as both a prototype and a flight model and thus is subjected to tests that combine the elements of both flight acceptance and prototype testing. Protoflight tests serve to qualify the design and method of manufacture of an item and accept it for flight operations. Qualification of the design is accomplished by imposing environmental levels more severe than those expected during ground and flight operations. Hardware fatigue is avoided by limiting the exposure time so as not to expend a significant portion of the useful life of the hardware. These tests also serve to detect latent defects in material or workmanship and to provide experience with the test item's performance under conditions similar to the mission environment.

Flight hardware, like protoflight hardware, is equipment that is designed to be used operationally. Testing for this equipment, called flight acceptance testing, is conducted at the maximum expected flight level and is intended to demonstrate the flightworthiness of the test item under simulated conditions expected from ground handling, launch, and orbital operations. The purpose is to locate latent material and workmanship defects in a proven design. A secondary purpose is to provide experience with the item's performance under conditions similar to the mission environment. In general, programs that use the flight acceptance test approach are preceded by engineering model or prototype hardware which, in many circumstances, is limited to the primary structural components. Inert masses are used to simulate electronic packages on the prototype hardware in order to obtain the proper mass and thus generate the appropriate loads during testing.

Regardless of the approach taken, the goals of the structural designer/analyst are the same: to establish well-defined load paths, avoid or eliminate offsets which cause undesirable bending moments, and keep the design as simple as possible. Doing so not only makes the design and analysis much easier and more straightforward, but also makes it easier to fabricate the hardware, thus creating a part which is cost effective with minimum mass.

8.7.1 Spacecraft Design Criteria

Launch loads (or limit loads) obtained from the launch vehicle user's manual are used to evaluate the capability of any given design. These limit loads are typically expressed in terms of the gravitational constant g and produce forces in each of the structural elements. Representative values for several different expendable vehicles are presented in table 8.4. These values should be applied simultaneously to the spacecraft center of gravity in determining the forces generated in the primary structure of the spacecraft

Table 8.4 Spacecraft system limit loads for expendable launch vehicles (gs) (+ = compression, − = tension)

Vehicle	Lateral	Axial
Delta II		
Liftoff	±4.0	+2.8/ −0.2
MECO	±0.2	7.5 ± 0.6
TECO	±0.1	8.4 to 10.3
Delta III		
Liftoff	±2.5	+2.7/ −0.2
MECO	±0.5	3.7+1.5
Delta IV		
Max Lateral	±2.0	+2.5/ − 0.2
Max Axial	±0.5	+6.5/ −1.0
Atlas IIAS		
Liftoff	±1.3	+1.2 ± 1.1
Flight winds	+0.4 ± 1.6	+2.7 ± 0.8
BECO (max. axial)	±0.5	+5.0 ± 0.5
BECO (max. lateral)	±2.0	+2.5 ± 1.0
SECO	±0.3	+2.0 ± 0.4
MECO (max. axial)	±0.3	+4.5 ± 1.0
MECO (max. lateral)	±0.6	±2.0
Atlas III		
Liftoff	±1.3	+1.2 ± 1.1
Flight winds	±0.4 ± 1.6	+2.7 ± 0.8
BECO (max. axial)	±0.5	+5.5 ± 0.5
BECO (max. lateral)	±1.5	+2.5 ± 1.0
MECO (max. axial)	±0.3	+4.5 ± 1.0
MECO (max. lateral)	±0.6	±2.0
Atlas V		
Liftoff	±2.0	+1.8 ± 2.0
Flight winds	±0.4 ± 1.6	+2.8 ± 0.5
Strap-on separation	±0.5	+3.3 ± 0.5
BECO (max. axial)	±0.5	+5.5 ± 0.5
BECO (max. lateral)	±1.5	+3.0 ± 1.0
MECO (axial)	±0.3	+4.5 ± 0.5
MECO (lateral)	±0.6	±2.0
Proton		
Liftoff	+2.6	±2.1
Max. dynamic pressure	±1.2	+2.2
1st/2nd stage pre-separation	±0.9	+4.65
1st/2nd stage post-sep. (max. compression)	±1.7	+3.0
1st/2nd stage post-sep. (max. tension)	±1.7/ ±1.2	−2.8/ −3.2
2nd/3rd stage separation	±0.3	+3.0
3rd/4th stage separation	±0.3	+2.8

MECO = Main engine cutoff BECO = Boost engine cutoff
TECO = Third stage burnout SECO = Second engine cutoff

(Continued)

Table 8.4 (Continued)

Vehicle	Lateral	Axial
Pegasus		
Captive flight	±0.7 (Y)	
	+3.6/ −1.0 (Z)	±1.0
Drop transient	±0.5 (Y)	
(payload interface)	±3.85 (Z)	±0.5
Aero pull-up	±0.2 ± 1.0 (Y)	+3.7 ± 1.0
	±2.33 ± 1.0 (Z)	
Stage burnout	±0.2 ± 1.0 (Y&Z)	+9.5 ± 1.0
Post-stage burnout	±0.2 ± 2.0 (Y&Z)	±0.2 ± 1.0
Taurus (3 *stage*)		
Liftoff	±2.0	+6.0/ −1.0
Transonic	±2.5	+8.0/ −1.0
Supersonic gust	±2.5	+4.5/ + 3.5
S1 resonant burn	±1.5	+6.0/ +2.0
Stage burnout	±0.5	+7.2
Titan II		
Liftoff	±2.0	+3.0/ −1.0
Max. airloads	±2.5	+3.0/ + 1.0
Stage I burnout	±4.0	+7.0/ + 4.0
Stage II shutdown	±2.0	+9.0/ + 2.5
Ariane 4		
Max. dynamic pressure	±1.5	+3.0
Pre-thrust termination	±1.0	+5.5
Thrust tail-off	±1.0	−2.5
Ariane 5		
Liftoff	±1.5	+3.2
Max. dynamic pressure	±2.0	+3.2
P230 burnout	±0.5	+4.5
H155 burnout	±0.1	+3.6
H155 thrust tail-off	–	+2.1/ −0.7
Athena 1		
Launch/1st stage ignition	±1.5	+3.0/ −1.0
1st stage acceleration	±2.0	+2.0
2nd stage resonance	±2.5	+4.0 ± 3.0
2nd stage max.	±2.0	+8.0
3rd stage max.	±1.0	+7.0
Athena 2		
Launch/1st stage ignition	±2.4	−8.7 to 9.2
1st stage resonance	±2.9	+3.6 ± 2.7
1st stage max. acceleration	±2.0	+8.1
2nd stage ignition	±2.0	−2.0 to 5.0
2nd stage max. acceleration	±1.0	+7.0

Table 8.5 Spacecraft system stiffness requirements

Vehicle	Thrust (Hz)	Lateral (Hz)
Delta II	35	20
Delta III (2 stage)	27	10
Delta IV Med/Med+	27	10
Delta IV Heavy	30	8
Atlas IIAS	15	8
Atlas III	15	8
Atlas V	15	8
Proton	25	10
Pegasus	–	20
Taurus	35–45, > 75[1]	25
Titan II	24	10
Ariane 4	31	10
Ariane 5	18[2]	8[3]
Athena 1	30, $\neq 45 - 70$[4]	15
Athena 2	30, $\neq 45 - 70$[4]	12

[1] Coupled vehicle/payload system requirement.
[2] Sum of effective mass at a given frequency.
[3] Worst case—payload mass dependent.
[4] $\neq$ applies to all values within the range.

system. For programs where the launch vehicle is unknown at the beginning of the design, the analyst must use this available information to make an "educated guess" for design loads.

The limit loads specified in table 8.4 are valid only if the spacecraft meets certain stiffness, or natural frequency, requirements. The stiffness requirements are based on the launch vehicle's dynamic characteristics and are designed to prevent an amplification of the loading. The spacecraft stiffness requirements, assuming the spacecraft is rigidly mounted at the interface plane, are presented in table 8.5. For systems that do not meet or exceed these natural frequency requirements, a coupled analysis of the integrated analytical models of the spacecraft and launch vehicle is necessary to verify the limit loads used in the design.

8.7.2 Component Design Criteria

In the case of secondary structure or component design such as that of electronic packages, design loads may not be available initially even if the launch vehicle is known. When such is the case, various methodologies have been developed for determining the design loads.

Design methods for components can be broken down into two categories: loads estimation and loads prediction. Loads estimation assumes the worst tuning between the launch vehicle and spacecraft and yields an upper bound of the expected flight loads. This allows early definition of the loads (with little or no knowledge of the launch vehicle) with little sensitivity to a change in loads. Through past experience, these loads have been shown to be only moderately conservative. Thus, when these values are used in the design process, a configuration is produced which yields a favorable trade off

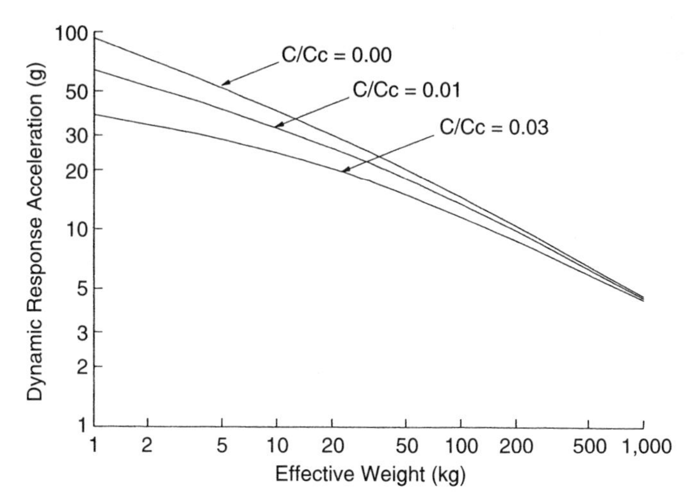

Figure 8.14 Component design limit loads.

between weight and design margin. Loads prediction aims at accurately predicting the time history responses of loads in members and the accelerations of major masses.

The most basic method of loads estimation will be presented here. This method is primarily used for preliminary analysis and is based on the mass acceleration method. This method gives the analyst a "rule of thumb" for sizing a preliminary structure and can be used for the design of secondary structure for which the loads analysis approach is not applicable.

Figure 8.14 illustrates a general loads estimation curve that can be used for structural analysis of *components and secondary structure*. The curves presented here are shown for three damping values (c/c_c) which represent the range of typical values for secondary structures and components. The damping ratio c/c_C represents the actual damping, c, compared to the critical damping, c_C. The critical damping value, defined in section 8.14.1, is the value of damping where a structure or item returns to its equilibrium conditon without a vibratory oscillation after being subjected to a forced disturbance. Given an effective weight or mass of a secondary structure or component, the design limit load can be obtained from these curves for a particular value of damping. This design limit load is applied as quasistatic accelerations separately in each of three orthogonal directions to determine whether the structure is capable of withstanding the worst-case expected loading conditions. Note that for masses greater than 200 kg, all response curves converge to values that approximate the limit loads for spaceacraft systems (as shown in table 8.4).

Component stiffness requirements, are not as easily defined. For components the stiffness or dynamic requirement is strongly influenced by the mounting location of the hardware and the supporting structure. The primary goal in this situation is to reduce or minimize the dynamic interaction between the interacting elements (that is, between the component natural frequency and the natural frequency of the mounting structure). To accomplish this dynamic separation between systems, it is best to apply what is known as the octave rule. This rule states that

Component natural frequency ≥
2 x *Natural frequency of the mounting or support structure*

Using this basic relationship between interacting elements will ensure that proper protection occurs; in other words, one item will be isolated from the other dynamically and any input loading will not be amplified. Later on in this chapter the foundation for this rule will be discussed.

Example 8.1

Given that the system illustrated below will be configured to launch on an Atlas IIAS launch vehicle and that the spacecraft primary structure just meets the minimum stiffness requirements for the axial or thrust direction condition, what are the axial stiffness requirements for units A and B if the natural frequency of the support deck is 50 Hz?

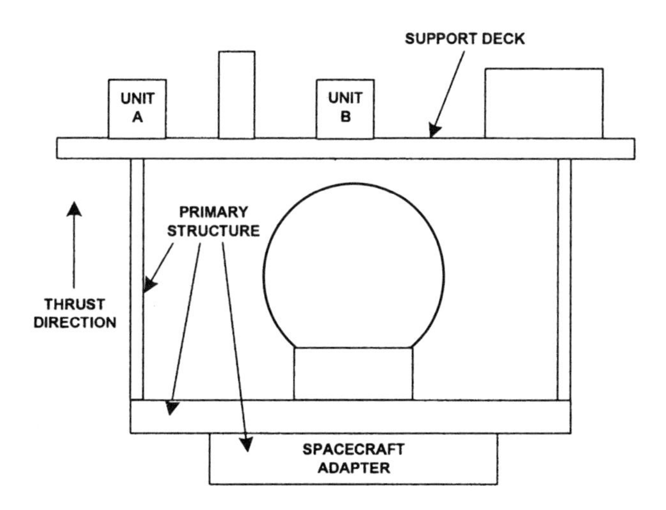

Solution: From table 8.5 for the Atlas IIAS launch vehicle, the minimum stiffness requirement in the thrust direction is 15 Hz. Unit A is mounted such that its dynamic characteristics will be dictated by the primary structure. Applying the octave rule, to prevent any loads amplification or coupling between unit A and the primary structure, unit A should be designed to a minimum of twice the natural frequency of the primary structure, or 30 Hz. Unit B, however, is mounted in the middle of the support deck and thus its requirements will be based on the design of this deck. Applying the octave rule, unit B should be designed for a minimum natural frequency of 100 Hz. Note that the relationship between the dynamic characteristics of the support deck and primary structure also satisfies the octave rule.

8.7.3 Factor/Margin of Safety

Launch loads used by the analyst for preliminary design are derived from information in the user's manual for the selected launch vehicle, from past experience, and from any additional documents that will help formulate a reasonable design criterion. While the magnitude of the launch loads can account for any uncertainty in the environment, the

Table 8.6 Test factors of safety (FS_1) and durations

	NASA			USAF	
	Strength	Random	Sine	Random	Sine
Flight	1.0	+ 0 dB 60 s/axis	1.0 4 oct/min	+0 dB 60 s min.	+ 0 dB 4oct/min
Protoflight	1.25	+3 dB 60 s/axis	1.25 4 oct/min	+3 dB 60 s min.	+3 dB 4 oct/min
Qualification, prototype, or engineering model	1.25	+3 dB 120 s/axis	1.25 2 oct/min	+6 dB 3 × max flt or 120 s min.	
Reference	General environmental verification specification for STS & ELV payloads, subsystems, and components			MIL-STD-1540D (or later) Test Requirements for Space Vehicles	

structural engineer must account for any design uncertainties as well as limitations in the analytical techniques. This is accomplished through the use of factors of safety. The *factor of safety (FS)* represents the ratio of the load (stress) that would cause yielding or failure of a member or structure to the load that is imposed upon it in service. With a given hardware category as defined in section 8.7 (flight, protoflight, etc.), structural design criteria are defined by the following:

- Limit loads = maximum expected flight loads.
- Test levels = $(FS)_1 \times$ limit loads.
- Material yield condition = $(FS)_2 \times$ limit loads.
- Material ultimate condition = $(FS)_3 \times$ limit loads.

where $(FS)_i$ are the factors of safety.

Table 8.6 provides guidelines for test factors of safety $(FS)_1$ for the hardware categories previously defined. These methodologies are provided for NASA and Air Force space systems and use experience gained from launch vehicles to determine cost-effective methods for estimating design loads. The cost-effectiveness is aimed at reducing the cost and schedule for payload loads analysis by decoupling the payload analysis from the launch vehicle as much as possible.

The flight yield condition, f_y, is defined as the worst-case loading condition during flight which must not exceed the material yield strength. Similarly the flight ultimate condition, f_u, must not exceed the material ultimate strength. The material yield is the maximum strength that can be developed without exceeding a specified permanent deformation; a permissible set of 0.1 to 0.2% is commonly used. The material ultimate strength is the maximum stress the material can handle before failing. Typical factors of safety used in the design of metallic structures and components are shown in table 8.7.

Some special areas, such as propulsion systems, follow slightly different approaches than those presented in table 8.7 because of the unique nature of the component or subsystem. As an example, table 8.8 provides typical factors of safety used in the design of propulsion system components as per MSFC-HDBK-505, Revision A.

In general, factors of safety may vary slightly from program to program and from organization to organization. Whatever factors of safety are used, the intent of all

Table 8.7 Typical material factors of safety (FS_2 and FS_3)

	Yield	Ultimate Unmanned	Ultimate Manned	Static Test
Flight preceded by dedicated test article	1.00	1.25	1.40	f_y = no DD f_u = no failure
Test one flight (protoflight)	1.25	1.40	1.40	f_y = no DD
Proof test each flight	1.10	1.25	1.40	f_y = no DD
No static qualification	1.60	2.00	2.25	—
Mechanical ground support equipment	3.00	5.00	5.00	>2.5

DD = detrimental deformation

structural design criteria is the same—to establish sufficient design margin without introducing ultra-conservative approaches to the program. For example, an overly conservative design FS causes the structural members to be much heavier than they need to be, thus causing excessive spacecraft weight.

A standard practice of indicating the design capability when considering limit loads and factors of safety is via the *margin of safety*. The margin of safety (MS) is defined as the amount of margin that exists above the material allowables for the applied loading condition (with the factor of safety included). For space applications the applied loading condition is the worst-case flight condition that exists, that is , typically limit loads. The general equation for margin of safety is

$$\text{MS(margin of safety)} = \frac{\text{Allowable load (stress)}}{\text{Actual load (stress)} \times \text{Design factor of safety}} - 1$$

Table 8.8 Propulsion system component factors of safety

Propellant Tanks		
Manned	Proof pressure*	1.05 × MEOP**
	Yield pressure	1.10 × MEOP
	Ultimate pressure	1.40 × MEOP
Unmanned	Proof pressure	1.05 × MEOP
	Yield pressure	1.10 × MEOP
	Ultimate pressure	1.25 × MEOP
Hydraulic and Pneumatic Systems, including reservoirs		
Lines and fittings less than 1.5 in. (38 mm) diameter	Proof pressure	2.0 × MEOP
	Ultimate pressure	4.0 × MEOP
Reservoirs, actuating cylinders, valves, filters, switches	Proof pressure	1.5 × MEOP
	Ultimate pressure	2.0 × MEOP

*Proof factor, determined from fracture mechanics service life analysis, must be used if greater than those shown.
**MEOP = maximum expected operating pressure.

Margin of safety calculations are typically made for material yield and ultimate strength (stress) conditions. For circumstances where FS = 1 or where the FS for the yield condition is equal to the factor of safety for testing, a positive margin of safety must exist (MS > 0). For all other conditions, MS ≥ 0.

Example 8.2

A structural member on a NASA protoflight program experiences a stress of 30,000 psi when subjected to the expected limit loading condition for the mission. Determine the margin of safety if the yield strength of the material is 42,000 psi.

Solution: From table 8.7 for protoflight hardware, the FS = 1.25. Thus

$$MS = \frac{\text{Allowable}}{\text{Actual} \times \text{FS}} - 1 = \frac{42{,}000}{30{,}000 \times 1.25} - 1 = 0.12$$

8.8 Stress Analysis

The environments experienced during the launch sequence can introduce various loading conditions in the structural members. Given that the structural elements can have various cross-sections and be made from a variety of materials, when the member is subjected to these external forces, distortion occurs and the force is transmitted through the structure. Accounting for the geometry, the transmitted force is expressed as *stress,* or force per unit area, while the distortion is expressed as *strain,* or displacement per unit length. The next several sections focus on the stresses and strains generated for the various types of loading conditions that occur during launch. A simple bar-type member is used to illustrate the basic relationships that must be satisfied for these loading scenarios. While designs always involve a greater level of detail than is shown in the upcoming sections, these basic relationships must still be applicable. Approaches presented here can be used in performing preliminary analysis.

Given a particular design, the structural engineer must determine whether the elements in the configuration have adequate strength to survive the launch events. Using the loads design criteria, the analyst proceeds to evaluate the capability of the structural members in the configuration, with a goal of optimizing the capability of the structural members and thus minimizing the mass of the structural subsystem. From the different launch environments the structure can be subjected to axial, bending, buckling, and/or torsional loads. The stresses generated from these loading conditions are compared to the stress, or strength, of the various materials used in the application. Figure 8.15 presents a general relationship between stress and strain for any given material.

An important area for all designs is the linear part of this curve up to the yield strength point. For the linear part, the following relationship, known as Hooke's Law, exists and is expressed by the equation

$$\sigma = E\varepsilon \tag{8.8.1}$$

where σ = stress, E = modulus of elasticity or Young's modulus, and ε = strain.

The modulus of elasticity is a measured mechanical property for materials and will be shown in later sections to be an important parameter in the structural evaluation

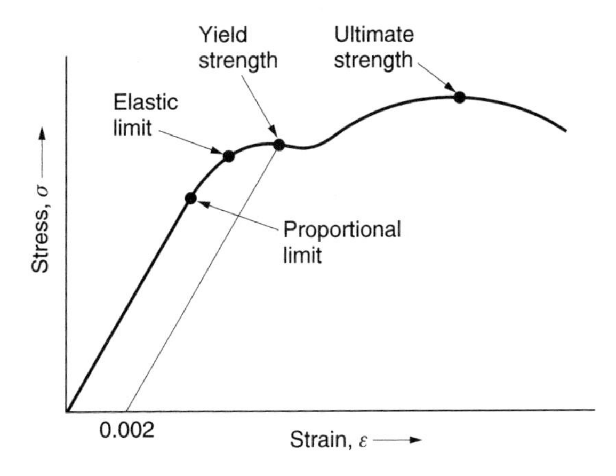

Figure 8.15 Typical stress–strain relationship.

process. The point up to which this relationship applies is known as the proportional limit. For practical applications, the proportional limit is difficult to define. As a result, a point close to the proportional limit, called the yield point, is most commonly used. The stress at this point, the yield strength, is defined to be the maximum stress that can be developed in the material without exceeding 0.2% strain. The yield strength and the maximum allowed stress (ultimate strength) are used in the structural design criteria presented in section 8.7. To ensure uniformity in structural design, yield and ultimate strengths for many of the commonly used aerospace materials have been compiled in a single document: MIL-HDBK-5, compiled by the Department of Defense and accepted as the standard for the aerospace industry.

8.8.1 Normal Stress Conditions

Loading due to the vehicle thrusting introduces uniaxial loads which produce normal stresses in the members. For a simple bar member of length L and cross-sectional area A loaded in the axial direction due to a thrusting event, the axial stress and strain from this force are

$$\left.\begin{aligned} \text{Axial stress} &= \sigma = P/A \\ \text{Axial strain} &= \varepsilon = \Delta/L \end{aligned}\right\} \tag{8.8.2}$$

where P is the force acting on the member and Δ is the total elongation due to the load. The member shortens uniformly under compression loading, giving rise to the uniform stress distribution shown in figure 8.16. Substituting the above expressions into Hooke's Law gives

$$\left.\begin{aligned} \sigma = E\varepsilon \Rightarrow \frac{P}{A} &= E\frac{\Delta}{L} \\ \text{or } P &= \frac{AE}{L}\Delta \end{aligned}\right\} \tag{8.8.3}$$

Comparing this last equation with the equivalent expression for the force acting on a spring ($F = kx$) gives

$$K_{\text{eq}} = \frac{AE}{L} \tag{8.8.4}$$

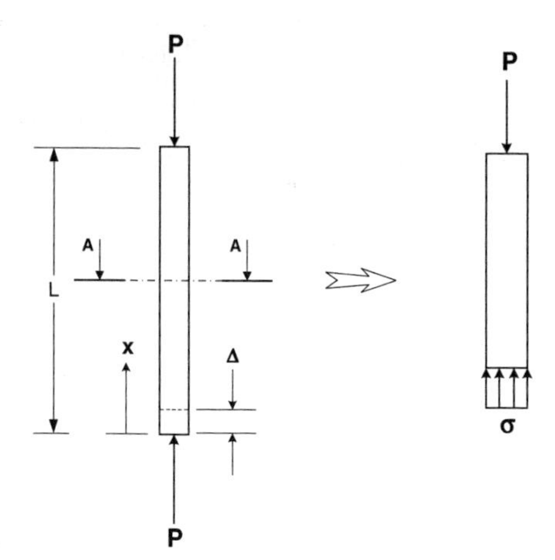

Figure 8.16 Uniform stress distribution.

This relationship defines an equivalent spring stiffness analogy for the member loaded in this manner. Note that to determine the equivalent stiffness one only needs to know the geometry and material modulus of elasticity for the member. This equivalent stiffness parameter is important in the later determination of the hardware's natural frequency of vibration.

Additionally, for the bar subjected to an axial load as shown in figure 8.16, a lateral deformation or strain occurs. The ratio of the lateral strain to the axial strain is a constant for a particular material and is defined as

$$\nu = \frac{\varepsilon_{\text{lateral}}}{\varepsilon_{\text{axial}}} \tag{8.8.5}$$

where ν = Poisson's ratio. This mechanical property is also defined in material handbooks, with typical values ranging between 0.1 and 0.35. For tensile loading a lateral contraction occurs, whereas for compressive loading a lateral expansion occurs.

A structural element will, in general, be subjected not only to axial loads but also to lateral or transverse loading that introduces bending of the member. For the simple beam shown in figure 8.17, equilibrium requires the reactive forces, shear force V and bending moment M, to exist on any general cross-section at $x = x_1$. In this figure F_i represent the applied loads and R_i the reactive loads from supports.

The bending of this member introduces a normal stress distribution on the section. The magnitude of stress produced by the moment M varies across the cross-section at $x = x_1$ according to

$$\sigma = \frac{My}{I} \tag{8.8.6}$$

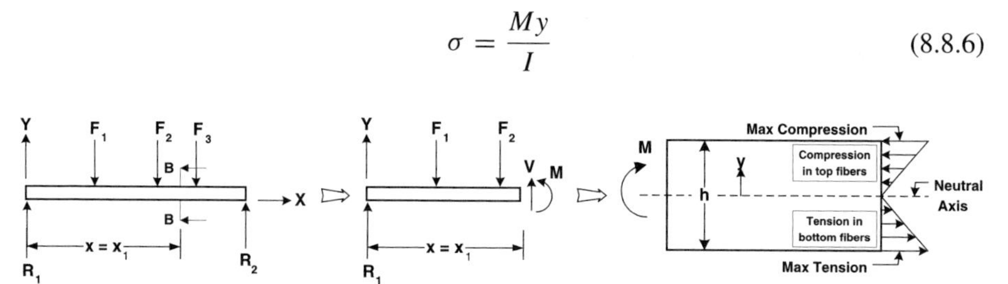

Figure 8.17 Simple beam: F_i represent the applied loads and R_i the reactive loads from supports.

where $I\,(=\int y^2\,dA)$ is the area moment of inertia of the cross-section and y is the distance from the neutral axis or surface. Figure 8.17 illustrates the bending stress distribution on this section of the element. The neutral axis is defined as the plane where no stress (or elongation) exists and can be shown to contain the centroid of the cross-sectional area.

From figure 8.17 and the above equation, the maximum stress occurs at the outermost fibers and has a magnitude of

$$\sigma = \frac{M(h/2)}{I} \tag{8.8.7}$$

where h is the beam thickness.

In considering the structural capability of the hardware, the structural engineer needs to address the worst case scenario. Given that axial and transverse (that is, bending in this situation) loading can exist simultaneously, the worst-case normal stress induced is represented as a combination of stresses due to each loading condition:

$$\sigma = \pm\frac{P}{A} \pm \frac{M(h/2)}{I} \tag{8.8.8}$$

The sign of each term is determined by the direction of the applied load. For the above equation, the worst-case stress levels are defined as

$$\left.\begin{aligned} \sigma &= \frac{P}{A} + \frac{M(h/2)}{I} \quad \text{(tensile stress)} \\ \sigma &= -\frac{P}{A} - \frac{M(h/2)}{I} \quad \text{compressive stress} \end{aligned}\right\} \tag{8.8.9}$$

Table 8.9a provides a tabulation of moments which may be substituted into the above equations for several of the cases more commonly used. Similar analyses, although rather more complicated, can be conducted to arrive at solutions of simple plate configurations with various boundary conditions. Some of the more common cases are presented in Table 8.9b.

Example 8.3

Given the general spacecraft configuration shown below with a spacecraft weight of 500 lb , if this system is launched on an Atlas IIAS launch vehicle, assuming a factor of safety 1.25, what is the maximum stress in the spacecraft adapter?

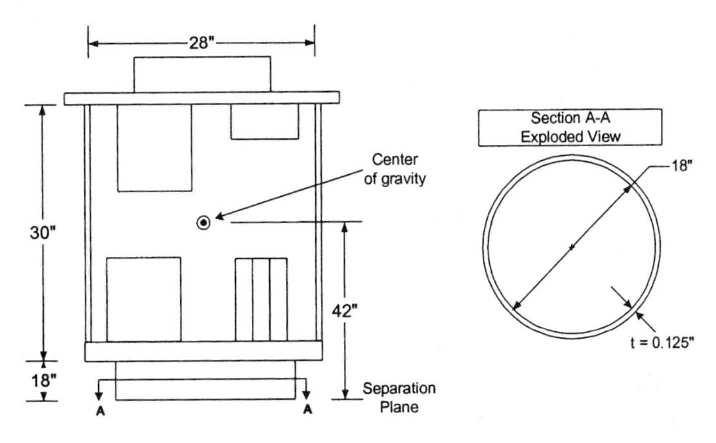

Solution: From table 8.4, the Atlas IIAS limit loads are:

Event	Lateral	Axial
Liftoff	± 1.3	$+1.2 \pm 1.1$
Flight winds	$+0.4 \pm 1.6$	$+2.7 \pm 0.8$
BECO (axial)	± 0.5	$+5.0 \pm 0.5$
BECO (lateral)	± 2.0	$+2.5 \pm 1.0$
SECO	± 0.3	$+2.0 \pm 0.4$
MECO (axial)	± 0.3	$+4.5 \pm 1.0$
MECO (lateral)	± 0.6	± 2.0

From the above load cases the worst-case axial loading occurs at BECO (axial) whereas the worst-case lateral loading occurs at flight winds or BECO (lateral). In considering the worst-case scenario we have

$$\sigma = \frac{P}{A} + \frac{Mc}{I}$$

For the configuration shown

$$A = 2\pi r\, t = 2\pi(9)(0.125) = 7.1\ \text{in}^2$$
$$I = \pi r^3\, t = \pi(9)^3(0.125) = 286\ \text{in}^4$$

For BECO (axial)

$$\sigma = \left[\frac{500(5.5)}{7.1} + \frac{500(0.5)(42)(9)}{286}\right](1.25) = [387.3 + 330.4](1.25)$$
$$= 897.1\ \text{psi}$$

For flight winds and BECO (lateral)

$$\sigma = \left[\frac{500(3.5)}{7.1} + \frac{500(2)(42)(9)}{286}\right](1.25) = [246.5 + 1321.7](1.25)$$
$$= 1960\ \text{psi}$$

The maximum stress in the adapter is thus 1960 psi.

8.8.2 Shear Stress Conditions

The transverse loading on the structure shown in figure 8.17 is reacted by a force V at any section. This force represents a shearing-type action and translates into a shear stress acting at this location.

Similarly to bending stress, the shear stress also varies over the cross-sectional area but as shown in figure 8.18. For this loading the maximum occurs at the neutral axis and is zero at the outermost fibers. This shear stress, τ_S, is defined as

$$\tau_S = \frac{VQ(y)}{It} \tag{8.8.10}$$

where $Q(y)$ is the first moment of the area above the point (y) where shear stress is being determined, y is the distance from the neutral axis, t is the thickness, and I is the

Table 8.9a Beam deflection/moment equation summary

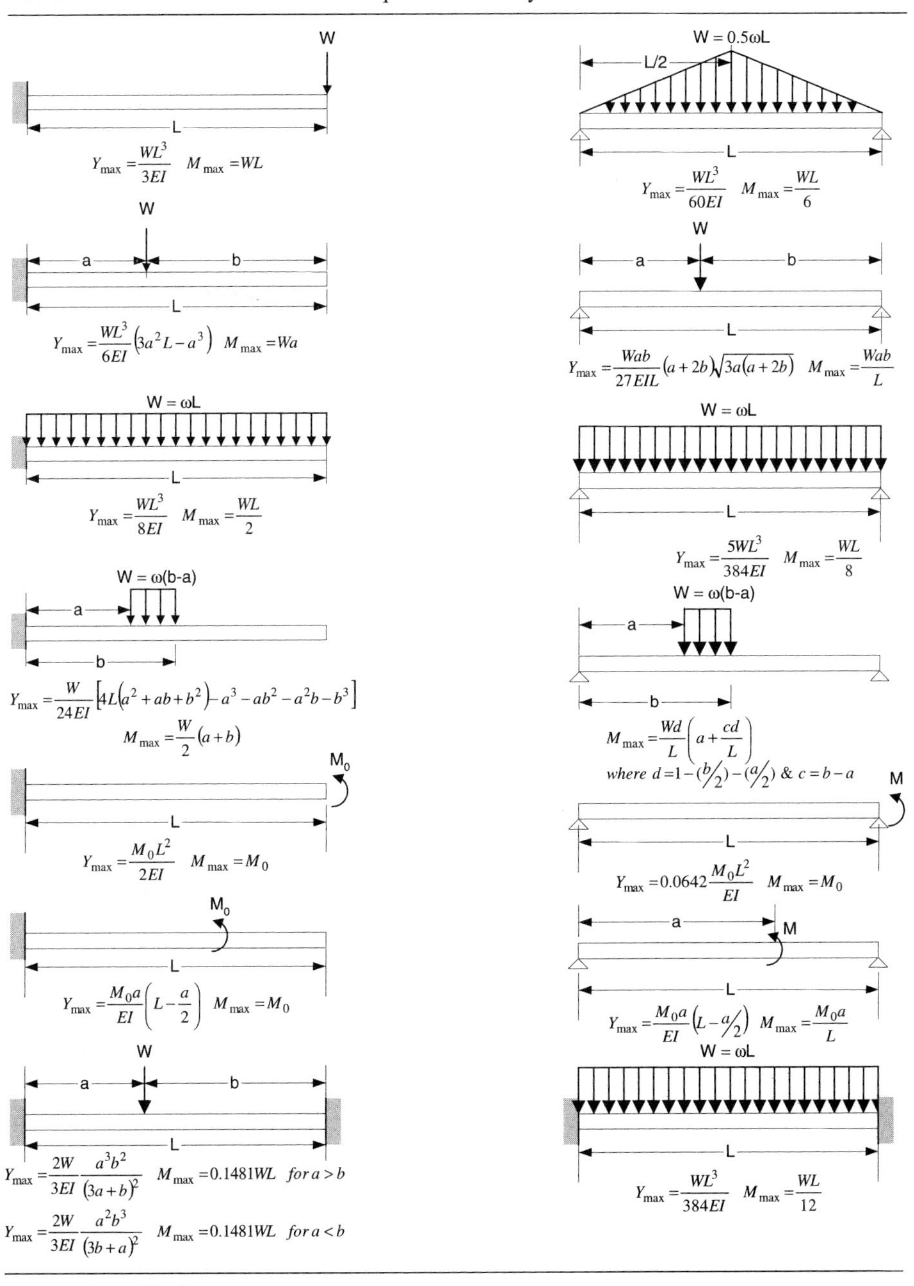

Y_{max} = maximum deflection
M_{max} = maximum bending moment

Table 8.9b Plate deflection/moment equation summary

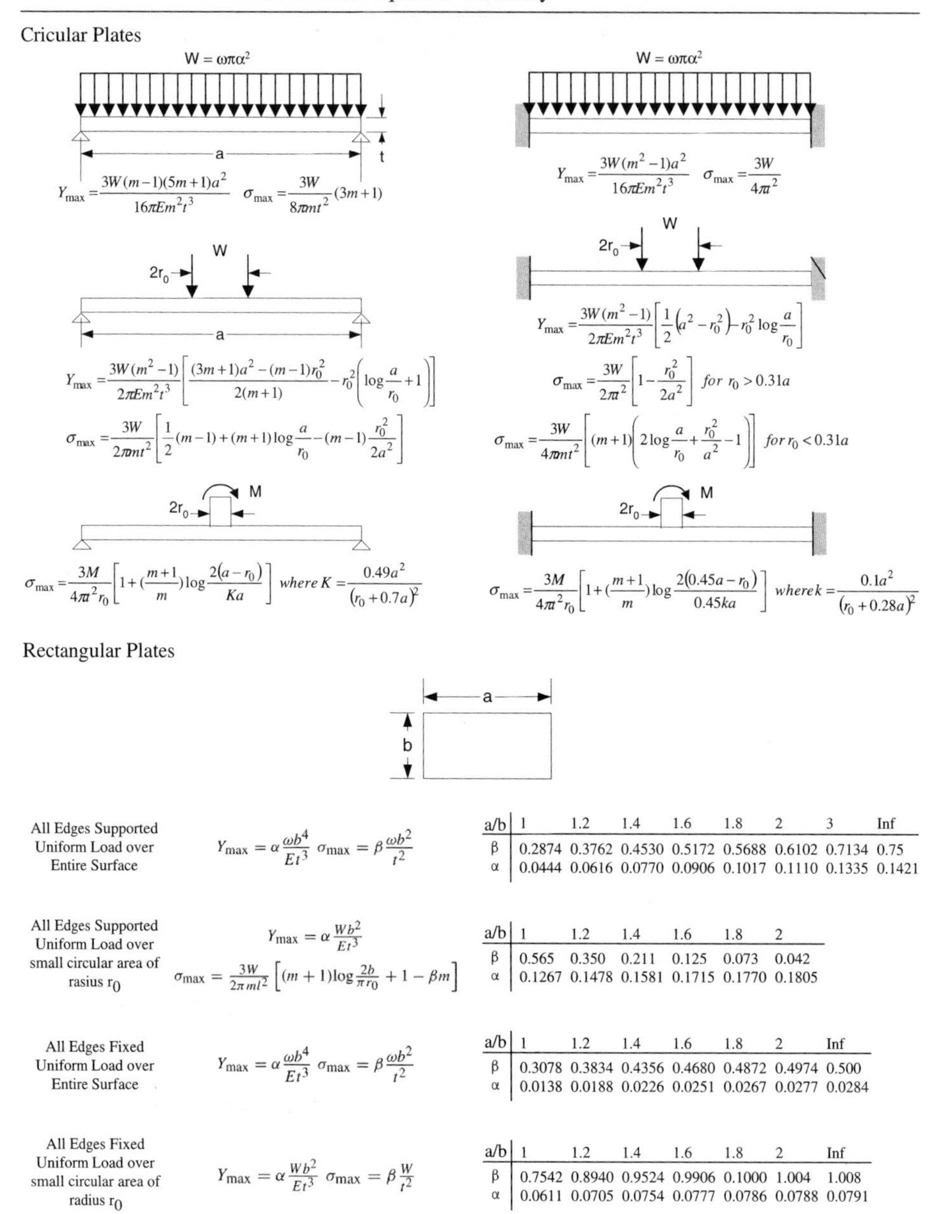

Cricular Plates

$W = \omega\pi\alpha^2$

$$Y_{max} = \frac{3W(m-1)(5m+1)a^2}{16\pi Em^2t^3} \quad \sigma_{max} = \frac{3W}{8\pi mt^2}(3m+1)$$

$W = \omega\pi\alpha^2$

$$Y_{max} = \frac{3W(m^2-1)a^2}{16\pi Em^2t^3} \quad \sigma_{max} = \frac{3W}{4\pi t^2}$$

$$Y_{max} = \frac{3W(m^2-1)}{2\pi Em^2t^3}\left[\frac{(3m+1)a^2-(m-1)r_0^2}{2(m+1)} - r_0^2\left(\log\frac{a}{r_0}+1\right)\right]$$

$$\sigma_{max} = \frac{3W}{2\pi mt^2}\left[\frac{1}{2}(m-1)+(m+1)\log\frac{a}{r_0}-(m-1)\frac{r_0^2}{2a^2}\right]$$

$$Y_{max} = \frac{3W(m^2-1)}{2\pi Em^2t^3}\left[\frac{1}{2}\left(a^2-r_0^2\right)-r_0^2\log\frac{a}{r_0}\right]$$

$$\sigma_{max} = \frac{3W}{2\pi t^2}\left[1-\frac{r_0^2}{2a^2}\right] \quad for\ r_0 > 0.31a$$

$$\sigma_{max} = \frac{3W}{4\pi mt^2}\left[(m+1)\left(2\log\frac{a}{r_0}+\frac{r_0^2}{a^2}-1\right)\right] \quad for\ r_0 < 0.31a$$

$$\sigma_{max} = \frac{3M}{4\pi a^2 r_0}\left[1+\left(\frac{m+1}{m}\right)\log\frac{2(a-r_0)}{Ka}\right] \quad where\ K = \frac{0.49a^2}{(r_0+0.7a)^2}$$

$$\sigma_{max} = \frac{3M}{4\pi a^2 r_0}\left[1+\left(\frac{m+1}{m}\right)\log\frac{2(0.45a-r_0)}{0.45ka}\right] \quad where\ k = \frac{0.1a^2}{(r_0+0.28a)^2}$$

Rectangular Plates

All Edges Supported Uniform Load over Entire Surface

$$Y_{max} = \alpha\frac{\omega b^4}{Et^3} \quad \sigma_{max} = \beta\frac{\omega b^2}{t^2}$$

a/b	1	1.2	1.4	1.6	1.8	2	3	Inf
β	0.2874	0.3762	0.4530	0.5172	0.5688	0.6102	0.7134	0.75
α	0.0444	0.0616	0.0770	0.0906	0.1017	0.1110	0.1335	0.1421

All Edges Supported Uniform Load over small circular area of rasius r_0

$$Y_{max} = \alpha\frac{Wb^2}{Et^3}$$

$$\sigma_{max} = \frac{3W}{2\pi ml^2}\left[(m+1)\log\frac{2b}{\pi r_0}+1-\beta m\right]$$

a/b	1	1.2	1.4	1.6	1.8	2
β	0.565	0.350	0.211	0.125	0.073	0.042
α	0.1267	0.1478	0.1581	0.1715	0.1770	0.1805

All Edges Fixed Uniform Load over Entire Surface

$$Y_{max} = \alpha\frac{\omega b^4}{Et^3} \quad \sigma_{max} = \beta\frac{\omega b^2}{t^2}$$

a/b	1	1.2	1.4	1.6	1.8	2	Inf
β	0.3078	0.3834	0.4356	0.4680	0.4872	0.4974	0.500
α	0.0138	0.0188	0.0226	0.0251	0.0267	0.0277	0.0284

All Edges Fixed Uniform Load over small circular area of radius r_0

$$Y_{max} = \alpha\frac{Wb^2}{Et^3} \quad \sigma_{max} = \beta\frac{W}{t^2}$$

a/b	1	1.2	1.4	1.6	1.8	2	Inf
β	0.7542	0.8940	0.9524	0.9906	0.1000	1.004	1.008
α	0.0611	0.0705	0.0754	0.0777	0.0786	0.0788	0.0791

area moment of inertia about the neutral axis. Again, the structural analyst is interested in the worst case or maximum. The maximum shear can be expressed in terms of the average shear stress acting on the cross-section according to

$$\tau_{max} = \alpha\frac{V}{A} = \alpha\tau_{avg} \tag{8.8.11}$$

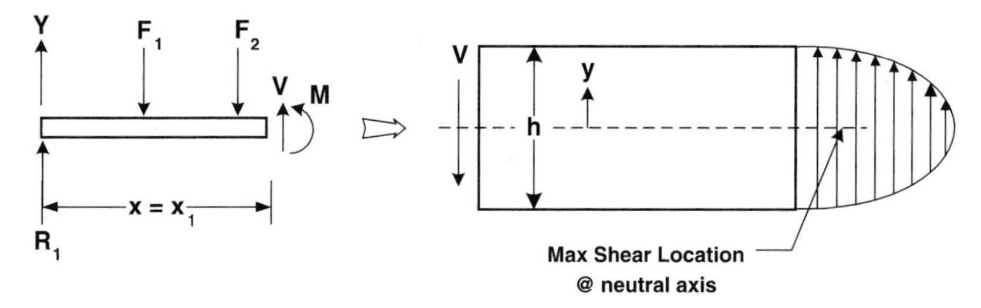

Figure 8.18 Shear stress distribution.

where V/A is the average shear stress and α is a factor that depends on the shape of the cross-section. Note that the maximum shear stress at any section occurs at the neutral axis provided the net width of the beam is as small there as anywhere else. (For this case, $y = 0$ in equation 8.8.10 and the entire area above the neutral axis is used in calculating $Q(y)$.) If the section is narrower elsewhere, the maximum shear stress may not occur at the neutral axis. Values of α for some simple beam cross-sections are given in table 8.10.

Shear stresses can also be produced by torsional moments acting on the member, as shown in figure 8.19. The torques produce a twisting of the member, giving rise to a shear strain γ. For the elastic range, the relationship between shear stress and shear strain is

$$\tau_s = G\gamma \tag{8.8.12}$$

where the shear modulus G is given by

$$G = \frac{E}{2(1+\mu)} \tag{8.8.13}$$

The angle of twist associated with this loading can be determined according to

$$\phi = \frac{TL}{JG} \tag{8.8.14}$$

where L is the length, T is the applied torque, and $J(=\int x^2\, dA)$ is the polar moment of inertia.

Table 8.10 Maximum shear stress for various cross-sections

Section	α	τ_{max}	Location
Rectangular	3/2	$\frac{3}{2}\tau_{avg}$	Neutral axis
Solid circular	4/3	$\frac{4}{3}\tau_{avg}$	Neutral axis
Hollow circular	2	$2\tau_{avg}$	Neutral axis
Triangular	3/2	$\frac{3}{2}\tau_{avg}$	Midway between top and bottom
Diamond shaped	9/8	$\frac{9}{8}\tau_{avg}$	Points $(1/8)d$ above and below neutral axis
I-beam or C-channel*	—	$\frac{Vbh}{2I}(1+\frac{h}{4b})$	Neutral axis

*b is the flange width and h is the height of the cross section

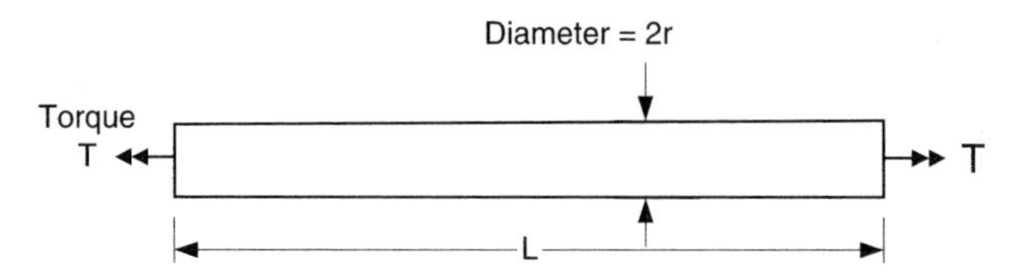

Figure 8.19 Member loaded in torsion.

For designs that use a solid circular shaft or tubular member with hollow circular cross-section, the shear stress τ_s varies across the cross-section linearly in proportion to the distance from the center of the shaft according to

$$\tau_s = \frac{Tr}{J} \tag{8.8.15}$$

For the circular member of radius R, $J = \pi R^4/2$ and the above equation reduces to $\tau_s = 2T/\pi R^3$.

While closed sections such as the solid circular or tubular section are preferred, in many cases structural members with non-circular cross-sections are used. Members with these cross sections are, in general, inefficient in reacting to shear conditions. However, they must be able to adequately resist the shear stresses that develop. To estimate the shear stress for members with these cross-sections, a rectangular cross-section must first be examined. For a rectangular section of width b and thickness t, as shown in figure 8.20a, the shear stress must act parallel to the boundary over the section. If the width b is large compared to the thickness t, the shear stress and angle of twist are

$$\tau_s = \frac{3T}{bt^2} \text{ and } \phi = \frac{3TL}{bt^3 G} \tag{8.8.16}$$

For any general rectangular section, these expressions can be written as

$$\tau_s = \frac{T}{\alpha bt^2} \text{ and } \phi = \frac{TL}{\beta bt^3 G} \tag{8.8.17}$$

where α and β are constants which are dependent on the cross-sectional dimensions. A comparison of the general expression of the twist angle ϕ (equation 8.8.14) with that from above gives

$$J = \beta bt^3 \tag{8.8.18}$$

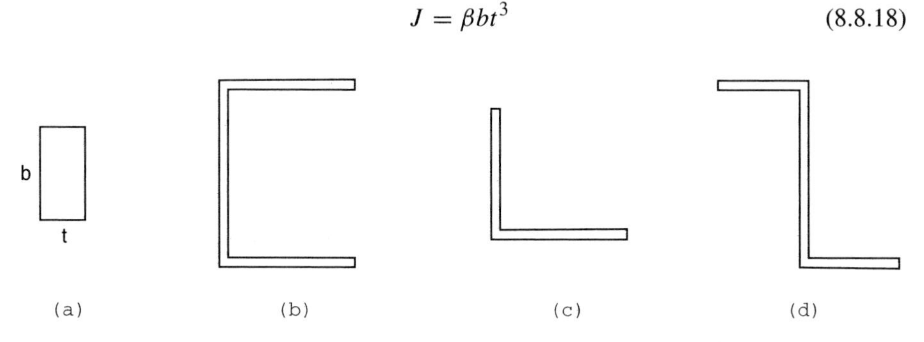

Figure 8.20 Typical non-circular cross-sections.

Table 8.11

b/t	1.00	1.50	1.75	2.00	2.50	3.00	4	6	8	10	∞
α	0.208	0.231	0.239	0.246	0.258	0.267	0.282	0.299	0.307	0.313	0.333
β	0.141	0.196	0.214	0.229	0.249	0.263	0.281	0.299	0.307	0.313	0.333

for the single rectangular element. Values for the constants α and β for ratios of the width and thickness are given in table 8.11.

The polar moment of inertia J for other sections shown in figure 8.20 can be viewed as a cross-section made up of several rectangular elements with dimensions b_1 and t_1, b_2 and t_2, etc., where b_i and t_i are the width and thickness, respectively, of each element. For this case, the polar moment of inertia becomes

$$J \equiv \sum \beta_i b_i t_i^3 \tag{8.8.18}$$

The use of this relationship in the equations above provides an estimate of the shear stress and angle of twist for the non-circular members. As mentioned previously, the thickness t is typically small, making the polar moment of inertia J even smaller because of the smallness of t^3. From the general equation for shear stress in equation 8.8.15, for small J, τ_s is large. From this it is obvious that these kinds of sections for structural members are only capable of resisting small values of torque.

8.8.3 Buckling Conditions

8.8.3.1 Beams

The stress–strain analysis presented previously inherently assumed that the structural member remained stable throughout the loading condition. In some circumstances, certain members become unstable when loaded in compression. Since the predominant loading during major launch vehicle thrusting events is compression, this condition can prove catastrophic if not properly addressed. This elastic instability, which can occur below the elastic limit, is commonly referred to as *buckling*.

The buckling phenomenon for beam-type applications occurs primarily for columns which are long in comparison with their other dimensions. This critical buckling stress will depend on the end boundary conditions of the column and can be determined from the following expression

$$\sigma_{\mathrm{cr}} = \frac{C\pi^2 E}{(L/\rho)^2} = \frac{\pi^2 E}{(L'/\rho)^2} \tag{8.8.19}$$

where $\sigma_{\mathrm{cr}} = P_{\mathrm{cr}}/A$, the critical buckling stress, and P_{cr} is the critical buckling load. In the above expression

A = cross-sectional area
E = modulus of elasticity
L = column length
$\rho = \sqrt{(I/A)}$, radius of gyration of the cross-section

I = second moment of inertia of the cross-section about the buckling axis
C = a constant, dependent on the end boundary conditions
$L' = L/\sqrt{C}$, effective length of the column
L'/ρ = slenderness ratio

The above equation is commonly referred to as the Euler equation and applies to beams satisfying the following conditions:

(1) columns whose slenderness ratio (L'/ρ) is on the order of 110 or greater,
(2) columns whose stress does not exceed the elastic limit of the material,
(3) columns with a "stable" cross-section (that is, closed section with relatively heavy wall thickness)

Critical loads for beams of various loading and end boundary conditions are presented in table 8.12.

Figure 8.21 illustrates a beam-type structure for the Midcourse Space Experiment (MSX) program that was buckling critical. This truss structure, made from composite materials, was approximately 6 ft tall by 5 ft square and constructed from individual members that had an "H" cross-section measuring approximately 2 in. by 2 in.

As indicated previously, the design of many aerospace structures involves columns with thin walls and open cross-sections. Because of the thin-walled sections, these columns may also fail by a local collapse of the wall at a stress much lower than those given by the Euler equation. This local failure, which may combine with the primary failure defined by Euler's equation, is called *crippling*. To calculate the crippling stress, the unstable cross-section is viewed as a single angle or combination of angles

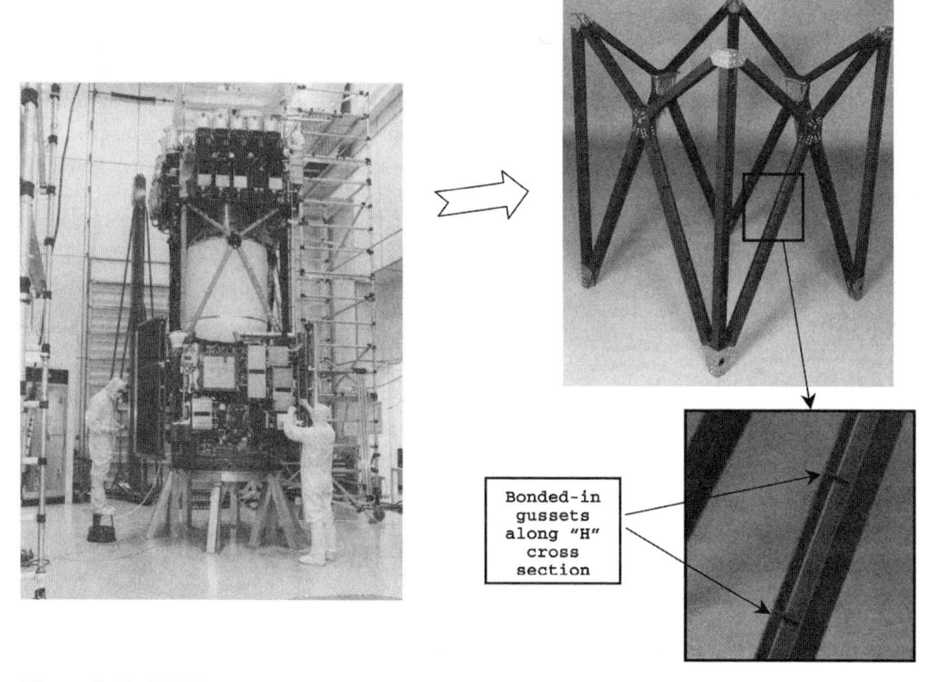

Figure 8.21 MSX truss structure.

Table 8.12 Beam bucking equation summary

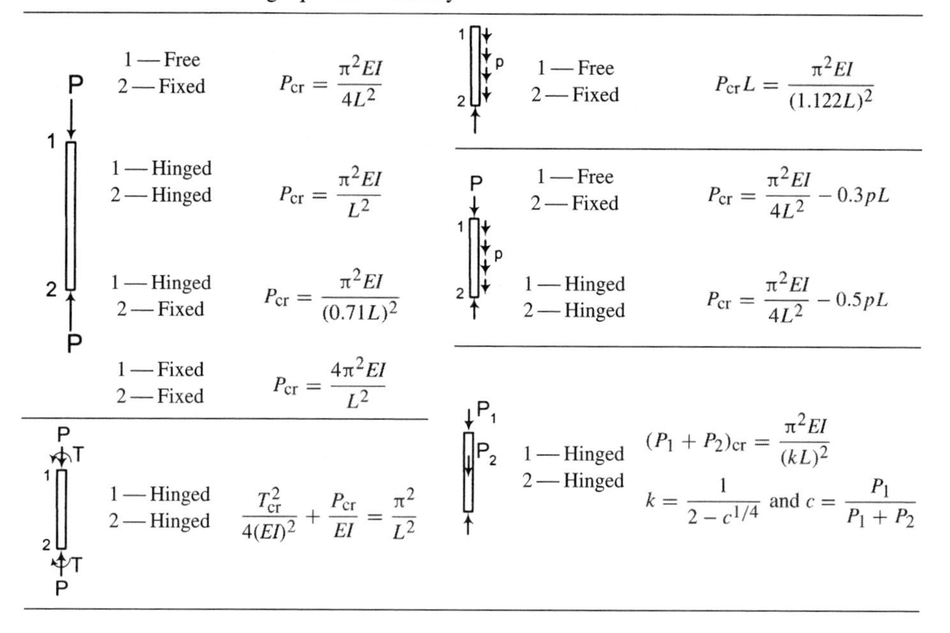

End conditions	Critical load
1—Free, 2—Fixed	$P_{cr} = \frac{\pi^2 EI}{4L^2}$
1—Hinged, 2—Hinged	$P_{cr} = \frac{\pi^2 EI}{L^2}$
1—Hinged, 2—Fixed	$P_{cr} = \frac{\pi^2 EI}{(0.71L)^2}$
1—Fixed, 2—Fixed	$P_{cr} = \frac{4\pi^2 EI}{L^2}$
1—Hinged, 2—Hinged (with torque T)	$\frac{T_{cr}^2}{4(EI)^2} + \frac{P_{cr}}{EI} = \frac{\pi^2}{L^2}$
1—Free, 2—Fixed (distributed p)	$P_{cr}L = \frac{\pi^2 EI}{(1.122L)^2}$
1—Free, 2—Fixed (P and distributed p)	$P_{cr} = \frac{\pi^2 EI}{4L^2} - 0.3pL$
1—Hinged, 2—Hinged (P and distributed p)	$P_{cr} = \frac{\pi^2 EI}{4L^2} - 0.5pL$
1—Hinged, 2—Hinged (P_1, P_2)	$(P_1 + P_2)_{cr} = \frac{\pi^2 EI}{(kL)^2}$; $k = \frac{1}{2 - c^{1/4}}$ and $c = \frac{P_1}{P_1 + P_2}$

with various edge conditions. A few examples of common cross-sections are shown in figure 8.22.

The crippling stress for the angle section (figure 8.22a) can be determined from the following expression

$$\sigma_c = k_e \frac{(E_c \sigma_{cy})^{0.5}}{(b'/t)^{0.75}} \tag{8.8.20}$$

Where

σ_c = crippling stress
E_c = compressive modulus of elasticity
σ_{cy} = material compressive yield strength
$b'/t = (a + b)/2t$
k_e = coefficient that depends on angle edge condition
= 0.316 (two edges free)
= 0.342 (one edge free)
= 0.366 (no edge free)

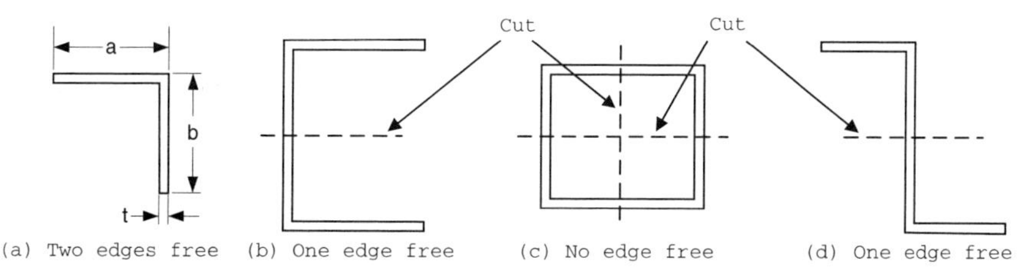

Figure 8.22 Thin-walled crippling cross-section.

The crippling load, P_c, on the angle is $\sigma_c A$. For the cross-sections shown in figure 8.22b, c, and d, the crippling stress is determined by dividing the shape into a series of angles as shown and computing the crippling loads for each section. The average crippling stress for the cross-section is then defined by

$$\sigma_{cs} = \frac{\sum \sigma_{ci} A_i}{\sum A_i} \tag{8.8.21}$$

where

σ_{cs} = member section average crippling stress
σ_{ci} = i^{th} angle crippling stress
A_i = i^{th} angle cross-sectional area

Referring again to the MSX truss structure in figure 8.21, the "H" cross-section was made of thin flanges and a web with a "free" end that would extend from the bottom to the top of the structure. To preclude the effects of crippling, small stiffening gussets were added along each of the open ends of the individual truss members. The gussets are those indicated in figure 8.21 by arrows.

8.8.3.2 Flat Plates and Cylinders

In the case of flat plates, buckling can occur when these members are loaded in compression, shear, or bending. For flat plates in compression, the buckling stress is defined as

$$\sigma_{cr} = K \frac{E}{12(1-\nu^2)} \left(\frac{t}{b}\right)^2 \tag{8.8.22}$$

where

$\sigma_{cr} = P_{cr}/A$
K = buckling coefficient, depending on the edge boundary conditions and aspect ratio (a/b)
E = modulus of elasticity
ν = Poisson's ratio
b = short dimension of plate or loaded edge
t = plate thickness

The buckling stress of a flat plate can also be written in a more general form

$$\sigma_{cr} = K_c E \left(\frac{t}{b}\right)^2 \tag{8.8.23}$$

where K_c is a function of a/b. For plates loaded in shear and bending, the above general equation can be used by substituting K_s (shear buckling coefficient) and K_c (bending buckling coefficient) for K_c. Buckling stress allowables for plates with various edge conditions are given in table 8.13a. Note that for plates with combined loading conditions the equations become considerably more complicated than the simple expression above.

In general, the flat sheet is inefficient for carrying compressive loads. Stiffeners are therefore added to reinforce the plate. These stiffeners assume many shapes (such as those indicated in figure 8.20) and can be bent-up sheet metal, formed, extruded, or machined parts. To determine the compressive load-carrying capability of these structural elements,

Table 8.13a Plate bucking equation summary (F = fixed, SS = simply supported)

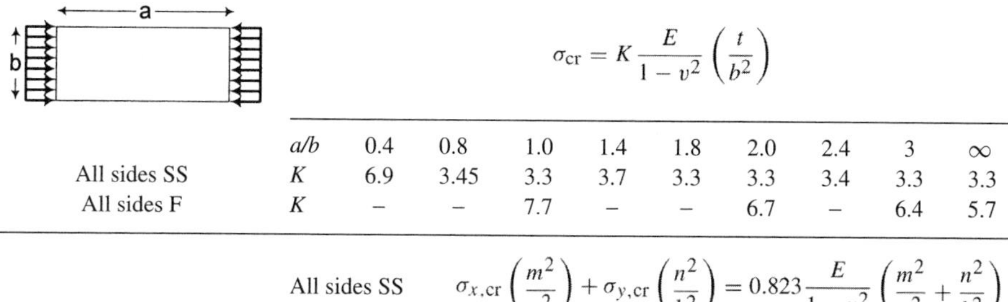

$$\sigma_{cr} = K \frac{E}{1-v^2}\left(\frac{t}{b^2}\right)$$

	a/b	0.4	0.8	1.0	1.4	1.8	2.0	2.4	3	∞
All sides SS	K	6.9	3.45	3.3	3.7	3.3	3.3	3.4	3.3	3.3
All sides F	K	–	–	7.7	–	–	6.7	–	6.4	5.7

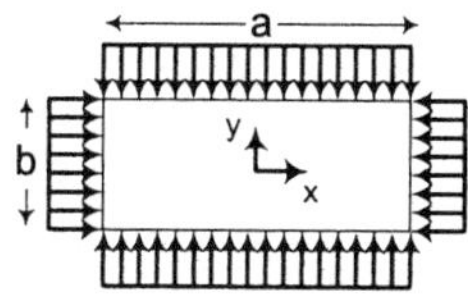

All sides SS
$$\sigma_{x,cr}\left(\frac{m^2}{a^2}\right) + \sigma_{y,cr}\left(\frac{n^2}{b^2}\right) = 0.823\frac{E}{1-v^2}\left(\frac{m^2}{a^2}+\frac{n^2}{b^2}\right)$$

All sides F
$$\sigma_{x,cr} + \frac{a^2}{b^2}\,\sigma_{y,cr} = 1.1\frac{Et^2a^2}{1-v^2}(3/a^4 + 3/b^4 + 2/a^2b^2)$$

where m, n refer to the modes in the x & y directions, respectively, and with m, n = 1,2,3,...

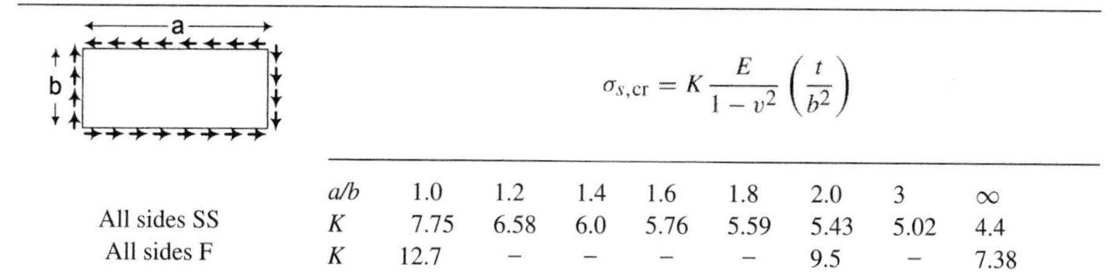

$$\sigma_{s,cr} = K\frac{E}{1-v^2}\left(\frac{t}{b^2}\right)$$

	a/b	1.0	1.2	1.4	1.6	1.8	2.0	3	∞
All sides SS	K	7.75	6.58	6.0	5.76	5.59	5.43	5.02	4.4
All sides F	K	12.7	–	–	–	–	9.5	–	7.38

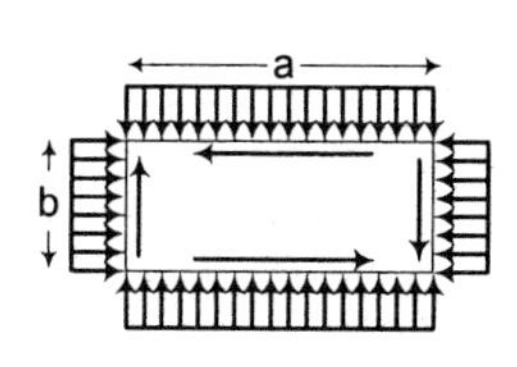

All sides SS
$$\sigma_{cr} = \sqrt{C^2\left(2\sqrt{1-\frac{\sigma_y}{C}}+2-\frac{\sigma_x}{C}\right)\left(2\sqrt{1-\frac{\sigma_y}{C}}+6-\frac{\sigma_x}{C}\right)}$$

where $C = \frac{0.823}{1-v^2}\left(\frac{t}{b^2}\right)$

All sides F
$$\sigma_{cr} = \sqrt{C^2\left(2.31\sqrt{4-\frac{\sigma_y}{C}}+\frac{4}{3}-\frac{\sigma_x}{C}\right)\left(2.31\sqrt{4-\frac{\sigma_y}{C}}+8-\frac{\sigma_x}{C}\right)}$$

where $C = \frac{0.823}{1-v^2}\left(\frac{t}{b^2}\right)$

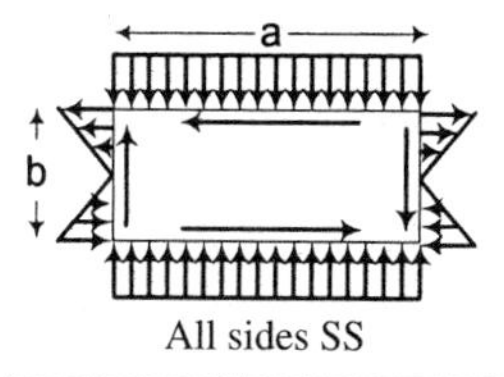

$$\sigma_{cr} = K\frac{E}{1-v^2}\left(\frac{t}{b^2}\right)$$

All sides SS						
$\sigma_s/\sigma_{s,cr}$	0	0.2	0.4	0.6	0.8	1.0
K	21.1	20.4	18.5	3.7	16.0	0

Table 8.13b Cylinder Buckling Equation Summary

	Condition	Equation
	Ends not constrained	$\sigma_{cr} = \dfrac{\gamma E}{\sqrt{3(1-\upsilon^2)}}\dfrac{t}{r}$
$\phi = \frac{1}{16}\sqrt{\frac{r}{t}}$ for $\frac{r}{t} < 1500$	Axial	$\gamma = 1 - 0.901(1 - e^{\phi})$
	Bending	$\gamma = 1 - 0.731(1 - e^{\phi})$
M, M	No constraint	$M_{cr} = K\dfrac{E}{1-\upsilon^2}rt^2$ where K for pure bending and long tube = 0.99, K_{avg} = 1.14 and K_{min} = 0.72
T, T	Ends hinged (simply supported)	$\sigma_{s,cr} = E\left(\frac{t}{L}\right)^2\left[1.8 + \sqrt{1.2 + 0.201\left(\frac{L}{\sqrt{tr}}\right)^2}\right]$ for $\upsilon = 0.3$
	Ends fixed	$\sigma_{s,cr} = KE\left(\frac{t}{r}\right)^{1.35}$
2α, 2r₁, 2r₂, L	Axial Compression	$P_{cr} = \gamma\dfrac{2\pi Et^2\cos^2\alpha}{\sqrt{3(1-\upsilon^2)}}$ where $\gamma = 0.33$ for $10° < \alpha < 75°$
	Bending	$M_{cr} = \gamma\dfrac{\pi Et^2 r_1\cos^2\alpha}{\sqrt{3(1-\upsilon^2)}}$ where $\gamma = 0.41$ for $10° < \alpha < 75°$
	Torsion	$T_{cr} = 52.8\,\gamma D\left(\frac{t}{L}\right)^{1/2}\left(\frac{r}{a}\right)^{5/4}$ where $r = r_1\cos\alpha(1 + A^{1/2} - A^{-1/2})$, $D = \dfrac{Et^3}{12(1-\upsilon^2)}$, $A = \dfrac{1}{2}\left(1 + \dfrac{r_2}{r_1}\right)$ and $\tau_{cr} = \dfrac{T_{cr}}{2\pi r_1^2 t}$

Ends fixed, values of K:

L/r	0.2	0.3	0.4	0.5	0.75	1.0	2	4
K	3.3	2.45	2.02	1.78	1.45	1.27	0.94	0.68

the results derived from plate buckling analysis can be used. As an example, the simple angle section shown previously in figure 8.20 can be treated as a group of two flanges with the common side acting as simply supported. If the flanges are of equal size, each will buckle at the same stress. For this condition, the buckling stress is determined to be

$$\sigma_{cr} = 0.385E\left(\frac{t}{b}\right)^2 \tag{8.8.24}$$

Utilizing this approach, the buckling capability of other stiffener cross-sections can be determined.

Similar buckling conditions also arise for cylindrical and conical structures. Table 8.13b provides some fundamental critical loading conditions for these types of members.

8.9 Types of Structure

The most common and basic forms of beam-type members are thin- or thick-walled tubing (circular or square), I-beams, H-beams, or angle construction. These members are typically used for support frames, stiffening structures, or boom configurations. Thin or thick flat plate construction may be used for mounting platforms for hardware but are usually mass-prohibitive for such applications. Flat plates are efficient for use in panel closeout configurations where only shear capability is desired and there is no need to mount hardware. Thin plates are also frequently used as radiator surfaces by the thermal engineer.

A special kind of structure used commonly in aerospace applications is the sandwich construction, sometimes referred to as honeycomb panels or bonded panel assemblies. Sandwich construction consists of a lightweight core material "sandwiched" between two thin face sheets as shown in figure 8.23. In many circumstances the core used in the construction of the panel has a hexcel or honeycomb cross-section, thus giving rise to the common expression honeycomb panel. An analogy of general sandwich construction can be made with the more common I-beam configuration. For the I-beam, when loaded with a transverse load, the flanges will provide the reaction to bending while the web will resist the transverse shear. For sandwich construction, the face sheets are analogous to the flanges and resist bending while the core, similarly to the web, will resist the transverse shear. The adhesive holds the two elements together and serves to transfer the load between the face sheets and the core. This technique thus takes advantage of the strengths of the individual components which, when joined together, produce a structure that is both strong and stiff while being lightweight.

This sandwich structure can be made from conventional materials or composite materials such as graphite epoxy. Face sheet thicknesses typically range in size from 0.005 in. to 0.030 in. Typical panel thicknesses are 0.25 in. to 4.00 in., although smaller and larger sizes are possible. Core sizes can vary as well as the thickness of the material used to make up the core; however, for most space applications, the core material thickness is usually relatively thin (0.0007 in. or 0.0010 in). The particular configuration used

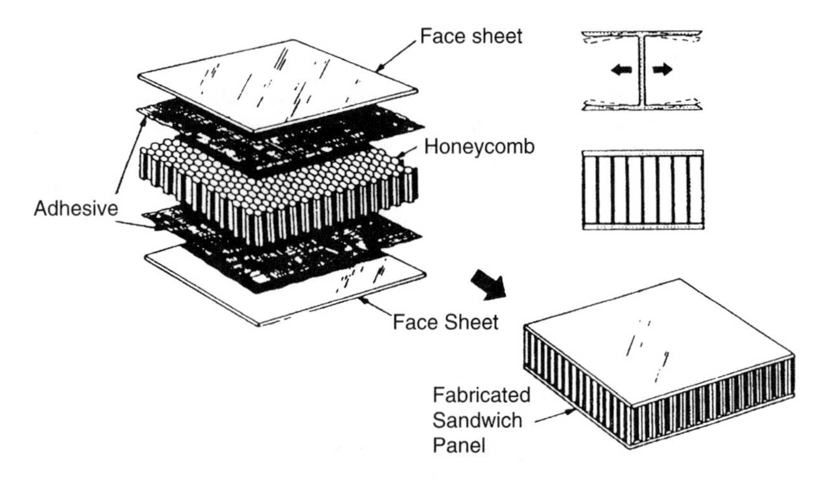

Figure 8.23 Sandwich construction.

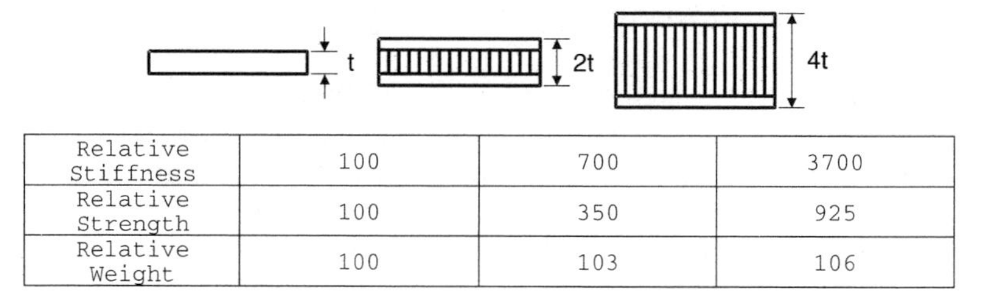

Relative Stiffness	100	700	3700
Relative Strength	100	350	925
Relative Weight	100	103	106

Figure 8.24 Comparison of sandwich construction to that of a typical flat plate.

is based on the desired application and the mass, strength, and stiffness requirements. Most applications use face sheets of similar materials. The cores used can be of the same material as the face sheets, although it is not uncommon to have a dissimilar core material. A common configuration which uses dissimilar materials is aluminum core with graphite epoxy face sheets. A comparison of sandwich construction with that of a typical flat plate is shown in figure 8.24.

Sandwich construction has various failure mechanisms which must be considered when examining these configurations analytically. The types of failure that can occur, and analytical equations to check these various conditions, are contained in table 8.14. MIL-HDBK-23 provides a good reference for analysis of plate structures made from sandwich construction.

8.10 Structural Load Path

During the development of any structural configuration, one key aspect in the process is to define an efficient means of support and arrangement of the structural elements. By carefully considering how the load is transmitted through the structure, in other words, the load path, the structural engineer can optimize the structural elements. To illustrate this consider the following example.

Example 8.4

Consider the beam with two parallel members separated by an inner core as shown below. The beam is loaded at its end with force W. Determine how much of this load is carried by member 1 and member 2, given that they are both made of the same material with modulus E and that member 1 is much thicker than member 2, i.e., $I_1 \gg I_2$. The combined elements are assumed to act as a single member.

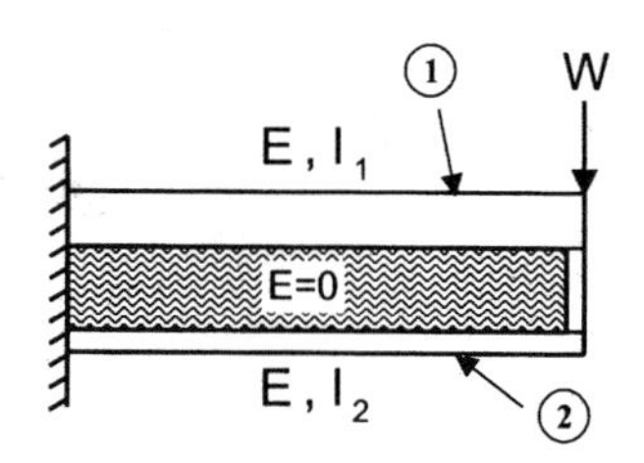

Solution: For proper design it is important to define how much of the load is distributed through the structure for each of the corresponding members or elements. Since $E = 0$ for the inner material, this has little stiffness in resisting the applied load by itself (as is the case with honeycomb core) but is used as the interconnecting structure for the outer two members.

Assuming W_1 is the amount carried by the upper member and W_2 is the amount of loading for the lower member, then the total loading can be represented as

$$W = W_1 + W_2$$

Since the member acts as a single unit, the deflection of each of the components is the same: thus $y = y_1 = y_2$. From table 8.9a,

$$y_1 = \frac{W_1 L^3}{3EI_1} \text{ and } y_2 = \frac{W_2 L^3}{3EI_2}$$

Equating gives

$$W_2 = \frac{W_1 I_2}{I_1}$$

Substituting into the first equation gives

$$W = W_1 + W_2 = W_1 + \frac{W_1 I_2}{I_1}$$

Since $I_1 \gg I_2$, $I_2/I_1 \approx 0$ then $\rightarrow W = W_1$.

Rearranging $y = WL^3/(3EI)$ into the form $F = k_{eq}X$ gives $k_{eq} = 3EI/L^3$. Comparing the individual members of the beam indicates that member 1 is stiffer than member 2 and thus

the stiffer member carries the greater part of the loading.

A more realistic example for the above concept is shown in figure 8.25. For this particular example, the launch vehicle generated a significant thrust loading condition during launch. The upper sections (instrument and instrument electronics) were supported by the thin lightweight honeycomb outer structure and the rectangular center structure. The edge of the honeycomb panel (penthouse deck) was attached to the outer structure via the 0.03 in. flat plate. Considering the thrust loading, the flat plate offers little resistance to any normal applied load (try bending a thin flat plate). From the previous example, to transfer any significant loading, stiffness is required. Thus the load path, or load distribution, for support of the upper sections will be via the penthouse honeycomb deck and through the center support structure.

This principle is an important one and provides the basis for many designs. Another such example where the relative stiffness plays an important role is the use of hardware called flexures. These flexures are blade-type structures that are typically used for mounting optical devices. By controlling the stiffness in various directions, the structural engineer can control how the load is transferred from one structure to another. By controlling the load and how it is transferred, the deformation of the structure is controlled.

8.11 Thermal Stress

The previous sections addressed loading from events such as those that occur during launch. Once through the launch sequence and on orbit, other conditions such as a bulk

Table 8.14 Sandwich Construction Beam Equations

Beam Type	Max Shear Force V	Max Bending Moment M	Bending Deflection Constant K_b	Shear Deflection Constant K_s
W = ωL, L	0.5 W	0.125 WL	0.0132	0.125
W = ωL, L	0.5 W	0.0833 WL	0.0026	0.125
W, L/2, L	0.5 W	0.25 WL	0.0208	0.25
W, a, L	0.5 W	0.125 WL	0.0052	0.25
W = ωL, L	W	0.5 WL	0.125	0.5
W, L	W	WL	0.333	1
W = ωL/2, L	W	0.3333 WL	0.0667	0.3333
W = ωL, L	0.625 W	0.125 WL	0.0054	0.0704

Condition	Governing Equation
Bending stress in facings	$\sigma_f = \dfrac{M}{t_f h b}$ where $i = 1, 2$
Core Shear Stress	$\tau_{cs} = \dfrac{V}{hb}$
Deflection	$\Delta = \dfrac{2K_b P L^3 \lambda}{E_f t_f h^2 b} + \dfrac{K_s P L}{h G_c b}$ (for same skin matcrials)
Face Dimpling	$\sigma_{cr} = \dfrac{2E_f}{\lambda}\left(\dfrac{t_f}{s}\right)^2$
Face Wrinkling	$\sigma_{cr} = 0.82 E_f \left(\dfrac{E_c t_f}{E_f t_c}\right)^{1/2}$

$$D = \frac{E_1 t_1 E_2 t_2 h^2 b}{E_1 t_1 \lambda_2 + E_2 t_2 \lambda_1}$$

Terminology

P = total load
b = width
L = span
σ_{cr} = critical facing stress
σ_f = facing stress
D = panel stiffness
t_f = skin thickness
$\lambda = 1 - \nu^2$
h = centroid distance
ν = facing poisson's ratio
τ_{cs} = core shear stress
S = cell size
Δ = deflection
E_c = core compression modulus
E_f = facing modulus
t_c = core thickness
G_c = core shear modulus
I = moment of inertia

Sandwich construction failure modes

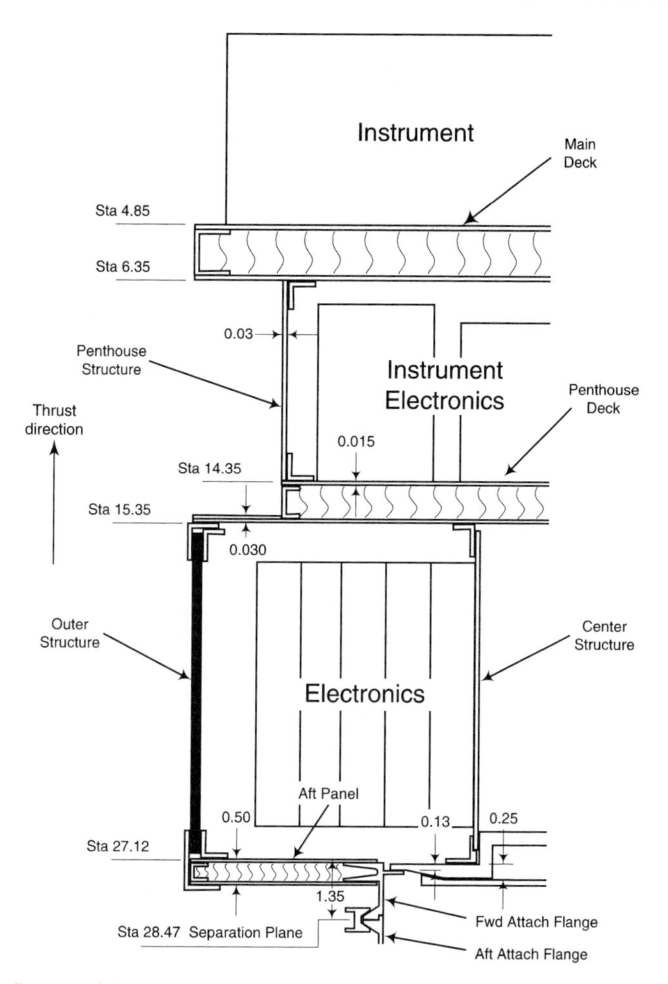

Figure 8.25 Spacecraft load path example.

temperature change or temperature gradient can produce significant mechanical stresses. These stresses, sometimes referred to as thermal stresses, are of particular importance when materials with dissimilar mechanical properties are used, or in applications where stable mounting platforms are required. These stresses are developed as a result of the constraints that exist on the structural member. Such constraints occur, for instance, when the ends of a structural member are attached to an adjacent structure. To investigate these effects, consider the simple beam shown below.

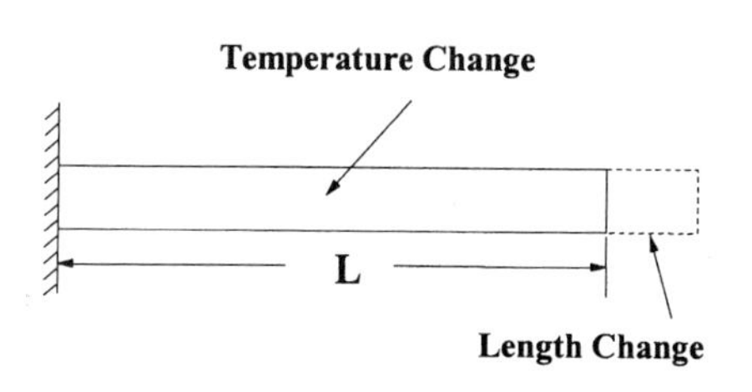

If this beam experiences a bulk temperature change ΔT, the length of the beam will increase according to the following relation

$$\delta_T = \alpha L(\Delta T) \tag{8.11.1}$$

where δ_T is the deformation, α is the coefficient of linear thermal expansion, L is the original length, and ΔT is the change in temperature. Under this condition, the beam is not constrained and thus no stress will occur; the member will just grow to a new length $L+\delta_T$. If, however, the beam is constrained as shown below, deformation is not permitted when the member is subjected to a change in temperature.

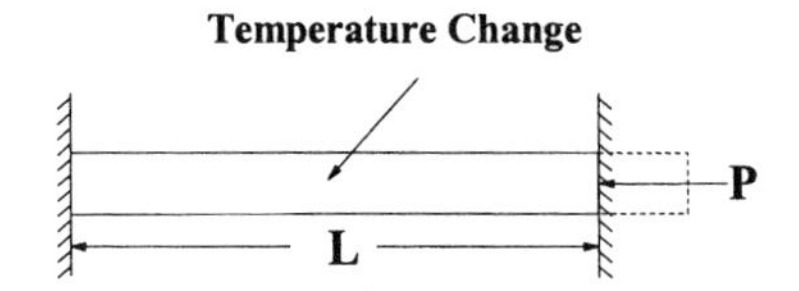

Because of this constraint, an internal force is generated which resists this deformation, causing an internal stress. The mechanical (or "thermal") stress that is generated can be determined using the fundamental definition of strain and Hooke's Law. The definition of strain gives

$$\varepsilon = \frac{\delta_T}{L} = \frac{\alpha L(\Delta T)}{L} \quad \text{or } \varepsilon = \alpha \Delta T \tag{8.11.2}$$

Substituting this into Hooke's Law gives an expression for the stress σ that is developed in the structural beam:

$$\sigma = E\varepsilon = E\alpha\Delta T \tag{8.11.3}$$

When the temperature of the member is increased, the member will want to expand thereby giving rise to a compressive stress. Alternatively, when the temperature is decreased, a tensile stress will develop.

As can be seen from the above equation, determining the effects of the temperature change requires knowledge of the temperature distribution and the mechanical properties of the material. For this example, the modulus of elasticity and coefficient of thermal expansion are shown to be constant. This assumption is generally adequate for most situations encountered, but for larger temperature ranges the variation of material properties over the temperature range must be considered. As an example, from MIL-HDBK-5H, for 6061-T6 aluminum, the coefficient of thermal expansion is seen to vary with temperature as shown in figure 8.26. This variation must be considered for hardware applications such as cryogenic instruments, where temperatures change from room temperature (20°C) to temperatures as low as 10–20K (or approximately −250°C)!

The above approach obviously represents a very simple view of the thermal stress problem. Even so, in many practical situations, simple calculations such as these produce results that will give the structural engineer an idea of the magnitude of the impact. This kind of approach usually bounds the problem and is extremely helpful early in the design phase.

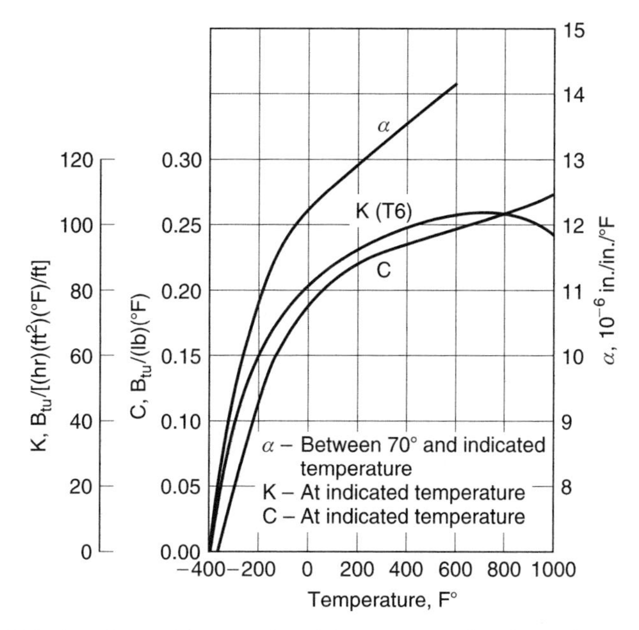

Figure 8.26 Variation of material properties with temperature for 6061-T6 aluminum.

The discussion in the first part of this section addresses the situation where the entire member experiences a uniform change in temperature. For many applications, the change in temperature may vary along the length or through the thickness of a particular member. This circumstance will not only introduce the stress described above but can also introduce a distortion which is undesirable. Consider, for example, the structure in the example of section 8.10 with members on the top and bottom having the same geometry and free at both ends. If the top member experiences a temperature that is 5° cooler than the bottom, the contraction of the top member will be greater and thus cause the complete structure to bend. This introduces a curvature in the structure which may have a detrimental impact for any hardware mounted on it that has stringent pointing constraints. Examples of such hardware include optical instruments and guidance and control hardware such as star trackers or gyros. Table 8.15 presents some fundamental equations that can be used for examining the impact of temperature changes on common structural members.

8.12 Combined Stresses and Stress Ratios

The structural theory presented previously can accurately predict the ultimate strength of a member when subjected to a single loading. In most circumstances, however, the structural member is simultaneously subjected to various combinations of axial, bending, and torsional loading. Calculation of the ultimate strength of members under combined loadings has best been handled by the use of *stress ratios*. For a single load condition, the stress ratio is defined as

$$R = \frac{f}{F} \tag{8.12.4}$$

Table 8.15

Member/Edge Condition	Calculated Parameter	
Uniform plate, edges fixed	Maximum bending stress	$\sigma = \dfrac{\alpha E \Delta T}{(1-v)}$
Bar, rectangular cross section	Free ends, radius of curvature	$R = \dfrac{h}{\alpha \Delta T}$
	Fixed end moments	$M_0 = \dfrac{\alpha EI \Delta T}{h}$
	Fixed end maximum bending stress	$\sigma = \dfrac{\alpha E \Delta T}{2}$
Plate of thickness h (same radius of curvature as above case)	Maximum bending stress	$\sigma = \dfrac{\alpha E \Delta T}{2(1-v)}$
Square plate fixed at ends	Maximum bending stress	$\sigma = \dfrac{\alpha E \Delta T}{2}$

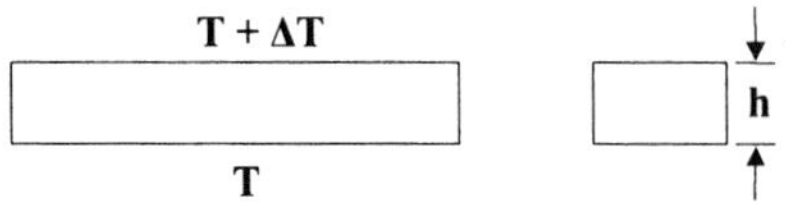

where f is the applied load, moment, or stress, and F the allowable, or critical, load, moment, or stress. For a combined loading condition, the general condition for most failures is

$$R_{\mathrm{c}}^{a} + R_{\mathrm{t}}^{b} + R_{\mathrm{b}}^{c} + R_{\mathrm{st}}^{d} + R_{\mathrm{s}}^{e} + R_{\mathrm{p}}^{f} + \cdots = 1.0 \qquad (8.12.5)$$

where R_{c}, R_{t}, R_{b}, R_{st}, R_{s}, and R_{p} are stress ratios for the compressive, tensile, bending, torsional, shear, and pressure loadings, respectively, and the exponents a, b, c, d, e, $f, \ldots,$ which give the relationship for the combined stresses, are determined by various yield and failure theories or by actual failure tests of combined load systems. The implication of this equation is that failure of a structural member under combined loading will occur only when the sum of the stress ratios (raised to the appropriate power) is equal to or greater than 1.0. Stress ratios and factors of safety for several loading conditions are presented in table 8.16.

Table 8.16a Interaction formulas for columns under combined loading

Loading	Interaction Equation	Factor of Safety
Biaxial tension or compression	$R_x = \frac{f_x}{F}, R_y = \frac{f_y}{F}$	$\dfrac{1}{R_{\max}}$
Axial and bending stress	$R_{\mathrm{a}} + R_{\mathrm{b}} = 1$	$\dfrac{1}{R_a + R_b}$
Normal and shear stress	$R_f^2 + R_{\mathrm{s}}^2 = 1$ $R_f = R_a + R_b$	$\dfrac{1}{\sqrt{R_f^2 + R_{\mathrm{s}}^2}}$
Bending, torsion, and compression	$R_b^2 + R_{\mathrm{st}}^2 = (1 - R_c)^2$	$\dfrac{1}{R_c + \sqrt{R_b^2 + R_{\mathrm{st}}^2}}$
Bending and torsion	$R_b + R_{\mathrm{st}} = 1$	$\dfrac{1}{R_b + R_{\mathrm{st}}}$
Tension and Shear	$R_{\mathrm{t}}^2 + R_{\mathrm{s}}^3 = 1$	

Table 8.16b Interaction formulas for plates under combined loading

Loading	Interaction Equation	Factor of Safety
Biaxial compression	$R_x + R_y = 1$	$\frac{1}{R_x + R_y}$
Longitudinal compression and shear	$R_c + R_s^2 = 1$	$\frac{2}{R_c + \sqrt{R_c^2 + 4R_{st}^2}}$
Longitudinal compression and bending	*Astronautic Structures Manual*, NASA Technical Memorandum X-73305, August 1975	
Bending and Shear	$R_b^2 + R_s^2$	$\frac{1}{\sqrt{R_b^2 + R_{st}^2}}$
Bending, shear, and transverse compression	*Astronautic Structures Manual*, NASA Technical Memorandum X-73305, August 1975	
Longitudinal compression, bending, and transverse compression	*Astronautic Structures Manual*, NASA Technical Memorandum X-73305, August 1975	

Table 8.16c Interaction formulas for cylinders under combined loading

Loading	Interaction Equation	Factor of Safety
Longitudinal compression and pure bending	$R_c + R_b = 1$	$\frac{1}{R_c + R_b}$
Longitudinal compression and torison	$R_c + R_s^2 = 1$	$\frac{1}{R_c + \sqrt{R_c^2 + 4R_{st}^2}}$
Torison and longitudinal tension	$R_{st}^3 - \lvert R_t \rvert = 1$ where $R_t = \frac{\text{applied tensile stress}}{\text{compression buckling stress}}$ and $R_t < 0.8$	—
Pure bending and torison	$R_b^{1.5} + R_t = 1$	—
Pure bending and transverse shear	$R_b^3 + R_s^3 = 1$	—
Longitudinal compression, pure bending, and transverse shear	$R_c + \sqrt[3]{R_s^3 + R_b^3} = 1$	—
Longitudinal compression, pure bending, transverse shear, and torison	$R_c + R_{st}^2 + \sqrt[3]{R_s^3 + R_b^3} = 1$	—
Longitudinal compression, pure bending, and torison	$R_c + R_b + R_{st}^2 = 1$	$R_c + R_b + \sqrt{(R_c + R_b)^2 + 4R_{st}^2}$

Example 8.5

A telescope is mounted on a rigid platform via the three support struts equally spaced as shown. For a weight of 145 lb with the center of gravity indicated, determine whether the strut with the cross-section shown is acceptable if the system is simultaneously subjected to load factors of $\pm 7g$ in the x-direction and $\pm 10.8g$ in the z-direction. Assume a factor of safety of 1.4 and a material ultimate strength of 42,000 psi in both tension and compression.

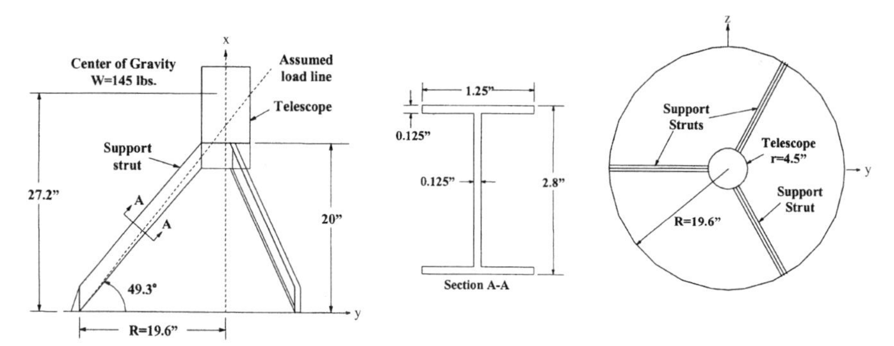

Solution: The design is considered acceptable if the stress levels (including the factor of safety) are within the material allowables. To solve the problem, the maximum loads acting on cross-section A–A must be determined. This is accomplished by determining the reaction loads at the base from which the maximum strut loads can be obtained. The desired solution is obtained by resolving the maximum loads into an axial force, a shear force and a bending moment acting on the cross-section.

The applied forces and moments based on the information given are:

Force in x-direction	$F_x = 145[(1.4)(7)] = 1420\ lb$
Force in z-direction	$F_z = 145[(1.4)(10.8)] = 2190\ lb$
Moment at apex	$M_{\text{apex}} = 145[(1.4)(10.8)](27.2 - 22.8) = 9865\ in\text{-}lb$

Note: The apex is defined as the point of intersection of three assumed load lines.

To determine loads in the struts, the reaction loads at the base due to the above applied loads must be defined. This operation can be broken into the two following categories:

Reaction forces due to F_x

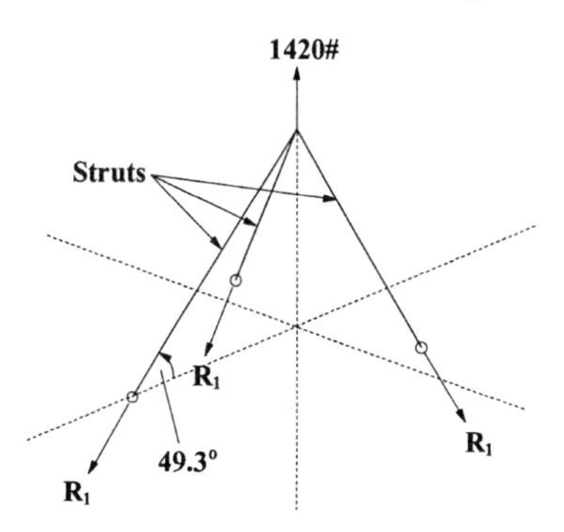

$$3(R_1)(\sin 49.3°) = 1420$$
$$R_1 = 624\,lb$$

Reaction forces due to F_z and M_{apex}

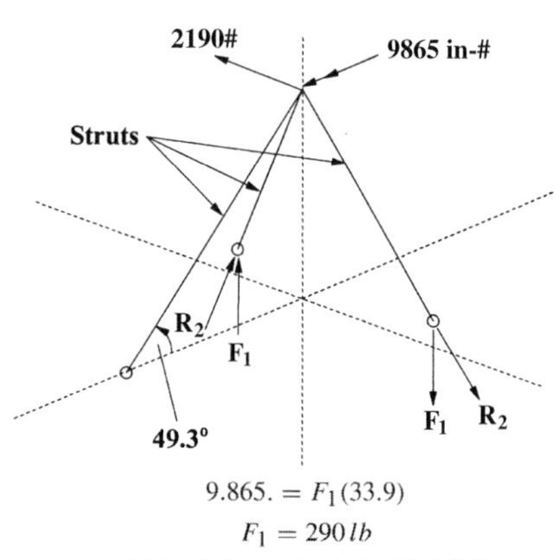

$$9.865. = F_1(33.9)$$
$$F_1 = 290\,lb$$
$$2190(22.5) = (R_2 \sin 49.3°)(33.9)$$
$$R_2 = 1934\,lb$$

Combining results from above, the worst-case loading in the struts is

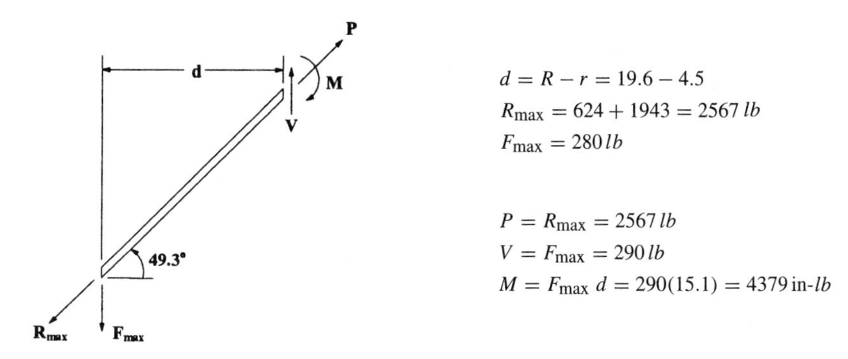

$$d = R - r = 19.6 - 4.5$$
$$R_{\max} = 624 + 1943 = 2567\, lb$$
$$F_{\max} = 280\, lb$$

$$P = R_{\max} = 2567\, lb$$
$$V = F_{\max} = 290\, lb$$
$$M = F_{\max}\, d = 290(15.1) = 4379 \text{ in-}lb$$

Resolving the loads into axial, shear, and bending moments acting on section A–A:

$$P' = P + V \cos 49.3^\circ$$
$$= 2567 + 290 \cos 49.3^\circ$$
$$P' = 2756\, lb$$
$$V' = V \sin 49.3^\circ = 220\, lb$$

For section A–A, the area A and the moment of inertia I can be calculated. The results are $A = 0.95$ in^2 and $I = 0.904$ in^4 (this exercise is left to the student). The maximum stress in the member due to the axial force and bending moment can be calculated using equation 8.8.8.

$$\sigma_{\max} = \frac{2756}{0.95} + \frac{(4379)(1.4)}{0.904} = 2901 + 6782$$
$$\sigma_{\max} = 9863 \text{ psi}$$

The acceptability of the strut design is evaluated by calculating the margin of safety under the given loading condition. For an allowable of 42,000 psi, the margin of safety is

$$\text{MS} = \frac{42{,}000}{9683} - 1 \text{ or MS} = 3.34$$

Thus the strut design given is acceptable.

8.13 Materials

The selection of materials for the spacecraft structure is dependent on many factors which are specific to each program. Mass, stiffness, strength, and stability are primary factors in determining the proper material for the structural elements. The technical requirements, combined with the cost and schedule, must be considered in each application and the choice between the use of more conventional materials like aluminum and more exotic materials like composites is one that must be made early in the design process. To properly make these decisions, it is important for the structural engineer to have a basic understanding of the properties of the various materials. For the more common materials, much of this information is available in numerous reference texts or handbooks.

8.13.1 Metals

Conventional metals such as aluminum, magnesium, and titanium will be sufficient for use on many space applications. One of the major advantages of using these materials is that their properties are well known, homogeneous, and isotropic. Responses and behaviors are therefore predictable and represent little risk for most applications. Mechanical and physical properties for most commonly used aerospace materials can be obtained from a variety of texts or manuals; however, throughout the years, MIL-HDBK-5, Military Standardization Handbook, Metallic Materials and Elements for Aerospace Vehicle Structures, has been the most accepted reference for use in aerospace applications. The current version, MIL-HDBK-5H (Department of Defense, 1983), provides information on a number of common metals such as aluminum, magnesium, and titanium. Mechanical and physical properties are provided for various alloys and a variety of conditions such as varying thickness. Since many of these properties vary somewhat with temperature, data is provided which indicates how the properties will change with temperature. Examples of applications where varying mechanical properties can play a significant role are cryogenically cooled instruments or items such as solar arrays which are exposed to the sun with large areas and small thermal mass. For the latter example, the solar arrays could reach temperatures in excess of 250°C for a mission to orbit the planet Mercury. These kinds of temperatures will not only significantly alter the mechanical properties of metals but have a serious effect on materials such as adhesives. Figure 8.27 provides information taken from MIL-HDBK-5H for 6061 aluminum alloy plate. A definition of the symbols used in specifying material allowables is given in table 8.17.

Some mechanical properties for a number of more commonly used metals are presented in table 8.18. For aluminum and magnesium two different alloys are included to illustrate how the properties can be affected by different composition or processing. The use of any of these materials for a given application involves a tradeoff between the properties, based on the desired need. Note that the alloy of titanium in the chart has a much lower coefficient of thermal expansion (α) than the other materials. Because of this low α, titanium is commonly used for joints in structures using composite materials (see next section). The stainless steel (SS) alloy in the table is not typically used for structural applications but is more commonly used for attachment hardware.

8.13.2 Composite Materials

With many systems in current missions having a greater sensitivity to mass limits and changes that occur on orbit due to temperature differences, conventional materials may not be adequate to satisfy the requirements. For these situations, composite materials

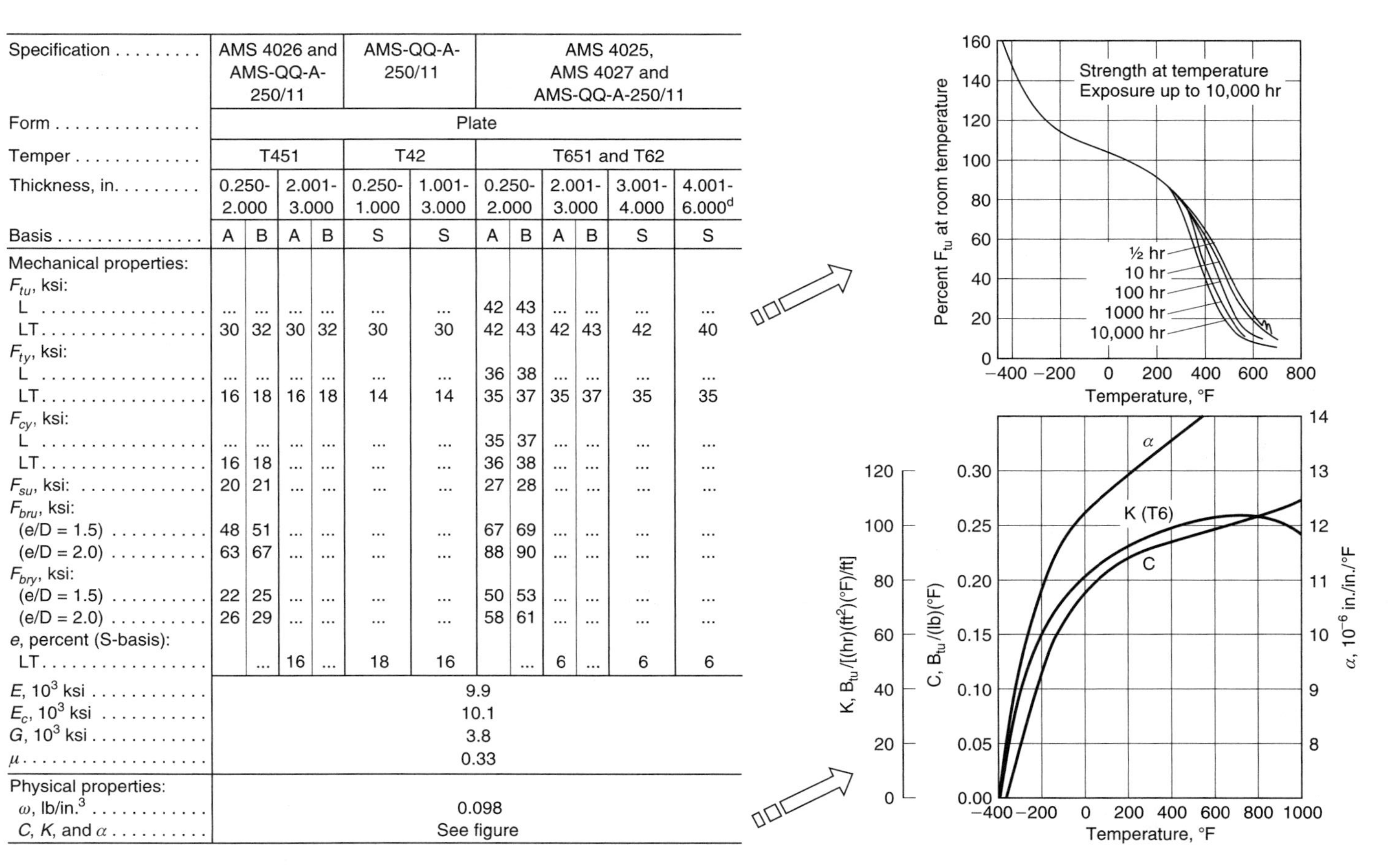

Specification	AMS 4026 and AMS-QQ-A-250/11				AMS-QQ-A-250/11		AMS 4025, AMS 4027 and AMS-QQ-A-250/11					
Form	Plate											
Temper	T451				T42		T651 and T62					
Thickness, in.	0.250-2.000		2.001-3.000		0.250-1.000	1.001-3.000	0.250-2.000		2.001-3.000		3.001-4.000	4.001-6.000[d]
Basis	A	B	A	B	S	S	A	B	A	B	S	S
Mechanical properties:												
F_{tu}, ksi:												
L	...	...	...	...	...	...	42	43	...	...	...	...
LT	30	32	30	32	30	30	42	43	42	43	42	40
F_{ty}, ksi:												
L	...	...	...	...	...	...	36	38	...	...	...	...
LT	16	18	16	18	14	14	35	37	35	37	35	35
F_{cy}, ksi:												
L	...	...	...	...	...	...	35	37	...	...	...	...
LT	16	18	...	...	...	...	36	38	...	...	...	...
F_{su}, ksi:	20	21	...	...	...	...	27	28	...	...	...	...
F_{bru}, ksi:												
(e/D = 1.5)	48	51	...	...	...	...	67	69	...	...	...	...
(e/D = 2.0)	63	67	...	...	...	...	88	90	...	...	...	...
F_{bry}, ksi:												
(e/D = 1.5)	22	25	...	...	...	...	50	53	...	...	...	...
(e/D = 2.0)	26	29	...	...	...	...	58	61	...	...	...	...
e, percent (S-basis):												
LT		...	16	...	18	16		...	6	...	6	6
E, 10^3 ksi	9.9											
E_c, 10^3 ksi	10.1											
G, 10^3 ksi	3.8											
μ	0.33											
Physical properties:												
ω, lb/in.^{3}	0.098											
C, K, and α	See figure											

Figure 8.27 Design mechanical and physical properties of 6061 aluminium alloy plate.

Table 8.17 Symbols used in material specifications

Temper = condition of a material defining its hardness or toughness
A basis = at least 99% of the population will equal or exceed this mechanical property allowable with a 95% confidence level
B basis = at least 90% of the population will equal or exceed this mechanical property allowable with a 95% confidence level
S basis = minimum mechanical property allowable specified by the appropriate specification

L = longitudinal grain direction	LT = long transverse grain direction
F = stress	e/D = ratio of edge distance to hole diameter
t = tension	
c = compression	e = elongation
u = ultimate	E = modulus of elasticity
y = yield	G = shear modulus
s = shear	μ = Poisson's ratio
br = bearing	ω = density
ksi = kilopounds per square inch.	α = coefficient of thermal expansion

have helped to serve this need. Composite materials generally consist of directionally reinforced components such as fibers or whiskers embedded in a homogeneous and isotropic matrix material. Graphite fibers and epoxies, or cyanate ester resins, are commonly used as fibers and matrices, respectively, in many aerospace applications.

A composite material structural element is obtained when several individual layers, or plies, are stacked together to form a structural element or laminate. Each ply of the laminate consists of fibers oriented in a particular direction embedded in a matrix to bond the fibers together. The nomenclature for defining the stacking sequence of a representative laminate is

$$[0_3/90_2/45/-45_3/Z_c]_s$$

For this particular example, the 0, 90, 45 and −45 represent the orientation of the fibers, the subscripts 3, 2, blank (indicating a 1) and 3 represent the number of each of the

Table 8.18 Summary comparison of metal Properties

Material	Al	Al	Mg	Mg	Ti	SS
Alloy	6061	7075	AZ31B	ZK60	6Al/4V	AISI 301
E (psi)	10×10^6	10×10^6	6.5×10^6	6.5×10^6	16×10^6	30×10^6
F_{tu} (ksi)	30–40	70–80	30–40	40–46	130–150	75–185
F_S (ksi)	20–28	40–48	17	22	80–90	50–100
ω (lb/in^3)	0.100	0.100	0.064	0.064	0.160	0.283
α (in/in/°F)	12.5×10^{-6}	12.5×10^{-6}	14×10^{-6}	14×10^{-6}	5×10^{-6}	8.5×10^{-6}
Corrosion resistance	High	Low–High	Mod.	Mod.	High	High

Notes:
(1) E = modulus of elasticity, F_{tu} = tensile ultimate stress, F_S = shear stress, ω = density, and α = coefficient of thermal expansion.
(2) Properties in general are temperature dependent.
(3) Stress corrosion is the combined action of sustained tensile stress and corrosion to cause premature failure of materials.

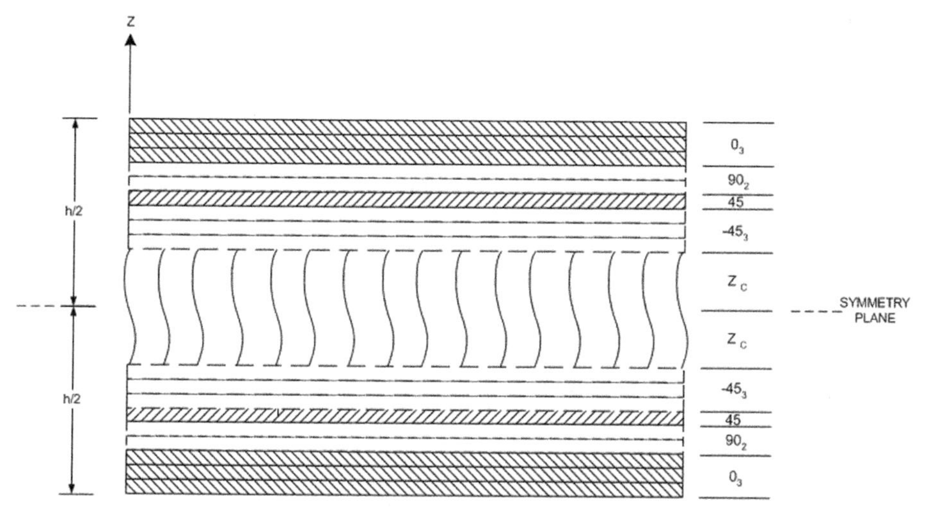

Figure 8.28 Example of a symmetric laminate.

plies, Z_C indicates the laminate has a honeycomb core, and the S indicates the laminate is symmetric. Illustratively, this is shown in figure 8.28. (Note that the ply stacking sequence starts from the bottom of the laminated plate). Individual plies are typically 2.5 to 5 mils thick and the number of plies, ply orientation, and stacking sequence vary, depending on the application.

Now that one has a better understanding of a composite material or plate, it is worth understanding some of the advantages and disadvantages of using these materials. The primary advantage of using composite materials is that the mechanical properties of the laminate can be tailored for the application. By manipulating the fiber orientation, number of plies, and stacking sequence, laminates can be tailored to enhance such properties as stiffness, strength, and/or stability. An additional benefit of using composites is to help in minimizing weight (composites have densities which are typically on the order of 60–70% of that of aluminum). Stability is particularly useful in the design of optical bench structures. In these applications, requirements dictate that the expansion or contraction of a mounting surface must be very small (sometimes on the order of arc-second level) when undergoing a bulk temperature change or thermal gradient. If the coefficient of thermal expansion of the structure can be tailored to near zero, the mounting platform will remain stable during temperature changes. This is especially desirable for instruments or hardware which need to remain pointing or oriented in a specific direction during the mission.

While composite structures have a number of advantages, they unfortunately do not come without shortcomings. In general, composites have poor electrical and thermal conductivity. While recent advances have produced fibers which have greater thermal conductivity, the problem still exists of conducting heat through the matrix into the fiber. From the electrical viewpoint, special measures are often required to ensure that there are no electrostatic discharge (ESD) issues. The resin matrix in a composite will absorb and desorb moisture. Taking on moisture can induce a swelling which more than likely

Table 8.19

	6061-T6 Aluminium	P75/ERL1962
Tensile strength (0°)	42 ksi	132.4 ksi
Tensile strength (90°)	42 ksi	4.27 ksi
Tensile modulus (0°)	10.1 msi	43.3 msi
Tensile modulus (90°)	10.1 msi	1.06 msi
Compressive strength (0°)	35 ksi	64 ksi
Compressive strength (90°)	36 ksi	8.3 ksi
Compressive modulus (0°)	10.1 msi	41.1 msi
Compressive modulus (90°)	10.1 msi	0.9 msi
Shear strength	27 ksi	9.8 ksi
Shear modulus	3.8 msi	0.6 msi
Poisson's ratio	0.33	0.28
CTE (0°)	11.5 ppm/°F	−0.6 ppm/°F
CTE (90°)	11.5 ppm/°F	15.5 ppm/°F

ksi = kilopounds per square inch, msi = million pounds per square inch, ppm = parts per million.

will have a negative impact on a stable structure. Furthermore, if the material desorbs, or gives off, moisture on orbit, this can deposit on sensitive optics or components and greatly impact the performance of the hardware. Because composites normally have coefficients of thermal expansion lower than those of most metals, mounting a composite structure on a metallic structure introduces an incompatibility between adjacent parts of the structure. If both parts of the structure undergo the same change in temperature, one will want to expand or contract a greater amount than the other, thus causing an internal stress to develop. In some situations, the stresses developed can be significant. Finally, composites have higher outgassing characteristics, different failure mechanisms which must be addressed, and are usually more expensive than metal structures for one-of-a-kind applications. Over the past ten years, though, the cost has reduced considerably, enough to make composite materials a more cost-effective alternative in many applications. Although it may seem to appear that the disadvantages outweigh the advantages, with proper precautions and good design the negative impacts can be minimized, thus making the "pluses" outweigh the "minuses."

Conventional materials such as metals are isotropic, that is, they have the same material properties in all directions. For composites, the material properties of an individual ply are highly directional. For a unidirectional ply or laminate, the fibers dictate the properties along the fiber direction while the matrix dictates the cross-wise and through-the-thickness properties. Properties for a unidirectional laminate of P75 carbon fibers in an ERL1962 epoxy matrix are illustrated in table 8.19. A typical aluminum alloy used in space applications is shown for comparison.

The unilateral laminate would be ideal if loading only occurred in a single direction. However, as shown in previous sections, loading on space structures during and after launch can occur in a variety of directions. Since the worst case must always be considered, the composite structure must be designed to carry multiple loading conditions. To deal with this, composite structures are more often configured in what is called a quasi-isotropic configuration. For this configuration, each of the individual plies are

Table 8.20

	Quasi-isotropic P75/ERL1962	Unilateral P75/ERL1962
Tensile strength (0°)	42.8 ksi	132.4 ksi
Tensile strength (90°)	(*)	4.27 ksi
Tensile modulus (0°)	15.2 msi	43.3 msi
Tensile modulus (90°)	15.0 msi	1.06 msi
Compressive strength (0°)	24.3 ksi	64 ksi
Compressive strength (90°)	25.8 ksi	8.3 ksi
Compressive modulus (0°)	13.8 msi	41.1 msi
Compressive modulus (90°)	15.1 msi	0.9 msi
Shear strength	26.2 ksi	9.8 ksi
Shear modulus	(*)	0.6 msi
Poisson's ratio	0.32	0.28
CTE (0°)	−0.12 ppm/°F	−0.6 ppm/°F
CTE (90°)	−0.12 ppm/°F	15.5 ppm/°F

(*) Value unavailable.

oriented in such a way that the properties in the plane of the laminate are very similar. Properties through-the-thickness are still greatly different from the in-plane properties and are dominated by the matrix. Table 8.20 provides a comparison of the properties for a unidirectional and quasi-isotropic configuration made of the same P75/ERL1962 material mentioned above.

The P75 fibers are referred to as high-stiffness fibers in that, comparing the properties with 6061 aluminum, an advantage in material stiffness is realized. Higher strength fibers also exist which provide enhanced strength over some of the more conventional materials. A general comparison of a number of conventional and quasi-isotropic composite material configurations is shown in table 8.21. To provide a more direct comparison, the properties are represented in terms of specific strength and specific stiffness (that is., relative to mass).

Table 8.21

Material	Specific Modulus, GPa (Msi)	Specific Strength, GPa (Msi)
Steel (AISI 4340)	25 (3.7)	230 (33)
Aluminium (7075-T6)	25 (3.7)	180 (26)
Titanium (Ti-6Al-4V)	25 (3.7)	250 (36)
Beryllium	42 (6.2)	260 (38)
E-glass/epoxy	11.2 (1.6)	260 (37)
S-glass/epoxy	15 (0.2)	430 (62)
Kevlar 49/epoxy	21 (3.0)	340 (49)
HS graphite/epoxy	35 (4.1)	300 (44)
HM graphite/epoxy	49 (7.1)	170 (24)
UHM graphite/epoxy	61 (8.8)	130 (19)
Boron epoxy	39 (5.7)	240 (34)

HS = high strength, HM = high modulus, UHM = ultra-high modulus.

8.14 Structural Dynamics

8.14.1 Single Degree of Freedom

In section 8.8, the effect of static (or steady-state) loads on a structure was examined. As mentioned in section 8.5, various dynamic loads are generated on the spacecraft by the launch vehicle. The degree to which these loads affect the spacecraft depend not only on the dynamic characteristics of the spacecraft but also on those of the launch vehicle. It is thus an important part of analysis to determine these dynamic characteristics in order that the launch loads can be accurately predicted. Before one can deal with analyzing the effects of the vibration environment on a complex system such as a spacecraft, an understanding of the effects on a simple system must be achieved.

A brief review of the vibratory environment for the much simpler single-degree-of-freedom system is presented here. Consider the system shown in figure 8.29, which consists of a mass (m), spring (k) and dashpot (c) connected to an immovable support. The position of the mass at any time can be described by the displacement, x, measured from the static equilibrium position. Using Newton's second law, the equation of motion for mass m is

$$m\ddot{x} + c\dot{x} + kx = F(t) \tag{8.14.1}$$

where

$$\left.\begin{aligned} F_m &= m\ddot{x} = \text{inertia force}, \\ F_k &= kx = \text{spring force}, \\ F_c &= c\dot{x} = \text{dissipated force}, \\ F(t) &= \text{applied force}. \end{aligned}\right\} \tag{8.14.2}$$

In many situations a structure is excited by transient forces over a short duration. When the transient force is eliminated, the structure is left to vibrate freely at a certain frequency. For this state of free vibration, and neglecting damping ($c = 0$), the equation of motion reduces to

$$m\ddot{x} + kx = 0 \tag{8.14.3}$$

The general solution to this equation is

$$x = A\cos\omega t + B\sin\omega t \tag{8.14.3}$$

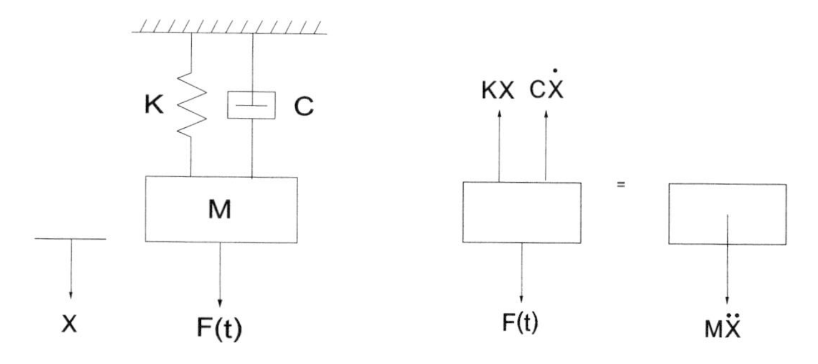

Figure 8.29 Single-degree-of-freedom system.

where A and B are constants of integration and are dependent on the initial conditions. When no forces are applied to the system ($F(t) = 0$) and the system is free to oscillate, $\omega = \omega_n =$ natural frequency, where

$$\omega_\mathrm{n} = (\mathrm{rad/s}) = \sqrt{\frac{k}{m}} \quad \text{and} \quad f_\mathrm{n}(\mathrm{Hz}) = \frac{1}{2\pi}\sqrt{\frac{k}{m}} \tag{8.14.4}$$

In some structural applications, damping cannot be ignored. For this situation the general equation of motion becomes

$$m\ddot{x} + c\dot{x} + kx = 0 \tag{8.14.5}$$

The general solution to this equation is

$$x = Ce^{pt} \tag{8.14.6}$$

Substituting the general solution into the equation of motion gives two values of p for this solution. These are

$$p_{1,2} = -\left(\frac{c}{2m}\right) \pm \sqrt{\left(\frac{c}{2m}\right)^2 - \left(\frac{k}{m}\right)} \tag{8.14.7}$$

The final solution is thus

$$x = Ae^{p_1 t} + Be^{p_2 t}.$$

Three possible conditions exist, depending on the value of the expression under the radical (negative, zero, or positive). The value of c at which the expression under the radical is zero is called *critical damping*, c_c, where

$$c_\mathrm{c} = 2\sqrt{mk} = 2m\,\omega_\mathrm{n} \tag{8.14.8}$$

For structural applications, damping is typically expressed in ratio form in terms of the critical damping. This ratio, ξ, is defined as

$$\xi = \frac{c}{c_\mathrm{c}} = \frac{c}{2\sqrt{mk}} = \frac{c}{2m\omega_\mathrm{n}} \tag{8.14.9}$$

The primary solution of interest in structures is where

$$\left(\frac{c}{2m}\right)^2 \ll \frac{k}{m} \quad \text{or} \quad \frac{c}{c_\mathrm{c}} \ll 1 \tag{8.14.10}$$

The solution with this condition indicates that p is a complex number. Using substitutions such as $e^{ja} = \cos a + j \sin a$, where $j = \sqrt{-1}$, and the expression

$$\omega_\mathrm{d} = \sqrt{\frac{k}{m} - \left(\frac{c}{2m}\right)^2} = \omega_\mathrm{n}\sqrt{1 - \xi^2} \tag{8.14.11}$$

for the damped natural frequency gives the following solution

$$x = \frac{x_0}{1 - \left(\frac{c}{c_\mathrm{c}}\right)} e^{-\left(\frac{c}{c_\mathrm{c}}\right)\omega_\mathrm{n} t} \sin(\omega_\mathrm{d} t + \theta) \tag{8.14.12}$$

This solution is a sinusoidal motion with exponentially decreasing amplitude. Examining the damping ratio for typical values of structural damping ξ:

$$\left.\begin{aligned} \text{for } \xi &= \frac{c}{c_c} = 0.05 \rightarrow\rightarrow\rightarrow\rightarrow \omega_d = 0.09987\omega_n \\ \text{and } \xi &= \frac{c}{c_c} = 0.02 \rightarrow\rightarrow\rightarrow\rightarrow \omega_d = 0.09988\omega_n \end{aligned}\right\} \tag{8.14.13}$$

Thus the damped and natural frequencies are essentially the same for the majority of structural applications.

For a system with a sinusoidal external force, the equation of motion is

$$m\ddot{x} + c\dot{x} + kx = F_0 \sin \omega t \tag{8.14.14}$$

Because the governing equation is linear, the solution for $\xi < 1$ is the sum of the free vibration and a forced vibration at the forcing frequency. The free vibration will damp out, leaving the steady-state solution

$$x_s = x_0 \sin(\omega t - \theta)$$

where, from substitution into the governing equation,

$$x_0 = \frac{F_0/k}{\sqrt{(1 - \frac{\omega^2}{\omega_n^2})^2 + [2\xi(\frac{\omega}{\omega_n})]^2}} = \frac{x_{st}}{\sqrt{(1 - \frac{\omega^2}{\omega_n^2})^2 + [2\xi(\frac{\omega}{\omega_n})]^2}} \tag{8.14.15}$$

where $x_{st} = F_0/k$ is the static deflection due to the force.

This last equation can be rewritten as an amplitude ratio which is referred to as the *transmissibility*, T, and is defined as

$$T = \frac{x_0}{x_{st}} = \frac{1}{\sqrt{(1 - \frac{\omega^2}{\omega_n^2})^2 + [2\xi(\frac{\omega}{\omega_n})]^2}} \tag{8.14.16}$$

Transmissibility, which defines the amplification of an input, is shown as a function of frequency ratio for various values of damping in figure 8.30. Note that the system response is maximum when the forcing frequency is close to the natural frequency. When the system experiences this condition, it is said to be at *resonance*. Note also that resonant peaks occur at approximately the same frequency ratio for low to mid damping ratios. While damping has the effect of reducing the peak it flattens out the part of the curve where $T < 1$. Also note that all curves cross the $T = 1.0$ axis at the same point, where the frequency ratio is $\sqrt{2}$. Since transmissibility represents the ratio of the output response over the input, above the frequency ratio of $\sqrt{2}$ the output response will be lower than the input and thus a degree of isolation is achieved relative to any input force. This is a key aspect in the proper dynamic design of systems.

From the transmissibility equation, assuming a forcing function at resonance $(\omega = \omega_n)$ with typical structural damping of $\xi = 0.05$,

$$T = \sqrt{\frac{1}{(2\xi)^2}} = \frac{1}{2\xi} = 10 \tag{8.14.17}$$

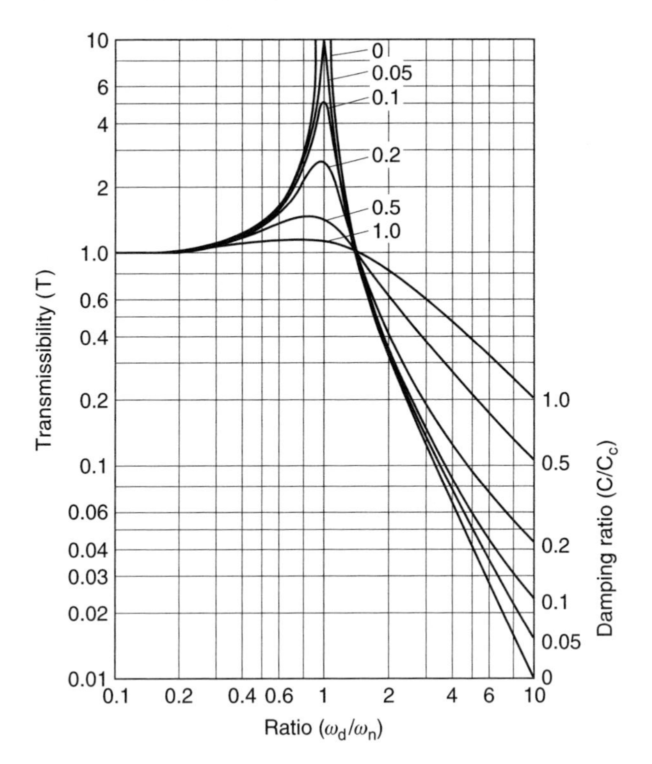

Figure 8.30 Transmissibility for single-degree-of-freedom system.

For many structural applications, this value is typical. Considering the "octave rule" mentioned in section 8.7.2, for $\omega = 2\omega_\text{n}$ and with $\xi = 0.05$,

$$T = \frac{1}{\sqrt{(1-4)^2 + [2(0.05)(2)]^2}} = \frac{1}{\sqrt{9+0.04}} = 0.33 \tag{8.14.18}$$

Thus changing ω/ω_n by 2 decreases the response by a factor of 30! This is the underlying premise of the benefit of using the octave rule in the design of structures.

8.14.2 Continuous Systems

While the single-degree-of-freedom analyses previously presented provide rough-order-magnitude (ROM) estimates, very few spacecraft can be accurately analyzed as single-degree-of-freedom systems. To determine more accurate results as well as to account for the numerous degrees of freedom that exist, the finite element method can be used to provide a more in-depth analysis. Using this method, a spacecraft system, which realistically has an infinite number of degrees of freedom, can be divided into many different elements each having at most six degrees of freedom. This finite number of elements is combined to represent the entire spacecraft structure.

Prior to discussing the application of the finite element method to a dynamics problem, a brief discussion of the dynamics of continuous systems is in order. In many circumstances, the solution to a simple continuous system can be used to estimate the fundamental characteristics of a complicated structure. Although the discussion in this section will be limited to beams, results for other structural configurations will be included in tabular form.

Consider the beam shown below

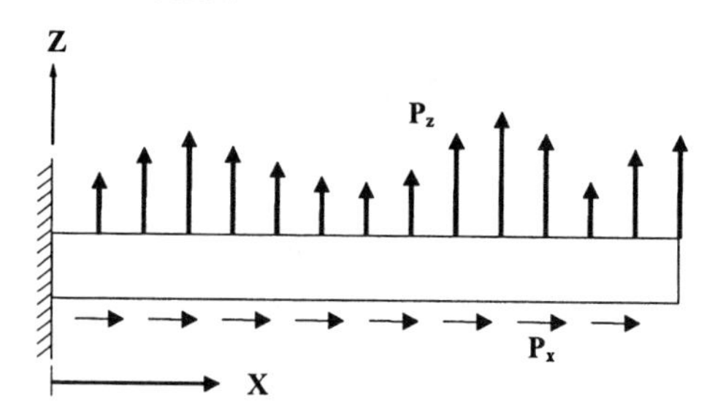

where

$$
\begin{aligned}
m(x) &= \text{mass/unit length of beam} \\
EI(x) &= \text{bending stiffness} \\
E &= \text{modulus of elasticity} \\
I &= \text{area moment of inertia} \\
p_z &= \text{vertical force per unit length} \\
p_x &= \text{horizontal force per unit length}
\end{aligned}
$$

To solve this problem one must consider an elemental section in the undeformed and deformed states. Considering the displacement field for small rotations and the corresponding strain–displacement relations, equilibrium conditions on the deformed element yield (after some manipulation) the fourth-order differential equation determining the beam deformation

$$\frac{\partial}{\partial x}\left(F\frac{\partial w}{\partial x}\right) - \frac{\partial^2}{\partial x^2}\left(EI\frac{\partial^2 w}{\partial x^2}\right) + p_z = 0 \tag{8.14.19}$$

where w = displacement of the beam in the z-direction and F = total axial force acting on the elemental section ($\partial F/\partial x = p_x$). To solve this equation, two boundary conditions are needed at each end of the beam. Some of the more common boundary conditions are presented in table 8.22.

Table 8.22 Element boundary conditions

Boundary Condition	Governing Equation
Hinged end (or pinned end)	$M = EI\frac{\partial^2 w}{\partial x^2} = 0$ and $w = 0$
Clamped or fixed end	$w = \frac{\partial w}{\partial x} = 0$
Free end	$M = EI\frac{\partial^2 w}{\partial x^2} = 0$ and $Q = \frac{\partial}{\partial x}\left(EI\frac{\partial^2 w}{\partial x^2}\right) = 0$
Force P_0 applied at free end	$M = EI\frac{\partial^2 w}{\partial x^2} = 0$ and $Q = -P_0\left(\frac{\partial w}{\partial x}\right)$
Spring (k) at right end (positive w upward)	$M = EI\frac{\partial^2 w}{\partial x^2} = 0$ and $Q = kw$
Spring (k) at left end (positive w upward)	$M = EI\frac{\partial^2 w}{\partial x^2} = 0$ and $Q = -kw$

w = displacement, M = bending moment, Q = shear force.

To formulate the dynamics equation, the vertical applied force p_z can be written as a combination of the applied force f_z and the inertia force $m\ddot{w}$:

$$p_z = f_z - m\ddot{w} \tag{8.14.20}$$

Substituting in the equation above gives

$$(E''Iw)'' - (F'w)' + m\ddot{w} = f_z \tag{8.14.21}$$

where

$$w'' = \frac{\partial^2 w}{\partial x^2} \quad w' = \frac{\partial w}{\partial x} \quad \ddot{w} = \frac{\partial^2 w}{\partial t^2} \quad E'' = \frac{\partial^2 E}{\partial x^2} \quad F' = \frac{\partial F}{\partial x}$$

This equation is the general form of the differential equation that governs the motion of any beam. For no externally applied force ($f_z = 0$) and no axial force ($F_z = 0$), the equation reduces to that of a free vibration problem,

$$(EIw'')'' + m\ddot{w} = 0$$

Using the separation of variables approach and for a constant *EI*, the solution to this equation is

$$w(x, t) = W(x)T(t) \tag{8.14.22}$$

where $W(x) = C \sinh \lambda x + D \cosh \lambda x + E \sin \lambda x + F \cos \lambda x$ and $\lambda = \sqrt[4]{mw^2/EI}$ with $\omega =$ natural frequency.

For a beam with constant *EI* which is clamped (fixed) at the left end and free at the right end, the boundary conditions from table 8.22 are, at $x = 0$,

$$\left.\begin{aligned} w = 0 \rightarrow\rightarrow\rightarrow\rightarrow W = 0 \\ w' = 0 \rightarrow\rightarrow\rightarrow\rightarrow W' = 0 \end{aligned}\right\} \tag{8.14.23}$$

and, at $x = L$,

$$\left.\begin{aligned} M = EIw'' = 0 \rightarrow\rightarrow\rightarrow\rightarrow W'' = 0 \\ Q = EIw''' = 0 \rightarrow\rightarrow\rightarrow\rightarrow W'' = 0 \end{aligned}\right\} \tag{8.14.24}$$

Substituting the boundary conditions and rearranging gives

$$\begin{bmatrix} 0 & 1 & 0 & 1 \\ 1 & 0 & 1 & 0 \\ \sinh \lambda L & \cosh \lambda L & -\sin \lambda L & -\cos \lambda L \\ \cosh \lambda L & \sinh \lambda L & -\cos \lambda L & \sin \lambda L \end{bmatrix} \begin{Bmatrix} C \\ D \\ E \\ F \end{Bmatrix} = 0 \tag{8.14.25}$$

For a non-trivial solution, det [ ] = 0. This gives what is known as the *eigenvalue or characteristic equation*

$$\cos \lambda L = -\frac{1}{\cosh \lambda L} \tag{8.14.26}$$

This equation defines values for $\lambda_r L$ for equation 8.14.22 so that

$$\omega_r = (\lambda_r L)^2 \sqrt{\frac{EI}{mL^4}} \quad (r = 1, 2, 3, \ldots,) \tag{8.14.27}$$

For equation 8.14.25 to be satisfied, det [ ] = 0. The four individual equations used in solving this above are not independent. Using the first three with $F = 1$ (thus producing three independent equations) gives

$$W(x) = \left(\frac{\cosh \lambda L + \cos \lambda L}{\sinh \lambda L + \sin \lambda L}\right)(\sinh \lambda x - \sin \lambda x) - (\cosh \lambda x - \cos \lambda x) \tag{8.14.28}$$

These last two equations define the natural frequencies ω_r for the beam and the relative shape $W(x)$, or mode shape, for the beam at these frequencies.

The above method can be used to solve many beam configurations with various boundary conditions. As previously mentioned, approximate results for somewhat complex structures can be determined by the use of these simple configurations. Similar approaches to that described above can be used to determine the dynamic characteristics of other structures. Table 8.23 contains a number of common cases for beam, plates, and cylinders that can be used to calculate the fundamental natural frequency f.

In the solution of the complex spacecraft many components, each with different dynamic characteristics, interact with each other. A goal of the analyst is to minimize this interaction so that loads are not amplified significantly and that disastrous results do not occur. This can usually be accomplished by careful separation of the critical system frequencies. When conventional methods fail to sufficiently minimize the environment, alternative approaches must be taken. Vibration isolators offer such an alternative.

The basic concept of vibration isolation can be shown by examining the response of a single-degree-of-freedom system. As indicated previously in figure 8.30, the response of such a system is a function of the ratio of the forcing frequency to the system's natural frequency. The measure of response, represented as the transmissibility, or variation of the dynamic output to the dynamic input, was previously defined for a forced vibration problem. For a problem involving structures which are driven at the base by time-varying forces (such as launch vibration), the transmissibility is defined as

$$T = \sqrt{\frac{1 + \{2(\frac{f_d}{f_n})(\frac{c}{c_c})\}^2}{(1 - \frac{f_d^2}{f_n^2})^2 + \{2(\frac{f_d}{f_n})(\frac{c}{c_c})\}^2}} \tag{8.14.29}$$

where f_d = damped natural frequency and f_n = natural frequency (no damping). For small damping ($c/c_c \ll 1$) and at resonance ($f_d/f_n = 1$)

$$T = \frac{1}{2(\frac{c}{c_c})} = \frac{1}{2\xi} \tag{8.14.30}$$

From figure 8.30, it is seen that isolation is attained primarily by maintaining proper frequency relations. When the excitation frequency is greater than the natural frequency

Table 8.23a Beam Fundamental Frequencies

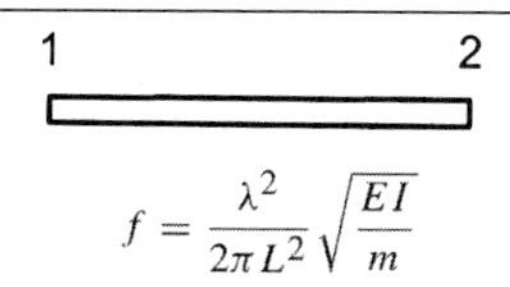

$$f = \frac{\lambda^2}{2\pi L^2}\sqrt{\frac{EI}{m}}$$

where m = mass/unit length

End		λ
1	2	
Free	Free	4.730
Free	Sliding	2.365
Clamped	Free	1.875
Free	Pinned	3.927
Pinned	Pinned	3.142
Clamped	Pinned	3.927
Clamped	Clamped	4.730
Clamped	Sliding	2.365
Sliding	Pinned	1.571
Sliding	Sliding	3.142

Description	Fundamental Natural Frequency
	$\frac{1}{2\pi}\sqrt{\frac{k}{M + 0.33M_s}}$
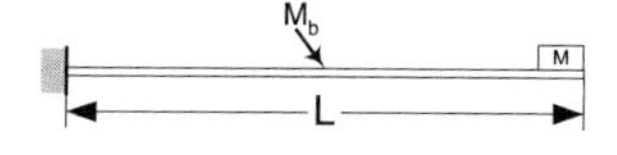	$\frac{1}{2\pi}\sqrt{\frac{3EI}{L^3(M + 0.24M_b)}}$
	$\frac{1}{2\pi}\sqrt{\frac{3EI}{L^3 M_b}\left(1 + \frac{5.45}{1 - 77.4(M/M_b)^2}\right)}$
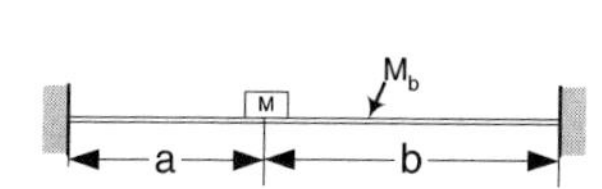	$\frac{4}{\pi}\sqrt{\frac{3EI}{L^3(M + (\alpha+\beta)M_b)}}$ $\alpha = \frac{a}{a+b}\left[\frac{(3a+b)^2}{28b^2} + \frac{9(a+b)^2}{20b^2} - \frac{(a+b)(3a+b)}{4b^2}\right]$ $\beta = \frac{b}{a+b}\left[\frac{(3b+a)^2}{28a^2} + \frac{9(a+b)^2}{20a^2} - \frac{(a+b)(3b+a)}{4a^2}\right]$
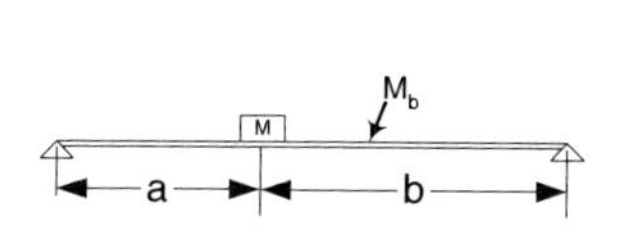	$\frac{1}{2\pi}\sqrt{\frac{3EI(a+b)}{a^2b^2(M + (\alpha+\beta)M_b)}}$ $\alpha = \frac{a}{a+b}\left[\frac{(2b+a)^2}{12b^2} + \frac{a^2}{28b^2} - \frac{a(2b+a)}{10b^2}\right]$ $\beta = \frac{b}{a+b}\left[\frac{(2a+b)^2}{12a^2} + \frac{b^2}{28a^2} - \frac{b(2a+b)}{10a^2}\right]$

Table 8.23b Plate fundamental frequencies

$$f = \frac{\lambda^2}{2\pi a^2}\sqrt{\frac{Eh^3}{12\gamma(1-\upsilon^2)}}$$

where h = plate thickness, γ = mass/unit area, and ν = Poisson's ratio

Description/End Conditions	λ^2 For Fundamental Frequency
Free edge	5.253
Simply supported (SS) edge	4.977
Clamped edge	10.22
Clamped at center, Free along edge	3.752
Clamped at center, SS along edge	14.8
Clamped at center, clamped along edge	22.7
Free edge point mass at center	9.0
SS edge, point mass (M) at center	$\frac{4}{\sqrt{3+4\upsilon+\upsilon^2}}\sqrt{\frac{\gamma\pi a^2}{M}}$ where $M > \gamma\pi a^2$
Clamped edge, point mass at center	10.2

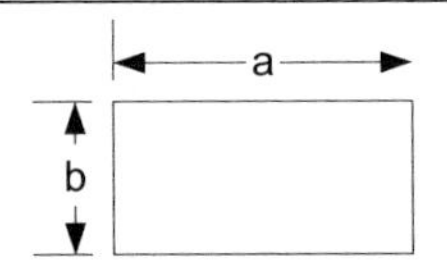

$$f = \frac{\lambda^2}{2\pi a^2}\sqrt{\frac{Eh^3}{12\gamma(1-\upsilon^2)}}$$

where h = plate thickness, γ = mass/unit area, and ν = Poisson's ratio

	a/b				
Boundary Conditions	0.4	0.667	1.0	1.5	2.5
All edges free	3.463	8.946	13.49	20.13	21.64
Long edges free short edges SS	9.76	9.698	9.631	9.558	9.484
Long edges free, short edges clamped	22.35	22.31	22.27	22.21	22.13
All edges SS	11.45	14.26	19.74	32.08	71.56
Long edges clamped, short edges SS	12.13	17.37	28.95	56.35	145.5
All edges clamped	23.65	27.01	35.99	60.77	147.80

Table 8.23c Cylinder fundamental frequencies

Simply Supported Cylindrical Shells

$$f = \frac{\lambda}{2\pi R}\sqrt{\frac{E}{\gamma(1-\upsilon^2)}}$$

where R = cylinder radius to the midsurface, γ = density of cylinder material, ν = Poisson's ratio and L = length of the cylinder

Mode	λ	Mode	λ
Torsion	$\frac{\pi R}{L}\sqrt{\frac{1-\upsilon}{2}}$	Radial	1 for long cylinders, $(L/R) > 8$
Axial	$\frac{\pi R}{L}\sqrt{1-\upsilon^2}$	Bending	$\left(\frac{\pi R}{L}\right)^2\sqrt{\frac{1-\upsilon^2}{2}}$

by a factor of $\sqrt{2}$, the input is significantly reduced or isolated. As the frequency ratio becomes larger, the degree of isolation increases significantly. As indicated in the above equation, the degree of isolation is dependent on damping. Figure 8.30 graphically illustrates the effect that damping has on this relationship.

The function of an isolator is thus to reduce the force transmitted from the vibrating support structure to the mounted equipment. To accomplish this, the isolator stores the incoming energy and dissipates this energy so that the motion to the equipment being supported is reduced, thus providing a level of protection. Properly designed isolators will provide a more comfortable environment which in turn will allow the equipment and systems to function as intended for extended periods of time.

There are two essential features of each isolator: (1) a resilient means of supporting the hardware (spring); and (2) a means by which energy is dissipated (damper). In some types of isolator the two functions are performed by a single element whereas in others a separate element is needed for each function. For analysis purposes, it is assumed that the spring and damper are separate elements. The type of isolator is based on the material from which it is made. Several of the types used are:

(1) *Elastomer isolator* (synthetic and natural rubber)—can be molded into many shapes and stiffnesses, has more internal damping than a metal spring, usually requires a minimum of space and weight.
(2) *Plastic isolator*—characteristics similar to rubber-to-metal type. Low cost, outstanding uniformity, but operating temperature range is limited.
(3) *Metal spring*—commonly used where static deflection required is large, or where temperature or other environmental conditions make elastomer unsuitable.
(4) Others include felt, cork, sponge rubber, and some compound materials—generally cut from large slabs and used flat. They lack ready adaptability of elastomeric parts and are often bonded or cemented in place.

Many factors enter into the selection of an isolator, the primary ones being stiffness and damping. These properties, as well as other information, are generally found in literature supplied by the isolator manufacturer. Other important factors which must also be considered include:

(1) Source of dynamic disturbance—isolate the source or the item?
(2) Type of dynamic disturbance—vibration, shock, or both?
(3) Direction of dynamic disturbance—isolation in one or more directions?
(4) Allowable response of a system to dynamic disturbance—maximum allowable transmitted levels?
(5) Space and location available for the isolator,
(6) Weight and center of gravity of supported equipment—load requirement for the isolator?
(7) Space available for equipment motion,
(8) Ambient environment,
(9) Desired service life,
(10) Fail-safe installation—what happens in the event of an isolator failure?
(11) Cost and availability.

Example 8.6

At different times during the launch scenario, the antenna assembly shown will be subjected to a maximum sinusoidal vibration input of 8.4g in the x-direction and the random vibration environments given in both the x- and z-directions (defined at the base of the assembly). If the structure is made from 6061-T4 drawn tube and the 3/32-in. rivets have an allowable shear load of 217 lb, determine (a) the design load factors in both the x- and z-directions and (b) the design margin that exists for both the tube section and the rivets. An amplification factor of 12 can be used for loads applied at the center of gravity of the antenna. The desired factors of safety for yield and ultimate material allowables are 1.15 and 1.25, respectively.

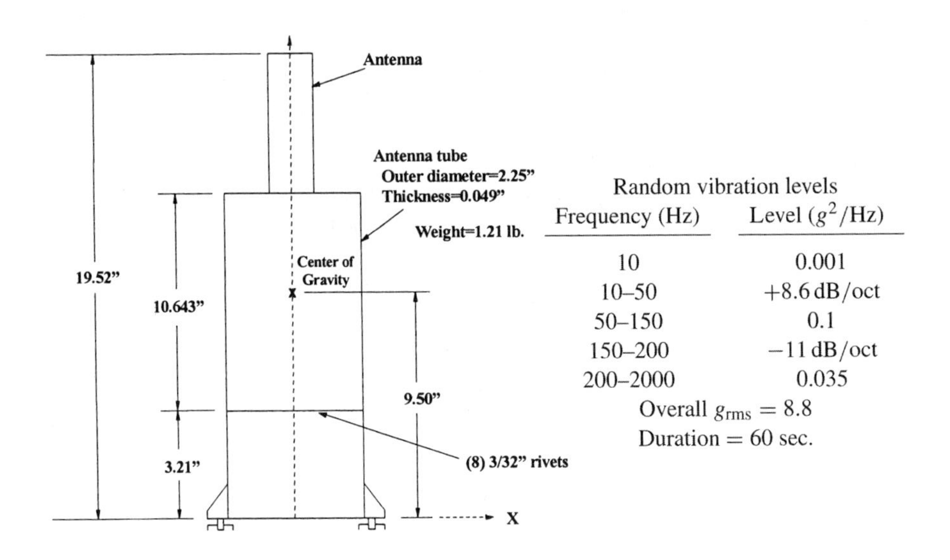

Random vibration levels

Frequency (Hz)	Level (g^2/Hz)
10	0.001
10–50	+8.6 dB/oct
50–150	0.1
150–200	−11 dB/oct
200–2000	0.035

Overall $g_{rms} = 8.8$
Duration = 60 sec.

Frequency determination

To determine the effects due to vibration, the natural frequencies in both directions must be determined.

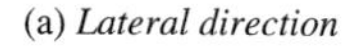

(b) *Axial direction*

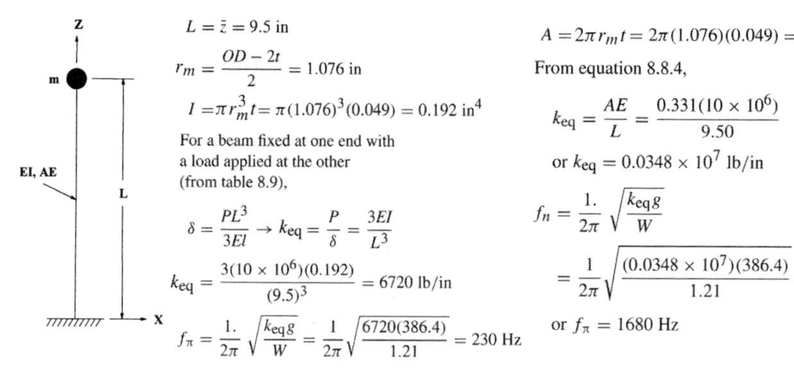

Frequency response

Determine the response at the center of gravity due to the sinusoidal and random inputs at the base, using an amplification factor of 12. These load factors will be used for design parameters.

(a) Sinusoidal (lateral)

$$\ddot{x} = \ddot{y} = (8.4g)(12) = 100g$$

(b) Random (axial and lateral). For both the axial (1680 Hz) and lateral (230 Hz) fundamental frequencies, the spectral input is 0.035 g^2/Hz. The response can be determined by the use of equation 8.17.1 in section 8.17.8. Thus

$$\ddot{x} = 3\sqrt{\frac{\pi Q s_0 f}{2}} = 3\sqrt{\frac{\pi(12)(0.035)(230)}{2}} \quad \text{or} \quad \ddot{x} = 37g$$

and

$$\ddot{z} = 3\sqrt{\frac{\pi(12)(0.035)(1680)}{2}} \quad \text{or} \quad \ddot{z} = 100g$$

The design must be examined for the worst-case environments to preclude any failure. Examining the responses calculated above indicates a design load factor of $100g$ should be used in both the axial and lateral directions. These will be applied separately since the problem states that the maximum sinusoidal and random environments occur at different times.

Stress calculations

(a) Tube structure. The bending moments at the base of the antenna assembly, taking into account the factor of safety, are

$$M = 1.15(1.21)(100)(9.50) = 1322 \text{ in. lb (yield)}$$
$$M = 1.25(1.21)(100)(9.50) = 1435 \text{ in. lb (ult)}$$

The stress can be determined using the general bending equation from equation 8.8.7 with $c = h/2$. Thus

$$\sigma_b = \frac{Mc}{I} = \frac{1435(1.125)}{0.192}$$
$$= 8410 \text{ psi}(ult)$$
$$= 7735 \text{ psi}(yield)$$

The allowable strengths for 6061-T4 drawn tube from MIL-HDBK-5H are

$$F_{cy} = 14{,}000 \text{ psi}$$
$$F_{ty} = 16{,}000 \text{ psi}$$

Using the minimum value, the margin of safety is

$$\text{MS} = \frac{14{,}000}{7735} - 1 = 0.81$$

(b) Rivets. The bending moment at the rivet joint is

$$M = 1.25(1.21)(100)(9.50 - 3.21) = 950 \text{ in. lb(ult)}$$

The shear force acting at the rivet joint is

$$V = 1.25(1.21)(100) = 1500 \text{ lb(ult)}$$

To determine the adequacy of the rivets, the worst-case shear load acting on a rivet must be determined. To do this, the concept of shear flow must be introduced. Simplistically, shear flow can be viewed as a distributed loading which develops on a cross-section of a structural member or between two structural members when they are subjected to a force or bending moment. Shear flow is the incremental loading developed to ensure equilibrium of an infinitesimal element and is defined in terms of force per unit length. When multiplied by the distance over which this loading acts, shear flow produces the total force acting on the section. In this example, for a circular cross-section subjected to bending and shear, the shear flow for each loading situation is defined as

$$q_{zM} = \frac{M}{\pi R^2} \quad \text{and} \quad q_{zV} = \frac{V}{\pi R}$$

where q_{zM} is the shear flow due to bending moment M, q_{zV} is the shear flow due to shear force V, and R is the mean radius of the circular cross-section. Thus

$$q_{zM} = \frac{950}{\pi(1.706)^2} = 260 \text{ lb/in.}$$

and

$$q_{zV} = \frac{150}{\pi(1.706)} = 44.4 \text{ lb/in.}$$

To determine the shear force component due to each loading condition, the distance over which the shear flow acts must be determined. For eight equally spaced rivets, this distance is

$$s = \frac{\pi}{4}R = \frac{\pi}{4}(1.706) = 0.845 \text{ in.}$$

The two shear force components are

$$(F_s)_M = 260(0.845) = 220 \text{ lb(ult)}$$

and

$$(F_s)_V = 44.4(0.845) = 38 \text{ lb(ult)}$$

The two components are then combined to give the resultant shear load which must be reacted by the rivet.

$$F_{sR} = \sqrt{(F_s)_M^2 + (F_s)_V^2} = \sqrt{(220)^2 + (38)^2} = 223 \text{ lb(ult)}$$

As stated in the problem, $(F_s)_{\text{allowable}} = 217$ lb; thus

$$\text{MS} = \frac{217}{223} - 1 = -0.03$$

The design of the rivet spacing is not acceptable. If the number of rivets were doubled, the distance s and thus the resultant shear load would be one-half the previous value. The margin of safety then becomes

$$\text{MS} = \frac{217}{\frac{1}{2}(223)} - 1 = 0.95$$

8.15 Finite-Element Analysis

While the single-degree-of-freedom analyses previously presented provide rough-order-magnitude (ROM) estimates, very few spacecraft can be accurately analyzed as single-degree-of freedom systems. As mentioned earlier, to determine more accurate results as well as to account for the numerous degrees of freedom that exist, finite element analysis (FEA) is used to provide a more in-depth analysis. Using this method, a spacecraft system having an infinite number of degrees of freedom is divided into a finite number of different elements. Each of these elements can be approximated by a finite number of coordinates and degrees of freedom. These finite elements are combined to represent the entire spacecraft structure. An example of a finite element model (FEM) created for NASA's TIMED spacecraft is illustrated in figure 8.31.

In finite element modeling, a structure is defined as an array of locations called grid points, connected by elements having elastic, mass, and damping properties. Each grid point can have six degrees of freedom—three translations and three rotations. Loading conditions may be applied to the structure as forces and moments at these grid points, as well as thermal or enforced deformations to the elements. The model is usually constructed with a knowledge of the anticipated loading. The connections between the grid points are the "finite elements." The complexity of the elements range from the simple one-dimensional rod element to the more complex three-dimensional solid elements.

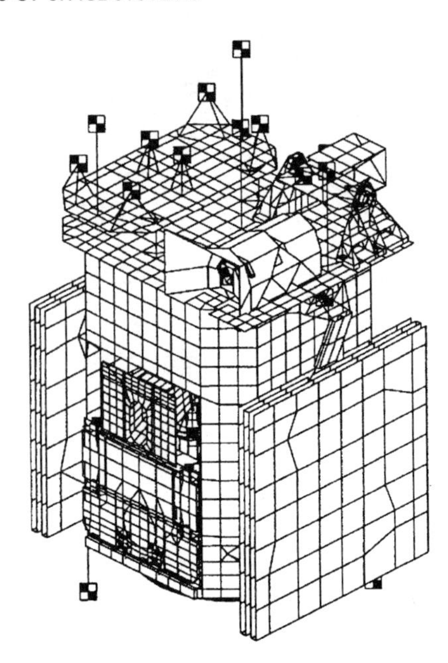

Figure 8.31 TIMED spacecraft finite-element model.

8.15.1 Static Analysis

For the finite element approach a structural member, such as a bar or plate, is modeled as a number of smaller, or differential, elements. For quasi-static analysis, equilibrium is applied and the forces are equated on each of these differential elements. Using the vectoral form of the stress/strain relationship, $\sigma = E\varepsilon$, and applying boundary conditions, an equilibrium equation in vectoral form is created. For many applications this may be the final equation we want to solve. However, for the finite element formulation it is desirable to work in the scalar form. This form is important in finite element analysis because the representation for each element does not vary with any coordinate transformation. Since various local coordinate systems are commonly used in the FEM, element invariance is highly desirable.

To develop the scalar form of the equilibrium equation, the concept of virtual displacement is introduced into the vectoral equation. This virtual displacement is purely fictitious (that is, not related to the actual displacement of the element) and is consistent with the boundary conditions of the element. Applying the virtual displacement to the vectoral equilibrium equation (force equilibrium) and summing the differential elements over the total length, area, or volume leads to an equation of the following general form

$$\delta\pi = \delta U + \delta W = 0 \tag{8.15.1}$$

The terms of this equation involve the actual and virtual displacements and their first derivatives. δU is called the *internal virtual work* and δW is the *external virtual work*. This equation is sometimes referred to as the *weak form* since it involves lower-order derivatives than that of the differential equation formulation.

As an example, the analysis described above for a simple bar element of length L, consisting of two nodes or grid points, one at each end, would yield the following scalar equation

$$\delta\pi = \int_0^L EA \frac{\partial u}{\partial x}\frac{\partial \delta u}{\partial x} dx + \int_0^L f\delta u dx = 0 \tag{8.15.2}$$

where

$$\delta U = \int_0^L EA \frac{\partial u}{\partial x}\frac{\partial \delta u}{\partial x} dx$$

and

$$\delta W = \int_0^L f\delta u dx$$

In this equation x is the coordinate along the length of the bar, u is the axial displacement, and f is the applied load per unit length, that is, $f(x)$. The finite element formulation involves solving this equation for the displacement u for each finite element. For this solution, one must look at the geometry of each element. As shown in the figure below, a local coordinate system is used to define the element geometry that allows one to represent the physical coordinates by a set of local coordinates, or nodes.

$x=x_1$ Node 1 — $x=x_2$ Node 2 → $\xi=-1$ Node 1 — $\xi=1$ Node 2

The use of the local coordinates establishes a relationship to the physical coordinates by what is referred to as "shape functions." For the two-noded bar element, the relationship between the physical and local coordinate is

$$x = \tfrac{1}{2}(1-\xi)x_1 + \tfrac{1}{2}(1+\xi)x_2 \quad \text{or} \quad x = N_1x_1 + N_2x_2 \tag{8.15.3}$$

where x is the physical coordinate, x_1 and x_2 are the coordinates of the two nodes of the element, ξ is the local coordinate, and N_1 and N_2 are the shape functions of the bar element. Note that these shape functions are defined such that they are equal to 1 for the corresponding node, that is, $N_1 = 1$ for node 1, and thus $x = x_1$ at node 1. In most finite element applications, these same shape functions are also used for mapping the displacement to the local coordinates. Using the same mapping function for the element coordinates and displacements is referred to as the *isoparametric* formulation. Again, for the two-noded bar element, the displacement is

$$u = N_1u_1 + N_2u_2 \tag{8.15.4}$$

Using these definitions for the element coordinates and displacements in the first part of equation 8.15.2 (the internal virtual work) yields the stiffness matrix for the individual elements. For the bar example shown, the element stiffness can be shown to be

$$\widetilde{K} = \frac{EA}{L}\begin{bmatrix} 1 & -1 \\ -1 & 1 \end{bmatrix} \tag{8.15.5}$$

Similarly, from the second part of equation 8.15.2 (the external virtual work), the element nodal load vector is determined. This process is conducted for each of the finite elements that make up the structural system. Once all of the individual element stiffness and load vectors are determined, they are combined together to provide a representation for the stiffness and loading on the complete structure. The system-level parameters are referred to as the "global" stiffness and load vectors. In finite-element terminology, this summing process is called "assembly" of the element stiffness matrices and element nodal load vectors.

From equation 8.15.2, with displacement $u = q$, and using equations 8.15.4 and 8.15.5, a scalar equation of the following general form results:

$$\delta\pi = \delta\tilde{q}^{\mathrm{T}}(\widetilde{K}\tilde{q} - \widetilde{F}) = 0 \tag{8.15.6}$$

In this equation, K is an $n \times n$ matrix, with n being the number of nodes or grid points, and both $\tilde{q}$ and $\widetilde{F}$ are matrices of size $n \times 1$. For this equation to be satisfied,

$$\widetilde{K}\tilde{q} - \widetilde{F} = 0 \tag{8.15.7}$$

This represents the equation solved by the FEM. To do so, boundary conditions are first applied to this equation, thus reducing the overall size of the matrices. The resulting expressions, called the *discretized equilibrium equations*, are solved for the displacement q. Having determined the displacement, one can compute the element strains and stresses according to the strain–displacement and stress–strain relations.

For the more general case, the scalar equation takes the following form:

$$\begin{aligned}\delta\pi &= \{\delta U\} + [\delta W] \\ &= \left\{\int \delta\tilde{\varepsilon}^{\mathrm{T}}\widetilde{C}\tilde{\varepsilon}\,dV - \int_{\mathrm{V}} \delta\tilde{\varepsilon}^{\mathrm{T}}\widetilde{C}\tilde{\varepsilon}^{0}\,dV\right\} + \left[-\int_{\mathrm{V}} \delta\tilde{u}^{\mathrm{T}}\overline{F}_{B}\,dV - \int_{\mathrm{S}} \delta\tilde{u}^{\mathrm{T}}\overline{T}\,dS\right] = 0\end{aligned} \tag{8.15.8}$$

where $\tilde{\varepsilon}^{0} \equiv$ initial strain (for example, thermally induced strains), $\overline{F}_{\mathrm{B}} \equiv$ body forces (for example, gravitational forces), $\overline{T} \equiv$ traction forces (that is, forces applied to the boundary), and the volume V can be partitioned into ΣV_i for i distinct elements. This general equation can be applied to any problem in one, two, or three dimensions. The first integral of this equation yields the stiffness matrix while the last three integrals have no unknown displacement and represent force terms; therefore, these produce the nodal load vector. Note that the scalar equation presented earlier is a simplified form of the above general equation in one dimension. For the above general equation, the internal virtual work, or work done by internal forces, is determined from the first two terms whereas the external virtual work, or work done by external forces, is obtained from the last two terms.

As an example of the more general case, consider the two-dimensional plate problem. The local coordinate system used to define the geometry and displacements similar to the above example is

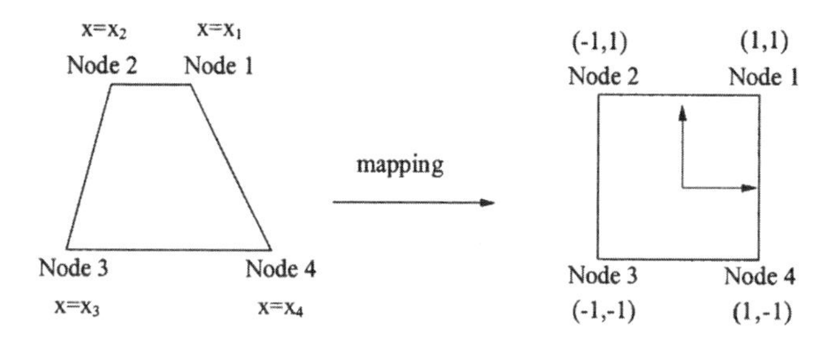

For this situation, each node has two degrees of freedom and two local coordinates, ξ and η, to represent the coordinates of each node. Using this transformation, the element coordinate (x, y) and displacement functions (u, v) can then be expressed as follows:

$$x = \sum_{1}^{4} N_i x_i \ \& \ y = \sum_{i}^{4} N_i y_i \rightarrow$$

$$\begin{Bmatrix} x \\ y \end{Bmatrix} = \begin{bmatrix} N_1 & N_2 & N_3 & N_4 & 0 & 0 & 0 & 0 \\ 0 & 0 & 0 & 0 & N_1 & N_2 & N_3 & N_4 \end{bmatrix} \begin{Bmatrix} x_1 \\ x_2 \\ x_3 \\ x_4 \\ y_1 \\ y_2 \\ y_3 \\ y_4 \end{Bmatrix} = \tilde{N}\tilde{x} \tag{8.15.9}$$

$$u = \sum_{1}^{4} N_i u_i \ \& \ v = \sum_{1}^{4} N_i v_i \rightarrow$$

$$\begin{Bmatrix} u \\ v \end{Bmatrix} = \begin{bmatrix} N_1 & N_2 & N_3 & N_4 & 0 & 0 & 0 & 0 \\ 0 & 0 & 0 & 0 & N_1 & N_2 & N_3 & N_4 \end{bmatrix} \begin{Bmatrix} u_1 \\ u_2 \\ u_3 \\ u_4 \\ v_1 \\ v_2 \\ v_3 \\ v_4 \end{Bmatrix} = \tilde{N}\tilde{q} \tag{8.15.10}$$

where the shape functions are

$$\left. \begin{aligned} N_1 &= \frac{1}{4}(1+\xi)(1+\eta), \quad N_2 = \frac{1}{4}(1-\xi)(1+\eta) \\ N_3 &= \frac{1}{4}(1-\xi)(1-\eta), \quad N_4 = \frac{1}{4}(1+\xi)(1-\eta) \end{aligned} \right\} \tag{8.15.11}$$

From equation 8.15.8, using the first term of the internal virtual work, the two-dimensioal strain—displacement relations, and equations 8.15.9 through 8.15.11, the stiffness matrix for the two-dimensional 4-noded element is

$$\widetilde{K}^i = \int_{-1}^{1}\int_{-1}^{1} \widetilde{B}^{\mathrm{T}}\widetilde{C}\widetilde{B}|\widetilde{J}|t \; d\xi \; d\eta \tag{8.15.12}$$

where $\widetilde{B}$ is the matrix that relates strains to displacements ($\tilde{\varepsilon} = \widetilde{B}\tilde{q}_i$) in two dimensions (defined in local coordinates) and, $|\widetilde{J}|$ the Jacobian matrix, is

$$\widetilde{J} = \begin{bmatrix} \dfrac{\partial x}{\partial \xi} & \dfrac{\partial y}{\partial \xi} \\ \dfrac{\partial x}{\partial \eta} & \dfrac{\partial y}{\partial \eta} \end{bmatrix} \tag{8.15.13}$$

which is a function of geometry only. The definition of the $\widetilde{B}$ matrix is beyond the scope of this text but can be found in any reference on the finite element method.

From the first term of the internal virtual work in equation 8.15.8, with no initial strain and applied force, the nodal load vector $\{Q^i\}$, is

$$\widetilde{Q}^i = \int_{-1}^{1}\int_{-1}^{1} \widetilde{N}^{\mathrm{T}}\widetilde{F}_B|\widetilde{J}|t \; d\xi d\eta \tag{8.15.14}$$

As indicated before, the individual element stiffness and nodal load vectors are combined to form the global stiffness and load vectors, respectively. Once the global matrices are assembled and the boundary conditions applied, the displacements can be determined from $[K]\{q\} = \{F\}$. Again, the displacements, strains, and stresses will be computed by using the strain–displacement and stress–strain equations.

In summary, the general steps taken in the solution of a problem by the finite element method are:

(1) Representation of the structure by a finite number of nodes and elements.
(2) Computation of the element stiffness matrices $[K^i]$ and the element nodal load vectors $\{Q^i\}$ via the use of local coordinate representation.
(3) Assembly of the element matrices to form the global matrices $[K]$ and $\{F\}$ for the complete system.
(4) Application of the boundary conditions to reduce the size of $[K]$ and $\{F\}$.
(5) Solution of $\{K\}\{q\} = \{F\}$ for $\{q\}$.
(6) Calculation of strain from the strain–displacement equations.
(7) Calculation of stress from the stress–strain equations.

The stresses resulting from this FE solution are due to the applied loading conditions such as those that would occur during the launch sequence. These stresses are then compared to material allowables (with appropriate factors of safety) to determine compliance of the design with the requirements.

8.15.2 Dynamic Analysis

The formulation of a dynamics problem in the finite element approach requires the introduction of mass and acceleration into the problem. This involves the determination of a mass matrix for each element; these mass matrices are in turn combined to give the global mass matrix. To determine the mass matrix, examine the third integral of equation 8.15.8. From this, with $F = ma$ and recalling that a is the second derivative of displacement with respect to time, one gets

$$-\int_V \delta\tilde{u}^{\mathrm{T}}\bar{F}_{\mathrm{B}}dV = \int_V (X_{\mathrm{B}}\delta u + Y_{\mathrm{B}}\delta v + Z_{\mathrm{B}}\delta w)dV$$
$$= \int_V \rho(\ddot{u}\delta u + \ddot{v}\delta v + \ddot{w}\delta w)dV \qquad (8.15.15)$$

where

$$X_{\mathrm{B}} = -\rho\ddot{u}, \quad Y_{\mathrm{B}} = -\rho\ddot{v}, \quad Z_{\mathrm{B}} = -\rho\ddot{w} \qquad (8.15.16)$$

represents the inertia effect with ρ = mass per unit volume and

$$F_{\mathrm{B}} \equiv \begin{pmatrix} X_{\mathrm{B}} \\ Y_{\mathrm{B}} \\ Z_{\mathrm{B}} \end{pmatrix} \quad \text{and} \quad u = \begin{pmatrix} u \\ v \\ w \end{pmatrix} \qquad (8.15.17)$$

are the components of the body force/volume and the displacement components, respectively.

For the finite element approach, similarly to the previous analyses, it can be shown that

$$\ddot{u} = [N]\{\ddot{q}_u\}, \quad \ddot{v} = [N]\{\ddot{q}_v\}, \quad \ddot{w} = [N]\{\ddot{q}_w\} \qquad (8.15.18)$$

Similar expressions can be defined for δu, δv, and δw, which, upon substitution into equation 8.15.15 and with

$$\widetilde{m} = \int_V \rho\widetilde{N}^{\mathrm{T}}\widetilde{N}dV, \qquad (8.15.19)$$

give the following expression

$$\int_V \rho(\ddot{u}\delta u + \ddot{v}\delta v + \ddot{w}\delta w)dV = \{\delta\widetilde{q}_u^{\mathrm{T}} \quad \delta\widetilde{q}_v^{\mathrm{T}} \quad \delta\widetilde{q}_w^{\mathrm{T}}\} \begin{bmatrix} \widetilde{m} & 0 & 0 \\ 0 & \widetilde{m} & 0 \\ 0 & 0 & \widetilde{m} \end{bmatrix} \begin{Bmatrix} \widetilde{q}_u \\ \widetilde{q}_v \\ \widetilde{q}_w \end{Bmatrix} \qquad (8.15.20)$$
$$= \delta\widetilde{q}_i^{\mathrm{T}}\widetilde{M}^i\widetilde{q}_i$$

where $[M^i]$ is referred to as the "consistent" mass matrix of the ith element.

For the two-noded bar element previously examined with $v = w = 0$,

$$\widetilde{m} = \widetilde{M}^i = \int_V \rho\widetilde{N}^{\mathrm{T}}\widetilde{N}\,dx\,dy\,dz = \int_{x_1}^{x_2} m\widetilde{N}^{\mathrm{T}}\widetilde{N}\,dx \qquad (8.15.21)$$

where $m = \int \rho \, dz \, dy$ = mass per unit length. Solving this equation for the mass matrix gives

$$\widetilde{M}^i = \int_{-1}^{1} m \widetilde{N}^{\mathrm{T}} \widetilde{N} \left(\frac{L}{2}\right) d\xi = \frac{L}{2} \int_{-1}^{1} m \begin{Bmatrix} \frac{1}{2}(1-\xi) \\ \frac{1}{2}(1+\xi) \end{Bmatrix} \left[\tfrac{1}{2}(1-\xi) \quad \tfrac{1}{2}(1+\xi)\right] d\xi$$

$$= \frac{L}{2} \int_{-1}^{1} m \begin{Bmatrix} \frac{1}{4}(1-\xi)^2 & \frac{1}{4}(1-\xi^2) \\ \frac{1}{4}(1-\xi^2) & \frac{1}{4}(1+\xi)^2 \end{Bmatrix} d\xi \tag{8.15.22}$$

If m is constant, the bar element mass matrix reduces to

$$\widetilde{M}^i = mL \begin{bmatrix} \frac{1}{3} & \frac{1}{6} \\ \frac{1}{6} & \frac{1}{3} \end{bmatrix} \tag{8.15.23}$$

Similarly, for the four-noded plate element,

$$m = \int_V \rho \widetilde{N}^{\mathrm{T}} \widetilde{N} dV = \int_A m \begin{Bmatrix} N_1 \\ N_2 \\ N_3 \\ N_4 \end{Bmatrix} [N_1 \quad N_2 \quad N_3 \quad N_4] |\widetilde{J}| d\xi \, d\eta$$

$$= \int_A \begin{bmatrix} N_1 N_1 & N_1 N_2 & N_1 N_3 & N_1 N_4 \\ N_2 N_1 & N_2 N_2 & N_2 N_3 & N_2 N_4 \\ N_3 N_1 & N_3 N_2 & N_3 N_3 & N_3 N_4 \\ N_4 N_1 & N_4 N_2 & N_4 N_3 & N_4 N_4 \end{bmatrix} |\widetilde{J}| \, d\xi \tag{8.15.24}$$

where $m = \int \rho \, dz$ = mass per unit area and the shape functions are as defined earlier.

Having now formed the mass matrices along with the stiffness matrices and load vectors, the finite element solution to the dynamics problem is obtained by solving

$$[M]\{\ddot{q}\} + [K]\{q\} = F(t) \tag{8.15.25}$$

with given initial conditions

$$q(0) = q_0 \quad \text{and} \quad \dot{q}(0) = \dot{q}_0 \tag{8.15.26}$$

The solution of the dynamics problem gives the system's natural frequencies and corresponding mode shapes.

8.16 Launch Loads — Dynamic Coupling Analysis

The main objective of the coupling analysis is to determine the spacecraft's dynamic response to a simulated launch environment. To accomplish this, the spacecraft is represented in terms of its eigenvalues (frequencies) and eigenvectors (characteristic shapes). This model is combined with the model of the launch vehicle and subjected to the analytic representation of critical forcing functions. These forcing functions simulate the loading imposed during the launch sequence. The responses of the spacecraft (load, deflection, acceleration) are determined and used to update the structural design parameters that were previously imposed (for example, the load factors described in table 8.4 of section 8.7.1). A verification cycle is usually conducted several times so that the loads are updated using the most current information available. A final verification is sometimes performed after environmental testing of the spacecraft has been completed and the model updated to reflect test results.

8.17 Structural Test Verification

Although the analysis efforts are essentially complete with the fabrication of hardware, the worries of the structural engineer are not completely over. Once the hardware is assembled (either at component, subsystem, or assembly level), it is subjected to a series of tests that attempt to simulate the environment that the structure will experience during launch or subsequent mission operations. These are the same environments the engineer has used for design purposes. If the engineer has done the job thoroughly, this series of tests may be a very uneventful formality. In some cases, even though the analytical efforts have been performed well, "surprises" may occur. When such is the case, the structural engineer becomes involved in determining a solution to the problem.

The environmental testing conducted during the latter phases of the program is performed to demonstrate that the hardware can withstand the structural loading due to the predicted environments, to provide a screening for the hardware (which can detect workmanship and other problems), and to demonstrate interaction of components as is done at the subsystem and system levels. Testing is usually conducted at all levels—from component to spacecraft system—with a desire to perform as much testing early on in order to preclude problems at a higher level of assembly.

At component-level testing, one of the primary goals of testing is to screen the equipment in order to minimize problems that might occur after integration with the spacecraft. Discovery of problems at this stage of the program can serve to avoid unnecessary delays which would have occurred if this anomaly were discovered at the system level. Spacecraft-level tests are designed to demonstrate that the entire system, once assembled, can withstand the predicted environments. At this level, the interaction between components and spacecraft structure is more realistically tested.

The common types of test that are typically performed on space hardware are: (1) strength testing; (2) sinusoidal vibration, both at a lower input or survey level and at a higher amplitude; (3) sinusoidal dwell or sinusoidal burst; (4) random vibration, both at full level and at workmanship level; (5) acoustic; (6) shock; (7) modal survey; and (8) spin balance. Each of these types of test is described in detail in sections 8.17.1 through 8.17.14. The exact tests to be conducted for each program can vary and are tailored to suit the program requirements and objectives. In general, major structures are usually strength tested separately, components are usually exposed to sinusoidal and random vibration (and on occasion shock testing), while complete spacecraft systems are exposed to sinusoidal vibration and acoustics testing. In some cases random vibration testing is substituted for acoustic testing at the spacecraft level since both tests serve to accomplish the same purpose. Each of these tests will be discussed further in the subsequent sections.

8.17.1 Strength Testing

Strength testing is typically performed at a component or subsystem level and is conducted to demonstrate that the test article can withstand the worst-case expected load conditions that the item will experience during launch. For primary structures, static loads testing is a common way of accomplishing this goal. During these tests, the structure is attached to a somewhat rigid fixture and pulled or pushed with hydraulic actuators and load cells. Instrumentation typically consists of displacement and strain gages mounted in numerous places to ensure the article is tested to the proper loads while not exceeding the allowable limits.

Strength testing of the structure can also be accomplished using an electrodynamic shaker. In this situation, mass simulators representing the components mounted to the structure are used. The shaker imparts a high level of acceleration which in turn generates the required loading into the structure. Accelerometers are used to measure the responses at numerous locations to verify the correct loading has been achieved.

8.17.2 Vibration Testing

Vibration testing is conducted by use of an electrodynamic shaker and, for most organizations, a digital control system. The input signal which drives the shaker is measured by an accelerometer, commonly referred to as the control accelerometer. Most vibration systems use a closed-loop system for controlling the tests. In this setup, a signal is generated by the digital control system and sent to the control accelerometer. The response of this accelerometer is, in turn, fed back into the control system to determine whether the test is being performed at the appropriate level. If the input is not at the proper level the control system adjusts the signal sent to the control accelerometer to bring the test input within the specification. In all testing, a separate accelerometer, located next to the control accelerometer, is used as an independent monitor of the control input. This system has the capability of automatically aborting the test should the level of the control accelerometer exceed the test specification. During most tests, additional instrumentation is located on the test article to measure the response of the hardware to the input. From this, the natural frequency and amplification can be measured. Most vibration testing is conducted in three orthogonal axes of the test article.

During vibration testing of the hardware, system responses may exceed the worst-case expected flight loads at the primary structure natural frequencies. To preclude structural failures due to unrealistic test levels that may occur at the natural frequencies, the input levels are limited at appropriate locations so that loads in critical structural members do not exceed the maximum expected flight loads. When limiting of the test occurs, control is temporarily transferred from the control accelerometer (typically mounted at the base of the test article) to this limiting accelerometer during the time in which the maximum loads would be exceeded. Once the loading reduces below the maximum allowed level, the control of the test is transferred back to the original control accelerometer. The maximum expected flight loads on which this testing is based are determined from the launch vehicle/spacecraft coupling analysis. A photograph of the NEAR spacecraft undergoing vibration testing is shown in figure 8.32.

Figure 8.32 NEAR spacecraft vibration test.

8.17.3 Sinusoidal Vibration

Sinusoidal vibration (simple harmonic motion) is periodically varying motion specified by amplitude (displacement, velocity, or acceleration), frequency, and phase angle. This testing is conducted over a specified frequency range with input levels defined at each specific frequency. Types of sinusoidal vibration tests include the sinusoidal survey test, the sinusoidal sweep test, the sinusoidal burst or pulse test, and the sinusoidal dwell test.

8.17.4 Sinusoidal Survey

The sinusoidal survey test is a low-amplitude test conducted through a specified frequency range at a prescribed rate. The main purpose of this test is to identify the primary structural resonances of the test article and amplification factors that occur at these resonances. The test is performed by sweeping at the specified rate from the beginning (lower) frequency and level through the final (higher) frequency and level. In some circumstances, the sweep is also conducted starting from the higher frequency region and sweeping to the lower frequency region because of the slightly different dynamic characteristics that exist in approaching a resonant condition from the higher frequency region. The sinusoidal survey is usually conducted at the beginning and end of major tests and serves as a means for detecting changes in dynamic characteristics of the system. Differences between pre- and post-survey testing imply that some aspect of the system

Table 8.24 Typical sinusoidal survey test levels

Frequency (Hz)	Level
10–2000	$0.5g$
All axes	
Rate = 4 oct/min	

has changed as a result of the higher-level testing and should be investigated before any further testing. The changes could be as minor as the loosening of some attachment hardware or as significant as the failure of a structural member. Test levels are typically in the range of $0.25g$ to $0.5g$ and sweep rates are normally either 2 or 4 octaves per minute. While the test frequency range can vary, many survey tests are conducted between 10 and 2000 Hz. A typical specification for a sinusoidal survey test is presented in table 8.24.

8.17.5 Sinusoidal Sweep

The sinusoidal sweep test is similar in nature to the sine survey except that the amplitudes are greater in magnitude and generally vary with different frequency regions. These tests are normally conducted over a smaller frequency region than the sinusoidal survey. The primary function of this test is to demonstrate that the hardware will endure the worst-case loading which may develop. During this test, the test article is subjected to higher acceleration input levels (and thus higher forces) in spectral regions of primary structural or vibration concern. This amplified level may be the result of a launch vehicle input, a primary structural resonance, or a locally induced perturbation. Resonant responses of the structure are monitored and if deemed unrealistic are limited as indicated in section 8.17.2 so as not to exceed the worst-case predicted flight loads. As with the sinusoidal survey tests, these tests may also be performed by a forward or backward sweep. A typical input for the swept sine vibration test is shown in table 8.25.

8.17.6 Sinusoidal Burst

The sinusoidal burst, or pulse, test is one that provides a convenient way to simulate a static load or steady-state condition on the test item in a laboratory environment. This

Table 8.25 Representative sinusoidal sweep test levels

Frequency (Hz)	Level
5–23	0.4 in. DA
23–100	$11.3g$
100–200	$6.4g$
All axes	
Rate = 4 oct/min	
DA = double amplitude	

Table 8.26 Representative sinusoidal burst test specification

Frequency (Hz)	Level
15	14.0g
Test duration = 6½ cycles at max. level	

test is short in duration and is conducted at a specified frequency for a specified number of cycles. The frequency at which the test is performed is chosen to be well below the fundamental frequency of the test item so as to avoid any dynamic amplification during the test. As a rule of thumb, this test should be performed at a frequency that is less than or equal to 1/3 of the resonant frequency of the test article. The primary function of this test is basically the same as that of the high-amplitude sinusoidal sweep test in that it demonstrates the test article can withstand the high loads environment. In this test, however, the test article is not exposed to the larger number of vibratory cycles, thus minimizing any concern over fatigue. Unlike other sinusoidal testing, this testing is conducted without the benefit of closed-loop control. This is because the signal that drives the shaker is a simple short-duration pulse input signal that cannot be updated via feedback because of its nature and duration. A specification for a sine pulse test must include the test level, frequency, and duration. For a system with a natural frequency greater than 45 Hz, a representative sinusoidal burst test specification is shown in table 8.26.

8.17.7 Sinusoidal Dwell

The sinusoidal dwell test is a high-amplitude, constant-frequency sinusoid applied at the resonant frequency of the test article over a prescribed time interval. The primary function of this test is to provide a thorough and rigorous design test of the test article. This type of test usually introduces unnecessary risks and is seldom considered a realistic representation of the actual environment because it is performed at the resonant frequency of the test article. A more realistic approach to the sinusoidal dwell test would be if the test frequency were based on resonant frequencies of the mounting structure instead of the test article. Similarly to the sinusoidal burst test, the dwell test is also performed with an open control loop. Specifications for the sinusoidal dwell test are similar to those of the sinusoidal burst test except that the duration is usually specified as an actual time versus a number of cycles.

8.17.8 Random Vibration

Random vibration is a vibration test whose instantaneous magnitude cannot be explicitly defined and one which theoretically has all frequencies present during the complete test. Random vibration contains non-periodic or quasi-periodic components and thus an exact value at a future time cannot be predicted. The instantaneous magnitude is specified by probability distribution functions giving the mean-square value that lies within a specified frequency range. Random vibration tests, typically performed over a specified time period, are viewed as energy inputs to the system and usually provide greater power inputs in the higher frequency ranges. As a result, electronics and secondary structure, which generally have higher resonant frequencies, are usually exercised to a greater extent than the primary structure. The response of an item to a specific random vibration input can be estimated by the Miles equation

$$g = 3\sqrt{\frac{\pi f Q s_0}{2}} \tag{8.17.1}$$

where g = estimated response in g units, f = frequency in Hz, Q = amplification factor, and s_0 = power spectral density input in g^2/Hz at the frequency f. Use of this equation assumes the system can be approximated as a single-degree-of-freedom system at the frequency of concern. The factor of three in the equation gives the statistical 3σ value which could occur at any given time.

8.17.9 Full Level Random Vibration

Full level random vibration tests are high energy input tests intended to envelope the random vibration environments generated by acoustics or any mechanical random vibration generated by the launch vehicle. These tests are normally conducted over the frequency range of 10–2000 Hz and for a time of 60 s. Responses are monitored as in other testing; should these responses exceed worst-case design limits, the inputs are "notched" to preclude overtesting of the test article. Notching is a method in which test input levels are reduced over a finite frequency range, thus effectively limiting system responses to test levels that do not exceed the worst-case expected environment. Full level random vibration testing is normally conducted in three orthogonal axes of the test article. A representative input for a random vibration test is shown in table 8.27. The rms value of a random vibration input is obtained by integrating the power spectral density over the frequency range and taking the square root.

8.17.10 Workmanship Random Vibration

Workmanship random vibration tests are typically performed on hardware that has already gone through full level acceptance testing and has required some rework due

Table 8.27 Random vibration test levels

Frequency (Hz)	Level
20	0.027 g^2/Hz
20–60	+3 dB/octave
60–170	0.08 g^2/Hz
170–400	+4 dB/octave
400–800	0.25 g^2/Hz
800–2000	−5 dB/octave
2000	0.055 g^2/Hz

Overall = 16.9g rms
Duration = 1 minute/axis
Three mutually perpendicular directions

to a problem not related to vibration. The extent and amount of workmanship testing is based on the amount and significance of the rework. For example, if a piece of electronic hardware had to replace a specific component that did not require much disassembly or rework then a full set of random vibration tests would probably not be required. In this circumstance, a test or tests designed to best exercise the mounting structure on which the component is mounted would likely be specified. In most cases workmanship testing is a subset of full level testing and is conducted at lower test levels than the full level tests. The specification for the workmanship vibration test is similar to the full level test except that the level is most likely lower and the number of testing axes is usually less then three.

8.17.11 Shock

Shock testing involves exposing the hardware to a non-periodic excitation that is characterized by its suddenness and severity. Shock tests are high-frequency, high-level impulsive loading inputs that are typically generated by devices like the pyrotechnic actuators and bolt cutters which are used at the spacecraft/launch vehicle separation plane. The acceleration generated from a shock event can be represented in terms of a shock response spectrum. The shock response spectrum is defined as the maximum response of a single-degree-of-freedom system for which the natural frequency of the oscillation is variable. The shock spectrum will depend on the assumed damping and is usually expressed in terms of the amplification Q. For shock response specifications, a Q of 10 is typically used. Although the shock level may appear to be significant, the energy generated from these events quickly dissipates through each joint and with increasing distance from the shock source. As a result, hardware that is not mounted in the immediate vicinity of the shock-generating event will be unlikely to experience any significant environmental loading.

Shock testing can be performed via shaker simulation or through the actuation of the device generating the input. For shaker tests, inputs are specified as a shock response spectrum and performed in three orthogonal axes. When testing is performed through device actuation, the event producing the environment is conducted at least twice to generate scatter in the environment. A representative input for a shock response spectrum with an amplification (Q) of 10 is shown in table 8.28.

8.17.12 Acoustic

Acoustic testing is generally a system-level test that subjects the hardware to direct acoustic noise in an acoustic chamber. The intent of this test is to subject the spacecraft to the predicted acoustic environment that will be generated by the launch vehicle (usually

Table 8.28 Separation shock response spectrum ($Q = 10$)

Frequency (Hz)	Level
50	$40g$
600	$2000g$
3000	$2000g$
Three orthogonal axes	

Table 8.29 Acoustic test levels

Octave Band Center Frequency (Hz)	Sound Pressure Level (dB) (ref. 2×10^{-5} N/m^2)
31.5	125
63	128
125	134
250	137
500	141
1000	134
2000	131
4000	127
8000	124
Overall SPL = 144 dB	
Duration = 60 s	

SPL = sound pressure level.

at liftoff and during transonic flight). The acoustic environment generates a random vibration environment due to the noise pressure acting on the surfaces of the spacecraft. This environmental test provides a better test for exterior structures of the spacecraft (such as solar arrays and antennae) which have a lighter density. Because the loading of the acoustic environment is generated from sound pressure, structures of this type may not normally be subjected to adequate excitation during random vibration testing. Acoustic testing is generally preferred over a random vibration test for most spacecraft because of the relatively high acoustic levels and low mechanically transmitted vibrations encountered during launch. The estimated response of an item from an acoustic input is determined by calculating the equivalent sound pressure power spectral density and applying equation 8.17.1. The equivalent sound pressure spectral density is given by

$$s = \frac{\sqrt{2} P_0^2 10^{(K/10)}}{f} \tag{8.17.2}$$

where

s = sound pressure power spectral density
P_0 = reference pressure level (2×10^{-5} N/m^2)
K = sound pressure level (SPL) in dB relative to P_0
f = frequency in Hz

Acoustic testing is conducted in a reverberant chamber equipped with an array of high-frequency noise generators with horns. Microphones are located in the vicinity of the spacecraft and measure the intensity of the acoustic field. Typically a minimum of four microphones are used to develop an average input spectrum at each of the octave band center frequencies. As with vibration testing, accelerometers are used to measure the response of the test article at key locations. Table 8.29 provides a representative test specification for a spacecraft-level acoustic test.

8.17.13 Modal Survey

Modal survey testing is a non-destructive test conducted for evaluating analytical models. It is typically conducted with very low level vibration levels and is used to characterize the dynamic response of the test article. Numerous accelerometers are located throughout the structure to define in great detail the natural frequencies and mode shapes of the structure.

8.17.14 Spin Balance

Spin balance testing is conducted to dynamically balance the system to ensure optimal performance of the launch vehicle during the launch sequence. This sequence of testing will decide where to properly place the required weights so as to accurately balance the spacecraft. During this testing components on the perimeter of the spacecraft are exposed to high acceleration levels.

Problems

1. A combined spacecraft ($W_{SC} = 1000$ lb) and upper stage ($W_u = 200$ lb) are spinning at 100 rpm while being thrust at 7500 lb. At the end of the burn, the weight of the upper stage, W_u, is still 200 lb. A specially developed meter is located 40 in. from the spin axis and is cantilevered as shown from point A. The cantilevered graphite mast is negligible in weight compared to the meter. In addition to the prescribed thrust and spin an engine anomaly has been noted on recent flights of the upper stage. Unfortunately, this anomaly lasts for the duration of the burn and produces a sinusoidal input in the thrust direction at the base of the cantilevered meter, point A, of 0.8 g. More unfortunately, the input is at the same frequency as that of the cantilevered mass. Recent tests on the mast indicated a damping ratio of $c/c_c = 0.05$. Calculate

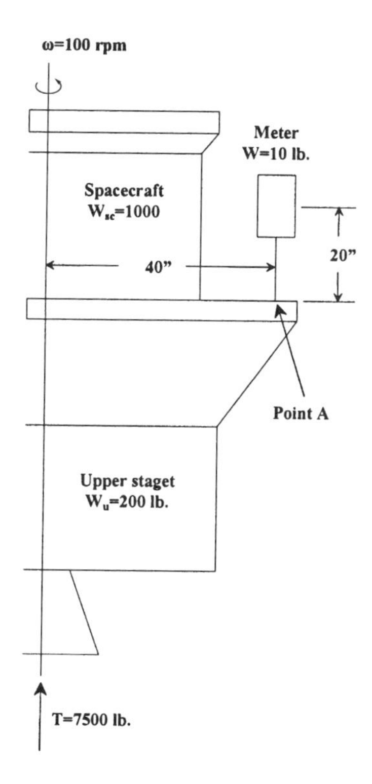

the axial load P_z and the bending moment M_a *in the plane of the diagram* at point A for the combined spin, thrust, and vibration.
Answer: $F = 142$ lb, $M = 2270$ in.lb.

2. An antenna element weighing 2 lb is shown in the deployed position. A vertical pin is inserted into the shaft (via a spring mechanism) at the completion of the deployment scenario to prevent the antenna arm from freely moving about the z-axis during any subsequent spacecraft motions. At some point during the mission (after antenna deployment), the spacecraft's propulsion system is used to maneuver the spacecraft. During this maneuver, it is expected that the antenna will experience a steady-state acceleration of 20 g in the y-direction. What size circular pin (i.e., radius) is needed to resist this loading condition? Assume a shaft radius $R = 1$ in., $L = 36$ in., $a = 0.25$, a factor of safety of 1.6, $\sigma_{tu} = \sigma_{cu} = 40,000$ psi and $\tau_{su} = 50,000$ psi. Neglect the weight of the deployment arm.

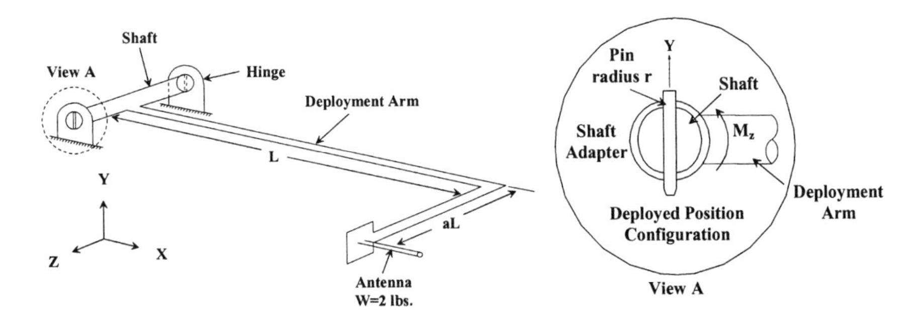

Answer: $r_{min} = 0.1$ in.

3. The spacecraft shown below weighs 2125 lb and is supported on the launch vehicle by the spacecraft attach fitting (SPAF). Assuming the SPAF is essentially a thin

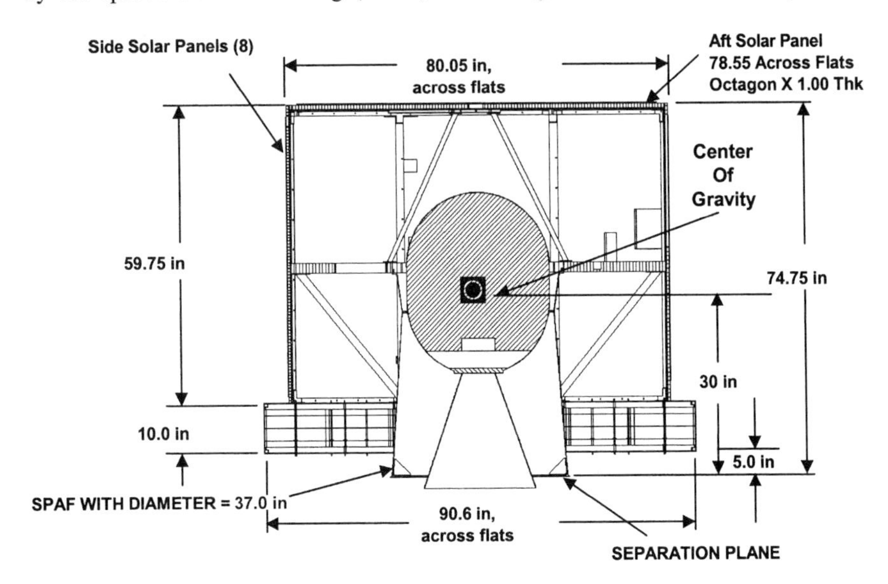

cylindrical support structure with a diameter of 37 in. and a wall thickness of 0.1 in., what is the maximum compressive and tensile stress in the adapter if the spacecraft is launched on a Delta II launch vehicle. The center of mass is located on the centerline of the cross section shown and at a distance of 30 in. above the separation plane. *Answer*: Max. tensile stress = 2197 psi, max. compressive stress = 1830 psi.

4. Compare the allowable axial compressive loads for a 3 in. × 2 in. × 1/4 in. aluminum alloy angle 43.5 in. long if it acts as a pin-ended column, or if it is so restrained that its effective length L' is $0.9L$. Assume a factor of safety of 2.5 and a radius of gyration of 0.435 in. Calculate the margins of safety if the actual loading were 5000 lb.

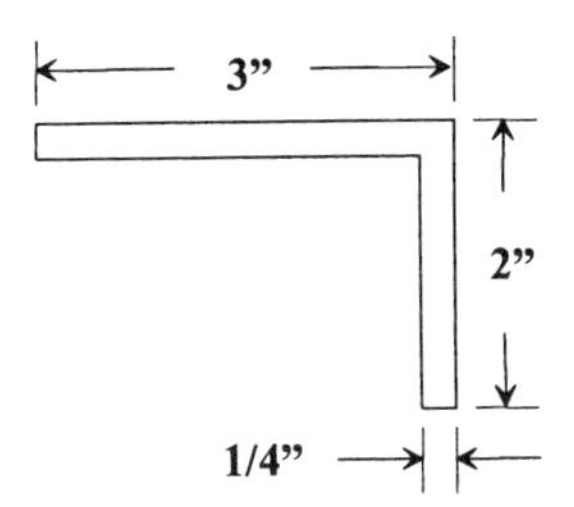

Answer: (a) MS = −0.06, (b) MS = 0.16.

5. The MSX truss structure in figure 8.21 is composed of three different truss member designs as shown in the table below. Given that each member is loaded in compression at each end with the loads in the table, what is the minimum margin of safety for buckling of this structure? The material of the truss members is P75/ERL1962 quasi-isotropic composite material with properties as given in table 8.20; assume a factor of safety of 2.0. Note that the truss structure is attached to other spacecraft structure at both ends.

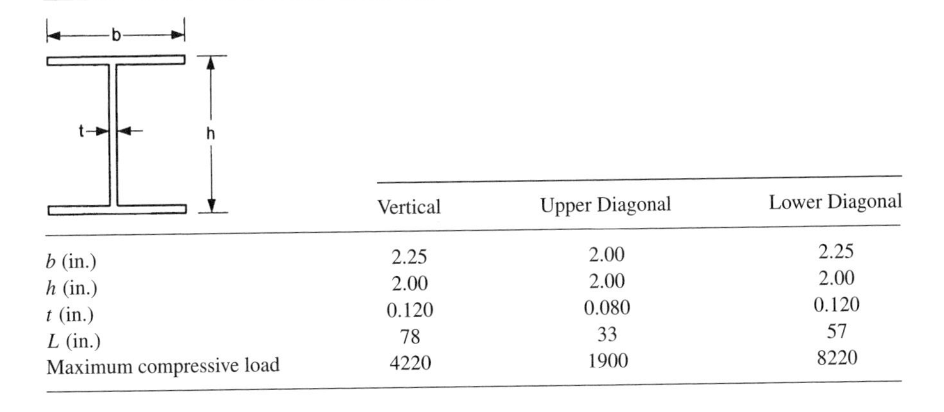

	Vertical	Upper Diagonal	Lower Diagonal
b (in.)	2.25	2.00	2.25
h (in.)	2.00	2.00	2.00
t (in.)	0.120	0.080	0.120
L (in.)	78	33	57
Maximum compressive load	4220	1900	8220

Answer: $\text{MS}_{\text{min}} = 0.67$

6. A criterion commonly used in the selection of materials for structural members is weight. For a structural column pinned at each end and loaded in uniaxial compression (that is, buckling critical), determine which one of the materials indicated

below provides the optimum weight design for the compressive loading condition. The thickness t is chosen as the design variable for this problem.

Material	Aluminium	Magnesium
Alloy	7075	AZ31B–H24
Ultimate strength (psi)	77,000	40,000
Density (lb/in^3)	0.100	0.065
Modulus of elasticity (psi)	10×10^6	6.5×10^6

Answer: Magnesium—aluminum would be 33% heavier.

7. During an on-orbit maneuver, a small sensor ($w = 5$ lb) that is cantilevered ($L = 60$ in.) from the primary spacecraft structure is subjected to a quasi-static load of $11.6g$ in the z-direction. At the same time, the base of the deployment arm is excited by the random vibration environment in the z-direction as shown below. Assume the arm has negligible mass and is made of aluminum ($E = 10 \times 10^6$ psi). (a) If the deployment arm has a tubular cross-section with diameter, d = 2 in, and with a natural frequency of 50 Hz, determine the thickness t of the tube if the structural damping is 2.5% and the allowable stress level is 42,000 psi. (b) If, in addition, a temperature gradient exists that causes an elongation of 0.005 in. at the end of the arm, what effect would this have on your answer in part (a)?

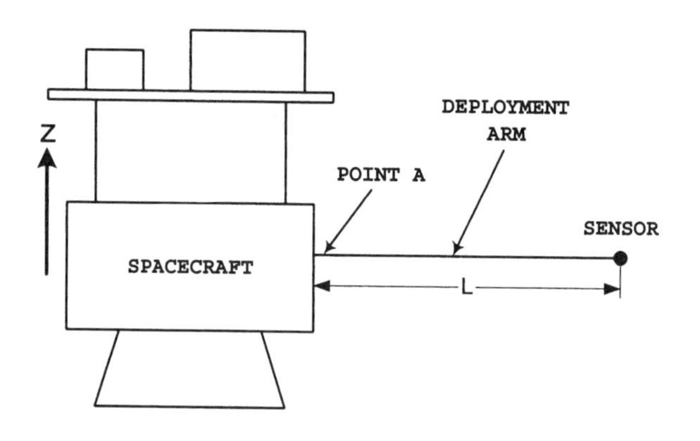

Frequency	Level
10–80	0.005 g^2/Hz
80–200	+3.6 dB/octave
200–1000	0.015 g^2/Hz
1000–2000	−9 dB/octave

Overall $g_{rms} = 4.4$
Duration = 15 s

Answer: $t = 0.046$ in.

8. The MSX truss structure of figure 8.21, represented as a 'stick' figure below, was designed for stability as well as its light weight. Given that all members of the truss are identical and the structure was assembled at room temperature (70°F or 21°C), what will be the estimated tilt, θ, of plane BbcC if the on-orbit temperature of side ABCD is −4°F (−20°C) and that of side abcd is 86°F (30°C)? The quasi-isotropic properties of the composite material used in the truss members are given in table 8.20. All end fittings are made of titanium with properties as given in table 8.18.

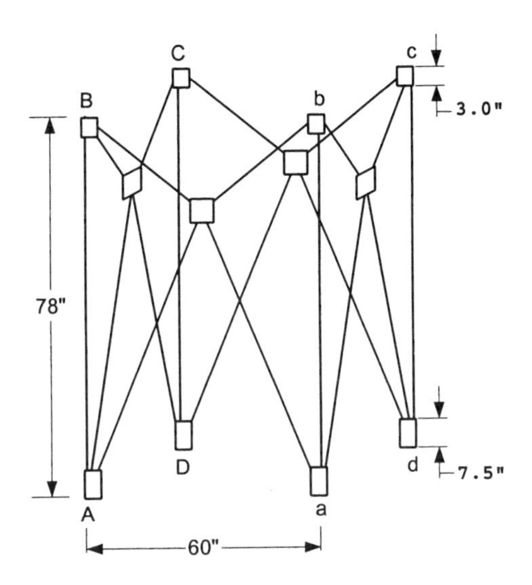

Answer: $\theta = 0.004°$.

9. The fundamental frequency of the support structure shown (without any end mass) has been experimentally determined to be 5 Hz. When a 2000 lb secondary spacecraft is added at the end of this structure, the frequency was found to be 4 Hz. If the secondary spacecraft is replaced by an experiment weighing 1000 lb, calculate the new fundamental frequency. (Assume the support structure acts as a single-degree-of-freedom system.)

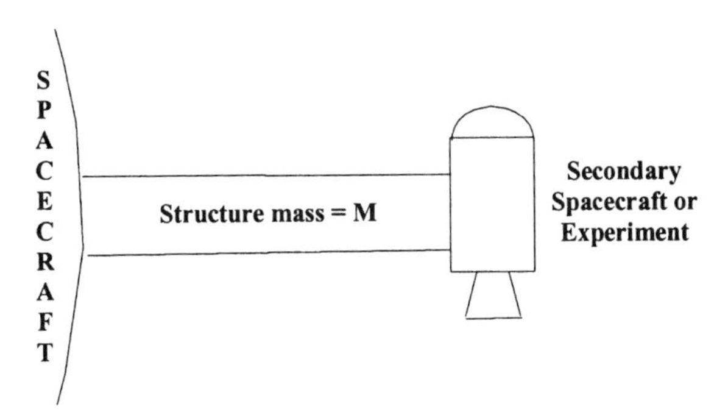

Answer: $f = 4.42$ Hz.

10. Write the equation of motion for the mass M if the base is given a prescribed displacement $Y_b(t)$, using an electrodynamic shaker.

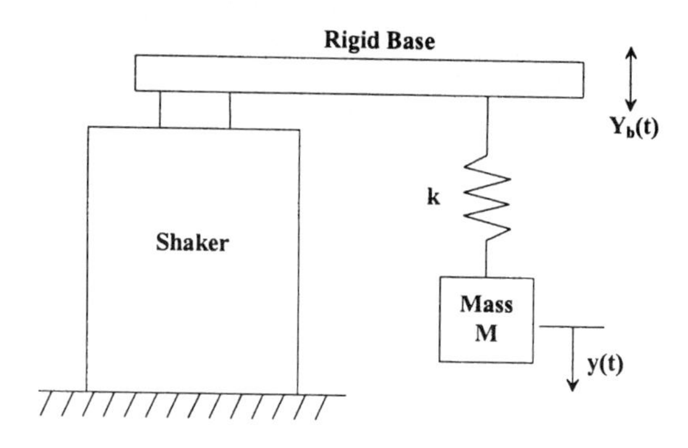

Answer: $My + k(y - y_b) = 0$

11. The top deck of the NEAR spacecraft structure is a 58 inch square aluminum sandwich honeycomb structure which supports 275 lb of hardware. The deck, with a top and bottom face sheet thickness of 0.015 in., can be assumed to be simply supported on all four sides. To prevent coupling with the primary spacecraft thrust mode, the deck must be designed to a minimum fundamental natural frequency of 30 Hz (in the z direction). (a) Given that Poisson's ratio (ν) is 0.33, determine the overall thickness, h, of the deck which will satisfy this requirement. (b) If, during a thrusting event, the sinusoidal vibration levels shown in the table are generated in the z-direction at the deck's mounting interface, what is the maximum response force generated on the structure during this event? Assume a deck natural frequency of 30 Hz and a structural damping of 5%.

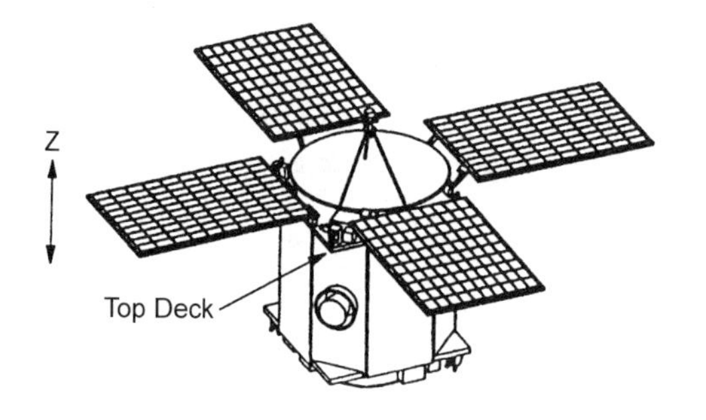

Frequency (Hz)	Level (g)
10–45	1.5
50–60	12.5
75–100	0.5

Answer: (a) $h = 1.61$ in., (b) $F_{max} = 4125$ lb.

12. The instrument (with weight $W = 24$ lb) is supported by the aluminum support structure and stowed as shown prior to deployment. In this configuration the arm experiences a transient condition in the z-direction which generates the response shown. During this time, the overall temperature of the support structure and instrument is 40°F less (that is, colder) than the spacecraft structure to which it is mounted. Given these two conditions acting simultaneously, and neglecting the weight of the support structure, what is the maximum longitudinal or axial stress (z-direction) in the support structure arm?

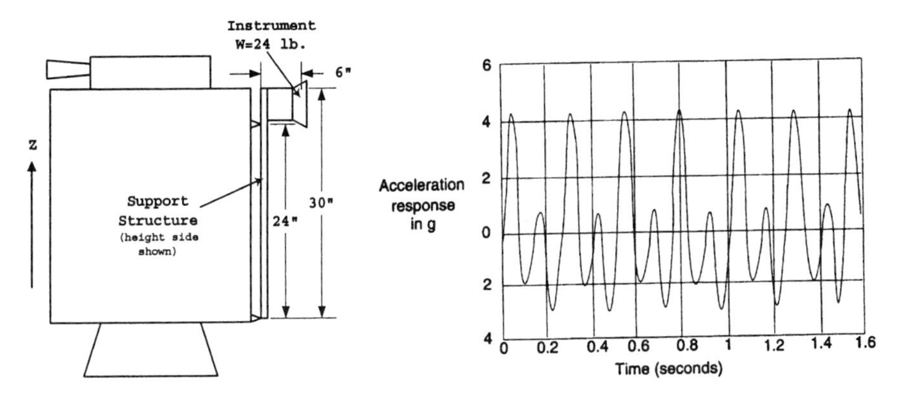

Support structure configuration with base $b = 0.2$ in. and height $h = 3$ in. $E = 10 \times 10^6$ psi, $A = 0.6$ in^2, $\alpha = 12 \times 10^{-6}$ in/(in°F) $\Delta T = -40$°F.

Answer: $\sigma = 6880$ psi.

13. An electronic housing contains several printed circuit boards. Each board is made of fiberglass and can be considered to be simply supported on all four sides. The following information is known about the configuration:

    Board dimensions 5 in. × 7 in.
    Uniformly distributed weight of 1 lb
    $E = 2.4 \times 10^6$ psi $\quad t = 0.125$ in.
    Poisson's ratio = 0.125 $\quad$ Damping factor = 0.033

    If the board is subjected to the random vibration input given below, what are the maximum stress levels in the board? If the allowable stress is 30,000 psi, what is the margin of safety?

Random Vibration Input

Frequency (Hz)	Level
20–100	0.001 g^2/Hz
100–200	+21.0 dB/octave
200–600	0.13 g^2/Hz
600–2000	−5.4 dB/octave
Overall $g_{rms} = 10.8$	
Duration = 60 s	

Answer: MS = 16.4. (Typically, printed circuit boards are designed for stiffness and thus usually exhibit large strength margins.)

14. Spacecraft are usually attached to the launch vehicle with mechanisms that upon completion of the boost phase allow separation either from a command stored on

the spacecraft or by ground command. These mechanisms are typically explosive bolts, explosive nuts, or V-band clamp straps. They do not, in general, provide any separation velocity to the spacecraft or upper stage of the launch vehcile. A separation velocity of two to three feet per second between these two bodies is generally desirable. Such a relative velocity allows for configuring a spacecraft for orbit (i.e., deploying booms, solar panels, etc.) without concern for impact with the spent booster. If the spacecraft is to remain in this orbit it also provides for a slightly different orbit for the two bodies. The separation velocity is typically provided by linear compression springs that provide a force between the two bodies until separation allows the extension of the springs and the resultant relative velocity. Assume that we have a spacecraft that weighs 400 lb and a spent booster that weighs 600 lb. We desire to have 3 ft/s relative velocity between the two bodies at the 3 in. full extension of a spring. Derive the stiffness of the spring (k) required to perform this task and the velocity imparted to each body. The problem is illustrated below:

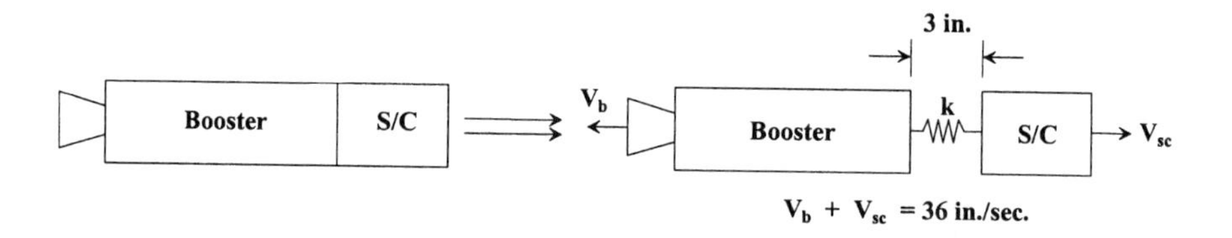

Answer: Stiffness $k = 89.5$ lb./in. Velocities, $V_{\text{booster}} = 14.4$ in./s, $V_{\text{spacecraft}} = 21.6$ in./s. (Note: velocities in opposite direction.)

15. A three-noded element has nodal displacements u_1, u_2, and u_3, as shown in the figure. The assumed displacement can be written as $u = a_1 + a_2 s + a_3 s^2$ or $u = N_1 u_1 + N_2 u_2 + N_3 u_3$.
   (a) Express N_1, N_2, and N_3 explicitly as functions of s.
   (b) Determine the element stiffness matrix.
   (c) For a constant uniaxial force per unit length ($f = c$), find the element nodal load vector Q^i_j.
   (d) Using only one three-node element, solve for the displacements of the vertical column.

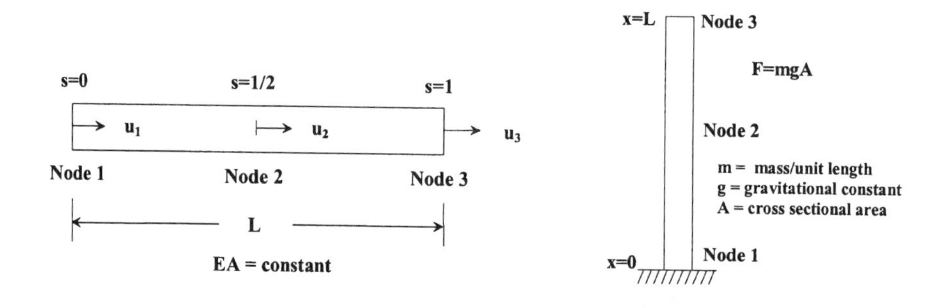

Answer:

$$\text{(a)}\quad \begin{aligned} N_1 &= 1 - 3s + 2s^2 \\ N_2 &= 4s(1 - s) \\ N_3 &= s(2s - 1) \end{aligned}$$

$$\text{(b)}\quad [k] = \frac{EA}{3L}\begin{bmatrix} 7 & -8 & 1 \\ -8 & 16 & -8 \\ 1 & -8 & 7 \end{bmatrix}$$

$$\text{(c)}\quad \{Q\} = \frac{cl}{6}\begin{Bmatrix} 1 \\ 4 \\ 1 \end{Bmatrix}$$

$$\text{(d)}\quad \begin{Bmatrix} u_1 \\ u_2 \\ u_3 \end{Bmatrix} = \frac{mgL^2}{8E}\begin{Bmatrix} 0 \\ -3 \\ -4 \end{Bmatrix}$$

Definitions

"A" basis allowables: The minimum mechanical property allowable (strength values) which 99% of the population will meet or exceed, with a statistical confidence level of 95%.

Acceptance tests: The required formal tests conducted on flight hardware to ascertain that the materials, manufacturing processes, and workmanship meet specifications and that the hardware is acceptable for intended use.

Allowable load (stress): The maximum load (stress) that can be accommodated by a material (structure) without rupture, collapse, or detrimental deformation in a given environment. Allowable stresses are commonly the statistically based ultimate strength, buckling strength, and yield strength.

Applied load (stress): The actual load (stress) imposed on the structure in the service environment.

"B" basis allowable: The minimum mechanical property allowable which 90% of the population will meet or exceed, with a statistical confidence level of 95%.

Critical condition: The most severe environmental condition in terms of loads, pressures, and temperatures, or combination thereof, imposed on the structure, systems, subsystems, and components during service life.

Design safety factor: A factor used to account for uncertainties in material properties and analysis procedures. Design safety factor is often called design factor of safety or simply factor of safety (FS).

Detrimental deformation: Any structural deformation, deflection, or displacement that prevents any portion of the structure from performing its intended function, or that reduces the probability of successful completion of the mission.

Development test: A test to provide design information that may be used to check the validity of analytic techniques and assumed design parameters, to uncover unexpected system response characteristics, to evaluate design changes, to determine interface compatibility, to prove qualification and acceptance procedures and techniques, or to establish accept/reject criteria.

Dynamic envelope: The space or volume allocated to a component, which includes allowance for all displacements and deflections associated with the worst combination of dynamic loading.

Environments: The combination of static loads, dynamic excitations with associated durations, and environmental exposures (such as humidity, temperature, radiation levels) that the systems are subjected to after completion of manufacture and final inspection.

Limit load stress: The highest predicted load or combination of loads that a structure may experience during its service life in association with the applicable operating environments. When a statistical estimate is applicable, the limit load is that load not expected to be exceeded at 99% probability with 90% confidence. The corresponding stress is called the limit stress.

Margin of safety (MS):

$$\text{MS} = \frac{\text{Allowable load (strees)}}{\text{Limit load (strees)} \times \text{Design safety factor}} - 1$$

Qualification tests: The required formal contractual tests used to demonstrate that the design, manufacturing, and assembly have resulted in hardware designs conforming to specification requirements.

Quasi-static acceleration: The combination of sustained and transient acceleration applied statically to determine the loading in structures. Transient environments occur in the low-frequency range where structures are susceptible to higher loading.

Residual strength: The maximum value of nominal load (stress) that a cracked or damaged body is capable of sustaining without unstable growth or collapse.

Residual stress: The stress that remains in a structure as a result of processing, fabrication, assembly, testing, or operation. A typical example is the welding-induced residual stress.

"S" basis allowable: The minimum mechanical design allowable specified by the governing industry specification. The statistical confidence on this value is unknown.

Static envelope: The space of volume allocated to a component with no allowance for displacements and deflections included.

Stress corrosion cracking: A mechanical-environmental induced failure process in which sustained tensile stress and chemical attack combine to initiate and propagate a crack or crack-like flaw in a metal part.

Ultimate load: The product of the limit load and the design ultimate load factor of safety. It is the load which the structure must withstand without rupture or collapse in the expected operating environments.

Yield load: The product of the limit load and the design yield load factor of safety. It is the load beyond which the structure will experience significant permanent deformation.

Acronyms

CAD: Computer-aided drafting
CBE: Current best estimate
CDR: Critical design review
CoDR: Conceptual design review
CTE: Coefficient of thermal expansion
ELV: Expendable launch vehicle
FEA: Finite-element analysis
FEM: Finite-element model
FOV: Field of view
FS: Factor of safety
MEOP: Maximum expected operating pressure
MS: Margin of safety
PAF: Payload attach fitting
PDR: Preliminary design review
PLF: Payload fairing
PSD: Power spectral density

References

Agrawal, Brij N., 1986, *Design of Geosynchronous Spacecraft*, chapters 1, 4, Prentice-Hall.

Barry Controls, Barry Wright Corporation, 1988, *Shock, Vibration and Noise Control—Application Selection Guide*, Bulletin C11-188.

Blake, A., 1990, *Practical Stress Analysis in Engineering Design*, second edition, Marcel Dekker.

Blevins, R. D., 1987, *Formulas for Natural Frequency and Mode Shape*, Krieger.

Bruhn, E. F., 1973, *Analysis and Design of Flight Vehicle Structures*, Jacobs.

Chen, J. C., J. A. Garba, M. A. Salama, and M. R. Trubert, 1984, *Methods for Estimating Payload/Vehicle Design Loads*, Jet Propulsion Laboratory, NASA Tech Brief, Vol. 7, No. 4, Item #50, April.

Department of the Air Force, 1982, MIL-STD-1540D, *Test Requirements for Space Vehicles*, October.

Department of Defense, 1983, *Military Standardization Handbook, Metallic Materials and Elements for Aerospace Vehicle Structures (MIL-HDBK-5H)*, Project No. 1560-0126, June.

George C. Marshall Space Flight Center, 1981, *Structural Strength Program Requirements*, MSFC-HDBK-505, Revision A, January.

Goddard Space Flight Center, 1996, *General Environmental Test Specification for Spacecraft and Components (Expendable Launch Vehicles)*, Change 3, October.

Harris, C. M., 1988, *Shock and Vibration Handbook*, third edition, McGraw-Hill.

Lord Industrial Products, 1988, Industrial Products Catalog PC-2201H.

Marshall Space Flight Center, 1975, *Astronautic Structures Manual*, NASA Technical Memorandum NASA TM X-73305, August.

Peery, D. J., and J. J. Azar, 1982, *Aircraft Structures*, McGraw-Hill.

Roark, R. J., 1995, *Formulas for Stress and Strain*, fourth edition, McGraw-Hill.

Silverman, E., 1996, *Composite Spacecraft Structures Design Guide*, NASA Contractor Report 4708, March.

Websites:

www.boeing.com/defense-space/space/delta/guides.htm—Delta launch vehicle planners' guides.

www.ilslaunch.com/missionplanner—Atlas and Proton vehicle planners' guides.

www.orbital.com/LaunchVehicle/SpaceLaunchVehicles/index.htm—Pegasus and Taurus launch vehicles.

9

Space Communications

ERIC J. HOFFMAN

9.1 Introduction and Overview

Almost without exception, spacecraft must receive operating commands and data from the ground and return ("telemeter") data to the Earth. Indeed, communications is usually so mission critical that it is the first subsystem to be made redundant. Figure 9.1 shows the block diagram of a typical spacecraft communications subsystem. Its "baseband" elements collect, assemble, and store data, and interpret commands. Radio

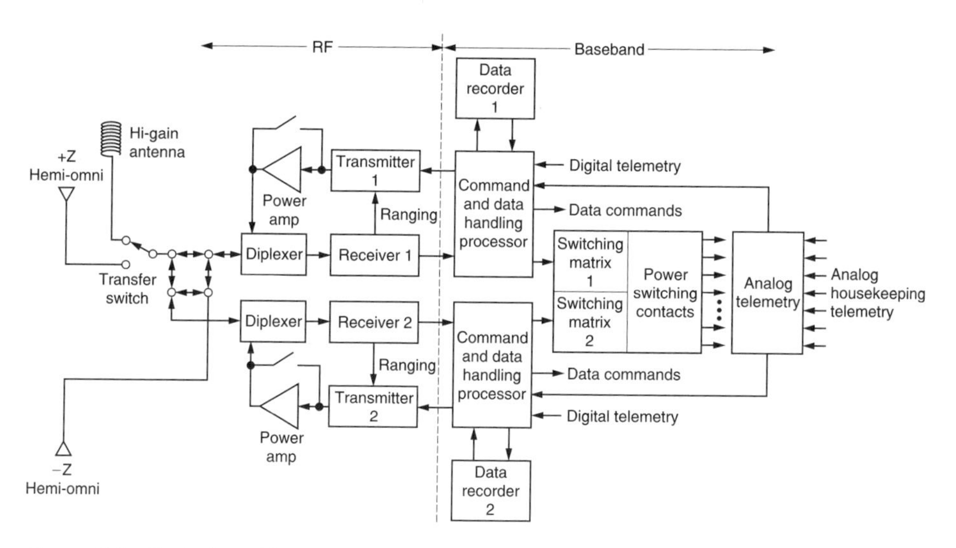

Figure 9.1 Typical satellite communications block diagram.

frequency (RF) elements link the spacecraft with the Earth. The baseband portion will be described in detail in chapter 10. This chapter focuses on the RF and communication systems aspects.

Exercise

Can you think of any spacecraft for which command or telemetry systems would be either unnecessary or undesirable?

9.1.1 Space versus Terrestrial Communications

Space communications links share many characteristics with their terrestrial counterparts, but there are also some important differences. "Slant ranges" between the spacecraft and the ground are typically large, ranging from a few hundred or a few thousand kilometers for low Earth orbiters up to 13,000,000,000 km for planetary probes. Figure 9.2 shows how to compute the typical LEO communications geometry in terms of the satellite altitude h and the worst-case elevation angle ε. Received signals are often incredibly weak, and time delays due to the finite speed of light must sometimes be considered.

The spacecraft is usually moving at high speed (up to 8 km/s) relative to its ground station. This causes the received frequency to differ from the transmitted frequency f by a significant (and time-varying) Doppler shift Δf that is proportional to the rate of change of slant range,

$$\Delta f = \frac{-f}{c}\left(\frac{dR}{dt}\right), \tag{9.1.1}$$

where $c \approx 3 \times 10^8$ m/s.

Figure 9.3 gives a bound on typical Doppler shift for circular orbits; it can exceed $\pm$ 25 ppm for low-altitude satellites. On an S-band (2 GHz) signal, that can mean as

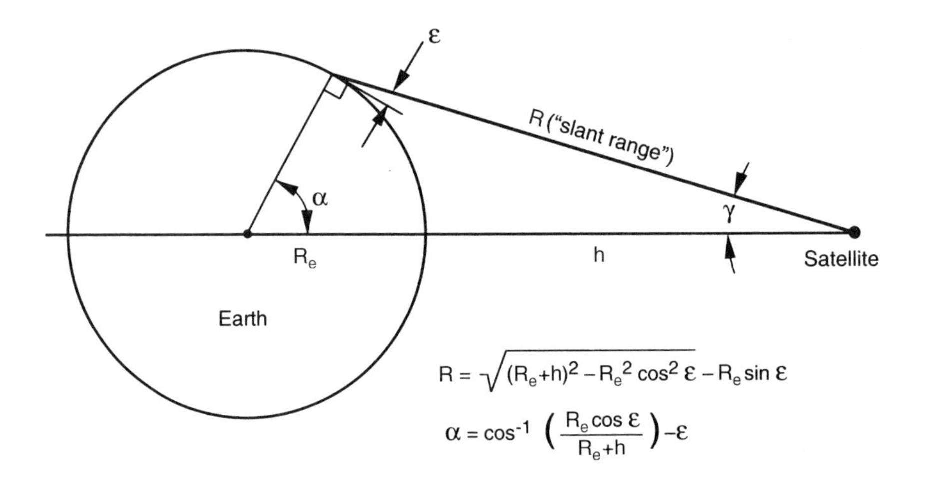

Figure 9.2 Typical satellite communications geometry.

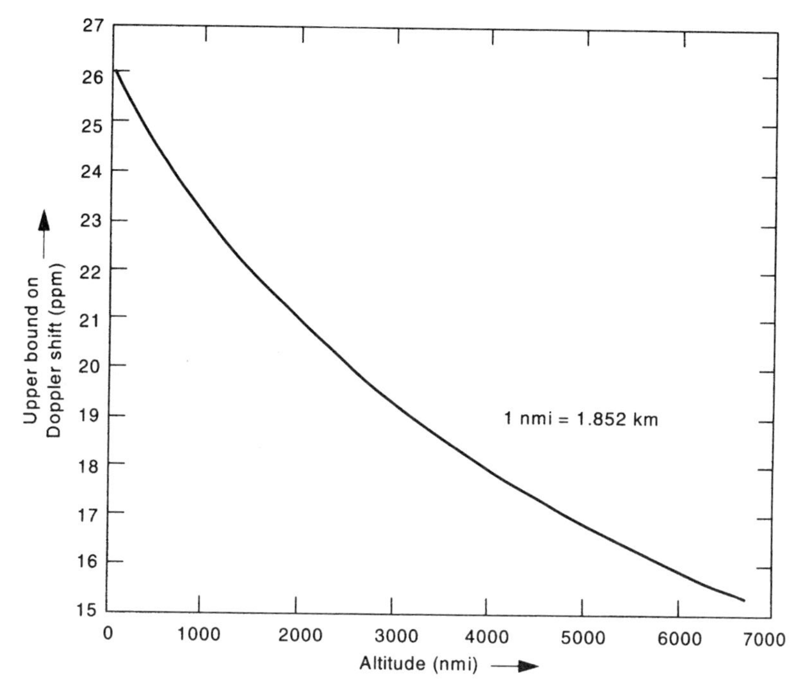

Figure 9.3 An upper bound on satellite Doppler shift for circular orbits.

much as ± 50,000 Hz of frequency uncertainty that must be searched during acquisition and tracked during reception. Doppler shift complicates receiver design, widens bandwidths, and lowers signal-to-noise ratios. On the plus side, Doppler shift has formed the basis for several different space-based navigation and tracking schemes (see section 9.6.6).

The orbital motion of the typical satellite means that any one ground station can see it only during short "passes." For low-altitude satellites, passes may last only a few minutes. The limited radio horizon of low-altitude satellites means that a hundred or more ground stations would be needed if continuous real-time coverage were required (figure 9.4). The impracticality of that approach has given rise to three alternatives, as detailed below.

Onboard data storage. Data recorders—originally mechanical tape recorders but today almost always their solid-state equivalent—can accumulate data slowly throughout the orbit and dump it quickly when over a ground station. In deep space missions the recorder is often used the opposite way, to capture high-rate data and play it back slowly over data-rate-limited links.

Special orbits. Certain specially designed orbits can cause the satellite to appear to hover over a ground station. A satellite launched into an equatorial orbit with a period of precisely 24 sidereal hours will appear to hang motionless over one point on the equator. The altitude to achieve this "geostationary" orbit is about 35,800 km, high enough that

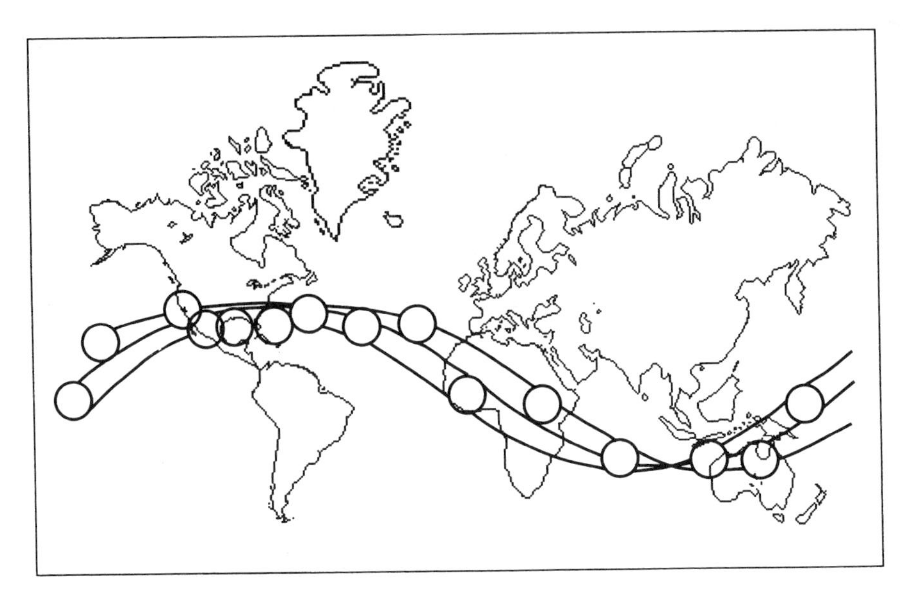

Figure 9.4 Coverage (visibility) circles for low-altitude spacecraft. These stations supported the early manned U.S. flights.

three such satellites can cover almost the entire Earth (excluding the polar regions; see figures 9.5 and 9.6).

The Molniya orbit provides a quasistationary alternative to the geostationary orbit, and is useful at high latitudes. This orbit is highly eccentric, with a period of 12 sidereal hours and an inclination of 63.4° (figure 9.5). Recall from chapter 3 that this particular

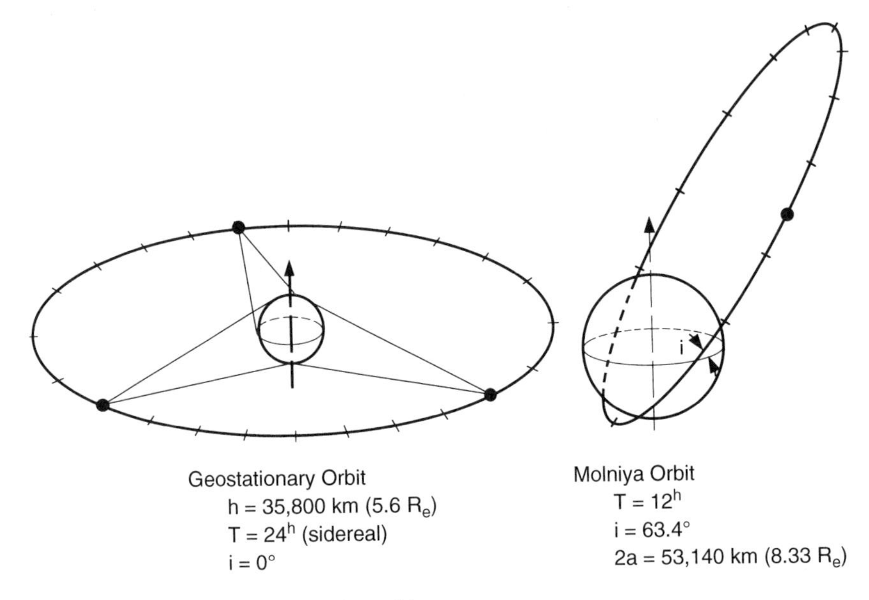

Figure 9.5 Geostationary and Molniya orbits.

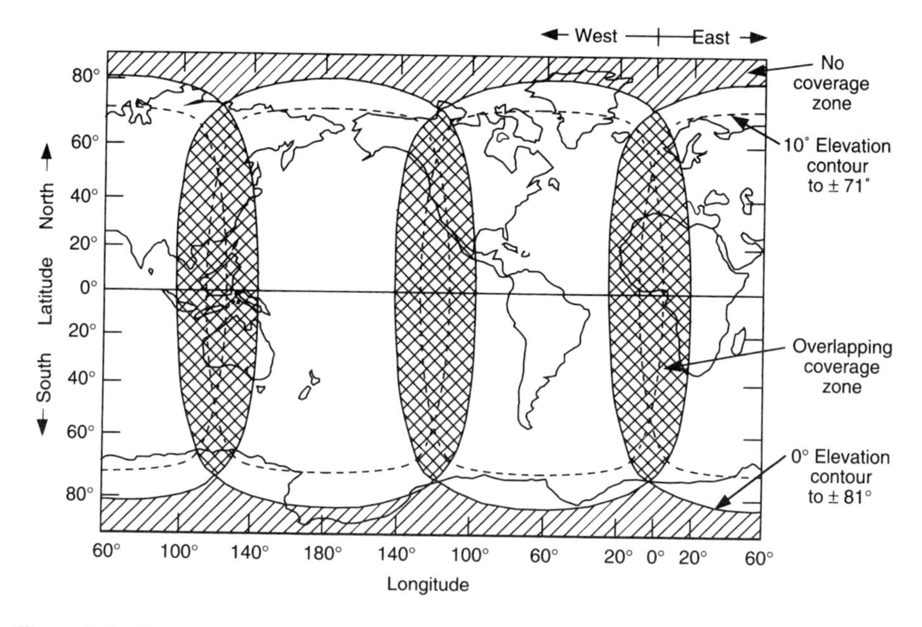

Figure 9.6 Ground coverage by three geostationary satellites.

inclination (namely, where $5\cos^2(i) = 1$) drives perigee precession to zero. The geosynchronous period ensures a repeating ground track. A Molniya satellite launched with its apogee at maximum latitude will therefore spend long periods of time over a fixed location in the Northern hemisphere. It is clear why the Russians find this orbit attractive for their communication satellites.

Data relay satellites. NASA's Tracking and Data Relay Satellites (TDRS), in geostationary orbit, relay commands to and telemetry from low-altitude satellites. They also provide nearly continuous satellite tracking. Prior to TDRS, astronauts could communicate with the Earth for only about 15% of the orbit, using 14 NASA ground stations around the world. With TDRS, coverage of low-altitude satellites is global, from a single ground station at White Sands, New Mexico.

From its high vantage point a satellite looks down on much more of the Earth than is visible to terrestrial stations. This increases its exposure to RF interference, possibly from intentional jamming. In fact, an effective strategy to defeat military space systems is to try to sever the communication links. Military spacecraft must often be designed with "robust" communications and high levels of onboard autonomy to counter this threat.

The satellite communication signal must also pass through the ionosphere and troposphere, where it experiences various disturbances that are important to the design of the RF links (discussed in section 9.2.6).

Finally, spaceborne communications equipment must be designed to be small, low power, low mass, extremely reliable, able to survive launch, and hardened to space radiation. And, increasingly, cost effectiveness is becoming a driving consideration.

9.2 Propagation of Radio Waves

Consider an infinitesimally small transmitter in free space that is radiating an electromagnetic wave with power P_t watts. The wave front, coming from a point source, must be spherically symmetric. When the wave front arrives at a distance R from the source, the power flux density is simply the total power P_t divided by the area of a sphere of radius R,

$$\text{flux density} = \frac{P_t}{4\pi R^2}. \tag{9.2.1}$$

A receiving aperture of physical area A and "effective area" $A_e = \eta A$ (where η represents aperture efficiency, typically around 55%) intercepts a fraction P_r of the total radiated power,

$$P_r = \frac{P_t A_e}{4\pi R^2}. \tag{9.2.2}$$

This is a simple form of the "one-way radar range equation" used in RF link design. For a typical geostationary link, R could be 40,000 km. A_e for a typical shipboard antenna might be 0.01 m^2. That gives a received power of only 0.0000000000000000005 of the transmitted power!

9.2.1 Antennas and Directive Gain

Clearly part of the reason for the low received power is that our hypothetical satellite is radiating its RF power uniformly in every direction of space ("isotropic" radiation) instead of concentrating it in the direction of the receiver. An "antenna" is a device that couples electromagnetic radiation between free space and a transmitter or receiver and, at the same time, increases the intensity of radiation in a preferred direction. Suppose we put a transmitting antenna on the spacecraft to concentrate the radiation in the direction of the receiver by a factor called the transmitting gain, G_t. (The product of G_t and P_t is the "effective" isotropic radiated power, EIRP.) Of course the power radiated in other directions must go down, to conserve energy. The one-way radar range equation then becomes

$$P_r = \frac{P_t G_t A_e}{4\pi R^2}. \tag{9.2.3}$$

As one would expect, there is a relationship between an antenna's gain and its area. High-gain antennas will invariably be many wavelengths long or have an area of many square wavelengths. The exact relationship is

$$G = \frac{4\pi A_e}{\lambda^2} = \eta \left(\frac{\pi D}{\lambda}\right)^2, \tag{9.2.4}$$

where η is the efficiency, D is the diameter (of a circular aperture), and λ is the wavelength, given by

$$\lambda = \frac{c}{f}. \tag{9.2.5}$$

9.2.2 Gain versus Area of an Antenna

Equation 9.2.4, which relates the gain and the area of antenna, is an important and fundamental equation. Yet it can be derived solely by conducting two *gedanken* experiments (thought experiments) that utilize two important properties of antennas. The first is that, at a given frequency, all antennas exhibit the same characteristics (gain, beamwidth, efficiency, pattern, polarization, etc.) whether used for transmitting or receiving. This convenient property is referred to as *reciprocity*. Secondly, antennas "scale", that is, they preserve their properties under change of dimension (of course, the wavelength or frequency must change accordingly). Any good radar textbook (see References) can provide a derivation of equation 9.2.4.

9.2.3 The One-way Radar Range Equation

By using equation 9.2.4 that relates G to A_e we can recast the one-way radar equation into three different forms, depending on whether the antennas at the transmitter and receiver are both fixed in gain, both fixed in area, or are fixed in gain at one end and area at the other:

$$\text{gain–gain}: P_r = \frac{P_t G_t \lambda^2 G_r}{4\pi R^2 4\pi} = \frac{P_t G_t G_r}{(4\pi)^2 R^2} \times \frac{c^2}{f^2} \tag{9.2.6}$$

$$\text{area–area}: P_r = \frac{P_t 4\pi A_{e_t} A_{e_r}}{\lambda^2 4\pi R^2} = \frac{P_t A_{e_t} A_{e_r}}{R^2} \times \frac{f^2}{c^2} \tag{9.2.7}$$

$$\text{gain–area}: P_r = \frac{P_t G_t A_{e_r}}{4\pi R^2}. \tag{9.2.8}$$

Note that the frequency dependence of the fundamental link design equation depends totally on the nature of the constraints on the antennas at each end of the link. If the antennas are constrained in *gain* at both ends, received power goes *down* with frequency. If the antennas are constrained in *area* at both ends, received power goes *up* with frequency. If one antenna is gain-limited and the other is area-limited, the received power is *independent* of frequency. It is important to recognize at the outset of any RF link design which case we are dealing with, as a guide to selecting operating frequencies.

9.2.4 The Decibel

Link design involves multiplying and dividing a large number of factors, ranging over many orders of magnitude. In the days B.C. (before calculators), it helped to express each term logarithmically using decibels; this tradition has continued. A power ratio expressed in decibel form is simply

$$\text{decibels} = 10 \log_{10}(\text{power ratio}). \tag{9.2.9}$$

A factor of 100 thus becomes 20 dB, a factor of 2 is 3.01 dB, and so on. It is also possible to express absolute quantities in dB notation by using "dB with respect to x" or dB(x).

Thus power can be expressed in milliwatts or in dBm (dB above 1 milliwatt). Antenna gain is often expressed in dBi (dB above an ideal isotropic radiator).

In link design we use the term *spreading loss* or *path loss* to define the loss between two ideal isotropic (unity gain) antennas at a distance R and frequency f. By combining the terms $c^2/(4\pi)^2 R^2 f^2$ in equation 9.2.6, we can express the path loss in dB by

$$\text{path loss (dB)} = 32.5 + 20\ \log(R) + 20\ \log(f), \tag{9.2.10}$$

where R is the slant range in kilometers and f is the frequency in megahertz.

Exercise

What constant substitutes for 32.5 if the range is in nautical miles? In astronomical units?

9.2.5 Some Properties of Antennas

Antenna gain varies with angle, usually achieving a maximum along the physical "boresight" (figure 9.7). Standard spherical coordinates (θ, φ) are used. The 3 dB beamwidth is defined as the angle between the points where gain has dropped to half its maximum. High-gain antennas must necessarily have narrow beamwidths, and in a perfectly efficient antenna the product of gain times the vertical and horizontal beamwidths (θ_v, θ_h) would equal the total number of square degrees in a sphere,

$$4\pi \text{ steradians} \times \left(\frac{180}{\pi}\right)^2 = 41{,}253\ \deg^2, \tag{9.2.11}$$

or

$$\text{gain} = \frac{41{,}253}{\theta_v \theta_h}. \tag{9.2.12}$$

When $\theta_v = \theta_h = \theta$ we have a so-called "pencil beam" antenna, for which $G\theta^2$ is constant. After taking efficiency into account, constants of 23,000–27,000 are

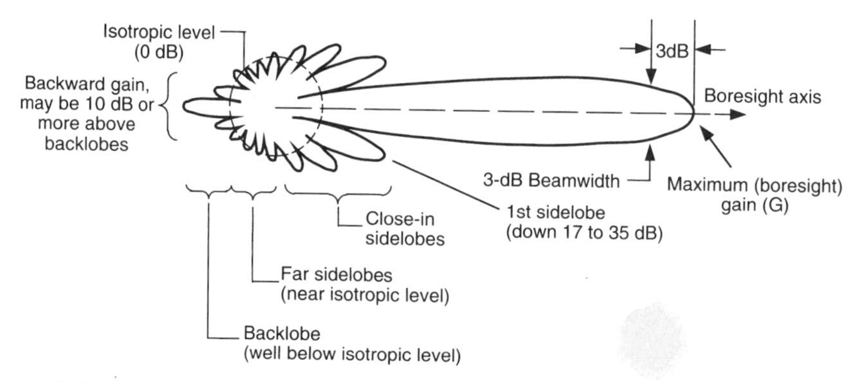

Figure 9.7 Typical radiation pattern for a high-gain antenna.

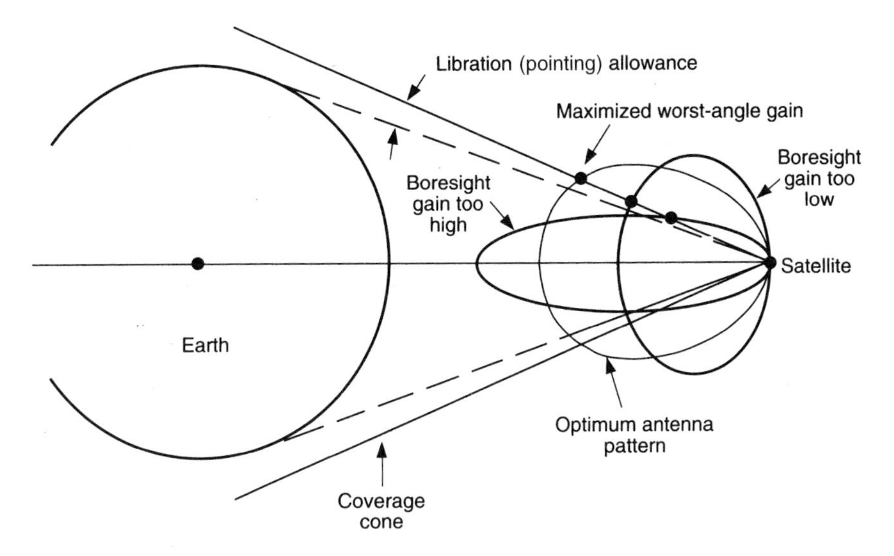

Figure 9.8 Optimizing the beamwidth to cover a given cone.

more typical for real antennas. A high-gain antenna might not illuminate every area of interest and must be carefully aimed, because gain falls off rapidly beyond the 3 dB points.

For every required cone of coverage, there is an optimum beamwidth that maximizes the worst-case antenna gain (that is, the gain at the worst viewing angle, see figure 9.8). For the most commonly used antenna feed illumination, the optimum antenna places the -3.9 dB ($= 2/\pi$) down points on the edge of the coverage cone. The optimum 3 dB beamwidth in this case is roughly 88% of the required full coverage cone. Any pointing errors (libration) of the satellite must be included in the required cone of coverage.

An important property of an antenna is its polarization, the orientation of the electric field vector in the wave launched from (or received by) the antenna. Polarization can be linear (either horizontal or vertical) or circular. In circular polarization, the electric field direction rotates 360° (in either a right- or left-hand sense) as the wave traverses one wavelength. Circularly polarized antenna gains are expressed as dBic. Circular polarization is sometimes achieved by combining two linear components in phase quadrature. That allows a linearly polarized wave to be received circularly (or vice versa) with never more than 3 dB polarization loss, a useful technique for dealing with Faraday rotation (section 9.2.6.1). A circularly polarized wave received with the opposite handedness (or a vertically polarized wave received horizontally) may be suppressed 25 dB or more. These cross-polarization effects can form the basis for frequency reuse schemes to share a single antenna and frequency, particularly for commercial communications satellites. Circular polarization is often used with low-Earth-orbit satellites to prevent dropouts due to cross-polarization as viewing angles change.

The selection of the antennas at both ends of the link is a crucial step in the link design, and many consequences for the system will flow from this decision. A wide variety of spacecraft antennas has evolved over the years to provide omnidirectional coverage,

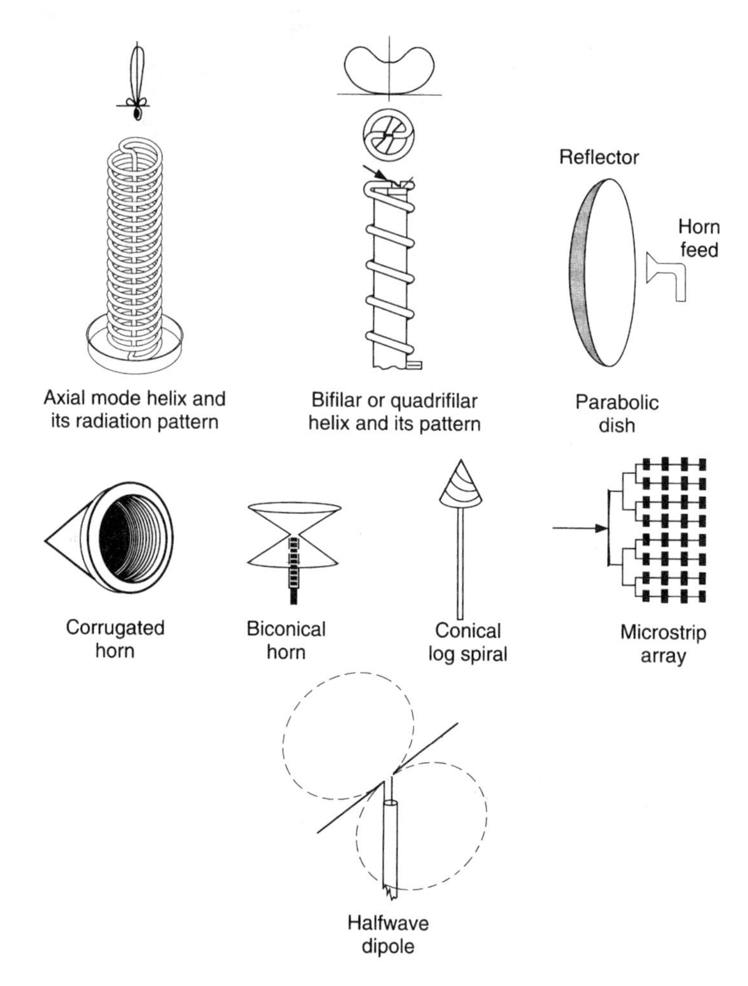

Figure 9.9 Representative space communications antennas.

hemispherical coverage, toroidal beam shapes for spinning spacecraft, high-gain pencil beams for stabilized spacecraft, and so on (figure 9.9, table 9.1). Electronically and mechanically despun antennas have been designed to let spin-stabilized spacecraft take advantage of pencil beam performance. Spacecraft antennas are often large, delicate structures that must be stowed for launch and then erected on orbit. Being outside the spacecraft, they are exposed to extreme heat and cold and to the damaging effects of atomic oxygen and ultraviolet radiation.

Ground-station antennas are typically large parabolic dishes, although various types of helices and linear arrays have been employed for lower gain (broader beamwidth) applications. To achieve high gain from a large dish requires that the surface should not depart from a parabola by more than about 1/20 of a wavelength rms. That presents a major design challenge for a large, steerable dish in strong winds at high frequencies.

Table 9.1 Gain and effective area of several antennas

Type of Antenna	Typical Gain	Effective Area
Isotropic (reference)	1	$\frac{\lambda^2}{4\pi}$
Halfwave dipole	1.64	$\frac{1.64\lambda^2}{4\pi}$
Horm	$\frac{10A}{\lambda^2}$	0.81 A
Axial mode helix ($\pi D \approx \lambda$)	$\frac{6L}{\lambda^2}$	$\frac{L\lambda}{2}$
Parabolic reflector	$\frac{6.9A}{\lambda^2}$	0.55 A
Broadside array (ideal)	$\frac{4\pi A}{\lambda^2}$	A

9.2.6 Propagation Effects in the Space Channel

In the typical space communications geometry the line-of-sight ray passes through the ionosphere and the atmosphere, whose properties were discussed in chapter 2. Their effects can be important in selecting communication frequencies, antennas, and polarization. Figure 9.10 shows the attenuation experienced by passage through the atmosphere. Rainfall can cause additional attenuation, particularly at frequencies above a few gigahertz.

9.2.6.1 Ionospheric Effects

The presence of free electrons in the ionosphere (figure 9.11) causes refraction (bending of the ray and increased delay), dispersion (frequency-dependent delay), and scintillation (rapidly fluctuating amplitude and phase). Fortunately, all ionospheric effects decrease as the square of frequency, so we can minimize them by using higher frequencies or measure them by transmitting two frequencies simultaneously. Additional refraction arises from the troposphere. Unfortunately, tropo refraction is frequency independent, and models are required to correct for it. Passage through the ionosphere, in conjunction with the Earth's magnetic field, also rotates the direction of the electric field vector ("Faraday rotation," figure 9.12). Faraday rotation usually dictates that at least one end of a low-frequency link be circularly polarized.

A more severe ionic effect is that caused by the plasma sheath that surrounds a re-entering space vehicle. Hot plasma attenuation can reach 80 dB or more and act as a completely conductive "shield" around the spacecraft. This can cause a total communications "blackout" on re-entry.

9.2.6.2 Multipath

Rays arriving at an antenna can arrive from both the direct (shortest) path and from an indirect (reflected) path (figure 9.13). The two signals can sum destructively, producing "multipath" fading. Multipath can also shift the times of zero crossings of the signal, which can become the design driver for navigation and high-speed communication systems that require nanosecond-level timing.

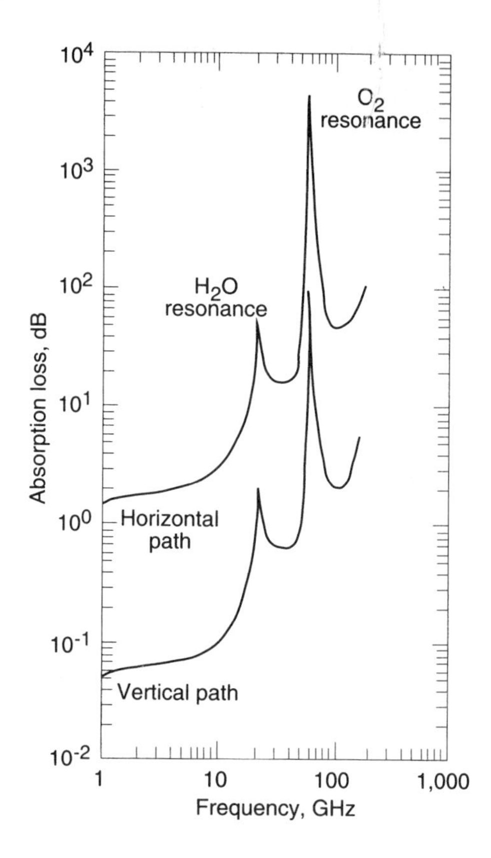

Figure 9.10
Atmospheric attenuation.

9.2.6.3 *Equipment Delays*

Propagation delays in the receiving equipment itself—and particularly variations in these delays—can affect the performance of navigation and tracking systems, altimeters, rendezvous radars, etc. For such systems equipment delays must be kept small and stable, and possibly even calibrated continually during operation.

9.3 Modulation of Radio-Frequency Carriers

For an RF carrier wave to transmit information, some property of the wave must be "modulated" (varied) by that information. For an unmodulated carrier signal, expressed as

$$s(t) = A\cos(2\pi f t + \theta), \tag{9.3.1}$$

it is possible to modulate either the amplitude A or the angle of the cosine, $2\pi ft + \theta$. Frequency modulation (FM) and phase modulation (PM) are two related types of angle modulation.

In analog modulation, some parameter of the carrier is made to vary in direct proportion to the signal being measured. In digital modulation, periodic samples of the signal

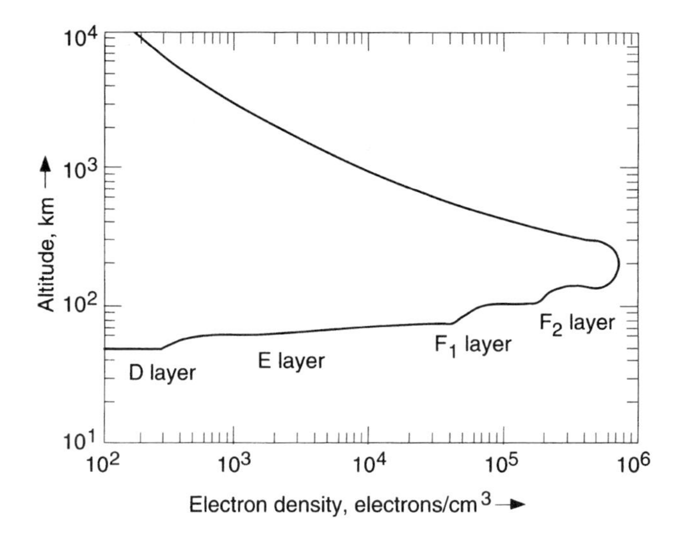

Figure 9.11 Typical daylight ionosphere, mid-latitudes.

to be measured are quantized into discrete amplitude steps and then expressed as an integer number ("digitized"). The carrier is then modulated between discrete (usually two) states.

9.3.1 Analog Modulation

Amplitude modulation (AM, figure 9.14) was historically the first type of analog modulation used and remains the simplest to detect. It thus finds use where receivers must be simple for reasons of cost (car radios) or reliability (early spacecraft command receivers). AM is bandwidth conservative (the RF bandwidth is only twice the information band-

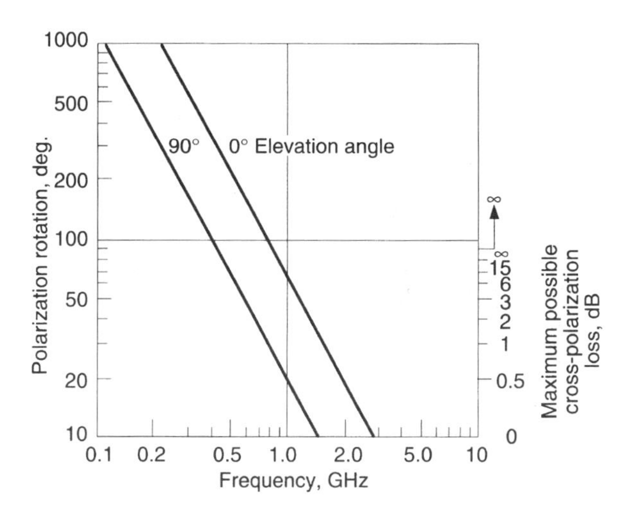

Figure 9.12 Faraday rotation of linearly polarized waves.

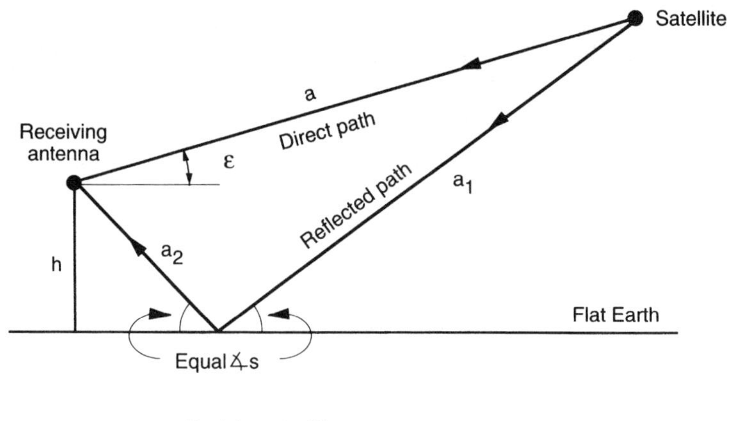

Figure 9.13 Multipath situation (flat Earth approximation).

width) and is linear in its signal-to-noise response (a small decrease in $(S/N)_{in}$ results in a small decrease in $(S/N)_{out}$; see figure 9.15). AM therefore still finds some use when wideband, high-resolution data (e.g., video data) must be sent over band-limited channels.

In FM, the ratio of carrier frequency deviation to modulating frequency is called the *modulation index*, β,

$$\beta = \frac{\Delta f_c}{f_m} \tag{9.3.2}$$

High-performance analog FM usually operates with large modulation indices that "spread" the RF bandwidth to be much larger than the information bandwidth,

$$BW_{RF} \approx 2(\Delta f_c + f_m). \tag{9.3.3}$$

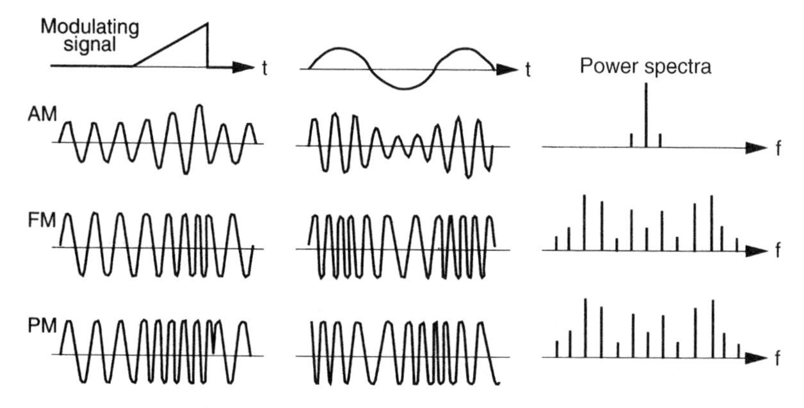

Figure 9.14 AM, FM, and PM analog modulation.

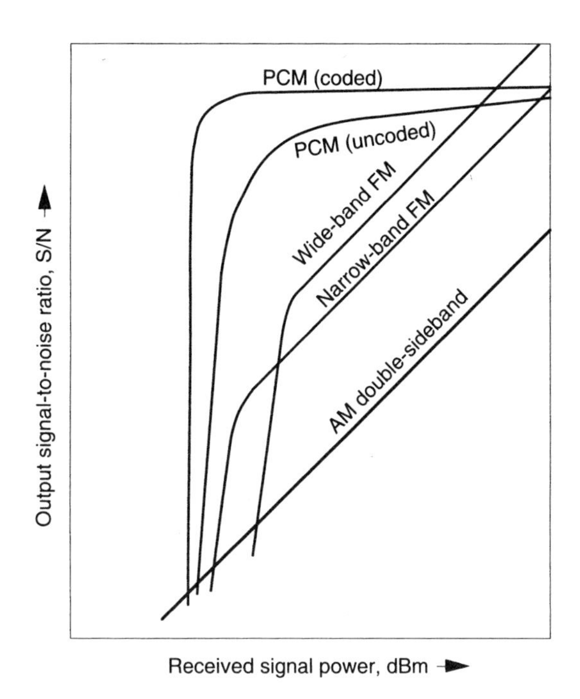

Figure 9.15 Signal-to-noise behavior of analog and digital modulations.

This leads to improved signal-to-noise performance, *provided* the input signal-to-noise ratio exceeds some "threshold." For $\beta \gg 1$, the FM performance improvement over AM can approach

$$\frac{(\mathrm{S/N})_{\mathrm{FM}}}{(\mathrm{S/N})_{\mathrm{AM}}} \approx 3\beta^2. \tag{9.3.4}$$

Below threshold, performance deteriorates rapidly and can be worse than AM. FM thresholds are typically in the range of 9–13 dB S/N.

The performance improvement of high-index FM makes it a good choice for analog data. FM is also sometimes used for digital data when the simplicity of a noncoherent detector is desired.

Because PM performance improvement is limited by the maximum phase deviation of $\pm 180°$, PM is not typically used for analog data. PM requires a receiver that can recover phase ("coherent receiver"); this complication is reserved for digital communication, where phase-shift keying (PSK) is in fact an optimal modulation.

9.3.2 Multiplexing and Digital Modulation

It is impractical to provide a separate RF carrier or baseband subcarrier (frequency multiplexing) for each of the hundreds of measurements on a typical spacecraft. This has spurred the development of time-division multiplexing (TDM, figure 9.16) and various types of pulse and digital modulation (figure 9.17). In analog TDM, multiple data sources are sampled periodically and the samples are interleaved (commutated). If each data source is sampled at a rate of at least twice its highest frequency content (that is, above

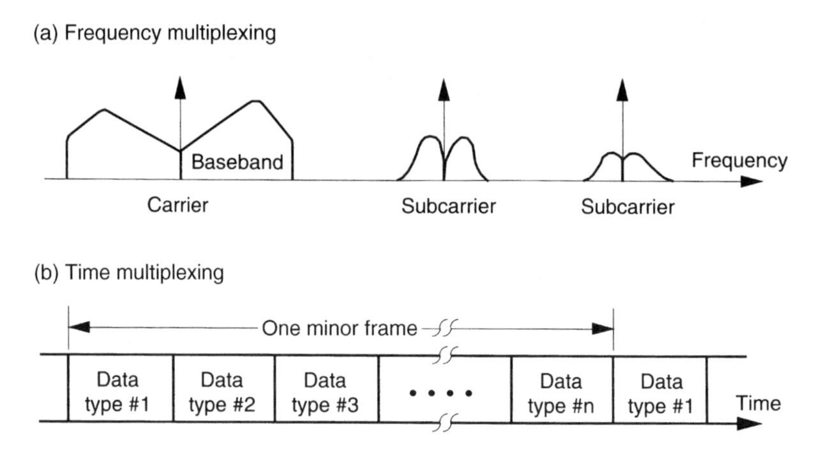

Figure 9.16 (a) Frequency- and (b) time-division multiplexing of data.

the so-called "Nyquist rate"), the samples can be decommutated and the original signal reconstructed without error.

The most important form of digital modulation today is pulse-code modulation (PCM, figures 9.18–9.21). In PCM the sampled data are converted to digital words ("digitized"), which are then assembled into a single serial bit stream (more about this in chapter 10). The bit stream then modulates the RF carrier by either amplitude, frequency, or phase-shift keying (ASK, FSK, or PSK). Note that the ratio of RF bandwidth to bit rate can vary considerably, depending on the modulation format chosen. In digital communications, the bit error rate (BER), rather than noise or distortion, becomes the measure of data

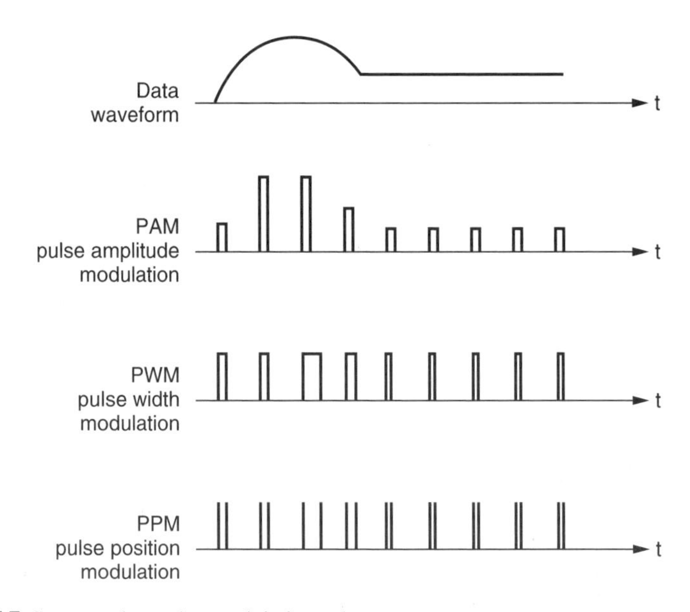

Figure 9.17 Some analog pulse modulation schemes.

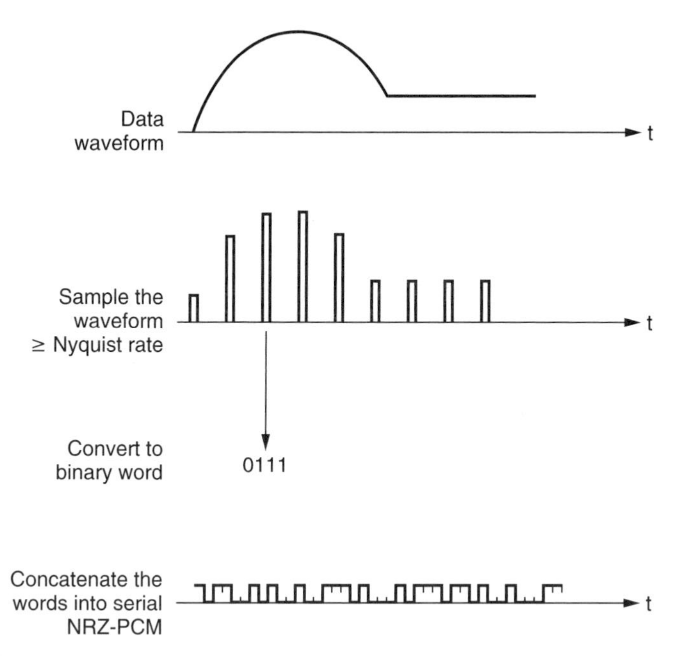

Figure 9.18 Pulse code modulation (PCM).

quality. The job of the RF system is to return the assembled bit stream with an acceptable error rate, using minimum EIRP.

Most space communication channels are "nondispersive" (that is, the time delay is constant with frequency over the bandwidth of interest) and are perturbed by additive, white, Gaussian noise (described in the next section). The theory of optimal detection of digital signals for this case has been highly developed. The theory shows that the most power-efficient modulation is to phase-shift key the carrier phase to be 0° for a binary "0" and 180° for a "1." The matched optimum receiver cross-correlates the received noisy signal against noise-free replicas of the "0" and "1" signals and selects the correlation with the highest output. This PCM-PSK with coherent detection is widely used today. Like other high-performance spectrum-spreading modulations, PSK experiences a sharp threshold effect in its bit error rate (BER) performance.

In phase-dispersive channels, the signal wave shape becomes too distorted for coherent cross-correlation reception to be used. For this type of channel, frequency-shift keyed (FSK) modulation and non-coherent (that is, simple power measurement) reception are often used. In FSK a carrier frequency of f_0 is sent for a "0" and a nearby frequency f_1 is sent for a "1." FSK is more tolerant of amplitude and phase anomalies in the channel and, if noncoherently detected, uses simpler hardware, but exacts a performance penalty of 3–4 dB. Figures 9.21 and 9.26 show the bit error rate performance of several modulation types.

One reason pure PSK (0° and 180°) performs so well is that it converts *all* the RF power into useful data sideband power. But sometimes residual power might intentionally be left in the carrier for ground receivers to lock onto. The ratio of data sideband power to total transmitted power is called *modulation loss*, and must

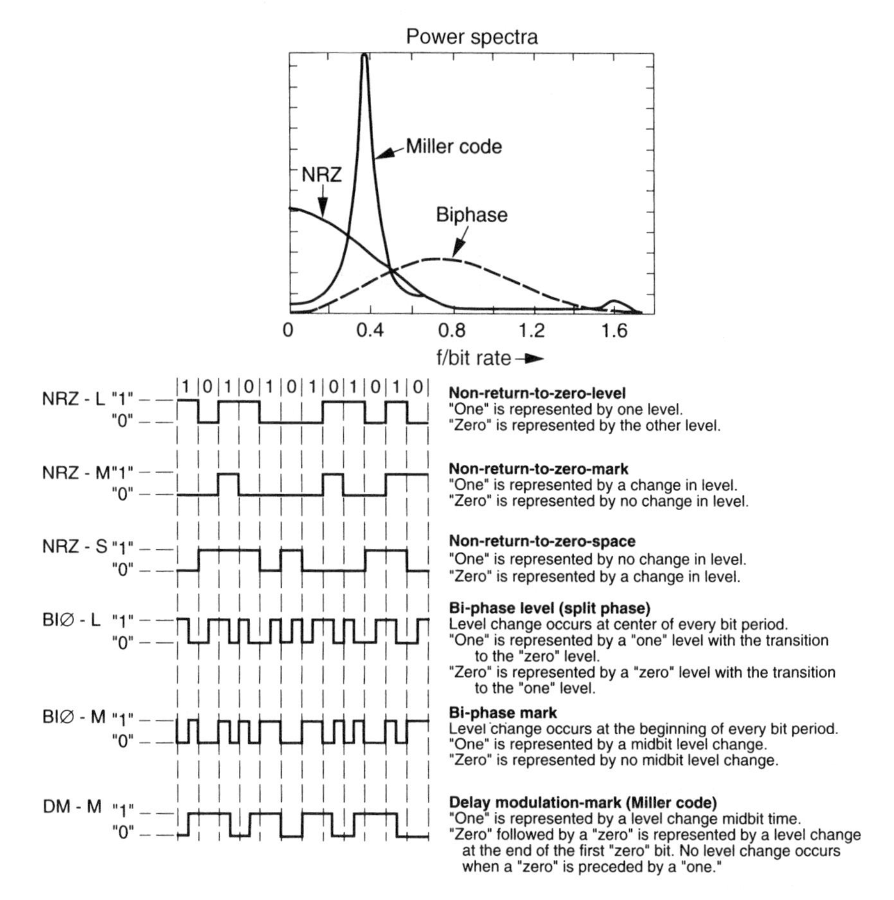

Figure 9.19 Standard PCM code definitions.

be carried in the link design. The problem of apportioning total radiated power among carrier, subcarriers, and data sidebands is a mathematically complex task, especially when modulation index tolerances are considered. We usually select modulation indices such that various portions of the link reach threshold in a desired order as signal level decreases.

Sophisticated pulse modulation schemes require complex receivers, with much of the complexity focused on solving the "synchronization problem." A modern system may require antenna tracking, carrier lock, subcarrier lock, bit sync, error correcting code sync, decryption equipment sync, word sync, and frame sync *all* to be present before information transfer can take place.

9.4 Noise, the Enemy of Communications

Noise is any unwanted interference with the signal. Some noise sources are manmade and predictable, such as PCM quantization noise (discussed in chapter 10) or interference from nearby transmitters. Random noise is unpredictable except in an average sense.

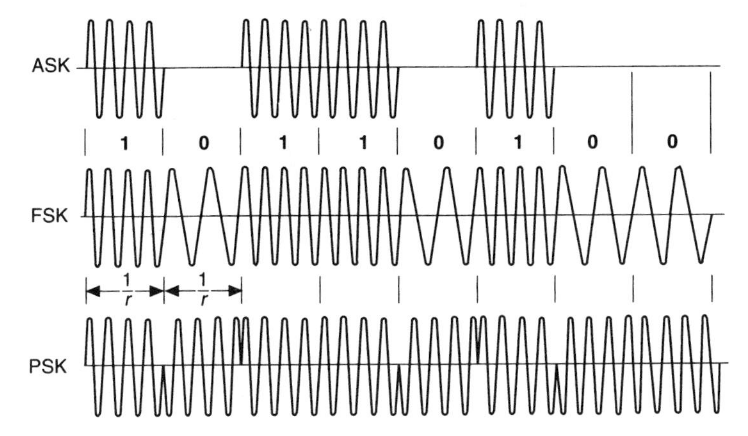

Figure 9.20 ASK, FSK, and PSK digital modulation.

It arises from various natural sources: from resistances; from amplifying devices; from the galaxy, the Sun, and the atmosphere.

Much of the natural noise we deal with has a Gaussian probability density of amplitude. The time behavior of a noise signal can be expressed by how quickly it becomes decorrelated with itself, as measured by the autocorrelation function $R_v(\tau)$ of a noise voltage $v(t)$,

$$R_v(\tau) = \int_{-\infty}^{\infty} v(t)v(t+\tau)dt, \tag{9.4.1}$$

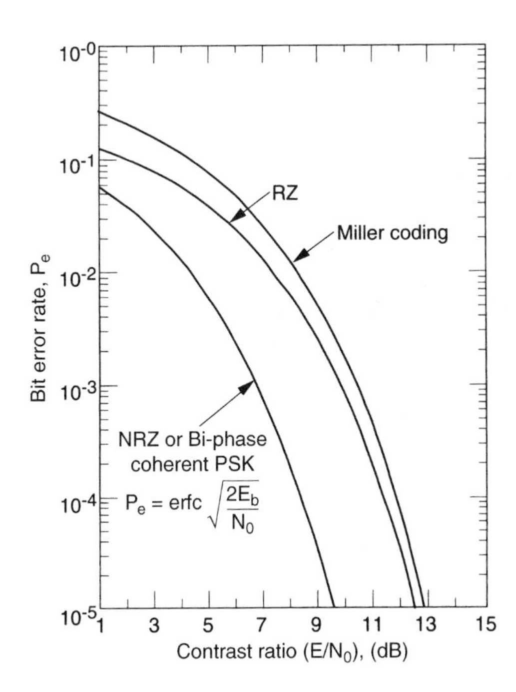

Figure 9.21 Bit error rate versus contrast ratio for three types of PCM (perfect bit synchronization assumed).

which integrates $v(t)$ times a replica of itself delayed by τ. It is usually more convenient to use the Fourier transform of the autocorrelation function, the power spectral density (PSD) $G_v(f)$,

$$G_v(f) = \int_{-\infty}^{\infty} R_v(\tau) e^{-j2\pi f \tau} d\tau, \tag{9.4.2}$$

which describes how the noise power is distributed in the frequency domain. Much of the natural noise we encounter has a power per unit bandwidth, N_0, which is essentially constant with frequency ("white noise"). (Note: N_0 is taken here to be "single sided," that is, to apply only to positive frequencies.) PSD is also useful for showing how *signals* (both deterministic and random) spread their power in frequency or occupy bandwidth. For example, the split-phase PCM signals in figure 9.19 occupy twice the bandwidth of the NRZ signals.

The intensity of a noise source can also be expressed through its noise "temperature" in kelvins,

$$T = \frac{N_0}{k} \frac{(\mathrm{W/Hz})}{(\mathrm{W/Hz\ K})} \tag{9.4.3}$$

where k is Boltzmann's constant, 1.38×10^{-23} W/(Hz K).

All resistors, and lossy attenuators of any type, exhibit thermal noise proportional to their resistance and physical temperature. The noise is "white," so its total power is proportional to bandwidth. Attenuation through the atmosphere and through cables, filters, and connectors is a common source of this noise.

Figure 9.22 shows the effect of additive white Gaussian noise on a PCM signal. The signal-to-noise ratio is 9 dB, in a bandwidth equal to the bit rate.

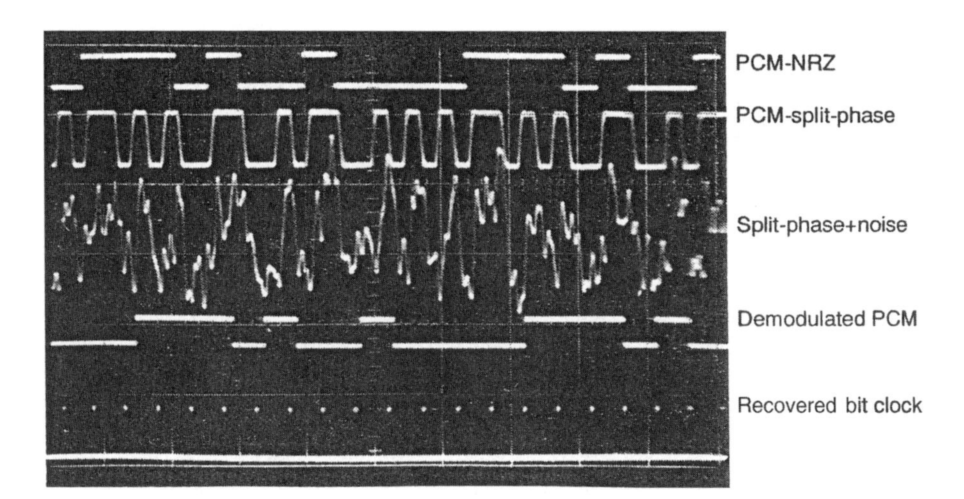

Figure 9.22 PCM signal plus noise ($E/N_0 = 9$ dB, measured BER $= 0.001$).

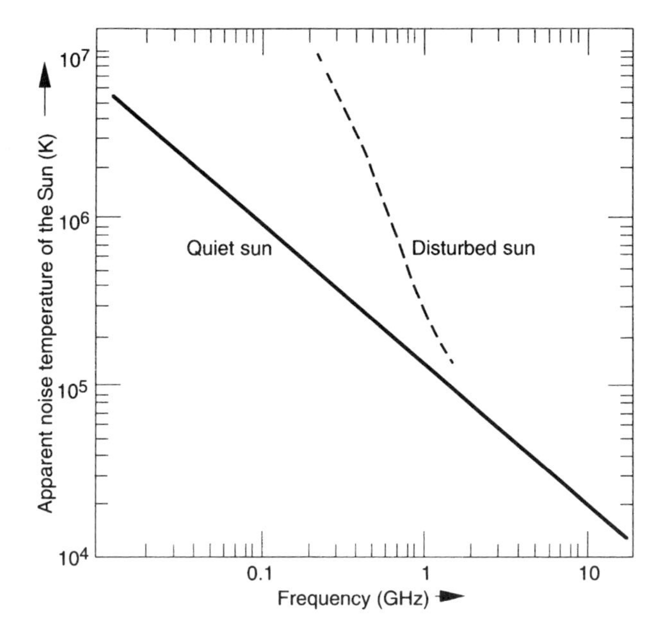

Figure 9.23 Sun noise.

9.4.1 Sources of Noise

The Sun is a powerful radio noise source whose noise temperature can range up into the millions of kelvins. Solar flares and sunspot activity can cause order-of-magnitude increases in RF noise (figure 9.23). Missions must usually be designed to keep from pointing the ground antenna main lobe (and often side-lobes as well) at the Sun when critical communications are needed.

The Earth is modeled as a 290 K blackbody radiator. Therefore, that portion of an antenna looking at the Earth will "see" a noise input of 290 K.

Galactic and cosmic noise arises from radio stars within our galaxy, from ionized interstellar gas, and from diffuse sources outside our galaxy. The noise power falls with frequency roughly as f^{-2}, becoming negligible above 2 GHz (figure 9.24). It can be ten times more intense from the direction of our galactic center than from the galactic poles.

The attenuation of the atmosphere and troposphere rises with both frequency and path length (as determined by antenna elevation angle). These losses contribute to the total sky noise. Where molecular absorption by oxygen and water vapor causes attenuation peaks, the noise also peaks. The combination of high attenuation (figure 9.10) and noise at the oxygen resonance makes the atmosphere essentially opaque to communications at 60 GHz. This permits satellite–satellite communication without fear of interception or jamming from the Earth.

A major and often dominant source of noise is the receiver electronics itself. Deep-space communications in particular has stimulated the development of ultra-low-noise receivers using cryogenically cooled maser front ends. Less demanding requirements use cooled parametric amplifiers, uncooled paramps, GaAs FET amplifiers,

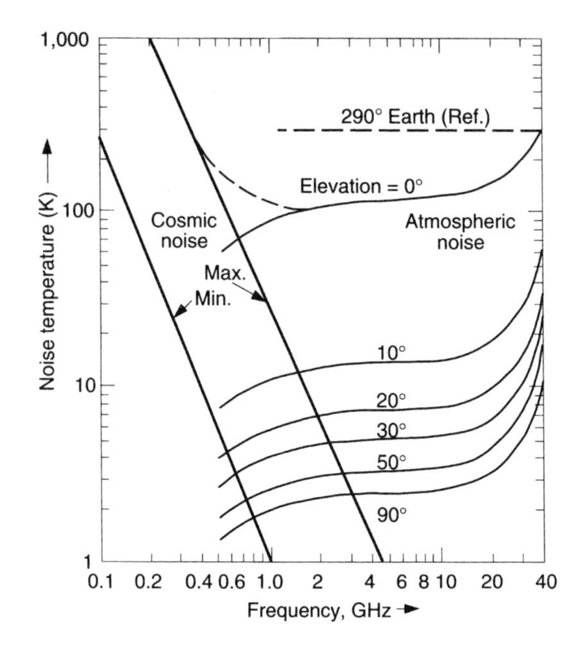

Figure 9.24 Sun noise.

and so on (see figure 9.25). A receiver's noise performance is sometimes given by its "noise figure" (NF) in dB or "noise factor" (F), which are related to the noise temperature by

$$T = 290\left(10^{\frac{\mathrm{NF(dB)}}{10}} - 1\right) = 290(F - 1). \tag{9.4.4}$$

9.4.2 System Noise Temperature

When a receiver is made up of several cascaded stages, each with its own gain G_i and noise temperature T_i, the composite noise temperature at the input of the cascade is given by the Friis equation

$$T_{\text{total}} = T_1 + \frac{T_2}{G_1} + \frac{T_3}{G_1 G_2} + \cdots \tag{9.4.5}$$

and the composite noise factor by

$$F_{\text{total}} = F_1 + \frac{F_2 - 1}{G_1} + \frac{F_3 - 1}{G_1 G_2} + \cdots. \tag{9.4.6}$$

The Friis equations show that the noise figure of a passive loss ($G < 1$) is equal to the loss itself. From equation 9.4.5 we see that to achieve lowest receiver noise it is essential that the *first* stage should have low noise and high gain, and that any losses preceding the first active stage should be held to a minimum. For this reason, the low-noise preamp stage is often mounted right at the antenna feed point. At the receiving station all these diverse noise sources must be combined to arrive at the overall "system operating noise temperature," T_s.

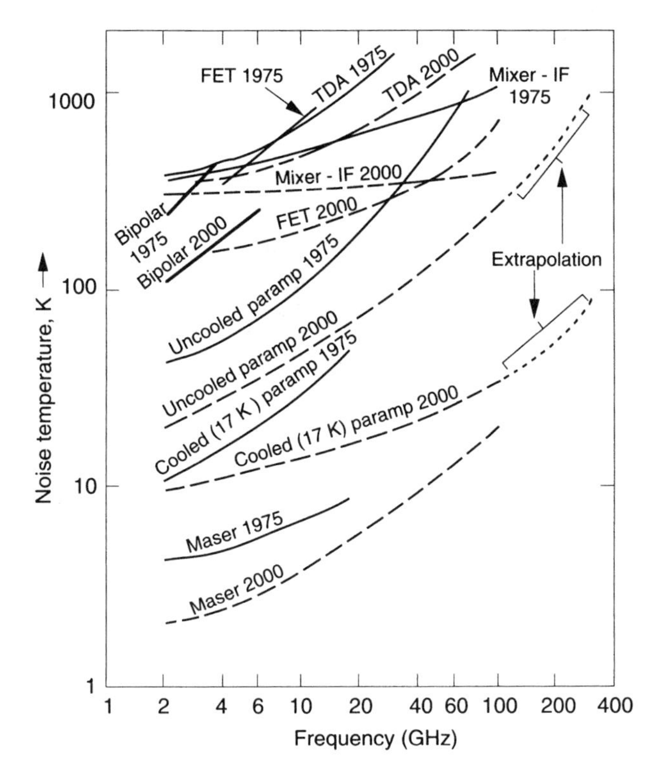

Figure 9.25 Typical receiver noise performance. (Courtesy of NASA GSFC.)

The antenna itself has many contributors to its antenna noise temperature, T_a. For example, the antenna main lobe may see galactic and atmospheric noise, some Sun noise may sneak in on a sidelobe, and the back lobe may pick up a contribution from the 290 K Earth. T_a is found by integrating the product of antenna gain and the noise source temperature over the sphere surrounding the antenna,

$$T_a = \frac{1}{4\pi} \int_{\phi=0}^{2\pi} \int_{\theta=0}^{\pi} G(\theta, \phi) T(\theta, \phi) \sin\theta \, d\theta \, d\phi. \tag{9.4.7}$$

In practice, antenna gain and noise source temperature are often constant over given portions of the sphere, so equation 9.4.7 can be approximated by

$$T_a \approx a_1 G_1 T_1 + a_2 G_2 T_2 + a_3 G_3 T_3 + \cdots, \tag{9.4.8}$$

where a_i is the fraction of the sphere over which G_i and T_i are constant, and

$$a_1 + a_2 + a_3 + \cdots = 1. \tag{9.4.9}$$

Cable, feed, and filter losses add more noise, and the receiver itself always adds noise.

T_s and received signal calculations must be referenced to the same point for consistency; usually the antenna feed is selected. The ideal receiving station has high antenna gain and low T_s. A useful "figure of merit" for receiving stations is G/T, expressed in dB/K.

9.4.3 Signal-to-Noise Ratio

An important design parameter is the signal-to-noise ratio, S/N. It is the ratio of signal power to noise power in a given bandwidth. For digital communications, we often use the bandwidth-independent parameter E_b/N_0, which is the energy per bit divided by the noise PSD. E_b, of course, is simply signal power divided by the bit rate, so that E_b/N_0 is equivalent to S/N in a bandwidth equal to the bit rate.

Noise and bandwidth combine ultimately to limit the rate and accuracy of information transfer. In 1948 Claude Shannon proved that a channel W Hz wide with a signal-to-noise ratio of S/N could transmit no more than

$$C = W \log_2(1 + S/N) \tag{9.4.10}$$

bits/second, error free. In fact we do much worse than this in real systems, and the past half century has been spent devising modulation and coding schemes that try to approach the "Shannon capacity." By setting $N = N_0 W$ and letting W go to infinity, we can show that error-free digital communication cannot take place below an E_b/N_0 of -1.6 dB ($= \ln 2$).

Exercise

Prove this statement.

Figure 9.26 shows the current state of the art. As a rule, the price paid for high performance is bandwidth spreading, abrupt thresholds, complex coding and decoding equipment, and computational delays during decoding. (Coding is discussed in section 9.6.2.)

9.4.4 Link Margin

To find the E_b/N_0 required for a particular link, we began with the theoretical E_b/N_0 for the particular modulation type and desired bit error rate. For example, the bit error rate for PSK modulation follows the complementary error function shown in figure 9.21, and for a bit error rate of 10^{-5} an E_b/N_0 of 9.6 dB is required. To this we add an "implementation loss" that quantifies the difference between practical detection hardware and theoretical performance. Depending on link complexity, this can range from 0.5 to 2.5 dB. The difference between $(E_b/N_0)_{\text{reqd}}$ and the actual $(E_b/N_0)_{\text{rcvd}}$ is called the *link margin*. Margins are usually specified on the basis of link complexity and the consequences of link failure. Typical required minimum margins are 3 dB for spacecraft downlinks, 6 dB for uplinks, and 12 dB for range safety (for example, command destruct) links.

9.5 Space Communication Link Design Example

A satellite at geostationary altitude (35,800 km) must communicate 1000 bps to shipboard terminals with a bit error rate $<10^{-5}$. It must cover every point on the Earth for which it is above 10° elevation. Spacecraft pointing error is $\pm 1^\circ$. Because of the ship's roll and pitch, the receiving antenna cannot be aimed more accurately than $\pm 15^\circ$. Assume

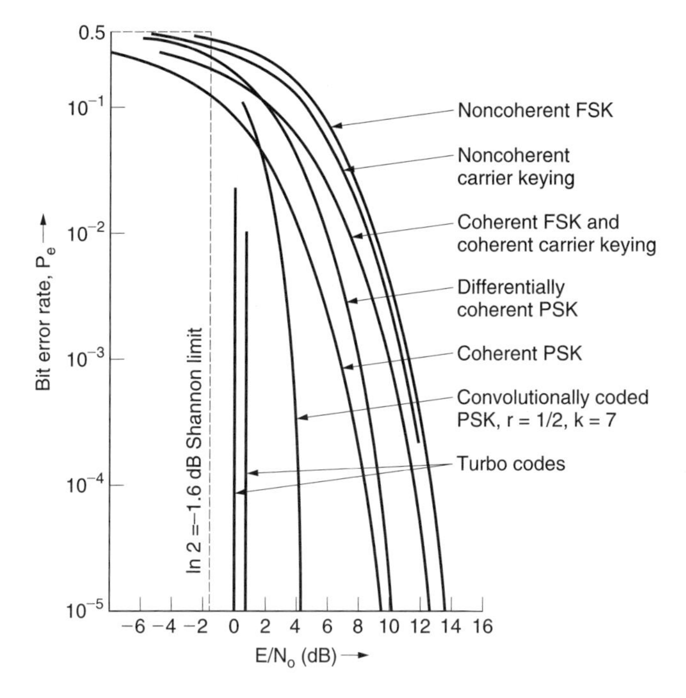

Figure 9.26 Performance of digital modulation techniques.

that frequency allocations are available at UHF (400 MHz), L-band (1500 MHz), and S-band (2200 MHz). Which is the best frequency to use? What antennas are needed? How much transmitter power is needed for a 3 dB link margin?

Solution

Begin by using figure 9.2 to establish the basic geometry:

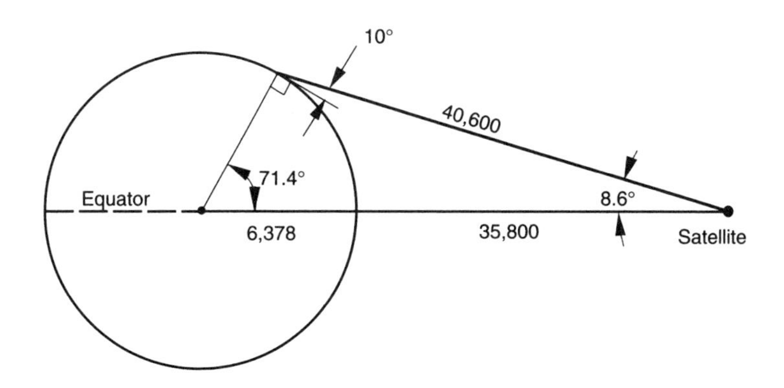

The full coverage cone is $2 \times (8.6° + 1°$ pointing error$) = 19.2°$. A uniformly fed, single spacecraft antenna to cover this cone would have a

$$\text{beam width}_{\text{opt}} \approx 0.88 \times 19.2° = 16.9°.$$

The maximum spacecraft antenna gain depends on whether a pencil beam or a toroidal beam is selected:

$$\left.\begin{aligned} \overline{G} &= \frac{27{,}000}{(16.9)^2} = 94.5 = 19.8 \text{ dBic} \\ G \text{ at edge} &= 19.8 - 3.9 = 15.9 \text{ dBic} \end{aligned}\right\} \text{pencil beam antenna}$$

$$\left.\begin{aligned} \overline{G} &= \frac{27{,}000}{360^\circ \times 16.9^\circ} = 4.4 = 6.5 \text{ dBic} \\ G \text{ at edge} &= 6.5 - 3.9 = 2.6 \text{ dBic} \end{aligned}\right\} \text{toroidal or pancake beam antenna}$$

The difference between the two antenna options is more than 13 dB. This could mean a 20:1 difference in transmitter power, so let us temporarily assume that the system will need the higher gain of the pencil beam antenna. This establishes a system requirement for a stabilized spacecraft antenna. The antenna situation on the ship is pictured here:

The ship's motion and view angle requirements dictate that the shipboard antenna cover at least 5° below its deck horizontal. A single antenna to provide this coverage would have a pattern of

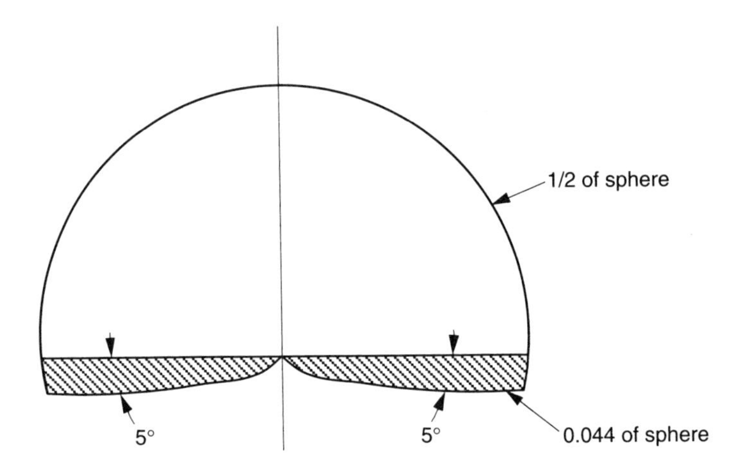

The gain of such an antenna, if 50% efficient, could not exceed

$$G_r \tilde{\leq} 0.5 \times \frac{1}{0.5 + 0.044} \approx -0.4 \text{ dB}.$$

	Cosmic noise		Atmospheric noise		Backlobes 0.05 × 290K		T_a
at 400 MHz	200 K	+	7 K	+	15 K	=	222 K
at 1500 MHz	12 K	+	12 K	+	15 K	=	39 K

We will use a working value of $G_r = -1$ dBic. If we want simple, unsteered antennas at each end of this link, we see that they are each constrained in beamwidth, or, what is the same thing, constrained in gain. Therefore this is a gain–gain type link, and selection of the lowest frequency (400 MHz) will minimize transmitter power. This assumes that the required spacecraft antenna is not impractically large at this frequency. If the spacecraft antenna is too large to build, it would become a gain–area type link. So the next step is to test the feasibility of the spacecraft antenna.

We want to estimate the size of, say, a dish antenna at 400 MHz ($\lambda = 0.75$ m) that has a maximum gain of 19.8 dB. Assuming 55% efficiency ($\eta = 0.55$), equation. 9.2.4 gives

$$\overline{G} = \eta \left(\frac{\pi D}{\lambda}\right)^2 = 19.8 \text{ dB},$$

Then $D = 3.15$ m (10.3 ft). This is somewhat large for a spaceborne dish, so we might have to settle for a spacecraft antenna that is smaller than optimum. But since the path loss at the next highest frequency is $(1500/400)^2 = 11.5$ dB higher, 400 MHz would remain the best choice until the spacecraft dish became so small (2.75 ft diameter) that its boresight gain dropped below $19.8 - 11.5 = 8.3$ dB.

We can compute the spreading or path loss from equation 9.2.10:

$$\begin{aligned}\text{Path loss (400 MHz)} &= 32.5 + 20\log(40{,}600) + 20\log(400)\\ &= 176.7 \text{ dB}.\end{aligned}$$

We next need to examine shipboard antenna noise and total system noise temperatures at the two frequencies we are trying to compare. Assume that the upper hemisphere of the shipboard antenna receives cosmic and atmospheric noise in accordance with figure 9.24, and that the backlobe, representing 0.044 of a sphere, intersects the 290 K Earth. The upper hemisphere represents $0.5/(0.5+0.044) = 92\%$ of the pattern, and the backlobes the remaining 8%. Then, using the lumped sum approximation equation 9.4.8, we have:

	Cosmic noise		Atmospheric noise		Backlobes at 290 K		T_a
at 400 MHz	0.92 × 209 K	+	0.92 × 7 K	+	0.08 × 290 K	=	222 K
at 1500 MHz	0.92 × 7 K	+	0.92 × 10 K	+	0.08 × 290 K	=	39 K

The large difference in natural noise at these two frequencies will have to be considered. But, before jumping to any conclusions, we must compute *total* system noise temperatures, T_s. Making typical assumptions about receiving equipment gains, losses, and noise figures (and using figure 9.25), we have:

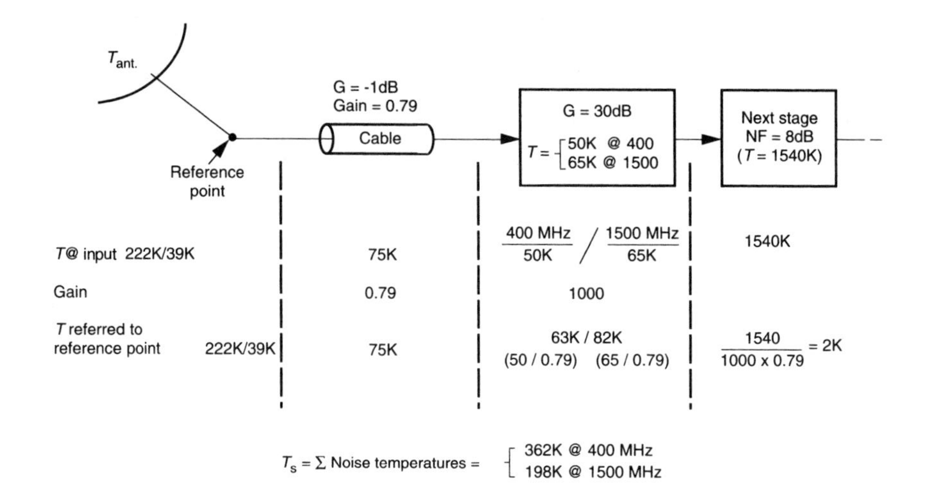

The ratio of total noise (T_s) is 2.6 dB in favor of 1500 MHz; so the decision to stick with a choice of 400 MHz is valid only if we can get a spacecraft gain (on boresight) of at least $19.8 - 11.5 + 2.6$ dB $= 10.9$ dB. This would require a 3.7 ft dish (or equivalent) at 400 MHz.

The total noise density N_0 at 400 MHz can now be calculated:

$$N_0 = kT_s = 1.38 \times 10^{-23} \times 362 \text{ K} = 5 \times 10^{-21} \text{J/K} \times \text{K}$$

$$\begin{aligned} N_0 &= 5 \times 10^{-21} \text{ W} - \text{s} \\ &= 5 \times 10^{-21} \text{ W/Hz} \\ &= 5 \times 10^{-18} \text{ mW/Hz} \\ &= -173.0 \text{ dBm/Hz}. \end{aligned}$$

We are now ready to compute the required spacecraft transmitter power, P_t. An easy way to do this without using a spreadsheet is to run a trial link calculation with 1.0 W transmitter power and adjust it after comparing it with the desired result. Here

Trial RF Link (400 MHz)	
Transmitter power, 1 W	+30.0 dBm
Modulation loss	−1.0 dB
Spacecraft cable & filter losses	−0.5 dB
Spacecraft antenna gain (worst angle)	+15.9 dB
Path loss, 10° elevation	−176.7 dB
Polarization loss, circ/circ	−0.5 dB
Shipboard antenna gain	−1.0 dB
Received power	−133.8 dBm
Bit rate, 1000 bps	+30.0 dB-Hz
Energy/bit, E_b	−163.8 dBmJ
Noise density, $N_0 (T_s = 362 \text{ K})$	−173.0 dB m/Hz
Received E_b/N_0	+9.2 dB

the designer has chosen to leave some unmodulated power in the carrier to aid signal acquisition, reducing data sidebands by 1 dB. There is also a 0.5 dB polarization loss between the two nominally circularly polarized antennas, representing their imperfect axial ratio.

From figure 9.26, the theoretical E_b/N_0 required for a 10^{-5} bit error rate is 9.6 dB. Adding 1.4 dB typical implementation loss, we need $9.6 + 1.4 = 11.0$ dB. We are therefore short by $11.0 - 9.2 = 1.8$ dB, and we must add the 3 dB required margin. We therefore need 4.8 dB more than the assumed 1.0 W, or

$$P_t = 3.0 \text{ W}.$$

9.6 Special Communications Topics

9.6.1 Optical Communications

As we go up in frequency, antenna beams can be made very narrow, making interception difficult. K_a-band (27–40 GHz) is used for military communications for just this reason. However, the ultimate in high-frequency, narrow-beam satellite communications is optical communications using a laser transmitter. The beam can be made so narrow that interception is virtually impossible. Optical links can use antennas that are just a few percent of the size and mass of RF antennas, yet offer the possibility of 10–100 × higher data return because the very narrow beam concentrates a larger fraction of the transmited power onto the ground receiving aperture.

Of course, working with narrow optical beams requires ultraprecise pointing. This might require gimbaling the optical antenna portion of the spacecraft, and possibly an RF "beacon" to aid optical acquisition. However, laser communications is attractive for robust satellite–satellite links if the satellites must be well stabilized for other reasons. Attenuation and dispersion in the atmosphere (and by clouds) are also significant problems, and might require geographical diversity in the ground stations. Even better (for deep space optical communications) would be an optical relay satellite in Earth orbit, well above the clouds.

Deep space optical communications is receiving increased attention by the military (for wideband and secure communications), by NASA (for high data rates at interplanetary ranges), and by commercial satellite providers (particularly for satellite–satellite links). NASA's Galileo Optical Experiment (1992) transmitted digital data optically to an experimental receiver aboard the Galileo spacecraft as it swung by Earth on its way to Jupiter. Communications was successfully demonstrated out to a range of 6 million km. With industry now beginning to supply flightworthy optical components, optical communications in space has a promising future.

9.6.2 Error-Correcting Codes

By adding extra "check" bits to a data stream it is possible to detect, and even to correct, bit errors. Coding can be as simple as appending a single parity check bit to each word, or as complex as concatenating two powerful codes in series, each designed to attack a certain kind of error. Modern codes can detect and correct multiple errors, even those

Table 9.2 Performance of various coding/decoding methods

Code	Decoding Method	Coding Gain at $P_e = 10^{-5}$	Complexity
Block	Majority—hard decision	1.5–3.5 dB	Simple
Block BCH	Algebraic	1.5–4 dB	Complex
Convolutional	Threshold—hard decision	1.5–3 dB	Fairly simple
Convolutional	Viterbi—soft decision	4.5–5.5 dB	Fairly complex
Convolutional	Sequential—soft decision	5–7 dB	Fairly complex
Concatenated block-convolutional	Viterbi + algebraic	6.7–7.5 dB	Very complex
Turbo	Maximum à posteriori	8.8–9.4 dB	Fairly complex

Note: Theoretical BPSK requires $E_b/N_0 = 9.6$ dB for $P_e = 10^{-5}$.

occurring in long bursts. In fact, a very sophisticated coding scheme is used in audio compact discs, a consumer product.

The addition of check bits means that the overall bit rate must increase to preserve the original data rate. This reduces the E_b/N_0 of the channel bits, but the coding more than compensates, resulting in a net "coding gain." The coding gain can be substantial, as shown by table 9.2 and figure 9.26, thus allowing the EIRP to be reduced accordingly. Most coding techniques are asymmetric; that is, they are almost trivially easy to *en*code but far more difficult to *de*code. This typically limits their use to spacecraft downlinks, which are usually more power-starved than the uplinks anyway. An example is convolutional coding, often used for satellite downlinks because it is simple to encode and the complex decoding is kept on the ground. A typical implementation uses as many check bits as there are information bits, making the "symbol" rate twice the data rate but in return yielding 5 dB net coding gain.

Figure 9.27 shows three simple examples of coding. In (a) a single parity check bit, chosen to maintain an odd number of ones, can detect (but not correct) any odd number of errors. The concept is expanded in (b) to a "block code" with row and column parity that can detect and correct a single error in the block of data words. A completely different coding concept—convolutional coding—is shown in (c). Here, a pair of carefully computed "symbols" is transmitted for each data bit. An advanced derivative of this code—Turbo code—represents today's state-of-the-art, less than 2 dB away from Shannon's limit. Coding is another example of trading bandwidth spreading for higher performance, as predicted by Shannon's equation (9.4.10).

9.6.3 Encryption and Authentication

When communications must be kept secret they are encrypted. Typically this is done by adding (modulo-2) a random, secret key stream to the "plaintext" bit stream to generate a "ciphertext" stream. (Analog communications can also be scrambled, but the techniques are less secure.) The receiver must have an identical and synchronized copy of the key. If the key is truly random (generated by coin tosses, for example) and is used only once without repeats, the encryption is provably unbreakable. This "one-time pad" is not practical for large quantities of data, however.

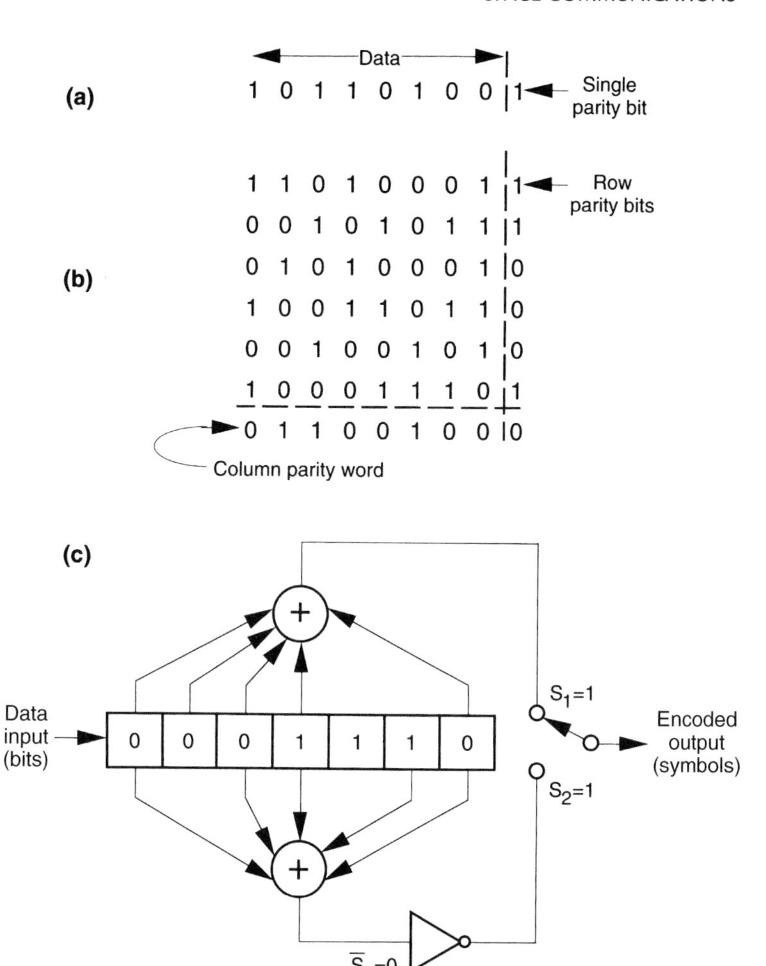

Figure 9.27 Three simple coding schemes: (a) single parity bit per word; (b) row-column parity; (c) convolutional code (rate = 1/2, constraint length = 7).

Practical systems generate a key stream using pseudorandom logic whose exact principles are highly classified. The key streams are designed not to repeat for a *very* long time, to appear to be truly random, and to have no structural fault that might reveal the method of generation. Some forms of encryption exhibit "error extension," so that a single error in the ciphertext input to the decryptor may cause a long burst of errors in the plaintext output. Some additional power may have to be added to the link to compensate.

In uplink communications, we usually do not care about secrecy but rather want to guarantee that commands have come from a legitimate source. Authentication can be achieved by using pseudorandom "challenge" and "reply" word pairs or by augmented encryption techniques. Modern authentication systems are resistant even to playback of previous legitimate commands intercepted by an enemy ("anti-spoof" feature).

9.6.4 Low-Probability-of-Intercept (LPI) Communications

Sometimes the goal is to hide the fact that communication is even taking place. Ultra-narrow beamwidths and optical frequencies can play a role here, but another technique is to spread the transmitted power over such a wide bandwidth that the signal is completely buried in the noise (that is, its peak value falls below the noise floor, N_0).

One "spread spectrum" technique is to add modulo-2 to the information bit stream a pseudorandom spreading sequence with a "chipping rate" much higher than the bit rate (see figure 9.28). This widens the bandwidth by the ratio of the chip rate to the data rate (the "processing gain"), which can be 40–50 dB or more. The friendly receiver has an exact, synchronized replica of the spreading sequence, allowing it to despread the spectrum back to the data rate. But an observer without the sequence or the synchronizing information might never even suspect there is a signal buried in the noise. The principal problem in spread spectrum communications is to obtain sync in a timely fashion (and to hang on for dear life once obtained!). The presence of large Doppler shifts makes spread spectrum acquisition for satellites even more difficult, turning it into a two-dimensional search in time and frequency.

A related spread spectrum technique is to "frequency hop" the RF carrier every few bits over a wide bandwidth according to a pseudorandom pattern known to only the transmitter and receiver. Performance is similar to the "direct sequence" spread spectrum mentioned, but with faster sync acquisition and better tolerance for some types of fading, dispersion, and multipath.

For satellite-to-satellite covert communications, it is possible to take advantage of the oxygen absorption attenuation peak in the atmosphere at 60 GHz. "Hiding behind the oxygen line" can provide an additional 100–200 dB of shielding from inquisitive receivers on the Earth. Optical and 60 GHz links are two competing satellite–satellite technologies being studied by both military and commercial users.

9.6.5 Antijam Techniques

Often in military communications it is necessary to communicate in the presence of intentional enemy jamming. The jammer may even be closer to the receiver than the transmitter is and may have a substantial power advantage. Again, spread spectrum comes to the rescue. The received signal consists of the wideband spread spectrum signal plus the jamming carrier (a single spectral line). The receiver despreads the received signal by multiplying it by a synchronized replica of the spreading signal. This collapses the incoming spectrum to recover the desired information signal and simultaneously *spreads* the jamming signal (because it was not premultiplied by the secret spreading sequence). The jamming power is now spread over a wide bandwidth, at a correspondingly reduced amplitude. Subsequent bandpass filtering rejects all the jamming "noise" except for the small amount falling in the information bandwidth. A narrowband jammer is thus suppressed by an amount equal to the processing gain (53 dB in the case of the Global Positioning System signals).

Another effective antijam technique is to steer a null in the receiving antenna pattern in the direction of the jammer at the same time that the main beam is kept aimed toward the friendly transmitter. Null-steering phased arrays are challenging to design but can provide an additional 15–30 dB of jammer rejection.

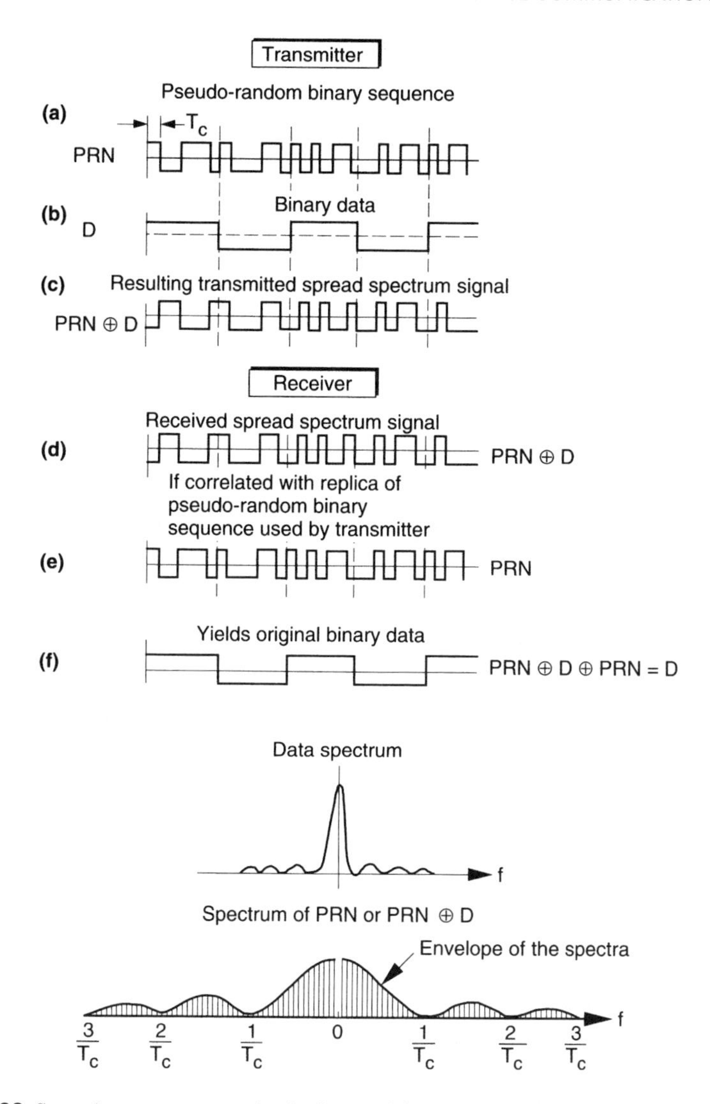

Figure 9.28 Spread spectrum operation in time and frequency.

9.6.6 Range and Range-rate Tracking

An important system design consideration is what level of spacecraft tracking accuracy is required and how it will be achieved. A number of techniques are available. For least impact on the satellite itself, radar or optical tracking from the ground can be employed (an obvious choice for non-cooperating satellites as well). Satellite cross-sections can be enhanced with radar corner reflectors, radar transponders, laser cubes, or flashing lights. The best "skin track" radars (no transponder) can measure range to well under 1 m and angle to a few tens of microradians.

More common is RF tracking by Doppler shift. The satellite can transmit a very stable beacon frequency whose one-way Doppler shift can be tracked by a network of ground stations. A precise orbit can be calculated from surprisingly small samples of Doppler shift. Dual-frequency tracking is often used to calculate out ionospheric delays. Another approach is to measure two-way Doppler shift by transponding the uplink frequency received at the spacecraft back to the ground. Of course, the downlink frequency cannot be the same as the received frequency, but is made "coherent" with it (that is, made a rational fraction of the uplink frequency). This removes spacecraft oscillator errors and simplifies the ground station hardware. Spaceborne transponders are somewhat more complex than separate receivers and transmitters, but may eliminate the need for an ultrastable oscillator.

For slowly moving or geostationary satellites we must measure ranges, rather than range rates, by transponding tones or digital pseudonoise codes from the ground to the satellite and back to the ground station. Pseudonoise tracking can measure spacecraft range to a meter or better, and works out to planetary distances.

For low-Earth-orbit satellites, both Doppler and ranging provide good tracking data. For geostationary satellites the Doppler signal is virtually nonexistent, so ranging and angle tracking are usually used.

Modern satellites can also do their own tracking onboard, without depending on ground stations. For example, the satellite can measure range and Doppler signals from a subset of navigation satellites in the Global Positioning System. This was first demonstrated on Landsat-D in 1982 using the GPSPAC (GPS PACkage) spaceborne navigation set. In addition to being autonomous, this approach is passive and emits no signals that could reveal the satellite's location. The ultimate in autonomy is perhaps the spaceborne star sextant, but this system is too complex for most spacecraft.

Problems

1. Do the exercise in section 9.1.
2. Do the exercise in section 9.2.4.
   Answer: 37.85 dB; 196.0 dB.
3. A satellite in a 1000 km circular, polar orbit must communicate 20,000 bps to a ground station whenever the satellite is above 20° elevation. The satellite is vertically stabilized with negligible attitude error and transmits at 2200 MHz.
   (a) What is the maximum slant range?
   (b) How many minutes will the longest pass last?
   (c) How many minutes will the shortest pass last?
   (d) By how many dB will the signal strength vary between minimum and maximum slant range?
   Answers: (a) 2121 km; (b) 9.15 min; (c) zero; (d) 6.5 dB.
4. Assume the ground station has a 55% efficient 15 ft diameter dish antenna and a system noise temperature of 400 K. Assume an E_b/N_0 of 12 dB is required, allow 3 dB for all miscellaneous losses, and carry a 3 dB link margin.
   (a) What is the wavelength?
   (b) What is the gain of the ground-station dish?
   (c) What is the maximum path loss in dB?
   (d) What is the gain at worst-view angle of the optimum spacecraft antenna?
   (e) Set up the link budget.
   (f) How much transmitter output power does the spacecraft require?
   Answers: (a) 0.136 m; (b) +37.9 dBi; (c) −165.9 dB; (d) +0.8 dBi; (f) 36 mW.

References

Bokulic, R. S., 1991. Use basic concepts to determine antenna noise temperature. *Microwaves & RF*, pp. 107–116.

Hoffman, E. J., and W. Birmingham, 1978. GPSPAC: A spaceborne GPS navigation set. *IEEE Position Location and Navigation Symposium. Record*, Nov. 1978.

Holmes, J. K., 1982. *Coherent Spread Spectrum Systems.* New York: Wiley.

Interplanetary Network Progress Reports. JPL Deep Space Network: *http://tda.jpl.nasa.gov/progress_report/index.html*

Levanon, Nadav, 1988. *Radar Principles.* Broderick, CA: Wiley.

Lin, S., 1970. *An Introduction to Error-Correcting codes*, Englewood Cliffs, NJ: Prentice-Hall.

Roddy, Dennis, 1996. *Satellite Communications.* New York: McGraw-Hill.

Sklar, B., 2001. *Digital Communications: Fundamentals and Applications.* Englewood Cliffs, NJ: Prentice-Hall. Revised edition with CD-ROM.

Spilker, J. J., 1977. *Digital Communications by Satellite*, Englewood Cliffs, NJ: Prentice-Hall.

Yuen, Joseph H. (ed.), 1983. *Deep Space Telecommunication Systems Engineering.* New York and London: Plenum Press.

10

Spacecraft Command and Telemetry

ROBERT C. MOORE

The command and telemetry systems of a spacecraft provide communication to and from the spacecraft. The command system is responsible for communication(s) from the ground to the spacecraft; the telemetry system is responsible for communication(s) from the spacecraft to the ground. Commands are sent to the spacecraft to tell it what we desire it to do; telemetry data are sent from the spacecraft to report back what it has measured and done, plus engineering status data that indicate the health and status of the spacecraft itself. It is common for spacecraft command and telemetry systems to process the command and telemetry data using custom, onboard, embedded microprocessors.

10.1 Command Systems

The purpose of a spacecraft command system is to permit the spacecraft or its subsystems to be reconfigured in response to radio signals that are sent up to the spacecraft from the ground. The command system must receive the radio signals, decide what they mean, and then respond accordingly so that the desired reconfiguration or operation takes place at the proper time. Obviously the command system has a vital role in the overall operation of the spacecraft.

A wide variety of commands may be sent to a spacecraft. Some of these commands will require immediate execution; others may specify particular delay times that must elapse prior to their execution, an absolute time at which the command must be executed, or an event or combination of events that must occur before the command is executed. There may be still other commands that have no strict timing requirements; they simply need to be executed as soon as it is convenient. These alternatives are differences in priority of execution.

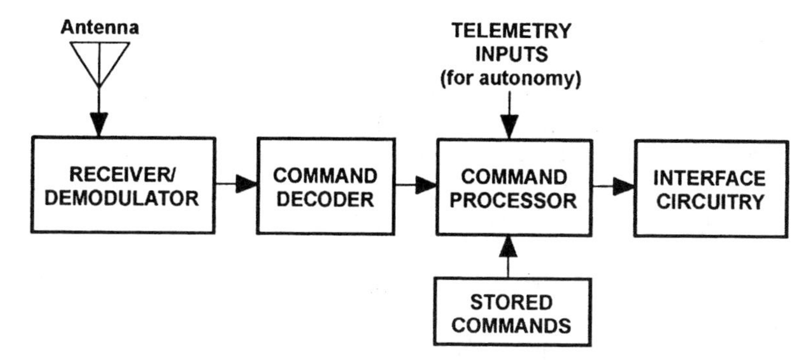

Figure 10.1 Spacecraft command system.

Commands also have a variety of functions they are expected to perform. Some will cause power to be applied to or removed from a spacecraft subsystem or experiment. Others will be used to alter the operating modes of subsystems. Still others will be used to control the various functions of the spacecraft guidance and attitude control systems, and thereby control the spacecraft orbit and orientation. There are commands to deploy booms, antennas, solar cell arrays, and protective covers. There are even commands to upload complete computer programs into the random access memories of programmable, microprocessor-based, onboard subsystems.

10.1.1 Generalized Spacecraft Command System

A block diagram of a generalized spacecraft command system is shown in figure 10.1. The radio-frequency (RF) signal that is sent up from the ground is received by the command receiver/demodulator. Typically this command *up-link* carrier frequency is S-band (1.6–2.2 GHz), C-band (5.9–6.5 GHz), or Ku-band (14.0–14.5 GHz), although VHF, UHF, L, and X bands are also used (Pratt and Bostian, 1985, p. 5). This RF signal has been *modulated* on the ground so that it carries an encoded *command message* up to the spacecraft. The task of the receiver/demodulator is to capture this signal, amplify it, *demodulate* the command message, and then deliver the message to the command decoder. The message is delivered in the form of an encoded *subcarrier* signal. The subcarrier frequency is usually in the range 14–16 kHz.

The command decoder looks at the subcarrier that is produced by the receiver/demodulator and "undoes" the command encoding to reproduce the original command message. This message is in the form of a serial digital binary bit stream; that is, the command message that comes out of the decoder is a sequence of binary digits: logic ones and zeros. Most decoders also will provide two other outputs. These are:

(1) a *lock* or *enable* signal, and
(2) a clock or timing signal that is synchronized to the serial bit stream (typically 100–2000 bps).

With regard to spacecraft command and telemetry systems it is very important to recall a powerful concept from digital communication theory: any message that can be communicated using written language also may be communicated using binary ones and zeros (Shannon, 1948).

The command logic checks the command message to see whether the command is valid. Several checks are implemented to validate each command. It is more important for the command system to reject invalid commands than to accept and execute valid commands, because invalid commands—for example, commands that have been contaminated by electrical noise—may wreak havoc with the spacecraft and its subsystems. A valid command that is missed usually can be sent again from the ground. Sometimes the retransmission occurs automatically as part of the ground station's command operations protocol. Command validation is therefore of critical concern in the design of a spacecraft command system, and it has a large impact on the choices that are made as the design proceeds.

Once a valid command has been detected, the command logic drives the appropriate interface circuitry so that the command is executed. If necessary, the command execution is delayed by a specified amount of time or postponed until a specified time. Sometimes the execution of a command will be conditional; that is, the command will not be executed unless certain conditions are first met. For example, commands that control a stepping motor that points a telescope may be conditioned on the presence of a signal that tells the spacecraft that the telescope's protective cover has been opened. *Event-driven* commands are initiated by detection of an event, or combination of events, on board the spacecraft. Event-driven commands and command sequences are used to implement spacecraft autonomy (discussed below).

The interface circuitry for some commands is quite simple; for others it is complex. Some commands simply change the voltage level on a single wire; others transmit thousands of bits of data to a subsystem. It is this variety of circuitry and function that makes command system design so interesting and challenging.

10.1.2 Command System Components

Figure 10.2 shows another block diagram of a full spacecraft command system. This diagram includes the system components that are necessary to generate and transmit the commands to the spacecraft system. The additional system components are:

- Ground support equipment (GSE).
- Modulation.
- Radio-frequency (RF) link.

The ground support equipment has many tasks, among which is the job of generating the commands for the spacecraft. A database of potential command messages is stored in

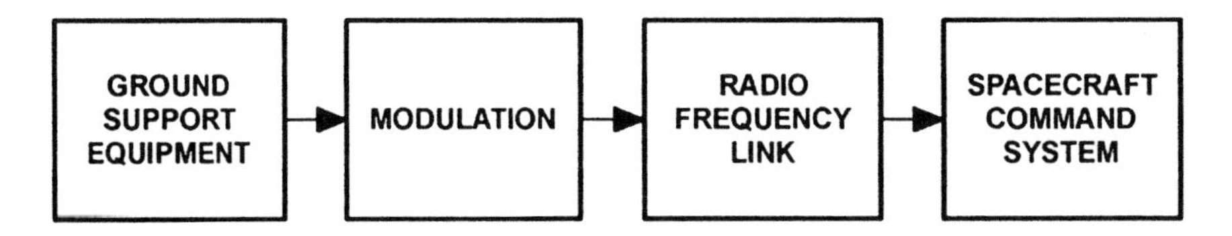

Figure 10.2 Complete command system.

the GSE memory. For a given command, the database contains the *command mnemonic* (command operation code) and a format for the binary command message. The specific instance of the command message format may be supplied by the operator at the time the command is entered, or it may be selected automatically from a stored *dictionary* or *look-up table* of standard command templates. In either event the resulting command message consists of a string of binary digits (*bits* for short, each a logic one or a logic zero) to which a specific meaning has been assigned. The meaning of the command message is said to be *denoted* by that particular string of bits.

The ways by which meaning may be denoted by a sequence of binary ones and zeros are myriad. Each method of assigning meaning to bit patterns is called a scheme of *encoding* that meaning. The *code* is an agreed-upon *a priori* mapping of meaning onto bit patterns. To take the desired meaning and generate the corresponding bit pattern is called *encoding*; to take a bit pattern and reconstruct the intended meaning is called *decoding*. Encoding and decoding always take place when two digital electronic systems communicate.

10.1.2.1 Sending a Command

When a particular command needs to be sent, the GSE operator types in the command mnemonic at the GSE console. The GSE computer then accesses its stored dictionary file for that command mnemonic and looks up the message format or template. If necessary, the operator is prompted for the specific data to be included within the command message. Then the GSE transmits the complete message to the command modulator. The mnemonic code is designed for use by humans only; it does not have to be transmitted to the spacecraft.

Some command mnemonics activate an entire sequence of pre-programmed commands. Each of these sequences then may be transmitted as a *batch* or group. The mnemonic for such a command sequence is called a *macro*; the command sequence itself is a *macro command.*

As received by the command modulator, the message is a string of bits that has been encoded using a pulse code modulation (PCM) scheme. PCM encoding represents each binary digit, using a pulse or a pulse pattern. For example, a logic zero may be represented by a pulse followed by no pulse, and a logic one may be represented by no pulse followed by a pulse. The particular PCM scheme that is chosen will be determined by the designer after evaluation of the system losses and signal-to-noise (S/N) ratio. More elaborate encoding schemes can increase the probability that the command message will be detected and decoded successfully at the spacecraft.

The command message is then used to modulate an RF carrier so that the message will be *carried*, or broadcast, to the spacecraft. In a typical command system the RF carrier is not modulated directly; rather, the command message is first encoded onto a *subcarrier*. There are several ways by which this may be done. One is called *frequency shift keying* (FSK), in which one of two separate subcarrier frequencies is selected, dependent on whether the message bit is a logic one or a logic zero. Another common subcarrier encoding scheme is called *phase shift keying* (PSK), in which the phase of a single-frequency subcarrier is modulated by the bits in the command message. Other types of modulation are described in chapter 9.

10.1.2.2 Modulation of the Carrier

Once the subcarrier has been encoded with the command message, it is used to modulate the main RF carrier. The most common modulation schemes are amplitude modulation (AM), frequency modulation (FM), and phase modulation (PM).

AM signals are the easiest to detect, but in general they also are the most vulnerable to corruption by electrical noise. The subcarrier is used to control the amplitude of the main carrier. At the spacecraft, using AM modulation, a simple measurement of the RF amplitude yields the subcarrier.

FM and PM signals require that the receiver measure the frequency or phase of the RF carrier to detect the subcarrier. This is more difficult to do, but FM and PM systems are much more immune to electrical noise than AM systems are, and therefore provide more reliable communication.

10.1.2.3 RF Link

The RF link includes everything from the command transmitter on the ground to the command receiver in the spacecraft. Detailed link calculations, or link analyses, are performed to determine the characteristics that are needed by the antennas on the ground and in the spacecraft. Usually the characteristics of the ground station antennas are well known, and their locations are fixed. This means that, so far as the antennas are concerned, the only real variable is the design of the spacecraft command receiver antenna (and the associated receiver sensitivity; for example, the noise figure).

Because spacecraft command systems must be able to receive commands even when the desired spacecraft attitude has not yet been established, the command receiver antennas are normally omnidirectional; that is, they can receive RF signals equally well from all directions.

10.1.2.4 Command Receivers

The command receiver amplifies and demodulates the RF carrier that was transmitted from the ground. Amplification is necessary because the signal strength is quite small by the time it has traversed the RF uplink. The command receiver takes the RF energy from the command antenna and generates a replica of the original subcarrier signal. At this point the command message is still modulated onto the subcarrier.

The most basic characteristic of the command receiver is the kind of demodulation that it performs. The modulation (AM, FM, PM) used on the ground determines the kind of demodulation that must be used by the command receiver. The system requirements that are placed on the command receiver frequently are the driving forces that determine what type of modulation is used for the RF carrier.

AM receivers are much less complex than FM or PM receivers, but the performance of an AM receiver is not as good as that of an FM or PM receiver. The goal is to obtain maximum signal-to-noise ratio for a given (usually small) received signal power. For most reasonable levels of received power the FM modulation scheme outperforms the AM scheme. (Refer to chapter 9 for additional details regarding modulation methods.)

Classes of command receiver. There are two major classes of command receiver: tuned RF and superheterodyne. In a tuned RF (TRF) command receiver the tuning is performed

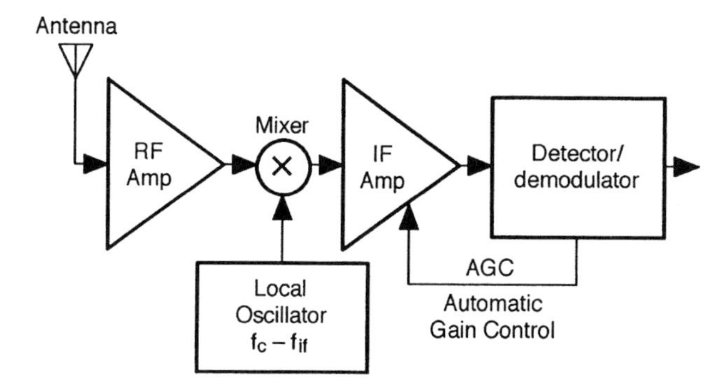

Figure 10.3 Block diagram of a superheterodyne receiver. (Adapted from Carlson, 1975.)

by a crystal filter. A crystal filter uses a crystal that resonates at the particular frequency of interest; thus it responds only to signals at the carrier frequency. Amplification and detection (demodulation) are performed directly on the RF signal.

In a superheterodyne receiver a signal from a local oscillator (LO) is mixed with the RF signal to form sum and difference signals. The lower of these two frequencies is filtered to become the intermediate frequency (IF) for the receiver. Amplification and detection (demodulation) then take place at the IF instead of the RF. This is easier to do because the IF is lower than the RF, and amplifiers and detectors are easier to build at lower frequencies (Malvino, 1984, p. 720).

Figure 10.3 is a block diagram of a superheterodyne receiver. The RF amplifier is tuned to amplify signals near the RF carrier frequency f_c. This RF amplifier has a bandwidth greater than that of the transmitted signal, but less than twice the IF; that is,

$$B_{\text{transmitted}} < B_{\text{RF}} < 2f_{\text{IF}}$$

The output of the mixer is at the IF, which is $f_{\text{LO}} - f_c$. The IF amplifier has a bandwidth approximately equal to that of the transmitted signal $B_{\text{transmitted}}$. The gain of the IF amplifier may be adjusted as required by the detector/demodulator, using automatic gain control feedback.

Design considerations. Other considerations in the design of a command receiver are the center frequency and the bandwidth. The center frequency of the receiver must be the same as the RF carrier frequency. This frequency must be allocated and approved by the appropriate international government regulatory agencies. There are certain bands of frequencies that are allocated for use by spacecraft command systems.[1]

When the RF carrier is modulated by the subcarrier (which itself has been modulated by the command message) the transmitted frequency spectrum is "broadened out." The *bandwidth* of the command receiver must be large enough to pass the entire frequency spectrum of the modulated command RF signal, so that none of the modulated information is lost or compromised. This bandwidth is a measurement of the range of RF

[1] See, for example, U.S. Department of Commerce (1989). Note especially the allocation of the following frequency bands, all of which pertain to spacecraft and space research: 1.62–1.66 GHz, 1.67–1.71 GHz, 2.29–2.30 GHz, 2.50–2.69 GHz, 3.40–4.20 GHz, 5.725–7.075 GHz, 7.25–7.75 GHz, 7.90–8.50 GHz, 10.6–13.2 GHz, 13.4–15.4 GHz, 17.3–21.4 GHz, 27.0–31.0 GHz, 36.0–37.0 GHz, 37.5–47.0 GHz, 47.2–76.0 GHz.

frequencies that it can amplify without causing excessive distortion. If the bandwidth is too narrow, portions of the transmitted signal will be lost; if it is too wide, excessive and unwanted noise will be received and the signal-to-noise ratio (S/N) of the receiver will be degraded. As the bit rate (bits per second) of the command message increases, the necessary bandwidth of the command receiver also must increase.

Power dissipation is another major consideration in the design of a spacecraft command receiver. Normally the command receiver is powered on all of the time, so any design reduction in its power dissipation will be integrated over the life of the spacecraft. This can result in considerable power savings that will impact the design of the power system.

Local interference. Spacecraft do more than receive command signals from the ground. Most spacecraft also transmit signals. These transmitted signals include spacecraft telemetry and navigational beacons. It is not unusual for a spacecraft to have several different transmitting antennas, each of which is transmitting a different carrier frequency. The transmitted power (energy transmission rate) of these signals is much greater than the received power of the command signal.

The command receiver and its antenna must be designed so that their performance is not degraded by the presence of the onboard transmitters; that is, the command receiver must reject the signals that are transmitted by its own spacecraft. The frequency spectrum of the transmitted signals therefore is another primary consideration when the designer chooses the carrier frequency for the command link.

Command decoders. A command decoder examines the subcarrier signal and detects the command message that it is carrying. Any PCM encoding also is detected so that the output of the command decoder is a simple binary bit stream. The normal output is called *non-return-to-zero* or NRZ data. A binary zero is represented by one particular logic level, and a binary one is represented by the other logic level. Detailed descriptions of various binary digit representations are given in chapter 9.

The command decoder also provides a clock to the command logic. This clock tells the command logic when a bit is valid on the serial data line. The clock is used to shift the command message into the command logic bit-by-bit (bit-serial) or word-by-word (bit-parallel, word-serial). Because the bit rates usually are low for commands (approximately 1 kbps), serial transmission is almost always used.

The command decoder also provides a *lock* signal. This signal becomes active (that is, it is asserted as *true*) when any command has been detected. The lock signal indicates to the command logic that a command is about to be shifted out of the decoder on the serial data line.

Like the command receiver, the command decoder usually is fully redundant. Either of the two redundant command decoders can decode commands. The two receivers and the two decoders are *cross-strapped* so that all four subsystems are active simultaneously. In the event that one command decoder fails, the other can be used in its place. Frequently there is a unique bit pattern in the command message that is used to select the command decoder to be used. Sometimes this command decoder address is part of the spacecraft address field, which is a group of bits that identifies the particular spacecraft to which the command is directed.

Those portions of the command system that are downstream from the command decoder, as well as the telemetry system, may be unpowered until activated by receipt of a valid synchronization word and spacecraft address. This activation usually follows spacecraft orbit insertion. To conserve stored energy, some spacecraft turn off the downstream portions of the command system when no command has been received for a sufficiently long period of time. The full command system is reactivated when a valid synchronization word and spacecraft address are received.

Some commands may be decoded further by the destination subsystem. This type of *decentralized decoding* can significantly reduce the mass of the wiring harness, which otherwise contains at least one wire for each switching command.

Synchronization words. Command systems, like most digital communication systems, normally use a pseudo-random synchronization word to mark the beginning of a command (Gruenberg, 1967, pp. 14–99). These frame-synchronization words are sometimes called *Barker sync* words (U.S. Naval Ordnance Laboratory, 1961). The Barker word is a bit pattern that may seldom or never occur in the rest of the command messages. This Barker word will appear at or near the beginning of each command message. It is used to acquire, and verify that the receiver and decoder have properly "locked on" to, the transmitted command signal. Barker sync words can reduce the probability that false commands will be received (Barker, 1953, pp. 273–287).

When a Barker synchronization word is used to synchronize fixed-length data frames, the probability of false synchronization can be given an upper bound:

$$P_{fs} < 1 - (1 - 2^{-n})^{m-1},$$

where P_{fs} is the probability of false synchronization, n is the length of the synchronization (Barker) word in bits, and m is the size of the data frame in bits (Masching, 1964, p. 1). A P_{fs} of 1×10^{-7} can be obtained by using synchronization words of reasonable length (say, 32 bits). For a 32-bit Barker word some typical values for P_{fs} are given in table 10.1. This table assumes that a fixed 32-bit sync word is used. From the table it is clear why, for a given sync word length, smaller frames are used where the data are more critical.

A typical 32-bit Barker word might look like this:

1111 1110 0110 1011 0010 1000 0100 0000

In hexadecimal (radix-16) notation, this synchronization word is

F E 6 B 2 8 4 0

The Barker sync word is chosen so that it has a sharply peaked autocorrelation function; that is, it does not correlate well with a replica of itself unless it is in exact

Table 10.1 Typical values for probability of false synchronization (P_{fs}) for a 32-bit sync word

Frame size, m	P_{fs}
1024	3×10^{-7}
2048	5×10^{-7}
4096	9×10^{-7}
8192	2×10^{-6}
16384	3×10^{-6}
32768	5×10^{-6}
65536	1×10^{-5}

synchronization with that replica (Wu, 1984, vol. 1, pp. 323–334). This means that the absolute value of its *autocorrelation sidelobes* (the value of the correlation of a sync word with a time-shifted version of itself) is small (Sklar, 1988, p. 461). This characteristic makes the sync word easy to detect and reduces the probability of false synchronization. The autocorrelation function is computed by treating the two code symbols (usually **0** and **1**) as the values $+\mathbf{1}$ and $-\mathbf{1}$, then adding the bit-wise products of the sync word and a shifted version of itself. A related function, the *agreement function*, is computed by shifting the Barker word relative to itself and tallying (counting) the number of bit locations where the shifted bit matches the unshifted bit. For example, if the above 32-bit Barker word is received in a stream of logic zeroes, the agreement for a relative shift of two (as shown below) is 18; that is, for this particular alignment of the received data with the Barker word, eighteen of the Barker bits match the received data bits.

0 0 1 1 1 1 1 1 1 0 0 1 1 0 1 0 1 1 0 0 1 0 1 0 0 0 0 1 0 0 0 0 0 0

1 1 1 1 1 1 1 0 0 1 1 0 1 0 1 1 0 0 1 0 1 0 0 0 0 1 0 0 0 0 0 0

Obviously, when the received Barker synchronization pattern is perfectly aligned with the Barker test string, the agreement will be equal to the length of the Barker word (+32 in the example).

Bit detector. The design of the command decoder is determined by the kind of encoding scheme that is used by the encoder on the ground. Typically the decoder circuit is called a *bit detector*. There are two basic types of bit detector. A *slow bit detector* (SBD) works at bit rates that are lower than 1000 bits per second (bps), and usually is associated with FSK encoding. A *fast bit detector* (FBD) works at bit rates of 1000 bps and higher, and usually is associated with PSK encoding. The bit detector provides bit synchronization (timing recovery) (Messerschmitt, 1991, p. 134).

"Off-the-shelf" decoders. Sometimes a "standard" command decoder is used. The most common of these is the NASA Standard Command Detector Unit (CDU). This unit is part of the NASA Standard Transponder, which is used to communicate with various NASA ground stations and their data networks. The CDU may be configured to work with several encoding schemes and at several data rates. Command data rates may range from 7.8125 bps to 2000 bps. These rates are typical for a spacecraft command receiver.

Bit error rate. The performance of the command receiver and its decoder is measured by computing a *bit error rate* (BER) for the system. The bit error rate is the statistical probability that a given bit will not be decoded correctly; that is, it is the probability that a transmitted *logic one* will be incorrectly decoded as a *logic zero*, or vice versa. The bit error rate is always a decreasing function of the signal-to-noise ratio (S/N). A more thorough treatment of the bit error rate for an RF link (command or telemetry) is given in chapter 9. The command system BER is usually smaller than 10^{-6}.[2]

[2] A bit error rate (BER) of 10^{-9} (an average of one error every 10^9 bits) is considered "acceptable" for a ground-to-ground communications link. See, for example, Saleh and Teich (1991), p. 894.

10.1.2.6 Secure (Encrypted) Links

For certain missions it is necessary to have a secure command link; that is, a link that makes it difficult for an unwanted or unauthorized listener to decode the command messages that are being transmitted to the spacecraft. Without encryption it is relatively easy for people with a receiver and a little knowledge of the carrier frequency and encoding method to decode the command messages, then to transmit commands of their own. If the message is put into a special code prior to modulation, the unwanted listeners must go to the additional trouble of *breaking* or *cracking* the code before they can determine what the original command message was.

Carefully crafted codes that are very difficult to crack are called encryption codes. By using one or more of these codes, eavesdropping may be made arbitrarily difficult (Wu, 1984, vol. 2, pp. 102–103). This provides additional security for the command link. Encoding a message using one of these codes is called encryption; decoding the message is called decryption. Encryption also may be employed by the spacecraft telemetry system to provide an increased margin of security for the telemetry data.

To prevent an unauthorized user from recording a certain encrypted command, observing the result, then retransmitting that command (without having to decrypt the command message) while wishing to cause the observed result to occur again, *authentication* bits can be included in the beginning of the command. These bits will be known only to the spacecraft command system and authentic sources of commands. For example, the number of commands executed since launch could be included as part of each command. This means that every time a particular command is sent, the encrypted form of the command is unique. If an unauthorized source of commands attempts to record a command, then retransmit it, the unauthorized command will be rejected by the spacecraft command system.

Of course, if the command messages are encrypted on the ground they must also be decrypted by the command system in the spacecraft. This is an additional hardware function for the command system. The decryption device is usually placed between the output of the command decoder and the input to the command logic. Typical inputs to the decrypter are *encrypted data*, *data clock*, and *lock detect*. The outputs are the decrypted command message and a bit clock.

10.1.2.7 Command Messages

The command message is the digital stream of bits to which the command logic of figure 10.1 must respond. It is the job of the command logic to interpret the meaning of each command message and react accordingly. Some of the components of a typical command message are

- Input checkerboard bits.
- Synchronization (Barker word) bits.
- Command bits.
- Error detection bits.

The input checkerboard is the first component of the command message. Not all command systems use an input checkerboard; but if a system does use it, the checkerboard is always the first component of each message. The checkerboard is a sequence of alternating logic ones and zeros. Its purpose is, at the beginning of each command, to provide the bit detector and command decoder time to *acquire*, or lock on to, the modulated subcarrier. If some of the checkerboard bits are missed or improperly decoded during acquisition of the subcarrier, no command data are lost.

The input checkerboard field must have enough bits to make its duration sufficiently large to permit the bit detector to acquire lock. Acquisition of lock is indicated by the assertion of the *lock detect* signal at the output of the detector. This informs the command logic that a new command message is imminent.

The synchronization pattern (Barker word) is the next portion of the command message. This pattern helps to prevent false commands from being executed by the command logic. Synchronization patterns vary widely in length; longer patterns usually provide better immunity to false commands. (See the section on synchronization words above.)

When synchronization patterns are used, the command logic will insist on seeing a valid synchronization pattern prior to accepting the rest of the command message. If the synchronization word is not detected, the remainder of the command message will be ignored. By choosing the synchronization patterns carefully, a very significant measure of error immunity can be achieved.

Command bit types. Once a correct synchronization pattern has been found, the command logic will begin to process the bits that follow. These are the bits that denote the actual operation to be performed. The length and format of command bits varies widely from system to system; a typical format usually contains the following types of bits:

- Spacecraft address bits.
- Command type bits.
- Command select bits.

In all command messages there are *spacecraft address bits* that denote a spacecraft-specific identification code. This code identifies the exact spacecraft for which the command is destined. The presence of these spacecraft address bits prevents the command logic of one spacecraft from attempting to execute commands that were destined for another spacecraft.

This spacecraft identification code is necessary because it is possible to have several spacecraft that use the same network of earth stations and communicate using the same carrier frequency and modulation type. Again, these address bits must match the correct pattern or the spacecraft command logic will not execute the rest of the command message. This provides additional error protection for the spacecraft.

The next field of bits in the command message—the *command type* bits—contains information about the type of command that is being sent. Common command types include:

- *Relay commands* that set or reset electromagnetic relays in the power switching unit of the command system.
- *Pulse commands* that provide short pulses (typically 50–100 ms in duration) to appropriate subsystems on the spacecraft.
- *Level commands*, each of which toggles the logic level on a dedicated discrete wire that goes to one of the spacecraft subsystems.
- *Data commands* that transfer binary data words to a designated subsystem or instrument (these are sometimes called *value commands*) (Chetty, 1991, p. 242).

The number of bits in the *command type* field is a logarithmic function of the number of unique command types that the system uses:

$$\mathbf{N}_{\text{command\_type\_bits}} = \lceil \log_2(\mathbf{N}_{\text{command\_types}}) \rceil$$

The next field in the command message specifies which instance of a given command type is to be executed. These *command select* bits act as an *operation code* or *opcode* to specify a given command from the set of commands that match the specified type.

Error detection. Another field that is usually specified in a command or data message contains error detection and correction (EDAC) bits. These bits are used to detect, and possibly correct, bit errors in the message, thereby avoiding execution of erroneous or corrupted commands. Using a sophisticated algorithm, the EDAC bits are computed from the rest of the bits in the command message. This computation is performed on the ground, before the command message has had any opportunity to be corrupted by noise or enemy jamming. When the command logic processes the message in the spacecraft it recomputes the EDAC bits, using the same algorithm that was used on the ground. If there is a discrepancy, the command message is known to have been corrupted; then it is either corrected or rejected.

Perhaps the most common error detection bit is the *parity* bit. The parity bit is selected on the ground so that the entire command message—parity bit included—always has an odd (even) number of *logic ones.* This is called *odd (even) parity.* If the spacecraft command logic receives a command message that has an even (odd) number of *logic ones,* then it knows that an odd number of errors has occurred. The command is therefore rejected.

Notice that this simple parity scheme cannot detect an even number of bit errors. If it is highly unlikely that any one message will sustain more than a single bit error, a single parity bit will suffice for error detection. More parity bits are required if it is desired to detect more than one error in a message, or to correct errors.

Most EDAC codes are very complex, but the basic principles they employ are the same as that used by the simple parity bit (Carlson, 1975, pp. 407–417; Lin, 1970, pp. 43–47). One of the most common EDAC codes is described in detail in chapter 11, in connection with correcting single–event errors in a microprocessor memory. A typical EDAC code for a near-Earth spacecraft may employ seven EDAC bits in each 64-bit command message (57 data bits, 7 EDAC bits). This permits double error detection and single error correction. The use of single error correction, double error detection (SEC–DED) codes is very common.

In chapter 9 we showed that as the distance of the spacecraft from the ground command transmitter increases, the susceptibility of the command receiver to noise increases also. So, for a spacecraft that will travel far from its ground stations (high or eccentric orbit, or interplanetary mission), the benefit of using EDAC codes quickly outweighs the costs of the increased hardware complexity that is required to implement them.

Multiple commands. Sometimes multiple commands may share a command message; that is, more than one command follows a common checkerboard and synchronization pattern. In this case the leading command will have a bit field that indicates how many commands will be present in the subsequent message.

If any of the commands is a data command it will indicate how many data words will follow. These data words may then be loaded into the destination system or subsystem of the spacecraft before the next command is executed. Macro commands frequently arrive at the spacecraft as multiple commands.

Command logic. The function of the command logic in figure 10.1 is to decode the command (that is, determine the command type and the specific action that is to be taken), verify that the command is valid, and then activate the appropriate interface circuits so that the command is executed (a *receive–verify–execute* sequence in which there is no attempt to execute without prior verification). In the case of delayed commands, the command is not executed immediately. Instead, the command is tagged with an execution time or delay time and priority, and then is placed in a *command pending queue*.

The checks for command validity that are performed by the command logic will include some or all of the following:

- Check for the correct spacecraft address code.
- Apply error detection and correction to the command message.
- Check the command operation code (command type and command select) bits. Certain combinations of these bits may not represent valid commands. Frequently these codes are designed so that any two valid codes are different from each other in more than one bit position; that is, the so-called *Hamming distance* between any pair of valid operation codes is at least two. This means that it takes more than one bit error to convert one valid command code into another. Also, this check can prevent false commanding resulting from errors in encoding the command on the ground.
- Verify that the command was received during a valid time period, and that the command did not require too short or too long a time to be received.
- Verify that the command was sent by someone who is authorized to send it. This is called "authenticating" the command. Commands that are received over an encrypted link usually are self-authenticating, because only an authentic source knows how to properly encrypt the command message.

The command validation process is needed to ensure that false commands are never executed. The probability of accepting a false command needs to be very low, typically 10^{-18} to 10^{-22}. Sometimes critical commands receive special treatment; for example, longer or multiple operation codes. Some critical commands are held in a register following validation, while the operation code is telemetered back to the ground. Then, following ground validation, an "execute" command is transmitted to the spacecraft.

Once the command has been decoded and checked, the appropriate interface circuitry must be driven to execute the command. For example, if the command is a relay command, the appropriate relay coil must be driven with a current pulse. In a subsequent section we will discuss the interface circuitry in more detail.

Microprocessor implementation. Because there is such a wide variety of command types and command messages, most command systems (that is, the *command logic* function that is indicated in figure 10.1) are implemented using programmable microprocessors. The microprocessor receives inputs from the command decoder, operates on these inputs in accordance with a program that is stored in read-only memory (ROM) or random-access memory (RAM), and then outputs the results to the interface circuitry. Usually the inputs are the decoded and partially checked command messages; the microprocessor performs the rest of the checks. Sometimes the microprocessor performs some of the decoding as well.

The crucial measure of a microprocessor-based command system design is its reliability. The command system must remain immune to errors and be fault tolerant. If the command system is unreliable or degraded, the entire spacecraft may be lost. This special emphasis on system reliability must extend beyond the hardware design to include the software design as well. Because it is often critical to the mission, command system software must be entirely *error-free* so that anomalous behavior never occurs.

Delayed commands. Some command systems have the capability to execute commands on a delayed basis. Delayed commands are decoded and checked, then loaded into a *command pending queue* in memory. At the specified time (*time-tagged* command), or after a specified delay (*delayed* command), or when a particular event occurs (*event-driven* command), the command is pulled from the queue and executed. The command pending queue may be part of the command system memory, or it may be located in one or more of the onboard spacecraft processors. For example, simple instruments and subsystems may have their delayed commands stored in the command system processor queue, whereas complex instruments that have their own microprocessor controllers may *queue up* their own delayed commands locally, within the instruments themselves.

Delayed commands are required when it is necessary to execute commands at very precise times or time intervals, or when commands must be executed during a period of time when ground station coverage is not available. They also may be used as part of a sequence of commands for which the relative time delays are critical. The time delay between commands is determined by timing bits in the command messages. The resolution of the delay time is typically ± 1 s; finer resolution is seldom needed. The maximum time delay between successive delayed commands can be in the tens of hours.

When a particular delayed command sequence is decoded by the command logic, the sequence will automatically be transmitted back to the ground through the telemetry system. This *delayed command memory readout* enables the ground operators to verify the contents of the command pending queue. Usually a separate command, called an *epoch start* command, is used to initiate timing for the command pending queue once it has been loaded.

Final command decoding and error checking are performed at the time of command execution. This is yet another check on the command data integrity.

The need for delayed and event-driven commands is a very important factor in designing a command system, because delayed and event-driven commanding capability adds significantly to the complexity of the design. The time required to test a command system with delayed or event-driven command capability is significantly greater than that required to test a simpler command system. In addition, the need to verify the contents of the command pending queue requires that an interface be designed between the command system and the telemetry system. Of course, once this interface has been included in the design, many new capabilities become available. For example, certain commands may be made conditional, based on the values of critical onboard data. It might be inadvisable to turn on a certain subsystem if its temperature is outside certain specified limits.

10.1.2.9 Interface Circuitry

The type of interface circuitry that is driven by the command logic varies depending on the type of command that is being executed. Every spacecraft command system has relay commands. Other command types include pulse commands, level commands, and data commands. Each type has its own kind of interface circuitry, as described below.

Relay commands. Relay commands activate the coils of electromagnetic relays in the central power switching unit. Typically, each latching relay has two coils: one coil pulls the relay switch contacts into one position, and the other coil pulls the contacts into the other position. The command logic activates one or the other of the coils by forcing a current pulse through the coil. The amplitude of these current pulses lies typically in the range from 50 to 300 mA. For relay commands, the ultimate outputs from the command system are the relay contacts. For purposes of reliability these contacts are usually redundant.

The primary use for relay commands is to switch spacecraft bus power, typically +28 V direct current (DC), to the various subsystems and instruments on the spacecraft. This means that power distribution wires for almost all of the spacecraft subsystems must be routed through the command system central power switching unit. Even the design of this wiring harness must be taken into consideration by the power switching subsystem designer.

One special set of contacts on each relay generates a discrete (or *telltale*) signal for the telemetry system. These telltales are used to telemeter the state of each relay. The telltale is another way to validate proper execution of relay commands.

Because the integrated circuits (ICs) in the command logic are not capable of driving relay coil currents directly (typical ICs can drive only a few tens of milliamperes, at most), power switching driver hardware must be included in the interface circuits for relay commands. These drivers are discrete *source* and *sink* bipolar or metal-oxide semiconductor (MOS) transistors arranged in a two-dimensional array. The *source drivers* drive columns of the array and the *sink drivers* drive the rows. At the intersection of each column and row there is a relay coil. Steering diodes are used to guarantee that the coil currents always flow in the intended direction.

Much protection is built into the power-switching relay driver circuits to guard against erroneous or undesired relay activation. Obviously, one source driver or one sink driver alone cannot drive any of the relays. One source *and* one sink driver must be activated synchronously (that is, simultaneously) before the coil receives a current pulse. In practice, additional *enable signals* must be present before any relay coil may be activated.

Pulse commands. Pulse commands are short pulses of voltage or current that are sent by the command logic to the appropriate subsystem. These low-energy pulses have typical durations in the range from 1 ms to 100 ms. The pulses are used to drive small relays or logic latches in the destination subsystem. If the pulse command is used to drive a small relay in the subsystem, the command is frequently called a *remote relay command.* Sometimes the remote relay coil is connected electrically to the power-switching driver matrix just as the large power relay coils are.

A logic pulse command sends to the destination subsystem a logic pulse, usually a voltage in the range from 5 to 10 V, that is compatible with the subsystem logic technology. These pulses are used to set or clear flip-flops or latches that control the operating mode of the commanded subsystem. For example, logic pulse commands to the telemetry system could be used to control the type and output bit rate of the telemetry data.

Level commands. A level command is exactly like a logic pulse command except that a logic level is delivered instead of a logic pulse. Each time the command logic decodes the level command it toggles the level (from a logic one to a logic zero, or vice versa). Alternatively, logic level commands may be associated with two commands, one to set the level to a logic one and the other to clear the level to a logic zero. This latter method of generating level commands is usually preferred, because the set/clear command always forces the level to a known state (that is, the outcome of command execution is not dependent on the previous state of the level being controlled).

Data commands. Data commands transfer data words to the destination subsystem. Each data command may transfer from eight bits to 65,536 (64 K) bits of data or more. The data usually go into a memory located in the destination subsystem.

Serial data transfers usually occur over a three-wire interface that is similar to the one between the bit detector and the command logic. The three wires carry the following signals:

- *Enable*, an *envelope* signal (asserted prior to transmission of the first data bit, then de-asserted following transmission of the last data bit, thus *enveloping* the data) that indicates when valid data are present on the data line.
- *Data*, the data (usually in NRZ-L form) to be transferred.
- *Clock*, a periodic pulse stream that indicates when each bit of the data is valid.

The three-wire interface provides a degree of error protection because all three lines must be active and properly synchronized for the transfer to take place.

Sometimes the data from a data command are transmitted over a bus (bit-parallel, word-serial) to a bus interface unit or bus adapter that is located in the destination subsystem. Bit-parallel transmission is used only in situations for which a high data transfer rate is required.

Data commands may be used to reconfigure subsystems, load programs, or *program patches* into subsystem memories, and to modify stored lookup tables and parameter blocks.

10.1.3 System Requirements

A spacecraft mission is planned with various scientific, commercial, or military objectives in view. These objectives cause an appropriate spacecraft orbit or trajectory to be chosen. The orbit is one of the primary system factors that determine which ground stations will be used. Other factors are:

- Cost.
- Ground station capability (for example, if delayed command capability is required, can the ground stations handle this?).
- Program sponsor's preference, based upon other system criteria.

Often there is little choice in selecting a ground station network; frequently the designer must work within the capabilities and limitations of an existing network.

Among the ground station capabilities to be evaluated are antenna gain, types of modulation available, carrier frequency bands that are supported, and the amount of coverage that can be provided. Other issues include the uplink bit rate and the maximum message length. If there is a choice of ground stations, the effect of these parameters on the design of the command and telemetry systems will determine which ground station network is chosen. Of course, if there is no choice of ground station networks, then the command and telemetry system designs will have to accommodate the specifications of whatever ground station network is available.

Once the orbit and ground station network have been selected, a detailed RF link analysis is performed as discussed in detail in chapter 9, and then the design of the antennas, receivers, and decoders can proceed. The orbit will also determine the radiation environment of the spacecraft, which may require the command and telemetry system designer to choose radiation-hardened electronic components for the design. The requirement for radiation hardness (total ionizing dose radiation tolerance, and immunity to single-event upsets and latch-up) is a very important one, and is discussed in chapter 2 (why radiation is a problem) and chapter 13 (how radiation affects solid-state electronic devices, and how to deal with the problem).

The amount of ground station coverage is also an important factor in the design of the command system, because it can cause a need for delayed command capability. If coverage is available for only a small fraction of the orbit period, there is an increased likelihood that delayed command capability will be necessary. Low ground station coverage may also require significantly more spacecraft autonomy; that is, the ability of the spacecraft to monitor its own functions and take corrective actions without intervention from the ground.

The uplink bit rate and total message length are also very important in determining the design of the command logic. Requirements for radiation tolerance, immunity to single-event upsets (SEUs), and fault tolerance through the use of redundancy (to ensure graceful degradation) also impact the design.

10.1.4 Resources Required

The spacecraft resources that are required by a typical command system (command receiver and command logic) for a small, radiation-tolerant, low-Earth-orbit spacecraft are approximately:

Mass: 1–10 lb (0.45–4.5 kg)
Size: 7 × 7 × 2 in. (98 in^3 or 1600 cm^3)
Power: 0.5–3 W

The *power switching* subsystem, which contains the relays and interface electronics, consumes additional resources:

Mass: 10–12 lb (4.5–5.4 kg)
Size: 6 × 6 × 10 in. (360 in^3 or 5900 cm^3)
Power: almost zero (average) if relays are switched infrequently

10.2 Telemetry Systems

The word *telemeter* means "to measure from a distance" (Gruenberg, 1967, p. 1). When direct measurements are impossible, telemetry is used to provide remote indication of

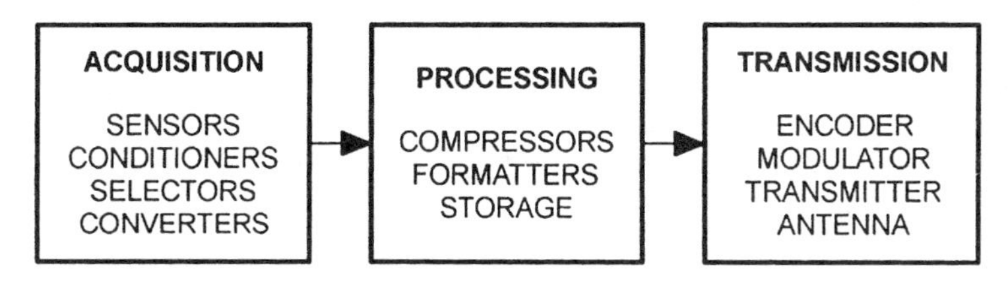

Figure 10.4 Spacecraft telemetry system.

what the desired measurements are. In the case of a spacecraft, accurate and timely telemetry is essential if we are to know in detail what is going on in the spacecraft and its environment. Kinds of information that may be telemetered include:

- Status data concerning spacecraft resources, health, attitude, and mode of operation.
- Scientific data gathered by onboard sensors: telescopes, spectrometers, magnetometers, accelerometers, electrometers, thermometers, etc..
- Specific spacecraft orbit and timing data that may be used for guidance and navigation by ground, sea, or air vehicles.
- Images captured by onboard cameras (visible or infrared).
- Locations of other objects, either on the Earth or in space, that are being *tracked* by the spacecraft.
- Telemetry data that have been *relayed* from the ground or from another spacecraft in a satellite constellation.

A telemetry system must acquire, process, and transmit data of many different kinds. A simple block diagram of the spacecraft portion of such a system is shown as figure 10.4. The acquisition of data is accomplished using sensors, signal conditioners, data selectors, and (usually) analog-to-digital converters. The data are processed in the telemetry system processor or, with certain "smart" sensor instruments, in the sensor's resident processor. Acquisition and processing of telemetry are described in the following sections; transmission of telemetry (modulation and the RF link) is discussed in chapter 9.

The processed data are transmitted from the spacecraft to the ground in much the same way that command message data are transmitted from the ground to the spacecraft, except that for telemetry transmissions the transmitter power and antenna size are more restricted than for command transmissions. Of course, for most ground stations the telemetry receiving antenna may be quite large; this tends to compensate for the reduced transmitter power at the spacecraft. Detailed RF link analyses, as described in chapter 9, must be performed to guarantee that the telemetry system will have a sufficiently high signal-to-noise ratio (S/N) and a sufficiently low bit error rate. Typically the telemetry *downlink* carrier frequency is S-band (2.2–2.3 GHz), C-band (3.7–4.2 GHz), or Ku-band (11.7–12.2 GHz).

Convolutional coding or Turbo coding (see chapter 9) may be used on the downlink to improve the bit error rate of the telemetry communications. In addition, an outer code, typically a *Reed–Solomon code*, is also used to provide the capability to correct bursts of bit errors in the received message. The NASA standard (255, 223) Reed–Solomon code, using an interleave depth of 5, adds 160 code bytes to each 1115 bytes of data and is capable of correcting an error burst 80 bytes in length (Fortescue and Stark, 1991, p. 356).

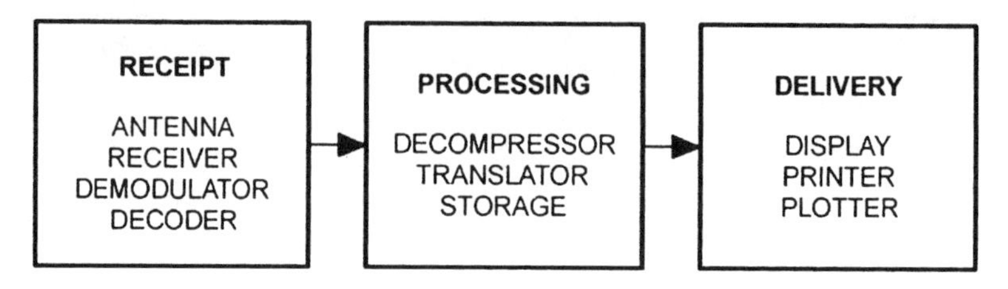

Figure 10.5 Spacecraft telemetry system (ground segment).

On the ground the telemetry data are demodulated and decoded, then stored or transmitted to the final users (see figure 10.5). A complete telemetry system comprises all of the basic components that are shown in figures 10.4 and 10.5. The following sections describe each major telemetry system component.

10.2.1 Sensors

Any device that changes its state in response to an external event or stimulus is called a sensor (Pastor and Miller, 1967). For example, the electrical resistance of certain metallic conductors will change with variations in the external temperature. A spacecraft temperature sensor may employ such a metallic conductor. The resistance of the conductor can be measured, and then that measurement can be telemetered. The measurement of the resistance becomes an indirect indication of the temperature of the sensor.

A transducer is a device that converts one form of energy into another. For example, a pressure working against a diaphragm may be converted into a voltage that works into an electrical load. Transducers may be used as sensors in a spacecraft.

Because the spacecraft telemetry system uses electronic signals to modulate the transmitted RF energy, the outputs of all sensors must eventually be electrical in nature. These electrical outputs include resistance, capacitance, inductance, frequency, voltage, and current.

Certain phenomena are more difficult to measure than others. Sometimes the sensors for these phenomena are complex. For example, magnetometers, altimeters, and radiometers are complex sensors that frequently are designed as entire spacecraft subsystems.

The parameters that are measured most frequently in a spacecraft are voltage, current, and temperature. These are relatively easy to measure. Integrated silicon sensors are available that produce output voltage or current in direct proportion to absolute temperature. These are called *proportional to absolute temperature* (PTAT) devices. Operational amplifiers may be configured as current-to-voltage converters. The result is that all of the most common signals that must be telemetered can be made available as analog voltages.

10.2.2 Signal conditioning

The process of taking raw sensor outputs and converting them to voltages in a standard range is called signal conditioning. Several important design factors must be considered when designing signal conditioning amplifiers:

- *Frequency response.* Does the conditioning circuitry accept the desired range of signal frequencies and reject unwanted frequencies?
- *Impedance.* Is the input impedance matched to the output impedance of the sensor?
- *Ground reference.* Are the sensor and the conditioning amplifier operating with respect to a common reference potential?
- *Common mode range.* Is there a bias voltage or current that must be removed from the sensor output? (Sometimes the actual sensor output signal is riding on a larger *common mode* signal, which must be rejected.)

10.2.3 Signal Selection

Usually, the telemetry system telemeters data one at a time in a *time-division multiple access* scheme. Signals that have been conditioned are selected one at a time by a multiplexer that examines only one particular voltage out of an array of many. The multiplexer usually steps through the conditioned voltages in a fixed sequence of rotation; such a multiplexer is called a *commutator.* One commutation cycle forms a *frame* of telemetry data (Pratt and Bostian, 1985, p. 238). Each datum is sampled at least once per frame.

If a data type is sampled by the commutator more than once per frame, that data type is said to be *supercommutated.* Supercommutated data are sampled at a rate that exceeds the frame rate. A second commutator may be connected to one of the inputs of the main commutator. Data sampled by this second commutator are said to be *subcommutated.* Subcommutated data are sampled at a rate that is less than the frame rate. Figure 10.6 illustrates commutation, supercommutation, and subcommutation.

The telemetry commutation sequence is determined by *filter tables* or *sequencing tables* that drive the software that controls the hardware multiplexers. A filter table determines which telemetry data will appear on the downlink (or which data are recorded on the solid-state data recorder) and at what rate each data type will be sampled.

The commutation system must be designed so that each data signal is sampled sufficiently often that no significant information is lost. The *Nyquist sampling theorem* (see chapter 9) indicates that a minimum sampling rate of twice the highest frequency of interest should be used; in practice the sampling frequency is typically five to ten times the highest frequency of interest.

Digital signals may also be commutated. A multiplexer that is designed to select analog signals usually may be used to select digital signals also; the converse is seldom true.

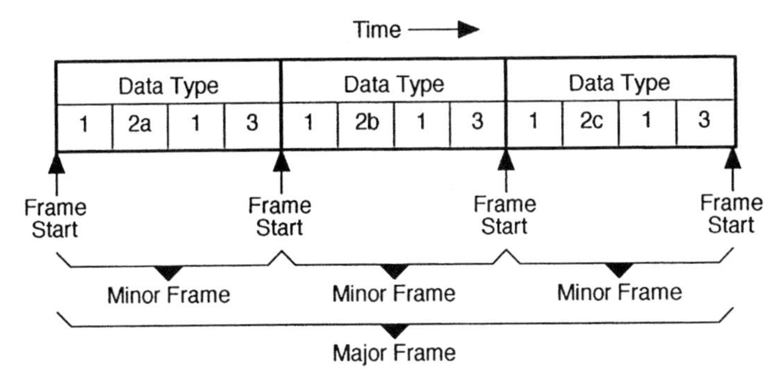

Figure 10.6 Subcommutation and supercommutation.

10.2.4 Conversion to Digital Form

Because most telemetry systems use digital input data to modulate the RF transmitter, the conditioned and commutated analog voltages must be converted into encoded digital numbers prior to modulation. The device that converts an analog voltage or current into a digital number is called an analog-to-digital converter (ADC).

The ADC accepts as input a voltage that has been conditioned to be in a specified range. This range of input voltage typically is 0 to 5.10 V (unipolar) or −2.56 to +2.54 V (bipolar). The ADC produces as output a binary word that encodes a digital value N. If the word has n bits, then, for a unipolar converter, the output integer N will be in the range from 0 to $2^n - 1$. When N equals $2^n - 1$ the converter output is said to be at *full scale*.

Generally, the transfer function of the ADC is linear; that is, each incremental change in input voltage causes identical increments in output numerical value. The conversion sensitivity of an ADC is usually measured in counts per millivolt; a typical sensitivity for an 8-bit unipolar ADC is 1 count per 20 millivolts, with a full-scale output ($N = 255$) corresponding to an input of 5.10 V.

Figure 10.7 shows a transfer function for a 3-bit ADC. This design produces a full-scale output for input voltages that exceed 3.5 V. Notice that the transfer function resembles a staircase; the number of steps is 2^n, where n is the number of bits in the ADC output word.

The staircase transfer function introduces error into the measurement because the output is *quantized* to one of 2^n discrete values. Consider what happens when the ADC of figure 10.7 is given an input voltage of 1.25 V. The output integer N will be 2, which corresponds to 1.00 V. (Assume that the decoding on the ground will compute V_{in} from N using the equation, $V_{\text{in}} = N/2$.) So the *quantization error* for this particular measurement is 0.25 V. The worst-case error is usually equal to the step size, or plus or minus one-half the step size.

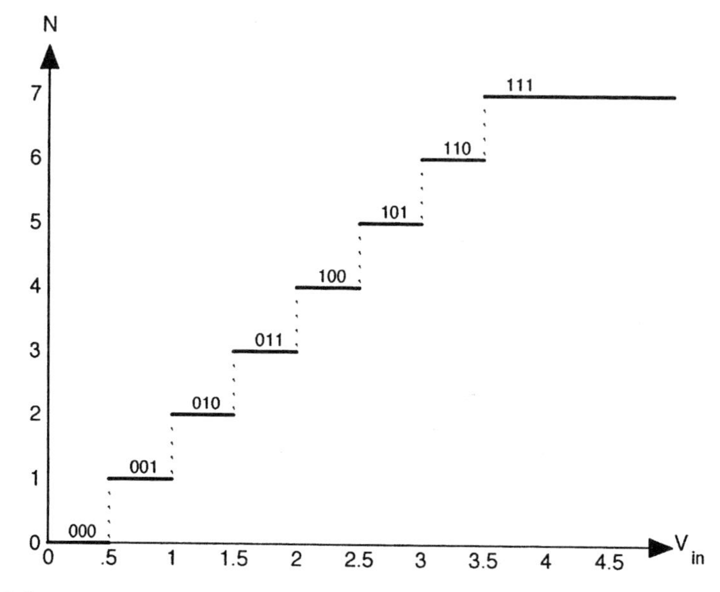

Figure 10.7 Transfer function for a 3-bit ADC.

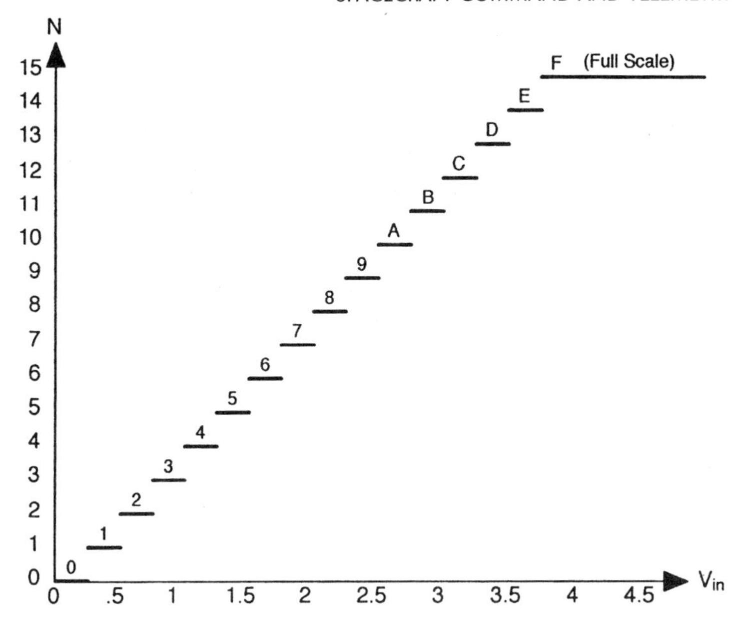

Figure 10.8 Transfer function for a 4-bit ADC.

This quantization error may be reduced by using more bits in the output word. Figure 10.8 shows an improved ADC design that employs four bits in the output word. With this design the full-scale output ($N = 1111$ binary, or "F" in hexadecimal) will be produced for inputs that exceed 3.75 V, and the steps in the staircase are smaller in size by a factor of two. Now the ground computation is $V_{in} = N/4$, and the worst-case quantization error is only 0.25 V—an improvement over the 3-bit design by a factor of two. This converter is said to *resolve* 0.25 V changes in its input voltage.

The maximum quantization error for an ADC that has $R = 2^n$ steps in its transfer function is given by the expression

$$(V_{max} - V_{min})/2R,$$

where V_{max} and V_{min} are the maximum and minimum analog values. For the example shown in figure 10.8 the maximum quantization error is $(4\ V - 0\ V)/2 \times 2^4 = 4\ V/32 =$ 125 mV.

The hardware for ADC designs varies considerably with the number of bits in the output word, desired power dissipation, conversion speed, and so on. Some typical ADC conversion specifications are:

High-Speed ADC:	50,000,000 conversions/second 8-bit output word 2.5 W power dissipation
High-Resolution ADC:	100,000 conversions/second 16-bit output word 1.5 W power dissipation
Low-Power ADC:	25,000 conversions/second 8-bit output word 0.005 W power dissipation

10.2.5 Processing

Once all of the telemetry data have been converted into digital form, they may be processed prior to transmission to the ground. This processing can reduce the number of data bits that must be transmitted, and thus reduce the average telemetry bit rate. This reduces the required bandwidth of the telemetry transmitter. Processing that reduces the number of transmitted bits in a telemetry system is called *data compression*. This and other forms of onboard spacecraft data processing are described in chapter 11.

Onboard telemetry data processing can also help to provide spacecraft autonomy. The processor can analyze the telemetry data and automatically inform the command system so that anomalies may be detected and necessary corrective actions may be initiated by command.

10.2.6 Formatting

After the data have been collected and processed, the telemetry system must add the following information:

- *Synchronization word* to indicate which bit is the first in the frame.
- *Frame count* to indicate which frame is currently being transmitted.
- *Spacecraft identification* to indicate which spacecraft is the source of the telemetry data.
- *Error detection/correction bits* so that errors in each *block* of transmitted data may be detected and corrected.
- *Frame format identification* to indicate which of several formats is in use.
- *Spacecraft time* when the telemetry frame or packet was assembled.

The addition of these data fields causes the telemetry data to appear in a standard frame or packet format. The process of adding these data is called *formatting*. The function of these fields is similar to that of the same fields in command messages.

The Consultative Committee for Space Data Systems (CCSDS) publishes standards for "wrapping" or packaging variable-length command and telemetry data. Most standard ground stations employ these standards, the software and hardware for which may be re-used on subsequent missions.

10.2.7 Data Storage

Continuous ground station coverage is not always possible. For example, Earth satellites in low orbits may be over a given ground station for only ten or twenty minutes out of each 100-minute orbit. In cases such as this, onboard data storage is required. Data that are accumulated in a mass storage device over a long period (say, twelve hours) can be played back quickly (say, in ten or fifteen minutes) during an RF contact as the satellite makes a *pass* over the ground station. Media for mass data storage are described in chapter 11.

10.2.8 Modulation and Transmission

Telemetry data are sent to the ground station over an RF link similar to that used by the command system. The most common type of modulation is phase shift keyed pulse code modulation (PSK–PCM). The use of PCM data facilitates data encryption and the implementation of error detection and correction (EDAC) codes.

The transmission of telemetry data can require significant amounts of spacecraft energy. To transmit energy at a rate of 12 W using a carrier RF of 2 GHz may draw energy from the power system at a rate of 60 W; that is, for each watt transmitted, four watts are wasted. For this reason it is desirable to keep the spacecraft transmitter power low. Use of high-gain, low-noise antennas at the ground station is mandatory.

10.2.9 Autonomy

Autonomy in a spacecraft is the ability of that spacecraft to monitor its internal functions and take whatever actions are appropriate, without any direct intervention from the ground. The requirement for autonomy in a given spacecraft may be caused by insufficient ground station coverage, communications geometry (for example, spacecraft too near the Earth–Sun line so that the solar noise prohibits RF communications), or special security requirements. Delayed command sequences may be used to react to or implement planned events during periods of no ground coverage, but full autonomy is necessary if the spacecraft is to deal with unplanned events (anomalies).

An unplanned or unexpected spacecraft event is called a *spacecraft anomaly*. An autonomous spacecraft is designed to detect these anomalies and respond correctly to them. Autonomy is generally implemented using the command and telemetry systems. The telemetry system already has the capability to monitor spacecraft functions, and the command system already has the ability to reconfigure the spacecraft by executing commands. Three additional capabilities are necessary for autonomous spacecraft operation:

- *Detect the abnormal*: the telemetry system must know when the functions that it monitors have deviated beyond normal ranges.
- *Determine the response*: the command system must know how to interpret abnormal functions, so that it can generate a proper command response.
- *Communicate*: the command and telemetry systems must be capable of communicating with each other.

Spacecraft autonomy is often implemented using a *rule-based* approach. With this approach, autonomy rules (often numbering in the hundreds for a given spacecraft) are generated for each potential fault or anomalous behavior that can be detected. Each rule is a conditional statement that comprises a premise (the condition), a persistence (usually implemented as m "true" evaluations of the premise out of n sample intervals), a corrective action to be taken if the rule "fires" (usually this is a macro command sequence), and a maximum "fire count" for the rule, which limits the number of times the rule may fire. Autonomy rules are assigned priority levels, so that, when a rule of a given priority fires, its macro can temporarily disable all rules of lower priority to prevent unwanted rule interactions.

The premise of each autonomy rule is an expression that logically or arithmetically combines several telemetry values with certain state variables that indicate mission phase, spacecraft state, subsystem health, data validity, and so on. An example of an autonomy rule is:

IF mission_phase = "orbit" AND
(time_to_next_eclipse < 4 hours) AND
(battery_state_of_charge < 85%) FOR 58 OF 60 samples, THEN
execute low_state_of_charge macro

For errors or failures caused by transient effects, such as radiation-induced single-event upsets in a microprocessor, the autonomous corrective action may be to reset the subsystem's microprocessor. For "hard" failures, in which the hardware is permanently damaged in some way, the autonomous corrective action may be to switch over to a backup (redundant) hardware unit.

Some anomalies, such as an autonomous *fail-over* to a backup processor, will result in an autonomous *mode demotion* of the spacecraft. Usually there are two or more *safe modes* to which the autonomy system can demote. These modes typically place the spacecraft into a known state with a pre-defined attitude. For interplanetary missions there is often a *Sun-safe* mode, in which the spacecraft solar arrays face the Sun in an attempt to restore charge to the battery, and an *Earth-safe* mode, in which the RF communications antennas are aimed toward Earth in an attempt to restore RF communications.

If the mission involves a constellation of multiple spacecraft, rather than just one satellite, then autonomy becomes even more important. Each spacecraft in the constellation must be capable of detecting anomalous behavior not just within itself but also within other members of the constellation. At any given time several members of the constellation may be out of view of a ground station. If one of those spacecraft develops a problem, or if it detects some kind of external problem or threat, then its nearest neighbors—or even the whole constellation—may need to be informed. If the appropriate response must be generated immediately, there is no time to confer with ground station personnel; the constellation must react in an autonomous fashion.

The design of an autonomous constellation of spacecraft is one of the exciting challenges that face spacecraft designers today.

10.2.10 Resources Required

The spacecraft resources that are required by a typical radiation-tolerant telemetry system are approximately:

- Mass: 1–5 lb (0.45–2.3 kg).
- Size: $7 \times 7 \times 1$ in. (49 in^3 or 800 cm^3).
- Power: 0.5–2 W.

Problems

1. A spacecraft in a near-earth polar orbit must be reconfigured each time it passes over Moscow. Unfortunately, there are no ground stations that can issue commands to the spacecraft at that time. Is there a way by which the satellite may be commanded to reconfigure itself as it passes over Moscow? Describe in your own words how this may be accomplished.
2. A spacecraft telemetry system uses an analog-to-digital converter (ADC) to measure the baseplate temperature of an onboard data processor (refer to figure 10.9.) The expected range for the processor baseplate temperature is $-30°$ to $+77°$C. We would like the telemetered digital output word, **N**, to be able to represent temperatures in the range from $-40°$C to $+87.5°$C. That is, "$\mathbf{N} = 0$" means $-40°$C, and "$\mathbf{N} =$ maximum (full scale)" means $+87.5°$C.

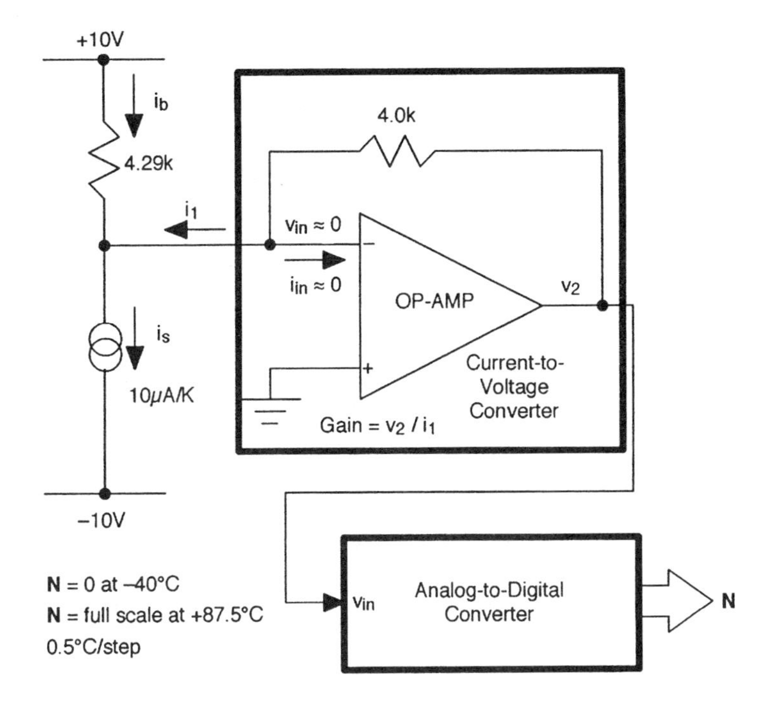

Figure 10.9 Temperature telemetry circuit.

The temperature sensor is a current source that passes a current of 10 μA/K; thus, at 0°C the sensor current will be 2.73 mA:

$$i_s = 2.73 \text{ mA at } 0^\circ\text{C, and}$$
$$i_s = 2.74 \text{ mA at } +1^\circ\text{C.}$$

A current-to-voltage converter (transimpedance amplifier) is used to generate a voltage input for the ADC, which has a conversion sensitivity of fifty steps per volt; that is, the digital output **N** increases by 1 if the input voltage increases by 1 V/50 = 20 mV.

$$\text{If } \mathbf{N} = 130 \text{ for } V_{\text{in}} = 2.60 \text{ V, then}$$
$$\mathbf{N} = 131 \text{ for } V_{\text{in}} = 2.62 \text{ V.}$$

The desired output resolution is 0.5°C per step:

$$\text{If } \mathbf{N} = 130 \text{ corresponds to } +25.0^\circ\text{C, then}$$
$$\mathbf{N} = 131 \text{ should correspond to } +25.5^\circ\text{C.}$$

For this particular temperature telemetry design, answer the following four design questions:

(a) How many binary digits (bits) must be in the output integer **N**?

(b) What should be the transimpedance (gain), in V/mA, of the current-to-voltage converter? Hint: the input voltage of the current-to-voltage converter is always zero.)

(c) If the processor baseplate temperature is +25.25°C, what will be the temperature that is recorded on the ground?

(d) What will be the worst-case (maximum, in absolute value) error between the actual baseplate temperature and the temperature that is reported on the ground?

(e) What is the maximum quantization error for this ADC?

References

Barker, R. H., 1953. Group synchronizing of binary digital systems. In: Jackson, W. (ed.), *Communication Theory.* New York: Academic Press.

Carlson, A. B., 1975. *Communication Systems: An Introduction to Signals and Noise in Electrical Communication,* 2nd ed. New York: McGraw-Hill.

Chetty, P. R. K., 1991. *Satellite Technology and its Applications.* Blue Ridge Summit, PA: Tab Professional and Reference Books.

Fortescue, P., and Stark, J. (eds.), 1991. *Spacecraft Systems Engineering,* New York: John Wiley & Sons.

Gruenberg, E. L., 1967. Space command/telemetry sychronizaion. In: *Handbook of Telemery and Remote Control.* New York: McGraw-Hill.

Lin, S., 1970. *An introduction Error-Correcting Codes.* Englewood Cliffs, NJ: Prentice-Hall.

Malvino, A. P., 1984. *Electronic Principles,* 3rd ed. New York: McGraw-Hill.

Masching, R. G., 1964. A simplified approch to optimal PCM frame synchronization formats. *Proceedings of the 1964 National Telemetering Conference,* 2–4 June.

Messerschmitt, D. G., 1991. Isochronous interconnect. In: Meng, T. H (ed.), *Synchronization Design for Digital Systems.* Boston: Kluwer Academic Publishers.

Pastor, G. J., and Miller, V. L., 1967. Sensing of information. In: *Handbook of Telemetry and Remote Control.* New York: McGraw-Hill, p. 3–3.

Pratt, T., and Bostian, C. W., 1985. *Satellite Communications.* New York: John Wiley & Sons.

Saleh, B. E. A., and Teich, M. C., 1991. *Fundamentals of Photonics.* New York: John Wiley & Sons.

Shannon, C. E., 1948. A mathematic theory of communications. *Bell System Tech. J.,* 379–423, 623–656.

Sklar, B., 1988. *Digital Communications: Fundamentals and Applications.* Englewood Cliffs, NJ: Prentice-Hall.

U.S. Department of Commerce, 1989. Allocations, allotments and plans. In: *Manual of Regulations and Procedures for Federal Radio Frequency Management.* Washington, DC: U.S. Department of Commerce National Telecommunications and Information Administration [NTIA], pp. 4-1–4-95.

U.S. Naval Ordinance Laboratory, 1961. *Synchronization Methods for PCM Telemetry (Report no. 542).* Instrumentation Division, for Aeronautical Systems Division, Air Force Systems Command, Wright-Patterson Air Force Base, Report ASD-TR-61-9.

Wu, W. W., 1984. *Elements of Digital Satellite Communication,* vol. I. Rockville, MD: Computer Science Press.

11

Spacecraft Computer Systems

ROBERT C. MOORE

Spacecraft computer systems require space-qualified microcomputers, memories, and interface devices. Functionally, these are not unlike the devices that are used in desktop business computers and personal computers for the home. However, spacecraft applications impose design constraints that are much more severe than those of the commercial market. For example, low power dissipation, volume, and mass must be achieved in a spacecraft without sacrificing overall performance. In addition, faults and failures in a spaceborne data processing system cannot usually be repaired. Therefore, space flight systems must exhibit excellent reliability and should be able to tolerate many kinds of faults.

In this chapter we shall examine the design of spacecraft computer system architectures with a view toward satisfying the unique requirements imposed on the design by the space environment. We shall be concerned with embedded real-time processors, memory, mass storage, input/output, fault tolerance, custom (special-purpose) processors and peripherals, and the methods by which such complex functions are designed into packages that have maximum reliability and minimal mass, volume, and power.

The processors described here are very similar to the processors used in the spacecraft command and telemetry systems; however, the processors under discussion here are typically assigned tasks that are more general than those of the latter systems. For example, one of these processors might be the intelligent controller in the spacecraft attitude control system or power-monitoring system. Another might be the controller and data processor for a complex scientific experiment. These processors therefore must be more general-purpose than their command and telemetry system counterparts. In addition, they frequently require custom, special-purpose peripheral devices and interfaces that add to the challenge of the computer system design.

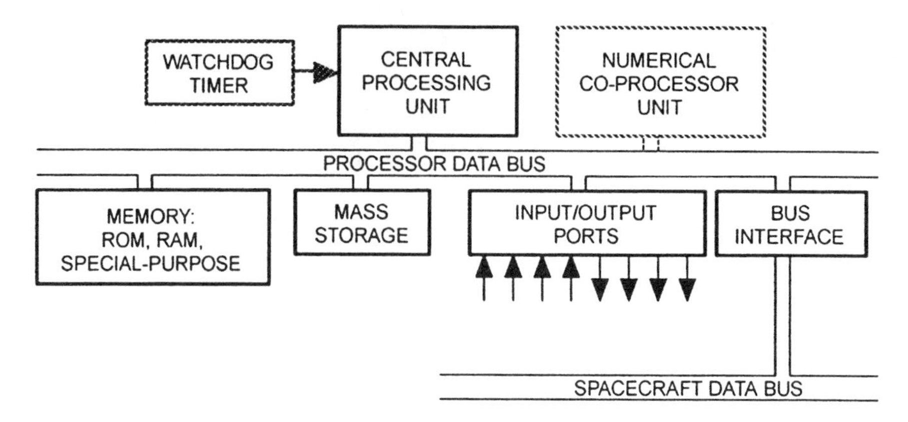

Figure 11.1 Spacecraft data-processing system.

Figure 11.1 is a simplified block diagram of a typical spacecraft computer system. One or more processing units have access to various kinds of memory, mass storage, and input/output devices. Command and memory "upload" data are transferred to the processor from the spacecraft command system; telemetry and memory "download" data are transferred out of the processor to the spacecraft telemetry system. The central processing unit (CPU) receives its instructions by fetching them from an application program stored in memory. The CPU communicates with its data sensors and other processors in the spacecraft through various kinds of input/output channel or over the spacecraft data bus. Often the CPU will have a coprocessor of some kind to which it can offload certain tasks such as floating-point arithmetic, high-speed image processing, fast Fourier transform, and so on.

Each processor in the spacecraft is programmed to do a unique set of tasks with a minimum of ground monitoring and supervision; that is, spacecraft data processors must be relatively autonomous. Spacecraft processors must operate flawlessly for years without a human operator or repair, and they must not require frequent attention from the ground. If the main processor "hangs" or "freezes," a *watchdog timer* (WDT), which is normally reset periodically by a top-level flight software task, will time-out and autonomously reboot the entire processor system.

11.1 Central Processing Units

The data-processing units used in spacecraft are usually microprocessors. These are integrated circuits having up to 64 million transistors each. It was the development of these processors for space and military applications during the 1970s, 1980s, and 1990s that led to the wide availability of personal computers today.

If a particular data-processing system has only one processing element, or if one of the processor elements is a *master* that coordinates all the others, then that processor is called a *central processing unit* (CPU). Systems having more than one processing unit are called *multiprocessor systems*.

Multiprocessor systems in which any processor may assume the role of *master*, or in which the executive tasks normally associated with the master are shared by all processors in the system, are called distributed processing systems. Distributed systems are very difficult to design and test; however, they have definite advantages that make them particularly attractive for use in spaceborne designs. A distributed system can be designed to tolerate faults by enabling designated processors to take over the workload of any failed processor, once that failure has been detected.

The job of the processing unit is to execute the program that is stored in memory, interpret and execute commands that are received from the spacecraft command system, maintain system status and health data (commonly referred to as *housekeeping* and *tell-tale* data), and format subsystem telemetry for transmission to the spacecraft telemetry system or the solid-state data recorder. The processing unit may elect to delegate some of these tasks to special-purpose peripheral processors that then execute the delegated subtasks in parallel (concurrently) with the execution of the main processor's own tasks.

11.2 Memory

Spacecraft computer systems use a wide variety of memory devices and architectures. These include read-only memory (ROM), random access memory (RAM), and special-purpose memories such as multiport memory (where more than one user has unambiguous access to the memory), first-in-first-out (FIFO) memory, cache memory, content-addressable memory (CAM), associative memory, and high-speed table-lookup memory. Memory types and architectures are tailored to the specific processor application.

11.2.1 Read-Only Memory

Read-only memory (ROM) stores data that are permanent. The systems engineer or application software designer places programs, constants, lookup tables, and so on, in ROM and they remain there for the duration of the mission. The stored data may be read but not altered. ROMs are *nonvolatile*: when electrical power is removed from the devices they do not lose the data that are stored in them (that is, the data do not "evaporate" or "fly away").

Normally ROM is used to store programs and subsystem parameters. The programs are read out of ROM into (usually faster) random access memory (RAM) from which individual instructions are fetched for execution. Sometimes instruction data are fetched and executed immediately out of ROM. In the latter case, no program may be self-modifying, because the physical program storage is not alterable. Parameters are loaded into RAM from ROM and are then used to control the functions of the software. The functions of the software may then be modified during the mission simply by uploading to the ROM a new set of parameters (if the ROM is fabricated using an erasable technology) and then commanding the software to reload the parameters from ROM into RAM.

Other forms of ROM include electrically alterable read-only memory (EAROM) and electrically erasable read-only memory (EEROM). These types of memory are referred to under the generic category *read mostly memory*. Both are nonvolatile; however, individual words of EAROM may be changed, while with EEROM an entire integrated circuit, or a significant block of it, must be erased (all words cleared to zero) at once before new data may be stored. Electrically erasable, programmable read-only memory (EEPROM, or E^2PROM) is EEROM that may be programmed by the user, not just the vendor, even after the spacecraft has been launched. EEPROM memory is commonly re-programmed in flight, although special care must be taken because an EEPROM is significantly more vulnerable to radiation-induced single-event upsets (SEUs) during its write cycle. A typical spacecraft computer subsystem will have from 1M to 64M bytes (8-bit words) of ROM, usually in the form of E^2PROM.

11.2.2 Random-Access Memory

Random access memory (RAM), when not implemented with magnetic cores or "flash" memory, is volatile; that is, if power to the devices is interrupted then the data stored in memory are lost. RAM is often the fastest kind of memory, and is read/write memory in which any word may be addressed at any time. Words of data stored in RAM do not have to be accessed sequentially or in blocks or tracks; the user has fast random access to the memory contents. A typical spacecraft subsystem will have from 4M to 256M bytes of RAM. One rule of thumb is that a single-CPU spacecraft computer system will need to have approximately 512 kB (524,288 bytes) of local RAM for each MIPS (million instructions per second) of throughput that is used. For missions that have a requirement for high immunity to radiation effects, it is often quite difficult to meet this requirement for RAM capacity.

11.2.3 Special-Purpose Memory

Some memory devices are used for unusual or special purposes. Multiport memory is used for inter-processor communication; cache memory is used for high-speed table lookup and data accessing, particularly by CPUs; fast multiply–accumulate memory is used to hold partial results in digital signal processing operations (for example, digital Fourier transforms); content-addressable memory is used for cache directories, translation look-aside buffers, statistical computation, and data type or format matching. Other special-purpose memories include energy spectrum accumulators, auto-correlators, cross-correlators, first-in-first-out buffers, and video frame-grabbers for image processing.

Because memory circuits are very regular and repetitive, they tend to be used more than any other single component type in spacecraft data processing systems. The space systems designer generally tries to use memory to solve problems wherever possible.

11.3 Mass Storage

As is the case with personal desktop computers, spacecraft data processors frequently need to store large amounts of data. The system RAM capacity is not large (at most a few hundred megabytes), so some form of mass data storage must be provided. The main kinds of mass storage for spacecraft are disks and integrated circuits (ICs).

11.3.1 Disk

Most users of personal computers are familiar with magnetic disks because they are used for the computer's internal "hard drive." Digital data are stored as magnetic flux changes in a thin film on a disk-shaped medium that rotates under moveable read/write heads. Random access to any datum can be achieved in milliseconds. A disk storage unit that occupies $4 \times 6 \times 1.25$ in. (30 cm^3 or 490 cm^3) may store as much as 200 gigabytes (1.6 trillion bits) of data.

One problem with disk storage is that the mechanical devices necessary to drive the disks and read/write heads are not as reliable as the rest of the components in the data processing system. In particular, the read/write heads are so fragile that even when they are "parked" properly they are unlikely to survive the vibration of launch. Another problem with disks is that the mechanical motion of the medium introduces momentum and vibration that may upset the spacecraft attitude control. There are also undesirable torques on the spacecraft when the rotation of the disk assembly is started or stopped. For these reasons, disk storage is seldom used in small spacecraft.

In the near future we may see optical, laser-written disks used for archival write-only mass storage in spacecraft with short mission durations. Perhaps a flight-qualified read/write optical or magneto-optical disk will become available by the end of this decade (the 2000s).

11.3.2 Solid-State Memory

An alternative to magnetic surface memory is solid-state mass storage. With this approach, large banks of random-access memory (RAM) integrated circuits (ICs) are interconnected to implement the mass storage. This method requires higher power, but offers high speed and density. For reasonable volume and power, the storage capacity of RAM mass memory may be as large as 4 gigabits. The cost of RAM memory is relatively high compared to that of magnetic disks.

Memory technology used for solid-state data recorders includes synchronous dynamic random-access memory (SDRAM) and "flash" memory (high-density, high-speed, low-power, nonvolatile RAM). Flash memory has limited total-dose radiation immunity—approximately 20 krad(Si) maximum—but offers the advantage of nonvolatility.

Use of phase changes in amorphous semiconducting materials (to represent stored logic 1s and 0s) may soon provide a high-capacity, nonvolatile, very radiation-immune random access storage medium for mass storage of space flight computers (solid-state data recorders).

11.4 Input/Output

For a data processing system to be useful, it is necessary to transfer data into and out of the system. A wide variety of methods is used to accomplish this input/output (I/O) activity. Specific I/O devices and techniques are tailored to the processor application.

11.4.1 Ports

An I/O port is a hardware device, driven by software as necessary, that transfers data to and from the processing system data bus. Typically, data are transferred to and from the bus in parallel, with the number of bits per transfer equal to the bus width. The I/O port serves as an interface between the processing system and some kind of external device.

In general there are two kinds of data transfer to and from the external device: serial and parallel. Also, there are two basic ways in which an I/O port may appear to the host processor: I/O mapped and memory mapped.

11.4.1.1 Serial I/O Ports

Serial ports transfer data to and from the system one bit at a time. They are frequently used when data transfer rates are low; for example, to transfer commands and (sometimes) housekeeping telemetry. A serial interface can consist of clock, data, and envelope signals as described in chapter 10; or it can be an asynchronous system that uses a universal asynchronous receiver–transmitter (UART) at each end of the transmission path. The latter method is similar to the familiar RS232-C standard interface used by many personal computers and engineering workstations. In a spacecraft, data transfer rates for serial I/O ports seldom exceed fifty million bits per second.

Certain serial data ports are controlled by dedicated hardware that implements and enforces the transfer protocol for a standard bus. Often these protocol controllers are designated as *master* or *slave* to indicate the degree of authority they may have over the bus.

11.4.1.2 Parallel I/O Ports

Parallel ports transfer data to and from the system one word at a time. A *word* is almost always a multiple of eight bits, and usually has as many bits as the processor data bus. For example, a 32-bit processor would normally use parallel I/O ports that are 32 bits wide. Parallel ports are used to transfer data between two bus-oriented systems. They are especially attractive when data transfer rates must be in the range from 20 million to 500 million bits per second.

11.4.1.3 I/O-Mapped Ports

Serial and parallel data ports are accessed by the processor in different ways. One way is to have special instructions in the machine language for inputting and outputting data from and to the ports. Each input or output instruction then carries an identification tag by which the specific port may be addressed. This kind of port addressing is done independently from the memory addressing, and is called I/O *mapped* because the port location addresses are unique to the input/output functions.

11.4.1.4 Memory-Mapped Ports

The I/O ports may also be accessed by addressing them as if they were locations in memory. Instead of using special input and output instructions to initiate data transfer, ordinary load and store operations are used, with the effective physical address specifying the particular port to be accessed. Input ports generally appear as read-only locations in the memory address space; output ports appear as write-only locations.

Sometimes memory-mapped output ports can be used as read/write memory; that is, data written to the output port may be read back for verification, or an acknowledgement (*handshake*) signal from the destination device may be read there.

With memory-mapped I/O the input/output ports usually occupy one or two contiguous ranges in the memory address space. These ranges of memory locations then receive special attention when memory address decoding is done by the processor hardware.

Regardless of whether I/O mapping or memory mapping is used, the input/output ports are usually sophisticated enough to perform their functions, once initiated by the processor, without any intervention by the processor. For example, a UART that is memory mapped will receive a byte of output data during the execution of a *store accumulator* instruction. It then will proceed to generate the appropriate start bit (synchronization pattern), data bits, parity bit, and stop bit while the processor continues on with its programmed instructions.

Large-scale and ultra-large-scale integration of transistors equips the system designer with "intelligent" peripheral devices that are very sophisticated. This enables the processor to delegate more tasks to its peripherals, thereby reducing the workload of the CPU. Input/output, memory management, and floating-point arithmetic are just three of the many *intelligent peripheral* functions. Sometimes these intelligent peripherals are called *coprocessors*, because their processing capabilities are comparable to that of the CPU.

11.4.2 Direct Memory Access (DMA)

When large blocks of data must be transferred, or when data transfer rates are high, data may be moved directly into or out of the processor memory. This method of input and output is called *direct memory access* (DMA). A DMA controller synchronizes the data transfers with the ongoing microprocessor activity.

Cycle-stealing DMA is a method by which processor cycles (such as an instruction fetch cycle) in which no RAM access will be made by the CPU (assuming instructions are always fetched from ROM or cache memory) are used simultaneously by the DMA controller to move data words into or out of RAM. Usually the DMA controller is intelligent enough to cycle-steal during the execute cycles of instructions that make no memory references (for example, register operations, flag checking).

If DMA data transfers are large, or have very high priority, the DMA controller may actually introduce *wait states* into the microprocessor activity so that the DMA takes priority over all other CPU activity. Data transfer rates ranging from 1 million to 400 million bits per second are possible with DMA in a typical spacecraft computer system.

11.4.3 Multiport Memory

When high data transfer rates are needed and wait states cannot be tolerated, multiport memory can be used for I/O. A multiport memory is one in which two or more users may access the memory simultaneously. The memory ICs themselves (or a very intelligent, fast memory controller IC) arbitrate so that no conflicts arise. An external device can be writing data into RAM at the same time as the CPU is reading data out of RAM, and so on. Multiport memory eliminates many of the timing hassles normally associated with DMA controller design.

Multiport memory may also be used for interprocessor communication. Certain regions of the system memory map can be made common to several processors. Data that are to be transferred from one processor to another are simply stored in these locations, which act as electronic *mailboxes*. When another processor needs to reference those data, it simply reads the predefined locations. The data do not have to be transferred or copied from one processor's address space into that of the other processor.

For fault tolerance, access to these multiport mailboxes may be deliberately limited. For example, only certain processors in a multiprocessor system may have access to write to some location, while other processors may enjoy only read-only access or perhaps no access at all. Execution privileges for stored code images are especially controlled, so that a particular processor never executes another processor's program. This is one way in which undesirable or unexpected interprocessor interactions are substantially reduced.

11.4.4 Interrupts

One form of input/output operation is the processor interrupt. An interrupt may be generated by a timer or an event. When the interrupt occurs, the processor stops whatever it is doing (as soon as that is convenient) and begins to execute an *interrupt service routine* (ISR). This routine locates the source of the interrupt and responds accordingly. If, for example, a sensor on a scientific experiment detects an event (say, a Sun sensor on a rotating spacecraft "sees" the sun at an unacceptable angle relative to the spacecraft body), it signals the processor with an interrupt. This can synchronize the processor activity to external events as they occur in real time.

When the interrupt occurs, a *vector* address is generated by the interrupt controller. This vector address steers the CPU to the appropriate ISR in memory.

When an instrument receives a command from the command processor, a "command ready" interrupt could be generated so that the instrument would know that a command is sitting in the command input buffer, awaiting execution by the instrument processor. Similarly, the telemetry system can signal an instrument processor with an interrupt when it is time for that processor to format a new subframe or packet of telemetry data.

Interrupts are also a convenient way of synchronizing the activity of multiple processors in a spacecraft. Periodic *real-time interrupts* (RTIs), which occur at regular time intervals, are used to synchronize tasks and functions within a single CPU. For example, a one-second RTI might be used to mark the beginning of a new frame of telemetry data.

A particular processor may have many interrupts, some requiring more immediate attention than others. For this reason, multiple interrupts are assigned priorities so that in the event that multiple interrupts occur simultaneously the highest priority interrupt is serviced first. Some spacecraft processors even have a complex *rotating priority* scheme in which priority is a function of time, or a function of recent processor interrupt activity.

11.4.4.1 Context Switching

Managing interrupts is a major task for the typical spacecraft processor. A processor that must respond to multiple interrupts needs to have an architecture that permits rapid *context switching* for its software procedures. A context switch occurs when the processor stops doing one task, stores the status of that task as a *context* that is pushed onto a stack, and then restores the context of another task so that work may proceed on the new task. Real-time interrupts frequently require context switching when they occur. If the processor cannot execute a context switch quickly, it may find itself spending a large portion of its time pushing and popping contexts to and from the stack. This would be counterproductive.

11.4.5 Timers

A timer is a piece of hardware that generates periodic interrupts or makes a real-time clock available to a processor. Timers are used to measure elapsed time between events, synchronize processors, and measure periods (and therefore rates). Special-purpose timers may be used to verify proper operation of the processor and to interrupt *infinite loops* (sometimes referred to as *do forever* loops) that might result from hardware upsets or software bugs. These are called *watchdog timers* or *improper sequence detectors*. See sections 11.5.9 and 11.5.10 below.

11.4.6 Bus Interface

If the spacecraft has a data bus that is common among many processors, some kind of bus interface is required. Typically this is done with a bus adapter. The bus adapter detects and corrects bus errors, arbitrates for bus access, and acts as a very intelligent DMA controller for the processor. The bus adapter stands as a hardware "bridge" between the local processor bus and the common spacecraft bus. For example, the local processor bus might be 64 bits in width, and the spacecraft bus (usually a standard bus, such as the compact peripheral computer interconnect bus, CPCI, or the IEEE-1394 standard bus) might be 32 bits in width. The "bridge" function would resolve the differences in bus width, timing, protocol, etc., to make data transfers between the two buses occur seamlessly.

Commands, telemetry, memory loads, and memory dumps all can occur over the common bus. The bus interface makes all of these data transfers happen reliably, without placing any burden or overhead on the individual processors.

Because the bus adapter's task is so important to the instruments and processors within the spacecraft, great care must be taken in designing the bus interface hardware and software. Frequently fully redundant busses and bus interface units are used to guarantee reliable spacecraft operation (Chetty, 1991, pp. 378–380).

11.5 Fault Tolerance

There are many ways in which spaceborne data processing hardware and software may fail. Some failures are *hard* failures, meaning that once the failure has occurred it continues to have effect for the remainder of the mission. Other failures are called *soft* failures, meaning that the failure causes its effect only once; after that the system reverts to normal operation. Whenever possible, spacecraft systems must be capable of tolerating both hard and soft failures. The design techniques by which this capability is implemented are grouped under the generic heading of *fault tolerance*.

The design of fault-tolerant systems is very complex. The scope of this chapter does not permit even a comprehensive overview of the subject. However, certain fault tolerance methods will be discussed to give the general idea of how spacecraft systems may be designed to deal with soft and hard failures.

11.5.1 Radiation Hardness

Radiation encountered in space, particularly on lengthy deep-space missions, can cause cumulative damage to semiconductor devices so that their performance degrades over time. These cumulative radiation effects are a function of the total radiation dose (*total ionizing dose*, TID) encountered by each semiconductor chip. As the cumulative ionizing dose increases, transistor beta goes down, threshold voltages change, metal oxide semiconductor (MOS) channel mobility degrades, leakage currents increase, and so on. This means, inter alia, that digital logic slows down, operational amplifier offset voltages change, current drive capability is reduced, and power dissipation goes up. The details of these cumulative TID effects are given in chapter 13.

Several different types of radiation are encountered in space. Most of them emanate from the Sun or from space itself (remnants of the "big bang") and are called cosmic radiation. Other types are trapped radiation (captured by the magnetic field of a planet) and man-made radiation (for example, emanations from nuclear explosions or nuclear reactors).

Very high energy photons, called *gamma rays*, can pass through an integrated circuit package with very little energy loss. The effect of a gamma ray is to leave a trail of electrical charge in its wake. Most of these charges recombine with units of opposite charge and disappear; however, some of the charges get "stuck" on the surfaces and boundary layers of the integrated devices. With sufficient exposure to gamma radiation, these trapped charges can accumulate and alter the operating characteristics of the device. This undesirable effect is a function of the *total accumulated dose* of gamma radiation, which is measured in *rads*. Other forms of radiation, most notably charged or neutral particles that impinge on orbiting space flight hardware, cause similar cumulative effects.

A rad is the quantity of any form of ionizing radiation that will impart 100 ergs of energy to a gram of material. With integrated circuits the material of interest is generally silicon; therefore, the unit of radiation dose is *rad silicon*, abbreviated rad (Si), to represent the energy deposited in silicon. For a magnetic bubble memory chip fabricated on a garnet substrate the unit of interest would be *rad (garnet)*; for an astronaut in space the unit of interest would be *rad (tissue)*.

Gamma rays are the result of a reconfiguration of nuclear energy levels in atoms. In space, only small amounts of gamma radiation appear, usually from galactic sources.

Gamma radiation is also caused locally when cosmic rays or highly energetic protons activate atoms of the spacecraft.

X-rays also occur in space; they result from solar processes and *bremsstrahlung* generated by the interaction of electrons with the spacecraft.

Essentially, the TID received by electronic components in a spacecraft is due to neutrons and charged particles that impinge on the vulnerable areas of those components. Neutrons are uncharged high-energy nuclear particles that behave as miniature projectiles and can cause structural damage to the solid-state materials from which integrated circuits are fabricated. Charged particles are energetic electrons, protons, alpha particles, or ions that can pass through an electronic device (for example, an integrated circuit) and generate a cloud of electrical charge. If this charge appears on the gate of a metal-oxide semiconductor (MOS) transistor, it can induce a single-event upset (SEU) that may have serious consequences to the operation of the system.

To guard against these radiation effects, conservative design rules are followed carefully. Integrated circuit technology is developed so that integrated circuits are able to survive large total doses of radiation. Every possible failure must be anticipated and then either (1) the component that might fail is designed out of the system, (2) dense material such as tantalum or tungsten is wrapped around the "soft" component to act as a shield against the radiation, or (3) the effects of the failure are mitigated by some other means. With complementary metal-oxide semiconductor (CMOS) device technology, for example, TID tolerance (also called *radiation hardening*) is achieved through meticulous care in design and processing. Guard rings, very thin oxides, radiation-hardened field dielectrics, and carefully controlled processing temperatures during integrated circuit fabrication are all used to accomplish this.

11.5.2 Single-Event Upsets

Another annoying kind of radiation-induced failure is the single-event upset (SEU). Because of the extremely small dimensions used in the design of modern integrated circuitry, one high-energy particle or ion can pass through an integrated circuit and deposit enough charge to change the state of a stored binary digit. When this happens, one bit in a RAM containing 1,073,741,824 bits (1 Gb) can toggle. This does not sound very significant until you realize that the changed bit could convert a stored program instruction into some other instruction by altering the operation code. This "bit flip" would certainly cause almost instant chaos in a data processing system.

Because it is impossible to predict where or when a SEU will occur, the system designer must assume that it will occur anywhere, and at the worst possible moment, and must therefore design the system to tolerate it. SEUs and their effects can be detected or corrected using error detection and correction (EDAC) codes, watchdog timers, fault rollback, watchdog processors, and other mitigation methods that are discussed below.

11.5.3 CMOS Latch-up

With CMOS logic a phenomenon called *latch-up* can occur. If noise currents of sufficient magnitude occur in the CMOS device, it will latch up across the power supply. This causes a very large current to flow in the device, which will usually destroy the device within a few tens of milliseconds.

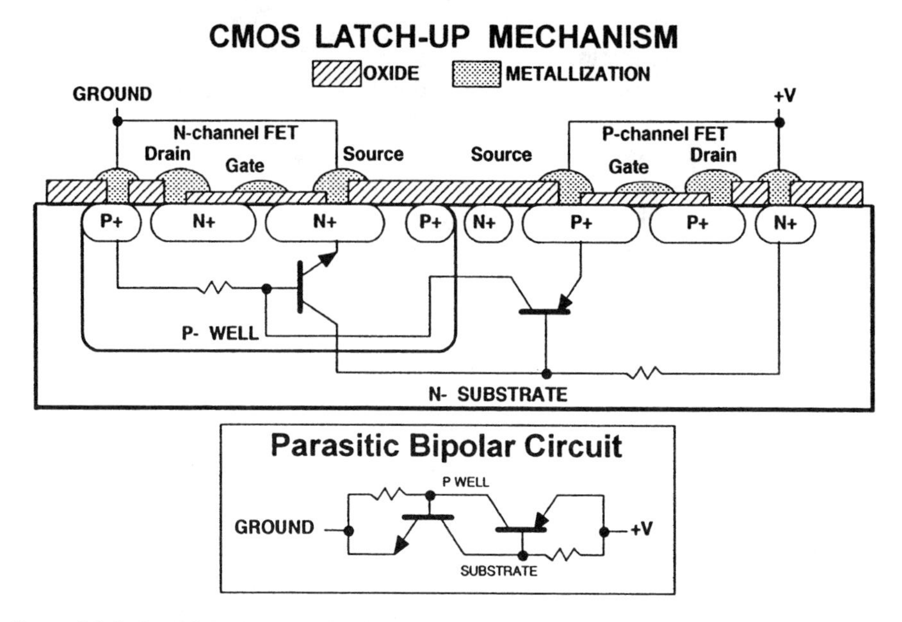

Figure 11.2 CMOS latch-up mechanism, showing parasitic bipolar transistors.

Figure 11.2 shows a cross-section of a typical CMOS device. The shaded regions represent the oxide that is grown on the surface of the silicon substrate. In the diagram, a p-channel field-effect transistor (FET) is fabricated in the n^- substrate material by diffusing a p^+ source and drain around the gate. (A metal gate is shown in the figure; self-aligned polysilicon gates are more common today.) An n^+ diffusion is used to make contact with the substrate. The n-channel FET is fabricated in a *well* or *tub* of p^- diffusion. (The wells are often called tubs because in three dimensions they have the appearance of a bathtub. n^+ diffusions on either side of the gate make the drain and source for this transistor. Contact with the p-well is made using a *plug* of p^+ diffusion in the well. The diode formed by the p^-well in the n^- substrate is reverse-biased by connecting the substrate to the positive supply rail and the p-well to the negative supply (ground) rail.

Notice the two bipolar transistor symbols that are shown in figure 11.2. The source of the p-channel FET forms a diode with the n^- substrate material; the p-well forms another diode with the substrate. These two bipolar junctions constitute a pnp bipolar transistor. Similarly, the source of the n-channel FET forms a diode with the p-well; with the p-well/substrate diode this makes an npn bipolar transistor. A similar situation arises if the p-channel FET is diffused into a well of n^- diffusion.

These two bipolar transistors are an inherent part of the CMOS processing. Their interconnection (figure 11.2) is that of a bipolar four-layer diode structure. Normally the two bipolar transistors are biased off. However, if sufficient noise current is generated, flowing into the base of the npn bipolar transistor or out of the base of the pnp bipolar transistor, then the corresponding transistor will begin to conduct. The collector current from each transistor adds to the base current of the other. This positive-feedback condition is called *regeneration*; if it begins, the parasitic bipolar circuit quickly *latches up* across the supply rails. The p-well rises above ground potential by approximately 0.76 V, and the source of the p-channel FET is pulled down to approximately 1.0 V. This latched-up CMOS device thereby presents a 1.0 V short circuit to the logic power supply, and the supply will attempt to dump its short-circuit output current into the CMOS device that has "latched up."

The latched CMOS device looks much like a turned-on silicon-controlled rectifier (SCR). It is capable of conducting tens of amperes (!) from the $+V$ supply rail to the ground rail, if the logic power supply can supply that much current. Whatever current the supply is capable of sourcing flows through the power rail wire bonds for the latched chip. These bonds, which typically are redundant gold wires 0.001 in. (one mil or 25 μm) in diameter, are not capable of carrying that much current for long. With a couple of amperes flowing, the wires self-heat to destruction in a few tens of milliseconds. This, of course, is catastrophic to the CMOS component.

If the flow of current to the latched CMOS device is interrupted, or reduced below some *extinction* value, the latch condition will go away. Provided the large latch-up current has not caused any irreversible damage, the part will then function normally. This suggests an architecture in which latch-up is detected and then the latch condition is extinguished by interrupting the supply of electrical power to the latched device.

The most common method for preventing damage caused by CMOS latch-up is to limit the current that can flow into any one CMOS device if latch-up does occur. If the current limit is low enough in value, latch-up cannot destroy the device. Commands from the ground may then extinguish the latch condition by turning the power to the subsystem off and then back on.

With many subsystems, such as the spacecraft command system, cycling the power to the subsystem is not feasible. For those systems, an automatic latch-up protection circuit is required. Figure 11.3 shows a typical latch-up protection circuit. The circuit continuously monitors the current that is flowing into the supply lead(s) of the device(s) that are to be protected. If that current ever exceeds a certain threshold value, the protection circuit "trips" and opens up a solid-state switch in the power supply line to the device(s). This automatically extinguishes the latch-up condition within a few milliseconds. The protection circuit may then be automatically restarted by a periodic power-on reset (POR) pulse, as shown, or by another appropriate signal.

11.5.4 Parity

One way to look for a SEU in a memory is to keep track of the evenness or oddness of the number of zero or nonzero bits in each word stored. This is called parity checking. For example, when the word is stored, an extra bit can be stored along with the data bits. The value of the extra bit is computed so that the number of logic ones in the word is always even. Then, when the word is read out of memory, the number of logic ones can be counted. If the number is odd, the processor can assume that an odd number of SEUs has occurred.

Complex systems of multidimensional parity checks can be implemented that will locate the erroneous bit so that it may be toggled back to its original state. In general, however, these parity-checking schemes require extensive hardware. If multiple errors must be detected and corrected, simple parity checking quickly becomes awkward.

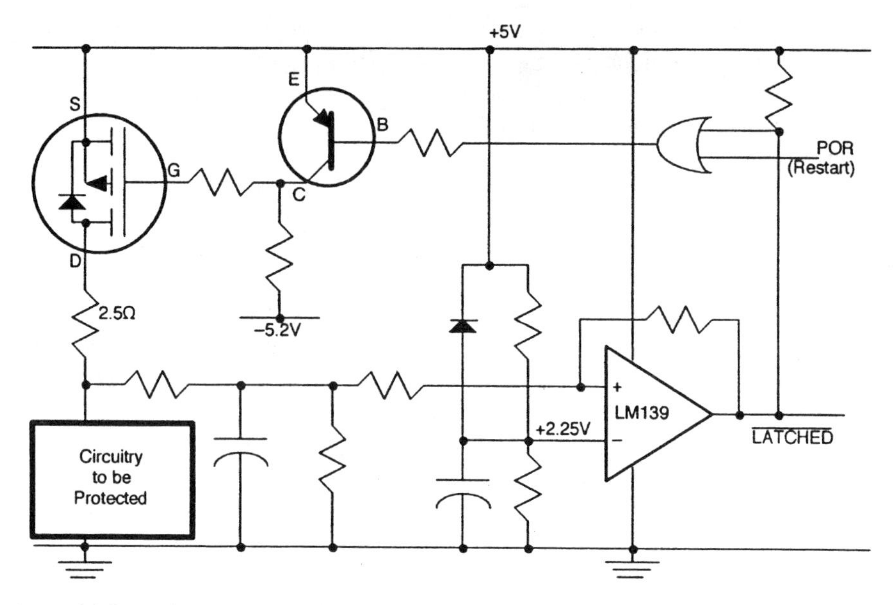

Figure 11.3 Typical CMOS latch-up protection circuit (POR is a power-on reset pulse).

Another form of parity check may be performed over many words, with the logic ones being summed modulo 2^n to form an n-bit memory checksum. This checksum is computed prior to transmitting a block of data (say, a memory load) and then recomputed by the receiving processor. The receiving processor compares the computed checksum with the transmitted checksum to verify that the data were transmitted without errors. If errors are detected, the receiving processor can request a retransmission of the data.

11.5.5 Error Detection and Correction

Other sophisticated techniques exist that can detect and correct multiple errors in data. Cyclic redundancy checks may be devised to detect and correct almost any number of simultaneous errors. These methods are especially attractive because errors introduced by the error-checking circuitry itself are also detected and corrected. These are called self-checking checkers.

11.5.5.1 Hamming Codes

A method for detecting and correcting errors was devised by Richard Hamming at Bell Telephone Laboratories in 1950, and is known as the *Hamming error correcting code* (Baer, 1980, pp. 235–237). In this scheme, an n-bit data word is encoded into an $(n+k)$-bit composite word by adding k parity bits at special positions as shown below.

1	2	3	4	5	6	7	8	9	10	11	12	13	…	n+k

Bits 1, 2, 4, 8, 16, 32, 64, 128,… are parity bits.

If the bits are numbered starting with 1, then the parity bits in the encoded word are those whose positions are a power of two, beginning with $2^0 = 1$. The parity bits are chosen to make the parity even or odd for a specific subset of the bits in the encoded word. The scheme is to make each bit position in the composite word be part of the parity check for those parity bits whose position numbers add up to the bit position. This causes each bit position to be checked by a unique set of the parity bits, because their positions make up a binary numbering system. The pattern of checks is said to be *orthogonal* over the bits of the data word.

The table below shows the subset of bits (from the *composite* word, not the data word) over which each parity bit is computed:

$\mathbf{P}_1$ XORS bits 3, 5, 7, 9, 11, 13, 15, 17, 19, 21, 23, ...
$\mathbf{P}_2$ XORS bits 3, 6, 7, 10, 11, 14, 15, 18, 19, 22, 23, ...
$\mathbf{P}_4$ XORS bits 5, 6, 7, 12, 13, 14, 15, 20, 21, 22, 23, ...
$\mathbf{P}_8$ XORS bits 9, 10, 11, 12, 13, 14, 15, 24, 25, 26, ...
$\mathbf{P}_{16}$ XORS bits 17, 18, 19, 20, 21, ..., 30, 31, 48, 49, ...

Notice that composite bit 5 is checked by parity bits 1 and 4 because $1 + 4 = 5$. Composite bit 11 is similarly checked by parity bits 1, 2, and 8. Notice also that for this scheme to work we must have

$$2^k - 1 = n + k.$$

For example, if we are working with seven-bit data such as simple ASCII characters, $n = 7$, and we need a k of 4.

It is easy to work out the value of k that is needed to correct single errors in data words of various lengths. For at most one bit error, there may be $n + k$ locations for the error. Add to this the indication of "no error," and we find that the k bits must be able to assume one of $n + k + 1$ unique combinations; that is, $n + k + 1 = 2^k$. This means that $\log_2 (n + k + 1) = k$, and it can be shown that $\log_2 (n + 1) < k$.

Table 11.1 shows the values of k for some standard data lengths.

A useful concept is the *Hamming distance* between two code words. This is the number of bits that must be toggled in the first word to transform it into the second word. If the minimum Hamming distance between any pair of code words is at least 2, no single-bit error (for example, a radiation-induced SEU) will be able to transform one legal code word into another. If the minimum Hamming distance is at least 3, it is possible to detect double-bit errors.

Table 11.1

n	k
8	4
16	5
32	6
64	7
128	8

11.5.5.2 Hamming Code Example

As a concrete example of the Hamming scheme, we show in figure 11.4 the even-parity encoding for a 7-bit ASCII character, the lower-case letter "c" (Tanenbaum, 1984, pp. 38–40).

The method for finding the incorrect bit in the received composite word is to compute the parity of each of the subsets defined above. Then these new parity bits are compared with those that were stored as part of the composite word. The bits that result from this comparison are called the *syndrome* bits. If all of the newly computed parity bits match those of the composite word (that is, if the syndrome is zero), then there is no detectable error. By adding up the position numbers of the check bits giving incorrect parity, the position in the composite word of the incorrect bit is obtained (Sklar, 1988, p. 298; Odenwalder, 1976). That bit then is toggled to correct the error. For example, if the parity determined by bits S2 and S8 is incorrect and all others are correct, then composite bit 10 ($2 + 8 = 10$) must be toggled to correct the error. This is true because the parity bits and the data bits are *orthogonal* in the bit-numbering space (Sklar, 1988, p. 39).

11.5.5.3 Implementation

One way this Hamming code may be implemented in a computer system is illustrated by the block diagram of figure 11.5 (Hayes, 1978, p. 151). The n data bits (that is, the data to be transferred or stored) arrive at the left side of figure 11.5. A block of combinatorial logic generates, in parallel, all k of the required parity bits.

At the receiving end of the transmission path, or when the data are read out of the memory, all $n + k$ bits are received. Then another parity-bit generator, called the *syndrome generator*, computes parity again over the received data and compares the results with the parity bits that were transmitted or stored in the composite word. If there are no discrepancies (that is, if all of the newly computed syndrome bits are zero), then there were no errors. If one and only one of the syndrome bits is a logic one, then the parity bit itself is the error (assuming at most one error has occurred), and no correction is required.

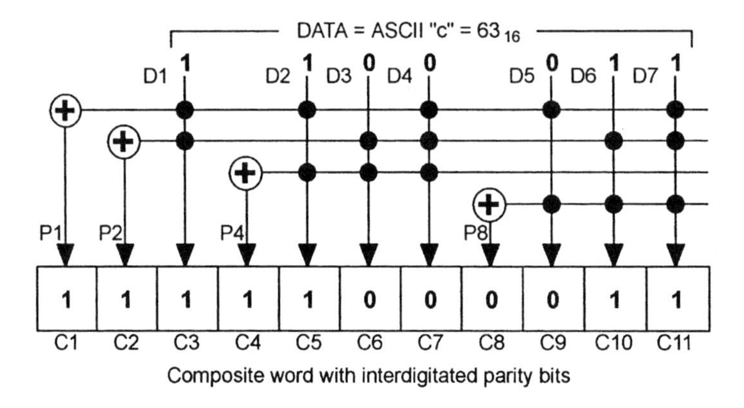

Figure 11.4 Example of error-correcting encoding.

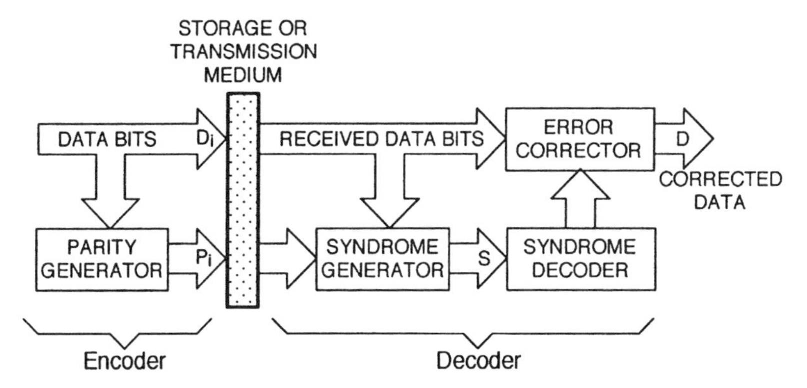

Figure 11.5 Block diagram of error-correcting logic.

If more than one of the syndrome bits are nonzero, then the new parity bits are assembled into a binary word, called the *syndrome*, that identifies the location in the composite word of the data bit that must be toggled. An error corrector decodes this syndrome and exclusive-ORs the result with the bad bit. The corrected data are then used by the receiving subsystem.

Imagine, for example, that the encoder of figure 11.4 is used when writing data into a memory. The decoder hardware would then be employed when data are read out of the memory. Any single-bit errors that occurred in a given data word while it was resident in that memory would be corrected automatically during the readout; the errors would never be seen by the rest of the system. This automatic correction of stored data is *transparent* to the user of that memory. The users of the memory "see through" the error correction mechanism to view the uncorrupted data.

Figure 11.6 shows the details of a hardware parity encoder for the $n = 7, k = 4$ case. The input to the encoder is the 7-bit data word that is to be transmitted or stored. The output of the encoder is the transmitted or stored (composite) word, which includes both the data bits and the parity bits, and is labeled as $\mathbf{C}_1$ through $\mathbf{C}_{11}$ in figure 11.6.

A shorthand notation has been employed in figure 11.6 to simplify the diagram, and to make clear the binary pattern by which the encoder may be expanded to encode more data bits. The bold lines in the figure indicate multiple inputs to the exclusive-OR (XOR) gates, whose function can be represented mathematically as

$$\begin{aligned}
\mathbf{P}_1 &= \mathbf{D}_1 \oplus \mathbf{D}_2 \oplus \mathbf{D}_4 \oplus \mathbf{D}_5 \oplus \mathbf{D}_7 \\
\mathbf{P}_2 &= \mathbf{D}_1 \oplus \mathbf{D}_3 \oplus \mathbf{D}_4 \oplus \mathbf{D}_6 \oplus \mathbf{D}_7 \\
\mathbf{P}_4 &= \mathbf{D}_2 \oplus \mathbf{D}_3 \oplus \mathbf{D}_4 \\
\mathbf{P}_8 &= \mathbf{D}_5 \oplus \mathbf{D}_6 \oplus \mathbf{D}_7
\end{aligned}$$

where $\oplus$ denotes the exclusive-OR logic function. The exclusive-OR of multiple binary digits is identical with the sum, modulo 2. A multiple-input exclusive-OR gate asserts its output if (and only if) the number of logic ones at its inputs is odd.

Similarly, figure 11.7 shows the syndrome generator that is located at the receiver (or active during memory readout). Notice the similarity between figure 11.6 and figure 11.7.

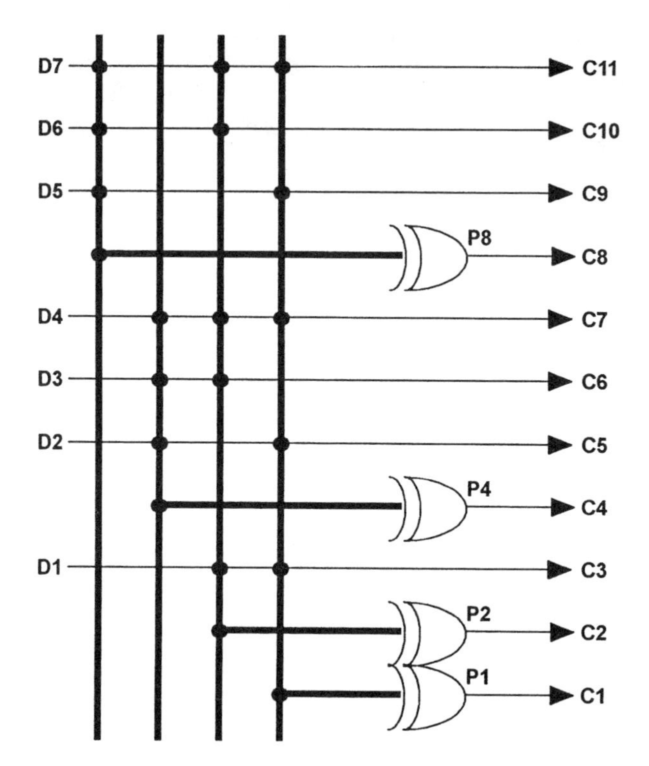

Figure 11.6 Parity encoder for seven data bits.

The parity generator examines the data bits only; the syndrome generator also examines the parity bits themselves.

To see how the parity generator and syndrome generator may be expanded to handle a larger number of data bits, look closely at the pattern of connection dots in the matrix to the left of the exclusive-OR gates. The dots form a binary pattern that may easily be expanded to any desired number of data bits. The key to this is the careful numbering of the bits, and the interdigitation of the parity bits with the data bits in the composite (transmitted, stored) word.

Figure 11.8 shows how the error detection and correction are implemented. Once the syndrome has been generated, a simple binary decoder may be used to generate a 1-of-n code, in which one and only one of n lines is asserted. The number of the asserted line corresponds to the value of the syndrome. These lines are then used to toggle the erroneous bit, and the corrected output data are delivered to their destination.

The entire encode/decode operation requires only combinatorial logic: exclusive-OR gates to generate the parity bits, more exclusive-OR gates to generate the syndrome, a 1-of-n decoder to detect the location of the error in the composite word, and a final level of two-input exclusive-OR gates to perform the correction.

High-speed, ultra-large-scale integrated circuits can be designed to perform error detection and correction *on the fly*; that is, in real time as the data are stored into and read out of memory. Because these self-checking checkers can be implemented in

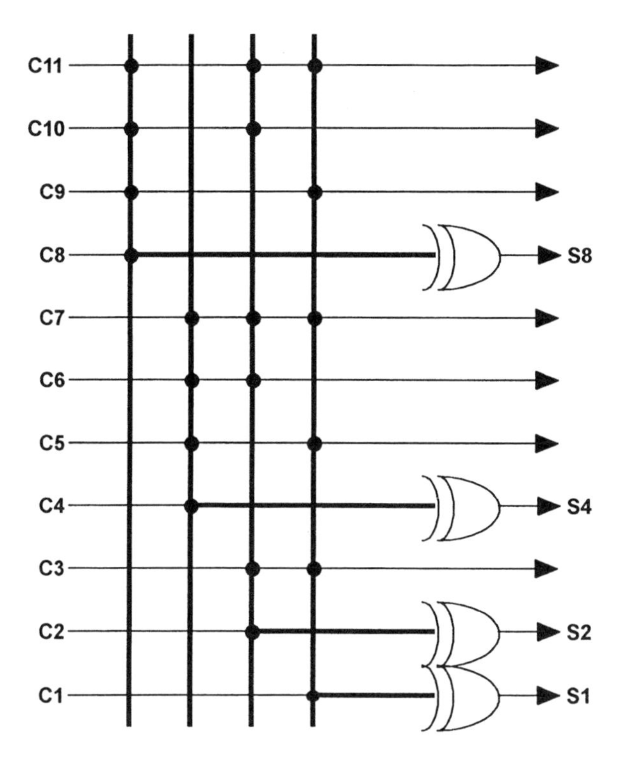

Figure 11.7 Syndrome generator.

single integrated circuits, they add very little to the circuitry, mass, volume, and power dissipation of the overall system.

Future processors and memory chips will have error detection and correction built right in. Vulnerability to SEUs and even multiple-bit upsets will then be greatly reduced. (Multiple-bit errors will be corrected so long as each of the erroneous bits is located in a different word of the memory.) All hardware and software modules that must handle data transfers need to be designed carefully so that data integrity is never lost.

Today's ultra-high-density memory chips have data storage transistors integrated so tightly together that a single radiation event can upset a neighborhood of stored bits, not just a single bit. Correcting these multiple-bit errors requires a more powerful error-correcting code than the one just discussed; however, the system architecture remains the same.

11.5.6 Triple Modular Redundancy

What if a subsystem develops a *hard* fault; that is, one that does not go away? If the fault is due to a modified piece of software, it can be fixed by reloading the software from onboard ROM object code images or from the ground. But what if the hard fault is in the hardware? In this case the affected subsystem is either "dead" or performing some kind of defective function. How can this situation be detected so that a spare subsystem may be used in place of the failed one?

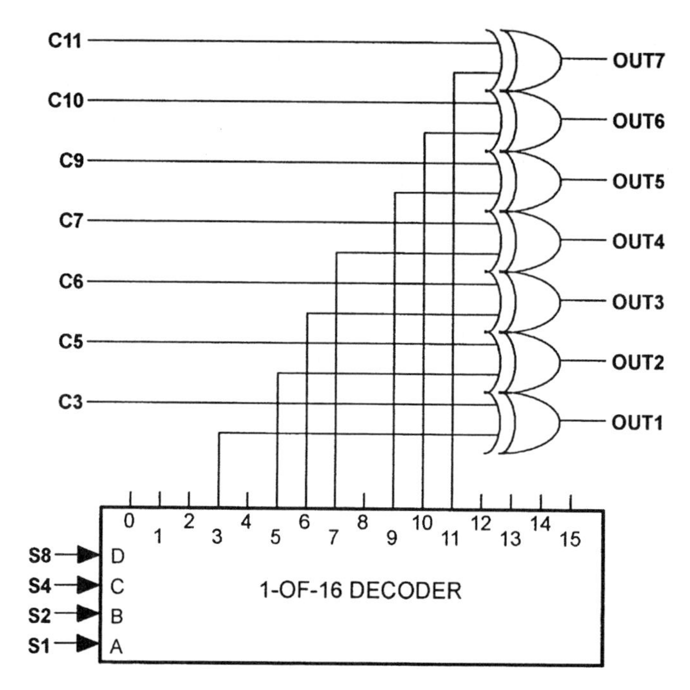

Figure 11.8 Error detector/corrector.

One very popular method is called triple modular redundancy (TMR). Three identical subsystems are included in the spacecraft, so that all three are simultaneously performing identical tasks. Then a simple majority voting circuit looks at the outputs of all three modules to decide what the "correct" result should be. If one of the three modules fails, the majority voting overrules its error.

Notice that TMR can correct multiple *soft* errors provided that they affect only one of the three subsystems. It also can correct *hard* errors, again provided that they affect only one of the three subsystems. The obvious disadvantage of TMR is that it requires three times as much hardware as a single nonredundant system, plus the hardware required for the majority voter. Its advantage is that errors are corrected immediately and autonomously, without interaction from the ground by machines or human operators.

If redundant components are built into the CPU integrated circuit, it is possible for the CPU to repair itself "on the fly." By periodically running self-test diagnostic software, defective hardware components can be identified, and then redundant components are switched in to take the place of the failed ones. At the subsystem level, a *watchdog processor* can perform this diagnostic function and autonomously switch in the cold spare subsystem when a fault in the primary subsystem is detected. This autonomous *fail-over* is controlled by a watchdog processor that is itself redundant.

11.5.7 Multiple Execution

Another method of achieving redundancy is to have more than one processor execute each critical task and then compare the results. Any discrepancy indicates that at least one of the processors was in error, and error correction may then be initiated (Avizienis,

1981), or execution of the task deferred until the problem has been diagnosed further on the ground. Critical tasks usually cannot afford to wait for ground intervention.

One very convenient way to implement multiple execution is to add peripheral circuitry to the CPU integrated circuit so that it may be either a *master* or a *slave*. The *master* executes normally; the *slave* also executes, and observes the outputs of the *master*. When the master's outputs differ from those of the slave, an error has been detected and an interrupt is generated. Because of its function, the slave processor is called a *watchdog processor* (Mahmood and McCluskey, 1988).

Only one pin of the integrated circuit is necessary to designate that circuit as *master* or *slave*. Pins that are outputs on the master are used as inputs on the slave, and bit-for-bit comparison is used within the slave to detect discrepancies between its outputs and those of the master. The master and slave integrated circuits therefore may be wired in parallel, pin for pin, with the single exception of the master/slave select pin. The two integrated circuits may even be physically mounted together, piggyback fashion, so that the impact on packaging volume is minimized. This is an especially effective way to achieve multiple execution.

Another form of multiple execution is to assign two processors identical tasks, but to give each processor its own unique program or algorithm by which the task is to be accomplished (Avizienis, 1985). Preferably, the two flight software tasks would be developed by two different teams of software designers. Significant discrepancies in the results may then be flagged so that software "bugs" can be identified. The bug can be patched out of the offending version of the software, and then normal system operation may continue. This post-launch method of eliminating software bugs assumes that two independent teams of software engineers will not make identical mistakes as they develop and test their codes.

Still another implementation of multiple execution involves a *master* procedure that leaves behind a carefully specified *certification trail* that may be used by the *slave* process. The certification trail contains a history of the master's procedure, and *hints* about the decisions the master used (such as initial conditions) as it executed the procedure. Because it has access to the master's certification trail data, the slave process then executes much more quickly than the master. When the slave has finished, the two results are compared. If they agree, the computation is accepted as correct; otherwise, an error is generated. With many algorithms the slave process can run concurrently with that of the master (Blough and Masson, 1990).

11.5.8 Fault Rollback

In spacecraft, as on the ground, the best software is well structured and modular; that is, it has been partitioned into clearly defined segments that are relatively independent from one another. These segments are given names such as tasks, procedures, and packages. One method of fault tolerance is to save the context whenever control is transferred from one segment to another. This intersegment gap is called a *rollback point*. If an error is detected (say, by one of the methods outlined above) during or immediately following the execution of a segment, a rollback can be initiated. The context at the most recent rollback point is restored, and execution is restarted from that point. If the error was a soft error caused by, say, a radiation-induced SEU, then the error will not recur during the second pass through the code segment.

Fault rollback is an excellent method of fault tolerance, provided that rollback points can be defined frequently enough so that no one rollback significantly reduces the timing margin of the task. If the rollback requires so much extra execution time that the task misses a critical real-time milestone or deadline, then the fault rollback has not really corrected the fault.

Fault rollback is usually easy to implement in a system that is heavily interrupt-driven, because such a system is designed to be very efficient at context switching.

11.5.9 Watchdog Timers

A watchdog looks for anything out of the ordinary. A *watchdog timer* (WDT) is a hardware timer that looks for program execution that is out of the ordinary. SEUs can cause the processor to exit its normal flow of execution and jump into an abnormal mode. If the normal mode is designed so that the WDT is reset periodically, then it is likely that an abnormal mode will not successfully accomplish this function.

When the WDT times out (because it was not reset sufficiently often), a hardware restart is initiated. This automatically resets the processor, reloads the software (if necessary), and restarts normal program execution.

The WDT is a simple way to prevent SEUs from causing endless loops. However, failures that continue to reset the WDT will not be detected. Also, a failure in the WDT itself may render the entire processor useless.

The software that resets the WDT is called the *watchdog manager*. The watchdog manager is a top-level executive task that periodically receives health status from all other active tasks in the system. So long as all active tasks report that they are healthy (during, say, a one-second real-time interval), then the watchdog manager resets the WDT.

11.5.10 Improper Sequence Detection

A refinement on the watchdog timer is the *improper sequence detector*. This is a device that looks for the program to do a number of different, independent things in the proper sequence. If any of these things do not occur, or if they do not occur in the proper sequence, then an error in program flow has been detected and a rollback, software reset, or hardware restart is initiated.

Few single-event upsets will fool a well-designed improper sequence detector; however, improper sequence detectors are more complicated than simple watchdog timers and are not self-checking. A hard failure in the improper sequence detector may render the processor useless.

To avoid having hardware that can be a single-point failure for a system, systems with multiple processors can implement improper sequence detectors in software, with one processor programmed to detect improper sequences in the flow of another processor. This can be effective, but it is very difficult to implement and test.

11.6 Custom, Special-Purpose Peripherals

Because of the wide variety of data processing functions that must be performed in a spacecraft, custom-designed peripherals are often used. These are intelligent pre- or post-processors that can perform important functions in parallel with the main processor(s). These functions include preprocessing of sensor and command data, compression of telemetry data, signal processing, and arithmetic coprocessing. Sometimes memory management, interrupt handling, and other similar functions are delegated to custom peripherals. For example, a DMA management function might be incorporated into a *bridge chip* that enforces proper protocol for two standard buses that must communicate with each other.

11.6.1 Data Acquisition

Data acquisition is a term that refers to gathering data from multiple sources and making those data available to the processor in a standard format. The data may come from many different sources and originate in many different forms.

Digital and analog sources must be accommodated. For this reason, a data acquisition peripheral must contain variable gain blocks and at least one analog-to-digital converter (A/D or ADC). An analog multiplexer is used to select the data source and gain; the output from the ADC is made available to the processor through an input port that is either I/O-mapped or memory-mapped. Details of analog-to-digital conversion are given in chapter 10.

Data from digital sources require digital multiplexing, possible gain multiplication, and offset addition before they are ready to be ported into the processor.

11.6.2 Logarithmic and Data Compression

Data that may cover a wide dynamic range will sometimes require logarithmic data compression so that the number of bits used to represent each datum is not too large. Logarithmic compression is very similar to the conversion of a fixed-point number to floating point representation. Instead of representing a number as an integer, the compressor represents it as a mantissa (*significand*) and an exponent (*characteristic*).

For example, if the value 1,000,000 has been measured to an accuracy of one part in one thousand, then there is no loss in precision if we represent that number as 1.000E6. If the decimal point and the exponent delimiter ("E") are implied, the latter representation uses only five decimal digits instead of the seven required by the former representation.

In binary, the fixed point number 11110100001001000000 could be compressed to .1111010000E10100 for a saving of five bits. If the source of these data needed to be sampled ten thousand times each second, the data transfer rate would need to be 200,000 bits per second without data compression and only 150,000 bits per second with data compression. This can yield a very significant saving in transmission bandwidth and a decrease in the bit error rate.

Other forms of data compression may be used to reduce bandwidth. For example, for slowly varying functions many bits can be saved by transmitting the differences between sample amplitudes instead of the actual amplitudes. This is called *delta compression.*

11.6.3 Frequency Domain Transformation

Sometimes data are more meaningful if they are observed and analyzed in the frequency domain. Frequency spectra can be analyzed quickly to determine the data type or source. For example, a "chirped" radar modifies the transmitted frequency during transmission, and in the return signal this "chirp" can be identified easily in the frequency domain.

A very useful special-purpose peripheral is the *digital fast Fourier transformer* (DFT). This device takes sampled data and then computes the discrete Fourier transformation of those data. The frequency domain data may then be analyzed by the spacecraft processor and all uninteresting data discarded. Only interesting spectra (or statistics of those spectra) need to be telemetered to the ground.

With very-large-scale integration techniques, the multiplications and additions necessary to implement a discrete Fourier transform can be performed by one or two integrated circuits. A complete 8192-point DFT can be computed in less than 5 μs using a *fast Fourier transform* (FFT) algorithm implemented in dedicated hardware.

11.6.4 Spectrum Accumulation

A frequency spectrum is not the only kind of spectrum that is of interest. Frequently, energy spectra are also useful. An energy spectrum may be used to determine the energetic particle environment in the magnetosphere of a planet, for example. Special-purpose peripherals can be designed to accumulate energy spectra automatically and then make those data available to the central processor. With ultra-large-scale integrated circuits, a 1024-bin energy spectrum accumulator with 32-bit bins can be implemented in a single special-purpose integrated circuit.

11.6.5 Image Processing

Spacecraft may have to process image data. Sometimes a moving target must be recognized; at other times the centroid of an object must be identified. Perhaps the edges of objects in the image must be detected, or objects that exceed a certain size must be counted. All of these functions can be performed by special-purpose image-processing peripherals. One such peripheral is a cellular logic transformation integrated circuit that can perform any of these transformations on a 256×256 pixel image in less than 1/30 s. If a special-purpose image-processing peripheral is used, the central processor is not then burdened with the image-processing task; it needs only to interpret the results.

11.7 Ultra-Large Scale Integration

Ultra-large-scale integration (ULSI) of transistors is a revolutionary technological development that makes possible all kinds of intelligent processors and peripherals for spacecraft. Today, 64,000,000 transistors or more may be integrated on a single silicon chip measuring 0.6×0.6 in. This gives potential processing power per cubic centimeter that is unprecedented. We have only just begun to explore this potential.

One obvious application of this integration technology is to include redundancy (for fault tolerance) on the chip with the CPU, peripheral controller, memory management unit, and so on. Another application is to implement sophisticated parallel processing architectures that exhibit graceful degradation as failures occur during a mission. The increasing density of transistor integration has permitted IC memory to replace magnetic tape recorders for flight processor mass storage.

Several methods are available to the system designer for applying ULSI. One is to use off-the-shelf devices manufactured by commercial integrated circuit suppliers. These *commercial off-the-shelf* (COTS) components must be screened and qualified for space flight application. Another method is to design an *application-specific* circuit using a gate array or field-programmable gate array (FPGA). A third method is to design a full-custom ULSI integrated circuit in which every transistor and interconnection is optimized for the application at hand. Modern computer-aided design (CAD) tools have been developed to facilitate the full-custom design process, so that a full-custom design can be implemented and simulated on a personal computer.

The capability of ULSI device technology far exceeds our present ability to apply it; that is, we currently have more capability than we know what to do with! A new generation of spacecraft design engineers is being handed this unprecedented potential for innovation. What that generation accomplishes over the next decade will no doubt far exceed even our most optimistic predictions.

Problems

1. A spacecraft instrument has a flight computer that requires 7.68 s out of each 8.192-s real-time interval to complete its data processing, housekeeping, and telemetry tasks. If commands to that instrument are all 32 bits in length, and if the instrument processor requires 64 ms to process each command after it is received, what is the maximum average rate at which commands may be sent to the processor without impacting its data processing, housekeeping, and telemetry tasks? You may assume that the 32-bit command data are transmitted from the command system to the instrument at 2000 bits per second. Briefly explain your answer. (Hint: Not all of the data given in this problem are necessary for its solution.)
2. The designer of a spacecraft data-processing system has chosen a microcomputer architecture that allows software subroutines to be called from any part of the main program. The return address of the subroutine is stored in the first two bytes of the subroutine itself, so that when the subroutine has completed its task it can return program control to the point from which it was most recently called. The designer also has chosen to place all program instructions, including subroutines, in read-only memory (ROM).

   Why will this design fail to work? What could the designer do to remedy the problem?
3. You are assigned the task of designing a data-processing system for an energetic particle detection instrument on board the Galileo deep-space mission to the planet Jupiter. Power and mass are especially critical for such a mission, and it is imperative that the instrument survive the trip so that meaningful data may be gathered as the spacecraft orbits Jupiter.

   The mission will begin when the Space Shuttle launches the spacecraft into Earth orbit. After orbiting the earth 48 times, a cruise rocket will take the spacecraft out of Earth orbit and send it on its way toward Jupiter. At an average velocity of 14,000 miles per hour, the spacecraft will travel the 403,200,000 miles to Jupiter in 1200 days. Once orbit is established at Jupiter, scientists desire to gather as much data as possible.

Integrated circuits in the spacecraft will receive 50 rad (Si) of gamma radiation per orbit while in Earth orbit. During the long cruise to Jupiter they will receive an average of 50 rad (Si) per day. Once in orbit about Jupiter, the circuits will receive 150,000 rad (Si) per orbit, due to the high-energy radiation environment of the Jupiter magnetosphere.

Which of the following microprocessor integrated circuits is best suited to this mission?

(a) The 2901 bit-slice microprocessor, which can tolerate 1×10^6 rad (Si). A 2901 system dissipates 2.9 W.

(b) The commercial 80C86 CMOS microprocessor, which can tolerate 1×10^4 rad (Si). An 80C86 system dissipates 0.4 W.

(c) A custom silicon-on-sapphire (SOS) microprocessor, which can tolerate 1×10^5 rad (Si). The SOS system dissipates 0.8 W.

(d) A radiation-hardened 80C86 CMOS microprocessor, which can tolerate 1×10^6 rad (Si). The rad-hard 80C86 system dissipates 0.6 W.

For the system you choose, how many orbits around Jupiter will provide scientific data before the microprocessor fails? (You may assume that the microprocessor is the one integrated circuit most vulnerable to radiation damage.)

4. A system is using an *even parity* Hamming code to transmit 24-bit data words. Five parity bits are used for error detection and correction. The assumption is that there will be no more than one bit error per word transmitted. The following composite word, which contains at most one bit error, is received:

C_1	C_2	C_3	C_4	C_5	C_6	C_7	C_8	C_9	C_{10}	C_{11}	C_{12}	C_{13}	C_{14}
1	1	1	1	0	1	1	1	0	1	1	0	1	1
P_1	P_2	D_1	P_4	D_2	D_3	D_4	P_8	D_5	D_6	D_7	D_8	D_9	D_{10}

C_{15}	C_{16}	C_{17}	C_{18}	C_{19}	C_{20}	C_{21}	C_{22}	C_{23}	C_{24}	C_{25}	C_{26}	C_{27}	C_{28}	C_{29}
1	0	1	1	0	1	0	1	1	1	1	0	0	0	0
D_{11}	P_{16}	D_{12}	D_{13}	D_{14}	D_{15}	D_{16}	D_{17}	D_{18}	D_{19}	D_{20}	D_{21}	D_{22}	D_{23}	D_{24}

The $\mathbf{C}_i$ are the bits of the composite word, the $\mathbf{P}_i$ are the received 'Parity bits, and the $\mathbf{D}_i$ are the received data bits. What is the *syndrome* that is generated by the receiver? What is the corrected composite word? What was the original (correct) data word?

5. For each of the following three situations, describe briefly what steps you would take (in hardware and/or software) to protect your spaceborne microprocessor-based data-processing system. If you add hardware or software resources to the design, explain what they are and what they do.

(a) Your complementary-symmetry metal-oxide semiconductor (CMOS) static random-access memory (SRAM) integrated circuits are found to be vulnerable to latch-up. For the entire local RAM the predicted latch-up rate is two latch-ups per month, worst case, and your mission duration is five years.

(b) Your CMOS SRAM chips are found to be vulnerable to single-event upsets (SEUs). The expected SEU rate for the entire 2^{24}-bit SRAM (16,777,216 bits, organized as 524,288 32-bit words) is 2×10^{-6} errors per bit per day. Your microprocessor has at least 60 ms of idle time in each 1.024 s real-time interval, and can execute one 32-bit load or store instruction in 15 ns.

(c) Your microprocessor central processing unit (CPU) chip is found to be vulnerable to single-event upsets (SEUs). The SEUs can affect data stored in any of the CPU's registers, including the program counter, memory address register, instruction register (which holds the operation code for the instruction that is currently being executed), and the translation look-aside buffer (which translates virtual addresses into physical memory addresses). The predicted SEU rate for the CPU is one upset per month.

References

Avizienis, A., 1981. Fauit tolerance by means of external monitoring of computer systems. *Proceedings of the National Computer Conference.* pp. 27–40.

Avizienis, A., 1985. The N-version approach to fault tolerant software. *IEEE Transactions on software Engineering,* SE-11(12): 149–501.

Baer, J.-L., 1980. *Computer Systems Architecture.* Rockville, MD: Computer Science Press.

Blough, D. M., and G. M. Masson, 1990. Performance analysis of a generalized concurrent error detection procedure. *IEEE Transactions on Computers*, 39(1): 47–62.

Chetty, P. R. K., 1991. *Satellite Technology and its Applications.* Blue Ridge Summit, PA: Tab Professional and Reference Books.

Hayes, J. P., 1978 *Computer Architecture and Organization.* New York: McGraw-Hill.

Mahmood, A., and E. J. McCluskey, 1988. Concurrent error detection using watchdog processors. *IEEE Transactions on Computers*, 37(2): 160–174

Odenwalder, J. P., 1976. *Error Control Coding Handbook.* San Diego, CA: Linkabit.

Sklar, B., 1988. *Digital Communications: Fundamentals and Applications.* Englewood Cliffs, NJ: Prentice-Hall.

Tanenbaum, A. S., 1984. *Structured Computer Organization.* Englewood Cliffs, NJ: Prentice-Hall.

12

Embedded Software Systems

HARRY K. UTTERBACK

12.1 Introduction

In this chapter we will explore the world of embedded software systems, including a brief overview of past systems, current systems, and what we might expect in the future. A discussion of the benefits of embedded systems to overall spacecraft design and the attendant drawbacks will lead into the application of software engineering to the problems of developing embedded software systems. Development environments, simulation of developing hardware interfaces, and changing requirements will be discussed. Software development issues will be examined, including language selection and the ramifications of a bad choice, the reuse problem, and testing, testing, testing. Automatic code generation from various design sources will also be discussed. We will focus on the software aspects of these embedded systems, looking at the planning and development processes necessary to produce reliable and trusted spacecraft software.

Embedded software systems, albeit primitive, have been a part of space flight systems from the 1960s. Embedded software systems are those software systems which are used to control one or more aspects of a larger system. In our case, the larger system is a spacecraft, or a spacecraft subsystem; for example, the telemetry system, command system, attitude control system, or an onboard instrument (Malcom and Utterback, 1999).

One of the first, if not the first, spacecraft to employ a general-purpose digital computer was Triad, developed by the Johns Hopkins University Applied Physics Laboratory. The primary significance of this accomplishment was that the computer was re-programmable from the ground. This gave the spacecraft operators the ability to run diagnostics, fill the computer's memory with operational programs and data, and, if necessary, carry out a whole new mission on the fly (Whisnant et al., 1981), based on modified software if necessary. The computer was interfaced with the command system and the telemetry system with which there was two-way communication. Triad, the first of the Transit Improvement Program and part of the Navy Navigation Satellite Program, had a custom-built 16-bit processor with 32k words of memory. Prior to Triad, onboard subsystems were controlled by hard-wired logic circuits that responded to commands transmitted from the ground. Post-Triad satellites have become dependent on embedded processors to control the satellite subsystems.

As a measure of the current technology the Midcourse Space Experiment, MSX, was a spacecraft launched in the mid 1980s. It had 54 embedded processors on board, serving 19 subsystems, and software that exceeded 275,000 lines of code (Huebschman, 1996). Today virtually every instrument and sensor on board a spacecraft will have an embedded programmable computer as a part of the subsystem.

12.1.1 Benefits

Prior to the use of re-programmable software systems, if one of a satellite's subsystems malfunctioned, the ground controllers had very little choice in trying to regain control or otherwise rescue the mission. Re-programmable systems allow a wide choice in possible "repairs" and new functions that the satellite or instrument may be able to perform (Whisnant et al., 1981).

12.1.2 Drawbacks

Software takes time. Spacecraft design takes time. The problem is that the software is strongly affected by the hardware design and a good preliminary design for the software cannot be done until the hardware design has "settled down" so that the software requirements can be identified, documented, and reviewed. A second and larger problem that confronts the software development team is complexity. The tasks of managing a command system, a telemetry system, a power system, an attitude control system, and the various scientific instruments that may be on board the spacecraft rapidly become very complex. "The flight software must respond in real time to events that usually occur asynchronously so that it can affect the progress of the process under its control. The real time response is achieved through the use of hardware-generated interrupts. These interrupt signals indicate, for example, the arrival of new commands, the expiration of timers, the extraction of telemetry data, the production of sensor data, the existence of error conditions, etc." (Malcom and Utterback, 1999). Added to this is the question of how to test the software once it has been written.

12.2 Engineering Flight Software

The software development process is more than writing code, putting it through a compiler, and seeing if it runs. Software development takes place in five gross phases: requirements gathering, preliminary design, detailed design, implementation, and operation. There are many activities that take place in each of the phases. Table 12.1 illustrates the process. Other activities would include independent verification and validation if the implementation required such assurances.

Table 12.1 Software development process

	Requirements Identification	Preliminary Design	Detailed Design	Implementation
Management	1—Cost estimation 2—Software development plan 3—Coordinate software requirements review (SSR)	1—Cost estimation refinement 2—Size estimation 3—Prepare configuration management plan 4—Coordinate preliminary design review (PDR)	1—Refine cost and size estimates 2—Coordinate critical design review (CDR) 3—Implement configuration management plan	1—Refine cost and size estimates 2—Coordinate software acceptance review
Development	1—Software requirements 2—Prepare software requirements specification (SRS) 3—Software requirements review (SRR)	1—Prepare functional design document (FDD) 2— Prepare interface control documents (ICD) 3—Preliminary design review (PDR)	1—Prepare detailed program design (DDD) 2—Critical design review (CDR) 3—Update FDD and ICD	1—Translate design into code 2—Update DDD 3—Conduct inspections 4—Software acceptance review
Testing	1—Review system requirements 2—Review SRS 3—Prepare testing specification	1—Prepare test plan 2—Review FDD 3—Review ICD 4—Attend PDR	1—Prepare test procedures 2—Review DDD 3—Attend CDR	1—Review test procedures 2—Execute tests 3—Prepare test reports 4—Software acceptance review
Quality Assurance	1—Develop quality assurance plan 2—Review SRS 3—Attend SRR	1—Review FDD 2—Review ICD 3—Review test plan 4—Attend PDR	1—Review DDD 2—Review test procedures 3—Attend CDR	1—Review test reports 2—Software acceptance review

Early decisions need to be made about which programming language is to be used to develop the software. The choice may involve training classes for the development staff, purchase of compilers, debugging aids such as inline circuit emulators, bus activity recorders, and so on, together with development platforms.

The choice of language is an interesting problem. Programmers like to use the language that they used in school: C or possibly C++. Old timers may prefer Fortran or Pascal or some other obscure language. The program manager, therefore, has his or her hands full in making a selection that makes sense for the long haul. Modern languages like Ada are to be preferred because of the portability of previously written packages; the validated compilers; the commitment of the Department of Defense to the language; the availability of validated, reusable components; the enforced type checking (and other safety features); the multitasking features that are built into the language; and the real time features of the language. That sounds like a commercial and in some ways it is, but the points made about Ada are just the language features that a project leader needs to complete a successful embedded software project. The question of reuse also comes into play at this point. Do we have a body of software already written that could be used for the mission? Are there tools to aid in the testing process for this language/development platform? This task is further complicated by the choice of an onboard processor or processors. The development platform must be able to produce code that will execute on the selected onboard processor(s). The choice of processors is often limited. There are only a few processors that are suitable for use in the space radiation environment and the software development tools that are available are also limited. Furthermore, the space-qualified processors that are available are usually not "state of the art"; that is, they are several generations behind the latest and greatest. Material for the memories of the processors may also add to the developer's woes. Memory-scrubbing routines may have to be written to keep the memory free of single-bit upsets and double-bit errors. Flash memories cause problems in loading and changing the memory because of the special protocols involved.

In any case, a highly structured language is to be preferred. Programs written in such a language are easier to read, easier to test, and easier to modify without introducing new errors in the process.

A technique that may be of use in some applications—attitude control comes to mind—is the use of code generation tools to not only construct the flight code itself but provide useful test code and simulations. The NEAR–Shumacher team used this method to produce code for the mission's attitude control system and orbit transfer system.

12.3 Software Configuration Management

Software configuration management (CM) is the means by which the software development team knows that the software product being built and tested is the one that is required (Humphery, 1990). Software configuration management is used to control the entropy that is present in software development projects. Every aspect of the software under development should be placed under control. This includes the requirements specifications, the design documents, the test data and scaffolding necessary to conduct the test(s), and of course the code.

Care must be taken in deciding when to place an item under configuration control, establishing a baseline without strangling early development excessively and letting things go until they have "settled down." If done too early, the developers feel they are hampered by the paperwork and, if done too late, the managers feel that the development is out of control. Usually a baseline is established when the item in question has been approved. For example, the requirements baseline is established when the specification is reviewed and approved. This baseline includes the operational concept document as well as the approved requirements documents. When new requirements appear and are approved, a new requirements baseline is established. Baselines are also established for design documents, code units (modules), the integration phase, and the operational phase of the program.

Being able to identify which version of a program is being executed is very important. In the case of a spacecraft the version identification is often reported in telemetry so that operations can verify that the final, complete version is on board and not some test version with one or more of the instrument interfaces stubbed out! Such things have happened, causing missions to fail or at least giving spacecraft controllers many anxious hours until the proper program can be loaded and executed.

Configuration management tools (Whitegift, 1991) are available to aid in the implementation of a configuration control system. A configuration management plan should be made and agreed to by the various software development parties (project management, software design, software implementation, the testing team, and others.) The configuration management plan (ANSI/IEEE Std 828-1990; Buckley, 1993) consists of:

- *Introduction*. The introduction identifies the document and its scope, that is, the items to be controlled.
- *Management*. The management section follows and sets forth the allocation of responsibilities and authorities for the CM activities. This includes the project organization, the CM responsibilities, and any applicable policies and procedures.
- *Activities*. The tasks or activities that the CM is to perform are listed. This section would normally contain processes to be used for configuration identification, change control, status accounting, and audits, and would include the problem-reporting system. If an organization has established procedures for accounting, problem reporting, etc., then references to these procedures would be sufficient.
- *Schedules*. This section ties the CM tasks to the overall project schedules and milestones.
- *Resources*. A section on resources is included to describe the tools and personnel needed to implement the plan.
- *Maintenance*. The plan maintenance section concludes the configuration management plan.

12.4 Organization

Ideally, the software organization would consist of a development team, a testing team, a quality assurance team, and the operations team. Each of these teams would be organized in the same fashion as is the parent organization.

The development team consists of the team leader(s) or manager(s) with (ideally) experts in requirements specification, software design, and programming.

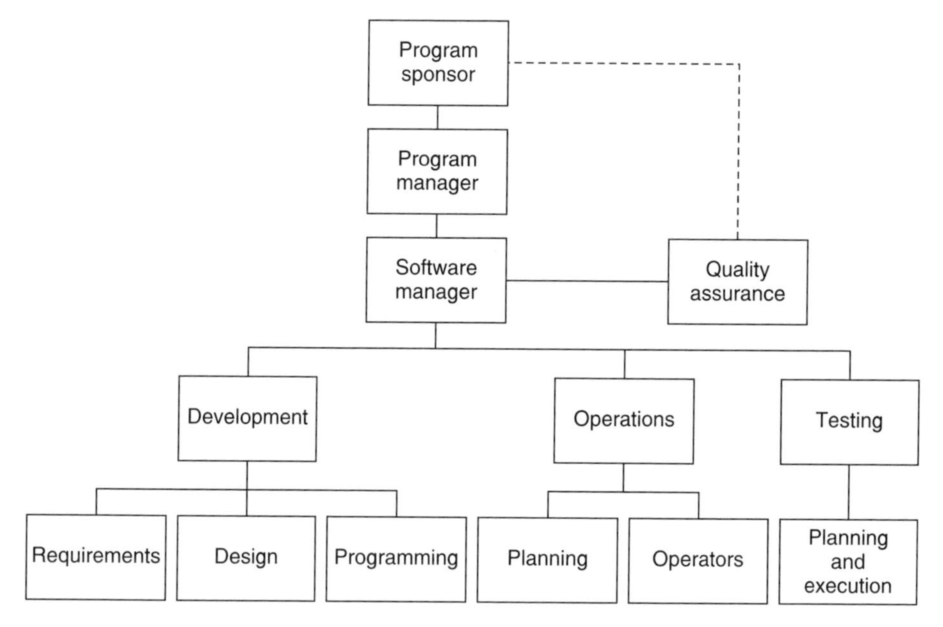

Figure 12.1 Software organization chart.

The testing team consists of people with experience in test development (experience in program development is a must) and a knack for discovering holes in requirements specifications, design documents, and, of course, errors in program code.

The quality assurance team is probably the smallest group. The team needs a direct line to the overall manager of the product being produced and the authority to stop production if things are not being done according to the written and approved quality assurance plan.

The operations team needs planners and operators who understand the product and its intended use. They need to be able to understand the requests of the community using the spacecraft and the capabilities of the software they are expected to operate. Typically these people have their programs in two places: on board the spacecraft, and in the ground control software.

A typical organization might look something like figure 12.1.

A software configuration control board (SCCB) should be appointed, consisting of members from the development team, the testing team, quality assurance, and operations. The SCCB decides whether a change request should be implemented or not. This includes judgment on the merits of the change, the consequences of no change, and schedule impact, that is, the cost of the change. The SCCB's responsibility continues after the change is (dis)approved. The SCCB usually has no say as to how a change is to be implemented but its responsibility continues until the updated item is finally approved for inclusion in the product. As changes are approved, it is imperative that the change is documented at all levels. If a new requirement is requested, then the change is to be reflected in the requirements document, the design document, the test plan, and, of course, the code. While this is time consuming and may be viewed as tedious, it is imperative that the documentation reflects the product as built, with a record of the modifications. Trying to track down an error without adequate and up-to-date documentation can be very costly.

Configuration management also helps in eliminating the problems of errors showing up that have been fixed earlier; in other words, it keeps the tester and the developers informed on which version of the program is being tested or modified.

Excellent treatment and guidance in software configuration management may be found in Humphery (1990) and Babich (1986).

12.5 Sizing and Estimation

How big will the sub-program, module package program, system, etc., be? How long will it take to produce? How many instructions or locations in memory will be required? Why do we care?

We care, first, because the hardware engineers have to construct a computer capable of handling the software system we are devising and, secondly, because we need to estimate how many people we will need to construct the software in the allotted time.

Typically the estimate is in SLOCs, source lines of code. The main advantage to using SLOCs is that a compiler can easily and objectively count them. The disadvantages are numerous: defining the rules for counting lines of code for a particular language; defining which lines of code to include, such as declaratives, macros, system service functions, and so forth. What about those cases where the system uses different languages?

One of the best tools to use in sizing and estimation is, of course, experience. But even experience does not always help, especially if we are dealing with a totally new problem or a new machine or new personnel or new management. The best teacher of these skills is your own personal record of time and sizing estimates and the actual outcomes. To this end it is recommended that each software engineer pursue Watts S. Humphery's book *Introduction to the Personal Software Process* (Humphery, 1997). It takes discipline and rigor to follow it, but the experience will pay off.

Another tool is the use of a historical database. Today's corporate memory is often short, so the establishment of a database containing the size of programs, the number of people who worked on the project and their experience level, the amount of time spent in development and testing, is very important. The database can give a good starting point for the estimation of size, time, and effort on a new project if it is somewhat similar to the planned program.

Function point analysis (Symons, 1991; Garmus and Herron, 1996) is another method that may be used for sizing and estimation. Allan Albrecht of IBM needed a method of estimating and measuring performance in an environment that was using several programming languages. He was able to develop a method of counting inputs, outputs, master files, etc., that, when added up using a weighted sum, gave an index of the size of the system. Albrecht's method has undergone three revisions and now another variation has been devised, called MkII Function Point Analysis, which purports to overcome some of the perceived weaknesses in Albrecht's method. Whichever method you choose to use, it is most important to be consistent and clear about what you are measuring.

Even when the estimation of size and time is accurate, the software developer is subject to the pressures to reduce the cost and shorten the schedule. This is often done without consulting the software developer and consequently the software is too expensive and late to boot! This error is difficult to overcome. Independent estimations are one solution, as long as the independent estimators are knowledgeable about software and not involved in the project management issues (Anderson and Dorfman, 1991).

Scheduling of the project after one has an idea of the size is another variable to be considered. How much time will be allocated to design, detailed design, code and unit testing, integration testing, and the final qualification tests? Record keeping is necessary to build up a database of size versus time to complete. Two tables of experience, one from TRW and one from Martin Marietta, are to be found in Humphery (1990).

Planning begins with an estimate of the size. That begins with a work breakdown structure (WBS) as described in chapter 1. There are two parts to a WBS: the product structure and the software process structure. The product structure reflects the overall design. It is refined and the process tasks to produce each element are defined. The more detail, the better the estimate, the better the plan, the better the tracking.

12.6 Reuse

Software development for space systems is a very expensive proposition and usually under very tight time constraints. To mitigate both the expense and the time pressure, the software development team needs to be able to reuse designs and code from previous projects. Completely reinventing the software for each project or insisting that the loader for subsystem A is not suitable for subsystem B (when both subsystems are using the exact same processor) has to become a thing of the past. As teams of developers work on a particular project they should think of future projects and how a particular module could be constructed to be of use in the future. Libraries of subprograms, packages, and so on, should be established and used. Management of such a library of routines can save many hours of development time on a project, not to mention the fact that the routines have been thoroughly tested before they were placed in the library. The hardware for spacecraft systems is, very often, "off the shelf" and the software should be too.

In this vein, a library of constants to use in the production of the software is very important. All developers on a team need to use the same units and constants to maintain uniformity of results and to keep surprises to a minimum.

Not to be excluded from reuse are the architecture and designs of the software systems, the test cases (especially those which had a high error yield), and the documentation. In an environment where the underlying hardware remains the same or the same language is employed, reuse can be of great benefit to a project.

Software reuse must be planned. It should be kept in mind from the outset of the project. Developing reusable components is, generally, more expensive but the extra cost is amortized with each later use of the component.

There are some repositories of reusable code sponsored by the United States Department of Defense (DoD). Access to these sources is limited to those entities doing business with DoD and using the Ada programming language. One of these repositories is the Common Ada Missile Packages (CAMP) at Eglin Air Force Base (Anderson and Dorfman, 1991). Commercial Ada components are available as "Booch components" which are general and fairly low level.

Ongoing activity at the Software Engineering Institute and the Software Productivity Consortium gives hope for more widespread and useful library functions for reusable components.

12.7 Process

A well-defined and executed software development process is paramount for the successful completion of a software effort in any situation and even more so in the development of software for a spacecraft. One should produce a development plan that includes requirements gathering and review, preliminary design and review, detailed design and review, software inspections or walkthroughs or the equivalent, and a comprehensive testing plan. Reviews at each step of the process are used to ferret out any misconceptions that the software developers have *vis-à-vis* the hardware developers and the subsystem leads. The early discovery of these misconceptions or errors can save countless hours of rework down the road when doing integration testing. This means that the design of the software must be carefully done and not put together on the fly during the coding. The reviews previously mentioned are to be viewed as part of the testing of the software, with the intent to find errors. The usual approach to the design review process is to have the review encompass both hardware and software. It is possible to do this with a well-defined set of hardware and mission requirements, but if the hardware or the mission has not settled down a good software review is unlikely. Interfaces must be defined and agreed to early in the development process for an effective design to be accomplished.

Quality in the design leads to success in the execution. Conceptual errors made during the design phase of a project and not caught will cost many times more to correct when they are discovered in later phases of the development (Boehm, 1976). Software developers are faced with the task of gathering the requirements for the software while the hardware requirements and designs are still being collected and executed. This process of concurrent development is, perhaps, the most difficult part of the design phases of embedded software development. As new ideas appear in the hardware, such as new instruments on the spacecraft, new interfaces for existing hardware, a new manager, and so forth, the software requirements change rapidly. Ideally, one could wait to begin the design of the software until all of the requirements were known. Reality, in the guise of schedule demands, sets in early and it becomes necessary to begin the design process before all of the requirements are known. While this is risky, careful planning can mitigate the risk by the use of structured programming methods and by delaying as many of the implementation details in the design as long as possible.

The Software Engineering Institute (SEI) was established at Carnegie Mellon University to aid the Department of Defense to define and refine the software development process. It was referred to as the "software crisis" at that time. The problem was that very few DoD software projects were on time and within budget and very often were in error (Paulk et al., 1993).

After studies, surveys, and discussions, the SEI produced a framework for improving the software development process. It is named the Capability Maturity Model (CMM). This model describes five levels of maturity, level 1 being the least capable and level 5 the most mature. Levels 2–5 describe the activities or processes that a software development

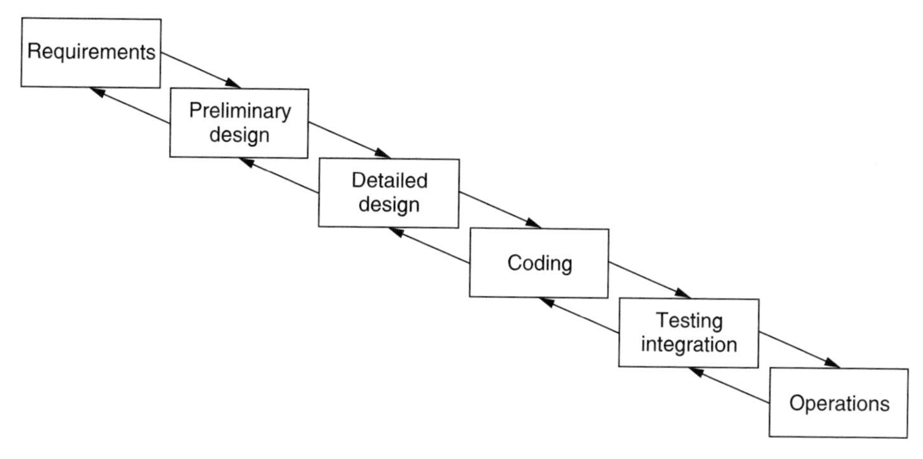

Figure 12.2 Simplified waterfall model.

organization goes through to produce software. Level 1 organizations have few, if any, processes and are, at best, chaotic in their work. Organizations at level 5 are few and far between but there are several now in existence. The IBM Houston's Space Shuttle project is at level 5. In general, all you have to do is look around yourself to see a level 1 organization (Paulk et al., 1993). The CMM is best used as a guide to better software development. Small organizations would have difficulty in staffing a full-blown implementation, but you should embrace the elements that fit your need, with particular attention to the goals defined for each level.

Selection of a life-cycle model gives the framework for your software process. There are several to choose from including the waterfall model, the spiral model, rapid, throwaway prototyping, and evolutionary prototyping. Each of these is different from the others and requires some thought as to which one is to be selected for a project. Consideration should be given to: (1) requirements stability; (2) lifetime of the system; and (3) availability of resources for the implementation (Anderson and Dorfman, 1991).

The waterfall model, developed by Winston Royce (1970) and illustrated in figure 12.2, is perhaps the best known and most widely used life-cycle model. As the name implies, the steps in the development can be viewed as a waterfall, with each product piece falling down to be used by the next cycle.

The spiral model was developed by Barry Boehm and published in *IEEE Computer* in May 1988. The model emphasizes a continuum of risk analysis, prototyping, and planning until an operational prototype is built. It is from that prototype that the final software is designed, coded, and tested.

Rapid, throwaway prototyping is a useful model to follow, especially if the requirements are rapidly changing and are not well known. The customer's needs are rapidly put into code (with as much scaffolding as necessary) to see how the "program" fits the need. This method brings the customer and the developer together quickly, arriving at a better understanding of the user needs for the developing system. However, overall time to the final product is no better than with the waterfall method (Anderson and Dorfman, 1991). As the name implies, the prototypes are thrown away once they have served their purpose.

Incremental development and evolutionary prototyping follow the spiral model by their nature; in other words, prototyping with the idea of building up to a finished product. Software is built to satisfy a few requirements, with the understanding that new requirements will be added and enhancements made as time progresses. These methods produce a product faster, but with limited capability (reduced functionality). The products have a potentially longer life because they are constructed with modification in mind. The suitability of any of these models to embedded software is purely in the eyes of the beholder. Rapid prototyping may be necessary so that system users can see the consequences of their decisions. Ground control software may be a candidate for use of the incremental development or evolutionary prototyping as a model as a spacecraft system is defined and subsystems and instruments are modified or added and deleted.

12.8 Software Inspections

The purpose of software inspections is to improve quality and productivity. It is not to evaluate personnel! A software inspection is a peer review to find problems and improve quality. It provides a method of finding errors early in the process when they are easier to correct.

Software inspections were the idea of Michael E. Fagan of International Business Machines (IBM) in the early 1970s (Fagan, 1976). He was looking for a method to improve programming quality and productivity. An inspection team is selected to examine each product. The inspection team is made up of a moderator, the author, a scribe, and two or three other inspectors. Here, small is good. The scribe is selected from the inspection team as a whole. Usually the moderator and the author are exempt from this task but the author may be chosen. The team should be made up of knowledgeable people; that is, those who are experienced in the task at hand—requirements, design, coding, testing, and so forth.

The method Fagan developed comprises six steps:

(1) *Planning*. In the planning step the inspection is organized. The author of the product to be inspected gathers the materials to be inspected as well as the supporting documentation. The moderator, or inspection leader, selects the other participants and reviews the materials to be sure that the product is ready to be inspected. The meeting times are selected and the others are notified of the schedule and given the materials.

(2) *Overview*. The overview or kickoff meeting is held to educate the other members of the inspection team about the product to be inspected.

(3) *Preparation*. This is the step where the inspection team studies the material. Check lists are used to guide the members in their preparation. Defects found during this process are recorded. The time spent in preparation is also noted.

(4) *Inspection meeting*. This meeting is used to detect defects. Everyone involved in the inspection attends the meeting: moderator, author, and the rest of the inspection team.

(5) *Rework*. This step is used by the author to fix the defects found in the meeting.

(6) *Follow-up*. This step is used to verify that the defects were fixed in a satisfactory manner and that the inspection report was completed.

A word about the inspection report is in order. The report consists of data about the amount of preparation effort, examination rate, product size, and the type of defects found (wrong, missing, or extra). These defects are also classified by severity, that is, major or minor. This information is used to improve the development process and not to punish.

The reader is encouraged to review Fagan (1976), Strauss and Ebenau (1994), and Gilb and Graham (1993) for further guidance on software inspections.

12.9 Software Testing

Approximately 50% of the total development time for software is taken up with testing. It is not an activity to be crammed in at the last minute before launch (Myers, 1979).

Software testing is the process used to detect errors in software. Restated: *The goal of software testing is to find errors*! (Note: this is not the same as determining whether the software performs correctly.) The process begins, not with the running of code, but with the very first documentation. So in fact it begins with the very first review that the software undergoes, usually the requirements review. Every review, including the inspections and/or walkthroughs, is a form of testing.

Ideally, someone other than the author does testing beyond the module level or first program tests. It is very hard to look for errors in one's own work (Humphery, 1990). Additionally, testing is hard creative work, and it is done most effectively by a team of people who are independent from the development team. They start with the specification of the program, are present at all reviews and construct tests to find as many errors as possible. Remember, a good test case is one that finds errors.

As the development progresses through the various stages of generating systems requirements, gathering and reviewing software requirements, conceptual design, preliminary design, and detailed design, one must also develop a plan for testing the software that will eventually be written. Planning for testing is, perhaps, one of the most neglected phases of software development. We proceed through our designs and design reviews, implement the design, do module testing, and all at once we are faced with the task of testing the system with the real hardware on the real satellite. Problems arise here because the various instruments and subsystems are either not ready for integration or cannot be turned on in a non-space environment. The task of rapidly developing a simulator in either software or hardware or a combination of both is, at the least, daunting. As the spacecraft systems come on line, the complexity of data handling increases manifold and the testing becomes harder to accomplish.

The parallel development of a spacecraft simulator is often the solution. This is a combination of hardware and software that is developed to simulate the spacecraft-bus activity, and ideally includes a provision to place real hardware into the simulation.

The system test is the point where the program's objectives are reviewed and tested. The end users are involved to ensure that the program is, in fact, the one that is desired. The test cases should include the possibilities of operational errors and intentional misuse. The system should be able to defend itself from such problems.

Test planning is as important as the development of the program, and should begin at the requirements phase of the project. Independent testers are of prime importance. The testers are a part of every review: requirements, preliminary and final design reviews, and inspections. The insights that they gain will help them in planning their tests of the system and as a consequence help field a better product. Test planning is described in great detail in Hetzel (1984).

12.9.1 Embedded Systems

Embedded systems pose special problems to the tester (Anderson and Dorfman, 1991). The systems are usually connected to specialized instruments or hardware outside of the computer. For example, attitude control systems are connected to gyros, star sensors, sun sensors, and magnetometers. The computer usually has no peripherals so it becomes difficult to assess what is going on during testing. The actual computer may not be available because it is being integrated onto the spacecraft. All of this is multiplied by the fact that these systems must operate in real time. The real-time aspect of these systems makes testing very difficult until the full system is implemented.

Testing of real-time systems should be done in two stages. One stage is used to test the functional correctness and the second stage to test the timing aspects. When all of the hardware and software come together at the spacecraft a series of tests is run to ensure the integrity of the system as a whole. The spacecraft controllers devise scenarios to exercise the spacecraft and its instruments as they will be used in space. Again, because some of the instruments cannot be turned on in a non-space environment, an instrument simulator may have to be used so that commands may be vetted and telemetry received. If an instrument simulator is not available the flight software may have to be "stubbed out" to prevent inadvertent exercise of the instrument in question. In the space environment the arrival of data is unpredictable and the testing must reflect the various possibilities that these conditions present. Timing conflicts are of great concern and so the systems must be tested rigorously and *any* glitch must be followed up completely. Many spacecraft missions have been jeopardized because a small glitch was not seen as mission threatening and later proved to be excruciatingly painful to the operators of the systems who had to address the problem—the Mars Lander comes to mind.

Test files should be maintained and retained as part of the testing process. The test file contains the test cases; that is, the driver, if there is one, the input data, the output, expected and realized, and the errors found. A record of the version of the program is also kept. This information is particularly helpful in selecting regression test cases. Regression testing is brought about when new modules are added to the existing program or when previously discovered errors have been fixed. This very often causes new errors, so the regression testing is very important. Selection of test cases for regression testing is "more art than science" (Humphery, 1990), but the selection of tests that have uncovered errors before is a prime candidate.

12.10 Software Independent Validation and Verification (SIV&V)

Errors are a result of complex software systems and the different ways software developers interpret specifications. Large teams of developers, however carefully managed, are not going to produce error-free designs and code. Because of the expense of failure in the usually high profile space systems, it behooves the software teams to employ an independent verification and validation team. By this we mean an independent third party, who answers to the person or entity who is paying for the software system. Definitions:

- "Software verification is an iterative process aimed at determining whether the product of each step in the development cycle (a) fulfills all the requirements levied on it by the previous step and (b) is internally complete, consistent, and correct enough to support the next phase."
- "Software validation is the process of executing the software to exercise the hardware and comparing the test results to the required performance" (Lewis, 1992).

SIV&V should start as early in the program as the regular software development, that is, during the requirements gathering phase. It is not, as generally thought, just a process to look at executing code. The SIV&V life cycle is very similar to the software life cycle, but with the word verification added to the cycle name. As such, a plan should be prepared that contains most of the elements of a software development plan (ANSI/IEEE Std 1012–1986).

Validation is the final phase of SIV&V. It is the equivalent of the software integration and testing phase of the software life cycle.

Lewis (1992) states six rules for judging the authenticity of an IV&V effort. Paraphrased, these are:

(1) the effort must be independent;
(2) it must be an overlay, not an integral part of the development cycle;
(3) it must report to the customer;
(4) data and tools may be shared with the developers, but IV&V must provide its own methods and tools that are complementary to those at the developing organization;
(5) as noted in the definition, each phase must be verified relative to itself and the preceding and succeeding phases; and
(6) SIV&V must be able to validate all testable software performance requirements.

All critical software should be subjected to SIV&V. By critical, we mean software which, if it fails, jeopardizes the mission. All manned flight software is critical. Real-time programs, programs having unverifiable results after every run, such as software that controls attitude control systems, are critical. Programs where the economic consequences of failure outweigh the cost of SIV&V are candidates for an SIV&V scrutiny.

Ego-less programming (Weinberg, 1971) is a great asset when producing this sort of software. The goals are to find errors so that the software can perform in the way in which it was envisioned. A good quality assurance program is not a substitute for IV&V. The internal system does not represent the customer and often is not able even to enforce the published standards of the organization. The internal organization is less likely to discover sins of omission in requirements specifications and the like.

12.11 Software Quality Assurance

Software quality assurance is governed by the following four basic principles:

(1) Unless you establish aggressive quality goals, nothing will change.
(2) If these goals are not numerical, the quality program will remain just talk.

(3) Without quality plans, only you are committed to quality.
(4) Quality plans are just paper unless you track and review them.

What these principles mean is that the senior management needs to establish aggressive and explicit numerical quality goals. The quality measures are objective, not subjective; in other words, "the software shall be good" is subjective but "numbers of errors per module" is something that can be measured. The measures need to be precisely defined and documented. A quality plan must be produced which commits to specific numerical targets and complies with the management goals. The quality performance is tracked and publicized. Quality measures are treated as indicators of overall performance.

A software quality assurance plan should be produced during the initial planning cycle. The plan should be documented, reviewed, tracked, and compared to actual performance, just like every other plan for the project. A responsible authority is named to own the quality data and the tracking and reporting system. The quality performance is tracked and reported to this authority during both development and maintenance. Resources are established for validating the reported data and retaining it in the process database. Actual product and organizational performance data are periodically reviewed against the plan. Results are reviewed with the responsible line management and discrepancies resolved. Performance against the targets is determined. If performance is bad, action plans are made and executed. If performance is much better than the targets, a more aggressive target is established. The quality performance is published, with a highlight report going to senior management.

Exercises

1. Write a paper describing a documented "horror story" about software development. Add an anecdote from your own experience.
2. Prepare an outline of a software test plan.
3. Prepare an outline of a software development plan.
4. Prepare an outline for an independent verification and validation plan.
5. Design a change request form for the design phase of a project.
6. List and describe the five levels of software maturity, as defined by the SEI.
7. Describe any two life-cycle models for software.
8. Write an in-depth paper on one of the following topics:
    - Configuration management: why is it needed?
    - Benefits of domain analysis in establishing a software reuse library.
    - Evolution of life-cycle models, with emphasis on current practice.
    - Estimation: what is needed, where is the industry?
    - Process: what are the current best practices?
    - Embedded software systems in spacecraft applications.
9. Describe your company's/division's/project's software process. Assess this process *vis-à-vis* the SEI software capability maturity model.
10. What can be done at your workplace to improve your company's/division's/project's software process?
11. Outline a software quality assurance plan.

References

Anderson, C., and Dorfman, M., eds. (1991). *Aerospace Software Engineering*. Washington, DC: American Institute of Aeronautics and Astrophysics.

ANSI/IEEE Std 828–1990 (1993). *Standard for Configuration Management Plans*. New York: IEEE.

ANSI/IEEE Std 1012–1986, reaff. 1992 (1993). *Standard for Software Verification and Validation Plans*. New York: IEEE.

Babich, W. A. (1986). *Software Configuration Management, Coordination for Team Productivity*. Reading, MA: Addison-Wesley.

Boehm, B. W. (1976). Software Engineering. *IEEE Transactions on Computers*, **C-25**(12): 1228.

Buckley, F. J. (1993). *Implementing Configuration Management*. New York: IEEE Press.

Fagan, M. E. (1976). Design and code inspections to reduce errors in program development. *IBM Systems Journal*, **15**(3).

Garmus, D., and Herron, D. (1996). *Measuring the Software Process, A Practical Guide to Functional Measurements*. Englewood Cliffs, NJ: Prentice-Hall.

Gilb, T., and Graham, D. (1993). *Software Inspection*. Reading, MA:Addison-Wesley.

Hetzel, W. (1984). *The Complete Guide to Software Testing*. Wellesley, MA: QED Information Sciences.

Huebschman, R. K. (1996). The MSX Spacecraft System Design. *Johns Hopkins University APL Technical Digest*, **17**(1).

Humphery, W. S. (1990). *Managing the Software Process*. Reading, MA: Addison-Wesley.

Humphery, W. S. (1997) *Introduction to the Personal Software Process*. Reading, MA: Addison Wesley Longman.

Lewis, R. O. (1992). *Independent Verification and Validation*. New York: John Wiley & Sons.

Malcom, H., and Utterback, H. K. (1999). Flight software in the Space Department: a look at the past and a view toward the future. *Johns Hopkins University APL Technical Digest*, **20**(4).

Myers G, J. (1979). *The Art of Software Testing*. New York: John Wiley & Sons.

Paulk, M.C., et al. (1993). *The Capability Maturity Model. Guidelines for Improving the Software Process*. Reading, MA: Addison-Wesley.

Royce, W. W. (1970). Managing the development of large software systems. *Proceedings of IEEE WESCON*. New York: IEEE.

Strauss, S. H., and Ebenau, R. G. (1994). *Software Inspection Process*. New York: McGraw-Hill.

Symons, C. R. (1991). *Software Sizing and Estimating MkII FPA*. Chichester, England: John Wiley & Sons.

Weinberg, G. M. (1971). *The Psychology of Computer Programming*. New York: Van Nostrand Reinhold.

Whisnant, J. M., Jenkins, R. E., and Utterback, H. K. (1981). On-board processing for the NOVA spacecraft. Presented at the *International Telemetering Conference*, San Diego, CA.

Whitegift, D. (1991). *Methods and Tools for Software Configuration Management*. Chichester, England: John Wiley & Sons.

13

Spacecraft Reliability, Quality Assurance, and Radiation Effects

R. H. MAURER

13.1 Introduction

In addition to being an engineering activity, reliability and quality assurance are disciplines. In terms of the institution or facility that builds hardware, the internal reliability function represents self-discipline that is often in conflict with the creative design functions.

13.1.1 Reliability

The reliability of a system is the probability that, when operating under stated environmental conditions, the system will perform its intended function adequately for a specified interval of time. Implicit in this statement are three assumptions:

(1) the acceptance of the probabilistic notion of reliability that admits the possibility of failure;
(2) the concept of adequate or acceptable performance for system parameters that deteriorate slowly with time;
(3) the judgment necessary to determine the proper statement of environmental conditions.

13.1.2 Quality Assurance

The quality control function assures final product acceptance. This function includes the documentation and certification of activities such as procurement, inspection, production, testing, and packaging/shipping.

The quality or performance assurance engineer serves as the program historian, keeping inspection, test, and failure records, and documenting the changes to drawings, the resolutions of problems, and waiver requests. This engineer is responsible for presenting the required documentation that supports the pedigree of the hardware to the sponsor or customer at the point of product acceptance or buy-off.

Obviously, space hardware is unique in that it usually cannot be repaired after being put into use. Thus, any significant failure in orbit may result in the partial or total loss of a spacecraft at considerable cost (tens to hundreds of millions of dollars) and schedule delay (some interplanetary science missions have windows for investigation that may not recur for a decade). Therefore, the objective for space hardware is to do it right the first time.

To ensure success many opinions are solicited, experts are polled, and design review action items are originated. The philosophy behind this approach is that the accumulated wisdom and experience of the past will serve as a heritage for present and future space system development.

From an organizational point of view, the reliability function should report directly to the head of the department or company so that decisions and considerations related to reliability and quality are not compromised by design or manufacturing activities or program office schedule pressures. Likewise, engineers who carry out quality, reliability, component engineering, and performance assurance functions should have an independent review board and sign-off authority.

13.2 Reliability

13.2.1 Systems Reliability

A system can be *simple*, such as a two-piece mechanical system (a jar and a lid), or *complex,* such as a spacecraft. Systems have subsystems arranged in series, parallel, or both. Some parallel subsystems may have been designed to be redundant.

If the probability of success or survival at time t is defined as $R(t)$, then for a system in series the individual reliabilities of the subsystems must be multiplied and

$$R_{\mathrm{AB}}(t) = R_{\mathrm{A}}(t)R_{\mathrm{B}}(t) \tag{13.2.1}$$

or, in general,

$$R_{\mathrm{AB}\ldots N}(t) = R_{\mathrm{A}}(t)R_{\mathrm{B}}(t)\ldots R_N(t). \tag{13.2.2}$$

For a system with parallel subsystems, the probability of failure of any one branch is independent of the failure of any other branch. The only way that the system can fail is if all subsystems have failed simultaneously by the same time, t. Thus, we are forced to define a probability of failure $P(t)$ which, in general, is related to the reliability by

$$R(t) = 1 - P(t) \tag{13.2.3}$$

If a system has two parallel subsystems, they must both fail simultaneously by time t and then

$$P_{\mathrm{AB}}(t) = P_{\mathrm{A}}(t)P_{\mathrm{B}}(t) \tag{13.2.4}$$

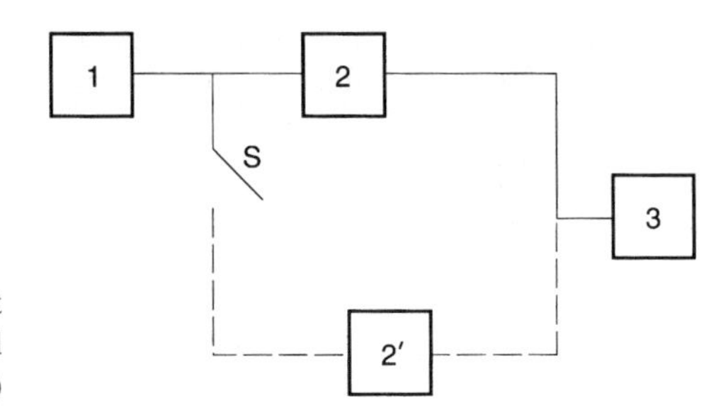

Figure 13.1 A simple redundant system. (Source: Kapur and Lamberson, 1977.)

leading to

$$R_{AB}(t) = 1 - P_{AB}(t) = 1 - P_A(t)P_B(t) \tag{13.2.5}$$

or, again in general,

$$R_{AB...N}(t) = 1 - P_A(t)P_B(t)\dots P_N(t) \tag{13.2.6}$$

Of course, most complex systems are a combination of series and parallel subsystems design.

The oldest known method of reducing the probability of failure is redundancy. Redundancy provides spare components that remain dormant (or in standby) until the original components cease to function adequately. An example of such a situation is figure 13.1, which could be representative of a spare tire on an automobile, a standby spare pump in a hydraulic system, or a redundant microprocessor in a spacecraft system. If subsystem 2 does not fail, the operation path is 1 to 2 to 3. However, if subsystem 2 does fail, then the path is 1 to S to 2′ to 3, where S represents a switch or sensor, for example. Since subsystem 2 is in parallel with S and 2′ we have, using the above results,

$$\begin{aligned} R_{2,S2'}(t) &= 1 - P_2(t)P_{S2'}(t) \\ &= 1 - \{[1 - R_2(t)][1 - R_{S2'}(t)]\} \\ &= 1 - \{[1 - R_2(t)][1 - R_S(t)R_{2'}(t)]\}. \end{aligned} \tag{13.2.7}$$

For the complete system in figure 13.1,

$$R_{system}(t) = R_1(t)R_3(t)(1 - \{[1 - R_2(t)][1 - R_S(t)R_2(t)]\}) \tag{13.2.8}$$

We comment that $R_{2'}(t)$ and $R_S(t)$ must be computed at time t, recognizing that they have been dormant until $t_0 (0 < t_0 < t)$ and active from t_0 until t. Thus, $R_S(t)$ and $R_{2'}(t)$ would be larger values, in general, than if S and 2′ had been operating from time 0 until t.

Another form of redundancy is an r out of n system that has n parallel components of which r must survive for the system to continue operating. Examples would be cables of a bridge of which a certain number are necessary to support the load, or several parallel current paths where a critical total path area is necessary to handle the power. The reliability for such a system is

$$R_{sys} = \sum_{x=r}^{n} \binom{n}{x} R^x (1 - R)^{n-x}, \tag{13.2.9}$$

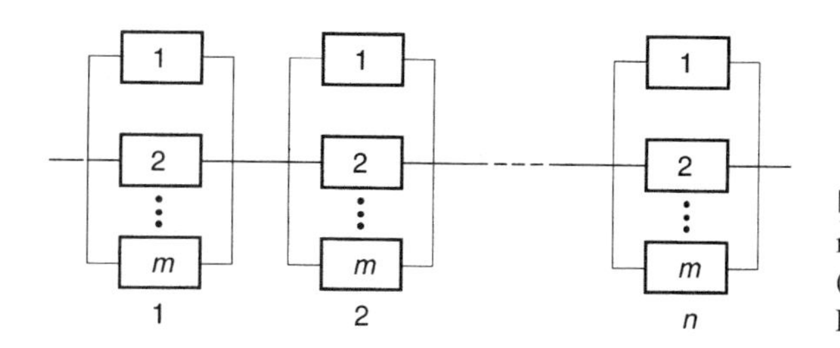

Figure 13.2 Low-level redundancy of components. (Source: Kapur and Lamberson, 1977.)

where R is the subsystem reliability, assumed equal for all subsystems, and

$$\binom{n}{x} = n!/x!(n-x)! \tag{13.2.10}$$

With just this simple mathematics we can already do an illustrative example of significant practical importance.

Example 1 (after Kapur and Lamberson, 1977)

Assume we have an n-component system. We usually have a choice of providing redundant components (figure 13.2) or of providing a totally redundant system (figure 13.3). How do these two levels of redundancy compare? Assume each component has the same level of reliability, R.

For figure 13.2 we have m components in parallel with each other, with n of these networks in series.

$$R_{\mathrm{S,low}} = [1-(1-R)^m]^n \tag{13.2.11}$$

For figure 13.3, we have n components in series with each other and m such branches in parallel with one another.

$$R_{\mathrm{S,high}} = 1-(1-R^n)^m \tag{13.2.12}$$

For example, if $R = 0.98$ (a typical value for a military specification component) and $n = 6$, $m = 2$, we find

$$R_{\mathrm{S,low}} = 0.998 \text{ (redundant components)},$$
$$R_{\mathrm{S,high}} = 0.987 \text{ (redundant subsystem)}.$$

13.2.1.1 Fault Trees and Reliability Prediction

In the past, predictions of reliability for complex space systems were based on individual component data from sources such as *MIL Handbook 217*. However, the values presented

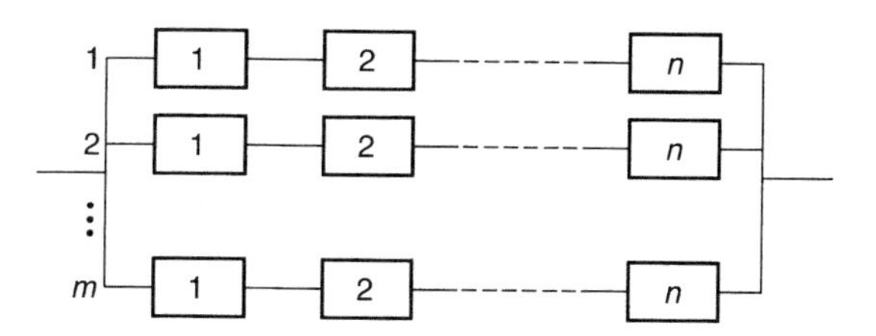

Figure 13.3 High-level redundancy of systems.

in this handbook have origins that are obscure. Some may be derived from field data, some from laboratory reliability testing, and some from military systems development qualification programs. Another important consideration is that the values given for number of failures per million hours of operation or some other similar parameter are determined at a 60% confidence level. Thus, if one collected the data again, one would have a 40% chance of obtaining a significantly different result. The main usefulness of *MIL Handbook 217* was, and remains, in comparing similar designs or different versions of the same design with the same methodology. The ability to make meaningful absolute predictions of lifetimes or failure probability with *MIL Handbook 217* was debatable.

Since the 1994 decision by the Department of Defense (DOD) to use commercial off-the-shelf (COTS) integrated circuit (IC) technology produced by best commercial practice, the reliability prediction and analysis software tools have been significantly improved. DOD made this decision to take advantage of the rapid improvements in the speed and capability of commercial IC technology to more rapidly advance military systems. The commercial products to be inserted into new military designs include plastic encapsulated microcircuits (PEMs). The IC industry had proved during the mid and late 1980s that PEMs were as reliable as hermetically sealed ceramic-packaged parts—that is, one could have long-term reliability without hermeticity.

Software package suites currently used for failure modes and effects analysis (FMEA), fault tree construction, and probabilistic risk assessment (PRA) include SAPHIRE (Systems Analysis Programs for Hands-on Integrated Reliability Evaluations), Relex, and PRISM.

SAPHIRE is useful for constructing event and fault trees and producing accompanying graphics. Once the set of basic events for the fault tree is defined, the initialization of the PRA is facilitated. The fault tree has value itself because the causality branches leading to the failure of a subsystem or system are logically linked in its construction. The complement to a fault tree is a success tree, which may have probabilities inserted into branches to yield a probability of success for the system at the top of the tree—ergo, a PRA. A trial version of SAPHIRE can be downloaded from www.nec.gov. SAPHIRE was originally developed by the Idaho Nuclear Engineering Laboratory for the old Nuclear Energy Commission. SAPHIRE also enables the FMEA, but a simpler method employing Microsoft Excel spreadsheets with MIL-STD 1629A and SAE (Society of Automotive Engineers) J1739 formats has proved more efficient. An FMEA looks at the consequences of a component or subsystem failure on associated components and systems and is often critical with respect to the incorporation of redundancy or fail-safe modes into the design.

Relex 7 contains part failure rate libraries and can undertake reliability block diagram analysis (often a starting point for complex space systems) and fault/event trees. Relex is available from the Relex Software Corporation, Greensburg, PA, and is currently at version seven. Relex includes the capability of doing FMEA, life cycle cost, and maintainability analyses. Maintainability involves the mean time interval needed for repair or refresh of a complex system. The cost incurred for maintainability over the life of the system must be estimated for its upkeep and operation; thus, it is called the life-cycle cost.

PRISM is available from the Reliability Analysis Center in Rome, NY, and is the improved response to the obsolescence and criticisms of *MIL Handbook 217*. PRISM allows the reliability engineer to tailor individual electronic part failure rates based on

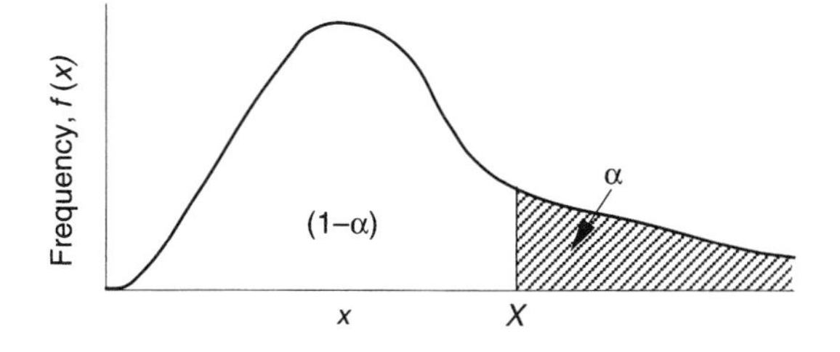

Figure 13.4 Frequency distribution plot showing the definition of the concept of confidence level.

one's own reliability testing and part screening, thus enabling a much more accurate specific failure rate. PEMs can be successfully evaluated. Other failure rates given in PRISM are mainly based on actual field experience or manufacturers' test data. The electronic part data from PRISM can be combined with the event tree of SAPHIRE and the block diagram from Relex to carry out a PRA for a complete spacecraft. Usually, in the concept and preliminary design stages, a parts count method is used in PRISM.

13.2.1.2 Confidence Level or Interval

With any population of things it is usually impractical or impossible to test every member of the population. Thus, we must determine the important parameters of the population (for example, the mean life or mean tensile strength) from a small sample of the whole population. The mean value of the sample, x, will usually not be identical to the mean value of the population, μ, but should be near the true population mean. Therefore, a confidence interval or level is chosen that will contain μ. If we designate the interval as $[(x - C_{\alpha/2})$ to $(x + C_{\alpha/2})]$, then the degree of confidence that this interval contains μ is $1 - \alpha$, where α is the risk we are willing to take of being wrong. That is, if $1 - \alpha = 95\%$, then $\alpha = 5\%$ is the risk we are taking of being in error. The greater the confidence level and hence the more certain we are, the wider the interval becomes. Figure 13.4 shows the case in which the confidence level that x will be larger than X is α, or, conversely, that it is smaller or equal to X is $1 - \alpha$.

Therefore, confidence is defined as the probability that a given interval determined from test data of a sample will contain the true parameters of the population.

Hence, using the older *MIL Handbook 217* with a 60% confidence level we have a 40% chance of obtaining a significantly different result if we repeat the experiment a second time. It is noted here that most decisions involving statistics use 90% and 95% confidence levels in order to be as certain as possible that specification or performance requirements are being met.

Confidence is also to be distinguished from reliability. Reliability applies to the hardware, component, system, and so on. Confidence applies to the test or experiment itself.

Suppose a piece of hardware has 90% reliability at 95% confidence. This statement means that when a single sample of 20 such parts is tested, two will fail (10% failure or 90% reliability). When 20 such tests of 20 parts each are carried out, 19 of the 20 experiments will have two failures or fewer (95% confidence).

Obviously, we desire to express our reliability predictions with the greatest degree of confidence. However, in many instances we must use the data available in a limited time frame or at reasonable cost. We are usually forced to couple our limited knowledge about potential space hardware with the heritage that it has from previously flown systems.

Heritage can give one a comfortable feeling due to the fact that some subsystem has performed successfully on a previous space mission. However, one must be careful that changes due to new technology or manufacturing processes, especially commercial ones, do not alter the pedigree of the previous subsystem significantly. In any case, the second subsystem should go through the same acceptance and qualification as the first.

An important consideration in any system's reliability evaluation is to consider the types of failure that may occur (for example, switch stuck open or closed, diode shorted or open) and whether any such single point of failure could cause the complete spacecraft to fail catastrophically. The best approach, as already discussed, is parallel redundancy. However, if the redundancy approach is not possible, extensive derating (using a component at less than 100% of its maximum rated parametric values and allowing for degradation by the end of the mission), testing, and protection may be necessary for the critical part. One may even pay a vendor for a special highest reliability design, perhaps an overdesign, to ensure mission success. Design of fault-tolerant systems has become an important goal in recent years.

When discussing failure and failure analysis, either before or after the fact, one must consider the physics of failure for a component or subsystem. Derating guidelines are based on such facts as the shortened mean time to failure for high temperatures at semiconductor material junctions or for high voltages on capacitors. Knowledge of the most likely failure mechanisms in key parts or subsystems help one to derate the appropriate operating parameter or environment for the hardware.

The designer should attempt to anticipate the weak links in the design by performing both worst-case operating and failure mode/fault tree analyses. What can be done to counteract or mitigate the weaknesses in the design? Can a more reliable component, even with reduced performance, satisfy the requirements? Will special qualification tests be necessary to gain confidence in a device? Is redundancy an option? Do fault-tolerant measures need to be employed? Are components being pushed to the limit of their performance (hence, the design is not conservative)?

We will say more about reliability of systems when discussing mathematical tools.

13.2.2 Tools for Reliability Assessment: Environments

How does one make a quantitative assessment of reliability? First, one must apply some stress to a system or component and continue the stress until significant changes or failure occur. These changes may be parametric or catastrophic. In the latter case one tests the object for its useful life (cycling relays until operational failure). In the former case, one, defines a priori the failure level (e.g., a change in DC resistance equal to ten times the initial theoretical constriction resistance for an electrical contact; a 20% degradation from the original drain–source saturation current of a field effect transistor (FET)). Table 13.1 presents commonly used environmental stresses or causes, together with the effects being investigated.

Since one cannot wait 10 to 20 years for results of an environmental test under field conditions, one must accelerate the stress by making the environment more severe.

Table 13.1 Commonly used environmental stresses

Stresses to determine	Responses
Tensile	Tensile strength
Electro/mechanical cycling	Life, wear
Mating/unmating	Insertion/withdrawal forces
Dielectric breakdown	Dielectric strength
Vibration	Mechanical durability; coupled loads
Thermal shock/cycling	Effects of differential thermal expansion
Elevated temperature	Accelerated aging
High humidity/moisture resistance	Insulation resistance; effects of corrosion; hermeticity
Thermal vacuum	Effects of outgassing; integrity of seals, efficacy of conductive heat sinks and radiators

Thus, we raise the temperature in an aging test or increase the applied voltage stress for dielectric breakdown. Questions then arise about such things as acceleration factors and failure modes.

The primary caution in any design of an accelerated environmental screen is that increasing the stress level should not introduce any potential modes of failure that do not exist in the field or mission environment. Thus, one does not want to increase aging temperatures above the glass transition temperature of G10 epoxy or above the outgassing temperature of polyvinyl chloride (PVC) insulated wire (both temperatures are in the 80 to 90°C range). For PVC the outgassing of chlorine from the wire insulation can combine with the moisture in a heat convection oven to form dilute HCl which will attack gold-plated copper alloy connector terminals at 120°C—a condition that is never approached in the field.

The best-known relationship for determining acceleration factors is that of Arrhenius, a Swedish physical chemist. His relationship is

$$\frac{t_2}{t_1} = \exp\left[\frac{E_a}{k}\left(\frac{1}{T_2} - \frac{1}{T_1}\right)\right] \tag{13.2.13a}$$

where t_2 is the time in field at T_2, the lower absolute temperature; t_1 the time in test at T_1, the higher absolute temperature; k is Boltzmann's constant; and E_a is the activation energy. The latter is that threshold necessary to activate a failure mechanism without causing a phase transition. Different modes of mechanical and electrical failure have different activation energies. Since the logarithm of the acceleration factor in the Arrhenius relationship is directly proportional to E_a, processes with larger activation energies yield larger acceleration factors for life tests. The ratio t_2/t_1 is the acceleration factor (>1) achieved by the environmental test.

Experimentally, this relationship is usually plotted with a log time plot on the ordinate and the inverse of the absolute temperature on the abscissa (figure 13.5). By carrying out two separate life tests at different temperatures which determine time to failure, one can plot a straight line and extrapolate to a field temperature of interest. The slope

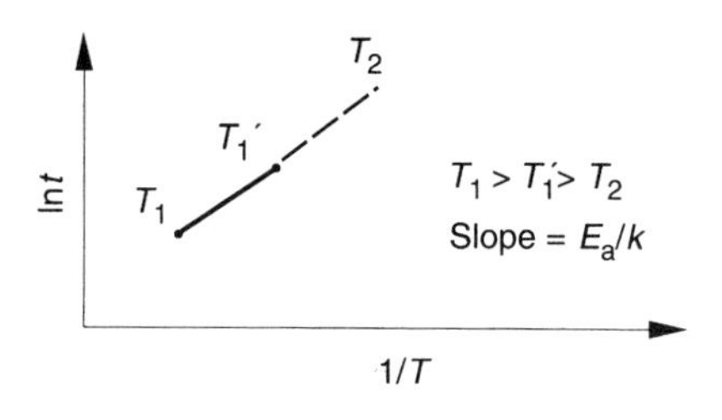

Figure 13.5 Typical Arrhenius plot of logarithm of time versus inverse of absolute temperature.

of this line can be used to determine the activation energy for the process causing the failure.

The Arrhenius relationship holds provided the system does not undergo a phase change between T_1 and T_2. Equation 13.2.13a shows that $\exp\,[(E_a/k)(1/T_2 - 1/T_1)]$ is actually an acceleration factor whose magnitude is determined by the activation energy of the process being studied and the difference between the test temperature and the field temperature. Thus, the temptation to increase the test temperature as high as possible emerges, often causing its own problems.

The activation energy is usually expressed in units of electron volts (eV). The higher the activation energy, the higher the threshold for a given process to be activated; but, once activated, the faster one can accelerate a test environment. Many different activation energies exist; for example, the energy to cause failure as a result of lack of tensile strength in an insulating material will not be the same as the energy to cause failure as a result of lack of dielectric strength in the same material.

In semiconductor materials, activation energies for diffusion processes that cause significant performance degradation are in the range of 0.6 to 1.0 eV for silicon devices and 1.3 to 1.8 eV for gallium arsenide devices. Consequently, acceleration factors are greater for gallium arsenide devices.

Equation 13.2.13a can be generalized to include temperature cycling and humidity environmental stresses as factors.

$$AF = \left[\frac{\Delta T_1}{\Delta T_2}\right]^q \left[\frac{RH_1}{RH_2}\right]^n \exp\left\{\frac{E_a}{k}\left(\frac{1}{T_2} - \frac{1}{T_1}\right)\right\} \tag{13.2.13b}$$

where

ΔT = range ($T_{max} - T_{min}$) of the temperature cycle
RH = relative humidity
T = absolute temperature at relative humidity RH in degrees Kelvin
E_a = activation energy for common models of failure, determined empirically, in eV
q, n = exponents resulting from log normal distribution of cumulative time to failure data at several levels of stress
k = Boltzmann's constant, equal to 8.62×10^{-5} eV/C

subscripts 1 and 2 refer to the test and application conditions, respectively.

For ground-based applications, all environments are relevant; for space flight, humidity is usually not a concern if ground processing facilities are environmentally controlled. The third or aging factor in equation 13.2.13b due to Arrhenius has already been described. The second factor on humidity exposure is due to D.S. Peck of Bell Laboratories who surveyed all published median life data on electrolytic corrosive failures of aluminum metalization during the 1980s. Peck's model fits a plot of 61 median lifetime data points. Correlation coefficients of > 0.98 are achieved with $E_a = 0.77 - 0.81$ eV and $n = 2.5$–3.0. The first, or temperature cycling, factor has been discussed by Harris Semiconductor investigators. Low-cycle (<10,000 cycles) fatigue failures are created by repeated mechanical stresses of thermal origin governed by the Coffin–Manson Law that relates plastic strain range to the temperature range. For bond wire fatigue failure, a value of $q = 4$ is most appropriate.

Example 2

Consider a 1000-cycle temperature cycling test between −55°C and +125°C, a test range of 180°C. Setting the aging and humidity factors to one in equation 13.2.13b for a space application in which the maximum temperature is +55°C and the minimum temperature is −30°C, a field range of 85°C, we have

$$AF = (180/85)^4 = 20.$$

Thus a 1000-cycle test has simulated 20,000 cycles in space. For a low Earth orbit mission with a 90–110 minute period this test represents 3.4–4.2 years.

The aim of life testing is to determine time until failure. Other environmental stress screens are designed to find weak points in hardware, whether from the components themselves or from workmanship such as solder joints. That is, the philosophy is to create failure in a short time at a high stress level so that a latent defect will not cause a field failure at a much later time. Temperature cycling is extensively used for plastic-encapsulated microcircuits (PEMs) since it produces stresses due to the differential thermal expansion between plastic package and silicon die with metal lead frames. One of the primary modes of failure being stressed is the lamination of the plastic package to the integrated circuit.

Life tests often run for hundreds to thousands of hours. Power burn-in tests (tests conducted under increased temperature with DC or AC electrical bias applied), on the other hand, usually run only for a week to ten days (168–240 h). The stress or bias conditions are often very much the same as during life testing. However, the object of the shortened duration is to precipitate failure only in those samples of the product lot so weak that they will fail after short service life—infant mortality. A mature product technology that is under control in manufacture should have little, if any, fallout due to burn-in.

13.2.2.1 Destructive Physical Analysis

For components, it is valuable to de-lid devices and actually observe, either optically or with the help of a scanning electron microscope (SEM), what is inside a component package. Thus, a small number of parts are removed from an incoming lot and subjected to internal inspection and analysis. This evaluation concentrates on the quality and workmanship of the part's physical structure. Chart A lists the procedures followed during destructive physical analysis (DPA) for integrated circuits. Figure 13.6 shows the kinds of defect that one inspects for when carrying out DPA on a relay.

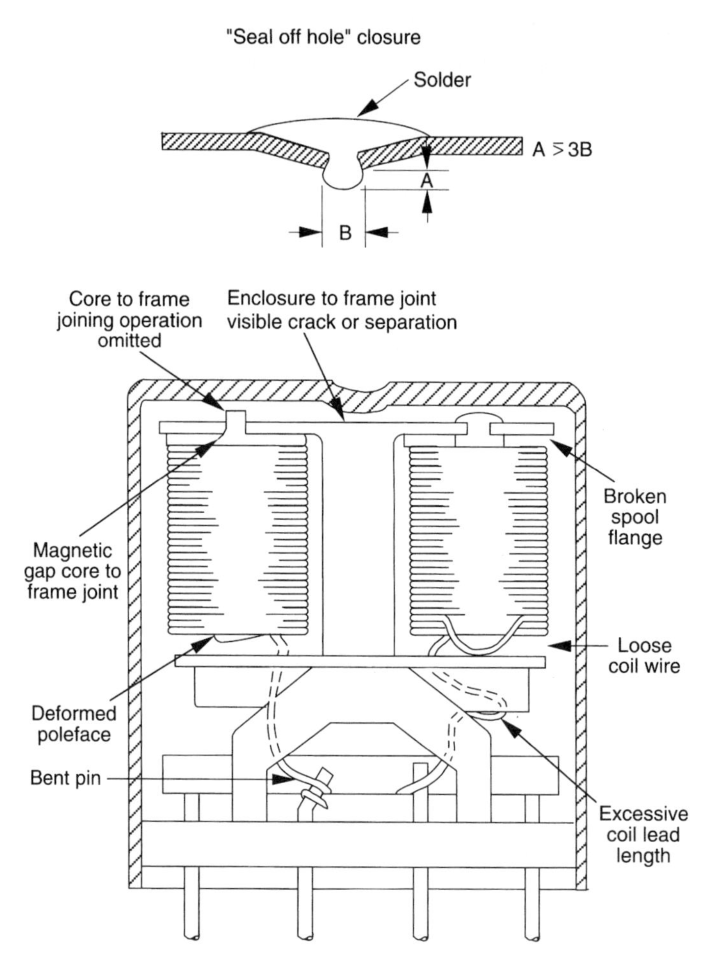

Figure 13.6 Relay defects.

If a component has some marginal performance in other tests or screens (that is, a few but not many parts are failing or marginal), the results of the DPA scrutiny can strongly influence the final decision about acceptance or rejection. Experience of the inspector is important in making useful judgments of a particular lot's quality versus the level of quality generally accepted for flight use.

13.2.3 Tools for Reliability Assessment: Mathematics

Reliability engineers must know and use some mathematics, particularly with respect to probability and statistics.

The probability density function can be defined as the function that yields the probability that a random variable takes on any one of its admissible values. This function is

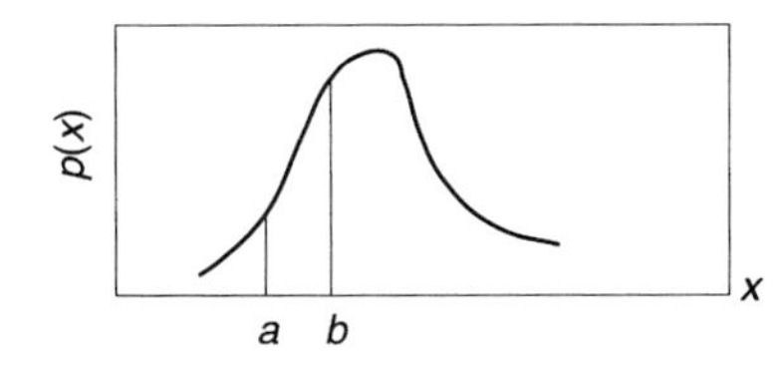

Figure 13.7 Density function for a continuous random variable.

usually designated $p(x)$, which is the probability of obtaining the value x with x as the random variable.

The cumulative distribution function is designated $F(x)$ and is defined as

$$F(x_i < a) = \sum_i^a p(x_i), \text{ for } x_i \text{ discrete and } x_i \leq a \tag{13.2.14}$$

or, if $p(x)$ is continuous,

$$F(x < a) = F(a) = \int_{-\infty}^{a} p(x)dx. \tag{13.2.15}$$

Figure 13.7 shows a general density function, while figure 13.8 shows a cumulative distribution function for a continuous random variable. In figure 13.8 the random variable takes on only positive definite values so that the domain of the integral is from 0 to a instead of from $-\infty$ to a.

We can now discuss measures of a sample or population of items. A sample is defined as a group of items under test, selected randomly from a population of infinite size containing similar items. The most common measure of the central tendency of a population or sample is the mean,

$$\mu = \int_{-\infty}^{\infty} x\ p(x)dx \quad \text{for a population of a continuous random variable} \tag{13.2.16}$$

or

$$\overline{x} = \frac{\sum_{i=1}^{n} x_i}{n} \quad \text{for the arithmetic mean of a discrete sample} \tag{13.2.17}$$

where n is the number of items in the sample.

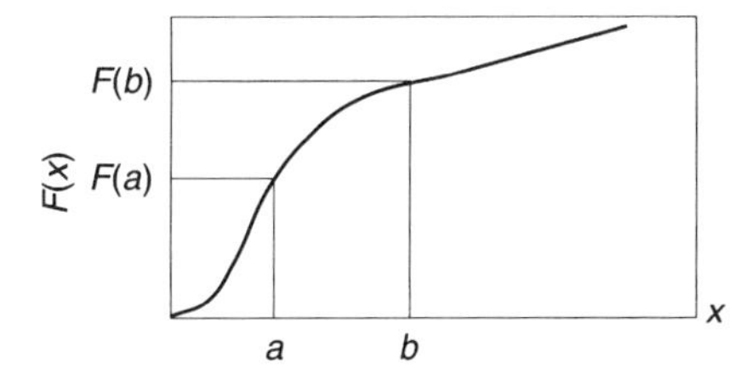

Figure 13.8 Cumulative distribution function for a continuous random variable.

CHART A. DPA PROCEDURE FLOW, INTEGRATED CIRCUITS

1. External visual
   a. Contamination
   b. Mechanical damage
   c. Heat or electrical damage
   d. Seal integrity
   e. Lead integrity and plating
   f. Correct marking
   g. Correct pin designation
   h. Dimensions
2. Opening or cross-sectioning
3. Internal visual inspection
   a. Optical or electron microscope looking for contamination and mechanical or electrical damage of semiconductor die
   b. Worst-case oxide steps
   c. Worst-case metalization (voids, cracking, bridging, etc.)
   d. Typical metalization pre- and post-glassivation removal
4. Bond pull (normally before SEM)
5. Die shear

The most common measure of the variability of a population or sample is known as the variance or standard deviation: for a continuous random variable, the variance

$$\sigma^2 = \int_{-\infty}^{\infty} (x - \mu)^2 p(x) dx \tag{13.2.18}$$

and the standard deviation

$$\sigma = \sqrt{\sigma^2}; \tag{13.2.19}$$

and for the standard deviation of a discrete sample

$$s = \sqrt{\sum_{i=1}^{n} (x_i - \overline{x})^2 / (n - 1)} \tag{13.2.20}$$

Equation 13.2.20 is generally referred to as an unbiased form of the standard deviation.

If a constant is added to or subtracted from each of the observations, the mean will be increased or decreased by that same constant but the variance and standard deviation will not be affected. If each observation is multiplied by a constant c, the new mean is c times the old mean, the new variance is c^2 times the old variance, and the new standard deviation is c times the old standard deviation.

13.2.3.1 Statistical Distributions

Table 13.2, taken from Lipson and Sheth (1973), summarizes the applications of the most commonly used statistical distributions. The most practical method of determining

Table 13.2 Fields of application of statistical distributions

Statistical Distribution	Fields of Application	Examples
Normal	Various physical, mechanical, electrical, chemical properties	Capacity variation of electrical condensers; tensile strength of aluminium alloy sheet; monthly temperature variation; penetration depth of steel specimens; rivet-head diameters; electrical power consumption in a given area; electrical resistance; gas molecule velocities; wear; noise generator output voltage; wind velocity; hardness; chamber pressure from firing ammunition
Log-normal	Life phenomena; asymmetric situations where occurrences are concentrated at the tail end of the range, where differences in observations are of a large order of magnitude	Automotive mileage accumulation by different customers; amount of electricity used by different customers; downtime of a large number of electrical systems; light intensities of bulbs; concentration of chemical process residues.
Weibull (two-parameter)	Same as log-normal cases. Also situations where the percent occurrences (say, failure rates) may decrease, increase, or remain constant with increase in the characteristic measured, for parts at debug, wearout, and chance failure stages of product's life.	Life of electronic tubes, antifriction bearings, transmission gears, and many other mechanical and electrical components; corrosion life; wearout life
Weibull (three-parameter)	Same as two-parameter Weilbull and, in addition, various physical, mechanical, electrical, chemical, etc., properties, but less common than in the case of normal distribution	Same cases as two-parameter Weibull. In addition, electrical resistance, capacitance, fatigue strength, etc.
Exponential	The life of systems, assemblies, etc. For components, situations where failures occur by chance alone and do not depend on time in service, frequently applied when the design is completely debugged for production errors	Vacuum-tube failure life; expected cost to detect bad equipment during reliability testing; expected life of indicator tubes used in radar sets; life to failure of light bulbs, dishwashers, water heaters, clothes wasthers, aircraft pumps, electric generators, automobile transmissions

(Continued)

Table 13.2 (Continued)

Statistical Distribution	Fields of Application	Examples
Binomial	Number of defectives in π sample size drawn from a large lot having p fraction defectives; probability of x occurrences in a group of y occurrences, that is, situations involving "go-no-go," "OK-defective," "good-bad" types of observations. Proportion of lot does not change appreciably as a result of sample drawn	Inspection for defectives in a shipment of steel parts; inspection of defective tires in a production lot; determination of defective weld joints; probability of obtaining electrical power of a certain wattage from a source; probability that a production machine will perform its function
Hypergeometric	Inspection of mechanical, electrical, etc., parts from a small lot having known percent defectives. Same as in binomial cases, except the proportion of lot may change as a result of sample drawn	Probability of obtaining 10 satisfactory resistors from a lot of 100 resistors having 2% defectives; similar cases involving light bulbs, piston rings, transistors
Poisson	Situations where the number of times an event occurs can be observed but not the number of times the event does not occur. Applies to events randomly distributed in time	Number of machine breakdowns in a plant; automobiles arriving simultaneously at an intersection; number of times dust particles found in atmosphere in some number of spot checks; industrial plant personnel injury accidents; dimensional errors in engineering drawings; automotive accidents in a given location per unit time; automotive traffic; hospital emergencies; telephone, circuit traffic; a defect along a long tape, wire, chain, bar, etc; tire punctures; stones hitting windshield; number of defective rivets in an airplane wing; radioactive decay; number of engine detonations; number of flaws per yard in sheet metal

*Source: Lipson and Sheth (1973).

a distribution from experimental data is to plot the data on various probability papers (normal, log-normal or Weibull), and determine the best linearity. More rigorously, one can use a goodness-of-fit test such as chi-squared. We will describe only six of the distributions specifically.

13.2.3.1.1 Normal Distribution. The probability density function

$$p(x) = \left(\frac{1}{\sigma\sqrt{2\pi}}\right)\exp\left\{-\frac{(x-\mu)^2}{2\sigma^2}\right\}, \text{ for all } x. \tag{13.2.21}$$

where μ and σ are the population mean and standard deviation respectively. The probability that x is greater than some value b can be designated $P(x > b)$ and equals the ordinate of the cumulative distribution function

$$F(b) = P(x > b) = \int_b^\infty p(x)dx. \tag{13.2.22}$$

This integral cannot be evaluated in closed form and thus it becomes necessary to resort to a tabulation in evaluating it. This is done by defining the standard normalized variate, z, where

$$z = \frac{x-\mu}{\sigma} \tag{13.2.23}$$

Using the transformation of 13.2.23, we find

$$P(z > z_\alpha) = \int_{z_\alpha}^\infty \left(\frac{1}{\sqrt{2\pi}}\right)\exp\left(-\frac{z^2}{2}\right)dz = \alpha, \tag{13.2.24}$$

where z_α is known as the upper α percentage point, α being the area under the probability curve between z_α and infinity. Tables of the integral in equation 13.2.24 can be found in every text on statistics.

13.2.3.1.2 Log-Normal Distribution. As figure 13.9 shows, the log-normal distribution is not symmetric about the mean value. It is applicable to life data or to parametric measurements that yield only positive definite values. The mathematical expressions for the log-normal distribution are the same as those for the normal distribution, equations 13.2.21 and 13.2.22, with x being replaced by $\log x$, sometimes designated

$$x_\ell \equiv \log x \tag{13.2.25}$$

with μ_ℓ the log population mean and σ_ℓ^2 the population variance. The probability of occurrence of x_ℓ can be found in the same manner as the probability of occurrence of x in the normal distribution case.

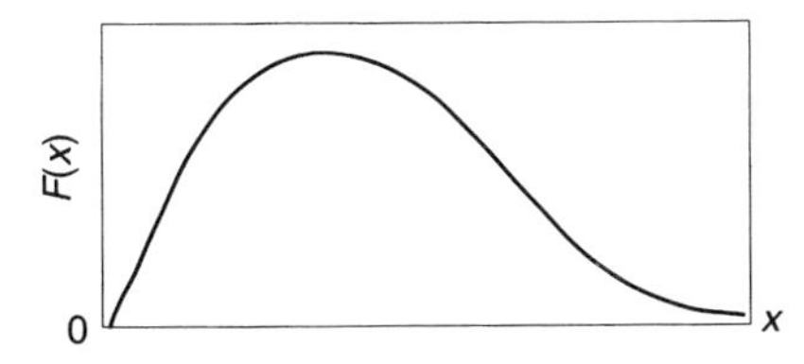

Figure 13.9 Log-normal distribution.

13.2.3.1.3 Weibull Distribution. Generally, the Weibull distribution describes the life characteristics of parts or components. The exponential distribution more adequately describes the life of systems, which is a superposition of the Weibull distributions of many parts or components. However, in some cases a system will also follow a Weibull distribution. This usually is indicative of immature design and/or quality problems (for instance, the first production versions of a missile with a new guidance set).

The Weibull distribution density function

$$p(x) = \left[\left\{\frac{b}{(\theta - x_0)}\right\}\left\{\frac{(x - x_0)}{(\theta - x_0)}\right\}^{b-1}\right]\left[\exp\left\{-\frac{(x - x_0)^b}{(\theta - x_0)^b}\right\}\right] \tag{13.2.26}$$

where x_0 is the expected minimum value of x (often taken as 0), b is the Weibull slope or shape parameter, and θ is the characteristic life or scale parameter. The Weibull slope b is the measure of the uniformity of the product: the larger the slope, the more uniform the product.

The three-parameter Weibull function (θ, b, x_0) occurs when one cannot assume $x_0 = 0$. Such cases would be represented by strength or wear data. Now the cumulative failure distribution function is

$$F(x) = 1 - \exp\left[-\frac{(x - x_0)^b}{(\theta - x_0)^b}\right]. \tag{13.2.27}$$

13.2.3.1.4 Exponential Distribution. The exponential distribution is really just a special case of the Weibull distribution in which $b = 1$, $x_0 = 0$, and $x = t$, where t usually represents time. Thus the probability density function is

$$p(t) = \left(\frac{1}{\theta}\right)\exp\left\{-\frac{t}{\theta}\right\} \tag{13.2.28}$$

and the cumulative distribution function is

$$F(t) = 1 - \exp\left\{-\frac{t}{\theta}\right\}. \tag{13.2.29}$$

For the exponential distribution a quantity called the hazard rate is important and is defined as $\lambda = 1/\theta =$ constant. The hazard rate is the probability that a given item will fail between t and dt when it has already survived until t. The simplicity of the exponential distribution is that this hazard rate is constant.

Thus, if a population has a given characteristic life $\theta = 100$ h, a given member of the population has a $\lambda = 1/100$ h or 1% chance per hour of failing, no matter how many hours it has already survived.

We can use the general definition of the hazard rate

$$\lambda = \frac{p(x)}{(1 - F(x))} \tag{13.2.30}$$

to find a hazard rate for the more inclusive Weibull distribution. The result is

$$\lambda = \left(\frac{b}{\theta^b}\right)t^{b-1}. \tag{13.2.31}$$

The Weibull distribution hazard rate depends on the Weibull slope b. Note that if $b = 1$, the Weibull hazard rate becomes the exponential hazard rate. When b is less than one (greater than one), the hazard rate decreases (increases) with time t. Hence, if a system follows a Weibull distribution, it is possible for it to be expected to last a very long time provided that its hazard rate is decreasing. One study has shown that this is indeed the case for spacecraft built with more mature technologies since 1977 (Hecht and Hecht, 1985). Consequently, predictions based on the exponential distribution have been too pessimistic.

13.2.3.1.5 Binomial Distribution. The binomial distribution is used when the random variable is discrete, when we are interested in analyzing attributes or pass/fail data. The conventional manner of defining the binomial distribution is to consider an experiment of n trials (independent) in which the probability of success is p and the probability of failure q, with $p + q = 1$. Then the probability that there will be r successes in n trials is

$$p(r) = ({}_r{}^n) p^r q^{n-r}, \quad r = 0, 1, 2, \ldots, n, \tag{13.2.32}$$

where

$$({}^n{}_r) \equiv \frac{n!}{r!(n-r)!} \tag{13.2.33}$$

The binomial distribution function has a mean value

$$E(r) = np \tag{13.2.34}$$

and a standard deviation

$$\sigma(r) = \sqrt{npq}. \tag{13.2.35}$$

Example 3

Space shuttle reliability. The catastrophic accident of the Columbia motivated this reliability calculation. A binomial distribution was used since the many and complex battery of tests, screens, checks, and processes used for Shuttle preparation mean that many parameters are measured. Thus, prior to flight, the Shuttle is judged as go/no go and the flight/mission is success/failure. In this case failure unfortunately means loss of vehicle and crew. The binomial distribution applies in such situations and the random variable x_i is discrete.

Defining p as the probability of failure and q as the probability of success, we have

$$p + q = 1.$$

The probability of obtaining r_ℓ defective units from a sample size n is the binomial:

$$P(r = r_\ell) = \left[\frac{n!}{r_\ell!(n - r_\ell)!}\right] p^{r_\ell} q^{n-r_\ell}.$$

For the Shuttle history: $p = 2/113 = 0.018$ and $q = 111/113 = 0.982$. The probability of success can be shown to have upper and lower limits (Lipson and Sheth, 1973): let

$$g = n - r$$

where g = the number of successes (111), n = the number of trials (113), and r = the number of failures (2). Then

$$q_{\rm L}\,(\text{lower limit}) = \frac{1}{1 + [(n - g + 1)/g]F_{\rm L}}$$

with degrees of freedom for the F distribution of

$$v_1 = 2(n - g + 1) \quad \text{and} \quad v_2 = 2g.$$

Similarly,

$$q_{\rm U}\ (\text{upper limit}) = \frac{1}{1 + (n - g)/(g + 1)(1/F_{\rm U})}$$

with degrees of freedom for F of

$$v_1 = 2(g + 1) \text{ and } v_2 = 2(n - g).$$

Tables of the F distribution can be found in any statistics book.

The upper and lower limits for success can be calculated for several levels of confidence, generally 90%, 95%, 99% ($F_{0.10,v_1,v_2}$, $F_{0.05,v_1,v_2}$, $F_{0.01,v_1,v_2}$). Using the formulae for $q_{\rm L}$ and $q_{\rm U}$, one obtains:

	$q_{\rm L}$ < true reliability < $q_{\rm U}$,
or at 95% confidence	0.95 < true reliability < 1.0,
and for 99% confidence	0.93 < true reliability < 1.0.

That is, the success history of 111 in 113 flights, which is equal to 0.982, gives us a point estimate of the Shuttle reliability. The statistical inference to be drawn from this relatively large sample size for future operation of the Shuttle is that the true reliability of this system and its support/maintenance is at worst 0.95, at best 1.0, with 95% confidence.

It should be noted that Nobel Laureate Richard Feynman stated in his comments on the Challenger accident investigation in 1986 that "...mature rocket systems have a failure rate of 1 in 50 (0.02)." The Shuttle has achieved 0.018, which includes reentry as well as launch—really pretty impressive.

13.2.3.1.6 Poisson Distribution. For the binomial distribution we can determine the probability of success and the probability of failure or the probability that an event occurs or does not occur. There are situations in which the numbers of times an event occurs are known, but the number of times it does not occur is not known. Examples are stones hitting windshields, nails puncturing tires or, if we include private aviation, plane crashes. One may know the number of plane crashes during a given time period, but usually does not know how many planes did not crash.

The Poisson distribution can be used in these circumstances to describe the occurrence of isolated events in a continuum

$$p(r = r_i) = \frac{e^{-y} y^{r_i}}{r_i!}, \quad r_i = 0, 1, 2, 3, \ldots \tag{13.2.36}$$

where y is the average number of occurrences; thus

$$E(r) = y \tag{13.2.37}$$

and

$$\sigma(r) = vy \tag{13.2.38}$$

for the mean and the standard deviation respectively. The Poisson distribution is a simple one that requires knowledge only of the average number of occurrences of an event during a given time period.

13.2.3.2 Statistics, Regression, and Inference

The whole point of finding and fitting statistical distributions to data is to be able to make predictions or inferences about a population from the behavior of a few samples. Obviously, this cannot be done with certainty unless every item of the population is tested, which is both impractical and undesirable. Thus, we must talk about degrees of confidence and risk. Even a one-of-a-kind spacecraft is composed of components that have come from a larger population. Although considerable qualification testing may be done using vibration and thermal vacuum environments, this approach only verifies that the spacecraft systems perform throughout the expected temperature and acceleration range, not that they can successfully survive thousands of thermal cycles of the actual spacecraft mission.

Regression analysis is used to determine the functional relationship between variables, while correlation determines the degree of association between variables. Regression methods usually assume that the error in the dependent variable is normally distributed and makes no assumption about the error of the independent variable. For correlation analysis, the errors of both variables should be normally distributed.

An important point is that correlation does not imply causation or a cause- and-effect relationship. An example that emphasizes this point is that during the 1960s people studying air pollution in cities recorded the fact that there was a significant correlation between high levels of air pollution and high levels of traffic noise. Had one believed that there was a cause-and-effect relationship, one could have gone to a good deal of trouble to make automobiles quieter without decreasing levels of air pollution.

Regression analysis establishes the functional relationship between two variables. When this relationship is a straight-line one, we have a special situation called linear regression analysis. The best fit to experimental data is usually determined by the method of least squares, which minimizes the sum of the squares of the deviations in the y direction of the data points (x_i, y_i) from the most probable curve.

We can give an example of the method of least squares for the linear case. Let

$$y = a_0 + a_1 x + \varepsilon \tag{13.2.39}$$

be the straight line that best fits the data, where ε is the measure of the deviation of the data points in the y direction.

The total sum of squares of the deviation, E, is given by

$$E = \sum e_i{}^2 = \sum_i^n (y_i - a_0 - a_1 x_i)^2 \tag{13.2.40}$$

with (x_i, y_i) the data points. The goal is to determine the regression parameters a_0, a_1 which minimize E and yield the best fit to the data.

How well a given regression line fits the data is measured by the correlation coefficient r. A value of $r = 1.0$ indicates perfect correlation; $r = 0$ indicates no correlation.

$$r = \frac{\text{Covariance of } x \text{ and } y}{\text{Square root of the product of variances of } x \text{ and } y}. \tag{13.2.41}$$

After r is determined, the square of r is a measure of the physical variation in the dependent variable that is accounted for by the variation in the independent variable. For example, if $r = 0.95$, $r^2 = 0.90$ and 90% of the variation in y is accounted for by variation in x with the remaining 10% being due to random variation not accounted for by the fit to the data.

13.2.4 Safety

The primary point of consideration about safety is to protect against the adverse impact of necessary operations, which often involve hazardous materials, ordnances, and pressurized systems. Another safety problem can be created by poor grounding techniques, in which the unexpected path to ground damages electrical circuitry. This latter case is an extreme example of an unwanted sneak circuit.

Safety is enhanced by limiting the number of hazardous materials to be used in a spacecraft system and by making sure that squibs and explosive bolts contain the debris produced by their firing.

Pressurized systems must usually be analyzed to determine what set or sets of conditions can cause a catastrophic event (usually an explosion, but sometimes just the venting of toxic material). The most intensive and complete analysis can be done by constructing a safety fault tree.

A safety fault tree has at its top the top event, whose occurrence is potentially catastrophic, leading to loss of life or mission failure, such as explosion or structural fragmentation. Basic events that either initiate the top event or enable it to occur are at the base of the various branches of the fault tree. AND gates and OR gates connect various branches of the fault tree and produce intermediate events.

Figure 13.10 (from Uy and Maurer, 1987) shows a safety fault tree for the explosion of a single lithium thionyl chloride battery cell. Chart B shows the ten sets of basic events that result in a single cell exploding. These ten sets of basic events were determined from a literature search and from discussion with experts involved in the manufacture and use of lithium batteries for both commercial and military applications.

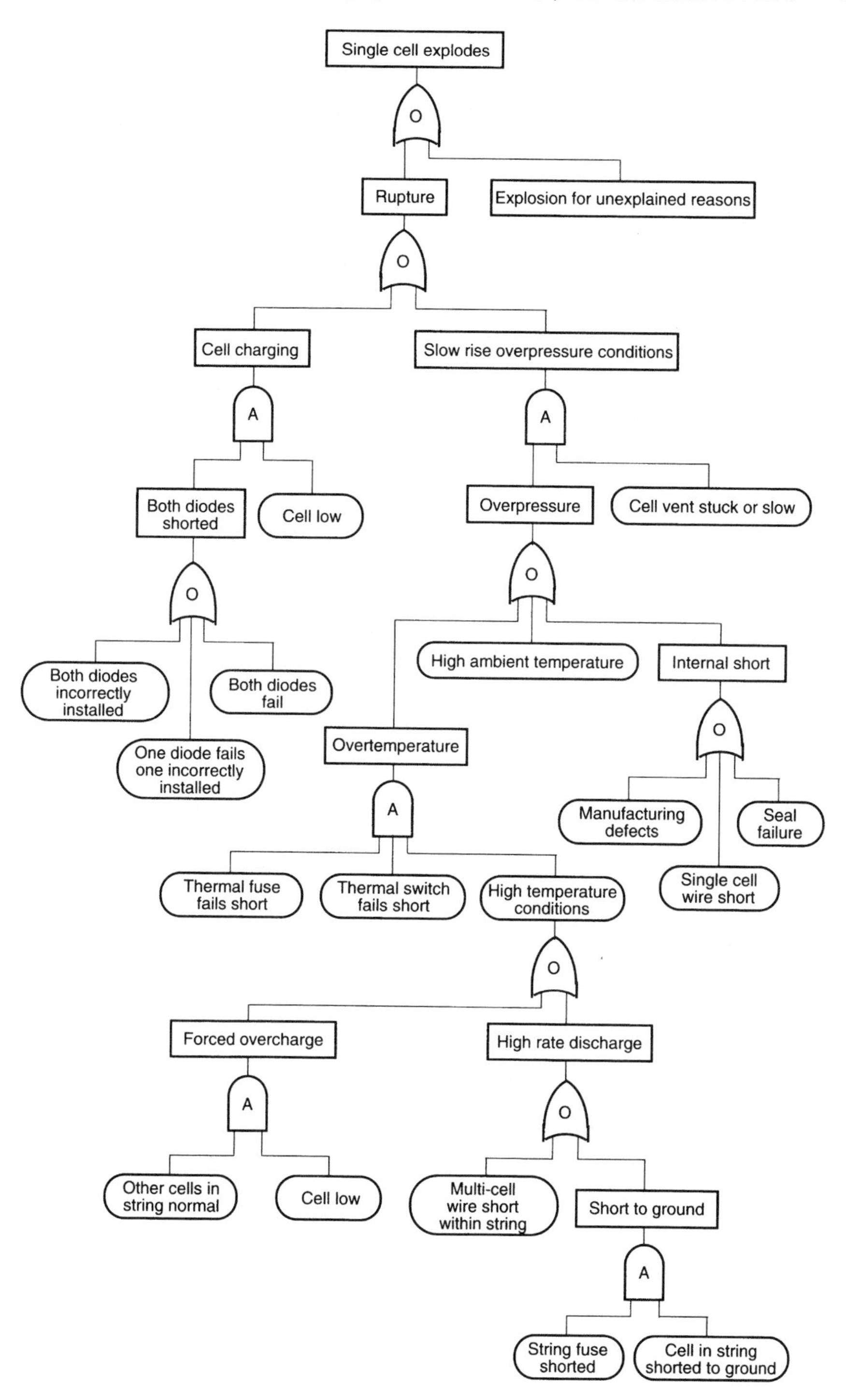

Figure 13.10 Lithium thionyl chloride single cell safety fault tree. (Source: Uy and Maurer, 1987.)

CHART B. MINIMUM SETS FOR CRITICAL SYSTEM STATES FOR THE EVENT "SINGLE CELL EXPLODES"

A. Cell charging

1. Cell low and diodes installed backwards
or 2. Cell low and diodes fail shorted
or 3. Cell low and one diode fails and the other is installed backwards

B. Overtemperature

1. High ambient temperature and cell vent stuck or slow

C. Internal short (leading to overtemperature)

1. Seal failure leading to shorting condition and cell vent stuck or slow
or 2. Single cell shorted by external wire or conductive debris and cell vent stuck or slow

D. High rate discharge (leading to overtemperature)

1. Multi-cell short due to external wire or debris and thermal fuse shorted and thermal switch shorted and cell vent stuck or slow
or 2. One or more cells shorted to ground and fuse shorted and thermal fuse shorted and thermal switch shorted and cell vent stuck or slow

E. Forced overdischarge (the rate may not be very high)

1. Cell within string with low capacity and other cells in string with normal capacity and thermal fuse shorted and thermal switch shorted and cell vent stuck or slow

To determine the relative importance of the various branches in the fault tree, estimates must be made of the probabilities of occurrence of all basic events. These are then propagated through the fault tree by addition at OR gates and multiplication at AND gates.

The main usefulness of a fault tree is in determining the relative importance of its various branches and the sensitivity of the top event occurrence frequency to significant changes in any of the basic event probabilities. The fault tree will show which factors are most important to be improved or closely controlled, so as to make the top event probability as low as possible.

Generally, each spacecraft program requires a safety plan at the launching site and/or test range. Such plans spell out procedures to be used in handling explosives, ordnances, fuels, and so on. Care must be taken that signals or communications normally used at the launch site do not inadvertently trigger some device on the spacecraft. The certification of the spacecraft hardware for the safety requirements of the test range usually requires testing of pressure vessels and verification of the fidelity of the electrical paths for ordnance.

13.3 Quality Assurance and Parts Selection

As previously mentioned, the quality assurance function assures final product acceptance and supplies the necessary documentation that must accompany the flight hardware to meet contractual requirements. Seemingly identical parts from different manufacturers fabricated at different times are not necessarily equivalent.

13.3.1 Systematic Controls

13.3.1.1 Component Level

The component level (resistors, capacitors, integrated circuits) is the level at which most flight hardware is purchased. If standard parts cannot be bought according to a Defense Electronics Supply Center (DESC) specification, then a source control drawing or purchase instruction must be written so that the parts received will meet the requirements of flight hardware. These documents are drafted by component and reliability engineers, who include the necessary performance and environmental requirements in the specification and recommend one or more vendors to the buyers. The source control drawing should also indicate whether the spacecraft contractor desires to make inspections at the vendor's manufacturing facility. Usually, pre-cap visual inspections are made for integrated circuits.

Generally, there are three ways to purchase flight components:

(1) Buy the part to an acceptable DESC or NASA qualification level;
(2) Pay the manufacturer to introduce additional steps or screens into his normal part-processing flow so that the parts produced for a particular program can be accepted as the equivalent of a standard part without additional screening;
(3) Obtain the best part the manufacturer can produce and implement the appropriate inspections, tests, and screens to qualify the part for flight.

Figure 13.11 and Chart C show a typical process flow with which an integrated circuit manufactured on a vendor's high-reliability line and tested with one or more of the test methods of MIL-STD 883 would be screened upon receipt. To be able to upgrade such an integrated circuit, the device must be hermetically sealed and operational over the usual military temperature range of −55°C to +125°C.

The steps in the incoming inspection flow shown in figure 13.11 are governed by the various method numbers (e.g., 1015, 2012) of MIL-STD 883, *Test Methods and Procedures for Microelectronics*. CMOS digital parts require a second static burn-in of 24 h in addition to the first dynamic burn-in of 168 to 240 h as specified in the NASA Goddard Space Flight Center Preferred Parts List 21. Lot qualification means a life test or extended burn-in, which is a special requirement for some programs. In addition to the tests shown on the flow card, radiation testing is often necessary. This topic will be discussed later.

The Goddard Space Flight Center (GSFC) Preferred Parts List (PPL) is a compilation of high-reliability electronic parts suitable for use in space flight applications. It combines and updates information from MIL-STD-975 (*NASA Standard Electrical,*

STEP (883 METHOD)	REJ	INSPR	DATE	IIDMR #
VISUAL (2009)				
DIMENSIONS (2016)				
INIT ELECT[1]				
STAB BAKE (1008/C)				
TEMP CYCLE(1010/C)				
CONST ACC (2001/E)				
PIND (2020/A)				
GROSS LEAK(1014/C)				
FINE LEAK(1014/A/B)				
INIT ELECT[1]				
BURN-IN (1015)				
ELECTRICAL[2]				
BURN-IN[2] (1015)				
FINAL ELECT				
X-RAY TV (2012)				
FINAL VISUAL				
TOTAL REJECTS		XXXXXXXXX	XXXXXXXX	
DATA REVIEW				
LOT QUAL (ATTACHED)				
DPA				
LOT ACCEPTANCE	XXXXXXXXX			

Note: 1. Electrical parameters specified in ____________.
2. Required for CMOS digital parts only, see PPL-18.

Figure 13.11 Flow card for incoming inspection of "/883" integrated circuit.

Electronic and Electromechanical (EEE) Parts List) and incorporates screening requirements from NASA GSFC Parts Branch Instruction 311-INST-001 (*Instruction for EEE Parts Selection, Screening and Qualification*) to provide detailed Grade 1 and Grade 2 screening requirements for all part categories.

The parts grades are defined below. Standard procedure is to stock Grade 1 and Grade 2 parts in the flight stockroom. Grades 3 and 4 require considerable upgrade screening while only the JANTXV and JANTX diodes and transistors of Grade 2 need power burn-in. Parts are subjected to 100% incoming inspection, with the exception of destructive tests such as radiation and destructive physical analysis (DPA) for which small samples are drawn from each lot. In some cases passive parts such as resistors and capacitors will just have flight lots sampled. For conservative design the parts are derated according to GSFC PPL 21 in addition to undergoing the screening process.

CHART C. PARTS SELECTION

APL Grades

Grade 1—preferred

As defined and specified in MIL-STD-975D, GSFC PPL-18, and MIL-STD-1547:

MIL-M-38510 class S integrated circuits
MIL-S-19500 JANS semiconductors
S failure rate level (0.001% per 1000 h) passive and electromechanical parts

Grade 2—acceptable

As defined in MIL-STD-975D and GSFC PPL-18:

MIL-M-38510 class B integrated circuits
MIL-S-19500 JANTXV and JANTX semiconductors
P failure rate level (0.1% per 1000 h) passive and electromechanical parts

Grade 3—conditional

MIL-STD-883 class B integrated circuits
MIL-S-19500 JAN semiconductors
M failure rate level (1% per 1000 h) passive and electromechanical parts

Grade 4—nonstandard

Commercial parts

Since many high-performance and large scale integrated (LSI) or very large scale integrated (VLSI) circuits are not on qualified parts lists, the task is often to ensure that these key devices, necessary for the performance of a state-of-the-art system, will function reliably for the duration of the mission. A great deal of challenging work goes into this type of part qualification.

Flight parts are stored in a bonded stockroom and are identifiable by both lot and part serial numbers for ease of traceability. Purchase orders are reviewed when parts are received for correct part number and count data. Parts are issued from stock in kits to the fabrication shops as dictated by system parts lists.

A very significant development took place in 1994 when the Department of Defense mandated the use of commercial-off-the-shelf (COTS) equipment, including plastic-encapsulated microcircuits (PEMs). The intent of the mandate was to procure the latest IC designs and generations into military equipment for maximum advantage in the US weapons systems.

Advantages of PEMs (which command 97% of the worldwide market for microcircuits) are

- Size (small outline packages, thin small outline packages)
- Weight (generally half as much as ceramic)
- Performance (lower dielectric constant of epoxy, smaller copper lead frame inductance)
- Availability (shorter acquisition times, 30% more part types available)
- Cost

Disadvantages of PEMs are

- Rapid obsolescence
- Non-hermetic package (reliability without hermeticity)
- Cannot readily de-lid for inspection, analysis, or test

Technology improvements over last 20 years by main-line manufacturers have made PEMS reliable. Among the major improvements are

- Molding compound purity
- Improved fillers, encapsulants, die passivation, metalization technology
- Significant process parameter improvement (assembly automation)

Low-cost PEMS now exhibit high quality, high reliability, and high performance. The major question is whether reliability without hermeticity is now sufficient for space missions, which raises additional concerns for microcircuit screening. Systems builders such as AT&T, Honeywell, and General Dynamics found that the performance, reliability, average failure rate, and failure modes of systems employing PEMs were better than or equal to those constructed with hermetically sealed ICs.

In 1999 APL performed an independent study of PEMs for the US Army. The study included 16 part types and 6 package types from 10 manufacturers for Army missile systems. The parts were purchased by APL space parts and reliability group in the same manner as spacecraft program parts are normally obtained to ensure run-of-mill product. All PEMs were subjected to a sequence of HAST (highly accelerated stress testing at 120–130°C, 85% RH) and temperature cycling (−55°C to 125°C at 10°C/min) environmental tests. Temperature cycling was performed first to attempt to separate or crack packages and then HAST was executed to drive moisture into the packages if possible. The only functional failure observed was for a 32-pin DIP hybrid memory stack package. These devices were from a third-party packaging house that combined memory die to produce a more highly integrated memory module. When inspected visually it appeared flimsy to the naked eye. All other PEM part and package types successfully passed the sequence of environmental tests. Since acoustic imaging showed substantial voids and cavities in some of the PEMs passing the environmental evaluation, the question that occurs is "how do these PEMs still function electrically within specifications?". We believe the answer lies with the excellent die passivation inside these plastic packages that serves as the last line of defense even if the package is breached by moisture.

In summary, the collective field experience from both commercial and military sectors finds PEMs manufactured with best commercial practice to be as reliable as their ceramic, hermetic counterparts. The major remaining question about the use of the plastic parts in military applications is the long-term dormant storage behavior of non-hermetic parts. With space systems, the period of dormancy is usually quite short compared with that for missiles, tanks, etc., and the environment during dormancy is well controlled.

13.3.1.2 Quality Control at Board and Box Level

The hardware fabrication controls include:

- Maintaining item identity by drawing number and serial number
- Monitoring fabrication processes

- Training and certification of personnel (e.g., NASA soldering standards are required)
- Stamp or signoff control system for inspectors
- Use of workmanship standards

When hardware is not being fabricated or tested, the assemblies are kept in controlled storage. It is desirable that completed assemblies be kept under Class 100,000 clean room conditions or better.

All flight hardware is inspected and tested, using calibrated test and measurement equipment and fixtures where special tooling is required. Specific procedures are written for system-level testing, and records of test and inspection are maintained. If anomalies occur, a system of discrepancy reporting and resolution is documented as described in chapter 1. If design changes are necessary, the appropriate engineering change notices must be generated. In some cases action by the material review board may be required. A material review board is composed of several colleagues who review test and screening data containing anomalies and reach agreement on the disposition of the components in question.

The goal of all this activity is to have the flight hardware fabrication, inspection, and testing governed by a system of controlling documents that can withstand an audit by representatives of the sponsor. The quality assurance personnel may conduct their own audits, particularly of subcontractors, and carry out vendor source inspections. One must ensure a quality product.

An activity that is allied with quality assurance is sometimes called performance assurance, which involves following the progress of the flight hardware and seeing that all significant fabrication activities are conducted according to pre-existing plans or standards, and that adequate documentation is being produced to deliver to the sponsor in support of the hardware's pedigree. Performance assurance activities include:

- Reviewing design specifications
- Reviewing test plans and procedures
- Supporting design reviews
- Gathering information for the acceptance data package
- Monitoring flowdown of requirements to subcontractors
- Participating in review board activities
- Monitoring procurement controls
- Preparing monthly status reports

The performance assurance engineer also serves as the configuration manager and participates in configuration review board activities. Chart D and table 13.3 define the different levels of configuration control and the types of hardware to which they

Table 13.3 Summary of configuration requirements

Hardware Unit	Drawing Level	Hardware Type
Flight model	2	A
Breadboard	1	C
Safety critical GSE	2	A
Selected GSE	2a or 2	B or A
Other GSE	1	C

GSE, ground support equipment.

apply. Most flight hardware is treated with Level 2 documentation for one-of-a-kind developments but with assured capability to reproduce the design, if necessary.

CHART D. CONFIGURATION CONTROL OUTLINE

Drawing Levels

- Level 1: Breadboard/brassboard development
  Informal drawings allowed
  Configuration control not possible
- Level 2a: Uses redlined control prints
  Limited capability to reproduce the design
  Will not support a configuration audit
- Level 2: Assured capability to reproduce the design
  Can provide spare parts to support the design
  Can verify correctness of hardware by documentation
  Prepared in accordance with DOD-STD-100
  Drawings stored in drawing vault after release
- Level 3: For quantity production
  Provides data to permit competitive procurement of items
  Allows for outside manufacture of hardware

Hardware Types (see table 13.3)

- Type A
- Type B
- Type C

Note: Principal differences between hardware types are in the activities of documentation preparation and verification, inspection and testing, and other quality. assurance activities

Finally, the quality/performance assurance activity produces

- Failure reports
- Failure analysis reports
- Parts and materials lists
- Material review board actions
- Safety and contamination plans
- Final configuration summaries

Some general guidelines for a contamination plan are shown in chart E. Requirements relating to volatility and mass loss of materials should be applied to ground support equipment as well as the spacecraft.

In general, a Class 100,000 clean room is used for spacecraft assembly and integration. However, in practice we have found that with appropriate ceiling to floor air flow and high-efficiency particle air (HEPA) filters we can achieve Class 10,000 without major modifications to the facility. This improved level of cleanliness meets the requirements of ISO 14644-1, Class 7, and FED STD 209. Typically, the science payload drives the contamination requirements, which should be defined at the concept stage, for each mission. For science instruments, especially imagers or cameras and UV spectrometers, a smaller Class 100 clean room is used for assembly and calibration. When integrated onto the spacecraft these instrument types are bagged and purged with dry nitrogen.

Filters are changed quarterly, floors are cleaned weekly, particulates are monitored continuously, and hydrocarbons are monitored periodically. Environmental monitoring, feedback, and maintenance create a semi-automatic clean room system. The clean room garb is laundered and tested weekly. For a "relaxed" Class 10,000 environment, personnel don, in this order, inner shoe covers, caps, coveralls, second outer shoe covers, wrist straps (to protect spacecraft electronics against electrostatic discharge), facemasks (for facial hair), and nitrile gloves.

CHART E. CONTAMINATION/CLEANLINESS

Spacecraft materials in accordance with NASA Publication 1124 on outgassing data

Total mass loss <1.0%
Collected volatile condensable materials <0.1%

Mirrors need Class 100 clean-room conditions
Bag and purge or seal necessary for mirrors and sensors during spacecraft integration
Start as clean as possible and bake out subsystems before integration

An important aspect of contamination control is clean room training and discipline. Formal presentations, including videotapes, give the rationale and background for certification of personnel. A walk through the clean room entrance and exit serves as a practical introduction at the start of the project.

A new and important issue in this arena is software quality assurance. Basically, one is trying to implement the same kind of discipline already present for flight hardware. Issues are maintenance of software engineering logbooks, firmware traceability, review of test results, use of check lists for task requirements, and configuration control to track changes.

13.3.2 Reviews

Probably, the most important vehicle for critiquing space system design is the sequence of design reviews that take place. Review boards are composed of experienced space system experts who know many of the pitfalls of designing flight hardware. Experts from all specialty areas, such as thermal and attitude control, compose the audience. Reviews (also discussed in chapter 1) usually occur in the following sequence:

A. Conceptual design review
B. Preliminary design review
C. Subsystem hardware design review
D. Critical design review
E. Pre-ship review

The conceptual design review (CoDR) presents the concept of how the mission requirements, often poorly defined at the time, will be accomplished. Useful information obtained from the review can often change the approach being used and usually leads to enough interchange among participants to cause better definition of requirements. The primary outcome of a CoDR is a better understanding of what is needed to accomplish the mission.

The preliminary design review (PDR) shows design detail down to the structure and circuit level. Layouts, schematics, and diagrams, showing details of the space system hardware, are presented. Block and box diagrams are expected; however, details down to the individual part level are not presented. Requirements should be quantitatively defined at this time. Weight and power estimates for each subsystem and the complete system are necessary. At PDR certain decisions about reliability, such as redundancy or fault tolerance, are made after review and discussion. Reviewers generally search for design details that have not been considered. Action items require designers to consider important consequences of these situations.

The subsystem hardware design review is usually an in-house review of the detailed hardware design by other members of the mission design team. The purpose of this review is to determine the readiness of a particular design for the critical design review (CDR). Complete details of the design down to individual electrical parts and screws are expected. The critiquing by peers is a good preparation for the program level CDR.

Critical design reviews determine whether the design team is ready to build hardware. The designs are presented in complete detail. Parts and materials lists should be available. This review is the last chance to make significant changes at minimum cost; that is, before existing hardware must be altered. Performance assurance plans and workmanship standards manuals should be available.

The pre-ship review usually concentrates on the performance of the hardware during environmental qualification testing. Test plans and results are presented which indicate that the hardware, fabricated according to the previously reviewed design, is ready to fly in space and meet the mission requirements. Both performance and housekeeping data are presented at the review.

Following this last review, the space system hardware is shipped to the launch site.

13.3.3 Launch Site Support

At the launch site the performance/quality assurance engineer is very concerned about safety, contamination, and adherence to the previously developed integration procedures. Here integration means interfacing the space system payload to the launch vehicle. Any discrepancies or last-minute changes must be thoroughly documented and approved. Last-minute anomalies or failure analysis must be expeditiously resolved. Environmental monitoring for contamination and safety purposes is carried out continuously. Difficult decisions about replacing a given subsystem with its spare must sometimes be made. If substantial repair is made to any subsystem, it must be environmentally requalified. Emphasis is placed on experience at this point, since there is no time left to gather large amounts of additional data. Additional evaluation time just delays the launch, which often has critical time windows.

13.3.4 Space System Reliability—NEAR Mission Case History

The Near Earth Asteroid Rendezvous (NEAR) spacecraft was the first mission under NASA's Discovery program. It had a twenty-seven month development time, a four-year cruise to the asteroid, and spent one year in orbit about the asteroid EROS. The spacecraft was successfully landed on EROS in February 2001 after its one year in orbit. Reliability was maximized by limiting the number of movable and deployable mechanical and electromechanical systems. This decision meant

- No deployable instrument booms
- Fixed RF communications antenna
- Deployable but fixed solar arrays
- Fixed instruments with movable filter wheels and deployable covers with viewing ports

New technology was used only when necessary to execute the mission. The four new developments were

- Gallium arsenide solar cell arrays
- Solid-state recorder
- Gamma ray spectrometer with sodium iodide detector crystal inside a BGO crystalline shield
- System of software autonomy rules

Purchasing and receiving functions were accelerated for parts procurement and the number of specialized purchase instructions was minimized. Incoming inspection and testing of flight parts was restricted to activities which added value. Weekly status reports from the fabrication department project leader tracked board design, drawing sign-off, board fabrication, and assembly for 118 flight boards. There were weekly small shop focus meetings with the program office, and a packaging engineer was assigned for each subsystem. A fabrication feasibility review (FFR) was held at the end of the board design phase.

Subcontractor technical problems experienced and solved were

1. DC/DC converters
   - Clamping diodes used to eliminate overshoot
   - Vendor's fabrication process changed to eliminate capacitor damage
2. 4 MEG X 4 DRAMS
   - Life tests successfully run to address high standby current concerns
3. Oscillators
   - Replacement of telemetry control unit (TCU) oscillator due to excessive noise and instability
   - Correction of unterminated CMOS input in the command and telemetry processor (CTP) oscillator
4. Magnetometer electronics
   - Replacement of more than twenty FETs due to flux contamination in original fabrication
5. Infrared spectrometer pin puller
   - Refurbishment of mating surfaces
6. High-gain antenna
   - Stripped, cleaned, and repainted
7. Laser range finder
   - Rework of optics train mechanical mounts that used epoxy adhesive for attachment

Thirteen waivers had to be issued to meet the accelerated schedule. The details of two major subsystems are described. The inertial management unit (IMU) had four problems:

- Four gyros not practically useable simultaneously; NEAR only used three
- High vibration sensitivity; IMU remained off during launch and was not disturbed by onboard thrusters
- Six of twenty performance parameters; two were compensated by onboard software; the other four were marginal and accepted
- Post-vibration shift in gyro alignment axes; latest values placed in IMU EEPROM

The qualification solar panel experienced an interconnect breakage problem. The silver web rigid metal lattice interconnect system had low resistivity, but was a limited life item, as demonstrated by temperature cycling (in air and vacuum) of test coupons and the qualification panel. Therefore, the range of qualification temperature profile was reduced by 10°C and the number of thermal cycles with excursions exceeding $\Delta T = 30°C$ was limited during the entire mission.

Upon the start of spacecraft integration, problems are controlled by a problem/failure report (PFR) system; 227 such documents were created before launch. Results from a statistical analysis of the PFRs indicated that the main effects, significant at a 95% confidence level, were

- The Guidance and control (G&C) subsystem (60) had a significantly higher number of anomalies but the propulsion subsystem (5) had fewer.
- There were significantly more software anomalies (129) than hardware (98) ones, particularly for the G&C software that was designed very late in the spacecraft development.
- More problems were caused by design (128) and fewer by workmanship (37) and parts (9), vindicating the parts screening philosophy.
- The less mature design of doors with opening mechanisms for the science instruments also created several PFRs.

The lessons learned from the NEAR development were

(1) Autonomy rules should be allowed to operate on computed values; CTP (command and telemetry processor) autonomy rules in EEPROM should be reprogrammable.
(2) Maximum uplink bit rate and high-gain antenna should be used to expedite massive software uploads; for spacecraft integration and test on ground, all units with significant amounts of reprogrammable software should have their own test port.
(3) The spacecraft should have switched power on redundant command receivers so that both can never be off simultaneously; autonomy can be used to switch receivers if necessary.
(4) Length of telemetry transfer frames should be variable so that shorter ones with only necessary information can be used in emergency modes.
(5) A complete spacecraft simulator is needed to test and debug safing software; tests need to simulate true flight conditions with respect to downlink data rates and round trip light time delays; there is a need for enough trained test personnel to staff several shifts.

The main things to do better next time are to follow a much more disciplined approach to software development, including its quality control, and to fund and develop the tools and platforms to provide accurate simulation of integrated flight tasks in the laboratory, concurrent with hardware fabrication.

13.4 Radiation

The one outstanding element that distinguishes the space environment is the presence of radiation. For the purposes of this discussion we will primarily confine ourselves to the natural space radiation. The enhanced or weapons environment is usually a short-lived one in space but some of the considerations are similar with respect to radiation effects on electronics.

The natural environment consists of electrons and protons trapped by planetary magnetic fields (Earth, Jupiter), protons and a very small fraction of heavier nuclei produced in energetic solar events, and cosmic rays (very energetic atomic nuclei) produced in supernova explosions within and without our galaxy. Inside large spacecraft structures such as the International Space Station (ISS), the primary cosmic beam of approximately 85% protons and 15% heavy nuclei is partially converted into secondary neutrons by collisions with the tens of grams per square centimeter (g/cm^2) of material. These secondary neutrons can present an additional threat via single-event effects in electronics. Weapons environments add neutrons, gamma rays, and X-rays to the natural background.

Some of these charged particles have sufficient energy to penetrate spacecraft and their electronics. Semiconductors are also semi-insulators from another point of view. When charged particles produce ionization tracks of electrons and holes in insulating material, this deposited ionization charge causes changes in the electrical conditions under which the solid-state device is operating. When enough charge accumulates, the threshold voltage shift or leakage current may cause the device to cease functioning under normal design conditions. Most of the damage from charged particles is due to ionization; however, when electrons and protons hit solar cells, displacement damage (dislodging atoms from their normal lattice sites) is the dominant mechanism.

Table 13.4 shows the terminology and units important to radiation exposure. Table 13.5 shows the number of electron–hole pairs produced in common semiconductor materials by 1 rad of ionizing radiation. Chart F summarizes the characteristics of the space environment.

Table 13.4 Important terminology and units of radiation exposure

Type of Radiation Exposure	Units of Measure
Flux	$particles/cm^2 s$
Fluence	$particles/cm^2$
Energy spectrum	$particles/cm^2 MeV$
Neutrons	
Fluence	n/cm^2
1 MeV equivalent fluence	n/cm^2
Ionizing radiation	
Stopping power (linear energy transfer function $(1/p)(dE/dx)$)	$MeV/(g/cm^2)$
Total radiation absorbed dose	rad[a]
Ionizing dose rate	rad(Si)/s

Source: McLean (1987).
[a] 1 rad(Si) = 100 ergs/g(Si) = 0.01 J/kg(Si); SI unit: I gray (Gy) = 100 rad = 1 J/kg.

Table 13.5 Electron–hole pair generation energies and pair densities generated by 1 rad

Material	Pair Generation Energy, E(eV)	Pair Density Generated per rad (pairs/cm^3)
Silicon	3.6	4.0×10^{13}
Silicon dioxide	17	8.1×10^{12}
Gallium arsenide	~4.8	$\sim 7 \times 10^{13}$
Germanium	2.8	1.2×10^{14}

Source: McLean (1987).

Models of the space radiation environment are used by the radiation effects engineer to establish the requirements for each spacecraft or instrument mission. For low Earth orbit (LEO) or geosynchronous orbit (GEO) missions, the protons and electrons trapped in the earth's magnetic belts are estimated using Goddard models AP-8 and AE-8 respectively. Solar proton fluence models developed at JPL in the early 1990s, and their successors, are used to predict levels of energetic protons during solar cycle maximum years. An important outcome of the JPL study is that there are no significant fluxes of energetic solar protons during the four minimum years of the 11-year solar cycle. CRÈME 86 and its update CRÈME 96 are used to assess the cosmic ray fluxes for both solar maximum and minimum periods and for disturbed and quiet epochs. A very useful web site for space radiation environment models is http://www.spenvis.oma.be/spenvis/, which is maintained and updated by ESA (European Space Agency).

CHART F. CHARACTERISTICS OF SPACE RADIATION ENVIRONMENT

Environment
- Low ionization dose rate ($\ll 1$ rad/s)
- Total dose ($\geqslant 10^5$ rad)
- High-energy electrons and protons trapped in Earth's magnetosphere
- Cosmic rays (electrons, protons, alphas, heavy ions)
- Solar protons (dose rate increases by 10^4 during large solar event)

Primary failure machanisms
- Total-dose-induced charge buildup
- Single-event upset

Test simulators
- Low-dose-rate ionization sources: ^{60}Co, low-energy X-ray tester
- High-energy particle sources: proton and heavy ion beams

(Source: McLean, 1987.)

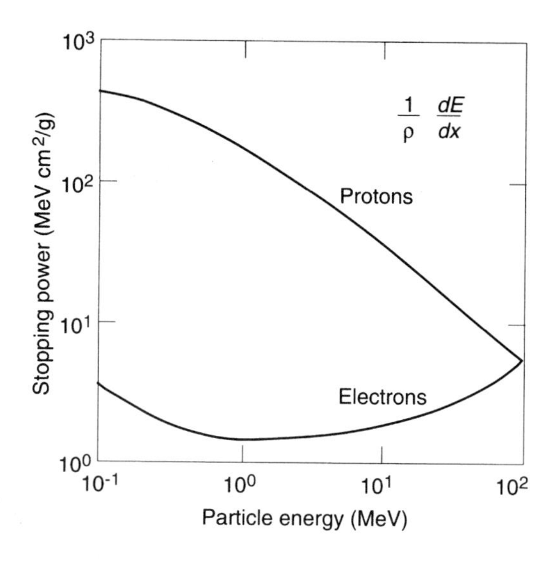

Figure 13.12 Stopping power as a function of particle energy for electrons and protons incident on silicon. (Source: McLean, 1987.)

13.4.1 Ionization Effects

13.4.1.1 Total Dose

For charged particles, the amount of energy that goes into ionization is given by the stopping power or linear energy transfer function, commonly expressed in units of MeV cm^2/g. Stopping powers for protons and electrons in numerous materials have been tabulated. Figure 13.12 shows the stopping power for electrons and protons incident on silicon. The absorbed ionizing dose is the integral of the product of the particle energy spectrum and the stopping power. The commonly used unit of absorbed ionizing dose is the rad, which is equal to an absorbed energy of 100 ergs per gram of material. Because the energy loss per unit mass differs from one material to another, the material in which the dose is deposited is always specified, hence rad (S) or rad (GaAs). The standard international (SI) unit for dose is the gray, equivalent to 100 rads.

When an electron–hole pair is created by incident radiation, the electron in the valence band is excited across the bandgap into a conduction band state, leaving a hole behind in the valence band. If an electric field is present, the electrons are readily swept away as their mobility in silicon is much greater than that of the holes. Except for some small fraction of pairs that undergo recombination immediately, the created electrons and holes are free to drift and diffuse in the material until they undergo recombination or are trapped. Insulators such as silicon dioxide (gate and field oxides in transistors and integrated circuits) contain large densities of trapping centers in which the radiation-induced charge can reside for long periods of time. These trapped charges generate internal space-charge electric fields that lead to voltage offsets or shifts in device operating characteristics. Sufficient space charge fields may cause device failure. Another site for trapping of unwanted charge in integrated circuits is at the interface or discontinuity between different materials such as silicon and silicon dioxide.

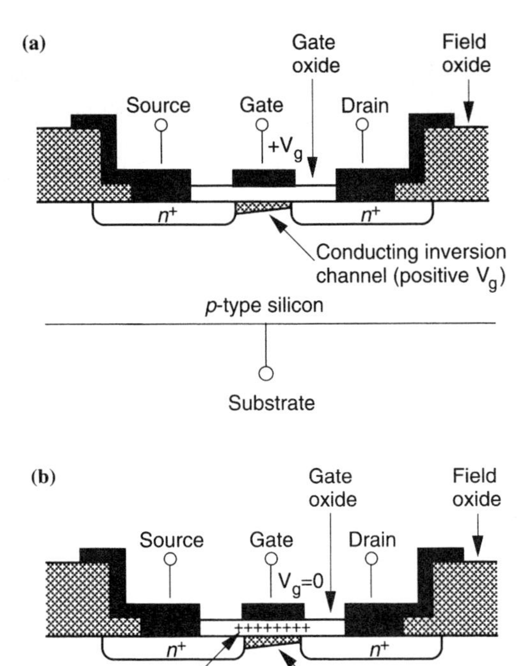

Figure 13.13 Schematic of n-channel MOSFET, illustrating the basic effect of total-dose-ionization-induced charging of gate oxide: (a) normal operation; (b) post-irradiation. (Source: McLean, 1987.)

Figure 13.13a shows a simple schematic of a MOSFET, in this case an n-channel device using a p-type Si substrate. When a bias potential is applied to the gate contact, there will be an electric field across the gate oxide region and into the Si surface region immediately below the gate region. If the gate bias is sufficiently large and positive (for n-channel operation), the majority carriers (holes in p-type SiO_2) will be repelled from or depleted in this surface region, and minority carriers (electrons) will be attracted to this region, forming what is called an inversion layer. If a potential difference is applied between the source and drain contacts (n^+-doped regions in figure 13.13), the inversion layer provides a low-resistance current channel for electrons to flow from the source to the drain. The device is then said to be turned on (figure 13.13a), and the control gate bias potential at which the channel just begins to conduct appreciable current is called the turn-on voltage or threshold voltage of the device.

The total-dose ionization problem that occurs in this structure is then due to the radiation-induced charging (normally positive) of the thin gate oxide region, which generates additional space-charge fields at the Si surface. These additional induced fields result in voltage offsets or shifts in the turn-on voltages of the devices, which in turn lead to circuit degradation and failure. For example, for sufficiently large amounts of trapped positive charge for the device schematically shown in figure 13.13, the device

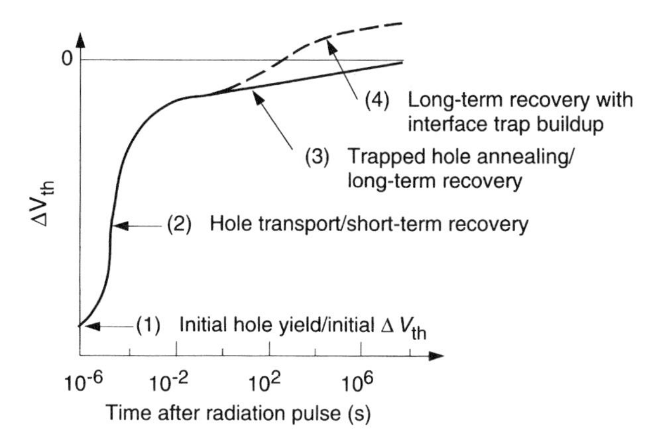

Figure 13.14 Schematic time-dependent threshold-voltage recovery of n-channel MOSFET following pulse irradiation, relating major features of response to underlying physical processes. (Source: McLean, 1987.)

may be turned on even for zero applied gate bias (figure 13.13b). The trapped positive holes have effectively shifted the applied voltage threshold in the MOSFET from about 3.3 V to zero. Functionally, if such a device were used as a switch with a square-wave pulse between 0 and +10 V applied to the gate, the MOSFET would no longer be a switch. Although this effect seems quite simple from this overview, the time history of the individual events combining to produce the result is quite complex.

Figure 13.14 is representative of the composite response of an actual hardened n-channel device. The initial maximum negative shift (10^{-6} s) is determined mainly by the electron–hole pair creation in the SiO_2 bulk and by initial recombination processes. The short-term annealing shown out to 10^{-2} s is due to the hole transport process, with the remaining negative shift at this time due to deep hole trapping near the Si/SiO_2 interface. The trapped holes anneal very slowly out to 10^8 s.

The solid curve in figure 13.14 corresponds to transport, trapping, and annealing of holes alone. In addition to long-term annealing of trapped holes, however, a buildup of radiation-induced interface traps may occur, typically in the time regime between 10^{-2} and 10^3 s, which is indicated by the dashed curve in figure 13.14. (For an n-channel device, the interface-state contribution to V_{th} is positive, corresponding to net negative interface trap charge at the threshold voltage.) If the interface trap contribution is relatively large, the threshold voltage may actually recover past its pre-irradiation value (going positive), giving rise to what is called super-recovery or rebound. This effect can also lead to circuit failure, if sufficiently large. If there is a relatively large component of prompt interface trap production, then there would simply be an additional upward translation of the solid and/or dashed curves for all time. Again, because of the annealing of trapped positive charge, super-recovery or rebound could occur.

The time history of the MOS device response to total doses of irradiation has important implications for sensible testing procedures, hardness assurance considerations, and predictions of device performance in the space environment. Since it is impractical to test at typical space environment dose rates, and since the annealing of the devices is quite important in determining the value of measured electrical parameters at the possible times of measurement, a new method of total dose testing has been devised.

Older methods just involved irradiating test samples in steps or continuously at a high dose rate to the appropriate dose level and reading electrical responses within 30 to 60 min of removal from the radiation chamber. These methods emphasize the short-term effects of oxide-trapped charge but do not adequately simulate the effects of interface states which may be dominant at long times (days). Various annealing techniques are being tried to adequately reproduce both trapped charge annealing and interface state production. Space dose rates for low Earth orbit (LEO), geosynchronous orbit (GEO), and interplanetary space missions are in the range of 10^{-3} to 10^{-7} rad/s, ignoring the contribution of any solar flares.

Since the early 1990s an enhanced low dose rate sensitivity (ELDRS) has been observed to cause increased total dose degradation in newly designed bipolar integrated circuits. Certain bipolar junction transistors degrade more when exposed to ionizing radiation at a low dose rate than at a high dose rate for the same accumulated dose level. ELDRS is caused by space charge effects in bipolar device emitter/base oxides (inserted into bipolar devices to reduce power consumption and improve performance) and in passivating and spacing or isolating oxides in MOS devices. Actually, the effect illuminates the fact that a high dose rate creates a significant space charge that suppresses radiation sensitivity to large amounts of localized and deposited charge in finite time intervals. The space charge initially trapped suppresses carrier and ion transport in the device due to the later deposited dose. The oxides or films used for passivation are typically 1 μm thick and have a higher level of impurities to act as trapping sites than the oxides required for gate dielectrics. To observe the peak effect of ELDRS, one should test devices for the space environment at a dose rate of 0.01 rad/s. However, this dose rate is impractical if many devices must be evaluated for a space program to a total dose of 10,000 or more rads. Thus, recent investigations into test methodology recommend a dose rate of 10 rad (Si)/s, with a design margin of a factor of three on the total dose, followed by a week of annealing at 100°C to assess recovery (Holmes-Siedle and Adams, 2002).

13.4.2 Displacement Damage

Displacement damage occurs in the space environment when protons or electrons with high enough energy (>1 MeV for electrons, >10^2 keV for protons) interact with a target atom in a crystalline lattice. Damage occurs in such lattices due to elastic two-body collisions, electronic interactions with atomic electrons (for charged particles), and nuclear interactions. Since silicon and many other materials have thresholds of about 25 eV for the creation of vacancy/interstitial pairs (see figure 13.15), considerable damage can be done by both primary particles and recoiling target atoms.

The billiard ball-type collisions (really elastic nuclear scattering) and electronic excitations take place for incident charged particles (electrons and protons) at energies of 1 MeV or less. Above 8 MeV (the average binding energy per nucleon for elements of mass number greater than 20—silicon has mass number 28), protons can also cause inelastic nuclear interactions in which energies considerably above 25 eV can be transferred to the lattice. Neutrons, carrying no electronic charge, can only cause damage by elastic or inelastic (above 8 MeV) nuclear collisions. The charged particles such as protons prefer to lose their energy through electronic processes. For example, a 10 MeV proton loses about 1000 times more energy due to electronic processes than energy due to nuclear processes.

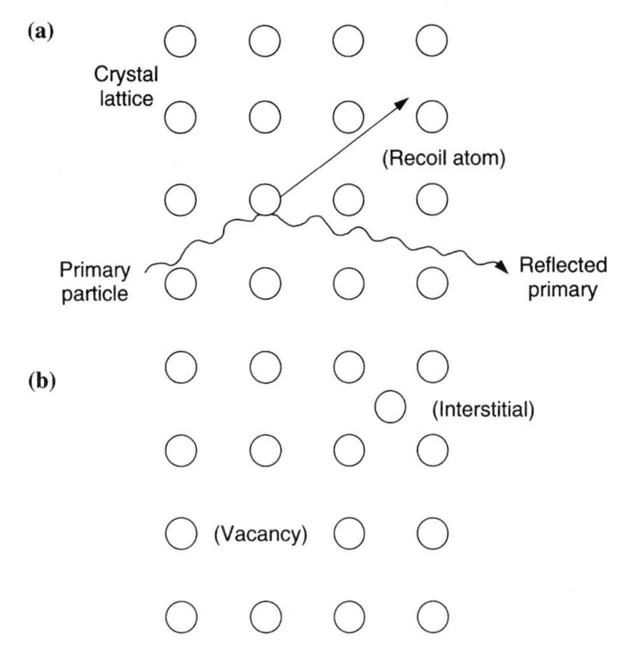

Figure 13.15 Schematic of atomic displacement damage in crystalline solid: (a) atomic displacement event; (b) simple radiation-induced defects (vacancy and interstitial). Atom displacements produce lattice defects that result in localized trap states–energy levels within bandgap. Measure of damage is the effect on a pertinent electrical parameter (e.g. minority carrier lifetime or transistor gain). (Source: McLean, 1987.)

Because of its preference for losing energy in electronic interactions, a proton and associated charged secondary particles can produce more damage in a localized area of a lattice. However, a neutron has much greater range in a material (primarily because it is not charged) and will ultimately cause more damage throughout the bulk of the material.

A discussion of the kinds of defects and defect clusters produced in semiconductor material is beyond the scope of this chapter. We will mention briefly how these defects affect the electrical properties of semiconductors. In general, any defect disrupts the periodicity of a crystalline lattice, producing localized "corrupting" states in the band-gaps of the semiconductor. Following McLean (1987), we will list the six main effects of the band-gap "corruption":

(1) Levels near midgap serve as thermal generation centers for electron–hole pairs, leading to increased dark current (noise with no stimulus).
(2) Localized states near midgap serve as recombination centers that shorten minority carrier lifetime, thereby reducing gain in bipolar transistors.
(3) Shallow traps near band edges can temporarily trap and re-emit charge, reducing charge transfer efficiency in CCDs.
(4) Deep radiation-induced traps compensate the majority carriers, causing carrier removal.

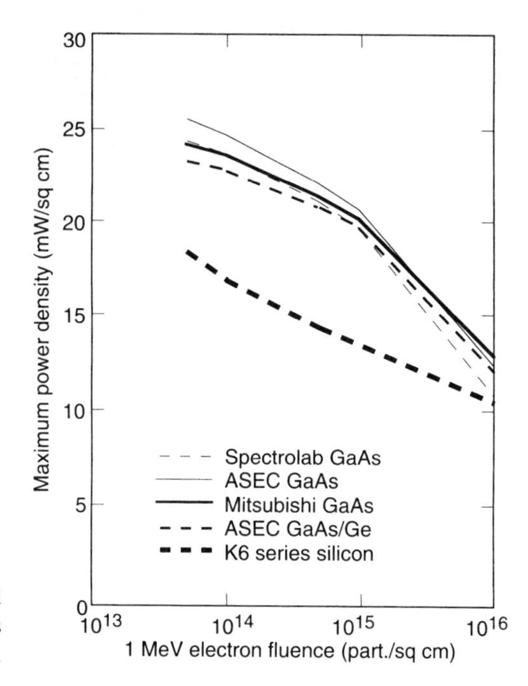

Figure 13.16 Solar cell maximum power density versus electron fluence.

(5) Trap-assisted tunneling leads to increased junction leakage current.
(6) Defects themselves act as scattering centers, decreasing carrier mobility.

Space systems most subject to permanent damage effects are the solar cells that produce power via photovoltaic processes, and the particle detectors and electromagnetic spectrum sensors such as charge-coupled devices (CCDs). These last two types of device make up the front ends of instruments designed for scientific or military missions as well as star trackers.

Figures 13.16 and 13.17 show the degradation of maximum power density versus 1 MeV electron fluence and 10 MeV proton fluence, respectively. Standard silicon cells degrade quasi-linearly on this semi-log plot, whereas gallium arsenide cells undergo a change in slope, degrading faster at higher fluence levels. Figure 13.18 shows the influence of junction depth and cell thickness on the degradation in relative maximum power for gallium arsenide solar cells. The effect of junction depth proves to be dominant (Herbert et al., 1989).

Subsequently, we made a thorough analysis (Maurer et al., 1989) of electron- and proton damaged GaAs solar cells. We found that although some electrical parametric and spectral response data were quite similar, the type of damage due to the two types of radiation was different. A current–voltage characteristic analysis model showed that electrons damaged the bulk of the cell and its currents relatively more, whereas protons damaged the junction of the cell and its voltages more.

To quote Janesick et al. (1989), "the radiation Achilles heel for present scientific CCDs appears to be bulk damage, especially in low signal applications." The onset of charge transfer efficiency (CTE) degradation due to high-energy electrons can occur at

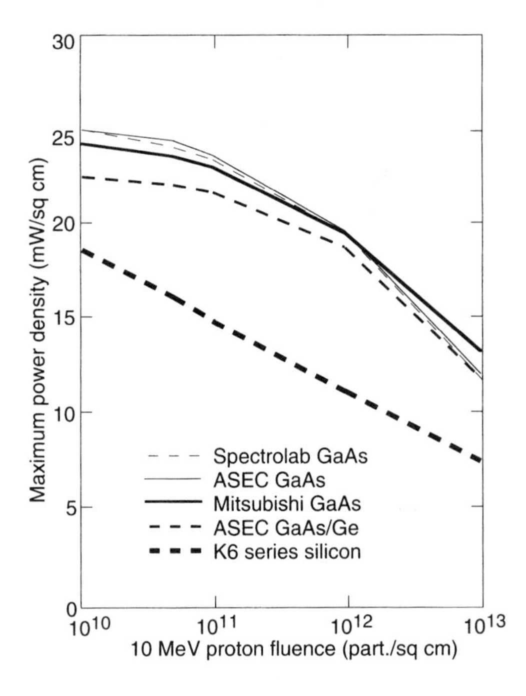

Figure 13.17 Solar cell maximum power density versus proton fluence.

fluences that produce doses as low as 1 krad (Si). Dark current spikes are produced in virtual phase CCDs exposed to 65 MeV protons. The levels of these dark current spikes are quantized, with some pixels having more damage sites than others; the maximum spike amplitudes are $> 3 \times$ pre-irradiation levels.

Srour et al. (1986) did very extensive experimental work on a Texas Instruments TC104 3456-element linear image sensor. The average pre-irradiation dark current density for individual cells was 0.1 nA/cm^2 at 303 K. The highest fluence level used for 99 and 147 MeV protons was 3×10^9 protons/cm^2, which produced 1432 and 884 damaged cells in the CCD, respectively. Cells exhibiting very large changes in dark current density are assumed to have experienced multiple interactions. Increasing proton fluence increases the number of damaged pixels, and cells with relatively large amounts of damage become more prominent. At the highest fluences for both 99 and 147 MeV protons, several cells appear to have been damaged by heavy fission fragments. Changes in dark current as great as 24–28 nA/cm^2 were observed after 3×10^9 p/cm^2 at both energies. The authors calculated the mean increase in dark current density per proton-silicon interaction, using a Poisson distribution. The results were 0.26 nA/cm^2 at 99 MeV (1849 interactions) and 0.34 nA/cm^2 at 147 MeV (1021 interactions). Previous work by the same authors had determined an average dark current increase of 1.17 nA/cm^2 per interaction for 14 MeV neutrons.

A practical method of evaluating CCDs for flight application is to damage them in steps at a proton accelerator facility and monitor dark current and image quality periodically outside the accelerator cave. Once the test hardware and procedures are developed, it is a simple matter to repeat the evaluations on flight lots when they are received. The best space systems mitigation of the displacement damage effects on

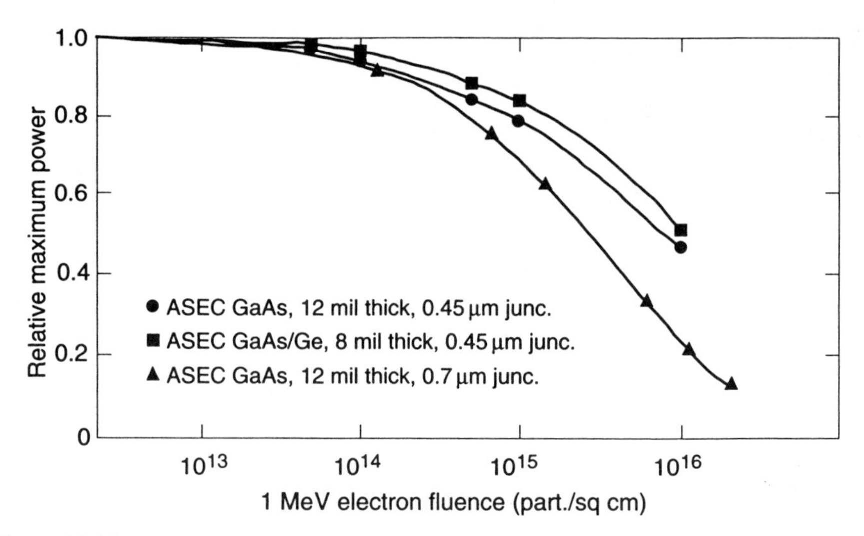

Figure 13.18 Relative maximum power versus electron fluence for GaAs solar cells of different thickness and junction depth. (Source: Herbert et al., 1989.)

CCDs is to operate them at a cold temperature. The CCDs can be kept cold by attaching a thermoelectric cooler.

13.4.3 Single-Event Effects

A single-event upset (SEU) is a temporary or permanent disturbance of an integrated circuit caused by the passage of a single high-energy particle, usually a heavy ion, through an electrical node of the die. A transient upset is a change of the logic state, as a 0 becomes a 1 or vice versa, that is ordinarily called a bit flip or soft error. This type of error is temporary and disappears when the location of the event is reused.

Latchup can be a permanent and destructive consequence of a single-event upset. Figure 13.19 illustrates the latchup mechanism in CMOS memory devices. The charge or current injected by single particle ionization yields a lower breakover voltage in analogy to a silicon-controlled rectifier (SCR) in which a high current latched state results. When this current path connects Vdd to ground in a CMOS device, burnout can result if no current limiting exists. The parasitic bipolar path, pnpn, results from CMOS fabrication in which a p-well resides on a n-substrate and has both n and p channel structures embedded in it. If the product of the gains of the back-to-back pnp and npn transistors is greater than one, the latched state is self-sustaining until the device burns out or until the power is turned off. If a device to be flown is susceptible to latchup, protection must be provided in the form of current-limiting resistors so that burnout does not occur, and one must be able to turn off the power to the device to erase the latched state. Obviously, this latter remedy is undesirable for something as critical as a satellite command system.

A third single-event effect is the burnout of power MOS transistors caused by the traversal of a heavy ion through the high-field drain depletion region of the FET. The charge collected from the particle ionization track can often produce sufficient voltage to forward bias the base–emitter junction of the parasitic bipolar transistor occurring in

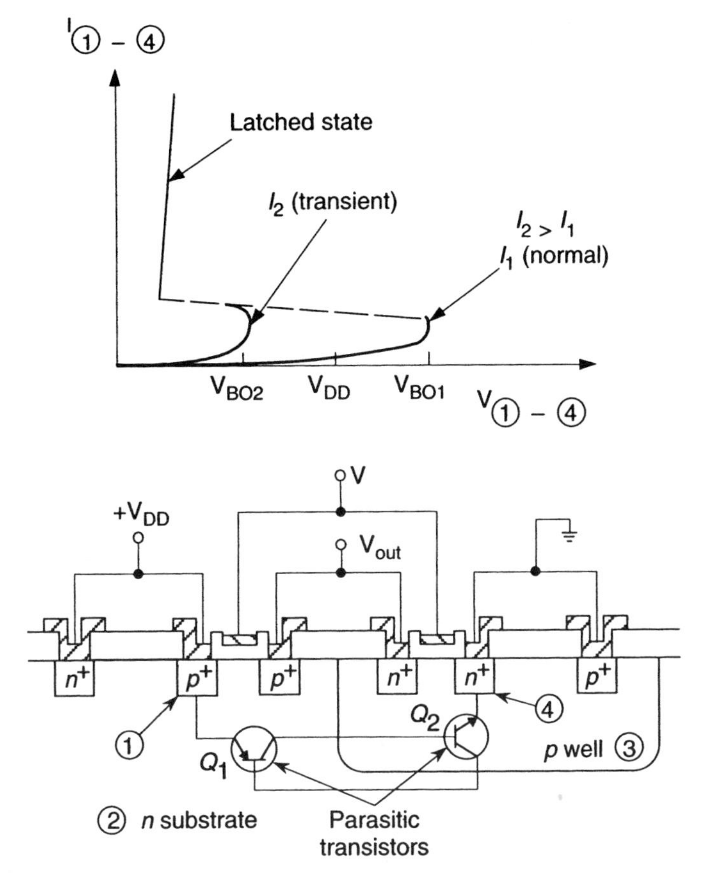

Figure 13.19 Illustration of CMOS latchup mechanism and analogy to SCR. Parasitic a path results from CMOS fabrication. When $\alpha(Q_1) + \alpha(Q_2) \geq 1$ a stable low impedance state occurs. α can be increased by current injected from single particle ionization, yielding smaller V_{BO}. When V_{BO} is reduced to less than V_{DD}, latchup occurs. (Source: Pickel, 1983.)

power MOSFET structures. If the applied drain–source voltage of the MOSFET is greater than the collector–emitter breakdown voltage of the parasitic transistor, an avalanche condition occurs, leading to high currents and local power dissipation and, finally, device burnout (Waskiewicz et al., 1986; Hohl and Galloway, 1987). The vertical structure of an n-channel power MOSFET introduces a parasitic bipolar transistor that is normally off but can be turned on by the charge introduced by a heavy ion strike. In an analog to latchup in bulk CMOS ICs, heavy current conduction leads to device burnout. A related phenomenon is single event gate rupture (SEGR), in which just the gate oxide is struck by a heavy ion and the resulting surge of current through the gate ruptures the oxide.

A concise history of single-event effects starts with the first prediction of upsets by Wallmark and Marcus (1962), a work ahead of its time in pointing out that upsets would occur as digital devices became smaller. Binder et al. (1975) identified upsets in flip-flop circuits in the space environment. In 1978, May and Woods (1979) proved that upsets in newly developed 4K memories were caused by alpha particles emanating from traces of

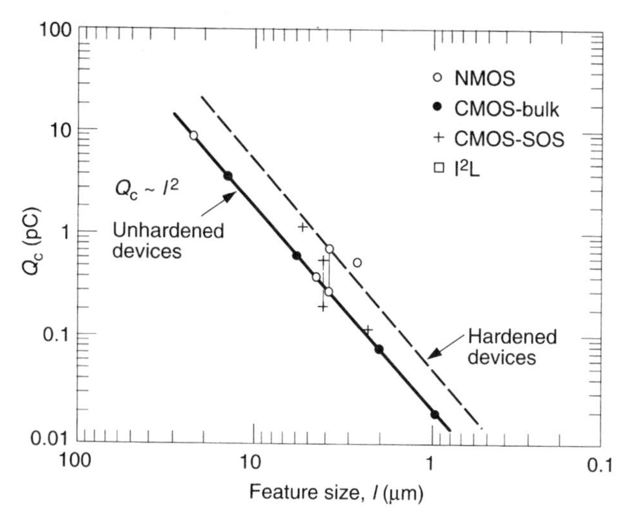

Figure 13.20 Critical charge for single-event upsets. (Source: Peterson, 1983.)

elements such as thorium in the device's ceramic packaging. Pickel and Blandford (1978) introduced the concept of critical charge and examined the sensitive nodes susceptible to upset in an analysis of dynamic memory upsets. Kolasinski et al. (1979) then succeeded in producing single-event effects, upsets, and latch-up, by using a high-energy particle accelerator to simulate cosmic-ray ions. During the 1980s upsets have been experienced on many satellites in a large number of device technologies: TTL, NMOS, CMOS, advanced bipolar. In addition, bulk CMOS devices have an architecture that makes them susceptible to single-particle induced latchup.

The recent miniaturization of integrated circuits (Peterson, 1983) has resulted in a crossing of a threshold at which single particles can deposit enough charge to cause upset. High speed and complexity in integrated circuits have been accomplished by decreasing both cell size and bit energy. As the number of gates per device increases and the total power on the chip is minimized, the energy per gate must decrease. In 1970, switching energy was 50 pJ, corresponding to 300 MeV. In 1975 it was down to 5 pJ or 30 MeV, an amount easily deposited by even relatively low LET particles.

For regular structures such as memories, in which a given cell is repeated many times, it is useful to define a concept called the critical charge, which is the charge necessary in the minimum pulse to cause an error in a device. This charge, Q_{c}, depends on the energy loss of the particle (linear energy transfer, LET, or linear charge deposition rate) and the pathlength in the sensitive volume of the memory cell. The reduced feature size of modern integrated circuits causes the critical charge to diminish proportionately. However, the combination of reduced dimensions and reduced critical charge indicates that the single-event error rate may remain fairly constant for very highly integrated circuits since the new smaller sensitive memory nodes are also harder to hit. Figure 13.20 shows the critical charge plotted as a function of feature size for a number of different semiconductor technologies. The critical charge appears to follow the simple scaling rule $Q_{\mathrm{c}} \approx \ell^2$, where ℓ is the feature size. The vertical bars connected to the points show roughly the range of critical charges that occur due to a process change that produces a hardened device.

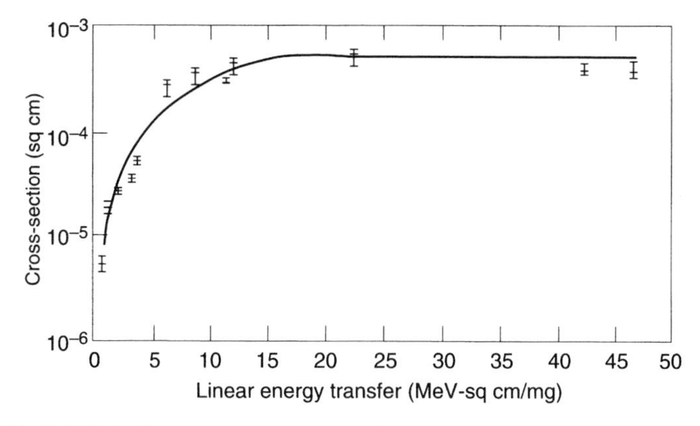

Figure 13.21 80186 single-event upset cross-section.

The convolution integral of a given space environment—geosynchronous is often used as a benchmark—with the upset cross-section, with both a function of linear energy transfer, gives a bit error rate (errors/bit day) or a device error rate (errors/chip day) for an integrated circuit in the specified environment. The calculation is done by evaluating the integral

$$N = \int_{L_{\min}}^{L_{\max}} \frac{dF(L)}{dL} \quad S(L)dL, \tag{13.4.1}$$

where N is the number of upsets per day, $dF/\,dL$ is the omnidirectional differential LET flux, S is the upset cross-section, also a function of LET, L is shorthand for LET, $L_{\min}$ is the threshold for upset for a device, and $L_{\max}$ is the effective cutoff LET of the natural environment. This integral is done numerically because simple analytic functions for $S(L)$ and $F(L)$ are difficult to define over the complete range of LET.

Figure 13.21 shows the upset cross-section for an Intel 80186 16-bit microprocessor as a function of LET. Figure 13.22 shows the heavy ion flux expected to be experienced by

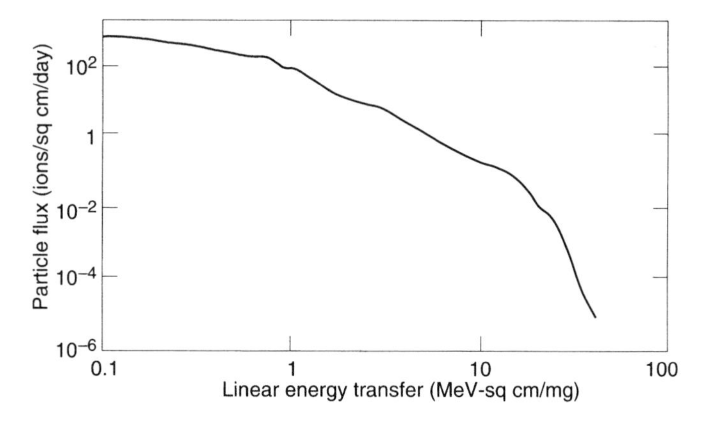

Figure 13.22 Heavy ion flux for 200 mils of aluminum.

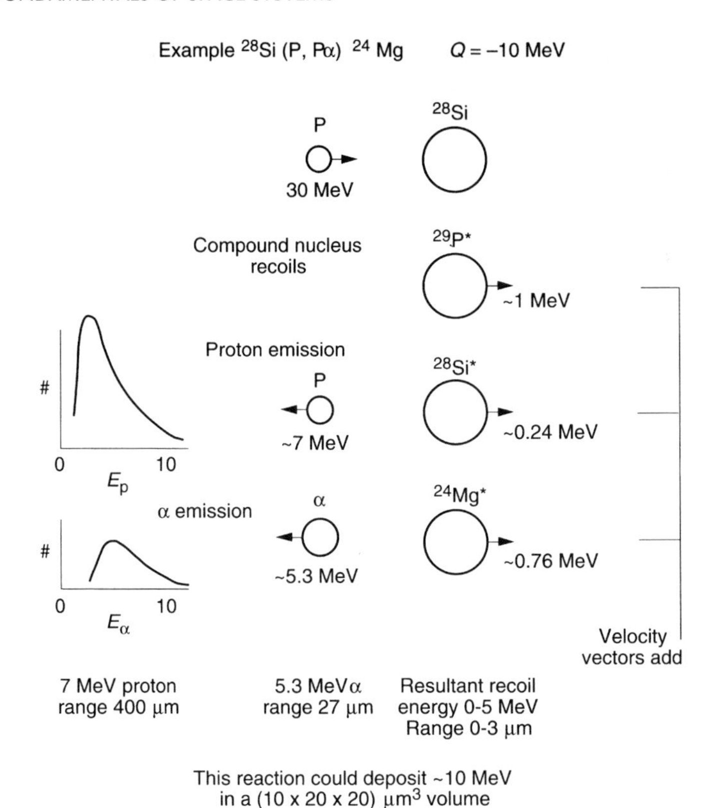

Figure 13.23 Nuclear reactions in silicon. (Source: Peterson, 1983.)

this microprocessor in a low Earth orbit (LEO) mission at a depth of 200 mils aluminum or approximately the location near the center of the spacecraft for such a sensitive device (Maurer et al., 1988).

Another means of depositing charge in sensitive regions of electronic devices, in addition to direct ionization by charged particles, is by the nuclear reaction in silicon induced by protons, which are quite numerous in the near-Earth space environment but which do not have high enough LET values to cause upsets by direct ionization, or by neutrons created by a thick space structure. Figure 13.23 shows a common nuclear reaction created by protons hitting silicon nuclei. The recoil magnesium nucleus has a very short range and high LET in silicon, depositing all of its energy as electron–hole pairs in 3 μm or less. Neutrons create similar collisions and recoils and penetrate the complete bulk of the semiconductor.

A rule of thumb is that devices that have LET thresholds for single-event upset by heavy ions at 6 MeV cm ^{2}/mg or less will also be susceptible to upsets induced by proton nuclear reactions in silicon. If a device falls into this category the upset rates for the two basic mechanisms must be combined in calculating the total upset rate estimate.

One way to reduce upset rates in the space environment is to choose an integrated circuit technology that has smaller charge collection regions. Single-event upsets are produced by the ionization of a single particle, either from primary heavy cosmic ray ions or from secondary recoil ions created by incident protons. As the particles pass through the semiconductor material they lose energy by ionization; this energy loss has units of MeV/μm or MeV/(g/cm^2), with the rate of loss designated as dE/dx. The ionization creates electron–hole pairs, with 3.6 eV needed to produce a single such pair in silicon. As an example, 3.6 MeV energy loss in a device will produce 10^6 electrons or 0.16 pC of charge. Modern semiconductor memories store information using charge in the range from 0.001 to 1.0 pC; heavy ions can easily generate such charge in silicon.

The stopping power or rate of energy loss, dE/dx, for a charged particle passing through matter can be expressed approximately by

$$dE/dx = f(E)MZ^2/E, \tag{13.4.2}$$

where x is the distance traveled in units of mass per unit area, f(E) is a very slowly varying function of energy, M is the mass of the ionizing particle, and Z is the charge of the ionizing particle.

Thus, for a given energy, the greater the mass and charge of the incident particle, the greater the amount of deposited charge or energy produced over a path length inside the IC. The intensity of heavy cosmic rays as a function of Z peaks at iron ($Z = 26$), abruptly decreasing thereafter. A very energetic 1 GeV iron nucleus will deposit about 0.14 pC in each micrometer of silicon traversed (Peterson, 1983).

Charges deposited by primary or secondary ions are collected in the high electric field depletion region and possibly in a temporary extension of the depletion region, called a funnel. When this charge is collected at a sensitive node, for example an n-channel transistor currently in an "off" state in a CMOS static RAM, the current pulse is transmitted to the other half of the flip-flop, in this example the "off" state p-channel transistor, appearing like a normal pulse and causing the flip-flop to change state if the charge is greater than the critical charge for such a change. If the bit so affected was originally zero, it is now one, and information has been changed. This situation is the basic single-event device physics for all CMOS static circuits.

Figure 13.24 (Pickel, 1983) shows a comparison of the charge collection regions for transistors in both bulk CMOS and CMOS/SOS technologies. The reduced charge collection volumes for an n-channel FET in the silicon-on-sapphire (SOS) technology are evident. Memories fabricated in CMOS/SOS technology are generally immune to latchup and less sensitive to transient upsets by one or two orders of magnitude than a comparable bulk CMOS memory. A more modern utilization of this concept is in silicon-on-insulator (SOI) technology.

Although gallium arsenide has proved to be quite resistant to total dose effects (compound semiconductor technologies have no insulating oxides to trap charge), it is not a solution for single-event upset effects. The reason for this is that the small node capacitance of highly integrated GaAs devices and the small voltage swings in digital devices, which permit the high switching speeds for GaAs, combine to yield small critical charges ($Q = CV$) at memory storage registers. Thus, nodes without charge can be charged by heavy ions, causing bit flips.

13.4.4 Solutions to Radiation Effects Problems

As already indicated, the best solutions to radiation effects problems caused by the natural environment are to design and fabricate radiation-hardened or radiation-tolerant

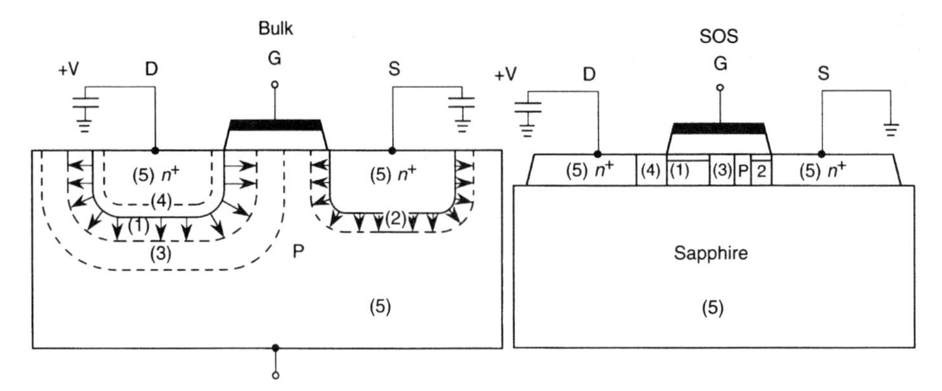

Region 1* Electrons swept to drain, holes swept to body – depletion region at drain.

Region 2* Electrons swept to source, holes swept to body – depletion region at source.

Region 3* Electrons may diffuse to drain.

Region 4* Holes may diffuse to body.

Region 5 Electrons and holes recombine – no net electric field.

*Results in charge disturbances on drain capacitance.

Figure 13.24 Charge collection regions for an n-channel FET in bulk and SOS technologies. (Source: Pickel, 1983.)

integrated circuits. Some technologies are inherently hard, while others, particularly MOS devices, must be hardened by employing thinner and cleaner oxides.

If the minimum hardness level of a given device or technology exceeds the total dose to be expected from the natural environment on a given mission by a factor of 2, the device can be considered hard for that mission. The factor of 2 is a design margin that allows for variability in hardness for the device and for the dynamic interplanetary or near-Earth radiation environment of the mission.

Devices with total dose hardness levels that do not exceed the mission requirement by a factor of 2 need to be derated for end-of-life performance or shielded with a high-density material or both. In some cases gradual degradation of device parameters in laboratory tests shows that a device need not be shielded if, for example, a higher leakage current or lower output voltage can be tolerated in circuit design. In these cases a derating of the device specification is all that is needed.

Other cases readily occur in which degradation of devices is so rapid, leading to catastrophic failure, that the total dose reaching the silicon die must be reduced. Heavy shielding materials such as lead, tantalum, or tungsten are used to surround or cover the part as completely as possible. One must consider the complete solid angle of access to the part and the part's hardness level in determining the shape and thickness of the shield. Additional design margin is sometimes introduced to account for sections of solid geometry that are impractical to shield. A practical advantage of tantalum is that it can be machined into desired shapes; tungsten is also an excellent heat sink for larger board applications.

Figure 13.25 shows a dose–depth curve for the TOPEX mission. The total dose becomes almost asymptotic at 10^4 rad (Si) with little reduction from 4 to 10 g/cm^2 areal density. On most light spacecraft the center is represented by an areal density of about 3.5–4.0 g/cm^2, and only an order of magnitude more shielding will significantly reduce the 10^4 rad (Si). Since huge amounts of shielding are impractical, a minimum acceptable device hardness level for the TOPEX mission is 10^4 rad (Si). Devices must be at least this hard so that we can shield them.

In contrast, the minimum areal density to be expected is usually about 0.5–1.0 g/cm^2. At 1.0 g/cm^2 figure 13.25 yields a total dose of 36 krad (Si) that, when multiplied by a design margin of 2, gives the mission requirement for a hard part that needs no shielding: 72 krad (Si). Between 10 and 72 krad (Si) parts will need shielding, as calculated from figure 13.25. This is usually done by ignoring incidental shielding, assuming 72 krad (Si) is incident at the surface of the part's package and using the difference in shield depth between the 72 krad (Si) requirement and the part's hardness level to determine the additional areal density needed to reduce 72 krad (Si) at the package surface to, say, 10 krad (Si) at the device die. This areal density can be converted to inches of aluminum and, by using mass density ratios, to inches of thickness of tantalum or tungsten needed for the radiation shield. If mass for the mission is at a premium, then a more sophisticated ray trace analysis is performed that takes incidental and inherent shielding inside electronics boxes into account. As a result, specific doses can be estimated at specific locations. The ray trace analysis usually produces lower dose estimates than the simplified generic geometry shielding routines.

Contending with single-event upsets is another matter. Shielding does not help here because energetic cosmic rays easily penetrate and pass through the whole spacecraft. Thus, again, the best approach (as discussed previously) is technology that is not readily upset—hence the interest in SOS and SOI. However, if devices do upset or latchup other measures can be taken. Current-limiting resistors and the means to turn off the power are necessary to correct latchup. For transient bit flips, error detection and correction schemes and watchdog timers are used and periodic memory scrubs can be performed. If upsets are occurring infrequently in science data streams, one may not care about a few bit flips. However, one cannot tolerate upsets in a command system microprocessor. Finally, if the upset rate expected is on the order of a few per year we again may decide that this rate is tolerable and that nothing need be done.

As an example of contending with SEU-induced latchup we discuss the Analog Devices 2100A digital signal processor. Samples tested at Brookhaven National Laboratory latched upon exposure to heavy ions with an LET threshold of 13 MeV cm^2/mg and an asymptotic cross-section of 10^{-4} cm^2. The ADSP2100A supply current before irradiation was approximately 2 mA. When the processor latched, power supply currents increased to as much as 300 mA. Holding current (that is, the minimum current required to sustain latch) was observed to range between 2 mA and 20 mA. Variation occurred when different internal latch paths were formed.

In spite of the progress made in removing the latchup susceptibility of the ADSP2100A, latchup detection and correction circuitry (figure 13.26) was designed to allow the use of the commercial processor in the event that an epitaxial version under development should fail some other tests associated with space qualification. A small resistor is used to limit the current to the device. When a latchup occurs, the voltage drop across the resistor changes. This is sensed by a comparator that produces a latch

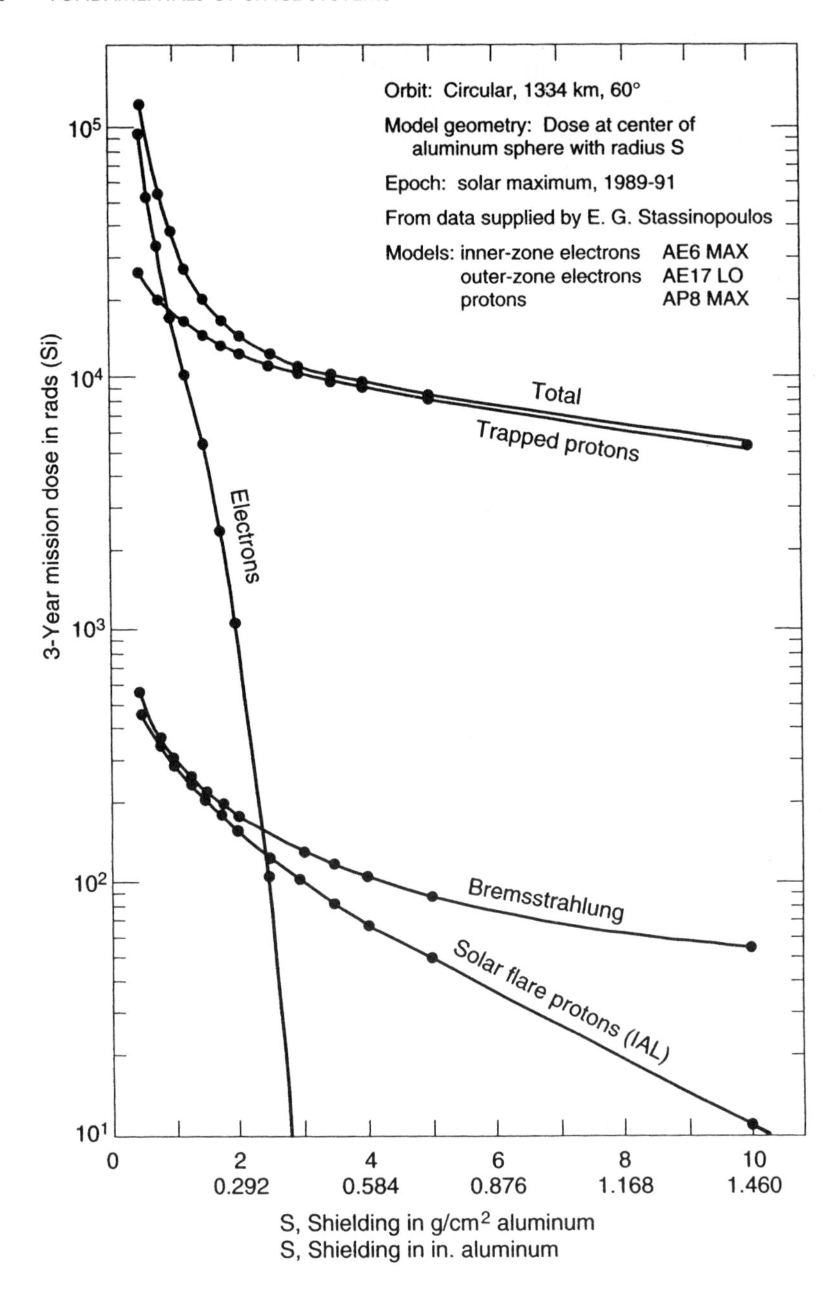

Figure 13.25 TOPEX radiation study.

detection signal. This signal is used to remove power to the ADSP2100A. Since each supply pin can latch individually, separate resistors and transistors are used for each pin. In addition, components connected to the signal lines of the processor are disabled by the latch detection signal to prevent damage to them by a latch condition in the ADSP2100A. After a specified delay, the device is powered up and reset. This circuit

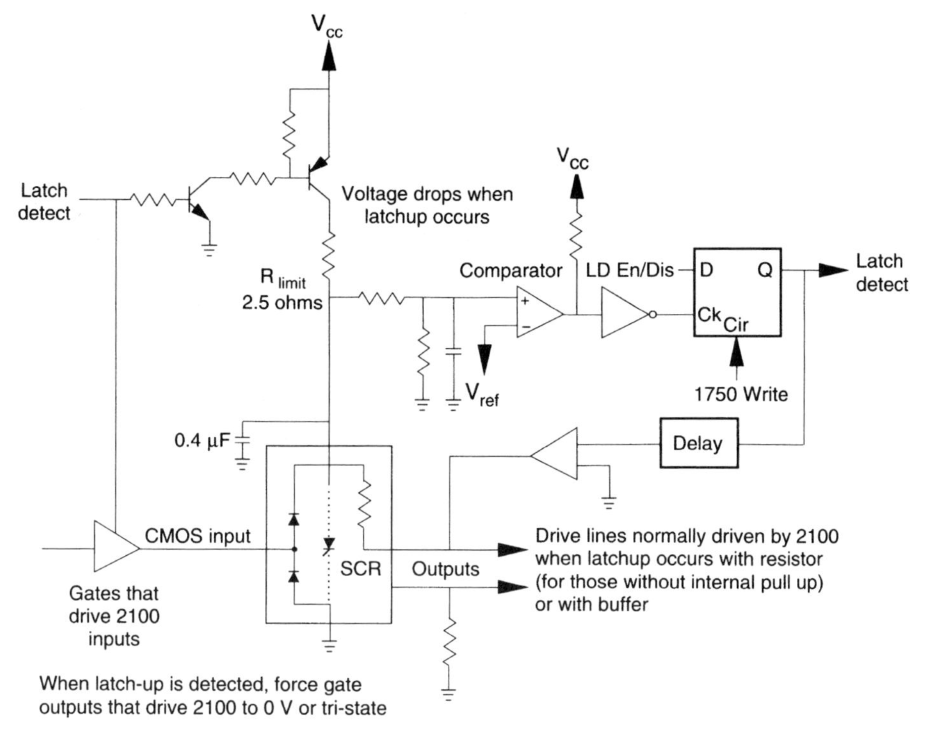

Figure 13.26 ADSP-2100 latchup protection circuit.

was tested numerous times with five samples of the commercial ADSP2100A. In every case, latchup was detected and the sample protected.

The first step in combating the soft error problem is to make an estimate of the upset rate using the cosmic ray spectrum to be seen by a device on a given mission and the experimental upset cross-section due to heavy ions (figures 13.21 and 13.22). Adams (1986) and Tylka (1997) have developed a suite of computer programs called CREME, which transforms interplanetary energy spectra into incident LET spectra for a particular spacecraft orbit and shielding thickness. CREME takes into account geomagnetic shielding and Earth shadowing and will calculate device upset rates if given device upset cross-section data. Kinnison et al. (1990) discuss another method of calculating upset rates that uses a Monte Carlo technique.

After the upset rate has been estimated, we decide whether to fly the device and, if so, whether error detection and correction or watchdog timer schemes are necessary.

Problems

1. For component reliabilities of $R = 0.8$ and $R = 0.98$, compare the system reliabilities for the ranges of $n = 1-10$ and $m = 2, 3$. Discuss your results with respect to whether it is better to provide spare parts or a spare system. Contrast your results for low (commercial) and high (military grade) reliability components.

2. A GaAs FET MMIC switch is determined to have failed when its insertion loss has increased by 50% over its initial value. At a test temperature of 225°C the mean time to failure is 1.5×10^4 h; at a test temperature of 250°C the mean time to failure is 3.4×10^3 h. What is the activation energy for this failure mode, and what mean time to failure at a maximum spacecraft device junction temperature of 100°C would be predicted? Auger electron spectroscopy determined that the dominant failure mechanism is the interdiffusion of gate metal and GaAs in the FET channel under the gate.

   Any of the environments listed in table 13.1 can be used to induce failure. The strength of the environmental stress screens can be increased by increasing the length of the test (hours of aging, number of thermal cycles), or by increasing the intensity of the environment, such as the temperature or rate of temperature change or acceleration amplitude during vibration. The advantage of step-stress testing is that, being carried out in steps or intervals, the degradation of parameters is usually gradual and failure analysis can be carried out on the devices after completion of the environmental test. Rapid catastrophic destruction yields ashes, rubble, and little useful information and should be avoided.
3. Let x_0 equal zero and, for the domain of x from 0 to T, find the cumulative distribution function $F(x) = F(T)$. This is the two-parameter (θ, b) Weibull function. Compare it to a simple exponential distribution.
4. A population of standard diodes is guaranteed to have 2% defectives or less. A certain board of electronics in a spacecraft low-voltage power supply (LVPS) uses 50 of these diodes. If no additional screening is performed by the spacecraft system fabricator, what is the probability of finding 4, 3, 2, 1, and 0 defectives on the LVPS board? Comment on your results.
5. The following observations on traffic flow indicate the number of times two vehicles approach an intersection simultaneously in any one minute. The total time for survey was 2 h.

Number of events in any one minute	0	1	2	3	4	5
Number of minutes having occurrences of above number of events	25	26	20	27	15	7

   Determine the probability of vehicles' approaching simultaneously 2, 3, or 4 times in any one minute. Compare the observed and predicted frequencies.
6. In a certain application the excessive amount of wear on a bearing was the result of excessive operating temperature, controlled by an oil bath. The amount of wear should not exceed 8 mg/100 hours of operation.

   An experiment was designed in which 10 bearings were tested for wear, each being run under a different operating temperature controlled by the oil bath. The following test data were obtained.

Operating temperature °F (x)	200	250	300	400	450	500	550	600	650	700
Amount of wear, mg/100 h (y)	3	4	5	5.5	6	7.5	8.8	10	11.1	12

Find the regression parameters a_0 and a_1 for the least-squares fit $y = a_0 + a_1 x$. Above what temperature does the wear exceed 8 mg/100 h?

7. Use the information in figures 13.21 and 13.22 to compute an upset rate for the 80186 microprocessor from equation 13.4.1. Note that figure 13.22 is an integral spectrum. Equation 13.4.1 should be evaluated numerically using Simpson's or the trapezoidal rule. For this problem, do not consider upsets due to proton-induced nuclear reactions but only those due to direct ionization.

References

Adams, J. H. Jr., 1986. Cosmic ray effects on microelectronics, Part IV. NRL Memorandum Report 5901.

Binder, D., E. C. Smith, and A. B. Holman, 1975. Satellite anomalies from galactic cosmic rays. *IEEE Trans. Nucl. Sci.* **NS-22**:2675.

Fleetwood, D. M., P. S. Winokur, L. C. Riewe, and R. L. Pease, 1989. An improved standard total does test for CMOS space electronics. *IEEE Trans. Nucl. Sci.* **NS-36**:1963.

Hecht, H., and M. Hecht, 1985. *Reliability prediction for spacecraft.* Rome Air Development Center Technical Report RADC-TR-85-229.

Herbert, G. A., R. H. Maurer, J. D. Kinnison, and A. Meulenberg, 1989. Radiation damage experiment on gallium arsenide solar cells. *Proc. of the IECEC*, August, Arlington. VA.

Hohl, J. H., and K. F. Galloway, 1987. Analytical model for the single event burnout of power MOSFETs, *IEEE Trans. Nucl. Sci.* **NS-34**:1275.

Holmes-Siedle, A., and L. Adams, 2002. *Handbook of Radiation Effects*, 2nd ed. New York: Oxford University Press.

Janesick, J., T. Elliott, and F. Pool, 1989. Radiation damage in scientific charge-coupled devices. *IEEE Trans. Nucl. Sci.* **NS-36**:572.

Kapur, K. C., and L. R. Lamberson, 1977. *Reliability in engineering design.* New York: John Wiley & Sons.

Kinnison, J. D., R. H. Maurer, and T. M. Jordan, 1990. Estimation of the charged particle environment for Earth orbits. *Johns Hopkins APL Tech. Dig.* **11**:300.

Kolasinski, W. A., J. B. Blake, J. K. Anthony, W. E. Price, and E. C. Smith, 1979. Simulation of cosmic ray induced soft errors and latchup in integrated circuit computer memories. *IEEE Trans. Nucl. Sci.* **NS-26**:5087.

Lipson, C., and N. J. Sheth, 1973. *Statistical design and analysis of engineering experiments.* New York: McGraw-Hill.

McLean, F. B., 1987. Interactions of hazardous environments with electronic devices. In IEEE Tutorial Short Course on Nuclear and Space Radiation Effects, Snowmass, CO.

Maurer, R. H., and J. J. Suter, 1987. Total dose hardness assurance for low Earth orbit. *IEEE Trans. Nucl. Sci.* **NS-34**:1757.

Maurer, R. H., and J. D. Kinnison, B. M. Romenesko, B. G. Carkhuff, and R. B. King, 1988. Space-radiation qualification of a microprocessor implemented for the Intel 80186. *Proc. 2nd Annual Conf. on Small Satellites*, Utah State University, Sept.

Maurer, R. H., G. A. Herbert, J. D. Kinnison, and A. Meulenberg, 1989. Gallium arsenide solar cell radiation damage study. *IEEE Trans. Nucl. Sci.* **NS-36**:2083.

May, T. C., and M. H. Woods, 1979. Alpha-particle-induced soft errors in dynamic memories. *IEEE Trans. Electron Devices* **ED-26**:2.

Messenger, G. C., and M. S. Ash, 1986. *The effects of radiation on electronic systems.* New York: Van Nostrand Reinhold.

Peterson, E. L., 1983. Single event upsets in space: basic concepts. In IEEE Tutorial Short Course on Nuclear and Space Radiation Effects, Gatlinburg, TN.

Pickel, J. C., 1983. Single event upset mechanisms and predictions. In IEEE Tutorial Short Course on Nuclear and Sapce Radiation Effects, Gatlinburg, TN.

Pickel, J. C., and J. T. Blandford, Jr., 1978. Cosmic ray induced errors in MOS memory cells. *IEEE Trans. Nucl. Sci.* **NS-25**:1166.

Srour, J. R., R. A. Hartmann, and K. S. Kitazaki, 1986. Permanent damage produced by single proton interactions in silicon. *IEEE Trans. Nucl. Sci.* **NS-33**:1597.

Tylka, A. J., J. H. Adams, Jr., P. R. Boberg, B. Brownstein, W. F. Dietrich, E. O. Flueckiger, E. L. Peterson, M. A. Shea, D. F. Smart and E. C. Smith, 1997. CRÈME 96: A revision of the Cosmic Ray Effects on Microelectronics code. *IEEE Trans. Nucl. Sci.* **NS-44**: 2150.

Uy, O. M., and R. H. Maurer, 1987. Fault tree safety analysis of a large $Li/SOCl_2$ spacecraft battery. *J. Power Sources* **21**:207–225.

Wallmark, J. T., and S. M. Marcus, 1962. Minimum size and maximum packing density nonredundant semiconductor devices. *Proc. IRE*, p. 286.

Waskiewicz, A. E., J. A. Groninger, V. H. Strahan, and D. M. Long, 1986. Burnout of power MOS transistors with heavy ions of californium 252. *IEEE Trans. Nucl. Sci.* **NS-33**:1710.

14

Spacecraft Integration and Test

MAX R. PETERSON AND ELLIOT H. RODBERG

14.1 Introduction

14.1.1 Spacecraft Design

The design of a spacecraft is divided into several phases, as described in chapter 1: conceptual design, preliminary design, detailed design, fabrication and assembly, and integration and test. This chapter will address the integration and test phase, which includes assembling the various mechanical, electrical, and thermal subsystems into an integrated spacecraft and performing tests on the integrated spacecraft to ensure that it will operate properly in the environment in which it must function. This process includes transport to a launch site where the spacecraft will be mated to the launch vehicle and again tested to ensure that all spacecraft systems are functioning correctly and that the spacecraft–launch vehicle combination is ready for launch.

Although the specific task of integration and test occurs at the end of the program, the actual planning for this important phase of the program starts with the conceptual phase. If one word could describe the integration and test phase it would be *logistics*, which deals with all aspects of the procurement, movement, maintenance and disposition of supplies, equipment, facilities, and personnel and the provision of services. To appreciate and understand the integration and test phase of a spacecraft, one must understand, to some extent, the other phases of the program. It is therefore prudent to review some of the concepts introduced and discussed in chapter 1 and discuss terminology, before we begin the discussion of the integration and test process.

14.1.2 Terminology

When discussing the integration, test, and qualification of any system, it is necessary to clearly understand the terminology which is used to describe the processes that occur.

Toward that end, a few important items are defined which should make the discussions in this chapter more concise.

A *part* is a hardware element that cannot normally be disassembled without destruction of its intended design function; examples are transistors, integrated circuits, circuit boards, housings, structure members, thermostats, and so on.

A *component* consists of parts which are assembled to perform a higher level function. Examples of components include

- a battery in a power subsystem, which provides stored energy to power the spacecraft during launch and eclipse periods;
- thermal blankets or louvers in a thermal control subsystem which are used to provide the desired thermal environment for the spacecraft;
- a reaction wheel in an attitude control subsystem which is used to orient the spacecraft to a desired position;
- a particle counter in an instrument that measures flux in the space environment.

A *subsystem* or *instrument* consists of components which are assembled to perform a specific function, such as a power subsystem which generates, stores, and conditions power for a spacecraft, or an instrument intended to measure a range of particle fluxes and energies. The term subsystem usually connotes an entity that performs a support function, whereas the term instrument usually connotes an entity that makes a specific scientific measurement or observation.

A *system* is composed of subsystems whose overall function is to execute the mission requirements. Specific terms used to describe a collection of subsystems and instruments intended for use in space are spacecraft, observatory, or payload.

A *unit* is a generic reference to a component, subsystem, instrument, or system.

Package design is the process of arranging electrical and mechanical parts into a component, subsystem, or instrument.

Integration is the process of assembling components into a subsystem (or subsystem and instruments together into a system) and ensuring that the individual units perform properly when interconnected.

Verification is the process of determining that the part, component, subsystem, instrument, or system meets specified requirements and will operate properly under the environmental conditions expected to be encountered during the mission.

- *Performance verification* may be accomplished by analysis, test, or a combination of both. An example of performance verification by test is the determination that an analog-to-digital converter will transform a voltage analog of some measured parameter into a digital number with the required accuracy while it is being subjected to thermal (or thermal vacuum) variations as well as power supply variations. As an example of performance verification by test and analysis, consider that it will most likely be necessary to perform significant analysis on a pressure vessel as well as to subject it to proof pressure testing.
- *Functional verification* is a subset of performance verification and is intended to demonstrate that the unit under test (that is, the part, component, etc.) continues to operate within specification by measuring only a few well chosen parameters. Functional verification can be used to shorten the total test time. For example, if it is necessary that a unit be subjected to several thermal cycles for purposes of environmental stress screening it may be appropriate to execute the

performance verification test during the first and last thermal cycles and to execute the functional verification test during all other cycles, just to ensure that time is conserved by not continuing the test if a failure occurs between the first and last cycle. Continuing the analog-to-digital converter example, a functional verification test might consist of measuring the transfer function of the converter at three points with nominal power supply voltage settings, whereas the performance verification test might require that the transfer function be measured for each possible output number of the converter for high, nominal, and low power supply voltage values.

14.1.3 Design Process Overview

14.1.3.1 Program Phases

Spacecraft development programs are organized into phases. At the end of each phase there is a major program review to ensure that designs are consistent and adequately support the program objectives and mission requirements. Figure 14.1 shows a typical block diagram for a spacecraft system, while figure 14.2 shows the programs phases and related reviews.

Phase A is the conceptual design phase, initiated at the beginning of the program. For many scientific missions, this phase is accomplished while developing a detailed proposal that defines the mission objectives, high-level mission requirements, and functional descriptions of the spacecraft subsystems and instruments. This conceptual design phase (or proposal phase) can be as short as a few months or as long as a year.

During phase A, the mission requirements are developed including mission success criteria and instrument performance requirements. Major system and subsystem functional requirements are determined along with initial system interface definitions. The system and subsystem designs are initiated. Phase A concludes with a conceptual design review (CoDR), which is often combined with the system requirements review (SRR).

The second phase of a spacecraft program, phase B, is used to complete the preliminary design of the spacecraft subsystems, instruments, and associated engineering disciplines such as stress analysis, thermal analysis, electromagnetic compatibility, contamination, ground systems, and mission operations. Flight and ground software architectures are defined and preliminary designs of the required software applications are completed. Phase B typically lasts eight to twelve months.

The implementation phase of a program, phase C/D, includes the detailed design of hardware and software subsystems. Breadboards and engineering models of selected subsystems are built and tested to confirm that subsystem designs meet the subsystem and system requirements. Software designs are developed and reviewed. About twelve months into phase C/D, the mission critical design review (CDR) is held. Detailed system and subsystem designs are presented along with resource margins, interface definitions, breadboard test results, and plans for testing, qualification, and operations. Subsystem fabrication generally begins following the mission CDR and subsystem detailed design.

Integration and test constitutes the last part of phase C/D. It includes installation and checkout of all components onto the flight structure, environmental testing of the spacecraft as a system, and testing at the launch site. Phase C/D typically lasts 24 to 36 months.

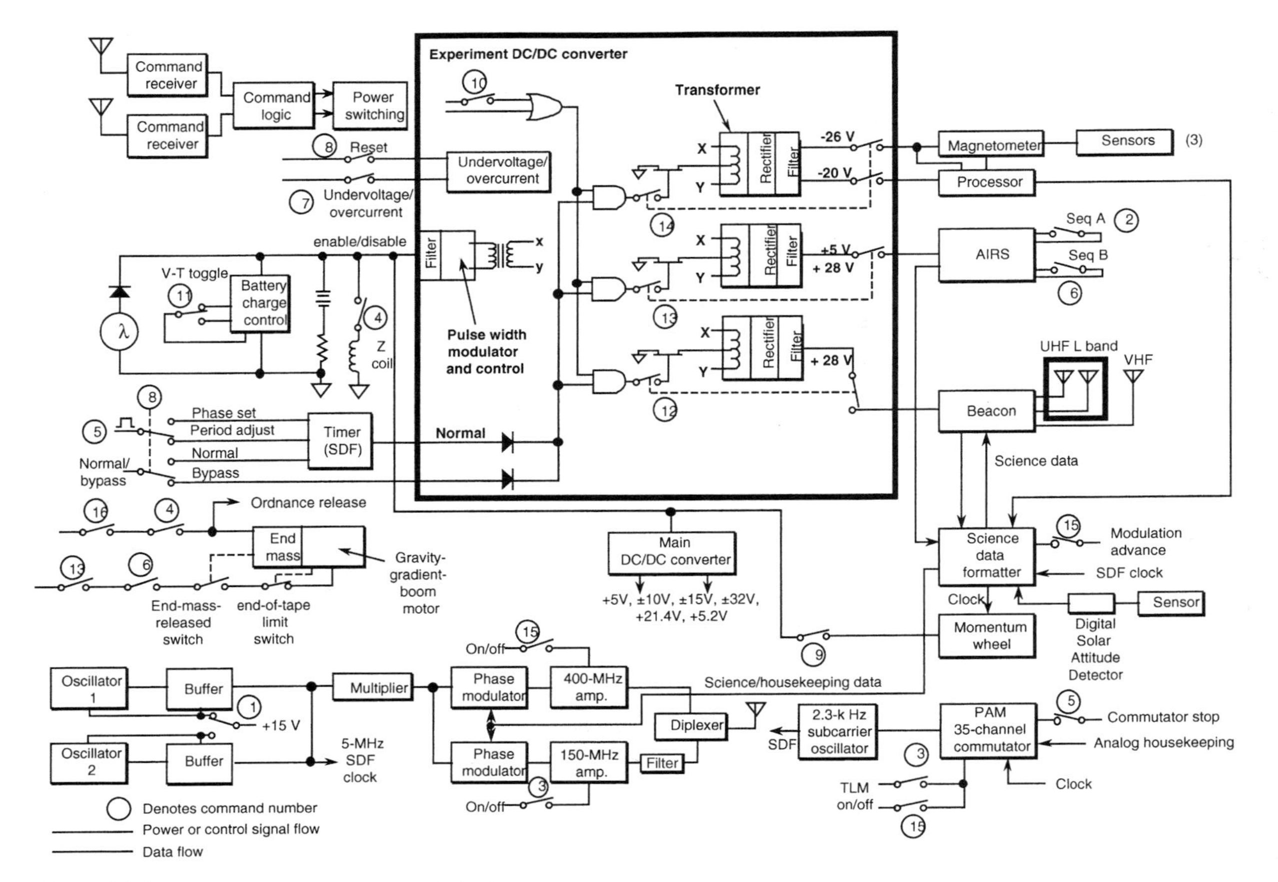

Figure 14.1 Typical spacecraft system block diagram.

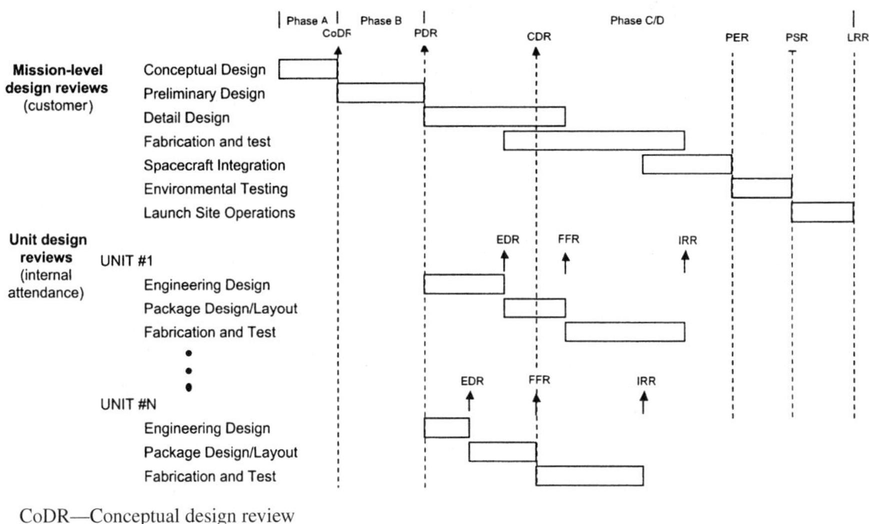

CoDR—Conceptual design review
PDR—Preliminary design review
CDR—Critical design review
PER—Pre-environmental review
PSR—Pre-ship review
LRR—Launch readiness review

Figure 14.2 Typical program phases and design reviews.

14.1.3.2 System and Subsystem Design

The system and subsystem design process is illustrated in figure 14.3. During the system design phase of the program, detailed electrical, mechanical, thermal, and software interface definitions are developed. Parts and materials must be carefully selected for use in the space environment. Several characteristics that must be considered in the selection include appropriate radiation tolerance for the expected environment,

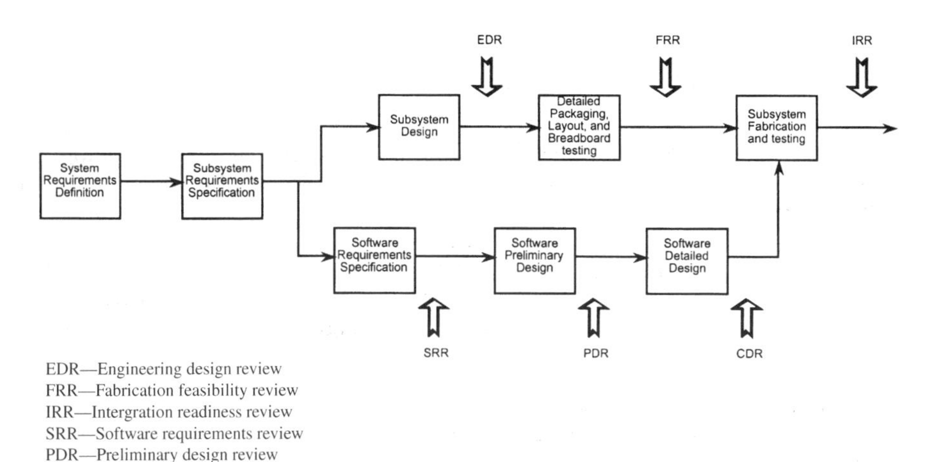

EDR—Engineering design review
FRR—Fabrication feasibility review
IRR—Intergration readiness review
SRR—Software requirements review
PDR—Preliminary design review
CDR—Critical design review

Figure 14.3 Subsystem design flow.

low out-gassing properties if sensitive surfaces of instrument sensors are not to be contaminated, light weight, strength, and appropriate thermal properties. Subsystem and instrument components are positioned on the spacecraft structure to minimize volume and harness interconnects, to avoid mechanical interferences, and to optimize, to the extent possible, the mass properties of the developing spacecraft. Instrument and RF fields of view must be considered as well as thermal design constraints. The integration and test team are an important part of this system design effort and it is their responsibility to help ensure that all subsystem and instrument interfaces are properly defined so that when the spacecraft enters the integration phase there will be few problems as the mechanical, thermal, and electrical subsystems are integrated and testing begins.

The subsystem design phase uses the system requirements and interface definitions agreed to during the preliminary design phase. Work begins on the detailed design, fabrication, and testing of the various subsystems, instruments, and software. While the electrical design of the subsystems and instruments progresses, the mechanical, structural, and thermal design of the spacecraft proceeds in parallel, so that when the electrical design is complete the mechanical and thermal package design can proceed. The flight software design must occur in phase with the subsystem design and deliver flight-qualified software to the spacecraft during the integration phase. The subsystem design process must be interactive; that is, there must be frequent meetings between the various subsystem designers, parts and materials experts, package designers, and the integration and test team to discuss, further develop, and refine the electrical, mechanical, and thermal interface agreements reached at the beginning of the program. It is also important to include the mission operations team in the subsystem design process. The operation of the subsystem (that is, the different modes or states in which it can function) must be well defined so that planning for on-orbit spacecraft operations by the mission operations team can proceed with assurance that the spacecraft will be capable of carrying out its intended mission.

Each subsystem must undergo a series of reviews and tests intended to detect flaws in the design or fabrication. It is important to understand that it is not just the reviews per se that correct design flaws, but, in addition, the process of preparing for the reviews forces interfaces and designs to be analyzed and documented, highlighting potential problems.

The terminology used to describe the design review process in this chapter differs from that used in chapter 1, since this chapter is intended to emphasize the review process specifically with respect to a subsystem or instrument that will be a part of a spacecraft system. This relationship is illustrated, for a typical program, in figure 14.3. Additionally, terminology may vary with different organizations; however, it is the process that should be emphasized, rather than terminology.

The key review for a subsystem or instrument should be an engineering design review (EDR), in which engineering details of the subsystem design are discussed. An EDR should be held for both electrical and mechanical subsystems when the designs are at or near completion. Although the following is not an exhaustive list, this review should include as a minimum

- presentation of subsystem requirements;
- for electrical designs—discussion of electrical circuit design, electrical part stress analysis, radiation tolerance of the design, circuit design simulations,

software/firmware requirements, results of breadboard testing performed to verify operation of the design over voltage and temperature variations, any unique or special part placement requirements, timing analyses, identification of high dissipative parts, operating modes, interfaces, and EMC design constraints;
- for mechanical designs—discussion of device function, preliminary stress analysis, material properties, deployment requirements, field-of-view analyses, "keep-out" zones of other subsystems, preliminary environmental or structural tests, radiation tolerance of optical components or materials, properties of seals or bearings, and thermal requirements;
- results of preliminary area or layout studies;
- safety issues;
- trade-offs considered during the design (i.e., "make–buy" decisions, alternate designs, etc.);
- test plans and procedures.

The second review of importance for a subsystem is a fabrication feasibility review (FFR) of the design prior to releasing the drawing package. Due to the nature of mechanical designs it is prudent, in most cases, to hold the FFR concurrently with the EDR. In the case of electrical designs, the drawing package should be ready for review and should include assembly, layout, detail drawings, and parts lists. At this review the designer and the organization which will manufacture the design are given an opportunity to review its manufacturability with respect to parts, materials, processes, and fabrication techniques. In addition, mechanical stress analysis, thermal analysis, and shielding analysis should be reviewed.

After manufacture, each subsystem must undergo a series of qualification tests before it reaches the spacecraft integration and test phase. (The types of tests performed are discussed in further detail later in this chapter.) At the conclusion of subsystem testing, an integration readiness review (IRR) should be held to discuss performance and status of the subsystem in relation to its readiness for integration onto the spacecraft. Any special operating constraints which must be observed by the integration and test team while the subsystem is on the ground should be identified at this review.

While somewhat repetitious of the material covered in chapter 1, an understanding of the concepts discussed above is mandatory for the proper execution of the integration and test phase. It must be emphasized that detection and correction of errors and flaws in the design before integration into the spacecraft will save time and reduce program cost. The program cost of a properly executed series of design reviews early in the program and comprehensive testing prior to integration is low with respect to the cost of the "marching army" which is in place to support the integration and test phase of a program. To amplify, the cost of any delays in the integration and testing of the spacecraft due to the necessity to correct errors in the subsystem design will far exceed the cost of the design reviews and prior testing, since it is not only spacecraft personnel who must be supported during these delays but other mission-related personnel such as the mission operations team, the launch vehicle team, and the science analysis team (in the case of an R&D program). Slips in the launch schedule can also affect the launch of other entirely unrelated programs, causing increased cost to them.

14.2 Planning for Integration and Test

14.2.1 Electrical Design Interactions

While subsystem design is proceeding, the integration and test team must develop more detailed block diagrams (such as power flow, signal flow, and grounding diagrams) and start designing the wiring harness and terminal boards which will interconnect all the subsystems electrically. Commands and engineering telemetry values are defined, and these will be used to operate the spacecraft and evaluate its health during ground testing and on-orbit operations. Detailed subsystem and instrument integration procedures must be developed that include electrical interface verification tests of both the spacecraft harness and the unit being integrated.

In addition, the design and fabrication of ground support equipment must be started. Test equipment needed for the safe operation of the spacecraft during integration and environmental testing must be designed and fabricated. Power subsystem test equipment includes such items as laboratory power supplies, fuse boxes, a non-flight "work" battery, battery-conditioning equipment, and a solar array simulator. Flight connector breakout boxes must be designed and fabricated for every type of connector used on the spacecraft. They provide a means to probe or stimulate individual pins on any connector. These boxes are used to verify proper characteristics of interface signals and to diagnose anomalous system behavior. It is also necessary to design and build ordnance simulators, instrument load simulators, and various "safing" and "arming" plugs. Umbilical ground support equipment must be designed, fabricated, and tested. It is used to monitor spacecraft health and status and transfer baseband commands and telemetry via the spacecraft–launch vehicle umbilical connector.

14.2.2 Thermal Design Interactions

The thermal design begins in parallel with the design of the subsystems with an assumed thermal analytical model of the spacecraft. The information from this model must be provided to the various subsystem package designers to ensure that the proper design is applied to their components in order to keep them and the spacecraft at the correct operating temperature on orbit. Conversely, as information from power dissipative components becomes better defined, this information must be fed back to the thermal engineer for refinement of the thermal analytical model. Details of the thermal design process are more completely discussed in chapter 7.

The thermal analytical model is the basis used by the thermal engineer to plan for the thermal ground support equipment required for individual package testing and to establish the thermal test requirements for each package. Thermal ground support equipment is not generally needed during integration because the thermal environment can be more easily controlled on the ground than on orbit. Occasionally there may be cases where portable cooling units are required to allow full testing of a particular subsystem which will have a low operating duty cycle on orbit, but must be operated at a higher duty cycle during testing. Planning for the special-purpose equipment necessary to simulate the orbital environment during subsystem and system thermal vacuum testing also proceeds in parallel with the spacecraft design effort. These tests verify that the thermal analytical model will accurately represent spacecraft thermal performance in orbit.

14.2.3 Mechanical Design Interactions

The environment to be encountered during launch is determined by the launch vehicle to be used. Using an initial structural analytical model of the spacecraft, a coupling analysis is initiated with the launch vehicle team to determine the dynamic response of the spacecraft structure to the vibration and acoustic forces expected during launch. Stress analysis of the spacecraft structures must be performed on the basis of the information determined by the launch vehicle coupling analysis to ensure that design limits are not exceeded. This information is used, along with previous test data, to determine mechanical package design requirements, as well as the levels to be used for testing the spacecraft and subsystems in random and sinusoidal vibration, and shock tests. Mass properties (mass, center-of-gravity, and moments-of-inertia) of the developing spacecraft must be closely tracked during the design phase to ensure that the mass delivery capability of the launch vehicle for the desired orbit is not exceeded. If a mass problem is recognized early enough in the package design of the subsystems and instruments, it may be possible to use lighter-mass materials to build the subsystems. (It should be noted, however, that lighter-mass materials are generally more expensive in terms of material and manufacturing costs.) More information on spacecraft mechanical design and analysis, as well as on the rationale and levels used for verification testing, is presented in chapter 8.

14.2.4 Technical Interchange Meetings

Technical interchange meetings (also referred to as working group meetings) must be held with the launch site team, the launch vehicle team, and the instrument teams. Meetings with the launch site team and the vehicle team are used to negotiate mechanical and electrical vehicle interfaces and to coordinate plans for launch site testing and launch operations.

Meetings with the instrument developers are required to plan for instrument integration and ensure information such as command and telemetry definitions and instrument operations and test procedures is conveyed to the integration team. The teams walk through the steps involved in integrating each instrument. Any special handling or test requirements are discussed. Operational considerations are worked out such as definition of operational constraints and electronic interfaces for transferring command and telemetry data.

14.2.5 Ground Support System

The ground support system (GSS) comprises the equipment used to operate the spacecraft through subsystem integration, environmental testing, and launch site operations. The GSS is used to verify the spacecraft operates as expected and meets its performance specifications. The design of the GSS to be used during the integration and testing of the spacecraft starts with a thorough understanding of the mission of the spacecraft and the design of its subsystems and instruments. This means that it is necessary for the integration and test team (and GSS system designers) to be extremely familiar with the various subsystems that make up the spacecraft and the operation of the flight software to be embedded in the spacecraft.

A cost-effective way to design and build the ground support system is to use the same equipment and software for subsystem testing (including flight software development), spacecraft testing, and post-launch operations. Usually, multiple instances of the same system design are used. Software maintenance systems may be maintained intact, but other subsystem test sets, when no longer needed, become part of the integration and test or mission operations ground system. Some of the equipment needed for subsystem testing is unique to that environment, but some of it migrates to spacecraft-level testing. This is usually the simulators and stimulators required to thoroughly test the spacecraft.

Mechanical ground support equipment (MGSE) is an important part of a spacecraft integration program. All MGSE that is used to lift or hold flight hardware must be inspected, tested, and certified for use. Special-purpose MGSE must frequently be developed to support the assembly, handling, and transport of the spacecraft during the integration, test, and launch phases. It includes unique subsystem and instrument mounting or handling fixtures, spacecraft support stands, turn-over fixtures, lifting slings, propellant loading/pressurizing units, cryogenic handling units, gaseous purge units, and shipping containers. General-purpose MGSE includes cranes, forklifts, scaffolding, work stands, and ladders.

Electrical ground support equipment (EGSE) must be developed to test the electromechanical and electrical subsystems of the spacecraft. Examples of subsystems which usually have individual EGSE include power, attitude determination and control, command and data handling, radio-frequency communications, propulsion, and, in the case of a scientific R&D spacecraft, the scientific instruments. Any EGSE that is used for testing that is repeated throughout the integration and test phase must be designed to be remotely controlled by the GSS command and control system. This is currently implemented using TCP/IP socket protocols over ethernet interfaces.

Figure 14.4 illustrates the function of the GSS in its most basic form. Here we see that the four primary functions are to

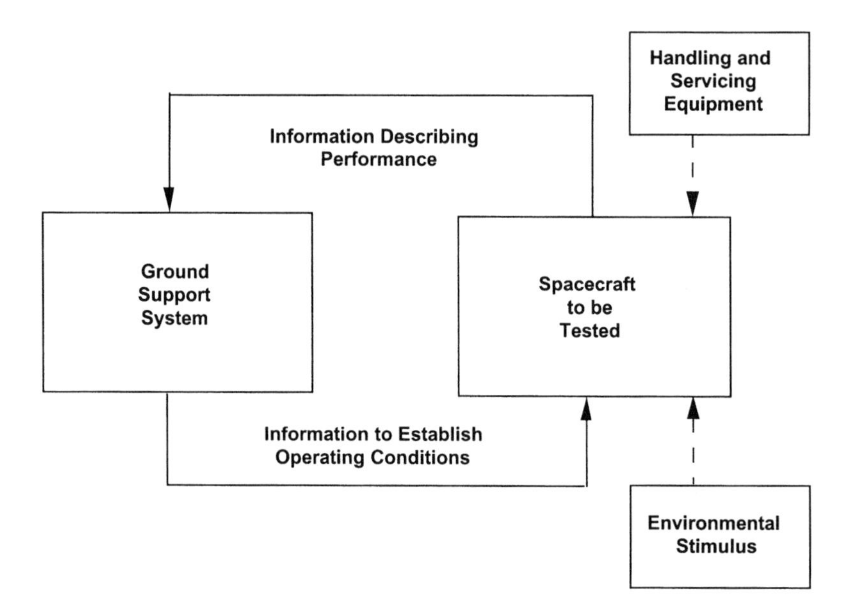

Figure 14.4 Ground support system (GSS) functionality diagram.

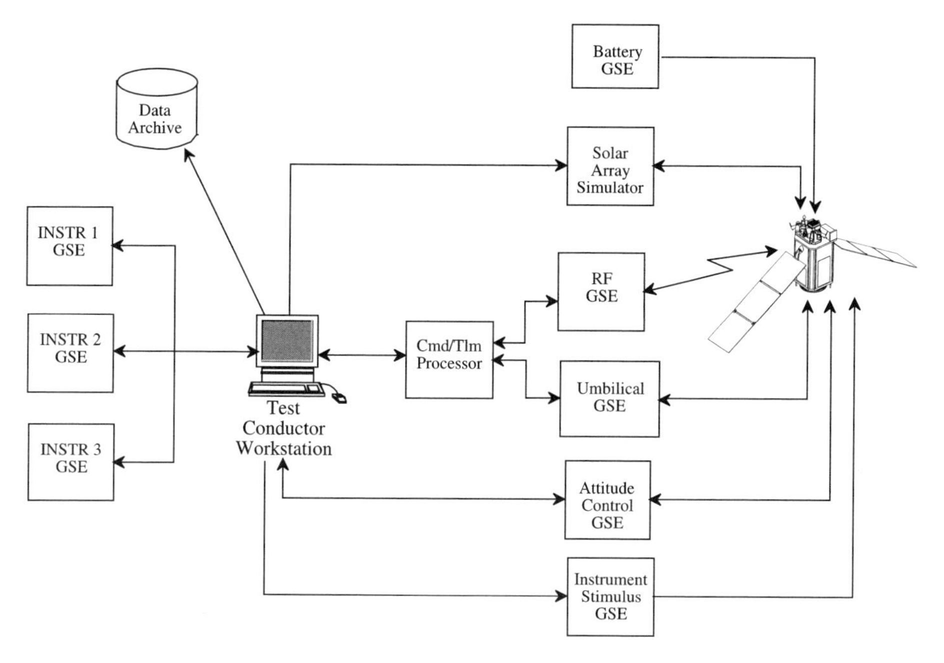

Figure 14.5 GSS block diagram.

(1) establish the operating conditions of the spacecraft subsystems;
(2) collect and process data from the spacecraft which indicates its performance in response to various stimuli;
(3) provide the simulated environment under which the spacecraft must be evaluated;
(4) provide safe handling of the spacecraft.

This figure does not indicate whether the GSS must be operated in a manual or automatic mode, although it does imply that the testing is done in a "closed loop" manner; that is, a stimulus is applied, a response is observed and, if satisfactory, another stimulus is applied and so on. As spacecraft complexity increases, however, it is essential that the GSS be operated automatically to make efficient use of critical schedule time.

Figure 14.5 shows a typical ground support system block diagram. At its center is the test conductor workstation, which is the controller of the whole system. All of the subsystem and instrument ground support equipment is electronically controlled in order to automate testing. The master workstation communicates with all of the other systems over ethernet. The equipment near the spacecraft can be commanded to the necessary modes and configurations. The central workstation forwards spacecraft telemetry data to a long-term archive and to any instrument ground support equipment computers for immediate or off-line analysis.

In a typical system the operator can enter a spacecraft or GSE instruction into the computer graphical user interface and view the spacecraft's response on customized display pages for immediate analysis. The ground system software provides the functionality to convert text versions of commands into their binary equivalents, then to format and transmit the command data to the spacecraft. Additionally, the ground system collects, displays, limit checks, and archives the telemetry data. Tools for converting raw telemetry to engineering units and other special processing such as memory load, dump, and compare are provided.

The operator can automate tests by using a scripting language to create a sequence of instructions to be executed by the command and control computer. Responses from the unit under test are automatically checked by the computer and logged to the screen and archived to a text file. The test sequence can continue with succeeding instructions, pausing only upon detection of an error or telemetry out of limit condition. This method of operation allows a complex system to be efficiently tested. More importantly, the tests are repeatable and provide consistent results that can be compared throughout the environmental test program. Trend data may then be analyzed for system faults or gradual degradation throughout the test program.

14.2.6 Mechanical Mock-ups

Mechanical mock-ups of various spacecraft structures must be designed and built. The specifics of the program will dictate the fidelity and number that will be required. Obviously, an attempt should be made to use as few mock-ups as possible, for as many different purposes as possible, commensurate with the spacecraft schedule. Several types of mock-up and their intended use are:

- *Wiring harness mock-up*—Used to determine cable routing and cable lengths during fabrication of the wiring harness. Also used to ensure there are no interferences between subsystems or instruments and the harness.
- *Radio frequency (RF) mock-up*—Used to verify the design of the spacecraft antennas. It is generally possible to build a scaled version of the spacecraft from sheet metal instead of duplicating it exactly, but this depends upon the RF frequencies used and other characteristics of the spacecraft.
- *Thermal mock-up*—Used by thermal engineering to design and fabricate thermal blankets, to estimate and plan for the amount of thermal coating material required to give the spacecraft the proper thermal characteristics, and to design and fabricate the thermal simulator to be used in thermal vacuum testing.
- *Launch vehicle interface mock-up*—Used to ensure that the spacecraft will mate properly with the upper stage of the launch vehicle.
- *Mechanical test prototype(s)*—Used to verify stress analysis and determine that adequate margin in the mechanical design exists. It may be necessary to fabricate high-fidelity prototypes of the spacecraft, subsystem, or instrument structure and to subject the unit to vibration or acoustic levels above the expected launch or orbital environment. These models might differ from the flight structure components to be launched in that dummy masses may be used in place of actual flight subsystems and instruments. Additionally, these models may be used to verify predictions of vibration levels at locations that are difficult to predict via modeling techniques.

14.2.7 Order of Integration

There are two basic considerations that determine the order of integration: function and accessibility. Generally, the mechanical flight structure and associated handling fixtures are required first. If the spacecraft includes an onboard propulsion system, the fuel tank, flow valves, thrusters, and fuel pipes should be the first subsystem installed. This cannot be fully tested until the flight electronics are integrated later, but the mechanical

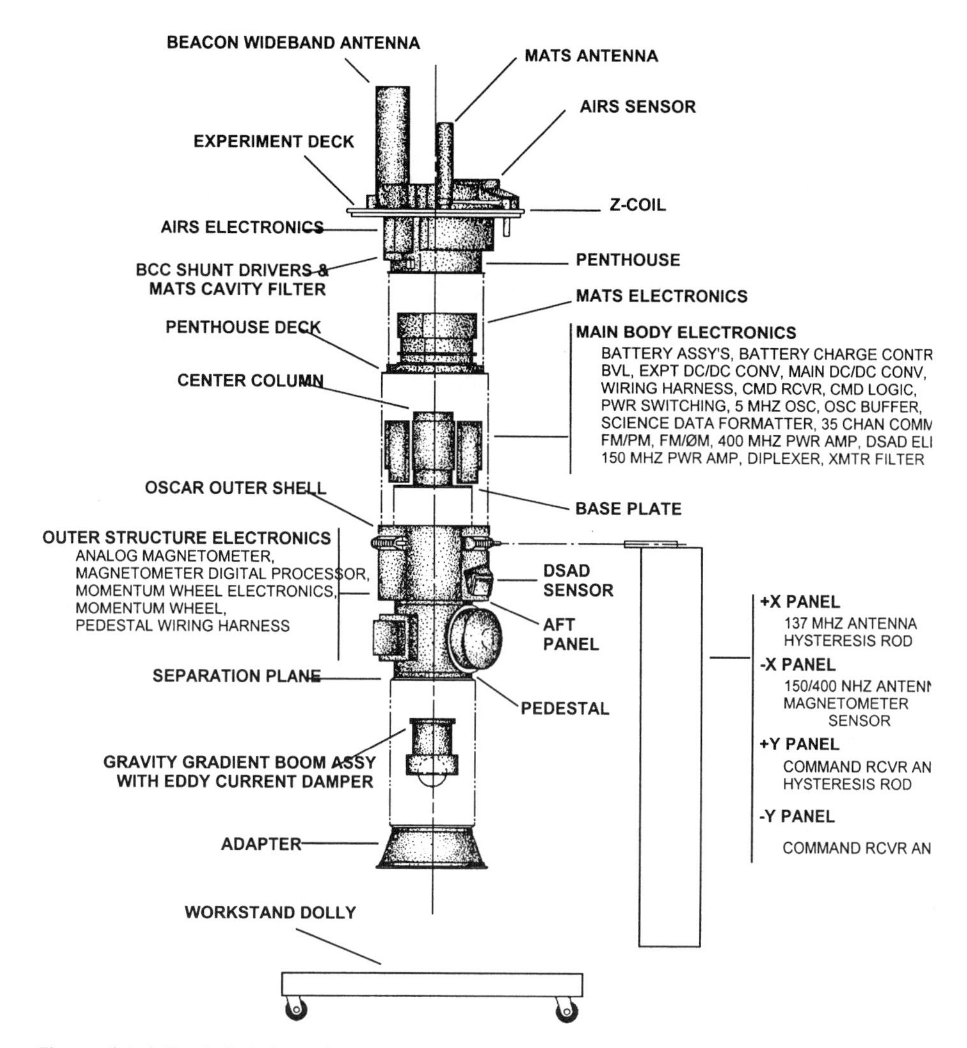

Figure 14.6 Exploded view of a typical spacecraft.

integration must be done first. Then the wiring harness and terminal boards can be mounted and tested in preparation for delivery of the subsystems and instruments. In terms of function, it is generally necessary to integrate the power subsystem first, followed by the command and data-handling subsystem. Then the RF subsystem and attitude determination and control subsystem can be integrated. For a scientific spacecraft, the instruments are normally integrated last, giving the scientists more time to complete instrument testing and calibration. With regard to accessibility, the order of integration may be determined by examining an "exploded view" of the spacecraft as shown in figure 14.6, and integrating systems from the bottom up, or, more typically, inner systems first, outer systems later.

Once the order of integration is established, the required delivery dates must be fed back to the master program schedule. Estimates of the time required to design, fabricate, and test the various subsystems and instruments must also be entered into the master schedule. Then schedule conflicts may be resolved so as to allow manpower and fiscal resources to be allocated in order to eliminate the conflicts and maintain the spacecraft integration, test, and launch schedule.

This feedback of schedule information often results in a decision to build a non-flight prototype of a unit that is on the critical path. This allows integration to proceed while the flight subsystem is fabricated and tested.

In the case of electrical subsystems and instruments, this non-flight prototype may take the form of a "packaged" breadboard or engineering model. A packaged breadboard will have electrical characteristics identical to those of the actual flight subsystem, but will have a different physical form. This particular option should only be considered for higher risk, low cost missions. A better solution is to build an engineering model. This is usually form, fit, and function like the flight model, except that high reliability, high cost, electronic parts are not used and it will not undergo all the rigorous environmental tests and inspections that the flight unit will receive.

Another technique that can reduce schedule risk is interface compatibility testing (both mechanical and electrical) with instruments and subsystems. The intent of the interface compatibility tests is to uncover inconsistencies and errors that can be corrected before the instruments or subsystems have completed fabrication and flight qualification testing. For this to be a viable alternative at the spacecraft level, integration must have progressed to the point where the spacecraft structure and/or electrical subsystems are available for these tests. A mechanical fit check can be used to confirm that bolt holes, connector clearances, and harness runs have been properly designed and implemented. For mechanical interface verification, drill templates are sometimes used to verify the spacecraft and the component have the same hole pattern.

An important consideration in the decision to perform interface compatibility tests and/or build packaged breadboards and engineering models is a tradeoff between cost and time. As mentioned earlier, it will generally cost less to the overall program to try and maintain schedule than to allow the launch date to slip, since there are many other variables associated with the launch date which are not directly controllable.

It is highly recommended that subsystem interface tests be performed as early as possible to detect design inconsistencies. Even if this is done with breadboards or engineering models, it is an extremely valuable check of the interface designs. This is particularly useful between the power subsystem or the command and data handling system and the other subsystems and instruments.

14.3 Integration and Test Facilities

14.3.1 Facility Scheduling

In addition to the activities described above, planning for integration and test facilities to be used must take place. Generally, facilities which can integrate and test large space systems are not abundant, due to the unique aspects of these systems, low usage, and potentially large maintenance costs. In the case of launch facilities, all are under the cognizance of some government agency. Scheduling of these facilities is but one of the logistic problems encountered by the integration and test team. The detailed schedule

Table 14.1 Class limits for Clean room facilities

	Measured Particle Size (μm)				
Class	0.1	0.2	0.3	0.5	5.0
1	35	7.5	3	1	NA
10	350	75	30	10	NA
100	NA	750	300	100	NA
1,000	NA	NA	NA	1,000	7
10,000	NA	NA	NA	10,000	70
100,000	NA	NA	NA	100,000	700

Notes:
1. Number of particles allowed cubic foot of room size equal to or greater than particle sizes shown.
2. The class limit particle concentrations shown in this Table are defined for class purpose only and do not necessarily represent the size distribution to be found in any particular situation.

showing the order in which the subsystems and instruments are required for integration and the order in which the tests will be performed is the basis for reserving the facilities at the proper time. It is important that all participants in the integration and test process be aware of this information and that they clearly understand the effects of a delay in delivery of a subsystem or instrument. Delay in reaching a facility as scheduled can cause an even greater delay due to conflicts with other systems competing for use of the facility and the necessity to reschedule its use.

14.3.2 Facility Cleanliness

In situations where it is necessary to avoid contamination of sensitive instruments, the spacecraft must be assembled and tested in an environment in which the humidity, temperature, particulate matter, and hydrocarbon content are maintained within limits set by the instrument manufacturers. Table 14.1, excerpted from FED-STD-209D, defines allowable levels of cleanliness for clean room facilities.

Rigid adherence to strict cleanliness requirements is necessary in certain situations, but for other situations it may be appropriate to examine cleanliness with respect to cost and realism; in other words, it may be prohibitively expensive to assemble and test all subsystems and instruments in an environment that may be essential only for unique instruments. Portable facilities (discussed below) may be a viable alternative. The bottom line is that cleanliness requirements for a spacecraft must be examined carefully and designed to meet the mission needs within the program constraints of cost and schedule.

14.3.3 Portable Facilities

In the event that facilities intended for use do not conform to the characteristics required to support the integration and test activities, provisions must be made to provide a portable environment while operating in that facility. Examples of portable environments that are often used are:

(1) a gaseous dry nitrogen purge of the instruments to prevent the adverse effects of humidity or outgassing products on optics or other sensitive surfaces;
(2) a portable clean room tent (and/or bagging) to enclose sensitive portions of the spacecraft, thereby keeping particulate contamination levels within required limits;
(3) portable air-conditioning equipment to provide convective spot cooling of instruments or subsystems that will be temperature controlled on orbit by conduction and radiation.

14.3.4 Test Facilities

A comprehensive spacecraft test program requires the use of several different types of facilities. These are required to fulfill the system testing requirements and may include some or all of the following facilities:

- Clean room
- Vibration test facility
- Acoustic test facility
- Mass properties test facility
- Thermal vacuum chamber
- EMC test facility
- Magnetics test facility
- RF compatibility test van or facility
- Launch site payload processing facility
- Launch site hazardous processing (fueling) facility

14.3.5 Transportation

Depending upon the size of the spacecraft and the test requirements, it is often necessary to transport the spacecraft between facilities listed in the previous section. If such is the case, the integration and test team must address such mundane questions as: "Will the spacecraft in transit fit under all bridges between destinations, and will it be within the load limit of any bridge it must cross?" And, at the completion of the test program, the spacecraft must be shipped to the launch site for final preparations and mating with the launch vehicle. In nearly all cases a shipping container is required and shipping arrangements must be made such that the spacecraft resides in a safe and clean environment at all times during transit between facilities and to the launch site.

14.4 Verification Program Considerations

14.4.1 Sequence of Testing

Verification and test at the component, subsystem, and system level is an extension of environmental stress screening (ESS) initiated at the part level (see chapter 13). Care must be taken to ensure that the units undergoing testing are not overstressed, while at the same time the environmental stress applied must be sufficient to uncover design flaws and deficiencies in manufacturing. In order to ensure that weak units are identified as early as possible and to provide margin in the design, the ESS levels are generally more severe in the earlier levels of verification; that is, ESS levels applied

Table 14.2 Summary of verification tests

Electrical	*Structural and mechanical*
Safe-to-mate	modal survey
Functional verification	Static loads
Performance verification	Acceleration
(including power supply voltage variation)	Sine vibration
Timekeeping	Random vibration
Phasing (polarity)	Acoustics
	Mechanical shock
Radiation	Pressure profile
	Mechanical function (deployments)
Total dose	Life test
Single event upset	Mass properties
EMC and magnetics	*Thermal*
Conducted emissions	Bakeout
Radiated emissions	Leak
Conducted susceptibility	Thermal balance
Radiated susceptibility	Thermal cycling
Magnetic properties	Thermal-vacuum cycling

to parts are more severe than those applied to components and subsystems, and those applied to components and subsystems are more severe than those applied at the system level.

In cases where extremely high reliability is required or the spacecraft is to be launched with a manned space vehicle, a prototype and a flight spacecraft will be built and tested to different levels. The prototype will receive the *same tests* as the flight spacecraft, the difference in the testing being that the severity of the levels and exposure time will be greater on the prototype to gain assurance that design margin exists in the flight spacecraft. More often, and in accordance with current philosophy, neither time nor money permits the fabrication and test of both a prototype and a flight spacecraft and a reasonable compromise is made by building and testing a protoflight spacecraft, with selected critical subsystem components built as spares. The protoflight spacecraft is tested to levels and exposure times that fall between those expected for flight and those defined for the prototype, to provide verification of design margin.

Generally, the sequence of testing is governed by the following rationale. Those tests that are judged most likely to cause the spacecraft to be disassembled if a malfunction occurs during the test are performed first. Vibration, acoustic, and thermal vacuum testing are performed last, since the spacecraft will be exposed to these environments (in that order) on its trip to orbit. An additional benefit associated with this sequence is that it aids in the detection of vibration and/or acoustic problems that do not fully manifest themselves until induced by the further stress of thermal cycling. Table 14.2 presents a list of verification tests, summarized by discipline.

14.4.2 Test Plans and Procedures

Test plans and procedures are an important method of communication and must be prepared and reviewed so that the tests the spacecraft will experience are clearly

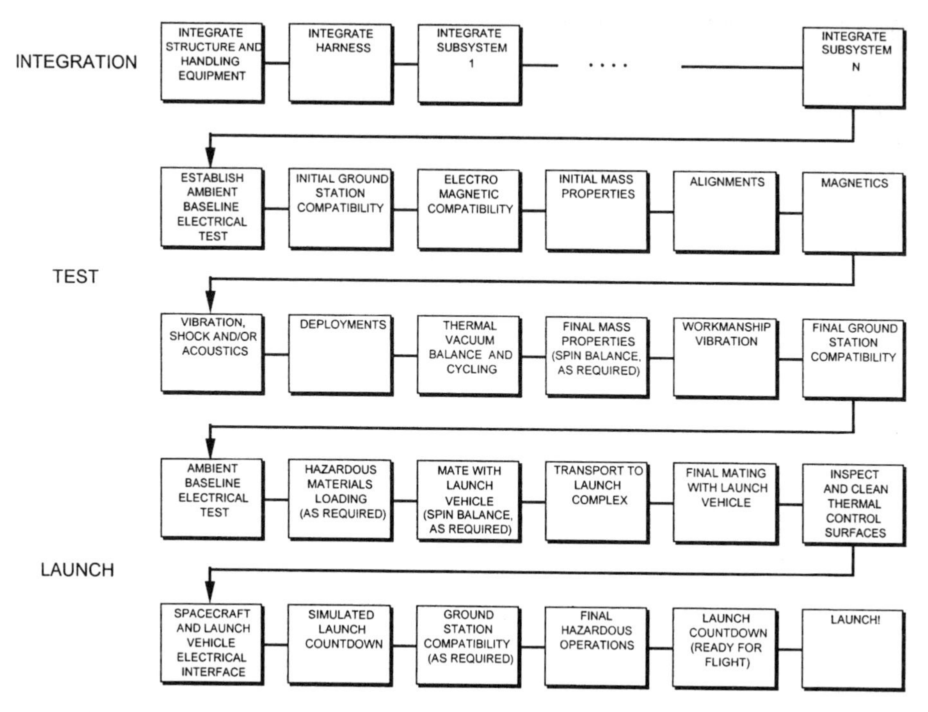

Figure 14.7 Integration and test flow.

understood and the spacecraft and its subsystems will be operated in a safe manner while undergoing the required sequence of tests.

The activities that comprise integration and test include hazardous operations (for example, cryogenic handling operations, radio-frequency radiation operations) and communicating when these activities will take place to participating subsystem and instrument engineers is extremely important from both safety and schedule aspects. To ensure that all personnel involved are clearly aware of the sequence of events associated with the integration and test activities it may be appropriate to include an activity flow diagram, similar to that shown in figure 14.7, in the system test plan.

14.4.3 Subsystem-Level Tests

Subsystems and/or their components must be tested to ensure that any design or manufacturing defects are uncovered and corrected before they are integrated onto the spacecraft. Failures detected during system-level integration and test often introduce schedule delays and, as a consequence, increase the cost of the program. The tests discussed below are listed in a recommended order which has been shown to uncover problems as early as possible to avoid costly and time-consuming rework late in a program. The actual requirement for the test, and the test limits, are determined during the mission definition phase as a result of the launch and orbital characteristics.

14.4.3.1 Radiation

Spacecraft intended to operate in orbits that will subject them to significant radiation dosage must be fabricated using parts that have been specifically designed to operate over the estimated orbital lifetime accumulated dose. In the case of certain parts (for example, digital integrated circuits) it may be necessary to perform single-event upset (SEU) testing to ensure that the part is immune to latchup and tolerant of soft errors. Refer to chapter 13 for a more significant discussion of the effects of radiation on parts. Additionally, in the case of spacecraft that must be "hardened" to the effects of nuclear blasts, it may be necessary to expose the subsystems and/or their components to radiation that simulates the expected blast environment.

14.4.3.2 Initial Magnetic Field

Spacecraft that carry sensitive magnetic field measuring instruments require that components and/or subsystems be tested to ensure that stray magnetic fields do not interfere with any magnetic field instrument's ability to correctly measure the on-orbit magnetic field. (Often this is accomplished by specifying that electronic parts and materials be non-magnetic.) Where necessary, early testing of other units must be performed to determine the effects of that unit on the spacecraft magnetic dipole. In other cases where the magnetic field instrument on the spacecraft is not particularly sensitive, it may be appropriate to perform analyses to determine whether other units will affect its operation.

14.4.3.3 Leakage

Hermetically sealed components and/or subsystems must be tested to ensure that they will remain sealed during vibration, acoustic, and thermal vacuum testing, launch vehicle ascent, and the vacuum of space.

14.4.3.4 Ambient Baseline Electrical

After initial electrical testing has been performed to debug the hardware, software, and the test procedure, a complete electrical performance verification test must be performed at ambient pressure, temperature, and humidity. The unit must be tested at the highest, lowest, and nominal input power supply voltages expected to be generated by the spacecraft power subsystem. These data become the baseline for all other testing which will be performed before, during, and after thermal vacuum, vibration, and acoustic environmental tests.

14.4.3.5 Electromagnetic Compatibility (EMC)

EMC testing of the subsystem and/or its components must be performed as early as possible to identify any problem areas that will keep the subsystem or component from being compatible with others and to correct those problems prior to system-level integration. MIL-STD-461 and -462 with tailored requirements are the generally accepted documents that control the conduct of EMC testing.

14.4.3.6 Optical and Mechanical Alignment

All subsystems with self-contained optics must be subjected to optical alignment at ambient pressure, temperature, and humidity. Components used as references for attitude determination and control must have reference surfaces established.

14.4.3.7 Mass Properties

In most cases the component and/or subsystem need only be weighed. However, for complex mechanical subsystems it may be necessary to measure moments of inertia and the location of the center of mass.

14.4.3.8 Temperature

The component and/or subsystem must be subjected to several hot–cold cycles in a thermal chamber (no vacuum) to induce thermal working of electrical connections via differential thermal expansion between dissimilar materials and to measure performance under the combined variables of temperature and input power supply voltage. Components with a high thermal mass must be allowed sufficient time to stabilize at the temperature extremes so as to ensure that the testing measures its performance properly. The baseline electrical performance verification procedure is performed at both hot and cold temperature plateaus and the results are compared to the ambient baseline electrical performance test. During temperature cycling, the shorter functional verification procedure may be executed. For components and/or subsystems that will experience rapid temperature changes it is important that they be subjected to thermal shock testing.

14.4.3.9 Vibration

Each component and/or subsystem is typically subjected to three-axis sine and random vibration at the levels defined for the program, with power applied. The requirement to expose the unit to vibration with power applied is based on previous experience that has shown this to be a valuable technique in identifying manufacturing defects. The functional verification procedure should be performed between each axis of the vibration exposure. In order to disclose potentially marginal situations, power to the unit should be monitored during the actual periods of vibration. If the unit is unpowered during the launch ascent and application of power presents an undue risk, it is prudent to reevaluate the requirement for vibration with power applied. Refer to chapter 8 for additional discussion of structural test verification.

14.4.3.10 Mechanical Shock and Acoustics

Mechanical shock and acoustics testing is performed on a case-by-case basis, depending upon the component and/or subsystem and the program requirements. Acoustic tests are primarily required of units that have large areas and low mass or thin-film windows and membranes. Shock testing is particularly appropriate if the component will be exposed to self-generated shock events or if it is located where it will receive significant shock from another event and is judged to be susceptible to that event.

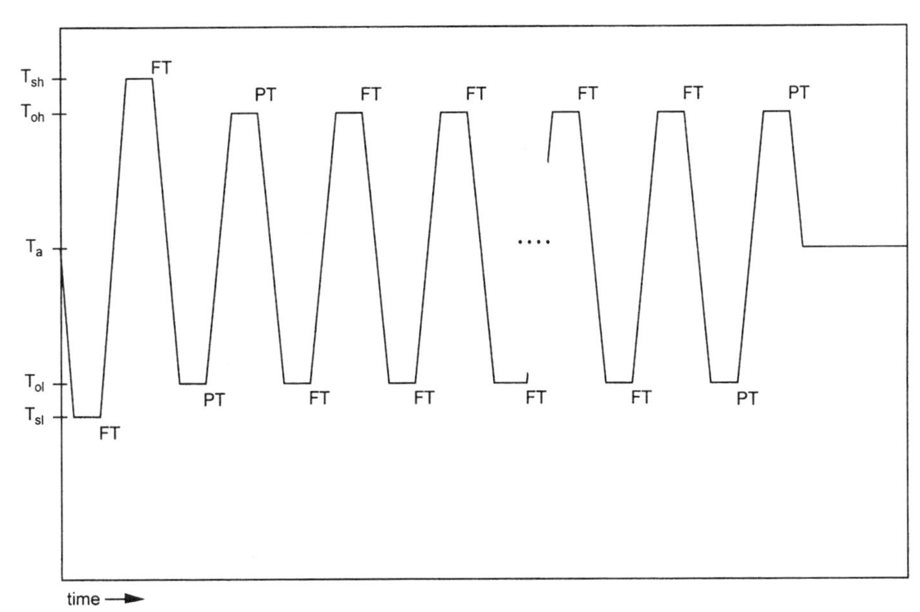

T_a = ambient temperature
$T_{sh,sl}$ = survival temperature, high or low
$T_{oh,\ ol}$ = operational temperature, high or low
FT = functional test at end of soak
PT = performance test at end of soak

Figure 14.8 Typical component thermal/thermal-vacuum profile.

14.4.3.11 Deployments

Deployment tests must be performed before and after exposure to the mechanical stresses listed above to demonstrate that the design is adequate and that there will be no "hangups." Examples of units that must undergo deployment testing include instrument sensor booms, antennae, solar arrays, instrument covers, and spacecraft–launch vehicle separation mechanisms.

14.4.3.12 Thermal Vacuum

Thermal vacuum tests demonstrate the ability of a component and/or subsystem to operate in the vacuum of space and, for components and/or subsystems that have high power dissipative "hot spots," that the thermal design is proper. If the unit has high-voltage power supplies that will be on during launch, they must be powered and monitored during pumpdown to ensure that there is no corona discharge, that is, no electrical arcing. If the unit has high-voltage power supplies but will not operate during launch, they must be left in an unpowered state until sufficient time has elapsed for adequate outgassing before power is applied. It is sometimes acceptable to substitute an increased number of thermal cycles over an extended temperature range in place of the thermal vacuum cycling if the component and/or subsystem has low power dissipation and no "hot spots." Figure 14.8 presents a typical component or subsystem thermal-vacuum test profile showing the use of the electrical functional and performance verification procedures at the end of each soak period.

14.4.3.13 Thermal Vacuum Bakeout

On spacecraft programs requiring above-normal cleanliness, thermal vacuum bake-out of components and/or subsystems is performed at the highest allowable temperature (that is, one that will not damage the unit in an unpowered state) to remove outgassing contaminants left over from the fabrication and test operations. The bake-out is normally performed during subsystem testing since there are components and/or subsystems on any spacecraft that cannot withstand high temperature bake-out (for instance., batteries) and a thermal vacuum bake-out at lowered temperature requires considerable time to remove outgassing contaminants. Normally, the test chamber is instrumented with a thermoelectric quartz crystal microbalance (TQCM) and a residual gas analyzer (RGA) to determine when all outgassing products have been removed. Figure 14.9 presents a profile of temperature and pressure versus time, which may be considered if it is not practical or possible to use facilities that are properly instrumented.

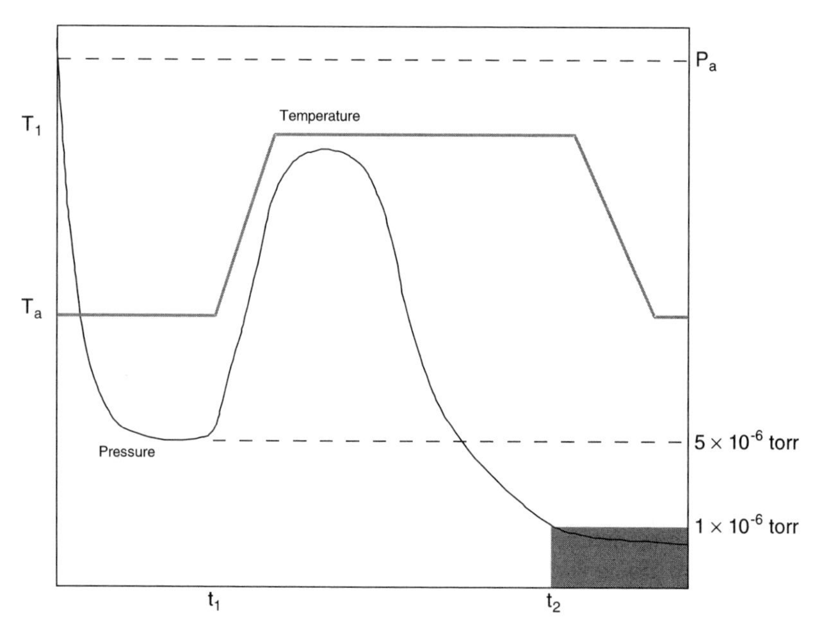

Note: Non-instrumented thermal vaccum bakeout. When a vaccum of 5×10^{-6} torr is achieved at the initial temperature T_a (usually ambient temperature), the chamber contents are heated to the bakeout temperature T_1 (limited by maximum component temperature), beginning at time t_1. When the temperature is increased, the outgassing rate of the chamber contents, and therefore the toal chamber pressure, increases. The elevated temperature is maintained until the total chamber pressure is reduced to 1×10^{-6} torr or less, shown in this plot at time t_2.

Figure 14.9 General time–temperature–pressure profile.

14.4.4 Subsystem Tests at the Spacecraft Level

Each subsystem and instrument that is delivered for integration with the spacecraft must first undergo an integration readiness review. The integration team will confirm that all required flight qualification tests have been executed and passed. Any outstanding issues will be discussed, drawings reviewed, and electrical interface definitions verified. Each electronics box installed on the spacecraft must have the following tests performed: safe-to-mate, aliveness, functional, and performance.

14.4.4.1 Safe-to-Mate

A safe-to-mate test is performed prior to mating any powered flight components. It consists of an unpowered test and a powered test. The unpowered test is used to verify that all power lines are isolated from return lines and ground. The powered test is used to verify that the proper voltages are provided on the proper pins, and not on any other pins in the harness connectors. The unit under test is also checked to ensure no circuits are shorted together that can cause damage to the spacecraft.

14.4.4.2 Aliveness Test

An aliveness test is a quick state-of-health verification that is used during portions of the environmental test program. Its main intent is to verify that the component can be powered on, receive commands, and generate telemetry. This test should take less than one hour to execute.

14.4.4.3 Functional Test

A functional test is designed to verify all power and data paths to and from the component under test. The test should exercise as many commands, as much telemetry, and as much functionality as the subsystem can provide without requiring external stimulus or GSE. If possible, this test should verify cross-strapping and other redundant interfaces. A functional test can take from one to three hours to complete.

14.4.4.4 Performance Test

A performance test is intended to be a detailed verification of the subsystem under test including as much testing of performance specifications as possible. GSE or external stimulus should be used, as necessary, to fully exercise the unit under test. Redundant interfaces and internal circuitry must be exercised and all modes of operation should be tested. This test should be as quantitative as possible and is the basis for the system baseline electrical test used throughout the environmental test program.

14.4.5 System Tests at the Spacecraft Level

Once the spacecraft has completed the integration phase, it is subjected to several tests to determine its readiness for launch. The spacecraft must be in the flight configuration with all subsystems integrated. Exceptions to this policy must be considered carefully on a case-by-case basis. For example, it is appropriate to use a solar array simulator during thermal vacuum testing to simulate solar array input to the power subsystem. As stated for subsystem and/or component testing, the requirement for the test, and the test limits, are typically unique for each program.

14.4.5.1 Ambient Baseline Electrical

This test is the reference point for determining proper performance of the spacecraft throughout all the other testing. It is performed at ambient pressure, temperature, and humidity. If the spacecraft is solar powered, a solar array simulator is used to simulate the output of the solar array(s) to the spacecraft battery charge control subsystem. Using a solar array simulator, the battery charge control subsystem is exercised to produce the minimum, maximum, and nominal bus voltages expected on orbit, simulating the range from 100% solar illumination to 100% eclipse. The functional test and performance test for each subsystem and instrument are executed under each condition of the battery charge control subsystem. This test specifically includes mission simulations run by the mission operations team.

14.4.5.2 Ground Station Compatibility

This test must be conducted to verify that the operational ground control station(s) can interface properly with the spacecraft and also with the mission operations ground system. RF compatibility must be verified between the ground station and the spacecraft. Often, a test van that is outfitted with identical RF equipment to that of the operational ground stations is used to conduct this test. RF performance is measured and all modes of command and telemetry encoding and decoding must be exercised. This is an excellent opportunity to exercise the software interfaces between the mission operations ground system and the ground station systems. This often constitutes the best end-to-end system test, involving all operational entities.

14.4.5.3 Mission Simulations

The mission operations team is provided opportunities to run the spacecraft in the manner they expect to operate on orbit. The launch-day simulation should be run many times to ensure pre- and post-launch configurations are properly activated and all operational personnel are familiar with the expected (and possibly unexpected) system performance. Mission-specific simulations should also be run; these often include rehearsal of early operations activities, rehearsal of propulsive maneuvers, a day-in-the-life, a week-in-the-life, and typical science data collection and download activities.

14.4.5.4 Electromagnetic Compatibility (EMC)

In order to determine compatibility between subsystems and satisfactory design margins, the spacecraft may undergo an EMC test. In certain cases a spacecraft may not require a formal MIL SPEC EMC test, but rather may rely on self-compatibility tests, where the primary interest is to verify that the spacecraft will not electrically interfere with itself.

The self-compatibility test is performed with the spacecraft in both the launch and orbital configurations. An important part of EMC is a clear understanding of the environment at the launch site, and it is appropriate to consider exposure of the spacecraft to the levels and RF frequencies which may be expected while in the processing facility or on the launch pad.

14.4.5.5 Initial Mass Properties

A set of initial mass property measurements of the spacecraft must be made to obtain an early indication of the amount and placement of balance weights that may be required to ensure that the spacecraft attitude control subsystem will be able to perform as required. This is also the first opportunity to evaluate predictions which have been used for mass properties up to this point, and confirm their accuracy. In the case of launch vehicles that are designed to be spin stabilized at some point in the launch sequence, the determination of balance weights and their location is a precursor to spin-balance with the launch vehicle.

14.4.5.6 Optical and Mechanical Alignment

In order to know the relationship between instrument sensors, antennas, and the spacecraft attitude subsystem components, an alignment test must be performed. This knowledge is necessary for determining the precise attitude of the spacecraft on orbit and later for processing science data from the instrument sensors. A full system alignment is performed prior to environmental testing, then verified between the vibration, acoustic, and thermal vacuum tests. It is verified again following transport to the launch site.

14.4.5.7 Magnetics

The spacecraft magnetic residual dipole must be measured (and in some cases trimmed to near zero) so that it will not affect the magnetic sensors on the spacecraft. Spacecraft containing magnetic torquing subsystems must have the characteristics of the torquing subsystem measured so that the effects on the magnetometer can be removed on orbit. Finally, the magnetometer subsystem itself must be calibrated.

14.4.5.8 Vibration, Mechanical Shock, and Acoustics

In order to validate the coupling analysis and simulate the environment during ascent, the spacecraft must be exposed to appropriate stress levels and durations, with all subsystems operating as they will during the actual launch. The spacecraft should be exposed to sinusoidal and random vibration testing. Large spacecraft should also have an acoustic test. These tests validate the ability of the spacecraft and its subsystems and instruments to survive the launch environment. Shock testing must be performed to validate that the system can survive launch vehicle engine ignition and burnout, fairing ejection, and separation of the spacecraft from the launch vehicle. The spacecraft subsystem and instrument performance should be verified, using the baseline electrical test procedure, between each major mechanical test. The shorter system functional test is generally used to verify the system performance between axes of the vibration test. Refer to chapter 8 for additional information on structural verification testing.

14.4.5.9 Deployments

Deployment tests must be performed after the complete sequence of vibration, mechanical shock, and acoustics testing to ensure that the devices still function properly after exposure to the simulated launch environment. Gravity compensation is required in some cases to validate the deployment. In certain situations the deployment tests also serve to validate the response of the spacecraft and its components to shock, since pyrotechnic devices will have been fired to perform the deployments. It should be recognized that some deployments are impractical or prohibited by instrument or facility constraints in a 1g ambient environment. In such a case, thorough analysis and testing must be conducted on a prototype of all or a part of that unit to ensure deployment will occur reliably when the spacecraft reaches orbit.

14.4.5.10 Thermal Vacuum

The thermal vacuum test consists of two parts: (1) thermal balance and (2) thermal cycling. Figure 14.10 illustrates some of the elements that must be considered as a part of the test, such as exposure to a survival temperature cycle, operation of the spacecraft

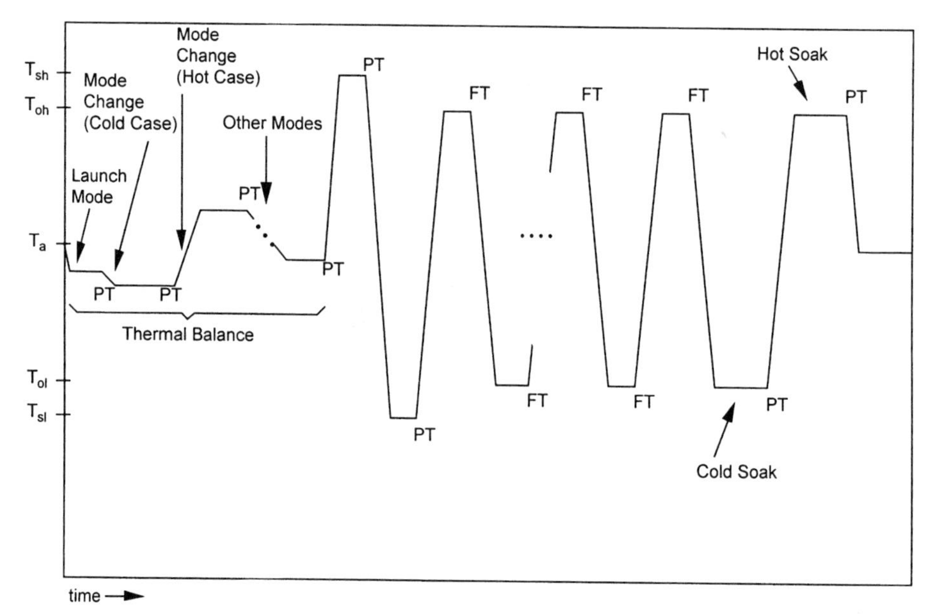

T_a = ambient temperature
$T_{sh,sl}$ = survival temperature, high or low
$T_{oh,ol}$ = operational temperature, high or low
FT = functional test at end of soak
PT = performance test at end of soak

Figure 14.10 Illustration, of spacecraft thermal-vacuum profile.

with respect to when it should be powered off and powered on, temperature levels, rate of change, dwell times, and number of cycles.

The purpose of the thermal balance test is to verify that the thermal design, based on the thermal analytical model, will keep the spacecraft operating within the required temperature limits. It is essential that temperature sensors be placed near points that can be correlated with the thermal analytical model to confirm the thermal design. The spacecraft must be placed in various operating modes that will occur on orbit and must be allowed to reach thermal equilibrium for various thermal input conditions, based on attitude and percent sunlight that are predicted to exist on orbit.

The thermal vacuum cycling test subjects the spacecraft to temperature extremes wider than it will experience on orbit. All subsystems that are to be on during the launch ascent must be on and operating during pumpdown. The spacecraft is subjected to a long cold soak and a long hot soak, with a number of cycles between the two extremes. The cycling tests are undertaken first, to detect malfunctions, followed by the long soaks which are intended to uncover malfunctions which might occur only after long exposure to the vacuum of space. The baseline electrical procedure is run at both hot and cold soak temperatures. Shorter functional tests are run at periodic intervals during temperature cycling to verify that the spacecraft subsystems continue to function properly.

14.4.5.11 Final Mass Properties and Spin Balance

Weight, center of gravity and moments of inertia must be measured. If the spacecraft contains a liquid propulsion system, the fuel tanks are normally filled with water to simulate the effects of fuel in the tanks, prior to the measurement of mass properties. Based on the final mass properties, balance (and despin weights, if any) must be calculated and installed and, if the launch vehicle is spin stabilized during the launch sequence, spin-balance is performed to ensure that the spacecraft will function properly during ascent and on orbit.

14.4.5.12 Workmanship Vibration

The inclusion of workmanship vibration in the verification program is optional and typically depends on the size of the spacecraft and unique program requirements. Its incorporation is intended to identify any defects which might have occurred during the various integration and test activities following exposure to earlier environments—for example, if a unit was removed, repaired, and reinstalled after vibration or acoustic tests. Levels and exposure times for this test will be lower than those required for initial verification. It is performed just prior to shipment to the launch site and, if performed, it must be followed by a complete baseline electrical test.

14.4.6 Launch Site Tests

After arrival at the launch site, the spacecraft is removed from the shipping container and thoroughly tested, using the baseline electrical procedure. Mating to the launch vehicle and required balancing operations (for example, final spin balance, if applicable) must be performed. All electrical interfaces between the launch vehicle and the spacecraft (such as umbilical and separation) must be checked, and flight ordnance must be installed. Radio-frequency compatibility tests between the spacecraft and the launch vehicle must be performed. Hazardous materials (cryogens, liquid fuel, etc.) must be loaded. The payload fairing (heatshield) must be installed. Finally, a simulated countdown must be performed, using the actual procedure that will be run on launch day.

Although this bounded discussion of launch site testing would seem to imply that it is a relatively simple operation, it is not! Each of the actions described above is a time-consuming and potentially hazardous operation and must be performed with care and caution.

Prior to launch day, non-flight protective covers and deployment or ordnance safing devices must be removed. Final electrical verification should be performed to ensure that the spacecraft is ready for its journey to a successful mission in space.

Problems

1. Using the exploded view of figure 14.6, develop a CPM ("bubble chart") schedule that indicates the order of integrating the subsystems onto the spacecraft. Assume the subsystems are qualified and ready for integration. You are not required to include dates or durations on your schedule.
2. You discover that the flight science data formatter (see main body electronics, figure 14.6) will be delivered later than required by your schedule of problem 1. Briefly discuss what, if anything, can or should be done to minimize a schedule delay.
3. Find an article in an engineering journal on the topic of integration of spacecraft or instruments. Turn in a copy of the article and a one-page, single-spaced summary of the important points.
4. Find an article in an engineering journal on the topic of testing of spacecraft or instruments. Turn in a copy and a one-page, single-spaced summary of the important points.
5. Find an article in an engineering journal on the topic of automated testing of spacecraft or instruments. Turn in a copy and a one-page, single-spaced summary of the important points.

References

Barney, R. D., 1990. Contamination control program for the cosmic background explorer: an overview. *Proc. 16th Space Simulation Conference.*, Albuquerque, NM, November.

Chu, M. L., and W. L. Mitnick, 1999. Lessons learned from the TIMED mini-moc. *Proc. R&D Symposium* 5, Space Satellite Technology Session, Laurel, MD, November, pp. 35–36.

Cranmer, J. H., et al., 1990. Thermal-vacuum bakeout procedure. JHU/APL private communication, October.

Dassoulas J., D. L. Margolies, and M. R. Peterson, 1985. The AMPTE/CCE spacecraft *IEEE Trans. Geoscience and Remote Sensing*, **GE-23**(3): 182.

FED-STD-209D, 1998. *Federal standard, clean room and work station requirements. controlled environment.* Commissioner, Federal Supply Service, General Service Administration, June 15.

GEVS-SE, 1990. *General environment verification specification for STS and ELV payloads, subsystems, and components.* NASA/GSFC, January.

Mehoke, D. S., 1991. Environmental test levels. JHU/APL private communication, January.

MESSENGER system test plan, 2002. APL Document No. 7384–9013, November 8.

MIL-STD-1540B, 1982. *Test requirements for space vehicles*. Department of the Air Force, October.

Mitnick, W. L., 1999. The TIMED mini-moc grows up: a successful strategy for reducing the cost of spacecraft testing and operations. *Proc. 18th Aerospace Testing Seminar*, Los Angeles, CA, March, pp. 449–454.

Oakes, J. B., 1998. Typical prototype development life cycle. JHU/APL private communication, January.

Peterson M. R., and D. G. Grant, 1987. The Polar BEAR spacecraft. *Johns Hopkins University APL Tech. Dig.,* **8**(3): 295.

SDO 11225, 2003. Space Department space flight system test requirements. JHU/APL, May.

SDO 11245, 2003. Space Department technical review requirements. JHU/APL, April.

15

Space Mission Operations

MARK E. HOLDRIDGE

15.1 Introduction

There comes a point in every space program when the development efforts are completed and the resulting flight hardware and software are ready for launch. The mission operations phase of the mission begins with the liftoff of the launch vehicle and continues through termination of spacecraft and payload operations. It is during this phase that a space program realizes its mission objectives and a return on the program's investment is finally realized.

But unmanned space operations come at significant risk as the space environment is a very unforgiving environment for robotics. Risks, if not identified and properly dealt with beforehand, can lead to only partial achievement of a mission's objectives or, in some unfortunate situations, complete mission failure. Hence, it is critical to start preparing early for a mission's operation. Because of the need to prepare in advance, and to constructively influence the architecture of successful flight and ground systems that will result, the mission operation team's participation should begin at program inception. Early involvement of mission operations personnel in the design process will add an element of realism to the process, as would the early involvement of a pilot in the design of an aircraft. Consequences to mission risk and operability are assessed as a "concept to operation" is developed and iterated with the mission objectives in mind.

In this day of widely varying space mission designs with diverse technologies and multiple objectives, successful missions are still built on very basic and proven building blocks of successful operations techniques. Mission objectives are driven by payloads that can be anything from a communications transponder to an imager on a planetary flyby trajectory. Because of this diversity in mission objectives, the concepts for conducting the primary phase of the mission can vary widely. But the day-to-day spacecraft maintenance or housekeeping operations must each be performed in the most disciplined manner. Rigorous operations approaches must also be applied to the payload operations.

This chapter will first describe space missions and the framework in which they are conducted. This will be followed by a description of the various activities performed by members of a space mission operation team over the course of a mission, from conceiving to preparing for and then, finally, conducting the mission at acceptable risks. Finally, proven operations techniques frequently used in successful space operations are summarized.

15.2 Supporting Ground System Architecture and Team Interfaces

While space operations take on many forms, it is possible to depict a highly general form applicable to all. Most space operations include the following key elements:

- Flight systems, including spacecraft and its payload.
- Tracking systems, including antennas and networks for two-way communications with flight systems.
- Ground systems, including the mission operations center (MOC), payload operations center (POC), data archive and distribution systems as required, and network infrastructure.
- Operations personnel, including core operations team in the MOC and supporting personnel in the POC, spacecraft engineering support, and management team.

Figure 15.1 depicts a generic version of a resulting space operation. Many variations are possible. These variations could include: backup MOCs required for system robustness, use of space-based relay satellites (TDRSS) for extended communications coverage, variations in flight systems architectures including surface lander or rover operations, and co-location or geographic separation between operational components and personnel.

15.2.1 Mission Operations Center (MOC)

The MOC is home to the core team responsible for the planning, assessment, and control of the spacecraft and sometimes its payload. Commands to the flight systems originate here or are piped through this facility to the tracking network and on to the flight systems. The MOC serves as the hub of operations and must therefore include the systems and tools required for the mission operations team to conduct its operation. MOC operational tools typically include

- Real-time spacecraft commanding and monitoring system
- Flight system performance and status assessment systems
- Mission planning and command scheduling tools
- Engineering data archive
- Flight simulators

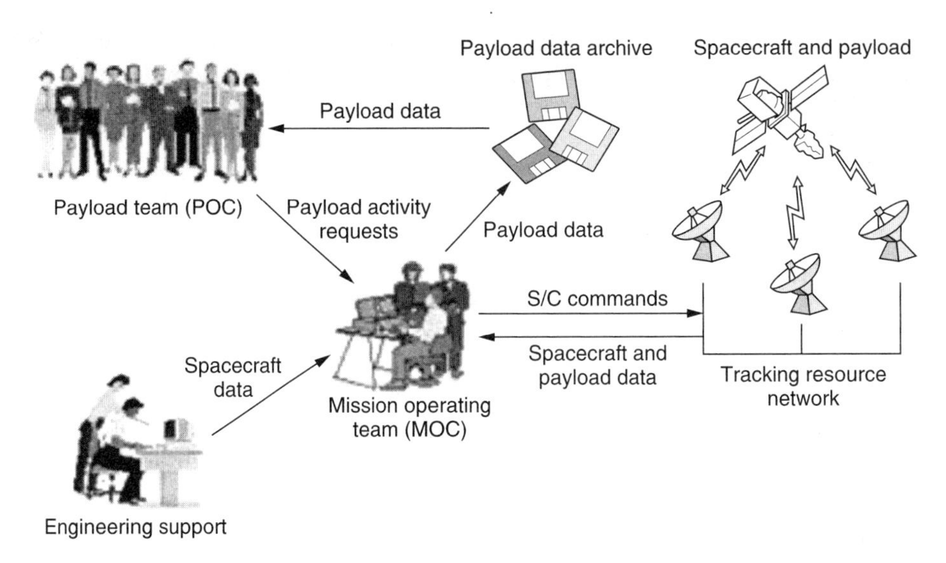

Figure 15.1 High-level space mission operations architecture.

Depending on the frequency and duration of spacecraft contacts, routine and contingency MOC operations may be round-the-clock. The MOC is generally the first receipient of spacecraft telemetry and many times serves as the "bent pipe" of payload data to the end user and must coordinate any data loss issues with the tracking network.

Because of its critical role in any space operation, the MOC architecture will typically incorporate a number of duplicate or backup systems to ensure critical operations and related data flows are not interrupted. These systems can be incorporated inside the MOC alongside existing or "primary" systems, forming primary and backup "strings" of equipment that support parallel data flows. A backup string might be implemented as a hot backup capable of taking over command and control functions immediately if a primary component fails. To protect against catastrophic facility failures (fire, flood, earthquake, and so on), backup systems and/or software are geographically displaced. When situations warrant, a backup MOC with fully redundant capabilities may be instituted.

15.2.2 Tracking Network

The tracking network includes the antenna systems and associated receivers and transmitters required to communicate with the spacecraft and compute range and range-rate measurements with the spacecraft. Antenna aperture sizes and transmitting power must be sufficient to provide the necessary uplink and downlink signal margins consistent with the mission's design including the spacecraft's RF system and Earth distances expected throughout the mission. Communications frequencies in the 1980s were typically S-band, but today X-band systems and Ka-band are more common because of requirements for increased data rates and improvements in RF technologies. The tracking network also includes those networks required for two-way data flow to and from the MOC systems and must be capable of handling expected data volumes and rates. Because of the large investment required to develop and operate a tracking network, these systems are typically used for multiple missions and hence tracking time must often be scheduled in advance.

15.2.3 Payload Operations Center (POC) and Payload Data Archive

The POC is where payload instrument planning and assessment functions are performed. For any given mission there may be one or more POCs (one for each instrument), or this function may be all or partly included in the MOC operations, depending on the level of coupling between the spacecraft and payload operations. The payload data archive serves as a repository and data distribution system for the payload data. The MOC may also have copies of this data for backup purposes but typically does not serve as the distributor to multiple instrumentors and scientists; the POC serves this function.

15.2.4 Spacecraft and Payload

This includes all orbiting flight systems. Typically the spacecraft includes subsystems for

- Power generation and distribution
- Onboard communications and data handling
- RF communications and tracking support
- Guidance and control (passive or active)
- Propulsion (optional)
- Mechanical structures

The payload could be a complement of science instruments for a science mission, a transponder for a communications spacecraft, visible or infrared imagers for Earth reconnaissance or weather, or one of many other possibilities. The payload is the reason for the mission, but its operation is typically not required to maintain spacecraft health and hence it may not always be operated. Instruments are often powered off during periods of inactivity or to conserve onboard power resources during special mission phases (launch, selected maneuvers, hibernation, and so on).

15.2.5 Operational Engineering Support

Engineering support to the operation includes members of the spacecraft and instrument development and test teams. Additional engineering support is required for launch and checkout activities and to be "on call" for the rest of the mission for special operations or contingency situations. Also included is support required for flight dynamics analysis of attitude and orbit maneuvers including guidance, control, and navigation functions. This includes the generation of actual and predicted spacecraft trajectories and related orbit products used in the operation.

15.2.6 Core Mission Operations team

Depending on the mission type and the payload, the mission operations team can be organized in many different ways; however, there is usually a core team responsible for the day-to-day operation of the spacecraft and planning its future operations. This team is located in the MOC with ready access to spacecraft engineering data and command capability. The team is composed of a multidisciplined set of operators and engineers/analysts, typically with education in aerospace engineering, computer science, and physics. Functions performed by this team include

- Flight and ground problem reporting and resolving
- Real-time command and monitoring
- Mission planning
- Spacecraft activity scheduling
- Payload activity scheduling (as required)
- Command scheduling and command load preparations
- Spacecraft performance assessment

Representatives of the mission operations team are brought on early in the mission concept development stage and are required for the duration of the operation. The distribution of these activities over time is discussed next.

15.3 Mission Phases and Core Operations Team Responsibilities

Figure 15.2 depicts a generic timeline for pre-launch activities and milestones, with typical activities conducted by members of the mission operations team shown below the timeline. Many reviews, both peer and external, are conducted over the spacecraft development and test cycle. It is critical to have members of the mission operations team actively participate in the development and reviews so as to expose concepts of operation to external review and report status. It is important to ensure adequate levels of resources are available to the operations. What is discussed here represents the generally accepted

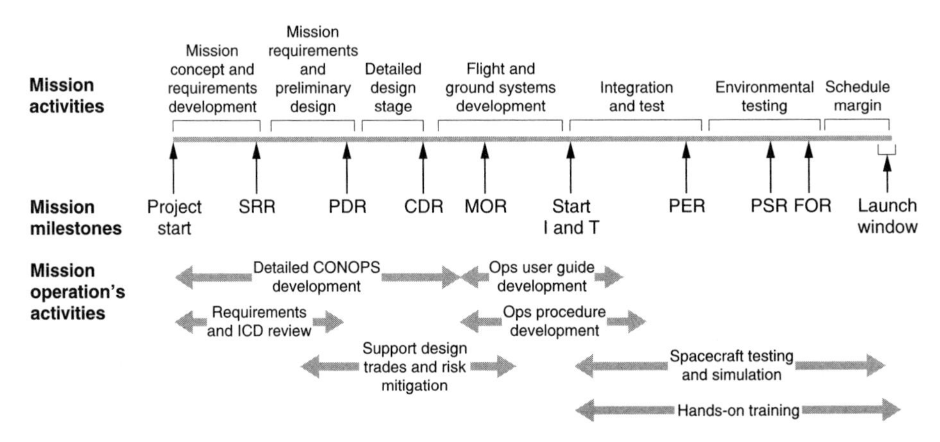

Figure 15.2 Mission timeline: pre-launch.

minimum level of preparations of a low-cost mission and not the limit to what can be done given sufficient time and funding.

15.3.1 Mission Concepts Development Phase

Space missions, while very diverse in their objectives, have a common set of phases they go through. From concept to completion, individuals, preferably with "hands on" mission experience, are involved in developing a concept of operation that minimizes total systems cost, evaluating risk consequences in design trades, and suggesting ways to improve overall operability. However, operations staffing levels for low-cost missions are at lowest numbers during the early phases of the mission. Hence it is important to carefully select the most experienced mission operations personnel during the start of the development when critical decisions are made, concepts are developed, and preparation plans finalized.

Mission operations personnel work with mission designers, flight system development, and ground system development team members to iterate on the system's designs and requirements to improve overall mission operability and reduce risk levels. Estimates of staffing levels required to perform the various mission scenarios are prepared, as well as ground system costs including communications or tracking resources that will be required to conduct the mission. The mission's design is evaluated for feasibility of implementation, such as the placement and frequency of orbital and attitude maneuvers required. A preliminary mission *Concept of Operations*, otherwise known as a CONOPS, is developed to flush out the best approaches for conducting the mission and to inform others less involved on how the systems they are developing will be used. The mission CONOPS maps out the following:

- Mission operations planning, control, and assessment strategies
- Spacecraft tracking resources utilized
- Ground systems and networks required, and plan for maintenance
- Data flows and interfaces between all mission elements
- Staffing level distribution over time
- Expected operational scenarios for key mission events including a "day in the life" of the mission

Mission planning activities include planning for all mission activities, both those related to spacecraft maintenance and those driven by payload requirements. Planning often includes coordination with all mission elements including payload operations, spacecraft engineering and analysis, and the production and use of products related to flight dynamics.

Control activities include commanding and monitoring of flight systems via the mission's ground systems. Control activities may be conducted with commands sent to the spacecraft and executed during real-time contacts with the spacecraft (via the tracking network) or they could be uplinked and stored on board the spacecraft for later execution at precise times as required. Real-time activities are typically those that require interactive control with verification steps. These include system checkouts and other off-nominal operations. Stored commands are typically used for those operations that can be safely conducted or simply need to occur at times when no real-time communications are available.

Assessment is the process of analyzing the flight system's performance either during real-time activities or after the fact, using recorded engineering telemetry. Flight systems activities are evaluated to see whether actual behavior is in accordance with the planned state of the flight systems over time. For those operations conducted out of sight of ground systems, engineering data recorded on board the spacecraft is transmitted to the ground for processing and analysis. The amount of assessment data available varies a great deal, depending on the design of the flight communications systems (antenna designs, frequencies, data rates), the type of tracking networks used (antenna aperture size, receiver design, and so on), tracking time available, and the "competition" for downlink bandwidth with payload operations.

Approaches for conducting all of these operations are developed during the concept phase.

15.3.2 Mission Requirements and Design Phase

During this phase, concepts are translated into specific requirements for flight and ground systems and then into specific designs for these systems. Members of the mission operations team develop and review requirements for flight and ground system components to ensure the systems that result are functionally complete and operable. This effort focuses on the flight (spacecraft) ground system (MOC) requirements. Details of designs are reviewed for compliance with requirements for general operability. Mission operations supports the design trades required to weigh the cost versus benefits of alternative implementations. Interface control documents (ICDs) for each external operational interface are reviewed and finalized.

Operations leadership prepares detailed schedules and staffing plans and detailed mission operations planning begins. This plan takes into account the CONOPS and the anticipated flight and ground system architecture based on requirements development. As designs mature, operations personnel generally assist in the make–buy–reuse decisions for flight and ground system components. Reuse is the key to reducing costs of development but is also very important in reducing operations-related costs as mission personnel can more efficiently operate systems with which they have experience.

15.3.3 Flight and Ground System Development Phase

During this phase, mission operations personnel continue to refine their detailed plans for mission operations. In addition, members of the mission operations team focus on developing the following items for later use during the integration and test phase and the eventual flight operation:

- Operations procedures and checklist development
- Operations user guide documentation
- Telemetry and command database development support
- Spacecraft and instrument command sequence development

15.3.4 Integration and Environmental Testing

The mission operations and payload operations team members continue with their procedure and command sequence development efforts during the integration and test period when flight components are delivered to the integration team and the spacecraft and its payload begin to take shape. The mission operations team will begin participation in coordinated spacecraft testing (the sooner the better) and actually run its own tests with the flight hardware. The emphasis will be on executing operation procedures and command sequences developed to date, to confirm their proper operation in as realistic an operational environment as possible.

Two-way data flow tests between the MOC and the spacecraft will be conducted to validate the ground system components. These same interfaces will be exercised during operations-directed mission simulations and dress rehearsals. Its is extremely important that these mission simulations be executed through the use of controlled flight procedures and scripts that will later be reused for flight operations. Since these simulations can later be conducted during spacecraft environmental testing, it is desirable to conduct these tests under realistic spacecraft conditions while the spacecraft is in the thermal vacuum chamber. This testing is the best "hands-on" training the team will receive prior to launch. The operation begins to look and feel like the actual flight operation, serving as a valuable training exercise that flushes out remaining problems with the end-to-end system. RF spacecraft communications are often routed through the satellite tracking service providers to test the space-to-ground interface but also to verify proper telemetry, command, tracking data, and antenna systems to monitor data flows with flight controllers in the loop.

As operational elements (telemetry and command databases, flight procedures, command blocks, and so on) are tested successfully, they are placed under configuration control. Once fully validated and under configuration control, these operational elements are considered ready for flight operations use.

15.3.5 Launch and Commissioning

"Mission operations" officially begin at launch vehicle liftoff. The first operations that take place (as shown in figure 15.3) are those surrounding the separation between the spacecraft and its launch vehicle. These separation-related activities are usually implemented using "canned" sequences loaded on board the spacecraft prior to launch and initiated automatically by the spacecraft if critical to the early operation. Separation activities typically include placing the spacecraft in a power-safe configuration and establishing communications with the ground.

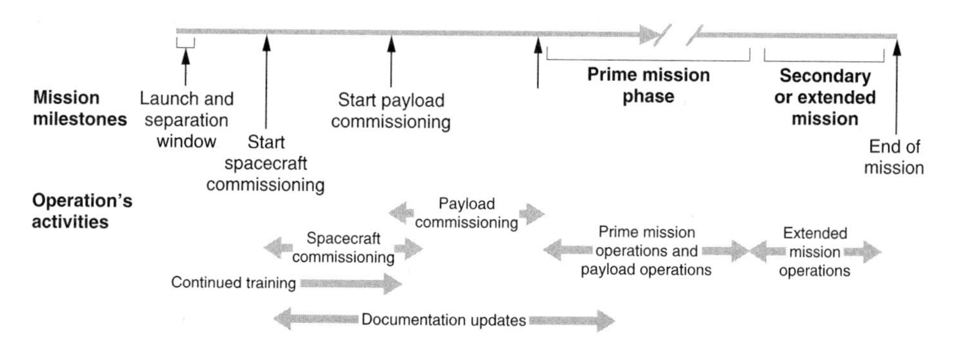

Figure 15.3 Mission timeline: post-launch.

After two-way communications are established between the MOC and the spacecraft, a full assessment of the spacecraft can begin as well as the conduct of any time-critical activities. These will be followed by spacecraft and payload checkout activities that can last for 30 to 60 days. An increased presence of spacecraft subsystem and payload engineering support is required to fully assess subsystem performance during this phase. All operations team members shown in figure 15.1 are utilized during launch and commissioning activities. After the flight systems have been commissioned, the primary mission begins.

15.3.6 Prime Mission Phase

During the prime mission phase, operation team members carry out the primary mission objectives. Payload operations are carried out in the MOC and the related payload operations centers, both taking center stage. For a planetary scientific mission, a cruise period of one or more years may be required before payload operations can commence. During this cruise period, in-flight testing of future science operations is conducted to ensure future readiness. For Earth-orbiting spacecraft, primary payload results might begin immediately after the commissioning. During this phase of the mission the operations team executes its tested procedures in the day-to-day conduct of the mission. All operations team members and facilities shown in figure 15.1 are involved in executing the primary mission.

The frequency of contacts with the spacecraft is entirely mission specific, with some missions (primarily geostationary spacecraft) maintaining constant communications while others, such as spacecraft operated in deep space, may go days or more between regular contacts. Mission operations activities during the mission typically can be thought of in two categories, real-time and offline activities.

Real-time flight operations are those conducted during the spacecraft contacts, and typically include the following activities:

- Real-time commanding and stored command loading
- Ephemeris (orbit prediction) updates for spacecraft pointing control
- Flight recorder data playbacks of engineering and payload data
- Real-time health and status assessments of flight systems
- Monitoring of critical operations activities
- Spacecraft tracking as required for orbit determination
- Contingency handling

Offline activities include the mission planning function and long-term assessment of spacecraft performance. These non-real-time activities include:

- Detailed assessment of spacecraft systems
- Planning of future spacecraft and payload events
- Command scheduling of future stored command loads
- Testing of command loads and updated command scripts
- Archiving and distribution of mission payloads data
- Evaluation of payload results and adjustments to mission plan as required
- Attitude and orbit maneuver planning to meet mission objectives

Mission operations vary widely in their duration but typically are one to five years long. Some missions go well beyond five years by design or if an extended mission is approved. During periods of decreased levels of intensity or during extended missions, the mission operations team is normally downsized and additional cost-saving measures are sought.

15.4 Mission Diversity

There is much diversity in space mission types and objectives. Mission characteristics can be defined in many ways; however, from a mission operations perspective it is useful to think of the following:

- Spacecraft's orbit or trajectory
- Mission sponsor and payload type
- Spacecraft attitude control approach

Each of these characterizations greatly affects mission operating characteristics and the manner in which the operation is performed. Table 15.1 summarizes these relationships. Mission operating characteristics include such things as expected real-time contact durations, spacecraft communications data rates, round-trip-light-times (communications delays), intensity and complexity of payload operations, commanding practices, use of autonomy, staffing levels, risk levels and level of "forgiveness" of mission objectives, and security practices. Higher security practices tend to increase

Table 15.1 Summary of operations characteristics versus mission characteristics

Spacecraft orbit	LEO	Short contacts, high data rates, negligible round-trip-light-times (RTLTs), repetitive operation, ground and flight systems automation candidate
	MEO/GEO	Longer contacts up to continuous coverage, moderate data rates, minimal RTLTs (< 1 sec), candidate for ground-based autonomy, low risk, steady state operation
	Deep-space	Periodic contacts, low data rates, long RTLTs, unique event-driven operations, flight-based autonomy, large aperture antennas, higher risk levels, science payloads.
Spacecraft type	3-axis	Flexible but complex operation, higher risk, can be highly automated
	Spin	Simpler implementation and low risk, less flexible, manual intensive
	Gravity gradient	Simple passive approach for orbiters, little flexibility, simple operation.
Mission sponsor (all U.S.)	NASA	Science and technology driven missions, moderate security, moderate risk/moderate cost.
	DOD	National defense and technology driven, highest security, moderate risk/moderate cost
	Commercial	Communications remote imagery applications, lower security, low risk/low cost

complexity and cost along with other aspects. A brief discussion, with examples of how mission characteristics vary depending on mission orbital characteristics, follows.

15.4.1 Spacecraft Orbit

Spacecraft orbits are typically classified as one of three types for mission operations. They are:

- Low to medium altitude Earth orbiting (LEO)
- Medium or geosynchronous Earth orbiting (MEO or GEO)
- Deep space (heliocentric)

While many spacecraft orbits do not fit directly into one of these categories, they can be described as having characteristics similar to those above. For instance, a highly eccentric Earth-orbiting spacecraft may exhibit operating characteristics similar to those of a GEO spacecraft at apogee and a LEO spacecraft at perigee.

LEO missions tend to have the most diverse possibilities for payloads, including communications, science, weather, and instrumentation related to Earth resources. Hence, LEO mission sponsors are just as diverse in nature. If a ground-based tracking network is being used for an LEO spacecraft operation, spacecraft contacts are typically 10–15 minutes in duration with 4–6 contacts per day visible from a single tracking site. Exact numbers of contacts are dependent on the LEO orbit inclinations, altitudes, and the number of tracking antennas/sites. The contacts have comparatively high communications data rates, lasting for short periods separated by long delays between.

LEO missions utilizing space-based communications relay satellites such as NASA's TDRS satellites are not bound to short contact durations, but that added flexibility comes at a price to develop and operate the relay spacecraft. LEO missions tend to be more repetitive in nature, with similar operations from one orbit to another, especially for Earth-observing satellites, hence they are excellent candidates for automation and extensive use of procedures, checklists, and hence less experienced operations personnel (each reduces mission operations costs). Autonomous monitoring functions (to reduce operating costs or decrease reaction times) can be placed on board or even on the ground (less costly) in the MOC, given the short communications delays typical of LEO spacecraft.

A GEO spacecraft "sits" over one Earth point and can "see" a very large footprint around that point. These missions typically have communications or weather-monitoring payloads with continuous real-time communications possible from a single antenna complex. A continuous operation with a somewhat slower pace and lower risk is possible. Engineering data rates for GEO missions tend to be lower as the data transfer is continuous and the reception of engineering data is spread over a much longer time period than for LEO missions that must burst data back during relatively short contacts. GEO mission operations can become steady state in nature (except for contingencies!) and hence are even more attractive to ground-based automation and operations cost reduction.

Spacecraft on Earth escape heliocentric trajectories travel great distances from Earth and are scientific in nature with a complement of scientific instruments. These missions are typically sponsored by governments in the spirit of discovery or exploration. This translates into much lower data rates to and from the spacecraft and delays in the signal transmissions, otherwise referred to as light-time delays or round-trip-light-times (RTLT). For instance, the speed of light is ~8.338 min/AU, so for a spacecraft at Jupiter distance (~5 AU) the RTLT, the minimum delay between sending a command and seeing its response, is approximately 84 minutes. Delays in communications can translate into delays in reaction to contingency situations. This leads to a greater dependency on responsive autonomous onboard corrective actions. Long cruise periods and the high cost of deep space tracking resources often lead to limited communications and tracking (typically 1–3/week) for deep space missions, hence the spacecraft must be able to take care of itself for days, weeks, or in some cases months without immediate assistance from the ground. This strengthens the need for autonomous decision making on board the spacecraft and its instruments, all of which must be thoroughly tested prior to launch.

Some planetary missions start off on a heliocentric trajectory and then are placed into orbit around anther body. These missions then behave more as a LEO mission (repetitive) but with long-distance communications characteristics. Other heliocentric missions can be characterized as flyby or encounter missions with a particular targeted body only viewable for a short time after years of cruise. These missions tend to be very unforgiving in that there is usually only one opportunity for success. Operations costs for planetary missions are difficult to reduce as the "payoff" is typically at the end of the operation and experienced personnel must be available after cruise periods that may last years. During cruise it is possible to reduce costs through the incorporation of hibernation or beacon-mode operations with only periodic "check-ins" necessary. However, a plan for mitigating knowledge loss by the operations team must be developed to compensate for long periods of inactivity and likely turnover in staffing.

15.5 Standard Operations Practices

Forty-plus years of operating unmanned space probes has led to many lessons learned as flight operations mature with time. Those who paved the way are leaving the workforce, so there is greater need than ever to document successful approaches for those that follow. Despite the many types of missions now being flown, as discussed earlier, standard practices can be developed and applied to the operation of these missions regardless of the orbit or payload type. Just as standard practices are used in the operations and maintenance of aircraft, so too are standard practices applied to space operations. These standards of conduct can greatly improve the predictability of mission outcome and the overall efficiency for a given mission, if applied correctly. If misused, standards can become a hurdle that complicates a mission operation with processes and bureaucracy that get in the way of the mission at hand. Not all practices are useful for all missions. The goal is to strike a balance between those practices that improve and those that impede the operation. An experienced operations leader or mission operations manager plays a key role in making those key judgment calls.

15.5.1 Concept of Mission Operations

It is important for experienced mission operations representatives to work closely with the design team to formulate operational scenarios for each phase of the mission.

This should translate early on to a concept of operations (CONOPS) that drives the requirements and design effort towards improved operability and lower risk. As actual design details emerge from the design efforts, concepts will likely be adjusted to account for the as-built design. As scenarios are taken to greater detail, detailed mission operations plans are developed that make these concepts known to all members of the operations team for continued iteration and improvement. Specific mission characteristics (as discussed previously) determine the necessary contents for the CONOPS and mission plans as resulting operations characteristics vary. For instance, spacecraft without a propulsion system need not have scenarios for conducting orbital maneuvers. However, despite differences in missions, there is commonality to all space missions, including the standard practices discussed next.

15.5.2 Configuration Management

The success of mission operations depends greatly on proper control of operationally critical elements. These elements include

- Scripts and procedures
- Operational databases (telemetry and command)
- Flight and ground software
- Spacecraft command loads
- Ground system hardware and networks
- Spacecraft simulation tools
- Critical documentation
- Spacecraft autonomy rules and onboard parameters
- Flight constraints

Prior to actual flight use of the above elements, each must be properly reviewed and tested in a manner as "flight like" as possible. After verification, each must be controlled in a manner that permits use of only authorized versions. The means for enforcing these controls can and should be tailored to available mission resources and appropriate risk levels without adding unnecessary workloads and delays.

15.5.3 Application of Flight Constraints

As the design of the spacecraft and its instruments matures, constraints or limitations in its operation surface. Some of these constraints will not be violated in any normal operating situation; others could easily be violated, depending on how a particular activity is conducted. Hence flight constraints or flight rules need to be documented and made available to the entire operations team. Mission operations plans and procedures must be developed in a manner that is consistent with these known operating limitations. Examples of operating limitations include

- Operating spacecraft pointing relative to Sun bounds (for power or thermal control)
- Sensor pointing constraints
- Command ordering (i.e., "command y must always follow command x")
- Command timing (i.e., "command y must always be separated by 2 seconds from command x")
- Power and heater management

- Configuration guidelines (i.e., "transmitter 1 shall always be connected to antenna A or B")
- Never do's (i.e., "never turn off communications and data handling processor")
- In the event (i.e, "in the event of under voltage, switch regulators")
- On-time limitation ("never leave calibration lamp on for more than 1 minute")

These are just examples of the types of rules that mission operations personnel utilize. As the complexity of spacecraft and instrument designs increases so does the number of rules. Each rule requires a means of operational implementation or enforcement. Depending on the level of autonomy built into the flight systems, many rules or constraints may be checked continuously on board via autonomy rules. Other rules may be checked by ground-based procedures or software in the command planning systems; others in standard and contingency procedures.

Regardless of how a rule or constraint is enforced, there needs to be a means to map each rule with the method or methods for implementing to ensure none are ever violated unintentionally. Also, rules can be divided into two groups: those that can never be violated, and those that can be violated under special circumstances such as significant operational events that warrant softening of operating bounds.

15.5.4 Training and Certification

Given the unforgiving nature of space operations, it is critical that all members of the team understand how to properly perform their duties and also back up other members of the team when needed. This is especially important for long-duration missions (one year +) where turnover in team members is likely. Standards for training are most often applied to the real-time operations as those operations are closest to the spacecraft and most likely to directly affect the spacecraft in a positive or negative way. For this reason rigorous training guidelines are required to certify each real-time operator in the conduct of both routine and contingency operations, using approved operational procedures. Typically this training is conducted as "on the job training" where a new team member works with more experienced personnel and, after having demonstrated competency in handling routine and contingency operations, is certified by that same experienced operator. Other means for training can include viewing taped video training exercises, reading operations user's manuals, and participating in simulator operations and rehearsals.

15.5.5 Real-Time Operations

It is important that real-time operations be conducted in a procedure-driven and highly disciplined manner to ensure predictable and consistent results. Spacecraft commanding activities are at the center of these activities and a controlled process is required to ensure only those commands planned and approved are sent to the flight systems. One way to ensure this is through the use of an uplink approval process, which usually utilizes a standard form that describes the commands to be used, the window to be sent in, the rationale for sending them, and approval signatures. Usually the mission operations manager or a designee serves as the gatekeeper, signing off on all non-routine commands. This ensures that a person responsible for the entire operation and knowledgeable on how one activity can affect another is kept informed of all real-time activities. Another technique commonly used to safeguard real-time commanding is the "two person rule." This rule requires at least two individuals participate in any commanding exercise and hence back each other up by reviewing each step before proceeding. This same technique can, of course, be applied to any and all aspects of the operation including the behind-the-scenes planning activities.

15.5.6 Documentation

Operations documentation is used to permanently preserve operations plans for review, to describe how operations systems (flight and ground) operate, and serves to define and control changes to procedures and approaches. It is most useful for those who will later use the documentation to develop it. Documents most commonly developed in support of space missions include

- Concepts of operations
- Spacecraft and ground system user guides
- Text procedures for standard and contingency operations
- Plans for training and configuration management
- Schedules for work plans
- Operating constraints (flight rules)
- Operating logs

15.5.7 Contingency Planning

Anticipating potential problems and preparing for them will reduce response times and could save a mission. Many contingencies are foreseen and included in onboard autonomy status checks and built-in responses to minimize reaction times. However, there are some contingencies that must be dealt with on the ground, including recovering from "safe modes" transitions and other configuration changes initiated by the onboard autonomy system. Other contingencies are ground system related, such as problems with the tracking systems and related networks that result in "negative acquisition of signal" or, in other words, unexpected loss of communications. Ground systems typically include backup systems and hence procedures are needed to fail-over to these redundant systems. There is never enough time to develop contingency procedures for all possible contingencies, so the focus is usually on those deemed either as most likely or most immediately threatening to the mission.

15.5.8 Spacecraft Performance Assessment

Spacecraft performance and status assessment is usually performed in two phases. The first is during real-time flight operations while engineering telemetry is flowing to the ground. Limits and status checks are normally performed by the real-time systems controlling the spacecraft. Alarms are displayed to user screens on event pages. This permits automatic and thorough checking of all critical telemetry points in real time and minimizes workloads on real-time operations personnel who would otherwise have to look at thousands of telemetry values, an impossible task. Contingency procedures provide the proper response even if it is simply to contact the experts. It is during these real-time contacts that engineering data recorded since the last contact is played back for offline assessment. In addition to limit checking, actual spacecraft status is compared to the expected status to verify the spacecraft is in the expected state. Care is taken to focus on those parameters deemed most likely to have problems or most critical to the operation. These typically include moving mechanical components, high-voltage items, consumables, onboard autonomy status, and critical operating limits (temperatures, voltages, currents, etc.).

More in-depth assessment is performed after the real-time contacts. This form of assessment usually involves plotting key telemetry over hours, days, or weeks to characterize behavior and compare actual behavior to that expected for current circumstances. Longer term behavior analysis is performed either by plotting data over longer time periods (CPU intensive) or by plotting samples or averages of each critical telemetry point over periods as long as several years to pull out trends in performance. This is usually referred to as "trending."

Whether by trending or real-time monitoring, the key goal of performance assessment is to proactively detect problems *before* they become mission threatening.

Problems

1. Assuming the predicted spacecraft signal strengths shown in the figure below (note logarithmic ordinate):

   (a) What is the maximum downlink bit rate possible at 0.6 AU, and which antenna should be used (parabolic high-gain antenna or omnidirectional low-gain antenna)? *Answer*: 5.5 kbps on HGA.

   (b) How long would it take to downlink a 1 megabit image at this data rate? *Answer*: 181.8 s or ~3 min.

   (c) What is the maximum uplink bit rate and the preferred antenna at the same distance? *Answer*: 500 bps on HGA.

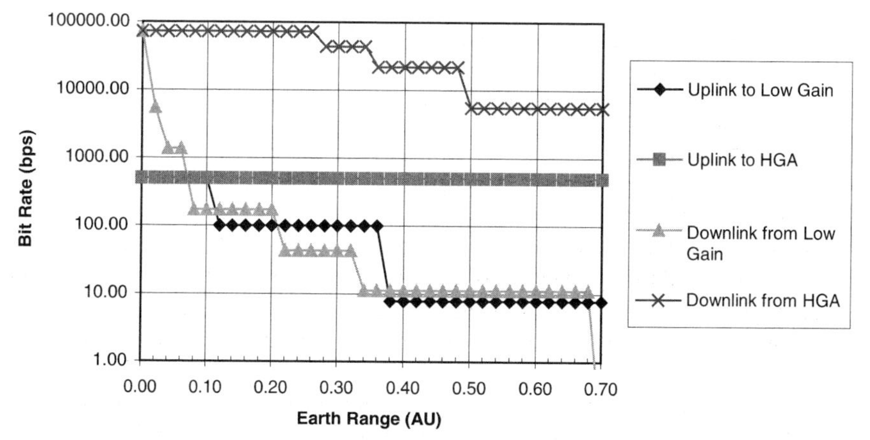

Bill rate capability versus Earth range.

(d) How quickly then could a 500 kilobit command load be uplinked? How long would it take on the low-gain antenna?
Answer: 1000 s on the HGA, or 13.8 h on the LGA.

(e) Under what circumstances would you expect the low-gain antenna to be used, given these results?
Answer: The LGA has omnidirectional capability so it can provide communications when the parabolic antenna cannot be directed at Earth. However the LGA provides significantly lower data rates, hence it would be used only when HGA communications are not possible or practical, or when higher data rates simply are not needed.

2. Assume 346 W is required to power a spacecraft and its instrument in a particular configuration. The spacecraft's solar panels can generate 500 W when the Sun's angle of incidence to the solar panels (α) equals zero (see figure).

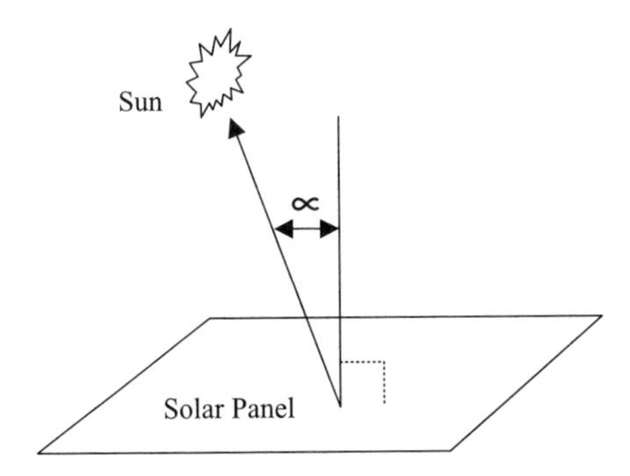

(a) Assuming the power generated drops off according to the cosine law, how great can α be (α_{max}) to maintain a positive power balance?
Answer: $\alpha_{max} = 46.21°$.

(b) How does this α_{max} angle change if a 10% power margin is desired?
Answer: $\alpha_{max} = 40.43°$ with 10% power margin.

(c) What is α later in the mission if power generation from the solar array is reduced (say from long-term degradation) by 10% and the 10% margin is still desired?
Answer: $\alpha_{max} = 32.24°$ with 10% power margin and 10% reduction in solar panel output.

3. A spacecraft's communications and data handling processor is capable of storing one megabit's worth of commands.

(a) Assuming each telecommand packet is 1280 bits long, how many commands can be stored onboard?
Answer: 781 commands.

(b) With an uplink bit rate of 7.8 bps, how long would a 781 command uplink take?
Answer: 35.6 h.

4. Find a relevant article on automation of space mission operations. Write a summary of methods used to automate the operation. Potential sources would be annual proceedings from American Astronautical Society (AAS), International Academy of Astronautics (IAA), or American Institute of Aeronautics and Astronautics (AIAA) conferences.
5. Find a relevant article on low-cost space mission operations. Write a summary of the methods used to reduce mission operations costs. Consider using the same sources as in problem 4.

6. Find a relevant article on a space mission ground system development. Describe the technologies utilized and the backup strategy. Consider using the same sources as in problem 4.
7. Find a relevant article on a particularly challenging or unique space mission operation. Describe the challenges and how they were overcome. Consider using the same sources as in problem 4.

16

Nanosatellite Conceptual Design

VINCENT L. PISACANE

16.1 Introduction

This chapter integrates the information presented earlier by illustrating the fundamentals of the conceptual design of a simplified spacecraft mission. A conceptual design is the initial step in the description of a space system in which the overall mission requirements are defined, the mission is described, a concept of operations defined, subsystem specifications determined, and initial characteristics and performance of each subsystem assessed. In addition, schedules, costs, and mission and development risks are identified. The simplified mission addressed here is one that engineering students in a senior design class could design and build over a period of about two years if supported by knowledgeable faculty and/or professionals from the space industry. It should be cautioned that the description as presented should not be construed as anything more than the first steps in a conceptual design that must go through the rigors of the preliminary and critical design phases. The emphasis in the following description is on the design of the spacecraft.

16.2 Mission Statement

The proposed mission is to deploy a nanosatellite to provide an amateur radio communications repeater node in space employing the Federal Communications Commission approved AX.25 amateur packet-radio protocol at 145.825 MHz. A secondary goal is to carry an instrument to support communications research consistent with the goals of the amateur radio community. The instrument is a radiation dosimeter to record and store radiation spectra and transmit the data to the control station. The ability to collect and store time-correlated spectra will permit the time dependency of the radiation environment to be determined. The spacecraft is to have a useful life of at least one year. Access to space will be obtained as a secondary payload on launches of opportunity. The Delta IV launch vehicle is used to characterize a typical launch environment. An existing satellite control station will monitor and control the spacecraft.

16.3 Background

Packet radio communication is a digital mode of amateur radio communications, similar to computer telecommunications. Data from a computer is sent by radio utilizing a terminal node controller (TNC) and is received by a site with similar equipment. During typical terrestrial transmissions the TNC formats the message into packets, keys the transmitter, and sends the packets to the transmitter. On reception at a station, the TNC receives the packet from the receiver, decodes the message, carries out error detection and correction, and sends the message to the terminal of a computer. In the spacecraft implementation, the spacecraft telecommunication system, with the TNC as a key component, acts as a repeater to retransmit messages. This will permit amateur radio operations over a wider area since communication at VHF frequencies is line of sight, limiting geographic coverage to less than about 30 km.

Knowledge of the environmental radiation intensity and distribution is important to terrestrial and space radio communications. Increased radiation levels, especially from enhanced solar activity, can disrupt terrestrial communications, provide a hazard to astronauts (Dicello et al., 1994), and in severe circumstances adversely affect power grids. This information would be a valuable contribution to the development of space weather models. A digital radiation dosimeter, similar to the type proposed by Rosenfeld and colleagues (1999, 2000, 2001) is employed.

16.4 System Requirements

System level requirements can be summarized as follows:

- Provide a spacecraft repeater to extend reliable amateur digital radio communications over a broader geographical region than afforded by direct line-of-sight VHF communications using the AX.25 packet-radio protocol at 145.825 MHz
- Make observations of the space radiation environment characterized by 8 spectra between transmissions to the ground station, with each spectrum up to 1024 energy levels and a maximum 10^9 counts per channel
- For data reduction, determine the ephemeris a posteriori to an accuracy of 10 km
- Operate the spacecraft from an existing ground station
- Design for a minimum mission duration of one year
- Satisfy the requirements for mass and size for classification as a nanosatellite
- Minimize cost to the extreme
- Have the capability to be launched as a secondary payload on a wide variety of launch opportunities

As a result, the mission design parameters selected are:

- System
  - —Mission duration: $\geq$1 year
  - —Clock accuracy: 1 s
- Spacecraft
  - —Mass: $\leq$ 20 kg
  - —Volume: $\leq$ 45 cm $\times$ 45 cm $\times$ 45 cm
- Orbit
  - —Altitude $\approx$ 500 km (nominal)
  - —Inclination $\approx$ 51° (nominal)
  - —Eccentricity: near zero (nominal)
  - —Orbit determination:
    Source: USSTRATCOM two-line elements
    Accuracy: 10 km a posteriori, 50 km real time
- Launch vehicle
  - —Delta IV with the Evolved Expendable Launch Vehicle (EELV) Secondary Payload Adapter (ESPA) and the Lightband® separation system
- Spacecraft telemetry
  - —Frequency: 145.825 MHz
- Ground station
  - —Frequency: 145.825 MHz
  - —G/T: 0.00125/K
  - —Operations about once per week
- Amateur radio telecommunications
  - —Frequency: 145.825 MHz
  - —AX.25 (level 2, version 2) link-layer protocol specification
  - —Bit error rate (BER) = 10^{-7}
  - —Coverage area: 1950 km for minimum elevation of 5°

The mission design life is one year but, since no expendables are anticipated, the spacecraft should perform for an extended period. The design orbit is not restrictive in terms of inclination and eccentricity and, to some extent, altitude. Consequently, this will permit the spacecraft to be placed in a variety of low-altitude orbit configurations to take advantage of a broad range of secondary launch opportunities. A non-equatorial inclination is desired to near-globally sample the radiation environment, especially the South Atlantic anomaly, and to provide a communications node for amateur radio operators in the middle latitudes. The launch vehicle is unspecified but the Delta IV launch vehicle is used to establish a representative launch environment. An existing ground station is used with the capabilities identified above, to minimize cost. A spacecraft system block diagram is given in figure 16.1, and the deployed configuration is illustrated in figure 16.2.

16.5 Concept of Operations

The spacecraft will be launched into an orbit of opportunity as a secondary payload on a Delta IV launch vehicle utilizing the Evolved Expendable Launch Vehicle (EELV) Secondary Payload Adapter (ESPA). This system provides for up to six secondary payloads with masses up to 180 kg, each fitting within an envelope of 61 cm $\times$ 61 cm $\times$ 96 cm as illustrated in figure 16.3. The altitude selected for the mission is 500 km and the inclination is 51° in order to support amateur radio communications over a broad

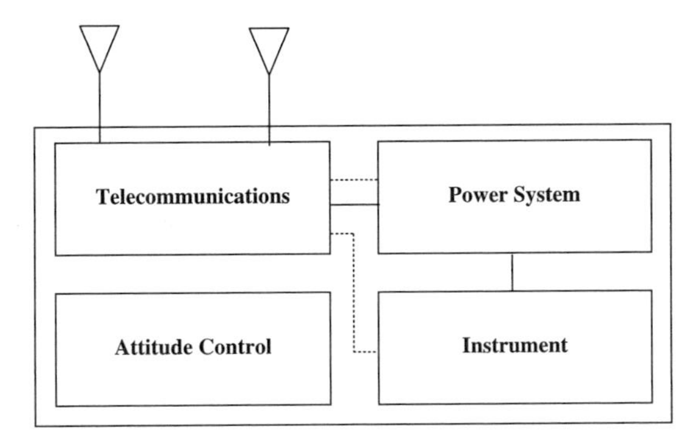

Figure 16.1 Spacecraft block diagram: ——, power; ...., signal.

geographic region. The spacecraft will be operated from a single ground station at 39° latitude. The amateur radio communications node and the instrument will be on continuously. Ground station operations will be minimized by storing the instrument data on the spacecraft with sufficient memory to transmit the data at arbitrary intervals. The spacecraft trajectory and ground station alerts will be obtained from the two-line elements provided by the United States Strategic Command, formerly NORAD.

16.6 Risk Assessment

There is no significant technical risk for this mission. Except for the instrument, the spacecraft subsystems and components have previous space flight experience. A breadboard model of the instrument has been developed and tested to ensure minimum risk to the development. No hazardous materials are used.

16.7 Astrodynamics

The orbit requirements are to:

1. Provide a lifetime of at least one year.
2. Provide a node so that amateur radio operators in a broad geographic area can extend the range of their packet communication using line-of-sight VHF communication.

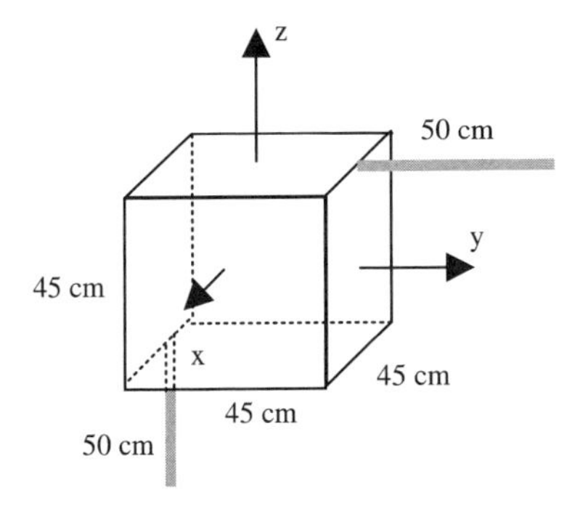

Figure 16.2 Deployed configuration.

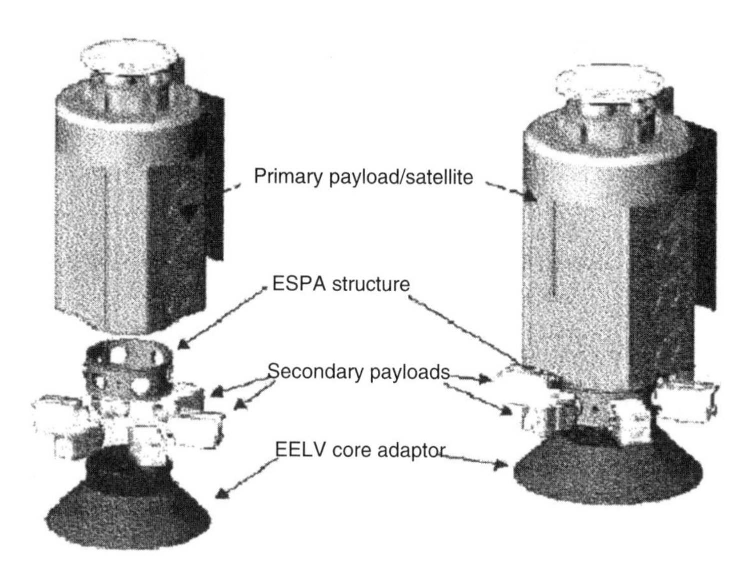

Figure 16.3 EELV Secondary Payload Adapter (ESPA). (Courtesy of the United States Air Force Space Test Program.)

3. Sample the broad radiation environment, especially the South Atlantic anomaly (SAA).
4. Determine the ephemeris of the spacecraft to an accuracy of 50 km in real time for alerting purposes and 10 km a posteriori to identify the location of the radiation observations.

The design orbit selected is:

- Altitude = 500 km.
- Inclination = 51°.
- Eccentricity: near zero.
- Orbit determination:
  — Source: USSTRATCOM two-line elements.

The orbit has an orbital period of 94.62 min and 15.2 rev/day. The precession of the orbit's line of nodes or rate of change of the right ascension of the ascending node is −8.1°/day. The maximum eclipse time is 35.8 min with the minimum eclipse time dependent on the declination of the Sun. If the longitude of the Sun is 90° or 270° from the spacecraft's line of nodes and the declination of the Sun is greater than 17.0° south or 17.0° north respectively, the spacecraft will experience intervals of 100% sunlight. The maximum duration of a pass over a ground station, from horizon to horizon, is 11.6 min.

Knowledge of the orbital position for data evaluation is required to 10 km so that the ephemerides provided by the US Strategic Command's (USSTRATCOM) two-line elements are adequate. Software provided by the USSTRATCOM will be used to produce station alerts with a maximum error of about 50 km and ephemerides for use in data reduction with position errors of less than 10 km.

16.8 Configuration and Structure

The configuration and structural requirements are:

1. Size and mass must satisfy the constraints of a nanosatellite measuring 45 cm × 45 cm × 45 cm with a mass of 20 kg or less.
2. Maintain the structural integrity of the spacecraft during launch and deployment by the Delta IV launch vehicle using the EELV Secondary Payload Adapter (ESPA).
3. Deploy two pairs of quarter-wave monopole antennas, each 50 cm in length.
4. Use a safety factor of 2.00 for yield strength to eliminate the requirement of a static test, to reduce cost and schedule.
5. Design a structure with minimum mass.

As specified in the Delta IV payload manual, the launch loads are (Boeing Company, 2000)

Minimum axial frequency:	27 Hz
Minimum lateral frequency:	10 Hz
Axial acceleration:	6.5g
Transverse acceleration:	2.0g
Sinusoidal accelerations:	
Axial	5.0–6.2 Hz at 1.27 cm double amplitude
	6.2–100 Hz at 1g, zero to peak
Transverse	5–100 Hz at 0.7g, zero to peak

Because of the cantilever positioning of the spacecraft on the ESPA, a simple, mass efficient, box-frame configuration with diagonal supports is employed that is consistent with the restrictions of the nanosatellite constraints on mass and size. The structure consists of 16 frame elements (12 to form the cube and 4 to support the electronics shelf) that have the cross-section of a right angle of the same widths and thicknesses while the two diagonal supports are flat bars. The structure without solar arrays has an outside dimension of 44 cm × 44 cm × 44 cm. The two antennas will be made of flexible copper strips that will be folded within the prescribed envelope during launch and released by a timer mechanism after the spacecraft is deployed in orbit. When extended, they will be oriented in the $+y$ and $-z$ axes, as illustrated in figure 16.2. Solar cells are mounted on all six sides, including the side attached to the ESPA.

The structural frame is illustrated in figure 16.4, with the primary load path illustrated in figure 16.5. The primary path is the stiffest and most direct path for the load to be transmitted from the support structure to the payloads. The structural material for the frame and sides is 7075 plate and sheet aluminum respectively. The load transfer will be from the attach flange of the ESPA assembly to the two vertical side supports and to the shelf on which the electronics are attached by the diagonal and horizontal supports. The specifications of the materials are given in table 16.1.

The maximum acceleration factors in the axial acc_{a} and lateral acc_{ℓ} directions are obtained by adding the steady-state and sinusoidal accelerations

$$\left.\begin{aligned} acc_{\mathrm{a}} &= 6.5 + 1.0 = 7.5, \\ acc_{\ell} &= 2.0 + 0.7 = 2.7 \approx 3.0. \end{aligned}\right\} \tag{16.8.1}$$

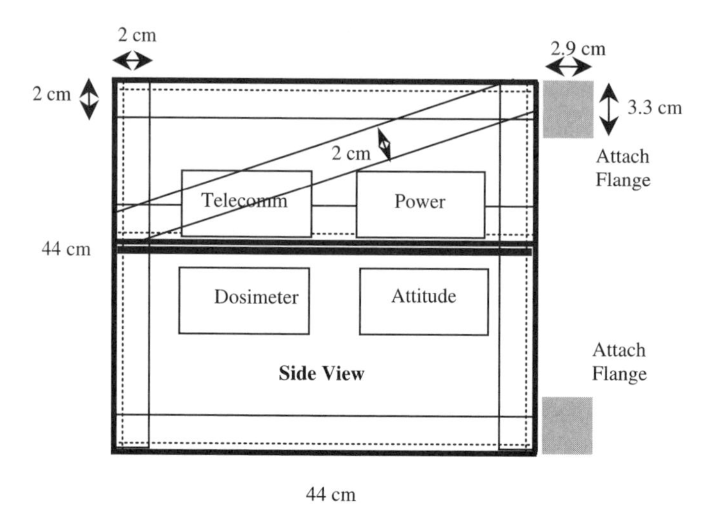

Figure 16.4 Side view of structural frame.

It is necessary to then estimate the payload mass. A typical a priori estimate of the structural mass fraction is 15 to 20% of the total spacecraft mass. The solar arrays, distributed over all six sides, are estimated to be about 5% of the total mass. As a result, exclusive of the structure and solar arrays, the total payload is estimated to be about 75% of the total mass or 15 kg, distributed fairly uniformly on the horizontal deck. The horizontal deck is supported by the identical combination of horizontal and diagonal supports on two sides so that each need support $M = 7.5$ kg. Imposing a safety factor in yield of $F_{sy} = 2$ and the dynamic load factors above gives the axial design load F_a to be

$$F_a = F_{sy} \times acc_a \times M = 2.0 \times 7.5g \times 7.5 = 1103 \text{ N}, \tag{16.8.2}$$

and the F_l lateral design load to be

$$F_l = F_{sy} \times acc_l \times M = 2.0 \times 3.0g \times 7.5 = 441 \text{ N}. \tag{16.8.3}$$

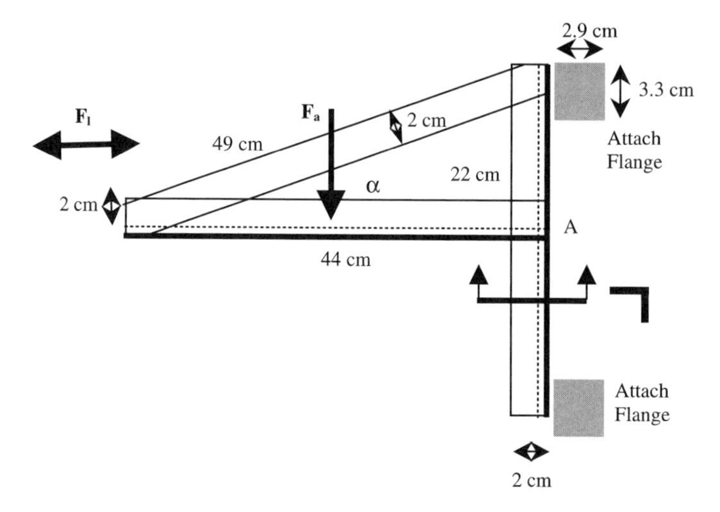

Figure 16.5 Primary load path.

Table 16.1 Material properties and acceleration loads

	Symbol	7075 T6 Bar	7 075 T73 Sheet	Units
Factor of safety, yield	F_{sy}	2.00	2.00	
Factor of safety, ultimate	F_{su}	2.00	2.00	
Ultimate stress	σ_u	5.70×10^8	4.60×10^8	$N/4m^2$
Yield stress	σ_y	5.00×10^8	3.80×10^8	N/m^2
Shear strength	τ_y	3.30×10^8	3.30×10^8	N/m^2
Modulus of elasticity	E	7.00×10^{10}	7.10×10^{10}	N/m^2
Poisson's ratio	e	0.33	0.33	
Density	ρ	2.71×10^3	2.80×10^3	kg/m^3
Specific heat	c	960	960	J/(kg K)

From figure 16.5, the force acting along the diagonal support, F_d, is obtained by summing moments of the forces about the point A, where the horizontal member attaches to the support wall. This gives

$$F_d l \sin\alpha = F_a \frac{l}{2},$$

$$F_d = \frac{F_a}{2\sin\alpha} = \frac{\sqrt{5}F_a}{2} = 1233\,\text{N}, \qquad (16.8.4)$$

where $l = 0.44$ m is the length of the side and the angle α between the diagonal and horizontal supports is given by

$$\alpha = \tan^{-1}\frac{l/2}{l} = 26.56^{\circ} \qquad (16.8.5)$$

The structure is characterized by assuming the widths of the angle members and the diagonal supports so that the task remains to determine the appropriate thicknesses. First the horizontal and then the diagonal supports that constitute the primary load paths are considered.

16.8.1 Horizontal Support

The horizontal member is an angle of equal-width sides $w = 0.02$ m, length $l_h = 0.44$ m, and thickness t to be determined. The maximum vertical shear stress τ in the horizontal member at the wall due to the axial force on the mass is given by

$$\tau = \frac{F_a}{A} = \frac{F_a}{2wt}, \qquad (16.8.6)$$

where A is the cross-sectional area, so the required thickness t is given in terms of the shear yield stress τ_y by

$$t = \frac{F_a}{2w\tau_y} = 8.36 \times 10^{-5}\text{m}. \qquad (16.8.7)$$

The required thickness for normal tension or compression at the wall in the figure above is determined from equations 16.8.4 and 16.8.5 as

$$\sigma = \frac{F}{A} = \frac{F_{\mathrm{d}} \cos \alpha + F_l}{2wt} = \frac{F_{\mathrm{a}} + F_l}{2wt}, \tag{16.8.8}$$

so that

$$t = \frac{F_{\mathrm{a}} + F_l}{2w\sigma_{\mathrm{y}}} = 7.72 \times 10^{-5} \text{ m}. \tag{16.8.9}$$

The crippling stress σ_{cr} for the horizontal angle is given in chapter 8 by equation 8.8.20 as

$$\sigma_{\mathrm{cr}} = 0.316 \frac{(E\sigma_{\mathrm{y}})^{1/2}}{(2w/t)^{3/4}}, \tag{16.8.10}$$

where E is the modulus of elasticity and the constant is for the ends treated as fixed–fixed because of the end constraints. The crippling compressive stress follows from equation 16.8.8 as

$$\sigma_{\mathrm{cr}} = \frac{F_{\mathrm{a}} + F_l}{2wt}, \tag{16.8.11}$$

consisting of the lateral load at the end due to the diagonal support and of the lateral load of the 7.5 kg distributed mass. Substituting equation 16.8.11 into 16.8.10 gives for the thickness t

$$t = \left[\frac{F_{\mathrm{a}} + F_l}{0.316\,(2w)^{1/4}\left(E\sigma_{\mathrm{y}}\right)^{1/2}} \right]^{4/7} = 5.29 \times 10^{-4} \text{ m}. \tag{16.8.12}$$

The fundamental frequency f_{i} of the horizontal angle with the mass m assumed distributed over the length l_{h} is given in table 8.21 as

$$f_{\mathrm{i}} = \frac{\lambda_{\mathrm{i}}^2}{2\pi l_{\mathrm{h}}^2} \sqrt{\frac{EI}{M/l_{\mathrm{h}}}}, \tag{16.8.13}$$

where I is the area moment of inertia about the bending axis, the constant $\lambda_{\mathrm{i}} = 3.927$ for a pinned–fixed beam, and the load supported is $M = 7.5$ kg. Ignoring the top part of the flange, a conservative estimate of the moment of inertia is

$$I = \frac{tw^3}{12} \tag{16.8.14}$$

and substituting into equation 16.8.13 with $l_{\mathrm{h}} = 0.44$ m and $E = 7 \times 10^{10}$ N/m^2 gives for the thickness

$$t = \frac{48\pi^2 f_{\mathrm{i}}^2 M l_{\mathrm{h}}^3}{E\lambda_{\mathrm{i}}^4 w^3} = 0.0017 \text{ m}, \tag{16.8.15}$$

for the design axial frequency of $f_{\mathrm{i}} = 27$ Hz.

To provide an additional margin of safety, to be able to take advantage of other launch opportunities, the thickness of the angle is selected to be 0.002 m.

16.8.2 Diagonal Support

The diagonal member has a rectangular cross-section, a length of $l_d = 0.492$ m, an assumed width $w = 0.02$ m, and undetermined thickness t. The required thickness for normal tension or compression is determined from F_d, the force along the diagonal, and the cross-sectional area A by

$$\sigma = \frac{F_d}{A} = \frac{F_d}{wt}, \tag{16.8.16}$$

so that

$$t = \frac{F_d}{w\sigma_y} = 1.23 \times 10^{-4} \text{ m}. \tag{16.8.17}$$

Treating the member as a fixed–fixed support, the appropriate equation for buckling, given by table 8.12 of chapter 8, is

$$P_{cr} = \frac{4\pi^2 EI}{l_d^2}, \tag{16.8.18}$$

where P_{cr} is the buckling load, E is the modulus of elasticity, and I is the cross-sectional area moment of inertia. For

$$I = \frac{wt^3}{12}, \quad \sigma_{cr} = \frac{P_{cr}}{wt}, \quad \text{and} \quad P_{cr} = F_d, \tag{16.8.19}$$

it follows that the required thickness is

$$t = \left(\frac{3F_d l_d^2}{\pi^2 E w}\right)^{1/3} = 0.0040 \text{ m}. \tag{16.8.20}$$

The required thickness for a lateral fundamental natural frequency of $f_i = 10$ Hz follows from equation 16.8.13 to be

$$t = \left(\frac{48\pi^2 f_i \rho l_d^4}{E\lambda_i^4}\right)^{1/2} = 6.72 \times 10^{-4} \text{ m}. \tag{16.8.21}$$

where $\rho = 2.71 \times 10^3$ kg/m^3 is the mass density and $\lambda_i = 4.730$ for a pinned–fixed beam.

As a consequence of the above estimates, the thickness of the diagonal support is selected to be 0.004 m.

16.8.3 Summary

In summary, the characteristics of the structure are that the angles have a width of 0.020 m and a thickness of 0.002 m and the diagonal support has a width of 0.020 m and a thickness of 0.004 m. From this, the mass of the structure is estimated in table 16.2 to be 6.200 kg. The total mass allocations with margin are given in table 16.3. Note from table 6.2 that the mass distributed on the horizontal shelf was assumed initially to be 15 kg, but is now estimated to be about 13.8 kg less the mass of the distributed solar cells. Thus the design of the structure is somewhat conservative.

Table 16.2 Mass estimate of structure

Quantity		Size (m)	Mass (kg)
2	Diagonal supports	$0.500 \times 0.020 \times 0.004$	0.217
16	Frame angles	$0.440 \times 0.020 + 0.020 \times 0.002$	1.526
6	Faceplates	$0.440 \times 0.440 \times 0.001$	3.253
1	Shelf	$0.440 \times 0.440 \times 0.002$	1.084
	Fittings		0.120
		Total	6.200

16.9 Attitude Determination and Control Subsystem

The attitude determination and control requirements are:

1. Ensure the spacecraft does not maintain the same orientation in space that would exacerbate the thermal control subsystem of the spacecraft.
2. Develop an attitude control system with minimum mass.
3. There is no requirement to determine the attitude.

Attitude stabilization is employed primarily to ensure that the satellite will not orient one face toward the sun for long periods of time that would exacerbate the thermal control subsystem. Attitude control is achieved by passive magnetic stabilization, using a ferromagnetic rod. Damping of librations about the geomagnetic field lines is by lossy magnetic hysteresis rods that dissipate energy through heat. Attitude determination is not required for this mission. However, it is expected that the attitude can be recovered by using differential power measurements from the six solar panels and a Kalman filter employing Euler's equation of motion with known mass and geometric properties.

Estimates for the performance of the attitude control system and masses of the magnet and the magnetic hysteresis rods can be made by utilizing the simplified centered dipole geomagnetic field model illustrated in figure 16.6. The radial and tangential components of the geomagnetic induction are

$$\left.\begin{aligned} B_r &= -\frac{2B_0 \sin\phi}{r^3}, \\ B_\phi &= +\frac{B_0 \cos\phi}{r^3}, \end{aligned}\right\} \tag{16.9.1}$$

where r is the ratio of the orbit radius to the radius of the Earth, ϕ is the geomagnetic latitude, and B_0 is $3.011\,53 \times 10^{-5}$ T. For the orbital radius of 500 km, $r = 6{,}878{,}137/6{,}378{,}137 = 1.078\,394$, and the magnetic field has a minimum value of 2.401×10^{-5} T at the geomagnetic equator and an upper bound of 4.803×10^{-5} T, which is the value at the geomagnetic poles. In the vicinity of the ground station, at the geographic latitude of 39° and geomagnetic latitude of $\phi = 49.6°$, the deviation of the orientation of the geomagnetic field line from the local vertical is given by equation 16.9.1 as

$$\alpha = \tan^{-1}(0.5 \cot\phi) = 23.0°. \tag{16.9.2}$$

The mass of the spacecraft's magnetic dipole moment can be determined as follows. The oscillation of the spacecraft about the field line can be approximated by

$$I\ddot{\theta} + MB \sin\theta = 0, \tag{16.9.3}$$

Table 16.3 Mass table

Subsystem		Mass (kg)
Structure		6.200
Frame	1.526	
Diagonals	0.217	
Sides	3.253	
Shelf	1.084	
Fasteners	0.120	
Power systems		3.150
Solar arrays	1.200	
Batteries	1.450	
Electronics	0.500	
Thermal control		0.500
Mylar insulation	0.500	
Telecommunications		1.000
Transmitters (2)	0.070	
Receivers (2)	0.070	
TNC (1)	0.500	
T/R switches (2)	0.050	
Electronics	0.300	
Antennas (2)	0.100	
Attitude control		0.200
Permanent magnet	0.085	
Hysteresis rods	0.065	
Fittings	0.050	
Instrument		5.950
Sensor and electronics	0.350	
Multichannel analyzers	0.300	
Electronics	0.300	
Radiation absorber	5.000	
Incidentals		1.000
Fasteners, harness, etc.	1.000	
Margin		2.000
	Total	20.000

where I is the mass moment of inertia, M is the magnetic moment, B is the geomagnetic induction, and θ is the angular deviation from the field line. The magnetic moment required to produce a frequency of oscillation about the field line of magnitude ω is given by

$$M = \frac{\omega^2 I}{B} = \frac{4k^2\pi^2 I}{B\tau^2}, \tag{16.9.4}$$

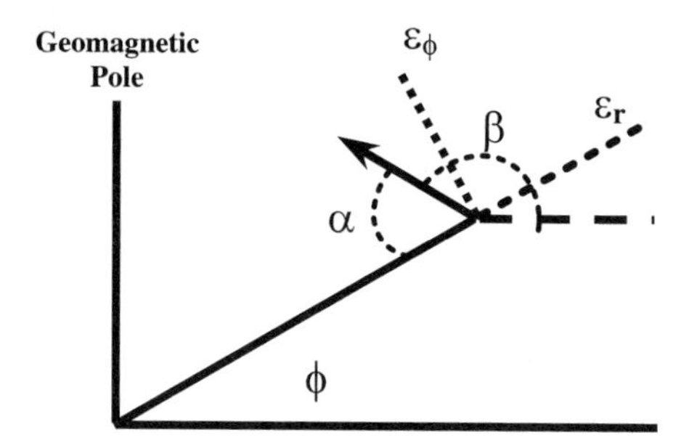

Figure 16.6 Dipole magnetic field.

where, for convenience, $\omega \equiv 2\pi k/\tau$ where k is an arbitrary constant and τ is the orbital period. The constant k represents the number of times per orbital period that the spacecraft would oscillate about the field line for a given geomagnetic induction B. To ensure adequate damping of excess librational energy, a value of $k = 20$ is selected. For the minimum magnetic field $B_{\text{min}} = 2.401 \times 10^{-5}$ T, an orbital period of 94.61 min, a mass moment of inertia for a uniform mass distribution

$$I = \frac{ml^2}{6} = \frac{20 \times 0.45^2}{6} = 0.675 \text{ kg m}^2, \tag{16.9.5}$$

it follows that the required magnetic dipole moment is $M = 13.8$ A m^2. The volume of the magnet can be determined from

$$M = \frac{B_{\text{i}} V}{\mu_0}, \tag{16.9.6}$$

where B_{i} is the intrinsic induction of the magnet in tesla that can be approximated by the residual flux density B_{r}, V is the volume in m^3, and μ_0 is the permeability constant of $4\pi \times 10^{-7}$ H/m. To achieve the dipole moment specified above, the volume of an Alnico magnet with $B_{\text{r}} = 1.5$ T is 1.16×10^{-5} m (11.6 cm^3). Alnico has a density of 7.3×10^{-6} kg/m^3 (7.3 g/cm^3), so the mass of the magnet is 0.085 kg.

The orientation of the geomagnetic field line with respect to the equator in a geomagnetic meridian, as illustrated in figure 16.6, is given by

$$\beta = \phi + \pi - \alpha, \tag{16.9.7}$$

where α is the orientation of the field with respect to the local vertical, ϕ is the geomagnetic latitude, and $\tan \alpha = (\cot \phi)/2$. The angular acceleration of the field line follows directly as

$$\ddot{\beta} = -\ddot{\alpha} = 12n^2 \sin\phi \cos\phi (1 + 3\sin^2\phi)^{-2}, \tag{16.9.8}$$

where $\dot{\phi} = n$ is the mean motion of the spacecraft, assumed to be in a near-circular orbit. The maximum value of $\ddot{\beta}$ can be determined from equation 16.9.8 to occur at approximately $\phi = 18°$ from the geomagnetic equator. The offset between the axis of the magnetic moment and the field line γ is given by

$$\theta = \sin^{-1}(I\ddot{\beta}/BM), \tag{16.9.9}$$

where B is the magnitude of the geomagnetic induction, I is the mass moment of inertia, and M is the dipole moment. For $I = 0.675$ kg m^2, $M = 13.8$ A m^2, $r = 1.078394$, and $\phi = 18°$,

$$\begin{aligned} B &= (B_0/r^3)(1 + 3\sin^2\phi)^{1/2} \\ &= (3.011\,53 \times 10^{-5}/1.078\,4^3)(1 + 3\sin^2 18°)^{1/2} \\ &= 2.724 \times 10^{-5}\ \text{T}, \end{aligned} \tag{16.9.10}$$

$$\begin{aligned} \ddot{\beta} = -\ddot{\alpha} &= 12n^2 \sin\phi \cos\phi (1 + 3\sin^2\phi)^{-2} \\ &= 12\left(\frac{2\pi}{94.61 \times 60}\right)^2 \sin 18° \cos 18° (1 + 3\sin^2 18°)^{-2} \\ &= 2.638 \times 10^{-6}\ \text{rad/s}^2, \end{aligned} \tag{16.9.11}$$

$$\begin{aligned} \theta &= \sin^{-1}(I\ddot{\beta}/BM) \\ &= \sin^{-1}(0.675 \times 2.638 \times 10^{-6}/2.724 \times 10^{-5} \times 13.8) \\ &= 0.27°, \end{aligned} \tag{16.9.12}$$

so the maximum deviation between the equilibrium orientation of the spacecraft and the geomagnetic field line is 0.27°.

Damping of the motion about the field line will utilize rods of High Permeability Alloy 49, an easily magnetized nickel–iron alloy that dissipates energy by heat through magnetic hysteresis. The estimate of the mass required is based on the Transit 3B spacecraft that had an altitude of approximately 1000 km, a moment of inertia of 11 kg m^2, and employed four hysteresis rods, each with a volume of 6.2×10^{-6} m^3. The volume of hysteresis rods required can be estimated by scaling with the ratio of the inertias and the square of the ratio of the geomagnetic fields encountered. For the nanosatellite, two sets of orthogonal rods with a total volume of 2.00×10^{-6} m^3 (2 cm^3) will be required, orthogonal to each other in a plane that is orthogonal to the dipole magnet. For Alloy 49, with a specific mass of 8180 kg/m^3 (8.18 g/cm^3), the estimated mass of the hysteresis rods is 0.016 kg for each pair for a total of 0.032 kg. To ensure adequate damping, this estimate is increased by a factor of 2 to a total mass of 0.065 kg. In addition to the magnetic hysteresis damping, additional energy will be dissipated by eddy currents induced in the aluminum structure by motion in the geomagnetic field.

In summary, the spacecraft will be magnetically stabilized so that it will rotate about twice per orbit to average out the thermal environment, if in a high inclination orbit. This is accomplished by a permanent magnet with dipole moment of 13.8 A m^2 and mass of 0.085 kg and magnetic hysteresis rods with total mass of 0.065 kg. A sketch of the anticipated orientation is given in figure 16.7 when the spacecraft is in a geomagnetic meridian. The actual orientation will depend strongly on the inclination of the orbit.

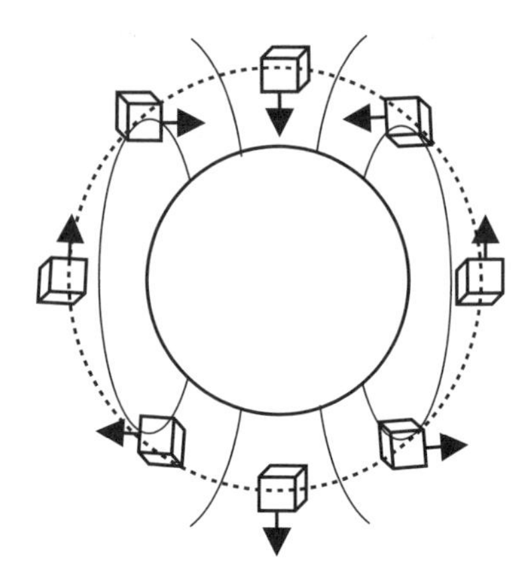

Figure 16.7 Spacecraft orientation in plane through magnetic poles

16.10 Power Subsystem

The power subsystem requirements are:

1. Provide sufficient power to ensure amateur radio communications anywhere in the orbit up to a maximum average of 25% of the orbit with a transmitter power of 2 W for each of the two antennas for an orbit average power of 1 W.
2. Provide sufficient power to operate the instrument with an orbit average power of 6 W.
3. Provide heater power with an orbit average power of 2 W.

The power required for the instrument is 6 W, with the telecommunications and command and data systems hardware consuming a fraction of a watt in standby mode. If power is limited during maximum eclipse, the instrument can be unpowered if necessary for selected periods of time without affecting the mission objectives. The transmitter will require 2 W of power for each antenna when transmitting, which equates to 1 W orbit average power assuming a conservative duty cycle of 25%. A summary of the average orbit power requirements is given in table 16.4; with margin, this totals 13 W.

The design is based on space-qualified solar cells that have an efficiency of 15.2%, a dimension of 2×6 cm, and in full Sun produce 0.5 A at 0.5 V for a power output of 0.25 W. Assuming a minimum packing factor of 75%, 108 cells can be placed on each side of the spacecraft for a power output of 27 W in full Sun. During the maximum eclipse orbit, the excess energy generated by the solar arrays is 13.7 W h, while the energy required during the eclipse is 7.8 W h, so that the storage and recovery efficiency need be only about 50%. Energy storage will use NiCd batteries, each with a nominal voltage of 1.2 V, capacity of 5.2 W h, mass of 0.145 kg, and an allowed depth of discharge of 30%. One pack of 10 cells at 12 V has a mass of 1.45 kg and a total capacity of 52 W h, equating to a maximum depth of discharge of 15%.

Power transfer to the bus will be by a direct energy transfer (DET) configuration with the solar arrays connected directly to the batteries and loads as illustrated in figure 16.8. A blocking diode is used to prevent reverse current through the solar panels that would cause damage. The bus voltage will be an unregulated nominal 12 V. A shunt circuit with a Zener diode will be used to limit battery overcharging. The dosimeter instrument

Table 16.4 Power budget

Subsystem	Orbit Average Power (W)
Telecommunications	1.0
Transmitters (2)	
Receivers (2)	
TNC (2)	
T/R switch	
Electronics	
Instrument	6.0
Sensor and electronics	
Multichannel analyzers	
Electronics	
Power system	1.0
Batteries	
Electronics	
Thermal system	2.0
Heater	
Electronics	
Margin	3.0
Total	13.0

will be connected to the bus by a low-voltage sensing switch (LVSS) to turn it off should a low power condition be encountered. The LVSS is reset automatically when sufficient power is available. The receiver will be continuously powered. While a Zener diode may be used, it will be prudent to consider the voltage–temperature control of battery charging in the preliminary design.

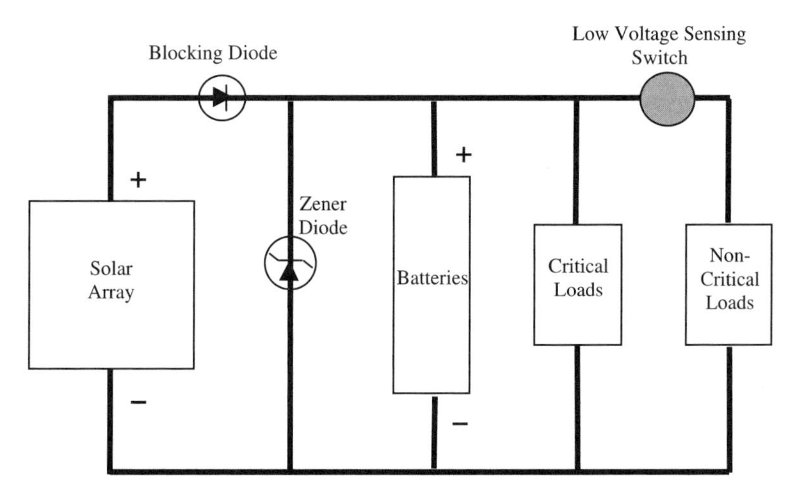

Figure 16.8 Power system schematic.

16.11 Thermal Subsystem

The thermal subsystem requirements are to:

1. Maintain the temperature of the electronics between 0 and 35°C.
2. Minimize the mass and power required.

The spacecraft temperature will be controlled by selection of surface properties and insulation, augmented by a thermostatically controlled electrical heater. The infrared emissivity and solar absorptivity of areas around the solar panels will be selected to ensure that the temperatures remain between the operating limits of the electronics, nominally 0 to 35°C. A thermal lumped-parameter or finite-element model will be developed to assist in the placement of the components to ensure uniform power dissipation. A thermal balance test will be used to confirm the model.

Upper and lower bounds to the temperature excursions can be determined as follows by evaluating the steady-state temperatures in full sunlight and in full eclipse. The equilibrium temperature can be obtained by equating the heat rate inputs and losses, where

$$\left.\begin{aligned}
&\dot{Q}_{\text{albedo}} = a\alpha f_e A_s S, \text{ heat rate input from Earth albedo}\\
&\dot{Q}_{\text{dissipation}} = \text{heat rate dissipated by internal equipment}\\
&\dot{Q}_{\text{EarthIR}} = \varepsilon f_e q_e A_s, \text{ heat rate input from Earth IR}\\
&\dot{Q}_{\text{heater}} = \text{heat rate dissipated by electrical heaters}\\
&\dot{Q}_{\text{loss}} = -\sigma\varepsilon A_s T^4_{\text{spacecraft}}, \text{ heat rate lost from the spacecraft by radiation}\\
&\dot{Q}_{\text{space}} = \sigma\varepsilon f_s A_s T^4_{\text{space}}, \text{ heat rate input from space}\\
&\dot{Q}_{\text{Sun}} = \alpha A_c S, \text{ heat rate input due to direct Sun}
\end{aligned}\right\} \quad (16.11.1)$$

where

$$\left.\begin{aligned}
a &= \text{albedo factor}\\
\alpha &= \text{spacecraft solar absorptivity}\\
\varepsilon &= \text{spacecraft emissivity}\\
\sigma &= \text{Stefan–oltzmann constant, } 5.669 \times 10^{-8}\ \text{W/m}^2\ \text{K}^4\\
\eta &= \text{solar cell efficiency}\\
A_c &= \text{cross-sectional area, approximated by a sphere with surface}\\
&\quad\ \text{area of cube}\\
A_s &= \text{surface area}\\
f_e &= \text{Earth view factor, } f_e = \tfrac{1}{2}[1 - \cos\rho] \text{ where } \rho = \sin^{-1}\left(\frac{R}{R+h}\right)\\
q_e &= \text{Earth IR, W/m}^2\\
R &= \text{mean radius of the Earth}\\
h &= \text{altitude of the spacecraft}\\
f_s &= \text{space view factor, given by } 1 - f_e\\
\dot{Q}_{\text{dissipation}} &= \text{internal heat dissipation rate, W}\\
\dot{Q}_{\text{heater}} &= \text{heat rate generated by the heater, W}\\
S &= \text{solar constant, W}
\end{aligned}\right\} \quad (16.11.2)$$

The upper bound on the maximum temperature and the lower bound to the minimum temperature can be determined from

$$T_{\text{spacecraft upper bound}} = [(\dot{Q}_{\text{Sun}} + \dot{Q}_{\text{space}} + \dot{Q}_{\text{Earth IR}} + \dot{Q}_{\text{albedo}} + \dot{Q}_{\text{dissipation}})/\sigma\varepsilon A_s]^{1/4}, \quad (16.11.3)$$

$$T_{\text{spacecraft lower bound}} = [(\dot{Q}_{\text{space}} + \dot{Q}_{\text{Earth IR}} + \dot{Q}_{\text{dissipation}} + \dot{Q}_{\text{heater}})/\sigma\varepsilon A_s]^{1/4}. \tag{16.11.4}$$

For the following conditions

$$\left.\begin{aligned}
a &= 0.3\text{, albedo factor}\\
\alpha - \eta &= 0.75\text{, absorptivity of solar cells}\\
\varepsilon &= 0.82\text{, emissivity of solar cells}\\
A_c &= 0.3038\ \text{m}^2\text{, cross-section of sphere with equivalent surface area}\\
A_s &= 1.215\ \text{m}^2\text{, surface area of cube}\\
S &= 1368\ \text{W/m}^2\text{, the solar constant}\\
f_e &= 0.3128\text{, Earth view factor for altitude of 500 km}\\
f_s &= 0.6872\text{, space view factor}\\
T_{\text{Earth}} &= 290\ \text{K}\\
-q_e &= 227 \pm 21\ \text{W}
\end{aligned}\right\} \tag{16.11.5}$$

the heat rate inputs are

$$\left.\begin{aligned}
\dot{Q}_{\text{albedo}} &= a\alpha f_e A_s S = 117.0\ \text{W}\\
\dot{Q}_{\text{dissipation}} &= 13.0\ \text{W}\\
\dot{Q}_{\text{EarthIR}} &= \varepsilon f_e q_e A_s = 70.7\ \text{W},\\
\dot{Q}_{\text{heater}} &= 5.3\ \text{W (orbit average power} = 2\ \text{W)}\\
\dot{Q}_{\text{space}} &= \sigma\varepsilon f_s A_s T_{\text{space}}^4 = 9.8770\times10^{-6}\ \text{W}\\
\dot{Q}_{\text{Sun}} &= \alpha A_c S = 311.7\ \text{W}
\end{aligned}\right\} \tag{16.11.6}$$

which yield $T_{\text{spacecraft upper bound}} = 36.3°\text{C}$ and $T_{\text{spacecraft lower bound}} = -53.7°\text{C}$. Since solar cells cover most of the surfaces, the opportunity to select emissivity and absorptivity is limited. In the eclipse, the temperature is independent of absorptivity and is weakly dependent on emissivity since the primary source of heat is the Earth's infrared radiation and internal dissipation. Consequently, varying the emissivity or absorptivity has little effect. Reducing the emissivity to increase the lower-bound temperature directly increases the upper bound to the temperature when the spacecraft is in the Sun. Thus, a more detailed analysis is required taking into account the temperature distribution in spacecraft and the use of interior thermal blankets.

In general, spacecraft have significant thermal capacity and insulation to store heat in order to reduce the extreme temperature determined by steady-state analyses. A first-order assessment of the transient behavior can be made by taking into account the thermal capacity of the spacecraft, where

$$\Delta Q = cm\Delta T, \tag{16.11.7}$$

where ΔQ is the change in heat, c is the specific heat of the composite spacecraft, m is the spacecraft mass, and ΔT is the change in temperature. This is a conservative analysis as it assumes no radial insulation so that the whole spacecraft has the same temperature. Under this assumption, the spacecraft temperature as a function of time can be approximated by

$$T_{\text{spacecraft}} = T_{\text{initial}} + \Delta T_{\text{spacecraft}}(t) = T_{\text{initial}} + \frac{\dot{Q}(T_{\text{initial}})}{cm}\Delta t, \tag{16.11.8}$$

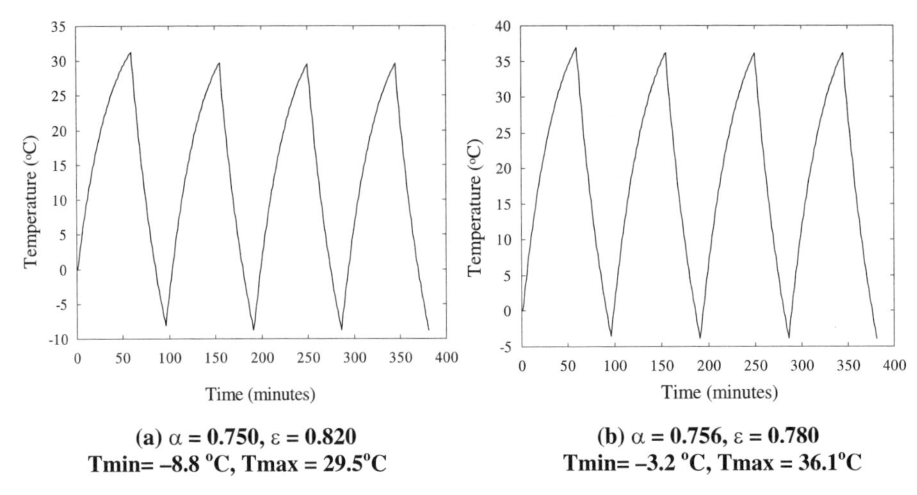

(a) α = 0.750, ε = 0.820
Tmin= –8.8 °C, Tmax = 29.5°C

(b) α = 0.756, ε = 0.780
Tmin= –3.2 °C, Tmax = 36.1°C

Figure 16.9 Transient spacecraft temperature.

where

$$\left.\begin{aligned} \dot{Q}\Big|_{\text{in eclipse}} &= \dot{Q}_{\text{space}} + \dot{Q}_{\text{EarthIR}} + \dot{Q}_{\text{dissipation}} + \dot{Q}_{\text{heater}} + \dot{Q}_{\text{loss}}, \\ \dot{Q}\Big|_{\text{in Sun}} &= \dot{Q}_{\text{Sun}} + \dot{Q}_{\text{space}} + \dot{Q}_{\text{EarthIR}} + \frac{2}{\pi}\dot{Q}_{\text{albedo}} + \dot{Q}_{\text{dissipation}} \\ &\quad + \dot{Q}_{\text{EarthIR}} + \dot{Q}_{\text{loss}}, \end{aligned}\right\} \quad (16.11.9)$$

$$\dot{Q}_{\text{loss}} = -\sigma\varepsilon A_s T^4, \quad (16.11.10)$$

where T_{initial} = initial temperature, and the $2/\pi$ factor is introduced to average the Earth albedo over the sunlit side of the orbit. The first T_{initial} is arbitrary as the temperature profile will converge. Since the spacecraft mass is principally aluminum, at least 30–40%, the specific heat of the spacecraft is assumed to be 600 J/(kg K) since AL 7075 has a specific heat of 960 J/(kg K). Assuming an initial temperature $T_{\text{initial}} = 0$°C gives the result illustrated in figure 16.9, where (a) relates to an absorptivity of 0.750 and an emissivity of 0.820, the values for the solar cells, while (b) relates to 20% of the surface coated with a material with an absorptivity of 0.9 and an emissivity of 0.5 that yields a composite absorptivity of 0.756 and a composite emissivity of 0.780. The temperature range reaches steady state after one to two orbital periods. In both cases, the temperature variation is close to the desired range 0 to 35°C, appropriate for most electronics. Insulation between the exterior and the instruments should provide an even more suitable temperature environment.

16.12 Telecommunications Subsystem

The telecommunications subsystem requirements are to:

(1) Provide a spacecraft repeater for digital traffic between amateur radio operators at 145.825 MHz using the AX.25 amateur radio protocol for digital packets.
(2) Send commands from the ground station to the spacecraft.
(3) On command from the operating ground station, have the spacecraft transmit real-time housekeeping data and the data stored by the instrument.

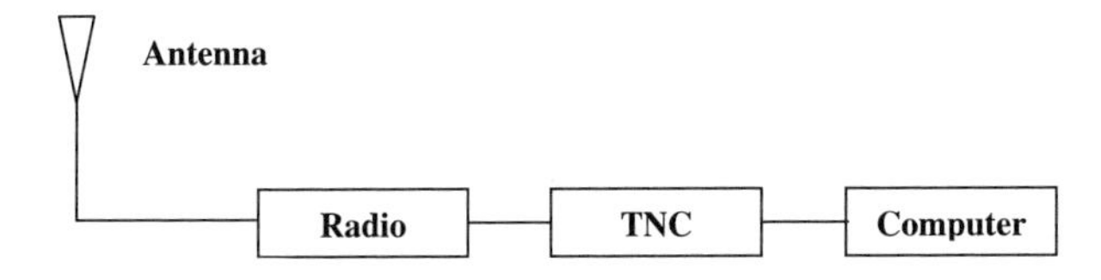

Figure 16.10 Ground station telecommunications system of schematic.

(4) Provide a timer to deploy two monopole antennas.
(5) Provide a bit error rate of one part in ten million for the uplink and downlink.
(6) There is no need for delayed commands.

The ground station equipment consists of a digital computer connected by a cable to an RS-232C serial modem port of a terminal node controller (TNC), similar to the one in the spacecraft, that is in turn connected to the ground station radio system with a circularly polarized antenna as illustrated in figure 16.10. The schematic of the spacecraft telecommunications system is illustrated in figure 16.11 and consists of the TNC and dual transmitters, receivers, transmit/receive switches, and an analog multiplexer. Both receivers will be on at all times to assure connectivity no matter what the orientation of the spacecraft. The downlink telemetry is initiated by an uplink command from the ground station that instructs the instrument to send the contents of its memory to the TNC and the telecommunications system to transmit. This will provide the scientific data, status of the instrument, and real-time analog housekeeping data consisting of temperatures, battery voltage, and solar panel voltages. The real-time housekeeping data is multiplexed into the five analog channels available in the TNC. The TNC digitizes the analog data and formats the data into packets and then keys the transmitter. The TNC, measuring 2 cm × 17 cm × 17.5 cm, has a mass of 500 g, and is powered at 5.5–25.0 V DC with a maximum current of 45 mA. The transmitter, receiver, and terminal node controller are existing, off-the-shelf technology. The antenna system is a pair of monopole antennas, mounted on opposite edges of the spacecraft, that transmit and receive. The AX.25 (level 2, version 2) amateur radio protocol will format and transmit the data at a rate of 1200 bits per second.

The TNC is a microprocessor-controlled multi-port packet communicator usually connected between a source of digital data, such as a computer or microprocessor, and radio by an RS-232 DCE (data communications equipment) connector as illustrated in figure 16.11. This permits the TNC to act similarly to a conventional telephone

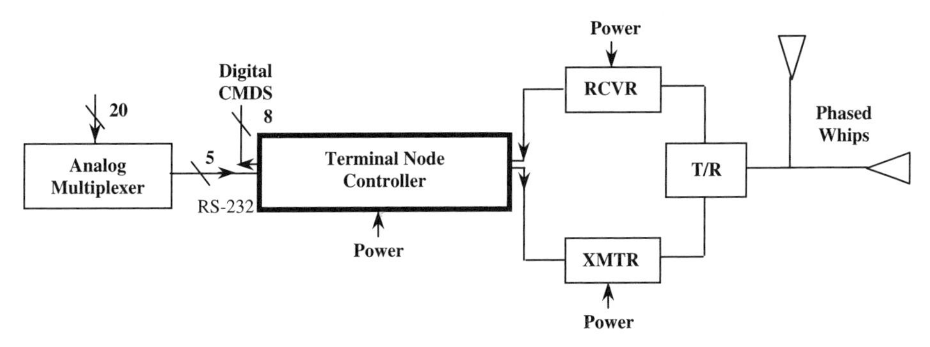

Figure 16.11 Spacecraft telecommunications system schematic.

Table 16.5 Typical AX.25 frame format

FLAG	ADDRESS	CONTROL	PID	INFORMATION	FCS	FLAG
01111110	112/256 bits	8 bits	8 bits	N(0 to 256) × 8 bits	16 bits	01111110

FLAG field defines the beginning of the frame.
ADDRESS field identifies the source and destination address for the packet.
CONTROL field identifies whether the frame is a supervisory frame used to manage the connection or an informational frame containing data; if the latter, it provides the sequence number.
PROTOCOL IDENTIFIER FIELD (PID) appears only in information frames and identifies the protocol.
INFORMATION field contains the data with a maximum of 256 bytes, with bit stuffing to prevent flags from accidentally appearing in this field by inserting a 0 after every five sequences of ones.
FRAME CHECK SEQUENCE (FCS) field provides error detection.
FLAG field defines the end of the frame.

modem. In the spacecraft, the RS-232 connector is used to send analog data to the TNC, which digitizes it and also relays commands to component subsystems in the spacecraft. AX.25, the amateur X.25 protocol, is a standard link-layer protocol popular with amateur radio operators to carry out point-to-point packet communications. Each amateur radio communicator must have the radio connected to a terminal node controller (TNC) that consists of a modem and the AX.25 protocol logic. In the AX.25 protocol, each message contains the source, destination address, control bits, and information. The protocol is capable of detecting transmission error and retransmitting packets if necessary. Information is transmitted using blocks of data called frames, with each frame consisting of fields with an integral number of bytes as illustrated in table 16.5.

The frequencies and signal levels used are comparable to those of typical single sideband (SSB) radio transceivers used by both the amateur radio and commercial services. Assuming carrier suppression and opposite sideband suppression in the receiver and transmitter are sufficient, selection of the lower (mark) or upper (space) audio tone is equivalent to frequency-modulating the carrier of an ordinary frequency shift keying transmitter. This binary frequency shift keying (BFSK) technique is commonly called audio frequency shift keying or AFSK, as the TNC uses audio frequency tones of 1300/2100 Hz.

Data rates can be estimated from the required capacity of the multichannel analyzer to store up to 8 spectra at 1024 energy levels at 32 bits each for a data rate of 262,144 bits per transmission. Housekeeping data will be real-time data only of 20 channels of 32 bits each for a total of 640 bits, giving a data rate of 263 kbits per transmission. Since the AX.25 format can have a maximum overhead of about 30%, this yields a total data rate of 342 kbits per transmission. At a data rate of 1200 bps, it will require less than 5 min of transmission time to download the required information. A timer in the spacecraft will terminate the transmission after initiation by ground control.

The required transmitter power is determined as follows. For BFSK signaling, the bit-error-rate (BER) is given by

$$\mathrm{BER} = \frac{1}{2}\exp\left(-\frac{E_\mathrm{b}}{2N_0}\right), \tag{16.12.1}$$

so that the required E_b/N_0 is

$$E_\mathrm{b}/N_0 = -2\mathrm{Log_e}(2\mathrm{BER}), \tag{16.12.2}$$

Table 16.6 Link budget

Identifier	Symbol	Units	Value	dB
Constants				
Speed of light	c	m/s	3.000E + 08	
Boltzmann's constant	k	J/K	1.381E − 23	−228.6
Frequency	f	Hz	1.45825E + 08	
Bit rate	R_b	b/s	1.200E + 03	+30.8
Maximum range (horizon)	r	m		
Bit error rate	BER	–	1.000E − 07	−70.0
Design bit energy/ Noise energy density	E_b/N_0	–	3.085E + 01	+14.9
Transmitter				
Modulation attenuation	L_m	–	0.794E + 00	−1.0
Line attenuation	L_t	–	0.794E + 00	−1.0
Antenna gain	G_t	–	0.501E + 00	−3.0
Propagation path				
Space attenuation	L_s	–	3.820E − 15	−144.2
Polarization atten (circ-linear)	L_p	–	0.794E + 00	−1.0
Atmospheric attenuation	L_a	–	0.794E + 00	−1.0
Receiver				
Antenna gain	G_r	–	0.501E + 00	−3.0
System noise temperature	T_s	K	4.000E + 02	+26.0
Required transmitter power	P_{tr}	W	0.535E + 00	−3.4
Transmit power margin factor			3.465E + 00	+ 5.7
Transmitter power per antenna	P_t	W	2.000E + 00	+ 4.8
Total transmitter power		W	4.000E + 00	

which for a BER of 1.0×10^{-7} gives $E_b/N_0 = 30.85$. The received channel signal power C to noise spectral density N_0 is given by

$$\frac{C}{N_0} = \frac{E_b R_b}{N_0} = \frac{P_{tr} L_t L_m G_t L_s L_p L_a G_r}{k T_s}, \tag{16.12.3}$$

where the transmitter power to achieve the BER follows as

$$P_{tr} = \left(\frac{E_b}{N_0}\right) \frac{R_b k T_s}{L_m L_t G_t L_s L_p L_a G_r}, \tag{16.12.4}$$

where the symbols are defined in the link budget in table 16.6. The antennas on the spacecraft and on the ground are monopole quarter-wave (half meter) whip antennas with a conservative attenuation of 3 dB fed 180° out of phase. They will be deployed by a timer on the spacecraft subsequent to launch. The transmitter power required to achieve the BER of 1×10^{-7} is 0.535 W, and a margin of 3.465 yields a design transmitter power of 2 W as illustrated in the link budget in table 16.6.

The commands to the spacecraft will include the following eight commands: two transmitter commands (on, off) and six instrument commands (on, off, etc.). There will be no delayed command capability. The telemetry will include, in addition to instrument data, housekeeping data including battery temperature, battery voltages, and the six solar panel voltages.

16.13 Schedule

For this simplified mission, development is assumed to be carried out by students in an aerospace engineering design course led by a knowledgeable faculty.

Emphasis during the first year will be on the development of the flight-qualified instrument and the design of the spacecraft. The emphasis during the second year will shift to fabrication, integration, and test of the spacecraft subsystems; integration of the instrument; and carrying out of system-level tests. Both the instrument and spacecraft development will follow the standard space development process that includes a phase A (conceptual design), phase B (concept definition), phase C (design and development), and phase D (fabrication, integration, test, and evaluation). Each phase will be concluded by the appropriate reviews identified in chapter 1.

Structured status meetings will be used to provide coordination, and a review committee of local experts in the field of space system development will be established to monitor progress throughout the two-year program. At the end of the critical design phase, evidence will exist that each subsystem and the spacecraft will perform as required through simulation, breadboards, brassboards, and analytical studies. Two of the major deliverables of the preliminary and design definition phases are detailed schedules and budgets. The top-level schedule follows in table 16.7.

Costs are not addressed here as they are unique to the cost structure of the particular organization making the estimate.

Table 16.7 Development schedule

Milestones	Month from Start
Instrument	
Preliminary design review	3
Engineering model development completed	6
Critical design review	6
Instrument development complete	11
Instrument tests completed	12
Instrument spare components delivered and tested	14
Spacecraft	
Conceptual design review	3
Preliminary design review	6
Order of selected long-lead items	9
Critical design review	11
Subsystems completed	20
System level tests completed	23
Schedule margin (one month)	24

References

Boeing Company, 2000. *DELTA IV Payload Planners Guide.* Huntington Beach, CA.

Dicello, J. F. M. Zaider and M. N. Varma, 1994. An inductive assessment of radiation risks in space. *Adv. Space Res.*, **14**(10):899–910.

Rosenfeld, A. B., 1999. Semiconductor microdosimetry in mixed radiation field: present and future. *Rad. Prot. Dosim.,* **84**(1–4):385–388.

Rosenfeld, A., P. Bradley, I. Cornelius and J. Flanz, 2000. New silicon detector for microdosimetry applications in proton therapy. *IEEE Trans. Nucl. Sci.,* **47**(4):1386–1394.

Rosenfeld, A. B., P. D. Bradley and M. Zaider, 2001. Solid state microdosimetry. *Nucl. Instr. Meth. Phys. Res. B*, **184**:135–157.

Appendix: Units, Conversion Factors, and Constants

Basic SI Units

Quantity	Name	SI Base Units
length	meter	m
mass	kilogram	kg
time	second	s
electric current	ampere	A
thermodynamic temperature	kelvin	K
amount of substance	mole	mol
luminous intensity	candela	cd

SI-Derived Units

Quantity	Name	Symbol	Common Units	SI Base Units
absorbed dose	gray	G	$J\,kg^{-1}$	$m^2\,s^{-2}$
capacitance	farad	F	$C\,V^{-1}$	$kg^{-1}\,m^{-2}\,s^4\,A^2$
dose equivalent	sievert	Sv	$J\,kg^{-1}$	$m^2\,s^{-2}$
electric charge	coulomb	C		$A\,s$
electric potential, potential difference	volt	V	$W\,A^{-1}$	$kg\,m^2\,s^{-3}\,A^{-1}$
electrical conductance	siemens	S	$A\,V^{-1}$	$kg^{-1}\,m^{-2}\,s^3\,A^2$
electrical resistance	ohm	Ω	$V\,A^{-1}$	$kg\,m^2\,s^{-3}\,A^2$
energy, work, heat	joule	J	$N\,m$	$kg\,m^2\,s^{-2}$
force	newton	N		$kg\,m\,s^{-2}$

(Continued)

(Continued)

Quantity	Name	Symbol	Common Units	SI Base Units
frequency	hertz	Hz		s^{-1}
inductance	henry	H	$Wb\,A^{-1}$	$kg\,m^2\,s^{-2}\,A^{-2}$
magnetic flux density	tesla	T	$Wb\,m^{-2}$	$kg\,s^{-2}\,A^{-1}$
magnetic flux	weber	Wb	V s	$kg\,m^2\,s^{-2}\,A^{-1}$
plane angle	radian	rad		$m\,m^{-1}$ = unity
pressure, stress	pascal	Pa	$N\,m^{-2}$	$kg\,m^{-1}\,s^{-2}$
solid angle	steradian	sr		$m^2\,m^{-2}$ = unity

Derived From SI-Derived Units

Quantity	Symbol	SI Base Units
absorbed dose rate	$Gy\,s^{-1}$	$m^2\,s^{-3}$
angular acceleration	$rad\,s^{-2}$	s^{-2}
angular velocity	$rad\,s^{-1}$	s^{-1}
electric field strength	$V\,m^{-1}$	$kg\,m\,s^{-3}\,A^{-1}$
heat capacity, entropy	$J\,K^{-1}$	$kg\,m^2\,s^{-2}\,K^{-1}$
heat flux density	$W\,m^{-2}$	$kg\,s^{-3}$
moment of force	N m	$kg\,m^2 s^{-2}$
permeability	$H\,m^{-1}$	$kg\,m\,s^{-2}\,A^{-2}$
permittivity	$F\,m^{-1}$	$kg^{-1}\,m^{-3}\,s^4\,A^2$
specific energy	$J\,kg^{-1}$	$m^2\,s^{-2}$
specific heat capacity	$J\,kg^{-1}\,K^{-1}$	$m^2\,s^{-2}\,K^{-1}$
thermal conductivity	$W\,m^{-1}\,K^{-1}$	$kg\,m\,s^{-3}\,K^{-1}$

Conversion Factors

To convert from	To	Multiply by	
angstrom (Å)	meter (m)	1.0	E − 10
atmosphere, standard (atm)	pascal (Pa)	1.013 25	E + 05
bar (bar)	pascal (Pa)	1.0	E + 05
millimeter of mercury (mmHg 0°C)	pascal (Pa)	1.333 22	E + 02
degree celsius, temperature °C	kelvin (K)	t + 273.15	
degree Kelvin, temperature K	degree celsius (°C)	t − 273.15	
dyne (dyn)	newton (N)	1.0	E − 05
gram (gm)	kilogram (kg)	1.0	E − 03
international nautical mile	meter (m)	1.852	E + 03
knot (kt)	meter/second	5.144 444	E − 01
light-year (ly)	meter (m)	9.460 73	E + 15
liter (L)	cubic meter (m^3)	1.0	E − 03
millibar (mbar)	pascal (Pa)	1.0	E + 02
parsec (pc)	meter (m)	3.085 678	E + 16
pound force per square inch (psi)	pascal (Pa)	6.894 757	E + 03

(Continued)

(Continued)

To convert from	To	Multiply by	
ton metric (t, tonne)	kilogram (kg)	1.0	E + 03
torr (Torr)	pascal (Pa)	1.333 224	E + 02
watt-hour (W-h)	joule (J)	3.6	E + 03

Constants

atomic mass constant (m_u), kg	1.660 538 73	E − 27
Avogadro constant (N_A), $gmol^{-1}$,	6.022 141 99	E + 23
$kmol^{-1}$	6.022 141 99	E + 26
Boltzmann constant (k) $J K^{-1}$	1.380 650 3	E − 23
electron mass (m_e), kg	9.109 381 88	E − 31
electron volt (eV), J	1.602 176 462	E − 19
elementary charge (e), C	1.602 176 462	E − 19
Faraday constant (F), C $gmol^{-1}$	9.648 534 15	E + 04
gravitational constant (G), $kg^{-1}\ m^3\ s^{-2}$	6.672 59	E − 11
magnetic constant (μ_0), $N A^{-2}$	4π	E − 07
molecular mass of carbon-12, kg/kmol	1.2	E + 01
neutron mass (m_n), kg	1.674 927 16	E − 27
Planck's constant (h), J s	6.626 068 76	E − 34
proton mass (m_p), kg	1.672 621 58	E − 27
Rydberg constant (R_∞), m^{-1}	1.097 373 156 854 9	E + 07
second, sidereal, SI second	9.972 696	E − 01
speed of light in a vacuum (c) $m\ s^{-1}$ defined	2.997 924 58	E + 08
standard atmosphere (atm), Pa	1.013 25	E + 05
Stefan–Boltzmann constant (σ), $W\ m^{-2}\ K^{-4}$	5.670 400	E − 08
universal (molar) gas constant (R),		
$J\ gmol^{-1}\ K^{-1}$	8.314 472	E + 00
$J\ kmol^{-1}\ K^{-1}$	8.314 472	E + 03
Wien displacement law constant (b), m K	2.897 7686	E − 03

Astrophysical Parameters

General

Astronomical unit (AU), m	1.495 978 706 91	E + 11
Gravitational constant, $kg^{-1}\ m^3\ s^{-2}$	6.673	E − 11
Hubble constant (H_0), km/s-Mpc	1.00	E + 02
Julian day, s	8.640 0	E + 04
Julian year, d	3.652 5	E + 02
Julian century, d	3.652 5	E + 04
Mean sidereal day, s	8.616 409 054	E + 04
Obliquity of ecliptic (J2000), arcsec	8.438 141 2	E + 04
Sidereal year, d	3.652 56 36	E + 02

Sun

black-body temperature, K	5.778	E + 03
equatorial radius, m	6.960	E + 08
equatorial surface gravity, m/s^2	2.75	E + 02
escape velocity, m/s	6.177	E + 05
flattening	5	E − 05
GM, m^3/s^2	1.327 124 400 18	E + 20
mass, kg	1.988 9	E + 30
mean density, kg/m^3	1.416	E + 03
orbital obliquity, deg	7.25	E + 00
sidereal rotation period, h	6.091 2	E + 02
solar cycle, yr	1.14	E + 01
solar constant, W/m^2	1.368	E + 03

Mercury

black-body temperature, K	4.42 5	E + 02
equatorial radius, m	2.439 7	E + 13
equatorial surface gravity, m/s^2	3.7	E + 00
escape velocity, m/s	4.4	E + 03
flattening	0	E + 00
GM, m^3/s^2	2.204	E + 13
J2	6.0	E − 05
length of day, h	4.222 6	E + 03
mass, kg	3.303	E + 23
mean density, kg/m^3	5.427	E + 03
orbit eccentricity	0.205 6	E + 00
orbit inclination, deg	7.005	E + 00
orbital obliquity, deg	1.0	E − 02
polar radius, m	2.439 7	E + 06
semimajor axis, m	5.790 92	E + 10
sidereal orbit period, d	8.796 9	E + 01
sidereal rotation period, h	1.4076	E + 03
tropical orbital period, d	8.796 8	E + 01

Venus

black-body temperature, K	2.317	E + 02
equatorial radius, m	6.0518	E + 06
equatorial surface gravity, m/s^2	8.87	E + 00
escape velocity, m/s	1.036	E + 04
flattening	0	
GM, m^3/s^2	3.247 695	E + 14
J2	4.458	E − 06
length of day, h	2.802	E + 03
mass, kg	4.868 5	E + 24
mean density, kg/m^3	5.25	E + 03

(Continued)

(Continued)

orbit eccentricity	6.77	E − 03
orbit inclination, deg	3.395	E + 00
orbital obliquity, deg	1.774	E + 02
polar radius, m	6.051 8	E + 06
semimajor axis, m	1.082 1	E + 11
sidereal orbit period, d	2.247 01	E + 02
sidereal rotation period, h	−5.832 5	E + 03
tropical orbital period, d	2.246 95	E + 02

Earth

black-body temperature, K	2.543	E + 02
equatorial radius, m (WGS84)	6.378 137	E + 06
equatorial surface gravity, m/s^2	9.78	E + 00
escape velocity, m/s	1.118 6	E + 04
flattening (WGS84)	1/298.257 223 563	
GM, m^3/s^2 (includes atmosphere)	3.986 004 418	E + 14
J2 (WGS84)	−1.082626683	E − 03
length of day, h	2.4	E + 01
mass, kg	5.973 6	E + 24
mean density, kg/m^3	5.515	E + 03
orbit eccentricity	1.671	E − 02
orbit inclination, deg	0	E + 00
orbital obliquity, deg	2.345	E + 01
polar radius, m	6.356 752 3	E + 06
semimajor axis, m	1.495 978 90	E + 11
sidereal orbit period (year), d	3.652 564	E + 02
sidereal rotation period, h	2.393 447 2	E + 01
standard acceleration of gravity, m/s^2	9.806 65	E + 00
standard atmosphere (atm), Pa, N/m^2	1.013 25	E + 05
tropical orbital period (year), d	3.652 422	E + 02
vernal equinox orbital period (year), d	3.652 424	E + 02

Moon

black-body temperature, K	2.745	E + 02
equatorial radius, m	1.738 1	E + 06
equatorial surface gravity, m/s^2	1.62	E + 00
escape velocity, m/s	2.38	E + 03
flattening	1.2	E − 03
GM, m^3/s^2	4.9	E + 12
J2	2.027	E − 04
mass, kg	7.349	E + 22
mean density, kg/m^3	3.340	E + 03
orbit eccentricity	5.49	E − 02
orbit inclination, deg	5.145	E + 00
orbital obliquity, deg	6.68	E + 00

(Continued)

(Continued)

polar radius, m	1.735 0	E + 06
semimajor axis, m	3.844	E + 08
sidereal orbit period, d	2.732 2	E + 01
sidereal rotation period, h	6.557 28	E + 02

Mars

black-body temperature, K	2.166	E + 02
equatorial radius, m	3.397	E + 06
equatorial surface gravity, m/s^2	3.63	E + 00
escape velocity, m/s	5.03	E + 00
flattening	6.48	E − 03
GM, m^3/s^2	4.283	E + 13
J2	1.960 45	E − 03
length of day, h	2.465 97	E + 01
mass kg	6.418 5	E + 23
mean density, kg/m^3	3.933	E + 03
orbit eccentricity	9.35	E − 02
orbit inclination, deg	1.850	E + 00
orbital obliquity, deg	2.519	E + 01
polar radius, m	3.375	E + 06
semimajor axis, m	2.279 4	E + 11
sidereal orbit period, d	6.869 80	E + 02
sidereal rotation period, h	2.462 29	E + 01
tropical orbital period, d	6.869 30	E + 02

Jupiter

black-body temperature, K	9.06	E + 01
equatorial radius, m	7.14 92	E + 07
equatorial surface gravity, m/s^2	2.312	E + 01
escape velocity, m/s	5.954	E + 04
flattening	6.487	E − 02
GM, m^3/s^2	1.266 86	E + 17
J2	1.473 6.	E − 02
length of day, h	9.925 9	E + 00
mass, kg	1.898 6	E + 27
mean density, kg/m^3	1.326	E + 03
orbit eccentricity	4.89	E − 02
orbit inclination, deg	1.304	E + 00
orbital obliquity, deg	3.13	E + 00
polar radius, m	6.685 4	E + 07
semimajor axis, m	7.783	E + 08
sidereal orbit period, d	4.332 589	E + 03
sidereal rotation period, h	9.925	E + 00
tropical orbital period, d	4.330 595	E + 03

Saturn

black-body temperature, K	9.5	E + 01
equatorial radius, m	6.033 0	E + 07
equatorial surface gravity, m/s^2	8.96	E + 00
escape velocity, m/s	3.55	E + 04
flattening	9.796	E − 02
GM, m^3/s^2	3.793 1	E + 16
J2	1.63	E − 02
length of day, h	1.066	E + 01
mass, kg	5.685	E + 26
mean density, kg/m^3	6.88	E + 02
orbit obliquity, deg	2.67	E + 01
polar radius, m	5.436	E + 07
semimajor axis, m	1.433 5	E + 12
orbit eccentricity	5.65	E − 02
orbit inclination, deg	2.485 53	E + 00
sidereal orbit period, d	1.075 922	E + 04
sidereal rotation period, h	1.06 56	E + 01
tropical orbital period, d	1.074 694	E + 04

Uranus

black-body temperature, K	5.82	E + 01
equatorial radius, m	2.556	E + 07
equatorial surface gravity, m/s^2	8.69	E + 00
escape velocity, m/s	2.13	E + 04
flattening	2.293	E − 02
GM, m^3/s^2	5.794	E + 15
J2	3.343 43	E − 03
length of day, h	1.724	E + 01
mass, kg	8.683 2	E + 25
mean density, kg/m^3	1.30	E + 03
orbit eccentricity	4.57	E − 02
orbit inclination, deg	7.72	E − 01
orbital obliquity, deg	9.777	E + 01
polar radius, m	2.497 3	E + 07
semimajor axis, m	2.872 46	E + 12
sidereal orbit period, d	3.068 54	E + 04
sidereal rotation period, h	−1.724	E + 01
tropical orbital period, d	3.058 874	E + 04

Neptune

black-body temperature, K	3.52	E + 01
equatorial radius, m	2.476 6	E + 07

(Continued)

(Continued)

equatorial surface gravity, m/s^2	1.1	E + 01
escape velocity, m/s	2.35	E + 04
flattening	1.710	E − 02
GM, m^3/s^2	6.835 1	E + 15
J2	3.411	E − 03
length of day, h	1.611	E + 01
mass, kg	1.024	E + 26
mean density, kg/m^3	1.64	E + 03
orbit eccentricity	1.12	E − 02
orbit inclination, deg	1.77	E + 00
orbital obliquity, deg	2.832	E + 01
polar radius, m	2.434 1	E + 07
semimajor axis, m	4.498 3	E + 12
sidereal orbit period, d	6.018 9	E + 04
sidereal rotation period, h	1.611	E + 01
tropical orbital period, d	5.98	E + 04

Pluto

black-body temperature, K	4.27	E + 01
equatorial radius, m	1.195	E + 06
equatorial surface gravity, m/s^2	6.6	E − 01
escape velocity, m/s	1.1	E + 03
flattening	0	E + 00
GM, m^3/s^2	8.3	E + 11
J2	–	
length of day, h	1.532 82	E + 02
mass, kg	1.25	E + 22
mean density, kg/m^3	2.050	E + 03
orbit eccentricity	2.444	E − 01
orbit inclination, deg	1.714	E + 01
orbital obliquity, deg	1.225 3	E + 02
polar radius, m	1.195	E + 06
semimajor axis, m	5.906 4	E + 12
sidereal orbit period, d	9.046 5	E + 04
sidereal rotation period, h	−1.532928	E + 02
tropical orbital period, d	9.058 8	E + 04

Index

Page numbers for illustrations are shown in bold type